# 4th International Conference on Advanced Materials Science and Technology (ICAMST-2016)



IOP Conference Series: Materials Science and Engineering
Volume 202

Malang, Indonesia
27 - 28 September 2016

**Editors:**

**Markus Diantoro**

ISBN: 978-1-5108-4176-5
ISSN: 1757-8981

**Printed from e-media with permission by:**

Curran Associates, Inc.
57 Morehouse Lane
Red Hook, NY 12571

**Some format issues inherent in the e-media version may also appear in this print version.**



Printed by Curran Associates, Inc. (2017)

For permission requests, please contact the Institute of Physics
at the address below.

Institute of Physics
Dirac House, Temple Back
Bristol BS1 6BE UK

Phone: 44 1 17 929 7481
Fax: 44 1 17 920 0979

techtracking@iop.org

**Additional copies of this publication are available from:**

Curran Associates, Inc.
57 Morehouse Lane
Red Hook, NY 12571 USA
Phone: 845-758-0400
Fax: 845-758-2633
Email: curran@proceedings.com
Web: www.proceedings.com

# Table of contents

## Volume 202

**The 4th International Conference on Advanced Materials Science and Technology**

**27–28 September 2016, Universitas Negeri Malang, Indonesia**

**Accepted papers received: 12 May 2017**
**Published online: 31 May 2017**

### Preface



### Papers

IOP Conf. Series: Materials Science and Engineering **202** (2017) 011001 doi:10.1088/1757-899X/202/1/011001

# Preface

The 4$^{th}$ International Conference on Advanced Materials Science and Technology 2016 (ICAMST-2016).

International Conference on Advanced Materials Science and Technology 2016 (ICAMST-2016) is an international forum for sharing knowledge and results in theory, computation, synthesis, characterization in all aspects of advanced materials and its technological application. This ICAMST is designated to promote research interests, knowledge sharing, and transfer, as well as to improve our common ground of science and technology.

The conference has brought together researchers and academicians from both academia as well as industry to meet and share cutting-edge development in the field. The Conference welcomes significant contributions in all major fields of theoritical & analyses, synthesis & Characterization, nanoscience, Functional Materials. This volume is a collection of 100 selected from 212 manuscripts by 503 authors. These papers were presented at the International Conference on Advanced Materials Science and Technology 2016 (ICAMST-2016) which was held in Malang, Indonesia, September 27 - 28, 2016. The selection of papers included in this volume was based on an international peer review procedure. We feel the variety of topics will be of interest to researchers.

I would like to thank:

- The Scientific Committee of ICAMST 2016 for their precious contribution.
- The distinguished keynote speakers and invited speakers for their difficult task and unique lectures Prof. Dr. Takayuki Ishida (Japan); Prof Dong-Sing Wuu (Taiwan), Assoc. Prof. Dr. Graeme R. Blake (The Netherlands), Prof. Dr. Andrivo Rusydi (Singapore), Prof. Dr. Robert Jann (Taiwan), Prof. Hadi Nur, Universiti Teknologi Malaysia (Malaysia), Prof. Dr. Darminto, Indonesia, Dr. Edy Giri Rachman Putra (Indonesia), Dr. Atsushi Okazawa (Japan), Dr. Hendrik Lintang (Indonesia), on their respective fields of expertise.
- The Organizing Committee Prof. Arif Hidayat (UM), Prof. Khairurijal (ITB), Assoc. Dr. Nandang Mufti, Dr. Sunaryono, Dr. Ahmad Taufiq, Assoc. Prof. Kuwat Triyana (UGM), for their dedication.
- The Institue of Physics for the inclusion of these proceedings into the famous series of the IOP Conference Proceedings. Material Sciences and Engineering.
- PT. KGC-LJUHV Technology, Rigaku-Lab Sistematika, Vanadia, Eramitra Perdana, and PT Indotech for their sponsorship.
- My sincere thanks goes to Prof. Ahmad Rofi'udin, the Rector of Universitas Negeri Malang, for the valuable supports.

I would also like to express my gratitude to all authors and contributing reviewers. Last but not least, my special thanks goes to the respectable the president of KoSaTeM Consortia, Prof. Khairurijal, and his team.

**Assoc. Prof. Dr. Markus Diantoro**
Chairman of ICAMST 2016
Editor-in-Chief of ICAMST 2016
Department of Physics, Faculty of Mathematics and Natural Sciences,
Universitas Negeri Malang, Jl. Semarang 5 Malang 65145, Indonesia

IOP Conf. Series: Materials Science and Engineering **202** (2017) 011002 doi:10.1088/1757-899X/202/1/011002

# ORGANIZING COMMITTEE OF ICAMST 2016

| | | |
|---|---|---|
| Chairman | : | Dr. Markus Diantoro, UM |
| Co-Chairmen | : | Prof. Dr. Arif Hidayat, UM |
| | : | Prof. Dr. Khairurijal, ITB |
| | : | Prof. Dr. Heru Susanto, UNDIP |
| | : | Prof. Dr. Sutiko, UNNES |
| | : | Dr. Kuwat Triyana, UGM |
| | : | Dr. Risa Suryana, UNS |
| | : | Dr. Edy Giri Rachman Putra, BATAN |
| | : | Dr. Nandang Mufti, UM |
| | : | Dr. Djoko S, UB |
| International Advisory Board | : | Prof. Dr. Andrivo Rusydi NUS, Singapore |
| | : | Ass. Prof. Dr. Gareme Blake, RUG Netherlands |
| | : | Dr. Edy Giri Rachman Putra, BATAN Indonesia |
| | : | Prof. Dr. Hadi Nur, UTM Malaysia |
| | : | Prof. Dr. Eko Hadi Sudjiono, UNM Indonesia |
| Secretary | : | Dr. Eny Latifah |
| Administration Coord. | : | Dr. Hari Wisodo |
| | | Atsnaita Yasrina, M.Sc |
| Treasurer | : | Era Budi Pryekti, M.Si |
| Information and | : | Dr. Ahmad Taufiq |
| Documentation | : | Nugroho Adi Pramono, M.Sc |
| Paper Publication Division | : | Dr. Nandang Mufti |
| Sponsorship Division | : | Abdulloh Fuad, M.Si |
| Plenary Presentation | : | Dr. Sunaryono |
| Division | : | Nurul Hidayat, M.Si |
| | | Dr. Rr. Poppy Puspitasari |
| Supply Division | : | Daeng Achmad Suaidi, M.Kom |
| Transportation and Accommodation Division | : | Samsul Hidayat, M.T |

IOP Conf. Series: Materials Science and Engineering **202** (2017) 011002 doi:10.1088/1757-899X/202/1/011002

## EDITORIAL BOARDS

IOP Conf. Series: Materials Science and Engineering **202** (2017) 011003 doi:10.1088/1757-899X/202/1/011003

# MALANG REGENCY MAP

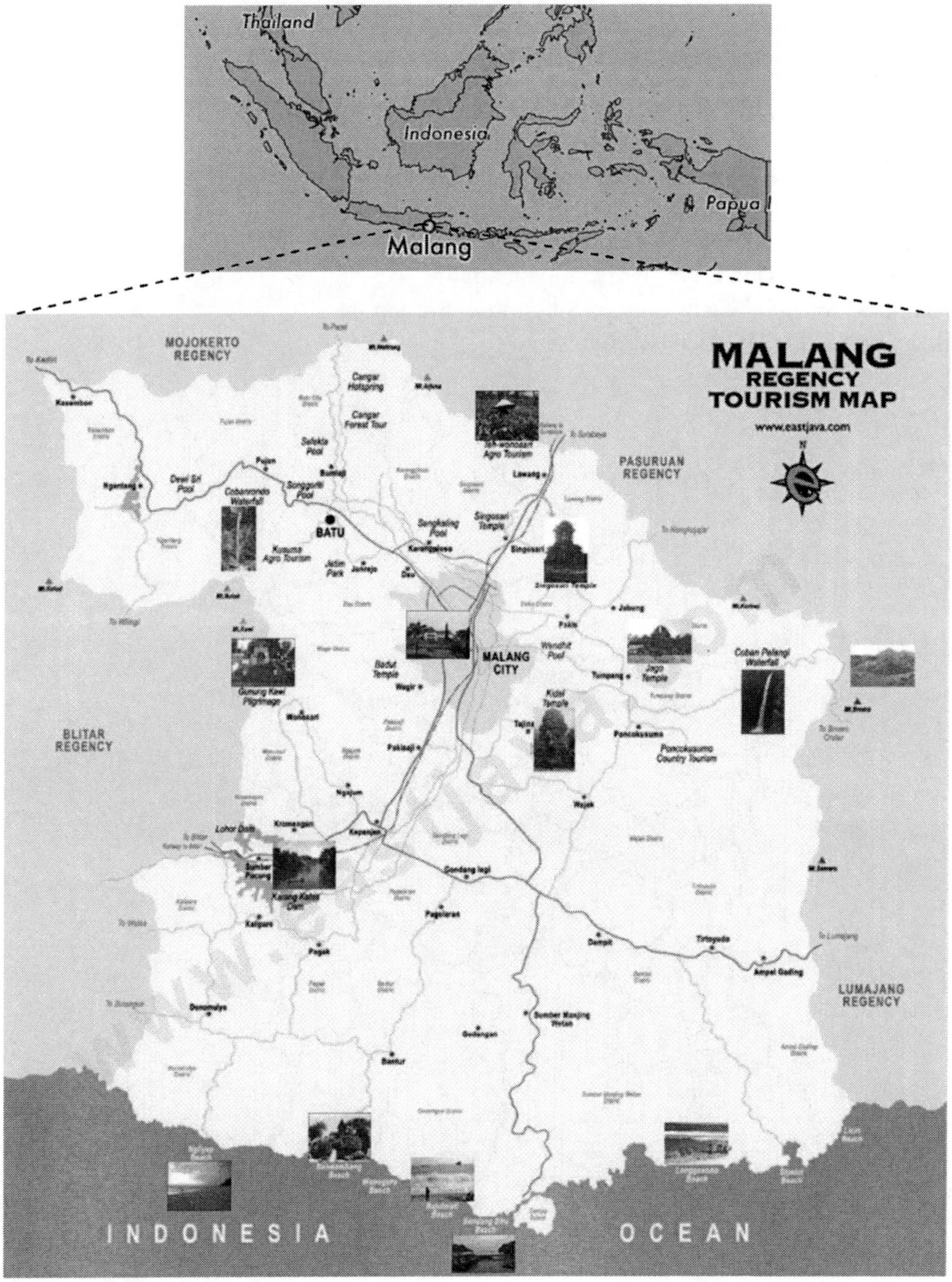

# MAP OF UM CAMPUS I

http://ft.um.ac.id/wp-content/uploads/2012/03/kampus-1.jpg

IOP Conf. Series: Materials Science and Engineering **202** (2017) 011003 doi:10.1088/1757-899X/202/1/011003

## Index Map of UM Campus I

| | |
|---|---|
| A1 | Gedung Rektorat |
| A2 | Gedung Administrasi Umum |
| A3 | Gedung Administrasi Akademik |
| B1 | Gedung Hotma |
| C1 | Gedung FPPPsi |
| C2 | Bank BNI Cabang Pembantu UM |
| C3 | UKM 1 |
| D1-D2 | Fakultas Ilmu Pendidikan |
| D3-D5 | Gedung Ekonomi |
| D7-D8 | Fakultas Sastra |
| D9 | Gedung Teknik 6 Jurusan Teknik Sipil |
| E1 | Gedung Ilmu Pendidikan 1 |
| E2 | Gedung Ilmu Pendidikan 2 |
| E3 | Gedung Ekonomi 1 |
| E4 | Gedung Ekonomi 2 |
| E5 | Gedung Ekonomi 3 |
| E6 | Gedung Sastra 1 |
| E7 | Gedung Sastra 2 |
| E8 | Gedung Sastra 3 |
| F1 | Perpustakaan |
| F2 | Masjid Al-Hikmah |
| F3 | Perpustakaan Masjid |
| G1 | Gedung WR4, HI, Pusat TIK, Kerjasama & Humas |
| G2 | Gedung Teknik 1, Jurusan Teknik Mesin |
| G3 | Gedung Teknik 2, Jurusan Teknik Sipil |
| G4 | Gedung Teknik 3, Jurusan Teknik Elektro |
| G5 | Gedung Teknik 4, Jurusan Teknolgi Industri |
| G6 | Gedung Teknik 5, Jurusan Teknologi Industri |
| H1 | Gedung Pascasarjana 1 |
| H2 | Gedung Pascasarjana 2 |
| H3 | Gedung Pascasarjana 3 |
| H4 | LP2M |
| H5 | Fakultas Teknik |
| H6 | KPRI |
| H7 | LP3M |
| H8 | Gedung Multi Guna 1 |
| I1 | Ukm 2 |
| I2 | Gedung Ekonomi 4 |
| I3 | Gedung Ekonomi 5 |
| I4 | Gedung Ekonomi 6 |
| I5 | Gedung Ekonomi 7 |
| I6 | Gedung Multi Guna 2 |
| I7 | UKM 3 |
| I8 | Kafetaria |
| I9 | Dharma Wanita |
| I10 | Sasana Budaya |
| I11 | LP2P |
| J | Graha Rektorat |
| K1 | SD Percobaan |
| K2 | SD Laboratorium UM |
| K3 | TK Laboratorium UM |
| L1 | Kantor Pos Cabang Pembantu UM |
| L2 | Poliklinik |
| M1-M5 | SMU Negeri 8 |
| M6 | SLTP Negeri 4 |
| M7 | Bengkel |
| O1 | Gedung Mipa 1 |
| O2 | Gedung Mipa 2 |
| O3 | Gedung Mipa 3 |
| O4 | Gedung Mipa 4 |
| O5 | Gedung Mipa 5 |
| O6 | Gedung Mipa 6 |
| O7 | Gedung Mipa 7 |
| Q1 | Asrama Mahasiswa Putra |
| Q2 | Sasana Krida |
| Q3 | Gedung Sastra |
| Q4 | Asrama Mahasiswa PGSD |
| R1 | Penerbit Percetakan UM |
| R2 | Garasi |
| S1-S3 | Wisma UM, BKBK,SPI,P2LP |
| S5-S10 | Wisma UM |
| S4 | Pusat Pengkajian Pancasila |
| T1,T3 | UPP PGSD 1 |
| T4,T6 | UPP PGSD 1 |
| T2 | Fakultas Ilmu Keolahragaan |
| T5 | SD Laboratorium UM |
| T7-T10 | SLTP Laboratorium UM |
| RD | Rumah Dinas |
| WTL | Warung Telekomunikasi |
| LT | Lapangan Tenis |
| LSB | Lapangan Sepakbola |
| LB | Lapangan Basket |
| LV | Lapangan Volly |

IOP Conf. Series: Materials Science and Engineering **202** (2017) 011003 doi:10.1088/1757-899X/202/1/011003

# CONFERENCE HALL MAP

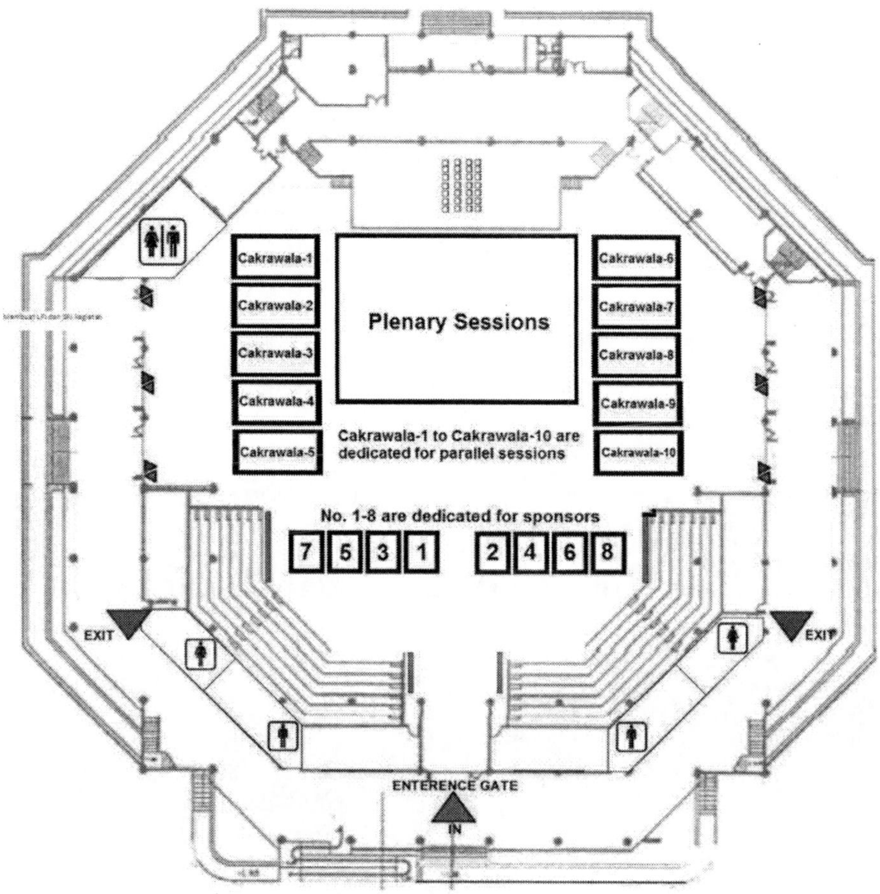

ICAMST
UNIVERSITAS NEGERI MALANG
The International Conference on Advanced Materials Science and Technology (ICAMST) 2016
Universitas Negeri Malang, 27-28 September 2016
http://icamst2016.um.ac.id

UNIVERSITAS NEGERI MALANG
The International Conference on Advanced Materials Science and Technology (ICAMST) 2016
Universitas Negeri Malang, 27-28 September 2016
http://icamst2016.um.ac.id

university of groningen
Thermoelect
Carrier concentration
The International Conference on Advanced Materials Science and Technology (ICAMST) 2016
Universitas Negeri Malang, 27-28 September 2016
http://icamst2016.um.ac.id

Chofu-city in Tokyo
I'm here !
The International Conference on Advanced Materials Science and Technology (ICAMST) 2016
Universitas Negeri Malang, 27-28 September 2016
http://icamst2016.um.ac.id

IOP Conf. Series: Materials Science and Engineering **202** (2017) 011003 doi:10.1088/1757-899X/202/1/011003

UNIVERSITAS NEGERI MALANG
UM
The International Conference on Advanced Materials Science and Technology (ICAMST) 2016
Universitas Negeri Malang, 27-28 September 2016
http://icamst2016.um.ac.id

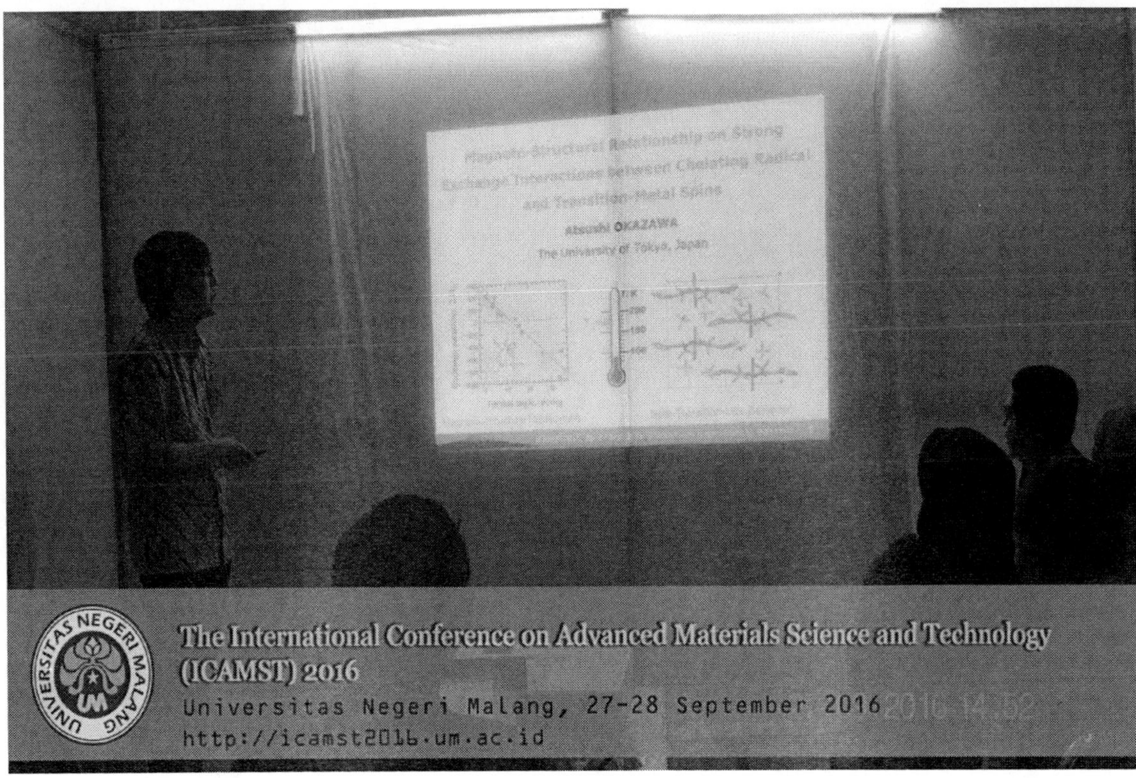
Magneto-Structural Relationship on Strong Exchange Interactions between Chelating Radical and Transition-Metal Spins
Atsushi OKAZAWA
The University of Tokyo, Japan
UNIVERSITAS NEGERI MALANG
UM
The International Conference on Advanced Materials Science and Technology (ICAMST) 2016
Universitas Negeri Malang, 27-28 September 2016
http://icamst2016.um.ac.id

IOP Conf. Series: Materials Science and Engineering **202** (2017) 011003 doi:10.1088/1757-899X/202/1/011003

IOP Conf. Series: Materials Science and Engineering **202** (2017) 011003 doi:10.1088/1757-899X/202/1/011003

# ICAMST 2016

**The 4th International Conference on Advanced Materials Science and Technology**


## Funding Acknowledgement

On Behalf of Organizing Committee We Greatly Appreciate to The Following Partners for Sponsorship Support

**UNIVERSITAS NEGERI MALANG**

**PT. KGC (LJ-UHV) TECHNOLOGY**

**PT. LAB SISTEMATIKA (RIGAKU)**

**PT. VANADIA**

**PT. INDOTECH**

**PT. ERA MITRA PERDANA**

September 28, 2016
Universitas Negeri Malang, Indonesia

**Dr. Markus Diantoro**
Chairman

IOP Conf. Series: Materials Science and Engineering **202** (2017) 011004 doi:10.1088/1757-899X/202/1/011004

# Peer review statement

All papers published in this volume of *IOP Conference Series: Materials Science and Engineering* have been peer reviewed through processes administered by the proceedings Editors. Reviews were conducted by expert referees to the professional and scientific standards expected of a proceedings journal published by IOP Publishing.

# Spin-Parity Behavior in the Exchange-Coupled Lanthanoid-Nitroxide Molecular Magnets

**T Ishida**

Department of Engineering Science, The University of Electro-Communications, Chofu, Tokyo 182-8585, Japan

Email: takayuki.ishida@uec.ac.jp

**Abstract**. To develop lanthanoid-based magnetic materials and relevant devices, reliable prescriptions for molecular/crystal design have long been desired. $Ln^{3+}$-ion dependence on the molecular magnetism was investigated in the isomorphous series [Ln(hfac)$_3$(2pyNO)] (Ln = Tb, Dy, Ho, Er), where 2pyNO stands for *tert*-butyl 2-pyridyl nitroxide as a paramagnetic ligand, and hfac for 1,1,1,5,5,5-hexafluoropentane-2,4-dionate. The slow magnetization reversal was evaluated as an indication of single-molecule magnets (SMMs) by out-of-phase ac magnetic susceptibility $\chi$". Whereas the $Tb^{3+}$ ($4f^8$) and $Ho^{3+}$ ($4f^{10}$) derivatives exhibited frequency-dependent $\chi$", practically null $\chi$" was recorded for the $Dy^{3+}$($4f^9$) and $Er^{3+}$ ($4f^{11}$) derivatives. As for another series with Ln/radical = 1/2, [Ln(hfac)$_3$(TEMPO)$_2$] complexes were prepared (Ln = Tb, Dy, Ho, Er, Tm; TEMPO = 2,2,6,6-tetramethylpiperidin-1-oxyl). The $Dy^{3+}$ and $Er^{3+}$ derivatives showed appreciable $\chi$", but the $Tb^{3+}$, $Ho^{3+}$, and $Tm^{3+}$ derivatives did not. Thus, the $S$ = 1/2 paramagnetic ligands play a role of a spin-parity switch to regulate whether the compound behaves as an SMM. In the strongly exchange-coupled regime owing to the direct radical coordination bond, the whole molecular electron counting may provide a useful criterion to predict Kramers molecules and accordingly to explore potential SMM candidates.

**Keywords:** Spin parity, exchange coupling, lanthanoid, nitroxide, molecular, magnet

## 1. Introduction

Lanthanoid (Ln) elements are many-talented and utilized for commercial magnets and light emitters. To develop lanthanoid-based magnetic materials, media, refrigerants, and relevant devices, reliable prescriptions for molecular/crystal design have long been desired [1-7]. A spin-only $_{64}Gd^{3+}$ ion ($4f^7$) is often chosen as an initial attempt, because of facile comprehension in studies on structure-function relationship [8-14]. The isotropic $Gd^{3+}$ ion leads to simple analysis even with a powder sample. Compared with $Gd^{3+}$, heavy lanthanoid ions ($_{65}Tb^{3+}$ or heavier) have advantage with an unquenched orbital moment and large magnetic anisotropy. At the second stage of our research strategy, a chemical trend of $Ln^{3+}$ ion series ($Ln^{3+}$-ion dependence) in the corresponding isomorphous analogs is an important issue for designing various magnets. Single-molecule magnets (SMMs) [1], displaying magnetic hysteresis at the molecular level, have attracted much attention, owing to the paradigm shift in science and engineering of magnets and also to future application in high-density information storage media, quantum computing materials, and other nanodevices.

The mechanism of SMMs is based on the extremely slow magnetization relaxation, which is described by a double-well potential surface with a relatively high activation energy barrier ($\Delta$) (Figure 1). This feature is regulated with the uniaxial magnetic anisotropy and large magnetic moment. Heavy lanthanoid ions ($_{65}Tb^{3+}$, $_{66}Dy^{3+}$, $_{67}Ho^{3+}$, $_{68}Er^{3+}$, and $_{69}Tm^{3+}$ with 8, 9, 10, 11, and 12 electrons in the 4f orbitals, respectively) are promising candidates for the magnetic source. Furthermore, ions with a half-integer spin are known as a Kramers ion, which guarantees the ground doublet degeneracy. In this sense, it is reasonably accepted that various $Dy^{3+}$-based SMMs have been well developed [6,15,16]. The class of mononuclear SMMs was later named as single-ion magnets (SIMs) [17]. There have been many SIMs involving $Dy^{3+}$ ions reported to date [18-20]. On the other hand, $Tb^{3+}$ and $Ho^{3+}$-based magnets require appropriate crystal field effects suitable for the uniaxial magnetic anisotropy and double-well potential energy structure.

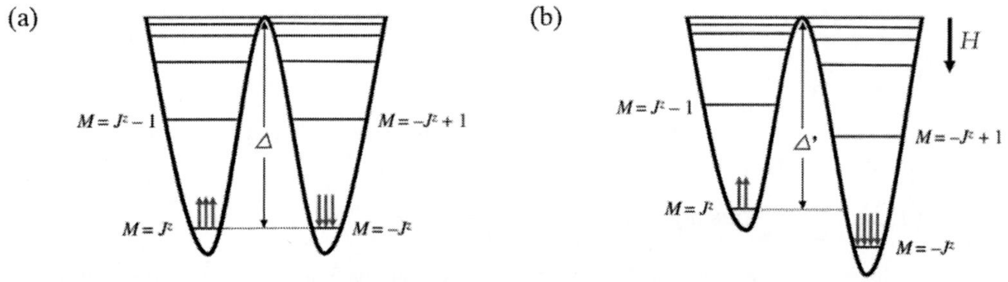


**Figure 1.** Schematic drawings of a double-well potential surface as a function of $J^z$ (a) without any dc applied field and (b) with a dc bias field.

We have concentrated our attention on 4f-2p heterospin systems and utilized persistent nitroxide radicals (>N-O•) as a paramagnetic ligand. Two kinds of nitroxide radicals will appear in this report; *tert*-butyl 2-pyridyl nitroxdie (2pyNO) [21] and a commercially available spin-probe agent, 2,2,6,6-tetramethylpiperidin-1-oxyl (TEMPO). The 4f spin sources have been selected from heavy-lanthanoid ions. Sterically bulky hfac anions (1,1,1,5,5,5-hexafluoropentane-2,4-dionate) were utilized as a cap and magnetic insulator, so that the resultant complexes are discrete and magnetic properties are ascribable to the single-molecule origin. The even-odd category of electron counting in a whole molecule is switched after an $S = 1/2$ paramagnetic ligand is incorporated. In the present study, we focus on the role of paramagnetic ligands, after overviewing new experimental findings together with some published results.

## 2. Experimental Method

### *2.1. Preparation of 2pyNO complexes*

Complexes [Ln(hfac)$_3$(2pyNO)] (abbreviated as **Ln-2pyNO**, hereafter) were prepared as orange platelets or needles from 2pyNO [21] and [Ln(hfac)$_3$($H_2O$)$_2$] [22] in a dichloromethane-heptane mixed solvent, according to the method known for **Gd-** and **Tb-2pyNO** [8,23,24]. **Dy-2pyNO**: Yield 50%. Mp. 96-98°C. The elemental analysis (C, H, N) of the complex on a PerkinElmer CHNS/O 2400 by a usual combustion method supported the chemical composition. Anal. Calcd.: C, 30.38; H, 1.70; N, 2.95% for $C_{24}H_{16}F_{18}DyN_2O_7$. Found: C, 30.61 H, 1.49; N, 2.97%. IR spectrum (neat; attenuated total reflection (ATR) method on a Nicolet FT-IR spectrometer) 773, 1001, 1059, 1134, 1250, 1648 cm$^{-1}$. The IR spectra for Ln = Ho, Er are practically the same, irrespective of Ln ions. **Ho-2pyNO**: Yield 53%. Mp. 84-85°C. Anal. Calcd.: C, 30.30; H, 1.70; N, 2.94% for $C_{24}H_{16}F_{18}HoN_2O_7$. Found: C, 30.31 H, 1.44; N, 3.12%. **Er-2pyNO**: Yield 80%. Mp. 85-87°C. Anal. Calcd.: C, 30.23; H, 1.69; N, 2.94% for $C_{24}H_{16}F_{18}ErN_2O_7$. Found: C, 30.13 H, 1.42; N, 3.10%.

### *2.2. Preparation of TEMPO complexes*

Complexes [Ln(hfac)$_3$(TEMPO)$_2$] (abbreviated as **Ln-TEMPO$_2$**, hereafter) were prepared as light yellow polycrystals from TEMPO (Tokyo Chemical Industry) and [Ln(hfac)$_3$($H_2O$)$_2$] [22] in a dichloromethane-heptane mixed solvent, according to the method known for **Gd-** and **Er-TEMPO$_2$** [25,26]. **Tb-TEMPO$_2$**: Yield 55%. Mp. 135–136°C. Anal. Calcd.: C, 36.33; H, 3.60; N, 2.57% for $C_{33}H_{39}F_{18}GdN_2O_8$. Found: C, 36.33; H, 3.55; N, 2.75%. IR spectrum (neat; ATR) 976, 1136, 1253,

1651, 2947 $cm^{-1}$. The IR spectra for Ln = Dy, Ho, Tm are practically the same, irrespective of Ln ions. **Dy-TEMPO$_2$**: Yield 62%. Mp. 132–133°C. Anal. Calcd.: C, 36.16; H, 3.59; N, 2.56% for $C_{33}H_{39}F_{18}DyN_2O_8$. Found: C, 36.21; H, 3.32; N, 2.67%. **Ho-TEMPO$_2$**: Yield 88%. Mp. 127–128°C. Anal. Calcd.: C, 36.08; H, 3.58; N, 2.55% for $C_{33}H_{39}F_{18}HoN_2O_8$. Found: C, 36.34; H, 3.55; N, 2.69%. **Tm-TEMPO$_2$**: Yield 68%. Mp. 119–120°C. Anal. Calcd.: C, 35.95; H, 3.57; N, 2.54% for $C_{33}H_{39}F_{18}TmN_2O_8$. Found: C, 35.79; H, 3.43; N, 2.69%.

### *2.3. X-Ray crystallographic analysis*
X-ray diffraction data of **Ln-2pyNO** and **Ln-TEMPO$_2$** were collected on a Rigaku Saturn70 CCD diffractometer with graphite monochromated MoKα radiation ($\lambda$ = 0.71073 Å). The structures were directly solved using a heavy atom method and expanded using Fourier techniques in the *CrystalStructure* program package [27]. All of the hydrogen atoms were refined as "riding." The thermal displacement parameters of the non-hydrogen atoms were refined anisotropically. Selected crystallographic data and geometrical parameters are listed in Tables 1 and 2. CCDC reference numbers are 912946, 1504137, 1504138, and 1504139 for **Tb-**, **Dy-**, **Ho-**, and **Er-2pyNO**, respectively, and 1488109 – 1488113 for **Tb-**, **Dy-**, **Ho-**, **Er-**, and **Tm-TEMPO$_2$**, respectively.

### *2.4. Magnetic measurements*
The alternating-current (ac) magnetic susceptibility of polycrystalline specimens were measured on a Quantum Design PPMS ac/dc magnetometer equipped with a 9 T coil in a temperature range 1.9 – 300 K. Samples were mounted and not fixed in a gelatin capsule. The direct-current (dc) susceptometry was conducted on a Quantum Design MPMS SQUID magnetometer equipped with a 7 T coil in a temperature range 1.8 – 300 K.

## 3. Results

### *3.1. 2pyNO complexes*
Complexation between [Ln(hfac)$_3$(H$_2$O)$_2$] and 2pyNO in a heptane-dichloromethane mixed solvent afforded [Ln(hfac)$_3$(2pyNO)] **(Ln-2pyNO)** as fine orange polycrystals in moderate yields. The elemental and spectral analyses support the chemical formula. The X-ray diffraction study on **Ln-2pyNO** clarified that the Tb through Er derivatives are completely isomorphous (Figure 2 and Table 1). The space group was triclinic $P\bar{1}$ with $Z$ = 4, and there are two crystallographically independent molecules in a unit cell. The two molecules are similar to each other. The Ln ions are eight-coordinate from one nitrogen and eight oxygen atoms, to form a square antiprism (SAPR), according to the *SHAPE* analysis [28]. The 4f and 2p spin centers are directly bonded in a five-membered chelate ring.

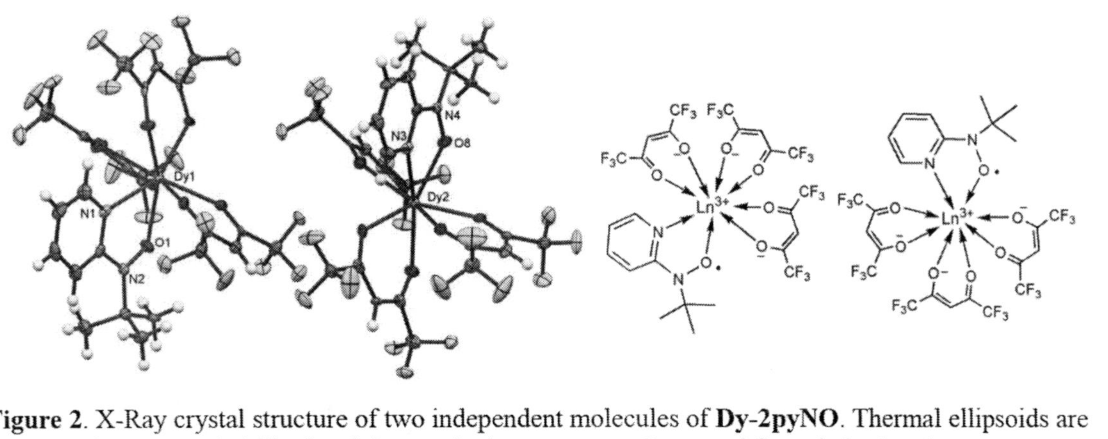


**Figure 2**. X-Ray crystal structure of two independent molecules of **Dy-2pyNO**. Thermal ellipsoids are drawn at the 50% probability level for non-hydrogen atoms. Structural formula is also shown.

**Table 1.** Selected crystallographic data of **Ln-2pyNO** (Ln = Tb, Dy, Ho, Er).

| | **Tb-2pyNO** | **Dy-2pyNO** | **Ho-2pyNO** | **Er-2pyNO** |
|---|---|---|---|---|
| Formula | $C_{24}H_{16}TbF_{18}N_2O_7$ | $C_{24}H_{16}DyF_{18}N_2O_7$ | $C_{24}H_{16}HoF_{18}N_2O_7$ | $C_{24}H_{16}ErF_{18}N_2O_7$ |
| Crystal system | triclinic | triclinic | triclinic | triclinic |
| Space group | $P\bar{1}$ | $P\bar{1}$ | $P\bar{1}$ | $P\bar{1}$ |
| $a$ / Å | 9.491(3) | 9.507(2) | 9.4973(9) | 9.5025(11) |
| $b$ / Å | 15.518(5) | 15.515(2) | 15.5375(13) | 15.540(2) |
| $c$ / Å | 22.240(5) | 22.174(3) | 22.173(2) | 22.144 (3) |
| $\alpha$ / deg | 73.07(2) | 72.943(7) | 72.817(5) | 72.806(5) |
| $\beta$ / deg | 87.91(2) | 87.740(7) | 87.723(6) | 87.840(6) |
| $\gamma$ / deg | 88.39(2) | 88.364(8) | 88.349(6) | 88.312(6) |
| $V$ / Å$^3$ | 3131(2) | 3123.9(7) | 3122.9(5) | 3121.0(6) |
| $Z$ | 4 | 4 | 4 | 4 |
| $R$ [a] $(I > 2\sigma(I))$ | 0.0480 | 0.0354 | 0.0662 | 0.0698 |
| $R_w$ [b] (all) | 0.0962 | 0.0376 | 0.0736 | 0.0755 |
| $T$ / K | 100 | 100 | 100 | 100 |
| Reference | c) | This work | This work | This work |

[a] $R = \Sigma \| F_o| - |F_c\| / \Sigma |F_o|$. [b] $R_w = [\Sigma w(F_o^2 - F_c^2)^2 / \Sigma w(F_o^2)^2]^{1/2}$. [c] Murakami R et al. 2013 *Dalton Trans.* **42** 13968.

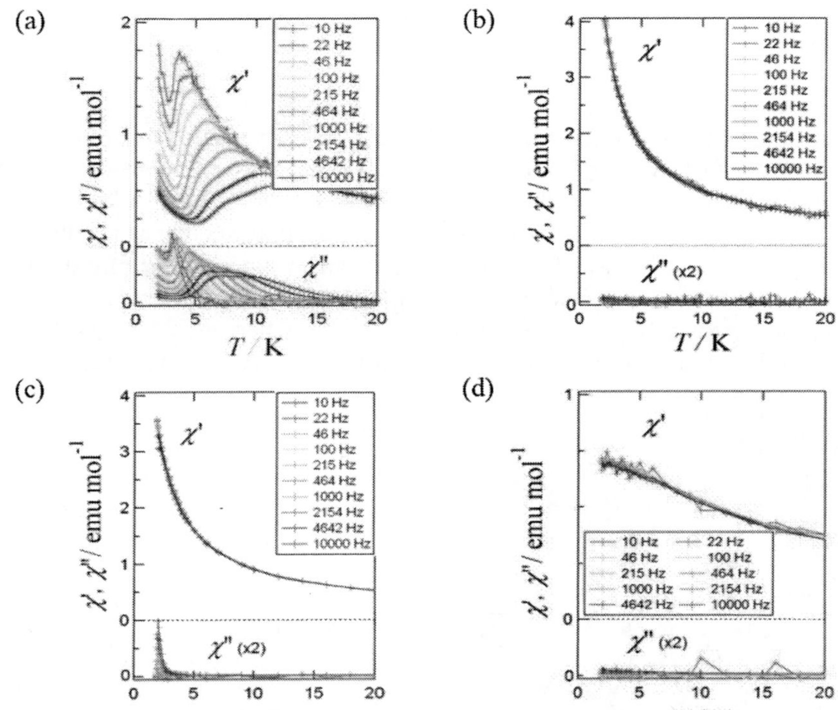


**Figure 3.** Ac magnetic susceptibilities (top: in-phase $\chi'$; bottom: out-of-phase $\chi''$) for (a) **Tb-**, (b) **Dy-**, (c) **Ho-**, and (d) **Er-2pyNO** without any bias field. Lines are shown as a guide to the eyes. Relatively large noises were recorded in the 10 Hz data.

Relatively strong antiferromagnetic coupling was found in **Gd-2pyNO**, originated in the direct Gd-radical coordination bond [8]. The $Gd^{3+}$-radical exchange coupling constant was determined from the magnetic susceptibility measurements, giving $J_{Gd\text{-}2pyNO}/k_B = -13.8(3)$ K, where the constant is formulated as $H = -J_{Gd\text{-}2pyNO}\, S_{Gd} \bullet S_{2pyNO}$. Study was extended to the heavier lanthanoid ions, but $J_{Ln\text{-}2pyNO}$ could hardly be obtained only from the susceptibility data. Instead, high-field and high-frequency electron paramagnetic resonance spectroscopy and inelastic neutron diffraction study can afford the information of the energy level scheme. Analysis on **Tb-2pyNO** gave $J_{Tb\text{-}2pyNO}/k_B = -9.52$ and $-7.19$ K [24]. The sign of exchange coupling is common among the heterospin molecules of $Gd^{3+}$ and heavy $Ln^{3+}$ ions coupled with 3d transition metal ions [29-33] and radicals [24]. The exchange coupling mechanism proposed for Gd compounds [10] seems to hold for the heavier Ln compounds. The intramolecular antiferromagnetic couplings in **Tb-**, **Dy-**, **Ho-**, and **Er-2pyNO** giving a ground ferrimagnetic state were confirmed by the saturation magnetization in the magnetization curves measured at 1.8 K.

The slow magnetization reversal was evaluated with ac magnetic susceptibility. Eventually, **Tb-2pyNO** was characterized as an SMM [23]. Two series of the out-of-phase ac susceptibility ($\chi$") signals appeared around 10 K even at zero direct-current (dc) bias field (Figure 3a), which is consistent with the presence of two independent molecules in a unit cell. The activation energies for the magnetization reversal were estimated as $\Delta/k_B = 39.2(3)$ K with $\tau_0 = 4.2(4)\times10^{-8}$ s for one molecule and $\Delta/k_B = 36(2)$ K with $\tau_0 = 4.1(12)\times10^{-7}$ s for the other, where $\tau_0$ stands for the pre-exponential factor in the Arrhenius analysis [34]. The Cole-Cole plot [35] supports this argument. On the other hand, unexpectedly, no $\chi$" was recorded for **Dy-2pyNO** (Figure 3b). A $Dy^{3+}$ ion has been supposed to be preferable for development of SMMs because of large magnetic moment and anisotropy, and moreover the Kramers theorem tells us that the half-integer spin ($S_{Dy^{3+}} = 5/2$) would bring about the double-well potential. As for the $Ho^{3+}$ analog, **Ho-2pyNO** exhibited appreciable $\chi$" with frequency dependence (Figure 3c), though a $Ho^{3+}$ ($S_{Ho^{3+}} = 2$) ion is not a Kramers ion. The $Er^{3+}$ ($S_{Er^{3+}} = 3/2$) derivative, **Er-2pyNO**, did not show any meaningful $\chi$" (Figure 3d). The present chemical trend indicates that an even-odd effect regulates the SMM characteristics and that a paramagnetic ligand with $S = 1/2$ switches the even-odd classification on electron counting.

### *3.2. TEMPO complexes*

As a series with a Ln/radical ratio of 1/2, [Ln(hfac)$_3$(TEMPO)$_2$] complexes have been prepared and characterized. Complexation between [Ln(hfac)$_3$(H$_2$O)$_2$] and TEMPO afforded **Ln-TEMPO$_2$** as light yellow crystalline products in moderate yields. The elemental and spectral analyses support the chemical composition. The X-ray diffraction study on **Ln-TEMPO$_2$** clarified that all the derivatives are isomorphous (Figure 4 and Table 2). The Ln ions are coordinated by eight oxygen atoms, consisting of two from TEMPO nitroxides and six from three diketonate groups. The *SHAPE* [28] analysis clarified that the coordination polyhedron belongs to a triangular dodecahedron (TDD). Two TEMPO groups are located at "*trans*" positions with respect to the Ln center, and the two TEMPO ends are well separated with three hfac groups. In a crystal, the peripheral $CF_3$ and TEMPO alkyl groups magnetically isolate adjacent molecules.

Relatively strong exchange coupling was observed for **Gd-TEMPO$_2$** with $J_{Gd\text{-}TEMPO}/k_B = -12.9(5)$ and $+8.0(6)$ K although TEMPO---TEMPO exchange coupling would disturb the analysis [25]. A chemical trend was investigated also in this series, and dominant antiferromagnetic $J_{Ln\text{-}TEMPO}$ couplings were proven from the saturation magnetization in the magnetization curve measurements [26]. As for the Kramers-ion analogs, **Dy-** and **Er-TEMPO$_2$** showed an indication of an SMM, as frequency-dependent $\chi$" was recorded (Figure 5). The crystal field parameters were evaluated for **Er-TEMPO$_2$**, and considerably large $Er^{3+}$-radical exchange couplings have been pointed out [26]. The $\chi$" upsurge observed for **Er-TEMPO$_2$** is only slight, but substantial as shown in the dc bias field experiments [26]. In contrast, practically null $\chi$" was recorded for **Tb-**, **Ho-**, and **Tm-TEMPO$_2$**.

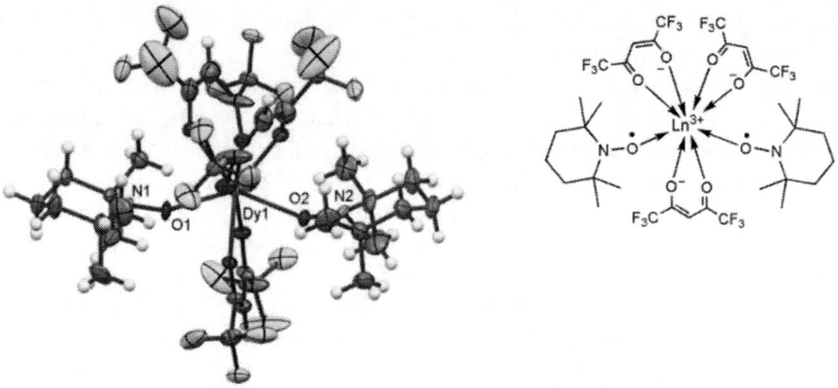


**Figure 4**. Molecular structure of (a) **Dy-TEMPO$_2$**. Thermal ellipsoids are drawn at the 50% probability level for non-hydrogen atoms. Structural formula is also shown.

**Table 2.** Selected crystallographic data of **Ln-TEMPO$_2$** (Ln = Tb, Dy, Ho, Er, Tm).

| | **Tb-TEMPO$_2$** | **Dy-TEMPO$_2$** | **Ho-TEMPO$_2$** | **Er-TEMPO$_2$** | **Tm-TEMPO$_2$** |
|---|---|---|---|---|---|
| Formula | $C_{33}H_{39}Tb$-$F_{18}N_2O_8$ | $C_{33}H_{39}Dy$-$F_{18}N_2O_8$ | $C_{33}H_{39}Ho$-$F_{18}N_2O_8$ | $C_{33}H_{39}Er$-$F_{18}N_2O_8$ | $C_{33}H_{39}Tm$-$F_{18}N_2O_8$ |
| Crystal system | momoclinic | momoclinic | momoclinic | momoclinic | momoclinic |
| Space group | $P2_1/c$ | $P2_1/c$ | $P2_1/c$ | $P2_1/c$ | $P2_1/c$ |
| $a$ / Å | 11.980(3) | 11.970(3) | 11.9648(13) | 11.943(3) | 11.937(2) |
| $b$ / Å | 16.158(4) | 16.150(4) | 16.152(2) | 16.145(4) | 16.133(3) |
| $c$ / Å | 22.288(6) | 22.249(5) | 22.255(3) | 22.015(4) | 22.195(4) |
| $\beta$ / deg | 90.98(1) | 91.04(1) | 90.9350(9) | 90.981(9) | 91.0296(10) |
| $V$ / Å$^3$ | 4314(2) | 4300(2) | 4300.4(8) | 4283(2) | 4273.6(12) |
| $Z$ | 4 | 4 | 4 | 4 | 4 |
| $R^{a)}$ ($I > 2\sigma(I)$) | 0.0607 | 0.0602 | 0.0686 | 0.0626 | 0.0640 |
| $R_w^{b)}$ (all) | 0.0854 | 0.1013 | 0.1248 | 0.0793 | 0.1099 |
| $T$ / K | 100 | 100 | 100 | 100 | 100 |
| Reference | This work | This work | This work | c) | This work |

$^{a)}$ $R = \Sigma\, ||F_o| - |F_c|| / \Sigma\, |F_o|$. $^{b)}$ $R_w = [\Sigma\, w(F_o^2 - F_c^2)^2 / \Sigma\, w(F_o^2)^2]^{1/2}$. $^{c)}$ Karbowiak M et al. 2016 *Chem. Phys. Lett.* **662** 163.

To compare the specification values of **Dy-** and **Er-TEMPO$_2$**, we performed the ac magnetic susceptibility measurements on **Dy-TEMPO$_2$** under the same conditions; namely, a dc bias field of 1000 Oe was applied to prevent possible quantum tunneling of the magnetization (Figure 1b). An obvious frequency dependence was recorded (Figure 6a). As the Arrhenius plot shows (Figure 6b), a downward deviation of the data points from the Arrhenius expectation is found below 2.5 K, suggesting the presence of a Raman-type relaxation and/or residual quantum tunneling process. In the Arrhenius regime, the activation energy for the magnetization reorientation for **Dy-TEMPO$_2$** was estimated as $\Delta/k_B = 24(1)$ K with $\tau_0 = 1.2(5) \times 10^{-8}$ s, while $\Delta/k_B = 18.2(19)$ K with $\tau_0 = 9(5) \times 10^{-8}$ s for **Er-TEMPO$_2$** [26]. The activation energy of **Dy-TEMPO$_2$** is larger than that of **Er-TEMPO$_2$**. The Cole-Cole plot on the $T = 2.0$ and 3.0 K data afforded $\alpha = 0.29(1)$ and 0.26(2), respectively (for the definition of $\alpha$, see [36]). The presence of single relaxation process is indicated, and it becomes clearer when $T = 3.0$ K than when $T = 2.0$ K (Figure 6c). We have to stress that **Dy-** and **Er-TEMPO$_2$** behaved as a typical SMM. The present chemical trend reveals an even-odd effect, and the even-odd classification of **Ln-TEMPO$_2$** is just the same as that of $Ln^{3+}$ single ions.

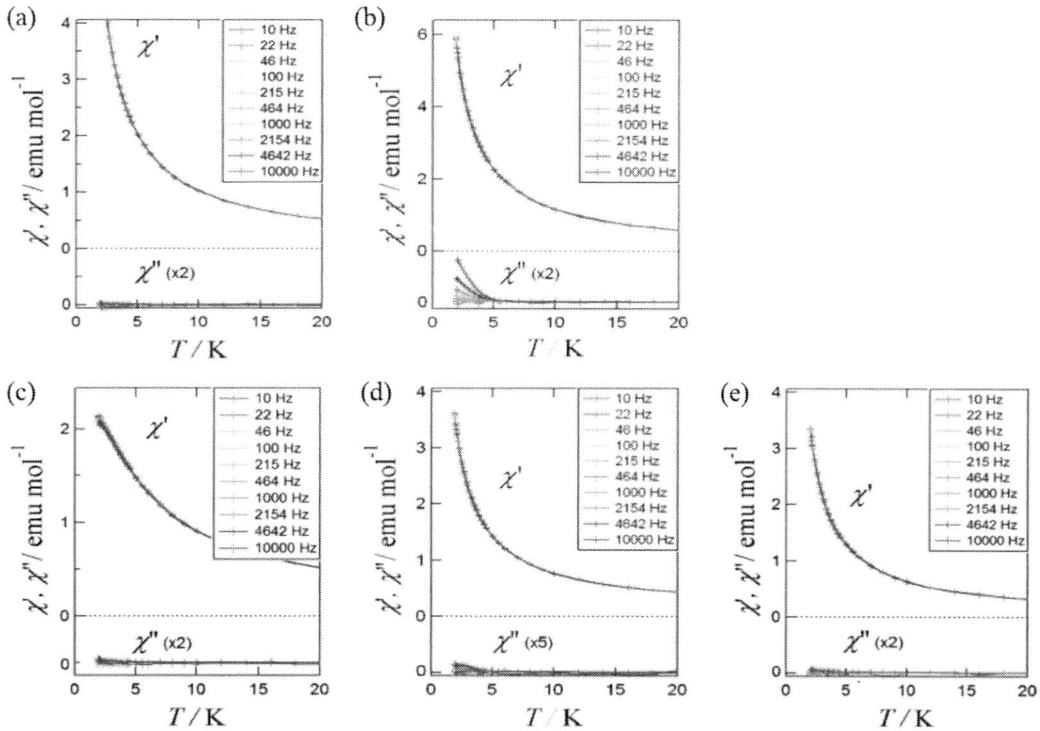


**Figure 5.** Ac magnetic susceptibilities (top: in-phase $\chi'$; bottom: out-of-phase $\chi''$) for (a) **Tb-**, (b) **Dy-**, (c) **Ho-**, (d) **Er-**, and (e) **Tm-TEMPO₂** without any bias field. Lines are shown as a guide to the eyes.

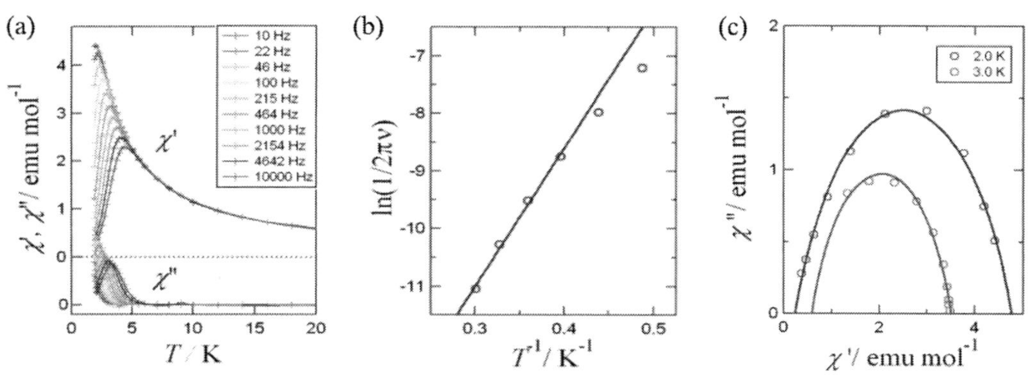


**Figure 6.** (a) Ac magnetic susceptibilities measured for **Dy-TEMPO$_2$** with a dc bias field of 1000 Oe. (b) The Arrhenius plot and (b) the Cole-Cole plot of the data taken from (a).

## 4. Discussion

All the compounds investigated here possess a ferrimagnetic ground state, and each molecule with the residual magnetic moment behaves as a paramagnet. During the screening of SMMs by means of ac susceptometry, a clear empirical rule emerges. The paramagnetic ligands play a role of a spin-parity switch to regulate whether the complex behaves as a potential SMM or not.

In the Ln/radical = 1/1 families, SMM behavior was observed for **Tb-2pyNO** [23] and **Tb-6bpyNO** [37] but not for **Dy-2pyNO** or **Dy-6bpyNO** (6bpyNO is a derivative of 2pyNO, 2,2'-bipyridin-6-yl *tert*-butyl nitroxide [38]; for the structural formula, see Figure 6). The electron countings of the whole molecule are odd for the Tb-radical compounds and even for the Dy-radical compounds. Similarly, the whole electron countings are odd for **Ho-2pyNO** and even for **Er-2pyNO**. Replacing the paramagnetic ligand with a diamagnetic one gave additional evidence [37]. Namely, **Tb-6bpyCO** is not an SMM despite having practically the same crystal field (6bpyCO = 2,2'-bipyridin-6-yl *tert*-butyl ketone; Figure 7). Since **Dy-2pyNO**, **Dy-6bpyNO**, and **Tb-6bpyCO** were not an SMM, the crystal field is unimportant to design SMMs in these systems.

The results on **Ln-TEMPO$_2$** with Ln/radical = 1/2 are completely different from those of **Ln-2pyNO**; **Dy**- and **Er-TEMPO$_2$** showed an indication of an SMM, whereas almost zero $\chi$" was recorded for **Tb-**, **Ho-**, and **Tm-TEMPO$_2$**. Supporting evidence was given from the study on **Ln-TMIO$_2$** [39] (TMIO = 1,1,3,3-tetramethylisoindolin-2-oxyl; Figure 7). Two paramagnetic ligands in a molecule are symmetrically related, and Tb- and Dy-radical antiferromagnetic couplings have been evaluated by means of electron paramagnetic resonance spectroscopy [40]. A Kramers molecule **Dy-TMIO$_2$** displayed distinct frequency-dependent $\chi$" with a 1000 Oe dc bias field, and a non-Kramers molecule **Tb-TMIO$_2$** did not [40]. The presence of two $S$ = 1/2 paramagnetic ligands maintains the even-odd classification of the whole molecular electron counting. Consequently the experimental results are totally compatible with the Kramers theorem. The bistability is not guaranteed for a non-Kramers $Tb^{3+}$ ion, and actually **Tb-TEMPO$_2$** and **Tb-TMIO$_2$** exhibited flat $\chi$". The present finding indicates that the crystal field effect is not a decisive factor to bestow an SMM character to these systems.

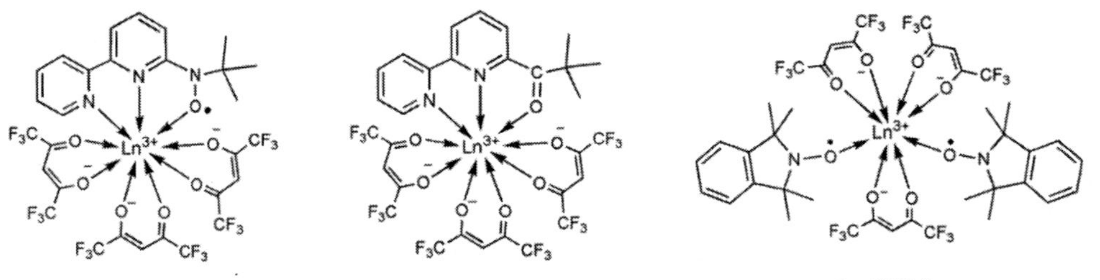


**Figure 7.** Structural formulas of **Ln-6bpyNO**, **Ln-6bpyCO**, and **Ln-TMIO$_2$**.

In the strongly exchange-coupled systems having the direct radical coordination, the total spin may be a useful criterion to predict Kramers molecules. It is noticed at the present stage that the integer or half-integer of the total spin is essential, regardless of the angular momentum. The $LS$ coupling is defined within a single ion whereas the spin-spin coupling works between two paramagnetic centers or more. When the latter is stronger than the former, the coupling scheme of the spin-spin coupling first and the $LS$ coupling later might be a possible explanation. In the heavy Ln group, $L$ and $S$ point to the same direction, i.e., $J = L + S$. After the spin-spin antiferromagnetic coupling is operative, the Ln total moment and radical moment are antiferromagnetically correlated. The Kramers theorem is a useful criterion to find SMM candidates, assuming that the spin-spin coupling is strong enough (typically in the order of 10 K), compared with cryogenic temperatures (e.g. 4.2 K). Such a discrimination method is not valid for 4f-3d or 4f-4f systems, because their spin-spin couplings are comparably small (typically in the order of 1 K or smaller).

Although a simple spin-parity behavior has already been reported for the DOTA-based SIMs [41] (DOTA = 1,4,7,10-tetraazacyclododecane-1,4,7,10-tetraacetic acid), an example of 4f-2p heterospin compounds is unprecedented. The establishment of the spin-parity effect on SMMs [42] needs detailed experiments on dynamics and theoretical consideration. If a factor of the tunneling probability at zero applied field is removed or focused on, experiments with dc bias fields would afford a clue, and each possible relaxation process should be separated and evaluated. So far, such measurements revealed a qualitatively unchanged trend for **Ln-2pyNO**, **Ln-TEMPO$_2$**, and **Ln-TMIO$_2$** at a bias field of 1000 Oe and for **Ln-6bpyNO** and **Ln-6bpyCO** at a bias field of 2000 Oe. Absorption and emission measurements will clarify the energy level structure with the crystal field splitting. Further studies are now underway.

## 5. Conclusion

The present study unveils the role of the paramagnetic ligand incorporated in 4f-2p heterospin systems. When 4f-2p heterospin compounds possess sizable exchange coupling due to the direct radical coordination, a quite unique situation is realized. In the strongly exchange coupling regime, the whole molecular electron counting provides a useful criterion to predict Kramers doublet molecules and accordingly to search for potential SMM candidates. The control of the relaxation time constants by means of introduction of a paramagnetic ligand would lead to new tactics in designing 4f-based molecules toward novel SMMs, data storage media, and quantum computing devices.

## 6. References

[1] Gatteschi D, Sessoli R and Villain J 2006 *Molecular Nanomagnets*, Oxford University Press, New York
[2] Demir S, Jeoh I–R, Long J R, Harris T D 2015 *Coord. Chem. Rev.* **289** 149
[3] Andruh M, Costes J P, Diaz C and Gao S 2009 *Inorg. Chem.* **48** 3342
[4] Liu K, Shi Wand Cheng P 2015 *Coord. Chem. Rev.* **289-290** 74
[5] Bogani L, Sangregorio C, Sessoli R and Gatteschi D 2005 *Angew. Chem. Int. Ed.* **44** 5817
[6] Reinhart J D and Long J R 2011 *Chem. Sci.* **2** 2078
[7] Suarez A I O, Lyaskovskyy V, Reek, J N H, van der Vlugt J I and de Bruin B 2013 *Angew. Chem. Int. Ed.* **52** 12510
[8] Ishida T, Murakami R, Kanetomo and T, Nojiri H 2013 *Polyhedron* **66** 183
[9] Kanetomo T and Ishida T 2014 *Inorg. Chem.* **53** 10794
[10] Kanetomo T, Yoshitake T and Ishida T 2016 *Inorg. Chem.* **55** 8140
[11] Kanetomo T and Ishida T 2014 *Chem. Commun.* **50** 2529
[12] Kanetomo T and Ishida T 2016 *AIP Conf. Proc.* **1709** 020015
[13] Nakamura T and Ishida T 2016 *AIP Conf. Proc.* **1709** 020016
[14] Ikegaya N, Kanetomo T, Murakami and R, Ishida T. 2012 *Chem. Lett.* **41** 82
[15] Zhang P, Guo Y-N and Tang J 2015 *Coord. Chem. Rev.* **257** 1728
[16] Mori F, Nyui T, Ishida T, Nogami T, Choi K-Y and Nojiri H 2006 *J. Am. Chem. Soc.* **128** 1441
[17] Ishikawa N, Sugita M, Ishikawa T, Koshihara S Y and Kaizu Y 2003 *J. Am. Chem. Soc.* **125** 8694
[18] Cucinotta G, Perfetti M, Luzon J, Etienne M, Car P-E, Caneschi A, Calvez G, Bernot K and Sessoli R 2012 *Angew. Chem. Int. Ed.* **51** 1606
[19] Jiang S-D, Wang B-W, Su G, Wang Z-M and Gao S 2010 *Angew. Chem. Int. Ed.* **41** 7448
[20] Watanabe A, Yamashita A, Nakano M, Yamamura T and Kajiwara T 2011 *Chem. Eur. J.* **17** 7428
[21] Okazawa A, Nogami T and Ishida T 2007 *Chem. Mater.* **19** 2733
[22] Richardson M F, Wagner W F and Sands D E 1968 *J. Inorg. Nucl. Chem.* **30** 1275
[23] Murakami R, Ishida T, Yoshii S and Nojiri H 2013 *Dalton Trans.* **42** 13968
[24] Baker M L, Tanaka T, Murakami R, Ohira-Kawamura S, Nakajima T, Ishida T and Nojiri H 2015 *Inorg. Chem.* **54** 5732
[25] Murakami R, Nakamura T and Ishida T 2014 *Dalton Trans.* **42** 5893
[26] Karbowiak M, Rudowicz C, Nakamura T, Murakami R and Ishida T 2016 *Chem. Phys. Lett.* **662** 163
[27] *CRYSTALSTRUCTURE* ver. 4.0. 2010 Rigaku Corp. Tokyo, Japan
[28] *SHAPE* V2.1. Lluncll M, Casanova D, Circra J, Bofill J M, Alcmany P, Alvarez S, Pinsky, M, Avnir D, University of Barcelona and The Hebrew University of Jerusalem, Barcelona, 2005
[29] Kobayashi Y, Ueki S, Ishida T and Nogami T 2003 *Chem. Phys. Lett.* **378** 337
[30] Ueki S, Ishida T, Nogami T, Choi K-Y and Nojiri H 2007 *Chem. Phys. Lett.* **440** 263
[31] Ueki S, Kobayashi Y, Ishida T and Nogami T 2005 *Chem. Commun.* 5223

[32] Okazawa A, Nogami T, Nojiri H and Ishida T 2008 *Chem. Mater.* **20** 3110
[33] Okazawa A, Watanabe R, Nojiri H, Nogami T and Ishida T 2009 *Polyhedron* **28** 1808
[34] Sessoli R, Tsai H-L, Schake A R, Wang S, Vincent J B, Folting K, Gatteschi D, Christou G and Hendrickson D N 1993 *J. Am. Chem. Soc.* **115** 1804
[35] Cole K S, Cole H R 1941 *J. Chem. Phys.* **9** 341
[36] Sessoli R and Powell A K 2009 *Coord. Chem. Rev.* **253** 2328
[37] Kanetomo T, Yoshii S, Nojiri H and Ishida T 2015 *Inorg. Chem. Front.* **2** 860
[38] Osanai K, Okazawa A, Nogami T and Ishida T 2006 *J. Am. Chem. Soc.* **128** 14008
[39] Nakamura T and Ishida T 2015 *Polyhedron* **87** 302
[40] Ishida T, Nakamura T, Kihara T and Nojiri H 2017 *Polyhedron in press* (doi:10.1016/j.poly.2017.03.008).
[41] Boulon M-E, Cucinotta G, Luzon J, Degl'Innocenti C, Perfetti N, Bernot K, Calvez G, Caneschi A and Sessoli R 2013 *Angew. Chem. Int. Ed.* **52** 350
[42] Wernsdorfer W, Chakov N E and Christou G 2005 *Phys. Rev. Lett.* **95** 037203

**Acknowledgments**
This work has been carried out in collaboration with Dr. Kanetomo T, Mr. Nakamura T, and Ms. Murakami R. The author thanks Prof. Nojiri H (IMR, Tohoku Univ.) for valuable discussion. This work was financially supported by KAKENHI (Grant Number JSPS/15H03793).

# Magneto-Structural Relationship on Strong Exchange Interactions between Chelating Nitroxide Radical and Transition-Metal Spins

**A Okazawa**

Department of Basic Science, The University of Tokyo, Komaba, Meguro-ku, Tokyo 153-8902, Japan.

Email: cokazawa@mail.ecc.u-tokyo.ac.jp

**Abstract.** To develop room-temperature molecule-based magnets, metal–radical approach is one of the most promising strategies. Strong magnetic interactions beyond a thermal energy of room temperature are essential for achieving such coveted materials. We have developed 2-pyridyl *tert*-butyl nitroxide (2pyNO) derivatives as a chelating radical ligand with a copper(II) or nickel(II) ion. The torsion angle ($\phi$) around M–O–N–$C_{2py}$ is a useful metric in describing correlation between the structure and metal–radical exchange coupling $2J$ (defined as $H = -2J\Sigma S_i \cdot S_{i+1}$). Although we previously reported such a magneto-structural correlation, in this paper we updated the plot of $2J$ versus $\phi$ by dividing into two categories: the hexacoordinatenickel(II) and copper(II) complexes. An empirical linear correlation was formulated for both Cu and Ni categories, as $2J/k_B = a + b|\phi|$ with $a$ = 503(112) K and $b$ = −36(3) K deg$^{-1}$ and with $a$ = 449(46) K and $b$ = −21.0(17) K deg$^{-1}$, respectively. In the latter plot we confirm that the relationship can be generalized for the oxoverdazyl and dithiadiazolyl systems as well as the 2pyNO systems. The results are supported by the calculations based on the density functional theory.

**Keywords:** magneto-structure, exchange coupling, chelating ligand, metal–radical complex

## 1. Introduction

Varied stable organic radicals have been devolved along with the prosperity of the field of molecule-based magnetic and conductive materials, and have remained attracting much attention to the researchers in materials chemistry [1]. In 1989, Gatteschi *et al.* exposited a synthetic strategy, "metal–radical approach" [2], toward molecule-based magnets by using appreciable magnetic exchange interaction between paramagnetic metal ions and directly-coordinating radical ligands, which has been recognized as one of the most fruitful approaches for developing coveted magnets. Actually, there are a few room-temperature molecular magnets reported up to the present, which consists of a transition-metal cation and an organic radical anion [3,4], except for a Prussian blue analog [5]. To achieve a molecule-based magnet, the enhancement of exchange coupling constants, $J$ (defined as $-2J\Sigma S_i \cdot S_{i+1}$), is indispensable and it correlates well with overlapping among magnetic orbitals. Therefore, the exchange interactions have investigated both experimentally and theoretically from the viewpoint of molecular structures [6].

Various 2-pyridyl derivatives substituted with a stable radical such as nitronyl nitroxide (NN) [7,8], imino nitroxide (IN) [9,10], oxoverdazyl (VD) [11-13], and dithiadiazolyl (DTDA) [14] groups have been developed, which are capable of chelating a metal ion to form metal–radical magnetic systems.

We have proposed *tert*-butyl 2-pyridyl nitroxide (2pyNO; NO stand for the *tert*-butyl nitroxide group) as a candidate of magnetic couplers promising to interact strongly with a chelated metal center. The NO derivatives have some advantage on magnetic properties compared with other stable radicals. The spin density of the ligating oxygen atom is assumed to be almost twice as large as that of the others, because the spin-delocalized area on NO is approximately a half of those of other conventional radicals as shown in Figure 1. Since we have developed the first *tert*-butyl 2-pyridyl nitroxides, e.g., 2pyNO [15], 6bpyNO [16], phpyNO [17], and so on [18,19], thanks to the stabilization by the complexation or the introduction of bulky substituents, the number of the reports on metal–radical complexes has been increased by many other researchers (see below). It would be meaningful if we revisit the magneto-structural relationship by adding new data to the proposed correlation plot.

We illustrate here magneto-structural relationship on dσ–π* exchange coupling constants for transition metal compounds coordinated with a paramagnetic chelating ligand, 2pyNO, its derivatives, and other related stable radicals. Some compounds of them show extremely strong exchange coupling beyond a thermal energy of room temperature ($2J > 300$ K).

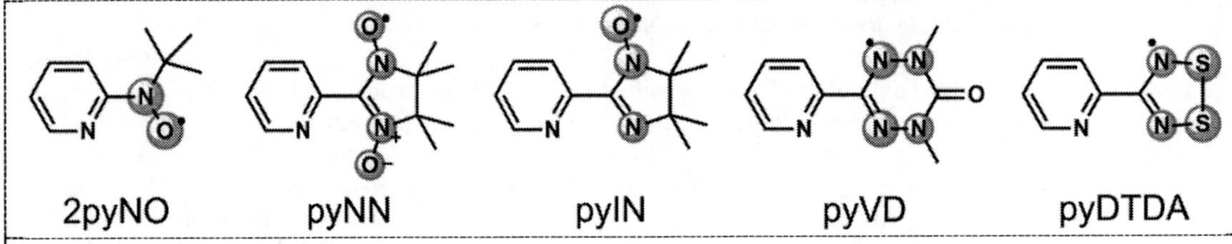


**Figure 1.** Structural formulas of 2-pyridyl-substituted stable radicals with schematic spin density distributions superimposed.

## 2. Magneto-Structural Relationship

We have focused our attention on the complexation with transition metal ions having dσ-type magnetic orbitals, *e.g.*, nickel(II) and copper(II) ions, because 2pyNO and its derivatives have a π*-character single-occupied molecular orbital (SOMO). Their complexes are anticipated to be arranged in the orthogonal manner between metal and radical magnetic orbitals, giving a ferromagnetic coupling. The six-coordinate octahedral geometry is the most common arrangement of metal complexes, especially for first-row transition metal ions. A $d^9$copper(II) ion frequently has an axial-elongated octahedral geometry due to the Jahn-Teller effect, where a magnetic orbital is well defined as $3d_{x2-y2}$. A $d^8$nickel(II) ion in an octahedral coordination sphere has two electron spins in SOMOs described as the $3d_{z2}$ orbital in addition to as the $3d_{x2-y2}$ one. These orbitals are of a dσ-character.

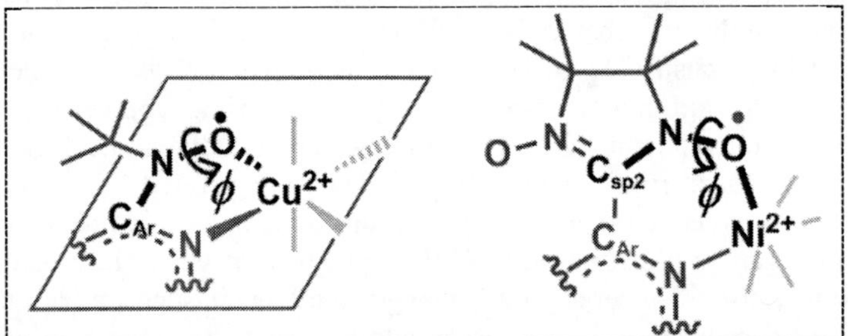

**Figure 2.** Schematic partial molecular structures of (left) the copper(II)–NO and (right) the nicke(II)–NN systems which have a six-coordinated geometry for the representative data listed in Tables 1 and 2. Each $\phi$ stands for the dihedral angle defined as Cu–O–N–$C_{Ar}$ for copper(II)–NO or Ni–O–N–$C_{sp2}$ for nickel(II)–NN systems.

**Table 1.** Exchange coupling and geometrical parameters on copper(II)–radical complexes.

| Compounds[a] | $2Jk_B^{-1}$/K [b] | $\lvert\phi\rvert$/° | Cu–X/Å [c] | Cu–X–Y/° [d] | Reference[e] |
|---|---|---|---|---|---|
| ***tert-Butyl nitroxides*** **(NO)** | | | | | |
| $[Cu(phpyNO)_2(H_2O)](ClO_4)_2{\cdot}C_6H_4Cl_2$ | +434(12) | 1.6(2) | 1.9481(13) | 114.66(11) | ref 17 |
| $[Cu(phpyNO)_2(ClO_4)_2]{\cdot}2CH_2Cl_2$ | +276(4) | 4.2(8) | 1.934(3) | 113.2(2) | ref 17 |
| $[Cu(meopyNO)_2(H_2O)_2](ClO_4)_2$ | +178(3) | 6.9(2) | 1.9688(13) | 113.71(10) | ref 19 |
| $[Cu(2pyNO)_2(ClO_4)_2]$ | −548(4) | 25.9(5) | 1.942(4) | 111.8(2) | ref 15 |
| $[Cu(phpyNO)_2(CF_3SO_3)_2]{\cdot}0.35H_2O$ | −940(30) | 33.4(2) | 1.9491(18) | 110.32(14) | ref 17 |
| ***Nitronyl nitroxides*** **(NN)** | | | | | |
| $[Cu(4ImNNH)(NO_3)_2]$ | −58(4) | 15.6(4) | 1.986(3) | 124.7(2) | ref 28 |
| $[Cu(4ImNNH)_2Br_2]$ | −372(5) | 34.4(6) | 2.032(3) | 119.4(3) | ref 28 |
| $[Cu(bpybNN)(ClO_4)_2]$ | −480 | 35.3(8) [f] | 1.908(6) [f] | 117.8(1) [f] | ref 29 |
| $[Cu(4MeImNNH)_2(NO_3)_2]$ | −443 | 39.6 | 1.975(4) | 118.9(3) | ref 30 |
| $[Cu(2,6\text{-}pybNN)_2](ClO_4)_2{\cdot}CH_3CN$ [g] | −920 | 47(7) [f] | 1.955(1) [f] | 112.2(12) [f] | ref 31 |
| $[Cu(4MeImNNH)(hfac)_2]$ | −1036 | 46.7 | 1.955(3) | 118.5(3) | ref 28 |
| $[\{Cu(hfac)_2\}_3(BrpyNN)_2]$ | −1266 | 54.2 | 1.9522(15) | 111.25(12) | ref 32 |
| $[Cu(phtrzNN)(hfac)_2]$ | −934 | 61.9(4) | 1.977(3) | 121.6(2) | ref 33 |
| ***Imino nitroxides*** **(IN)** | | | | | |
| $[Cu(pyIN)(hfac)_2]$ | +420 | 1.0 | 2.003(4) | 112.7 | ref 24 |

[a] See References for abbreviated ligands.
[b] The $2J$ is the experimental value itself, not normalized.
[c] The Cu–O and Cu–N bond lengths for NO/NN and IN, respectively.
[d] The Cu–O–N and Cu–N–$C_{sp2}$ angles for NO/NN and IN, respectively.
[e] See References in the main text.
[f] Average on the geometries of crystallographically independent units.
[g] Only the data on equatorially coordinated moieties are used.

For copper(II)–nitroxide complexes, it has frequently been mentioned that an equatorially-coordinated pair results in a strong antiferromagnetic metal–radical exchange coupling ($J \ll 0$), while an axially-coordinated in a moderate ferromagnetic coupling ($J > 0$) [2,21-23]. However, Luneau and co-workers reported obviously ferromagnetic couplings in the equatorial arrangements of copper(II)–nitroxide systems [24,25], and we have also found varied ferromagnetic copper(II)–nitroxide compounds [17,19,26,27]. Moreover, we formulated the magneto-structural relationship with respect to the orbital overlap between d$\sigma$ and $\pi^*$ magnetic orbitals on metal–radical complexes [17, 19, 26], which contain an NO or NN radical. The exchange coupling of d$\sigma$–$\pi^*$ spins range extensively in proportion to the change of a simple and convenient geometrical parameter, $\lvert\phi\rvert$, indicating the dihedral angle between the M–O–N (correlated with the $d_{x2-y2}$ magnetic orbital) and O–N-$\pi^*$-conjugated (correlated with the $\pi^*$ magnetic orbital) planes [17,19], where M represents Ni or Cu. The parameter, $\phi$, is here defined as the torsion angle around M–O–N–$C_{Ar}$($C_{sp2}$).

Here we updated the plot of $2J$ versus $\lvert\phi\rvert$ for the reported nickel(II) and copper(II) complexes of not only NO and NN but also other stable radicals such as IN, VD, and DTDA. Among a huge amount of radical and copper(II)/nickel(II) complexes reported, we picked up discrete hexacoordinate copper(II) and nickel(II) complexes [15-17,19,28-50] and listed in Tables 1 and 2, respectively, which are chelated by radical ligands at an heteroatom on radical moiety and at the N atom on the substituent heteroaromatic ring as shown Figure 2. Although a five-coordinate geometry around the copper(II) ion was found in some related compounds such as $[Cu(6bpyNO)Cl_2]$ [16], its data point did not obey the present relationship due to distorted basal plane where the magnetic orbital $3d_{x2-y2}$ is hardly defined to be located strictly on the $xy$ plane. As far as we investigated, there is no report on copper(II)–VD and –DTDA systems where the radical ligand occupies the equatorial sites, while some axially-coordinated complexes were found [13,51,52].

**Table 2.** Exchange coupling and geometrical parameters on nickel(II)–radical complexes.

| Compounds[a] | $2Jk_B^{-1}$/K [b] | $\lvert\phi\rvert$/° | Ni–X/Å [c] | Ni–X–Y/° [d] | Reference[e] |
|---|---|---|---|---|---|
| ***tert-Butyl nitroxides* (NO)** | | | | | |
| $[Ni(phpyNO)_2(H_2O)_2](ClO_4)_2$ | +409(10) | 1.5(3) | 1.981(2) | 114.56(18) | ref 17 |
| $[Ni(2pyNO)_2(H_2O)_2](ClO_4)_2$ | +252(6) | 3.9(3) | 2.008(2) | 115.6(2) | ref 15 |
| $[Ni(meopyNO)_2(H_2O)_2](ClO_4)_2$ | +288(5) | 4.2(3) | 2.020(2) | 114.27(15) | ref 19 |
| $[Ni(6bpyNO)_2](PF_6)_2$ | +384(4) | 5.0(5)[f] | 2.095(6)[f] | 114.1(3)[f] | ref 16 |
| ***Nitronyl nitroxides* (NN)** | | | | | |
| $[Ni(4ImNNH)_2(NO_3)_2]$ | +85(3) | 6.3(4) | 2.038(2) | 125.6(2) | ref 28 |
| $[Ni(bpybNN)(H_2O)(ClO_4)]ClO_4{\cdot}CH_3CN$ | +80.0 | 8.1 | 1.986 | 123.5 | ref 29 |
| | +19.9 | 14.0 | 1.993 | 122.1 | |
| $[Ni(thzNN)_3](ClO_4)_2$ | −204 | 34(7)[f] | 2.051(5) | 119(3)[f] | ref 34 |
| $[Ni_3(pzNN)_6]{\cdot}4CH_3CN$ | −80(2) | 35.0(6)[f] | 2.126(16)[f] | 117.8(11)[f] | ref 35 |
| $[Ni(2ImNNH)_3](ClO_4)_2$ | −319(20) | 38(9)[f] | 2.04(2)[f] | 118(3)[f] | ref 36 |
| $[Ni(bztfac)(temen)(pyNN)]PF_6$ | −287 | 39.5 | 2.056(2) | 117.8(2) | ref 37 |
| $[Ni(bzac)(temen)(pyNN)]PF_6$ | −317 | 39.8 | 2.055(4) | 117.9(3) | ref 37 |
| $[Ni(acac)(temen)(pyNN)]PF_6{\cdot}{}^1/_3CH_2Cl_2$ | −429 | 40.8 | 2.052(4) | 118.9(3) | ref 37 |
| $[Ni(pyNN)_2Cl_2]{\cdot}2CH_2Cl_2$ | −316(32) | 41.0 | 2.064(1) | 116.3(1) | ref 38 |
| $[Ni(MeBzImNN)_2(NO_3)(CH_3OH)]NO_3{\cdot}CH_3OH$ | −119 | 42(6)[f] | 2.07(3)[f] | 118(3)[f] | ref 39 |
| $[Ni(pyNN)(NTB)](ClO_4)_2{\cdot}CH_3OH$ | −55.9 | 42.7(7) | 2.098(4) | 114.1(3) | ref 40 |
| $[Ni(pyNN)(hfac)_2]$ | −480(17) | 43.0 | 2.029(2) | 115.4(2) | ref 38 |
| $[Ni(picNN)_2]{\cdot}3H_2O$ | −555 | 47.3 | 2.0842(14) | 114.78(11) | ref 41 |
| $[Ni(2,6\text{-}pyNN_2)_2](ClO_4)_2{\cdot}0.5H_2O{\cdot}0.5CH_3OH$ | −480 | 48(2)[f] | 2.030(7)[f] | 111.4(10)[f] | ref 31 |
| ***Imino nitroxides* (IN)** | | | | | |
| $[Ni(pyIN)(hfac)_2]$ | +273(11) | 0.7(3) | 2.050(5) | 113.8(4) | ref 25 |
| $[NiCl(2,6\text{-}pybIN)(H_2O)_2]Cl{\cdot}2H_2O$ | +345(6) | 1.6(2)[f] | 2.137(11)[f] | 112.3(2)[f] | ref 42 |
| $[Ni(NCS)_2(mepyIN)_2]$ | +193(4) | 5(4)[f] | 2.124(9)[f] | 112.8(2)[f] | ref 43 |
| $[Ni(acac)_2(pyIN)]$ | +119(6) | 12.0(4) | 2.151(3) | 111.4 | ref 44 |
| ***Oxoverdazyls* (VD)** | | | | | |
| $[Ni(bpyVD)_2](PF_6)_2{\cdot}C_3H_6O$ | +345(11) | 1.5(4)[f] | 2.211(2)[f] | 112.1(4)[f] | refs 13,45 |
| $[Ni(35dipyVD)_2](PF_6)_2$ | +432 | 3.7(2)[f] | 1.979(5)[f] | 120.3(3)[f] | ref 46 |
| $[Ni(hfac)_2(2ImVD)]$ | +278 | 6.4 | 2.229(3) | 111.8(2) | ref 47 |
| ***Dithiadiazolyls* (DTDA)** | | | | | |
| $[Ni(hfac)_2(pyDTDA)]$ | +264(2) | 1.8 | 2.082(15) | 116.5(13) | ref 48 |
| $[\{Ni(hfac)_2\}_2(pmDTDA)]$ | +130(4) | 2.6(19)[f] | 2.143(4)[f] | 111.1(3)[f] | ref 49 |
| $[Ni(hfac)_2(boaDTDA)]$ | +182(4) | 4.8(6) | 2.149(4) | 114.2(3) | ref 50 |

[a] See references for abbreviated ligands.
[b] The $2J$ is the experimental value itself, not normalized.
[c] The Ni–O and Ni–N bond lengths for NO/NN and IN/VD/DTDA, respectively.
[d] The Ni–O–N and Ni–N–$C_{sp2}$ angles for NO/NN and IN/VD/DTDA, respectively.
[e] See references in the main text.
[f] Average on the geometries of crystallographically independent units.

Figure 3 shows the magneto-structural correlation plot, when the normalized exchange parameter $J$ is plotted against the dihedral angle $\lvert\phi\rvert$. Each $J$ value was normalized based on the spin density at the ligating radical atom, because the exchange coupling is generally proportional to the spin density [53]. For the NO and NN systems, the spin density is assumed to be in proportion to the $a_N$ value determined in electron spin resonance (ESR) spectroscopy [54-56] ($a_N$ = 1.02 mT for 2pyNO [17] and 0.743 mT for PhNN [56]). The spin densities ofthe NN and IN have been directly measured by means of polarized neutron diffraction study (0.277(13) for the oxygen atom in phenyl NN [57] and 0.236(7) for the imino nitrogen atom in *m*-nitrophenyl IN [58]). Based on the spin density of NO, we applied the normalization factors of 1.4 and 1.6 to the plots of the NN and IN systems, respectively.

We can empirically find a plausible correlation (a dashed line) for Cu complexes. Using a simple linear formula, $2J/k_B = a + b|\phi|$, afforded $a$ = 493(110) K and $b$ = –36(3) K $deg^{-1}$. The critical $|\phi|$, which is the border between ferromagnetic and antiferromagnetic couplings, is 14(2)°. If limited among 2pyNO-type compounds, an excellent correlation was displayed, where $a$ = 476(27) K, $b$ = –41.4(14) K $deg^{-1}$, and the critical $|\phi|$ is 11.5(4)°.

For Ni complexes, the data widely vary compared with those of copper(II) complexes, and the slope was relatively small (Figure 3b). In this plot we confirm that the relationship can be generalized for the oxoverdazyl and dithiadiazolyl systems. The normalization factors of VD and DTDA are 1.6 and 2.0, respectively, which were estimated from the ESR hyperfine coupling constants ($a_N$ = 0.64 mT for pyVD [11] and 0.502 mT for pyDTDA [59]). As a result, the critical $|\phi|$ shifts to be 21.0(14)° and apparently larger than that of copper(II) compounds. The correlation can be roughly approximated by using $2J/k_B = a + b|\phi|$ with $a$ = 464(44) K and $b$ = –22.0(16) K $deg^{-1}$.

The nickel(II) ion has two paramagnetic spins at $3d_{z2}$ in addition to $3d_{x2-y2}$. The $d_{x2-y2}$–$\pi^*$ interaction mainly contributes the observed $J$, viewing from the overlapping and mutual lobe direction between interacting SOMO's. However, it is most likely that the interaction of $d_{z2}$–$\pi^*$ somewhat contributes to the entire exchange coupling and is insensitive to the geometrical change represented as $\phi$ from the aforementioned reason.

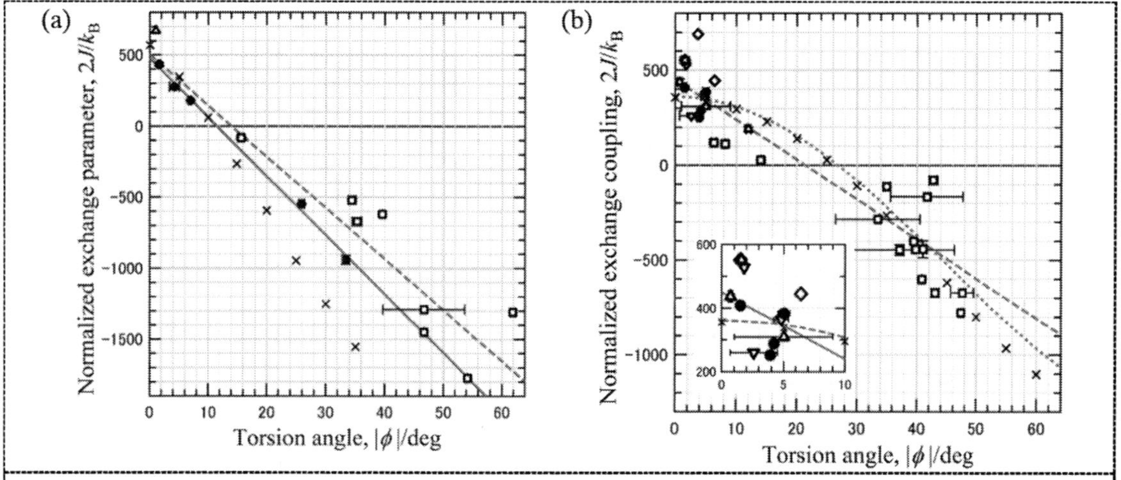


**Figure 3.** Plot of the observed exchange coupling parameter ($2J$) versus M–O–N–$C_{2Ar}$ torsion angle ($|\phi|$) as a magneto-structural relationship in the six-coordinate (a) copper(II) and (b) nickel(II) complexes with radical chelating ligands. Filled circles stand for the data points on NO, open squares NN, triangles IN, diamonds VD, and inverted triangles DTDA. The $2J$ values except for NO are normalized by a factor according to the spin density at the ligating atom (see text). The dashed line stands for a linear fit for all the data herein and the solid line for those limited among NO. The DFT calculation results (see ref 17 for copper(II) systems) for each geometry are superimposed as the cross marks. The dotted line represent the best fit for the data by using a cos $2\phi$ function.

## 3. DFT Calculations

To estimate a magneto-structural correlation, density functional theory (DFT) calculations were carried out on the Gaussian09 program [60]. for the model nickel(II)–radical diad, $[Ni(2pyNO)(H_2O)_3(NH_3)]$, capped with $H_2O$ and $NH_3$ molecules. The peripheral substituents were simplified because of reducing computational cost. The DFT study for copper(II)–radical system have already been reported in the previous paper [17]. The molecular geometries of nickel(II) complexes were fully optimized on the unrestricted B3LYP functional [61,62] with the 6-311G(d,p) basis set except for the constraint of the Cu–O–N–$C_{2py}$ dihedral angle ($\phi$). The calculation points were set from $\phi$ = 0 to 60° with an interval of 5°. Exchange coupling constants ($J$) were evaluated from the calculated energies ($E_{HS}$ and $E_{LS}$) of the high-spin quintet and low-spin doublet states, according to the following equation by using of the spin-projected method proposed by Yamaguchi *et al* [63].

$$J = \frac{E_{LS} - E_{HS}}{\langle S^2 \rangle_{HS} - \langle S^2 \rangle_{LS}}$$

We successfully terminated the optimization and calculated the energies of unrestricted quartet and doublet spin states. In the range of $\phi$ = 0° to 60°, the Ni–O bond length ranges from 2.007 to 2.085 Å and the Ni–O–N angle from 117.4° to 103.4°, respectively. The coordination distance is most likely to affect the metal–radical exchange coupling; however, their variation on the optimized structures fit within the experimental range (see Table 1). The $2J$ value obtained by DFT calculations is superimposed in Figure 3b as cross marks. They monotonically decreased with an increase of $\phi$, which is in good agreement with a tendency on the experimental results. The critical angle of $|\phi|$ is estimated to be 26.0°.

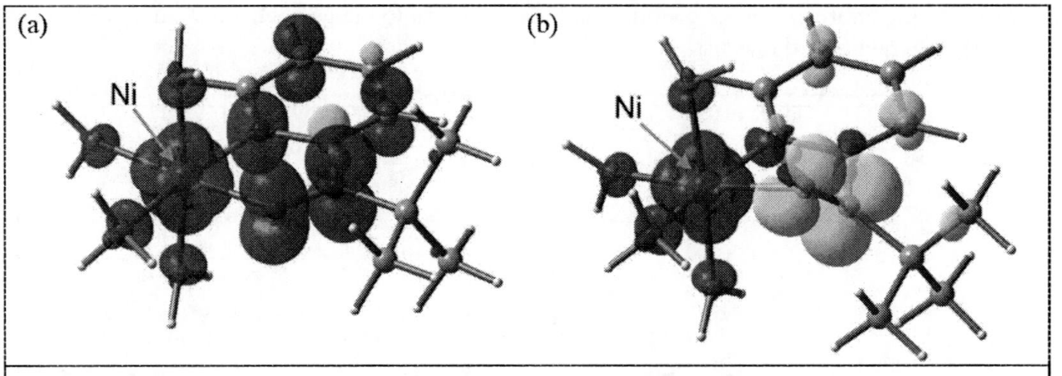


**Figure 4.** Spin density surfaces for (a) the ground quartet state of $\phi$ = 0° and (b) the doublet state at $\phi$ = 60° of [Ni(2pyNO)($H_2O$)$_3$($NH_3$)]. They were calculated on the UB3LYP/6-311G(d,p) level. Dark gray, positive spin; light gray, negative spin.

Figure 4 show the spin density distributions for the ground quartet state of the model molecule with $\phi$ = 0° and for the ground doublet one at $\phi$ = 60°, respectively. They are qualitatively represented as the SOMOs and we can confirm the σ-character made by combining $d_{x2-y2}$- and $d_{z2}$-type SOMOs on the nickel atom and the π-character derived from the pπ* nitroxide radical spin. Their orthogonal arrangement is found in the $\phi$ = 0° configuration, leading to a strong ferromagnetic coupling. In contrast, the distorted structure of $\phi$ = 60° indicates the SOMO–SOMO overlap giving a strong antiferromagnetic interaction.

On the nickel(II) complexes, the $2J$ value seems to obey cos $2\phi$ (or $\cos^2\phi$) dependence [64], in contrast to the calculation result reported for a copper(II)–nitroxidediad where the calculation curve does not exactly obey the cos $2\phi$ law. When the correlation is formulated by using a cos $2\phi$ function, $2J/k_B = A\cos(2|\phi|) + B$, we obtained $A$ = 939(76) K and $B$ = –566(58) K, and the critical $|\phi|$ is 26(3)°. Some reports claimed that DFT-calculated exchange parameters tend to be overestimated [65,66], and other reports mentioned that configuration interaction methods often afford a better result due to a credible contribution originated from the noninnocent nature of a radical ligand [67,68]. The relationship introduced in this work is out of consideration of their problems; however, we can obtain a semiquantitative plot of $2J$ versus $\phi$. It is concluded that the simple structural parameter represented as M–O–N–$C_{2py}$ is a convenient and useful metric for the exchange coupling constant between copper(II)/nickel(II) and radical spins on metal–radical approach.

## 4. Conclusion

We here introduced the magneto-structural relationship for nickel(II)– and copper(II)–radical complexes. The linear correlation is formulated as $2J/k_B = a + b|\phi|$ with $a$ = 464(44) K and $b$ = –22.0(16)

K deg$^{-1}$ for the Ni complexes and with $a$ = 493(110) K and $b$ = −36(3) K deg$^{-1}$ for the Cu complexes. If limited among 2pyNO-type compounds, the 2$J$ value is excellently correlated with the $\phi$ angle, where $a$ = 476(27) K, $b$ = −41.4(14) K deg$^{-1}$, and the critical $|\phi|$ is 11.5(4)°. On the nickel(II) complexes, the exchange coupling $J$ is relatively obtuse in decreasing itself according to the increment of $\phi$. This result may be explained as follows. The nickel(II) ion has two paramagnetic spins at $3d_{z^2}$ in addition to $3d_{x2\text{-}y2}$. Viewing from the overlapping and mutual lobe direction between interacting SOMO's, the interaction of $d_{z2}$–$\pi^*$ is insensitive to the geometrical change represented as $\phi$.

A useful semiquantitative correlation between 2$J$ and $\phi$ was obtained, and it also holds for the recently reported complexes including other paramagnetic ligand systems, VD and DTDA. Furthermore, this type of correlation scheme has been successfully applied to lanthanoid–nitroxide systems, though the slope is considerably different from that of 3d-nitroxide systems [69]. It is emphatically stressed that the simple structural parameter represented as M–O–N–$C_{2py}$ is a convenient and useful metric for the exchange coupling constant between copper(II)/nickel(II) and radical spins on metal–radical approach.

## 5. References

[1] Hicks R G ed 2010 *Stable Radicals: Fundamentals and Applied Aspects of Odd-Electron Compounds* (Chichester: Wiley)
[2] Caneschi A, Gatteschi D, Sessoli R and Rey P 1989 *Acc. Chem. Res.* **22** 392
[3] Manriquez J M, Yee G T, McLean R S, Epstein A J and Miller J S 1991 *Science* **252** 1415
[4] Jain R, Kabir K, Gilroy J B, Mitchell K A R, Wong K-C and Hicks R G 2007 *Nature* **445** 291
[5] Ferlay S, Mallah T, Ouahes R, Veillet P and Verdaguer M 1995 *Nature* **378** 701
[6] Kahn O 1993 *Molecular Magnetism* (Yew York: VCH)
[7] Davis M S, Morokuma K and Kreilick R W 1972 *J. Am. Chem. Soc.* **94** 5588
[8] Ulrich G, Ziessel R, Luneau D and Rey P 1994 *Tetrahedron Lett.* **35** 1211
[9] Ulrich G and Ziessel R 1994 *Tetrahedron Lett.* **35** 1215
[10] Helbert J N, Kopf P W, Poindexter E H and Wagner B E 1975 *J. Chem. Soc., Dalton Trans.* 998
[11] Brook D J, Fornell S, Stevens J E, Noll B, Koch T H and Eisfeld W 2000 *Inorg. Chem.* **39** 562
[12] Barr C L, Chase P A, Hicks R G, Lemaire M T and Stevens C L 1999 *J. Org. Chem.* **64** 8893
[13] Barclay T M, Hicks R G, Lemaire M T and Tompson L K 2003 *Inorg. Chem.* **42** 2261
[14] Hearns N G R, Preuss K E, Richardson J F and Bin-Salamon S 2004 *J. Am. Chem. Soc.* **126** 9942
[15] Okazawa A, Nogami T and Ishida T 2007 *Chem. Mater.* **19** 2733
[16] Osanai K, Okazawa A, Nogami T and Ishida T 2006 *J. Am. Chem. Soc.* **128** 14008
[17] Okazawa A, Nagaichi Y, Nogami T and Ishida T 2008 *Inorg. Chem.* **47** 8859
[18] Okazawa A, Nogami T and Ishida T 2007 *Polyhedron* **26** 1965
[19] Okazawa A, Nogami T and Ishida T 2009 *Polyhedron* **28** 1917
[20] Caneschi A, Gatteschi D and Rey P 1991 *Progr. Inorg. Chem.* **39** 331
[21] Caneschi A, Gatteschi D, Grand A, Laugier J, Pardi L and Rey P 1988 *Inorg. Chem.* **27** 1031
[22] Gatteschi D, Laugier J, Rey P and Zanchini C 1997 *Inorg. Chem.* **26** 938
[23] Bencini A, Benelli C, Gatteschi D and Zanchini C 1984 *J. Am. Chem. Soc.* **106** 5813
[24] Luneau D, Rey P, Laugier J, Fries P, Caneschi A, Gatteschi D and Sessoli R 1991 *J. Am. Chem. Soc.* **113** 1245
[25] Luneau D, Rey P, Laugier J, Belorizky E and Conge A 1992 *Inorg. Chem.* **31** 3578
[26] Okazawa A, Hashizume D and Ishida T 2010 *J. Am. Chem. Soc.* **132** 11516
[27] Okazawa A and Ishida T 2010 *Inorg. Chem.* **49** 10144
[28] Aoki C, Ishida T and Nogami T 2003 *Inorg. Chem.* **42** 7616
[29] Luneau D, Romero F M and Ziessel R 1998 *Inorg. Chem.* **37** 5078
[30] Fursova E, Romanenko G, Ikorskii V and Ovcharenko V 2003 *Polyhedron* **22** 1957
[31] Francese G, Romero F M, Neels A, Stoeckli-Evans H and Decurtins S 2000 *Inorg. Chem.* **39** 2087
[32] Rajadurai C, Ostrovsky S, Falk K, Enkelmann V, Haase W and Baumgarten M 2004 *Inorg.*

*Chim. Acta* **357** 581
[33] Zhang Y-J, Wang J-J, Huang Q and Chen J 2012 *Transition Met. Chem.* **37** 743
[34] Wang L-Y, Ma L-F, Jiang Z-H, Lio D-Z and Yan S-P 2005 *Inorg. Chim. Acta* **358** 820
[35] Yamada S, Yasui M, Nogami T and Ishida T 2006 *Dalton Trans.* 1622
[36] Fegy K, Sanz N, Luneau D, Belorizky E and Rey P 1998 *Inorg. Chem.* **37** 4518
[37] Yoshida T, Suzuki T, Kanamori K and Kaizaki S 1999 *Inorg. Chem.* **38** 1059
[38] Luneau D, Risoan G, Rey P, Grand A, Caneschi A, Gatteschi D and Laugier J 1993 *Inorg. Chem.* **32** 5616
[39] Wang L-Y, Sun X-Y, Yang R-H, Jian K and Wang Y-F 2010 *J. Mol. Struct.* **964** 97
[40] Zhang C-X, Zhang Y-Y, Zhang H-L, Zhao Y and Sun Y-Q 2009 *Inorg. Chim. Acta* **362** 5231
[41] Coronado E, Giménez-Saiz C, Romero F M and Tarazón A 2009 *Inorg. Chem.* **48** 2205
[42] Oshio H, Yamamoto M, Kawauchi H, Koga N and Ito T 2001 *J. Solid State Chem.***159** 455
[43] Yamamoto Y, Suzuki T and Kaizaki S 2001 *J. Chem. Soc., Dalton Trans.* 2943
[44] Tsukahara Y, Kamatani T, Iino A, Suzuki T and Kaizaki S 2002 *Inorg. Chem.* **41** 4363
[45] Barclay T M, Hicks R G, Lemaire M T and Thompson L K 2000 *Chem. Commun.* 2141
[46] Brook D J R, Richardson C J, Haller B C, Hundley M and Yee G T 2010 *Chem. Commun.* **46** 6590
[47] Norel L, Pointillart F, Train C, Charmoreau L-M, Boubekeur K, Journaux Y, Brieger A and Brook D J R 2008 *Inorg. Chem.* **47** 2396
[48] Hearns N G R, Fatila E M, Clérac R, Jennings M and Preuss K E 2008 *Inorg. Chem.* **47** 10330
[49] Wu J, MacDonald D J, Clérac R, Jeon I-R, Jennings M, Lough A J, Britten J, Robertson C, Dube P A and Preuss K E 2012 *Inorg. Chem.* **51** 3827
[50] Fatila E M, Clérac R, Jennings M and Preuss K E 2013 *Chem. Commun.* **49** 9431
[51] Britten J, Hearns N G R, Preuss K E, Richardson J F andBin-Salamon S 2007 *Inorg. Chem.* **46** 3934
[52] Gilroy J B, Koivisto B D, McDonald R, Fergusonb M J and Hicks R G 2006 *J. Mater. Chem.* **16** 2618
[53] McConnell H M 1963 *J. Chem. Phys.* **39** 1910
[54] Forrester A R, Hepburn S P and McConnachie G 1974 *J. Chem. Soc., Perkin Trans. I* 2213
[55] Goldman J, Petersen T E, Torssell K and Becher J 1973 *Tetrahedron* **29** 3833
[56] Ullman E F, Osiecki J H, Boocock D G B and Darcy R 1972 *J. Am. Chem. Soc.* **94** 7049
[57] Zheludev A, Barone V, Bonnet M, DeUey B, Grand A, Ressouche E, Rey P, Subra R and Schweizer J 1994 *J. Am. Chem. Soc.* **116** 2019
[58] Zheludev A, Bonnet M, Delley B, Grand A, Luneau D, Örström L, Ressouche E, Rey R and Schweizer J 1995 *J. Magn. Magn. Mater.* **145** 293
[59] Hearns N G, Preuss K E, Richardson J F and Bin-Salamon S 2004 *J. Am. Chem. Soc.* **126** 9942.
[60] Frisch M J, Trucks G W, Schlegel H B, Scuseria G E, Robb M A, Cheeseman J R, Scalmani G, Barone V, Mennucci B, Petersson G A, Nakatsuji H, Caricato M, Li X, Hratchian H P, Izmaylov A F, Bloino J, Zheng G, Sonnenberg J L, Hada M, Ehara M, Toyota K, Fukuda R, Hasegawa J, Ishida M, Nakajima T, Honda Y, Kitao O, Nakai H, Vreven T, Montgomery J A Jr, Peralta J E, Ogliaro F, Bearpark M, Heyd J J, Brothers E, Kudin K N, Staroverov V N, Kobayashi R, Normand J, Raghavachari K, Rendell A, Burant J C, Iyengar S S, Tomasi J, Cossi M, Rega N, Millam J M, Klene M, Knox J E, Cross J B, Bakken V, Adamo C, Jaramillo J, Gomperts R, Stratmann R E, Yazyev O, Austin A J, Cammi R, Pomelli C, Ochterski J W, Martin R L, Morokuma K, Zakrzewski V G, Voth G A, Salvador P, Dannenberg J J, Dapprich S, Daniels A D, Farkas Ö, Foresman J B, Ortiz J V, Cioslowski J and Fox D J 2009 *Gaussian 09, Revision B.01* (Wallingford CT : Gaussian, Inc.)
[61] Becke A D 1993 *J. Chem. Phys.* **98** 5648
[62] Lee C, Yang W, Parr R G 1998 *Phys. Rev. B* **37** 785
[63] Soda T, Kitagawa Y, Onishi T, Takano Y, Shigeta Y, Nagao H, Yoshioka Y and Yamaguchi K 2000 *Chem. Phys. Lett.* **319** 223
[64] Venkataraman L, Klare J E, Nuckolls C, Hybertsen M S and Steigerwald M L 2006 *Nature* **442** 904

[65] Dai D, Whangbo M-H, Koo H-J, Rocquefelte X, Jobic S and Villesuzanne A 2005 *Inorg. Chem.* **44** 2407
[66] Grau-Crespo R, de Leeuw N H and Catlow C R 2003 *J. Mater. Chem.* **13** 2848
[67] Rota J-B, Norel L, Train C, Amor N B, Maynau D and Robert V 2008 *J. Am. Chem. Soc.* **130** 10380
[68] Rota J-B, Calzado C J, Train C and Robert V 2010 *J. Chem. Phys.* **132** 154702
[69] Kanetomo T, Yoshitake T and Ishida T 2016 *Inorg. Chem.* **55** 8140

**Acknowledgments**
This work was partly supported by JSPS KAKENHI Grant Number JP15K17841. The author is thankful to Prof. Takayuki Ishida (The University of Electro-Communications) for helpful discussions.

# Fabrication of Mesoporous Silica/Alumina Hybrid Membrane Film Nanocomposites using Template Sol-Gel Synthesis of Amphiphilic Triphenylene

**H O Lintang[1,2], M A Jalani[3], L Yuliati[1,2] and M M Salleh[2]**

[1]Ma Chung Research Center for Photosynthetic Pigments, Universitas Ma Chung, Malang 65151, East Java, Indonesia
[2]Centre for Sustainable Nanomaterials, Ibnu Sina Institute for Scientific and Industrial Research, Universiti Teknologi Malaysia, 81310 UTM, Johor Bahru, Johor, Malaysia
[3]Department of Chemistry, Faculty of Science, Universiti Teknologi Malaysia, 81310 UTM, Johor Bahru, Johor, Malaysia

Email: hendrik.lintang@machung.ac.id

**Abstract.** Herein we reported that by introducing a one-dimensional (1D) substrate with a porous structure such as anodic aluminum oxide (AAO) membrane, mesoporous silica/alumina hybrid nanocomposites were successfully fabricated by using amphiphilic triphenylene ($TP_{C10TEG}$) as a template in sol-gel synthesis ($TP_{C10TEG}$/silica$_{hex}$). For the optical study of the nanocomposites, $TP_{C10TEG}$/silica$_{hex}$ showed absorption peak at 264 nm due to the ordered and long-range $\pi$-$\pi$ stacking of the disc-like aromatic triphenylene core. Moreover, the hexagonal arrangement of $TP_{C10TEG}$/silica$_{hex}$ was proven based on their diffraction peaks of $d_{100}$ and $d_{200}$ at $2\theta = 2.52°$ and 5.04° and images of transmission electron microscopy (TEM), respectively. For fabrication of mesoporous silica/alumina hybrid membrane, $TP_{C10TEG}$/silica$_{hex}$ was drop-casted onto AAO membrane for penetration into the porous structure via gravity. X-ray diffraction (XRD) analysis on the resulted hybrid nanocomposites showed that the diffraction peaks of $d_{100}$ and $d_{200}$ of $TP_{C10TEG}$/silica$_{hex}$ were still preserved, indicating that the hexagonal arrangements of mesoporous silica were maintained even on AAO substrate. The morphology study on the hybrid nanocomposites using TEM, scanning electron microscope (SEM) and field emission scanning electron microscope (FE-SEM) showed the successful filling of most AAO channels with the $TP_{C10TEG}$/silica$_{hex}$ nanocomposites.

**Keywords:** silica, alumina hybrid thin film, nanocomposite membrane, sol-gel, amphiphilic triphenylene.

## 1. Introduction

Ordered mesoporous silica (OMS) materials have gained much attention for the last few decades. In the early 1990s, Kuroda and his co-workers pioneered the synthesis of mesoporous silicate called FSM (Folded Sheet Materials)-16 [1] and then followed by the report of the mesoporous M41S family such as MCM (Mobil Composition of Matter)-41 by Kresge *et al.* in 1992 [2]. In particular, mesoporous silica is an ordered and well-defined pore structure with large surface areas, good thermal stability, and narrow pore size distribution [3]. Generally, mesoporous silica has been synthesized through silica formation catalyzed by either acid or base medium with the presence of amphiphilic template surfactants. Since most of OMS materials were produced in the powdered form [4, 5], it is interesting to investigate the properties of mesoporous silica synthesized as thin film nanocomposite membranes.

Recently, there are many works reported on the synthesis of thin film nanocomposites by utilizing the columnar assembly of functional amphiphilic organic moieties in the channels of mesoporous silica [6-14]. Also, in 2002 Okabe *et al.* showed that amphiphilic triphenylene derivatives could be used as a self-assembled template to synthesize mesostructured silica films [15]. However, their silica films depend on the donor-acceptor system and ethanol vapour-mediated maturing for the synthesis of nanocomposites. Theoretically, triphenylene is a disc-shaped molecule of a flat polycyclic aromatic hydrocarbon (PAH) consisting of four fused benzene rings with a planar structure and three-fold rotation axes. Moreover, triphenylene derivatives possessed graphite-like conducting pathway [16] that ideal to be used in energy devices like battery due to its good ionic conductivity properties. Nevertheless, previous studies reported that as thin films, the formed mesostructured silica tends to have channel direction oriented in parallel to the substrate [17] and this phenomenon is not optimal for energy storage purposes. Xie *et al.* stated that a one-dimensional (1D) arrangement is crucial in energy conversion and storage as it can help to minimize the path of carrier transport, increase the carrier collection ability, and act as an electron expressway in the axial axis [18]. Thus, the development of mesoporous silica thin film with 1D alignment is really important, and the formed silica nanochannels must be arranged perpendicularly to the substrate.

Anodized aluminum oxide (AAO) membrane was used as a template for controlling the perpendicular orientation of mesoporous silica to the substrate [19-23]. AAO was made up of porous cylindrical channels with ordered and vertical 1D arrangement [24, 25]. In 2003, the first work of filling the mesoporous silica inside the AAO was done by Yang *et al.* [26], followed by another outstanding work by Teramae *et al.* in 2004 [21]. However, in their work, they used the commercially available surfactants such as Pluronic F-127 ($EO_{106}PO_{70}EO_{106}$) and cetyltrimethylammonium bromide (CTAB) as the template for synthesis of mesoporous silica. Therefore, in our work, we have used stepwise-synthesized useful surfactants, which is amphiphilic triphenylene ($TP_{C10TEG}$) as a template in the sol-gel synthesis of mesoporous silica ($TP_{C10TEG}$/silica$_{hex}$) in the AAO. Generally, triphenylene derivatives were mainly applied for molecular electronics, field-effect transistors, and photovoltaic due to its excellent self-healing property as well as giving the anisotropic columnar assembly of liquid crystalline phase to produce 1D or charge and energy migration materials [16, 27]. However, although triphenylene derivatives have an excellent ionic conductivity, not many works have been done for investigating the potential of this nanomaterial as energy devices. Thus, the interesting utilization of this triphenylene derivative for potential application in energy aspects that mainly require fast charge transfer and / or transport that may be achieved when using triphenylene derivative. Hence, this report was done based on the idea of synthesizing a hybrid nanomaterial that has the potential to be used in energy conversion or storage. Likewise, sol-gel synthesis was utilized in this study due to its simple and easy method of synthesizing hybrid of mesoporous silica/alumina membrane nanocomposites consisting of organic functional groups with dense filling in the nanochannels [28].

## 2. Experimental Methods

### *2.1 General*

Proton and carbon nuclear magnetic resonance ($^{1}$H– and $^{13}$C–NMR) spectra were performed by a Bruker AVANCE 300 MHz NMR spectroscopy. The chemical shifts ($\delta$) were determined using deuterated chloroform ($CDCl_3$) at 7.24 and 77.0 ppm for $^{1}$H– and $^{13}$C–NMR as internal standards. XRD measurements at room temperature were carried out using D8 Advance diffractometer from Bruker with Cu$K\alpha$ radiation at 40 mA and 40 kV with a step size and scan speed of 0.02° and 2° min$^{-1}$. Thermo Scientific GENESYS 10S with 100 nm min$^{-1}$ as a scan speed was used as an ultraviolet-visible (UV-vis) spectroscopy. For monitoring chemical reactions, thin-layer chromatography (TLC) was used on pre-coated silica gel plates of 0.2 mm in thickness from Merck. AB Sciex MALDI-TOF/TOF™

IOP Conf. Series: Materials Science and Engineering **202** (2017) 012003 doi:10.1088/1757-899X/202/1/012003

5800 spectrometry was used for the measurement of mass spectra on reflection mode with ionization in ion positive mode. For crystallization, a matrix such as dithranol (1, 8, 9- antracenetriol) in chloroform was spotted onto a conductive plate containing the samples. The AAO membrane used was commercially purchased Whatman Anodisc 25 with 25 mm in diameter and 0.2 μm pore size. JEOL JEM-2100, JSM-6390LV, and JSM-6701F microscopies were used to visualize TEM, SEM and FE-SEM images at 200, 15 and 5 kV, respectively.

### *2.2 Synthesis of $TP_{C10TEG}$*

According to Okabe *et al.* [15], $TP_{C10TEG}$ was prepared by reacting hydroxy type of triphenylene and amphiphilic alkyl bromide. Typically, in dry N,N-dimethylformamide solution (10 mL), the mixture of 80.0 mg for 2,3,6,7,10,11-hexahydroxytriphenylene (0.247 mmol) and 1.0 g for 1-(2-(2-(2-methoxyethoxy)ethoxy)ethoxy)-10-bromodecane ($C_{10}TEGBr$; 2.61 mmol) with the presence of 600.0 mg of potassium carbonate (4.35 mmol) was refluxed at 90 °C for 66 hours under an inert condition. Insoluble substances from the reaction mixture were filtered off, and then the filtrate was evaporated under a reduced pressure to give residue as the desired product. Silica column chromatography in dichloromethane ($CH_2Cl_2$) and methanol (MeOH) with a ratio of 80:1 was used as an eluent for the purification procedure. The collected fraction as the desired product was evaporated until all solvents were completely removed. This desired product was dried under a reduced pressure and followed by freeze–drying in benzene until reached a formation of dried oily liquid with a brownish color with 30% yield (298.0 mg, 0.139 mmol).

### *2.3 Sol-gel synthesis of $TP_{C10TEG}$/silica$_{hex}$*

For confinement of 1D columnar assembly of $TP_{C10TEG}$ into the nanoscopic channels of mesoporous silica, the sol-gel synthesis was performed under an acidic solution of aqueous ethanol (EtOH, 595.6 mg, 0.01 $\mu$mol; HCl, 3.9 mg, 30.60 $\mu$mol; $H_2O$, 45.4 mg, 0.26 $\mu$mol) for a mixture of $TP_{C10TEG}$ (36.3 mg, 0.17 $\mu$mol) and tetrabutoxysilane (TBOS, 162.0 mg, 5.10 $\mu$mol). This sol-gel reaction with TBOS was carried out for partial oligomerization in 12 hours at room temperature with mole ratios of $TP_{C10TEG}$/ [TBOS]/ [$H_2O$]/ [EtOH/ [HCl] in 1:30:150:770:1.8. The mixture was completely wrapped with aluminum foil to avoid the interaction of the light. Spin-coater was then used to make a thin film on a quartz plate from the viscous solution with a speed of 3000 rpm for 15 s. The thin film provided colorless transparent material and was then dried for 24 h at room temperature. For the synthesis of $TP_{C10TEG}$/silica$_{hex}$ inside channels of AAO, the sol-gel solution was drop-casted onto the AAO membrane surface and allowed gravity to flow through the channels of the membrane for the formation of mesoporous nanocomposites. The resulting thin film was dried in open air for 24 h at room temperature.

## 3. Results and discussion

### *3.1 Structure elucidation of $TP_{C10TEG}$*

From Williamson ether substitution reaction of triphenylene and amphiphilic $C_{10}TEGBr$, structure elucidation of the desired product for purified $TP_{C10TEG}$ showed $^1H$–NMR spectrum with $\delta_H$ as illustrated in Figure 1(a) with special labels for the presence of protons according to their environments (**Ha-i**). Particularly, it gave $\delta$ at 7.86 (**Hi**, singlet, 5.61H), 4.19 (**Hh**, triplet, 11.50H), 3.64–3.51 (**Hb** = $CH_2CH_2OCH_2CH_2OCH_2CH_2$, multiplet, 72.43H), 3.42 (**Hc** = $OCH_2$, triplet, 12.42H), 3.35 (**Ha** = $OCH_3$, s, 18.00H), 1.91 (**Hg** = $CH_2$, triplet, 2.15H), 1.56 (**Hd** = $CH_2$, singlet, 5.62H) and 1.40-1.28 ppm (**He** = $CH_2$, **Hf** = $(CH_2)_4$, overlapped, 12.12H). For $^{13}C$–NMR, the spectrum gave $\delta_C$ as shown in Figure 1 (b) for the presence of carbon according to their environments (**Ca-m**) at 148.88 (**Ck**), 123.52 (**Cm**), 107.28 (**Cl**), 71.54 (**Cb**), 71.15 (**Cd**), 70.42 (**Cc**), 69.96 (**Ce**), 69.63 (**Cj**), 58.94 (**Ca**), 29.56-29.46 (**Cf**, **Cg**, **Ch**) and 26.12 ppm (**Ci**). Moreover, the mass spectrum of purified $TP_{C10TEG}$ observed a monoisotopic peak at 2137.2607 Dalton for $[M]^+$ where it was almost similar to the calculated value for a monoisotopic peak with a molecular formula of $C_{120}H_{217}O_{30}$ at 2137.5376

Da. This results indicated that amphiphilic version of $TP_{C10TEG}$ was successfully synthesized and then can be used as a template for the sol-gel synthesis of mesoporous silica nanocomposites containing columnar assembled in the nanoscopic channels.

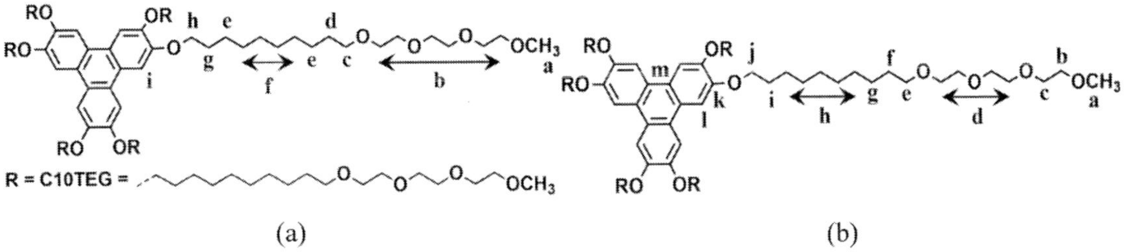


**Figure 1**. Representative molecular structures of $TP_{C10TEG}$ for (a) $^{1}$H– and (b) $^{13}$C–NMR spectra

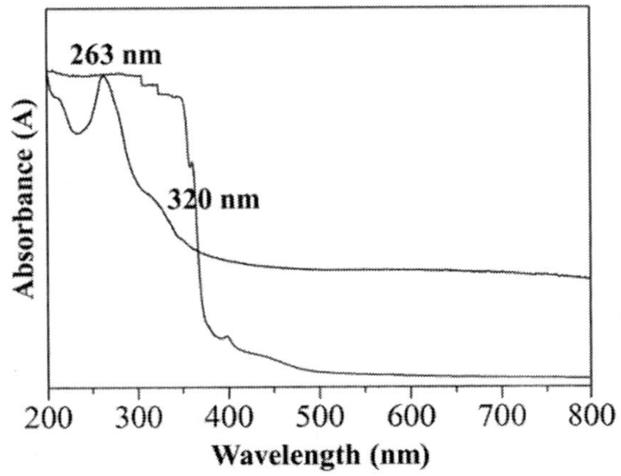


**Figure 2**. UV-vis spectra of the bulk $TP_{C10TEG}$ (blue line) and $TP_{C10TEG}$/silica$_{hex}$ fabricated onto quartz substrate (black line)

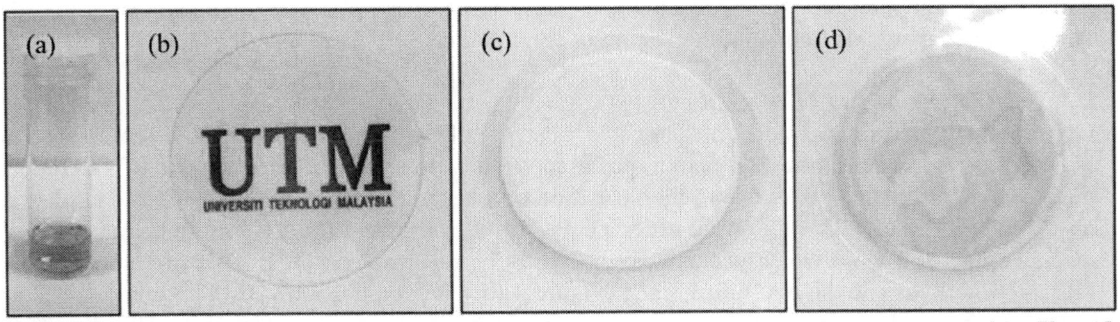


**Figure 3**. Photograph images of (a) sol-gel solution of $TP_{C10TEG}$/silica$_{hex}$, (b) transparent thin film of $TP_{C10TEG}$/silica$_{hex}$ fabricated onto quartz substrate, (c) blank AAO membrane and (d) $TP_{C10TEG}$/silica$_{hex}$ after drop-casted onto AAO membrane via a gravitational filtration

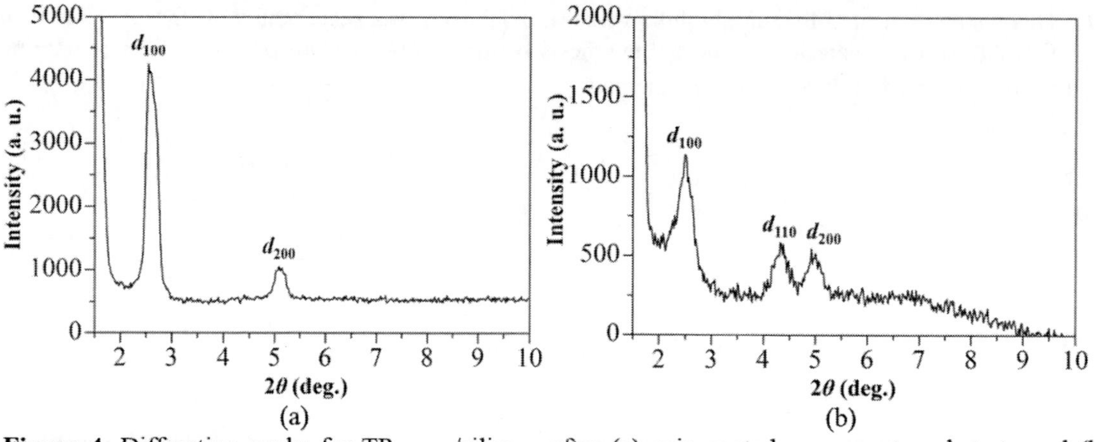


**Figure 4**. Diffraction peaks for $TP_{C10TEG}$/silica$_{hex}$ after (a) spin-coated on a quartz substrate and (b) drop-casted inside AAO channels via gravitational filtration

### *3.2 Optical properties of $TP_{C10TEG}$/silica$_{hex}$*

For identification of appearance of polycyclic aromatic hydrocarbons, a study on the absorption of organic chromophores can be evaluated using UV-vis spectroscopy [29]. Figure 2 shows the UV-vis spectrum of bulk $TP_{C10TEG}$ (blue line) and $TP_{C10TEG}$/silica$_{hex}$ that was spin-coated on a quartz substrate (black line), where this film was used as a reference throughout this study. In this figure, bulk $TP_{C10TEG}$ showed a broad peak from 200 to 270 nm that maybe due to a $\pi$-$\pi$ stacking between triphenylene molecules. Interestingly, $TP_{C10TEG}$/silica$_{hex}$ showed a strong absorption peak of 264 nm that may due to the ordered and long-range $\pi$-$\pi$ stacking of disc-like aromatic triphenylene core from their self-assembly in the silicate nanochannels, along with a shoulder at 320 nm for the charge-transfer of the organic molecules. Figure 3 (a) shows the photograph images of a sol-gel solution of $TP_{C10TEG}$/silica$_{hex}$ that has a yellow-brownish color. After the sol-gel solution was spin-coated by a quartz substrate, it gave a colorless thin film as shown in Figure 3 (b). Moreover, the sol-gel solution was also fabricated onto the blank AAO membrane as in Figure 3 (c) to produce a light brownish hybrid AAO membrane as observed in Figure 3 (d). Although, it is really difficult to check the absorption peak of $TP_{C10TEG}$/silica$_{hex}$ after being fabricated inside the AAO channels due to the opaqueness of the hybrid AAO nanocomposites, the images in Figure 3 (d) and its UV-vis spectrum provided a strong confirmation for the formation of columnar assembly with well self-organization inside the nanostructured materials from oligomerization of silica source.

### *3.3 Structural Analysis of $TP_{C10TEG}$/silica$_{hex}$*

The structural analysis was done on fabricated $TP_{C10TEG}$/silica$_{hex}$ nanocomposites and AAO substrate using XRD [15]. After fabrication onto a quartz substrate with a slow oligomerization of TBOS as a silica source, diffraction peaks due to the formation of a hexagonal structure for mesoporous silica at $2\theta = 2.59^o$ ($d_{100}$) and $5.08^o$ ($d_{200}$) were observed as shown in Figure 4 (a). This diffractogram gave $d$-spacing value of 3.4 nm which was almost closed to the reported one [15]. After the fabrication of $TP_{C10TEG}$/silica$_{hex}$ inside the AAO membrane (Figure 4 (b)), an additional $d_{110}$ diffraction peak at $2\theta = 4.40^o$ was observed and this peak was not present in the fabricated one using quartz. The absence of the $d_{110}$ was usually observed from the c-axis of a hexagonal unit cell that was oriented in parallel to the quartz [8]. Interestingly, although the intensity of the diffraction peaks after fabricated on the AAO membrane was slightly lower compared to the fabricated one onto a quartz substrate, the diffraction patterns for the typical hexagonal structure of mesoporous silica $TP_{C10TEG}$/silica$_{hex}$ were still preserved. Moreover, it was suggested that the lower intensity might be highly possible for the formation of most hexagonal channels of mesoporous silica that were already inside the AAO pores.

*3.4 Morphology Studies of $TP_{C10TEG}$/silica$_{hex}$*

The XRD results were also supported by morphology studies using the TEM images in Figure 5 (a) and (b). TEM images showed the presence of mesoporous silica materials with a hexagonal structure containing pore sizes ranging from 2.0 to 2.5 nm after fabricated on both a quartz substrate and an AAO membrane.

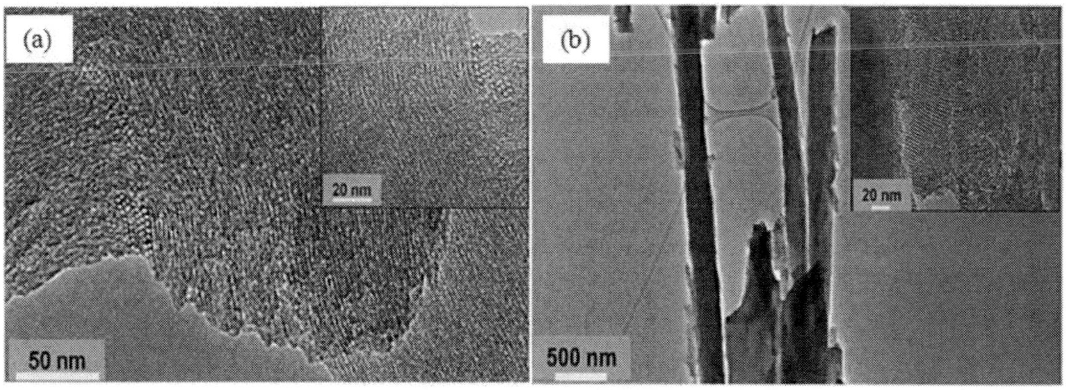


**Figure 5**. TEM images of (a) $TP_{C10TEG}$/silica$_{hex}$ and (b) $TP_{C10TEG}$/silica$_{hex}$ inside AAO channels. The inset Figures are to magnify the image with a scale bar of 20 nm

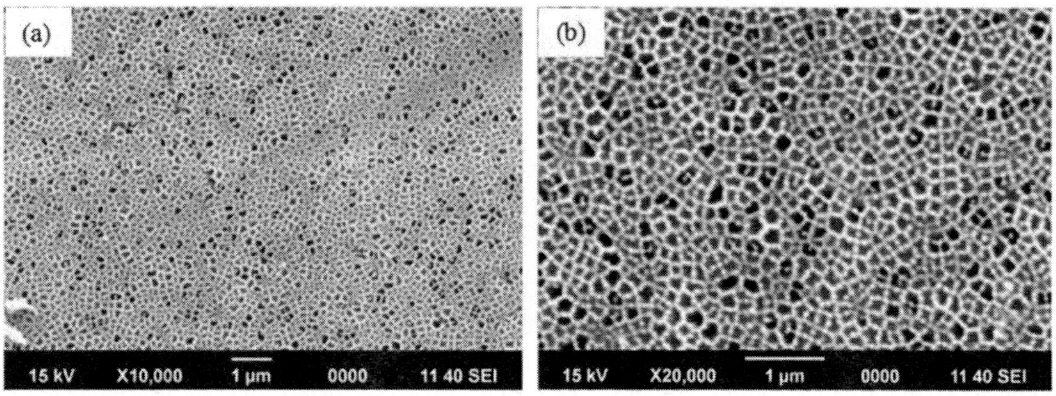


**Figure 6**. SEM images of $TP_{C10TEG}$/silica$_{hex}$ after fabricated inside AAO channels via gravitational filtration at (a) 10K and (b) 20K times magnification at 1 $\mu$m scale

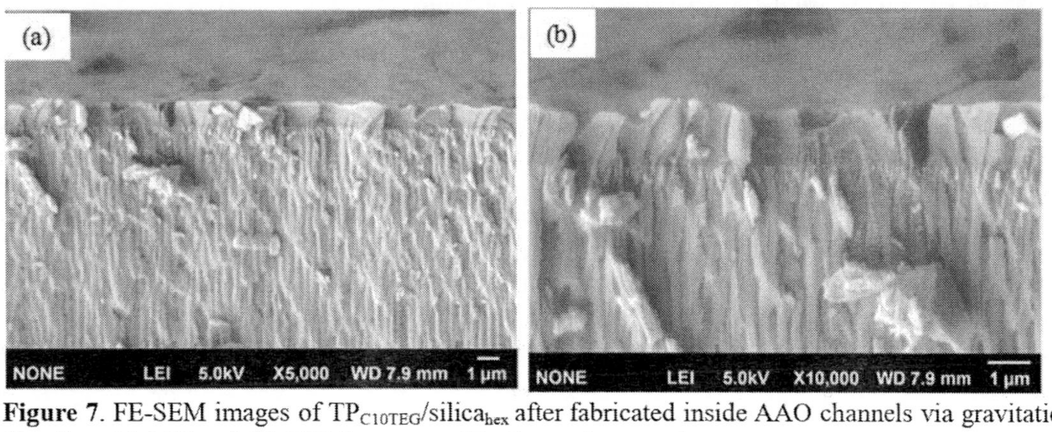


**Figure 7**. FE-SEM images of $TP_{C10TEG}$/silica$_{hex}$ after fabricated inside AAO channels via gravitational filtration at (a) 5K and (b) 10K times magnification at 1 $\mu$m scale

IOP Conf. Series: Materials Science and Engineering **202** (2017) 012003 doi:10.1088/1757-899X/202/1/012003

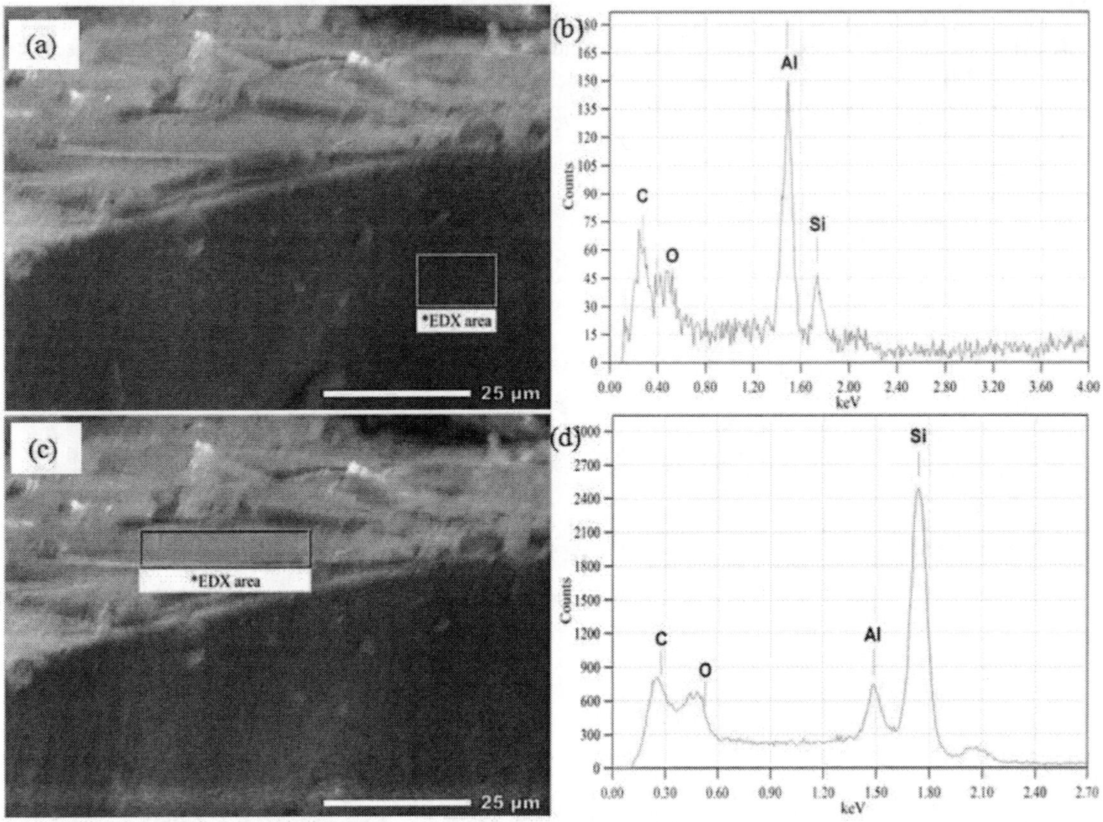


**Figure 8**. FE-SEM images of $TP_{C10TEG}$/silica$_{hex}$ after fabricated inside AAO channels showing the selected area indicated by * for EDX analysis in (a) the middle of the membrane with (b) its respective EDX and (c) the top area of the membrane with (d) its corresponding EDX

Based on SEM images in Figure 6 (a) and (b), it showed the successful filling of most AAO channels with the $TP_{C10TEG}$/silica$_{hex}$ composites, with the diameter of AAO channels ranging from 200 to 300 nm. These images also suggest that many $TP_{C10TEG}$/silica$_{hex}$ composites have been successfully penetrated to the AAO channels without a structure collapse. Such filling of the template for mesoporous silica was in good agreement with their X-ray diffractogram and TEM images. Hence, the 1D arrangement of $TP_{C10TEG}$ silica nanocomposites in the AAO membrane film was achieved via a simple template sol-gel synthesis and a gravitational filtration.

By characterization of the mesoporous silica/alumina hybrid membrane film using the cross-section of FE-SEM as shown in Figure 7 (a) and (b), it was confirmed that the mesoporous silica not only formed inside the AAO channels but also on top of AAO membrane. Energy-dispersive X-ray (EDX) analysis suggests the existence of Al, Si, C, and O elements inside AAO of the $TP_{C10TEG}$/silica$_{hex}$. Upon characterizing the area of the nanocomposites from Figure 8 (a), the Si element can be found along with Al as shown in Figure 8 (b), indicating that mesoporous silica nanocomposites were present inside the AAO pores. Further EDX analysis on the top layer as shown in Figure 8 (c) and (d), Si ratio was found to be higher than Al. This result indicated that the formation of mesoporous silica would cover the pores on top of the AAO membrane. Therefore, in order to prevent any formation of mesoporous silica on top of the membrane, Teramae *et al.* [17] proposed the usage of aspiration technique while doing the filtration other than using gravity. However, this method might be useful for silica source with faster oligomerization rate such as tetramethyl orthosilicate (TMOS).

## 4. Conclusion

Triphenylene bearing a $C_{10}$TEG as the amphiphilic chains ($TP_{C10TEG}$) in the form of light yellow oily liquid was successfully prepared in 30 % yield (298.0 mg, 0.139 mmol). The resulting $TP_{C10TEG}$ was used in the sol-gel synthesis as a template to form $TP_{C10TEG}$/silica$_{hex}$ that later was fabricated on a quartz substrate and inside the AAO membrane. After the fabrication of $TP_{C10TEG}$/silica$_{hex}$ onto a quartz substrate, the diffraction peaks due to a hexagonal structure of mesoporous silica nanocomposites were observed and it was in good agreement with their TEM images. Upon characterization of $TP_{C10TEG}$/silica$_{hex}$ inserted into the AAO channels using XRD, it was confirmed that the diffraction patterns for hexagonal channels of mesoporous silica were still preserved, accompanied by the presence of additional diffraction peak of $d_{110}$ at $2\theta$ = 4.40°. Based on the SEM, FE-SEM and EDX analysis, it was confirmed that 1D arrangement of mesoporous silica nanocomposites were achieved as the $TP_{C10TEG}$/silica$_{hex}$ grew one-dimensionally inside the porous AAO with a perpendicular orientation to the AAO substrate.

## 5. References

[1] Yanagisawa T, Shimizu T, Kuroda K and Kato C 1990 Bull. Chem. Soc. Jpn. 63 988
[2] Kresge C T, Leonowicz M E, Roth W J, Vartuli J C and Beck J S 1992 Nature 359 710
[3] Kato M, Shigeno T, Kimura T and Kuroda K 2005 Chem. Mater. 17 6416
[4] Zhao D, Feng J, Huo Q, Melosh N, Fredrickson G H, Chmelka B F and Stucky G D 1998 *Science* **279** 548
[5] Chen C Y, Li H X and Davis M E 1993 *Microporous Mater* **2** 17
[6] Otani W, Kinbara K and Aida T 2009 *Farad. Discuss.* **143** 335
[7] Lu Y, Yang Y, Sellinger A, Lu M, Huang J, Fan H, Haddad R, Lopez G, Burns A R, Sasaki D Y and Shelnutt J 2001 *Nature*, **410** 913
[8] Lintang H O, Kinbara K, Tanaka K, Yamashita T and Aida T 2010 *Angew. Chem. Int. Ed.* **49** 4241
[9] Lintang H O, Kinbara K, Yamashita T and Aida T 2010 Proceedings of International Conferences on Enabling Science and Nanotechnology (ESciNano) 5700970 ISBN: 978-1 4244-8853-7
[10] Lintang H O, Kinbara K, Tanaka K, Yamashita T and Aida T 2012 *Chem Asian J* **7** 2068
[11] Lintang H O, Kinbara K and Aida T 2012 Proceedings of International Conferences on Enabling Science and Nanotechnology (ESciNano) 6149684 ISBN: 978-1-4577-0799-5
[12] Lintang H O, Yuliati L and Endud S 2014 *Mater Res Innov* **18** S6 444
[13] Jalani M A, Yuliati L, Endud S and Lintang H O 2014 *Adv Mat Res* **925** 233
[14] Jalani M A, Yuliati L and Lintang H O 2014 *Jurnal Teknologi* **70** 131
[15] Okabe A, Fukushima T, Ariga K and Aida T 2002 *Angew. Chem. Int. Ed.* **41** 3414
[16] Bushby R J, Donovan K J, Kreouzis T and Lozman, O R *Opto-Electron. Rev.* **13** 269
[17] Yamaguchi A, Kaneda H, Fu W and Teramae N 2008 *Adv. Mater.* **20** 1034
[18] Le Xie J, Guo C X and Li C M 2014 *Energy Environ. Sci.* **7** 2559
[19] Platschek B, Petkov N and Bein T 2006 *Angew. Chem. Int. Ed.* **118** 1152
[20] Wu Y, Livneh T, Zhang Y X, Cheng G, Wang, J, Tang, J, Moskovits M and Stucky G D 2004 *Nano Lett.* **4** 2337
[21] Wu Y, Cheng G, Katsov K, Sides S W, Wang J, Tang J, Fredrickson G H, Moskovits M and Stucky G D 2004 *Nat. Mater.* **3** 816
[22] Xomeritakis G, Liu N G, Chen Z, Jiang Y B, Kohn R, Johnson P E, Tsai C Y, Shah P B, Khalil S, Singh S and Brinker C J 2007 *J. Membr. Sci.* **287** 157
[23] Yamaguchi A, Uejo F, Yoda T, Uchida T, Tanamura Y, Yamashita T and Teramae N 2004 *Nat. Mater.* **3** 337
[24] Luo H, Lin Q, Baber S and Naalla M 2010 *J. Nanomater.* 3
[25] Lu Q, Gao F, Komarneni S and Mallouk T E 2004 *J. Am. Chem. Soc.* **126** 8650
[26] Yang Z, Niu Z, Cao X, Yang Z, Lu Y, Hu Z and Han C C 2003 *Angew. Chem. Int. Ed.* **42** 4201

[27] Mulder F M, Stride J, Picken S J, Kouwer P H, de Haas M P, Siebbeles L D and Kearley G J 2003 *J. Am. Chem. Soc.* **125** 3860
[28] Yao B, Fleming D, Morris M A and Lawrence S E 2004 *Chem. Mater.* **16** 4851
[29] Giger W and Blumer M 1974 *Anal. Chem.* **46** 1663

**Acknowledgments**
The authors would like to thank to Universiti Teknologi Malaysia (UTM), Malaysia for characterization facilities. The authors also thank to Research University Grant Scheme (RUGS Tier 1) with cost center no. Q.J130000.2526.10H11 from Ministry of Higher Education (MOHE) through UTM for the financial supports.

# Preparation of Low fouling PolyethersulfoneMembranes by Simultaneously Phase Separation and Redox Polymerization

**A Roihatin**[1] **and H Susanto**[2]

[1]Conversion Energy Department, Polytechnic State Semarang
Jl. Prof. Soedarto-Tembalang, Semarang, Indonesia
[2]Department of Chemical Engineering, Diponegoro University
Jl. Prof. Soedarto-Tembalang, Semarang, Indonesia

Email: heru.susanto@undip.ac.id

**Abstract**. This paper presents preparation of low fouling PES membranes by non solvent induced phase separation (NIPS) coupled with redox polymerization.The membrane characterization included water permeability, morphology structure (by SEM) and surface chemistry (by FTIR).Water permeability measurements showed thatthe membranes have water permeability within the range 10-50 L/h.m$^2$.bar. Addition of PEG dan PEGMA intopolymer solution increased water permeability, whereas blending redox initiator and crosslinker, MBAA in polymer solution decreased water permeability.Surface morfology of membranes by SEM showed that unmodified PES membrane had smaller pore size than PEG or PEGMA modified PES membranes. Furthermore, PES-PEG or PES-PEGMA membranes modified by blending with redox initiator and MBAA as crosslinker showed smaller pore size than unmodified membrane. FTIR analysis showed that all membranes have typical spectraof PES polymer;however no additional peak was observed forthe membranes prepared with addition of PEG/PEGMA, initiator redox and also crosslinker. The addition of PEG/PEGMA, redox initiator and crosslinker resulted in membranes with high rejection and an acceptable flux as well as more stable due to relatively high fouling resistance.

**Keywords**: fouling, NIPS, redox polymerization, low fouling UF membranes

## 1. Introduction

UF membrane technology has been applied in various industries such as water and wastewater treatment, beverage and food processing, medicine and pharmaceutical industry [1-4]. But along with the increasing application of ultrafiltration, fouling which causing significant loss of performance with respect to flux and often selectivity is still the biggest problem. Furthermore, fouling can also shorten the membrane life.

Eventhough several methods to control fouling of the membrane have been developed [5,6], but in many cases fouling is determined by the membrane itself [7,8]. Therefore, preparation of low fouling membranes is urgently needed. The hydrophobic caharacter of PES polymer has driven modification of PES membrane. Three different approaches have been developed to increase the hydrophilicity of PES membranes, namely (i) modification of polymer membranes (pre-modification) [9] (ii) mixing the polymer membranes with additives [10-11] and (iii) modification of the surface after membrane preparation (post-modification) [12]. However, each approach has drawbacks such as takes a long time, stability of additives in the polymer matrix is low, and also some treatments can cause changes in membrane structure. Integrationof NIPS and chemical modification is the best solution becauseit is simple and effective way to make low fouling membranes with high stability. Development of NIPS modified by redox polymerization is expected to be able to synthesis low fouling UF membrane in single process with high stability of hydrophilic character.

In general, the objective of this research is to synthesis low fouling UF membranes by using NIPS integrated with redox polymerization. In particular, this research investigates the effects of additives, initiator redox and crosslinking agent on the characteristics and stability of the resulting membranes. In addition, fouling behaviour is also investigated.

## 2. Experimental Method

### *2.1. Materials*
PES polymer Ultrason E 6020P was purchased from BASF. N-methyl-2-pyrrolidone (NMP) was purchased from Merck. Polyethylene glycol (PEG) and N,N'-methylenbisacrylamide (MBAA) were purchased from Sigma-Aldrich. Poly(ethyleneglycol) methacrylate (PEGMA 400) was purchased from Polysciences. TEMED (N,N,N,N'-tetramethylenediamine), ammonium peroxydisulfate (APDS), pottasium dihydrogen phospate ($KH_2PO_4$)and disodium hydrogen phospate dihydrate ($Na_2HPO_4.2H_2O$)were purchased from Fluka Chemica.Bovine serum albumin (BSA) was purchased from ICN Biomedicals.

### *2.2. Membrane preparation*
PES polymer was dissolved in NMP. PEG and PEGMA were then added to the polymer solution. In addition, oxidizing agent (TEMED) and MBAA were also added into polymer solution. The homogenous polymer solution was left without stirring until no bubbles is observed. The polymer solution was cast with a thickness of 200 μm using a steel casting knife on a glass substrate. Thereafter, the proto-membrane was solidified by immersing in a coagulation bath containing water and reducting agent (APS) for one hour.The resulting membranes was washed and soaked in the water for 24 h before drying.

### *2.3. Membrane characterization*
Membrane characterization included water permeability, morphology structure (by SEM) and surface chemistry (by FTIR). Measurement of water permeability followed our previous publication [18]. The top surface morphology of the membrane was observed by using a Quanta 400 FEG (FEI) environmental scanning electron microscope (ESEM). Before observation, sputtering to coat the outer surface of the sample with gold/palladium should be conducted. The membrane surface chemistry was analyzed by using the Varian 3100 Fourier transform infrared spectroscopy (FTIR) Excalibur series.

## 3. Results and Discussion

### *3.1. Water permeability*
Measurement of water permeability was performed for unmodified and modified membranes.The effects of PEG, PEGMA, redox initiators and crosslinking agent were investigated. The results are depicted in Figure 1.

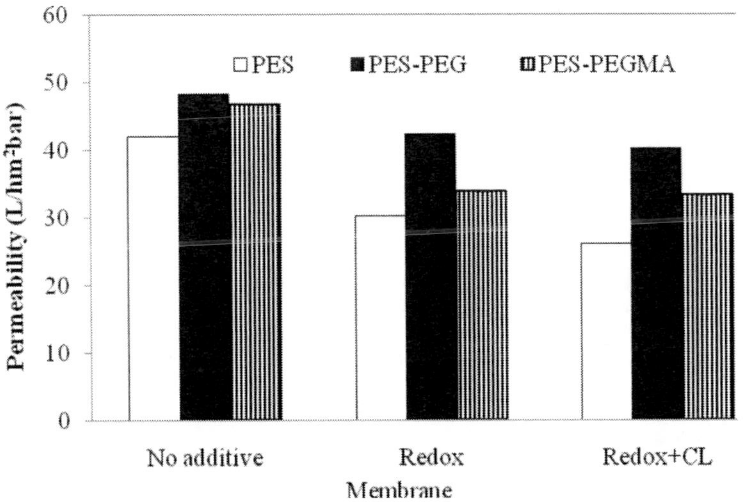


**Figure 1.** Permeability of unmodified membrane and modified membranes

The water permeability data presented in Figure 1 suggestthat the resulted membranes are in the range of UF membranes, which have water permeabilityof 10–50 L/h.m$^2$.bar [1]. Addition of PEG and PEGMA increased water permeability. PEG is a hydrophilic polymer,which will increase the hydrophilic character of PES membrane. I addition, the presence of PEG into polymer solution may favor the formation of larger pores and porosity, resulting an increase of water flux. This is in agreement with the results obtained by previous authors [13-16]. Similar phenomenon was obtained using PEGMA as the additive and in agrement with the result obtained by Susanto et al. [12].

Blending redox initiator in polymer solution decreased water permeability of modified membrane. The presence of initiator may stimulate the chemical reaction which may reduce pore size. Similar phenomenon is also seen for MBAA addition in the polymer solution. The addition ofcrosslinker may form a network structure. This result is in agreement with the experiment conducted by Peeva et al.[17]. Crosslinking with MBAA yielded denser hydrogel layers on the porous base membrane and consequently decrease water permeability.

*3.2. Surface morphology*

Figures 2, 3 and 4 show surface morphology of unmodified PES and PEG/PEGMA modified PES membranes. It is *observed that all membranes had* pore sizes within the nanometer range. Comparing the unmodified PES and modified PES membrane*s* by PEG or PEGMA addition, it is *seen* that all *membranes had* significant difference in pore size. Similar with previous authors, unmodified PES membrane had smaller pore size than PEG or PEGMA modified PES membrane*s* [14,18]. *These results support previous explanation about* water permeability of membranes.PES-PEG or PES-PEGMA membranes modified by blending with redox initiator and MBAA as crosslinker showed smaller pore size than unmodified membrane. Redox initiator generated radicals on the membrane surface to create new functional groups on the membrane surfaces [19].

IOP Conf. Series: Materials Science and Engineering **202** (2017) 012004 doi:10.1088/1757-899X/202/1/012004

**Figure 2.** Surface morphology of membrane: (a) PES only, (b) PES-PEG, (c) PES-PEGMA

**Figure 3.** Surface morphology of membrane: (a) PES-PEG, (b) PES-PEG-R, (c) PES-PEG-R-CL

**Figure 4.** Surface morphology of membrane: (a) PES-PEGMA, (b) PES-PEGMA-R, (c) PES-PEGMA-R-CL

### *3.3. Surface chemistry*

Surface chemistry of membrane was characterized by FTIR. All membranes showed typical spectra of PES, i.e., aromatic bands at 1578 and 1485cm$^{-1}$ from the benzene ring and C–C bond stretch and aromatic ether band around 1240 cm$^{-1}$ (IR spectra are not shown). However no additional peak was observed for the membranes prepared with addition of PEG, PEGMA and MBAA. The reason for this result would be overlapping bands of the strongest bands for PEG and PEGMA with bands for PES.Indeed, a significant increase in absorbance at ~1105cm$^{-1}$, due to additional intensity of C–O bond stretch from PEG was observed. This confirms the presence of additives in the membrane polymer matrix. Addition of PEGMA to the PES membrane showed a new band associated with the

C=O vibration in the ester molecule, which is located at 1714 $cm^{-1}$. Such overlapping band is also found in previous publications by Susanto and Ulbricht [18] and Belfer et al. [20].

## 4. Conclusion

Modification of PES UF membranes by redox polymerization using hydrophilic monomers (PEG/PEGMA), initiator redox and also MBAA as crosslinker has been successfully developed. Water permeability measurment suggested that all membranes should be UF membrane. Addition of PEG dan PEGMA increased water permeability. By contrastblending redox initiator and crosslinker into polymer solution decreased water permeability. Surface morfology of membranes indicated thatthe pore sizes are within the nanometer range. Unmodified PES membrane had smaller pore size than PEG or PEGMA modified PES membrane . Furthermore, PES-PEG or PES-PEGMA membranes modified by redox initiator and MBAA showed smaller pore size than unmodified membrane. FTIR analysis showed that no additional peak was observed for the modified membranes. The fouling resistance of unmodified and modified membranes are under evaluation.

## 5. References

[1] Mulder M 1996 Basic *Principle of Membrane Technology 2nd ed* (Dordrecht: Kluwer Academic Publishers)

[2] Baker R W 2004 *Membrane Technology and Applications 2nd ed* (Chichester: John Wiley & Sons, Ltd)

[3] Cheryan M 1998 *Ultrafiltration and microfiltration handbook* (Pennsylvania: Technomic Publishing Company Inc)

[4] Zeman, L J and Zydney A L 1996 *Microfiltration and ultrafiltration: principles and applications* (New York: Marcel Dekker Inc)

[5] Belfort G, Davis R H and Zydney A L 1994 The behavior of suspensions and macromolecular solutions in crossflow microfiltration *J. Membr. Sci.* **96** 1-9

[6] Hilal, N, Ogunbiyi O O, Miles N J and Nigmatullin R 2005 Methods employed for control of fouling in MF and UF membranes: A comprehensive review *Sep. Sci. Technol.* **40** 1957-64

[7] Marchese J, Ponce M, Ochoa N A., Pradanos P, Palacio L and Hernandez A 2003 Fouling behaviour of polyethersulfone UF membranes made with different PVP *J. Membr. Sci.* **211** 1-11

[8] Susanto H and Ulbricht M 2009 *Polymeric membranes for molecular separations, in E. Drioli, L. Giorno (Eds), Membrane Operations. Innovative Separations and Transformations* (Weinheim: Wiley – VCH) 34-45

[9] Chalida K, Ladewig B P, Max L G Q and Wang L 2011 Preparation and characterization of sulfonated polyethersulfone for cation-exchange membranes *J. Membr. Sci.* **368**(1-2) 48-53

[10] Liu Y, Koops G H and Strathmann H 2003 Characterization of morphology controlled polyethersulfone hollow fiber membrane by the addition of polyethylene glycol to the dope and bore liquid solution *J. Membr. Sci.* **223** 187-195

[11] Idris A, Zain N M and Noordin M Y 2007 Synthesis, characterization and performance of asymmetric polyetehrsulfone (PES) ultrafiltration membranes with polyethylene glycol of different molecular weights as additives *Desalination* **207** 324-333

[12] Susanto H, Balakrishnan M and Ulbricht M. 2007 Via surface functionalization by photograft copolymerization to low-fouling polyethersulfone-based ultrafiltration membranes *J. Membr. Sci.* **288**(1-2) 157-167

[13] Wang Y Q, Wang T, Su Y L, Peng F B, Wu H amd Jiang Z Y 2006 Protein-adsorptionresistance and permeation property ofpolyethersulfone and soybean phosphatidylcholine blend ultrafiltration membranes *J Membr. Sci.* **270** 108-141

[14] Arthanareeswarana G, Mohan D and Raajenthiren M 2010 Preparation, characterization and performance studies of ultrafiltration membranes with polymeric additive *J. Membr. Sci.* **350** 130-138

[15] Kim J H and Lee K H 1998 Effect of PEG additive on membrane formation by phase

separation *J. Membr. Sci.* **138** 153-161

[16] Su Y L, Cheng W, Li C and Jiang Z 2009 Preparation of antifouling ultrafiltration membranes with poly(ethylene glycol)-graft-polyacrylonitrile copolymers *J. Membr. Sci.* **329** 246-252

[17] Peeva P D, Pieper T and Ulbricht M 2010. Tuning the ultrafiltration properties of anti-fouling thin-layer hydrogel polyethersulfone composite membranes by suited crosslinker monomers and photo-grafting conditions *J. Membr. Sci.* **362** 560-568

[18] Susanto H and Ulbricht M 2009 Characteristics, performance and stability of polyethersulfone ultrafiltration membranes prepared by phase separation method using different macromolecular additives *J. Membr. Sci.* **327** 125-134

[19] Zhao C, Xue J, Ran F and Sun S 2013 Modification of polyethersulfone membranes–A review of methods *Progress in Materials Science*. **58** 76-150

[20] Belfer S, Fainchtain R, Purinson Y and Kedem O 2000 Surface characterization by FTIR-ATR spectroscopy of polyethersulfone membranes-unmodified, modified and protein fouled. *J Membr. Sci.* **172** 113-124

# Crystallinity and Electrical Conductivity of PANI-Ag/Ni Film: The Role of Ultrasonic and Silver Doped

**M Diantoro[1,2], I N Fitriana[1], F Parasmayanti[1], Nasikhudin[1], A Taufiq[1,2], Sunaryono[1,2], N Mufti[1,2] and H Nur[3]**

[1]Department of Physics, Faculty of Mathematics and Natural Sciences, Universitas Negeri Malang (State University of Malang), Jalan Semarang No 5, Malang 65145, Indonesia
[2]Center for Minerals and Advanced Materials, Faculty of Mathematics and Natural Sciences, Universitas Negeri Malang (State University of Malang), Jalan Semarang No 5, Malang 65145, Indonesia
[3]Centre for Sustainable Nanomaterials, Ibnu Sina Institute for Scientific and Industrial Research, Universiti Teknologi Malaysia, 81310 UTM Skudai, Johor, Malaysia

Email : markus.diantoro.fmipa@um.ac.id

**Abstract.** Polyaniline (PANI) is typically a conductive polymer which has a uniqueness in structure and physical properties. The physical properties generally can be controlled by introducing a specific dopant as well as other processing parameters. This polymer has been widely investigated and applied in various fields, especially in electronics. Some metallic elements have been introduced to modified its characteristic such as iron and copper. In order to improve the characteristic of this material, it is important to study the optimum process as well as the elemental substitution of PANI films. In this study, we report the influence of employing sonochemistry technic and the silver substitution on the crystallinity and electrical conductivity of PANI/Ni films. The samples have been prepared similarly to the previous report with silver doped in the range of $0.1 \leq x \leq 0.5$ M. Other series of the sample were also prepared at various irradiation time of ultrasonic exposure. FTIR spectra showed that EB and ES Polyanilines had been successfully synthesized. X-RD histogram also revealed that crystallinity of the films comparably increased with the increase ultrasonic irradiation time as well as the silver in dopant. Along with the crystallinity, the electrical conductivity of films also increased as an increase of the dopant and of ultrasonic irradiation time.

**Keywords:** polyaniline, crystallinity, conductivity, sonochemistry, silver ion.

## 1. Introduction

The development of polymer materials engineering technology has found a wide range of materials that have certain advantages, one of which is a conductive polymer. The polymer which having an intrinsic electron conductivity appears to be an attractive organic semiconductor. These polymers have semiconducting properties associated with the availability of electron conjugation bond -π which is a source of free charge in the polymer conjugation. The electron of unpaired -π bonds, can move in conjugated carbon-carbon bonds of the polymer chain [1-3].

One of the most useful conjugated polymers to be used are polyaniline (PANI), which is characterized by the simplicity of synthesis, non-toxic, high sensitivity to external factors and its potential as an optical material [4], high potential to be a good synthetic conductor [MacDiarmid]. Polyaniline is very promising for a variety of electronic applications, energy storage, pharmaceuticals and information technology. The basic form of polyaniline is the emeraldine base (EB), and emeraldine salt (ES) respectively the half-oxidized, and partially oxidized [1, 5].

Basically electrical conductivity or other physical properties of PANI can be improved by physical or chemical doping [5]. One way to do is adding a proton $H^+$ from HCl [2], CSA [6], iron, copper [7], lanthanum [8], nickel [9], cobalt [10] metallic ions, transition ions of arsenic [11], or other metal oxalates. So far the conductivity of PANI showed an increase with the increase of various dopants and other process parameters. However, the results showed that the range of conductivity was very large. There is also no detail information the relation between the dopant introduction, crystallinity and the electrical conductivity. The incomprehensive information shows no clear guidance of inferring microscopic mechanism in the material.

It is shown that the use of functional protonic acid such as CSA (camphor sulfonic acid) as a dopant, resulting PANI complex in conductive and non-conductive form soluble in the common solvents such as chloroform, xylene, m-cresol and others. Thus functional protonic acid allows conductivity of polyaniline can be processed as fiber, sheet, transparent thin films and others [5]. Some methods of coating films PANI-Ag / Ni is spin coating, screen printing, spray coating and so on [3]. The method used in this research is the spin coating method because of its advantages which can produce thin films with a thickness reaching nanometer scale. This process can also produce a homogeneous layer and relatively short time. The fabrication in the form of a thin film is very broad, because the properties of the material can be modified and close to application of devices. The combination of PANI and Ni as a substrate will give good results because it has a maximum conductivity, crystal quality, and high homogeneity because the Ni substrate has a high crystal orientation. So far there is no comprehensive study Ag dopants and ultrasonic irradiation with CSA on PANI on Ni substrate. Ultrasonic irradiation is not only intended as an enhancer of the PANI on conventional conductivity but also gives a chance to form nano and the crystalline phase of PANI-Ag/Ni. In this study, Ag nano has been used as a dopant into PANI/Ni film as well as the influence of ultrasonic irradiation to determine its conductivity.

## 2. Experimental Method

Two series of silver doped PANI-Ag/Ni and ultrasonic irradiation time films have been prepared using spin coating method. Each series was provided 5 samples. The basic configuration of PANI-EB and ANI ES were synthesized following the procedure of previous report [7]. The solutions of PANI-Ag had been prepared with synthesizing PANI-EB from aniline using the chemical oxidizing method. About 1.82 mL aniline (0.1 M) was dissolved into 50 mL HCl (0.2 M) liquid for about 1 hour. Along with this, a solution of 5.71 grams $(NH_4)_2S_2O_8$ (0.1 M) in 50 mL aquadest also prepared at the same time. After 1 hour, those two solutions just mixed and stirred a while then exposed at room temperature for about 24 hours for polymerization. The precipitated material was filtered using Whatman filter paper then washed using deionized water and aquadest until clean liquids observed. The obtained precipitated materials then mixed and homogenized using magnetic stirrer in NH4OH (0.5 M) for 4 hours and then let them about 24 hours and washed using aquadest to obtain a blue PANI-EB. The powder of PANI-EB can be obtained after annealing the material for 5 hours at 80 °C. PANI-EB then mixed with the Camphor sulfonic acid (CSA) and then mixed with the $AgNO_3$ solution in chloroform and employing ultrasonic irradiation for various irradiation time. The obtained solution was filtered to obtain a PANI-Ag solution for deposition on nickel substrate using spin coating method. A different series of various $AgNO_3$ PANI doped of 0.1 M, 0.2 M, 0.3 M, 0.4 M, and 0.5 M also prepared in the same manner. The films have been deposited onto 1 x 1 $cm^2$ Ni substrate using spin coating method at 1000 rpm for 1 minutes. The obtained films then annealed at 100 °C for 1 minutes.

To check the formation of PANI-EB and PANI-ES (II), both series of the samples were characterized using FTIR and X-RD respectively to evaluate the bonds formation and crystallinity. X-

RD diffraction was conducted under Cu-Kα at 0.02 degrees of 2θ. The electrical conductivity measurement was conducted using the 4-probe method with 0.2 cm distance separated of the probes. Conductivity (σ) of the film was obtained by taking the reciprocal of its resistivity ρ which were obtained using the following Equation 1.

$$\rho = 2\pi s \frac{V}{I} \tag{1}$$

Where $\rho$, $V$, $I$, and $s$ expressed respectively of resistivity (Ωcm), voltage (volt), current (ampere), and probes' distance.

**3. Results and Discussion**

Spectra of PANI-EB and PANI-ES(II) obtained from FTIR measurement is displayed in Figure 1. Basically, there are six main peaks showing the existing of functional groups of PANI-EB. The six peaks respectively appear at the wave number of 1581 cm$^{-1}$,1489 cm$^{-1}$, 1377 cm$^{-1}$, 1303 cm$^{-1}$, 1151 cm$^{-1}$, dan 820 cm$^{-1}$[12] which are appear in the sample indicated by red bars in Figure 1.

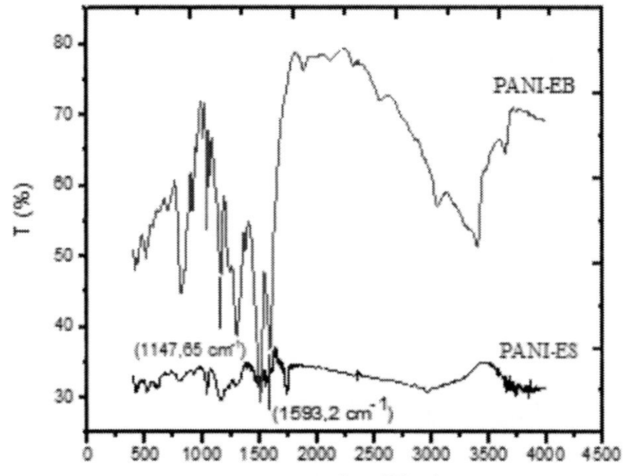


**Figure 1.** Transmittance (%) of (a) PANI EB and (b) PANI ES

A strong peak at a wave number of 1593.2 and 1492.9 cm$^{-1}$ express the existing of C=N and C=C vibrations of quinoid and benzoid rings. The peak observed at 1300.02 cm$^{-1}$ shows a secondary aromatic amino C-N (–N–Benzoid–N–). The peak at 1147.65 cm$^{-1}$ indicated stretching mode of benzoid unit. The next peak at 1147.65 cm$^{-1}$ explains the existing electronic band or vibration of nitrogen quinoid (C=N stretching –N=quinoid=N–). A peak at 815 cm$^{-1}$ is originated from C–C and C–unit benzoid units which are used for indicating the key appearance of substituted benzene type. A typical of PANI-EB is the appearance of a strong peak at 1140 cm$^{-1}$ which is showing a vibration of quinoid nitrogen [14]. It means that the PANI-EB has been successfully synthesized. FTIR peaks of PANI-ES are closely similar to the PANI-EB. The main difference of the ES and EB spectra is the peak of a wave number in the area of 1593.2 cm$^{-1}$ and 1147.65 cm$^{-1}$ in which the EB spectrum appears very strong peaks which are not visible in the ES spectrum. This indicates imine C = N group at PANI-EB has turned into a form of amine C-N and -NH- in the PANI-ES.

Motomichi *et al.* [13] reported that the diffraction pattern for polycrystalline PANI exhibit Bragg's peaks at $2\theta$ of 5.2°, 9.5°, 15°, 20.9°, 25°, 27.5°, and 30°. Compare to the result of our pattern showed at Figure 2 shows that a strong peak exist at 5.2°. This peaks consistently appear at series of silver doped as well as time exposure of irradiation. A very strong peak at about 44° for different time irradiation samples is belong to silver. A less strong peaks series shown by time irradiation PANI at 5° indicates that the film is more oriented than the silver doped PANI.

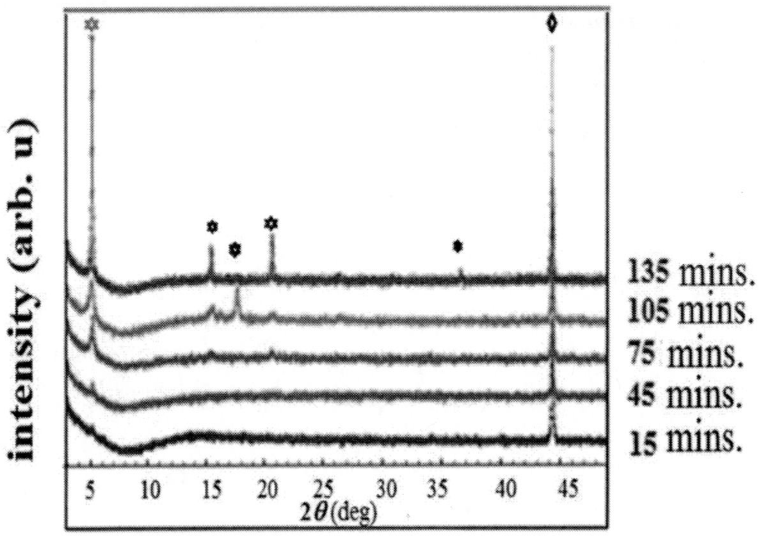


(a)

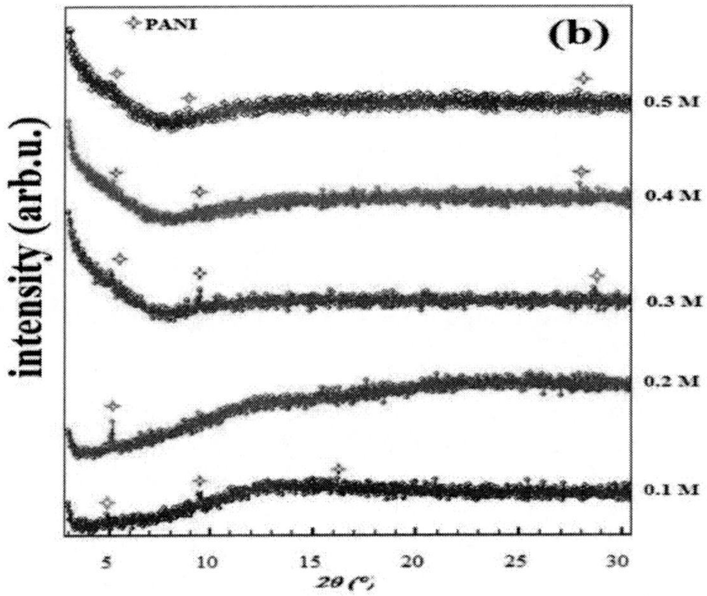


**Figure 2.** Diffraction pattern of ultrasonic time (a) and silver doped (b) of PANI-Ag/Ni films

Theoretically, silver crystalline will appear at $2\theta = 44.08°$. It is observed that the increase of silver ion doped give rise to a slightly shift to higher degress. The respective peaks of 0.1 M, 0.2, 0.3, 0.4 0.5 M correspod to the appearance of the peaks at 44.30°, 44.32°, 44.30°, 44.28°, and 44.27°. The shift of tho the higher shift is correspond to the slightly decrease of silver particles. The subtitution of silver ion is indicated also by the higher shift of the Bragg's peaks.

Further analyses of the first peaks explore the relation between the dopant and irradiation time relative to the crystallinity. The crystallinity shows the degree of the crystal fraction in PANI-Ag/Ni as well as the time irradiation PANI is shown in Figure 3. The crystallinity may play a key important role on the doped PANI and various time irradiation PANI for further examination of Figure 5.

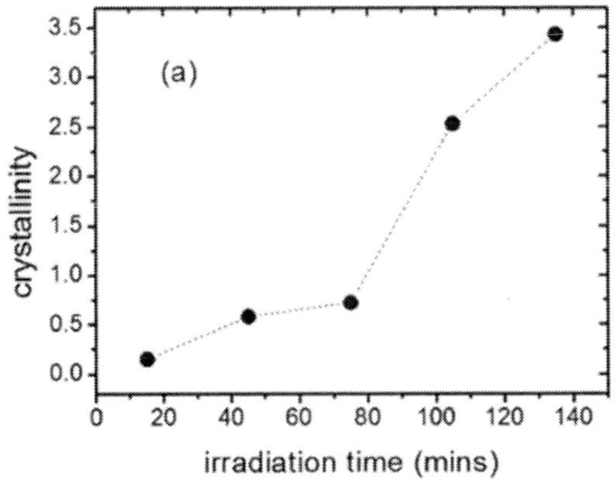


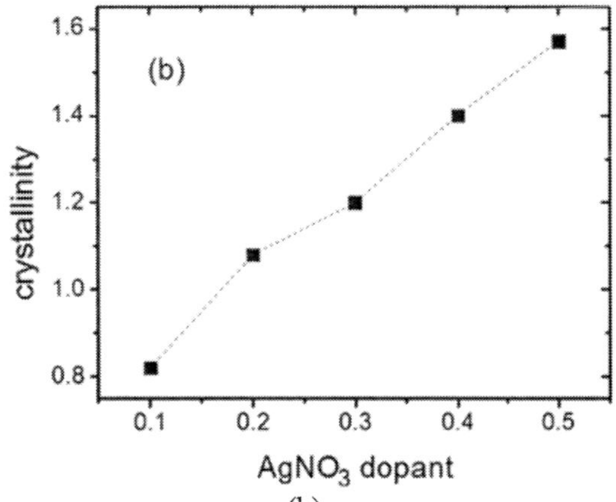


(a) (b)

**Figure 3**. Crystallinity PANI/Ni as a function of silver doped (a) and ultrasonic irradiation time (b).

SEM images of PANI-Ag films which were exposed under magnification of 10000 x showing a relatively similar morphology. At this magnification, the samples exhibit a homogenous surface. There is no crack, or another grain boundary fault appears on them. It looks that the grain or island are greater than 5 μm as shown by bar scale.

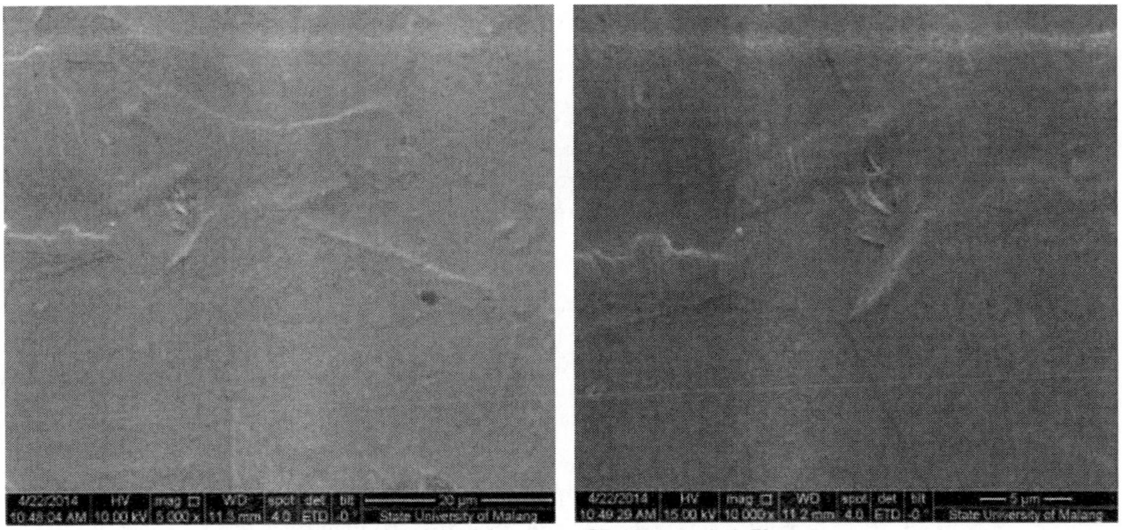

**Figure 4.** SEM images of PANI-Ag/Ni films.

The crystallinity of the films shows an increase by increase of the dopant as well as the time of irradiation. As shown in Figure 3 the crystallinity was calculated from X-RD data is relatively low. This crystallinity can not be observed by SEM as depicted in Figure 4. For detail examination may be can be observed by TEM diffraction. Other important physical properties related to the dopant and the irradiation time of PANI is electrical conductivity. The electrical conductivity of both of the two series of samples is displayed in Figure 5.

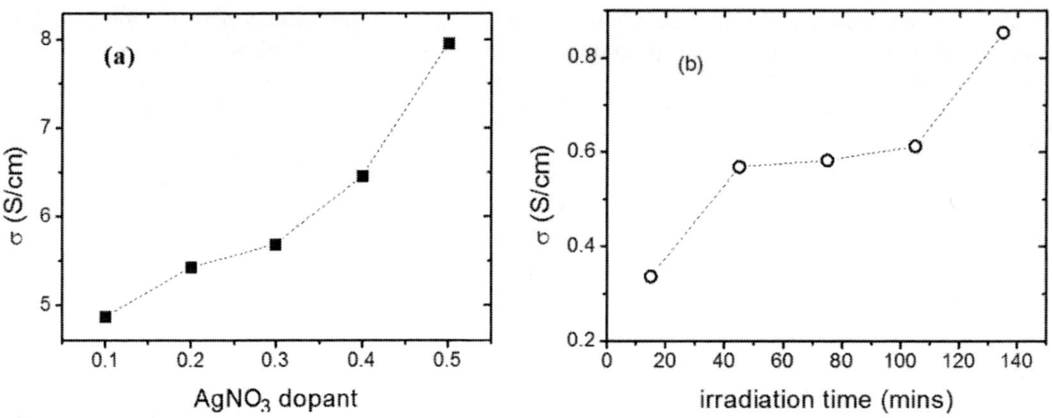


**Figure 5.** Electrical conductivity at different silver doped (a) and irradiation time of sonochemistry (b).

It is observed that silver ion concentration, Figure 5(a) as well as irradiation time of ultrasonic, Figure 5(b) showing a similar trend i.e. increase the electrical conductivity. Even though they do not have a simple relation. The influence of silver ions PANI-Ag/Ni generally resulted electrical conductivity compare to the effect of irradiation time. The highest electrical conductivity achieved by 0.5 M silver doped sample which reaches up to 7.96 S/cm. This conductivity is higher than oriented PANI sample reported by Mihai [15] which reach 1 S/cm. On the other hand, this silver doped PANI/Ni is still lower than $Ar^+$ and $Ga^+$ doped PANI which was reported by Aleshin et. al [11] of 150 S/cm. This experimental data suggest that the prediction of MacDiarmid [1, 2] is revealed and potential to be studied further. Both of the crystallinity and conductivity obtained from this work is closely similar to the silver doped natural polymers [16].

## 4. Conclusion

In conclusion, we confirmed that the role of silver dopant, as well as the ultrasonic exposure, give rise to increasing the crystallinity as weel as the electrical conductivity. The electrical conductivity of PANI/Ni is higher than that of ultrasonic exposure. This result suggests that the silver doped can be used to improved the electrical conductivity in combination with CSA as well as ultrasonic irradiation.

## 5. References

[1.] MacDiarmid A G 2001 Synthetic Metals: A Novel Role for Organic Polymers (Nobel Lecture) *Angew. Chem. Int. Ed.* **40** 2581 - 2590

[2.] Cristofolini L, Fontana M P, Konovalov O, Berzina T and Smerieri A 2009 Doping-Induced Conductivity Transitions in Molecular Layers of Polyaniline: Detailed Structural Study, *Langmuir* **25(21)** 12429–12434

[3.] Stakhira P, Cherpak V, Volynyuk D, Hotra Z, Belukh V, Aksimentyeva O, Tsizh B and Monastyrskyi L 2010 Growth and Properties Of Conducting Polyaniline Thin Films Obtained By Means Of Ionic Sputtering In Crossed Electrical And Magnetic Fields, *Rev.Adv.Mater.Sci.* **23**180-184

[4.] Saravanan S, Anantharaman M R, Ventachalam S, Avasthi D K 2008 Studies on the optical band gap and cluster size of the polyaniline thin films irradiated with swift heavy Si ions. Vacuum **82** 56-60.

[5.] Stejskal J, Gilbert R G, Polyaniline. Preparation of a conducting polymer *Pure Appl. Chem.* **74** 5 pp. 857–867, 2002

[6.] Mathew H, Punnackal V S, Kuriakose S, Kumari B S, Manuel A.cy, Synthesis and Electrical Characterization of Polyaniline-Multiwlled Carbon Nanotube Composite with Different Dopants *International Journal of Scientific Reserch Publication* **3(8)** 2013 1-10

[7.] Diantoro M, Purwaningtyas D, Muthoharoh N, Hidayat A, Taufiq A, Fuad A 2012 The Influence of Iron- and Copper- doped of PANi Thin Film on their Structure and Dielectric Properties *American Institute of Physics Proceedings* 1454. p268-271.2012, Melville New York

[8.] Krishna J B M, Saha A, Okram G S, Soni A, Purakayastha S and Ghosh G 2009 Electrical propertties of polyaniline doped with metal ions *Journal of Phyisics D Appl. Phys.* **42** (9)095404.

[9.] Tan Y, Zhang Y, Kan J 2009 Synthesis and properties on polyaniline in the presence of nickel chloride *eXPRESS Polymer Letters* **3** 6 333-339.

[10.] Aghazadeh M, Aghazadeh F 2013 Electrical Conductivity Property Study of Polyaniline, Cobalt *Nanocomposite Journal of Applied Chemical Research* **7(3)** 47-55

[11.] Aleshin A N, Mironkov N B, Suvorov A V, Usov I O, Co klin J A, Su T M and Kaner R B 1998 Conductivity and magnetoconductivity of Polyaniline films implanted with $Ar^+$ and $Ga^+$ ions near the critical regime of the metal insulator transition, *Journal of Physics: condensed matter* **10 (22)** pp. 4867

[12.] Wallace G G, Geoffrey M S, Maguire K, Peter R 2009 Conductive Electroactive polymers Iintelligent Polymer System *Perancis:CRC Pres.*

[13.] Motomichi I, Castillo-Ortega M, Michiko M, Inoue B 2011 Polyaniline Toluenesulfonates: X-Ray Diffraction and Electrical Conductivity. X-Ray Diffraction and Electrical Conductivity *Journal of Macromolecular Science* 10.1080/10601329708011058.

[14.] Nobuta T, Claire H et al 2009 Oriented Thin Film of PANI by Friction Transfer Method. *Mol. Cryst Liq Cryst* **505** pp 80-86

[15.] Mihai I, Addiego F, del Frari D, Bour J, Ball V 2013 Associating oriented polyaniline and eumelanin in a reactive layer by layer manner: composites with high electrical conductivity, *Colloids and Surface A: Physicochemical and Engineering Aspects* **434(2013)** 118-125.

[16.] Diantoro M, Hidayati N N S, Latifah R, Fuad A, Nasikhudin, Sujito, Hidayat A 2016 Electrical Conductivity Modification of Extracted Jatropha Multifida L. and Pterocarpus Indicus Wild Latex Films *AIP Conference Proceedings* 1719, 030034 (2016); doi: 10.1063/1.4943729.

## Acknowledgments

The authors would like to thank the DRPM, KEMENRISTEKDIKTI, Republic of Indonesia for providing the research grant for MD 2016-2017.

# The Effect of Thickness of ZnO Thin Films on Hydrophobic Self-Cleaning Properties

**N Mufti, D Arista, M Diantoro, A Fuad, A Taufiq and Sunaryono**

Department of Physics, Faculty of Mathematics and Natural Sciences, Universitas Negeri Malang (State University of Malang) Jl. Semarang 5, Malang 65145, Indonesia

Email: Nandang.mufti.fmipa@um.ac.id

**Abstract**. Glass coating can be conducted by using ZnO-photocatalyst based semiconductor material since it is preeminent in decomposing organics compound and dangerous bacteria which often contaminates the environment. If there are dirt containing organics compound on the glass, the ZnO photocatalyst coat can be applied as self-cleaning, usually called self-cleaning glass. It depends on the coating thickness which can be controlled by setting the speed of spin coating. In this research, the various rotating speeds of spin coating were conducted at 2000 rpm, 3000 rpm, and 4000 rpm to control the thickness. The raw materials used in this research were $Zn(CH_3COOH)_2.2H_2O$ (PA 99,5%), Ethylene glycol, Diethanolamine (PA 99%), Isopropanol Alkohol, Glycerol, and Ashton. Synthesis methods used were sol-gel prior to spin coating technic were applied. The results of the film were characterized by using SEM, XRD, and UV-Spectrophotometer. The crystal structure was analyzed by using Highscore plus and GSAS software, the size crystal was calculated by using Scherrer equation, a contact angle with ImageJ software. It was shown that ZnO thin film had been successfully synthesized with the crystal size around 21 nm up to 26 nm. The absorption value is higher due to the increasing of coat thickness with bandgap ± 3.2 eV. The test result of hydrophobic and hydrophilic characteristics show that all samples of ZnO thin film with the thickness ± 1.050 μm, ± 0.450 μm, ± 0.250 μm can be applied as self-cleaning glass. The best result was gained with the thickness of thin film ± 1.050 μm.

**Keywords:** Photocatalyst, ZnO, thin film, spin coating, self-cleaning property

## 1. Introduction

The use of glass has increased in the various field especially in building construction or household property. Glass also used as a decorator for the alternative of energy saving since it can help lighting of building in the day. One of the glass weaknesses is it is easily contaminated by dirt that gradually reduces the room lighting even they can damage the surface of material. Therefore, it needs other alternatives to solve such problem. One of them is surface engineering technology [1,2]. One kind of surface engineering technologies is coating. Glass coating can employ photocatalyst based semiconductor material. Semiconductor photocatalyst has drawn people's attention as it produces free radical (OH) that can mineralize dangerous substances into environment friendly-substance [1–3].

ZnO is one of the photocatalytic semiconductors. ZnO has been applied as the main material of light transmitter diode, cathode ray tube, electroluminescence thin film, and a gas sensor [4,5] ZnO gives better efficiency rather than $TiO_2$ as the photodegradation in water solution [5]. Besides, the combination of hydrophobic photocatalytic effectivity and super hydrophilic allows ZnO becoming a good candidate for self-cleaning glass application.[1,5]. ZnO has some such preeminences as stable to the light (it is not degraded when it does light absorption), non-toxic, photosensitive, cheap cost, and it is able to decompose organical compound and harmful bacteria that often contaminates [6,7].

The thin ZnO Films can be created through some methods such as vapor deposition, pulsed laser deposition, molecular beam epitaxy, metal organic chemical vapor deposition, sputtering, electron beam evaporation, spray pyrolysis and spin coating [4,8]. The sol-gel spin coating is a method to form a coat from photoresist polymer material which is positioned on the surface and other flat materials. After the sol-gel solution is dripped on the substrate, the rotary speed is set by centrifugal force to produce homogeneous thin coat. This sol-gel spin coating method combines physics and chemistry methods. This method is considerably easy and effective to make thin coat only by setting the time parameter, rotation speed as well as solution viscosity. Sol-gel spin coating method has some benefits, for examples, cheap cost, low annealing temperature, easy-controlled composition and structure, homogeneous composition, no need high vacuum, the coating thickness is controllable and shows good microstructures. This method, therefore, is frequently used in this recent years (Foo, 2013). This method is easy and effective to form thin coat only by setting the time, rotation speed as well as solution viscosity parameters [8,9]

There two types of the self-cleaning coating, i.e. hydrophobic and hydrophilic which has getting interested for daily application [1,5,5]. Hydrophilic is a condition when water touches the glass; the particle spreads out the water over the glass so that it cleans the glass automatically [5]. Hydrophobic is a condition in which the surface cannot hold the water and has big contact angle size when it can not do irradiation. The physical surface might influence the hydrophobic characteristic of the material, the substance coated by the material is in nano size and has wider surface rather than bulk material [2]. When the glass is touched by water, the glass surface will be brighter. The glass will look cleaner for a longer time, and the maintenance cost is cheaper. Self-cleaning glass helps to reduce the use of glass cleaner products which are not environmentally safe, so the environment pollution might reduce as well [1–3]. The photocatalytic coating of ZnO can be applied as self-cleaning. However, the appropriate of ZnO film thickness to self-cleaning properties is still rarely investigated. Therefore, in this study, we investigate the effect of ZnO film thickness to self-cleaning properties. The thickness ZnO Films controlled by setting the speed of spin coating. The crystal size, crystal structure, of ZnO films were characterized by XRD, while effectivity of self-cleaning glass of ZnO films was characterized by hydrophobic and hydrophilic behaviors.

## 2. Experimental Method

This sample has prepared by glass cleaning route before coating process. The glass was washed with soap, soaked in alcohol 96% for 15 minutes and soaked in acetone for 15 minutes. The glass, subsequently, was put into an oven at temperature of 105 $^0$C for one hour. Before using the glass, it was cooled to room temperature. After that, the making process of the solution with sol-gel method and solution of ZnO was prepared by using the raw material of zinc acetate dihydrate [$Zn(CH_3COOH)_2.2H_2O$]. Firstly, 5.488 g zinc acetate dihydrate was mixed into 3 mL of ethylene glycol. It was then heated at a temperature of 150 °C for 30 minutes. After that, 20 ml of isopropyl alcohol (IPA) was added and stirred at room temperature for 30 minutes. When the solution was whitish, 0.5 mL glycerol was added and following by adding 1.5 mL of diethanolamine (DEA) as a stabilizer. Subsequently, it was stirred for 60 minutes to obtain a homogeneous sol solution. The coating synthesis was, then, conducted by using spin coating method to get thin coat deposition on a substrate which subsequently. The ZnO created was then shaken and dripped on glass surface to cover the entire surface. The spin coating was set at rotation speed of 2000, 3000, and 4000 rpm for 30 seconds. The resulted film was then heated at a temperature of 300 °C for 10 minutes and subsequently annealed at a temperature of 550 °C for one hour. The ZnO thin films were characterized by using X-ray Diffractometer, SEM, and UV-VIS Spectrometer. The characteristics of hydrophobic and hydrophilic of ZnO thin films was conducted by dripping water on the coated surface. For hydrophilic, the sample was irradiated under sunlight for 30 minutes, and then the water was dripped on it.

## 3. Results and Discussion

The microstructure of ZnO thin films were analyzed by using SEM to obtain the samples' thickness and morphology. Table 1 shows the average coat thickness of each ZnO thin film from the different speed of spin coating.

**Table 1.** The Thickness of ZnO Thin Films

| Rotation speed (rpm) | Average film thickness (μm) |
|---|---|
| 2000 | 1.050 |
| 3000 | 0.450 |
| 4000 | 0.250 |

From Table 1, we can see that the lower speed of spin coating resulted in bigger film thickness and vice versa. The decrease of film thickness may caused by the high rotation speed which leads to higher centrifugal force. The force brings the liquid distribution at a higher rate and spread out more to cover a larger area on the glass substrate. Figure 1 shows the morphology of ZnO thin films with 50.000 times magnification.

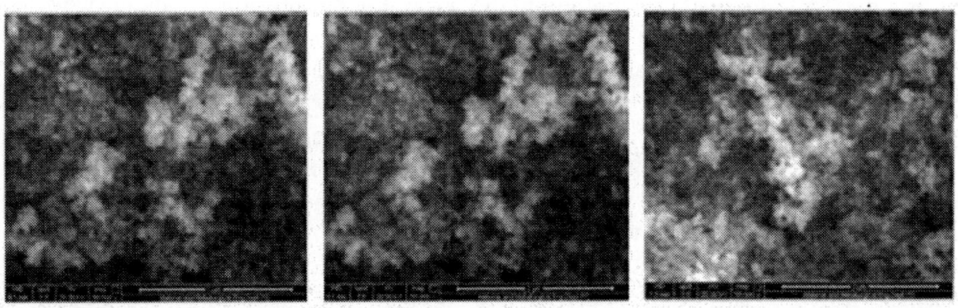

(a) (b) (c)

**Figure 1.** Morphology of ZnO thin films with the rotation speed (a) 2000 rpm (b) 3000 rpm (c) 4000 rpm

The diffraction pattern of ZnO thin film sample with an average thickness of 0.250, 0.450, and 1.050 μm are shown in Figure 2. From Figure 2, we can see that when a coat was being thicker, the crystal orientation of thin film will preferably grow on (002) plane. The films with lower thickness tend to grow with less preferred orientation. The later exhibits more randomly showing the existence of (100), (002) and (101) Bragg's planes.

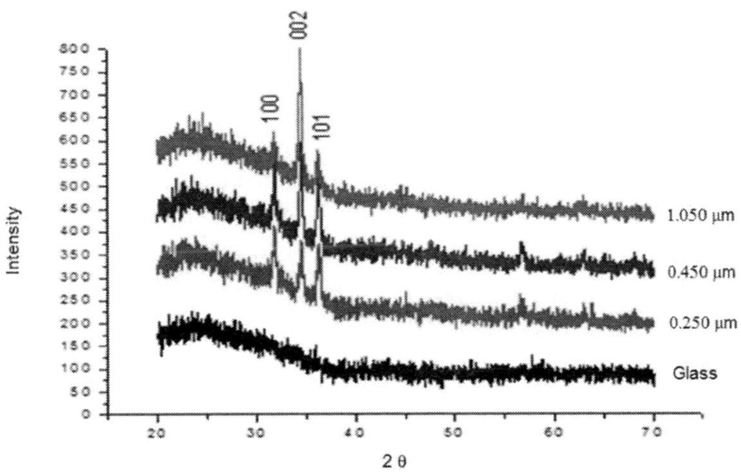


**Figure 2.** Diffraction patterns of ZnO thin films with various thicknesses

**Table 2.** The Analysis Result of Crystal Structure of Thin Film ZnO

| Parameter | Model | ± 1.050 µm | ±0.450 µm | ±0.250 µm |
|---|---|---|---|---|
| $a$ = b (Å) | 3.250 | 3.249 | 3.249 | 3.248 |
| $c$ (Å) | 5.250 | 5.206 | 5.203 | 5.203 |
| α = β (°) | 90 | 90 | 90 | 90 |
| γ (°) | 120 | 120 | 120 | 120 |
| $c/a$ | 1.6 | 1.6 | 1.6 | 1.6 |

From Table 2, it can be shown that there is almost no difference between lattice parameters at various thicknesses, where the a and c parameters are 3.250 and 5.250, respectively with c/a is around 1.6. These lattice parameters are similar as reported by Li et al. [8]. The crystal size was calculated by using Deye Scherer equation. We firstly inferred the FWHM and 2θ values of the X-ray diffraction data. The size of crystal size as a function of film thickness is summarized in Table 3.

**Table 3.** The Crystal Size of ZnO Thin Film on Various Speed of Spin Coating

| Average Film Thickness (µm) | Bragg planes (hkl) | Crystallite Size (nm) |
|---|---|---|
| 1.050 | 100 | 32.15 |
| | 002 | 24.30 |
| | 101 | 22.89 |
| | Average | 26.45 |
| 0.450 | 100 | 26.07 |
| | 002 | 27.99 |
| | 101 | 22.73 |
| | Average | 25.59 |
| 0.250 | 100 | 22.32 |
| | 002 | 21.62 |
| | 101 | 19.61 |
| | Average | 21.18 |

These results were in line with the early prediction related to the width of the peak in X-Ray diffraction pattern. The wider the peak, the higher the value of FWHM leads the smaller of crystal size. It is clearly seen that the thinner the film, the smaller the crystal size. The results were in line with the research conducted by Gayen (2011), the increase of crystal size was parallel with the increase of coat thickness.

Based on the UV-Vis measurement, we are able to know the transmission value of ZnO thin film as a function of different wavelengths with the various of thickness as shown in Figure 3.

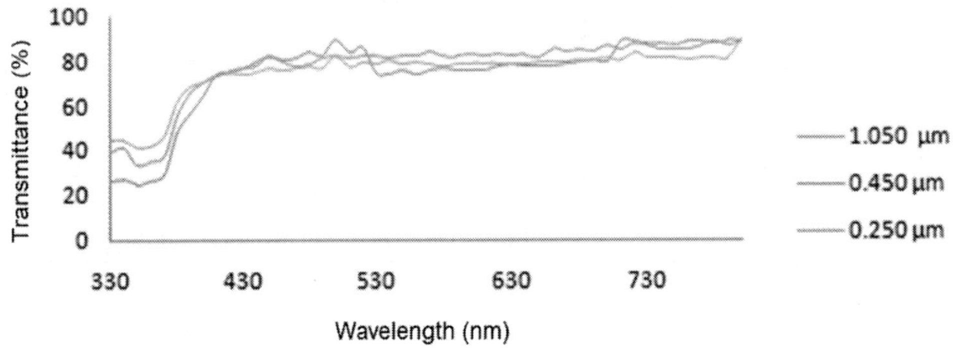


**Figure 3.** The percentage of transmittance of ZnO thin films as function of wavelength with various of the film thickness

Based on Figure 3, generally, the absorption values of all samples decreased for longer wavelength and this was the characteristic of ZnO semiconductor absorption. The thicker the film, the bigger the absorption value. If the sample was not gradually transparent, the more coats were formed and more molecules of ZnO were included in the process of visible light absorption. The calculation of optical bandgap was begun by deciding the maximum and minimum transmittance values, thickness, the coefficient of thin coat absorption. It was ended by the calculation applying Tauc method by doing extrapolation from the correlation graph (hυ) as the x-axis and $(\alpha\, h\upsilon)^n$ as y –axis so it cut the energy axis and the bandgap value was gained. Transmittance measurement was conducted at the wavelength 330 nm – 800 nm. Our results indicate that the coating thickness did not influence the bandgap value. In all variations of thickness, the results were similar, that was 3.2 eV. This result was suitable for the range of bandgap value of ZnO, that was 3.2-3.5 eV [4,8]. The small value of bandgap might allow electron was easily excited from valence band to the conduction band.

The picture of the contact angle of ZnO thin film with the different thickness was taken by using a camera and then it was measured by using ImageJ software. The result of contact angle was shown in Figure 4. Kenanakis et. al [5] found that self-cleaning glass application should be hydrophobic and superhydrophobic. In table 6, we can see that the contact angle resulted $\geq 90^0$ when it was in the room and $\leq 10^0$ after got sunlight in 30 minutes. The hydrophobic characteristic caused the water rolling down, so it did not stamp to the glass by bringing the dirt. With hydrophilic characteristics, the water would spread out the surface of glass cleaning the dirts and the gravitation forces caused the spread water dropped, so the glass remained clean and bright although it was affected by water.

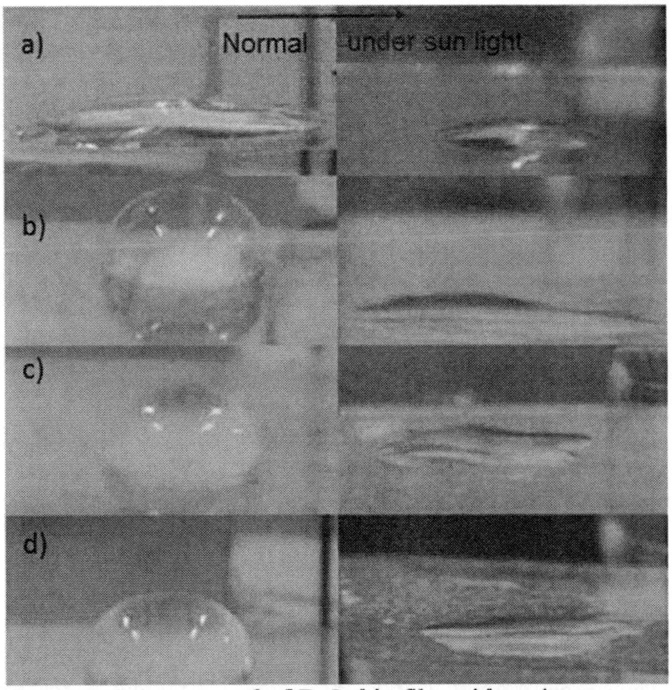


**Figure 4.** Contact angel of ZnO thin film with various thickness (a) uncoated, (b) 1.050 μm (c) 0.450 μm, (d) 0.250 μm

Table 4 shows the contact angle value of ZnO thin film with a various thickness of speed of spin coating. It shows that ZnO thin film with ±0.250 μm thickness had the biggest contact angle, that was 96.32° when it was inside a room. ZnO thin film had the smallest contact angle – 20.61° – when it was under sunlight in 30 minutes. On the contrary, ZnO thin film with ±1.050 μm thickness had the biggest contact angle, that was 99.99° when it was inside a room. It had the smallest angle – 6.03 ° – after it was lighted by the sun in 30 minutes. The variation shows good hydrophobic characteristic since it had contact angle above 90°. it also shows good super hydrophilic characteristic since it had contact angle under 10°.

**Table 4.** Contact Angel of ZnO Thin Film with Various Coat Thickness

| Average Thickness (μm) | Normal Contact Angle (°) | Sunlight Contact Angle (°) |
|---|---|---|
| Uncoated | 16.81 | 17.83 |
| 1.050 | 91.03 | 6.03 |
| 0.450 | 97.84 | 8.54 |
| 0.250 | 99.74 | 10.44 |

The contact angle increased based on the coating thickness. It was because the thicker the coat, the more particles of ZnO were deposited. The role of ZnO was detaining liquid on the surface, so there of material contact with liquid was narrower, so the number of ZnO deposition influenced the hydrophobic characteristic of material.

IOP Conf. Series: Materials Science and Engineering **202** (2017) 012006 doi:10.1088/1757-899X/202/1/012006

## 4. Conclusion

Based on data analysis and discussion, it can be concluded that coat thickness did not influence crystal structure of ZnO thin films. The thicker the coat, the bigger the coat crystallite size of ZnO thin film was. The thicker the coat, the higher the UV absorption value of ZnO thin film, however, it did not influence the bandgap value, that was 3.2 eV. Besides, the thicker the coat, the bigger the hydrophobic characteristic of ZnO thin film. It was because of the thicker the coat, the more particles were deposited and it influenced hydrophobic characteristic since the material contact area was narrower. Sun lighting caused hydrophilic characteristic of ZnO thin film smaller since sunlight could produce a pair of electron and hole on the surface of ZnO resulted by hydroxyl cluster that caused the contact angle decreased.

## 5. References

[1] Kim S M, In I and Park S Y 2016 Study of photo-induced hydrophilicity and self-cleaning property of glass surfaces immobilized with TiO2 nanoparticles using catechol chemistry *Surf. Coat. Technol.* **294** 75–82

[2] Sanderson K and Goer B 2016 Glass, Self-Cleaning *Reference Module in Materials Science and Materials Engineering* (Elsevier)

[3] Nosrati R, Olad A and Nofouzi K 2015 A self-cleaning coating based on commercial grade polyacrylic latex modified by TiO2/Ag-exchanged-zeolite-A nanocomposite *Appl. Surf. Sci.* **346** 543–53

[4] Kumar R, Kumar G, Al-Dossary O and Umar A 2015 ZnO nanostructured thin films: Depositions, properties and applications—A review *Mater. Express* **5** 3–23

[5] Kenanakis G, Vernardou D and Katsarakis N 2012 Light-induced self-cleaning properties of ZnO nanowires grown at low temperatures *Appl. Catal. Gen.* **411–412** 7–14

[6] Wang J, Li Y, Ge J, Zhang B-P and Wan W 2015 Improving photocatalytic performance of ZnO via synergistic effects of Ag nanoparticles and graphene quantum dots *Phys. Chem. Chem. Phys.* **17** 18645–52

[7] Yunus I S, Harwin, Kurniawan A, Adityawarman D and Indarto A 2012 Nanotechnologies in water and air pollution treatment *Environ. Technol. Rev.* **1** 136–48

[8] Li Y, Xu L, Li X, Shen X and Wang A 2010 Effect of aging time of ZnO sol on the structural and optical properties of ZnO thin films prepared by sol–gel method *Appl. Surf. Sci.* **256** 4543–7

[9] Tari O, Aronne A, Addonizio M L, Daliento S, Fanelli E and Pernice P 2012 Sol–gel synthesis of ZnO transparent and conductive films: A critical approach *Sol. Energy Mater. Sol. Cells* **105** 179–86

# Optimalization of Freezing-Thawing Process in Enhancing Magnetic Properties of $Fe_3O_4$/PAA/PVA Magnetic Hydrogel Composites

**Sunaryono[1,2], H Hifdziyah[1], A Taufiq[1,2], M Diantoro[1,2] and N Mufti[1,2]**

[1]Department of Physics, Faculty of Mathematics and Natural Sciences, Universitas Negeri Malang (State University of Malang) Jl. Semarang 5 Malang 65145, Indonesia
[2]Center for Minerals and Advanced Materials, Faculty of Mathematics and Natural Sciences, Universitas Negeri Malang (State University of Malang) Jl. Semarang 5 Malang 65145, Indonesia

Email: sunaryono.fmipa@um.ac.id

**Abstract.** Magnetic hydrogel composites of $Fe_3O_4$/polyacrylic acid (PAA)/polyvinyl alcohol (PVA) have been successfully fabricated through a freezing-thawing cycle method. The natural iron sand-based $Fe_3O_4$ filler nanoparticle has been successfully synthesized through coprecipitation at a room temperature. The crystal structure and the morphology of the $Fe_3O_4$ nanoparticle were investigated by using X-ray Diffractometer (XRD), and Transmission Electron Microscopy (TEM), respectively. The study of magnetic characteristics of the $Fe_3O_4$/PAA/PVA magnetic hydrogel composites were characterized by using Vibrating Sample Magnetometer (VSM). Based on the quantitative analysis, the morphology of $Fe_3O_4$ filler characterized by using TEM has a particle size around of 8 nm. This result was well confirmed with the analysis of the XRD and VSM data. Additionally, the study of the magnetic properties of the $Fe_3O_4$/PAA/PVA magnetic hydrogel composites decreased saturation magnetization significantly with the increase of the number of freezing-thawing cycles. This result indicated that the distribution pattern of the $Fe_3O_4$ filler in the magnetic hydrogel composites was more optimal when the process of freezing-thawing reached seven times cycles.

**Keywords:** $Fe_3O_4$, PAA, PVA, composite, magnetic hydrogel

## 1. Introduction

The magnetic iron oxide, or in the chemistry formulation it is called $Fe_3O_4$, is a magnetic material having the potential for some fields, especially nanotechnology and nanoscience. In the field study, $Fe_3O_4$ is a necessary material for fabrication of liquid magnetic [1], sensor [2], gel magnetic [3], and spintronic [4]. $Fe_3O_4$ shows the most attractive magnetic material characteristic, and it is commonly used due to its magnetic characteristic and low poisonous content [5]. $Fe_3O_4$ has attracted considerable attention as the magnetic filler in the hydrogel polymer especially in the research of magnetic gel. Hydrogel polymer is one of the hydrophilic macromolecule polymer types that is in the form of three-dimensional tissue, has swelling and deswelling abilities in the water [6]. Hydrogel has caught the researchers' attention due to its potential application in the biomedical field such as functioning as a contact lens, bandage, controlled medicine distribution system, scaffold-cell and so forth [7].

Polymer hydrogel, as matrix material for gel magnetic fabrication, has been reported by many previous researchers either in the variation of polymer type, synthesis method or the study on a physical characteristic. For example, Wu *et al.* [8] use 10-hydroxy camptothecin-based hydrogel (HCPT) for drug delivery application. The modification of acrylate PVA (Acr-PVA)-based hydrogel fabrication, and glycidyl methacrylate (GMA) was conducted to maximize the affinity and tissue viability [9]. Chitosan (CS) based biomaterial hybrid composite, hydroxyapatite (HA), and $Fe_3O_4$ are for the study on the mechanical characteristic [10]. Furthermore, Sivudu and Ree [11] investigate the sensitive act of pH to the polyacrylamide (PAM) based magnetic hydrogel composite. Such study has explained that the sensitive act of magnetic hydrogel pH is shown by the presence of the lower thermal stability at the initial decomposition temperature and the higher thermal stability at the end of its decomposition. Meanwhile, in the previous research, the researcher has successfully investigated the magneto-elasticity properties of polymer (PVA) based magnetic hydrogel composite [12] and the distribution pattern of magnetic filler in the PVA hydrogel [13]. However, the study on magnetic properties of polymer polyacrylic acid (PAA) and polyvinyl alcohol (PVA) based magnetic hydrogel composite is rarely reported.

This study investigated the magnetic characteristic of magnetic hydrogel composite by using a freezing-thawing method by varying some cycles and controlling the duration of the freezing-thawing process. The distribution filler of $Fe_3O_4$ in the magnetic hydrogel as the effect of the change of magnetic characteristic is reported as well.

## 2. Experimental Method

The materials used in this study were natural iron sand from Wedi Ireng Beach in Banyuwangi, HCl of 12.063 M (Sigma-Aldrich), $NH_4OH$ of 6.5 M (Sigma-Aldrich), PAA, PVA, and distilled water. In line with the previous report [14][15][16], the raw material of iron sand was separated from the impurity material by using a permanent magnet to get the high content of Fe. The Fe in the iron sand was characterized by using XRF to confirm the Fe content. The $Fe_3O_4$ nanoparticle was then synthesized by using coprecipitation method. The extracted natural iron was dissolved in HCl to produced $FeCl_2$ and $FeCl_3$ solutions. The solution then was titrated by using $NH_4OH$ so that the $Fe_3O_4$ precipitate. Those processes used hot plate magnetic stirrer. The precipitate of titration results was washed many times until reaching the normal pH (pH 7) and it was then filtered. The $Fe_3O_4$ nanoparticle of the filtering result was dried to be characterized by using XRD and TEM. The magnetic hydrogel composite was fabricated by using the combination of PAA and PVA hydrogels as well as a $Fe_3O_4$ filler. PAA and PVA were dissolved in the distilled water. The magnetic hydrogel was obtained by mixing the $Fe_3O_4$ nanoparticle with hydrogel through freezing-thawing (FT) method in three and seven cycles process. Finally, the magnetic hydrogel composite was characterized by using Vibrating Sample Magnetometer (VSM) to know the magnetic properties.

## 3. Results and Discussion

The crystal structure characterization result of iron sand before and after extraction is shown in Figure 1. Figure 1 shows that the iron sand without separation has many crystal peaks. This case indicated that there were many other compounds besides $Fe_3O_4$ contained in the sample. Meanwhile, in the diffraction pattern of iron sand, the separation result showed the particle crystal peak dominated by $Fe_3O_4$ phase.

The diffraction pattern of the $Fe_3O_4$ nanoparticle as the synthesis result is shown in Figure 2. All diffraction peaks indicated the $Fe_3O_4$ phase which was suitable for the Inorganic Crystal Structure Database (ICSD) with a code number of 9013529. The refinement result by using Rietica software showed that the size of $Fe_3O_4$ nanoparticle was 7.46 nm with a chi-squared $\chi^2$ was 0.043. The particle size ($D$) was determined as well by using Debye-Scherrer equation as following

$$D = \frac{k\lambda}{B\cos\theta} \tag{1}$$

where $k$ is constant of "Shape Factor", $\lambda$ is wavelength, $B$ is FWHM (Line broadening at half of the maximum intensity), and $\theta$ is Bragg's angle. The result of fitting the $Fe_3O_4$ particle size by using equation (1) through OriginPro software is shown in Figure 3. Based on the fitting result, the $Fe_3O_4$ particle size was 8.67 nm. This result was well confirmed by the characterization result of TEM (Figure 4) and XRD that showing the average value of particle size of 8.24 and 7.46 nm respectively. Furthermore, the average value of particle size obtained approached the previous report result that was around 8.9 nm [14]. The other research reports also obtained almost similar value. For example, Zhao *et al.* [17] report that the average value of magnetic particle size was about 11.4 nm. Sun *et al.* [18] obtained the average value of particle size was approximately 6 nm, and Girod e*t al.* [19] reported the average value of magnetic particle was about 7 nm. Furthermore, Saravanan *et al.* [20] and Cai & Wan [21] also published their study result stating that the primary particle magnetic was around 7 up to 9 nm.

Morphology of $Fe_3O_4$ nanoparticle (Figure 4) shows the presence of agglomeration as the effect of electrostatic force among particles. Agglomeration of $Fe_3O_4$ sample was difficult to avoid due to the magnetic nature of each particle. Giving surfactant material expected to be a solution to avoid agglomeration. The surfactant might played in reducing the electrostatic force among particles so that the particle might perfectly dispersed.

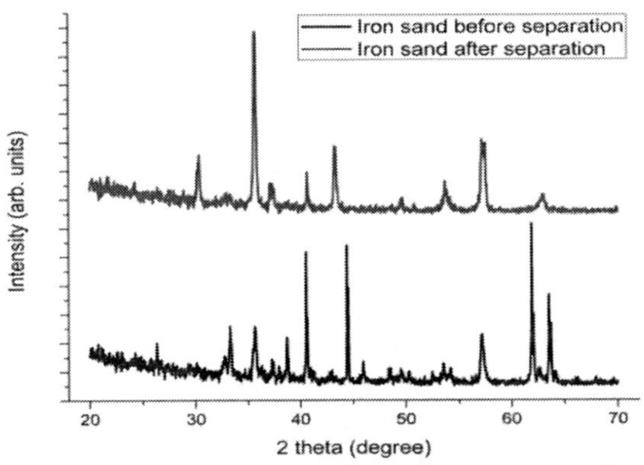


**Figure 1.** XRD patterns of the iron sands before and after separation process

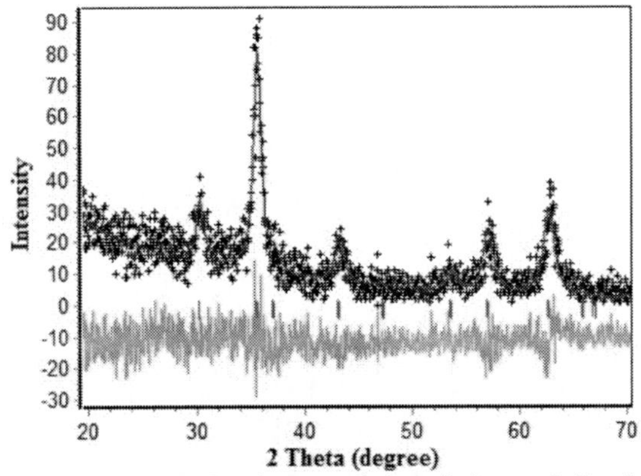


**Figure 2.** XRD data and its fitting of $Fe_3O_4$ particle

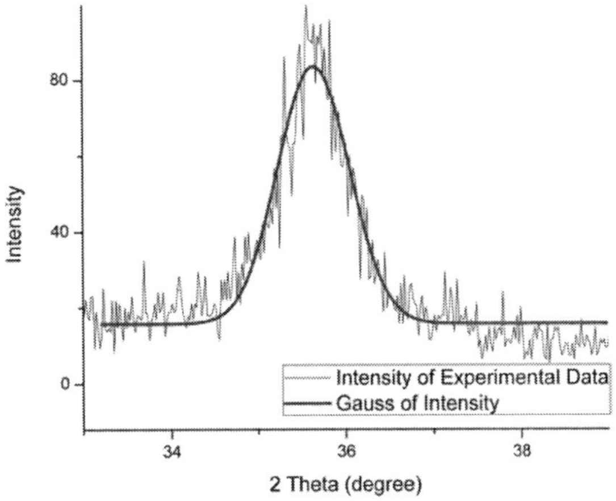


**Figure 3.** The fitting result of XRD data of $Fe_3O_4$ particle

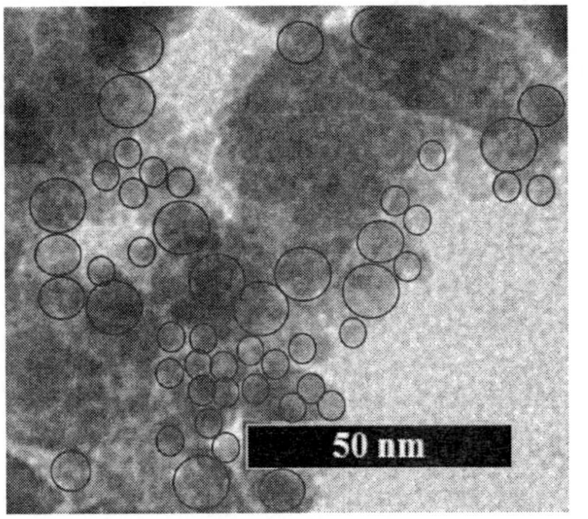


**Figure 4.** TEM image of $Fe_3O_4$ nanoparticle

IOP Conf. Series: Materials Science and Engineering **202** (2017) 012007 doi:10.1088/1757-899X/202/1/012007

**Table 1.** The Magnetization of magnetic hydrogel composite

| Sample | $M_r$ (emu/g) | $M_s$ (emu/g) | $H_c$ (T) |
|---|---|---|---|
| FT 3x | 0 | 35.676 | 0.009 |
| FT 7x | 0 | 28.073 | 0.009 |

**Table 2.** Magnetic properties and particle size of $Fe_3O_4$/PAA/PVA magnetic hydrogel composites

| Sample | $M_r$ (emu/g) | $M_s$ (emu/g) | $H_c$ (T) | D (nm) |
|---|---|---|---|---|
| FT 3x | 0.07 | 36.013 | 0.009 | 8.35 |
| FT 7x | 0.03 | 28.240 | 0.009 | 8.10 |

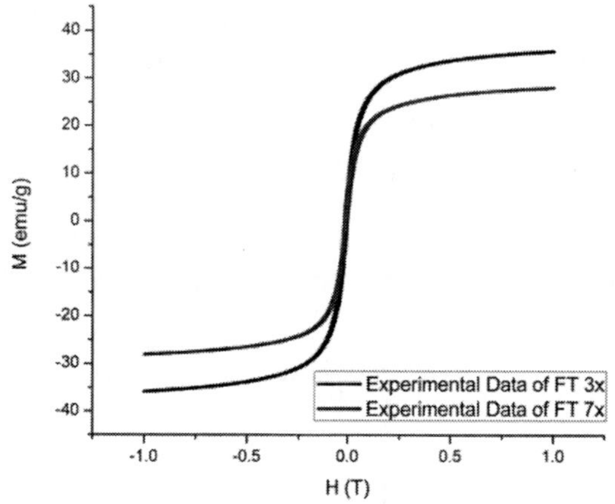


**Figure 5.** Magnetization hysteresis of $Fe_3O_4$/PAA/PVA magnetic hydrogel composite

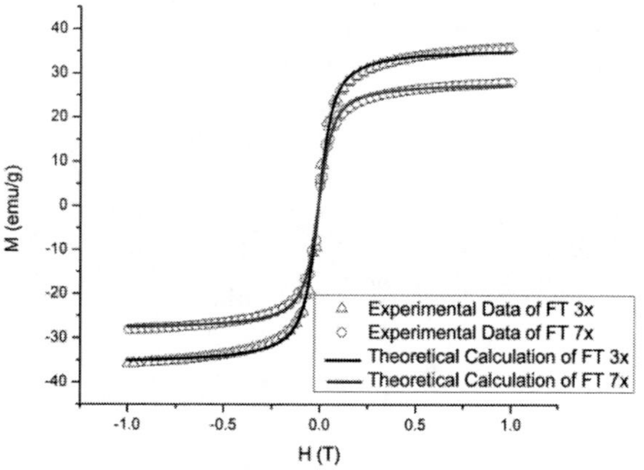


**Figure 6.** Fitting profile of magnetization hysteresis of $Fe_3O_4$/PAA/PVA magnetic hydrogel composite with cycle of (a) FT 3× and (b) FT 7×

The study on crystal structure and morphology of $Fe_3O_4$ filler has been discussed comprehensively. The next study is an investigation of the magnetic characteristic of magnetic hydrogel composite. The instrument used to characterize the magnetic characteristic of magnetic hydrogel composite was VSM. The data resulted from VSM instrument was hysteresis curve between magnetization ($M$) and magnetic field ($H$) as presented in Figure 5. The hysteresis value of certain material depends on the grain size, domain, pressure, and temperature. From the hysteresis curve, we can see that the other magnetic amount like remanent magnetization ($M_r$), saturation magnetization ($M_s$), and coercivity magnetic field ($H_c$) as shown in Table 1.

Furthermore, the hysteresis curve of $Fe_3O_4$/PAA/PVA magnetic hydrogel composite was fitted by using Lengevinmod model. The amount obtained will be used to calculate the sample magnetization and particle grain size. This result can be a subsequent confirmation from the value got by using XRD and TEM methods. The fitting result is shown in Figure 6 and the amounts got from fitting are shown in Table 2.

The correlation of value of magnetization ($M$), Bohr magneton ($\mu$), and Temperature ($T$) are written in Langevin equation as expressed in equation 2 [22],

$$M = M_s\left(\coth\left(\frac{\mu H}{k_B T}\right) - \frac{k_B T}{\mu H}\right) \quad (2)$$

where $k_B$ is Boltzmann constant ($1{,}38.10^{-23}$ $J/K$). The combination of linearity between Langevin equation and LangevinMod model was used to get the amount of sample magnetization based on the fitting result. LangevinMod model used was suitable with the equation 3,

$$y = y_0 + C\left(\coth\left(\frac{x-xc}{s}\right) - \frac{s}{x-xc}\right) \quad (3)$$

The elaboration of Langevin equation (equation 2) will result in correlation of particle grain size (*D*) and material magnetization as written in the equation 4 [23],

$$D = \sqrt[3]{\frac{18k_B T}{\pi}\frac{\chi}{\rho M_s}} \tag{4}$$

where $\chi$ is susceptibility of the initial sample, $\rho$ is density, and $M_s$ is saturation magnetization.

Based on the fitting result of magnetization hysteresis of $Fe_3O_4$/PAA/PVA magnetic hydrogel composite, the value of magnetization *Ms* and *Hc* obtained from the experiment data was in line with the model data. Based on the amount obtained both from VSM investigation (Table 1) and fitting process result (Table 2), the value of magnetization *Mr* and $H_c$ approached 0 and the size of $Fe_3O_4$ filler was around 8 nm. Thereby, it can be shown that $Fe_3O_4$/PAA/PVA magnetic hydrogel composite indicates superparamagnetic behavior. This phenomenon is in line with the previous report [13], and the study of Ramesbhabu *et al.* [24], as well as Sivudu *et al.* [11], revealed that particle is presenting superparamagnetic behavior when the coercivity value of the particle is relatively low. Besides referring to the value of magnetization *Mr* and *Hc*, the particle size can determine the magnetic material characteristic. Saravanan *et al.* [20] and Cai & Wan [21] reported that the primary particle of magnetite around 7 to 9 nm indicate superparamagnetic properties. Moreover, the nanoparticles can be exhibited having superparamagnetic characteristic if its size is less than 15 nm [25].

Furthermore, the value of saturation magnetization of $Fe_3O_4$/PAA/PVA magnetic hydrogel composite with FT 7x cycle process shows the lower value than FT 3x cycle process. This result shows that FT 7x cycle process is an optimal cycle for distribution process of $Fe_3O_4$ filler in the PAA/PVA matrix based magnetic hydrogel composite. This phenomenon can be explained by magnetic domain theory [26]. A magnetic domain is classified into such two areas as a single-domain and multi-domains restricted by critical point between both of them. The single-domain area consists of stable and unstable areas. The unstable area is more known as superparamagnetic area. The $Fe_3O_4$ filler used in this study was relatively small; it was around 8 nm so that it can be categorized existing in the single domain area. This phenomenon is in line with the studies of some researchers. Tong *et al.* [27] reveal that the magnetic particle of which size is less than 25 nm is categorized existing in the single domain area. Meanwhile, Morrish and Yu [28] reported that the magnetic particle categorized existing in the single domain area if the particle size is smaller than 50 nm. Cornell [29] categorized the single domain area positioned for a particle of which diameter is less than 100 nm. If the particle aggregation more increases, the saturation magnetization more increases as well. Thereby, the reduction of magnetic hydrogel composite magnetization in the FT 7x cycles is the representation of the decrease of particle aggregation in the cycle. Therefore, the distribution $Fe_3O_4$ filler in the magnetic hydrogel composite at FT 7x cycle becomes evenly distributed over the entire matrix network of PAA/PVA polymer hydrogel compared to FT 3x cycle.

Besides, PPA/PVA polymer matrix based magnetic hydrogel composite generally has saturation magnetization value greater if it is compared to PVA (*Ms* = 6.5 emu/g) polymer hydrogel matrix based magnetic hydrogel [13]. It can be concluded that PAA matrix can construct an effective crosslinking agent so it can increase the aggregation of $Fe_3O_4$ filler in the PAA/PVA polymer matrix network. This condition is strengthened as well by another researcher's report. They were devoted that the combination of PAA network will enhance the primary cyclization degree in the PAA/PVA polymer matrix network and increase the crosslinking agents of polymer chain so that it causes the increase of primary cyclization degree which is very sensitive to pH during polymerization process [30]. Thereby, the increase of crosslinking agents of polymer chain will increase the network structure in the PAA/PVA matrix based magnetic hydrogel composite. As a consequence, the aggregation of $Fe_3O_4$ filler in the matrix network of PAA/PVA hydrogel will enhance followed by the increase of the saturation magnetization value of magnetic hydrogel composite.

## 4. Conclusion

$Fe_3O_4$/PAA/PVA magnetic hydrogel composite has been successfully fabricated by using freezing-thawing method for 3 and 7 times cycle processes. The magnetic hydrogel of both samples presents a superparamagnetic behavior. This case is caused by two things, namely the magnetization value and the particle size. The remanent magnetization and coercivity field of both samples approaching 0 and the size of $Fe_3O_4$ filler (~ 8 nm) which is relatively small can be categorized existing in the single domain area. Based on the analysis result, the distribution of $Fe_3O_4$ filler and magnetization characteristic of magnetic hydrogel composite is influenced by freezing-thawing cycle and matrix polymer material used. The distribution of $Fe_3O_4$ filler in the PAA/PVA hydrogel matrix was optimally spread out when freezing-thawing 7x cycles compared to the 3x cycles. Generally, the additional PAA polymer matrix on the $Fe_3O_4$/PAA/PVA magnetic hydrogel composite is more effective in increasing its magnetization properties.

## 5. References

[1] Taufiq A, Sunaryono, Rachman Putra E, Okazawa A, Watanabe I, Kojima N, Pratapa S and Darminto 2015 Nanoscale Clustering and Magnetic Properties of $Mn_xFe_{3-x}O_4$ Particles Prepared from Natural Magnetite *J. Supercond. Nov. Magn.* **28** 2855–63

[2] Freire T M, Dutra L M U, Queiroz D C, Ricardo N M P S, Barreto K, Denardin J C, Wurm F R, Sousa C P, Correia A N, de Lima-Neto P and Fechine P B A 2016 Fast ultrasound assisted synthesis of chitosan-based magnetite nanocomposites as a modified electrode sensor *Carbohydr. Polym.* **151** 760–9

[3] Ramanujan R V and Lao L L 2006 The mechanical behavior of smart magnet–hydrogel composites *Smart Mater. Struct.* **15** 952

[4] Noh J, Osman O I, Aziz S G, Winget P and Brédas J-L 2015 Magnetite $Fe_3O_4$ (111) Surfaces: Impact of Defects on Structure, Stability, and Electronic Properties *Chem. Mater.* **27** 5856–67

[5] Häring M, Schiller J, Mayr J, Grijalvo S, Eritja R and Díaz D D 2015 Magnetic Gel Composites for Hyperthermia Cancer Therapy *Gels* **1**

[6] Lao L L and Ramanujan R V 2004 Magnetic and hydrogel composite materials for hyperthermia applications *J. Mater. Sci. Mater. Med.* **15** 1061–4

[7] Jiang F, Wang X, He C, Saricilar S and Wang H 2015 Mechanical properties of tough hydrogels synthesized with a facile simultaneous radiation polymerization and cross-linking method *Radiat. Phys. Chem.* **106** 7–15

[8] Wu C, Li R, Yin Y, Wang J, Zhang L and Zhong W 2017 Redox-responsive supramolecular hydrogel based on 10-hydroxy camptothecin-peptide covalent conjugates with high loading capacity for drug delivery *Mater. Sci. Eng. C* **76** 196–202

[9] Drozdova M G, Zaytseva-Zotova D S, Akasov R A, Golunova A S, Artyukhov A A, Udartseva O O, Andreeva E R, Lisovyy D E, Shtilman M I and Markvicheva E A 2017 Macroporous modified poly (vinyl alcohol) hydrogels with charged groups for tissue engineering: Preparation and in vitro evaluation *Mater. Sci. Eng. C* **75** 1075–82

[10] Heidari F, Razavi M, E.Bahrololoom M, Bazargan-Lari R, Vashaee D, Kotturi H and Tayebi L 2016 Mechanical properties of natural chitosan/hydroxyapatite/magnetite nanocomposites for tissue engineering applications *Mater. Sci. Eng. C* **65** 338–44

[11] Sivudu K S and Rhee K Y 2009 Preparation and characterization of pH-responsive hydrogel magnetite nanocomposite *Colloids Surf. Physicochem. Eng. Asp.* **349** 29–34

[12] Sunaryono, Taufiq A, Munaji, Indarto B, Triwikantoro, Zainuri M and Darminto 2013 Magneto-elasticity in hydrogels containing $Fe_3O_4$ nanoparticles and their potential applications *AIP Conf. Proc.* **1555** 53–6

[13] Sunaryono, Taufiq A, Putra E G R, Okazawa A, Watanabe I, Kojima N, Rugmai S, Soontaranon S, Zainuri M, Triwikantoro, Pratapa S and Darminto 2015 Small-Angle X-Ray Scattering Study on PVA/$Fe_3O_4$ Magnetic Hydrogels *Nano* **11** 1650027

[14] Sunaryono, Taufiq A, Mashuri, Pratapa S, Zainuri M, Triwikantoro and Darminto 2015 Various Magnetic Properties of Magnetite Nanoparticles Synthesized from Iron-sands by Coprecipitation Method at Room Temperature *Trans Tech Publ. Switz.* **827** 229–34

[15] Taufiq A, Sunaryono, Giri Rachman Putra E, Pratapa S and Darminto 2015 Nano-structural studies on $Fe_3O_4$ particles dispersing in a magnetic fluid using X-ray diffractometry and small-angle neutron scattering *Trans Tech Publ. Switz.* **827** 213–8

[16] Baqiya M A, Taufiq A, Sunaryono, Ayun K, Zainuri M, Pratapa S, Triwikantoro and Darminto 2017 Spinel-Structured Nanoparticles for Magnetic and Mechanical Applications *Magnetic Spinels - Synthesis, Properties and Applications* ed M S Seehra (InTech)

[17] Zhao Y, Qiu Z and Huang J 2008 Preparation and Analysis of $Fe_3O_4$ Magnetic Nanoparticles Used as Targeted-drug Carriers Supported by the Technology Project of Jiangxi Provincial Education Department and Jiangxi Provincial Science Department. *Chin. J. Chem. Eng.* **16** 451–5

[18] Sun J, Zhou S, Hou P, Yang Y, Weng J, Li X and Li M 2007 Synthesis and characterization of biocompatible $Fe_3O_4$ nanoparticles. *J. Biomed. Mater. Res. A* **80** 333–41

[19] Girod M, Vogel S, Szczerba W and Thünemann A F 2015 How temperature determines formation of maghemite nanoparticles *10th Int. Conf. Sci. Clin. Appl. Magn. Carr. 10-14 June 2014 Dresd. Ger.* **380** 163–7

[20] Saravanan P, Alam S, Kandpal L D and Mathur G N 2002 Effect of substitution of Mn ion on magnetic properties of $Fe_3O_4$ nanocrystallites *J. Mater. Sci. Lett.* **21** 1135–7

[21] Cai W and Wan J 2007 Facile synthesis of superparamagnetic magnetite nanoparticles in liquid polyols *J. Colloid Interface Sci.* **305** 366–70

[22] Chantrell R W, Popplewell J and Charles S 1978 Measurements of particle size distribution parameters in ferrofluids *Magn. IEEE Trans. On* **14** 975–7

[23] Liu Z L, Liu Y J, Yao K L, Ding Z H, Tao J and Wang X 2002 Synthesis and Magnetic Properties of $Fe_3O_4$ Nanoparticles *J. Mater. Synth. Process.* **10** 83–7

[24] Rameshbabu R, Ramesh R, Kanagesan S, Karthigeyan A and Ponnusamy S 2014 Synthesis and Study of Structural, Morphological and Magnetic Properties of $ZnFe_2O_4$ Nanoparticles *J. Supercond. Nov. Magn.* **27** 1499–502

[25] Wu W, He Q and Jiang C 2008 Magnetic Iron Oxide Nanoparticles: Synthesis and Surface Functionalization Strategies *Nanoscale Res. Lett.* **3** 397–415

[26] Cullity B D and Graham C D 2008 Fine Particles and Thin Films *Introduction to Magnetic Materials* (John Wiley & Sons, Inc.) pp 359–408

[27] Tong G, Liu Y, Cui T, Li Y, Zhao Y and Guan J 2016 Tunable dielectric properties and excellent microwave absorbing properties of elliptical $Fe_3O_4$ nanorings *Appl. Phys. Lett.* **108** 072905

[28] Morrish A H and Yu S P 1956 Magnetic Measurements on Individual Microscopic Ferrite Particles Near the Single-Domain Size *Phys Rev* **102** 670–673

[29] Cornell R M and Schwertmann U 2004 Electronic, Electrical and Magnetic Properties and Colour *The Iron Oxides* (Wiley-VCH Verlag GmbH & Co. KGaA) pp 111–37

[30] Elliott J E, Macdonald M, Nie J and Bowman C N 2004 Structure and swelling of poly(acrylic acid) hydrogels: effect of pH, ionic strength, and dilution on the crosslinked polymer structure *Polymer* **45** 1503–10

**Acknowledgements**
The authors would like to thank to DRPM the Ministry of Research, Technology, and Higher Education (DIKTI) for the financial support from the "*Penelitian Produk Terapan*", 2016-2017 for S.

# Fabrication of Magnetite Nanoparticles Dispersed in Olive Oil and Their Structural and Magnetic Investigations

**A Taufiq[1,2], R E Saputro[1], Sunaryono[1,2], N Hidayat[1,2], A Hidayat[1,2], N Mufti[1,2], M Diantoro[1,2], A Patriati[3], Mujamilah[4], E G R Putra[3,5] and H Nur[6]**

[1]Department of Physics, Faculty of Mathematics and Natural Sciences, Universitas Negeri Malang (State University of Malang), Jl. Semarang 5, Malang 65145, Indonesia
[2]Center for Minerals and Advanced Materials, Faculty of Mathematics and Natural Sciences, Universitas Negeri Malang (State University of Malang), Jl. Semarang 5, Malang 65145, Indonesia
[3]Centre for Science and Technology of Advanced Materials, National Nuclear Energy Agency of Indonesia, Kawasan Puspiptek Serpong, Tangerang 15314, Indonesia
[4]Centre for Technology of Nuclear Industry Materials, National Nuclear Energy Agency of Indonesia, Kawasan Puspiptek Serpong, Tangerang 15314, Indonesia
[5]Sekolah Tinggi Teknologi Nuklir, National Nuclear Energy Agency of Indonesia, Yogyakarta 55281, Indonesia
[6]Centre for Sustainable Nanomaterials, Ibnu Sina Institute for Scientific and Industrial Research Universiti Teknologi Malaysia, 81310 UTM Skudai, Johor, Malaysia

Email: ahmad.taufiq.fmipa@um.ac.id

**Abstract:** In this work, the iron sand taken from Wedi Ireng Beach in Banyuwangi, Indonesia, was employed as the main precursor in fabricating magnetite nanoparticles. The magnetite nanoparticles were then functionalized in preparing magnetic fluids coated by oleic acid as a surfactant and dispersed in olive oil as a liquid carrier. The phase purity, crystallite size and crystal structure of the dried magnetic fluids were characterized by using X-Ray Diffractometer. Meanwhile, the functional groups of the magnetic fluids were investigated by means of Fourier Transform Infra-Red (FTIR) spectroscopy. The particle size and morphology of the magnetite particles were also investigated by using Transmission Electron Microscopy (TEM). The magnetic behaviors of the magnetic fluids were determined by using Vibrating Sample Magnetometer (VSM). Based on the XRD data analysis, the magnetite particles crystallized in the spinel structure without the presence of any other phases. The FTIR spectra showed that the functional groups of the magnetic fluids were referring to the magnetite, oleic acid, and olive oil. The TEM image presented that the magnetite particle was formed in a nanometric size. Finally, the saturation magnetization of the magnetic fluids varied in the mass composition and particle size of the magnetite nanoparticles.

**Keywords:** Magnetite, magnetic fluid, olive oil, structure, magnetization

IOP Conf. Series: Materials Science and Engineering **202** (2017) 012008 doi:10.1088/1757-899X/202/1/012008

## 1. Introduction

Magnetic fluids or ferrofluids, colloidal suspensions containing magnetic nanoparticles in a single domain and dispersed in appropriate media, become a group of smart magnetic materials that play an essential role in many advanced applications. Due to their unique properties, the magnetic fluids can be applied in several fields, such as for rotating seal and rotating damper [1], as contrast agent in magnetic resonance imaging [2], for magnetic sensing [3], and so forth.

In 2016, Borin and co-workers successfully prepared magnetite-based magnetic fluids by employing alkenyl succinic anhydride, stearic acid, and oleic acid as combined surfactants [4]. In 2015, El-Boubbou and co-workers successfully synthesized the magnetite nanocolloids by using fatty acid as a stabilizer [5]. Furthermore, Lobato *et al.* successfully prepared magnetic fluids by employing oleic acid as appropriate surfactant and produced the magnetic with average particle size of about 10 nm [6]. Therefore, it can be pointed out that oleic acid becomes one of the best choices to be used as an essential surfactant in surfacing the magnetite particles in magnetic fluids

It was reported that the magnetic fluids surfaced by oleic acid as a surfactant were prepared using several methods to produce them in various sizes and morphologies. In general, the preparation of magnetic fluids was started by preparing the magnetic nanoparticles using several methods such as coprecipitation [7,8], sonochemistry [9], hydrothermal [10], one-pot synthesis [11], and so on. Moreover, the magnetic fluids were also prepared by dispersing the magnetic nanoparticles in various liquid carriers such as water [7], dialkyldiphenyl and polyethylsiloxane [4], transformer oil [12], and others. However, several problems of the utilization of the mentioned liquid carriers still appeared and needed to be overcome by suitable strategies. The strategies can be conducted such as by lowering the particle size of the magnetic and reducing the agglomeration using appropriate liquid carrier and surfactant.

In this work, we develop synthesis method of the magnetic fluids by introducing a combination of oleic acid and olive oil as a stabilizer and as a dispersing media of the magnetic nanoparticles, respectively. The oleic acid and olive oil were selected because they have good performances related to their similar physical properties. Therefore, we believed that olive oil could play an essential role as a suitable liquid carrier to produce the magnetic fluids with a homogeneous magnetic particle size. So far, based on our knowledge, there is no a report of the utilization of oleic acid as a surfactant and olive oil as a liquid carrier in preparing the magnetic fluids from iron sand becomes a novelty of this work. In order to cut off the preparation cost, we propose the use of an inexpensive raw material in preparing the magnetic fluids by exploring the natural iron sand from Indonesia. Furthermore, this paper also presents the structural and magnetic behaviors of the magnetic fluids.

## 2. Experimental Method

The materials used were iron sand taken from Wedi Ireng Beach in Banyuwangi, Indonesia, HCl, $NH_4OH$, distilled water, oleic acid, and olive oil. The iron sand was washed by using distilled water for several times and followed by natural drying process under sunlight. The iron sand was then extracted by using a magnetic separator to obtain the magnetite powder [13]. The powder was reacted with HCl as a dissolving mediator by using the method proposed in our previous work [14]. The product of the previous process was then reacted with $NH_4OH$ using a magnetic stirrer to achieve a black precipitate sample [15]. The process was continued by washing the precipitate sample for several times to get magnetite particles. The prepared magnetite particles were then coated with oleic acid as a surfactant and dispersed in olive oil as a liquid carrier. The three samples were prepared by varying the mass compositions of the magnetite particles in the magnetic fluids to 0.32, 0.38, and 0.41 g. The samples were coded by MF1, MF2, and MF3 for the magnetite particles' compositions of 0.32, 0.32, and 0.41, respectively.

For characterization processes, X-Ray Diffractometer (XRD) was employed to study the phase purity, crystallite size and crystal structure of the dried magnetite fluids. Fourier Transform Infra Red (FTIR) spectroscopy was used to characterize the functional groups of the magnetic fluids. Transmission Electron Microscopy (TEM) was utilized to investigate the particle size and morphology of the magnetite particles. Finally, Vibrating Sample Magnetometer (VSM) was employed to study the magnetic behaviors of the magnetic fluids.

## 3. Results and Discussion

The X-ray diffraction pattern of the dried magnetic fluid prepared from iron sand taken from Wedi Ireng Beach in Banyuwangi, Indonesia, is presented in Figure 1. In the figure, the circle and solid line represent the diffraction data and refinement model, respectively.

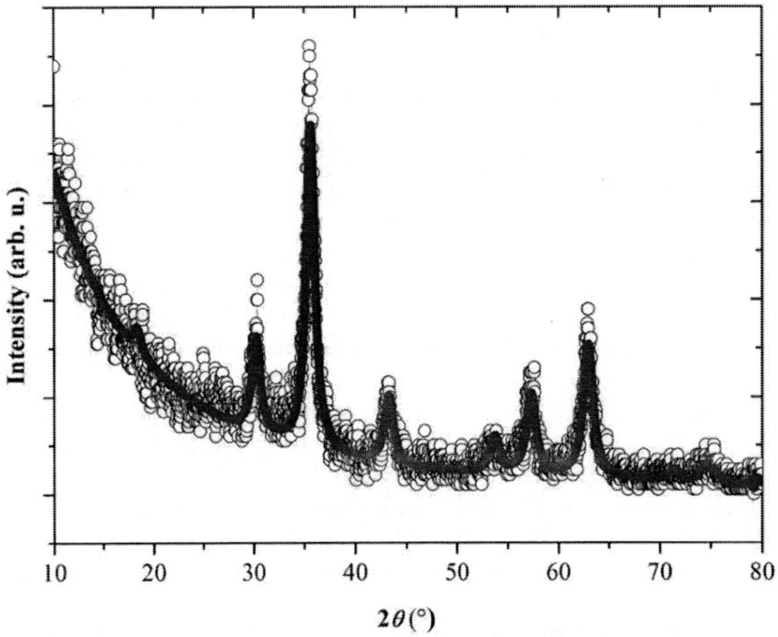


**Figure 1.** X-ray diffraction pattern of the dried magnetic fluid using Cu-Kα radiation 1.5406 Å.

The quantitative analysis for the X-ray diffraction pattern exhibited a single phase of magnetite without any impurities. The refinement parameters such as *Rp, Rwp,* and *GoF* were 13.64, 18.91, and 0.58, respectively. These values indicated that the experimental data were well refined by the theoretical model. Furthermore, the lattice parameter and crystal volume of the dried magnetic fluid obtained from Rietveld analysis were 8.354 Å and 583.02 Å$^3$, respectively. The sample was crystallized in a cubic crystal structure. In this structure, the $Fe^{2+}$ and $Fe^{3+}$ ions randomly placed tetrahedral and octahedral sites. Furthermore, the dried magnetic fluid formed on a nanometric scale with the size of about 9 nm. This result corresponds to the size of secondary particles constructed by primary particles as building blocks [13]. In order to confirm the particle size of the sample, we also characterized the dried magnetic fluid by using TEM as presented in the following figure.

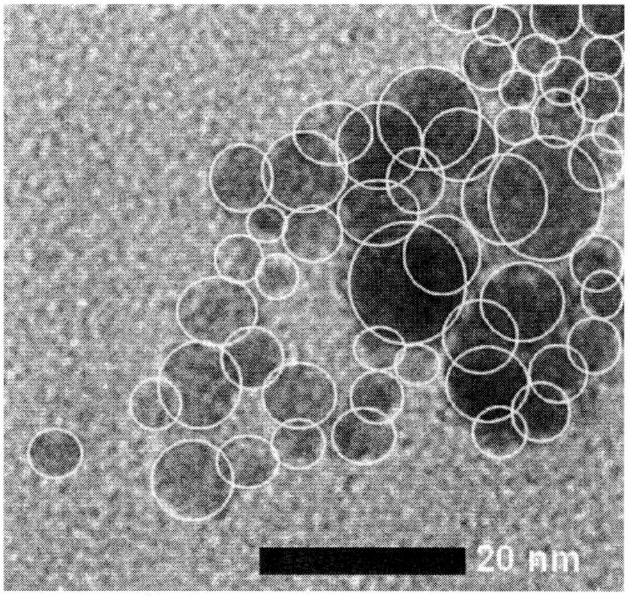


**Figure 2.** TEM image of the dried magnetic fluid.

The TEM image in Figure 2 shows that the dried magnetic fluid tended to agglomerate in a nanometric size. Theoretically, the agglomeration phenomenon in the magnetite occurs during nucleation process of the magnetic nanoparticles. Quantitatively, the average particle size of the sample was approximately 9.3 nm. This result is quite similar to the particle size obtained by a calculation using Debye Scherer's formula for the XRD data. Another work reported by Song and co-workers has successfully employed a co-precipitation method in preparing magnetite nanoparticles with the size of about 12 nm from commercial precursors [16]. Furthermore, Liu et al. reported that they successfully synthesized the magnetite nanoparticles by using a rapid microwave-assisted hydrothermal method with the particle size of about 20-25 nm, also from commercial precursors [17]. Therefore, the use of iron sand in this work was effective as a raw material in producing the magnetic fluids with the magnetite's particle size below 10 nm.

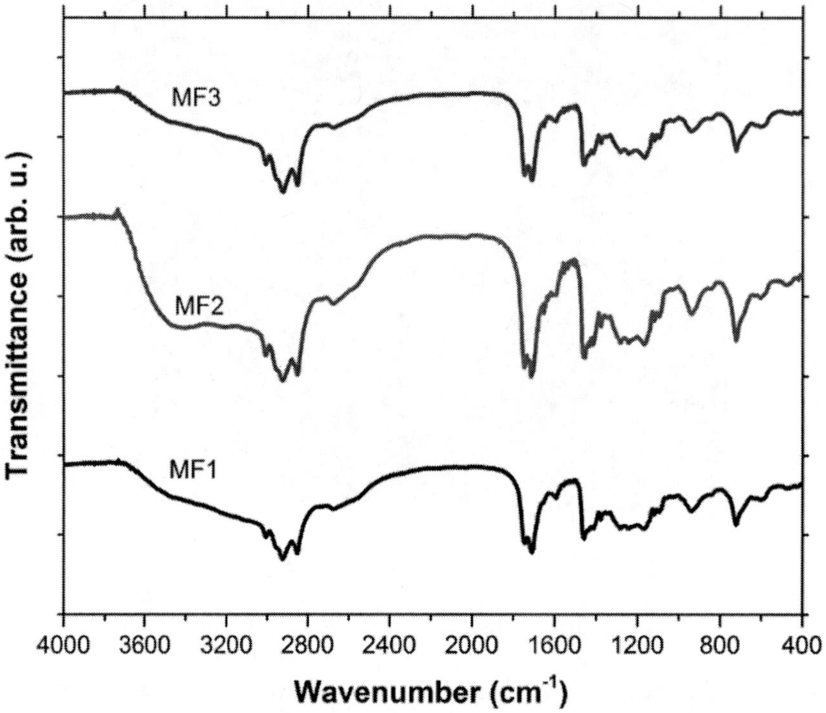


**Figure 3.** FTIR spectra of the magnetic fluids.

Based on Figure 3, it can be seen that there are several peaks indicating the functional groups of the samples. The magnetite particles can be identified at the wavenumbers of around 590 cm$^{-1}$ and 600 cm$^{-1}$. These peaks represented the intrinsic stretching vibrations of $Fe^{2+}$-O and $Fe^{3+}$-O that randomly placed in the octahedral and tetrahedral positions. In a good agreement with this experiment, other groups reported that vibrations of C-H could be detected at around 1112 cm$^{-1}$, stretching vibrations of C=C in oleic acid could be detected at around 1384 cm$^{-1}$, stretching vibrations at 3404 cm$^{-1}$ and 1636 cm$^{-1}$ were originated from -$CH_3$ and C=C groups, respectively [18]. Furthermore, the presence of oleic acid could be detected by the stretching vibrations of -$CH_2$ groups at the bands around 2921 cm$^{-1}$ and 1852 cm$^{-1}$ [19]. The bands at around 1430 cm$^{-1}$ and 1590 cm$^{-1}$ represented the vibrations of a oleat (-COO) [20]. Furthermore, the peak observed at around 1744 cm$^{-1}$ represented the vibration of C=O, while the peaks at around 3005 cm$^{-1}$ and 1171 cm$^{-1}$ represented C-H and C-O, respectively, as band characters of olive oil. These results are in good agreement with the results reported by another group [21]. Therefore, the FTIR data showed the present oleic acid that successfully covered the magnetite nanoparticles of the fluids dispersed in the olive oil. As a final remark, in this work, the olive oil plays an essential role as a good liquid carrier of the magnetic fluids.

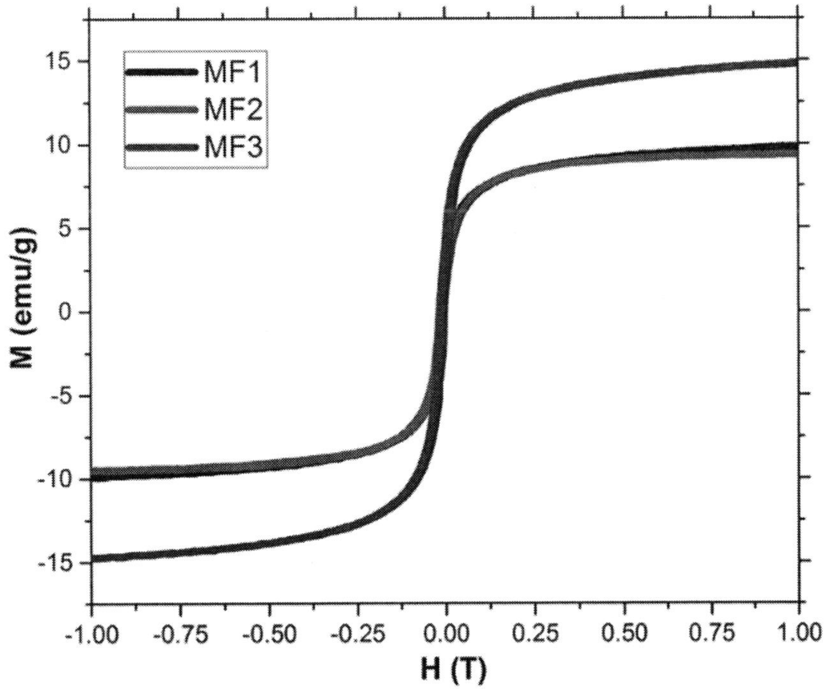


**Figure 4.** Magnetization curves of the magnetic fluids.

Figure 4 shows the magnetization curve of the magnetic fluids prepared from the iron sand. From the figure, MF1 and MF2 had similar saturation magnetization values, while MF3 had the highest one. The particle size obtained from the quantitative data analysis for magnetization curve and saturation magnetization of the magnetic fluids are presented in Table 1.

**Table 1.** Saturation magnetization ($M_s$), and particle size of the magnetic fluids

| Sample | $M_s$ (emu/g) | Particle size (nm) |
|---|---|---|
| MF1 | 9.81 | 8.0 |
| MF2 | 9.76 | 7.3 |
| MF3 | 14.81 | 8.4 |

Visually, all samples exhibited a superparamagnetic character at room temperature because they had an *S* shape and their coercive field and remanent magnetization were almost negligible. From Table 1, all of the samples had the particle size of less than 10 nm which indicated a single domain character of the magnetic fluids and exhibited a superparamagnetic character. In a single domain state, the magnetic particles are uniformly magnetized and alignment of all the spins are in the same direction [22]. It can be seen from Figure 4 and Table 1 that the MF1 and MF2 samples have a similar saturation magnetization value of around 9.8 emu/g, while MF3 sample has the highest one of around 14.8 emu/g. Quantitatively, the saturation magnetization ($M_s$) values of all samples were smaller than that of bulk magnetite [23]. The decreasing saturation magnetization was originated from the decreasing particle size of the sample. In more detail, the saturation magnetization values of MF1 and MF2 were quite similar. Theoretically, the saturation magnetization of MF2 should increase because its magnetite particle size was bigger than that of MF1. This phenomenon was originated from the increasing magnetite content in the magnetic fluid. Furthermore, MF3 has the highest saturation magnetization value than those of other samples originating from its increasing mass composition and particle size of the magnetite simultaneously. These results are in good agreement with the results reported by another group [24].

## 4. Conclusion

Based on the above discussion, we conclude that the magnetite particles from iron sand were successfully prepared as magnetic fluids using oleic acid as a surfactant and olive oil as a liquid carrier. All magnetic particles were crystallized in a cubic structure in a single phase. The magnetic particle sized in a nanometric scale of below 10 nm obtained by XRD, TEM, and VSM characterizations. The functional groups obtained by FTIR characterization showed the presence of the magnetite, oleic acid, and olive oil in the magnetic fluids. Moreover, the saturation magnetization of the magnetic fluids had a function of the mass composition and particle size of magnetite nanoparticles.

## 5. References

[1] Marinică O, Susan-Resiga D, Bălănean F, Vizman D, Socoliuc V and Vékás L 2016 Nano-micro composite magnetic fluids: Magnetic and magnetorheological evaluation for rotating seal and vibration damper applications *J. Magn. Magn. Mater.* **406** 134–43

[2] Casula M F, Corrias A, Arosio P, Lascialfari A, Sen T, Floris P and Bruce I J 2011 Design of water-based ferrofluids as contrast agents for magnetic resonance imaging *J. Colloid Interface Sci.* **357** 50–5

[3] Homa D and Pickrell G 2014 Magnetic Sensing with Ferrofluid and Fiber Optic Connectors *Sensors* **14** 3891–6

[4] Borin D Y, Korolev V V, Ramazanova A G, Odenbach S, Balmasova O V, Yashkova V I and Korolev D V 2016 Magnetoviscous effect in ferrofluids with different dispersion media *J. Magn. Magn. Mater.* **416** 110–6

[5] El-Boubbou K, Al-Kaysi R O, Al-Muhanna M K, Bahhari H M, Al-Romaeh A I, Darwish N, Al-Saad K O and Al-Suwaidan S D 2015 Ultra-Small Fatty Acid-Stabilized Magnetite Nanocolloids Synthesized by In Situ Hydrolytic Precipitation *J. Nanomater.* **2015** 1–11

[6] Lobato N C C, Ferreira Â de M and Mansur M B 2016 Evaluation of magnetic nanoparticles coated by oleic acid applied to solvent extraction processes *Sep. Purif. Technol.* **168** 93–100

[7] Taufiq A, Sunaryono, Rachman Putra E G, Pratapa S and Darminto 2015 Nano-Structural Studies on Fe3O4 Particles Dispersing in a Magnetic Fluid Using X-Ray Diffractometry and Small-Angle Neutron Scattering *Mater. Sci. Forum* **827** 213–8

[8] Sunaryono, Taufiq A, Putra E G R, Okazawa A, Watanabe I, Kojima N, Rugmai S, Soontaranon S, Zainuri M, Triwikantoro, Pratapa S and Darminto 2015 Small-Angle X-Ray Scattering Study on PVA/Fe3O4 Magnetic Hydrogels *Nano* **11** 1650027

[9] Wang Y, Nkurikiyimfura I and Pan Z 2015 Sonochemical Synthesis of Magnetic Nanoparticles *Chem. Eng. Commun.* **202** 616–21

[10] Rizzuti A, Dassisti M, Mastrorilli P, Sportelli M C, Cioffi N, Picca R A, Agostinelli E, Varvaro G and Caliandro R 2015 Shape-control by microwave-assisted hydrothermal method for the synthesis of magnetite nanoparticles using organic additives *J. Nanoparticle Res.* **17** 1–16

[11] Chen X, Organ M G and Pietro W J 2016 One-pot Synthesis of Size Controllable Amine-Functionalized Core-Shell Magnetic Nanoparticles for Use in Microfluidic Flow Separators *J Nanosci Adv Tech* **1** 25–31

[12] Rajnak M, Timko M, Kopcansky P, Paulovicova K, Tothova J, Kurimsky J, Dolnik B, Cimbala R, Avdeev M V, Petrenko V I and Feoktystov A 2017 Structure and viscosity of a transformer oil-based ferrofluid under an external electric field *J. Magn. Magn. Mater.* **431** 99–102

[13] Taufiq A, Sunaryono, Rachman Putra E G, Okazawa A, Watanabe I, Kojima N, Pratapa S and Darminto 2015 Nanoscale Clustering and Magnetic Properties of MnxFe3-xO4 Particles Prepared from Natural Magnetite *J. Supercond. Nov. Magn.* **28** 2855–63

[14] Sunaryono, Taufiq A, Mashuri, Pratapa S, Zainuri M, Triwikantoro and Darminto 2015 Various Magnetic Properties of Magnetite Nanoparticles Synthesized from Iron-Sands by Coprecipitation Method at Room Temperature *Mater. Sci. Forum* **827** 229–34

[15] Sunaryono, Taufiq A, Munaji, Indarto B, Triwikantoro, Zainuri M and Darminto 2013 Magneto-elasticity in hydrogels containing Fe3O4 nanoparticles and their potential applications *AIP Conf. Proc.* **1555** 53–6

[16] Song L N, Gu N and Zhang Y 2017 A moderate method for preparation DMSA coated Fe3O4 nanoparticles *IOP Conference Series: Materials Science and Engineering* vol 164(IOP Publishing)p 12026

[17] Liu Y, Cui T, Wu T, Li Y and Tong G 2016 Excellent microwave-absorbing properties of elliptical Fe3O4 nanorings made by a rapid microwave-assisted hydrothermal approach *Nanotechnology* **27** 165707

[18] Dhumal J, Bandgar S, Zipare K and Shahane G 2015 Fe 3 O 4 ferrofluid nanoparticles: synthesis and rheological behavior *Int J Mater Chem Phys* **1** 141–5

[19] Silva V A J, Andrade P L, Silva M P C, Bustamante D. A, De Los Santos Valladares L and Albino Aguiar J 2013 Synthesis and characterization of Fe3O4 nanoparticles coated with fucan polysaccharides *J. Magn. Magn. Mater.* **343** 138–43

[20] Darwish M S 2017 Effect of carriers on heating efficiency of oleic acid-stabilized magnetite nanoparticles *J. Mol. Liq.* **231** 80–5

[21] Laachari F, Maâtaoui H, El Bergadi F, Boukir A and Ibnsouda S K 2015 Improvement of a lipolytic activity assay protocol to increase the efficiency of the lipases activity measurement and validation by infrared spectroscopy *Afr. J. Biotechnol.* **14**

[22] Issa B, Obaidat I, Albiss B and Haik Y 2013 Magnetic Nanoparticles: Surface Effects and Properties Related to Biomedicine Applications *Int. J. Mol. Sci.* **14** 21266–305

[23] Cao S-W, Zhu Y-J and Chang J 2008 Fe3O4 polyhedral nanoparticles with a high magnetization synthesized in mixed solvent ethylene glycol–water system *New J. Chem.* **32** 1526

[24] Andrade Â L, Valente M A, Ferreira J M F and Fabris J D 2012 Preparation of size-controlled nanoparticles of magnetite *J. Magn. Magn. Mater.* **324** 1753–7

**Acknowledgements**
The authors would like to thank DRPM, Kemenristekdikti, Republic of Indonesia, for supporting a "research grant" for AT.

# Magnetically Separable $Fe_3O_4/SnO_2$/Graphene Adsorbent for Waste Water Removal

**V Paramarta[1,2], A Taufik[1,2] and R Saleh[1,2]**

[1]Departemen Fisika, Fakultas MIPA (Department of Physics, Faculty of Mathematics and Natural Sciences)-Universitas Indonesia, 16424 Depok, Indonesia
[2]Integrated Laboratory of Energy and Environment, Fakultas MIPA-Universitas Indonesia, 16424 Depok, Indonesia

Email: rosari.saleh@gmail.com

**Abstract**. Our previous study conducted the $SnO_2$ and $SnO_2$/graphene adsorption efficiency in Methylene Blue removal from aqueous solution, however, the difficulty of adsorbent separation from the methylene blue solution limits its efficiency. Therefore, in this work, $SnO_2$ and $SnO_2$/graphene was combined with $Fe_3O_4$ to improve the separation process and adsorption performance for removing the organic dyes. $Fe_3O_4/SnO_2$/grapheme were synthesized by using the co-precipitation method. The graphene content was varied from 1, 3, and 5 weight percent (wt%). The crystalline phase and thermal stability of the samples were characterized by using X- ray Diffraction (XRD) and Thermal Gravimetric Analysis (TGA). The adsorption ability of the samples was investigated by using significant adsorption degradation of MB observed when the graphene in $Fe_3O_4/SnO_2$ nanocomposite was added. The other parameters such as pH and initial concentration have also been investigated. The reusability was also investigated to study the stability of the samples. The fitting of equilibrium adsorption capacity result indicates that the adsorption mechanism of $Fe_3O_4/SnO_2$ nanocomposite with graphene tends to follow the Langmuir adsorption isotherm model.

**Keywords:** $SnO_2$, $Fe_3O_4/SnO_2$, graphene, adsorption, methylene blue.

## 1. Introduction

The environmental pollution has increased public concern nowadays because of its problem for aquatic life and environment. This problem must be decreased by using the suitable methods and technologies [1]. Methylene blue (MB) dye is one of the cationic dyes used as a material for dyeing wood, cotton, and silk [2]. A variety of methods such as adsorption, coagulation and flocculation, oxidation, chemical precipitation, and so forth, have been reported for wastewater treatment [3]. Among these, adsorption is an attractive and favorable technique because of their high efficiency, cost effectiveness, and simple operation [4].

A variety of sorbents have been tested to remove dyes in aqueous solution, and some metal oxides exhibit prominent potentials for adsorptive removal of dye waste [5]. One of metal oxide that is widely used in the degradation process of dye wastes is a tin oxide ($SnO_2$). $SnO_2$ has various advantages such as its sensitivity when it is exposed to the light and thermal energy, its numbers of an intrinsic defect in its structure caused by low formation energy and strong attraction between the tin sites and oxygen vacancies [6]. Graphene, a novel one-atom-thick two-dimensional graphitic carbon system with sp2-bonded carbon localized on the honeycomb lattice, has been under intense multidisciplinary study attribute to its extraordinary electrical and mechanical properties, large surface area, an excellent mobility of charge carriers, and good thermal conductivity [7]. Furthermore, combining the graphene material into the metal oxide can enhance the adsorption performance of the sample due to their functional groups and defect sites of graphene extending the lifetime of the adsorbent material by acting as support material which inhibits leaching of fine metal oxide nanoparticles into the treated water [8]. The coupling $SnO_2$ and graphene with magnetic adsorbent were investigated widely to overcome the disadvantage for difficult separation. Several researchers have reported on the use of $Fe_3O_4$ for removal and separation of dyes [9-10]. Therefore, it is believed that combining $SnO_2$, $Fe_3O_4$ and graphene will exhibit excellent adsorptive with highly effectively recovery by magnetic separation technique.

In this study, $Fe_3O_4/SnO_2$ nanoparticle will be combined with various weight percentages of graphene (1, 3, and 5 wt%) that are applied as adsorbents to degrade the methylene blue (MB) dye and the adsorption mechanism, pH, initial concentration, and temperature of these composites . The sample was characterized by using X-ray Diffraction (XRD) and Thermal Gravimetric Analysis (TGA).

## 2. Experimental Method

### *2.1 Chemicals*

All reagents used were analytical grade without further purification. Anhydrous tin chloride ($SnCl_2$), iron (II) sulfate heptahydrate ($FeSO_4.7H_2O$), sodium hydroxide (NaOH), methylene blue (MB), ethanol, ethylene glycol (EG) were purchased from Merck (Kenilworth, NJ, USA). Graphene was bought from Angstron Material.

### *2.2 Preparation of $SnO_2$ nanoparticle and $Fe_3O_4/SnO_2$ nanocomposites*

The $Fe_3O_4$ nanoparticles were synthesized by using the same method used in our previous study [11]. The $SnO_2$ nanoparticles were synthesized by using a modification of the method reported by Yue Li and co- workers [12]. First, $SnCl_2$ was dissolved in a mixture of ethanol and aqueous solution, which was then added into the NaOH solutions by using magnetic stirrer. Then, the mixed solutions were heated at 180 °C for 3 h, then chilled to room temperature. The precipitate was obtained by centrifugation and washed by using aqueous solutions and ethanol several times. The precipitated particles were dried under vacuum at 80 °C, and $SnO_2$ particles obtained by calcination for 3 h at 700 °C. The $Fe_3O_4/SnO_2$ nanocomposites were synthesized by using the sol-gel method. First, $SnO_2$ nanoparticles were mixed with $Fe_3O_4$ in a mixture of ethanol and aqueous solution. Each mixture was ultrasonicated for 2 h and then centrifuged to obtain the precipitate. The resulting product was then dried under vacuum at 80 °C for 12 h to obtain $Fe_3O_4/SnO_2$ nanocomposites.

### *2.3 Preparation of $Fe_3O_4/SnO_2$/graphene composites*

$Fe_3O_4/SnO_2$/graphene was synthesized by using the co-precipitation method. First, graphene with various weight percentage (1, 3, 5 wt%) was dissolved in distilled water and ethanol through the ultrasonic treatment for 2 hours, then $Fe_3O_4/SnO_2$ nanocomposites were poured into the solution and stirred magnetically. The mixed solution was then heated at 120 °C for 3 hours. The result of a solution was then centrifuged and dried at 70 °C under vacuum condition.

### *2.4 Characterization*

The sample was characterized by XRD using a Rigaku Miniflex 600 (Rigaku, Tokyo, Japan) with a Cu K-α radiation source ($\lambda$ =1.5406 Å) and thermal gravimetric analysis (TGA) of the sample obtained by using Rigaku TG8121 with the temperature from 27 °C until 900 °C.

IOP Conf. Series: Materials Science and Engineering **202** (2017) 012009 doi:10.1088/1757-899X/202/1/012009

*2.5 Adsorption processes*

Adsorption experiment was carried out by mixing $Fe_3O_4/SnO_2$/graphene into methylene blue (MB) solution used as pollutant model with the concentration of 20 mg/L, and the pH solution was adjusted By using NaOH. The concentration of dye was analyzed by using UV-Vis Spectrometer. The maximum degradation and the amount of MB adsorbed onto sample were calculated by using the following equation:

$$\text{Maximum degradation (\%)} = (C_t - C_0)\text{x } 100\%/C_0 \quad (1)$$

$$q_t = (C_0 - C_t)V/W;\ q_e = (C_0 - C_e)\ V/m \quad (2)$$

where $C_0$ and $C_t$ (mg $L^{-1}$) are initial and real-time concentration of organic pollutant, respectively. $q_t$ (mg $g^{-1}$) is the amount of adsorbed per unit mass of the absorbent at time $t$. $V$ (L) is the volume of the absorbed solution and $m$ (g) is the mass of the absorbent. Furthermore, adsorption experiment is carried out again by varying MB concentration for sample with best adsorption activity.

## 3. Results and Discussion

The crystalline structure of the $Fe_3O_4/SnO_2$/graphene composite with various weight percentages (1, 3, 5 wt% labeled as FS-g 1wt%, FS-g 2wt% and FS-g 5wt%, respectively) were analyzed by XRD and showed in Fig. 1. All of the $Fe_3O_4/SnO_2$/graphene composites displayed peaks corresponding to the tetragonal structure of $SnO_2$ and cubic spinel structure from $Fe_3O_4$. In Fig. 1, the diffraction peaks at $2\theta$ = 26,5°, 33.8°, 38°, 39°, 51.8°, 54.8°, 58°, 62°, 64.7°, 65.8°, 71.2°, 78.2°, 81.2°, 83.7° were assigned to the characteristic peak of $SnO_2$, which corresponded to the crystal planes of (110), (101), (200), (111), (211), (220), (002), (310), (112), (301), (202), (321), (400), and (222), respectively. The diffraction pattern of $Fe_3O_4$ was observed at the value of $2\theta$ = 30.14°, 35.49°, 43.28°, 57.20°, 62.83°, and 74.8° corresponding to each area (220), (311), (400), (511), (440) and (533), respectively. For graphene, there is only a broad peak at around of $2\theta$ of 25°, which shows the amorphous structure of graphene. As can be seen, the diffraction peak of the $SnO_2$ and graphene were overlapped at $2\theta$ around 26° at $Fe_3O_4/SnO_2$/graphene composites.

Figure 2 shows the weight loss curve for $Fe_3O_4/SnO_2$/graphene composites with different weight percentages (1, 3, and 5 wt%) obtained from TGA measurement. As can be seen, $Fe_3O_4/SnO_2$ nanoparticle did not exhibit the significant percentage loss in the result of TGA measurement which indicated that $Fe_3O_4/SnO_2$ nanoparticle without any addition of graphene was stable until 1000°C. The weight loss started around 100°C was due to the volatilization of adsorbed water [13]. The weight loss stage from 400°C decomposition took place thereafter [14]. The percentage loss of the samples could be identified as the total amount of graphene in the samples. Furthermore, the weight loss of graphene at TGA measurement could confirm the existence of the graphene in $Fe_3O_4/SnO_2$/graphene composite samples.

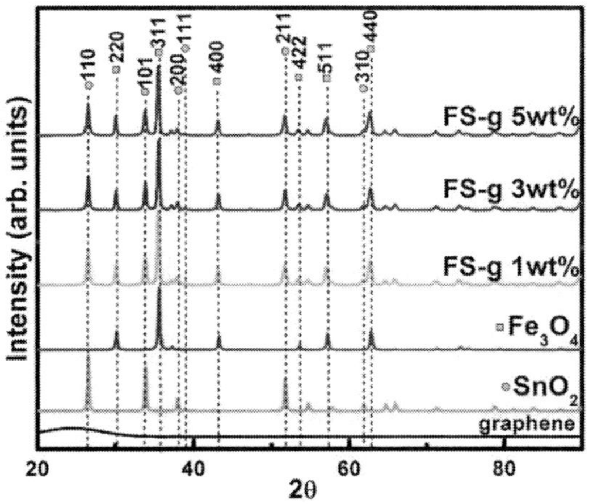


**Figure 1.** XRD patterns of $Fe_3O_4/SnO_2$ nanocomposite and $Fe_3O_4/SnO_2$/graphene with various weight percentages (1, 3, 5 wt%)

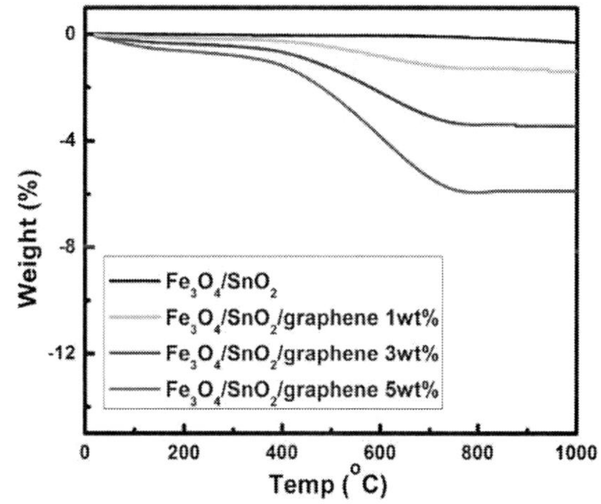


**Figure 2.** TGA curve of $Fe_3O_4/SnO_2$ nanocomposite and $Fe_3O_4/SnO_2$/graphene with various weight percentages (1, 3, 5 wt%)

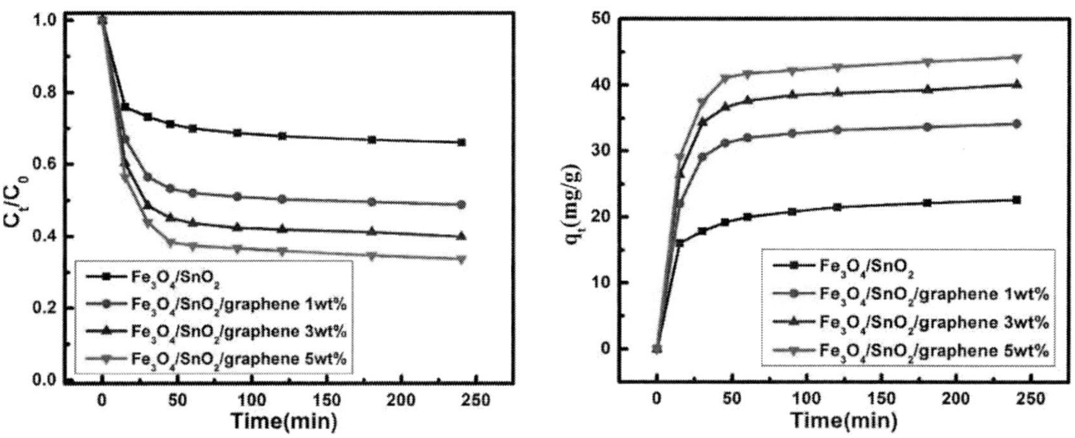


**Figure 3.** (a) Decolorization process of MB concentration and (b) adsorption capacity of $Fe_3O_4/SnO_2$/graphene 5 wt%

Figure 3 shows the adsorption ability of the $Fe_3O_4/SnO_2$/graphene composites with various weight percentages (1, 3, and 5 wt%). The results show that $Fe_3O_4/SnO_2$/graphene 5 wt% composites exhibited the highest adsorption ability than $Fe_3O_4/SnO_2$/graphene of 1 wt%, $Fe_3O_4/SnO_2$/graphene of 3 wt%, and $Fe_3O_4/SnO_2$ composite. Fig. 3a shows the color change (decolorization) of MB during the adsorption process. The decolorization process increased simultaneously with the addition of weight percentage of graphene at $Fe_3O_4/SnO_2$ from 33.94% for $Fe_3O_4/SnO_2$ nanoparticle until 51.25%, 60.13%, and 66.29% for $Fe_3O_4/SnO_2$/graphene of 1, 3, 5 wt%, respectively. The certain contact time during an adsorption process was performed to identify the adsorption capability of the adsorbent to adsorb the pollutants on its surface per unit of time [15]. Fig. 3b shows the adsorption capacity against time up to 250 minutes. As can be seen in Fig. 3b, $Fe_3O_4/SnO_2$/graphene 5 wt% showed the highest adsorption capacity (44.19 mg/g) then followed by $Fe_3O_4/SnO_2$/graphene 3 wt% (40.08 mg/g), $Fe_3O_4/SnO_2$/graphene 1 wt% (34.16 mg/g) and $Fe_3O_4/SnO_2$ (22.62 mg/g). The maximum adsorption capacity of the samples reached at amm ount graphene material in Fe3O4/SnO2/graphene was 5 wt%. It may due to the increase of the surface area because the addition of graphene was proportional to the mass of graphene in the solution[16].

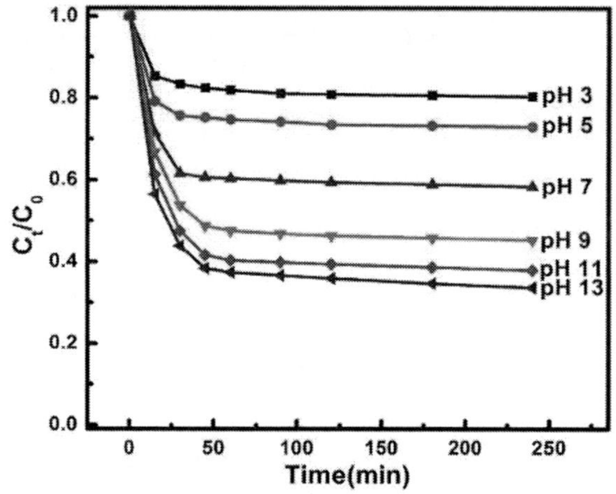


**Figure 4.** Effect of pH in adsorption capacity of $Fe_3O_4/SnO_2$/graphene of 5 wt%

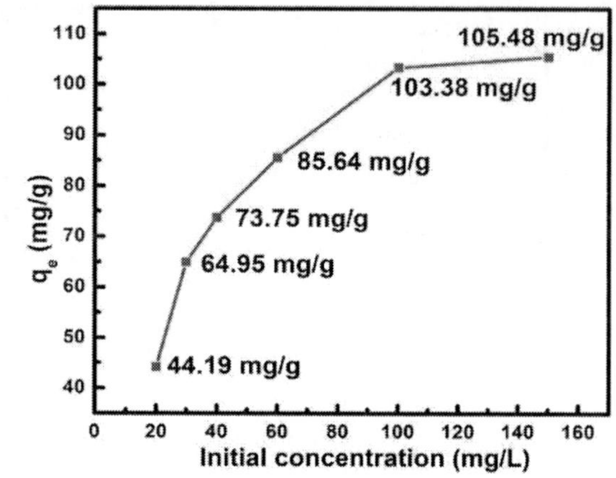


**Figure 5.** Adsorption capacity of $Fe_3O_4/SnO_2$/graphene of 5 wt% with different MB concentration

Figure 4 shows the influence of pH in the adsorption ability of $Fe_3O_4/SnO_2$/graphene of 5 wt% with broad range of pH from pH 3 until pH 13. All measurements were made at room temperature and for the contact time of 4 hours. The result showed the adsorption ability was found to increase with the increasing solution pH. The lower value of adsorption ability at acidic pH may be due to the presence of excess $H^+$ ions from the acid solution competing with the charge MB dye cations for vacant adsorption site [3,17]. That phenomenon indicated the maximum adsorption capacity was obtained at the pH value of 13 for $Fe_3O_4/SnO_2$/graphene 5 wt% composite.

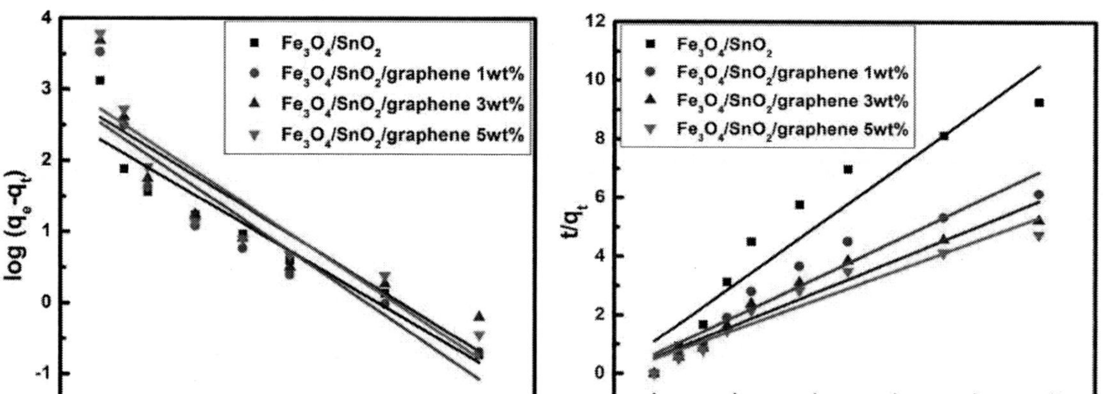


**Figure 6.** (a) Pseudo-first-order and (b) Pseudo-second-order plots for MB adsorption of $Fe_3O_4/SnO_2$ nanocomposite and $Fe_3O_4/SnO_2$/graphene with various weight percentages (1, 3, 5 wt%)

To investigate the influence of the initial concentration of MB dye, Figure 5 shows the adsorption capacity of $Fe_3O_4/SnO_2$/graphene of 5 wt% composite with the variation of MB concentration from 20 mg/L– 150 mg/L with the adsorbent dosage of 0.3 g/L. The result shows, the adsorption capacity of $Fe_3O_4/SnO_2$/graphene decreased with the increase of MB concentration with the equilibrium adsorption capacity reach its maximum of 105.48 mg/g.

The pseudo-first-order (Figure 6a) and pseudo- second-order (Figure 6b) models were examined in this study to analyze the kinetics of MB adsorption. The selection of the best applicable adsorption kinetics model for the adsorption of MB onto the samples was done on the basis of the values of the

IOP Conf. Series: Materials Science and Engineering **202** (2017) 012009 doi:10.1088/1757-899X/202/1/012009

correlation ($R^2$). The pseudo first-order model can be expressed as follows:

$$\mathrm{Log}(q_e - q_t) = \log q_e - (k_1/2.303)t \qquad (3)$$

where $q_e$ and $q_t$ are the adsorption capacity of $Fe_3O_4/SnO_2$ nanocomposite and $Fe_3O_4/SnO_2$/graphene with various weight percentages (1, 3, 5 wt%) for MB at equilibrium (mg $g^{-1}$) and at given time (min), respectively. While $k_1$ is the rate constant for the pseudo-first-order of the adsorption (g $mg^{-1}$ $min^{-1}$).

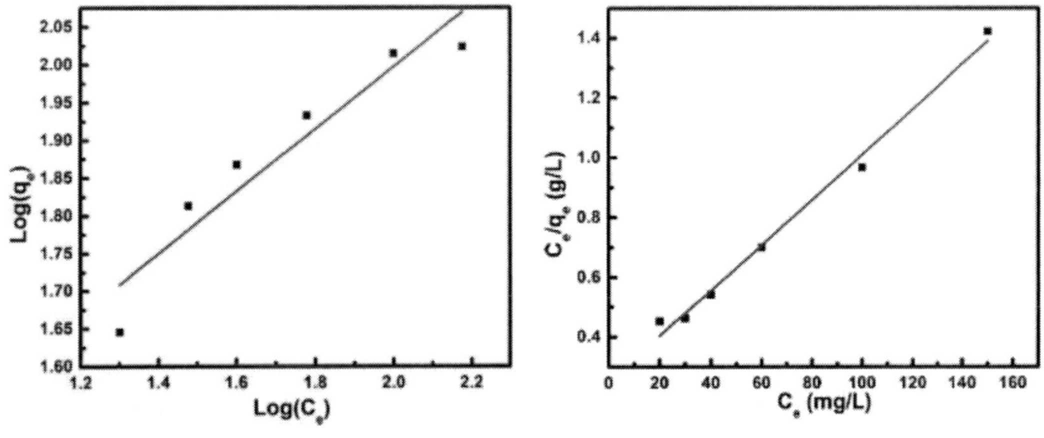


**Figure 7.** (a) Freudlich and (b) Langmuir adsorption-isotherm model fitting the experimental data of $Fe_3O_4/SnO_2$/graphene of 5 wt%

**Table 1.** Kinetic parameter of the pseudo-first-order and pseudo-second-order models for the adsorption of MB onto $Fe_3O_4/SnO_2$ nanocomposite and $Fe_3O_4/SnO_2$/graphene with various weight percentages (1, 3, 5 wt%)

| | | The pseudo-First-order rate | | | The pseudo-second-order rate | | |
|---|---|---|---|---|---|---|---|
| Sample | $q_e$ (mg/g) exp | $q_{1e}$ (mg/g) | $K_1$ | $R^2$ | $q_{2e}$ (mg/g) | $K_2$ | $R^2$ |
| Fe3O4/SO2 | 22.629 | 9.921 | 0.013 | 0.875 | 25.555 | 0.0013 | 0.90121 |
| Fe3O4/SnO2/graphene 1 wt% | 34.168 | 12.549 | 0.015 | 0.823 | 38.491 | 0.0010 | 0.9189 |
| Fe3O4/SnO2/graphene 3 wt% | 40.088 | 1.002 | 0.013 | 0.760 | 44.984 | 0.0008 | 0.91867 |
| Fe3O4/SnO2/graphene 5 wt% | 44.196 | 15.260 | 0.014 | 0.786 | 49.652 | 0.0008 | 0.91894 |

The pseudo-second-order model, can be expressed as follows:

$$t/q_t = 1/(k_2.q_e^2) + t/q_e \qquad (4)$$

where $k_2$ is the rate constant for the pseudo-second-order adsorption process [4]. The value of the $R^2$ for the pseudo-first-order and pseudo-second-order models are shown in Table 1. The low value of the correlation coefficient $R^2$ indicated that the adsorption of MB onto samples poorly fitted with pseudo first- order kinetic model. However, the adsorption of MB onto all samples shows a better fit to the correlation coefficient $R^2$ of the pseudo-second-order model. That case indicated the adsorption process were confirmed to the pseudo-second-order kinetics.Two adsorption kinetics models were applied to know the mechanism of the adsorption process. Figure 7 shows the Freundlich (Figure 7a) and Langmuir (Figure 7b) adsorption kinetics models from adsorption process of $Fe_3O_4/SnO_2$/graphene of 5 wt%. The figure shows the Langmuir isotherm is a better fit than the Freundlich isotherm kinetic model, with the value of regression coefficient ($R^2$) was 0.99 and 0.88 for Langmuir isotherm and Freundlich adsorption isotherm model, respectively.

## 4. Conclusion

The $Fe_3O_4/SnO_2$ nanocomposite and the different weight percentages of graphene (1, 3, 5 wt%) in $Fe_3O_4/SnO_2$/graphene have been successfully synthesized by using ultrasonic assisted and co-precipitation method, respectively. The structural characterization of the samples was studied by using XRD measurement, which revealed a combination phase from the tetragonal phase of $SnO_2$ and cubic spinel phase of $Fe_3O_4$. The TGA measurement confirmed the existence of graphene in $Fe_3O_4/SnO_2$/graphene composite by showing the weight loss at around 400oC which attributes to combustion effect from graphene. The maximum equilibrium adsorption capacity reached $Fe_3O_4/SnO_2$/graphene of 5wt% with the value of 44.19 mg/g. The maximum adsorption capacity was obtained at the pH value of 13 for $Fe_3O_4/SnO_2$/graphene of 5 wt% composite. The adsorption process of $Fe_3O_4/SnO_2$/graphene 5 wt% follows the Langmuir adsorption isotherm model.

## 5. References

[1] Zahra Khodami, Alireza Nazamzadeh-Ejhieh 2015 *Journal of Molecular Catalyst A: Chemical* **409** 59-68
[2] Sung Phil Kim, Myong Yong Choi, Hyun Chul Choi 2016 *Materials Research Bulletin* **74** 85-89 [3] Lukumoni Borah, Mridusmita Goswami, Prodeep Phukan 2015 *Journal of Environmental Chemical Engineering* **3** 1018-1028
[4] Jiali Chang, Jianchao Ma, Qingliang Ma, Duoduo Zhang, Nannan Qiao, Mengxiao Hu, Hongzhu Ma 2016 *Applied Clay Science* **119** 132-140
[5] Wenjuan Wang, Heipeng Zhang, Ling Zhang, Haiqin Wan, Shourong Zheng, Zhaoyi Xu 2015 *Colloids and Surface A: Physicochem. Eng. Aspects* **469** 100-106
[6] Haomiao Xu, Zan Qu, Songjian Zhao, Dongting Yue, Wenjun Huang, Naiqiang Yan 2016 *Chemical Engineering Journal* **292** 123-129
[7] Zhibin Wu, Hua Zhong, Xingzhong Yuan, Hou Wang, Lele Wang, Xiaohong Chen, Guangmin Zeng, Yan Wu 2014 *Water Research* **67** 330-344
[8] Huan Wang, Haihuan Gao, Mingxi Chen, Xiaoyang Xu, Xuefang Wang, Cheng Pan, Jianping Gao 2016 *Applied Surface Science* **360** 840-848
[9] Cao C, Xiao L, Chen C, Shi X, Cao Q, Gao L 2014 *Powder Technology* **260** 90-97
[10] Dalvand A, Nabizadeh R, Ganjali M R, Khoobi M, Nazmara S, Mahvi A H 2016 *Journal of Magnetism and Magnetic Materials* **404** 179-189
[11] Taufik A, Susanto I K, Saleh R 2015 *Materials Science Forum* **827** 37-42
[12] Yue Li, Yanqun Guo, Ruiqin Tan, Ping Cui, Yong Li, Weijie Song 2009 *Materials Letters* **63** 2085-2088
[13] Lihua Hu, Zhongping Yang, Limei Cui, Yan Li, Huu Hao Ngo, Yaoguang Wang, Qin Wei, Hongmin Ma, Liangguo Yan, Bin Du 2016 *Chemical Engineering Journal* **287** 545-556
[14] Loryuenyong V, Totepvimarn K, Eimburanapravat P, Boonchompoo W and Buasri A 2013 *Advances in Material Science and Engineering* **2013** 923403

IOP Conf. Series: Materials Science and Engineering **202** (2017) 012009 doi:10.1088/1757-899X/202/1/012009

[15] Singh D, Verma S, Gautam R K, Krishna V 2015 *Journal of Environmental Chemical Engineering* **3** 2161-2171
[16] Linson Lonappan, Tarek Rouissi, Ratul Kumar Das, Satinder K Brar, Antonio Avalos Ramirez, Mausam Verma, Rao Y Surampalli, Valero J R 2016 *Water Management* **49** 537-544
[17] Saleh R, Djaja N F 2014 *Spectrochim. Acta Part A* **130** 581-590

IOP Conf. Series: Materials Science and Engineering **202** (2017) 012010 doi:10.1088/1757-899X/202/1/012010

# Synthesis of High-Impact Polystyrene Fibers using Electrospinning

**A Zulfi[1,2], A Fauzi[1,2], D Edikresnha[1,2], M M Munir[1,2] and Khairurrijal[1,2]**

[1]Department of Physics, Faculty of Mathematics and Natural Sciences,
[2]Bioscience and Biotechnology Research Center, Institute for Research and Community Services, Institut Teknologi Bandung, Jalan Ganesha 10, Bandung 40132, Indonesia

**Abstract.** Synthesis of fibers from waste high-impact polystyrene (HIPS) have been successfully done using electrospinning method. The HIPS solutions were made with a single solvent (DMF or d-limonene), a mixed solvent (d-limonene/DMF), and with the addition of acetone to the previously stated solvents. The effects of HIPS concentration, a mix of solvent, and the addition of acetone on the morphology and the diameter of fibers were observed. The morphological change from particles to fibers took place along with the increasing concentration of HIPS in d-limonene. For other precursor solutions using DMF solvent, bead free fibers could be obtained even at low levels. The average diameter of fibers increased along with the increase of the HIPS concentration in DMF. At the concentrations of 15, 20, 25, 30, and 35 wt.%, the average diameters were 1.85, 2.09, 2.66, 3.59, and 7.38 μm, respectively. For the precursor solutions with the combination of different solvents (HIPS/DMF), the existence of beads was influenced by the ratio of solvents. When the ratio of d-limonene/DMF was 75:25, the obtained beaded fibers had a relatively large amount of beads. At the ratio of 50:50, fewer beads were found. Bead-free fibers were finally reached when the ratio of HIPS / DMF was 25:75. The addition of acetone reduced the diameter of the produced fibers. However, too much addition of acetone caused the fibers to be wet. Additionally, the diameter became larger if the addition of acetone surpassed a certain amount of volume.

## 1. Introduction

Nanotechnology has become a rapidly growing field in recent years. One area of great interest in nanotechnology is the synthesis and application of nanofibers. Nanofiber is a one-dimensional material which shaped like a fine thread [1]. It has a broad range of applications due to its unique physical properties, such as great ratio between the surface area and the volume. Nanofibers are elastic, and they have superior mechanical properties such as having high flexibility, high ultimate tensile strength and high porosity [2,3]. By acquiring these properties, nanofibers can widely be used in various applications e.g. as sensors, catalysts, and filters [4–6].

One technique to synthesize nanofibers is electrospinning. It is a simple approach to producing fibers with a relatively uniform diameter and various compositions using a high electric field. A high voltage source, a collector, a syringe pump, and a syringe with a metal needle are parts of the electrospinning apparatus [1]. The fibers produced are long and continuous while the diameters of fibers can be controlled by changing the polymer parameter (type and molecular weight), the solution parameter (polymer concentration, viscosity, conductivity and surface tension), the processing parameters (applied voltage, distance of needle tip to collector, flow rate, and diameter of needle tip), and the environmental parameters (temperature and humidity) [6].

Polystyrene is one of the commodity polymers that are widely used and an amorphous thermoplastic polymer that can easily be produced [7]. Recently, electrospun polystyrene fibers have been applied as biocatalysts and formalin sensors [8,9]. High-impact polystyrene (HIPS) and styrofoam are further developments of polystyrene. HIPS is a graft copolymer consisting of a small amount of butadiene rubber dispersed in a matrix of polystyrene [10]. Due to the special structure, HIPS has a series of advantages, such as excellent toughness, processability, and low production cost. Therefore, it is widely used in many fields, such as for furniture, household appliances, packaging materials, transportation, automobile, and so on [11].

The increasing use of HIPS does in turn increase the waste problem of HIPS. Usually, HIPS is recycled by transforming it into pellets using an extruder that will ultimately result in lower grade products [12]. Several studies have described the transformation of waste styrofoam into nanofibers using electrospinning and rotary force spinning methods [13–15]. In the case of electrospinning of HIPS, there is only one report in which HIPS was dissolved in a mixture of 1,2-dichloroethane (DCE) and N, N-dimethylformamide (DMF) [16].

The purpose of this present study was, therefore, to provide an alternative solution in the treatment of waste HIPS by converting it into fibers using electrospinning. The mixture of DMF and d-limonene was used as the main solvents, while acetone was employed as the additional solvent for d-limonene. The morphologies of produced fibers were observed using a digital microscope. Microscopic images of the fibers were analyzed for obtaining their average diameters and standard deviations so that their uniformity could be determined.

## 2. Experimental Method

Waste HIPS was collected from covers of broken compact discs. The main solvents used were N, N-dimethylformamide (DMF) and d-limonene, which were purchased from Sigma-Aldrich and Bratachem, Indonesia, respectively. Additionally, acetone, which was also obtained from Bratachem, Indonesia, was used as a supplementary solvent for the d-limonene solvent.

To observe the effect of HIPS polymer concentration on the produced fibers, the precursor solutions (15, 20, 25, 30, and 35 wt.%) were prepared by dissolving HIPS in a single solvent (DMF or d-limonene). Then, for the observation of the effect of solvent, the mixtures of solvents (d-limonene/DMF) with ratios of 75:25, 50:50 and 25:75 wt.% were prepared and then used to make HIPS solution with a concentration of 20 wt.%. Lastly, to observe the effect of acetone addition, a solution of HIPS/d-limonene with a concentration of 40 wt.% was made and then added by acetone forming 10, 20, and 30 wt.% solutions.

The schematic diagram depicting the synthesis process of fibers using electrospinning apparatus (Nachriebe 600, Indonesia) is shown in Figure 1. The precursor solution was placed in the plastic syringe with metal needle diameter of 0.8 mm. The syringe was mounted on the syringe pump to control the flow rate of the ejected solution. The metal tip of the needle was then connected to the positive pole of the high voltage source. The grounded collector plate was coated with aluminum foil to collect fibers and located as far as 20 cm from the metal tip.

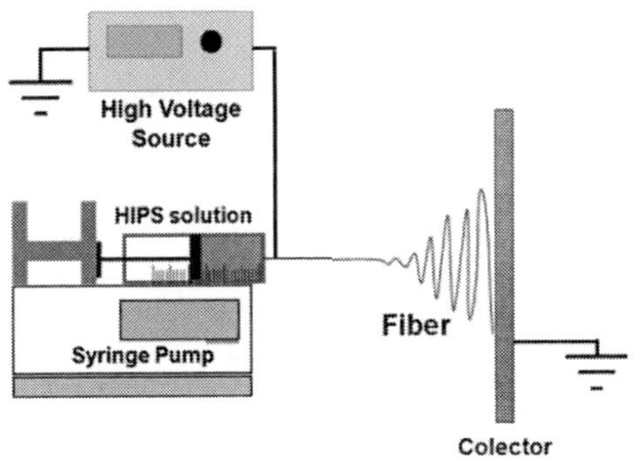


**Figure 1.** The schematic diagram of electrospinning

The characterization of the fibers was done by using a digital microscope (National DC3-163) with 200×, 400×, and 1000× magnifications. Then, the microscopic images were processed to obtain the average and distribution of the diameter of fibers. The observations were taken at different locations of up to 100 points. The determination of the uniformity of fibers was done by calculating the coefficient of variation (*CV*) using Equation (1) [17].

$$CV = \frac{\sigma}{\mu}, \quad (1)$$

where $\sigma$ represents the standard deviation and $\mu$ represents the average diameter of fibers.

## 3. Results and Discussion

By adjusting the electrospinning parameters, the morphology of nanofibers can be changed. It consists of the structures of simple beads (nanoparticles), beaded fibers, and bead-free fibers [18]. In this study, the effects of the concentration of HIPS polymer in a single solvent (d-limonene or DMF), in mixed solvents with certain ratios (d-limonene and DMF), and of acetone addition are given as follows.

### *3.1. Morphology of fibers from precursor solution using a single solvent (d-limonene or DMF)*

HIPS solutions with various concentrations of 15, 20, 25, 30 and 35 wt.% were created by using a single solvent (d-limonene or DMF). Figure 2 shows the microscopic images of the HIPS fibers with DMF solvent under a digital microscope with 400× and 1000× magnifications and their respected fiber size distribution. During the syntheses, the process parameters, such as the flow rate of 0.5 mL/h, the high voltage of 15 kV and the distance of the collector-the needle tip of 20 cm, were kept constant.

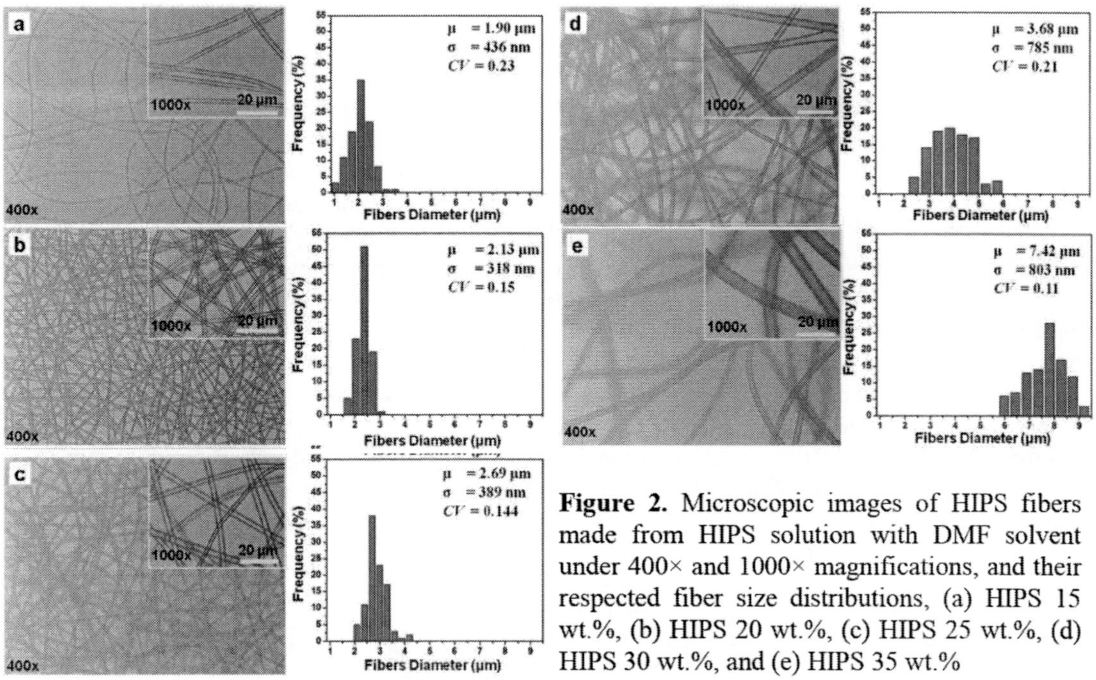


**Figure 2.** Microscopic images of HIPS fibers made from HIPS solution with DMF solvent under 400× and 1000× magnifications, and their respected fiber size distributions, (a) HIPS 15 wt.%, (b) HIPS 20 wt.%, (c) HIPS 25 wt.%, (d) HIPS 30 wt.%, and (e) HIPS 35 wt.%

As can be seen in Figure 2, the average diameter of fibers changed because of the change of HIPS concentration. For the concentrations of 15, 20, 25, 30, and 35 wt.%, the average diameter of fibers of were 1.85, 2.09, 2.66, 3.59, and 7.38 μm, respectively. The average diameter of fibers increased together with the increasing concentration of HIPS polymer since the viscosity became higher as a result of the enhancing HIPS polymer concentration. As the viscosity increased, it hindered the deformation process of the jets into fibers and a longer time was needed for the process to produce a perfect result. Therefore, by keeping all other parameters constant, the average diameter of fibers will be larger when the concentration of HIPS polymer is higher [19–21]. To determine the uniformity of fibers, the coefficient of variation (CV) was calculated using Equation (1). If the calculated CV was greater than 0.3, the diameters of fibers were then nonuniform [17]. From the fiber size distribution, it could be seen that the CVs were less than 0.3. Therefore, the HIPS fibers made from the precursor solution with DMF solvent produced fibers with a uniform diameter.

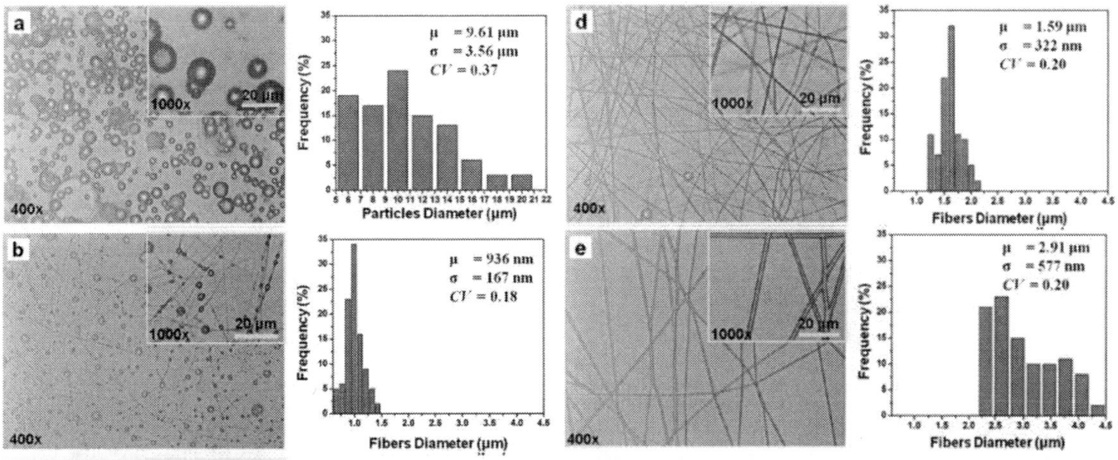


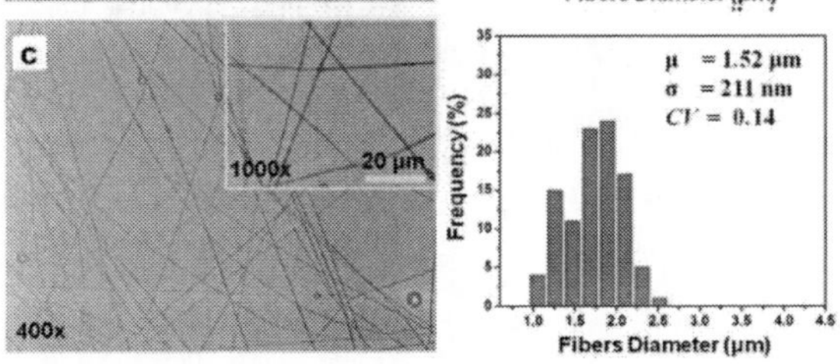


**Figure 3.** Microscopic images of HIPS fibers made from waste HIPS solution with d-limonene solvent under 400×and 1000× magnifications along with their respected fiber size distributions, (a) 15 wt.% HIPS, (b) 20 wt.% HIPS, (c) 25 wt.% HIPS, (d) 30 wt.% HIPS, and (e) 35 wt.% HIPS

Figure 3 shows the microscopic images at 400× and 1000× magnifications of HIPS fibers with d-limonene solvent along with their fiber size distribution. The process parameters were kept constant at the flow rate of 0.2 mL/m, the high voltage of 9.2 kV, and the distance between the collector and the needle tip of 20 cm. Unlike previous attempts when DMF was being used as the solvent, there was a lower limit of the HIPS concentration in the d-limonene solvent to produce bead-free fibers successfully. As depicted in Figure 3.(a), the microparticles with an average diameter of 9.61 μm were generated for the 15 wt.% HIPS. When the concentration of HIPS polymer is too little, the interactions among the molecules of the solvent were more dominant than the interactions among the molecules of polymer. Consequently, there was a tendency to form spheres to reduce the surface area [22]. Since the surface tension was much higher than the force between charges and the force due to the interactions between the molecular chains of polymers, particles were formed instead of fibers [23,24]. As shown in Figure 3.(b), beaded fibers were formed when the concentration of HIPS polymer was 20 wt.% proving that the interaction among the polymer chains had increased. However, the surface tension was still quite high that made beads still appeared along the fibers. A relatively similar result was achieved for the fibers made from the precursor solution of 25 wt.% HIPS. The images in Figures 3.(d) and 3.(e) illustrate the formation of fine and bead-free fibers after the concentration of polymer was further increased. The increase of polymer's concentration led to more interactions among the molecules of polymer and more interactions among the molecules of solvent and polymers, while the interaction between solvent and molecules became lower. Therefore, when the solution was being attracted on the collector under a high voltage, the molecules of the solvent would spread around the polymer molecule, which prevented the formation of beads [19]. The average diameters of the fibers made from the precursor solutions with the concentrations of 20, 25, 30, and 35 wt.% were 0.92, 1.52,

1.59, and 2.86 μm, respectively. From the fiber size distribution, it was found that the CVs for all the obtained fibers were below 0.3 and therefore, they were uniform

*3.2 Morphology of fibers with mixed solvent (d-limonene/DMF)*

The precursor solution with mixed solvents of 20 wt.% HIPS was prepared by using a mixture of solvents with various ratios of d-limonene/DMF to be 75:25, 50:50, and 25:75 wt.%. The viscosity and surface tension of the precursor solution were then measured, and the results are shown in Table 1. During the syntheses, the process parameters were kept constant. The flow rate was 0.2 mL/m, the applied high voltage was 9.2 kV, and the distance between the plate collector and the needle tip was 20 cm. Figure 4 shows the microscopic images of the HIPS fibers made from different ratios of solvent mixture (d-limonene/DMF)

**Table 1.** Properties of HIPS solution[a].

| Solution | Concentration (wt.%) | Viscosity (centi Poise) | Surface Tension (dyne/cm) |
|---|---|---|---|
| HIPS/d-limonene/DMF(75:25) | 20 | 309 | 34.2 |
| HIPS/d-limonene/DMF(50:50) | 20 | 312 | 33.5 |
| HIPS/d-limonene/DMF(25:75) | 20 | 325 | 32.8 |

[a]Values reported for 25 °C. The ratio represented the weight ratio of d-limonene to DMF

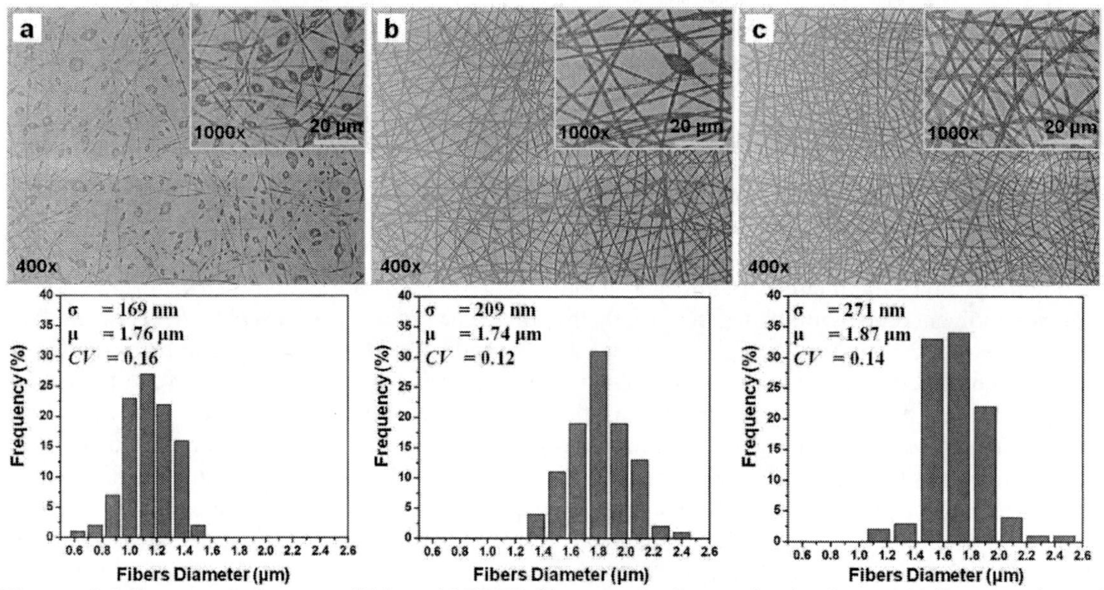


**Figure 4.** Microscopic images of 20 wt.% HIPS fibers made from mixed solvents (d-limonene/DMF) under a digital microscope with 400× and 1000× magnifications and their corresponding fiber size distributions, (a) d-limonene/DMF of 75:25%, (b) d-limonene/DMF of 50:50%, and (c) d-limonene/DMF of 25:75%.

As presented in Figure 4 (a), beaded fibers did appear with varying diameters and in an elliptical shape. By looking at this morphology, the beads were formed before the elongation while the elliptical shape was caused by the pulling effect of the jets during the elongation process [23]. When the percentages of d-limonene and DMF were equal, the diameter of fibers increased while the number of beads decreased significantly as depicted in Figure 4.(b). Lastly, in Figure 4.(c), as the DMF was increased further to 75%, the bead-free and straight fibers were formed. The average diameter of fibers increased as the percentage of DMF was increased. The average diameters were 1.06, 1.73, and 1.86 μm for the precursor solutions with the ratio of solvents (d-limonene: DMF) of 75:25, 50:50, and

25:75%, respectively. As seen from Table 1, along with the increase of the DMF concentration in the mixed solvents, the precursor solution of 20 wt.% HIPS experienced an increase in the viscosity and decrease in the surface tension. The increase in viscosity and the decline in surface tension thus caused the diameter to increase while the number of beads decreased or even disappeared [22]-[24].

From the fiber size distribution, all fibers were evenly distributed and almost all the fibers were uniform as they had below 0.3 CVs. Therefore, by adjusting the ratio of solvents (d-limonene and DMF) and setting a constant amount of polymer concentration, beaded fibers or bead-free straight fibers with uniform diameter can be achieved.

### *3.3. Addition of extra acetone into HIPS solution*

Figure 5 shows the microscopic images of the fibers as a result of acetone addition at 10, 20, and 30 wt.%, on the initial weight of the precursor solution of HIPS dissolved in d-limonene at a concentration of 40 wt.% as well as their fiber size distribution. The process parameters used in the form where the flow rate of 0.2 mL/m, the applied voltage of 9.2 kV, and the distance between the collector and the needle tip of 15 cm. All these parameters were held constant.

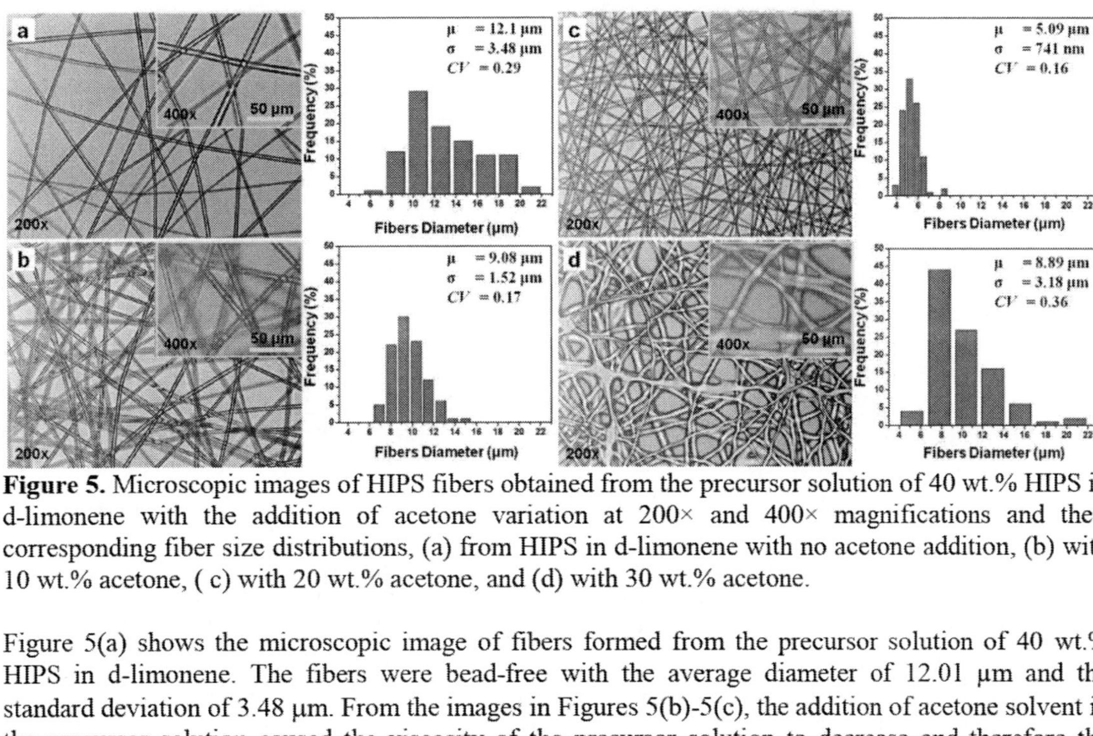


**Figure 5.** Microscopic images of HIPS fibers obtained from the precursor solution of 40 wt.% HIPS in d-limonene with the addition of acetone variation at 200× and 400× magnifications and their corresponding fiber size distributions, (a) from HIPS in d-limonene with no acetone addition, (b) with 10 wt.% acetone, ( c) with 20 wt.% acetone, and (d) with 30 wt.% acetone.

Figure 5(a) shows the microscopic image of fibers formed from the precursor solution of 40 wt.% HIPS in d-limonene. The fibers were bead-free with the average diameter of 12.01 μm and the standard deviation of 3.48 μm. From the images in Figures 5(b)-5(c), the addition of acetone solvent in the precursor solution caused the viscosity of the precursor solution to decrease and therefore the diameter of fibers also decreased proportionally [22]. Their average diameters were 9.08, 5.09, and 8.89 μm for the acetone additions of 10, 20, and 30 wt.%, respectively. Moreover, all fibers were evenly distributed and uniform as they had less than 0.3 CVs. However, when the acetone was added up to 30 wt.%, the fibers were wet, and the average diameter went larger as shown in Figure 5.(d). The addition of too much acetone made the molecules of solvent cannot reach the collector since they evaporated too soon [22].

## 4. Conclusion

High-impact polystyrene (HIPS) fibers have successfully been synthesized from waste HIPS dissolved in the DMF solvent, the d-limonene solvent, the mixture of DMF and d-limonene solvents as well as d-limonene and acetone solvents. Varying concentrations of HIPS polymer in the single solvent (DMF or d-limonene) caused morphological changes in the fibers. When DMF was used as the solvent, all fibers were free of beads. Meanwhile the diameter of the fibers increased as the concentration of the polymer increased. The average diameter of fibers for precursor solutions of 15, 20, 25, 30, and 35 wt.% were 1.85, 2.09, 2.66, 3.59, and 7.38 μm, respectively. When d-limonene was used as the solvent, there was an observable change in the morphology from particles, beaded fibers and up to beads-free fibers as the concentration of polymer increased. At a concentration of 15 wt.%, the particles were not uniform with an average diameter of 9.61 μm. When the concentration was increased to 20 %, beaded fibers were formed with an average diameter of fibers 0.92 μm and once the concentration was increased further to 25 wt.%, fewer beads were formed with an average diameter of fibers of 1.52 μm. The beads were totally disappeared for the concentrations of 30 and 35 wt.% with the average diameters of 1.59 and 2.86 μm, respectively. When mixed solvents of DMF/d-limonene were used, the higher volume ratio of DMF reduced the number of beads and increased the diameter of fibers. For the concentration of the mixture solvent of d-limonene / DMF at 75:25, 50:50 and 25:75, the average diameters of fibers were, respectively, 1.06, 1.73, and 1.86 μm. (c) Lastly, the addition of acetone was found to be able to decrease the diameter of the produced fibers. However, too much addition of acetone caused the fibers to be wet. For the additions of acetone of 0, 10, 20, and 30 %, the obtained average diameters of fibers were 12.01, 9.08, 5.09, and 8.89 μm, respectively.

## 5. References

[1] Munir M M, Iskandar F, Khairurrijal and Okuyama K 2008 A constant-current electrospinning system for production of high quality nanofibers *Rev. Sci. Instrum.* **79**

[2] Huang Z M, Zhang Y Z, Kotaki M and Ramakrishna S 2003 A review on polymer nanofibers by electrospinning and their applications in nanocomposites *Compos. Sci. Technol.* **63** 2223–53

[3] Yoon K, Kim K, Wang X, Fang D, Hsiao B S and Chu B 2006 High flux ultrafiltration membranes based on electrospun nanofibrous PAN scaffolds and chitosan coating *Polymer* **47** 2434–41

[4] Munir M M, Iskandar F, Djamal M and Okuyama K 2011 Morphology controlled electrospun nanofibers for humidity sensor application *AIP Conference Proceedings* vol 1415pp 223–6

[5] Widiyandari H, Munir M M, Iskandar F and Okuyama K 2009 Morphology-controlled synthesis of chromia-titania nanofibers via electrospinning followed by annealing *Mater. Chem. Phys.* **116** 169–74

[6] Khairurrijal, Munir M M, Nuryantini A Y, Rajak A, Prama Ekaputra M and Rahma A 2015 Fabrication of Poly(acrylonitrile)/PAN Nanofiber Using a Drum Collector Electrospinning System for Water Purification Application *Advanced Materials Science and Technology II* Advanced Materials Research vol 1123(Trans Tech Publications)pp 281–4

[7] Ebewele R O and Ebewele R O 2000 *Polymer science and technology* vol 1(CRC press Boca Raton, FL)

[8] An H, Jin B and Dai S 2015 Fabricating polystyrene fiber-dehydrogenase assemble as a functional biocatalyst *Enzyme Microb. Technol.* **68** 15–22

[9] Zhang C, Wang X, Lin J, Ding B, Yu J and Pan N 2011 Nanoporous polystyrene fibers functionalized by polyethyleneimine for enhanced formaldehyde sensing *Sens. Actuators B Chem.* **152** 316–23

[10] Liu J, Zhang Y, Peng S, Pan B, Lu C, Liu H, Ma J and Niu Q 2015 Fire property and charring behavior of high impact polystyrene containing expandable graphite and microencapsulated red phosphorus *Polym. Degrad. Stab.* **121** 261–70

[11] Hall W J and Williams P T 2006 Fast pyrolysis of halogenated plastics recovered from waste computers *Energy Fuels* **20** 1536–49

[12] Jung S H, Kim S J and Kim J S 2013 The influence of reaction parameters on characteristics of pyrolysis oils from waste high impact polystyrene and acrylonitrile-butadiene-styrene using a fluidized bed reactor *Fuel Process. Technol.* **116** 123–9

[13] Rajak A, Munir M M, Mikrajuddin A and Khairurrijal 2015 Photocatalytic Activities of Electrospun TiO<sub>2</sub>/Styrofoam Composite Nanofiber Membrane in Degradation of Waste Water *Mater. Sci. Forum* **827** 7–12
[14] Munir M M, Nuryantini A Y, Iskandar, Suciati T and Khairurrijal 2014 Mass Production of Stacked Styrofoam Nanofibers Using a Multinozzle and Drum Collector Electrospinning System *Adv. Mater. Res.* **896** 20–3
[15] Fauzi A, Edikresnha D, Munir M M and Khairurrijal 2015 Synthesis of Styrofoam Fibers Using Rotary Forcespinning Technique *Mater. Sci. Forum* **827** 279–84
[16] Choi E and Hwang T S 2011 전기방사에 의한 술폰화 HIPS 이온교환 나노섬유의 제조 및 특성 Preparation and Characteristics of Sulfonated HIPS ion Exchange Nanofiber by Electrospinning **49** 69–74
[17] Matulevicius J, Kliucininkas L, Prasauskas T, Buivydiene D and Martuzevicius D 2016 The comparative study of aerosol filtration by electrospun polyamide, polyvinyl acetate, polyacrylonitrile and cellulose acetate nanofiber media *J. Aerosol Sci.* **92** 27–37
[18] Nitanan T, Opanasopit P, Akkaramongkolporn P, Rojanarata T, Ngawhirunpat T and Supaphol P 2012 Effects of processing parameters on morphology of electrospun polystyrene nanofibers *Korean J. Chem. Eng.* **29** 173–81
[19] Sarkar K, Gomez C, Zambrano S, Ramirez M, De Hoyos E, Vasquez H and Lozano K 2010 Electrospinning to Forcespinning??? *Mater. Today* **13** 12–4
[20] Padron S, Fuentes A, Caruntu D and Lozano K 2013 Experimental study of nanofiber production through forcespinning *J. Appl. Phys.* **113**
[21] Badrossamay M R, McIlwee H A, Goss J A and Parker K K 2010 Nanofiber assembly by rotary jet-spinning *Nano Lett.* **10** 2257–61
[22] Ramakrishna S 2005 *An introduction to electrospinning and nanofibers* (World Scientific)
[23] Andrady A L 2008 *Science and technology of polymer nanofibers* (John Wiley & Sons)
[24] Munir M M, Suryamas A B, Iskandar F and Okuyama K 2009 Scaling law on particle-to-fiber formation during electrospinning *Polymer* **50** 4935–43

# Preparation of PVA/$TiO_2$ Composites Nanofibers by using Electrospinning Method for Photocatalytic Degradation

**Nasikhudin[1,3], E P Ismaya[1], M Diantoro[3], A Kusumaatmaja[1,2] and K Triyana[1,2]**

[1]Departement of Physics, Faculty of Mathematics and Natural Sciences, Universitas Gadjah Mada, Bulaksumur Yogyakarta 55281, Indonesia
[2]Nanomaterials Research Group, Universitas Gadjah Mada, Bulaksumur Yogyakarta 55281, Indonesia
[3]Department of Physics, Faculty of Mathematics and Natural Sciences, Universitas Negeri Malang (State University of Malang), Jl. Semarang 5, Malang 65145, Indonesia

Email: triyana@ugm.ac.id

**Abstract.** PVA/$TiO_2$ composites nanofibers have been prepared by using electrospinning method. Electrospinning was performed at a voltage of 15 kV, with a tip-collector distance of 14 cm. The PVA/$TiO_2$ composites nanofibers were characterized by FTIR, SEM, and EDAX. The results showed that PVA/$TiO_2$ composites nanofibers were successfully formed. SEM observation showed that the nanofibers has several hundred nanometers with a smooth surface and have $TiO_2$ nanoparticle distributed in the nanofibers. FTIR result shows a functional group of PVA and EDAX result demonstrates the presence of $TiO_2$ in the nanofibers. The photocatalytic activity of PVA/$TiO_2$ composites nanofibers were investigated for the degradation of methylene blue under UV light. The PVA/$TiO_2$ composite nanofibers suspending in the dye solution have dye removal of 70% dye was degraded in 5 h. As a comparison, the $TiO_2$ powder that has dye removal of 95% dye was degraded in 5 h. Photocatalytic activity of PVA/$TiO_2$ nanofiber was lower than $TiO_2$ powder, but PVA/$TiO_2$ nanofiber was more efficient than $TiO_2$ powder regarding process treatments photocatalytic.

**Keywords:** electrospinning, nanofibers, PVA, $TiO_2$, photocatalytic.

## 1. Introduction

Titanium dioxide has been widely studied for its ability to photocatalytic and applications have high performance for photovoltaic applications [1]. It also shows chemical stability, low toxicity, low pollutant load, and availability at low cost. $TiO_2$ size has been reduced to increase the reactive surface area to improve efficiency [2]. Wastewater treatment for environmental preservation based $TiO_2$ photocatalytic has been proven to degrade many organic pollutants effectively. This case, however, still leaves some problems when it is applied in environments such as rivers, lakes, and sea [3]. The main problem is the difficulty of recovery after photocatalytic treatment process, namely how to separate $TiO_2$ particles from the liquid waste that is already degraded. Several ways have been made to overcome this problem include the use of $TiO_2$ particle filtration system with ceramic materials, stainless steel, and activated carbon. It is less effective because it can cause new problems as the precipitation and cause further pollution [4].

IOP Conf. Series: Materials Science and Engineering **202** (2017) 012011 doi:10.1088/1757-899X/202/1/012011

The solution to this problem is the development of photocatalytic based nanofibers were synthesized by using electrospinning methods. $TiO_2$ has been widely used as photocatalytic due to its chemical stability and availability in the commercial. Many forms of $TiO_2$ nanomaterial have been prepared including nanoparticle, nanotube, nanofiber, and nanosheet [5]. Electrospinning is a unique spinning technique using electrostatic force to produce fine fibers from a polymer solution. Fibers produced from this electrospinning have a thinner diameter (in order of nanometers) and a surface area which is greater than that obtained from conventional spinning process [6]. The electrospinning process is a method to produce polymer fibers from polymer solutions, with diameters in the range of 10 - 500 nm and high surface areas per unit mass. Nanofibers were due to their high ratio of surface area to mass and extraordinary small diameters around 100 nm. Electrospinning has received considerable attention in recent years in a various applications due to the ease of forming fibers with different properties. Electrospinning offers some unique advantages such as a high ratio of surface to volume, porosity that can be adapted to the structure of electrospun, and flexibility to electrospinning into various shapes and sizes [7]. Nanofiber has excellent characteristics such as the ratio of surface area to volume ratio is vast and high porosity with small pore sizes. Therefore, the nanofibers are very promising materials for many biomedical applications such as tissue templates, medical prostheses, artificial organs, wound dressings, bone tissue engineering, drug delivery, pharmaceutical compositions, a formation of filter and sensing applications [8]. Compared to conventional $TiO_2$ particles, $TiO_2$ fibers have greater surface-to-volume ratios, and their porous structures provide a higher number of active surface sites for effective photocatalytic [9-10].

Many efforts are focused on the application of $TiO_2$ in the wastewater treatment system. However, it is hard to remove $TiO_2$ from wastewater after treatment using conventional separation methods due to the nanosized structures. This limitation worked as the main drawback for $TiO_2$ to be widely used as adsorbents and photocatalyst in the environmental protection. In this paper, $TiO_2$ nanoparticles have been bounded into nanofiber composites by using Polyvinyl alcohol as fiber template. The primary purpose why $TiO_2$ was bounded into a nanofiber wasto prevent $TiO_2$ particles not became a new pollutant in the water, as we know $TiO_2$ particles are very dangerous for an organism living in the water. In this paper, we study about electrospinning process of PVA/$TiO_2$ composites nanofibers and morphology of the fiber. The photocatalytic activity of PVA/$TiO_2$ composites nanofibers were investigated for the degradation of methylene blue under UV light and compared with the photocatalytic activity of $TiO_2$ nanopowder.

## 2. Experimental Method

The material used in the research were: PVA (Mw: 89.000-98.000 g/mol, 99% hydrolyzed), Titanium (IV) oxide, anatase, nanopowder < 25 nm particle size, 99.7% trace metal basis, Glutaraldehyde solution 50 wt% in $H_2O$, and methylene blue powder, all of them purchased from Sigma-Aldrich. Characterizations of PVA/$TiO_2$ nanofibers by using FTIR and SEM-EDAX. FTIR (Shimadzu, Type: IR Prestige21) were used to determine the functional group of PVA/Chitosan/$TiO_2$ nanofiber. The nanofibers surface morphology was studied by using SEM-EDAX (FEI, Type: Inspect-S50). Image processing software was used to measuring the diameter of the nanofibers captured in the SEM micrograph.

PVA/TiO2 electrospun solution was prepared by dissolving $TiO_2$ nanoparticle in deionized water at 90 °C for 1 hour in magnetic stirrer. $TiO_2$ dispersion was added with PVA powder and then the PVA and TiO were blended at a ratio of 4:1. They were then stirred for 1 hour at 90 °C. The prepared PVA/TiO2 solution was then loaded into a 12-mL plastic syringe equipped with a 0.4-mm needle diameter. A high voltage of 15 kV, with a tip-collector distance of 14 cm, was applied to the solution and fiber collected on the plane collector. Electrospun nanofibers were crosslinked by immersing in a solution of 2.5 wt% Glutaraldehyde solution (volume ratios of Glu (aq): acetone: HCl (aq) = 1:2:0.01 ) at 40 °C for 24 h.

Photocatalytic activities of nanofiber membrane PVA/$TiO_2$ were investigated by observing degradation of methylene blue (MB) under UV light. The experiment was set up with conditions of 4 x

10 W lamp as the UV light source (365 nm). The reaction was prepared by using 0.1 g of PVA/$TiO_2$ nanofibers mat into the beaker containing 10 ppm of Methylene blue solution (10mL). Degradation of methylene blue was observed by UV-vis spectroscopy each 1 hour during 5 hours which suggested PVA/$TiO_2$ composites nanofibers could degrade the methylene blue solution.

## 3. Results and Discussion

The SEM image of electrospun nanofibers and $TiO_2$ nanopowder is shown in Figure 1. The morphology of PVA nanofibers as depicted in Figure 1.(a) is clearly fine and smooth. Its average diameters are uniform about 355 nm. On the other hand, PVA nanofibers containing $TiO_2$ as in Figure 1.(b) has some $TiO_2$ nanoparticles in the fiber. $TiO_2$ in the nanofibers has agglomeration, same as $TiO_2$ nanopowder in Figure 1.(c). The SEM image showed that the fiber has several hundred nanometers with a smooth surface and have $TiO_2$ nanoparticle distributed in the fibers. The average diameters of PVA/$TiO_2$ nanofibers are about 160 nm. If we compare those two nanofibers, we find that adding $TiO_2$ gives rise to reduce the fiber's diameter. The presence of $TiO_2$ causes to decrease the solution's viscosity and increase the conductivity of the electrospun solution. A reduction in fiber diameter is very helpful to increase the active surface area to improve the ability of the photocatalytic.

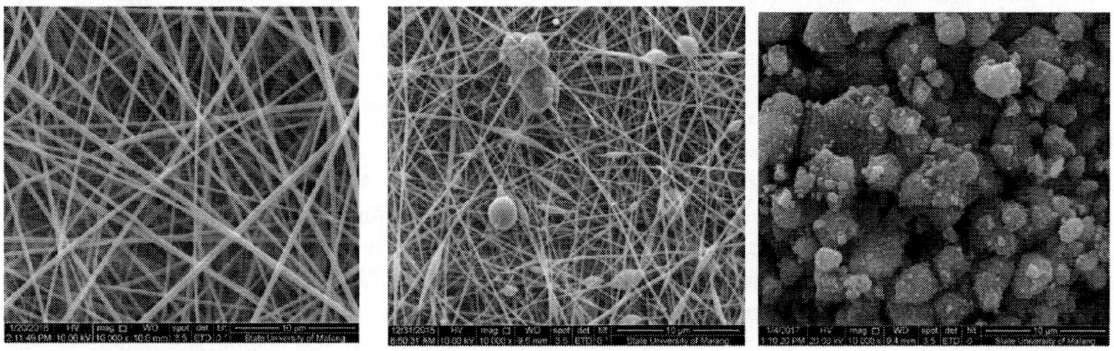

(a) (b) (c)

**Figure 1.** SEM image of (a) PVA/$TiO_2$ nanofibers, (b) PVA nanofibers, and (c) $TiO_2$ nanopowder.

The FTIR spectrum of PVA/$TiO_2$ nanofibers is shown in Figure 2. The Broadband at 3200-3550 $cm^{-1}$ indicates the O-H stretching. The peaks at 2910 $cm^{-1}$ indicate the C-H stretching. The peaks at 1330 and 1377 $cm^{-1}$ indicate the C-H group. The broad absorption bands at 1095 $cm^{-1}$ indicate the C-O stretching. C-H group vibration peaks were observed at 1230 and 1323 $cm^{-1}$. For the PVA/$TiO_2$ nanofibers adsorbents, a new broadband around 550–800 $cm^{-1}$ is attributed to the Ti-O-Ti band [11].

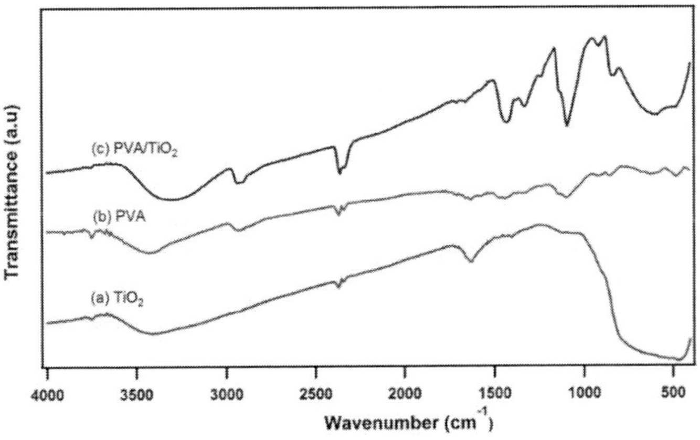


**Figure 2.** FTIR spectra of PVA/$TiO_2$ nanofibers.

The EDAX was used to investigate the content elements of electrospun fibers. EDAX result of the PVA/$TiO_2$ composite nanofibers is shown in Figure 3. The elemental analysis reveals the presence of $TiO_2$ nanoparticle on the fibers, the element ratio of C, O, and Ti are 68.03, 28.76 and 3.21% respectively in the entire area of PVA/$TiO_2$ composite nanofibers. It proves that the nanofiber contains $TiO_2$.

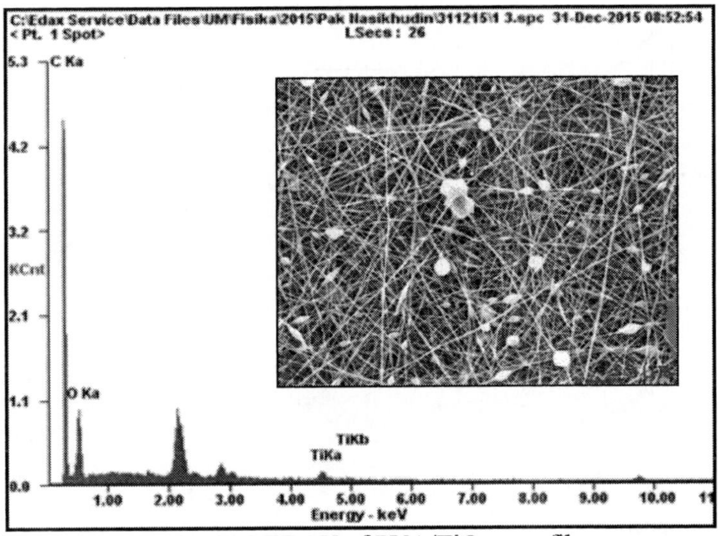


**Figure 3.** SEM-EDAX of PVA/$TiO_2$ nanofibers

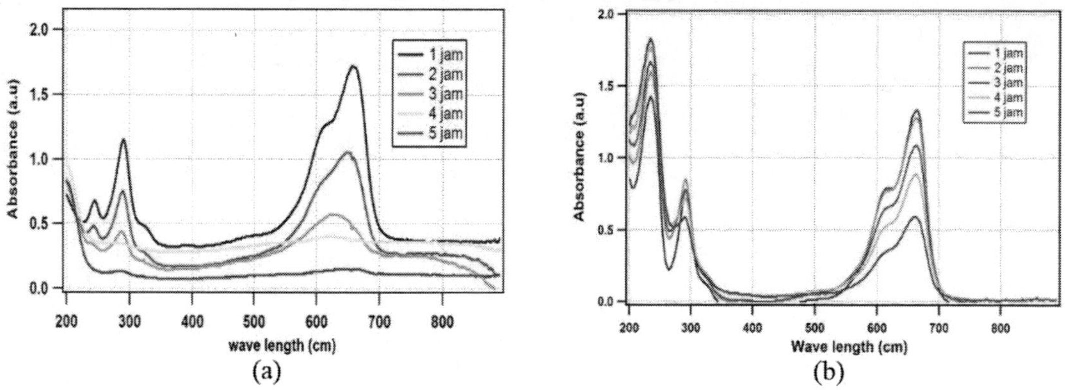


**Figure 4.** Uv-vis spectrum dye Methylene Blue solution after 1 – 5-hours degradation by (a) PVA/$TiO_2$ nanofibers (b) $TiO_2$ nanopowders.

The UV-Vis spectroscopy was used to observe the concentration of Methylene Blue solution after photocatalytic treatment. The UV-Vis spectroscopy observed every 1 hour during 5 hours. Figure 4 shows that the Methylene Blue solution are degraded by (a) PVA/$TiO_2$ nanofibers (b) $TiO_2$ nanopowder. Degradation of Methylene Blue solution by PVA/$TiO_2$ nanofibers was less than $TiO_2$ nanopowder. Comparison of photocatalytic activity to degrade Methylene Blue solution between PVA/$TiO_2$ nanofibers and $TiO_2$ nanopowder is shown in Figure 5.

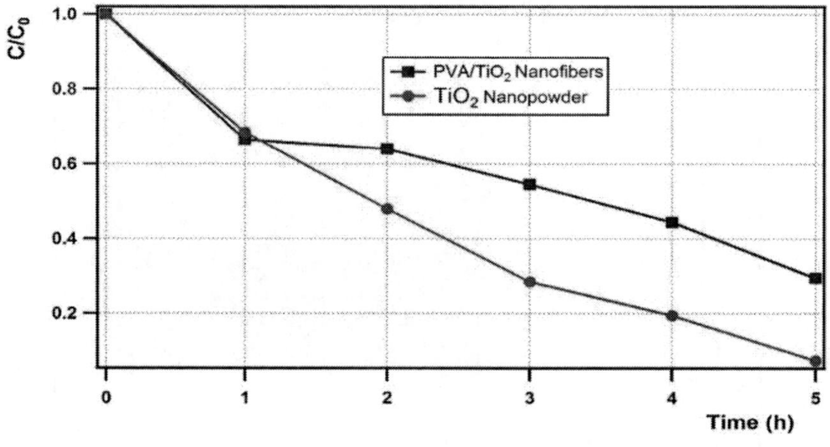


**Figure 5.** Removal of dye Methylene Blue by (a) PVA/$TiO_2$ nanofibers (b) $TiO_2$ nanopowder.

The plot of dye concentration on exposure time is shown in Figure 5. The results indicate that the PVA/$TiO_2$ composite nanofibers suspending in the dye solution and having dye removal of 70% dye was degraded in 5 h. In another research, PAN-based fibers contain $TiO_2$ particles removal 80% of the dye was degraded in 38 h [3]. PVA/PAAc/TiO2/CNTs composite nanofiber showed 30% of dye removal after 350 min at pH 7 and 80% of dye removal after 350 min at pH 10 [12]. $TiO_2$/PVA/PDMAPAAm composite nanofiber showed 50% of dye removal after 300 min at pH 7 and 90% of dye removal after 300 min at pH 2 [2]. As a comparison, the $TiO_2$ powder suspending in the dye solution and having dye removal of 95% dye was degraded in 5 h. Photocatalytic activity of PVA/$TiO_2$ nanofiber was lower than $TiO_2$ powder, but PVA/$TiO_2$ nanofiber was more efficient than $TiO_2$ powder regarding photocatalytic treatments. The treatments process of $TiO_2$ powder in photocatalytic should be done by using stirrer to mix the $TiO_2$ powder with methylene blue solution, after treatments process we must separate $TiO_2$ powder that has been dissolved in a methylene blue solution conducted by centrifuge and filtration. On the other hand, for nanofiber photocatalytic does not need to do it. This phenomenon proves that the process of photocatalytic with PVA/$TiO_2$ nanofiber is more efficient than $TiO_2$ powder. The result reveals that PVA/$TiO_2$ composite nanofibers could be applied for environmental field effectively.

## 4. Conclusion

The PVA/$TiO_2$ nanofibers were successfully fabricated by electrospinning method. The result of FTIR and EDAX shows that PVA and $TiO_2$ nanoparticles exist in PVA/$TiO_2$ nanofibers. SEM observation shows that the fiber has several hundred nanometers with a smooth surface and have $TiO_2$ nanoparticle distributed in the fibers. It also shows that by adding $TiO_2$ in the electrospun solution resulted in reducing the diameter of the fiber due to reducing the solution's viscosity and increasing the conductivity of the electrospun solution. Photocatalytic activity of PVA/$TiO_2$ nanofiber is lower than $TiO_2$ powder, but PVA/$TiO_2$ nanofiber is more efficient than $TiO_2$ powder regarding photocatalytic treatments. The result reveals that PVA/$TiO_2$ composite nanofibers could be applied for environmental field effectively.

## 5. References

[1] Gupta S M and Tripathi M 2011 A review of TiO2 nanoparticles *Chinese Sci. Bull.* **56** 1639–57
[2] Yun J, Jin D, Lee Y and Kim H 2010 Photocatalytic treatment of acidic waste water by electrospun composite nanofibers of pH-sensitive hydrogel and TiO 2 *Mater. Lett.* **64** 2431–4
[3] Im J S, Kim M Il and Lee Y-S 2008 Preparation of PAN-based electrospun nanofiber webs containing TiO2 for photocatalytic degradation *Mater. Lett.* **62** 3652–5
[4] Qu X, Alvarez P J J and Li Q 2013 Applications of nanotechnology in water and wastewater treatment *Water Res.* **47** 3931–46
[5] Bo Y I N, Ji-tong W, Wei X U, Dong-hui L and Wen-ming Q 2013 Preparation of TiO 2 / mesoporous carbon composites and their photocatalytic performance for methyl orange degradation *New Carbon Mater.* **28** 47–54
[6] Bhardwaj N and Kundu S C 2010 Electrospinning : A fascinating fiber fabrication technique **28** 325–47
[7] Garg K and Bowlin G L 2011 Electrospinning Jets and nanofibrous structures *Biomicrofluidics* **5** 1–19
[8] Elsabee M Z, Naguib H F and Morsi R E 2012 Chitosan-based nanofibers, review *Mater. Sci. Eng. C* **32** 1711–26
[9] Choi D, Hwang C, Lee J, Lee I, Oh I and Park J 2013 Characterization of TiO 2 fibers by combined electrospinning method and hydrothermal reaction **106** 41–4
[10] Prahsarn C, Klinsukhon W and Roungpaisan N 2011 Electrospinning of PAN / DMF / H 2 O containing TiO 2 and photocatalytic activity of their webs *Mater. Lett.* **65** 2498–501
[11] Abbasizadeh S, Keshtkar A R and Mousavian M A 2013 Preparation of a novel electrospun polyvinyl alcohol/titanium oxide nanofiber adsorbent modified with mercapto groups for uranium(VI) and thorium(IV) removal from aqueous solution *Chem. Eng. J.* **220** 161–71
[12] Jeon S, Yun J, Lee Y-S and Kim H-I 2012 Preparation of poly(vinyl alcohol)/poly(acrylic acid)/TiO2/carbon nanotube composite nanofibers and their photobleaching properties *J. Ind. Eng. Chem.* **18** 487–91

## Acknowledgement

This research was supported by a research grant of "PUPT (*Penelitian Unggulan Perguruan Tinggi*), Contract No:157/LPPM/2015" by the General Directorate of High Education (DIKTI), Ministry of Education and Culture, the Republic of Indonesia. The authors would also thank Mr. Kunto Wandono for maintaining the electrospinning machine.

# Synthesis, Single Crystal Structure, and Magnetic Properties of 3-D $Cu(NITpPy)_2[Cu(CN)_3].2CH_3OH.2H_2O$ Complexes

**I W Dasna[1], S Golhen[2], L Ouahab[2] and Subakti[3]**

[1]Jurusan Kimia, FMIPA, Universitas Negeri Malang, Jl. Semarang 5 Malang, 65145, Jawa Timur, Indonesia
[2]Laboratoire de Chimie du Solide et Moléculaire, Groupe Matériaux Moléculaires, UMR 6511, CNRS-Uniersité de Rennes 1, Institut de Chimie de Rennes, Avenue du Général Leclerc, 35042 Rennes cedex, France
[3]Jurusan Fisika, FMIPA, Universitas Negeri Malang, Jl. Semarang 5 Malang, 65145, Jawa Timur, Indonesia

Email: idasna@um.ac.id

**Abstract.** A new Cu(II) complexes containing nitroxide radicals $Cu(NITpPy)_2[Cu(CN)_3]$. $2CH_3OH.2H_2O$ has been successfully synthesized. The single crystal X-ray analysis revealed that the compound belongs to monoclinic crystal system (C2/c (#15)). The existence of the C≡N radical and water molecules are observed by using infra-red within the range of 4000-500 $cm^{-1}$ and 3500-3000 $cm^{-1}$, respectively. The $Cu(NITpPy)_2^{2+}$ units are connected through $N(CN)^{2-}$ bridging ligands in 2 coordination developing, respectively, into one-dimensional and two-dimensional polymeric networks. Paramagnetic interactions in the compound appear in the vicinity of room temperature. As the temperature decreased, low dimensional antiferromagnetic interactions established and Cu(II)-radicals antiferromagnetic interactions dominates down to 2 K with $\mu_{eff} = 0.8129\ \mu_B$

**Keywords :** Magnetism, Polycyanamide Complexes, Nitronyl Nitroxide

## 1. Introduction

There have been challenges to synthesize magnetic complexes compounds by using radical nitronyl nitroxide as ligand [1-5] [1],[2][3],[4][5] . The organic radical nitronyl nitroxide is organic stable at room temperature which possesses single unpaired electron. This unpaired electron could provide the possibility to obtain complexes compounds with unique magnetic properties. The magnetic properties of nitonyl nitroxide consisting complexes can be modified for example by forming bi-radical, or combining nitronyl nitroxide with the other functional groups, using the radical as a ligand to form transition metal organic frameworks [6], [7], [8], [9], [10],[11].

The latest method apparently provides interesting field to study the magnetic interaction between unpaired electrons coming from the nitronyl nitroxide as the ligand and transition metal as the atomic centers. These kinds of complexes having potential application in the field of molecular magnetic. Moreover, the possibilities of combining building blocks of transition metal-radical with a polydentate ligand can provide opportunities to obtain interesting polymeric magnetic complexes [7-11]. Previous work by Dasna *et al.* [7,8,11] shows that obtaining 1D and 2D complexes are possible which is very important crystallographically as well as to shed light on the search of functional magnetic complexes through structural modification. The reported complexes were consisting of 2+ transition metal and nitronyl nitroxide as the building block connected by dicyanamide polydentate ion $[N(CN)_2]^-$.

In this work, we aim to synthesize new building block of complexes $[Cu(NITpPy)_2]^{2+}$ bound to $[Cu(CN)_3]^{2-}$ ligand. This anion ligand is a polydentate ligand and is designated to provide the possibility of having 3D crystal structure by bridging the building block $[Cu(NITpPy)_2]^{2+}$ which would become an impressive system to explore the crystal structure magnetic properties interplay. Apart from that, the unique phenomenon in this synthesis process is the formation of $[Cu(CN)_3]^{2-}$ ligand during the crystal growing with slow liquid diffusion H-tube method. This ligand is neither commercially available nor almost impossible to obtain by additional $CN^-$ into CuCN compound. However, the $[Cu(CN)_3]^{2-}$ ligand was formed metathetically during the reaction between tetracyanoethylene (TCNE-) anion and $[Cu(NITpPy)_2]^{2+}$. That is a partial reaction between $[Cu(NITpPy)_2](NO_3)_2$ and Li[TCNQ] in methanol which results in $[Cu(CN)_3]^{2-}$ anion. It is suggested that there was possibly occurring reduction-oxidation reaction between $[TCNE]^-$ with a certain amount of $Cu^{2+}$ from reactant to form the $[Cu(CN)_3]^{2-}$ ion. Furthermore, the present of radical ligand in slow liquid diffusion process can accelerate the reduction-oxidation reaction.

The significant results in this contribution concerning the use of unpaired electrons in both nitronyl nitroxide radical and transition metal ions ($Cr^{2+}$, $Mn^{2+}$, $Co^{2+}$, $Ni^{2+}$, $Fe^{2+}$) connected by polydentate ligands to design and synthesize functional magnetic complexes compounds. The results have led us to extend the idea that besides using $[Cu(CN)_3]^{2-}$ and $[N(CN)_2]^-$ ions as ligands, it is also possible applying shorter bidentate ligands such as $CN^-$ and $SCN^-$ anions. These two latest ligands are expected to shorten the bridging atomic distances to strengthen the magnetic interaction at molecular level. The study of synthesis, structural and magnetic properties of 3D polymeric complexes $Cu(NITpPy)_2[Cu(CN)_3].2CH_3OH.2H_2O$ are described in this article.

## 2. Experimental Method

Compounds of $Cu(NO_3)_2.5H_2O$ (Aldrich), TCNE (Fluka) were used as purchased. NITpPy was prepared from the procedure described by Ulmanet *et al.* (1970) [12]. The Synthesis was carried out in argon environment and using fresh distilled methanol. The compound $Cu(NITpPy)_2[Cu(CN)_3]$. $2CH_3OH.2H_2O$ obtained by employing slow liquid diffusion in methanolic solvent using H-tube. About 200 mg of $Cu(NITpPy)_2(NO_3)_2$ was placed in one part of H-tube and 100 mg of LiTCNE placed in the other part of H-tube. The methanolic solvent was added with a very slowly rate into the tube. Single crystals were obtained after one-month growing process.

### *2.1 Crystallographic Data Collection and Structure Determination*

A black brown single crystal was mounted on an Enraf-Nonius CAD4 diffractometer equipped with a graphite monochromated Mo-Kα radiation source (λ = 0.71073Å). The air-sensitive crystals were protected with glue. The cell dimensions and the orientation matrix for data collection were obtained from least-squares refinement, using the setting angles of 25 centered reflections. The crystal data are given in Table 1. The intensities were collected by θ-2θ scans. There was no significant decay reflected on the three standard reflections measured every hour during data collection and correction. The crystal structures were solved by using SHELXS97 and refined with SHELXL97 [13] by the full-matrix least-square method.

### *2.2 Magnetic and Spectroscopic Measurements*

Magnetic measurements have been carried out down to 2 K employing a Quantum Design MPMS-5S SQUID magnetometer. All magnetic investigations were carried out for the polycrystalline sample. The molar susceptibility was corrected for the sample holder and for the diamagnetic contribution of all atoms by using Pascal's table [14]. The FTIR spectra were measured on a KBr pellet sample using IR Bomem-Michelson 100 spectrometer.

## 3. Result And Discussion

### *3.1 Crystal structure*

We have successfully grown enough green crystals with the form of needles to study their structures and magnetic properties. The FTIR study in a KBr matrix within the range of 4000-500 $cm^{-1}$ shown in Figure 1 reveals the peaks related to the anion and the C≡N radical. The peaks related to the water molecules are also observed between 3500-3000 $cm^{-1}$. Several significant observed peaks are : $\nu_{C\equiv N}$ : 127(s) ; $\nu_{N\text{-}O}$ : 1373(s), 1136(w) ; $\nu_{N\text{-}O}$ : 1371(s), 1136(w) ; $\nu_{py}$ : 1612(m), 1546(w), 1547(m), 1451(m), 1404(s) ; $\nu_{O\text{-}H}$ : 3440(m). The peaks at 459 and 542 $cm^{-1}$ are indicating the presence of Cu-N and Cu-O in the compound.

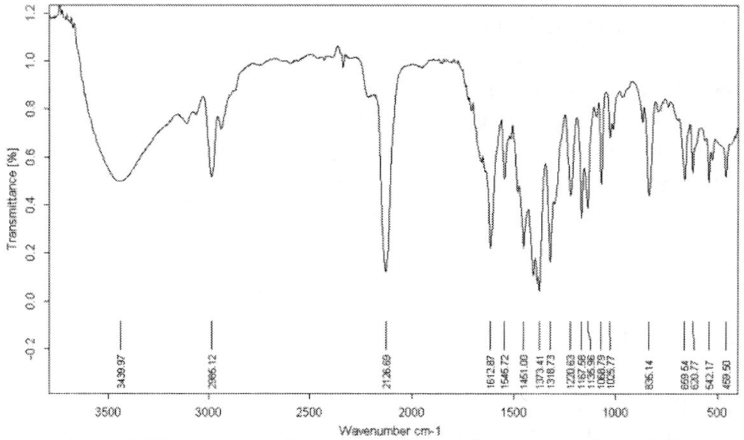


**Figure 1.** FTIR spectra in KBr pellet of compound $Cu(NITpPy)_2[Cu(CN)_3].2CH_3OH.2H_2O$

### 3.2 *Structural Analysis*

The crystallographic parameters of the $Cu(NITpPy)_2[Cu(CN)_3].2CH_3OH.2H_2O$ are summarized in Table 1. The crystal fits the space group C 2/c monoclinic crystal system. Cu1 and Cu2 the atoms are on the inversion centers (¼, ¼, 0) and (½, ½, 0), the third ion Cu3 is situated in general position.

**Table 1.** Crystallographic data for $Cu(NITpPy)_2[Cu(CN)_3].2CH_3OH.2H_2O$

| Compound | C57H86Cu4N17O15 |
|---|---|
| Molecular Mass/g.mol$^{-1}$ | 1503,59 |
| Crystal System | Monoclinic |
| Space Group | C2/c (#15) |
| Lattice Parameters: | |
| a/Å | 33,2524(11) |
| b/Å | 13,6270(4) |
| c/Å | 23,5211(6) |
| β/deg | 128,247(1) |
| Volume/Å$^3$ | 8370,3(4) |
| Z, V Mass (calculated)/ g.cm$^{-3}$ | 4, 1,193 |
| R value | R1 = 0,0706 |
| [I>2σ (I)] | wR2 = 0,2132 |

The structure ORTEP diagram given in Figure 2 shows that this compound contains two types of copper atoms with different valence state: copper (II ) and copper (I). Copper (II) ions (Cu1 and Cu2) are coordinated in transposition with two radicals through the nitrogen atom of the pyridine ring, two nitrogen atoms of the C≡N group, and two atoms of oxygen to methanol . Copper (I) (Cu3) is coordinated by three C≡N anions to yield a diamagnetic ligand tridentate CuI (CN) 32- which adopts a distorted trigonal geometry with angles C-Cu (3) -C ranging from 114.1 (2) ° to 124.1 (2) °.

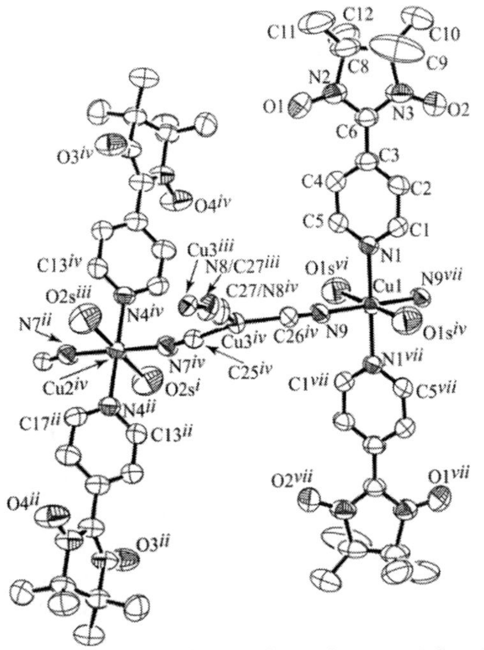


**Figure 2.** ORTEP view of $Cu(NITpPy)_2[Cu(CN)_3]$ $.2CH_3OH.2H_2O$ ( *i = x, 1+y, z-1,ii = 1-x, 1+y, ½+ z; iii: 1-x, -y, 2-z; iv: x, -y, z-0,5; vii = ½-x, -1/2-y, 1-z).*

Two nitrogen atoms of the anion $CuI((CN)_3)^{2-}$ are each connected to the copper atom of a Cu-group $(NITpPy)_2(CH_3OH)_2$, Cu1 ion on one side and an ion Cu2 the other. In this trans Cu1 or Cu2 ion include another group CuI (CN) 32-. The angles formed between the anion CuI (CN) 32- (N8iii plan N7iv, N9) and O1svi plan N9vii, O1siv, N9 on the one hand and the one formed between CuI (CN) 32- and O2siii, N7iv, O2Si, N7ii are 67.65(9) and 38.69° (11) , respectively. The ion Cu3 is also connected to a third group CN. The position of the nitrogen N8 derives from that of the C27 carbon atom by an axis of symmetry of order two. Thereby the position is statistically half occupied by an atom N8 and a C27 atom. The third nitrogen atom of the anion CuI (CN) 32- is then linked to an anionic group CuI (CN) 32-.

The structure of compound 18 is a three-dimensional network whereas Cu ion is bound to a -CN group in a spiral which develops along b axis. The CuII-radical groups adopt a distorted octahedral geometry with average bond lengths Cu-Cu-O and N of 1.9933 (4) Å and 2.4775 (49) Å, an average angle CuII N-N-90 , 0°, and O-Cu-N varying angles of 81.23 (12) 98.77° (12)°. The pyridine group plan and O1, N2, C6, N3, O2 form an angle of 28.09 (41)° similar to those observed in the foregoing. The angle of 75.25 (16)° formed between the square N1, N9, N1vii, N9vii, Cu1 and Cu plane (CN) 32- is similar (77.84 (15)°) that the angle of this anion with the second Cu-radical plan (N4iv, N7iv, N4ii, N7ii, Cuiv). Both Cu-radical groups are in different positions as shown in Figure 3. the O1svi plans N9vii, O1siv, N9 and O2Si, N7iv, O2siii, N7ii an angle of 82.23 (11)°. Distances and selected interatomic angles are summarized in Table 2.

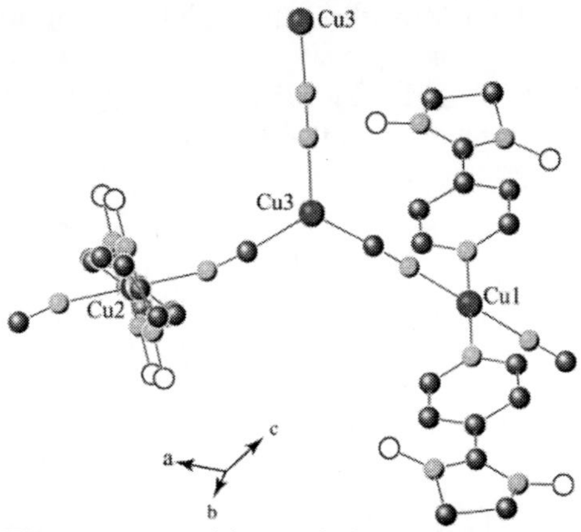


**Figure 3.** Position relative group *Cu-radicals* in compound Cu(NITpPy)$_2$ [Cu(CN)$_3$].2CH$_3$OH. 2H$_2$O.

**Table 2.** Selected Interatomic Distances (Å) and angles (deg) for Cu(NITpPy)$_2$[Cu(CN)$_3$] .2CH$_3$OH.2H$_2$O

| | | | |
|---|---|---|---|
| Cu(2)-N(7) | 1,951(4) | N(3)-O(2) | 1,267(7) |
| Cu(2)-N(4) | 2,036(4) | N(8)iii-C(27)iv | 1,162(9) |
| Cu(1)-N(9) | 1,943(4) | N(8)iii-N(8)iv | 1,162(9) |
| Cu(1)-N(1) | 2,043(4) | N(7)-C(25) | 1,144(6) |
| Cu(3)-C(26) | 1,907(5) | O(1)-N(2) | 1,250(7) |
| Cu(3)-C(25) | 1,928(5) | C(1S)-O(1S) | 1,417(10) |
| Cu(3)-N(8) | 1,932(5) | C(2S)-O(2S) | 1,403(10) |
| N(6)-O(4) | 1,284(6) | O(1S)iv-Cu1 | 2,4936(48) |
| O(3)-N(5) | 1,269(6) | O(2S)i-Cu2iv | 2,4613(50) |
| N(9)-C(26)iv | 1,152(6) | | |
| | | | |
| N(3)-O(2) | 1,267(7) | C(25)-Cu(3)-N(8) | 114,1(2) |
| N(8)iii-C(27)iv | 1,162(9) | C(1)-N(1)-Cu(1) | 121,8(3) |
| N(8)iii-N(8)iv | 1,162(9) | C(5)-N(1)-Cu(1) | 120,6(3) |
| N(7)-C(25) | 1,144(6) | C(13)-N(4)-Cu(2) | 121,5(4) |
| O(1)-N(2) | 1,250(7) | C(17)-N(4)-Cu(2) | 121,4(4) |
| C(1S)-O(1S) | 1,417(10) | C(26)iv-N(9)-Cu(1) | 176,8(5) |
| C(2S)-O(2S) | 1,403(10) | C(27)iv-N(8)iii-Cu(3)iii | 174,98(16) |
| O(1S)iv-Cu1 | 2,4936(48) | C(25)-N(7)-Cu(2) | 167,0(5) |
| O(2S)i-Cu2iv | 2,4613(50) | O(1)-N(2)-C(6) | 126,3(5) |
| N(7)-Cu(2)-N(4)I | 89,92(17) | O(1)-N(2)-C(8) | 121,8(5) |
| N(7)-Cu(2)-N(4) | 90,08(17) | N(7)-C(25)-Cu(3) | 174,7(5) |
| N(9)-Cu(1)-N(1) | 89,63(16) | N(9)v-C(26)-Cu(3) | 177,9(5) |
| N(9)-Cu(1)-N(1)ii | 90,37(16) | O(1S)iv-Cu(1)-N(1) | 98,77(12) |
| C(26)-Cu(3)-C(25) | 124,1(2) | O(1S)vi-Cu1-N(1)vii | 81,23(12) |
| C(26)-Cu(3)-N(8) | 121,8(2) | O(2S)iii-Cu(2)iv-N(4)iv | 89,99(2) |

*i : -x+1, -y-1, -z+2 ; ii : -x+1/2, -y+1/2, -z+1 ; iii : x, -y, z-1/2 ;*
*iv : -x+1, y, -z+5/2 ; v : x, -y, z+1/2*

The spiral-like crystal network develops along *b*-direction such that projection on the *ac* plane, a large hexagon is observed (Figure 4) containing four Cu-radical groups. Each hexagon contains ten Copper ions (four six CuII and CuI). The distances between Cu1..Cu3, Cu2 and Cu3 ... Cu3..Cu3 are respectively 4.9980 (6) Å, 4.9719 (5) Å and 5.011 (11) Å.

In a hexagon, the largest diagonal between two copper atoms (Cu 3 .... Cu3) equals 15.213 (1) Å and the distance ... N8 N8 is 17.9681 (5) Å (Figure 4). The water molecules are located inside the spiral in each hexagon without coordinates metal while the methanol molecules are coordinates the CuII atom. In the octahedron CuO2N4, the Cu-O distances are longer than the Cu-N distances indicating the existence of Jahn-Teller distortion for the cation CuII. Figure 5 shows the helical network $[Cu_2(CN)_3]_n$, in which for reasons of clarity, the radicals and solvent molecules are omitted. It shows that the structure consists of an assembly of left or right form spirals. The shortest interactions between groups N-O, 3.727 (11) Å and 3.691 (13) Å, are observed between O3 and O2 ... ... O3iii O2vii.

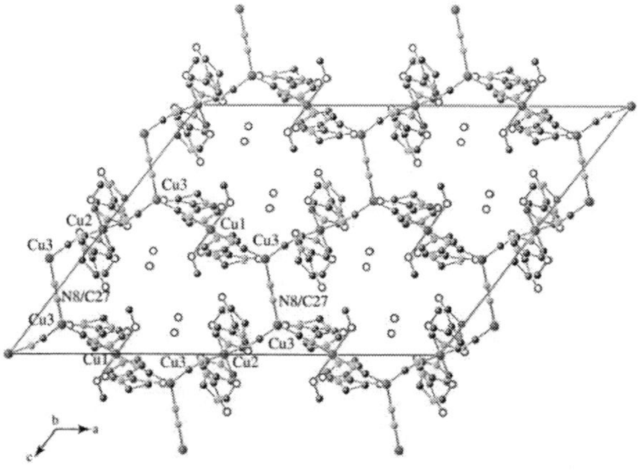


**Figure 4.** Projection on *ac* plan for compound $Cu(NITpPy)_2[Cu(CN)_3]$ $.2CH_3OH.2H_2O$.

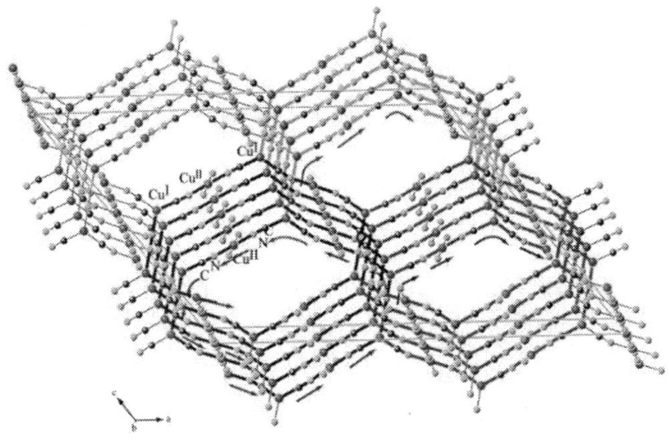

**Figure 5.** Network in 3-D for compound $Cu(NITpPy)_2[Cu(CN)_3]$. $2CH_3OH.2H_2O$ whereas radical and solvent are not viewed for crystal structure clarity.

*3.3 Magnetic Properties*

The $\chi_M T$ data are plotted against T in the range of 2 K and 300 K shown in Figure 6. At room temperature, the compound exhibits a paramagnetic behavior with a value of 0.826 $\chi_M T$ $cm^3.K.mol^{-1}$ which is lower to the expected value coming from the uncoupled spins of Cu(II) ion (S = 1/2) and two radicals (S = 1/2). When the temperature decreases down to 46 K, the $\chi_M T$ value decreases gradually and then decreases very fast approaching 2 K indicating the existence of antiferromagnetic interactions between the copper (II) ion and two radicals. This curve is modeled using the same Hamiltonian as for compound 16 pourconduire to an antiferromagnetic coupling constant between Cu-radical, J = -13.2 (3) $cm^{-1}$ and a weak intermolecular interaction antiferromagnetic (zJ = -0 74 (20) $cm^{-1}$) between the magnetic centers, for a value of g = 2.0 and reliability factor R = 0.99755.

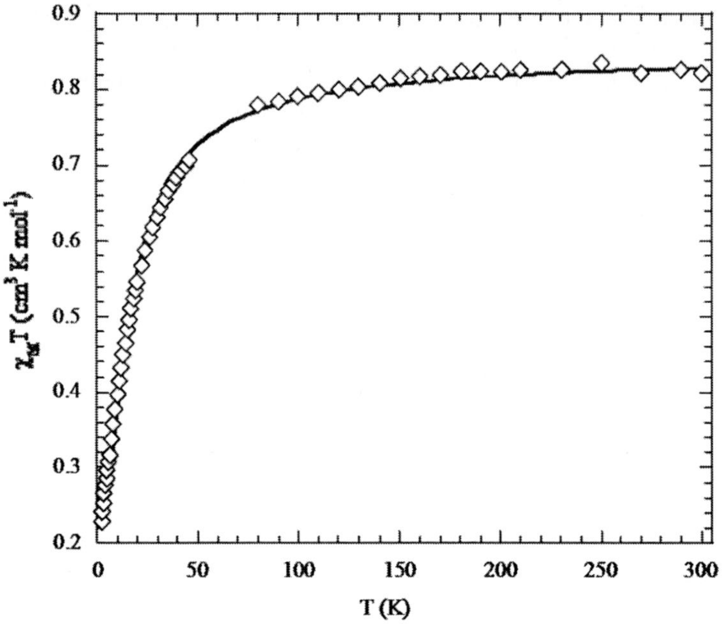


**Figure 6.** $\chi_M$T vs. T for $Cu(NITpPy)_2[Cu(CN)_3].2CH_3OH.2H_2O$

## 4. Conclusion

The coordination complexes with formula $Cu(NITpPy)_2[Cu(CN)_3].2CH_3OH.2H_2O$ has been successfully synthesized by means of slow liquid diffusion H-tube method in methanolic solution. The complexes obtained shows the 3D crystal behavior with polymeric chain. The building block of $[Cu(NITpPy)_2]^{2+}$ is connected by $[Cu(CN)_3]^{2-}$ anion ligand results in 3D chain network. The unpaired electrons contributed by $Cu^{2+}$ and radicals bound to $Cu^{2+}$ ion giving rise antiferromagnetic interaction. The results pave the way to design and synthesis controllable magnetic properties in coordination complexes by using different transition metal ions.

## 5. References

[1] Tretyakov E, Okada K, Suzuki S, Baumgarten M, Romanenko G, Bogomyakov A and Ovcharenko V 2016 Synthesis, structure and properties of nitronyl nitroxide diradicals with fused-thiophene couplers *J. Phys. Org. Chem.* **29** 725–734

[2] Lee J, Lee E, Kim S, Bang G S, Shultz D A, Schmidt R D, Forbes M D and Lee H 2011 Nitronyl Nitroxide Radicals as Organic Memory Elements with Both n-and p-Type Properties *Angew. Chem. Int. Ed.* **50** 4414–4418

[3] Sugano T, Blundell S J, Pratt F L, Hayes W, Uekusa H, Ohashi Y, Kurmoo M and Day P 1997 Magnetism in nitronyl nitroxide radicals and their ion radical salts *Mol. Cryst. Liq. Cryst.* **305** 435–444

[4] Suzuki S, Furui T, Kuratsu M, Kozaki M, Shiomi D, Sato K, Takui T and Okada K 2010 Nitroxide-substituted nitronyl nitroxide and iminonitroxide *J. Am. Chem. Soc.* **132** 15908–15910

[5] Yao R, Liao S, Chen X, Tang G, Wang G and Zheng F 2016 Effects of ZnO and NiO on material properties of microwave absorptive glass-ceramic tile derived from iron ore tailings *Ceram. Int.* **42** 8179–8189

[6] Wang X, Wen G, Wang F, Chen Y, Zhao X and Bao X 2016 Synthesis and crystal structure of an iron (II) complex with nitronyl nitroxides *Synth. React. Inorg. Met.-Org. Nano-Met. Chem.* **46** 206–209

IOP Conf. Series: Materials Science and Engineering **202** (2017) 012012 doi:10.1088/1757-899X/202/1/012012

[7] Dasna I, Golhen S, Ouahab L, Peña O, Daro N and Sutter J-P 2001 Ferromagnetic interactions in 0D and 2D Mn (II) coordination complexes containing nitronyl-nitroxide radicals and dicyanamide anions: Mn II (NITpPy) 2 [N (CN) 2] 2 (H 2 O) 2 and Mn II (NITpPy) 2 [N (CN) 2] 2· 2 CH 3 CN *Comptes Rendus Académie Sci.-Ser. IIC-Chem.* **4** 125–133

[8] Dasna I, Golhen S, Ouahab L, Fettouhi M, Pena O, Daro N and Sutter J-P 2001 Synthesis, X-ray crystal structures and magnetic properties of Cu (II)(NITpPy) 2 [N (CN) 2] 2· solv (NITpPy= nitronyl nitroxide radical, solv= H 2 O or CH 3 CN). From discrete molecules to 2-D polymeric coordination compounds *Inorganica Chim. Acta* **326** 37–46

[9] Dasna I, Golhen S, Ouahab L, Peña O, Guillevic J and Fettouhi M 2000 1-D mixed stack of coordinated and uncoordinated radicals in Mn II (NITpPy) 4 [N (CN) 2] 2 (NITpPy= nitronyl nitroxide radical) *J. Chem. Soc. Dalton Trans.* 129–132

[10] Dasna I, Golhen S, Ouahab L, Peña O, Daro N and Sutter J-P 2000 A dimeric Cu (II) acetate complex containing axially coordinated p-pyridyl nitronyl nitroxide radicals:[Cu II (CH 3 COO) 2 (NITpPy)] 2 *New J. Chem.* **24** 903–906

[11] Dasna I, Golhen S, Ouahab L, Daro N and Sutter J-P 2001 Ferromagnetic interactions in Mn (II) coordination complex containing nitronyl nitroxide radical and silver–dicyanide anion: structure and magnetic studies of Mn II (NITpPy) 2 [Ag (CN) 2] 2 *Polyhedron* **20** 1371–1374

[12] Ullman E F, Call L and Osiecki J H 1970 Stable free radicals. VIII. New imino, amidino, and carbamoyl nitroxides *J. Org. Chem.* **35** 3623–3631

[13] Sheldrick G M and Shelx97 A A program for the refinement of Crystal Structures, 1997 *Univ. Gött. Ger.*

[14] Mabbs, F.E. and Machin, D. J. 1973. Magnetism and Transition metals Complex. London: Chapman and Hall.

# Preparation of MWCNT-$Fe_3O_4$ Nanocomposites from Iron Sand Using Sonochemical Route

**R Rahmawati[1,2], A Melati[2], A Taufiq[3], Sunaryono[3], M Diantoro[3], B Yuliarto[1], S Suyatman[1], N Nugraha[1] and D Kurniadi[1]**

[1]Engineering Physics Study Program, Faculty of Industrial Technology, ITB Bandung 40132, Indonesia
[2] Departement of Physics, Faculty of Sains and Technology, UIN Sunan Kalijaga Yogyakarta 55281, Indonesia
[3] Department of Physics, Faculty of Mathematics and Natural Sciences, Universitas Negeri Malang (State University of Malang), Jl. Semarang 5, Malang 65145, Indonesia

Email: brian@tf.itb.ac.id

**Abstract:** The composites of multi-walled carbon nanotube (MWCNT) and magnetite ($Fe_3O_4$) nanoparticles from iron sand were successfully prepared via the sonochemical route. In this experiment, the MWCNT-$Fe_3O_4$ nanocomposites were prepared with different compositions of MWCNT (0.01%, 0.02%, and 0.04%) with the constant composition of $Fe_3O_4$ particles. The characterizations were performed by means of X-Ray Diffractometry (XRD), Fourier Transform Infra-Red (FTIR) Spectrometer and Scanning Electron Microscopy (SEM) integrated with Energy Dispersive X-Ray (EDX). The XRD data analysis showed that the $Fe_3O_4$ crystallize in spinel structure in nanometric size. Furthermore, the crystallinity of the samples tended to reduce by increasing the MWCNT compositions. The SEM images showed that $Fe_3O_4$ tend to agglomerate in nanometric size. The FTIR spectra detected the functional groups of Fe-O bonding that showed the existence of $Fe^{2+}$ and $Fe^{3+}$. In the composites, the $Fe_3O_4$ nanoparticles were physically mixed with the MWCNTs constructing a unique structure. The as prepared MWCNT-$Fe_3O_4$ nanocomposites have the potential for bio-applications.

**Keywords:** MWCNT, $Fe_3O_4$, iron sand, nanocomposite, sonochemical route.

## 1. Introduction

Nowadays, the development of nanomaterials was focused on improvement of properties and function as well as their promising application. So, the fabrication of nanocomposites has been extensively investigated by researchers for exploring the new type of materials with combine of two or more compounds with better performances. By definitions, a nanocomposite consists of a multiphase of solid materials where part(s) or all of the phases with dimensions less than 100 nanometers (nm) [1]. In recent years, nanocomposites of carbon nanotube (CNT) and magnetite ($Fe_3O_4$) become much attention due to their unique properties owned by these materials, especially in nanoscale. The CNT has great potential in many application since found in 1991 by Iijima [2] due to it has many advantages such as large surface area, high electric conductivity, superior in mechanical and thermal properties [3-4]. Meanwhile, the $Fe_3O_4$ also has standout properties particularly in magnetic properties, namely superparamagnetic behaviors in nanoscale [5]. Furthermore, the flexibility of $Fe_3O_4$ which can be arranged depending on their sizes and structures make it become the fascinating materials.

The nanocomposite of MWCNT-$Fe_3O_4$ have been applied in many fields such as for removal trace arsenic and chromium [4,6], for magnetic printing ink [7], as adsorbents [8], for enhancing electrochemical performance in Li-ion batteries [9], for immunoscreening [10], for immune sensor [11], for hyperthermia and drug delivery [12], and so forth. There are three general approaches for creating MWCNT-$Fe_3O_4$ composites i.e. the $Fe_3O_4$ can be encapsulated, incorporated within the walls and deposited on the outer surface of nanotubes [13]. Various techniques have been conducted by researchers for obtaining the nanocomposite of MWCNT-$Fe_3O_4$. Pourkhalil [14] successfully synthesized the MWCNT-$Fe_3O_4$ nanocomposite by using sol-gel methods. The results showed that the $Fe_3O_4$ distributed on the surface of the MWCNT with superparamagnetic behavior at room temperature and with saturation magnetization of 19 emu/g. Meanwhile, Balacianu et al. [15] reported that the synthesis of MWCNT-$Fe_3O_4$ was done by coprecipitation method with covalent functionalization of $Fe_3O_4$. However, this process was not significantly add the amount of MWCNT which attached on the $Fe_3O_4$. The coprecipitation method was also used by Qu et al. [16] resulting the $Fe_3O_4$ with average sizes of 10-30 nm found on the surface of MWCNT. Its electrochemical characterization revealed that the nanocomposite of MWCNT-$Fe_3O_4$ is very appropriate for biosensor applications. Furthermore, Zhou et al. [7] reported that the MWCNT-$Fe_3O_4$ nanocomposite was successfully prepared by hydrothermal method. The result showed that the $Fe_3O_4$ with cubic structure attaching to the surface of MWCNT. Its magnetic characterization showed that the nanocomposite has superparamagnetic behavior with saturation magnetization of 38 emu/g.

The various synthesis methods of the MWCNT- $Fe_3O_4$ nanocomposites and their application were reported by many researchers. However, the synthesis of the MWCNT-$Fe_3O_4$ with the sonochemical route using a starting material of $Fe_3O_4$ from iron sand have never been found in the literature. Therefore, this paper will report the effective and simple method to synthesize the MWCNT-$Fe_3O_4$ nanocomposites from iron sand via sonochemical route. Besides that, the influence of MWCNT composition on the morphology of nanocomposite MWCNT-$Fe_3O_4$ is also discussed.

## 2. Experimental Method

### *2.1. Materials*

In this work, the materials were multiwall carbon nanotubes (~ 50 nm in diameter, and ~ 1 μm in length), iron sand from the southern coast of Indonesia, HCl (molarity 12 M), $NH_4OH$ (molarity 6.5 M), $HNO_3$ (65 %), ethanol, and distilled water.

### *2.2. Functionalization of MWCNT*

The MWCNTs (1 g) were functionalized by treating nitric acid (1: 100 w/v) and sonicated at frequency of 40 kHz for 120 minutes at 50 °C. The functionalized MWCNTs were filtered by the paper filter and repeatedly washed with ethanol and distilled water until pH neutral. After that, the samples were dried in an oven at 100 °C for 5 hours.

### *2.3. MWCNT-$Fe_3O_4$ nanocomposite fabrication*

The first step, the iron sand that had been extracted by a magnet permanent was dissolved in HCl and stirred 900 rpm for 30 minutes until to obtain the iron salts solutions. Later, 10 mL the solutions were added into 0.1 g of MWCNTs (0.01 % of the total composite) and sonicated (40 kHz) for 10 minutes. The next process was the dropping of 12 mL $NH_4OH$ to get the black precipitate, sonication process continued for 20 minutes. The resulted precipitate was washed extensively with the distilled water, and the MWCNT-$Fe_3O_4$ was collected by a strong magnet. The as-prepared MWCNT-$Fe_3O_4$ was dried in an oven at 100 °C for 5 hours. This process was repeated with various MWCNT contents, i.e. 0.02% and 0.04%.

*2.4. Characterizations*

The samples were characterized by XRD for find out their crystalline size and crystal structure. Meanwhile, the functional groups of the nanocomposites were investigated by FTIR. Furthermore, the morphology and composition of the samples were observed by SEM-EDX.

**3. Results and discussion**

Figure 1 shows the XRD patterns of the samples. From the figure, the peaks of the $Fe_3O_4$ nanoparticles and MWCNT of all samples were clearly identified. The peak of MWCNT as shown in figure 1c for 0.04 % MWCNT at $2\theta = 25.9°$ on the plane (0 0 2), is in good agreement with the JCPDS 76-1651, and the MWCNT was identified as the hexagonal graphite structure [7]. The appearance peaks of MWCNT proves that the structure of MWCNT is not change due to the sonochemical process. Meanwhile, the peak of the $Fe_3O_4$ at 30.1°, 35.5°, 43.2°, 53.6°, 57.2°, and 62.7° well matched with the planes of (2 2 0), (3 1 1), (4 0 0), (4 2 2), (5 1 1), and (4 4 0) of the cubic spinel phase of $Fe_3O_4$ (JCPDS No 19-0629) [14]. Furthermore, the XRD pattern revealed that the peaks of MWCNT disappear along with the decreasing MWCNT composition in the composite. Related to the result, Zhou et al. [7], reported that the lowering of the MWCNT peaks for XRD patterns caused by the $Fe_3O_4$ particles with good crystalline structure that covered the e surface of the MWCNT. In this work, the crystallite size of the $Fe_3O_4$ particles was determined using Debye Scherer's formula [17]. The data analysis using the formula showed that the $Fe_3O_4$ particles have crystallite size in the range of 10-20 nm.

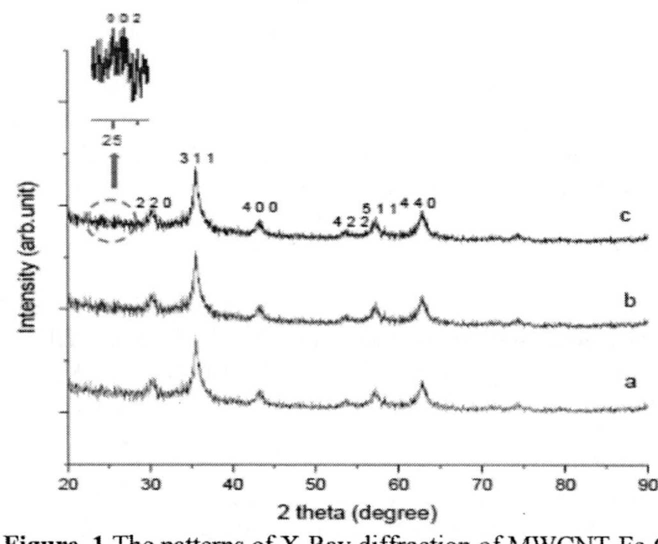


**Figure. 1** The patterns of X-Ray diffraction of MWCNT-$Fe_3O_4$, (a) 0.01% MWCNT, (b) 0.02% MWCNT, and (c) 0.04% MWCNT.

The FTIR spectra of the MWCNT-$Fe_3O_4$ nanocomposites were given in Figure 2. The spectra revealed that there are several peaks presenting the functional groups of the MWCNT and the $Fe_3O_4$. The peaks covered at 3321 $cm^{-1}$ (O-H stretching), 1720 $cm^{-1}$ (C=O stretching), 1574 $cm^{-1}$ (C=C stretching, 1310 $cm^{-1}$ (O-H bending) and 588 $cm^{-1}$ (F-O stretching and bending mode) [8,10]. Based on this characterization, it can be stated that the sonication process with nitric acid was successfully generated carboxyl groups on the surface of the MWCNTs. The carboxyl groups attached the $Fe_3O_4$ nanoparticles with occur electrostatic interaction between negative charge from carboxyl groups and positive charge from $Fe_3O_4$. Huang et al. [18] reported that the carboxyl groups could be attached to micro molecules that open potential opportunity for biomedical applications. According to Zhuo et al. [7], the functional groups of the MWCNT-$Fe_3O_4$ nanocomposites are easily to be homogeneous by dispersing in a liquid medium. The XRD result presented that $Fe_3O_4$ has a spinel cubic structure which corresponds to result from the FTIR spectra which confirm the existence of Fe-O stretching. Figure 2 also shows that the absorption bands of Fe-O become greater with increasing MWCNT composition in the nanocomposites. Moreover, Ghazanfari et al. [19] investigated that the addition of the amount of MWCNT in the composite increase the oxidation rate of $Fe^{2+}$ to $Fe^{3+}$.

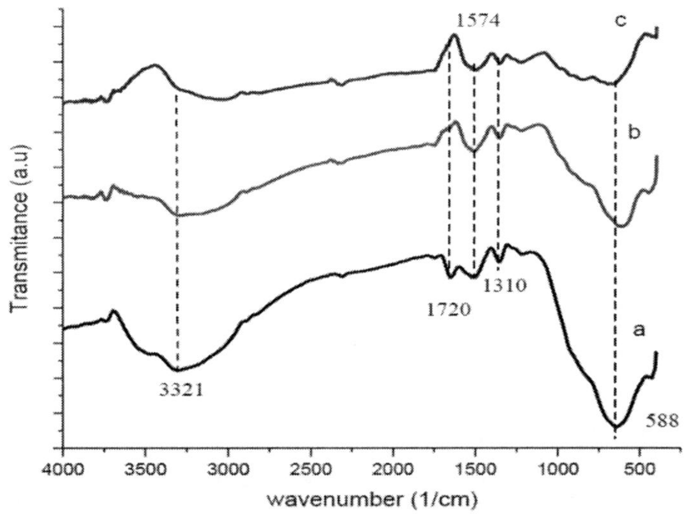


**Figure 2**. FTIR spectra of MWCNT-$Fe_3O_4$, (a) 0.01 % MWCNT, (b) 0.02 % MWCNT and (c) 0.04 % MWCNT

The morphology of the MWCNTs, $Fe_3O_4$ nanoparticles, and nanocomposites of MWCNT-$Fe_3O_4$ was investigated by SEM. As shown in Figure 3, the MWCNT was functionalized by nitric acid having particle size in diameter ranging from 39 to 63 nm. The decrease of the size of MWCNT if compared before functionalization process due to defect in the wall of MWCNT by the acid reaction. According to Samouhos et al. [13], the defect on the MWCNT surface become the nucleation sites for $Fe_3O_4$. As shown in Figure 1a, it can be seen that the MWCNT forms the bundles due to the strong Van der Waals attractions. Meanwhile, the morphology of the $Fe_3O_4$ synthesized by sonochemical methods (in Figure 3b) presents clusters of the $Fe_3O_4$ that cling together and form aggregates in spherical shapes with individual sizes ranging from 10 to 20 nm. The clusters of the $Fe_3O_4$ are originated from the magnetostatic coupling between particles [19]. The SEM images of the MWCNT-$Fe_3O_4$ nanocomposites are described in Figure 3c-3e. Based on the figure, it can be seen that network of MWCNTs is interwoven among the $Fe_3O_4$ nanoparticles. The morphology of the nanocomposite with 0.01 % MWCNT appears that the $Fe_3O_4$ adhered on the surface of MWCNT as well as form the large agglomeration as free particles. It can be explained that the amount of the MWCNT is less for nucleation and growth of $Fe_3O_4$. However, the agglomeration of the $Fe_3O_4$ decreased with increasing the MWCNT contents as shown in Figure 3d and 3e. Even, when utilization of 0.04 % MWCNT in the nanocomposite, the $Fe_3O_4$ attached uniformly on the surface of MWCNT. It can be indicated that the nanocomposite of MWCNT-$Fe_3O_4$ was successfully synthesized by sonochemical route.

IOP Conf. Series: Materials Science and Engineering **202** (2017) 012013 doi:10.1088/1757-899X/202/1/012013

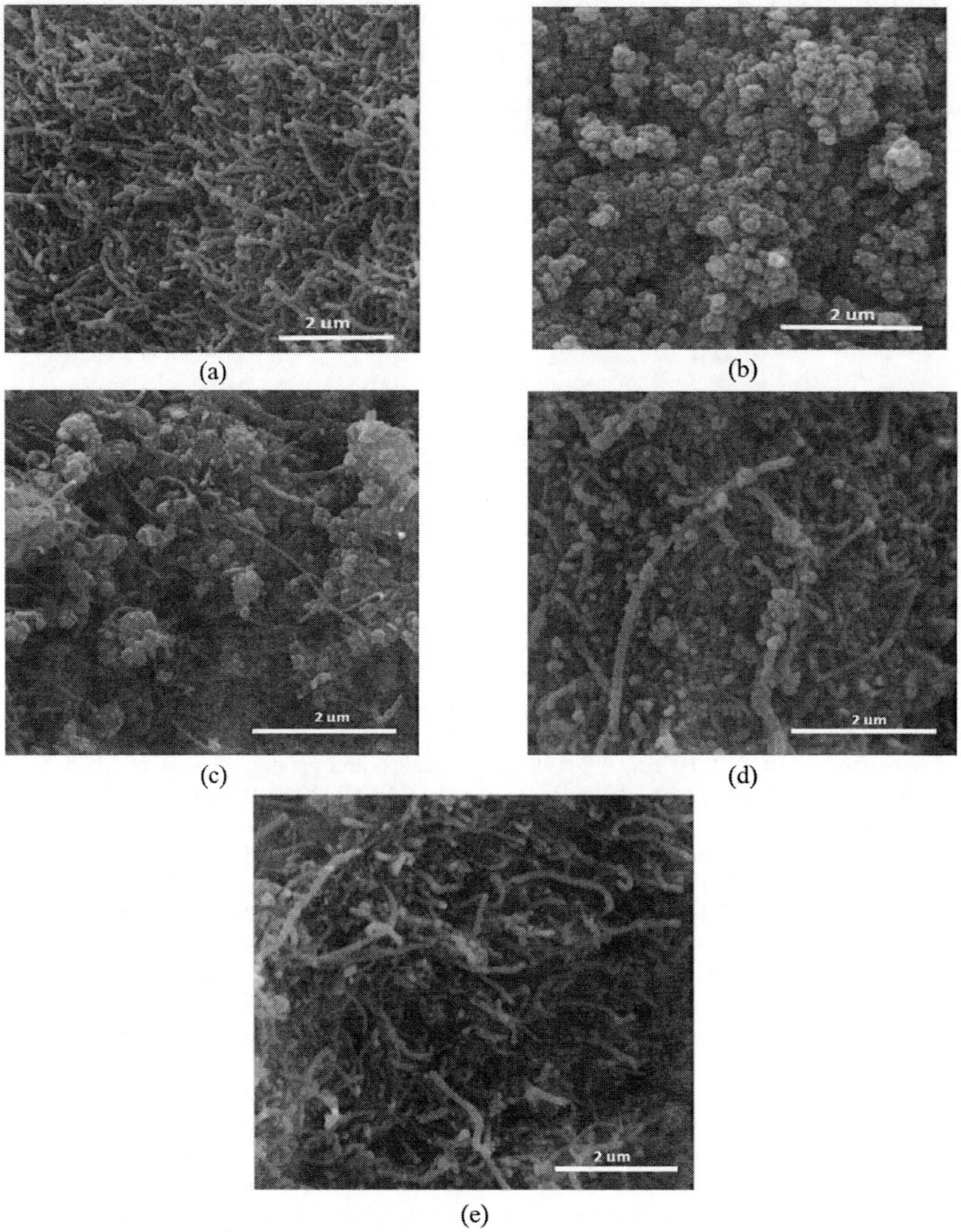


**Figure 3**. SEM images of (a) MWCNTs, (b) $Fe_3O_4$, (c) MWCNT-$Fe_3O_4$ with 0.01 % MWCNT, (d) MWCNT-$Fe_3O_4$ with 0.02 % MWCNT, (e) MWCNT-$Fe_3O_4$ with 0.04 % MWCNT.

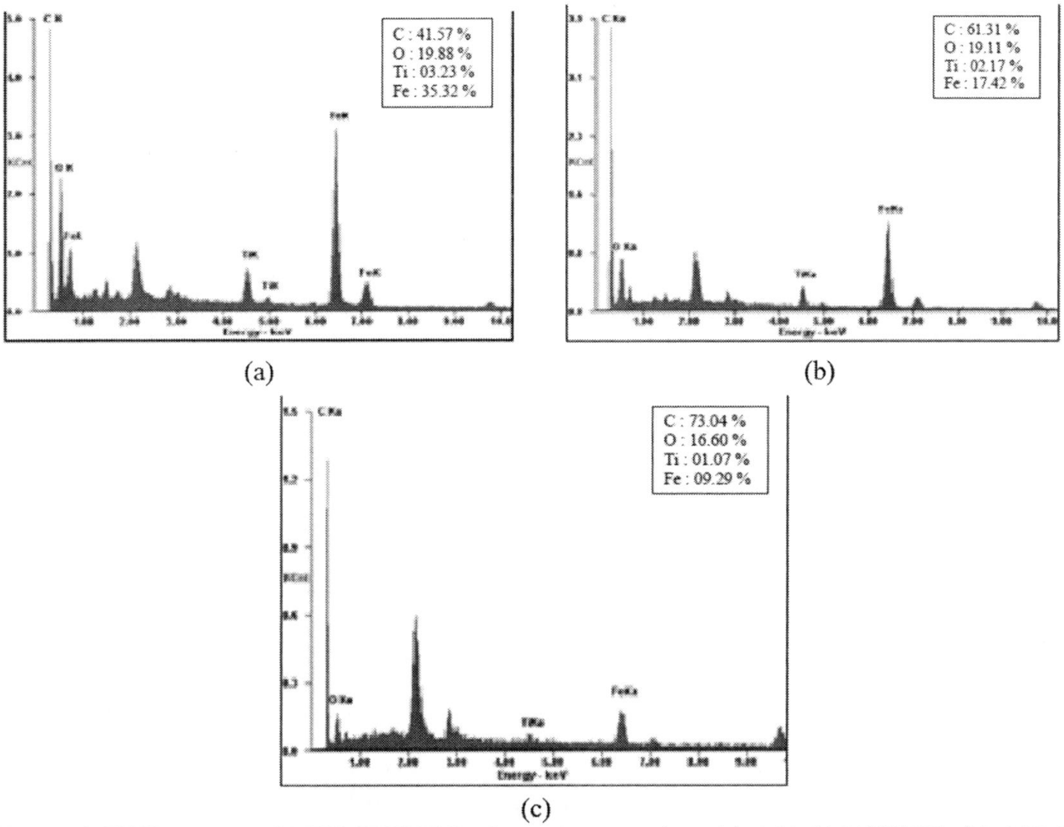


**Figure 4.** EDX spectra of MWMWCNT-$Fe_3O_4$ nanocomposites, (a) 0.01 % MWCNT, (b) 0.02 % MWCNT and (c) 0.04 % MWCNT

To compare the element content of nanocomposite, EDX characterization was carried out and the result is shown in Figure 4. The figure confirms the presence of C, O, and Fe element in the nanocomposites. These results are consistent with the result of XRD, FTIR, and SEM data. For more details, the C signal was generated by MWCNT. Meanwhile, the O signal appeared due to the functionalization process of MWCNT, and the Fe signal was generated by $Fe_3O_4$. However, in EDX appears the Ti element as impurity regarding to the iron sand as a starting material. From the EDX result. It also can be concluded that the percentage of Fe element (in wt%) decreases with increasing the amount of MWCNT.

## 4. Conclusion

A simple sonochemical route was employed to synthesize MWCNT-$Fe_3O_4$ nanocomposites. The $Fe_3O_4$ produced from iron sand as a starting material can be attached uniformly on the surface of the MWCNT. The agglomeration of the $Fe_3O_4$ in nanocomposite decreased with increasing the MWCNT composition. Based on the result of characterizations, the MWCNT-$Fe_3O_4$ nanocomposites are very recommended for many application, especially in bioapplication. This simple method can be extended for other metals, alloys or oxides nanoparticles which compatible with properties of CNT.

## 5. References

[1] Okpala C Chikwendu 2013 *International Journal of Engineering Research and Development* **817**
[2] Konga L, Lub X and Zhang W 2008 *J Solid State Chem* **181** 628
[3] Safari J and Ravandi G Soheila 2014 *J Mol Struct* **1074** 71
[4] Ntim A Susana and Mitra S 2011 *J. Chem. Eng. Data* **56** 2077

[5] Lv Z, Wang Q, Bin Y, Huang L, Zhang R, Zhang P and Matsuo M 2015 *J. Phys. Chem. C* **119** 26128
[6] Chen R, Chai L, Li Q, Shi Y, Wang Y and Mohammad A *J Environ Sci Pollut Res* DOI 10.1007/s11356-013-1671-4
[7] Zhou X, Fang C, Li Y, An N and Lei W 2016 *J Composites B* **89** 295
[8] Li S, Gong Y, Yang Y, He C, Hu L, Zhu L, Sun L and Shu D 2015 *Chem Eng J* **260** 231
[9] Yang L, Hu J, Dong A and Yang D 2014 *Electrochim Acta* **144** 235
[10] Ching C Liu, Subramaniam S, Sivasubramanian S and Huei F Ling 2016 *Mater Sci Eng B* **61** 422
[11] Zarei H, Ghourchian H, Eskandari K and Zeinali M 2012 *Anal Biochem* **421** 446
[12] Pistone A, Iannazzo D, Fazio M, Celegato M, Barrera G, Tiberto B, Giordano A, Azzerboni B and Galvagno S 2014 *Physica B* **435** 88
[13] Samouhos S and McKinley G, 2007 *J Fluid Eng* **129** 429
[14] Pourkhalil M 2016 Journal of Advanced Biotechnology and Research (IJBR) **7** 1310
[15] Balacianu F D, Nechifor A C, Bartos R, Voicu S I and Nechifor G 2009 *Optoelectron Adv M – Rapid Communications* **3** 219
[16] Qua S, Wang J, Kong J, Yang P, Chena G, 2007 *Talanta* **71** 1096
[17] Hua L J, Yu H R, Hua L G, Ying Z, Zhong L H and Guang W D 2010 *New Carbon Mater* **25** 192
[18] Huang Z, Li J, Chena Q and Wanga H 2009 *Mater Chem Phys* **114** 33
[19] Ghazanfari M and Yazdani A 2015 *Mat Sci Semicon Proc* **40** 152

**Acknowledgments**
This research was partially supported by scholarship dissertation of Lembaga Pengelola Dana Pendidikan (LPDP) and "Hibah Kompetensi" research grant of Ditlitabmas, Ditjen DIKTI, 2016. We would like to thank Advances Functional Materials Laboratory (AFM) Bandung Institute of Technology (ITB) and Central Laboratory for Minerals and Advanced Materials Faculty of Mathematics and Natural Sciences Universitas Negeri Malang (State University of Malang) which already provided the facilities for synthesis and characterizations.

# The Scale Formation of Barite ($BaSO_4$) from Laminar Flowing Water in The Presence of Tartaric Acid and $Ba^{2+}$ Concentration Variation of Solution

**F Fatra[1], G Ivanto[1], N S Dera[1], S Muryanto[1,2] and A P Bayuseno[1]**

[1] Department of Mechanical Engineering, University of Diponegoro, Tembalang Campus, Semarang, 50275, Indonesia
[2] Department of Chemical Engineering, UNTAG University, Bendhan Dhuwur Campus, Semarang, 50233, Indonesia

Email: fathrafahmi@gmail.com

**Abstract**. The barite ($BaSO_4$) scale is a mineral deposit that can be precipitated during the process of drilling oil and gas in the offshore. Deposite scale in pipes can cause a narrowing of the diameter of pipes, and can reduce water flowing in the pipe. The aim of this study is to investigation the effect of the tartaric acid additive and $Ba^{2+}$ concentration on the growth o the scale formation of barite in the laminar flow of the piping system. Solution forming barite crystal was prepared by mixing equimolar solutions of barium chloride ($BaCl_2$) and sodium sulfate ($Na_2SO_4$) with concentration variations of $Ba^{2+}$ of 3000, 3500, 4000, 4500, and 5000 ppm. The flow rate of solution is 40 ml /min at temperature of 50 °C. Various concentrations of tartaric acid ($C_4H_6O_6$) of 0 ppm, 5 ppm and 10 ppm were added to the solutions. The formation of barite from the solution was observed by ion conductivity measurement. The obtained barite crystals before and after adding tartaric acid were dried and characterized by using SEM/EDX for morphology and elemental analysis, and XRD for phase identification. The SEM results show that the morphology of the crystals are star-like particles, while XRD analysis confirmed that the barite crystals were produced during the experiments are high purity. Moreover, the tartaric acid can inhibit the crystal growth of barite.

**Keywords:** Barite, tartaric acid, laminar flowing water, morphology crystal

## 1. Introduction

Barium sulphate (barite) has been known as a scale deposit that is particularly prevalent in the pipeline of oil and gas industry [1, 2]. Scaling of barite in oil fields may be due to the presence of insoluble barium compounds in water. Once the soluble barium compounds exceed its solubility limit, they will crystallize from the solution. The precipitate and crystal growth may consequently reduce the diameter of the pipe and hence result in inhibition of fluid flow in the pipe system. The blocking flow of water may lead to increase temperature and pressure so that the pipeline may rupture [3]. Therefore, the scale formation may be prevented by softening and acquisition of mineral water. However, the use of demineralized water in the industry may improve the operating costs [4].

In general, the scale formation can be prevented by using acid as an inhibitor to lower the solution pH, control the inorganic compounds as well as the composition and the crystal morphology. Barite crystal growth can be controlled by the process of nucleation through the presence of chemical additives. Moreover, the removal of barite from the surface of pipes using high concentrations of acid may be not effective because the acid can increase the corrosion rate. Also, the use of chemical inhibitors containing phosphonate and carboxylate is quite dangerous, because these inhibitors will change the quality of water [5, 6]. Additionally, phosphonates are not environmentally friendly. This present research proposes a new scale inhibitor to prevent the barite scale formation, which is more effective and environmentally friendly such as the tartaric acid [7]. The addition of this additive concentration has shown to be a significant influence on the growth of the barite crystals [8, 9].

This is to investigate the effect of the tartaric acid additive on the growth of the scale barite by using a pilot scale of laboratory rig. Here, the flow rate of water in the pipe and temperature were fixed constant. In this study, parameter variations of the solution and additive concentration were examined. The barite crystal products were then characterized by using SEM/EDX for morphology and elemental analysis, and XRD analysis for confirming the mineral. Inhibiting growth rate of the barite scale by the tartaric acid may be applied to the piping systems in the petroleum and gas industries.

## 2. Experimental and Method

### *2.1. Raw materials for preparing crystalline forming solutions*

Materials, all with grade analytic powder (Merck$^{TM}$), were selected in this experimental research consisting of $BaCl_2$ and $Na_2SO_4$ crystals, distilled water and the tartaric acid ($C_4H_6O_6$) with the analytical grade as an additive.

### *2.2. Experimental method for barite crystallization and materials characterization*

In this present study, the mixing solution of $BaCl_2$ and $Na_2SO_4$ as scale formation was diluted in the water to form barite scale. Barite crystals stick to the wall of the pipe to form a crust, while the NaCl crystals are dissolved in water. The barite formation may be presented as follows:

$$BaCl_2\,(aq) + Na_2SO_4\,(aq) \rightarrow BaSO_4\,(s) + 2\,NaCl\,(aq) \qquad (1)$$

The experimental work was conducted according to the flow-chart diagram given in Figure 1. Variation of the $Ba^{2+}$ concentration in the amount of 3000, 3500, 4000, 4500, and 5000 ppm was prepared in the solution. The flow rate and temperature were fixed during the work (40 ml /min and 50 $^{o}$C). The tartaric acid in the amount of 0 ppm, 5 ppm, and 10 ppm was added in the solution during the experiments. The experiment was run by using the laboratory rig for scale formation described in elsewhere [10]. Barium scale formed in the absence and presence of additives was then examined by using XRD and SEM/EDX analysis.

XRD data were acquired by using a Bragg-Brentano diffractometer (D5005 SHIMATZU) operated on Cu-Kα radiation, and equipped by a graphite diffracted beam monochromatic, and a scintillation detector. The scan parameters (10-90 2θ$^{O}$, 0.020 steps, 15 seconds/step) were recorded. The x-ray source was conventional sealed by X-ray tube operated at 50 kV and 40 mA. For selecting high intensity, incident and diffracted beam slits were fixed at 1.0 and 0.2 $^{O}$ respectively. A PC-based search- match program (X'Pert plus Philips software) was used for identifying the mineral phases of precipitates through matching to the crystalline data included in the International Centre for Diffraction Data (ICDD)-PDF (powder diffraction file) database.

Furthermore, scanning electron microscopy (SEM) analysis on a JEOL DSM apparatus equipped with an EDX was conducted on the microstructure and morphology. The powder crystals were mounted on a circular metallic sample holder and subsequently sputtered with carbon for collecting imaging data.

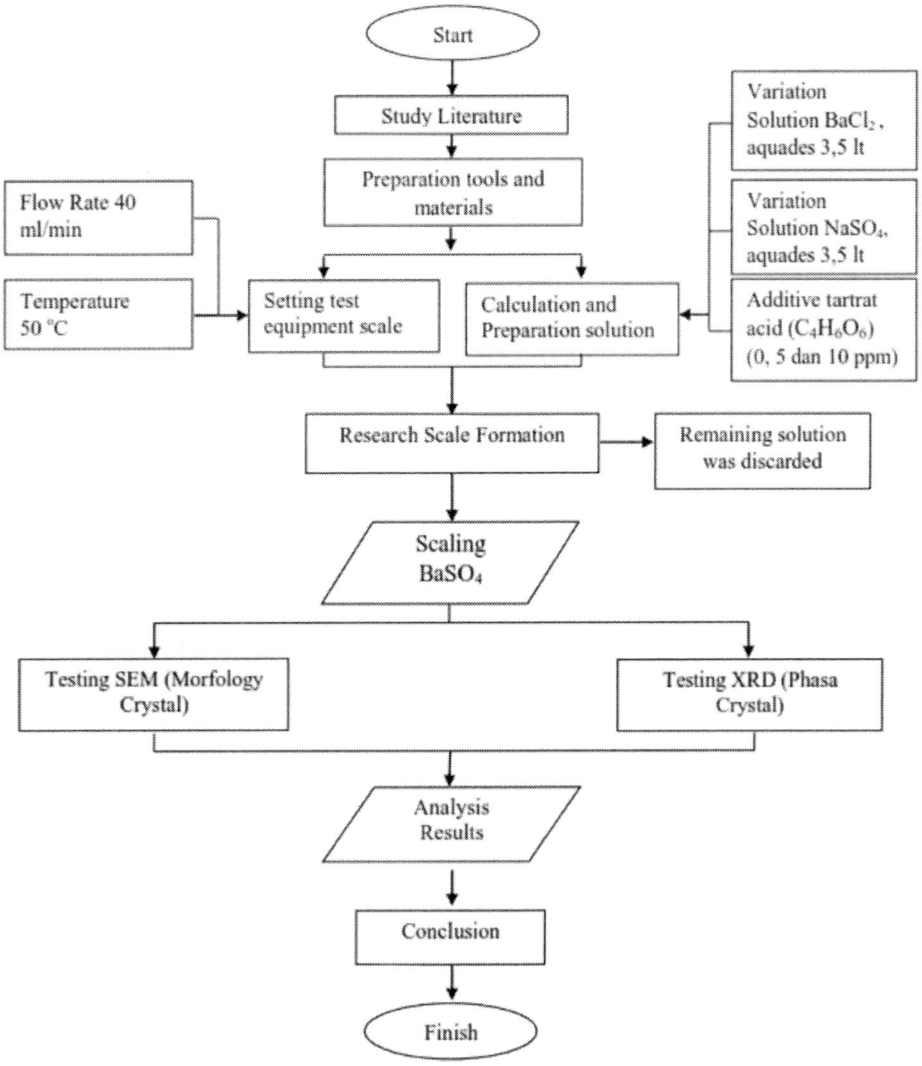


**Figure 1**. Research Flowchart Diagram

## 3. Result and Discussion

### *3.1. Induction Time*

The solution of $BaCl_2$ and $Na_2SO_4$ was met in the coupons to form barite scale. Once barite scale was formed, and then deposited on the wall of the coupon continuously for two hours. The scale mass results were obtained depending on the variation of testing parameters. This phenomenon can be seen from the measurement of conductivity for concentration variation of $Ba^{2+}$ in the presence of 10 ppm additive (Figure 2). At $Ba^{2+}$ concentration (5000 ppm), it shows that the crystallization reaction between $BaCl_2$ and $Na_2SO_4$ occurred for about 30 minutes. At this stage, drop in conductivity can be represented as the induction time. Moreover, the increasing $Ba^{2+}$ concentration in the solution yielded the increasing induction period (the respective 34, 38, 42, and 48 min.) for 3500, 4000, 4500, and 5000 ppm respectively.

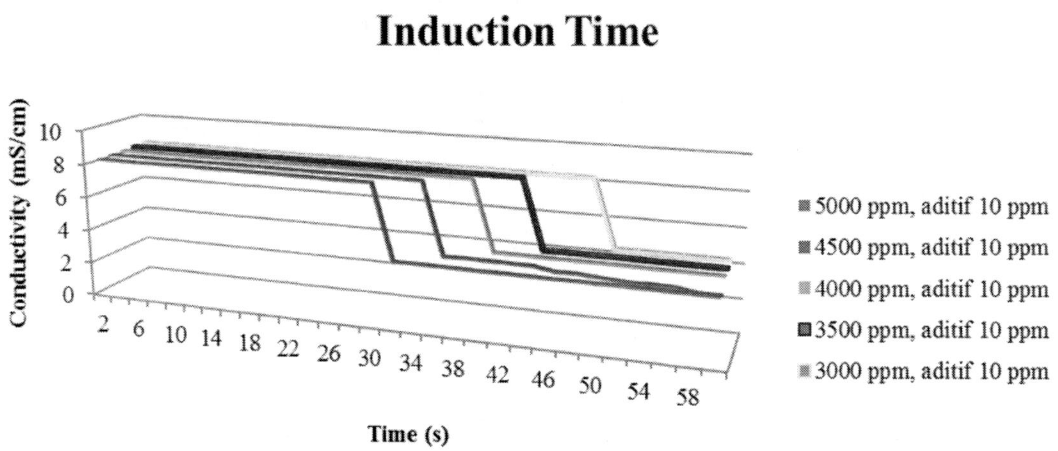


**Figure 2**. Plot of conductivity versus scaling time at additive 10 ppm

### *3.2. Mass scale produced from the solution*

This experiment employed five variations of the $Ba^{2+}$ concentration in solution (3000, 3500, 4000, 4500, and 5000 ppm), and three different concentrations of the tartaric acid (0, 5, and 10 ppm) were selected. Table 1 presents the mass scale of barite formed from the solution. It shows that the more $Ba^{2+}$ concentration was available in the solution, the more scale mass was produced in the piping system. Nevertheless, the more tartaric acid was added, the crystal growth of barite was inhibited. Consequently, the scale mass products were reduced in the presence of additives.

**Table 1.** Mass scale precipitated from the solution in the presence of tartaric acid

| Concentration $Ba^{2+}$ (ppm) | Flow Rate (ml/second) | Temperature ($^0C$) | Tartaric Acid (ppm) | Mass of scale (gram) |
|---|---|---|---|---|
| 3000 | 40 | 50 | 0 | 0.086 |
| 3000 | 40 | 50 | 5 | 0.082 |
| 3000 | 40 | 50 | 10 | 0.073 |
| 3500 | 40 | 50 | 0 | 0.103 |
| 3500 | 40 | 50 | 5 | 0.096 |
| 3500 | 40 | 50 | 10 | 0.090 |
| 4000 | 40 | 50 | 0 | 0.128 |
| 4000 | 40 | 50 | 5 | 0.120 |
| 4000 | 40 | 50 | 10 | 0.114 |
| 4500 | 40 | 50 | 0 | 0.158 |
| 4500 | 40 | 50 | 5 | 0.152 |
| 4500 | 40 | 50 | 10 | 0.144 |
| 5000 | 40 | 50 | 0 | 0.184 |
| 5000 | 40 | 50 | 5 | 0.176 |
| 5000 | 40 | 50 | 10 | 0.172 |

### *3.3. Microstructure of the scale*

Barite crystal morphology examined by using SEM is given in Figure 3. The elemental analysis of the scale by using EDX is also presented in Figure 4.

(a) (b)

**Figure 3**. Morphology Barite crystals, (a) the absence of additive and (b) the presence of additive for 10 ppm

It shows that the morphology of barite has a starlike crystal, which was obtained from the solution without additives and with additives. At all concentrations of $Ba^{2+}$ without and with additives, the same image of morphology was obtained, but the reduction of crystal size was noted in the presence of additives.

XRD analysis was also conducted to verify the mineral phase formed by comparing the measured data with the database of the International Center for Diffraction Data (ICDD). X-ray diffractogram results provided the peak intensity to be identified through the search-match procedure. Figure 4 shows the direct experimental evidence of barite crystals formed in the presence of 10 ppm tartaric acid and in the absence of the additive. The Powder Diffraction File (PDF) numbers of 04-012-5411 matched with all XRD patterns measured and in this study, and proved that the pure barite was precipitated from the solution.

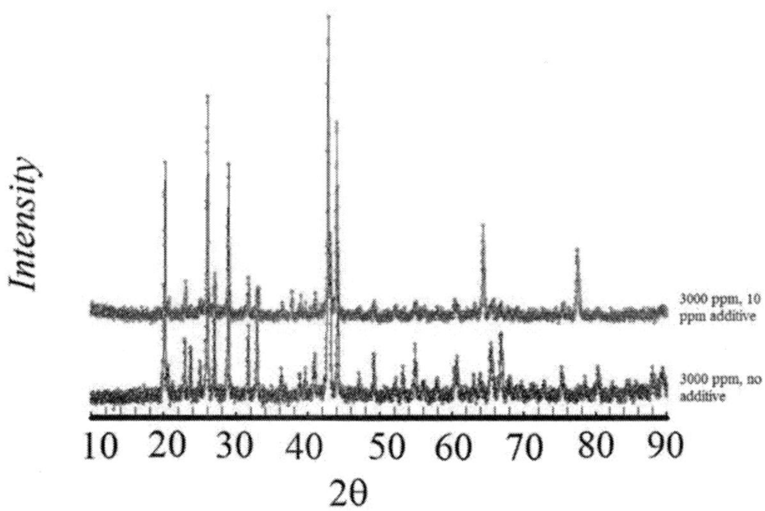


**Figure 4**. X-ray diffractogram of the crystal product obtained from the solution

## 4. Conclusion

Experimental scaling of barite with variations in concentration of $Ba^{2+}$ (3000, 3500, 4000, 4500, and 5000 ppm), temperature 50 $^0$C, a flow rate of 40 ml/min and the different concentrations of tartaric acid (0, 5, and 10 ppm) was performed. From the results of SEM, a visible star-like morphological for barite precipitated from the solution without and with additives can be observed. The presence of tartaric acid additive may inhibit the crystal growth of barite. It can also be explained that the decline

IOP Conf. Series: Materials Science and Engineering **202** (2017) 012014 doi:10.1088/1757-899X/202/1/012014

in the mass of barite may be due to the presence of tartaric acid. Although the scaling of barite was inhibited by tartaric acid additive affecting a little; the use of this additive from organic materials is environmentally friendly.

## 5. References

[1] Sorbie K S and Mackay E J 2000 *Petroleum Sci* Eng **27** 85.
[2] Benton W J, Collins I R, Grimsey I M, Parkinson G M, Rodger S A, and Chem J 1993 *Soc Faraday Disc* **95** 281.
[3] Asnawati 2001 *Effect of Temperature Reaction Against phosphonate in Inhibitor scale on Oil Well* Journal Basic Science **2** No 1 : 20.
[4] Sousa M F and Bertran C A (2014) *New methodology based on static light scattering measurements for evaluation of inhibitors for in bulk crystallization* Journal of Colloid and Interface Science **420** 57-64.
[5] Wang, Moo Y, Chen Y X, Gunawan C, and Xu P G R 2010 *Fast Preciptation of Uniform $CaCO_3$ Nanospheres and Their Transformation to Hollow Hydroxyapatite Nanospheres* Journal of Colloid and Interface Science **233** 65 – 72.
[6] Franca J, Philip J, and Mark I O 2007 *The Interaction of EDTA with Barite* Journal of Colloid and Interface Science **316** 553-561.
[7] Zahid A 1994 *Inhibition of Barium Sulfate Precipitation : Effects of additives, Solution pH, and supersaturation* China Ocean Press-Printed in Beijing 47 - 56
[8] Hennessy A J B and Graham G M 2002 *The effect of additives on the co-crystallisation of calcium with barium sulfate* J. Cryst Growth 237-239.
[9] Amer B M and Abu A M Y 2009 *Laboratory Study on Preciptation of Barium Sulphate in Malaysia Sandstone Cores* The Open Petroleum Engineering Journal 1 – 11.
[10] Muryanto S, Bayuseno A P, Sediono W, Mangestiyon W, and Sutrisno 2012 *Development of a versatile laboratory project for scale formation and control* Education for Chemical Engineers **7** e78–e84.

## Acknowledgments

The authors are grateful to the Laboratory of Physical Metallurgy Diponegoro University which provided the facility during the experiment. The authors would like also to appreciate the useful comments and the kind suggestions from the anonymous reviewers.

# Citric Acid Addition to Controlling Crystallization of Barium Sulphate ($BaSO_4$) in Pipes through $Ba^{2+}$ Concentration Variation in the Solution

**G Ivanto[1], F Fatra[1], N S Dera[1], S Muryanto[1, 2] and A P Bayuseno[1]**

[1]Department of Mechanical Engineering, University of Diponegoro, Tembalang Campus, Semarang, 50275, Indonesia
[2]Department of Chemical Engineering, UNTAG University, Bendhan Dhuwur Campus, Semarang, 50233, Indonesia

Email: ivanto.guntur@gmail.com

**Abstract**. The scale of barium sulphate ($BaSO_4$) is common scale for mineral deposit that found in the offshore oil and gas exploitation. This scale is related with precipitation and grown of mineral deposit on the pipelines surface. Therefore, it results in blockage at the pipe. This paper presents the experimental scaling of barium sulphate in the laminar flow. The barium sulphate solution was prepared by mixing an equimolar solution of barium chloride ($BaCl_2$) and sodium sulphate ($Na_2SO_4$). The flow rate is 40 ml/min at temperature of 50 °C. The solutions added by citric acid ($C_6H_8O_7$) with variation concentration of 0 ppm, 5 ppm, and 10 ppm. The crystallization of barium sulphate was measured by using the conductivity meters. The barite crytals were dried and characterized by using SEM/EDX and XRD. The SEM Results show that the morphology of Barite scale was change in the presence of citric acid. The mineral of barium sulphate was pure barite based on the XRD phase analysis. The presence of citric acid clearly inhibit the crystallization of barium sulphate.

**Keywords:** Barium sulphate, Mineral deposit, Citric acid, crystallization.

## 1. Introduction

Scaling of inorganic compounds in pipes are one of the major flow assurance concerns in offshore oil and gas production, and lead to significant reductions in productivity and increasing operating cost if the scale is allowed to form uncontrolled [1]. Barium sulphate (barite) is a known scale-producing precipitate in many industrial and oil drilling processes [2-4]. The formation of barium sulphate may follow the reaction below:

$$Ba^{2+} + SO_4^{-2} \rightarrow BaSO_4 \quad (1)$$

The scale of barite cannot be removed easily by acid, because this compound is highly insoluble (solubility in water only 2 mg / liter) [5, 6]. The relative low solubility in water makes barite is easily formed in the aqueous water system, when the soluble salt exceeds the solubility limit [7-9]. Barium sulphate precipitation can also occur in the oil production from brine that becomes saturated with a material due to a change in the flowing fluid conditions within a well. Therefore, the control of barium sulphate formation has become a great concerned through a chemical additive dissolution. The use of chemical additive such citric and tartaric acids for retarding crystal growth are believed to work effectively, hence it promotes the formation of small crystals, which act as nucleation sites, but it prevents them from becoming large enough to precipitate from solution [1, 10]. The initial nucleation is led to the reduction in the super saturation of the solution quickly. Consequently, the large molecules adhere to active crystal growth planes and hinder the crystal growth in one particular direction.

This present work was undertaken to examine the barium sulphate scaling of pipes with variation of $Ba^{2+}$ concentration in solution (3000, 3500, and 4000 ppm) and laminar flow rate of 40 ml/min. Other parameters examined were temperature of 50 °C and citric acid (0, 5, and 10 ppm) as additive. The objectives of this research are to prevent and overcome the scale formation of barite in the drilling pipes, which can be subsequently applied in the industry of offshore oil and gas production. This research is expected to provide some data or information on how to setup the parameters that influence the formation of barite so that the piping system will work on higher efficiencies.

## 2. Experimental Method

*2.1 Barium Sulphate Scale Formation*

This study used the crystal powders of $BaCl_2$ (p. a.) and $Na_2SO_4$ (p. a.) as the scale-forming solutions. Citric acid ($C_6H_8O_7$) as an additive, which was expected to inhibit scaling, was added into the solution in ppm amount. The reaction of the scaling process may follow:

$$BaCl_2\,(aq) + Na_2SO_4\,(aq) \rightarrow BaSO_4\,(s) + 2\,NaCl\,(aq) \tag{2}$$

The concentration of the $BaCl_2$ solution was determined based on the calculation of (3000 ppm, 3500 and 4000 ppm) $Ba^{2+}$ concentration in the solution. $BaCl_2$ solution was made by dissolving $BaCl_2$ crystals in distilled water at room temperature. Additionally, $Na_2SO_4$ solution was prepared in the same way. The citric acid ($C_6H_8O_7$) crystals (5 and 10 ppm) was added in the $BaCl_2$ solution.

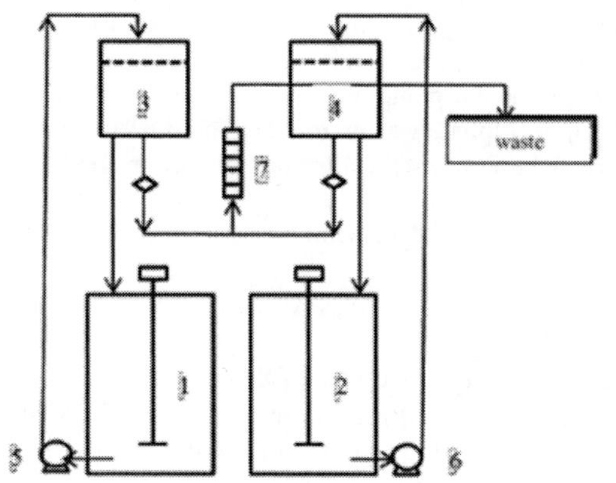


**Figure 1.** A pilot scale of the experimental rig for barium sulphate scale formation

The experimental rig employed in the study is shown in Figure 1. Five liters for each of $BaCl_2$ and $Na_2SO_2$ solutions were prepared in vessel (1) and (2) respectively. Two pumps (5) and (6) were operated to circulate the solutions into vessel (3) and (4). Both solutions were then simultaneously entered into the coupon house (7) which contained five copper coupons. Subsequently, those solutions met in the coupons (which were designed specifically for the growth of the barite scale). The conductivity of the solutions available in the coupons was continuously monitored by conductivity meter for up to two hours, i.e. until the experimental run ended. In this study, the induction time was determined by recording the conductivity of the solution coming out of coupons. Recording the conductivity of the solution was done every two minutes (the first hour) and every three minutes (the second hours). This method has been widely applied to monitor the nucleation and crystal growth of organic minerals, either in the absence or presence scale inhibitor. The end of the induction period was signed by a decrease in the value of conductivity of the solution of a sudden, it indicated that a large number of ions forming the crystals begin to leave the solution to form crystals.

At the end of the experiment, the coupons and the vessel (3 and 4) were disconnected and removed from the coupon house, followed by drying in the oven at 80 °C for one hour. After this, the coupon was weighed (difference weight of the coupon after and before testing was the mass of the scale).

This method was applied to monitor the nucleation and crystal growth of organic minerals progressing, either in the absence or the presence of scale inhibitor. At the end of induction period, a sudden decrease in the value of conductivity of the solution was noted [11]. This phenomenon indicates that a large number of ions for forming the crystals begin to leave out from the solution.

### *2.2 Characterization of barium sulphate*

Barium sulphate was characterized by SEM equipped by EDX and XRD. Before SEM examination, all samples were coated first by using platinum (Pt). SEM integrated with EDX was conducted to determine the morphology of crystals and the chemical elements of the scale respectively.

XRD analysis was conducted to confirm the crystalline phase of the scale formed. Scan parameters were chosen for observation to be 10-90° $2\theta$, 0.020 steps, 15 s/steps. Peak position and peaks intensity were identified by using the feature finder or on the screen in the software. The data obtained were compared with the ratio of PC based search data base program. Philips X'Pert Software (Philips Electronics NV) was employed to identify the crystalline phase of the sample. In this approach, the peak position was automated by using the diffraction minima in the second derivate of diffraction traces. The peak positions and heights were checked against XRD standard for barite available in the International Centre for Diffraction Data (ICDD)-PDF (powder diffraction file) database.

## 3. Results and Discussion

### *3.1. Induction Time*

The barium sulphate formation was due to the mixing process of a solution of $BaCl_2$ and $Na_2SO_4$ in the coupon (Figure 1 no. 7). The solutions of $BaCl_2$ and $Na_2SO_4$ in the coupons reacted to form the scale and then deposit to the wall of the coupon continuously for two hours of the experiment. The scale mass resulted variously depending on the variation of testing parameters. The experimental results showed that the reaction between $BaCl_2$ and $Na_2SO_4$ solutions resulted in the decline in conductivity of the solution coming out of the coupons. The conductivity of the solution was stabil at first and then fell suddenly.

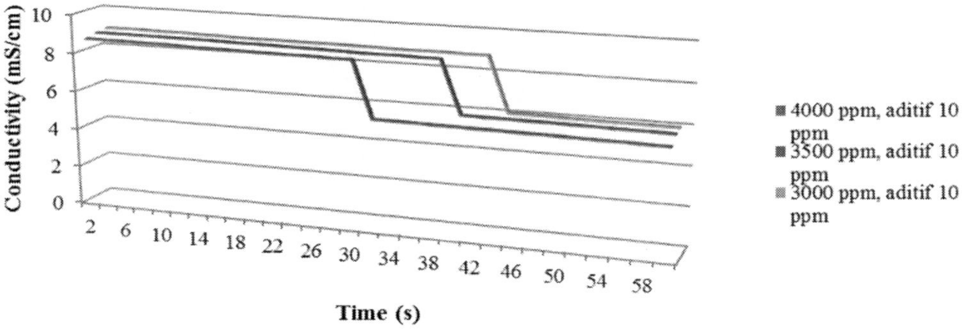


**Figure 2.** Plot of conductivity versus scaling time in the presence of 10 ppm citric acid

The time between mixing the solution to decrease the conductivity varied in accordance with the parameters investigated. This time interval is called as the induction time. Figure 2 shows the

IOP Conf. Series: Materials Science and Engineering **202** (2017) 012015 doi:10.1088/1757-899X/202/1/012015

induction time generated from the variation of $Ba^{2+}$ concentration in the solution and in the presence of 10 ppm additive. At the 4000 ppm of $Ba^{2+}$ concentration, the induction time of barium sulphate formation may start at around 30 minutes. The reducing $Ba^{2+}$ concentration from 4000 to 3000 ppm promoted the increasing induction time.

### *3.2. Effect of $Ba^{2+}$ concentration and citric acid on the mass scale produced*

This experiment selected five variations of the $Ba^{2+}$ concentration in the solution (3000, 3500, and 4000 ppm), and three variations of citric acid concentrations (0, 5, and 10 ppm) as an additive. Table 1 presents the variation of the research parameters and mass of the scale barium sulphate formed. It shows that more $Ba^{2+}$ concentration, the more the mass of scale produced in the piping system. In contrast, the more citric acid additive employed in the same concentration of solution, the mass of scale reduced. Obviously, the use of citric acids for inhibiting the scale formation is quite effective. The effect of barite concentration variation and mass additive citric acid to the weight of the scale mass contained in the pipe is shown in the Figure 3.

**Table 1.** The mass of scale produced from the solution with citric acid additives

| Concentration $Ba^{2+}$ (ppm) | Flow Rates (ml/second) | Temperature ($^0C$) | Citric Acid (ppm) | Mass of $BaSO_4$ (gram) |
|---|---|---|---|---|
| 3000 | 40 | 50 | 0 | 0.124 |
| 3000 | 40 | 50 | 5 | 0.120 |
| 3000 | 40 | 50 | 10 | 0.116 |
| 3500 | 40 | 50 | 0 | 0.138 |
| 3500 | 40 | 50 | 5 | 0.134 |
| 3500 | 40 | 50 | 10 | 0.130 |
| 4000 | 40 | 50 | 0 | 0.145 |
| 4000 | 40 | 50 | 5 | 0.142 |
| 4000 | 40 | 50 | 10 | 0.138 |

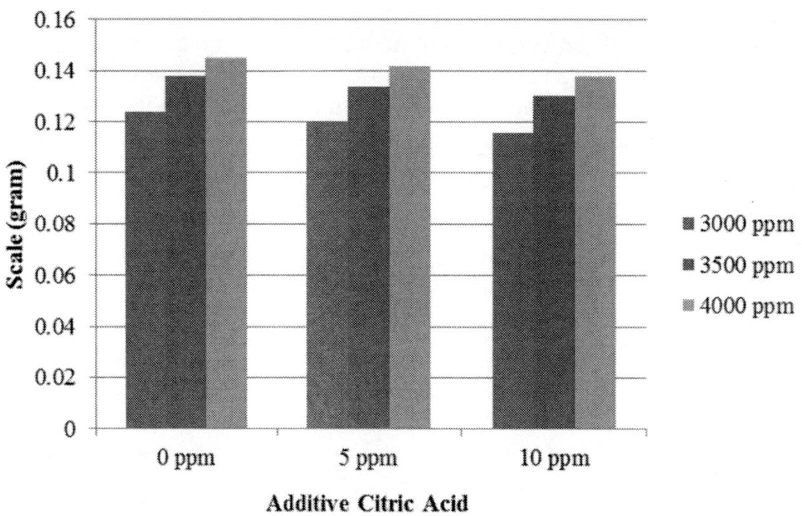


**Figure 3.** Mass of barite scale produced from the different variation of $Ba^{2+}$

### *3.3. Characterization of the scale*

Barite crystal morphology was examined by using SEM and it is given in Figure 4. The barium sulphate crystal has a starlike crystal morphology. It also shows that the morphological change was observed in the crystal produced from concentration of solution without and with additives. In the

IOP Conf. Series: Materials Science and Engineering **202** (2017) 012015 doi:10.1088/1757-899X/202/1/012015

solution without additives, the SEM image of irregular crystal with different shape and size was shown. At concentrations of $Ba^{2+}$ using the additive, it showed the star-like crystal with almost similar size.

**Figure 4**. Morphology of barite crystals precipitated in the absence of additive (left), and in the presence of 10 ppm citric acid (right).

XRD analysis was conducted to verify the crystal phase formed by comparing the data of experiment with the data base (ICDD-PDF). XRD measurement results in the form of intensity (peak). Figure 5 provides the direct experimental evidence of phase crystals formed in the presence of citric acid 10 ppm with and no additive, which matched with PDF number of 04-012-5411 for barium sulphate (barite). In fact, the barite crystal sample is the only mineral precipitated from the solution.

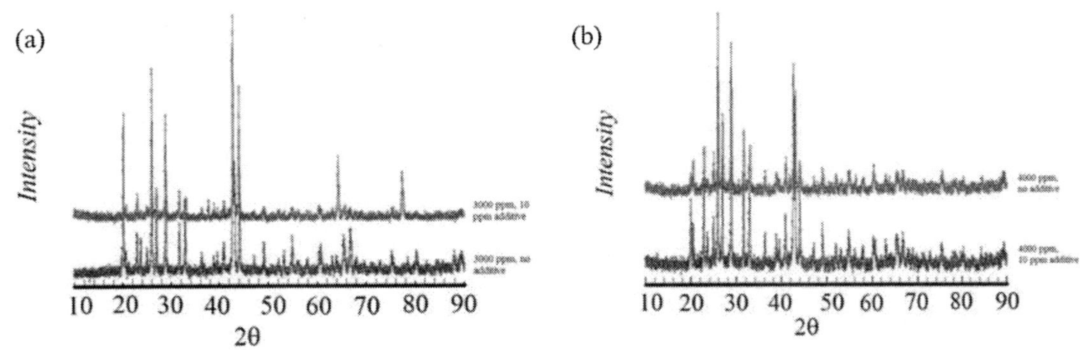


**Figure 5.** XRD analysis of the scale sample in the presence of citric acid at (a) 3000 and (b) 4000 ppm on $Ba^{2+}$ concentration

## 4. Conclusion

Experimental scaling of $BaSO_4$ with variations in $Ba^{2+}$ concentration (3000, 3500, and 4000 ppm), temperature of 50 $^0$C, a flow rate of 40 ml / and variations additive of citric acid (0, 5, and 10 ppm) were performed. The different mass of barite scale was produced as a function of $Ba^{2+}$ concentration in the solution and citric acid additives. At the time of using the citric acid, it yielded the reduction in mass of the scale. From the experimental results, it can be reported that the use of citric acid may reduce the size of the barite particles. Further research should be performed by using other additives for comparative analysis in term of mitigation strategy of scale formation.

## 5. References

[1] Amer B M 2012 *Inhibition of Barium Sulphate Scale at High-Barium Formation Water* Journal of Petroleum Science and Engineering 90 – 91 124 – 130.

[2] Sorbie K S, Mackay E J and Pet J 2000 *Sci Eng* **27** 85–106.

[3] Benton W J, Collins I R, Grimsey I M, Parkinson G M, Rodger S A and Chem J 1993 *Soc. Faraday Discuss* **95** 281–297.

[4] Breen P J, Downs H H and Diel B N 1991 *Spec. Publ. R. Soc. Chem* **97** 186–198.

[5] Asnawati 2001 *Effect of Temperature Reaction Against phosphonate in Inhibitor scale on Oil Well* Journal Basic Science **2** No1: 20.

[6] Amer B M and Abu A M Y 2009 *Laboratory Study on Preciptation of Barium Sulphate in Malaysia Sandstone Cores* The Open Petroleum Engineering Journal 1 – 11.

[7] Sousa M F and Bertran C A 2014 *New methodology based on static light scattering measurements for evaluation of inhibitors for in bulk crystallization.* Journal of Colloid and Interface Science **420** 57-64.

[8] Robert J Ferguson and Baron R Ferguson 2010 *The Chemistry of Strontium and Barium Scales* Assosiation of Water Technologies.

[9] Wang Y M, Chen Y X, Gunawan C and Xu P G R 2010 *Fast Preciptation of Uniform $CaCO_3$ Nanospheres and Their Transformation to Hollow Hydroxyapatite Nanospheres* Journal of Colloid and Interface Science **233** 65 – 72.

[10] Zahid A 1994 *Inhibition of Barium Sulphate Preception: Effects of Additives, Solution pH, and Supersaturation* China Ocean Press.

[11] Muryanto S, Bayuseno A. P, Ma'mun H, Usamah, and M Jotho 2014 *Procedia Chemistry* **9** 69 – 76.

**Acknowledgments**

The authors are grateful to the Laboratory of Physical Metallurgy, Diponegoro University which provided the facility during the experiment. The useful comments and kind suggestions from the anonymous reviewers are greatly appreciated.

# Phase Transformation of Limonite Nickel Ores with $Na_2SO_4$ Addition in Selective Reduction Process

**W Mayangsari and A B Prasetyo**

Research Center for Metallurgy and Material, Indonesian Institute of Sciences, South Tangerang, Banten, Indonesia

Email: wahy035@lipi.go.id

**Abstract.** This study aims to observe the transformations, both mineralogy and morphology of limonite nickel ore with coal as reducing agent and $Na_2SO_4$ as an additive in the selective reduction processes. A series of selective reduction processes were carried out at various temperatures in Carbolite muffle furnace for 1 hour. The range of reduction temperature was determined from Thermogravimetric-Derivative Thermal Analysis (TG-DTA). The mineralogy of raw and reduced ores was characterized by using XRD analysis. The morphology, particle size, and mapping were characterized by using SEM-EDS analysis. The results showed that goethite was the main component in the raw limonite nickel ore and it was decomposed by losing the crystalline water at 300 °C. The decompositions were continued to lizardite and trevorite at 600 °C and 800 °C respectively, then Fe and Fe-Ni were formed at 900 °C. Metallization continuously occurred by increasing reduction temperature up to 1100 °C. Moreover, the particle size and porosity of reduced ore were greater up to 2.5-times. The addition of $Na_2SO_4$ on the selective reduction processes decreased silica content; it also caused the formation of fayalite and FeS at 600 °C and 800 °C respectively. The intensity of fayalite and FeS decreased when the reduction temperature was more than 900 °C.

**Keywords**: Nickel, Limonite, $Na_2SO_4$, Reduction, Phase transformation

## 1. Introduction

Nickel ore is divided into two types, sulfide (high grade) and laterite (low grade). The world resources of laterite ore are up to 1.5 times higher than sulfide resources, but it does not reach a half of the total number of existing nickel resources used for world's nickel productions, sulfide ore dominates the raw material [1-4]. Utilizing the high-grade nickel ore is continued to rise, causing high-grade nickel resources decrease, so the low-grade nickel resources get much attention and leads to be used as raw material, especially for nickel containing less than 2% wt. of Ni [5,6].

Indonesia is one of countries that has the highest laterite nickel ore resources in the world, that is about 12% [1]. In addition, another advantage of utilizing laterite nickel ore as raw material is location for mining; the laterite nickel ore can be mined by open-cut mining as it is located closer to the soil surface [4]. The first layer near the surface is limonite, then alunite, and saprolite [7]. Based on that condition, Indonesia has highly potential to develop nickel processing of limonite nickel ore through this research with the raw limonite nickel ore spread in several areas of Indonesia, especially in the east part of Indonesia.

Halmahera, North Maluku is one of the areas in Indonesia which has laterite nickel ore resources [8]. Limonite ore is one of laterite nickel types that has low Ni content (<1.50%) and high Fe content (> 40%)[9]. In Halmahera, it is not used for raw material of nickel ore processing. In this study, it is used to investigate the potency for ore processing nickel by using limonite ores.

Several studies about laterite nickel ore were carried out. Valix and Cheung [7] found that reducing conditions on dehydroxylation reaction have a significant influence on the transformation of mineral during the reduction. The phase was formed at 800 °C, and it was irreversible during cooling. For the reduction of saprolite after calcination at temperatures more than 700 °C, the forsterite was formed. In the reduction process with $Na_2SO_4$ addition at temperatures more than 800 °C has performed by Guanghui Li [10]. Forsterite was formed and had not become inhibitor for further reduction process for the formation of FeS which can suppress the formation of metallic Fe. By this way, it could increase the Ni contain up to 9.48% [10]. On the other hand, nickel laterite from Indonesia has been studied by Subagja, et al. [11]. As much as 3.64% of Ni was obtained which is less than what resulted by Li in the selective reduction process with 20% addition of $Na_2SO_4$ [11]. Based on the research studied, it is known that selective reduction process with $Na_2SO_4$ addition is effective to increase nickel content due to the formation of FeS which can be physically separated by magnetic separation, which is related to the metallization process that leads transformation of metal and morphology. From those process, the product will be known from the composition of mixture in the reduction process at certain temperature, either as Fe-Ni, nickel matte or other products, but study which focuses on the phase transformation and its mechanism of reaction of limonite nickel ore from Indonesia by selective reduction process is still approximately, so it needs to identify the potency of limonite nickel ore from Indonesia, especially from Halmahera. The aims of this study are to examine the structural changes, both mineralogy and morphology of limonite ore in the selective reduction process by using coal as reducing agent and $Na_2SO_4$ as an additive.

## 2. Experimental Method

Limonite nickel ore from Halmahera, North Maluku was dried by using an electric oven at 110 °C for 12 hours, then it was crushed, grounded and screened to get -100 mesh of particle size. The morphology and composition of limonite nickel ore were examined by using SEM-EDS analysis, and the mineralogy was examined by using XRD (*X-Ray Diffraction*) analysis. Sub-bituminous coal was used as reducing agent. It was prepared by crushing, grounding and screening with a 100 mesh of a sieve. $Na_2SO_4$ was used as an additive, the specification was chemical grade (P.A) for analysis.

The prepared limonite nickel ore, reducing agent, and $Na_2SO_4$ were uniformly mixed by physical blending before reduction process. The addition of reducing agent and additive was 10% wt. of limonite ore used respectively. TG-DTA (Thermogravimetry - Derivative Thermal Analysis) Setaram Setsys-1750 was performed to the prepared mixture from the previously step to determine the percentage of mass reduction and decomposition temperature. Selective reduction processes were carried out to the other part of the prepared mixture by using Carbolite muffle furnace in the graphite crucible. The crucible was put into the furnace when the furnace was turned on until reached the particular temperature and restrained for 1 hour. The range of temperature for selective reduction process was determined from decomposition temperature of TG-DTA. The reduced mixture was then characterized through XRD analysis to determine its phase transformation and SEM-EDS analysis to determine the morphology, particle size, and distribution of metal formed when it reached the ambient temperature. XRD analysis was performed with Shimadzu XRD 7000 and Match software used to interpret the phase formed. SEM-EDS analysis was performed with SEM-EDS JEOL 6390A.

## 3. Results and Discussion

### 3.1. Characterization of raw limonite nickel ore

The compositions of raw limonite nickel ore are listed in Table 1. The EDS result shows the main components of limonite nickel ore are Fe and O elements. It means that the minerals in the ore are bound together with oxide as iron oxide wich can confirm with XRD analysis for its phases. Fe is the main mineral in the ore followed by Si, Cu, Ni, Mg, Al, and Cr.

Figure 1 shows the phases of limonite ore used are lizardite ($(Mg, Fe)_3Si_2O_5(OH)_4$), goethite ($FeO(OH)$), hematite ($Fe_2O_3$), trevorite ($NiFe_2O_4$), quartz ($SiO_2$), and wustite( $FeO$), the diffraction pattern in accordance with Match result. Quartz shows having the highest intensity, followed by goethite and hematite in the few points. It indicates that the dominant component in the limonite ore used are silica and iron-based compound. This is agree with EDS result in Table 1. Goethite is the main component in the limonite ore [7], and shows a higher intensity at some point than the other elements, the sharp diffraction peak intensity of goethite at 2θ at 21.25°, 36.5°, and 53°.

**Table 1.** Composition of raw limonite nickel ore

| Element | Mass (%) |
|---|---|
| O | 10.9 |
| Fe | 71.41 |
| Si | 5.07 |
| Mg | 2.24 |
| Ni | 3.76 |
| Al | 1.47 |
| Cr | 1.13 |
| Cu | 4.02 |

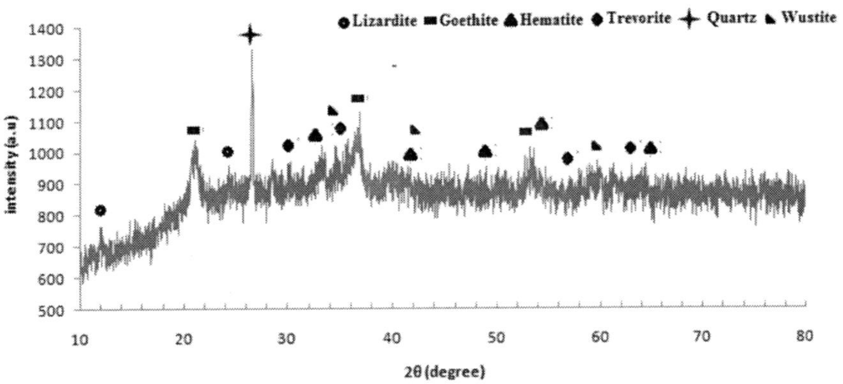


**Figure 1.** XRD analysis of limonite nickel ore

*3.2 TG-DTA Analysis*

The TG-DTA curve (Figure 2) shows that mass loss was started at 98 °C and 100 °C with 0.01% and 0.06% of mass loss respectively and represents the evaporation of water [12]. First endothermic peak occurred at 324 °C with 2.39% of mass loss; it represents the dehydroxylation of goethite to form hematite [13]. The low mass loss in TG indicates of poorly crystalline of goethite structure [13].The first endothermic temperature was lower than 337 °C which was raw limonite nickel ore without reducing agent and additive performed [14], it means that reducing agent and additive addition makes reaction faster. The second endothermic peak occurred at 625 °C with no significant change of mass loss, 4.21%. Decomposition at temperature higher than 600°C represents the decomposition of lizardite [15]. The third endothermic peak at 902 °C with mass loss 7.39% and the last endothermic peak, accompanied by mass loss of 8.89%. Finally, an exothermic peak at 1236 °C represents recrystallisation. The mass loss between 625 °C and 902 °C have high gap of mass loss, it is more than twice from prior decomposition temperature that represents the high crystallinity phase occurred between that temperature. Therefore the decomposition at 800 °C is required to determine phase transformation between that temperatures. Based on the TG-DTA results, a series of selective reduction processes were performed at 300, 600, 800, 900, 1000, and 1100 °C, then XRD and SEM-EDS analysis were performed to analyse the decomposition occurred to the reduced ores between those range of temperature, both, phase and morphology transformation.

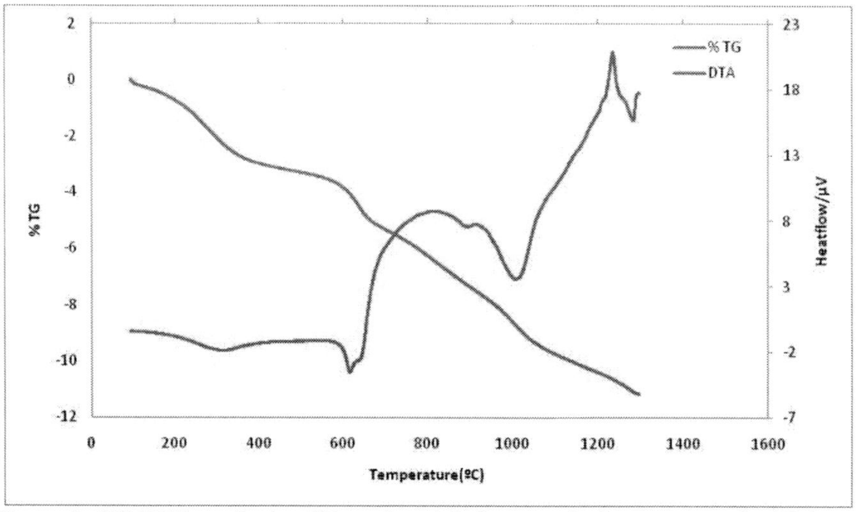


**Figure 2.** TG-DTA analysis of limonite nickel ore, reducing agent and additive mixture

*3.3 Phase Transformation of Reduced Ore*

Phase transformation of reduced limonite nickel ore with a reducing agent and $Na_2SO_4$ additive at various temperature is shown in Figure 3. It indicates that hematite and goethite were observed at XRD pattern at 300 °C, whereas lizardite, quarts, wustite and trevorite were not observed. The sharp diffraction peak intensity of goethite was at 2θ at 21.25° and 41.25°, while hematite was at 23°, 33°, 54.5°, 41°,45°, and 71°. At this temperature, the structure must be destroyed, as well goethite starts to lose most of the crystalline water to form hematite and lizardite and start to lose of crystalline water, but the transformation is not entirely completed [3] [16]. It caused XRD pattern of the sample treated at 300 °C is amorphous. Dehydroxylation of goethite and lizardite occurred such as reaction (1) and reaction (2), it can be confirmed with free energy diagrams, Figure 4., which shows that reaction (1) and reaction (2) have the lowest standard free energy at about 100 °C, and 300 °C respectively.

$$2FeO(OH) \rightarrow Fe_2O_3 + H_2O \qquad (1)$$
$$(Mg,Fe)_3Si_2O_5(OH)_4 \rightarrow (Mg,Fe)_2SiO_4 + (Mg,Fe)SiO_3 + 2H_2O \qquad (2)$$

For reduction temperature at 600 °C, the crystal with high intensity was observed. Hematite was reduced to magnetite with the presence of $CO_{(g)}$, Sharp diffraction peak intensity of magnetite at 2θ was at 30.25°, 35.3°, 58°, and 62.5°. Nickel mineral could not be liberated in the limonite ore, after the structure of trevorite damage was at 300 °C, it was incorporated with the hematite forming trevorite [13], as in reaction (6). Lizardite was continued to decompose to form $MgSiO_3$ and $Mg_2SiO_4$, as Lu Jie, et al. result [15]. $Na_2SO_4$ presence cause Fe liberates from Mg at lizardite [10-15], causing the formation of fayalite ($Fe_2SiO_4$), as in reaction (7). The sharp diffraction peak intensity of fayalite at 2θ was at 35.3° and 37.25°. Reaction (9) is the intermediate reaction that was reacted in further reduction temperature. From Figure 4, reaction (9) has the lowest free energy values at 800 °C.

$$C + O_2 \rightarrow CO_2 \qquad (3)$$
$$CO_2 + C \rightarrow 2CO \qquad (4)$$
$$3\ Fe_2O_3 + CO_{(g)} \rightarrow 2Fe_3O_4 + CO_{2(g)} \qquad (5)$$

$Fe_2O_3 + NiO \rightarrow NiFe_2O_4$ (6)
$(Mg,Fe)_2SiO_4 \rightarrow Mg_2SiO_4 + Fe_2SiO_4$ (7)
$(Mg,Fe)SiO_3 \rightarrow MgSiO_3 + FeSiO_3$ (8)
$FeSiO_3 + Na_2SO_4 + CO_{(g)} \rightarrow FeO + Na_2SiO_3 + SO_{2(g)} + CO_{2(g)}$ (9)

$Na_2SO_4$ attached quartz to form sodium metasilicate, so the presence of $Na_2SO_4$ at reduction process can decrease silica content, then the intensity of silica decreases. According to the standard free energy diagrams (Figure 4), $Na_2SO_4$ reacts spontaneously with quartz at 500 °C and more spontaneously with the increasing temperature.

$SiO_2 + Na_2SO_4 + CO \rightarrow Na_2SiO_3 + SO_2 + CO_2$ (10)

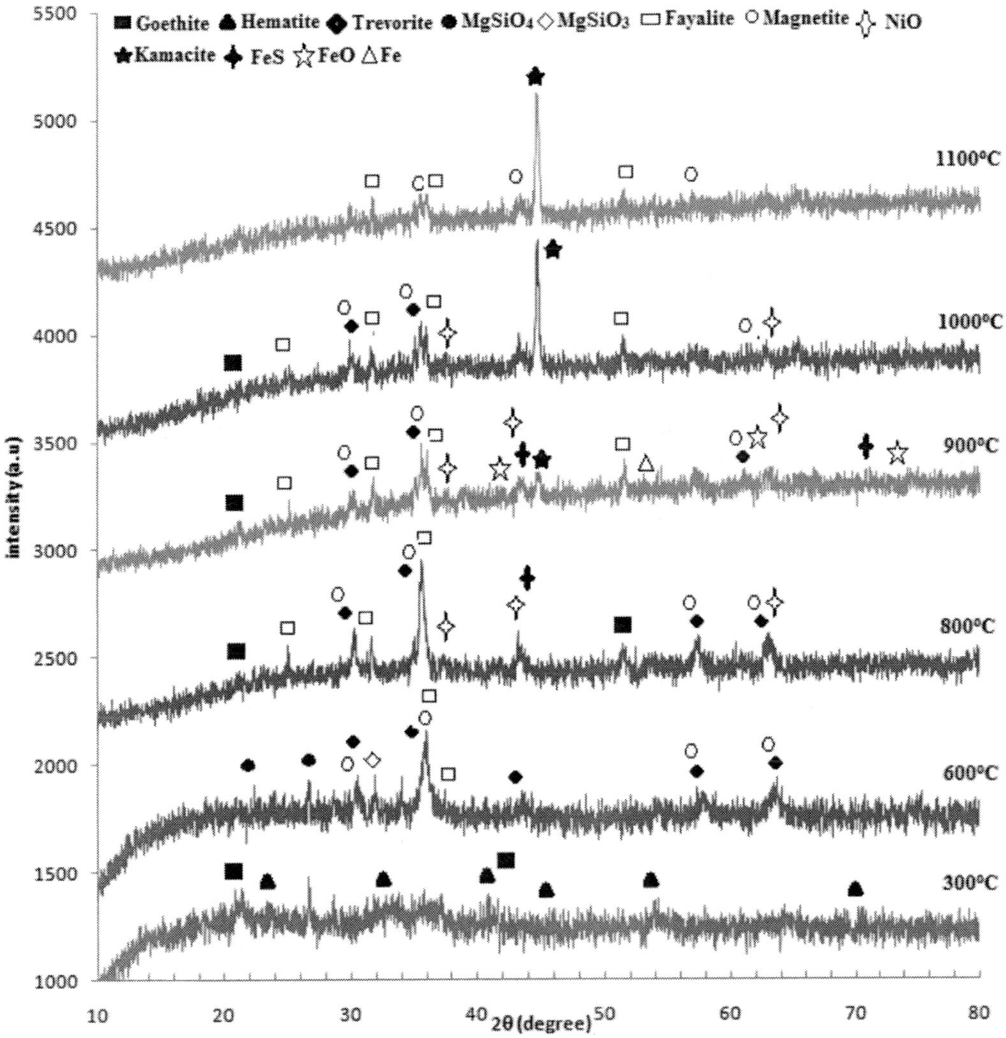


**Figure 3.** XRD analysis of reduced ores at 300° - 1100°C

At the XRD pattern of the reduction temperature of 800 °C, fayalite shows higher intensity than the previous one. Trevorite partially was decomposed into NiO and magnetite. The sharp diffraction peak intensity of NiO at 2θ at 37°, 43° and 63°. Instead, $MgSiO_3$ and $Mg_2SiO_4$ were not observed at peak with reduction temperature above 600 °C [10]. It changes because limonite nickel ore has small Mg content [9] so the peak was not apparent. FeS started to form at 800 °C, reaction (12). The sharp diffraction peak intensity of FeS at 2θ of 44°.

$1{,}5NiFe_2O_4 + \frac{1}{2} C \rightarrow 1{,}5\ NiO + Fe_3O_4 + \frac{1}{2} CO$ (11)
$FeO + SO_2 + CO \rightarrow FeS + CO_2$ (12)

Magnetite was reduced to form FeO at 900 °C, then with trevorite form (Fe,Ni)O.$Fe_2O_3$ [13] and led kamacite (Fe, Ni) phase and $Fe_2O_3$. It is appropriate with the XRD pattern that FeO and (Fe,Ni) were observed. The sharp diffraction peak intensity of FeO at 2θ was at 42°, 60.8°, and 73°, while Fe-Ni was at 44.5°. FeO partially reduced to Fe and also reacted with $SO_2$ to form FeS. Between 800° – 900°C, fayalite has highest intensity. It is appropriate with DTA curve that shows the increasing in heat flow. Instead, intensity of NiO decreased because partially incorporated to form kamacite.
$NiFe_2O_4 + Fe_3O_4 \rightarrow (Fe,Ni)O.Fe_2O_3 + Fe_2O_3$ (13)
$2Fe_2O_3 + 2CO \rightarrow 4FeO + CO_2$ (14)
$Fe_3O_4 + CO \rightarrow 3FeO + CO_2$ (15)
$FeO + CO \rightarrow Fe + CO_2$ (16)
$FeO + SO_2 + CO \rightarrow FeS + CO_2$ (17)

For reduction temperature at 1000°C, intensity of kamacite phase was higher than previous, instead trevorite and magnetite decrease. Fayalite and FeS decreased with reduction temperature more than 800°C, for further reduction temperature trevorite and goethite totally reduced and intensity kamacite was higher than previous.

XRD pattern shows the decomposition, either physical or chemical that is previously shown in the TG-DTA curve. Temperature affects the metallization process whereas additive $Na_2SO_4$ affects the Fe and silica content. Moreover, it increases the reaction rate. $Na_2SO_4$ addition provides beneficial for increasing selectivity of Ni by formation of fayalite and FeS that can be separated by magnetic separation.

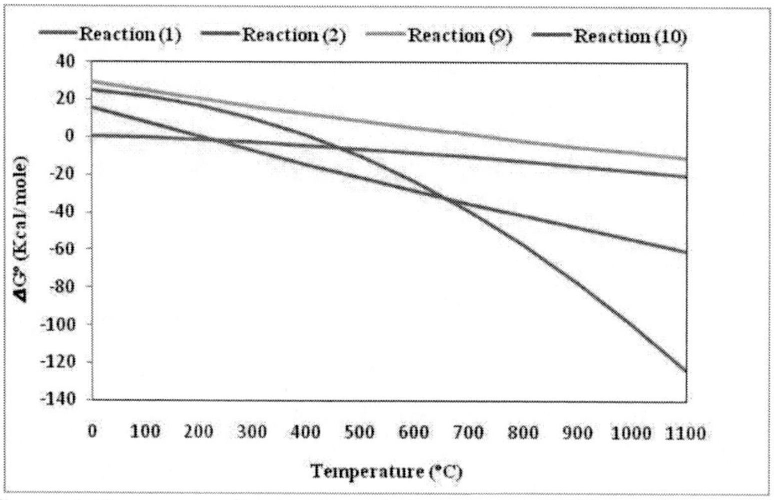


**Figure 4**. Free energy diagram in various temperature. Data calculated by HSC Chemistry 6

If we compare between TG-DTA curve and XRD pattern, it can determine the phase transformation of limonite nickel ore at reduction temperature up to 1100 °C with $Na_2SO_4$ addition and coal as reducing agent. At 324 °C, endothermic reaction occurred that caused the loss of crystalline water that restructuring limonite nickel ore. At 625 °C, endothermic reaction causes decomposition lizardite to form fayalite. At 902 °C FeO is formed as a result of reduction of magnetite and Fe-Ni was formed at a temperature 900 °C. At 1019 °C, goethite completely reduced into magnetite and at 1236 °C recrystallization of Fe-Ni occurred.

*3.4 Morphology Transformation of Reduced Ore*

Figure 5 shows the microstructure transformation of the reduced ores at 300 °C – 1100 °C. It can be seen that particle size was growing. The particle size of limonite ore is greater as the temperature reduction increases. This is also happening in the porosity. It can be seen in Figure 5 (a), and (b) that it still looks homogeneous, but in Figure 5 (c) it can be seen their separation because of the decomposition process at temperatures up to 900 °C, Figure 5 (d) and (e). In Figure 4 (f) and (g), aggregation occurred, so the size of metal particles formed is larger and better in porosity. Limonite ore size before reduced is about 27.41 μm and was continued to grow in line with increasing reduction temperature up to 60 μm at 1000 °C and 70.73 μm at 1100 °C. This is appropriate with the XRD analysis that showed an increasing intensity of metal formation in line with increasing the reduction temperature, it must be indicating the larger particle size.

**Figure 5.** SEM-EDS images of limonite nickel ore (a), and reduced ore for 1 hour of limonite nickel ore (b) at 300 °C, (c) at 600 °C, (d) at 800 °C, (e) at 900 °C, (f) at 1000 °C, (g) at 1100 °C

Figure 6 shows the mapping of SEM-EDS analysis of the reduced ore at 1000 °C for 1 hour. The elemental mapping illustrates the distribution of several elements, such as O (Figure 6 (b)), Fe (Figure 6 (c)), Ni (Figure 6 (d)), Si (Figure 6 (e)) and S (Figure 6 (f)). The distribution of element is indicated by the color indicator (left side) on the morphology of SEM analysis that refers to Figure 6 (a). The color indicator also shows the qualitative analysis. From the pictures it can be seen that the Fe has the highest concentration followed by O, Si, S, and Ni, it can be confirmed from EDS which shows the semi-quantitative analysis in Table 2. In mapping, Fe and S elements have similarities, but S distribution is not as high as Fe distribution, it shows that FeS formed is not in large amount. Mapping Fe and Si also have similarities. It indicates that fayalite has larger quantities than FeS, according to the XRD patterns, fayalite observed at some point even with weak intensity. It is also observed that Fe and O have the same distribution, which indicates the presence of iron oxides such as goethite and magnetite. Distribution of Fe and Ni in mapping also have similarities with the smaller amount of Ni indicating the Fe-Ni distributions.

**Table 2.** Composition of reduced ore at 1000 °C based on EDS result

| Element | Mass (%) | Element | Mass (%) |
|---|---|---|---|
| O | 6.31 | Fe | 74.94 |
| Mg | 0.56 | Ni | 4.49 |
| Al | 1.97 | Cu | 3.25 |
| Si | 3.43 | Zn | 1.27 |
| S | 1.11 | Cr | 2.67 |

**Figure 6**. Images of (a) reduced ore at 1000 °C, (b) mapping for O atom, (c) mapping for Fe atom, (d) mapping for Ni atom, (e) mapping for Si atom, (f) mapping for S atom

## 4. Conclusion

Phase transformation occurred to the reduced ore in the selective reduction temperature processes at 300 - 1100 °C for 1 hour. Four endothermic peaks were observed at TG-DTA that indicated the decomposition of goethite, lizardite, and trevorite, respectively. Moreover, one exothermic peak indicated recrystallization of Fe-Ni. The presence of $Na_2SO_4$ in the selective reduction process led the fayalite formation at 600 °C, and FeS formation at 800 °C, also decreased silica content with the $Na_2SiO_3$ formation. A higher reduction process at a temperature more than 900 °C gave an advantage for Fe and Fe-Ni formation. Otherwise, fayalite and FeS formation decreased. Phase transformation affected microstructure change, so the size and porosity also transformed. The size and porosity of the limonite ore particles were greater up to twice at 1000 °C and 2.5 times at 1100 °C with uniform distribution.

## 5. References

[1] Dalvi AD., Bacon, W.G., Osborne, R.C 2004 The past and the future of nickel laterite, PDAC 2004: International Convention. Trade show & investor exchange, North Carolina USA.

[2] Jinhui Li, Xinhai Li, Qiyang Hu, Zhixing Wang, et al 2009 Effect of pre-roasting on leaching of laterite, Elsevier: hydrometallurgy.

[3] Harris C.T., Peacey J.G., Pickles.C.A 2010 Selective sulfidation of a nickeliferous lateritic ore, Elsevier: Minerals Engineering Journal.

[4] Norgate, T., Jahanshahi, S 2011 Assessing the energy and greenhouse gas footprint of nickel laterite processing, Elsevier: Mineral Engineering.

[5] Hwa Young Lee, Sung Gyu Kim, Jong Kee Oh 2005 Electrochemical leaching of nickel from low-grade laterite, Elsevier: Hydrometallurgy.

[6] Kim Jungah, Dodbiba Gjergj et al 2010 calcination of low- grade laterite for concentration of Ni by magnetic separation, Elsevier: Mineral Engineering.

[7] Valix M., Cheung W.H 2002 Study of phase transformation of laterite ores at high temperature, Pergamon: Mineral Engineering.

[8] Usman, Ediar 2011 Prospek Pengembangan Sumber Daya Nikel Laterit di Kawasan Timur Indonesia *Bulletin ME* **8** 2: 60-68

[9] Mayangsari W, Prasetyo Agus B, Proses Reduksi Selektif Bijih Nikel Limonit Menggunakan Zat Aditif $CaSO_4$, Majalah Metalurgi **31** 1.2016

[10] Li Guanghui, dkk 2011 Beneficiation of nickeliferous laterite by reduction roasting in the presence of sodium sulfate, Elsevier: Mineral Engineering.

[11] Subagja Rudi, Prasetyo Agus B, Mayangsari W 2015 Pengaruh temperatur dan waktu, serta penambahan batubara dan $Na_2SO_4$ terhadap peningkatan kadar Ni dalam bijih nikel laterit, Prosiding Seminar Material dan Metalurgi 2015

[12] Zevgolis E N., Zografidis C., Perraki T., Devlin E 2009 Phase transformations of nickeliferous laterites during preheating and reduction with carbon monoxide, Springer, Hungary

[13] O'Connor F, Cheung W H., Valix M 2005 Reduction roasting of limonite ores: effect of dehydroxylation, Elsevier: International journal of mineral processing.

[14] Swamy, Y V., Kar, B B., Mohanty, J K 2003 Physico-chemical characterization and sulphatization roasting of low grade nickeliferous laterite, Hydrometallurgy 69, 89-98.

[15] Jie Lu, Shoujun Liu, Ju Shangguan, dkk 2013 The effect of sodium sulphate on the hydrogen reduction process of nickel laterite ore, Elsevier: Minerals Engineering.

[16] Ruan, H.D., Frost, R.L., Kloproggr, J.T., Duong, L 2002 Infrared spectroscopy of goethite dehydroxylation: III. Ft-IR microscopy of in situ study of the thermal transformation of goethite to hematite.

**Acknowledgments**
This research was performed at Research Center for Metallurgy and Materials - LIPI, which was funded by DIPA budget for 2016. Special thanks to Dr. Rudi Subagja and Dr. Yanni Sudiyani, M.Agr. who have provided valuable discussion for this research.

IOP Conf. Series: Materials Science and Engineering **202** (2017) 012017 doi:10.1088/1757-899X/202/1/012017

# The Use of Heterogeneous Catalysts of Chitosan Sulfonate Bead on the Esterification Reaction of Oleic Acid and Methanol

**H N Chamidy and Riniati**

Chemical Engineering Department, Politeknik Negeri Bandung, Bandung 40012 – Indonesia

Email : chamidy@gmail.com

**Abstract**. Biodiesel is one of the ester compounds with physical properties closer to a biodiesel which can be produced by the esterification reaction between methanol and oleic acid (one of major components present in Palm Fatty Acid Distillate, PFAD). The purpose of this study was to obtain an optimum condition of esterification reaction by using chitosan sulfonate bead as heterogeneous catalysts. Chitosan sulfonate bead was made from chitosan undergo sulfonation process using acidic reagents cross-linked with sulfosalicylic and glutaraldehyde with a high enough value of ion exchange capacity. The stage of esterification reactions was carried by varying the amount of catalyst being added (4, 6, 8, 10, 12% by oleic acid), the operating temperature was varied of 40, 50 and 60 °C, and the reaction time of 1, 2, 3, 4 and 5 hours. Conversion determination of the products was done by analysing the free fatty acids content in each sample. Having obtained from the optimum amount of catalyst being added, temperature, and time, it was found that the catalyst was at 8%, 50 °C, during 5 hours in operation. The maximum conversion of oleic acid into biodiesel was 73.12%.

**Keywords :** catalysts, chitosan, sulfonate, oleic acid, methanol

## 1. Introduction

Chitosan is a long chain of beta polymeric glucose constructed of (2-acetamido-2-deoxy-ß-D-glucose-(N-acetylglucosamine)) (Figure 1) derived from chitin, a natural polysaccharide usually obtained from carapaces of the marine crustaceans such as crabs and shrimps [19]. It is cheaper to use Chitosan since chitin is the second most abundant biopolymer in nature next to cellulose [10].

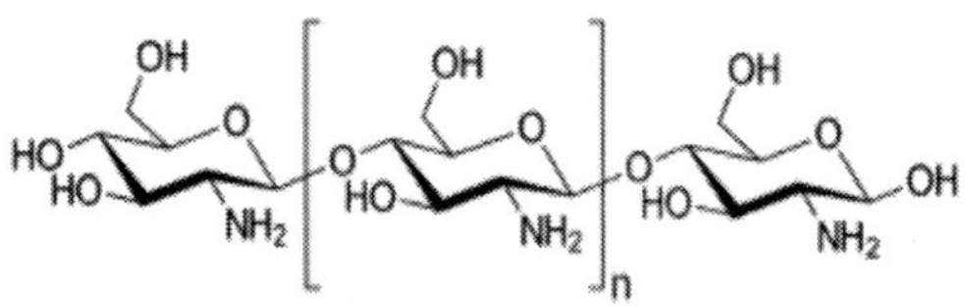


**Figure 1.** Chitosan structure (Thate, 2004)

Chitosan is the most attractive choice of immobilization since it offers several advantages such as versatility of available physical forms (flakes, porous beads, gel, fiber and membrane); low biodegradability; ease of handling; high affinity for proteins and, above all, non-toxicity (Felse and

Panda, 1999). Chitosan may have been applied in various fields such as biotechnology, pharmaceutical, waste water treatment, agriculture and food. To increase the use of chitosan, it was synthesized and sulfonated to produce an insoluble acid chitosan sulfonate bead [2].

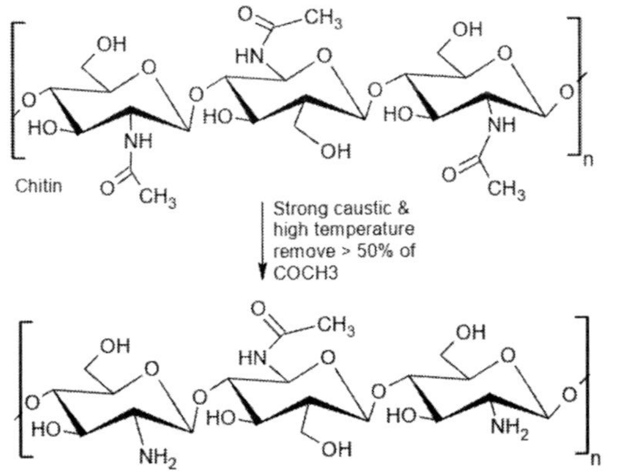


**Figure 2.** Chitosan derived from chitin (Thate, 2004)

Oleic acid, representing free fatty acid (FFA), is the second largest component present in the Palm Fatty Acid Distillate (PFAD) [1]. PFAD is a waste (a by-product of the refining process) produced during cooking oil production with a high free fatty acid content. The major components in PFAD are lauric acid (0.1%-1.7%), myristic acid (1.0%-1.8%), palmitic acid (46.4%-51.2%), stearic acid (3.7%-5.1%), oleic acid (33.0%-37.7%) and linoleic acid (7.8% -9.6%) [15]

Esterification of FFA with low-molecular-weight of alcohols using acid or base catalysts will produce biodiesel and water [18]. Homogeneous catalyst is a commonly used for acid catalyst [12]. However, homogeneous catalysts are corrosive; contaminate the product and causing environmental problems. The use of heterogeneous catalysts that are more environmentally friendly will partly solve these problems. It has an advantage of easily separated from its liquid with a low percentage loss of the catalyst.

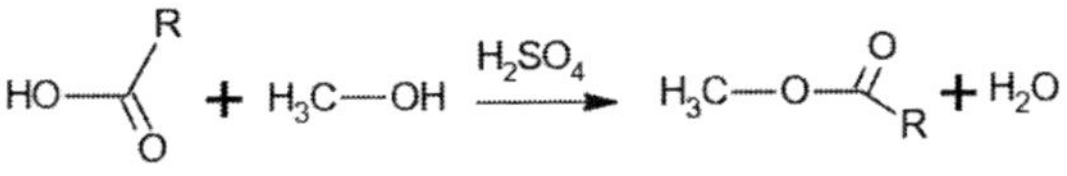

**Figure 3**. Esterification of FFA

The existence of -OH and $-NH_2$ functional groups in chitosan allows a wide variety of chemical modifications that are classified as highly functional chitosan biopolymer. These groups can be used as an acid catalyst through sulfonation reaction [3].

Sulfonation could occur by chemically attaching sulfonate groups to chitosan using various methods and conditions [9]. N-sulfonated chitosan is attaching sulfonate groups into NH site, whereas O-sulfonated is attaching sulfonate groups into OH site. Propane sultone [20] is the example of N-sulfonated reaction. 5-sulfosalicylic acid, and glutaraldehyde, $SO_3$ or chlorosulfonic acid reagents and dimethylformamide which has been patented by Hayashi (1993) are the example of O-sulfonated reaction. This method will improve the proton conductivity.

Chitosan which activated by $-SO_3H$ group called chitosan sulfonate, could be applied as a catalyst in the process of esterification for biodiesel production [3]. In 2013, Castanheiro *et al.* have done a heterogeneous catalyst preparation of sulfonate-based biopolymer chitosan by reacting chitosan and sulfosuccinic acid. He produces a heterogeneous catalyst in a sheet form. Heterogeneous catalyst could be prepared by dissolving chitosan in a 2% acetic acid to form poly-cation. Sulfosalicylic acid then be cross-linked to the functional groups -OH so that the chitosan has a sulfonate group ($SO_3H^+$).

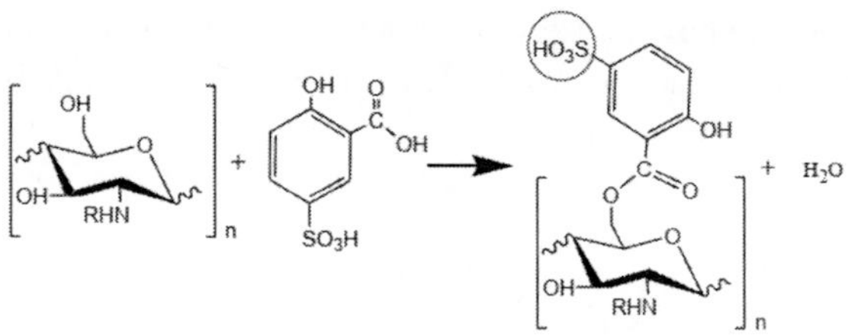


**Figure 4.** Cross-linked reaction of chitosan and sulfosalicylic acid

The -OH functional group is easier to remove proton ions (Figure 4) rather than –NH2. Sulfosalicylic acid as sulfonate group is a hydrophilic in which easier to remove proton ions at -OH groups. The production of chitosan sulfonate bead using sulfosalicylic acid was an electrophilic substitution, where $SO_2^+$ being attacked by a pair of free electrons of oxygen and nitrogen. The sulfonation can occur in two functional groups, namely the hydroxy and amine groups.

However, hydroxy group was the most likely sulfonated as the acidic atmosphere will protonated amine groups. The oxygen atom was more electronegative than nitrogen in the amine group so that when there was a free proton, the electron pair of the oxygen atom at the hydroxy group will attack it [2].

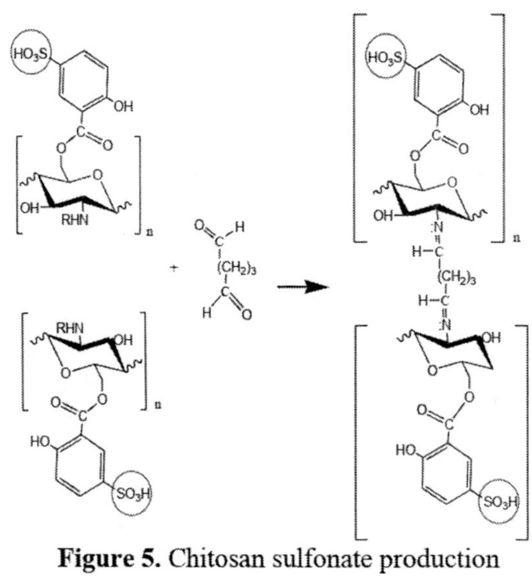


**Figure 5.** Chitosan sulfonate production

## 2. Experimental Method

Industrial grade chitosan with a degree of deacetylation of 84.39% was obtained from Biotech Surindo. 5-sulfosalicylic acid and glutaraldehyde were obtained from Merck. NaOH, $H_2SO_4$, a pure oleic acid, and methanol were obtained from commercial sources.

The chitosan sulfonate bead were prepared by dissolving 2 gram of chitosan in a 2 wt.% acetic acid solution (100 mL) to make a 2% in concentration and stirred in 30-40 °C for 2 hours. After complete dissolution of the chitosan flakes, the solution was added with sulfosalicylic varied at 17, 18, 20, and 22 mmol per liter of chitosan. Stirring was continued at ambient temperature for another 24 hours. The beads were formed by dripping the bubble-free mixture through a disposable plastic syringe with 2 mm outlet diameter into 200 mL, 8M NaOH solution for 30 minutes. The beads formed were then filtered and neutralized using aquadest. Cross-linking was done again by soaking the chitosan beads into 200 mL of 2.5% glutaraldehyde solution at ambient temperature for 24 hours. After filtering and washing, the beads were then soaked again in 200 mL of 2.5% acetic acid for another 24 hours. The chitosan sulfonate beads were then filtered, neutralized, and dried at 60 °C for 3 hours and stored in a desiccator for 1 hour.

On the esterification reaction, the amount of catalyst was varied at 4, 6, 8, 10, and 12 wt% of oleic acid. The reaction was carried at ambient temperature. The optimum result will then be used for further esterification reactions by varying the reaction time at 1, 2, 3, 4 and 5 hours and also varying the operating temperature at 40, 50 and 60 °C.

Characterisation was done by determining the ion-exchange capacity (IEC) of chitosan sulfonate beads. IEC indicates the number of mili-equivalents of ions in 1 g of dry polymer. In brief, required amount of chitosan sulfonate bead sample was soaked in 50 mL of 3M sodium chloride (NaCl) solution for 24 h. Nearly 10 mL of the above solution was titrated against 0.01N sodium hydroxide (NaOH) solution using phenolphthalein as the indicator. By measuring the amount of NaOH consumed in the titration, the molar quantity of the sulfonic acid groups contained in the chitosan sulfonate sample was determined. By using this value, IEC was estimated by Equation (1).

$$IEC=\frac{Vol.of\,NaOH\,consumed \times Normality\,of\,NaOH}{Dry\,weight\,of\,Chitosan\,sulfonate} \quad (meq/g) \qquad (1)$$

FTIR Shimadzu 8400 was used to analyze a known amount of chitosan sulfonate bead for sulfonate group existence. Degree of conversion of oleic acid to become biodiesel was done by analyzing the free fatty acids content in each sample.

## 3. Results and Discussion

Dried chitosan sulfonate bead which has been produced have a diameter of 1-1.5 mm (Figure 6). The important factor of the resulting catalyst quality was the value of Ion Exchange Capacity (IEC), particularly regarding with the existing number of $H^+$ ions. As the amount of sulfosalicylic acid rises, the IEC value rises. It was found that a harder bead indicating a higher IEC. In certain high amount of sulfosalicylic acid, the bead becomes soft, and will decrease the IEC.

**Figure 6.** Chitosan sulfonate beads

IEC could be used as proton conductivity estimation indirectly on the ion exchangeable groups present in a bead for proton conduction [13]. Figure 7 shows that IEC was directly related to degree of sulfonation. IEC of chitosan sulfonate bead was varied between 0.14 and 0.34 mmol/g. The IEC was low because of the low reactivity of the aromatic rings directly attached to the sulfone groups.

IOP Conf. Series: Materials Science and Engineering **202** (2017) 012017 doi:10.1088/1757-899X/202/1/012017

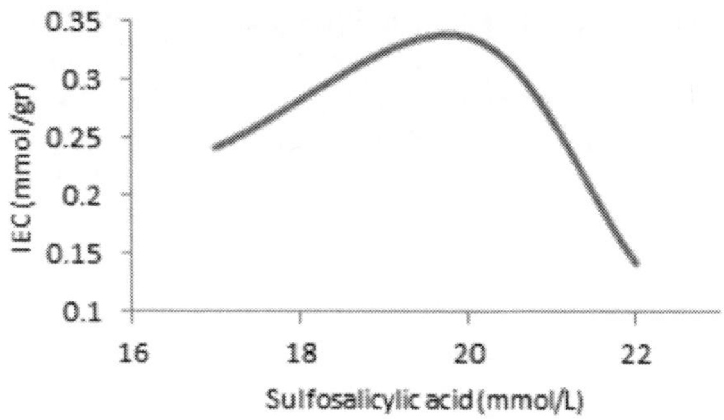


**Figure 7.** IEC of chitosan sulfonate Bead

The ability of the catalyst to perform an ion exchange was strongly influenced by the condition of the parent charged ions in the polymer chain, it was the existence of -$HSO_3$ group. A higher IEC value of sulfonate chitosan beads, indirectly indicating that cross-link was occurred.

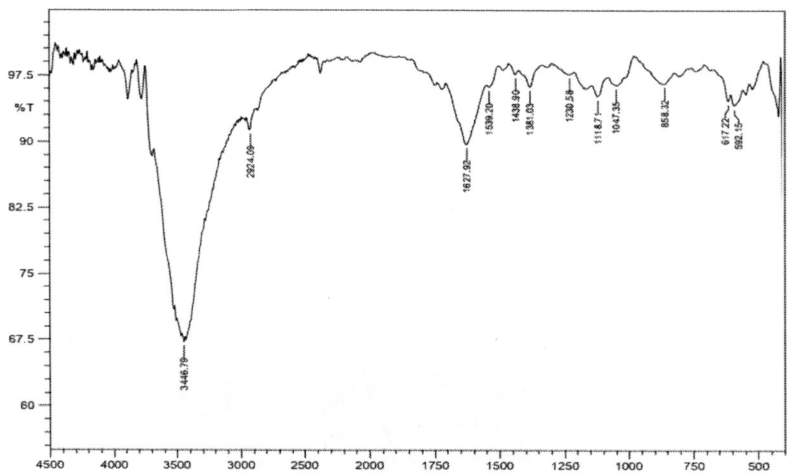


**Figure 8.** Chitosan sulfonate IR spectrum

The presence of sulfonate groups that resulted from crosslinking reaction in the sulfonation process can be observed from FTIR spectra of the chitosan sulfonate. The FTIR spectra of chitosan sulfonate shows in Figure 8. The peak at 1118 $cm^{-1}$ is the characteristic absorption bands of OH group in sulfonic acid. Meanwhile, the peak at 1124.50 $cm^{-1}$ could be ascribed as the stretching vibration of sulfoxide in sulfonic acid. The result suggests that the crosslinking between sulfosalicylic acid and chitosan was successful and chitosan was sulfonated. From the FTIR analysis, the results suggest that crosslinking reaction was successful and chitosan was sulfonated.

The chitosan sulfonate used as a heterogeneous catalyst in the esterification of oleic acid was the highest IEC achieved, which was using 20 mmol/g addition of sulfosalicylic acid. At ambient temperature, the conversion of oleic acid into biodiesel increases as the amount of catalyst increases from 6% into 8% by weight of oleic acid used. The highest oleic acid conversion was achieved during addition of 8% chitosan sulfonate bead.

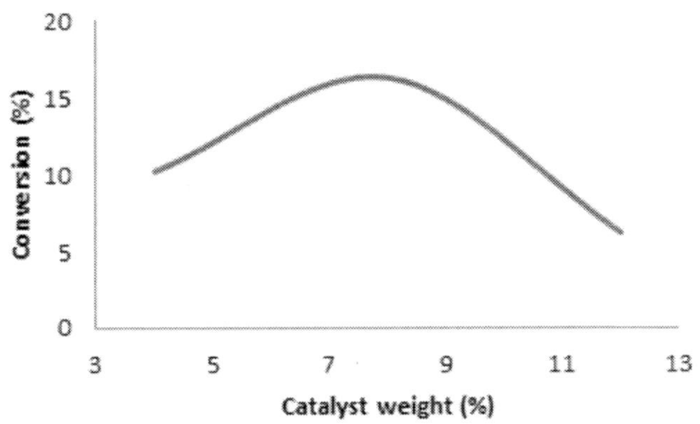


**Figure 9.** FFA Conversion by Catalyst Concentrations

To continue the esterification, the reaction condition was varied by varying the temperature and time taken using 8% addition of the catalyst. The highest value of conversions was achieved on hour-5 at the reaction temperature of 50 °C (Figure 10). The maximum conversion of oleic acid into biodiesel was 73.12%.

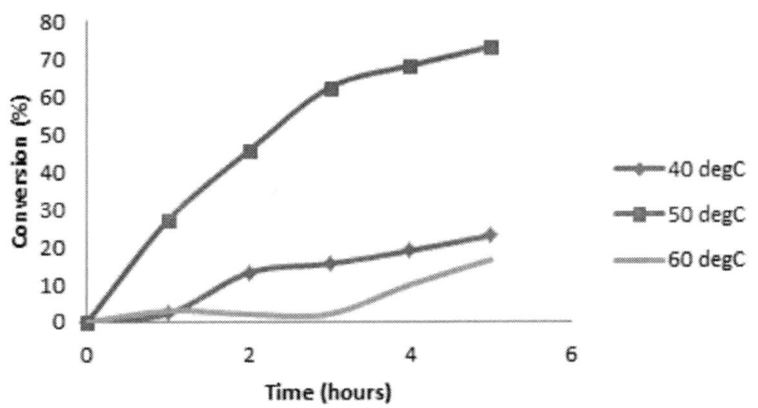


**Figure 10.** Conversion degree of FFA by varying time and temperature

## 4. Conclusion

The chitosan sulfonate bead as heterogeneous catalyst produced has the highest IEC value of 0.34 mmol/gram by addition 20 mmol/L sulfosalicylic acid. The optimum conditions were 8% of chitosan sulfonate concentration in esterification reaction of oleic acid and methanol with a temperature of 50 °C for 5 hours which resulted in the conversion of oleic acid to form biodiesel up to 73.10%. It is expected these conditions can be applied to the manufacture of biodiesel using oleic acid.

## 5. References

[1] Aranda D A G, Santos R T P, Tapanes N C O, Ramos A L D, Antunes O A C 2008 Acid-catalyzed homogeneous esterification reaction for biodiesel production from palm fatty acids *Catal. Lett.* **122** 20–25

[2] Basuki, Bagus R, and Sanjaya I G M 2009 Sintesis Ikat Silang Kitosan dengan Glutaraldehis serta Identifikasi Gugus Fungsi dan Derajat Deasetilasinya *Jurnal Ilmu Dasar* **10** 1 93 – 101

[3] Castanheiro et al 2013 Esterification of free fatty acids over chitosan with sulfonic acid groups *Chemical Engineering Journal* **230** 567–572

[4] Felse P A and Panda T 1999 Studies on Applications of Chitin and its Derivatives *Bioprocess*

*Engineering* **20** 505-512

[5] Furuta S, Matsuhashi H, Arata K 2004 Biodiesel fuel production with solid superacid catalysis in fixed bed reactor under atmospheric pressure *Catal.Commun* **5** 721–723

[6] Gotoh T, Matsushima K and Kikuchi K I 2004 Preparation of Alginate-Chitosan Hybrid Gel Beads and Adsorption of Divalent Metal Ions Chemosphere **55** 135

[7] Hassani M, Amini G, Najafpour G D and Rabiee M 2013 A Two-step Catalytic production of Biodiesel from Waste Cooking Oil *International Journal of Engineering Transaction C: Aspects* **26** 6 563-570

[8] Hayashi J 1993 Process for preparing a chitosan sulfonate *US patent 5299504*

[9] Jayakumar R, Nwe N, Tokura S and Tamura H 2007 Sulfated chitin and chitosan as novel biomaterials *International Journal of Biological Macromolecules* **40** 175–181

[10] Krajewska B 1991 Chitin and its Derivatives as Supports for Immobilization of Enzymes *Acta Biotechnologica* **11** (3) 269-277

[11] Ma F, Hanna M Biodiesel 1999 production: a review *Bioresour. Technol.* **70** 1–15

[12] Ma Jia, Sahai Y 2013 Review: Chitosan biopolymer for fuel cell applications *Journal of Carbohydrate Polymers* **92** 955-975

[13] Meenakshi S, Bhat S D, Sahu A K, Sridhar P, Pitchumani S, Shukla A K 2011 Chitosan-Polyvinyl Alcohol-Sulfonated Polyethersulfone Mixed-Matrix Membranes as Methanol-Barrier Electrolytes for DMFCs *Journal of Applied Polymer Science* E74-E82

[14] Mukoma P, Jooste B R and Vosloo H C M 2004 A comparison of methanol permeability in chitosan and Nafion 117 membranes at high to medium methanol concentrations *Journal of Membrane Science* **243** 293–299

[15] Ping B T Y, Yusof M 2009 Characteristics and Properties of Fatty Acid Distillates from Palm Oil *Oil Palm Bulletin* **59** 5-11

[16] Riniati, Chamidy H N and Widiastuti E 2015 Modified Chitosan Into Chitosan Sulfonate Bead For Heterogeneous Catalyst Applications *Australian Journal of Basic Sciences* **9**(33) p 224-228

[17] Rui Song, Rui Xue, Ling-hao He, Ying Liu and Qiao-ling Xiao 2008 The Structure And Properties Of Chitosan/ Polyethylene Glycol/Silica Ternary Hybrid Organic-Inorganic Films Chinese *Journal of Polymer Science* **26**(5) 621-630

[18] Sivasamy A, Cheah K Y, Fornasiero P, Kemausuor F, Zinoviev S, Miertus S N 2009 Catalytic applications in the production of biodiesel from vegetable oils *ChemSusChem* **2** 278–300

[19] Thate M R 2004 Synthesis and Antibacterial Assessment of Water-Soluble Hydrophobic Chitosan Derivates Bearing Quaternary Ammonium Functionality Louisiana

[20] Tsai H S, Wang Y Z, Lin J J, Lien W F 2010 Preparation and properties of sulfopropyl chitosan derivatives with various sulfonation degree *Journal of Applied Polymer Science* 116:1686-93

# $CaSO_4$ Scale Inhibition by a Trace Amount of Zinc Ion in Piping System

**W Mangestiyono and Sutrisno**

Mechanical Engineering, School of Vocational, Diponegoro University, Indonesia 50275

Email : rhombo_hedral@yahoo.co.id

**Abstract.** Usually, a small steam generator is not complemented by equipment such as demineralization and chlorination process apparatus since the economic aspect was a precedence. Such phenomenon was uncovered in a case study of green tea industrial process in which the boiler capacity was not more than 1 ton/hour. The operation of the small boiler affected the scaling process in its piping system. In a year operation, there was already a large scale of calcium attached to the inner surface of the pipe. Such large scale formed a layer and decreased the overall heat transfer coefficient, prolonged the process time and decreased the production. The aim of the current research was to solve the problem through a laboratory research to inhibit the $CaSO_4$ scale formation by the addition of trace amounts of zinc ion. This research was conducted through a built in-house experimental rig which consisted of a dosing pump for controlling the flow rate and a thermocouple to control the temperature. Synthesis solution was prepared with 3,500 ppm concentration of $CaCl_2$ and $Na_2SO_4$. The concentration of zinc was set at 0.00; 5.00 and 10.00 ppm. The data found were characterized by scanning electron microscopy (SEM) to analyze crystal polymorph as the influence of zinc ion addition. The induction time was also investigated to analyze the nucleation time, and it was found on the $9^{th}$, $13^{th}$, and $19^{th}$ minute of the zinc ion addition of 0.00, 5.00 and 10.00 ppm. After running for a four-hour duration, the scale grow-rate was found to be 5.799; 5.501 and 4.950 x $10^{-3}$ gr/min for 0.00; 5.00 and 10.00 ppm of zinc addition at 50 °C.

**Keywords**: $CaSO_4$, zinc, inhibition, scale, piping system.

## 1. Introduction

This research was conducted based on a case study of a small steam generator operation in a green tea industry where the product decreased extremely after a year operation. The main cause of the decreased product was identified due to the scaling process occurred on the inner surface of the boiler piping system. A small steam generator usually is not completed by any equipment such as demineralization and chlorination processed apparatus since the economic aspect was precedence. That is why the calcium scale can pile up easily in the inner surface of the pipe. When the scale was analyzed through SEM EDX, the results showed that scale composition consisted of $CaSO_4$ and $CaCO_3$ in majority and $MgSO_4$ in the minority. In the current research, the $CaSO_4$ scale would be the focus of this research since this type of scale was tough to clean.

The existence of $CaSO_4$ scale in piping system led to a disturbance of the system such as decreased overall heat transfer coefficient, prolonged process time and increased economic cost [1, 2, 3]. Many efforts to inhibit $CaSO_4$ scale have been conducted on the physical and chemical factors. The flow rate and temperature are classified as physical factors [4,5], otherwise impurities and additive classified as chemical factors [6,7]. Usually, the appropriate method to inhibit scale growth is performed on the chemical factors through the addition of the trace amount of a chemical substance [1].

In order to inhibit nucleation and crystal growth, the addition of the trace amount of a chemical substance in a supersaturated solution is a commonly employed technique [3]. Many additives have been used to inhibit those scale, such as citric acid, nitric acid, malic acid[1,2,8] and double-hydrophilic block copolymers [9] also phosphate component [3] and polyelectrolyte [10]. The presence of additives could affect the structure and shape of many crystals such as $CaSO_4$, $BaSO_4$, and ZnO [9] and also induce crystal morphology and crystal size [4].

In the current research, Zinc ion was used as the additive to inhibit the $CaSO_4$ scale. As an inhibitor, zinc ion must have some affinity with the host crystal to be adsorbed on its surface and to be able to modify its behavior [3]. The effects of zinc ion as an inhibitor of $CaSO_4$ scale formation from over-saturated solutions would be the focus of this research.

## 2. Experimental Method

### *2.1. Material*

$CaSO_4$ scaling process occurred synthetically by mixing $CaCl_2$ with a $Na_2SO_4$ solution, which was supplied by Merck®, according to the reaction as Equation (1).

$$CaCl_{2\,(aq)} + Na_2SO_{4\,(aq)} \rightarrow CaSO_{4(s)} + 2NaCl_{\,(aq)} \qquad (1)$$

The concentration of calcium solution was defined of as much as 3.500 ppm, prepared by dissolving 77.175 g of $CaCl_2$ to 6,000 mL of demineralized water. Sodium solution was prepared by dissolving 55.65 g of $Na_2CO_3$ to 6,000 mL of demineralized water, too. The straining process of the solution was done two times by 0.22 μm microspore® paper to ensure the unwanted particles remain in the solution. Each of the solutions was kept in each covered vessel to avoid dust and other contaminants. Weighing, mixing and straining processes must be done carefully to ascertain that there is not a bias in the research. The weighing process was done via Ohaus AR-2140 analytical balance in the accuracy of 0.1 mg. CHEMFEED Ca-92683 dosing pump did the mixing process.

### *2.2. Experimental design and characterization*

As the aim of the research was to uncover the influence of zinc ion on $CaSO_4$ scale formation, the research design was defined as described below. The chosen flow rate was at 35 mL/min. The selected temperatures were 27; 40 and 50 °C and the chosen Ca concentration was at 3,500 ppm for all experiments. Zinc ion was found from $ZnCl_2$ with the zinc concentration varied of 0.00; 5.00 and 10.00 ppm. The $CaSO_4$ scale grow-rate was measured when the experiment was affected by the additions of 5.00 and 10.00 ppm zinc ion and then was compared with the experiment without zinc ion. The different scale deposition was pointed as being influenced by an additive on the deposition rate. Induction time was also measured in the kinetic study to show the effect of the additive on the nucleation process. Crystal polymorph was characterized via scanning electron microscopy (SEM) to show the different crystal shape due to the zinc ion addition. Crystal component was investigated through SEM EDX to demonstrate the alteration of crystal structure along with the zinc ion addition.

### *2.3. Experimental rig*

To conduct the experiment, a built in-house experimental rig with the control of an electronic program was employed. The employment is shown schematically in Figure 1. The rig consisted of two vessels, each of which contained $CaCl_2$ and $Na_2CO_3$ solutions, respectively. A stirrer was attached in each vessel under computer program control and automatically set at 40 rpm to establish a homogeneous solution. All of the solutions in each vessel were pumped by CHEM-FEED Ca-92683 dosing pump (5) similarly in the flowrate of 35 mL/min, and the solutions met in the test pipe section (6). To set the temperature, an electrical heater (3) was immersed in the solution and controlled by a sensor (4) in order to keep stable. When the various temperatures were set,

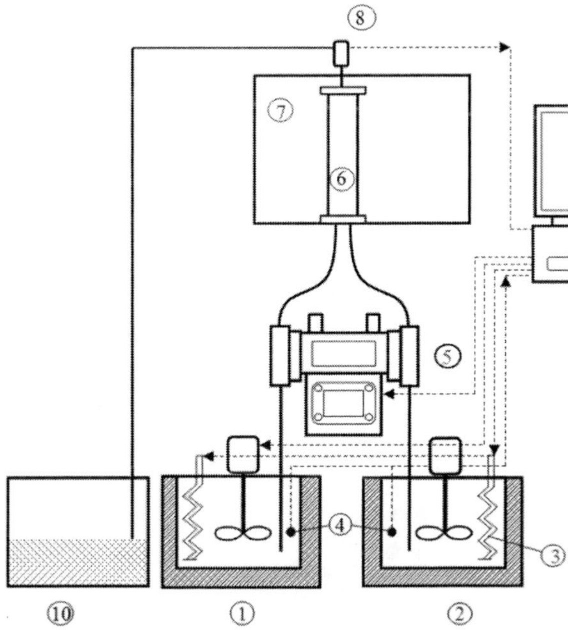


**Figure 1**. Experimental rig

the computer program sent a command to the pump to delay the operation for a couple of minutes until the predetermined temperature was achieved. In the test pipe section, four pairs of coupons were inserted to the parts where the precipitation process needed to form. The pipe was attached to a table (7) vertically. The conductivity of the solution after leaving the coupons was measured by TDS meter supplied by Komatsu (8). The data were collected through the computer program (9), and the solution was then sent to the vessel (10).

*2.4. Phreeqc prediction*

The Phreeqc prediction on the scaling potential showed that there were not any differences in the experiments with an additive and without an additive. The data input of the solution without Zinc ion was collected as follow, temp 25; pH 7; default units g/kgW; density 1; Ca 1.75; Cl 3.105; H(0) 0.175; O(0) 0.7; S(6) 4.2; Na 2.1 and water 1 kg. Otherwise, input data with 10 ppm of Zinc ion was done such as : temp 25; pH 7; units g/kgW; density 1; Ca 1.75; Cl 3.105; H(0) 0.175; O(0) 0.7; Na 2.1; S(6) 4.2; Cu 0.01 and water 1 kg. The specific conductance of the solutions in their initial condition showed differences i.e. 14.102 μS/Cm and 14.082 μS/Cm respectively in the solutions with and without 10 ppm Zinc ion.

The phreeqc program predicted the potential of crystal phase produced through the analysis of the amount of each saturation index. The prediction showed that there were not any differences between the experiments without and with zinc ion performed thermodynamically. Thus, the presence of 10 ppm zinc ion did not have any effects on the solution in its thermodynamic.

## 3. Results and discussion

The results of this experiment consisted of several forms of data such as induction time, the scale grow rate, inhibition rate, crystal polymorph and crystal structure. All data were measured in the absence and in the presence of zinc ion to show the influence of the zinc ion addition. Furthermore, the data were then compared with each other, and the differences found in those data stood as the effect of the zinc ion addition.

### *3.1. Induction time*

The time of firstly measurable change of conductivity is called as an induction time [11]. The measurable change of conductivity showed that some ions in the solution have been reacted and precipitated as scales. Induction time also describes the time needed to produce a nucleus which is also called as nucleation time [12].

To identify the induction time, the conductivity of the solution after leaving the coupon was measured by conductivity meter and acquired by a computer program in every two seconds. After being calculated in the mean value of every one minute, the data were plotted to show the measurable change of conductivity. To describe the influence of the addition of zinc ion to the solution, a conductivity versus time was plotted in a graph of the zinc ion concentration varied to 0.00, 5.00 and 10.00 ppm as shown in Figure 2.

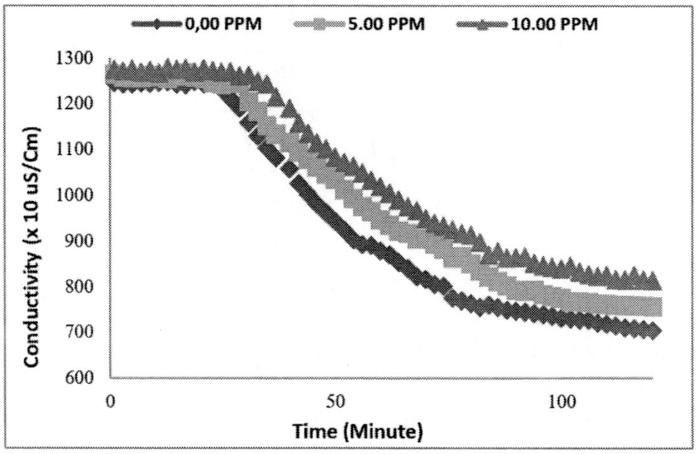


**Figure 2.** Plot of conductivity over the time

In this experiment, the induction time at 27 °C was found to be 21, 25 and 31 minutes of zinc ion addition in 0.00, 5.00 and 10.00 ppm. For the experiment at 40 °C the induction time was found in the 15$^{th}$, 19$^{th}$, and 23$^{rd}$ minute of zinc ion addition in 0.00, 5.00 and 10.00 ppm. When the temperature was set at 50 °C, the induction time was found to be at the 9$^{th}$, 13$^{th}$ and 19$^{th}$ minute of zinc ion addition in 0.00; 5.00 and 10.00 ppm. These data show close agreement with other research, such as the research published by Ang [7].

The extended nucleation time showed the evidence that the addition of zinc ion in 5.00 and 10.00 ppm has successfully inhibited the nucleation process of $CaSO_4$ scale formation. The data also showed the influence of zinc ion addition on induction time in various temperatures which successfully inhibited the nucleation process at higher temperatures. As the nucleation time was longer than normal, the scale formed could be fewer. The following section would discuss the scale grow rate and inhibition rate section in detail.

### *3.2. CaSO4 scale grow rate*

$CaSO_4$ scale, as the product of precipitation process, was found in the coupon. The assessment of the $CaSO_4$ scale grow as the influence of the zinc ion addition was conducted in three temperature conditions i.e. 27; 40 and 50 °C meanwhile the zinc ion concentration varied to 0.00, 5.00 and 10.00 ppm. The result was expressed in a graph such as shown in Figure 3.

The scale grow rate was calculated as the mass of the coupon which was formed in a 4-hour duration and was then converted in the unit of gr/min. When the temperature was set at 27 °C and zinc ion was as much as 0.00; 5.00

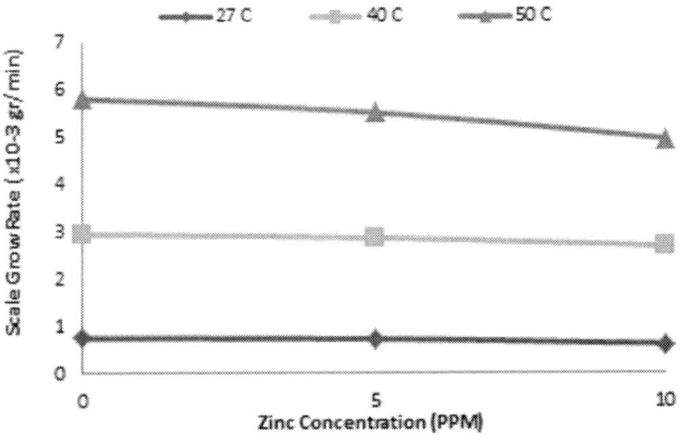


**Figure 3.** $CaSO_4$ scale grow rate in various zinc concentration and temperatures

and 10.00 ppm, the scale grow rates were found to be as much as 0.755; 0.728 and 0.618 x $10^{-3}$ g/min. The experiment with the temperature at 40 °C and zinc ion of as much as 0.00, 5.00, and 10.00 ppm resulted in the scale grow rates of as much as 2.942; 2.860 and 2.687 x $10^{-3}$ gr/min. the scale grow rates found in the experiment at the temperature of 50 °C and zinc ion varied of 0.00, 5.00, and 10.00 ppm were 5.799, 5.501 and 4.950 x $10^{-3}$ g/min.

The data showed that $CaSO_4$ scale grow rate was decreased by the addition of trace amount of zinc ion even at the amount of 5.00 and 10.00 ppm and in various temperatures. These inventions showed the evidence that the addition of the trace amount of zinc ion has successfully inhibited the $CaSO_4$ scale grow rate at different temperatures. The decrease of the $CaSO_4$ scale growth was in conformance with the research published by Ang [7].

### *3.3. Inhibition rate*

In this experiment, an inhibition rate was determined as the different scale growth between the experiment of 5.00 and 10.00 ppm and the experiment of 0.00 ppm of zinc ion of which result is presented in Figure 4.

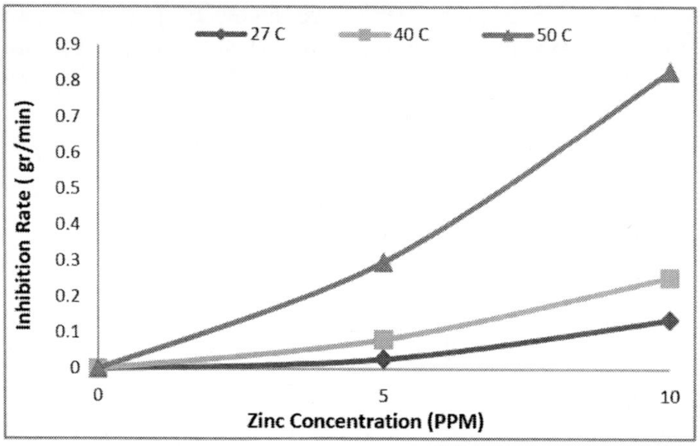


**Figure 4.** Inhibition rate of $CaSO_4$ scale grow in various zinc concentrations and temperatures

As for the measurement, the inhibition rates of $CaSO_4$ scale which were developed through the experiment at 27 °C were 0.00, 0.298 and 0.829 x$10^{-3}$ gr/min of zinc ion addition in the amount of 0.00, 5.00 and 10.00 ppm.

The inhibition rates of $CaSO_4$ scale grow of the experiment at 40 °C were 0.00, 0.082 and 0.255 x $10^{-3}$ g/min of zinc ion addition of as much as 0.00, 5.00 and 10.00 ppm. The inhibition rates of $CaSO_4$ scale grow of the experiment at 50 °C were 0.00, 0.027 and 0.137 x $10^{-3}$ g/min of zinc ion addition varied to 0.00, 5.00 and 10.00 ppm.

The graph of inhibition rate as the addition of zinc ion showed that the inhibition rate increased as the rising temperature. It has been broadly understood that calcium solubility will decrease when the temperature increase [7] as solubility is defined as the maximum amount of a solution that will dissolve in a given quantity of solvent at a specific temperature [13]. Thus, the rising temperature will be followed by the decreasing solubility. That is why the $CaSO_4$ scale-grow rate increases as the rising of temperature could be accepted. Moreover, when the inhibition rate increases as the rising $CaSO_4$ scale-grow rate could also be tolerated.

### *3.4. Crystal polymorph*

The polymorph of $CaSO_4$ crystal was found after a characterization by scanning electron microscopy (SEM) regarding investigating the influence of zinc ion addition to the crystal polymorph. Figure 5a) shows the $CaSO_4$ crystal polymorph in the experiment without zinc ion, meanwhile figure 5b) shows the $CaSO_4$ crystal polymorph in the experiment with the addition of 10.00 ppm of zinc ion.

IOP Conf. Series: Materials Science and Engineering **202** (2017) 012018 doi:10.1088/1757-899X/202/1/012018

(a)

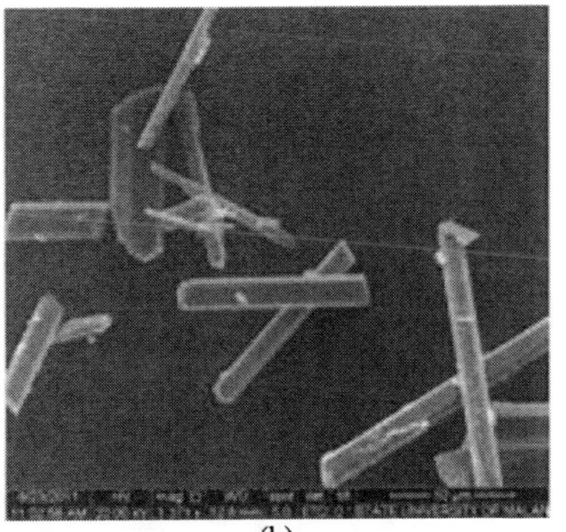
(b)

**Figure 5**. $CaSO_4$ crystal polymorph of a) 0.00 ppm and b) 10.00 ppm zinc ion

Both of the crystal polymorphs showed that they were not different in their shapes and their polymorphs. Refer to crystal system, SEM images in figure 5a and 5b indicate that the crystal polymorphs are rhombohedral [14] as well as the usual. However, it could be distinguished from their dimension that the crystal polymorph in the experiment with 10.00 ppm zinc ion has shorter needles and a smaller dimension.

The data show the mechanism of zinc ion in inhibiting the $CaSO_4$ scale growth. As the growth of crystal dimension was inhibited, $CaSO_4$ scale mass was decreasing. This phenomenon has a close agreement with the phreeqc prediction. The addition of zinc ion does not have any differences in the thermodynamic as predicted by phreeqc program. So, the inhibition mechanism was not based on its thermodynamic but through another mechanism i.e. through covering the crystal surface by a zinc ion. This phenomenon explained how the crystal growth was inhibited by the presence of zinc ion [7].

*3.5. Crystal composition*

The crystal composition was investigated through the characterized $CaSO_4$ scale by SEM EDX. The results consisted of crystal spectrum and composition in atomic percent (at %) also in weight percent (wt %). The crystal spectrum such as shown in figure 6 shows the atomic composition and its peak.

The spectrum shows the structure of the crystal components i.e. Ca, S and O. The existence of Na and Cl ions in the crystal composition could be caused by the trapping of those ions in the inner crystal. The composition of $CaSO_4$ shows that the crystal was pure and the process was done carefully [15]. The crystal was not found in its stoichiometry, which can be caused by the presence of Na and Cl ions in the crystal.

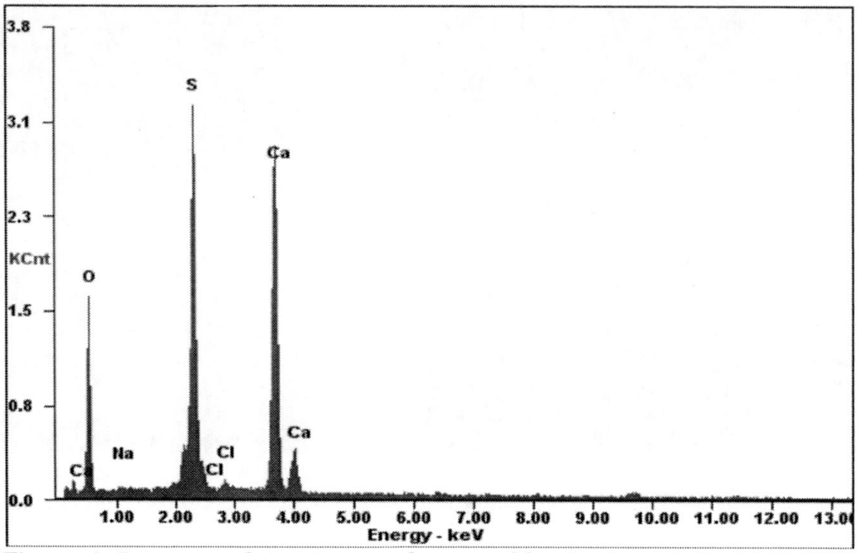


**Figure 6.** Spectrum of $CaSO_4$ crystal composition in the experiment without zinc ion

## 4. Conclusions

Based on a case study on green tea production of which steam generator piping system was hampered by the presence of calcium scale, this laboratory experiment had been employed. The experiment to inhibit $CaSO_4$ scale by a trace amount of zinc ion was successfully conducted. The addition of zinc ion of 5.00 and 10.00 ppm significantly decreased the $CaSO_4$ scale grow rate from 5.799 to 5.501 and 4.950 x $10^{-3}$ g/min. As the investigation, the inhibition rates of $CaSO_4$ scale growth of the experiment at 27 °C were 0.00, 0.298 and 0.829 x $10^{-3}$ g/min for zinc ion addition varied to 0.00, 5.00 and 10.00 ppm. The inhibition rates of $CaSO_4$ scale grow of the experiment at 40 °C were 0.00; 0.082 and 0.255 x $10^{-3}$ g/min for zinc ion addition of 0.00, 5.00 and 10.00 ppm. The inhibition rates of $CaSO_4$ scale grow in the experiment at 50 °C were 0.00, 0.027 and 0.137 x $10^{-3}$ g/min for zinc ion addition of 0.00, 5.00 and 10.00 ppm. The crystal polymorph remained in rhombohedral and showed that the zinc ion addition of 5.00 and 10.00 ppm does not have any effects on its polymorph, but their dimensions were changed. The microanalysis of the $CaSO_4$ crystal show that the crystal component was not in the stoichiometry because of the presence of Na and Cl ions in the inner part of the $CaSO_4$ crystal.

## 5. References

[1] Prisciandaro M, Olivieri E, Lancia A, Musmarra D 2006 Gypsum Precipitation From Aqueous Solution In The Presence Of Nitrilotrimethylnephosphonic Acid, *Ind. Eng.Chem.Res* **45**(2006)pp: 2070-20876

[2] Jamialahmadi M and Muller-Steinhagen H 2007 Heat Ex-changer Fouling And Cleaning In The Dehydrate Process For The Production Of The Phosporic Acid, *Chemical Engineering Research And Design* 2007 **85** pp: 245-255

[3] Dydo P, Turek M, Ciba J, Wandachowicz K, Misztal J 2004 The nucleation kinetic aspects of gypsum nanofiltration membrane scaling, *Desalination* **164** (2004) pp: 41-52

[4] Hernandez A, La Rocca A, Power H, Graupner U, Ziegenbalg G 2006 Modelling The Effect Of Precipitation Inhibitor On Crystalizat ion Process From Well Mixed Over Saturated Solutions In Gypsum Based On Langmuir Volmer Flux Correction *Journal Of Crystal Growth* **295** 2006 pp : 217-230

[5] Helalizadeh A, Müller-Steinhagen, Jamialahmadi M 2006 Application of fractal theory for characterisation of crystalline deposits *Chemical Engineering Science* **61** 2006 pp: 2069-2078

[6] Euvrard M, F Membrey, C Filiatre, A Foissy 2004 Crystallization of Calcium Carbonate *Journal of Crystall Growth* 2004 **265** pp: 322-330.

[7] Ang H, S.Muryanto S, T Hoang 2006 Gypsum scale Formation Control In Pipe Flow System: A Systemic Study On The Effect Of Process Parameters And Additives, Curtin *University Of Technology, Australia,***2006**

[8] Bosbach D, Hochella M F 1996 Gypsum growth in the presence of growth inhibitors: a scanning force microscopy study, *Chemical Geology* **132** pp: 227-236

[9] Lei M, Li P G, Sun Z B, Tang W H 2006 Effects of organic additives on the morphology of calcium carbonate particles in the presence of CTAB *Materials Letters* 60 (2006) 1261–1264

[10] Martinoda A, Euvrard M, Foissy A, Neville A 2008 Progressing the understanding of chemical inhibition of mineral scale by green inhibitors, *Desalination* **220** pp: 345–352

[11] Isopescu R, Mateescub C, Mihaia M, Dabijaa G 2009 The Effect Of Organic Additive On Induction Time And Characteristic Of Precipitated Calcium Carbonate, *Chemical Engineering And Research Design* **2009**.

[12] Petrou A L and Terzidaki A 2014 Calcium carbonate and calcium sulfate precipitation, crystalliza tion and dissolution: Evidence for the activated steps and the mechanisms from the enthalpy and entropy of activation values *Chem. Geo.* **381** (2014) pp 144–153

[13] Chang R and Overby J 2011 *General Chemistry: The Essential Concepts*, 6th edition, Mc. Graw Hill, New York.

[14] Callister WD 2001, Material Science And Engineering. An Introduction, 7th Edition, John Wiley & Sons Inc, USA

[15] Beck R, Hakkinen A, Sorensen D M, Andreassen J P 2009 The effect of crystalization condition, crystal morphology and size on pressure filtration of L-glutamic acid and an aromatic amine *Separation and Purification technology* 2009 **66** pp.549-558.

**Acknowledgements**
The funding of this research was supported by Diponegoro University.

# Low-Temperature Carbothermic Reduction of Indonesia Nickel Lateritic Ore with Sub-Bituminous Coal

**I Setiawan[1,2], S Harjanto[1] and R Subagja[2]**

[1]Department of Metallurgy and Material Engineering, University of Indonesia
[2]Research Center for Metallurgy and Material , LIPI

Email: iwan028@gmail.com

**Abstract.** In this work, the experiment was conducted to investigate the effect of temperature and sub-bituminous coal on upgrading nickel laterites ores by carbothermic reduction. The reduction was carried out in the muffle furnace at a temperature ranging of 800-1100 °C. The produced calcine from reduction step was grounded in vibrating mill, mixed with water and passed on a magnetic separator to separate nickel concentrates from the tailing. The experimental results showed that the concentrate with highest nickel grade was produced at temperature 1000 °C. The nickel grade on concentrate was 5.0 %, with the recovery of nickel was 80.6 %.

**Keywords**: laterite, carbothermic reduction, recovery, grade, concentrate

## 1. Introduction

Nickel laterite ore comprises 73% of the world's nickel resources and will be the dominant source of nickel in the future. With the continuous depletion of high-grade nickel ores such as millerite and niccolite, nickel laterite ores have become the major source for the production of nickel metal. However, currently, only 42% of the world's production of nickel comes from nickel laterite ore because the nickel content in the ore is low [1]. Nickel laterite ores are difficult to concentrate due to their low nickel content and isomorphic goethite or serpentine crystalline structure [2].

The presence of sulfur has a greater effect on the recovery achieved in weathered saprolite compared to fresh saprolite. The sulfur, in general, improves the recoveries of all the laterite species used in this study. Similar improvements in recoveries, however, area so achieved in the saprolite when reducing conditions are introduced prior to the dehydroxylation of the mineral phase [3]. Zhu et al [4] upgraded low-nickel content laterite ores using selective reduction with reduction agent is coal, followed by magnetic separation. These previous works investigated that coal-based reduction followed by magnetic separation is effective for the recovery of nickel and iron from nickel laterite ores. The reaction mechanism of the selective reduction of nickel laterite ore has also been characterized recently. A number of researchers suggested that the selectivity of reduction depends mainly on the reduction atmosphere and silica level; the presence of sulfur significantly promotes ferronickel particle growth to improve nickel enrichment in the metallic phases [1]. However, another researcher reported that the reduction with coal of ferrous minerals in the presence of $Na_2SO_4$ was not showed by the formation of fayalite or forsterite, whereas the decrease of iron metallization was exhibited to the formation of troilite (FeS) [5]. In their research, Jiang et al[2] revealed a ferronickel concentrate with a nickel grade about of 9.52% and a recovery of 84.04 % from high iron grade laterite ore using selective reduction and magnetic separation role, with the addition of 7% $Na_2SO_4$ at 1200°C for 50 min. Therefore, it is necessary to analyze the mechanisms of coal-based reduction of nickel laterite ore. In this study, the reduction roasting of low-grade nickel laterite ore was performed using coal-based reduction with sulfur addition and uses lignite as reducing agents. The effects of roasting reduction temperature on the growth of metallic particles were characterized. Moreover, the reduction mechanisms of nickel were investigated based on the properties of the crystal structure of the nickel laterite ore and the reduction route. The present study aims to provide a basis for the utilization of nickel laterite ores.

## 2. Experimental method

### *2.1. Material*

A low-grade nickel laterite ores from Sulawesi island, Indonesia, was selected as the raw material for experiment. The chemical composition of the nickel laterite ores is presented in Table 1, which shows the nickel content in the nickel laterite ores is 1,71 %. Another material exists in the laterites ores are iron oxide, cobalt oxide, magnesium oxide, chrome oxide, manganese oxide, silicate, and alumina.

**Table 1.** Chemical formulation of nickel laterite ore (mass fraction, %)

| Ni total | $Fe_2O_3$ | $Co_2O_3$ | MgO | $Cr_2O_3$ | $MnO_2$ | $SiO_2$ | $Al_2O_3$ | LOI | Others |
|---|---|---|---|---|---|---|---|---|---|
| 1.71 | 27.04 | 0.08 | 13.40 | 0.19 | 0.68 | 24.64 | 1.93 | 17.21 | 12.65 |

The sub-bituminous coal as reducing agents from Kalimantan province, Indonesia. The proximate analysis of the sub-bituminous coal is presented in Table 2. The coal is classified as a medium quality coal because of its medium fixed carbon content and low ash, volatile matter, and relatively low harmful element (S, P) content.

**Table 2**. Proximate analysis of sub-bituminous coal (mass fraction, %)

| Reducing agents | VM (%) | Ash (%) | Fixed Carbon (%) | S (%) | Moisture (%) | CV (MJ/kg) |
|---|---|---|---|---|---|---|
| sub-bituminous coal | 36.26 | 14.39 | 36.92 | 0.91 | 14.4 | 25.32 |

### *2.2. Carbothermic reduction and magnetic separation test*

The experiment to increase a nickel content in concentrates were carried out according to the following steps i.e. pelletizing step, reduction roasting step of nickel laterite pellet and magnetic separation step. At pelletizing step, nickel laterite ores were first dried at 100 °C in the oven for 24 hours. The dried ores were mixed with a certainties amount of coal (13% weight), and they were grounded in the ball mill. The fine particle of mixed nickel laterite coal was pelletized in the pelletizer to produce nickel laterite pellets. The amount of sub-bituminous coal was added as reduction agent are 2 times (13% weight) of the stoichiometry calculations (~6.5% weight).

At the reduction step, 200 grams of laterite pellets samples were put in a ceramic crucible, and it was heated with a heating rate 10 °C/min, until certainty desired temperature was achieved, after that, the pellet samples were kept heated at a desired temperature for 1 hours in the furnace. Once the reduction experiments were finished, the samples were taken out of from the furnace and cooled to room temperature by water quenching. The quenched materials were mixed with water and passed to the wet magnetic separator to separate nickel concentrates from the tailing. The nickel and iron content in the concentrates were analyzed with atomic Absorption spectrophotometers (AAS), their phases were analyzed with X-ray diffraction method (XRD 7000 Shimadzu), and to investigate the microstructures of reduced nickel laterite materials, those reduced nickel laterite pellets were analyzed by Scanning Electron Microscope (SEM JSM 6390A), in which the qualitative and mapping analysis were carried out by energy dispersive spectrometry (EDS) on an X-ray spectrometer combined with a SEM.

The Nickel and iron recovery in the concentrates were calculated by the following equations:

$$\text{Ni Recovery (\%)} = \frac{(\text{weight of Ni at concentrate})}{(\text{weight of Ni at reduced pellet})} \times 100\% \tag{1}$$

$$\text{Fe Recovery (\%)} = \frac{(\text{weight of Fe at concentrate})}{(\text{weight of Fe at reduced pellet})} \times 100\% \tag{2}$$

And the nickel and iron grade in the concentrate were calculated by the following equations:

$$\text{Ni Grade (\%)} = \frac{(\text{weight of Ni at concentrate})}{(\text{weight of concentrate})} \times 100\% \tag{3}$$

$$\text{Fe Grade (\%)} = \frac{(\text{weight of Fe at concentrate})}{(\text{weight of concentrate})} \times 100\% \tag{4}$$

## 3. Results

### *3.1. X-Ray Diffraction test result of nickel laterites ores*

Figure 1 shows the X-ray diffraction pattern of nickel laterite ores. The crystalline phases of the raw nickel laterite ore mainly consisted of lizardite [$(Mg,Fe)_3Si_2O_5(OH)_4$], Talc [$Mg_3Si_2O_5(OH)_4$], antigorite [$(Mg,Al)_3(Si,Al)_2O_5(OH)_4$], goethite [FeO.OH], quartz [$SiO_2$], and hematite [$Fe_2O_3$].

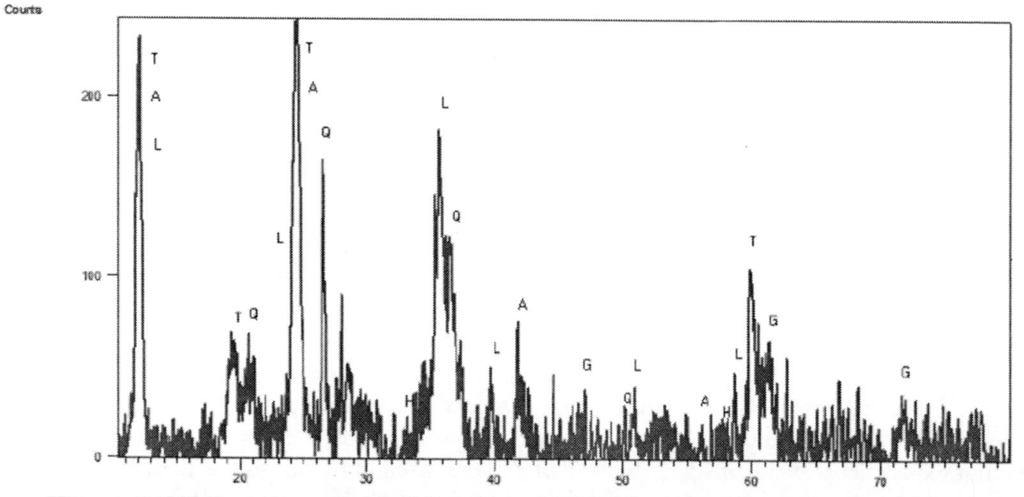


**Figure 1**. XRD pattern of nickel laterite ore (T: Talc, A:Antigorite, L:Lizardite, Q:Quartz, G:Goethite, H:Hematite)

*3.2. Thermal properties of nickel laterites ores and carbothermic reduction with DTA-TG*

In order to investigate the thermal properties of nickel laterites ores, a nickel laterites samples were analyzed by using Differential Thermal Analyzer (DTA). The result of DTA-TG analysis of nickel laterite ores is presented in Figure 2, which shows four steps of the weight loss process. First, the endothermic peak at 78.80 °C on the DTA curve, is attributed to the vaporization of free water. Second, an endothermic effect appears in 293.7 °C on the DTA curve, is attributed to the dehydroxylation of goethite and the desorption of crystalline water. Third, the endothermic peak center at 638.0 °C is caused by the dehydroxylation of lizardite or serpentine group minerals as clinochrysotile and antigorite. Finally, the exothermic peak at 828 °C does not have any associated mass loss resulting from there crystallization of forsterite.

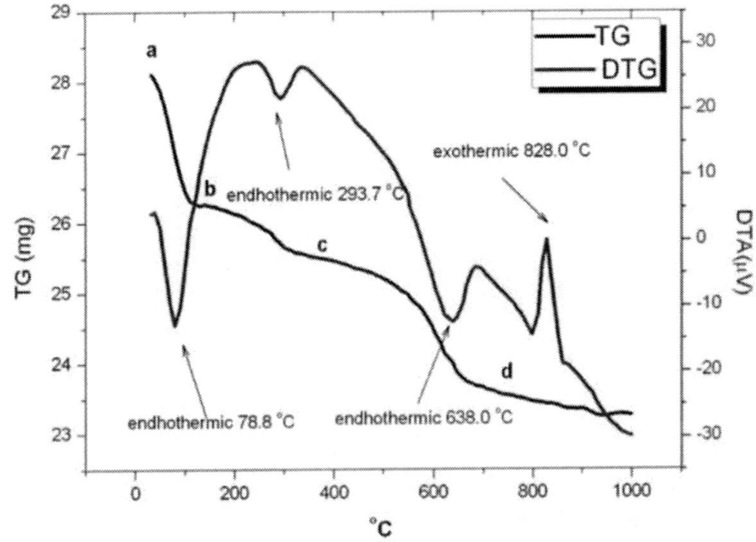


**Figure 2.** DTA-TG curves of the studied laterite ore (*Heating rate 10 °C/min, atmosphere 20% $CO_2$*)

In order to characterize the thermal properties of nickel laterites ores reduced by coal, a nickel laterites samples were analyzed by using Differential thermal Analyzer (DTA) as shown in Figure 3. Thermal events that occur when nickel laterite reduced the use of coal can be predicted using a DTA-TG. Approximately 20 mg sample of laterite ore is mixed with 20% of coal is then heated in a chamber DTA-TG with a heating rate of 10°C/min into 20% $CO_2$ atmosphere.

Based on the results shown in Figure 3, ore will undergo three endothermic reactions [6], namely;

1. At a temperature of 76.5 °C water began evaporation moisture / moisture
2. At a temperature of 344.9 °C $FeO.H_2O$ goethite dehydroxylation reaction occurs;
3. At a temperature of 624.8 °C occurred dehydroxylation reaction carrier magnesium silicate mineral compound into magnesium silicate hydrate;

While the exothermic reaction will undergo two reactions [6], namely;

1. At a temperature of 305 °C which is the burning of coal;
2. At a temperature of 820.4 °C associated with the crystallization of magnesium silicate into forsterite 2MgO. $SiO_2$.

At a temperature of 624.8 °C nickel mineral carrier began to damage of structure so that a transition temperature that can be used for the benefit of the reduction to nickel and iron[7]. Due to the conditions of the transition will be water evaporation magnesium silicate hydrate crystals by way of release of the crystal lattice. At the time of the release of the structure will be open to possible contact with surrounding conditions. The temperature of 820.4 °C is the crystallization of forsterite phase. At this temperature, phase is crystalline forsterite will be having a meeting and regular structure. In this condition everything was trapped in the crystal will be relatively inert to external influences. Similarly, if the element nickel or iron cannot be disassociated from this phase, it will be difficult to have suffered further treatment [8,9].

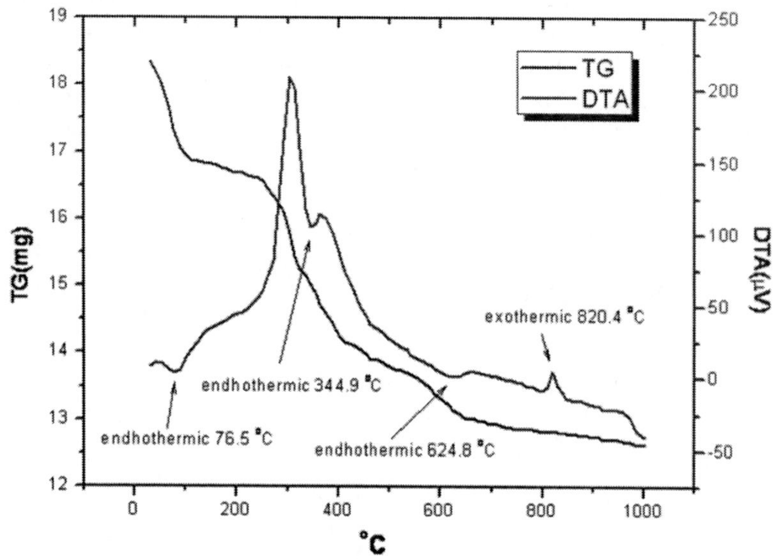


**Figure 3.** DTG–DTA curves of the studied laterite ore with 13% coal. (*heating rate 10°C/min, atmosphere 20% $CO_2$*)

*3.3. The effect of temperature on nickel–iron grade in the nickel concentrates*

In order to investigate the effect of reduction roasting temperatures to the nickel and iron content in the nickel concentrates, the nickel laterite pellets which containing 13 % sub-bituminous coal were roasted at temperatures 800, 900, 1000, and 1100 °C. The nickel and Iron grade which represents the nickel and iron content in the concentrate were determined by using AAS. Their results are presented in Figure 4. This figure shows that the nickel and iron grade in the concentrate increased when the reduction roasting temperatures were increased from 800 °C to 1100 °C. In the case of nickel recovery at concentrates, when the reduction roasting temperature was increased from 800 °C to 1000 °C, the nickel recovery increased, and it decreased when the reduction temperature was increased from 1000 °C to 1100 °C. The same tendencies are also found for iron recovery; it increased when the reduction temperature was increased from 800 °C to 1000 °C, and it decreased when the reduction temperature was increased from 1000 °C to 1100 °C. From this result, it is known that maximum recovery both for iron and nickel occurs at 1000 °C. The maximum grade percentages for Fe occur at 1000 °C and for Ni at 1100 °C. These results are in good agreement with the result of Zhu et al [1] which conducted an experiment for upgrading low nickel content laterite ores using selective reduction followed by magnetic separation. Their result of the experiment shows that the increase in temperature improved the degree of metallization of Ni and Fe. However in the case of Fe recovery, when the reduction roasting temperature was increased from 1000 °C to 1100 °C, Fe recovery decreased due to the formation of fayalite is more dominant than the formation of metallic iron from wustite.

IOP Conf. Series: Materials Science and Engineering **202** (2017) 012019 doi:10.1088/1757-899X/202/1/012019

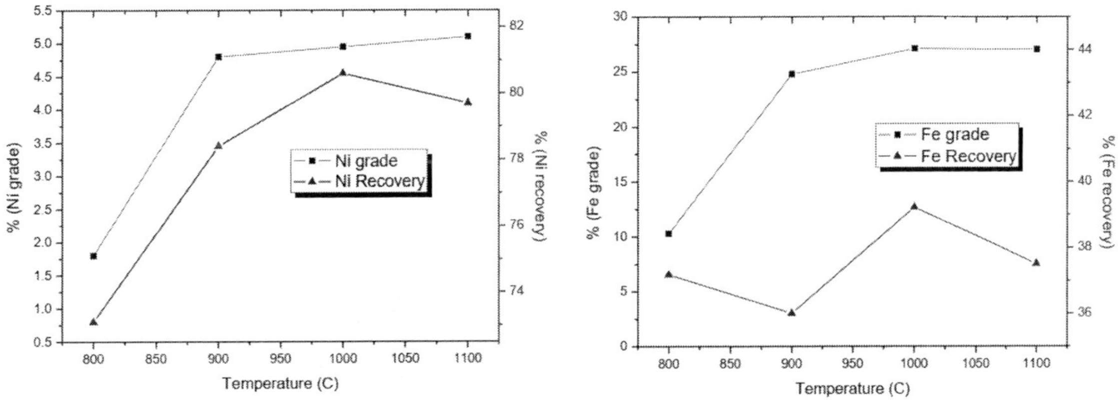


(a) (b)

**Figure 4.** The effect of the reduction temperature on nickel beneficiation (a) and iron beneficiation (b). (reduction with sub-bituminous coal 13% weight)

### *3.4. X-Ray diffraction of reduced pellet*

To determine the phases that form the calcine is cooled quickly with water, dried and then using magnetic separation performed manually and analyzed using XRD well to concentrate, tailings and ore early (Figure 5). The main compounds in the initial sample, namely; lizardite, goethite, and quartz, concentrates consisting of forsterite, quartz and taenit (FeNi), while in the tailings is dominated by quartz, forsterite and hematite.

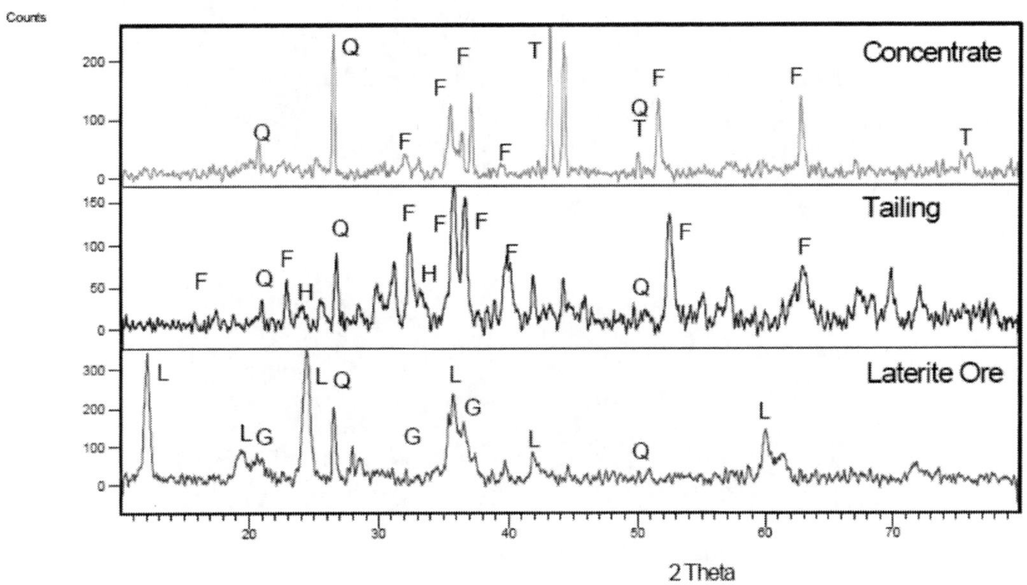


**Figure 5.** XRD graph sample concentrates, tailings and laterite (T = 1000 °C) (Q; quartz, F: forsterite, T; taenite, G: goethite, H: hematite, L; lizardite)

The main types of minerals contained in the ore are lizardite a compound of magnesium silicate hydrate. Another mineral contained quartz and goethite. After reduction and then separated magnetically, the concentrate contained tainite (FeNi) and forsterite phase that still brought when using magnetic separation. Goethite in ore will be transformed into hematite and will be integral as tailings.

IOP Conf. Series: Materials Science and Engineering **202** (2017) 012019 doi:10.1088/1757-899X/202/1/012019

*3.5. Morphology of reduced pellets*
To determine the morphology of the sample surface elemental mapping analysis using SEM-EDX of calcine after magnetic separation.

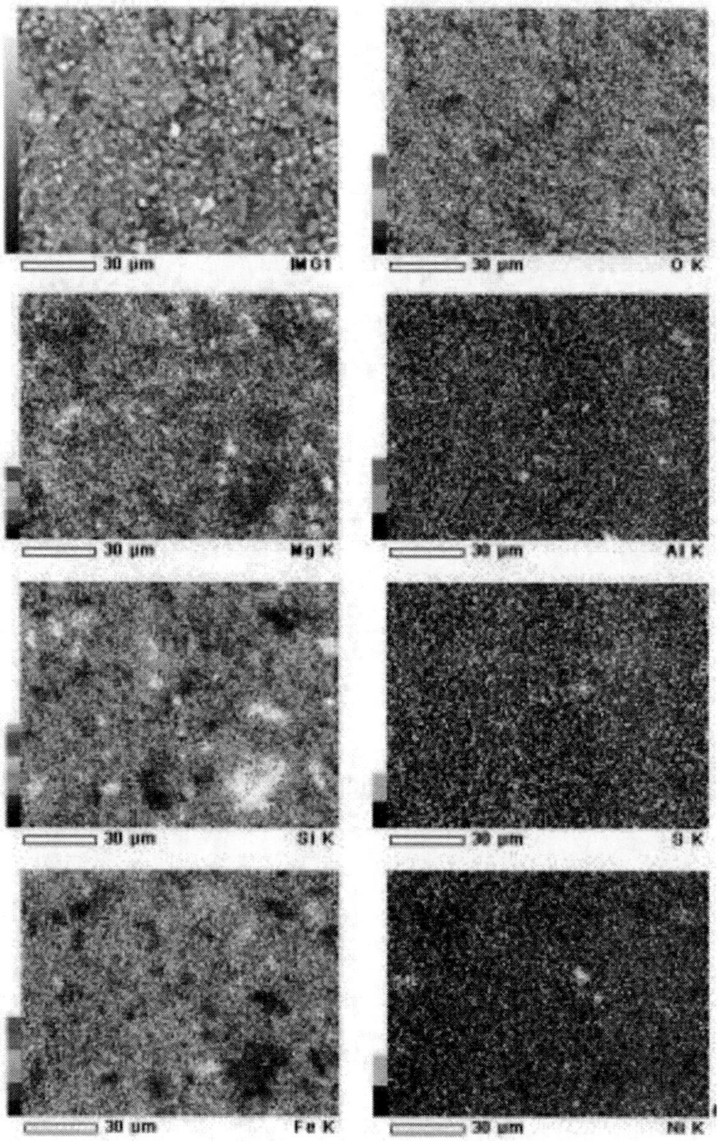


**Figure 6.** Elemental mapping of tailing

Figure 6 is a mapping of samples of concentrate. Based on these images, there are two groups of elements of nickel metal and iron came together to form a ferronickel, and a collection consisting of the elements Mg, Si and O which may compound forsterite or olivine mixed with fayalite. The size of a nickel metal look is still below 10 μm.

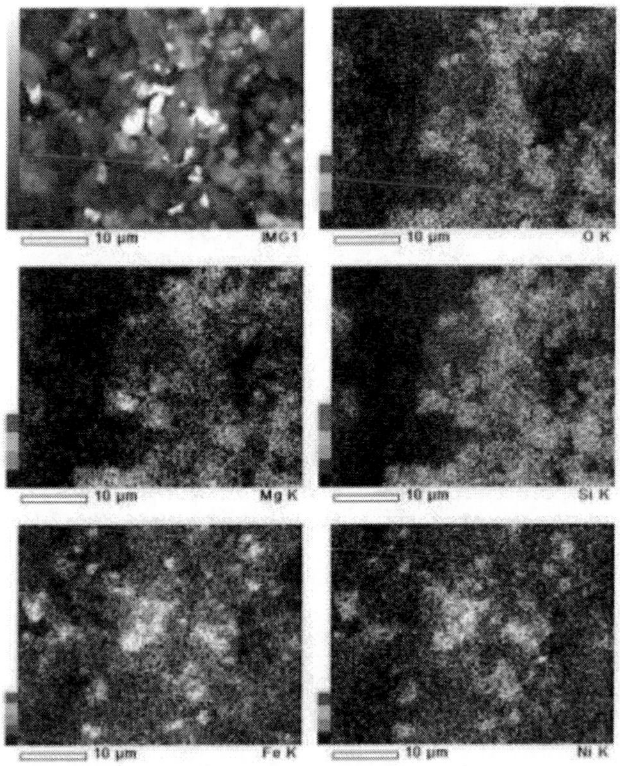


**Figure 7.** Elemental mapping of concentrate

Figure 7 shows an image of the tailings sample mapping; nickel metal found only a small point in the sense of a lower concentration of nickel concentrate. Set of elements consisting of elements Fe and oxygen with the possibility of iron oxide, and the other is a collection of Mg, Si and O form forsterite. The reduction reaction of nickel oxide with carbon in a reaction carbothermic can be estimated as follows:

$$NiO+C \rightarrow Ni+CO \quad (5)$$

$$2NiO+C \rightarrow 2Ni+CO_2 \quad (6)$$

The existence of CO and $CO_2$ is an equilibrium reaction following the Boudouard and equilibrium CO at atmospheric pressure. Both CO and $CO_2$ are in a position of equilibrium when the reduction is run under 1000 °C. But at temperatures above 1000 °C in the C-CO-$CO_2$ almost all in a state of CO so that the reduction reaction of nickel follow the following equations of reaction.

$$NiO + CO \rightarrow Ni + CO_2 \quad (7)$$

$$CO_2 + C \rightarrow 2CO \quad (8)$$

$$NiO + C \rightarrow CO \quad (9)$$

When plotted standard energy from reactions with a temperature above then it will look like in Figure 7.

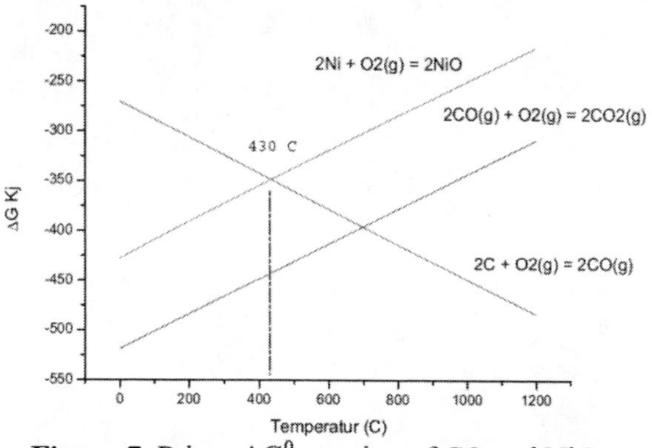


**Figure 7.** Prices $\Delta G^0$ reaction of CO and NiO

It appears that at standard atmospheric pressure state nickel oxide will readily reduce. A temperature of 430 °C is the minimum temperature reduction to metallic nickel.

## 4. Conclusion

The rise in temperature increases the degree of reduction of Fe and Ni metal. In specific circumstances, the reducibility of Ni decreases due to the formation of forsterite which protects the nickel ions from further reduction. At the low-temperature reduction (<1000 °C) carbothermic reduction with sub-bituminous coal as reductant produce an increasing grade of ore from 1.7% to 5%.

## 5. References

[1] Jungah K, Gjergj D, Hideaki T, Katsunori O, Seiji M,Toyohisa F, 2010 Calcination of low-grade laterite for concentration of Ni by magnetic separation. *Minerals Engineering* **23** 282–288.

[2] Mudd G, 2010 Global trends and issues in nickel mining: sulfides versus laterites. *Ore Geol. Rev.* **38** 9–26.

[3] Valix,M, Cheung, W H, 2002b. Effect of sulfur on the mineral phases of laterite ores at high temperature reduction. *Miner. Eng.* **15** 523–530

[4] Zhu D Q, Cui Y, Vining K, Hapugoda S, Douglas J, Pan J, Zheng G L 2012 Upgrading low nickel content laterite ores using selective reduction followed by magnetic separation [J]. *International Journal of Mineral Processing* 106109: 17.

[5] Li G H, Shi T M, Rao M J, Jian T, Zhang Y B 2012 Beneficiation of nickeliferous laterite by reduction roasting in the presence of sodium sulfate [J]. *Minerals Engineering* 32: 1926

[6] Bunjaku, Kekkonen M, Pietila K and Taskinen P 2012 Effect of mineralogy and reducing agent on reduction of saprolitic nickel ores, *Mineral Processing and Extractive Metallurgy (Trans. Inst. Min. Metall. C)* **121**

[7] Kim,J., Dodbiba,G., Tanno,H., Okaya,K., Matsuo,S., Fujita,T., 2010. Calcination of low-grade laterite for concentration of Ni by magnetic separation. *Miner.Eng.* **23** 282–288.

[8] Kawahara M, Toguri J M, Bergman R A 1988 Reducibility of laterite ores, Metallurgical transactions B. **19B** pp. 181-186.

[9] Kim,J., Dodbiba,G., Tanno,H., Okaya,K., Matsuo,S., Fujita,T., 2010. Calcination of low-grade laterite for concentration of Ni by magnetic separation. *Miner.Eng.* **23** 282–288.

# Preparation of Chitosan/Collagen Blend Membranes for Wound Dressing: A Study on FTIR Spectroscopy and Mechanical Properties

**D J Indrani[1,2], F Lukitowati[2,3] and Y Yulizar[4]**

[1]Department of Dental Materials Science, Faculty of Dentistry, Universitas Indonesia
[2]Biomedical Engineering, Multidisciplinary Studies, Universitas Indonesia
[3]Center for the Application of Isotopes and Radiation, National Nuclear Agency, Indonesia
[4]Department of Chemistry, Faculty of Mathematics and Natural Sciences, Universitas Indonesia.

Email: decky@ui.ac.id; deckyji@gmail.com

**Abstract.** The effect of different gamma-ray irradiations on the functional groups and the mechanical properties of chitosan, collagen, and chitosan/collagen blend have been studied. Commercially available chitosan and extract of collagen from bovine tendon by acid dissolution method were employed. Chitosan, collagen and chitosan/collagen blend membranes were prepared using the solvent evaporation method. The dried membranes were subjected to gamma-ray irradiation using 0, 15, and 25 kGy. The membranes then characterized by Infrared spectroscopy and measured in tensile and elongation at break. The irradiated chitosan membranes displayed reduced intensity of -OH, -CH, -NH and -C-O-C groups, whereas, the collagen membranes revealed amide I, II and III groups. The irradiated chitosan/collagen (50/50%) blend membranes demonstrated the decreased intensities in the –OH and amide groups. The application of gamma-ray irradiation had produced chitosan/collagen blend membranes with a tensile strength and elongation at break between the chitosan and collagen membranes. It was further determined that the decrease in the mechanical properties of the membranes was likely to be due to some changes of the functional groups.

**Keywords:** Chitosan, collagen, gamma-ray, tensile strength, elongation.

## 1. Introduction

Skin plays a major role in body protection. Severe damage to the skin may cause disruption or injury to skin structure and function, and gives the damage to intrinsic skin barriers [1]. In healing the wound, an inflammatory response occurred; the cells behind the dermis start building collagen, and it continues up to the end of establishing the epithelium. The healing process of skin damage may be sufficient with the promotion of wound dressing [1-3]. Wound dressings prepared from natural polymers have the capacity to uptake water [3-4]. The moist environment of wound dressings is aimed to maintain the existence of moisture around the wound, so it can accelerate tissue restoration and alleviates pain [1].

Natural polymers are widely acceptable for biomedical applications. Considerable research interest is focused on blend polymers that may produce new polymeric materials for which has little been existing as a blend of nature. Nowadays, it has become common to blend biopolymers to produce new materials with better properties [5-7]. The blend of biopolymers, such as chitosan and collagen (chitosan/collagen), have been widely applied as wound dressings about their potentially beneficial biological properties. As an addition, spectroscopic techniques have been used to better identify the interactions and structural changes in the biopolymer network [8-9].

Chitosan is a nitrogenous polysaccharide deacetylated from chitin, found in cell walls of crustaceans, such as crabs [10]. Chitosan contains aliphatic (–CH), eter (–C-O-C) and hydroxyl (–OH) groups of the primary alcoholic group and a number of basic amino amide (–NH) resulting in cationic polyelectrolyte properties and act as an ion exchanger [11-12]. Similar to many cationic polymers, chitosan can disrupt the outer membrane of bacteria that possesses anions, and subsequently hamper the growth of infectious substances [13]. Wound dressing from chitosan may prevent penetration of bacteria through the membranes [13-14]. However, the hard and inelastic linear polysaccharide structure of chitosan is known to cause the limitation of its application [10]. One way to overcome the poor mechanical strength of this biopolymer is to modify the physical properties of the membrane.

Collagen is the substance that holds the whole body together and found in the extracellular matrix. Collagen consists of a repeating amino acid that constructs the connective tissue throughout the body, acting as the triple helix structure in tissues. An H atom of the side chain of glycine, links the NH peptide bond (amide II) and CH (amide III) of a glycine to a peptide carbonyl (C=O; amide I) group in an adjacent polypeptide and helps hold the three chains together and, therefore, can be involved in all 3 phases of the wound-healing cascade [15-16]. Collagen is the most abundant protein found in mammals; parts of an animal that can be the main sources of collagen for biomedical uses are skin or tendon, such as from rat-tail or bovine [17-18]. Collagen builds a network of macromolecules that provide structural support, in relation to strength and degree of elasticity. For clinical applications, the use of non-sterilized biomedical materials is associated with the risk of infectious disease transmission. Wound dressing needs to be perfectly sterile to avoid the transmission of any illness [2]. The physical methods including UV or gamma irradiation have been shown to be effective in sterilizing material for biomedical applications. Gamma-ray irradiation is preferred for sterilization of biomaterials due to some advantages. Gamma irradiation, which has less atmospheric pollution, is an easy and cost-effective process, is regarded for sterilization and the application is without the introduction of chemical reagents [19-21]. Gamma irradiation doses required for sterilizing medical devices are 15 to 25 kGy (ISO (11137) [22]. Gamma irradiation at 25 kGy was found to be suitable for sterilization of the dressings, and infrared spectral scanning has shown that papain was stable on gamma irradiation at 25–35 kGy [21]. Studies have also shown that gamma-ray irradiation may alter the mechanical properties of some biopolymers [19, 23]. Growing irradiation doses up to 60 kGy applied to gelatin films have raised their tensile strength, although, the chitosan/gelatin blend films lowered the elongation at break [23].

In relation to their potentially beneficial biological properties for wound dressings, chitosan and collagen are mostly environmentally friendly because they do not produce harmful residues. Previous research has shown the extractions of chitosan from crabs and collagen fro bovine tendon to produce chitosan/collagen membranes [3, 14]. Here, chitosan/collagen blend membranes with or without gamma-irradiation up to 25 kGy have shown a capacity to uptake water [3] and also demonstrated the prevention of bacteria penetration power through the blend membranes [14]. On the other hand, the mechanical strength of wound dressings is of great importance because they give influence on the performance of the result. However, they have not been studied. Tensile strength and elongation at break are necessary to be investigated for unirradiated (initial) and irradiated wound membranes of single and blend components. Therefore, the purpose of the present study was to innovate wound dressing prepared with chitosan/collagen membranes by enhancing the properties of those obtained from chitosan or collagen membranes alone and analyze the effect of gamma-ray irradiation doses on their functional groups, tensile strength, and elongation at break.

## 2. Experimental Method

### *2.1. Materials*
Commercial medical grade chitosan, extracted from crabs having 90 % degree of deacetylation (DDA) and molecular weight of 170 kDa was obtained from the Biotech Surindo (Cirebon, Indonesia). The bovine tendon was available from the Research Laboratory of the National Nuclear Agency, Indonesia. Reagents, i.e. glacial acetic acid (100 % pure) and sodium chloride were purchased from Merck (Darmstadt, Germany) and used without any further purification.

### *2.2. Methods*
#### *2.2.1 Chitosan membrane preparation*
In the attempt to prepare chitosan, methods as provided in [5] and [7] were selected. Firstly, the chitosan powder was dissolved in 0.7 M acetic acid under a constant stirring of 750 rpm for 1 hour at 40 °C and continued without stirring at room temperature for 12 h to obtain a chitosan solution. This membrane-forming solution was cast in $7.5 \times 7.5$ mm$^2$ and subjected to solvent evaporation at 25 °C for 48 h in order to obtain chitosan membranes. Each dried chitosan membrane was peeled off manually using a spatula and then stored in a polyethylene bag before characterization.

#### *2.2.2.Collagen membranes preparation*
Collagen was obtained by enzymatic hydrolysis and chemical hydrolysis [17-18]. Collagen achieved by biological processes used the addition of enzymes [6]; which is more promising, but more expensive. This study used the chemical hydrolysis process, which is more commonly used. Collagen was extracted from bovine tendon using the acid-soluble method following the study of [5]. After all washes, clean collagen bundles were immersed in a 0.7 M acetic acid solution, then remained at -4 °C for 12 h and after that, they were folded until they gelled. The gel was reconstituted by adding 4M NaCl and then stirred at 750 rpm for approximately 1 hour, until it was white cotton-like. Next, the fibers were separated from the NaCl solution. The precipitated fibers were washed using a phosphate buffer solution to neutralize the pH. They were then subjected to freezing at -20 °C for 12 h and immediately freeze-dried at -109 °C at $5 \times 10^5$ bar for 24 h. The dried collagen was dissolved in 0.7 M acetic acid and then stirred at room temperature to produce the collagen solution. The preparation of collagen membrane was identical to the previously described preparation of the chitosan membranes.

#### *2.2.3. Chitosan/collagen blend membrane preparation*
Before irradiation, the chitosan, collagen-chitosan/collagen blend membranes were packaged in polyethylene film to prevent oxidation during the process and to maintain sterility after irradiation. Chitosan and collagen solutions obtained from the previous step were mixed in the ratio of 50:50 and the stirring was maintained at 750 rpm at room temperature for 24 h. The preparation of the chitosan/collagen membrane was identical as previously described preparation of the chitosan membranes.

#### *2.2.4. Irradiation Treatment*
Before irradiation, the chitosan, collagen-chitosan/collagen blend membranes were packaged in polyethylene film to prevent oxidation during the process and to maintain sterility after irradiation. The membranes were irradiated under a $^{60}$Co gamma-ray source using a linear electron accelerator of 7.9 kGy/h dose rate at room temperature. Doses required for irradiation in this study were 15 and 25 kGy. The membranes without irradiation were used as controls.

#### *2.2.5. FTIR spectroscopy*
The Fourier Transform Infrared (FTIR) spectrum from each membrane was obtained using a spectrometer UVmini-1240 (Shimadzu, Japan). The transmittance spectral profiles were determined at the wavenumbers ranging from 4000 cm$^{-1}$ to 400 cm$^{-1}$.

IOP Conf. Series: Materials Science and Engineering **202** (2017) 012020 doi:10.1088/1757-899X/202/1/012020

### *2.2.6. Mechanical properties*

The mechanical properties of chitosan, collagen and chitosan/collagen blend membranes were assessed by the measurements of tensile strength and elongation at break. The measurements were carried out according to ASTM Standard D882-91 [23]. The membranes were cut precisely to get rectangular (2.5 × 8 $cm^2$) test piece specimens. The specimens were placed in an extension grip of the Instron universal testing machine (Groove City, Pennsylvania, U.S) and stretched uniaxially with a 300 N load cell at a rate of 10 mm/min until breaking. The stress and strain curves obtained from the machine were used to determine the tensile strength and elongation at break [24]. The tensile strength was measured using the equation (1)

$$\sigma = F / A \quad (1)$$

Where $A$ is the area of the specimen, and the elongation at break was calculated using Equation (2).

$$E = \left( \frac{L_1 - L_0}{L_0} \right) 100\% \quad (2)$$

Where $L_0$ is the specimen area before the test and $L_1$ is the extension when broken up. The measurements were carried out at room temperature, and five specimens were tested for each membrane.

## 3. Results and Discussion

### *3.1. Data*

This study has shown chitosan, collagen and chitosan/collagen blend membranes regarding their functional groups and their mechanical properties. Visual inspection to the membranes showed that all membranes were in good homogeneity. FT-IR profiles for the non-irradiated and irradiated chitosan, collagen and the chitosan/collagen blend membranes in spectral ranged from 4000-400 $cm^{-1}$, $X$ axis, wavenumbers in $cm^{-1}$, $Y$ axis, the transmittance in % are presented in Figures 1, 2, and 3.

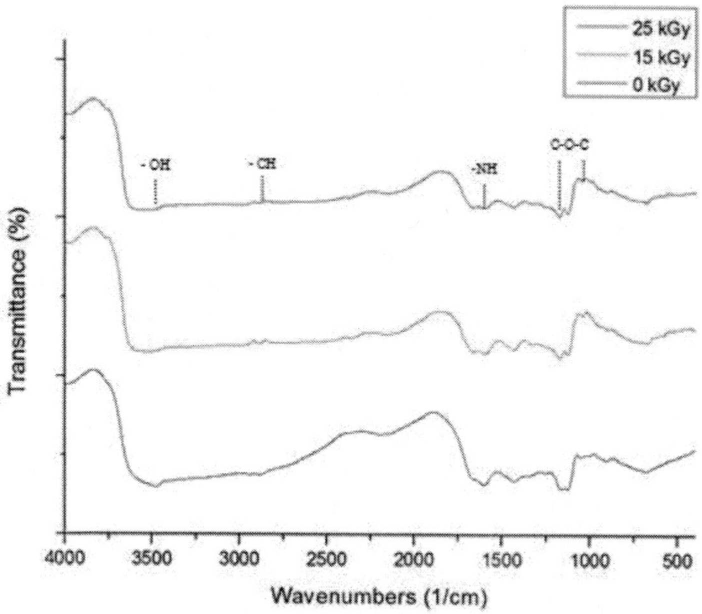


**Figure 1.** FTIR spectra for chitosan membranes with gamma-ray irradiation doses of 0,15 and 25 kGy

Figure 1 displays the spectra of characteristic peaks obtained from the chitosan membrane with and without gamma-ray irradiation that was similar. However, there were some changes in the peaks. The irradiated (0 kGy) chitosan membranes showed a peak of around 2993 $cm^{-1}$ which corresponded to –CH of the aliphatic groups, a broadband around 3458 was due to –OH groups, whereas, the characteristic peaks were located at 1664 and 1041 $cm^{-1}$ which represented the–NH (amide) and C–O–C groups, respectively. When gamma-ray irradiation was given in the doses of 15 or 25 kGy, the intensity of the peaks remained unchanged, with the exception of the –NH group that had its intensity decreased, as shown by visual inspection of the spectra.

The FTIR spectra in figure 2 showed characteristic peaks of the collagen membranes. Three different bands in the collagen membranes in position for amides I, II, and III characterizing the peptide bond. The peak of 1685 $cm^{-1}$ was mainly associated with C=O (amide I) groups, the peak of 1573 appears from C–N, and N–H (amide II) groups, and the peak of 1249 $cm^{-1}$ corresponded to $CH_2$ (amide III) groups of glycine backbone. After the implementation of 15 or 25 kGy gamma irradiation, they did not show any changes in their intensities.

The spectra of the chitosan/collagen blend membrane without gamma-ray irradiation can be seen in Figure 3. The chitosan/collagen blend membranes showed peaks from both the chitosan and collagen membranes that demonstrated changes. Three different bands showed a general shift in the positions of approximately 1692, 1597, and 1265 $cm^{-1}$ due to amide I, II and III, respectively, as compared to those in the collagen membrane alone. Visual inspection of the spectra of the blend membranes irradiated with 15 kGy indicated a lower intensity of –OH peak, for which the heights were less evident when considering a combination spectrum obtained from differences in absorbance values and higher irradiation at 25 kGy showed further reduced the intensity of the peaks.

In terms of mechanical strength, figure 4 summarized the tensile strength for the resulting chitosan, collagen and chitosan/collagen blend membranes with variations in the irradiation doses of 0, 15 and 25 kGy.

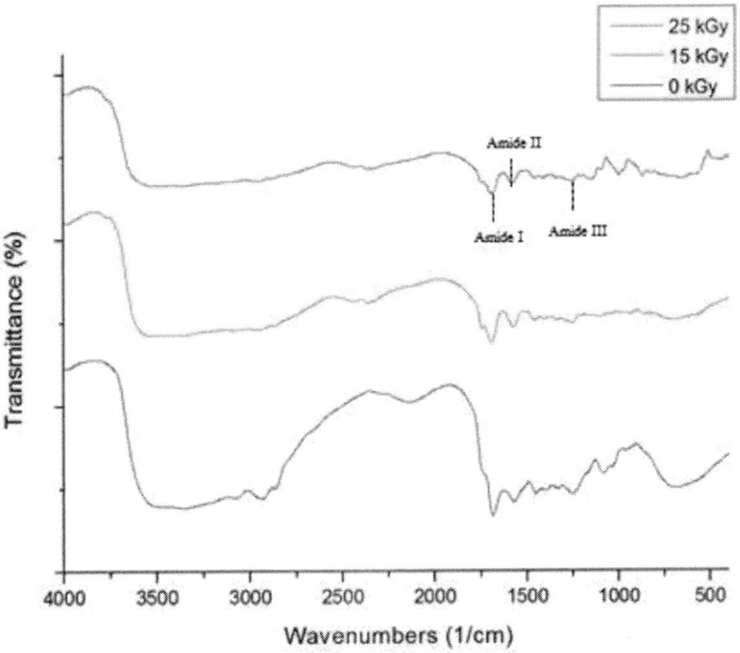


**Figure 2.** FTIR spectra for collagen membranes with gamma-ray irradiation doses of 0, 15, and 25 kGy

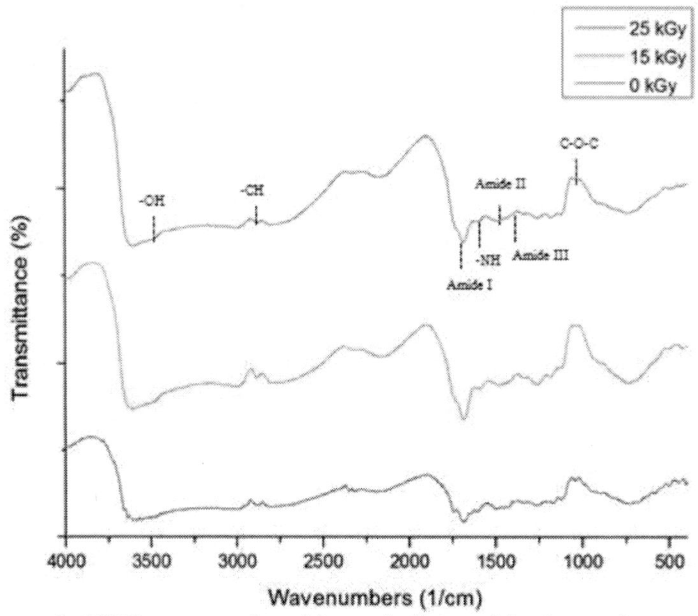


**Figure 3.** FTIR spectra for chitosan/collagen blend membranes with gamma-ray irradiation doses of 0, 15, and 25 kGy

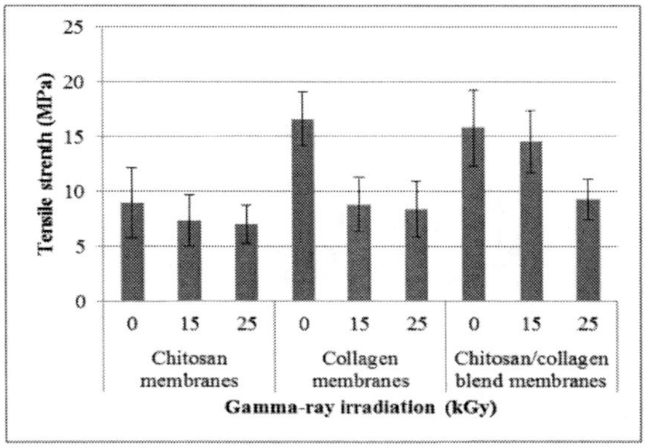


**Figure 4.** Tensile strength of chitosan, collagen and chitosan/collagen blend membranes irradiated with gamma-ray irradiation doses of 0, 15, and 25 kGy

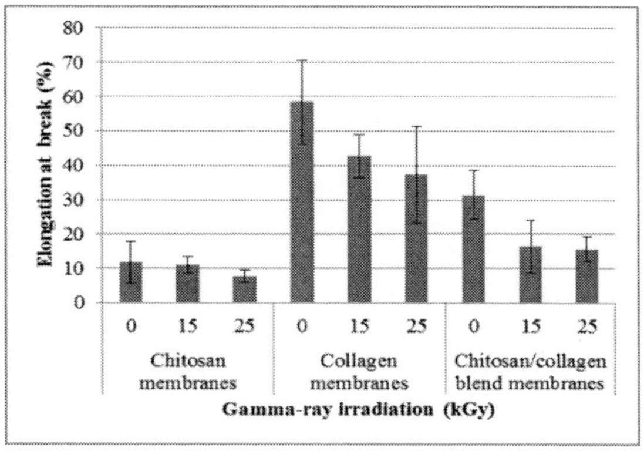


**Figure 5.** Elongation at break of chitosan, collagen and chitosan/collagen blend membranes irradiated with gamma-ray irradiation of 0, 15, and 25 kGy

As shown in Figure 4, the chitosan membranes without irradiation (0 kGy) showed the lowest tensile strength (8.97 ± 3.21MPa). In contrast, the collagen had the highest tensile strength (16.61 ± 2.45MPa), whereas, chitosan/collagen blend was between the chitosan and the chitosan/collagen (15.8 ± 3.48 MPa). With 15 kGy, the tensile strength of the chitosan, collagen and chitosan/collagen membranes decreased by 18.1%, 46.8% and 8.0% respectively, whereas, with 25 kGy, further decreased by 21.6%, 49.5% and 41.3%, respectively, when compared to that of each value without irradiation (0 kGy).

Regarding the elongation at break obtained from the chitosan, collagen and chitosan/collagen blend membranes against gamma irradiation doses, the values are plotted in Figure 5. In Figure 5, the chitosan membranes without irradiation demonstrated the lowest elongation at break (11.9±6.1 MPa), the collagen had the highest elongation at break (58.5±12.3 MPa), whereas, chitosan/collagen blend was between the chitosan and the chitosan/collagen membrane blend (31.4±7.1MPa). Similar to those revealed from the tensile strength of all membranes. With irradiation of 15 kGy, the value from the chitosan, collagen and chitosan/collagen membranes decreased by 6.7%, 27.0%, and 47.6%, respectively, and a higher irradiation of 25 kGy occurred with lowering values, i.e. 33.9%, 36.1%, and 49.8%, respectively, in comparison to those without irradiation.

### *3.2. Discussion*

Immersing the bovine tendon in an acidic solution has caused the solution to penetrate throughout the causing material swelling to two or three-time-bigger compared to its initial volume. The acetic acid used in this study was a common solvent to dissolve pure collagen partially. The non-cross-linked collagen, which was sensitive to the pH of the solvent, experienced dissociation. It may cause the non-covalent molecular bonds to disrupt.

The FTIR spectra of the chitosan/collagen blend membranes showed some peak changes in the characteristic peaks. The shifting of the amides I, II, and III groups of the blend membranes might indicate a new linkage between the chains contained the chitosan and collagen. A molecular interaction was likely to occur due to hydrogen bonding (H-O-H) and to electronic interaction between the cationic chitosan ($-NH_3^+$ groups) and neighboring anionic collagen ($-COO^-$ groups); these might produce a network of the blend membranes. Regarding the chitosan/collagen blend membranes irradiated with 15 kGy, the observed shifting in amides I, II, and III showed by lowering the intensity of each peak, indicating changes in the structure. With a higher irradiation of 25 kGy, further lowering of the peak intensities was likely to relate to further changes in the structure. This result was consistent with that of reported in [25] that utilized chitosan/starch films irradiated with gamma-ray.

The tensile strength or the elongation at break of the chitosan/collagen blend membranes were between each of the chitosan and the collagen membrane alone. The chitosan membranes showed a lower tensile strength or elongation at break when compared to those of the collagen membranes. It meant that the chitosan membranes were mechanically weaker than the collagen membranes. When chitosan was combined with collagen, the chitosan/collagen blend membranes could improve its tensile strength or its elongation at break. The improved values were probably related to the formation of attractive interactions between the chitosan and the collagen in the blend membranes. These attractive interactions were likely to cause additional interactions between chitosan and collagen, resulting in more stable network polymers, consequently, some enhancement in the strength. This meant that the collagen has a larger role in the blend membrane compared to that of chitosan. It agrees with the work reported [19] that showed the mechanical properties of chitosan membranes were improved by blending it with gelatin membranes. A previous study performed in [9] showed that there was also an increase in their tensile strength when collagen was added to chitosan to produce chitosan/gelatin blend membranes. Also, studies also showed that the mechanical properties of protein-based membranes are generally better than those of polysaccharide-based membranes [26].

The improvement of tensile strength and elongation at break of the blend membranes were likely to be due to an addition of a plasticizer. It is known that the addition of a plasticizer, in this case: collagen, can modify some of the physicomechanics of the blend membranes. In a previous study, [9] showed that there was also an increase in the tensile strength when collagen was added to chitosan to produce their blend membranes. Similarly, other researchers showed the mechanical properties of chitosan membranes raised by blending them with gelatin membranes [19].

When gamma-ray irradiation was applied, the chitosan, collagen or chitosan/collagen blend membranes demonstrated decreasing effects on their tensile strength and elongation at break. Regarding the characteristic peaks, this can be related to the lowering intensities of the amide I, II, and III peaks in the irradiated chitosan/collagen blend membranes. With irradiation of 15 kGy application, the main chains in each biopolymer might experience radiation degradation. After attaining the 25 kGy, the cross-linking probably showed further reduction and degradation increased gradually and the structure might have degraded into fragments. Observing the work reported in [9] of up to 5 kGy

doses, the tensile strength arose with the irradiation. However, at a higher dose of 10 kGy the tensile strength decreased. The authors proposed that the initial increase using the 5 kGy was due to a crosslink of both polymers. Then at a higher dose, the degradation of the chitosan chain (C-O-C) occurred. The irradiation might also destroy hydrogen bonds between those two macromolecules.
On the chitosan/collagen blend membranes, the values of the mechanical strength were between the chitosan and collagen membranes. The rise in the tensile strength and elongation at break of the blend membranes were likely to be due to the addition of a biodegradable plasticizer, in this case, the collagen. Therefore, it is required to overcome the relatively low mechanical strength of the chitosan membranes. The application of gamma-ray irradiation to the chitosan/collagen blend membranes, nevertheless, has lowered both tensile strength and elongation at break. However, when compared to the collagen membranes, the increase of both measurements in chitosan/collagen blend membranes may be an advantage. However, since chitosan has an antimicrobial property, it seemed that this added value might be produced from a chitosan/collagen blend membrane. Therefore, the chitosan/collagen blends may be applied as an alternative due to the combination of bacteriostatic of chitosan and the relatively high mechanical strength of collagen.

## 4. Conclusion

The addition of the collagen has improved both tensile strength and elongation at break of the chitosan/collagen blend membranes when compared to the chitosan membranes. It is supported by the infra-red spectroscopy results, which showed the shifting of the amide groups I, II, and III, in carbonyl and amino groups of collagen. The applications of gamma-ray irradiation with the doses of 15 or 25 kGy to the chitosan/collagen blend membranes has lowered the tensile strength and elongation at break. However, they were higher when compared to the chitosan membranes.

## 5. References

[1] Boateng J and Catanzano O 2015 Advanced therapeutic dressings for effective wound healing-a review *J. Pharm. Sci.* **104(11)** 3653–3680

[2] Laurel M, Tania J and Phillips T J 2016 Wound healing and treating wounds differential diagnosis and evaluation of chronic wounds *J. Am. Acad. Dermatol.* **74(4)** 589-605

[3] Lukitowati F and Indrani D J 2016 Water uptake of chitosan/collagen blend membranes for wound dressings with gamma-rays irradiations exposed to water *The 2nd Materials Research Society of Indonesia Meeting, Bandung*

[4] Jina S G, Yousafa A M, Kima K S, Kima D W, Kima D S, Kima J K, Yongc Y C, Yound Y S, Kimc J O and Choi H G 2015 Influence of hydrophilic polymers on functional properties and wound healing efficacy of hydrocolloid based wound dressings *Int. J. of Phar.* **501** 160-166

[5] Ramasamy P and Shanmugam A 2015 Characterization and wound healing property of collagen–chitosan film from Sepia kobiensis *Int. J. Biol Macromol.* **74** 93-102

[6] Huang R, Li W, Lu X, Lei Z, Bian Y, Deng H, Wang H, Li J and Li X 2015 Biomimetic LBL structured nanofibrous matrices assembled by chitosan/collagen for promoting wound healing *Biomaterials* **53** 58-75

[7] Montoya M, Moscoso J, Jatomea M, Ortega H S, Sandez O R, Lopez J L C, Rios E M and Brauer J M E 2010 Jumbo squid (Dosidicus gigas) mantle collagen: extraction, characterization, and potential application in preparation of chitosan-collagen biofilms *Journal Bioresource Technology* **101** 4212-4219

[8] Rimdusit S, Somsaeng K, Kewsuwan P, Jubsilp C and Tiptipakorn S 2012 Comparison of gamma radiation crosslinking and chemical crosslinking on properties of methylcellulose hydrogel *Engineering Journal* **6(4)** 15-28

[9] BenBettaieb N, Karbowiak T and Bornaz S 2014 Spectroscopic analyses of the influence of electron beam irradiation doses on mechanical, transport properties and microstructure of chitosan-fish gelatin blend films *Food Hydrocoloids* **48** 37-51

[10] Prabu K and Natarajan E 2012 Bioprospecting of shells of crustaceans *International Journal of Pharmacy and Pharmaceutical Sciences* **4(4)** 1-3

[11] Casettari L, Vllasaliu D, Castagnino E and Illum L 2012 PEGylated chitosan derivatives: synthesis, characterizations and pharmaceutical application *Polymer Scienc.* **37(5)** 659-685

[12] Li P, Poon Y F, Li W, Zhu H Y, Yeap S H, Cao Y, Qi X, Lamrani M, Beuerman R W, Kang E T, Mu Y, Li C M, Chang M W, Leong S S J and Chan-Park M B 2011 A polycationic antimicrobial and biocompatible hydrogel with microbe membrane suctioning ability *Nat Mater.* **10(2)** 149–156

[13] Dai T, Tanaka M, Huang Y Y and Hamblin M R 2011 Chitosan preparations for wounds and burns: antimicrobial and wound-healing effects *Expert Rev Anti Infect Ther.* **9(7)** 857–879

[14] Indrani D J, Lukitowati F and Darwis D 2016 Effect of gamma-ray irradiation on bacterial penetration power of chitosan/collagen blend membranes for wound dressing *Journal of International Cental and Medicine Research* **9(3)** 202-206

[15] Fleck C A and Simman R 2010 Modern collagen wound dressings: function and purpose *J Am Col Certif Wound Spec.* **2(3)** 50–54

[16] Rastogi S, Modi M and Sathian B 2009 The efficacy of collagen membrane as a biodegradable wound dressing material for surgical defects of oral mucosa: a prospective study *Journal of Oral and Maxillofacial Surgery* **67(8)** 1600-1606

[17] Silvipriya K S, Kumar K K, Bhat A R, Kumar B D, John A and Iakshmanan P 2015 Collagen: animal sources and biomedical application *Journal of Applied Pharmaceutical Science* **5(3)** 123-127

[18] Techatanawat S, Surarit R, Suddhasthira T, Siribang-on and Khovidhunkit P 2010 Type I collagen extracted from rat-tail and bovine achilles tendon for dental application: a comparative study *Asian Biomedicine* **5(6)** 787-798

[19] Huq T, Khan A, Dussault D, Salmieri S, Khan R A and Lacroix M 2011 *Effect of gamma radiation on the physico-chemical properties of alginate-based films and beads* Radiation Physics and Chemistry **81** 945-948

[20] Miller A 2011 *International Standards for Radiation Sterilization of Medical Devices* http://www.iaea.org/inis/collection/NCLCollectionStore/_Public/39/066.pdf (Accessed 10 October 2015)

[21] Singh D and Singh R 2012 Papain incorporated chitin dressings for wound debridement sterilized by gamma radiation *Radiation Physics and Chemistry* **81** 1781–1785

[22] ISO 11137-2:2013 Sterilization of health care products-radiation: establishing the sterilization dose

[23] ASTM D-882-91 Standard Test Methods for Tensile Properties of Thin Plastic Sheeting

[24] Sionkowska A, Wisniewski M, Skopinska J, Kennedy C J and Wess T J 2004 Molecular interactions in collagen chitosan blends *Biomaterials* **25(5)** 795-801

[25] Tuhin M, Rahman N, Haque M E, Khan R A, Dafader N C, Islam R, Nurnabi M and Tonny W 2012 Modification of mechanical and thermal property of chitosan-starch blend films *J Radiation Physics and Chemistry* **81(10)** 1659-1668

[26] Vieira M G A, Da Silva M A and Dos Santos L O 2014 Natural-based plasticizers and biopolymer films: a review *European Polymer Journal* **47(3)** 254-263

**Acknowledgements**
This study was supported by the Departement of Research, Technology and Higher Education, Indonesia, through Universitas Indonesia. The authors would also like to thank the National Nuclear Energy Agency, Indonesia, for the use of the gamma-ray irradiator plant, and Neil Trout of the School of Chemical Engineering, University of South Australia for the valuable discussions.

# Solid State Reaction Synthesis of Si-HA as Potential Biomedical Material: An Endeavor to Enhance the Added Value of Indonesian Mineral Resources

**Hartatiek, Yudyanto, S D Ratnasari, R Y Windari and N Hidayat**

[1]Department of Physics, Faculty of Mathematics and Natural Sciences, Universitas Negeri Malang (UM), Jl. Semarang 5, Malang 65145, Indonesia

Email: hartatiek.fmipa@um.ac.id

**Abstract**. In recent years, one of the most prominently investigated materials is hydroxyapatite (HA). It is because of its excellent properties for medical applications, essentially related to orthopedic. Also, the introduction of other materials to HA becomes another research focus of many leading scientists. In this present study, silicon with various concentrations was introduced, by means of solid state reaction route, to HA forming Si-HA. The crystal structure properties of the as-prepared samples were evaluated by X-ray diffractometer (XRD). Fourier Transform Infra Red (FTIR) spectroscopy data collection and analysis were done to investigate the functional groups within the samples. The microstructural characteristics as well as elemental mapping of the samples were captured by scanning electron microscopy and energy dispersive x-ray spectroscopy (SEM-EDX). Vickers hardness test was also conducted to investigate the hardness properties of the samples. Furthermore, *in vitro* characterization-based bio resorbability of the samples in a simulated body fluid were also described. This study revealed that Indonesian limestone can be utilized as the raw material for synthesizing HA. The silicon has been successfully incorporated into phosphate site of the HA crystal. Conclusively, the Si-HA reported in this study shows good bioresorbability characteristic.

**Keywords**: Si-HA, solid state reaction, biomedical material.

## 1. Introduction

Study of biomedical materials for bone implantation has been extensively made for the last decades. One of the most securing materials for bone repair and bone regeneration is the well-known hydroxyapatite ($Ca_{10}(PO_4)_6(OH)_2$, HA) [1,1] that is attributable to its resemblance to the human bone [2,3]. Many approaches have been conducted to produce HA, for instance by means of hydrothermal [4], sol-gel precipitation [5], ultrasound irradiation [6], and microwave-assisted approaches [7]. In addition, solid-state reaction technique is, in spite of everything, recognized as one of the most effective methods for HA production [8–11]. However, a study exclusively investigating the use of natural calcium-based materials to synthesize HA is rarely conducted. This breakthrough is important not only to reduce the cost of HA production but also to provide an added-value of the abundant natural resources, i.e. limestone.

Limestone is one of the most abundantly naturally-occurring minerals [12] consisting mainly of calcium carbonate ($CaCO_3$) in three different crystalline phases, i.e. calcite, vaterite, and aragonite [12,13]. It is captivating to know the way the nature adeptly models and accurately organizes the building block, constituent composition, polymorphism, and other remarkable characteristics of $CaCO_3$. Previous study showed that high-purity $CaCO_3$ could be found in many sites in Indonesia [14]. However, it becomes less valuable due to its unnecessary purposes of exploitation. This work, then, introduces the possibility of realizing a beneficial use of limestone as the calcium stand-in to fabricate calcium phosphate-based HA.

Eventhough HA offers outstanding osteoconductivity and biocompatibility [15], pure-HA, at the same time, performs low reactivity with the original (human) bone [16] leading to insufficient ability to create interfaces with the original bone and to stimulate the new bone formation [17]. So, having succeeded in producing pure-HA may not solve the challenge of creating an excellent biomedical material for bone grafting yet.

The best way to solve this problem is to modify HA crystal structure with insertion of small amount of silicon. Silicon is frequently used as dopant for bioceramics since it significantly enhances the ultimate properties of the biomaterials such as surface chemical structure, mechanical strength, bioactivity, and biocompatibility [17]. Particularly, silicon plays an important role at the early stages in stimulating normal growth of bone and soft tissue development [18]. Therefore, in this work, we report the use of limestone in HA production by means of solid state reaction and study the effect silicon on the characteristics of silicon-doped hydroxyapatite (Si-HA).

## 2. Experimental Method

A limestone from Malang, Indonesia, was selected prior to produce HA. The limestone was grounded and sieved. Only particles passing a mesh screen with 200 mesh were used in this study. The powder was calcined at1000 °C for 5 h to obtain CaO powder. The CaO was then dissolved in distilled water. $H_3PO_4$ (Sigma-Aldrich) was added to the CaO solution and stirred at 30 °C for 30 min. Filtration was made after stirring the solution. The precipitate was heated at 700 °C for 10 h to form calcium phosphate, $Ca_2P_2O_7$. The primary materials to synthesize Si-HA were prepared after the stoichiometric calculation, i.e. $CaCO_3$ from limestone, $Ca_2P_2O_7$, and $SiO_2$ (Sigma-Aldrich). These materials were mixed and grounded for 30 h followed by pressing to produce cylindrical-shaped samples. The as-prepared samples were sintered at 1100 °C for 20 h. The characterizations were performed by means of XRD for crystal structure and phase content study. The microstructural investigation and elemental mapping of the samples were conducted by using Scanning Electron Microscopy (SEM) and energy dispersive x-ray (SEM-EDX). Furthermore, Vickers hardness tes was used to study the mechanical strength od the samples. The functional groups of the samples were determination by using Fourier-Transform Infrared (FTIR).

## 3. Results and Discussion

The previous study [19] revealed that the content of calcium from the limestone was above 97%. The diffraction profile of the as-prepared calcium phosphate also depicted that the crystal structure was in good agreement with the crystallographic open database code 96-200-1133. It implies that the limestone can be well used as the raw material to produce calcium phosphate and for further used for the production of HA.

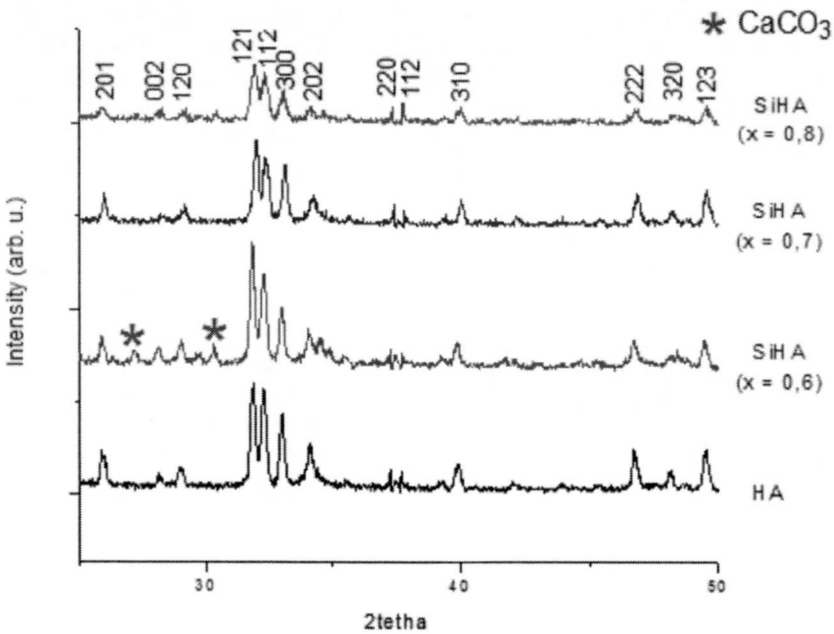


**Figure 1.** XRD profiles of pure-HA and Si-HA with various Si contents

The XRD data for Si-HA with various Si contents can be viewed in Figure 1. There is no peak found in which corresponding to silicon. The arrangement of 2-theta and intensities of the XRD profiles, particularly for pure-HA refers to the crystal structure of hydroxyapatite with the PDF reference code of 01-084-1998 without the presence of any minor phases. Introducing of silicon to HA does not change the XRD patterns, meaning that the crystal structures remain unchanged. Even if there is a lattice distortion, it will be very insignificant [15]. Only Si-HA with $x = 0.6$ containing additional peaks referring to aragonite $CaCO_3$ with the PDF code of 00-024-0025. The hydroxyapatite detected in the XRD patterns has hexagonal crystalline system with space group of P63/m. The absence of silicon-containing XRD peaks implies that the silicon might successfully incorporate to HA sites.

Further XRD data analysis was conducted to explore degree of crystallinity as well as crystalline size. The crystalline size was evaluated by means of Scherrer equation [20]. Since, in principle, the equation which based on the XRD peak broadening can be applied to predict the size of crystal [21]. The crystalline size of Si-HA with silicon contents of 0, 0.6, 0.7, and 0.8 are 26.85 nm, 34.85 nm, 40.06 nm, and 40.84 nm, respectively. These results are in a good agreement with another report [15]. In other words, there is an increasing size of crystal for silicon-containing HA with respect to the concentration of silicon. On the other side, crystallinity of the samples also varies with different silicon content. The crystallinities of Si-HA with silicon contents of 0, 0.6, 0.7, and 0.8 are 35.75%, 29.40%, 21.66%, and 20.99%, respectively. The reduced of the degree of crystallinity is strongly believed that it is due to the incorporation of $SiO_4^{4-}$ ions into HA sites by exchanging $PO_4^{3-}$ ions [22].

One of the most important characteristics of HA is the ratio of Ca/P or Ca/(P + Si) for Si-HA. Therefore, here we also report the ratio by means of SEM-EDAS measurement. Figure 2 shows the SEM-EDAS elemental analysis of Si-HA with $x = 0.6$ (data for pure-HA and other Si-HAs are not shown). From Figure 2, we can calculate the ratio of Ca/(P + Si) is 1.80. The ratio of Ca/P for pure-HA is 1.67 which is very close to the ratio reported by other researchers [20–24]. Figure 3 represents the microstructure of pure-HA and Si-HA. It informs us that there is a significant decrease of porosity as the increase of silicon contents. Quantitatively, the porosity of sample reduces from 10.27% (pure-HA) to 8.97% ($x = 0.6$), 7.69% ($x = 0.7$), and 7.60% ($x = 0.8$). The decrease of porosity reveals that the

IOP Conf. Series: Materials Science and Engineering **202** (2017) 012021 doi:10.1088/1757-899X/202/1/012021

incorporation of silicon into HA takes place the pore of the system. This case may assist dissolution of surface ions that stimulates the formations of apatite layer [17].

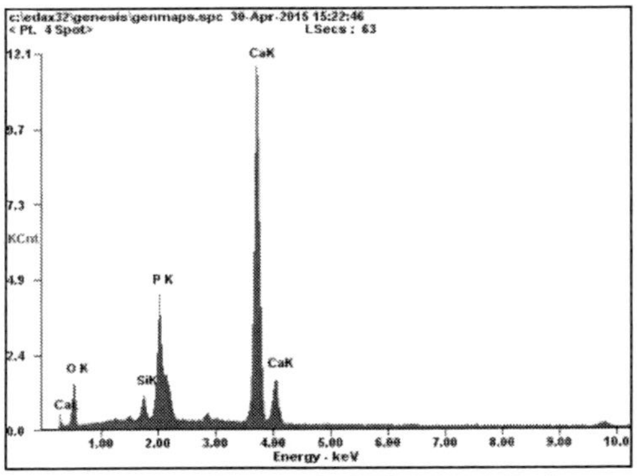


| *Element* | *Wt%* | *At%* |
|---|---|---|
| ***O*** | 30.58 | 50.84 |
| ***Si*** | 01.93 | 01.82 |
| ***P*** | 13.04 | 11.20 |
| ***Ca*** | 54.45 | 36.14 |

**Figure 2.** SEM-EDAS elemental analysis of Si-HA ($x = 0.6$)

**Figure 3.** SEM photographs of pure-HA and Si-HA

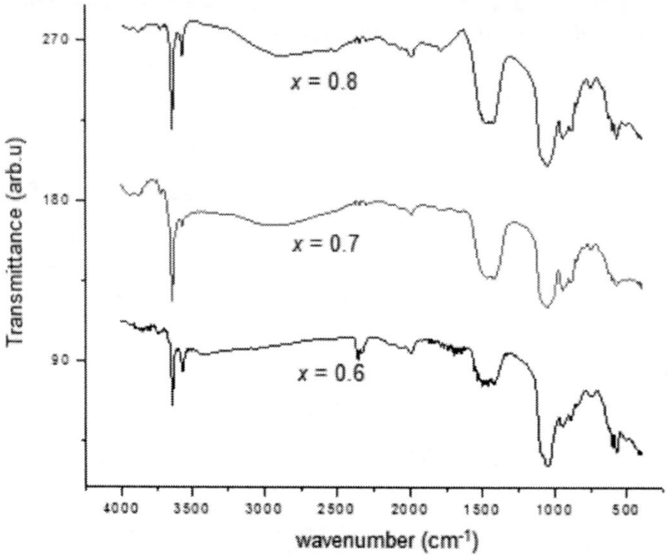


**Figure 4.** FTIR Spectra of Si-HA with different silicon concentrations

FTIR was applied to investigate the influence of the Si incorporation on the different functional groups of Si-HA. From Figure 4, Asymmetric bending vibration of $PO_4^{3-}$ groups is precieved at 570 $cm^{-1}$ [27]. Another phosphate bonding is also noted at 601 $cm^{-1}$ [28]. Hydroxyl group bending is found in 630 $cm^{-1}$ [5]. Si-O and P-O stretching, which decrease and become less resolved as the higher amount of silicate content [1], are shown by the FTIR bands in 950 $cm^{-1}$ and [1,15] and 1089 $cm^{-1}$ [15]. Hydroxyapatite broad absorption corresponding to $PO_4^{3-}$ stretching vibrational mode is discovered at 900-1100 $cm^{-1}$ [29]. In addition, silicate and phosphate tetrahedral molecular units share several similarities of asymmetrical stretching vibration modes in the range of 900-1100 $cm^{-1}$ [1]. Increase in broadening peak at 1060 $cm^{-1}$ is strongly believed as the incorporation of silicon and the decrease of crystallinity [22]. Asymmetric stretching of $CO_3^{2-}$ groups is detected at 1400 $cm^{-1}$ [27]. Hydroxyl group stretching in hexagonal channels and surface hydroxyl group are observed at 3572 $cm^{-1}$ [5] and 3643 $cm^{-1}$ [22].

The hardness characteristic of the Si-HA was characterized by means of Vickers hardness test. It is revealed that the addition of silicon into hydroxyapatite elevates the hardness. Figure 5 represents the graph that contains information on the interconnection between hardness and porosity of Si-HA.

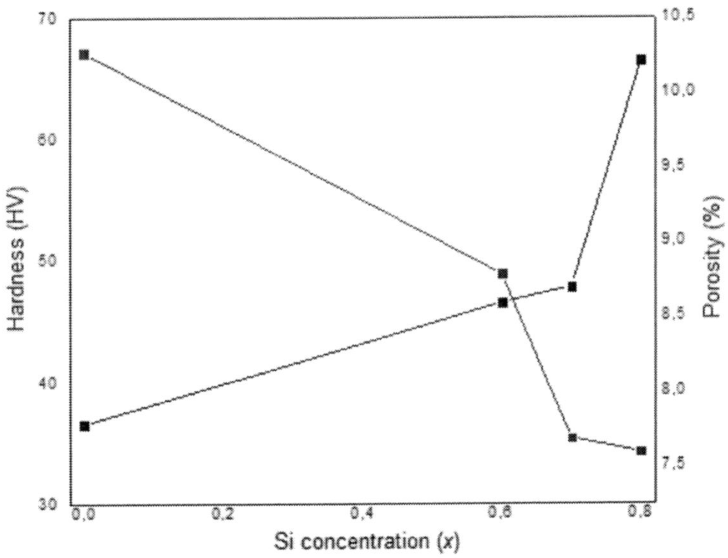


**Figure 5.** Hardness and porosity characteristics of Si-HA with different silicon concentrations

The high temperature sintering applied during the synthesis may give chance for the grains to grow due to the densification process [30]. Introducing silicon into hydroxyapatite yields the reduction of porosity, meaning that the Si-HA becomes denser as the quantity of silicon intensifies. Which in turns, it increases the hardness value of the prepared Si-HA.

Furthermore, the bioresorbability of the sample with the lowest silicon concentration in Si-HA prepared in the series of this experiment was evaluated by means of immersing the sample in simulated body fluid (SBF) solution for hours. Bioresorbance is one of biodegradation phenomena indicated by the decrease of mass during the immersion of sample in a fluid [31]. It can be evaluated by calculating the mass reduction of Si-HA in SBF solution. The average mass reduction of Si-HA per day was 0.003 g. It implies that the Si-HA is applicable for biomedical applications [31].

## 4. Conclusion

The present work describes the characteristics of Si-HA prepared by solid-state reaction approach. It is discovered that the limestone from Indonesian mineral resource can be effectively used as the raw-material for hydroxyapatite production. The XRD data revealed the absence of silicon-oxide-based diffraction peaks and FTIR spectra convincing the presence of Si-O functional groups. It implies that the silicon has been successfully inserted into phosphate site within the crystal of hydroxyapatite. The Vickers hardness of the samples increased as the decrease of porosity. Finally, the produced Si-HA shows the feasibility for biomedical applications particularly due to its bioresorbability characteristic.

## 5. References

[1] Padmanabhan S K, Ul Haq E and Licciulli A 2014 Rapid synthesis and characterization of silicon substituted nano hydroxyapatite using microwave irradiation *Curr. Appl. Phys.* **14** 87–92

[2] Rodríguez-Valencia C, López-Álvarez M, Cochón-Cores B, Pereiro I, Serra J and González P 2013 Novel selenium-doped hydroxyapatite coatings for biomedical applications *J. Biomed. Mater. Res. A* **101A** 853–61

[3] Marchat D, Bouët G, Lueckgen A, Zymelka M, Malaval L, Szenknect S, Dacheux N, Bernache-Assollant D and Chevalier J 2013 Physico-Chemical Characterization and In Vitro

Biological Evaluation of Pure SiHA for Bone Tissue Engineering Application *Key Eng. Mater.* **529–530** 351–6

[4] Zhang C, Yang J, Quan Z, Yang P, Li C, Hou Z and Lin J 2009 Hydroxyapatite Nano- and Microcrystals with Multiform Morphologies: Controllable Synthesis and Luminescence Properties *Cryst. Growth Des.* **9** 2725–33

[5] Sanosh K P, Chu M-C, Balakrishnan A, Lee Y-J, Kim T N and Cho S-J 2009 Synthesis of nano hydroxyapatite powder that simulate teeth particle morphology and composition *Curr. Appl. Phys.* **9** 1459–62

[6] Brundavanam S, Poinern G E J and Fawcett D 2015 Synthesis of a Hydroxyapatite Nanopowder via Ultrasound Irradiation from Calcium Hydroxide Powders for Potential Biomedical Applications *Nanosci. Nanoeng.* **3** 1–7

[7] Amer W, Abdelouahdi K, Ramananarivo H R, Zahouily M, Fihri A, Djessas K, Zahouily K, Varma R S and Solhy A 2013 Microwave-assisted synthesis of mesoporous nano-hydroxyapatite using surfactant templates *CrystEngComm* **16** 543–9

[8] Pramanik S, Agarwal A K, Rai K N and Garg A 2007 Development of high strength hydroxyapatite by solid-state-sintering process *Ceram. Int.* **33** 419–26

[9] Ingole V H, Hussein K H, Kashale A A, Gattu K P, Dhanayat S S, Vinchurkar A, Chang J-Y and Ghule A V 2016 Invitro Bioactivity and Osteogenic Activity Study of Solid State Synthesized Nano-Hydroxyapatite using Recycled Eggshell Bio–waste *ChemistrySelect* **1** 3901–8

[10] Guo X, Yan H, Zhao S, Li Z, Li Y and Liang X 2013 Effect of calcining temperature on particle size of hydroxyapatite synthesized by solid-state reaction at room temperature *Adv. Powder Technol.* **24** 1034–8

[11] Aina V, Cerrato G, Martra G, Bergandi L, Costamagna C, Ghigo D, Malavasi G, Lusvardi G and Menabue L 2013 Gold-containing bioactive glasses: a solid-state synthesis to produce alternative biomaterials for bone implantations *J. R. Soc. Interface* **10** 20121040

[12] Meddah M S, Lmbachiya M C and Dhir R K 2014 Potential use of binary and composite limestone cements in concrete production *Constr. Build. Mater.* **58** 193–205

[13] Fujiwara M, Shiokawa K, Araki M, Ashitaka N, Morigaki K, Kubota T and Nakahara Y 2010 Encapsulation of Proteins into CaCO3 by Phase Transition from Vaterite to Calcite *Cryst. Growth Des.* **10** 4030–7

[14] Hidayat N, Ramli I, Sunaryono S, Taufiq A, Zainuri M and Pratapa S 2015 Gradualisme Struktur Kristal dan Sifat Mekanik Material Fungsional Kalsit-Mg/Al Hasil Fabrikasi dengan Metode Infiltrasi Seminar Nasional Fisika dan Pembelajarannya 2015 (Malang: Department of Physics, UM) p FM19-FM24

[15] Rau J V, Fosca M, Cacciotti I, Laureti S, Bianco A and Teghil R 2013 Nanostructured Si-substituted hydroxyapatite coatings for biomedical applications *Thin Solid Films* **543** 167–70

[16] Ducheyne P, Radin S and King L 1993 The effect of calcium phosphate ceramic composition and structure on in vitro behavior. I. Dissolution *J. Biomed. Mater. Res.* **27** 25–34

[17] Khan A F, Saleem M, Afzal A, Ali A, Khan A and Khan A R 2014 Bioactive behavior of silicon substituted calcium phosphate based bioceramics for bone regeneration *Mater. Sci. Eng. C* **35** 245–52

[18] Hing K A, Revell P A, Smith N and Buckland T 2006 Effect of silicon level on rate, quality and progression of bone healing within silicate-substituted porous hydroxyapatite scaffolds *Biomaterials* **27** 5014–26

[19] Hartatiek H, Yudyanto Y, Windari R Y, Ratnasari S D and Hidayat N 2016 Karakteristik Hidroksiapatit Doping Silikon Si0,5HA Hasil Reaksi Keadaan Padat dan Potensinya sebagai Material Biomedis Seminar Fisika dan Pembelajarannya 2016 (Malang: Jurusan Fisika UM)

[20] Scherrer P 1918 Determination of the size and internal structure of colloidal particles using X-rays *Math-Phys Kl.* **2** 98–100

[21] Hidayat N, Sunaryono S and Taufiq A 2016 Komparasi Analisis Ukuran Kristal Partikel Nano Magnetit Berbasis Data Difraksi Sinar-X dengan Beragam Metode Seminar Fisika dan Pembelajarannya 2016 (Malang: Jurusan Fisika UM)
[22] Sutha S, Kavitha K, Karunakaran G and Rajendran V 2013 In-vitro bioactivity, biocorrosion and antibacterial activity of silicon integrated hydroxyapatite/chitosan composite coating on 316 L stainless steel implants *Mater. Sci. Eng. C* **33** 4046–54
[23] Jeong Y-H, Kim J-J and Choe H-C 2016 Surface Characteristics of Nano-Structured Silicon/Hydroxyapatite Deposition onto the Ti–Nb–Zr Alloy *J. Nanosci. Nanotechnol.* **16** 1783–6
[24] Harden F J, Gibson I R and Skakle J M S 2013 Simplification of the Synthesis Method for Silicon-Substituted Hydroxyapatite: A Raman Spectroscopy Study *Key Eng. Mater.* **529–530** 94–9
[25] Surmeneva M A, Surmenev R A, Chaikina M V, Kachaev A A, Pichugin V F and Epple M 2013 Phase and elemental composition of silicon-containing hydroxyapatite-based coatings fabricated by RF-magnetron sputtering for medical implants *Inorg. Mater. Appl. Res.* **4** 227–35
[26] Yu H, Liu K, Zhang F, Wei W, Chen C and Huang Q 2015 Microstructure and in vitro Bioactivity of Silicon-Substituted Hydroxyapatite *Silicon* 1–11
[27] López-Álvarez M, Solla E L, González P, Serra J, León B, Marques A P and Reis R L 2009 Silicon–hydroxyapatite bioactive coatings (Si–HA) from diatomaceous earth and silica. Study of adhesion and proliferation of osteoblast-like cells *J. Mater. Sci. Mater. Med.* **20** 1131–6
[28] Kannan S, Lemos A F and Ferreira J M F 2006 Synthesis and Mechanical Performance of Biological-like Hydroxyapatites *Chem. Mater.* **18** 2181–6
[29] Bianco A, Cacciotti I, Lombardi M and Montanaro L 2009 Si-substituted hydroxyapatite nanopowders: Synthesis, thermal stability and sinterability *Mater. Res. Bull.* **44** 345–54
[30] Barsoum M and Barsoum M W 2002 *Fundamentals of Ceramics* (CRC Press)
[31] Rameshbabu N and Rao K P 2009 Microwave synthesis, characterization and in-vitro evaluation of nanostructured biphasic calcium phosphates *Curr. Appl. Phys.* **9** S29–31

# Aging Time Effect on Porous Characteristics of Natural Mud-based Silica Prepared by Hydrothermal-Coprecipitation Route

**A Ubaid[1], N Hidayat[2] and Munasir[1]**

[1]Department of Physics, Faculty of Mathematics and Natural Sciences, Universitas Negeri Surabaya (UNESA) Jl. Ketintang, Surabaya 60231, Indonesia
[2]Department of Physics, Faculty of Mathematics and Natural Sciences, Universitas Negeri Malang (UM), Jl. Semarang 5, Malang 65145, Indonesia

Email: munasir_physics@unesa.ac.id

**Abstract:** In this present study, we report the influence of aging time on porous characteristics and chemical bonding within structurally related silicates of silica. The silica was well prepared by using a combination of hydrothermal and coprecipitation methods. Local natural mud from Sidoarjo, Indonesia was preferred as a starting material, rather than the expensive commercial ones. The aging time was set at 20, 24, 28, and 32 hours. The X-ray fluorescence (XRF) test revealed that the produced porous silica has a purity of 98.9%. The Fourier Transform of Infra-Red (FTIR) analysis showed that ≡Si-OH and ≡Si-O -Si≡ functional groups were found in the samples. The pores of the silica, after Brunauer–Emmett–Teller (BET) measurement, found to be macropore and mesopore. The prepared silica with different porous characteristics were strongly influenced by the chemical activities during the synthesis, known as syneresis and Ostwald processes.

**Keywords**: porous silica, aging time, natural mud, hydrothermal-coprecipitation

## 1. Introduction

Porosity is a vital parameter for adsorbent materials. Higher porosity will lead to better adsorption performance of an adsorbent. Regarding porosity, silica is one of the well-known materials that possess a high level of porosity which turns to meet the criteria of adsorbent materials [1]. Many experts and scientists have widely searched the new synthetic approaches of silica, particularly porous silica, as well as its advanced applications. To address some of its applications, a porous silica can be applied as an adsorbent [1], catalyst and water filter [2], and drug delivery system [3]. Porous characteristics of materials can influence other important desired properties of the materials [4]. The porous characteristics can be controlled by specific treatments, e.g. heating [5] and aging[6]. During an aging process, there will be a reaction that leads to better characteristics of the material in terms of its rigidity and strength [7].

Aging is expected to cause the formation of pores with controlled characteristics regarding radius, volume, and surface area of the pores [6,8]. Pores are formed in several shapes such as a cylinder, funnel, and flask [4,9,10]. The pore has an influence on the physical and chemical properties of the material, such as density, conductivity and surface area [11,12]. Since there are many influences of porous silica and the need for the specifications, it is, therefore, a challenging task to explore the effective route in synthesizing the porous silica. On top of that, we introduce the chance of using natural mud from Sidoarjo, which has been erupting for a decade, as the raw material to produce the porous silica.

## 2. Experimental Method

In this study, the natural mud from Sidoarjo, Indonesia, was extracted and immersed in 2 M HCl for 2 hours. The solution was then precipitated and washed with distilled water before obtaining the precipitates. Afterward, a hydrothermal process was carried out by stirring the precipitated solution in 7 M of NaOH for 2 hours at 80 °C. HCL was then wisely added to the stirred solution to reduce the pH to 4 before aging for 20, 24, 28, and 32 h. After aging, the silica was finally washed with distilled water and dried overnight. The produced silica was characterized by using Surface Area Analyzer (SAA), Brunauer–Emmett–Teller [13], Fourier Transform Infra-Red (FTIR), and X-ray fluorescence (XRF) Spectroscopy. Samples with aging time of 20, 24, 28, and 32 h were coded as SA20, SA24, SA28, and SA32, respectively.

## 3. Results and Discussion

The produced silica in this study had a purity of 98.9%. The result is closely to the purity of silica that was synthesized by using alkali fusion route was 98.9% [5,14], the continuous method was 98.8%, co-precipitation method was 97.6% , 96.1% [15], the drying process was 96.5% [14], and the extraction was 95.6% [16]. In this experiment, the mass of the obtained silica was approximately of 1.2 - 2.3 grams.

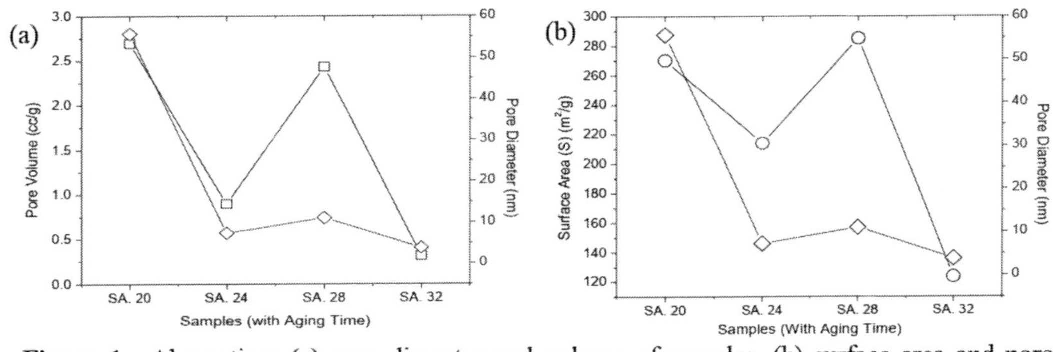


**Figure 1.** Absorption: (a) pore diameter and volume of samples, (b) surface area and pore diameter

The difference masses of the porous silica depend on the quantity of the sodium induced by NaOH solution. A previous study inferred that the greater the moles of NaOH, the greater the mass of the produced silica. Moreover, the reaction rate of the natural mud in the hydrothermal process and the final pH titration also affected different masses. It determined the quantity of the sodium induced by NaOH solution and the reaction rate of the natural mud in the hydrothermal process affecting the amount of the $Na_2SiO_3$ formation. The final pH affected the separation process of sodium from the $Na_2SiO_3$ and formed NaCl [14].

Figure 1 represents pore diameter and pore volume of the samples at absorption phenomena [13]. The ratio of pore diameter and pore volume of SA20, SA24, SA28, and SA32 were 55.7 nm and 2.7 cc/g, 7.3 nm and 0.9 cc/g, 11.1 nm and 2.4 cc/g, 3.8 nm and 0.3 cc/g, respectively. Fig. 1 b represents the surface area and pore diameter at adsorption phenomena. The ratio of pore diameter and surface area of SA20, SA24, SA28, and SA32 were 55.7 nm and 270 $m^2/g$, 7.3 nm and 214 $m^2/g$, 11.1 nm and 285 $m^2/g$, 3.8 nm and 124 $m^2/g$, respectively. The obtained surface area was relatively smaller compared to previous studies in which the aging for 24, 48, 72 hours produced an area of 380 $m^2/g$, 366 $m^2/g$, 111 $m^2/g$ [8,11].

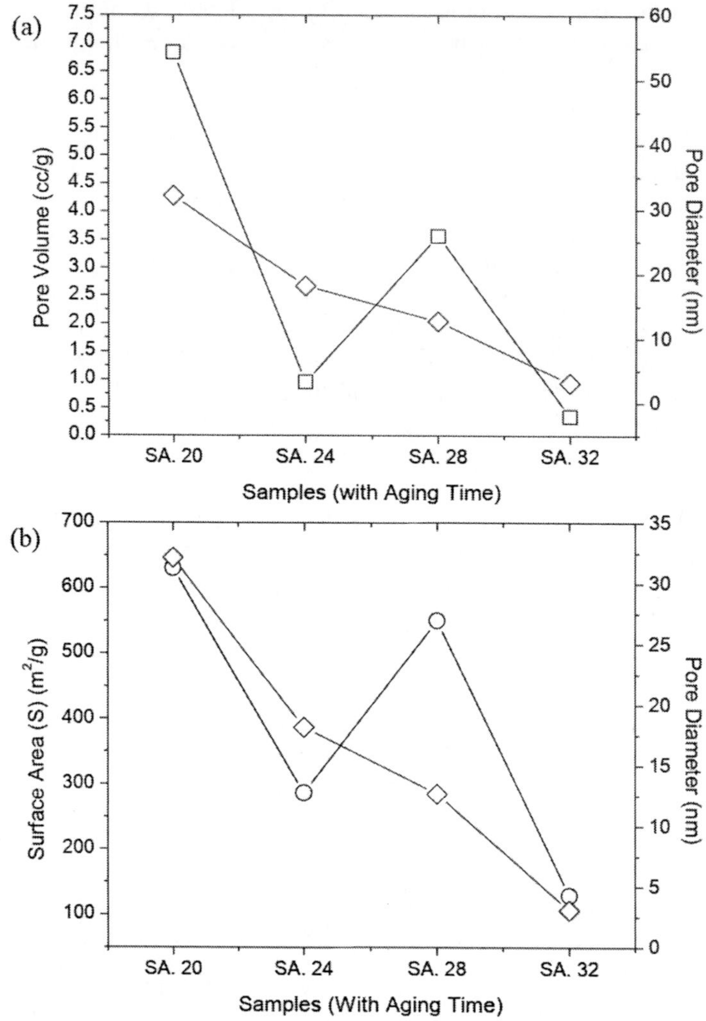


**Figure 2.** Desorption: (a) pores diameter and volume of samples, (b) surface area and pore diameter

There was a decrease in pore size on silica aging, but SA28 had a pore size that was larger than SA24 and SA32. Increased pore diameter, pore volume, and surface area on SA28 related to the formation of pores. The pore shapes are usually affected by the particle size when it is added wisely by HCl to obtain the final pH. If the size of silica particle was large, the phenomena of aging formed a large capacity of silica pore. The SA28 was known to have a small pore diameter but has a large pore volume and surface area. The inequality demonstrated different pore shapes of SA28 with SA20, SA24, and SA32.Thus it was possible to form pores in the forms of ink bottle pores or interconnected pores in SA28 [6].

Figure 2 represents pore diameter and pore volume of samples at desorption phenomena. The ratio of pore diameter and pore volume of SA20, SA24, SA28, and SA32 were respectively 32.1 nm and 6.8 cc/g, 18.1 nm and 1 cc/g, 12.7 nm and 3.6 cc/g, and 3.1 nm and 0.3 cc/g. Fig. 2 b represents surface area and pore diameter at desorption phenomena. The ratio of pore diameter and surface area of SA20, SA24, SA28, and SA32 were respectively 32.1 nm and 630 $m^2/g$, 18.1 nm and 287 $m^2/g$, 12.7 nm and 551 $m^2/g$, 3.1 nm and 130 $m^2/g$.The sizes of pore volume-pore diameter and surface area-pore diameter had the same pattern. Wherein the pore decreased at SA20, SA24, SA28, SA32 [10,17].

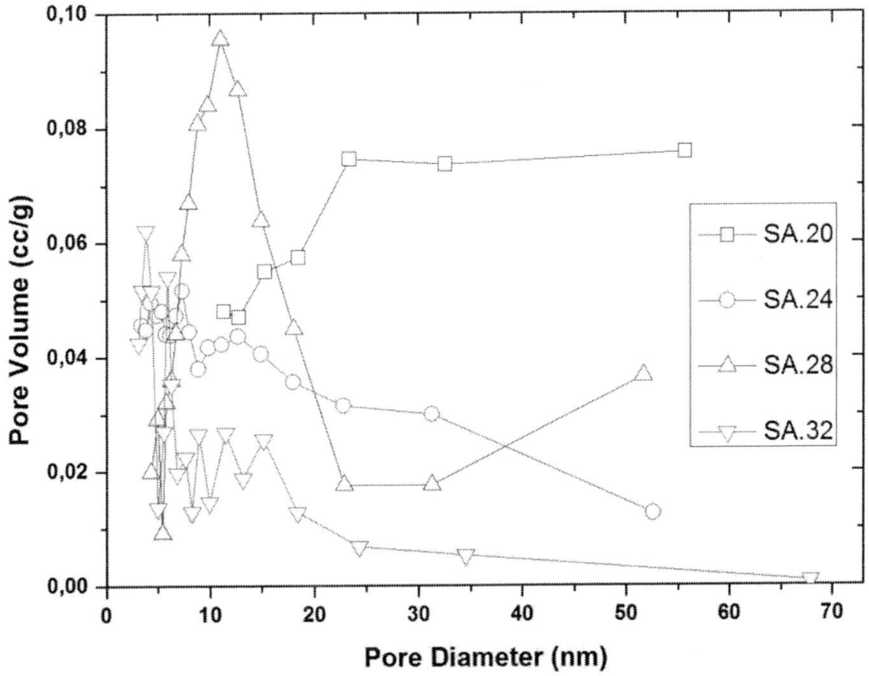


**Figure 3.** Pore size distribution of the produced silica

Figure 3 represents the distribution of as-prepared silica pore size taken from the results of surface area analyzer. It can be revealed that the silica pore size was macroporous and mesoporous. The smallest pore sizes of SA20, SA24, SA28, and SA32 are respectively 11.33 nm, 3.41 nm, 4.3348 nm, and 3.1113 nm; and the pore volume of each is of 0.04802 cc/g, 0.04563 cc/g, 0.01989 cc/g, and 0.04250 cc/g. On the SA20, the large pore volume was found by increasing the pore diameter, and the majority of the pore diameter is of >10 nm. At SA24, the majority of the pore diameter was <10 nm with the large pore volume that did not get much different and decreased by increasing the pore diameter. At SA28, the majority of the pore diameter was <10 nm, but it had the largest pore size in the pore diameter ±11 nm and then it was decreased by increasing the pore volume of the pore diameter. At SA32, the majority of pores <10 nm but the different pore volume was more striking than the SA24 [17].

From Figure 1, Figure 2 and Figure 3, it is clearly seen that there was a decreasing pattern of the pore diameter, pore volume, and surface area. It was due to the particle size and aging process with the phenomena known as syneresis and Ostwald ripening [3]. An Oswald ripening occurs when there is driving force to break the bond within the silicate. This phenomenon caused smaller silica particles that were able to fill pores formed from large silica particles that were firstly settled. Syneresis caused two groups ≡Si-OH adjacent the surface of silica that may be linked to each other and form ≡Si-O-Si≡ [3]. The change of ≡Si-OH groups into ≡Si-O-Si≡ may affect the morphology of the silica produced. There were some differences in pore size between the SA28 and SA24. This possibility occurred because of the effectiveness of the aging process and the effects of the circumstance such as the pH and the size of the current particle when it was synthesized to heat during the aging.

Figure 4 shows the Fourier transform Infrared spectrum. From Figure 4, identical groups of silica were found with the same results as other reports [3,18]. The assigned peaks for the Si-O vibrational mode, Si-O-Si vibrational group, Si-O-Si deformation, ≡Si-O vibrational bond, Si-O bonding, OH vibrational mode, and $H_2O$ molecule absorption were 463.6 $cm^{-1}$, 799.9 $cm^{-1}$, 947.3 $cm^{-1}$, 1102 $cm^{-1}$, 1638.8 $cm^{-1}$, and 3425 $cm^{-1}$, respectively. In addition, a functional group had widening wave numbers that bond the silica, ≡Si-O vibration, and the vibration of water molecules adsorbed. It can be identified that there was found the wave number that vibrated differently from one to others in a function group.

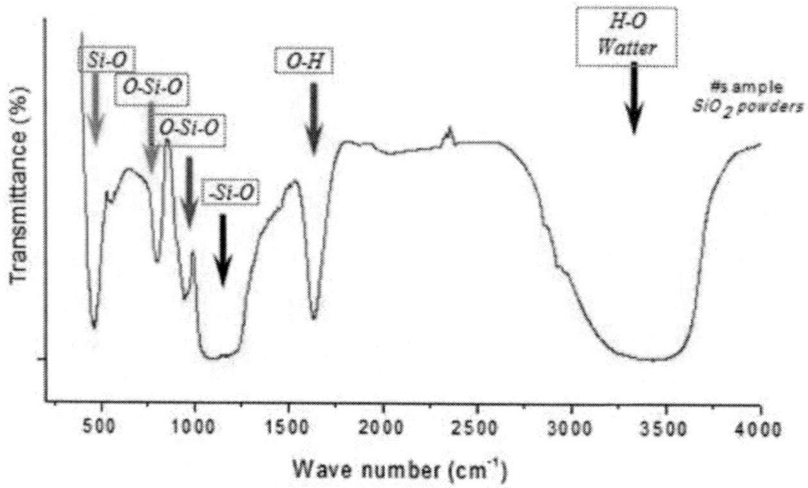


**Figure 4.** FTIR spectrum forming the porous silica produced by hydrothermal-coprecipitation process

There was an interesting effect of the aging on the wavenumbers that were detected by the FTIR, where the formation of bonds between the silica ≡Si-O on the surface of pores. It is also observed that the shift of peaks are due to the phenomenon of syneresis, i.e. the aging process that caused the energy between the atoms bind differently from one to others. The deformation of silica absorption's peak during the syneresis phenomenon is possible. However, the broadening of silica wavenumbers can be minimized by decreasing the pH and titration [6,19,20]. As a result, it will reduce the quantity of the produced silica as a result of the acid that degrades the sample with binding energy which shown by the deformation of the transmission peaks.

## 4. Conclusion

From our experiment, it can be concluded that the mass of the porous silica produced by a combination of hydrothermal-coprecipitation was found around 1.2 to 2.3 grams with the purity of 98.9%. The important functional groups of ≡Si-OH and ≡Si-O -Si≡ groups were observed. The pore size was in the form of the macropores and mesopores. The aging process affected the pore size of the silica through syneresis and Ostwald ripening phenomena that caused the pore size reduction that decreased the pore radius, pore volume, and pore surface area. From the pore size distribution of silica, it can be seen a shift in the dominant pores from a greater to a smaller range.

## 5. References

[1] Zamani C, Illa X, Abdollahzadeh-Ghom S, Morante J R and Romano Rodríguez A 2009 Mesoporous Silica: A Suitable Adsorbent for Amines *Nanoscale Res. Lett.* **4** 1303–8

[2] Priatama A, Abdullah M, Khairurrijal K and Mahfudz H 2010 Fabrication of Microporous Water Filter Using Titanium Dioxide Particles, Silica Particles, and Polyethylene Glycol *ITB J. Eng. Sci.* **42** 39–52

[3] aus Leningrad I S 2002 *Synthesis of silica aerogels and their application as a drug delivery system* (Germany: der Technischen Universität Berlin)

[4] Unger U 1979 *Porous silica : its properties and use as support in column liquid chromatography* (Amsterdam [etc.]: Elsevier Scientific Pub. Co.)

[5] Munasir, Sulton A, Triwikantoro, Zainuri M and Darminto 2013 Synthesis of silica nanopowder produced from Indonesian natural sand via alkalifussion route *AIP Conference Proceedings* INTERNATIONAL CONFERENCE ON THEORETICAL AND APPLIED PHYSICS (LCTAP 2012) vol 1555(AIP Publishing)pp 28–31

[6] Affandi S, Setyawan H, Winardi S, Purwanto A and Balgis R 2009 A facile method for production of high-purity silica xerogels from bagasse ash *Adv. Powder Technol.* **20** 468–72

[7] Buhrke V E, Jenkins R and Smith D K 1997 *A Practical Guide for the Preparation of Specimens for X-Ray Fluorescence and X-Ray Diffraction Analysis* (Wiley)

[8] Chen Q, Larismaa J, Keski-Honkola A, Vilonen K, SöDerberg O and Hannula S-P 2012 Effect of Synthesis Time on Morphology of Hollow Porous Silica Microspheres *Mater. Sci.* **18**

[9] Rouquerol J, Avnir D, Fairbridge C W, Everett D H, Haynes J M, Pernicone N, Ramsay J D F, Sing K S W and Unger K K 1994 Recommendations for the characterization of porous solids (Technical Report) *Pure Appl. Chem.* **66**

[10] Gonçalves M L, Dimitrov L D, Jordão M H, Wallau M and Urquieta-González E A 2008 Synthesis of mesoporous ZSM-5 by crystallisation of aged gels in the presence of cetyltrimethylammonium cations *Catal. Today* **133–135** 69–79

[11] Physical Chemistry Division 1994 Physical Chemistry Division, Recommendations For The Characterization Of Porous Solids *Pure Appl Chern* **66**1739–1758,

[12] Barrabino A 2011 *Synthesis of mesoporous silica particles with control of both pore diameter and particle size* (Sweden: Chalmers University Of Technology)

[13] Fagerlund G 1973 Determination of specific surface by the BET method *Matér. Constr.* **6** 239–45

[14] Munasir, Triwikantoro, Zainuri M and Darminto 2015 Synthesis of SiO2 nanopowders containing quartz and cristobalite phases from silica sands *Mater. Sci.-Pol.* **33** 47–55

[15] Larbi K K 2010 *Synthesis of High Purity Silicon from Rice Husks* (Toronto, USA: University of Toronto)

[16] Mourhly A, Khachani M, Hamidi A E, Kacimi M, Halim M and Arsalane S 2015 The Synthesis and Characterization of Low-Cost Mesoporous Silica $SiO_2$ from Local Pumice Rock *Nanomater. Nanotechnol.* **5** 35

[17] Gurav J L, Jung I-K, Park H-H, Kang E S and Nadargi D Y 2010 Silica Aerogel: Synthesis and Applications *J. Nanomater.* **2010** 1–11

[18] Zarezadeh-Mehrizi M and Badiei A 2014 Highly efficient removal of basic blue 41 with nanoporous silica *Water Resour. Ind.* **5** 49–57

[19] Kalapathy U, Proctor A and Shultz J 2000 A simple method for production of pure silica from rice hull ash *Bioresour. Technol.* **73** 257–62

[20] Waseem M, Mustafa S, Naeem A, Shah K H and Shah I 2009 Synthesis And Characterization Of Silica By Sol-Gel Method *J Pak Mater Soc* **3**

IOP Conf. Series: Materials Science and Engineering **202** (2017) 012023 doi:10.1088/1757-899X/202/1/012023

# Magnetic Susceptibility and Morphology of Natural Magnetic Mineral Deposit in Vicinity of Human's Living

**S Zulaikah, R Azzahro, Pranita S B, E S Mu'alimah, N Munfarikha, Dewiningsih, W L Fitria and H A Niarta**

Department of Physics, Faculty of Mathematics and Natural Sciences, Universitas Negeri Malang (UM), Jl. Semarang 5, Malang 65145, Indonesia

Email: siti.zulaikah.fmipa@um.ac.id

**Abstract**. Magnetic susceptibility and morphology of magnetic minerals have been explored to some samples from the different environment near the vicinity of human's living, such as hot spring, apple plantation, paddy plantation, and reservoir. Magnetic susceptibility ranged from -0.0009 × $10^{-6}$ $m^3$/kg (Peat in central Borneo) to 98.27 × $10^{-6}$ $m^3$/kg (Polluted Soil in Jalan Sukarno Hatta Malang). The grain size of magnetic mineral not more than that of 300 μm. Data analysis informs us that each environment where the magnetic minerals were deposited, influenced the two physical properties both of magnetic susceptibility and morphology of magnetic minerals. Regarding the environment process, magnetic susceptibility depends upon the grain size beside the kinds of magnetic minerals. So, it can be concluded that in every environment, the magnetic minerals have specific properties including magnetic susceptibility and the morphology of magnetic minerals.

**Keywords:** Magnetic susceptibility, morphology, magnetic minerals, specific environment.

## 1. Introduction

As the main carrier of magnetic properties in sediment, soil, or other natural samples, the magnetic mineral may contribute as a clue of special environment along it was growing. The change in kind, structure, shape, and size of magnetic minerals may change its magnetic properties [1]. The change on magnetic mineral represents a specific information related to environments and even a climatic system that has been influenced [2, 3]. It is also known that many natural magnetic minerals are valuable for industries. That is way we need to study further regarded to magnetic minerals.

In many cases, measuring of magnetic properties as a clue of environment and industrial studies, is very common in using magnetic susceptibility, natural remanent magnetization. In this study we use the two special techniques i.e., magnetic susceptibility and morphology using scanning electron microscope (SEM) to recognize the magnetic mineral in a specific environment clearly. The measurement of magnetic susceptibility has been used for many soil samples as an evaluation of polluted and non-polluted soils [4, 5, 6, 7]. Moreover, these measurements can be used to trace the existing of heavy metals dissolved in the samples [8]. Chemical composition of soils also influence on magnetic susceptibility [9, 10, 11]. The existing of magnetic mineral and analyzing of the size based on the SEM data and the connection to the magnetic susceptibility have been conducted on reservoir sediment in Selorejo [12]. However, the perfect relationship of these parameters has not formulated yet.

## 2. Experimental Method

Sediment or soil samples in this research were taken from some area, such as rivers, reservoir, plantation area, mangrove, peat area, and hot spring in the form of cores or a representative pick out a small portion. In some wet area, usually we picked out the sample using cores by covering the boat and in the area of plantation. Usually, we picked out a small portion. A sub-sample was then picked in the standard plastic holder for magnetic measurement, and some sample was extracted to get the magnetic mineral that carries of magnetic properties as the sample. A bulk sample was then measured the magnetic susceptibility by using magnetic susceptibility meter MS2B, and the extract of the magnetic mineral image was conducted by SEM (Scanning electron microscope). The composition of elements of extracted samples were characterized using EDAX. The measurements were performed in Central Laboratory of Universitas Negeri Malang, Indonesia.

## 3. Results and Discussion

**Table 1.** List of magnetic susceptibility of some samples in specific environment

| Sample | Sample Location | Magnetic susceptibility ($\times 10^{-6}$ m$^3$/kg) | Susceptibility dependence (%) |
|---|---|---|---|
| Soil Plantation | Apel orchard soil Pujon, East Java | 6.54 – 16.24 | Not calculating yet |
| | Apel orchard soil Poncokusumo, East Java | 9.39 - 18.49 | Not calculating yet |
| | Paddy plantation Araya, Malang East Java | 5.12 – 5.6 | Not measuring yet |
| | Paddy plantation Madiun, East Java | 0.58 – 2.96 | 0.2 – 7.2 |
| | Paddy plantation Malang, East Java | 0.85 – 3.82 | 0.3 – 9.4 |
| Polluted Soil | Polluted soil in Jalan Sukarno Hatta Malang, East Java | 5.93 – 98.27 | Not calculating yet |
| Reservoir | Selorejo, Ngantang, East Java | 4 – 47 | 1-4 |
| | Wlingi, Blitar, East Java | 2.59 – 38.82 | 0.6 – 4 |
| | Small pond Araya Malang, East Java | 12.36 – 13.48 | Not calculating yet |
| Rivers sediments | Metro river, Malang | 25.89 – 29.68 | 1.91 – 1.99 |
| | Small river in Araya | 11.83 – 16.16 | |
| | Kamp Wolker river, Papua | 11 - 25 | 0.031 – 0.37 |
| | Hubai river, Papua | 4 - 16 | 0.10 – 1.03 |
| Hot Spring sediment | Pacet, East Java | 1.50 – 4.63 | 2.11 |
| | Cangar, East Java | 1.40– 4.82 | 2.48 |
| Mangrove sediment | Prigi, Trenggalek, East Java | 2.13 – 34.61 | 0.29 – 1.36 |
| | Wonorejo, Surabaya, East Java | 0.44 – 22.47 | 0.14 – 2.88 |
| | Clungup beach, Malang, East Java | 0.05 – 10.58 | 0.70 – 3.47 |
| Peat | Central Borneo | - 0.0009 – 0.08 | - 37.85 – 42.9 |

IOP Conf. Series: Materials Science and Engineering **202** (2017) 012023 doi:10.1088/1757-899X/202/1/012023

Magnetic susceptibility of some magnetic mineral that deposited in any specific environment such as rivers, plantation area, hot springs and some other places are listed in Table 1. The magnetic susceptibility is also depended in what kind of magnetic minerals. Besides of magnetic susceptibility, we discuss the morphology of magnetic minerals. The image of magnetic minerals from scanning electron microscope (SEM) data was analyzing on four environment background, such as hot springs, paddy plantation area, apple plantation area and reservoir sediments.

*3.1 Hot Springs*

Some samples were taken from the two hot springs area, located in East Java Indonesia, called Pacet and Cangar. The magnetic susceptibility is ranged from $1.40 - 4.82 \times 10^{-6}$ $m^3/kg$. The morphology of the extracted magnetic minerals shows in Figure.1. The size of magnetic mineral is of around 200 μm and tend to have a rounded angle. In hot spring environment background, natural magnetic mineral grows with the smooth surface relatively, and some of them have a crystalline structure. Finer surface grain in hot spring environment occurs due to the flowing of warmer water that influenced of the minerals growing with a temperature of about 50 °C. Figureure 1 and Figureure 2 shows the distribution of magnetic minerals extracted from the sediment of hot spring area. In Figureure 1, we can find some crystalline magnetic minerals dissolved in C1.3 sample ID that taken from Cangar hot spring area. The magnetic susceptibility of the low frequency of this sample is 4.82 x 10 6 $m^3/kg^{-1}$ and susceptibility dependence frequency 0.392. The Fe content of this sample is about 42.26 %.

**Figure 1.** Image of magnetic mineral extracted from C1.3 sample, taken from Cangar hot spring with the size of around 200 μm.

**Figure 2.** One of another sample is P1.3 ID that taken from Pacet hot spring area. The magnetic susceptibility is about 1.496 and susceptibility dependence frequency 0.663. Fe element contained in the mineral assembly by EDAX is about 36.33 Wt%.

*3.2 Apple plantations*

The soil of apple plantation has magnetic susceptibility ranged from 6.54 to 18.49. Because of the existing of magnetic mineral have been deducted influenced the apple growing and fruit quality, we extracted the magnetic minerals in the apple orchard soil and imaging by SEM is presented in Figure. 3.

**Figure 3.** Sample AT6L1 with susceptibility 16.24 × $10^{-6}$ $m^3$/kg (left) BT1L8 with susceptibility 9.79 × $10^{-6}$ $m^3$/kg (right)

The size of the magnetic mineral is ranged from fine grain around 10 μm to a coarse grain that has a size up to 300 μm. The shape of magnetic minerals also very random, some are tended to spherical grains, and some are tens to elongated grains.

*3.3 Paddy Plantation*

Clay from paddy plantation has a lower susceptibility an order than that of soils apple orchard soils. The imaging of the magnetic mineral as shown as in Figureure 4. The magnetic susceptibility of the two area are ranged from 0.584 – 2.961 × $10^{-6}$ $m^3$/kg and the averaged is about 1.67 × $10^{-6}$ $m^3$/kg for Madiun and ranged from 0.847 – 3.817 × $10^{-6}$ $m^3$/kg with the average is about 2.1× $10^{-6}$ $m^3$/kg for Malang.

**Figure 4.** Sampel ID MDN.L2.S3(2) with the magnetic susceptibility 0.582 × $10^{-6}$ $m^3$/kg (left) and MLG.L1.S3(4) with susceptibility 3.802 × $10^{-6}$ $m^3$/kg (right)

The magnetic susceptibility of the two sample in deference order shows the different morphology of magnetic mineral. The sample which the magnetic susceptibility relatively low, has the smaller grain size and vice versa.

*3.4 Reservoirs*

Magnetic susceptibility of sediment from Selorejo reservoir has a same order of Apple plantation in Pujon and Poncokusumo Malang, as seen in Table 1. Magnetic minerals from the sediment of that reservoir shown in Figure. 5.

**Figure 5**. Sample ID TT7 with the magnetic susceptibility 2.28 □ 10-6 m3/kg (left) and crystalline magnetic mineral suggested titanomagnetite dissolved in TT7 (right).

The sediment reservoir still has a magnetic mineral that also has a random shape and size. For example, the sample that taken from the center of the reservoir has a magnetic mineral size of around 10 μm to 225 μm.

## 4. Conclusion

The magnetic susceptibility of some soil and sediment samples in general is ranged from 0.5 to 30 × $10^{-6}$ $m^3$/kg. In some natural samples, the size is not more than 300 μm and we still find a crystalline of magnetic mineral with has a more smooth or finer surface than that of other minerals. Some of the magnetic mineral has changed in size, and shape or its morphology was originated from the change of the environment. It was shown clearly in hot spring that influenced by the flowing warmer water were causing the rounded angle of crystal. The transformation in morphology of magnetic minerals informs us a good clue of environmental change.

## References

[1] Tarling D H and Hrouda F 1993 The Magnetic Anisotropy of Rocks Chapman & Hall
[2] Maher B A and Thompson R 2010 Quaternary Climates Environments and Magnetism New York Cambridge: University Press
[3] Evan M E and Heller F 2003 Environmental Magnetism: Principles and Applications of Enviromagnetics Academic Press
[4] Kanu O M *et al* 2013 Measurement of Magnetic Susceptibility of Soils in Jalingo N-E Nigeria: A Case Study of the Jalingo Mechanic Village World Applied *Sciences Journal*, **24 (2)** 178-187
[5] Dlouha S *et al* 2013 Investigation of Polluted Alluvial Soils by Magnetic Susceptibility Methods: a Case Study of the Litavka River. *Soil & Water Res.* **8(4)** 151–157

[6] Quijano L *et al* 2011 Magnetic Susceptibility in Topsoils and Bulk Cores of Cultivated Calcicols Proceedings Tandil Argentina

[7] Garbacea F G & Dumitru I 2010 Geophysical mapping of soils New data on romanian soils based on magnetic susceptibility *Rom. Geophys*. J. **54** 83–95

[8] Baghdadi E M *et al* 2011 Magnetic susceptibility and heavy metal contamination in agricultural soil of Tadla plain *J. Mater. Environ. Sci.* **2** 513-519

[9] Jivan C and Sala F 2014 Relationship between tree nutritional status and apple quality *Hort. Sci,* **1(41)** 1–9

[10] Murdock K J *et al* 2013 Rock magnetic properties, magnetic susceptibility, and organic geochemistry comparison in core LZ1029-7 Lake El'gygytgyn Russia Far East *Climate of the Past* **9** 467–479

[11] Lourenco A M *et al* 2012 Relationships between magnetic parameters, chemical composition and clay minerals of topsoils near Coimbra central Portugal *Nat. Hazards Earth Syst. Sci*. **12** 2545

[12] Zulaikah S *et al* 2015 Anisotropy of magnetic susceptibility (AMS) analysis for sedimentation tracing of Selorejo reservoir IOP Proceedings

[13] Dearing John 1999 Environmental Magnetic Susceptibility Using the Bartington MS2 System British Library Cataloguing in Publication Data

**Acknowledgements**
The authors would like to thanks to LP2M DIKTI for granted PUPT in 2016 on contract number 7.4.13/UN32.14/LT/2016 in which the first author is being the main researcher.

# Preparation of Superparamagnetic $Zn_{0.5}Mn_{0.5}Fe_2O_4$ Particle by Coprecipitation-Sonochemical Method for Radar Absorbing Material

**A Taufiq[1,2], S Bahtiar[1], Sunaryono[1,2], N Hidayat[1,2], A Hidayat[1,2], N Mufti[1,2], M Diantoro[1,2], A Fuad[1,2], Munasir[3], R Rahmawati[4], W A Adi[5], S Pratapa[3] and Darminto[6]**

[1]Department of Physics, Faculty of Mathematics and Natural Sciences, Universitas Negeri Malang (State University of Malang), Jl. Semarang 5, Malang 65145, Indonesia
[2]Center for Minerals and Advanced Materials, Faculty of Mathematics and Natural Sciences, Universitas Negeri Malang (State University of Malang), Jl. Semarang 5, Malang 65145, Indonesia
[3]Department of Physics, Faculty of Mathematics and Natural Sciences, Universitas Negeri Surabaya (UNESA), Jl. Ketintang, Surabaya 60231, Indonesia
[4]Engineering Physics Study Program, Faculty of Industrial Technology, Institut Teknologi Bandung (ITB), Bandung 40132, Indonesia
[5]Centre for Technology of Nuclear Industry Materials, National Nuclear Energy Agency (BATAN), Kawasan Puspiptek Serpong, Tangerang 15314, Indonesia
[5]Department of Physics, Faculty of Mathematics and Natural Sciences, Institut Teknologi Sepuluh Nopember (ITS), Jl. Arif Rahman Hakim, Surabaya 60111, Indonesia

Email: ahmad.taufiq.fmipa@um.ac.id

**Abstract.** One of many applications of spinel ferrite nanoparticles is related to their performance as radar absorbing materials. In this work, we report developing synthesis method through combined coprecipitation-sonochemical routes in preparing $Zn_{0.5}Mn_{0.5}Fe_2O_4$ nanoparticle from iron sand in Indonesia as a vital raw material. The structure, size, morphology, and elements of the $Zn_{0.5}Mn_{0.5}Fe_2O_4$ nanoparticle were investigated via X-Ray diffractometry and Transmission/Scanning Electron Microscopy (TEM/SEM) combining Energy Dispersive Spectroscopy (EDS). The magnetic properties of the $Zn_{0.5}Mn_{0.5}Fe_2O_4$ nanoparticle were characterized by using Vibrating Sample Magnetometer (VSM). Furthermore, the reflection loss character of the $Zn_{0.5}Mn_{0.5}Fe_2O_4$ nanoparticle was determined via Vector Network Analyzer (VNA). From the qualitative and quantitative analysis of the XRD data, it can be identified that the $Zn_{0.5}Mn_{0.5}Fe_2O_4$ particle formed a spinel cubic structure in a single phase with the lattice parameter of approximately 8.401 Å. It is known from the TEM image that the $Zn_{0.5}Mn_{0.5}Fe_2O_4$ particle had a size of about 9.7 nm and tended to agglomerate. Furthermore, the data analysis of the $M(H)$ curve presented that the $Zn_{0.5}Mn_{0.5}Fe_2O_4$ nanoparticle has a superparamagnetic behavior with the saturation magnetization of approximately 43 emu/g. Finally, the data analysis of the reflection loss as a function of frequency showed that the $Zn_{0.5}Mn_{0.5}Fe_2O_4$ nanoparticle performs as a radar absorbing material with the absorption performance of approximately -11.0 dB at the frequency of 10.8 GHz

**Keywords:** $Zn_{0.5}Mn_{0.5}Fe_2O_4$, nanoparticle, coprecipitation-sonochemical method, superparamagnetic, radar absorbing material.

## 1. Introduction

Spinel ferrite nanoparticles have general formula $AB_2O_4$ and form spinel crystal structure, where A represents the metallic cations located at the tetrahedral site, and B represents the metallic cations positioned at the octahedral site [1]. Spinel ferrite nanoparticles are a group of functional magnetic materials that have attracted significant attention due to their high performance in various applications. Nowadays, several researchers have extensively reported that the spinel ferrite nanoparticles have excellent application performances in many fields. The applications include for hyperthermia treatment [2], energy storage [3], core, switching and MLCI's [4], as a catalyst to eliminate organic pollutants [5], electromagnetic shielding [6], microwave absorption and high-frequency [7], and so forth.

Numerous efforts have been conducted intensively and continuously by many researchers to obtain the high-performance spinel ferrite nanoparticles as radar absorbing materials. It is imperative because radar absorbing materials have an essential role in fulfillment in civil and military purposes [8]. The efforts have been conducted through developing synthesis method, expanding characterization technique, and also developing a theoretical explanation to achieve the thin, flexible and light absorbers with high reflection loss performance.

For mass production, it is essential to develop the synthesis method efficiently and inexpensively in preparing spinel ferrite nanoparticles as radar absorbing materials. However, many papers reported that the ferrite nanoparticles as radar absorbing materials are prepared from commercial precursors which are relatively expensive. Therefore, to reduce production cost, in this experiment, we propose the utilization of iron sand from natural resources in Indonesia in preparing $Zn_{0.5}Mn_{0.5}O_4$ nanoparticle. Among various synthesis methods, the coprecipitation approach has advantages due to its simplicity, effectivity, low price, and ability in producing spinel ferrite nanoparticles [9]. Furthermore, the sonochemical approach also is also superior regarding its high performance in preparing the nanoparticles in various shapes and sizes with high homogeneity. Therefore, we also propose a combination of the two approaches in synthesizing the $Zn_{0.5}Mn_{0.5}O_4$ nanoparticle. Finally, we complete this works by discussing in detail the structural, magnetic, and microwave absorption characteristics of the prepared $Zn_{0.5}Mn_{0.5}Fe_2O_4$ nanoparticle.

## 2. Experimental Method

### *2.1 Sample Preparation*

The $Zn_{0.5}Mn_{0.5}Fe_2O_4$ particle was prepared by using a combined coprecipitation-sonochemical method using natural iron sand as a primary raw material. The iron sand, was taken from Sine Beach Tulungagung - East Java Indonesia, was extracted and purified by using a magnetic separator to obtain the $Fe_3O_4$ powder following our previous works [10,11]. The process of the powder was reacted with HCl (Merck, PA) to form a $FeCl_2$ and $FeCl_3$ solution tracking our previous experiments [12–14]. The solution was then mixed with $ZnCl_2$ (Merck, PA) and $MnCl_2.4H_2O$ (Merck, PA) by using a magnetic stirrer. The process was continued by wisely dropping $NH_4OH$ (Merck, PA) and combined with the sonochemical process by using an Ultrasonic Batch also at room temperature to obtain the black precipitated sample. The precipitated sample was washed several times by using distilled water until it reached a normal pH condition. Finally, the precipitated sample was filtered before drained at 100 °C to obtain the $Zn_{0.5}Mn_{0.5}Fe_2O_4$ particle.

### *2.2 Characterizations*

The crystal structure was investigated using X-ray Diffractometry (XRD) with Cu-Kα radiation in the range of 10° – 80°. The XRD data were analyzed by using Rietica program [15] based on Rietveld refinement approach [16] to obtain the phase purity, crystal structure, lattice parameters, and crystallite size of the sample. The particles size, morphology and elemental composition of the sample were investigated by means of Transmission/Scanning Electron Microscopy (TEM/SEM) combined with energy-dispersive spectroscopy (EDS). The magnetic properties of the sample, as well as saturation magnetization, coercive field, and remanent magnetization, were characterized by using Vibrating Sample Magnetometer (VSM) at room temperature. Finally, the performance of the sample as a radar absorbing material was investigated by using vector network analyzer (VNA) at a frequency in the range of 8 - 12 GHz.

## 3. Results and Discussion

The chemical reaction of the $Fe_3O_4$ from iron sand dissolved in HCl tracks the following equation of reaction [17].

$$Fe_3O_4 + 8HCl \rightarrow 2FeCl_3 + FeCl_2 + 4H_2O \quad (1)$$

From reaction (1), the metal salt solution containing $Fe^{3+}$ and $Fe^{2+}$ can result only through one step of reacting natural $Fe_3O_4$ with HCl. The prepared solution from the first reaction was then mixed with $ZnCl_2$ and $MnCl_2.4H_2O$, and then reacted with $NH_4OH$ by dropping process to form the $Zn_{0.5}Mn_{0.5}Fe_2O_4$ particle. The sonochemical route was also applied to reducing the agglomerated particles of the sample during the precipitating process. The detail of the chemical reaction is presented in equation of reaction (2). The remained or unwanted product in this reaction was removed by a washing process using distilled water for several times to achieve a normal pH condition and then followed by a heating process at 100 °C.

$$\begin{aligned} &4(2FeCl_3 + FeCl_2 + 4H_2O) + 3ZnCl_2 + 3MnCl_2.4H_2O + 44NH_4OH \rightarrow \\ &6Zn_{0.5}Mn_{0.5}Fe_2O_4 + 44NH_4Cl + 50H_2O \end{aligned} \quad (2)$$

The obtained $Zn_{0.5}Mn_{0.5}Fe_2O_4$ particle was characterized using XRD to investigate the phase purity, crystal structure, as well as lattice parameter and crystal volume, and also crystallite size. The XRD pattern of the $Zn_{0.5}Mn_{0.5}Fe_2O_4$ particle is presented in Figure 1.

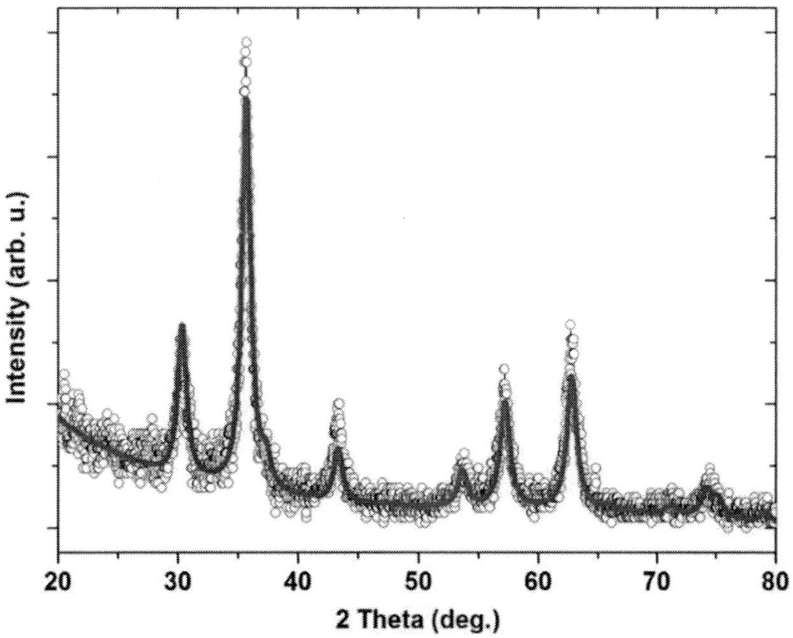


**Figure 1.** X-ray diffraction pattern and of the $Zn_{0.5}Mn_{0.5}Fe_2O_4$ particle synthesized by the coprecipitation-sonochemical method

The XRD data as illustrated in Figure 1 were analyzed qualitatively and quantitatively. From the qualitative analysis, the $Zn_{0.5}Mn_{0.5}Fe_2O_4$ particle formed a single phase without impurity. Meanwhile, from the quantitative analysis, the $Zn_{0.5}Mn_{0.5}Fe_2O_4$ particle built a spinel crystal structure in the cubic system. The measured XRD data and the Rietveld refinement plot are, respectively, shown by the circles and solid line. The lattice parameter of the crystal can be calculated using the following equation [18].

$$a = \frac{\lambda}{2}\sqrt{\left\{\frac{\left(h^2+k^2+l^2\right)}{\sin^2\theta}\right\}} \tag{3}$$

Where $a, \lambda$, *(hkl)*, and $\theta$ represent the lattice parameter, wavelength of X-ray, Miller indices, and Bragg angle, respectively. For this case, the lattice parameter is $a = b = c$, and the crystal volume is $V = a \times b \times c$ as the cubic structure with the space group of *F d* -3 *m Z*. From the quantitative analysis, the $Zn_{0.5}Mn_{0.5}Fe_2O_4$ particle has the lattice parameters and crystal volume of approximately 8.401 Å and 592.9 Å$^3$, respectively. These are in excellent agreement with the results reported by other researchers [19,20].

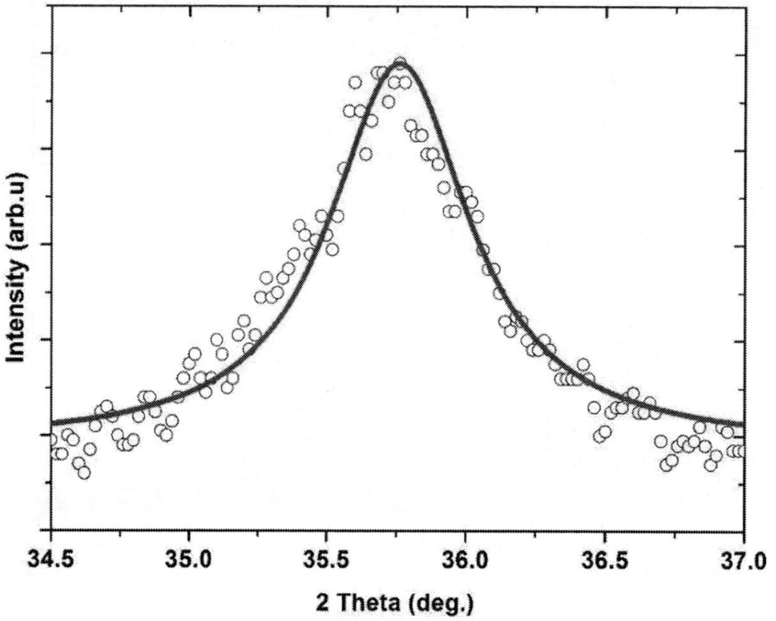


**Figure 2.** FWHM fitting (solid line) of the $Zn_{0.5}Mn_{0.5}Fe_2O_4$ XRD pattern (circle) synthesized by the coprecipitation-sonochemical method

Based on Figure 2, the crystallite size of the $Zn_{0.5}Mn_{0.5}Fe_2O_4$ particle was calculated by using Debye-Scherrer's formula to ensure that the particle formed in nanometric size. The fitting result of the FWHM for the highest peak is presented in Figure 2. The equation (4) represents the Debye-Scherrer's formula [21].

$$D = \frac{K.\lambda}{\beta \cos \theta} \tag{4}$$

Where $D$ is the crystallite size (nm), $K$ is the constant of 0.9, $\theta$ is the Bragg angle (degree), $\beta$ is the excessive line broadening (rad), and $\lambda$ is the wavelength of X-ray (0.15406 nm). From the data analysis in this experiment, the crystallite size of the $Zn_{0.5}Mn_{0.5}Fe_2O_4$ particle prepared by the coprecipitation-sonochemical method is 10.7 nm. This result is quite similar to the crystallite size of the $Zn_{0.5}Mn_{0.5}Fe_2O_4$ prepared by precursor-based combustion method [22]. It means that the combined method is effective and efficient to prepare $Zn_{0.5}Mn_{0.5}Fe_2O_4$ particle in nanometric size from local natural iron sand. Furthermore, the particle size is also compared to the TEM characterization of the $Zn_{0.5}Mn_{0.5}Fe_2O_4$ nanoparticle as shown in Figure 3.

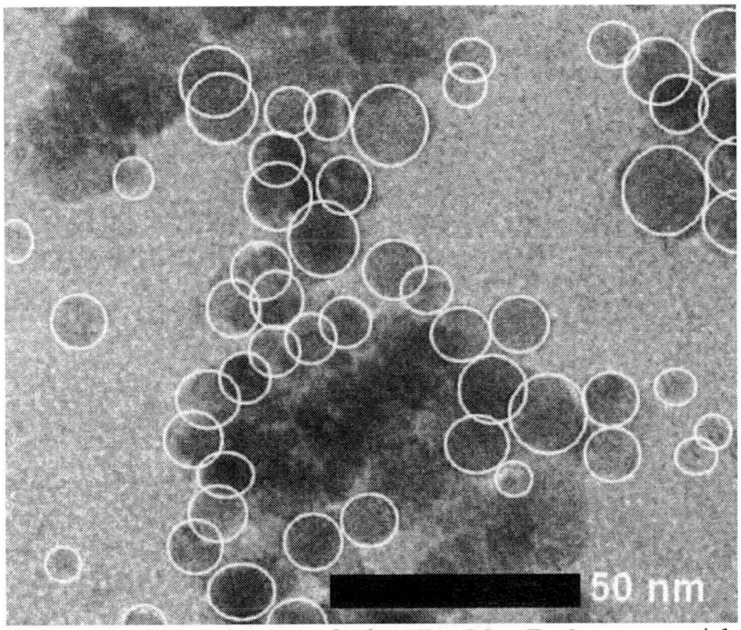


**Figure 3.** TEM image of the $Zn_{0.5}Mn_{0.5}Fe_2O_4$ nanoparticle synthesized by the coprecipitation-sonochemical method

Based on Figure 3, it is clear that the prepared $Zn_{0.5}Mn_{0.5}Fe_2O_4$ nanoparticle in this experiment has nanometric size. Quantitative analysis of the TEM image presented that the average particle size of the $Zn_{0.5}Mn_{0.5}Fe_2O_4$ nanoparticle is 8.7 nm with a standard deviation of 2.4 nm. This result is close to the size calculated from the XRD data. In this work, the $Zn_{0.5}Mn_{0.5}Fe_2O_4$ nanoparticle from iron sand has smaller particle size than that of $Zn_{0.5}Mn_{0.5}Fe_2O_4$ nanoparticle from commercial precursors prepared by microwave induced urea nitrate and coprecipitation processes [23,24]. The $Zn_{0.5}Mn_{0.5}Fe_2O_4$ nanoparticle had a tendency to form a spherical shape with aggregation. The small particle size can contribute the phenomenon of agglomeration in the magnetic nanoparticles.

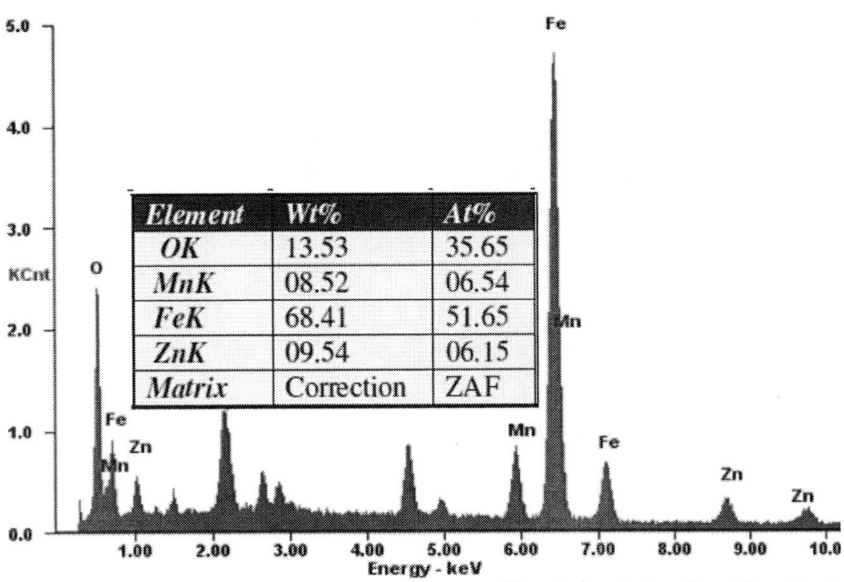


| *Element* | *Wt%* | *At%* |
|---|---|---|
| *OK* | 13.53 | 35.65 |
| *MnK* | 08.52 | 06.54 |
| *FeK* | 68.41 | 51.65 |
| *ZnK* | 09.54 | 06.15 |
| *Matrix* | Correction | ZAF |

**Figure 4.** EDS Spectra and elemental quantification of the $Zn_{0.5}Mn_{0.5}Fe_2O_4$

nanoparticle synthesized by the coprecipitation-sonochemical method

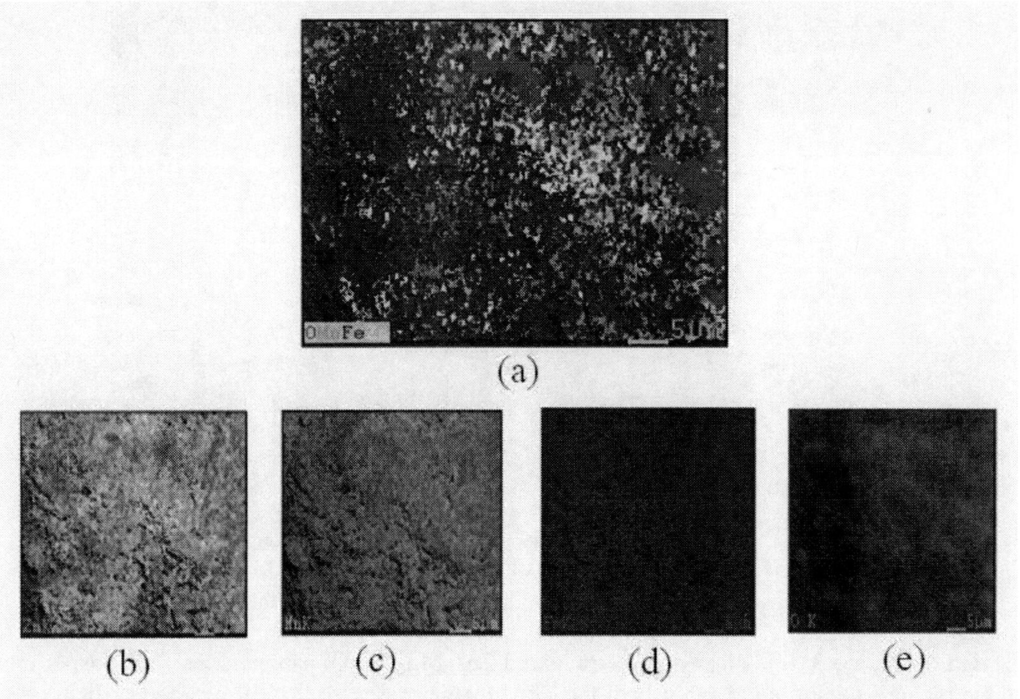


**Figure 5.** EDS Mappings of (a) all elements, (b) Zn, (c) Mn, (d) Fe, and (e) O for the $Zn_{0.5}Mn_{0.5}Fe_2O_4$ nanoparticle synthesized by the coprecipitation-sonochemical method

The result of Rietveld analysis of the XRD data for the $Zn_{0.5}Mn_{0.5}Fe_2O_4$ nanoparticle became physical evidence that the Mn and Zn ions successfully substituted Fe ion in the crystal structure. In order to enhance our knowledge about the composition and distribution of the metallic elements and oxygen, further investigation was also carried out by using SEM combined with the EDS. The investigation resulted in EDS spectra, elemental quantification as shown in Figure 4, and also elemental mapping for all elements and individual element as presented in Figure 5. From the data, the selected elements such as iron, zinc, manganese, and oxygen tended to distribute randomly in all space of the material. Furthermore, the distribution of iron was dominant because it has the highest composition. Interestingly, the atomic composition of zinc and manganese was quite similar representing that the theoretical calculation (Mn : Zn = 0.5 : 0.5) was appropriate for the experimental data.

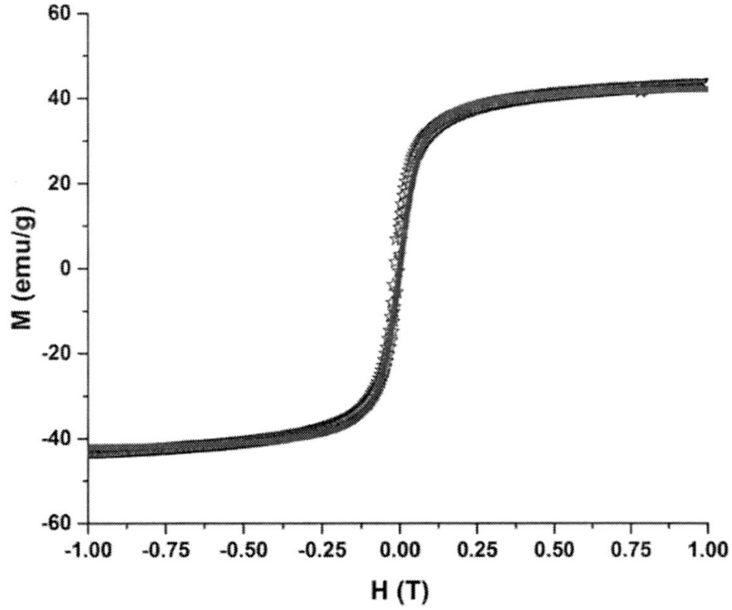


**Figure 6.** Magnetization curve of the $Zn_{0.5}Mn_{0.5}Fe_2O_4$ nanoparticle synthesized by the coprecipitation-sonochemical method

Figure 6 presents the magnetization curve *M(H)* of the $Zn_{0.5}Mn_{0.5}Fe_2O_4$ nanoparticle at room temperature with the magnetic field (*H*) in the range of -1 to 1 T. Qualitatively, the *M(H)* curve of the $Zn_{0.5}Mn_{0.5}Fe_2O_4$ nanoparticle displays *S* profile with the remanent magnetization ($M_r$) and the coercive force ($H_c$) values whih are nearly negligible. Therefore, these characters pointed out that the $Zn_{0.5}Mn_{0.5}Fe_2O_4$ nanoparticle demonstrates a superparamagnetic character [25]. Moreover, to study the magnetic properties comprehensively, the quantitative analysis of the *M(H)* curve was conducted via fitting analysis. The data analysis was done by employing Langevin function according to the following equation [26].

$$M = M_S\left(\coth\left(\frac{\mu H}{k_B T}\right) - \frac{k_B T}{\mu H}\right) \quad (5)$$

Where $M$ is the magnetization, $M_s$ is the saturation magnetization, $\mu$ is the magnetic moment, $H$ is the magnetic field, $k_B$ is the Boltzmann constant, and $T$ is the temperature. The solid line and star represent the fitting model and experimental data, respectively. The theoretical calculation fits the experimental data well with the *R*-square of 99.8%. From the result of data analysis, the $M_s$ and $\mu H/k_B T$ values were approximately 43 emu/g and 40, respectively. The saturation magnetization value of the $Zn_{0.5}Mn_{0.5}Fe_2O_4$ nanoparticle prepared in this work is higher than that of Mn-$ZnFe_2O_4$ nanoparticles reported by other researchers [23,27]. The saturation magnetization of the $Zn_{0.5}Mn_{0.5}Fe_2O_4$ in the nanoparticle is lower than that of the bulk as ascribed to the particle size. The saturation magnetization of the spinel-structured materials such as $Zn_{0.5}Mn_{0.5}Fe_2O_4$ nanoparticle can be determined from the magnetic moments of individual ion, the ions distribution in the spinel crystal structure, and the exchange interaction between tetrahedral and octahedral positions.

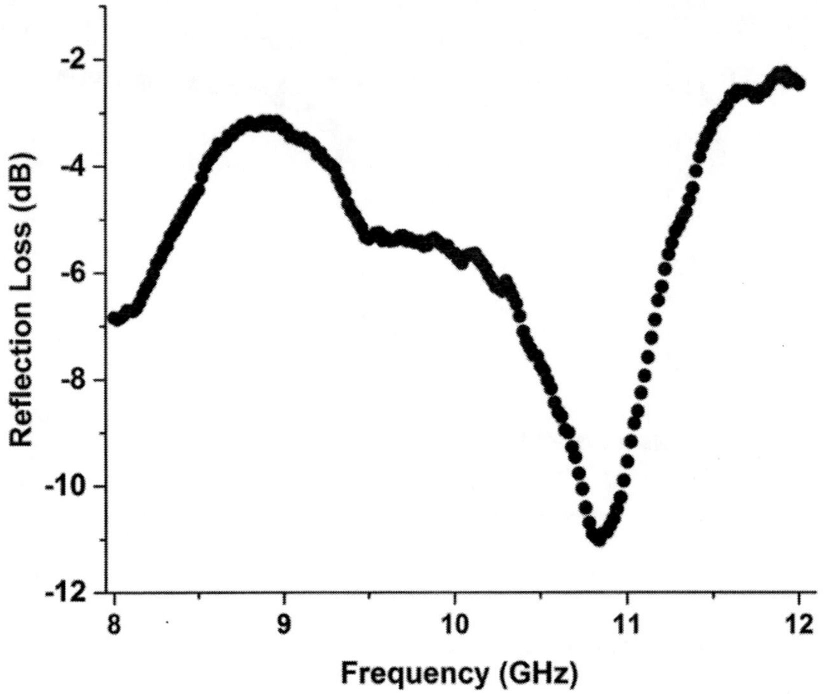


**Figure 7.** The reflection loss of the $Zn_{0.5}Mn_{0.5}Fe_2O_4$ nanoparticle synthesized by the coprecipitation-sonochemical method at the frequency ranged from 8 – 12 GHz

Figure 7 exhibits the reflection loss character of the $Zn_{0.5}Mn_{0.5}Fe_2O_4$ nanoparticle at a frequency ranging of 8 – 12 GHz. The reflection loss of the material which represents the ability to absorb electromagnetic wave can be determined by employing the following mathematical expression [28].

$$RL = 20\log\left|\frac{Z_{in} - Z_0}{Z_{in} + Z_0}\right| \tag{6}$$

Where $RL$, $Z_0$, and $Z_{in}$ are the reflection loss (dB), the characteristic impedance of free space, and input impedance, respectively. The characteristic impedance and the input impedance are presented in equation (7) and (8).

$$Z_0 = \left(\frac{\mu_0}{\varepsilon_0}\right)^{\frac{1}{2}} \tag{7}$$

$$Z_{in} = \left(\frac{\mu_r}{\varepsilon_r}\tanh\left\{j\frac{2\pi ft}{c}(\mu_r\varepsilon_r)\right\}\right)^{\frac{1}{2}} \tag{8}$$

Where $\mu_r$ and $\varepsilon_r$ represent the complex permeability and permittivity, f represents the frequency of operation, t represents the thickness of the sample, and c is the light velocity.

Based on Figure 7, there were two reflection losses of the absorption peaks at frequencies of approximately -7.0 dB and -11.0 dB at 8 GHz 10.8 GHz, respectively. The electromagnetic wave absorption after calculating reaches about 80 % at the highest peak frequency of 10.8 GHz with a thickness of 1.0 mm. Low reflectivity of the material makes excellent absorption performance. It means that the strongest performance of the $Zn_{0.5}Mn_{0.5}Fe_2O_4$ nanoparticle in term of absorbing electromagnetic wave is -11.0 dB at the frequency of 10.8 GHz. Therefore, the $Zn_{0.5}Mn_{0.5}Fe_2O_4$ nanoparticle prepared by a coprecipitation-sonochemical method in this experiment exhibits its sustainability as a radar absorbing material, especially for military purposes. Usually, the radar absorbing materials for military applications are maintained at a frequency in the range of 8 to 12 GHz [8]. The materials should have magnetic and dielectric characters with proper impedance and interaction with the electromagnetic energy to give a high performance in absorbing electromagnetic wave.

## 4. Conclusion

The $Zn_{0.5}Mn_{0.5}Fe_2O_4$ nanoparticle was successfully prepared by employing a mixed coprecipitation-sonochemical method using iron sand as the core precursor. From the XRD data analysis, the $Zn_{0.5}Mn_{0.5}Fe_2O_4$ particle formed a spinel cubic structure in a single phase with the particle size in nanometric scale and a lattice parameter of approximately 8.401 Å. Furthermore, the data analysis of the *M(H)* curve presented that the $Zn_{0.5}Mn_{0.5}Fe_2O_4$ nanoparticle has a superparamagnetic behavior with the saturation magnetization of approximately 43 emu/g. Finally, the data analysis of the reflection loss as a function of frequency showed that the $Zn_{0.5}Mn_{0.5}Fe_2O_4$ nanoparticle performs a radar absorbing material with the absorption performance of -11.0 dB at the frequency of 10.8 GHz.

## 5. References

[1] Kefeni K K, Msagati T A M and Mamba B B 2017 Ferrite nanoparticles: Synthesis, characterisation and applications in electronic device *Mater. Sci. Eng. B* **215** 37–55

[2] Jasso-Terán R A, Cortés-Hernández D A, Sánchez-Fuentes H J, Reyes-Rodríguez P Y, de-León-Prado L E, Escobedo-Bocardo J C and Almanza-Robles J M 2017 Synthesis, characterization and hemolysis studies of $Zn_{(1-x)}Ca_xFe_2O_4$ ferrites synthesized by sol-gel for hyperthermia treatment applications *J. Magn. Magn. Mater.* **427** 241–4

[3] Bashir B, Shaheen W, Asghar M, Warsi M F, Khan M A, Haider S, Shakir I and Shahid M 2017 Copper doped manganese ferrites nanoparticles anchored on graphene nano-sheets for high performance energy storage applications *J. Alloys Compd.* **695** 881–7

[4] Akhtar M N, Khan M A, Ahmad M, Nazir M S, Imran M, Ali A, Sattar A and Murtaza G 2017 Evaluation of structural, morphological and magnetic properties of CuZnNi ($Cu_xZn_{0.5-x}Ni_{0.5}Fe_2O_4$) nanocrystalline ferrites for core, switching and MLCI's applications *J. Magn. Magn. Mater.* **421** 260–8

[5] Zeng T, Yu M, Zhang H, He Z, Zhang X, Chen J and Song S 2017 In situ synthesis of cobalt ferrites-embedded hollow N-doped carbon as an outstanding catalyst for elimination of organic pollutants *Sci. Total Environ.* **593–594** 286–96

[6] Wu X, Yan S, Liu W, Feng Z, Chen Y and Harris V G 2016 Influence of particle size on the magnetic spectrum of NiCuZn ferrites for electromagnetic shielding applications *J. Magn. Magn. Mater.* **401** 1093–6

[7] Akhtar M N, Rahman A, Sulong A B and Khan M A 2017 Structural, spectral, dielectric and magnetic properties of $Ni_{0.5}Mg_xZn_{0.5-x}Fe_2O_4$ nanosized ferrites for microwave absorption and high frequency applications *Ceram. Int.* **43** 4357–65

[8] Folgueras L de C, Nohara E L, Faez R and Rezende M C 2007 Dielectric microwave absorbing material processed by impregnation of carbon fiber fabric with polyaniline *Mater. Res.* **10** 95–9

[9] Darminto, Cholishoh M N, Perdana F A, Baqiya M A, Mashuri, Cahyono Y and Triwikantoro 2011 Preparing $Fe_3O_4$ Nanoparticles from $Fe^{2+}$ Ions Source by Co-precipitation Process in

Various pH *AIP Conference Proceedings* The 4th Nanoscience And Nanotechnology Symposium (Nns2011): An International Symposium vol **1415**(AIP Publishing)pp 234–7

[10] Taufiq A, Sunaryono, Rachman Putra E G, Okazawa A, Watanabe I, Kojima N, Pratapa S and Darminto 2015 Nanoscale Clustering and Magnetic Properties of $Mn_xFe_{3-x}O_4$ Particles Prepared from Natural Magnetite *J. Supercond. Nov. Magn.* **28** 2855–63

[11] Sunaryono, Taufiq A, Munaji, Indarto B, Triwikantoro, Zainuri M and Darminto 2013 Magneto-elasticity in hydrogels containing Fe3O4 nanoparticles and their potential applications *AIP Conf. Proc.* **1555** 53–6

[12] Taufiq A, Sunaryono, Rachman Putra E G, Pratapa S and Darminto 2015 Nano-Structural Studies on $Fe_3O_4$ Particles Dispersing in a Magnetic Fluid Using X-Ray Diffractometry and Small-Angle Neutron Scattering *Mater. Sci. Forum* **827** 213–8

[13] Sunaryono, Taufiq A, Putra E G R, Okazawa A, Watanabe I, Kojima N, Rugmai S, Soontaranon S, Zainuri M, Triwikantoro, Pratapa S and Darminto 2015 Small-Angle X-Ray Scattering Study on PVA/Fe3O4 Magnetic Hydrogels *Nano* **11** 1650027

[14] Sunaryono, Taufiq A, Mashuri, Pratapa S, Zainuri M, Triwikantoro and Darminto 2015 Various Magnetic Properties of Magnetite Nanoparticles Synthesized from Iron-Sands by Coprecipitation Method at Room Temperature *Mater. Sci. Forum* **827** 229–34

[15] Hunter B A 1998 Rietica - A visual Rietveld program *Newsl. Int. Union Crystallogr. Comm. Powder Difraction*

[16] Rietveld H M 1969 A profile refinement method for nuclear and magnetic structures *J. Appl. Crystallogr.* **2** 65–71

[17] Malik Anjelh Baqiya, Ahmad Taufiq, Sunaryono, Khuroti Ayun, Mochamad Zainuri, Suminar Pratapa, Triwikantoro and Darminto 2017 Spinel-Structured Nanoparticles for Magnetic and Mechanical Applications *Magnetic Spinels: Synthesis, Properties and Applications* (InTechOpen) pp 253–72

[18] Cullity B D and Stock S R 2001 *Elements of X-ray diffraction* (Upper Saddle River, NJ: Pearson/Prentice Hall)

[19] Murugesan C and Chandrasekaran G 2016 Structural and Magnetic Properties of Mn1−xZnxFe2O4 Ferrite Nanoparticles *J. Supercond. Nov. Magn.* **29** 2887–97

[20] Li M, Fang H, Li H, Zhao Y, Li T, Pang H, Tang J and Liu X 2017 Synthesis and Characterization of MnZn Ferrite Nanoparticles with Improved Saturation Magnetization *J. Supercond. Nov. Magn.* 1–7

[21] Ahmad S, Riaz U, Kaushik A and Alam J 2009 Soft Template Synthesis of Super Paramagnetic Fe3O4 Nanoparticles a Novel Technique *J. Inorg. Organomet. Polym. Mater.* **19** 355–60

[22] Kundaikar G V S, Tangsali R B, Pissurlekar V J, Budkuley J S and Naik P P 2016 Synthesis of Uniform Size Superparamagnetic Grains of $Mn_xZn_{(1-x)}Fe_2O_4$ Ferrites by Precursor-Based Combustion Method *J. Supercond. Nov. Magn.* **29** 789–94

[23] Maleknejad Z, Gheisari K and Raouf A H 2016 Structure, Microstructure, Magnetic, Electromagnetic, and Dielectric Properties of Nanostructured Mn–Zn Ferrite Synthesized by Microwave-Induced Urea–Nitrate Process *J. Supercond. Nov. Magn.* **29** 2523–34

[24] Venkataraju C and Paulsingh R 2014 FTIR and EPR Studies of Nickel Substituted Nanostructured Mn Zn Ferrite *J. Nanosci.* **2014** 1–5

[25] Rameshbabu R, Ramesh R, Kanagesan S, Karthigeyan A and Ponnusamy S 2014 Synthesis and Study of Structural, Morphological and Magnetic Properties of $ZnFe_2O_4$ Nanoparticles *J. Supercond. Nov. Magn.* **27** 1499–502

[26] Cullity B D and Graham C D 2008 *Introduction to Magnetic Materials* (Hoboken, N.J: Wiley-IEEE Press)

[27] Kareem S H, Shamsuddin M and Lee S L 2016 Effect of Precipitation Temperature on Structural and Magnetic Features of Polyethylene Glycol-Coated $Mn_{0.8}Zn_{0.2}Fe_2O_4$ Nanoparticles *J. Supercond. Nov. Magn.* **29** 2691–7

[28] Mu G, Chen N, Pan X, Shen H and Gu M 2008 Preparation and microwave absorption properties of barium ferrite nanorods *Mater. Lett.* **62** 840–2

**Acknowledgments**
The authors would like to thank the DRPM, KEMENRISTEKDIKTI, Republic of Indonesia for providing the "research grant" for AT 2016-2017.

# The Effect of Sintering Temperature on The Rolled Silver-Sheathed Monofilament Bi,Pb-Sr-Ca-Cu-O Superconducting Wire

**Hendrik, P Sebleku, B Siswayanti and A W Pramono**

Research Center for Metallurgy and Materials-Indonesian Institute of Sciences (P2MM - LIPI), Kawasan Puspiptek, Serpong, Tangerang 15314, Indonesia

Email: hend027@lipi.go.id

**Abstract**. The manufacture of high critical temperature (*Tc*) Bi, Pb-Sr-Ca-Cu-O (HTS BPSCCO) superconductor wire fabricated by power-in-tube (PIT) is a multi-step process. The main difficulty is that the value of *Tc* superconductor wire determined by various factors for each step. The objective of this research is to investigate the effect of sintering parameters on the properties of final rolled material. The fabrication process of 1 m rolled-silver sheath monofilament superconductor BPSCCO wire using mechanical deformation process including rolling and drawing has been carried out. The pure silver powders were melted and formed into pure silver (Ag) tube. The tube was 10 mm in diameter with a sheath material: superconductor powders ratio of about 6 : 1. Starting powders, containing the nominal composition of $Bi_2$-$Sr_2$-$Ca_{m-1}$-$Cu_m$-$O_y$, were inserted into the pure silver tube and rolled until it reached a diameter of 4 mm. A typical area reduction ratio of about 5% per step has been proposed to prevent microcracking during the cold-drawing process. The process of rolling of the silver tube was subsequently repeated to obtain three samples and then followed by heat-treated at 820 °C, 840 °C, and 860 °C, respectively. The surface morphology was analyzed by using SEM; the crystal structure was studied by using X-RD, whereas the superconductivity was investigated by using temperature dependence resistivity measurement by using four-point probe technique. SEM images showed the porosity of the cross-sectional surface of the samples. The sample with low heating temperature showed porosity more than the one with high temperature. The value of critical temperature (Tc) of the sample with a dwelling time of heating of 8 hours is 70 K. At above 70 K, it shows the behavior of conductor properties. However, the porosity increased as the heating time increased up to 24 hours. The critical temperature was difficult to be identified due to its porosity. According to XRD results, the Bi-2212 phase is prominent in all samples.

**Keywords:** High-temperature superconductor, powder-in-tube, BPSCCO, Bi-2223, Bi-2212 phase

## 1. Introduction

Since conventional NbTi and $Nb_3Sn$ low-temperature superconductors dominated the commercial market which was applied to magnetic resonance imaging (MRI) and high-energy physics, the researchers began trying to apply Bi-2223 and REBCO high-temperature superconductors (HTS) for high-resolution NMR, maglev trains, MCZ, as well as magnetic separation of superconducting thin films such as electronic devices and transmission cables for power cables [1-3]. Many attempts in the last ten years have been done to observe the behavior of Bi-2223 / Ag superconducting tape [4-6], while these observations on the superconducting wire were not optimal due to the behavior of the wire drawing to produce porous wire simultaneously with the increase of sintering time. Therefore, in this paper, we will study the behavior of superconducting properties in the form of wire, especially in the Bi-2212/Ag.

The Ag-BPSCCO superconducting wire is very potential to be used to carry electricity for large-scale applications especially power cables. In Ag-BPSCCO superconducting wire, there are several phases system that can be formed. Whereas, the Bi-2223 phase is difficult to make due to its phase formation within a narrow temperature range, other superconducting phase such as Bi-2201, mixture of Bi-2201 and Bi-2212, and Bi-2212 can be easily formed.

To realize the full application of HTS, many issues should be overcome related to powder in tube (PIT) method and mechanical deformation process. To produce Bi-2212 phase/Ag superconducting wire, in-situ PIT method was chosen of which a powder with a nominal of composition of Bi-2223 was filled into Ag-tube and then mechanically deformed into the thin composite wire.

## 2. Experimental Method

Based on the stoichiometry, to get the phase of $Bi_{1.6}Pb_{0.4}Sr_2Ca_2Cu_2O_{8+\alpha}$ , the weight ratio of materials are $Bi_2O_3$ : $Pb(NO_3)_2$: $SrCO_3$: $CaCO_3$: CuO = 3.32: 1.18: 2.63 :1.0 :2.07. Weighing followed by mortal grinding was carried out. The powder mixture was then inserted into silver tube followed by rolling.

### *2.1. Mechanical deformation.*

A pure silver tube (ø inner tube = 4 mm; ø outer tube = 6 mm) was used for the rolling process. The tube was then filled with a mixture of $Bi_2O_3$, $Pb(NO_3)_2$, $SrCO_3$, $CaCO_3$ and CuO powders. An area reduction ratio of about 5% to 4 mm in diameter was carried out. The ratio of sheath material to superconductor fixed at 6:1.

### *2.2. Sintering.*

Three samples were rolled and sintered at 820, 840, and 860 °C for about 8 hrs (heating rate 5 °C/min), respectively. Other samples were heat-treated at 860°C, and then the temperature was held for 24 hours. Cooling to room temperature in the furnace was achieved in about 24 hours.

### *2.3. Resistivity measurement.*

To confirm the nature of the superconductivity of superconducting tape, the samples were measured by using a four-point probe (Figure 1). To expose the powder of superconducting wire to an electric current, the sample was mounted in a resin and cut for one side so that the surface of BPSCCO can be attached via four terminals by using Ag paste.

**Figure 1**. Image of sample Ag-BPSCCO exposed to four points probe

### *2.4. SEM JEOL model 6390A and XRD.*

The morphological analysis of the sample was performed by using scanning electron microscopy (SEM). X-ray diffraction analysis was conducted to investigate the crystal structure of superconducting wire.

## 3. Results and Discussion

The process of sintering at 840 to 860 °C temperature range forms two-phase systems which are Bi-2212 and Bi-2223 as also found in the literature [7]. However, there are possible mixture of two phases when the result of experimental research shows it is difficult to make Bi-2212 single phase material. It is indicated by two modes situated on the curve Bi-2201 at Tc zero (20 K) and *Tc* onset (70 K) as observed in Figure 2.

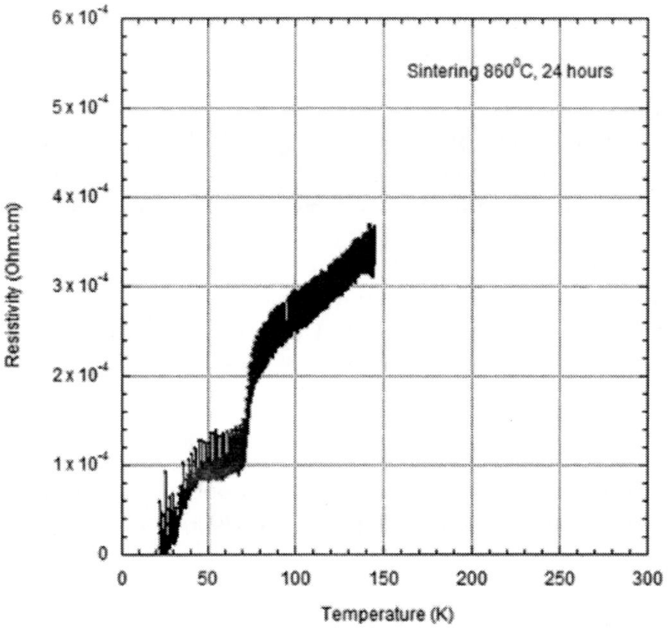


**Figure 2**. Resistivity dependence on temperature for the sample with sintering temperature at 860°C for 24 hours.

Figure 2 depicts our measurements of electrical resistance versus temperature. It points out that after long sintering time of 860°C sintering temperature there is a growing of Bi-2201 together with Bi-2212. This phenomenon suggests that it is likely Bi-2201 phase is a main phase. We also observed the data points in the $\rho$-$T$ curve of a sample sintered for 24 hours (Figure 2) is wider than the one in the curve $\rho$-$T$ of a sample sintered for 8 hours (Figure 3). Several factors can cause this discrepancy. Since the sample was cooled down from room temperature to 1.5 K, the surface cooled faster than the middle of superconductor, hence it is possible to get wider data points and the transition temperature.

Figure 3 depicts our measurements of electrical resistance versus temperature in the three sample wires. It is noted that with regard to 8 hours sintering time sample having a lower sintering temperature of 820°C exhibited no superconductivity behavior. In contrast to the one with lower sintering time, both samples with higher sintering temperature of 840°C and 860°C tend to shift toward superconducting phase having Tc onset 80 K and 85 K, respectively. The obtained value for the sample sintering temperature of 840°C and 860°C is in good agreement with literature data [8-10]. This phenomenon suggested conductor-superconductivity transition behavior occurred at 840°C.

Figures 4 and 5 revealed that the surface morphology of the cross section of wire samples after they have been cut. The image points out that the samples sintered for 8 hours (Figure 4) have less open channel than the sample sintered for 24 hours (Figure 5). The grains of the sample which was sintered for 24 hours are smaller and brighter in contrast to those sample sintered for 8 hours (Figure 4). Sintering in the air atmosphere can be accompanied by the formation of gas bubbles since the bubbles formed in the PIT process are intrinsic in the wire. Several authors claim that the use of overpressure may reduce the gas agglomeration as a source of current flow damage. Hence, it is proposed that the intermediate process like rolling and heat treatment is beneficial to produce the superconducting grains. In addition, sintering under vacuum environment can be used to prevent gas bubble formation.

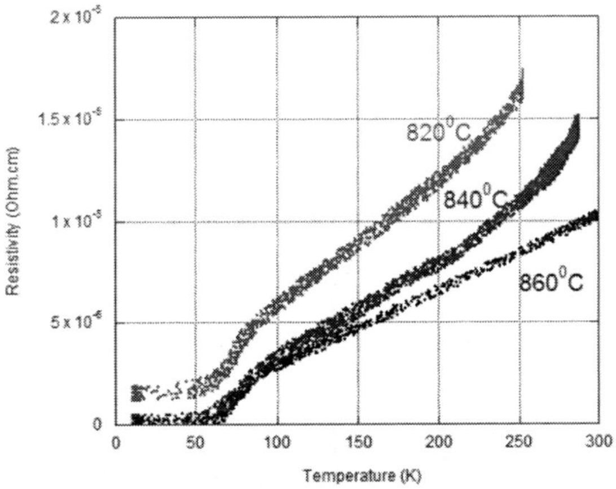


**Figure 3**. Resistivity dependence on temperature for samples with sintering temperature at 820°C, 840°C and 860°C for 8 hours.

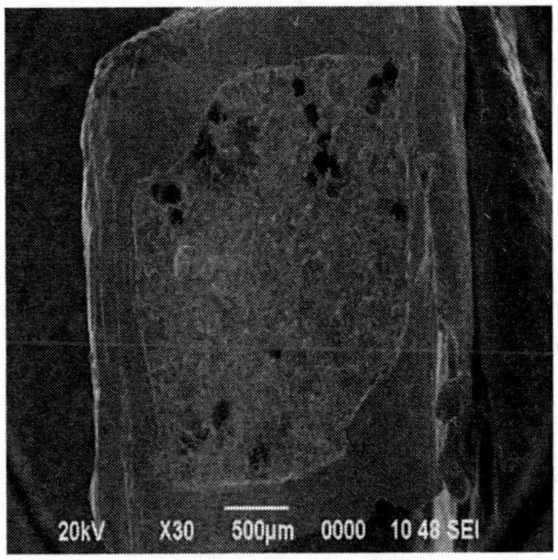


**Figure 4.** The surface morphology of cut sample of Ag-BPSCCO wire with 8-hour sintering time.

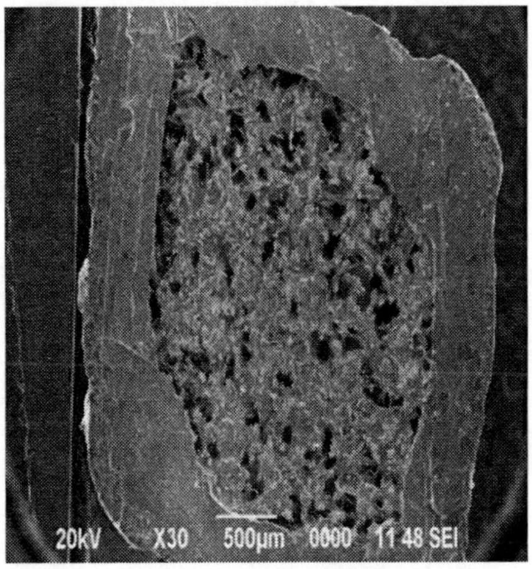


**Figure 5.** The surface morphology of cut sample of Ag-BPSCCO wire with 24-hour sintering time.

Figure 6 shows the XRD patterns of samples with a sintering temperature of 820, 840, and 860°C. Almost single phase of Bi-2212 compounds was detected for sample sintered at 820, 840, and 860°C, along with the small bump at 20° that related to resin during sample preparation.

It is interesting to note that there are various transitions of Bi-2212 with respect to the change in sintering temperature and sintering time. This phenomenon suggests the sample structure of Bi-2212 is disoriented and higher temperature leads to ordering of aligned structure in the copper-oxygen plane. By increasing temperature up to 860°C, upon melting and solidification of the typical of the superconducting grains of Bi-2212 are optimally formed and thus aligned crystallographically in the copper-oxygen plane so that its resistivity reached zero resistance.

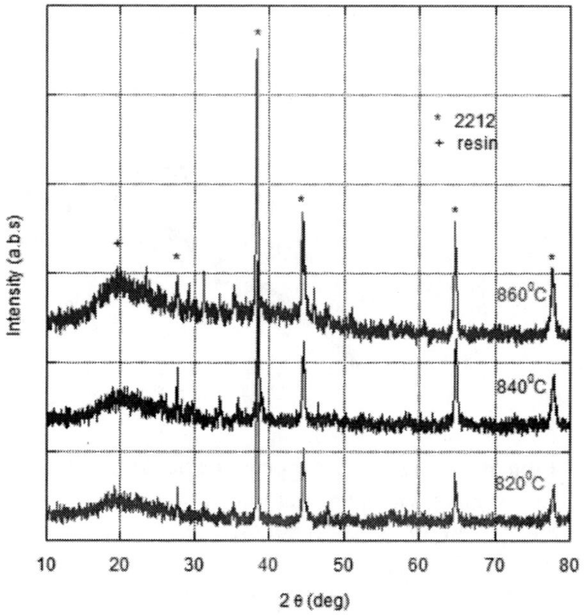


**Figure 6.** XRD result of sample of Ag-BPSCCO with a sintering temperature of 820°C, 840°C and 860°C for 8 hours.

## 4. Conclusion

Four samples of Bi,Pb-Sr-Ca-Cu-O silver-sheathed monofilament superconducting wire with the diameter of 4 mm and a length of 1 meter has been fabricated by the rolling process. Those samples were analyzed by the superconducting properties to provide a favorable path for further processing in the manufacture of superconducting wire. The effect of sintering temperature on crystalline phase formation and resistivity has been studied. The sample sintered at 860°C for 24 hours showed the dominant existence of Bi-2212 phase. Nevertheless, the sample sintered at 820°C for 8 hours has not yet formed a superconducting phase. It was clear that sample sintered at 840°C and 860°C showed that the critical temperature (Tc) shifted to zero resistance (0 ohms) indicated the Bi-2212 phase as also confirmed by X-RD data. All samples sintered for 8 hours are less porous than the one sintered for 24 hours.

## References

[1] Minervini J V 2014 *Natur Mater* **13**,326-327
[2] Matuccielo N, Giuileo F, Grimaldi G and Corat V 2015 *Supercond. Sci. Technol.* **28**
[3] Sato K 2015 *Superconductor in the Power Grid*,75-95
[4] Tjia M O, Darminto, Diantoro M, Rusydi A and Loeksmanto W 2000 *Advances in Superconductivity XII, Yamashita and Tanabe* (Eds) Springer-Verlag 679
[5] Diantoro M, Tjia M O, Kovac P, Husek I 2001 *Physica C* **357-360**, 1182-1185

[6] Oemry F, Diantoro M, Sutjahja I M, Tjia M O, Kopera L, Kovac P, Bonfait G J M 2005 *Physica C* **426-431**, 396-401
[7] Grigorashvili Y E 2012 *Superconductors-Properties, Technology, and Applications*
[8] Kametani F, Shen T, Jiang J, Scheuerlein C, Malagoli A, Michiel M D, Huang Y, Miao H, Parrell J A, Hellstrom E E and Larbalestier D C 2011 *Supercond. Sci. Technol.* **24**
[9] Darsono N, Imaduddin A, Raju K and K et al 2015 *Supercond Nov Magn* **28**,2259-2266
[10] Darsono N, Yoon D H and Raju K 2016 *Supercond Nov Magn* **29**, 1491-1497

# Low-Temperature Nitriding of Pure Titanium by using Hollow Cathode RF-DC Plasma

**J M Windajanti[1,2], D J Djoko H S[1] and Abdurrouf[1]**

[1]Physics Department, University of Brawijaya, Jl. Veteran Malang, 65145, Indonesia
[2]SMAK Santa Maria, Jl. Raya Langsep 41 Malang, 65116, Indonesia

Email: abdurrouf@ub.ac.id

**Abstract**. Pure titanium is widely used for the structures and mechanical parts due to its high strength, low density, and high corrosion resistance. Unfortunately, titanium products suffer from low hardness and low wear resistance. Titanium's surface can be modified by nitriding process to overcome such problems, which is commonly conducted at high temperature. Here, we report the low-temperature plasma nitriding process, where pure titanium was utilized by high-density RF-DC plasma combined with hollow cathode device. To this end, a pure titanium plate was set inside a hollow tube placed on the cathode plate. After heating to 450 °C, a pre-sputtering process was conducted for 1 hour to remove the oxide layer and activate the surface for nitriding. Plasma nitriding using $N_2/H_2$ gasses was performed in 4 and 8 hours with the RF voltage of 250 V, DC bias of -500 to -600 V, and gas pressure of 75 to 30 Pa. To study the nitriding mechanism as well as the role of hollow cathode, the nitrided specimen was characterized by SEM, EDX, XRD, and micro-hardness equipment. The TiN compound was obtained with the diffusion zone of nitrogen until 5 μm thickness for 4 hours nitriding process, and 8 μm for 8 hours process. The average hardness also increased from 300 HV in the untreated specimen to 624 HV and 792 HV for 4 and 8 hours nitriding, respectively.

**Keywords:** Low-temperature, nitriding, titanium, RF-DC plasma.

## 1. Introduction

Titanium is a lightweight metal representing some exceptional properties such as high strength, low density, high corrosion resistance, and stable properties at high temperature. Titanium and its alloys are widely used for structural and mechanical parts in the construction of industrial equipment such as heat exchanger, piping system in petrochemical and offshore industries. They can also be found in aircraft constructions and sports goods. They are also a bio-compatible material for medical implants [1-2]. However, the low surface hardness and the poor wear resistance of their products limit their usage life, especially if there is a friction on contact with other materials [3-4].

Recently, there has been an increasing interest in enhancing the surface properties of pure titanium for wider and more significant applications. To improve the wear resistance of pure titanium, the surface modification, such as nitriding process, is a crucial process in modern manufacturing. The nitriding process can increase the wear resistance of metals by surface hardening. The plasma nitriding is one of the promising nitriding processes, due to its shorter process time and reduced growth of the nitrided parts. In addition, it also saves gasses and keeps the environment clean. Titanium is a metal suitable for nitriding, but previously it needs high-temperature processing. Typically, the nitriding process of titanium has been conducted at a temperature above 700 °C. For example, plasma nitriding process for titanium alloy at a temperature of 1000 °C for 10 hours could be increased the hardness up to 1365 HV [4]. The increasing of surface hardness after nitriding is caused by the increase of nitrogen concentration as a solid solution and the formation of TiN and $Ti_2N$ on the metal surface [3].

The thermo-chemical process for titanium at high temperature causes a degradation of fatigue strength and a detrimental of microstructural changes in titanium substrates. Therefore, we need a method to perform the nitriding process at lower temperatures. In this present paper, a plasma nitriding process at a low temperature of 450 °C is introduced for increasing the surface hardness of pure titanium. A high-density RF-DC plasma with an addition of the hollow cathode tube was conducted for nitriding process. The use of hollow cathode device increases the electric field inside the tube, improves the number of ions and electrons trapped, and therefore increases the plasma density [5]. As a consequence, the specimen was effectively bombarded by the high plasma ion density, giving a higher nitrogen concentration in the surface of the nitrided specimen.

## 2. Experimental Methode

Figure 1 shows the RF-DC plasma system utilized for the present low-temperature plasma nitriding. This system consists of a vacuum chamber, a computerized control unit, a plasma generator, a gas supply sub-system and a vacuum pump system. The vacuum reactor consists of the RF dipole electrodes, the DC bias cathode plate, and the resistive heater is placed under the cathode plate where the temperature of the cathode and the samples on it can be control of the processing unit. At the low-pressure chamber, the glow discharge plasma can be generated independently by a DC bias generator and/or by RF generator with the frequency around of 2 MHz. The matching power condition is automatically adjusted around 2 MHz. The pressure processing and gas flow rate are controlled automatically. An addition of hollow cathode device made from stainless steel rectangular hollow tube was placed on the cathode plate. Setting up of the hollow cathode device is shown in Figure 2.

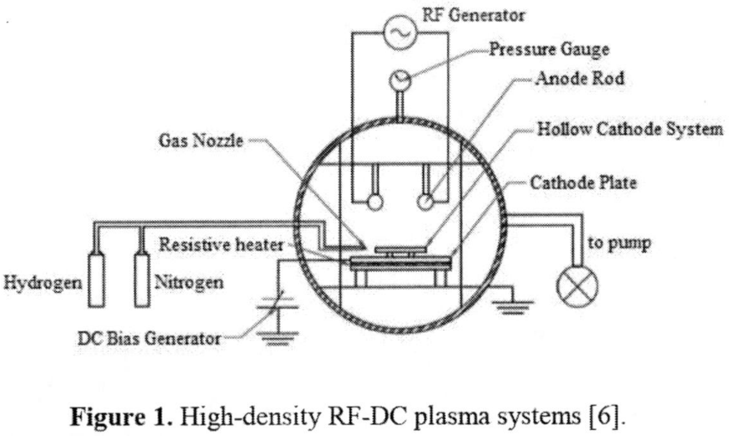


**Figure 1.** High-density RF-DC plasma systems [6].

In this study, a pure titanium plate with a dimension of 18 x 40 x 0.3 $mm^3$ was used as the substrate to investigate the nitriding process. In preparation, the specimen was cleaned with industrial soap (alkali solution) in the ultrasonic cleaner bath because it must be clean from oils and surface surfactants. After that, the titanium specimen was placed in the hollow cathode tube, and then the device was loaded into the chamber, and the system was pumped down to 0.7 Pa to remove oxygen gas from the chamber.

The series of the nitriding process begins by heating the specimen to 450 °C in a nitrogen atmosphere at a pressure of 300 Pa. After the target temperature had been reached, the chamber pressure was lowered back to the value of the set-up process pressure. Before the main nitriding process, a pre-sputtering was performed by applying a high voltage of -500 V DC plasma of nitrogen and hydrogen gas for 1 h. This process was used to remove the oxide layer and activate the surface. The main nitriding process followed the activation process by using both RF and DC plasma. The RF voltage was fixed at 250 V while the DC bias was varied from -500 V to -600 V. The nitriding process was conducted for 4 hours and 8 hours by using a mixed gas with the flow rate of 160 mL/min of $N_2$ gas and 40 mL/min of $H_2$. The effect of gas pressure on the hardening process was studied by varying the pressure of the system. The gas pressure was adjusted at 75 Pa, 50 Pa, and 30 Pa, respectively. After the main process, the specimen was cooled down in a nitrogen atmosphere until the temperature of 300 °C.

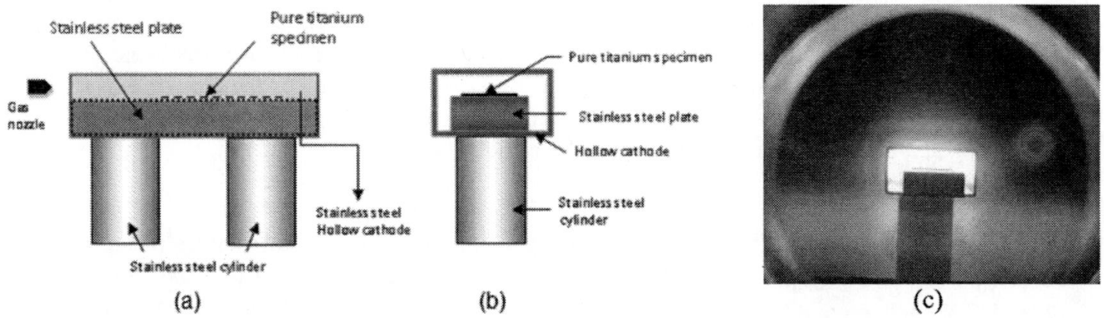


**Figure 2.** Setting up of the hollow cathode device, from a side view (a) and front view (b and c)

The surface of the nitrided specimen was observed by using SEM equipment, Shimadzu SSX-550M. The composition of elements of the nitrided subsurface of the specimen was obtained from EDX point analysis of the cross-sectional area. XRD measurements were carried out through Rigaku Ultima IV XRD system to identify the phases of nitrided materials with monochromated Cu K$\alpha$ radiation ($\lambda$ = 1.54598 Å). The diffraction angle (2$\theta$) was controlled to vary from 3° to 90°. The surface hardness of the specimens was tested by means of Akashi MVK – H1 tester under an applied load of 0.5 N or 50 gf for 10 s for five different positions of surface area.

## 3. Results and Discussion

A diffusion of nitrogen atoms into pure titanium by the nitriding process causes a significant change of its surface morphology. Before the nitriding process, the SEM image at 1600× magnification of the surface morphology (Figure 3, left panel) showed a striped line structure. After 4 hours nitriding process by high-density RF-DC plasma at a temperature of 450 °C, a gas pressure of 30 Pa, and DC bias voltage of -600 V, a top surface of nitrided specimens showed a pronounce of underlying granular structure morphology as seen in Figure 3, right panel.

**Figure 3.** Top surface morphology of a nitrided titanium before (left panel) and after subjecting to plasma nitriding (right panel).

IOP Conf. Series: Materials Science and Engineering **202** (2017) 012026 doi:10.1088/1757-899X/202/1/012026

Observation of cross section cutting by using SEM and EDX equipment showed a thin layer of the surface part with a thickness of few micrometers formed on the outermost surface of every nitrided specimen. Below the surface part, the darker parts were appeared, which is identified as nitride layers and followed by diffusion zone with a matrix of the pure titanium as shown in Figure 4. The increase of surface part thickness was obtained in the process with lower gas pressure and long-term process of 8 hours. At the outermost, a thin layer of the contaminant was identified. This element of Fe, Cr and Ni did not originally come from titanium alloys. These residues coated on the surface of pure titanium came from the stainless steel hollow via high-density plasma nitriding. The EDX point analysis result showed the maximum concentration of 18.55 at.% of nitrogen element detected at a depth of 2 μm for the nitrided sample after 4 hours treatment and amounted to 26.96 at.% detected at a depth of 3 μm on the nitrided sample for 8 hours. The nitrogen diffusion zone was detected up to a depth of 5 μm in the process for 4 hours and 8 μm in the process for 8 hours. Figure 5 shows nitrogen atom concentration from the surface to the nitriding depth of 8 μm in the specimen.

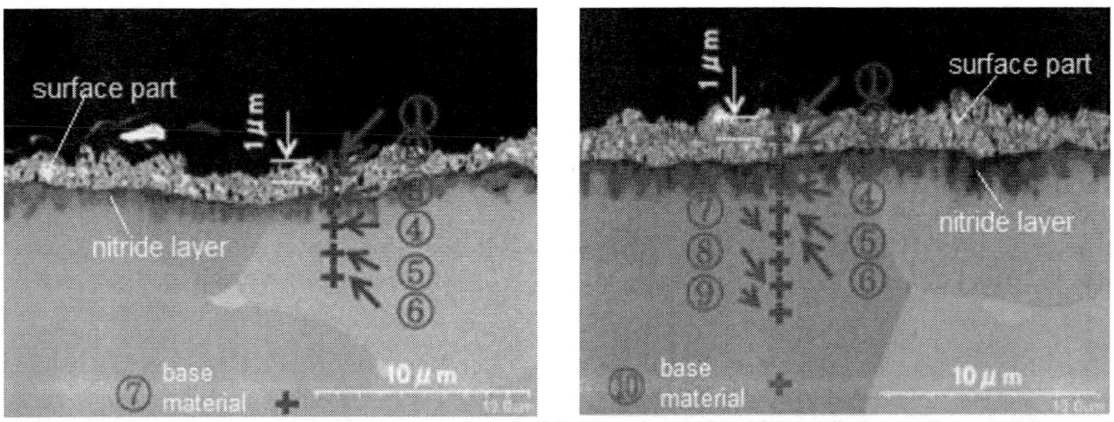

**Figure 4.** The microstructure of cross section nitrided specimen for 4 hours (left panel), and 8 hours (right panel). The nitriding process was conducted under DC bias of -600 V and at a gas pressure of 30 Pa.

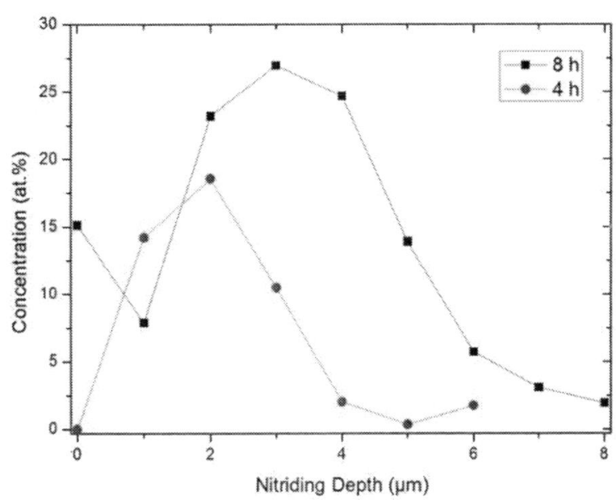


**Figure 5.** Nitrogen concentration versus nitriding depth.

The nitrogen diffusivity depends on energy and ion flux. Correspondingly, the nitrided layer thickness also depends on the energy and ion flux. Therefore, the higher the concentration of nitrogen

IOP Conf. Series: Materials Science and Engineering **202** (2017) 012026 doi:10.1088/1757-899X/202/1/012026

in the nitrided specimen leading to higher ion flux, so that the nitrogen diffusivity will increase as well. In this present study, the use of the hollow cathode device is proven to increase the plasma density, which is used to expose the titanium specimen. Then, the high-density plasma combined with the high energy of nitrogen ions causes the increase of nitrogen depth. The relationship between nitrogen concentration and hardness is an important subject of nitriding process, The incorporation of nitrogen increases, in general, the hardness of the material, so the physical result is an increase in the surface hardness of nitrided specimen.

In addition to changes in the surface morphology, the nitriding process using a hollow cathode device also causes a phase transformation of α-Ti to the metastable phase of ω-Ti accompanied by the formation of the TiN compound shown in the XRD spectroscopy results in Figure 6.

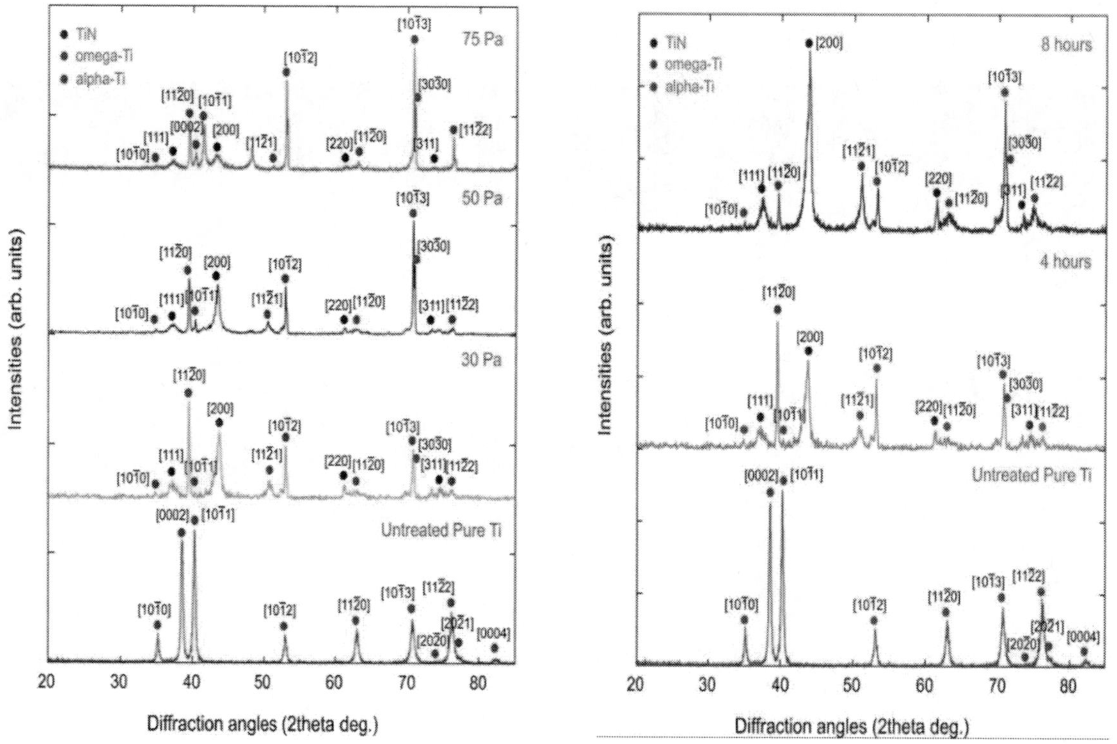


**Figure 6.** Left panel: Comparison of X-ray diffraction patterns of the nitrided specimen after subjecting for 4 hours to plasma nitriding by using a DC bias voltage of -600 V for different gas pressure at 75 Pa, 50 Pa, and 30 Pa. Right panel: Comparison of X-ray diffraction patterns of the nitrided specimen after subjecting to plasma nitriding by using a DC bias voltage of -600 V at a gas pressure of 30 Pa for different time processing of 4 hours and 8 hours.

In the nitrided specimen, there were 3 phases identified, i.e. α-Ti, ω-Ti, and TiN, respectively. The calculation results of lattice parameters for each phase of pure titanium and nitrided samples are summarized in Table 1. The c/a ratio of α-Ti phase around 1.58 and the c/a ratio of ω-Ti around 0.61 are obtained from the nitrided specimens. The phase transformation occurs because a high concentration of nitrogen was deposited in the interstitial lattice of pure titanium.

In the process of using lower pressure, there is an increasing formation of ω-Ti and TiN with the calculation of the interphase concentration of nitrided specimens were summarized in Table 2. Based on Table 2 it appears that the gas pressure reduction even triggers a phase transformation from α-Ti to ω-Ti, while the increasing concentration of ω-Ti also intensify the formation of TiN. In the long process of 8 hours, the interphase concentration of TiN increased significantly while the interphase concentration ω-Ti decreased. It can suggest that the formation of TiN as a compound layer which occurs more effectively after the saturation condition in the diffusion process. In this study, the saturation condition occurs faster in use of low gas pressure combined with high voltage DC bias.

**Table 1**. Lattice parameter of α-phase, ω-phase and TiN in the untreated pure Titanium and seven plasma nitrided specimens

| Sample Condition | α-Ti phase | | | ω-Ti phase | | | TiN |
|---|---|---|---|---|---|---|---|
| | *a* = *b* (Å) | *c* (Å) | *c*/*a* | *a* = *b* (Å) | *c* (Å) | *c*/*a* | *a*=*b*=*c* (Å) |
| Untreated pure Titanium | 2.9531 | 4.6826 | 1.5857 | | | | |
| Nitrided at 75 Pa, -500 V DC bias, 4 h | 2.9780 | 4.7150 | 1.5834 | 4.632 0 | 2.8300 | 0.6110 | 4.1582 |
| Nitrided at 50 Pa, -500 V DC bias, 4 h | 2.9510 | 4.6650 | 1.5808 | 4.657 2 | 2.8754 | 0.6174 | 4.3535 |
| Nitrided at 30 Pa, -500 V DC bias, 4 h | 2.9450 | 4.6650 | 1.5840 | 4.610 0 | 2.8100 | 0.6095 | 4.2520 |
| Nitrided at 75 Pa, -600 V DC bias, 4 h | 2.9527 | 4.6830 | 1.5860 | 4.651 7 | 2.8643 | 0.6158 | 4.3950 |
| Nitrided at 50 Pa, -600 V DC bias, 4 h | 2.9520 | 4.6680 | 1.5813 | 4.649 9 | 2.8580 | 0.6146 | 4.2870 |
| Nitrided at 30 Pa, -600 V DC bias, 4 h | 2.9450 | 4.6610 | 1.5827 | 4.632 0 | 2.8370 | 0.6125 | 4.2590 |
| Nitrided at 30 Pa, -600 V DC bias, 8 h | 2.9450 | 4.6560 | 1.5810 | 4.657 8 | 2.9179 | 0.6265 | 4.2460 |

**Table 2**. Phase concentration of α-phase, ω-phase and TiN in seven plasma nitrided specimens

| Sample Condition | Phase concentration (at.%) | | |
|---|---|---|---|
| | α-Ti | ω-Ti | TiN |
| Nitrided at 75 Pa, -500 V DC bias, 4 h | 66.45 | 33.46 | 0.46 |
| Nitrided at 50 Pa, -500 V DC bias, 4 h | 64.16 | 33.98 | 1.87 |
| Nitrided at 30 Pa, -500 V DC bias, 4 h | 53.18 | 40.73 | 6.09 |
| Nitrided at 75 Pa, -600 V DC bias, 4 h | 89.65 | 8.64 | 1.71 |
| Nitrided at 50 Pa, -600 V DC bias, 4 h | 62.49 | 30.38 | 7.13 |
| Nitrided at 30 Pa, -600 V DC bias, 4 h | 34.04 | 57.25 | 8.71 |
| Nitrided at 30 Pa, -600 V DC bias, 8 h | 24.79 | 37.11 | 38.1 |

One of the factors that can explain why the addition of hollow cathode can improve the efficiency of nitriding process is the high-density of the plasma inside the hollow pipe. In observation with the naked eye, the high-density plasma is characterized by higher emitted light intensities inside hollow tube compared to a planar cathode glow discharge as shown in Figure 2c because there was a high field current on the cathode surface [7]. The effect of the hollow cathode geometry is an increase in the density of plasma because of the electrostatic trapping of electrons between hollow cathode gaps. The trapping electrons oscillate inside the cavity, expanding all their energy for excitation and ionization of atoms [8]. These electrons cause a significant increase in the various collision processes between the plasma species (ions, electrons, atoms and neutral molecules), intensifying the ionization reactions and excitations [9]. As a result, there is also an increase in the quantity of the species (comprised basically of positive ions and neutral species) that impinge on the cathode surface (with a sample placed inside the cathode) that it is connected to the negative power source. Moreover, the result of this interaction is the bombardment of nitrided surface specimens by the plasma species which tend to activate surface diffusion of the nitrogen atoms that are deposited on the sample to achieve the formation of the TiN compound layer [10].

The increasing of plasma density inside the hollow cathode tube also depends on the pressure. In low pressure, the mean free path of electrons is large. Hence the electrons gain a lot of energy from applying an electric field to create more excitement and ionization through inelastic collisions with other plasma particles, and then the higher density of ion and electron was formed inside the hollow tube [5]. The effect of gas pressure on the hollow cathode discharge depends on the product gas pressure by the inter-cathode distance (p.d). Hence the decrease in p.d value by decreasing pressure for fixed inter-cathode distance in this study made the density of ions increases sharply. The other reason is it was connected with higher efficiency of ionization in the hollow cathode cavity, which is determined by the ratio of sheath width of gap lengths. The ionization efficiency is due to the ability of oscillating electrons to create secondary electrons in the sheath that caused an exponential increase of the discharge current when a substantial fraction of ionization events occur in the sheath [8, 11]. In this study, it was observed that the sheath width increased in the lower gas pressure, so the ratio of sheath width of gap lengths increased too. This condition shows that the higher ion density will be formed as the result of decreasing gas pressure in the nitriding process by using the hollow cathode device. The final result is expected on the increase of ion density is the possibility of the more effective diffusion process so that the surface hardening process can occur more optimally.

The increase of nitrogen concentration as interstitial elements leads to a drastic increase in surface hardness of nitrided specimen. In this study, the initial value of surface hardness of pure titanium increased from 300 HV to 624 HV after nitriding process by using DC bias of -600 V and gas pressure of 30 Pa for 4 hours, and then the hardness value increased again to 792 HV after 8 hours time processing. The trends of increasing hardness value against the lowering gas pressure after subjected to 4 hours plasma nitriding process are shown in Figure 7. It appeared that the hardness value increased in the process of using lower pressure, which is the higher hardness is found in the process by using DC bias of -600 V rather than -500 V with same gas pressure. It shows that the hardening process occurs more effectively in the nitriding process carried out at the lower gas pressure and the higher value of DC bias.

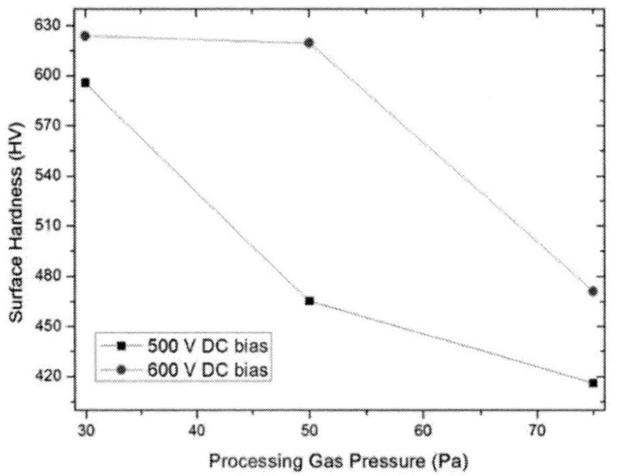


**Figure** 7. The correlation between increasing hardness value and lowering gas pressure.

## 4. Conclusion

The high-density plasma generated by RF-DC plasma system was applied to low-temperature nitriding of the pure titanium at 450 °C. The use of hollow cathode device with lowers gap distance increasing ion density in plasma, and greatly affects the success of the titanium nitriding process as a process at high temperature. The effective diffusion process of nitrogen atoms as a solid solution into a titanium matrix drives the increase of surface hardness. Variation of gas pressure and DC bias voltage as plasma nitriding parameters plays a major role in the hardening mechanism of pure titanium. The lower gas pressure combined with a high DC bias voltage can increase the surface hardness significantly because it causes the phase transformation of α-titanium to ω-titanium which drives a formation of fine TiN precipitates as the compound layer. The formation of ω-titanium can be a trigger for the formation of the compound layer because it facilitates the higher possibility of diffusion processes in the nitrided specimen. The use of lower pressure combined with a higher DC bias voltage also accelerates the saturation condition in the diffusion process of nitrogen atoms into a titanium matrix and then it will increase the concentration of TiN on the top surface of nitrides specimen.

## 5. References

[1] Buijs K and Stainless V L 2008 *Stainless World* 1-4
[2] Holz M and Walker G P 2011 *Ti Fact, International Titanium Association* (USA: Northglenn CO) p. 1-24
[3] Zhecheva A, Wei Sha, Malinov S and Long A 2005 *J. of Surface & Coating Technology* **200** pp 2192-2207
[4] Koyuncu E, Kahraman F and Karadeniz O 2009 *J. of Achievements in Material and Manufacturing Engineering* **37** (2) pp 434-441
[5] Yunata E E, Aizawa T, Santjojo D J H D 2014 *8$^{th}$ SEATUC Meeting and Symposium* (Malaysia: Universiti Teknologi Malaysia) **6** (14) pp1-4
[6] Aizawa T, private communication
[7] Awsi S K 2013 *American Journal of Modern Physics* **2** (6) pp 276-281
[8] Kolobov V I and Tsendin L D 1995 *Plasma Source Sci. Technol.* **4** pp 551-560
[9] Brunatto S F, Kühn I and Muzart J L R 2001 *J. Materials Research* **4** (4) pp 245-250
[10] Alves C Jr, Guerra Neto C L B, Morais G H S, da Silva C F and Hajek V 2006 *Surface & Coating Technology* **200** pp 3657-3663
[11] Barankova H and Bardos L 2001 *Surface and Coatings Technology* **146–147** pp 486-490

## Acknowledgements

The authors would like to thank Tatsuhiko Aizawa from Shibaura Institute of Technology, Japan for the research facilities provided as well as guidance for the implementation of the research in LLC Nanofilm & Coat Laboratory, Ota-ku, Tokyo. The authors are also thankful to ASMAT Research Collaboration, University of Brawijaya, which is funded through *Program Hibah Kerjasama Luar Negeri JSPS Dikti* with the contract number 033/SP2H /LT/DRPM/2/2016, so that this research can be accomplished.

# Optical Properties Characterization of Gamma Irradiated $CeO_2$ Nanoparticles Solution

**I Nurhasanah, A Luthfia and Z Arifin**

Physics Department, Diponegoro University, Jl. Prof. Sudharto, SH, Semarang 50275, Central Java, Indonesia

Email : nurhasanah.siregar@mail.ugm.ac.id

**Abstract.** Optical properties of $CeO_2$ nanoparticles solution have been studied to explore the possibility of $CeO_2$ nanoparticles as a liquid chemical dosimeter to measure gamma radiation in radiotherapy. $CeO_2$ nanoparticles were synthesized by ultrasound irradiation of cerium nitrate solution. The $CeO_2$ nanoparticles were characterized by X-ray diffraction (XRD) and scanning electron microscope (SEM). $CeO_2$ nanoparticles solution with concentration of 2.5 x $10^{-4}$ g/ml was irradiated by 1.3 MV $^{60}Co$ gamma rays with a dose of 1.0 to 2.5 Gy. The UV absorbance spectrum of unirradiated and irradiated $CeO_2$ nanoparticles solution was measured by using UV-Vis spectrophotometer. The UV absorbance of $CeO_2$ nanoparticles solution increased with increasing radiation dose and caused a change in the band gap. The change in optical properties of $CeO_2$ nanoparticles solution is correlated to chemical reactions induced by gamma radiation. The obtained chemical yield was in the order of $10^3$ mol/J indicates the sensitivity of $CeO_2$ nanoparticles to gamma radiation. The results show the possibility of $CeO_2$ nanoparticles for use as gamma radiation dosimeter.

**Keywords:** Optical properties, Gamma, irradiation, $CeO_2$, nanoparticle.

## 1. Introductions

Measurement of gamma rays is important in radiotherapy, medical diagnosis, and environmental monitoring. Gamma rays interact with materials through ionization and excitation. Ionization is the key process to detect gamma rays, where part or all of the gamma rays energy give up to an electron in materials. The materials used to detect gamma rays are gas, solid or liquid. Irradiation of gamma rays to the system of an aqueous solution of the solid materials generates free radicals by radiolysis process. The free radicals recombine or interact with other species present in the solution produce the chemical changes and induce optical property changes of the solution. That phenomenon is used as a basis for developing of the chemical dosimeter to determine the amount of energy absorbed from ionizing radiation. The degree of chemical change associated with the absorbed dose is called as *G*-value and determines the sensitivity of chemical dosimeter.

Dosimeters based aqueous chemical system (such as ferrous sulfate and iodide/iodate solution) have been used in radiation dosimetry. The most widely use of chemical dosimeter is ferrous sulfate or Fricke dosimeter due to high sensitivity and high accuracy to X-ray and γ-ray irradiation for high absorbed dose level [1, 2]. This level of detection is not sufficient to be used in radiotherapy. Consequently, some efforts have been made to improve sensitivity and develop standard Fricke dosimeter by adding various organic solutes including benzoic acid, xylenol and sorbitol [2-5]. Those efforts successfully improved the sensitivity of Fricke dosimeter. However, its capability to measure Co-60 γ-ray with the precision of 0,1% is limited for doses range of 5 - 25 Gy [6]. Meanwhile, the daily dose of γ-ray used in radiotherapy is in the range of 1.8 – 2.0 Gy. Therefore, there is interest to find dosimeter for dose level radiotherapy application. Nanoparticles can be used to detect the high radiation energy like as: X-rays, gamma rays, and other particles (β-particles, α-particles) with high sensitivity because of large surface area. $CeO_2$ nanoparticles have attracted attention due to two oxidation states of Ce ($Ce^{+4}$ and $Ce^{+3}$) and potential to be developed for medical application [7-10]. Both oxidation states are associated with optical properties and have two characteristics spectrophotometric UV absorbance peaks. In the 230–260 nm range corresponding to $Ce^{+3}$ absorbance and in the 300-400 nm range corresponding to $Ce^{+4}$ absorbance. Similar to ferric ion in Fricke dosimeter, the property of those $CeO_2$ nanoparticles can be explored as sensitive materials for chemical dosimeter.

In this paper, the optical property of $CeO_2$ nanoparticles solution was studied to explore the potential $CeO_2$ nanoparticles for chemical dosimeter. $CeO_2$ nanoparticles were used in this study produced by ultrasonic radiation as we have reported, previously [11]. The absorbance spectrum of $CeO_2$ nanoparticles solution unirradiated and irradiated by gamma rays with doses range 1 to 2.5 Gy was measured by using UV-Vis spectrophotometer.

## 2. Experimental Method

### *2.1. Synthesize of $CeO_2$ nanoparticles and preparation of $CeO_2$ solution*

To synthesize $CeO_2$ nanoparticles, Cerium (III) nitrate (Sigma Aldrich, Germany) was dissolved in a mixed solvent of aqua demineralized and isopropanol with a volume ratio of 1:6. Ammonium hydroxide (3M) was added drop by drop to solution until pH 10. The solution was irradiated by ultrasound wave with a frequency of 40 kHz for 60 minutes. The product of irradiated solution was then washed and calcined at 100°C for 3 hours. $CeO_2$ nanoparticles solution was prepared by dispersing $CeO_2$ nanoparticles in distilled water with a concentration of $0.25 \times 10^{-4}$ g/ml and then placed into the glass tube.

### *2.2. Material characterization*

X-ray Diffraction (*Phillips PW 1710*) measurement was used to analyze the structure of nanoparticles. Morphology of the synthesized nanoparticles was observed by scanning electron microscope (JEOL JSM-6360LA). Optical properties of nanoparticles were measured by using ultraviolet-visible (UV-Vis) spectrophotometer (Shimadzu 1420).

### *2.3. Irradiation of $CeO_2$ nanoparticles solution*

The $CeO_2$ nanoparticles solution in the glass tube was irradiated by Gamma rays by using Co-60 teletherapy air FCC8000F Series with the energy of 1.30 MV. The irradiation dose was varied at 1, 1.5, 2 Gy, and 2.5 Gy. The absorbance of unirradiated and irradiated $CeO_2$ nanoparticles solutions were measured by using UV-Vis spectrophotometer over the wavelength in the range of 200 nm – 400 nm to observe absorption peak. The absorption peak was further used to analyze response and sensitivity of $CeO_2$ nanoparticles solution.

## 3. Results and Discussion

### *3.1. Structure and morphology of $CeO_2$ nanoparticles*

The diffraction pattern of $CeO_2$ nanoparticle is shown in Figure 1(a). Certain of diffraction peaks were observed which indicate that $CeO_2$ nanoparticles are polycrystalline in nature. The (111), (200), (220), (311) dominant diffraction peaks corresponded to a pure fluorite cubic structure of $CeO_2$ (JPDS 43-1002). The sharp of those peaks is indicating that $CeO_2$ is well crystallized. To confirm the synthesized nanoparticles composed of a single phase with the cubic structure of $CeO_2$, the

IOP Conf. Series: Materials Science and Engineering **202** (2017) 012027 doi:10.1088/1757-899X/202/1/012027

Rietveld refinement yielded the agreement factors are $wRp$ = 13. 95% and $R_p$ = 10.82% with the goodness of fit ($\chi^2$) of 1.02 and the refined cell parameter is 5.4218 Å. Figure 1(b) shows the Rietveld refinement plot of XRD pattern of $CeO_2$ nanoparticles.

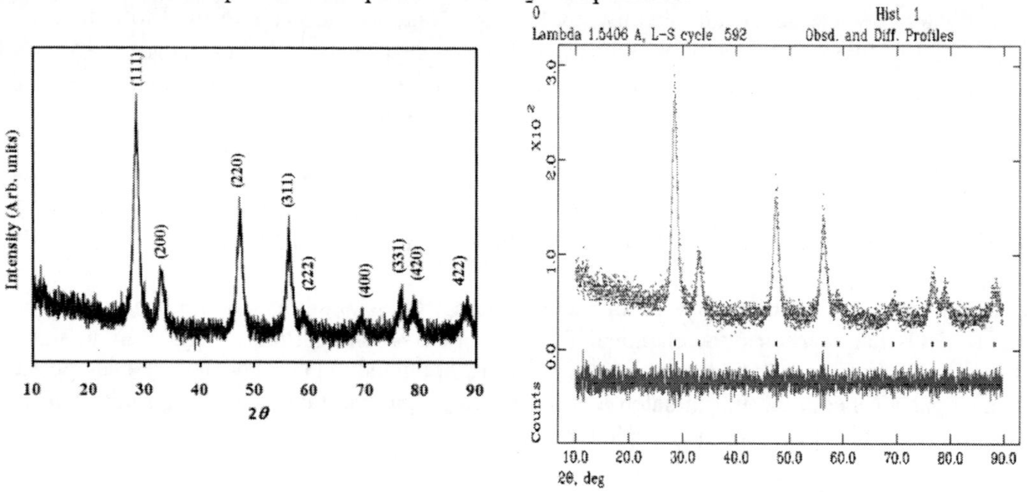


(a) (b)

**Figure 1.** (a) XRD pattern and (b) Rietveld refinement plot of $CeO_2$ nanoparticles. Experimental data is indicated by red tick marks, while calculated curve obtained after refinement is indicated by green solid line. The rose-colored of continuous curve represents the difference between the experimental data and the calculated curve.

The morphology and particles size of $CeO_2$ nanoparticles were analyzed by using SEM image as depicted in Figure 2. The surface morphology of $CeO_2$ appears to be rough and generally spherical in shape with estimated average particle size of 37 nm.

**Figure 2.** SEM image of $CeO_2$ nanoparticles

### 3.2. *Optical properties of $CeO_2$ nanoparticles solution*

The absorbance spectrum of un-irradiated and gamma irradiated $CeO_2$ nanoparticles solution under 1, 1.5, 2, and 2.5 Gy dose of gamma rays are shown in Figure 3. When $CeO_2$ nanoparticles solution was exposed to gamma radiation, the absorbance increased with increasing the gamma radiation dose and the optical absorption edge shifted to a shorter wavelength. Gamma irradiation with a dose of 1 to 2 Gy increased the absorbance slightly and the absorbance increased sharply at gamma irradiation dose of 2.5 Gy. The maximum absorbance was observed at a wavelength of 304 nm and became strong by increasing the dose. The wavelength of 304 nm was then used as a sensitive wavelength reference for further analysis of UV absorbance spectrum.

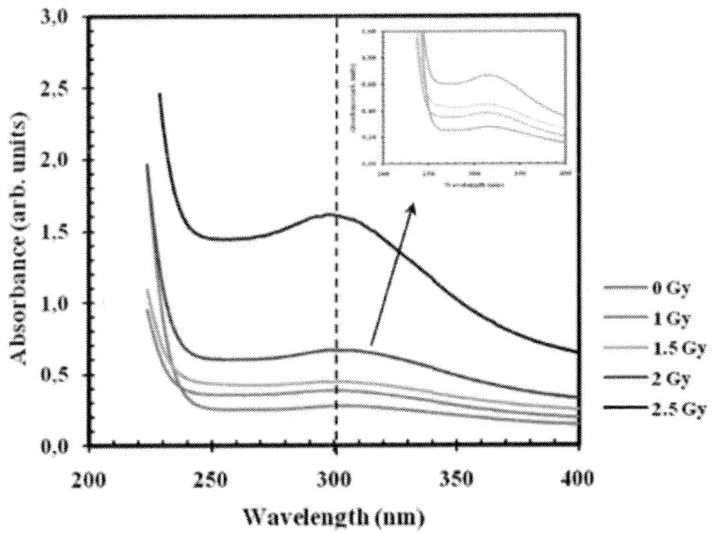


**Figure 3**. UV absorbance spectrum of the irradiated $CeO_2$ nanoparticles solution under gamma rays with different doses. The insert represents the UV absorbance spectrum of $CeO_2$ nanoparticles irradiated with dose 0, 1, 1.5, and 2 Gy.

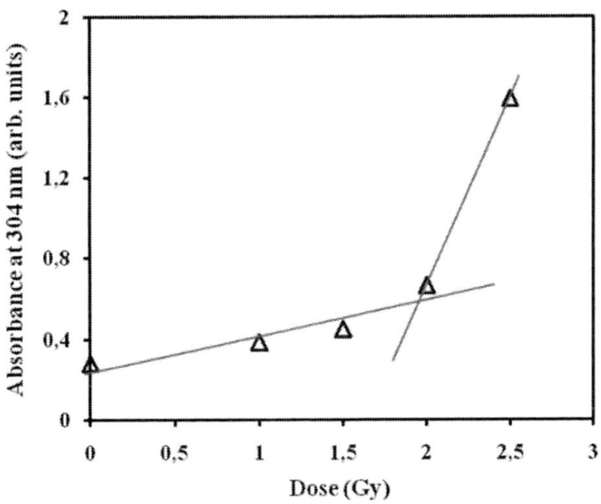


**Figure 4**. UV absorbance at 304 nm of irradiated $CeO_2$ nanoparticles solutions under gamma rays as function of radiation dose.

Figure 4 shows the absorbance at the wavelength of maximum absorption ($\lambda$ = 304 nm) of irradiated $CeO_2$ nanoparticles solution as function of radiation dose obtained from absorption spectra in Figure 3. The absorbance increased linearly with increasing radiation dose to 2 Gy and then the absorbance increased sharply at a dose of 2.5 Gy. It indicates that $CeO_2$ nanoparticles solution is capable of detecting gamma radiation. Capability of $CeO_2$ nanoparticles solution to detect gamma radiation can be explained through basic mechanism in Fricke solution dosimeter [12-13]. When the $CeO_2$ nanoparticles solution irradiated, the dominant reaction is water radiolysis. Radiolysis produces

$e_{aq}^{-}$,H*, OH, $H_2$, $H_2O_2$ that react with $CeO_2$ nanoparticles led to changes in the oxidation number of $Ce^{4+}$ into $Ce^{3+}$ and increases the absorbance [10].

The changes in absorbance of $CeO_2$ nanoparticles solution due to gamma radiation and the peak absorbance at 304 nm corresponded to the $CeO_2$ band gap. An increase in radiation dose causes an increase in number of radical and $Ce^{3+}$ ion formation at $CeO_2$ nanoparticles surface, and then change the band gap. The band gap of unirradiated and irradiated $CeO_2$ nanoparticles solution were determined by plotting $(ahv)^2$ vs $(hv)$ for direct transition as described by Equation (1).

$$(\alpha h\vartheta)^n = A(h\vartheta - E_g) \tag{1}$$

$\alpha$ is the optical absorption coefficient, $hv$ is a photon energy (eV), $E_g$ is a band gap, $A$ is a constant, and $n = ½$ for an indirect transition or $n = 2$ for direct transition. The calculated band gap is tabulated in Table 1. An increase in band gap with gamma radiation doses is mainly caused by changes in valence state and point defect formation due to radiation. The point defect may lead blueshift in the band gap [14-18].

**Table 1.** Band gap of $CeO_2$ nanoparticles for various gamma radiation doses.

| Dose (Gy) | Band gap (eV) |
|---|---|
| 0 | 3.163 |
| 1.0 | 3.164 |
| 1.5 | 3.166 |
| 2.0 | 3.208 |
| 2.5 | 3.356 |

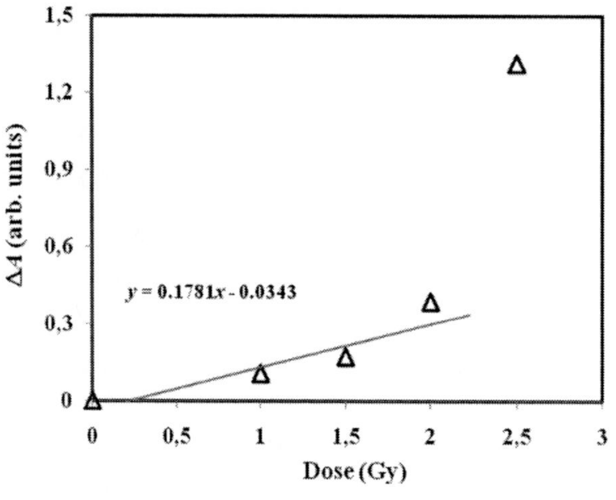


**Figure 5.** UV absorbance changes of $CeO_2$ nanoparticles solution for various doses

### *3.3. Sensitivity of $CeO_2$ nanoparticles*

The sensitivity of $CeO_2$ nanoparticles solution to gamma radiation can be determined from the change in absorbance ($\Delta A$) and the radiation chemical yield ($G$-value) as described in Equation (2) [18].

$$G = \frac{\Delta A}{D\varepsilon\rho b} \tag{2}$$

$\Delta A$ is the change in absorbance at $\lambda_{max}$, $D$ is a dose (Gy), $\varepsilon$ is the linear molar extinction coefficient ($L.mol^{-1}.cm^{-1}$), $\rho$ is a density ($gr.cm^{-3}$), and $b$ is the optical path length (1 cm). The change in absorbance of irradiated $CeO_2$ nanoparticles solution as function of irradiation dose is shown in Figure 5. The slope of plot of $\Delta A$ to dose in the range 0 – 2 Gy is used to calculate the *G*-value and the result was found to be 3.72 x $10^3$ mol/J. The *G*-value obtained in this study is greater than *G*-value of other chemical dosimeter that is ~ 1.6 x $10^{-6}$ mol/J for Fricke dosimeter [6] and ~14 mol/J for iodide dosimeter [19]. The result shows that $CeO_2$ nanoparticles solution possess a high sensitivity to gamma radiation.

## 4. Conclusion

The optical properties of gamma irradiated $CeO_2$ nanoparticles have been characterized by using UV-vis spectrophotometer. The UV absorbance spectrum shows the peak absorbance at 304 nm was corresponding to the band gap of $CeO_2$ nanoparticles. Gamma radiation causes an increase in UV absorbance and the band gap of $CeO_2$ nanoparticles. High sensitivity and linear response in the range of 0 – 2 Gy indicate the possibility of $CeO_2$ nanoparticles to be developed as a chemical dosimeter for radiotherapy application.

## 5. References

[1] Juarez-Calderon J M, Negron-Mendoza A and Ramos-Bernal S 2007 *Rad. Phys. Chem.* **76** 1829
[2] Gupta, B L 2003 *Radiat. Phys Chem.* **67** 737-743
[3] Bansal A K, Semwal M K, Arora D, Sharma D N, Julka P K and Rath G K 2013 *Phys. Med.* **29** 368-373
[4] Mhatre S G V and Adhikari S 2012 *Radiat. Meas.* **47** 430-433
[5] Moussous O, Medjadj T and Benguerba, M 2011 *Appl. Radiat Isot.* **69** 399-342
[6] Moussous O, Khoudri S and Benguerba, M 2011 *Australas. Phys. Eng. Sci. Med.* 34 523-528
[7] Trovarelli A 1996 *Catal. Rev. Sci. Eng.* **38** 439-459
[8] Arshak K and Korostynka O 2004 *Sensor and actuator A* **113** 319-323
[9] Das M, Patil S, Bhargava N, Kang J F, Riedel L M, Seal S, and Hickman J J 2007 *Biomaterials* **28** 1918-1925
[10] Heckert E G, Karakoti A S, Seal S and Self W T 2008 *Biomaterials* **29** 2705-2709
[11] Efendi A F and Nurhasanah I 2015 *Mater. Sci Forum* **827** 56-61
[12] Liu Y, Hu J and Wang J 2014 Radiat. Phys Chem 96:81-87.
[13] Costa C, Vandenborre J, Crumiere F, Blain G, Essehli R and Fatahhi M 2012 *Am. J Anal. Chem.* **3** 6-11
[14] Chen, M Y, Zu, X T, Xiang, X, and Zhang, H L 2007 *Phys. B* **389** 263-268
[15] Fan, S, Yu, C, He, D, Li, K., and Hu, L 2011 *Radiat. Meas.* **46** 46-50
[16] Kaur R, Singh S, and Pandey O P 2014 *J. Mol. Struc.* **1060** 251-255.
[17] Elbatal H A, Mandouh Z, Zayed H, Marzouk S Y, Elkomy G and Hosny, A 2010 *Phys. B* **405** 4755-4762
[18] Gafar S M, and El-Ahdal, ME 2014 *Dyes Pigments.* **109** 67-71
[19] Rahn R O, Gerstenberg H M and Vavrina M A 2003 *Appl. Radiat. Isot.* **56** 525-534

## Acknowledgments

This work was financially supported by Ministry of Research, Technology and High Education of The Republic of Indonesia as part of *Hibah Bersaing* project.

IOP Conf. Series: Materials Science and Engineering **202** (2017) 012028 doi:10.1088/1757-899X/202/1/012028

# Microstructures and Mechanical Study of Mg Alloy Foam Based on Mg-Zn-Ca-$CaCO_3$ System

**A Erryani, F Pramuji ,D Annur, M I Amal and I Kartika**

Research Center for Metallurgy and Materials, Indonesian Institute of Science (LIPI) Gedung 470, Kawasan Puspiptek Serpong Tangerang Selatan, 15314, Indonesia

Email: apri011@lipi.go.id

**Abstract**. Magnesium alloy, a material that has potential to use some applications such as aerospace components, computer parts, and mobile phones. Magnesium alloy can also be a popular candidate as an orthopedic implant material for biodegradability, non-toxicity, and mechanical and physical properties that are excellent. Magnesium, one of the main macro elements required for the proper functioning of the human organism, is used to test the materials for biodegradable implants. The main objective of this study was to find out the microstructure, and mechanical characteristics of the Mg-Ca-Zn-$CaCO_3$ alloy as porous implant materials are biodegradable. The presence of $CaCO_3$ on the alloy functions as a foaming agent expected to produce gas bubbles during manufacturing process taken place that will form pores in the alloy. Mg-Ca-Zn-$CaCO_3$ alloy was made by powder metallurgy method with three variations of composition (96Mg-Ca-3Zn-$CaCO_3$, 91Mg-Ca-3Zn-5$CaCO_3$, and 86Mg-Ca-3Zn-10$CaCO_3$ wt%). Milling process was by using a shaker mill for 2 hours to produce a powder size distribution which was more homogeneous. The mixed powder was uniaxially pressed at a pressure of 100 MPa for 2 minutes and 200 MPa for 3 minutes into green compacts with dimensions of 10 mm in diameter and 10 mm in length. The sintering process was carried out at 650°C with a variation of holding time of 10 and 15 hours, and then the specimens were cooled down at room temperature. Microstructural analysis was performed by using X-Ray diffraction technique and Scanning electron microscopy equipped with an energy disperse spectrometry (EDS). The mechanical characteristics were analyzed by using Universal Testing Machine. The density and porosity of specimen were further measured by using Archimedes method. The results show that the optimum microstructure and mechanical characteristics are the holding time of 10 hours. The value of compression was 208.398 $N/mm^2$, the density was 1.63 g/cc and a porosity was 18% on the composition of 86Mg-Ca-3Zn-10$CaCO_3$ wt%.

**Keywords**: Mg alloy, Microstructure, powder metallurgy method, mechanical characteristic, biodegradable

## 1. Introduction

For many years, people interact increasingly with metals. The pure or alloy metals have shown a significant role in many parts of daily life, especially in the medical field. The materials selection, as well as its composition and the correct process, are the key to be successful against many possible failures. One of metals that can be used for the biomedical application is magnesium, which has low density compared to other metals [1]. Magnesium and its alloys are widely used for structural applications in the aerospace, biomedical and automotive industries as well as in electronics, and other household equipment which needs light-weight property [2- 3].

Due to their excellent comprehensive performance compared with other materials, such as stainless steel, polymers, and ceramics, magnesium alloys become ideal implants [4]. The mechanical properties such as the fracture toughness are higher than hydroxyapatite, Young's modulus and compressive yield strength are closer to those of cortical bone [5- 6]. To improve its mechanical properties, people use alloying process. The choice of alloying elements and its composition are critical. Alloying elements determine the microstructure and therefore influence the properties of alloys [1,7].

The development of non-toxic and allergy-free biomaterials is one of the most important directions of material chemistry today[8]. Magnesium as one of the main macro elements required for the proper functioning of the human organism is used to test the materials for biomedical implants [2, 9]. In fact, pure magnesium has low mechanical properties, which in turn disqualifies magnesium as a material for a medical implant. Proper selection of alloying elements is to improve the mechanical properties of magnesium [10] effectively.

Alloying with Zn and Ca not only enhances the hardness of magnesium alloys, thus to offer adequate support to the injured tissue or bone[10], but also they are non-toxic. Zinc is one of the most abundant nutritionally essential elements and is a co-factor for specific enzymes in bone [11]. On the other hand, calcium is a major component in human bone and is also beneficial to bone growth and or healing [12]. It means that Zn, and Ca have good biocompatibilities when incorporating with magnesium.

Cellular metals and metallic foams are metals with pores deliberately integrated into their structure. The terms of cellular metals or porous metals are general expressions referring to metals having a large volume of porosities, while the terms of foamed metal or metallic foams apply to porous metals produced with processes where foaming takes place [13]. Cellular metals are a relatively new class of engineering materials that can be fabricated with either a random or controlled cellular structure. A controlled cellular structure allows the precise control of the pore geometry and hence following material properties that can be important for some applications such as orthopedic implants [14, 15].

Porous structure in Mg-Ca-Zn alloy can be obtained by using a foaming agent. Foaming process is performed with the gas-releasing agent and other additives that produce a foamed alloy, which differs greatly from the matrix alloy [16, 17]. Carbonate is one of the foaming agenst used in the manufacture of metal cellular. Carbonate releases $CO_2$ or CO on heating which produces porous in alloy[18, 19].

## 2. Experimental Method

Magnesium powder (purity 98.5%, particle size 0,06-0,3 mm), Zinc powder (purity 99%, particle size < 45 μm), Calcium granule (purity 98%) and $CaCO_3$ (purity 98.96%, particle size < 30 μm) were purchased from Merck. Magnesium powder, Zinc powder, and Calcium (Ca) granule were reduced by using a shaker mill. Mg, Zn and Ca as a base alloy were mixed with $CaCO_3$ as the foaming agent with a ratio: 96Mg-Ca-3Zn-$CaCO_3$, 91Mg-Ca-3Zn-5$CaCO_3$, and 86Mg-Ca-3Zn-10$CaCO_3$ wt%. Milling process was by using a shaker mill for 2 hours to produce a powder size distribution which was more homogeneous. The mixed powder was uniaxially pressed at pressure of 1 00 MPa for 2 minutes and 200 MPa for 3 minutes into green compacts with dimensions of 10 mm in diameter and 10mm in length.

The green compact was heat treated in a tube furnace under an argon atmosphere. The sintering process was carried out at 650°C with a variation of holding time of 10 and 15 hours and the specimens were cooled down at room temperature. Microstructural analysis was performed by using X-Ray diffractometer (Shimadzu XRD-7000), and Scanning electron microscopy (JEOL, JSM-6390A Japan) equipped with an energy disperse spectrometry (EDS). The mechanical characteristics were analyzed by using compression testing by using Universal Testing Machine (Shimadzu AGS-10 KN) with dimensions of Ø10 mm x 10 mm at room temperature and rate of 1.33 mm/min. The density was measured by using Archimedes method. Along with the mechanical analysis, we also measured the porosity of the specimens.

## 3. Results and Discussion

### *3.1. Microstructure Analysis*

*3.1.1. X-ray Diffraction Analysis.* The microstructural analysis was performed by using X-Ray diffraction technique used to investigate the existing phase in the Mg-Ca-Zn-$CaCO_3$ alloys. X-Ray Diffraction pattern of sample sintered with different Holding Time of 10 and 15 hours is shown in Figure 1 and 2.

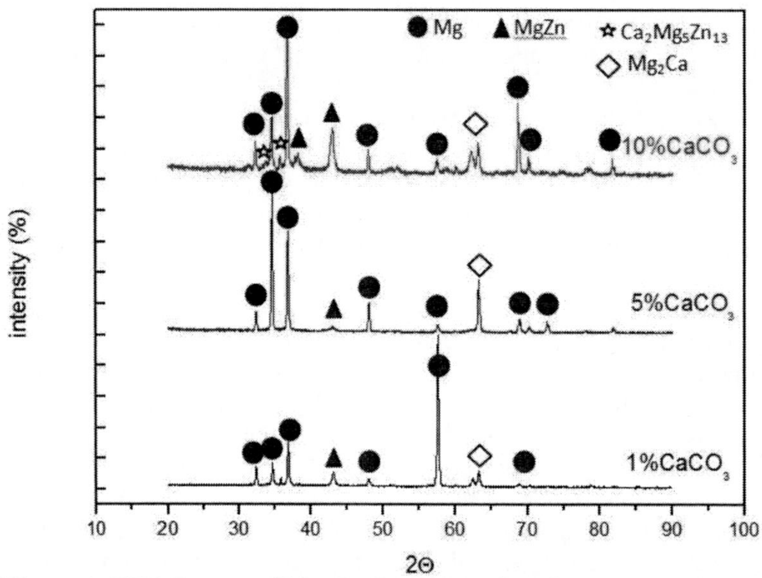


**Figure 1**. XRD Pattern of Mg-Ca-Zn-$CaCO_3$ sintering at 650°C Holding Time of 10 Hours.

In Figure 1, Mg peaks are detected in all of the samples. During the sintering process, theformed intermetallic phases were MgZn, $Mg_2Ca$, and $Ca_2Mg_5Zn_{13}$. MgZn and $Mg_2Ca$ have been formed in the alloy of 1% $CaCO_3$, while $Ca_2Mg_5Zn_{13}$ phase appeared in the alloy of 10% $CaCO_3$. MgZn phase can increase the mechanical properties of the alloy. However, $Mg_2Ca$ phase will make the alloy becomes brittle and rigid [5]. In alloys with a long holding time of 10 hours, the phase of $Mg_2Ca$ formed was less than MgZn phase. MgZn phase is the most phase formed on the alloy with 10% $CaCO_3$ that has the best mechanical properties of all alloys.

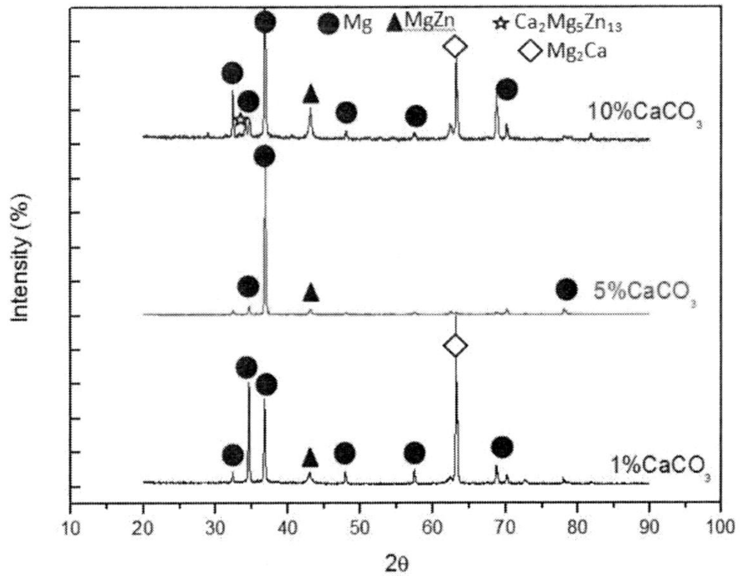


**Figure 2**. XRD Pattern of Mg-Ca-Zn-$CaCO_3$ sintering at 650°C Holding Time 15 hours.

Different from the holding time of 10 hours, sintering alloy at 650°C Holding time of 15 hours produced a phase of $Mg_2Ca$ quite a lot compared to the phase of MgZn (Figure 2). The amount of phase formed $Mg_2Ca$ resulting alloy becomes very brittle and rigid thereby lowering mechanical properties.

*3.1.2. SEM Analysis.* Microstructure analysis using SEM was to observe pore structure in the alloy and second phase which formed during sintering process. Figure 3 reveals the morphology magnesium alloy (Mg-Ca-Zn-$CaCO_3$) with three different compositions, sintering at 650°C, holding time of 10 and 15 hours.

The addition of foaming agent with 1, 5, and 10% $CaCO_3$ also shows pores formed have a greater size and tends to be shaped spherical. Pore size in the alloy composition 1, 5, and 10% $CaCO_3$ that is less than 100, 50-200, and 100-300 μm. Alloys with 1 and 5% $CaCO_3$ showed many pores with small size and enclosed generated from volume shrinkage during the sintering process takeing place, while the pore formed in the alloy composition with 10% $CaCO_3$ has a larger size and allow the formation of interconnected pores.

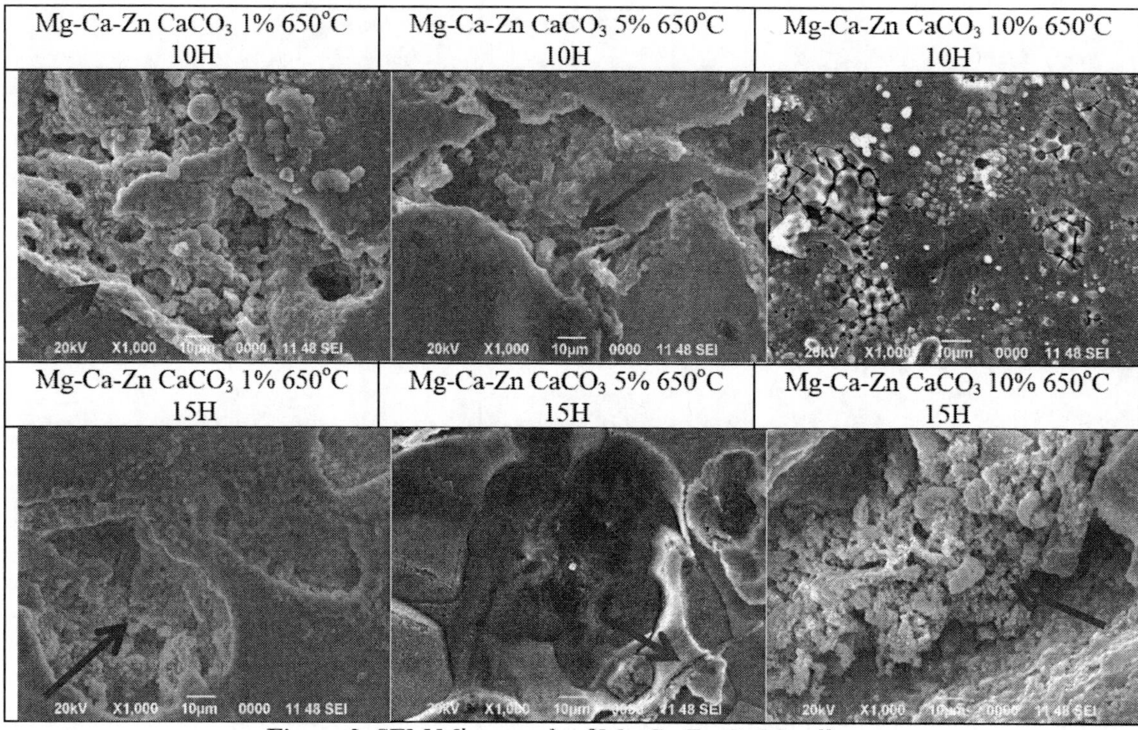


**Figure** 3. SEM Micrograph of Mg-Ca-Zn-$CaCO_3$ alloys.

### *3.2. Mechanical Properties*

Mechanical testing for the alloy is density, porosity and compression test. Actual density was obtained through Archimedes' method through the medium of water, and a theoretical density was obtained by calculating the density of each of the elements of the alloy composition. Base on the density of the sample, actual porosity was obtained based on the calculation of the ratio between actual density and the theoretical density of the specimen with the testing standard used was ASTM 378-88. The actual density is shown in Table 1.

**Table 1**. Density of Mg-Ca-Zn-$CaCO_3$ alloys.

| Composition | Temperature and Holding Time | Average Density (g/cc) |
|---|---|---|
| 96Mg-3Zn-1Ca-1$CaCO_3$ | 650°C 10H | 1.74052 |
| | 650°C 15H | 1.62929 |
| 91Mg-3Zn-1Ca-5$CaCO_3$ | 650°C 10H | 1.66324 |
| | 650°C 15H | 1.73733 |
| 86Mg-3Zn-1Ca-10$CaCO_3$ | 650°C 10H | 1.63743 |
| | 650°C 15H | 1. 65307 |

Table 1 shows that the alloy with 10% $CaCO_3$ which was sintered at 650°C with holding time of 10 h has the lowest density of 1.63743. Meanwhile,the alloy with the highest density is shown by the specimen with 1% $CaCO_3$ of 1.74052. Before sintering process, the density of Mg alloy (Mg-Ca-Zn-$CaCO_3$) was about 1.73 gr/cc. Increase in density on the sample as in alloys with 1% $CaCO_3$ sintered at 650 °C with holding time of 10H most likely occured due to the formation of the oxide layer during the sintering process. The presence of oxide will increase the mass of the sample so that the effect is on density. The decrease in density was caused by the formation of pores in the alloy. Pore was formed by the decomposition reaction of $CaCO_3$ as a foaming agent. The oxide layer is almost always the case in every sintering process. However, it was minimized by flowing inert gas which was argon during the heating process. The increase in density in the range of 1% $CaCO_3$ wasdue to the small pores formed so as not comparable with the existing oxide layer. The density reduction on the alloy was to signify the formation of pores. Changes in density values are presented in the following graphical.

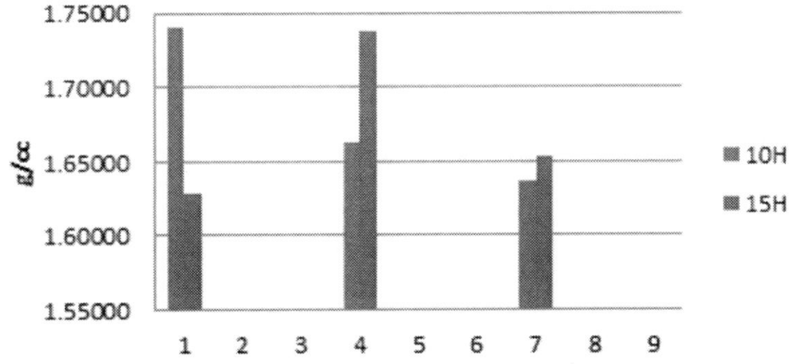


**Figure 4**. Chart Density of Mg-Ca-Zn-$CaCO_3$ alloys.

It can be seen from Figure 4, that the density of Mg alloy with holding time of 10 h decreased with increasing the amount of $CaCO_3$ on the alloy served as a foaming agent. A large number of pores formed resulting density decreases. However, this phenomenon did not apply to holding time of 15 h. The density obtained irregularly caused by sintering was too long so that the alloy became brittle and much apart.

**Table 2**. Porosity of Mg-Ca-Zn-$CaCO_3$ alloys.

| Composition | Temperature and Holding Time | Average Porosity (%) |
|---|---|---|
| 96Mg-3Zn-1Ca-1$CaCO_3$ | 650°C 10H | 8.56273 |
| | 650°C 15H | 14.35947 |
| 91Mg-3Zn-1Ca-5$CaCO_3$ | 650°C 10H | 14.63004 |
| | 650°C 15H | 10.95897 |
| 86Mg-3Zn-1Ca-10$CaCO_3$ | 650°C 10H | 18.52912 |
| | 650°C 15H | 17.76913 |

Table 2 shows that the highest porosity was 18.52912%, which belonged to the alloy with 10% $CaCO_3$ holding time of 10 H. as well as the density, porosity increases with increasing amount of $CaCO_3$ in the alloy. Obtaining porosity of 18.52912%, this means an increase of 5% compared to previous research that only earned a maximum of 13% that the Mg alloy composition at 10% $CaCO_3$ with sintering temperature 650 °C Holding time of 5 H [20]. However, porosity in the Holding Time of 15H became irregular due to the sintering process was too long resulting samples becoming brittle and making it difficult to analyze. 650 °C is the melting temperature of Magnesium so that the temperature limits allowed for sintering the powder metallurgy method. To obtain the pore in large quantities, we performed variations of the holding time. However, Holding time for 15 H resulted in Mg alloy became brittle and partly destroyed. With these conditions, the density and the number of pores could not be analyzed and did not produce accurate data.

IOP Conf. Series: Materials Science and Engineering **202** (2017) 012028 doi:10.1088/1757-899X/202/1/012028

Compression test was performed to determine the mechanical strength of the alloy. Based on Table 3, the mechanical strength of Mg was still weak, with the addition of Zn it can increase the mechanical strength. Proportional to the percentage of porosity which is owned by a sample, the result of compression strength test showed the higher the percentage of porosity, then the value of compression strength will be smaller. Therefore, the presence of Zn which improves the mechanical properties can be offset shaft formed by the foaming agent.

**Table 3**. Compression test of Mg-Ca-Zn-$CaCO_3$ alloys.

| Composition | Temperature and Holding Time | Compression Test ($N/mm^2$) |
|---|---|---|
| 96Mg-3Zn-1Ca-1$CaCO_3$ | 650°C 10H<br>650°C 15H | 233.389<br>201.644 |
| 91Mg-3Zn-1Ca-5$CaCO_3$ | 650°C 10H<br>650°C 15H | 211.301<br>213.788 |
| 86Mg-3Zn-1Ca-10$CaCO_3$ | 650°C 10H<br>650°C 15H | 208.398<br>210.792 |

## 4. Conclusion

Magnesium alloy foam has been successfully prepared through powder metallurgy method. From microstructure analysis, sintering process of Mg-Ca-Zn-$CaCO_3$ alloys produces intermetallic phase of MgZn, $Mg_2Ca$, and $Ca_2Mg_5Zn_{13}$. The existence of Zn which produces MgZn phase improves the mechanical properties of the alloy. In the present study, optimum microstructure and mechanical characteristics are the Mg-Ca-Zn-$CaCO_3$ alloys sintered at 650 °C with holding time of 10 hours. The value of compression of 208.398 $N/mm^2$, the density of 1.63 g/cc and the porosity of 18% on the composition of 86Mg-Ca-3Zn-10$CaCO_3$ wt%.

## 5. References

[1] Thesis M and Lekkala J 2013 Study Of Magnesium Metal And Its Alloys As A Biodegradable Material For Medical And Electrical March

[2] Wang H and Yang Z M S K 2009 Magnesium And Magnesium Alloys As Degradable Metallic Biomaterials **2008**

[3] 2010 The history of biodegradable magnesium implants: A review," *Acta Biomater* **6** 5 pp 1680–1692

[4] Y. Lu, "Microstructure and Degradation Behaviour of Mg- Zn(-Ca) Alloys," 2014.

[5] Zhang B P, Wang Y and Geng L Research on Mg-Zn-Ca Alloy as Degradable Biomaterial

[6] Zhao X, Shi L and J X Ã 2013 Biodegradable Mg–Zn–Y alloys with long-period stacking ordered structure : Optimization for mechanical properties *J. Mech. Behav. Biomed. Mater* **18** pp 181–190

[7] Zhang E, Yang L, Xu J and Chen H 2010 Microstructure, mechanical properties and bio-corrosion properties of Mg–Si(–Ca, Zn) alloy for biomedical application *Acta Biomater* **6** 5 pp 1756–1762

[8] Orsa Z, Orin A, Kupkova M, Hrubovc M and Morovska 2014 A Sintered metallic foams for biodegradable bone replacement materials pp 131–140

[9] Applications B 2013 The Potential Of Magnesium Alloys As Bioabsorbable / Biodegradable Implants For **3**

[10] Nowosielski R, Cesarz K and Babilas R 2013 Structure and corrosion properties of Mg 70-x Zn 30 Ca x(x=0.4) alloys for biomedical applications," *J. Achiev. Mater. Manuf. Eng* **58**

[11] 2001 Dietary reference intakes for vitamin A, vitamin K, arsenic, boron, chromium, copper,

IOP Conf. Series: Materials Science and Engineering **202** (2017) 012028 doi:10.1088/1757-899X/202/1/012028

iodine, iron, manganese, molybdenum, nickel, silicon, vanadium, and zinc in *Food and Nutrition Board IoM*, Washington DC National Academies Press pp 442–501

[12] 1997 Dietary reference intakes for calcium, phosphorus, magnesium, vitamin D, and fluoride in *Food and Nutrition Board IoM* Washington DC National Academies Press pp. 190–249.

[13] Lefebvre L-P, Banhart J and Dunand D C 2008 Porous Metals and Metallic Foams: Current Status and Recent Developments *Adv. Eng. Mater.* **10** 9 pp 775–787

[14] Ing B 2009 A Novel Processing Route for the Manufacture of Mg with Controlled Cellular Structure March

[15] Wendel J G H P 2014 Cytocompatibility evaluation of different biodegradable magnesium alloys with human mesenchymal stem cells pp 835–843

[16] Park H, Temenoff J S and Mikos A G Biodegradable Orthopedic Implants

[17] Seyedraoufi Z S and Mirdamadi S, "Synthesis, microstructure and mechanical properties of porous Mg – Zn scaffolds," *J. Mech. Behav. Biomed. Mater.*, vol. 21, pp. 1–8, 2013.

[18] Yang D, Hur B and Yang S 2008 Study on Fabrication and Foaming mechanism of Mg foam using $CaCO_3$ as blowing agent **461** pp 221–227

[19] Yang D, Seo C and Hur B 2008 Mg Alloy Foam Fabrication via Melt Foaming Method **24** 3 pp 302–304

[20] Kartika I 2015 Structural properties of Mg-Ca-Zn alloy with addition of $CaCO_3$ as foaming agent prepared by powder metallurgy method *Proceeding 2nd Int. Conf. Mater. Metall. Technol.* pp 423–434

**Acknowledgements**

This research was supported by Research Center for Metallurgy and Materials, Indonesian Institute of Science. Special thanks to biocompatible implant research group for support and teamwork. The authors are thankful to polymer technology centers of BPPT for help with Compression Strength test and also to Syarif Hidayatullah State Islamic University for help with XRD Testing.

# Phase Analysis and Crystal Morphology of Barium Sulphate Precipitated from The Laminar Flowing Water

**N S Dera[1], F Fatra[1], G Ivanto, S Muryanto[1,2] and A P Bayuseno[1]**

[1]Mechanical Engineering Graduate Program Diponegoro University, Tembalang Campus Semarang 50275, Indonesia.
[2]Department of Chemical Engineering, UNTAG University in Semarang, BendhanDhuwur Campus, Semarang 50233, Indonesia.

Email: santydera@gmail.com

**Abstract**. Barium sulphate ($BaSO_4$) is common scale deposits precipitated in pipes which can hinder the flow rate and lower heat transfer efficiency. Therefore, there is a need to address the understanding of scale formation in pipes. In this paper, the formation of $BaSO_4$ scale in the laminar flowing water was investigated in the laboratory rig of scale formation. The scale forming solution was prepared by $BaCl_2$ and $Na_2SO_4$ with $Ba^{2+}$ concentrations in equimolar ppm of 2500, 3000, and 3500. The pH solution was set up in the values of 6, 8, and 10. The crystals were deposited on the four coupons pipes made of copper inside the pipes. The scale deposited from the flowing water was then characterized by using SEM equipped by EDX for crystal morphology and elemental analysis. XRD method was used for the crystalline phase analysis. The results showed that $BaSO_4$ crystals with star-like morphology can be observed from SEM imaging. The pure crystal barite can be obtained from the experiments as can be confirmed by XRD analysis. It is obvious that the barite crystals can be easily formed in the basic solution.

**Keywords**: Barite, The laminar flowing water, Crystal morphology, Phase Analysis

## 1. Introduction

The formation of barium sulphate as called barite and other insoluble scale types (metal sulphate, calcium carbonate and silicates) is a very common case in the oilfield and frequently happen when incompatible types of water (injected sea water and formation water) are mixed in the reservoir. The control parameters in the scale formation may include the supersaturation index in the mixed solution [1]. Barite is one of the mineral deposits, becoming an extreme concern for the oil and gas industry because it can be deposited on the surface of the inside diameter of pipes. The existence of this mineral scale influences the oil production, hence leading an extreme economic impact. This scale can also be considered as one of factors that makes the production of oil and gas around the world declined. Now, the scale prevention strategy becomes a crucial issue in the fields of oil and gas production. Therefore, every year, the companies spend millions of dollars to mitigate the impact of the scale. Estimation shows that the industry spends around 5 billion USD every year. Considering from its operating costs, the scale prevention is the largest maintenance cost allocated for this field [2]. On the other hand, barium sulphate is the most difficult scale to be removed because of its low solubility in water (solubility in water only 2 mg/l). This scale is frequently difficult to remove by acid treatment.

Because of the relatively low solubility in water, barite can be easily precipitated from brine once its solubility limit exceeds. Barite formation typically occurs on the areas that are directly in contact with seawater during the oil production process, such as down-hole in the near well bore area and production tubular as well as in the topside, crude oil, processing equipment. In the oil production processes, deposition of the sparingly soluble salts is becoming great issues, because of the downtime consumed for cleaning the equipment and the increasing operating cost due to the possible loss of the product purity in the presence of unwanted minerals [3]. Barite is an unusual scale mineral not similar to calcium sulfate and calcium carbonate, which are typically found in water treatment. However, it is considered to be a common mineral formed into the marine environment and the area of oil and gas production [4]. This mineral can also occur in many processes ranging from desalination to oil production and hence resulting in great interest in cost reduction by the scale inhibitor. In comparison to the chemistry of calcium carbonate ($CaCO_3$), barite is more soluble in water with increasing temperature, while the solubility of calcium carbonate reduces as the temperature increases. In contrast, the formation of calcium carbonate depends highly on pH, while barite is not pH dependence. Additionally, calcium carbonate solubility is also higher than barite.

Furthermore, the solubility of barite at 25 and 50 $^{\circ}$C has the general increasing trends as temperature increases. Barite solubility product of $K_{sp}$ equals to 1.08 $\times$ $10^{-10}$ mg/l at 25 $^{O}$C and 2.16 x$10^{-10}$ mg/l at 50 $^{\circ}$C [1]. It can be easily found on the surface equipment and tubes in many industrial and oil drilling processes. Here the control of crystal growth and morphological changes of crystal barite needs a knowledge of mechanism crystallization [5]. In nature, the crystalline phase and morphology of barite may be related to a complex crystallization process; as a result of multiple masses transferred levels from supersaturated solutions. This process may be influenced by many factors such as solution temperature, supersaturation, mixing speed and the presence of additives. To change nucleation and crystal growth behavior, size and morphology of crystals may use chemical additives. For the selection of scale inhibitor, it can be made based on the results of spontaneous precipitation experiments conducted at the laboratory without and with additives.

In the present study, the formation of barite scale was examined by using a pilot scale of laboratory equipment consisting of a pipe system. The objective of this study was to analyze the crystalline phase and crystal morphology of barite precipitated from the laminar flow water. In particular, crystal growth of the piping surface was predicted through a kinetic study based on induction time. The scale sample was then analyzed by using X-ray diffraction (XRD) for phase composition and SEM/EDX for morphology and chemical elemental determination.

## 2. Experimental Method

### *2.1. Materials*

A feedstock powders with an analytical grade for crystal forming solution consisting of $BaCl_2$ and $Na_2SO_4$ (Merck$^{Tm}$), and 1 M KOH was added into the $BaCl_2$ solution to maintain pH solution. The tools used in the experiments involved conductivity meters, oven, pH meter, the glass beaker with volume of 1000, 250, and 50 ml, a pipette, and filter paper.

### *2.2. Crystal Growth Experiments*

$BaSO_4$ crystal can be formed as a result of a chemical reaction against $BaCl_2$ and $Na_2SO_4$ according to (Equation 1):

$$BaCl_2\,(aq) + Na_2SO_4\,(aq) \rightarrow BaSO_4\,(s) + 2NaCl\,(aq) \qquad (1)$$

Firstly, here an important step in making a supersaturated solution must be considered, based on a calculation of the concentration of barium ($Ba^{2+}$) in the solution. The concentration of $Ba^{2+}$ was determined to be 2500, 3000, and 3500 ppm, while a flow rate of 30 ml, temperature (30, 40, and

50°C), the time required for trial for 2.5 hours were selected. The need for the concentration is presented in Table 1. Meanwhile Figure 1 shows the pilot scale of laboratory equipment.

**Table 1.** Mass of $BaCl_2$ and $Na_2SO_4$

| Concentration $Ba^{2+}$ (ppm) | $BaCl_2$ (gram) | $Na_2SO_4$ (gram) |
|---|---|---|
| 2500 | 26.6874 | 15.51198 |
| 3000 | 32.02488 | 18.61664 |
| 3500 | 37.35522 | 21.71846 |

For the preparation of the solution, the weighing of powder feedstock was performed according to the calculations presented in Table 1. Then, the vessel was filled with 1 liter of distilled water equal to 1 liter of $BaCl_2$. The addition of distilled water continued until the desired volume reached. To remove particles carried in solution was filtered by using a filter paper. The same steps were also carried out in the preparation of $Na_2SO_4$ solution. For checking the pH value, $Na_2SO_4$ and $BaCl_2$ solutions were then entered into each vessel 1 and vessel 2. Afterward, both solutions were pumped into the coupons from the vessel 3 and vessel 4. Measurements of the conductivity of the solution were performed every two minutes for the first one hour experiment, every five minutes for the second one hour and every ten minutes for the third one hour, so that time for a trial was ended for two hours and thirty minutes.

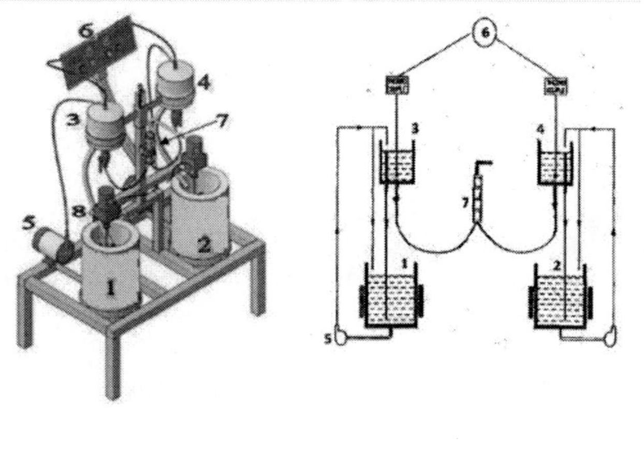

**Figure 1.** Experimental rig used for the study [6]

| | | | |
|---|---|---|---|
| 1. | Vessel under $BaCl_2$ | 4. | Water pump |
| 2. | Vessel under $Na_2SO_4$ | 5. | Temperature control |
| 3. | Vessel over $BaCl_2$ | 6. | Flow rate indicators |
| | | 7. | The house of coupon |

To perform conductivity measurements was carried out by inserting electrode's conductivity in the liquid coming out of the coupon. After two hours and thirty minutes, the pump was turned off, and the house of coupons was detached. Subsequently, the coupon was taken from the house coupons and dried by using an oven at a temperature of 100 °C for 1.5 hours. Weighing of the mass scale was initially done while the scale was still attached to the coupon. Furthermore, the masses difference between coupon mass with and without scale provided the mass of the resulting scale. After the experiments, the crystal scale product was subjected to the SEM-EDX and XRD analysis.

The scale deposited from the flowing water was then characterized by using SEM equipped by EDX for crystal morphology and elemental analysis. Characterization of the samples was performed by using SEM brand JEOL JED-2300 with EDX facility to determine chemical elemental composition in the sample. SEM image capture process was done by using the secondary electron detector (SE) on the acceleration energy of the primary electron-beam source tungsten wire at 20 keV with a working distance (WD), held constant at 10 mm. WD value was set up to get EDX spectrum in the best condition. EDX was taken on the average value of dead time between 20% and 40%. Additionally XRD method was used for the crystalline phase analysis. The XRD data for the crystal structure analysis was collected by using Shimadzu XRD brand XD610 with measurement angle 2θ of $10^0$-$90^0$. The X-ray was generated from Cu-Kα with a wavelength of 1.5405. The measured XRD data was analyzed by using the Full-Prop program for crystalline phase identification.

## 3. Results and discussion

### *3.1. The induction time for the variation of $Ba^{2+}$ concentration in the solution.*

In practice, the induction time is frequently used to represent crystal nucleation. The induction period, $t_{ind}$, can be explained as the time pass between the invention of supersaturation and the first showing of a new phase, ideally nuclei with the critical cluster size dimensions [7]. The induction time may be regular by following the change in concentration of one of the crystal ions over time (see Figure 2). However, the induction period can be determined experimentally and the precision is in agreement with nucleation turn on the analytical technique predicted. Therefore, a part at the induction time may also indicate the crystal growth into a detectable size [8].

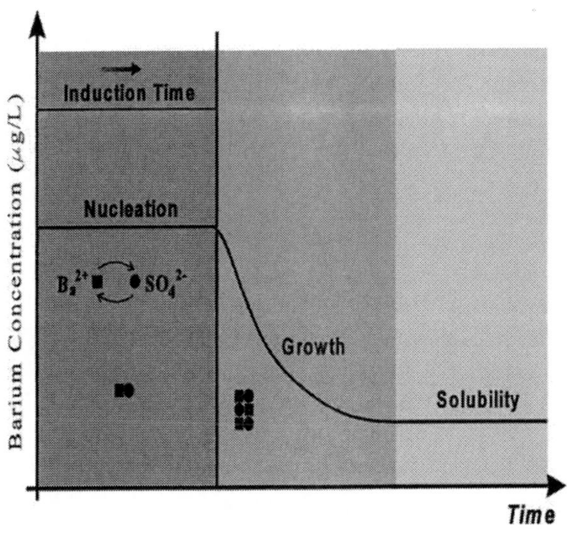


**Figure 2.** Barium concentration over time performance (i) a stable region where nucleation is measurable by the experimental determination of induction time followed by (ii) barium supersaturation due to the growth of nuclei, from spontaneous nucleation or constrained by seed crystal addition, to final equilibrium saturation [15].

Furthermore, the induction time can be considered as time taken when the ions in solution to, firstly, form a crystal nucleus. The induction time can be measured by looking at the conductivity of the solution coming out of the pipe specimen. The occurrence of a significant reduction in the conductivity of the solution indicates the end of the induction time. This phenomenon implies that the ions forming the crystal have left a new solution and form crystals that stick to the wall of the pipe. The induction time can be measured from the start reaction to ions forming barite until the conductivity drops significantly. In this present study, the induction time of barite crystal formation can be seen in the graph as shown in Figure 3.

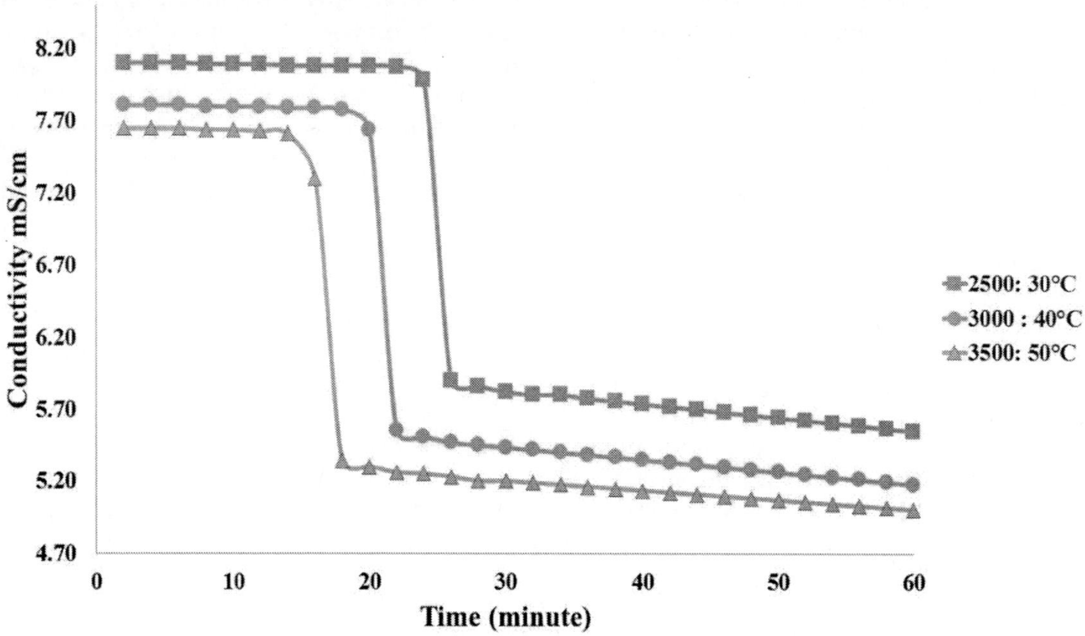


**Figure 3.** Effect of $Ba^{2+}$ concentration on the induction time

It shows that the higher concentration leads to crystallize faster than that for the lower concentration. If the concentration is high, then the molecules have a significant concentration that supports the formation of crystal nucleus. The higher the $Ba^{2+}$ concentration in the solution, the smaller critical nucleus size will get. Accordingly, the formation of crystal nucleus may increase. For this research, it is assumed that the nucleation time is much greater than that required for growth of crystal nuclei in the detectable size.

### *3.2. Effect of $Ba^{2+}$ concentration on the growth of barite scale*

Obviously, the increasing $Ba^{2+}$ concentration resulted in the increasing crystal growth rate of barite when the temperature also increases (Fig. 4). If the solution exceeds the saturation limit, there will be a reaction between ions in solution, and the scale deposit is produced [16]. Nucleation occurs when the saturated solution is achieved. As the solution exceeds the limit of the saturated condition, some molecules or ions combine to form clusters, which later grew into a crystal nucleus. Crystal's nucleus will dissolve when its size is smaller than the particle size (critical core), while the crystal nucleus subsequently develops when the particle size has become larger than the critical and crystalline structure is developed [17].

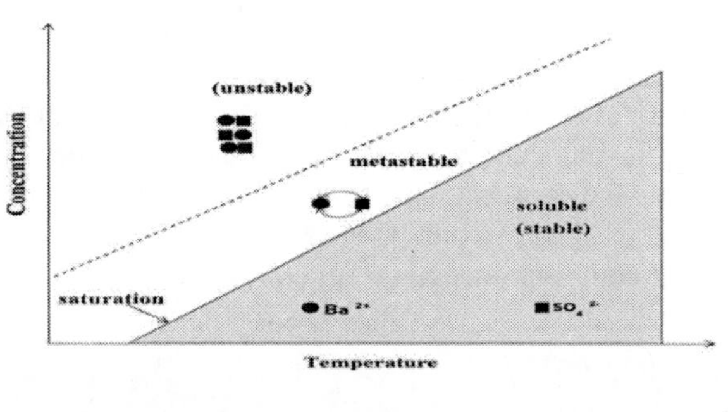


**Figure 4.** Solubility-super saturation zones for barium sulphate: (i) undersaturated zone where barium sulphate is soluble and no deposit is potential; (ii) metastable (supersaturated) zone where spontaneously deposit is not likely; (iii) unstable zone where spontaneously precipitation is extremely likely [18].

A metastable zone of supersaturation exists where the nucleation rate is so low that precipitation is impossible within the measured time frame. The breadth of this zone is given thermodynamically by the free energy of becoming the critical cluster size in homogenous nucleation. Homogeneous nucleation forms on the basis of some nucleation theories and can be comprehensive to heterogeneous nucleation. Two important parameters characterize the nucleation process: (i) the free-energy change, and (ii) the rate of nuclei forming [18].

In this present study, increasing the concentration of the solution at the same conditions will multiply to barite scale formed. The rate of growth at 2500 ppm $Ba^{2+}$ concentration providing 0.2689 grams / hour; the concentration of 3000 ppm providing 0.2702 grams / hour and 3500 ppm resulting in 0.2890 grams / hour. This case suggests that the increased concentration as the solution increased the average rate of scale growth.

### *3.3. Effect of temperature on the growth rate of barite scale*

Temperature is one of the main factors influencing the growth of barite scale. The higher temperatures result in increasing solubility of barite, in other words; the higher temperature decreases precipitated barite (Figure 4). According to Merdhah (2012), the barite scale formation decreases at low temperatures because the solubility increases with increasing temperature [13]. Barite precipitation rate and reduction in permeability intensity decreased with increasing temperature [14]. In this study the $Ba^{2+}$ concentration of 2500 ppm with temperatures rising from 30 $^0$C to 50 $^0$C yielded to reduction from 0.2767 gram scale mass to 0.2697 gram. This phenomenon is because the low temperatures causing the faster formation of barite scale.

### *3.4. Effect of pH value on the rate of growth of barite scale*

The pH control of cooling water may be used with the acid injection. It has long been applied to prevent scaling of calcium carbonate, calcium phosphate, and others. However, the pH value has the significant effect on the presence of inhibitors, where the higher pH will increase protonation inhibitors, and the lower pH generally reduced the level of protonation [3]. According to Akyol (2012), the pH solution affects the protonation degree of the carboxylic acid groups of PAA, and low pH generally reduces the protonation level. Therefore, the interaction between PAA and barite crystal was weakened at low pH; the inhibition effect of PAA on the crystallization of barite at low pH was impaired, and the morphology of the crystals was similar to that of the crystals ready in the absence of polymer. When the pH of the solution was increased to 9, the spherical particles were obtained. This result showed that the increase of pH of the solution led to the increase in the protonation level of the carboxylic acid groups of PAA [4]. The greater the change in pH will alter the degree of dissociation of the molecule inhibitors. The pH 8 with an inhibitor HEDP showed inhibition of the additive but increasing the pH from 8 to 12 result in a loss due to additive inhibition of the barite surface changes with PH [15]. In the present study, pH value did not have a significant effect on the growth of barite scale. Effect of pH on the crystal growth can be seen in Figures 3 in which no changes on the surface of barite crystals can be observed.

### *3.5. Morphological analysis*

Barite crystal structure is orthorhombic, has a unit cell comprising four units of barium ions. Barite growth by aggregation mechanism was followed by a recrystallization process [5], while particle morphology is strongly influenced by the level of saturation. The higher the saturation level will produce a particle size smaller crystals continuously until far below 100 μm [16] Crystallization reaction is the formation of a solid phase from a liquid as a result of appearance saturation followed by a chemical reaction [17].

According to Akyol (2014), the morphological characteristics precipitated barium sulfate in the absence of additives in tablet form rectangular as shown in Figure 5.ac. The crystal size is the mean length of 9.53 μm and a mean width of 5.79 μm. Barite crystals in the presence of additives CMI (carboxymethyl inulin) show products in the form of tablets or aggregates such as flowers. Spherical crystals were also formed in the presence of PAA additives. Crystal's morphology of barium sulfate in the presence of additives can be seen in Figure 5(b)-(d) [4].

**Figure 5.** Barite morphological image formation in the presence of PAA and CMI additives [4]

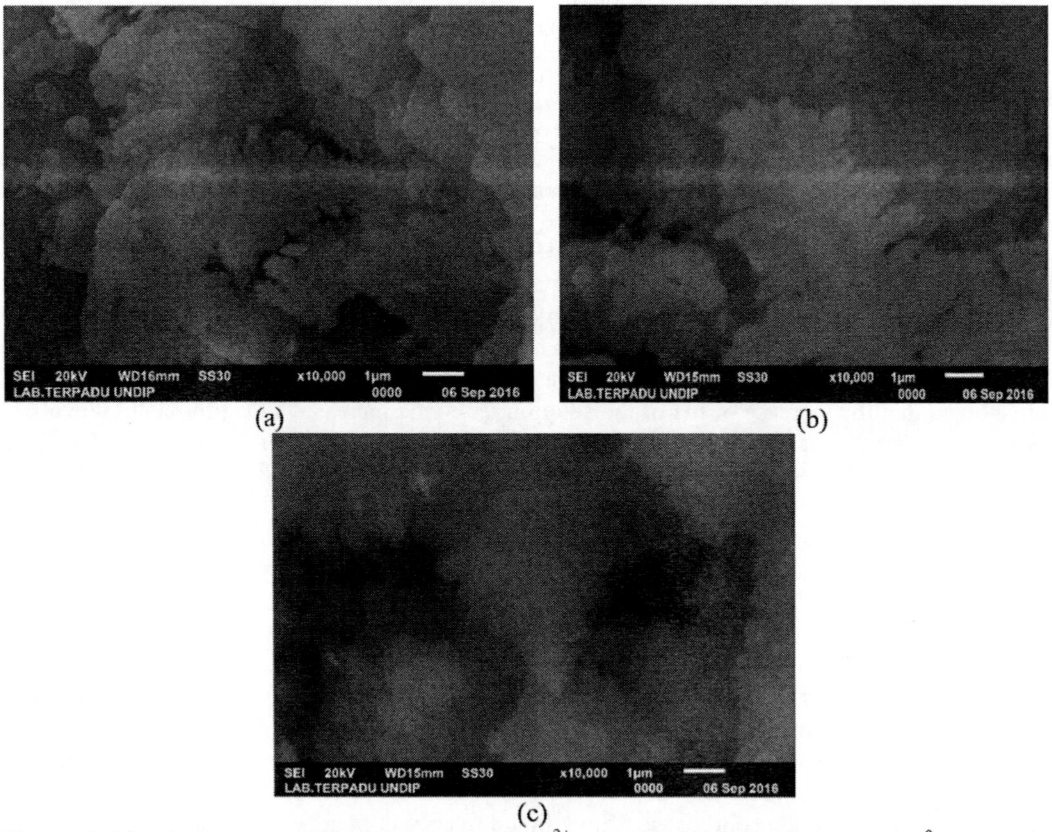


**Figure 6**. Morphology crystal barite obtained at $Ba^{2+}$ concentration a). 2500 ppm, 30 $^0$C, pH value of 8. b) 3000 ppm, 40 $^0$C, pH value of 6. c) 3500 ppm, 50 $^0$C, pH value of 10

In this study, morphological analysis may include the examination of shape and size of the crystals. The analysis was performed by using Scanning Electron Microscopy (SEM) to identify whether the crystal goes right shape as barite crystals. Morphology of barite crystals with the $Ba^{2+}$ concentration of 2500, 3000, and 3500 ppm can be seen in Figure 6.

From the result of SEM analysis, it can be seen that the form of barite crystals formed at a concentration of 2500, 3000, and 3500 ppm are shaped like a flower; it indicates that this crystalline form is characteristic of barite crystals. EDX analysis aims to look into the composition of barite crystals. EDX analysis on the results can be seen that the main elements in the scale formed barite is Ba (Barium), C (carbon), O (oxygen) and S (sulfate). This case indicates that the scale is generated for this study to be barite scale.

**Table 2.** EDX analysis of the result

| Parameters | Elements | Wt.% (actual) |
|---|---|---|
| 2500 ppm, 30 $^0$C, pH value of 8 | S | 21.79 |
| | Ba | 43.86 |
| | O | 17.64 |
| | C | 34.35 |
| 3000 ppm, 40 $^0$C, pH value of 6 | S | 21.38 |
| | Ba | 44.11 |
| | O | 17.42 |
| | C | 34.51 |
| 3500 ppm, 50 $^0$C, pH value of 10 | S | 16.93 |
| | Ba | 37.32 |
| | O | 14.04 |
| | C | 45.75 |

*3.6. XRD analysis*

XRD analysis was performed to determine the crystalline phase of the scale. X-ray diffractogram of the crystal product is presented in Figure 7. From XRD results of the different solution with various $Ba^{2+}$ concentration, each intensity peak was then matched with the data of the International Centre for Diffraction Data (ICDD)-Powder Diffraction File (PDF). It shows that all diffraction peak intensities of the crystal are matched with PDF numbers of 04-012-5411, indicating that the crystal of barite is present in the sample. All the diffraction intensity of the XRD pattern are also in good agreement with the reported data (JCPDS filed No: 24-1035) that can be indexed as barite with orthorhombic crystal structure [18].

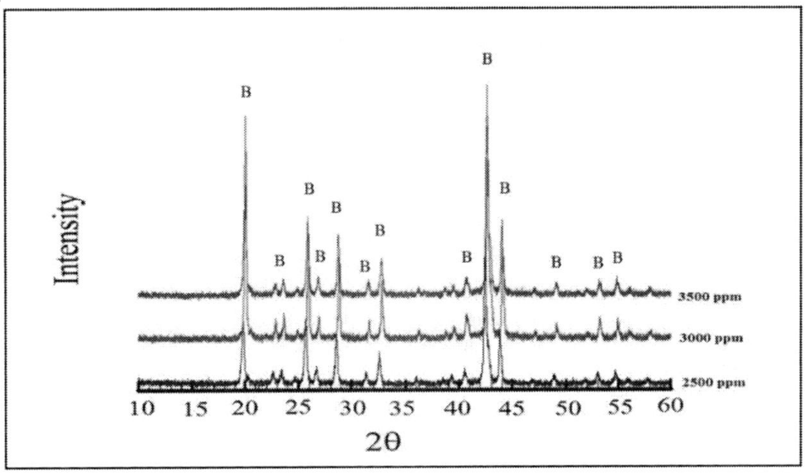


**Figure 7**. Graph XRD measurement results concentration of 2500, 3000, and 3500 ppm. Peaks are labeled B (barium) noted.

## 4. Conclusion

Varying the $Ba^{2+}$ concentration (2500, 3000, and 3500 ppm) in the solution was observed during the study by providing the induction time of 24, 20, and 16 minutes, respectively. This study suggested that the increased $Ba^{2+}$concentration in the solution reduced the induction time. This phenomenon is due to the increase of the level of solubility saturation limit, which results in the increasing rate of nucleation and crystal growth. The increasing temperature made the decreased mass of the scale, which is due to the more soluble barite in the solution at high temperatures. From the SEM analysis, barite crystal with starlike morphology was confirmed and the XRD analysis verifies that the crystal material is true barite.

## 5. References

[1] Mavredaki E, Neville A and Sorbie K 2011 Applied Surface Science Assessment of barium sulphate formation and inhibition at surfaces with synchrotron X-ray diffraction ( SXRD ) *Appl. Surf. Sci.* **257** 4264–71

[2] Akyol E, Aras O and Oner M Barium Sulfate Crystallization in the Presence of Additives 6–8

[3] Wang F, Xu G, Zhang Z, Song S and Dong S 2006 A systematic morphosynthesis of barium sulfate in the presence of phosphonate inhibitor *J. Colloid Interface Sci.* **293** 394–400

[4] Akyol E, Aras Ö and Öner M 2014 Control of barium sulfate crystallization in the presence of additives *Desalination Water Treat.* **52** 5965–73

[5] Judat B and Kind M 2004 Morphology and internal structure of barium sulfate - Derivation of a new growth mechanism *J. Colloid Interface Sci.* **269** 341–53

[6] Muryanto S, Bayuseno A P, Sediono W, Mangestiyono W, and Sutrisno W 2012 Development of a versatile laboratory project for scale formationand control education for chemical engineers, pp: 1-7,

[7] Van der Leeden M C 1991 The role of polyelectrolytes in barium sulfate precipitation Ph.D. Thesis TU Delft

[8] Söhnel O, Garside J 1992 Precipitation-Basic Principles and Industrial Applications *Butterworth- Heinemann*

[9] Boerlage F E, Kennedy M D, Bremere I, Witkamp G J, Hoek J P, Schippers J C 2000 Stable barium sulphate supersaturation in reverse osmosis *Journal of Membrane Science* **179** 53-68

[10] Sediono W, Bayuseno A P, Muryanto S 2011 Eksperimen Pembentukan Kerak Gipsum Dengan Konsentrasi $Ca^{2+}$ : 3500 Ppm Dan Aditif $Fe^{2+}$ Momentum **7** 2 41-44

[11] E. K. Ho. B. Zeiher, K. D. Williams. *"Novel antiscalant dosing control"*. Desalination, 157(1), 209-216, 2003.

[12] Mullin J W 1993 Crystallization 3rd Edition Butterworths Heinemann Oxford

[13] BinMerdhah A B 2012 Inhibition of barium sulfate scale at high-barium formation water *J. Pet. Sci. Eng.* **90–91** 124–30

[14] Naseri S, Moghadasi J and Jamialahmadi M 2015 Journal of Natural Gas Science and Engineering Effect of temperature and calcium ion concentration on permeability reduction due to composite barium and calcium sulfate precipitation in porous media *J. Nat. Gas Sci. Eng.* **22** 299–312

[15] Jones F, Oliveira A, Rohl A L, Parkinson G M, Ogden M I and Reyhani M M 2002 Investigation into the effect of phosphonate inhibitors on barium sulfate precipitation *J. Cryst. Growth* **237–239** 424–9

[16] Kucher M, Babic D and Kind M 2006 Precipitation of barium sulfate: Experimental investigation about the influence of supersaturation and free lattice ion ratio on particle formation *Chem. Eng. Process. Process Intensif.* **45** 900–7

[17] Matynia A, Piotrowski K and Koralewska J 2005 Barium sulphate crystallization kinetics in the process of barium ions precipitation by means of crystalline ammonium sulphate addition **44** 485–95

[18] Fang W, Guiying X, Zhiqing Z, Xia X 2005 Morphology control of barium sulfate by PEO-Ppo-PEO as crystal growth modifier *colloids and surface A: Phycochem. Eng. Aspects* **259** 151-154

**Acknowledgements**
Authors wish to thank the Gorontalo University for providing the scholarship and research grants. We are also grateful to the Laboratory of Physical Metallurgy Diponegoro University which provided the facility during the experiment. The useful comments and the kind suggestions from the anonymous reviewers would be also appreciated.

# Noninvasive and Painless Urine Glucose Detection by Using Computer-based Polarimeter

**Sutrisno, Y A Laksono and N Hidayat**

Department of Physics, Faculty of Mathematics and Natural Sciences, Universitas Negeri Malang, Jl. Semarang 5 Malang, Indonesia

Email: sutrisno.fmipa@um.ac.id

**Abstract.** Diabetes kills millions of people worldwide each year. It challenges us as researchers to give contribution in early diagnosis to ensure a healthy life. As a matter of fact, common glucose testing devices that have been widely used so far are, at least, glucose meter and urine glucose test strip. The glucose meter ordinarily requires blood taken from patient's finger. The glucose test strip uses patient's urine but records unspecific urine glucose level, since the strip only provides the glucose level in some particular ranges. Instead of detecting the glucose level in blood and using the non-specific technique, a noninvasive and painless technique that can detect glucose level accurately will provide a more feasible approach for diabetes diagnosis. The noninvasive and painless urine glucose level monitoring by means of computer-based polarimeter is presented in this paper. The instrument consisted of a power source, a sample box, a light sensor, a polarizer, an analyzer, an analog to digital converter (ADC), and a computer. The concentration of urine glucose concentration was evaluated from the curve of the change in detected optical rotation angle and output potential by the computer-based polarimeter. Statistical analyses by means of Gaussian fitting and linear regression were applied to investigate the rotation angle and urine glucose concentration, respectively. From our experiment, the urine glucose level, measured by glucose test strips, of the normal patient was 100 mg/dl, and the diabetic patient was 500 mg/dl. Our polarimeter even read more precise values for the urine glucose concentrations of those normal and diabetic of the same patients, i.e. 50.61 mg/dl and 502.41 mg/dl, respectively. In other words, the results showed that our polarimeter was able to quantitatively measure the urine glucose level more accurate than urine glucose test strips. Hence, this computer-based polarimeter could be used as an alternative for early detection of urine glucose with noninvasive and painless characteristics.

**Keywords:** glucose level, noninvasive and painless measurement, computer-based polarimeter.

## 1. Introduction

During the past years, one of the most critical issues on human health is diabetes mellitus [1]. Diabetes is a systemic disorder disturbing the body metabolism and may cause death [2]. In 2015, statistically speaking, more than 400 million grownup men and women were infected by diabetes mellitus, and it is predicted to rise by over 50% in the next 25 years [3]. Still, according to WHO, world health organization, it was reported that diabetes mellitus caused at least 1.5 million deaths in 2012. If the early detection of diabetes mellitus can be easily done, the risky effects of it can be significantly reduced. Since glucose levels are repeatedly used as an indicator of diabetes mellitus [3], then only can good instruments accurately detect the diabetes mellitus in terms of the glucose levels.

IOP Conf. Series: Materials Science and Engineering **202** (2017) 012030 doi:10.1088/1757-899X/202/1/012030

Furthermore, many researchers have developed instruments for measuring glucose levels. For instances, high-speed dual-wavelength optical polarimetries, near-infrared spectroscopies ([4-6]), multisensor noninvasive blood glucose monitoring systems [7], and blood glucose monitoring in the terahertz frequency range techniques [8]. However, those instruments are excessively complex and reasonably highly cost consuming. Thus, it is necessary to develop a simpler instrument for early detection of diabetes mellitus by measuring the glucose level. The measurement should be noninvasive and painless. Therefore, instead of measuring the glucose levels from blood, urine is more preferable. In this present study, we introduce a noninvasive and painless urine glucose levels computer-based polarimeter for diabetes early mellitus detection.

## 2. Experimental Method

A laser diode, a light dependent resistor (LDR), a polarizer, an analog to digital converter (ADC), and a computer were prepared to construct the computer-based polarimeter. The laser diode was chosen as a light source due to its coherence radiation characteristic. The laser diode was emited to the solutions and then detected by the LDR. The LDR is able to transform light intensities to be output voltages. The voltages as analog data were then converted into digital data by means of ADC and connected to the computer, so that the data can be recorded and analyzed as well. In principle, the optical activity property of the glucose solution rotates the polarization direction of the polarized light when passing through the glucose solution. The amount of rotation is defined as a rotation angle. The rotation angle depends on (1) the nature of the solution, (2) the concentration of the solution, (3) the length of solution that is exceeded by light, (5) the wavelength of source light, and (6) the temperature of the solution. The angle of rotation ($\alpha$) can be mathematically expressed as

$$\alpha = \alpha_t^D \frac{LC}{100} \tag{1}$$

Where $L$ is the length of solution that is exceeded by light and $C$ is the concentration of the solution. From the developed polarimeter, the data that can be read by the computer is the real time tabulation consisting of angel rotation polarimeter (in degree) and intensity light (in voltage). From those data, further analysis was done by executing Gaussian fitting to extract an angle that refers to the minimum value of output potential [9]. The concentrations glucose solutions were then calculated by Equation 1. The sensitivity of the instrument was evaluated by using linear regression analysis. To calibration correlate glucose solutions concentration and rotation angle is depicted in Table 1. while the instrumental set-up of the computer-based polarimeter is given in Figure 1.

**Table 1.** Glucose solution concentrations and rotation angle

| Cons. (mg/dL) | Rotation Angle (° ) | | | Change of rotation angle ( °) |
|---|---|---|---|---|
| | 1 | 2 | $\alpha_{average}$ (°) | |
| 0 | 86.990 | 86.830 | 86.990 | 0 |
| 25 | 87.120 | 86.890 | 87.005 | 0.015 |
| 150 | 87.160 | 87.020 | 87.090 | 0.100 |
| 500 | 87.090 | 87.150 | 87.120 | 0.130 |
| 5000 | 87.670 | 87.350 | 87.510 | 0.520 |

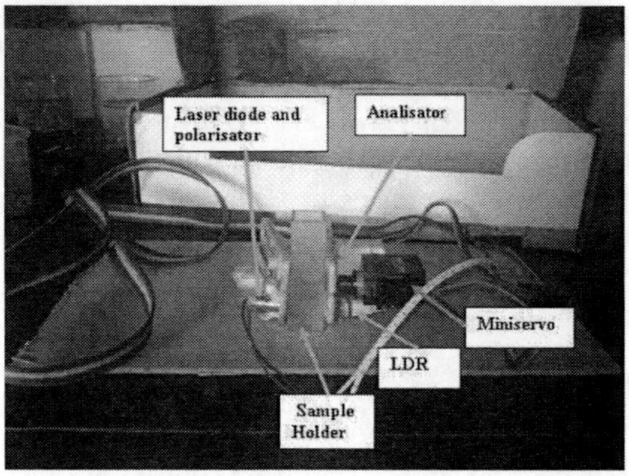


**Figure 1.** Instrumental set-up of the computer-based polarimeter

## 3. Results and Discussion

The collected data by means of the computer-based polarimeter is depicted in Figure 2. The lowest intensity represents that the polarizer and the analyzer are perpendiculars, so that the output potential is very low. This most moderate output potential is corresponding to the rotation angle of the measured solution. Figure 2 shows the relationship between rotation angle and output potential. As we can see from Figure 2, the graph can be approached by means of Gaussian fitting. The minimum value of the output potential is related to the rotation angle of 87.64°. Every measurement of the glucose solution concentration was repeated three times to obtain more precise and accurate measurement.

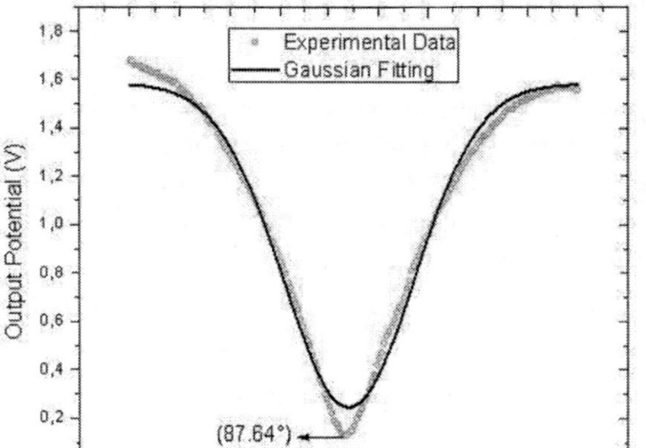


**Figure 2.** Graphical representation of rotation angle (α ) and output potential for water. The black solid line is the collection data from computer-based polarimeter. The green dot line is the Gaussian fitting.

The sensitivity of the computer-based polarimeter can be evaluated by linear regression analysis as given in Figure 3. The linear fitting equation for Figure 3 is $Y = 10040X - 401.19$, where $Y$ is the concentration of glucose solutions and $X$ is the change of rotation angle $\Delta\alpha$ (°) and correlation coefficient of $R^2 = 0.96$. with errors of $S_B = 0.34$ and $S_A = 0.10$. The value of the correlation coefficient of 0.96 indicates that (1) the experimental data is well fitted with the linear model and (2) the developed polarimeter is reliable for determining the concentration of glucose solutions.

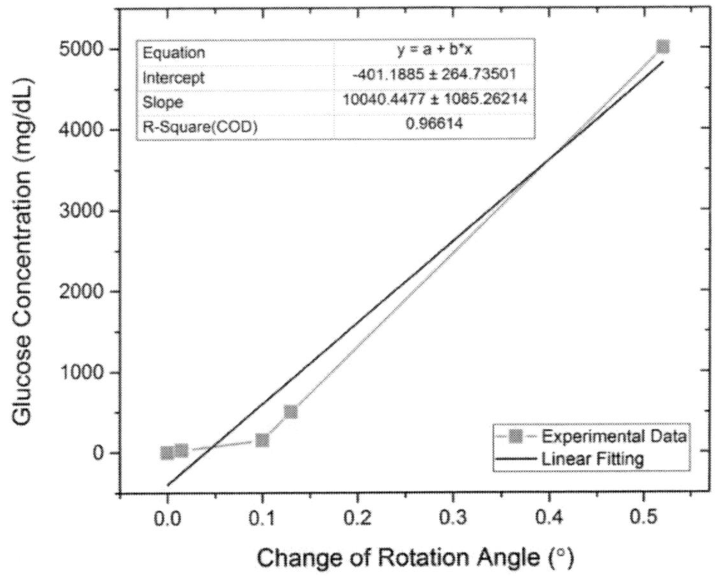


**Figure 3.** Linear fitting of glucose concentration as the function of the change of rotation angle.

The experiment to a human with sample one of the diabet and normal patient. The change of rotation angle for urine of patient was found from polarimeter measurement as 0.045° and 0.09°. Our polarimeter could even read more precise values for the urine glucose concentrations of those normal and diabetic patients, i.e. 50.61 and 502.41 mg/dl, respectively, depicted in Table 2. The urine glucose level, measured by glucose test strip, of the normal patient was 100 mg/dl and the diabetic patient was 500 mg/dL.

**Table 2.** The urine glucose level measurement by computer-based polarimeter

| Gender (M/F) | Age (Years) | Rotation angle α (°) 1 | 2 | average | Chang of rotation angle Δ α (°) |
|---|---|---|---|---|---|
| Male | 22 | 87.040 | 87.030 | 87.035 | 0.045 |
| Female | 61 | 86.890 | 87.270 | 87.080 | 0.090 |

## 4. Conclusion

A very simple computer-based polarimeter has been successfully built-up. The instrument can be used for urine glucose levels for early detection of diabetes mellitus. Our experiment showed that the polarimeter might read more precisely being compared with urine glucose test strip. Therefore, this polarimeter can be applied for noninvasive and painless urine glucose levels measurement.

## 5. References

[1] Zimmet P, Alberti K G, Magliano D J and Bennett P H 2016 Diabetes mellitus statistics on prevalence and mortality: facts and fallacies *Nat. Rev. Endocrinol* **12** 10 pp. 616–622

[2] Malik S, Gupta S, Khadgawat R and Anand S 2015 A novel non-invasive blood glucose monitoring approach using saliva," in *Signal Processing, Informatics, Communication and Energy Systems (SPICES), IEEE International Conference on* 2015 pp. 1–5.

[3] American Diabetes Association, 2012 Diagnosis and Classification of Diabetes Mellitus Diabetes Care 35 (Supplement 1) pp S64–71

[4] Li X and Li C 2016 Improved CEEMDAN and PSO-SVR Modeling for Near-Infrared Noninvasive Glucose Detection *Comput. Math. Methods Med* **2016**

[5] Song K, Ha U, Park S, Bae J and Yoo H-J 2015 An impedance and multi-wavelength near-infrared spectroscopy IC for non-invasive blood glucose estimation *IEEE J. Solid-State Circuits* **50** 4 pp 1025–1037

[6] Yadav J, Rani A, Singh V and Murari B M 2015 Prospects and limitations of non-invasive blood glucose monitoring using near-infrared spectroscopy *Biomed. Signal Process. Control*,**18** pp 214–227

[7] Samadarsinee S 2015 Multisensor noninvasive blood glucose monitoring system Wichita State University

[8] Cherkasova O, Nazarov M and Shkurinov A 2016 Noninvasive blood glucose monitoring in the terahertz frequency range *Opt. Quantum Electron* **48** 3 pp 1–12

[9] Sutrisno, Laksono Y A S and Hidayat N 2015 Dependensi sudut putar pada konsentrasi larutan gula dalam sistem polarimeter berbasis komputer *SNFP* **F-B-1-4**

**Acknowledgements**

We would like to acknowledge Ministry of Research, Technology and Higher Education of the Republic of Indonesia for providing Hibah Bersaing 2016 for SS under contract No. 7.4.9/UN32.14/LT/2016.

# Inexpensive Home-Made Single Wavelength Ellipsometer ($\lambda$ = 633 nm) for Measuring the Optical Constant of Nanostructured Materials

**L Z Maulana, K Megasari, E Suharyadi, R Anugraha, K Abraha and I Santoso**

Departemen Fisika, Universitas Gadjah Mada, Sekip Utara BLS 21 Yogyakarta, Indonesia

Email: iman.santoso@ugm.ac.id

**Abstract**. Inexpensive home-made Single wavelength Ellipsometry with RAE (Rotating Analyser Ellipsometer) configuration has been developed. Spectroscopic ellipsometry (SE) is an optical measurement technique which is based on the measurement of the change of the phase difference ($\Delta$) and the amplitude ratio ($\psi$) between p and s linear polarized of reflected (or transmitted) light. Our RAE configuration system composed of polarizer, sample, analyzer, detector, and He-Ne laser ($\lambda$ = 633 nm) that acted as the monochromatic light source. To test the reliability of our SE system, we measure the optical constant of Au bulk and Cr (30 nm thick) film. The optical constant and the thickness were extracted by employing the pseudo-dielectric function and numerical inversion which is based on the secant method, the $\psi$ and $\Delta$ of our SE data which is modelled by Fresnel equation. From the extraction using the secant method we obtain the optical constant of the Au bulk sample with $n$ = 0.11 to 0.22 and $k$ = 3.26 to 3.37 which is close to that of using pseudo-dielectric method. We obtain the same result for Cr film with $n$ = 3.66 to 3.81 and $k$ = 5.32 to 5.38 which is close to the result from reference. These results show that our inexpensive home-made Single wavelength Ellipsometry instrument and the extraction method are reliable for determining the optical constant of nanostructured materials.

**Keywords:** Secant methods, optical constant, Ellipsometry

## 1. Introduction

Recently spectroscopy has become a demand at physics, especially on nanostructured material analysis. In the nanoscale, some matters revealed the unique phenomenon compared to the large dimension [1]. Moreover, on some scales, they show the confinement effect [2]. The intrinsic properties of matter can be represented by the complex refractive index. There are several methods to calculate the optical constant of materials that has been developed such as reflectometer and UV-Vis spectroscopy [3]. The determination of both optical constant (refractive index and extinction coefficient) are difficult. This case is due to the correlation and the existence of the imaginary extinction coefficient on the optical constant. Ellipsometry is a technique that both optical constants can be calculated all at once. Ellipsometry is a reliable optical measurement technique that analyzes light reflection (or transmission) from the sample. The main feature of ellipsometry is measured the change in polarization of light upon reflection or transmission from the sample. It is well known as a

IOP Conf. Series: Materials Science and Engineering **202** (2017) 012031 doi:10.1088/1757-899X/202/1/012031

very sensitive and nondestructive technique in materials characterization methods. Moreover, it allows to measure accurately and high reproducibility the thicknesses and complex refractive index or dielectric constant of a given materials. In the Ellipsometry measurement, the optical model defining are necessary. That case means any dimension of the sample can be investigated according to our model, even in the nanostructured of the thin film. The precision of Ellipsometry measurement changes proportionally with the wavelength [4,5]. With the light source in this experiment, we can get less than 1 nm precision that leads to the precise result of the optical constant measurement. This paper is explaining the optical constant measurement and calculation of bulky and thin film in nanometer thick.

In the ellipsometry measurement, light are considered as the electromagnetic waves that compose of electrical field ***E*** and magnetic field ***B***. Ellipsometric measurement allows us to quantify the phase difference between ***Ep*** and ***Es*** as $\Delta$, and the ratio of their amplitudes given by tan $\psi$. If a sample undergoes a change especially in its thickness or the material composition, its reflection properties will change. For a reflecting surface, the forms of $\Delta$ and $\psi$ are defined:

$$\Delta = \delta_p - \delta_s, \qquad \tan\psi = \frac{|r_p|}{|r_s|}. \tag{1}$$

where $\delta p$ and $\delta s$, are the phase changes for the ***p*** and ***s*** components of light and $r_p$ and $r_s$ are the complex Fresnel reflection coefficients. Furthermore, in mathematical language, the ellipsometric analysis is based on Fresnel's equations for polarized light encountering boundaries in the planar sample [5]. The famous ellipsometric equation is conventionally written as:

$$\rho \equiv \frac{r_p}{r_s} \equiv \tan\psi e^{i\Delta}. \tag{2}$$

The general expressions for $r_p$ and $r_s$ for the interface between mediums are given by:

$$\begin{aligned} r_p &\equiv \frac{E_{rp}}{E_{ip}} = \frac{N_t\cos\theta_i - N_i\cos\theta_t}{N_t\cos\theta_i + N_i\cos\theta_t} \\ r_s &\equiv \frac{E_{rs}}{E_{is}} = \frac{N_i\cos\theta_i - N_t\cos\theta_t}{N_i\cos\theta_i + N_t\cos\theta_t}. \end{aligned} \tag{3}$$

where the ambient with complex refractive index $N_i$ and the air with complex refractive index $N_t$ with $\theta_0$ and $\theta_t$ are the angles of incidence and refraction. These expressions can be implied on single and multilayer medium. As long as $\psi$ and $\Delta$ are measured at a given wavelength, the optical parameters and thickness sample can be extracted [4].

The ellipsometer has been constructed in RAE (Rotating Analyser Ellipsometry) configuration. Ellipsometry can perform in reflection and transmission mode. Employing 2×2 Jones-matrix formalism on RAE configuration, the reflected electric field $\boldsymbol{E_r}$ regarding the incident field $\boldsymbol{E_i}$ can be expressed in matrix form:

$$\begin{aligned} \begin{bmatrix} E_A \\ 0 \end{bmatrix} &= \begin{bmatrix} 1 & 0 \\ 0 & 0 \end{bmatrix} \begin{bmatrix} \cos A & \sin A \\ -\sin A & \cos A \end{bmatrix} \begin{bmatrix} \sin\psi\exp(i\Delta) & 0 \\ 0 & \cos\psi \end{bmatrix} \\ &\times \begin{bmatrix} \cos P & -\sin P \\ \sin P & \cos P \end{bmatrix} \begin{bmatrix} 1 & 0 \\ 0 & 0 \end{bmatrix} \begin{bmatrix} 1 \\ 0 \end{bmatrix}. \end{aligned} \tag{4}$$

The above matrix are consisted by detector, Analyzer, sample, Polarizer and light source respectively. By simplifying this form we obtain Fourier equation with two constant consist in it as follows:

$$I = I_0[1 + \alpha\cos 2A + \beta\sin 2A] \tag{5}$$

From these two constant $\alpha$ and $\beta$, ellipsometer parameters $\Delta$ and $\psi$ can be written as:

$$\tan\psi = \sqrt{\frac{1+\alpha}{1-\alpha}}\left|\tan\beta\right|, \qquad \cos\Delta = \frac{\beta}{\sqrt{1-\alpha^2}}. \tag{6}$$

## 2. Experimental Method

We describe the instrument and calculation methods of inexpensive home-made single wavelength ellipsometer (cost below $8000). This ellipsometer composed by the light source, polarizer, sample holder, analyzer, and detector. The light source is a He-Ne laser tube emitting less than 5mW linearly polarized red spectrum ($\lambda$ = 633 nm). Rochon's prism (by Edmund Optic) configuration has been used as the polarizer and analyzer. Those prisms are mounted in vernier scales along 360° vertically with 1° precision. Sample holders were a form of acrylic substance that can be rotated easily along 360° horizontally to determine the incidence angle. The monochromatic light propagated from He-Ne laser polarized by quartz Rochon polarizer through the sample can be seen in Figure 1. The reflected beam passes through another quartz Rochon Analyzer caught by LP Power meter detector. A schematic diagram of these ellipsometers is shown in Figure 1.

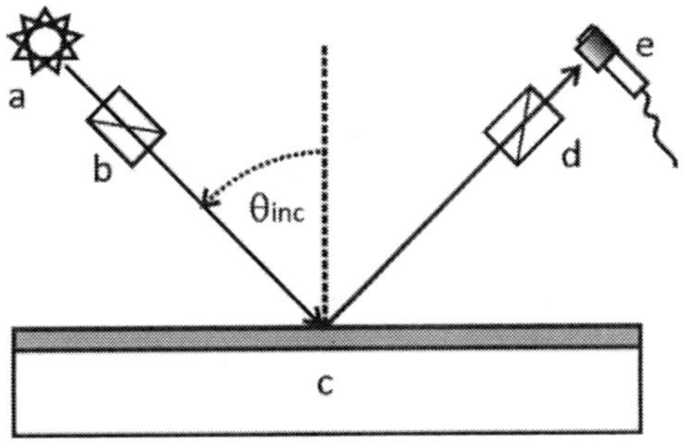


**Figure 1.** Schematic diagram of RAE configuration, (a) monochromatic light source (b) fixed linear polarizer (c) an isotropic sample (d) linear rotated analyzer and (e) photodetector.

In this paper two kinds of materials as the bulk and thin film are investigated. Purchased evaporated Au on a glass substrate is used as the bulk sample. Sputtered Cr with 30 nm thick on a glass substrate is used as the thin film sample. Thin films from deposition usually form on planar with cm dimension depend on the substrate. To fit the sample on the sample holder, the sample must be resized in 0.3 × 0.5 cm 1 × 1 cm maximum. The sample in out of this range must be fitted with other attachment. The first experimental section begins by setting up the Ellipsometer system. The important thing that should be done in the optical measurement is the instruments alignment. In this case, we made it sure that the light are trough the sample in 0° vertically and horizontally and continue to set the incident angle on 60, 65°, and 70°. An RAE's configuration analyzer are rotated in 0° to 180° with step of 5°. The initial and last intensity were recorded during the measurement and divided to get a normalized intensity versus analyzer angle.

Figure 2 shows intensity versus analyzer angle of Au, glass slide, and Cr 30 nm respectively. For each data analyzer is rotated from 0° to 180° with 5° step. The continuum line guides to light of 45° linear polarized light source, and the other line shows the ellipsometry curve indifference incident angle. As shown in those Figures, reflected light has changed both in amplitude and especially in that phases, that shows the interaction between electromagnetic waves and matter. This paper explains the measurement of the bulk Au (Figure 2a) and the Cr (Figure 2c) thin film. Since both samples are deposited on the glass slide substrate we also show glass slide measurement. Normally the intensity versus analyzer angle in the Ellipsometry measurement are shaped as follow equation 5. All those figures in Figure 2 show the same sinusoidal shape; they have difference peak and close position. Every kind of sample and the specific thickness has a characteristic on the intensity versus analyzer angle. The special sinusoidal shape that has specific $\alpha$ and $\beta$ will lead to one of the intrinsic properties of matter that called the optical constant.

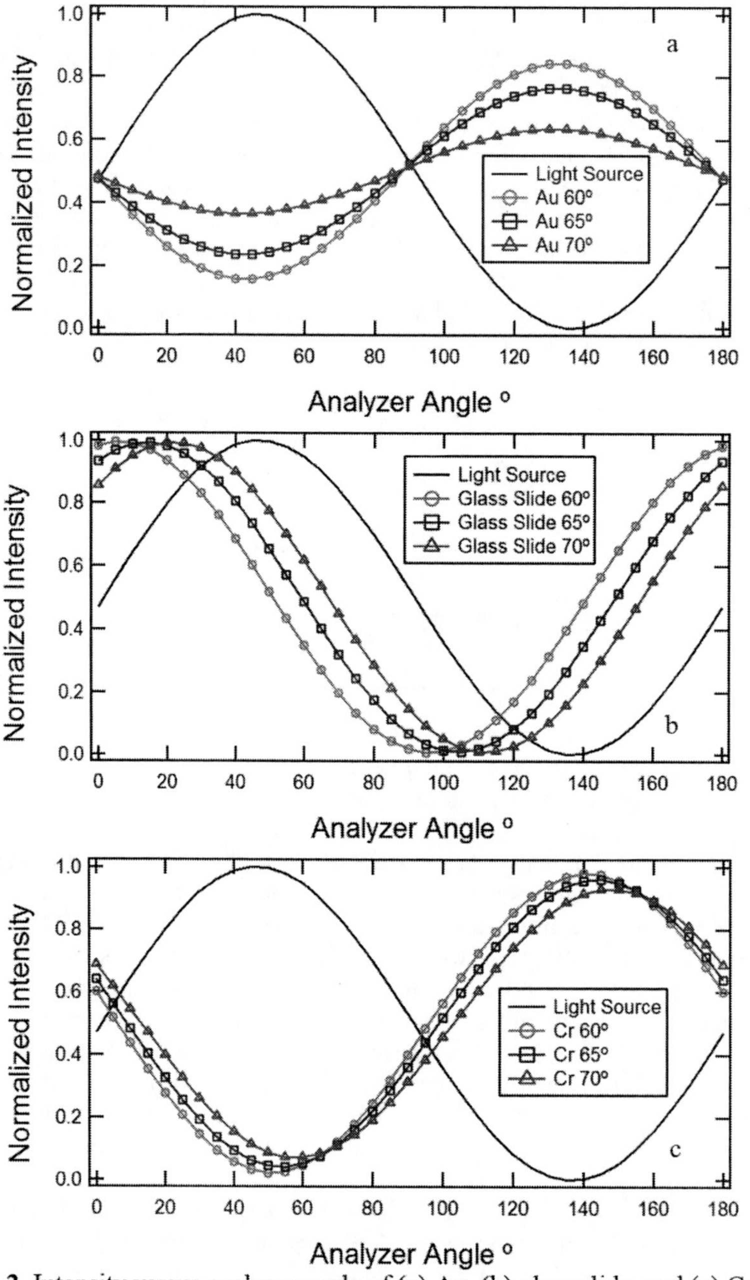


**Figure 2.** Intensity versus analyzer angle of (a) Au, (b) glass slide, and (c) Cr 30 nm.

## 3. Results and discussion

The intensity vs. analyzer angle that is modeled in Eq. 5 were curve fitted with that's equation to obtain $\alpha$ and $\beta$. Those parameters were used to calculate the ellipsometry parameters $\psi$ and $\Delta$ with equation 6 and of course $\rho$ from equation 2. The precision of $\psi$ and $\Delta$ can be expressed from the measurement errors of Fourier coefficients following from equation 7 [5].

$$\delta\psi = -\frac{(1-\cos 2P\cos 2\psi)^2}{2\sin^2 2P\sin 2\psi}\delta\alpha.$$
$$\delta\Delta = -\frac{\cos\Delta(\cos 2P-\cos 2\psi)(1-\cos 2P\cos 2\psi)}{2\sin^2 2P\sin\Delta\sin 2\psi}\delta\alpha - \frac{(1-\cos 2P\cos 2\psi)}{\sin 2P\sin\Delta\sin 2\psi}\delta\beta. \quad (7)$$

In this equation $P$ is polarization angle fixed in 45°. The change of the phase difference ($\Delta$) and the amplitude ratio ($\psi$) can be united in complex form $\rho$. The ellipsometry parameters are shown in the Table1 to 3.

**Table 1.** Ellipsometry parameters of Au sample.

| Inc. Angle | $A$ | $\beta$ | $\psi$° | Δ° |
|---|---|---|---|---|
| 60° | -0.050 ± 0.005 | -0.683 ± 0.005 | 43.7 ± 0.2 | 133.2 ± 0.4 |
| 65° | -0.0468 ± 0.0001 | -0.529 ± 0.007 | 43.757 ± 0.001 | 121.9 ± 0.5 |
| 70° | -0.0340 ± 0.0005 | -0.268 ± 0.002 | 44.13 ± 0.05 | 105.5 ± 0.1 |

**Table 2.** Ellipsometry parameters of glass slide.

| Inc. Angle | $I_0$ | $A$ | $\beta$ | $\psi$° | Δ° |
|---|---|---|---|---|---|
| 60° | 0.45 | 0.963 ± 0.004 | 0.206 ± 0.002 | 82.2 ± 0.1 | 40 ± 4 |
| 65° | 0.46 | 0.863 ± 0.004 | 0.467 ± 0.008 | 74.8 ± 0.1 | 22 ± 4 |
| 70° | 0.46 | 0.707 ± 0.002 | 0.683 ± 0.002 | 67 ± 2 | 14 ± 1 |

**Table 3.** Ellipsometry parameters of Cr thin film.

| Inc. Angle | $I_0$ | $A$ | $\beta$ | $\psi$° | Δ° |
|---|---|---|---|---|---|
| 60° | 0.46 | 0.201 ± 0.001 | -0.933 ± 0.001 | 50.90 ± 0.03 | 162.34 ± 0.04 |
| 65° | 0.46 | 0.279 ± 0.001 | -0.871 ± 0.002 | 53.21 ± 0.09 | 155.2 ± 0.2 |
| 70° | 0.46 | 0.375 ± 0.001 | -0.768 ± 0.001 | 56.10 ± 0.03 | 146.04 ± 0.06 |

Basically, the optical constants are calculated by solving the Fresnel equation (Eq. 3). In large shape two media are considered. They are ambient with optical constant $N_0$ and sample medium with optical constant $N_1$. These Fresnel equations are already shown in Eq. 3. The bulk shape sample optical constants are swimmingly calculated with famous pseudo-dielectric function below [4].

$$\hat{\varepsilon} = \hat{n}^2 = \sin^2\theta_0\left[1+\tan^2\theta_0\left(\frac{1-\rho}{1-\rho}\right)^2\right]. \quad (8)$$

More medium are involved more difficult and complicated equation formed. On the thin film sample, the light can propagate from ambient trough the substrate make three mediums are considered in thin film model. There are film thickness $d$, ambient with optical constant $N_0$, thin film with optical constant $N_1$, and the substrate with optical constant $N_2$ are involved. The Fresnel equation of thin film shapes are shown below [4]:

$$r_{012,p} = \frac{r_{01,p}+r_{12,p}\exp(-i2\beta)}{1+r_{01,p}r_{12,p}\exp(-i2\beta)},\quad r_{012,s} = \frac{r_{01,s}+r_{12,s}\exp(-i2\beta)}{1+r_{01,s}r_{12,s}\exp(-i2\beta)}, \quad (9)$$

$$\beta = \frac{2\pi d}{\lambda} N_1 \cos\theta_1 = \frac{2\pi d}{\lambda}\left(N_1^2 - N_0^2 \sin^2\theta_1\right)^{1/2}. \tag{10}$$

There are several ways to solve the Fresnel equation. In this paper Secant methods were used to calculate it. Secant methods are the development of Newton methods presented below [6]:

$$x_{k+1} = x_k - \frac{(x_k - x_{k-1}) f(x_k)}{f(x_k) - f(x_{k-1})}. \tag{11}$$

As mentioned above $\psi$ and $\Delta$ can be transformed to $\rho_{\exp}$, and we get $\rho_{calculation}$ from Fresnel equation. Secant methods are work on $\rho_{\exp}$ and $\rho_{calculation}$ to be same by putting an error on them. We divine error 0.01 to all of the Secant methods calculation. The function put in Secant methods are depend on the optical model of sample. Equation 12 shows bulky sample and eq. 13 for thin film sample.

$$f(x) = \rho_{\exp} - \frac{\left(\frac{N_1 \cos\theta_0 - N_0 \cos\theta_1}{N_1 \cos\theta_0 + N_0 \cos\theta_1}\right)}{\left(\frac{N_0 \cos\theta_0 - N_1 \cos\theta_1}{N_0 \cos\theta_0 + N_1 \cos\theta_1}\right)} = 0. \tag{12}$$

$$f(x) = \rho_{\exp} - \frac{\left\{\frac{r_{01,p} + r_{12,p}\exp(-i2\beta)}{1 + r_{01,p} r_{12,p}\exp(-i2\beta)}\right\}}{\left\{\frac{r_{01,s} + r_{12,s}\exp(-i2\beta)}{1 + r_{01,s} r_{12,s}\exp(-i2\beta)}\right\}} = 0. \tag{13}$$

As the bulk sample, both ways of secant methods and pseudo-dielectric function calculation of Au are reported. Both calculation methods show the same result that present in Table 4. Its proved that both calculation are reliable to calculated optical constant. We obtain the complex refractive index of the Au bulk sample with $n$ = 0.11to 0.22 and $k$ = 3.26 to 3.37 which is close to the result from reference [7,8,9,10]. Basically, the optical constant is the intrinsic properties of matter that shows same value even on the difference angle of incident. There are some discrepancies of our result and the reference. It may come from parallax error, sample tilting and surface inhomogeneity. These results are leading us to developed this algorithm and computational program to calculated another kind of sample.

**Table 4.** Optical constant of Au by secant methods and pseudo-dielectric function calculation.

| Inc. Angle | $P$ | $\hat{n}$ | $\hat{n}$ secant |
|---|---|---|---|
| 60° | -0.652804 - 0.696211 i | 0.23 + 3.34 i | 0.23 + 3.34 i |
| 65° | -0.507264 - 0.812135 i | 0.18 + 3.38 i | 0.18 + 3.38 i |
| 70° | -0.260148 - 0.934447 i | 0.11 + 3.26 i | 0.11 + 3.26 i |

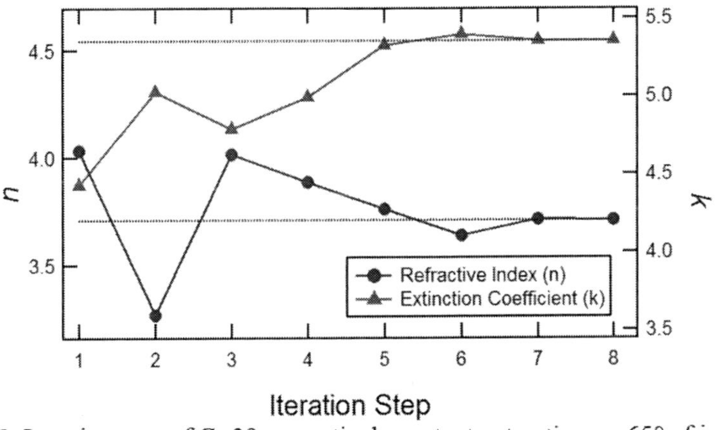


**Figure 3**. Iteration step of Cr 30 nm optical constant extraction on 65° of incident of angle.

Secant is the popular computational method to calculate non-linear or other complicated equation. In this paper, we perform these methods to extract the complex refractive index of the thin film sample. These methods successfully calculated the optical constant of the bulk sample. All of the secant methods calculation we choose 0.01 as the tolerance value. The reliability of the computational methods to solve some equations can be seen from how fast they reach the tolerance value, and the convergence of that iteration. Figure 3 shows the iteration step sample of our experiment. All of the secant calculation need less than 15 iterations to reach the final step. The convergence iteration in Figure 3 makes sure that these methods are reliable. Table 5 shows the complex refractive index of the Cr thin film with $n$ = 3.66 to 3.8 and $k$ = 5.33 to 5.38 which is close to the result from reference[10,11]. There are some discrepancies between our result and the reference result, especially on the extinction coefficient. The other reports are explain that deposition methods influence the optical properties of the film due to the crystal structures and the surface roughness [12]. Chromium is belonging to the wet metal that easily oxidized that lead to swallowing this coefficient [11].

**Tabel 5.** Optical constant of Cr 30nm thick by secant methods calculation.

| Inc. Angle | $\rho$ | $\hat{n}$ secant |
|---|---|---|
| 60° | -1.17292 – 0.373358 i | 3.81 + 5.33 i |
| 65° | -1.21429 – 0.560096 i | 3.71 + 5.35 i |
| 70° | -1.23451 – 0.831445 i | 3.66 + 5.39 i |

## 4. Conclusion

The construction and analysis methods of the inexpensive home-made Single wavelength Ellipsometry have been successfully developed. The instrument and the extraction method are reliable for determining the optical constant of bulk and nano-thin film sample. On the Ellipsometry, the light acted as an electromagnetic wave that interacted with matter. We obtained the same result of the Au optical constant that has been calculated by using both Pseudo Dielectric function and Secant method make sure that our Secant methods work properly to solve the two medium Fresnel equations. This capability leads the Secant methods been used to solved Cr nano-thin film that have 3 medium Fresnel Equation. These all systems are successfully to obtain the complex refractive index of Au as a bulky sample and Cr as a nano-thin film sample which is closed to the reference.

## 5. References

[1] Roco M C Mirkin C A and Hersam M C 2011 Nanotechnology research directions for societal needs in 2020: summary of international study. J Nanopart Res, DOI 10.1007/s11051-011-0275-5

[2] Mitin V, Sementsov D I and Vagidov N Z 2010 Quantum Mechanics for Nanostructures,*Cambridge University Press* UK.

[3] Potter R F, Reflectometer for Determining Optical Constants *Applied Optics* **4** 53-57

[4] Tompkins H G and Irene E A Handbook of Ellipsometry William Andrew New York

[5] Fujiwara H 2007 Spectroscopic Ellipsometry Principles and Applications John Wiley & Sons Ltd UK

[6] Klein A and Godunov A 2006 *Introductory Computational Physics* Cambridge University Press UK

[7] Babar S and Weaver J H 2015 Optical constants of Cu, Ag, and Au revisited *Optical Society of America* **53** 477-481

[8] McPeak K.M., Jayanti S.V., Kress S.J.P., Meyer S., Iotti S., Rossinelli A., and Norris D.J., 2015, Plasmonic Films Can Easily Be Better: Rules and Recipes. ACS Photonics 2, 326−333

[9] Olmon R L, Slovick B, Johnson T W, Shelton D, Oh S, Boreman G D and Raschke M B 2012 Optical dielectric function of gold *Physical Review B* **86** 235147

[10] Rakic´ D A, Djurisˇic´ A B, Elazar J M and Majewski M L 1998 Optical properties of metallic films for vertical-cavity optoelectronic devices *Journal of Optical Society of America* **37** 5271-5283.

[11] Johnson P B, and Christy RW 1973 Optical constants of transition metals: Ti, V, Cr, Mn, Fe, Co, Ni, and Pd *Physical Review B* **9** 5056

[12] Tompkins H G, Baker J, and Convey D 2000 Effect of process parameters on the optical constants of thin metal films *Surface and Interface Analysis* **29** 227–231

**Acknowledgments**

Grant of PUPT, Minister of Research, Technology, and Higher Education, Indonesia, Contract number: 733/UNI-P.III/LT/DIT-LIT/2016.

Nano-Fabrication of Nagoya University, Culture, Sport, Science, and Technology (MEXT) Nano-Project Platform, Japan, 2012-2017

# Effect of $Fe_3O_4$ Magnetic Nanoparticle Concentration on the Signal of Surface Plasmon Resonance (SPR) Spectroscopy

**M Oktivina, D T Nurrohman, Rinto A N Q Z, E Suharyadi and K Abraha**

Department of Physics, Universitas Gadjah Mada, Sekip Utara BLS 21 Yogyakarta Indonesia

Email: esuharyadi@ugm.ac.id

**Abstract**. Effect of $Fe_3O_4$ magnetic nanoparticle concentration on the signal of surface plasmon resonance (SPR) spectra has been successfully observed. The $Fe_3O_4$ nanoparticles with a particle size of about $10.5 \pm 0.2$ nm were used as active materials to increase the SPR response. X-ray diffraction (XRD) pattern showed that $Fe_3O_4$ nanoparticles have a high degree of crystallinity with spinel structure. The SPR system was successfully set up by using a glass prism coupler in a Kretschmann configuration in which gold (Au) thin film was thermally evaporated on the prism base. A green laser of wavelength 543 nm was used as light source. The angular scan in the attenuated total reflection (ATR) spectra showed a dropping intensity. Those things clearly indicated the appearance of SPR coupling phenomenon on the interface of Au thin film. The SPR spectra of fixed Au masses were also performed with same angular positions of dips. The $Fe_3O_4$ nanoparticles were deposited on gold thin film as a third layer which was synthesized via co-precipitation method. Hence, it was observed that the variation of $Fe_3O_4$ concentration affected the SPR spectra profile. The concentrations of $Fe_3O_4$ nanoparticles are 1, 3, 5, 7, 9, and 11 mg/ml which correspond to the angle shift of 0.1°, 0.3°, 0.5°, 0.7°, 0.9°, and 1.0°, respectively. The SPR angle of the dip was shifted to higher value due to change of refractive index of the medium as $Fe_3O_4$ nanoparticles concentration increases. Based on this result, we can conclude that the angle shift of SPR increases with increasing concentration of $Fe_3O_4$ nanoparticles.

Keywords: attenuated total reflection (ATR), $Fe_3O_4$, surface plasmon resonance.

## 1. Introduction

SPR is a resonant oscillation of conduction electrons at the interface between a negative and positive permittivity material stimulated by incident light. SPR-based biosensor can observe the interaction of biomolecules directly. The utilization of SPR-based biosensor as among others can measure the concentration of biomolecules, the thickness, and kinetic bond for certain biological analytes such as antigen/antibody, ligand/receptor, a protein reaction, and hybridization of DNA [1]. One of the observed responses in SPR-based biosensor is an SPR signal. For example, the trend of SPR signal against material concentration, in this case, is the polyclonal antibodies (PAb). The SPR signal increases with increasing PAb concentration [2]. The increase of SPR signal is observed as a phenomenon that occurs on the SPR system. The SPR phenomenon can be shown by the change in the angle of the SPR or SPR angle shift.

In general, the SPR angle shift occurs due to the difference in optical coating parameters, because of the angle depending on the optical parameter of the SPR system, such as refractive index from both sides of the metal, such as gold [3]. This case is the reason why SPR phenomenon can be used to a variety of biomolecules coated on top of the prism and metal. Given the characteristics of the various biomolecules, each of them that will be detected should then certainly be able to stick to the layer of metal. On the sample preparation process, there are several biomolecules that can be coated on the metal directly, but the obtained response is still low when detected by using SPR-based biosensor. To overcome this problem, SPR system needs an active material to increase the sensitivity, such as dielectric nanoparticles in detecting biomolecules.

One of the dielectric nanoparticles is $Fe_3O_4$. The $Fe_3O_4$ nanoparticles have a high dielectric constant that is very useful to enhance the input response. The polymer coated $Fe_3O_4$ can bind biomolecules by means of their ligand. For this purpose, the concentration of $Fe_3O_4$ nanoparticles might influence the SPR signal because the angle shift of SPR increases with increasing dielectric medium concentration [4]. Of course, varying the concentration of $Fe_3O_4$ is an interesting study that needs to be conducted in investigating their relation to the SPR signal. Therefore, this paper is focused on the SPR angle shift changing due to the concentration variation of $Fe_3O_4$ nanoparticles.

## 2. Experimental Method

### *2.1. Preparation of magnetic nanoparticles*

Magnetic nanoparticles of $Fe_3O_4$ were prepared by using co-precipitation method. The stoichiometric amount of 8.109 g $FeCl_3.6H_2O$ (Merck, Germany) was dissolved in 15 ml distilled water, 4.1703 g $FeSO_4.H_2O$ (Merck, Germany) was also dissolved in 15 ml distilled water. Then, the solution was added with dropwise of 24 ml $NH_4OH$ 10% (Merck, Germany) solution and also dissolved in 36 ml distilled water. The reaction was kept under constant stirring at 60°C for 90 minutes. Precipitated ferrite nanopowders were washed with distilled water for seven times. Finally, the sample was dried in a furnace at 80°C for 2 hours. The $Fe_3O_4$ nanoparticles of particle size about 10.5±0.2 nm were used as active materials to increase the SPR response. X-ray diffraction (XRD) pattern showed that $Fe_3O_4$ nanoparticles have a high degree of crystallinity with spinel structure [5].

### *2.2. Sample Preparation*

The $Fe_3O_4$ powder nanoparticles were dissolved in ethanol for analysis solution to make colloidal nanoparticles. The mixtures were placed in the ultrasonic cleaner about 30-60 minutes to form homogeneous colloids. Six samples of $Fe_3O_4$ nanoparticles were prepared in this study with a concentration range from 1% to 11% (Table 1). The $Fe_3O_4$ nanoparticle colloid was deposited on the Prism/Au system by using spray method to form a $Fe_3O_4$ film.

**Table 1.** The comparison of $Fe_3O_4$ powders and ethanol volume.

| Sample | $Fe_3O_4$ powders | Ethanol volume | Colloid concentration |
|---|---|---|---|
| A | 5 mg | 5 ml | 1 mg/ml |
| B | 15 mg | 5 ml | 3 mg/ml |
| C | 25 mg | 5 ml | 5 mg/ml |
| D | 35 mg | 5 ml | 7 mg/ml |
| E | 45 mg | 5 ml | 9 mg/ml |
| F | 55 mg | 5 ml | 11 mg/ml |

### *2.3. The observation of $Fe_3O_4$ nanoparticle concentration effect on the SPR-angle*

The SPR phenomenon was observed in the Prism/Air, Prism/Au/Air and Prism/Au/$Fe_3O_4$ nanoparticles/Air systems by using the method of attenuated total reflection (ATR) in the

IOP Conf. Series: Materials Science and Engineering **202** (2017) 012032 doi:10.1088/1757-899X/202/1/012032

Kretschmann configuration by using SPR set up device at Laboratory of Material Physics and Instrumentation, Physics Department of *Universitas Gadjah Mada*, Indonesia (Figure 1). Green laser (with wavelength $\lambda$=543 nm) was used here as the light source. The reflectance spectra were observed in range of 30°-80° with duplication to ensure the reliability of measurement. The BK7 type prisms were used in this study. 15 mg Au 99% was evaporated onto the clean prism by using vacuum evaporator. The evaporation was performed at pressure of 2 x $10^{-3}$ Pa and electric current of 50 A for 7 minutes, to make Au thin film. This chip was called Prism/Au system. Observation of the Prism/Air system was done to determine the refractive index value of each prism. Based on the value of critical angle $(\theta_c)$, by using the Snell's equation, the refractive index of prism $(n_p)$ can be calculated by using equation 1.

$$n_p = \frac{1}{\sin\theta_c}. \quad (1)$$

*2.4. Quantitative analysis*
The SPR angles against $Fe_3O_4$ concentration were plotted to built a model of sensor response for the prism/Au/$Fe_3O_4$/Air system. The model was obtained by using the standard linear curve fitting method facilitated by Igor Pro 6.36 for Windows software.

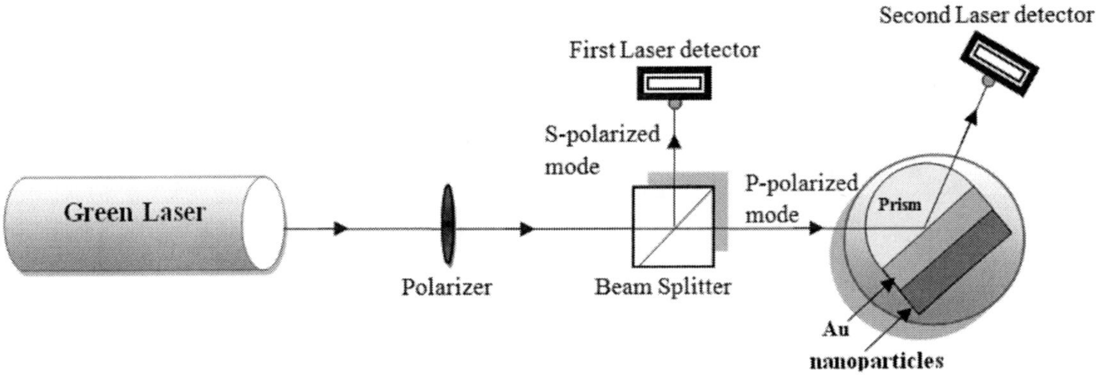


**Figure 1.** Set-up of SPR instruments

## 3. Results and Discussion

*3.1. Critical angle of prism*
Reflectance values are obtained from the comparison between the intensity of the reflected beam against the rays come. On each of the critical angle occurs optimally reflection produced by a prism. Figure 2 shows the ATR curves of Prism/Air system. Based on Figure 2, we know that the critical angles of prism A, B, C, D, E and F are 41.6°± 0.05°, 41.6°± 0.05°, 41.5°± 0.05°, 41.5°± 0.05°, 41.8°± 0.05° and 41.4°± 0.05°, respectively. The optimum reflection generated by the prism is due to the non-existence of a particle or other material over the surface of the prism that can excite the surface plasmons and causes a decrease in the value of the reflectance or the occurrence of vast rays absorption. We can retrieve the values of the refractive index of prism A, B, C, D, E and F, which are respectively 1.51, 1.51, 1.51, 1.51, 1.50 and 1.51 by using equation (1). Those figures indicate that the difference in refractive index of each prism is not too significant because we use six identical prisms in this study.

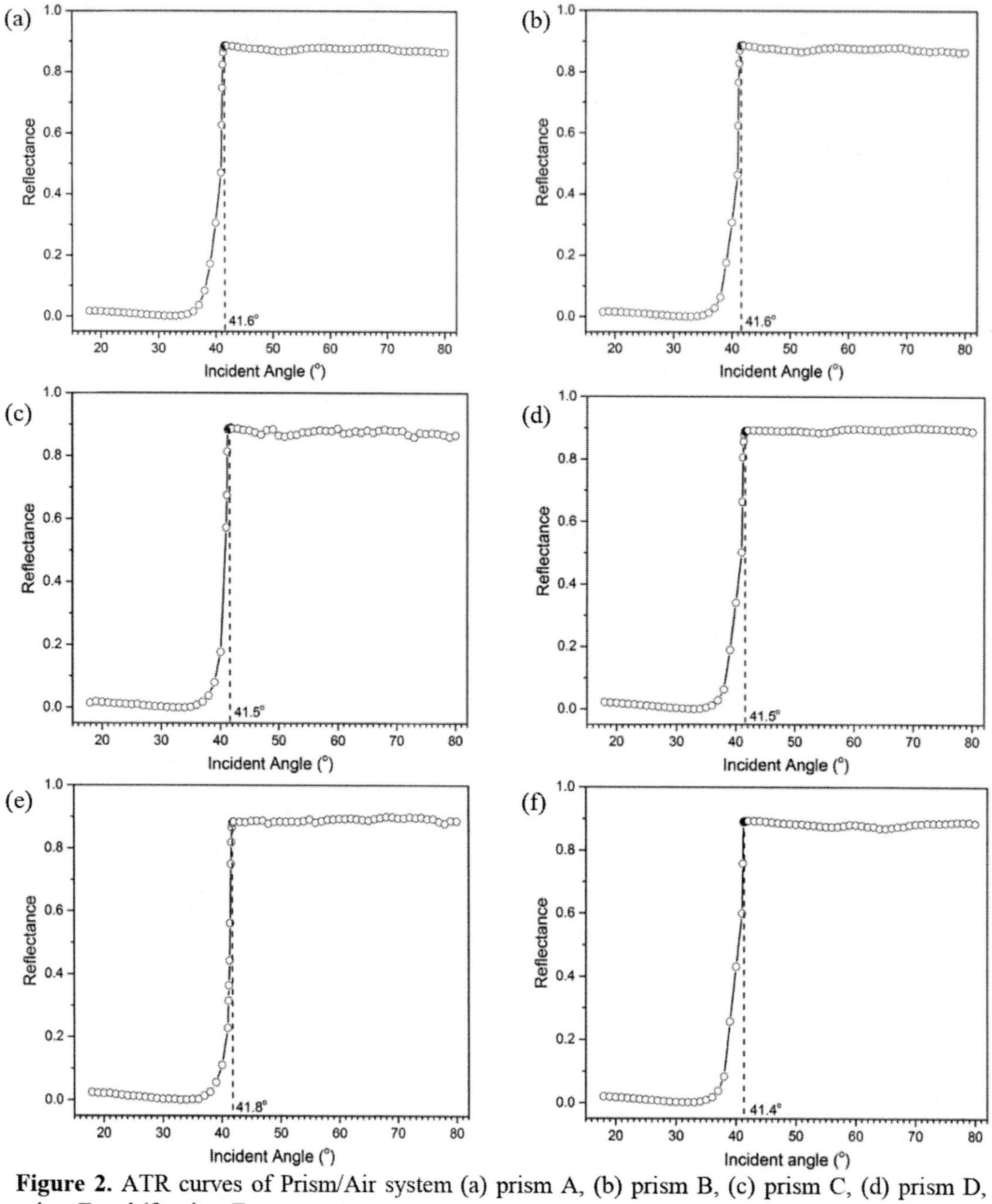


**Figure 2.** ATR curves of Prism/Air system (a) prism A, (b) prism B, (c) prism C, (d) prism D, (e) prism E and (f) prism F.

### *3.2. Effect of $Fe_3O_4$ concentration on the SPR-angle*

Figure 3 shows the ATR curves of Prism/Au/Air systems and Prism/Au/$Fe_3O_4$ nanoparticles/Air systems at different concentration. Curves of prism/Au are different because the difference of Au thickness result deposited on the prism. Based on Figure 3, it was observed that the variation of $Fe_3O_4$ concentration affects the SPR spectra profile. The dip of SPR angle shifts to higher value due to change of refractive index of medium as $Fe_3O_4$ nanoparticle concentration increases [4]. The concentrations of $Fe_3O_4$ nanoparticles are 1, 3, 5, 7, 9, and 11 mg/ml which correspond to the angle shifts of 0.1°, 0.3°, 0.5°, 0.7°, 0.9° and 1.0°, respectively.

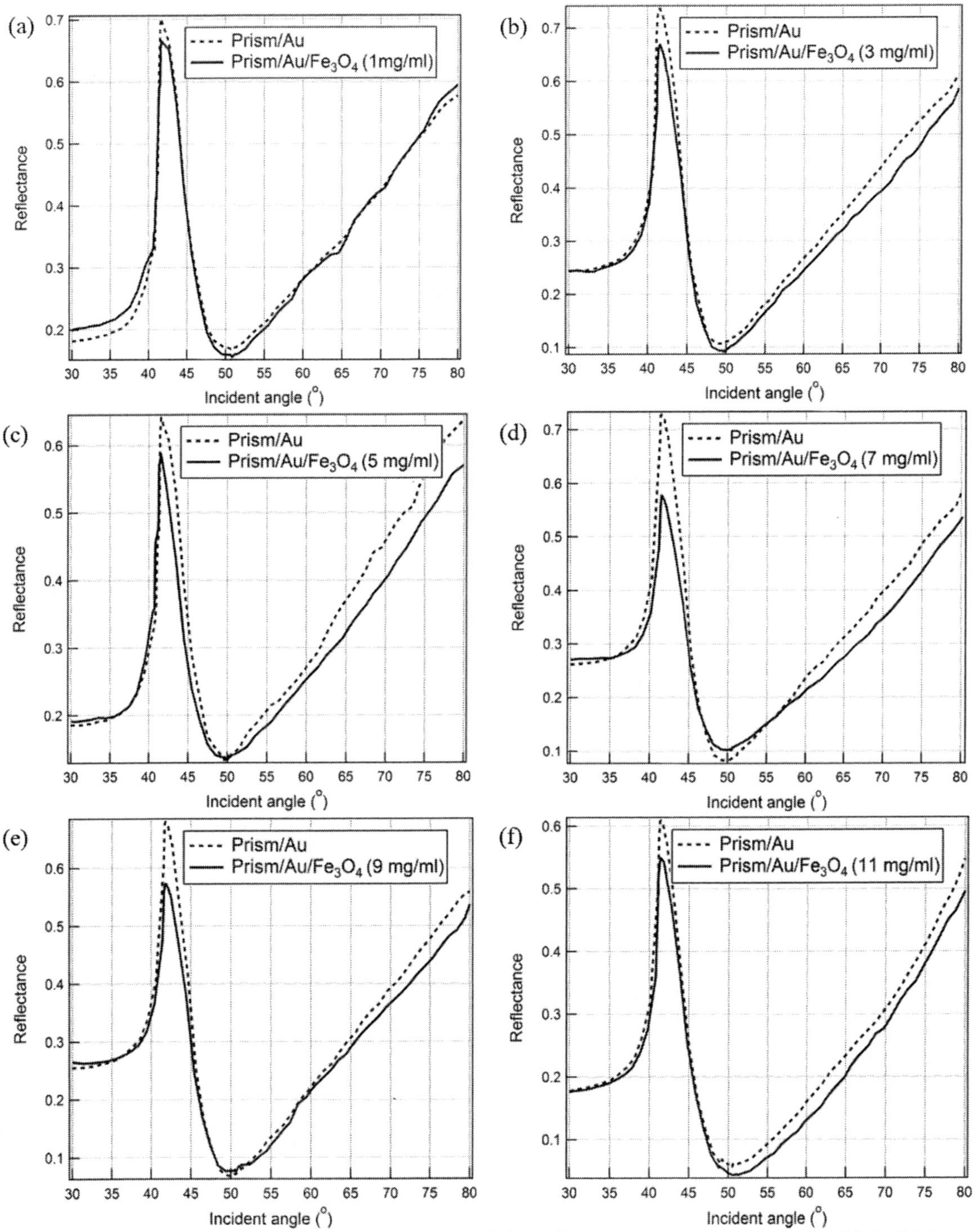


**Figure 3.** ATR curves of Prism/Au/$Fe_3O_4$ nanoparticles/Air systems at concentration of (a) 1, (b) 3, (c) 5, (d) 7, (e) 9, and (f) 11 mg/ml.

Figure 4 shows the relationship between the $Fe_3O_4$ concentration and SPR angle shift. The model in this section uses the SPR angle changes in determining the SPR response against the concentration variations of $Fe_3O_4$ nanoparticles. Based on the linear fitting model by using Igor Pro software for Window, the angle shift versus $Fe_3O_4$ concentration clearly shows a straight line relationship.

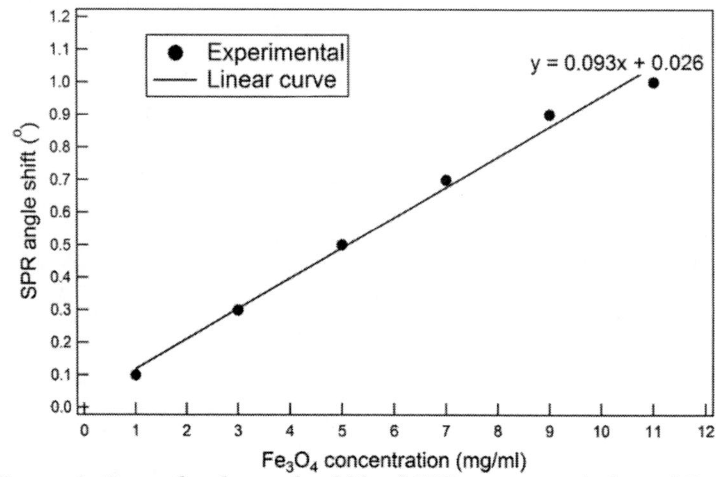


**Figure 4.** Curve for the angle shift of SPR versus variation of $Fe_3O_4$ nanoparticle concentration.

The equation of sensor response model is

$$\Delta\theta_{SPR} = 0.093c + 0.026 \quad (2)$$

where $\Delta\theta_{SPR}$ is a change of SPR dip angle and $c$ is $Fe_3O_4$ concentration. Based on equation (2), the sensitivity of the sensor is $0.093° \pm 0.004°$ to any change in the 1 mg/ml concentration of $Fe_3O_4$ nanoparticles. Based on Figure 4, the concentration of the $Fe_3O_4$ nanoparticles causes shifting in the SPR angle dip toward a greater value where both are linear in the range of 1-11 mg/ml concentration of $Fe_3O_4$ nanoparticles.The angle shift of SPR increases with increasing $Fe_3O_4$ concentration [4]. The SPR angle shifts due to the change of refractive index in each layer with different concentration. Indirectly, changes in the refractive index of $Fe_3O_4$ nanoparticle layers will cause a change in absorbance of such layers. The changing of concentration and absorbance of $Fe_3O_4$ nanoparticles change reflectance value. The absorbance value of the effect on the value of extinction coefficient signifies that the $Fe_3O_4$ nanoparticles can absorb light passing through it. The lower value of extinction coefficient which represents the imaginary part of the refractive index indicates that concentration of material decreases [6]. This phenomenon causes the refractive index is getting smaller.

## 4. Conclusion

$Fe_3O_4$ nanoparticles have been successfully synthesized through co-precipitation method and deposited on the Prism/Au system. The SPR angle of dip was shifted to higher value due to change of refractive index of the medium as $Fe_3O_4$ nanoparticle concentration increases. Based on this result, we can conclude that angle shift of SPR increases with increasing concentration of $Fe_3O_4$ nanoparticles.

## 5. References

[1] Choi S H, Kim Y L and Byun K M 2011 *Optics Express* **19** pp 458.
[2] Liu X Hu Y, Zheng S, Liu ., He Z, and Luo F *2016 Sensors and Actuators B: Chemical* **230** pp 191-198
[3] Tudos A J and Schasfoort R B M 2011 *Handbook of Surface Plasmon Resonance*, (United

Kingdom: RCS Publishing)
[4] Zhao Y, Deng Z, and Wang Qi 2013 *Sensor and Actuators B* **192** pp 229-233
[5] Riyanto A, Listiawati D, Suharyadi, E and Abraha K 2012 *Prosiding Pertemuan Ilmiah XXVI HFI Jateng & DIY*, ISSN : 0853-0823
[6] Fox M 2001 *Optical Properties of Solid* (New York: Oxford University Press)

**Acknowledgements**
The reported work was financially supported by *Program Penelitian Unggulan Perguruan Tinggi* (PUPT) from the Ministry of Research and Technology of High Education for period of 2015 – 2017 and thesis scholarship of Indonesia Endownment Fund for Eductaion (LPDP) batch III for period of 2016.

IOP Conf. Series: Materials Science and Engineering **202** (2017) 012033 doi:10.1088/1757-899X/202/1/012033

# Effects of Sintering Holding Time on the Structural, Electrical and Magnetic Properties of $Zn_{0.95}Ni_{0.05}O$

**M Ginting[1], D Aryanto[1], C Kurniawan[1], A Y Sari[1], A Subhan[1], T Sudiro[1], P Sebayang[1], E R Tarigan[2], Nasruddin M N[2] and K Sebayang[2]**

[1]Research Center for Physics, Indonesian Institute of Sciences, Serpong 15314, Tangerang Selatan, Banten, Indonesia
[2]Postgraduate Program, Universitas Sumatera Utara, Medan 20155, Indonesia

Email: masno.ginting@gmail.com

**Abstract**. $Zn_{0.95}Ni_{0.05}O$ has been synthesized by mixing 5% mol of NiO into ZnO using solid state reaction and high-speed shaker mill method. The samples were sintered at 900 °C with holding time for 2, 4 and 8 hours. Crystal structure, electrical and magnetic properties of $Zn_{0.95}Ni_{0.05}O$ were characterized by using XRD, I-V, C-V and VSM. XRD results showed that variation of holding time does not change the structure of ZnO and no other secondary phase observed. The value of lattice parameters (*a* and *c*) tends to decrease proportionally to the holding time. The Intensity value changes and the peak shifted to a higher $2\theta$ angle due to holding time variation. In general, the conductance of $Zn_{0.95}Ni_{0.05}O$ decreases and the magnetic properties decrease also as the holding time is increased.

**Keywords:** Sintering, holding time, $Zn_{0.95}Ni_{0.05}O$, structure, electrical and magnetic property

## 1. Introduction

Magnetic properties transition from paramagnetic to ferromagnetic in the semiconductor material is an interesting topic in solid-state physics. It is because of its advantage in spintronics applications, i.e. data processing speed, non-volatility and higher integration densities [1]. The change of semiconductor material magnetic properties was made by doping the transition-metal elements (TM), i.e. Sc, Ti, V, Cr, Mn, Fe, Co, Ni, and Cu [2]. Control and tuning of magnets properties in dilute magnetic semiconductor (DMS) such as ZnO, $TiO_2$, dan $SnO_2$ have been studied in the last decades. Among other materials, ZnO has advantages, such as non-toxic, low-cost and eco- friendly [3]. Moreover, ZnO as a semiconductor material has the chemical stability and high thermal, with an energy gap of 3.37 eV and exciton binding energy of 60 MeV [4]. It makes the ZnO high potential in many applications, such as solar cells, optoelectronic, magnetic tunnel junctions, photovoltaics, light-emitting diodes, spintronics-based devices, and sensors [5]. Ferromagnetic room-temperature in doped ZnO metal transition, particularly in Ni has been studied intensively in recent years [6]. However, the reported result is still far from the clear explanation on the origin of the magnetism.

Nickel (Ni) is a transition metal with a radius nearly equal to Zinc (Zn). This case is the reason why the $Ni^{2+}$ ions are more easily substituting $Zn^{2+}$ ions in the ZnO crystal [7]. Ni-doped ZnO has been studied through various synthesis methods, such as solid state reaction [5], sol-gel [4], co-precipitation [2,8,13], sputtering [9,10] and hydrothermal [11]. Solid state reaction method using a milling tool is an inexpensive and simple laboratory scale. Until today, only a few work reported on Ni-doped ZnO synthesized by using solid-state reaction method. In this study, we synthesized Ni-doped ZnO with a solid-state reaction method by using high-speed shaker mill. Then we discussed the effect of the Sintering holding time on the structure, electrical and magnetics properties of $Zn_{0.95}Ni_{0.05}O$.

## 2. Experimental Method

$Zn_{0.95}Ni_{0.05}O$ synthesis was done by using solidstate reaction method with ZnO and NiO powder as starting material. ZnO and NiO were mixed and milled in a vial cylindrical stainless steels by using a high-speed shaker mill with frequency oscillation of 700 / min. Steel ball diameter ($\phi$) of 4 mm was used with the powder weight ratio to the ball was 1:10. The process carried out for 3 hours in wet conditions by adding toluene as a solvent. The powder was then dried in an oven at a temperature of 100 °C for 3 hours; then powder was turned into the pellet shaped by using the automatic axial hydraulic pressure of 1500 kgf/cm$^2$. The resulted pellet samples have a diameter of 15 mm and thickness of 2 mm. In the final process, the samples were sintered at temperature of 900 °C with holding time of 2, 4 and 8 hours in the air atmosphere. The crystal structure of the $Zn_{0.95}Ni_{0.05}O$ of pellets was investigated by using X-ray diffraction (XRD)-Smartlab-Rigaku. Meanwhile, electrical and magnetics properties of the samples were analyzed by using the I-V meter (Fluke 8842A multimeter high impedance), C-V meters (Keithley 590 CV analizer) and a vibrating sample magnetometer (VSM 250, Dexing Magnet.ltd).

## 3. Results and Discussion

The detail results of XRD analysis of $Zn_{0.95}Ni_{0.05}O$ with holding time variation are shown in Figure 1. The XRD pattern of pure ZnO was used as a reference. The analysis result revealed that overall $Zn_{0.95}Ni_{0.05}O$ samples have the main hexagonal wurtzite structure (Figure 1-a). The presence of peaks can also be seen at an angle of 2θ, approximately at 31,7°; 34,4°; 36,3°; 47,6°; 56,6°; 62,9°; 66,4°; 67,9°; 69,1°; 72,6°; and 76,9°, that are related to planes of wurtzite structure (100) (002) (101) (102) (110) (103) (200) (112) (201) (004) and (202). Furthermore, a peak that indicates NiO phase appears at an angle of 2θ at 37.1° and 43.2° (Figure 1-b). Further analysis indicates that some of the $Ni^{2+}$ ions diffuse into the ZnO lattice, and some are still in the form of NiO compound. This phenomenon is confirmed by detailed analysis on ZnO plane (002) given in Figure 1-c. The peak intensity is decreased and shifted when ZnO doped with NiO. The change of intensity and peak position confirmed that the $Ni^{2+}$ ion is already penetrated into the ZnO lattice and substituting the $Zn^{2+}$ ion [8]. Then the increase of holding time causes the peak shifted to a higher $2\theta$ angle. Figure 1-c also reveals that the full width at half maximum (FWHM) decreases by increasing the holding time. The peak shift and FWHM changes, in this case, may be affected by the changes in other crystal parameters.

Crystal parameters such as lattice parameters (*a* and *c*) and crystal size were estimated based on the XRD data. Hexagonal lattice parameters of the phase weredetermined by using the equation:

$$a = \frac{\lambda}{\sqrt{3}\sin\theta}\sqrt{h^2 + hk + k^2} \tag{1}$$

and

$$c = \frac{\lambda}{2\sin\theta} l \tag{2}$$

where λ is the wavelength of X-rays, θ is the Bragg diffraction angle, and *hkl* is miller index, whereas the crystal size was determined according to ZnO plane (002):

$$D = \frac{0.9\lambda}{\beta\cos\theta} \tag{3}$$

where β is the FWHM of (002) diffraction peak. The calculation result of crystal parameters is shown in Table 1. The results of *a* and *c* of $Zn_{0.95}Ni_{0.05}O$ remain the same as the holding time was increased. As a result the *c/a* value is also constant to a change of holding time. This case is due to $Ni^{2+}$ ion has partially substituted $Zn^{2+}$ ions in ZnO lattice. However, the lattice parameter values obtained from this research is smaller than that of the theory value (ICDD No. 00-005-0664, a = 3.249Å and c = 5.205 Å). The Ionic radius of $Ni^{2+}$ (0.55Å) is smaller than the radius of $Zn^{2+}$ (0.60Å) which is the main factor of the change of the lattice parameter values [12].The lattice parameter changes were also reported by previous studies [4,5,8]. The $Zn_{0.95}Ni_{0.05}O$ crystal size increases with increasing holding time. This phenomenon can be explained that the longer the holding time causes the diffusion longer derived by the heat effect so that the small grains have enough time to grow into larger grains [13].

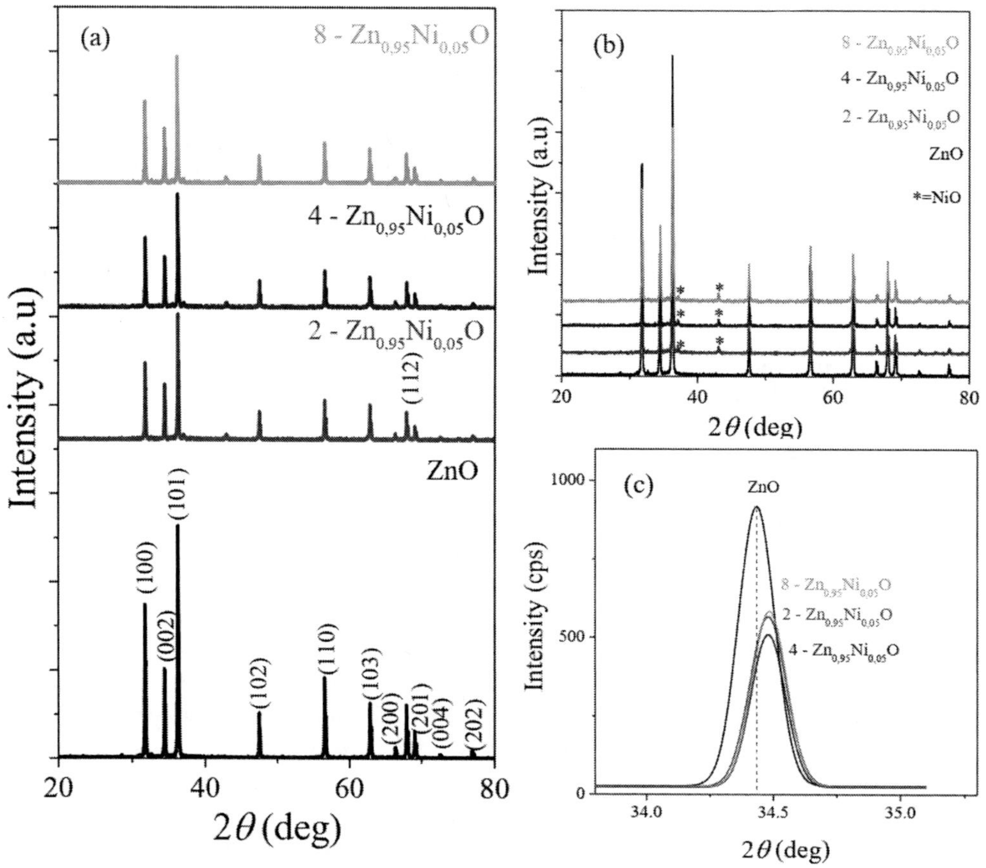


**Figure 1**. XRD Pattern of (a) $Zn_{0.95}Ni_{0.05}O$ with holding time of 2, 4 and 8 hours; (b) $Zn_{0.95}Ni_{0.05}O$ and (c) detailed observation at (002) plane.

**Table 1.** Calculation of lattice parameters (*a* and *c*) and average crystal size of $Zn_{0.95}Ni_{0.05}O$ at (002) plane as the effect of holding time.

| Sintering holding time (hour) | *a* (A) | *c* (A) | *c*/*a* | FWHM (deg.) | *D* (nm) |
|---|---|---|---|---|---|
| 0 | 3.249 | 5.205 | 1.602 | 0.154 | 50.6 |
| 2 | 3.245 | 5.198 | 1.601 | 0.112 | 74.6 |
| 4 | 3.246 | 5.199 | 1.601 | 0.106 | 75.0 |
| 8 | 3.246 | 5.198 | 1.601 | 0.102 | 77.9 |

The electrical properties of $Zn_{0.95}Ni_{0.05}O$ were observed in different holding time at room temperature. Conductivity value ($\sigma$), capacitance ($C$) and dielectric constant ($\varepsilon_r$) are listed in Table 2. It can be seen that the increase of the holding time leads to changes in conductivity and dielectric constant value of $Zn_{0.95}Ni_{0.05}O$. The capacitance value of $Zn_{0.95}Ni_{0.05}O$ increases as holding time is increased. The changes in the electrical conductivity on this study may be due to the Ni doping, which can also create crystal defects [14]. The difference value between the radius of $Zn^{2+}$ and $Ni^{2+}$ ions also causes the internal stress on the ZnO lattice [15] that also causes the defects on the grain boundaries. Then the charge movement will be blocked by the grain boundaries defect. Moreover, the absence of oxygen is also believed to contribute to the electrical conductivity values [9]. Meanwhile, the high obtained dielectric value of the $Zn_{0.95}Ni_{0.05}O$, since $Ni^{2+}$ ions occupy the center position in $Zn^{2+}$, which leads to the permanent electric dipole. The absence of oxygen is also another aspect that affects the dielectric properties of $Zn_{0.95}Ni_{0.05}O$ [11,15]. The sintering holding time influences very much on reducing defects at the grain boundaries. Thus, the holding time has an important role in improving the structure and electrical properties of $Zn_{0.95}Ni_{0.05}O$.

**Table 2** Conductivity, capacitance and dielectric constant of $Zn_{0.95}Ni_{0.05}O$ as the effect of holding time

| Sintering holding Time (Hour) | $\rho$ ( Ωcm) | $\sigma$ (S/cm) | $C$ (pF) | $\varepsilon_r$ |
|---|---|---|---|---|
| 0 | $3.25\times10^{7}$ | $3.06\times10^{-8}$ | 3.28 | 10.75 |
| 2 | $4.80\times10^{7}$ | $2.08\times10^{-8}$ | 25.29 | 47.55 |
| 4 | $5.99\times10^{7}$ | $1.67\times10^{-8}$ | 27.48 | 36.28 |
| 8 | $4.79\times10^{7}$ | $2.09\times10^{-8}$ | 35.35 | 58.33 |

Figure 2 shows the magnetic field versus the magnetization of $Zn_{0.95}Ni_{0.05}O$ sample with different Holding Time at room temperature below 20 kOe. As shown in Figure 2-a, the initial magnetic behavior of ZnO before Ni doped is diamagnetic. Doping with Ni causes the change in the magnetic properties of ZnO into ferromagnetic. This result is consistent with previous studies reported somewhere [4,5,7,8,9]. The ferromagnetic properties are formed due to part of $Ni^{2+}$ ions substituting $Zn^{2+}$ in ZnO lattice.

The exchange interaction between local spin-polarized electrons of $Ni^{2+}$ and conduction electrons or free carriers caused the ferromagnetic interaction [2]. Figure 2-a reveals that the increase of holding time leads to the change in the magnetic properties. Figure 2-b till 2-c with different scale on the *x*-axis (H value) are also shown to make the effect on the magnetics properties due to the holding time become clearer. The details of $Zn_{0.95}Ni_{0.05}O$ magnetic properties as the effect of holding time is shown in Table 3. It can be seen that, coercivity value (Hc) decreases with increasing of holding time. The coercivity value decreases from 147.78 to 117.99 (Oe), where the saturation magnetization (Ms) values are vary from 0.220 (emu/g) to 0.339 (emu/g). Holding time is close related to the grain growth during the sintering process as shown in the XRD results. The increase of holding time will increase the density value of $Zn_{0.95}Ni_{0.05}O$, and it will contribute to its magnetics behavior. Atoms have a spin in the same direction, which will increase its magnetics behavior. However, the presence of the crystal defects in the sample will also strongly affects the atomics spin rotation, so that the Ms value does not change linearly with the density value.

**Table 3**. Magnetic properties of the sintered $Zn_{0.95}Ni_{0.05}O$ at 900°C for different holding time

| Holding Time | Magnetic Properties | | |
|---|---|---|---|
| (hours) | Mr (emu/g) | Hc (Oe) | $M_s$ (emu/g) |
| 2 | 0.038 | 147.78 | 0.288 |
| 4 | 0.038 | 117.99 | 0.339 |
| 8 | 0.041 | 115.96 | 0.220 |

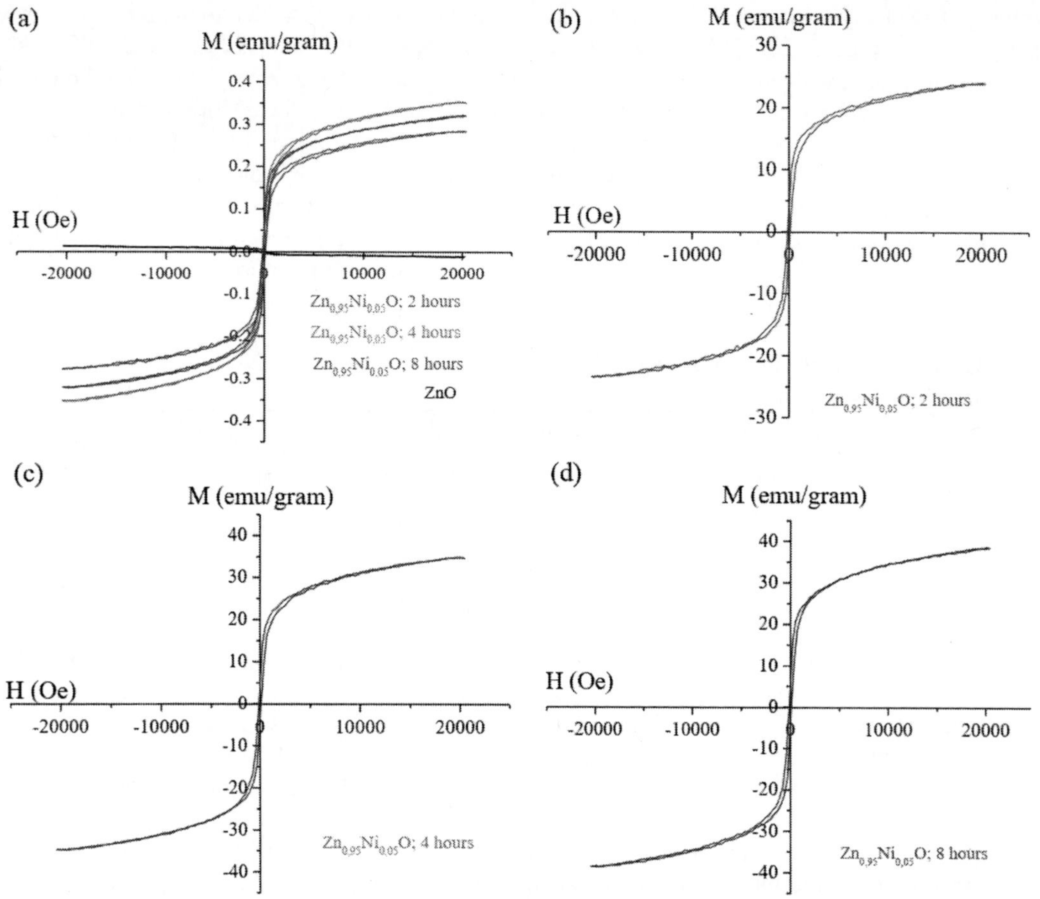


**Figure 2**. (a) The magnetic hysteresis loop of $Zn_{0.95}Ni_{0.05}O$ for Holding Time of 2, 4 and 8 hour at room temperature below 20 kOe, (b ) The magnetic hysteresis loop of $Zn_{0.95}Ni_{0.05}O$ with Holding Time of 2 hour (c ) The magnetic hysteresis loop $Zn_{0.95}Ni_{0.05}O$ with Holding Time of 4 hour and (d) The magnetic hysteresis loop of $Zn_{0.95}Ni_{0.05}O$ with Holding Time of 8 hour

## 4. Conclusion

$Zn_{0.95}Ni_{0.05}O$ was successfully synthesized by using solid-state reaction method with a holding time variation of 2, 4 and 8 hours. It is found that structural, electrical and magnetic properties change to the change of holding time. TheXRD results shown that overall $Zn_{0.95}Ni_{0.05}O$ samples have wurtzite hexagonal major structural. The $Zn_{0.95}Ni_{0.05}O$ sample resistivity and capacitance values increase with the increasing of the holding time. From VSM measurements shown that overall samples are ferromagnetic. The remanence value slightly increases with the increasing of the holding time, while the coercivity decreases significantly.

## 5. References

[1] Das J, Mishra D K, Srinivasu V V, Sahu D R and Roul B K 2015 *J. Magnetism and Magnetic Materials* **382** 111-116

[2] Vijayaprasath G, Murugan R, Asaithambi S, Sakthivel P, Mahalingam T, Hayakawa Y, and Ravi G 2016 *Ceramics International* **42** 2836-2845

[3] Aryanto D, Marwoto P, Sudiro T, Birowosuto M D, Sugianto and Sulhadi 2016 *AIP Conf. Proc.* **1729** 020039-(1-5)

[4] Srinet G, Kumar R and Sajal V 2013 *J. Applied Physics* **114** 033912(1-6)

[5] Pal B, Sarkar D and Giri P K 2015 *Applied Surface Science* **356** 804-811
[6] Doğan N, Bingölbali A and Arda L 2015 *J. Magnetism and Magnetic Materials* **373** 226-230
[7] Jadhav J, Patange M and Biswas S 2013 *Carbon-Sci. Tech.* **5** 269-274
[8] Shinde K P, Pawar R C, Sinha B B, Kim H S, Oh S S, and Chung K C 2014 *Ceramics International* **40** 16799-16804
[9] Liu X J, Zhu X Y, Song C, Zeng F and Pan F 2009 *J. Physics D: Applied Physics* **42** 035004(1-7)
[10] Abdel-Wahab M S, Jilani A, Yahia I S and Al-Ghamdi A A 2016 *Superlattices and Microstructures* 94 108-118
[11] Zhong M, Li Y, Tariq M, Hu Y, Li W, Zhu M, and Li Y 2016 *J. Alloys and Compounds* **675** 286-291
[12] Kakazey M, Vlasova M, Vorobiev Y, Leon I, Gonzalez M C and Urbiola E A C 2014 *Physica B: Condensed Matter* **453** 116-122
[13] Liu D, Gao Y, Liu J, Wang Y and An L 2016 *Ceramics International* **xx** yyy-yyy
[14] Huang J Q, Du P Y, Weng W J and Han G R 2008 *J. Electroceramics* **21** 394-397
[15] Srinet G, Kumar R and Sajal V 2013 *Ceramics International* **39** 7557-7561

**Acknowledgements**
The authors are grateful to Laboratory of Advance Material Magnetic, Research Center for Physics-LIPI, Indonesia.

IOP Conf. Series: Materials Science and Engineering **202** (2017) 012034 doi:10.1088/1757-899X/202/1/012034

# Synthesis and Characterization of Ti-6Al-6Mo Prepared by Arc Melting Process

**G Senopati, I N G Putrayasa A and C Sutowo**

Research Center for Metallurgy and Materials Indonesian Institute of Sciences Building 470, Puspiptek Technological Park Serpong, Tangerang Selatan, 15314, Banten, Indonesia

Email: gali002@lipi.go.id

**Abstract**. The Ti-6Al-6Mo alloy was arc-melted eight times from pure Ti, Al, and Mo. The Ti-6Al-6Mo were analyzed by using an optical microscope, SEM-EDS, XRD and Vickers Hardness Tester. Corrosion test was performed by using Hank's solution at 37 °C and pH 7.4. The optical micrograph showed a similar structure of this alloy. The microstructure of Ti 6Al-6Mo was basketweave with $\alpha$ lath. The map of Ti, Al and Mo showed a similar distribution of this alloy. The XRD analysis showed that the $\alpha$ and $\beta$ phase occurred. The Vickers hardness value of Ti6Al6Nb was 398.7 HV, higher than pure Ti. The corrosion rate of as cast Ti-6Al-6Mo was 0.0016 mmpy (mm/year). The result showed that this alloy can potentially be used for biomedical application.

**Keywords** : Ti-6Al-6Mo, arc melted, Biomedical application

## 1. Introduction

Ti (pure Ti) and its alloys have become attractive biomaterials due to their corrosion resistance, biocompatibility, specific strength, and lower young's modulus than other metallic biomaterials such as stainless steels and Co-Cr alloys [1]. The $\alpha$+$\beta$ Ti-6Al-4V alloy has been used widely because of better mechanical properties than pure titanium [2]. Although α+β Ti-6Al-4V alloy presents lower young's modulus than stainless steels and Co-Cr alloys, various studies have shown that the release of a small amount of V in the human body can induce cytotoxic effects [3]. Vanadium is known as $\beta$ isomorphous element. In this study, V element was substituted with Mo which is known as $\beta$ isomorphous element [4].

Various studies have been done by using Mo as a beta stabilizer in titanium alloy [5-7]. The $\beta$ grain boundary presented when Mo content above 5 wt.% although the X-ray diffraction does not confirm the occurrence of $\beta$ phase due to its small volume fraction [6]. Wen Fu Ho has studied the influence of Mo on Ti-5Cr system and reported that Ti-5Cr-9Mo alloy has the highest bending strength/modulus ratio 26.0 which is higher than pure Ti [7]. In the previous study, the influence of Mo content on the Ti-6Al system has been investigated but did not report the phase stability and mechanical properties in as-cast condition. Whereas $\alpha/\beta$ phase formation process begins in casting process. Cardozo report in developing titanium for implant application, the correlation of each step has an important role, start from determining chemical composition, microstructure characteristics, heat treatment process, and mechanical properties of alloy [6]. Another study reported that microstructure and phase formed in casting process have important role in the next process (thermomechanical process) [8-10]. The objective of this study is to investigate the phase stability, mechanical properties (hardness value), and electrochemical behavior of as cast Ti-6Al-6Mo alloy.

## 2. Experimental Method

### *2.1 Material Preparation*

Ti-6Al-6Mo was prepared from commercially available pure Ti (99,9 % pure), aluminum (99,9%pure) and Mo (99% pure). All materials were processed by arc melting with non-consumable tungsten electrode on a water-cooled copper hearth. The alloy was prepared in ultra-high purity argon atmosphere, and the ingot was melted eight times to improve chemical homogeneity. The obtained ingot was approximately 100 g.

### *2.2 Examination*

As cast Ti-6Al-6Mo was sectioned by using cutting wheel machine with dimension 1×1×1 cm. The block specimen mounted by using resin and mechanically polished via a metallographic procedure to a final level of 0.3 μm alumina powder and then etched in a solution of water, nitric acid, and hydrofluoric acid (80:15:5 in volume). The microstructure of the etched alloy was analyzed by using Meiji Japan optical microscope. X–ray diffraction (XRD) measurement for phase analysis was conducted by using Rigaku smart lab diffractometer, was performed at room temperature by using Cu K$\alpha$ radiation and operating at 30 kV/15 mA. Hardness measurements were measured by using Shimazu Vickers hardness tester with a load 500 gf for 30 s. Corrosion test in simulated body fluid were tested to confirm the biocompatibility properties of the Ti6Al6Nb alloy. The corrosion test was tested in Hank's Solution of pH 7.4 at 37 °C to simulate the body fluid environment and body temperature. A standard three electrodes cell used in this test. Graphite is the counter electrode; saturated calomel electrode is the reference electrode and sample as a work electrode. The total amount of Hank's Solution taken in the cell is about 500 ml. The potential scanning rate is 1 mV/s.

## 3. Results and Discussion

Figure 1 show pancake ingot obtained after eight times remelting process. The pancake weight was 100 g approximately with 100 mm diameter. The as cast appearance was shine which indicated there was no oxide layer on the as-cast surface. Figure 2 EDS energy spectrum analysis of as-cast Ti-6Al-6Mo. The energy of Ti, Al, and Mo was observed. The desired chemical compositions were Al 6 wt.%, Mo 6 wt.%, and Ti 88 wt.%. The actual chemical compositions were Al 4.69 wt.%, Mo 5.60 wt.% and Ti 89.71 wt.%.

**Figure 1.** Pancake As-Cast Ti-6Al-6Mo

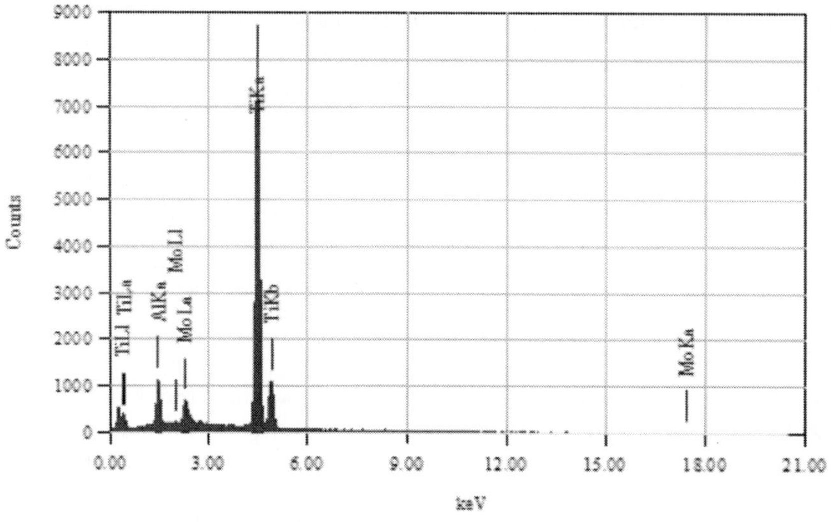


| Element | Wt.% |
|---|---|
| Al | 4.69 |
| Ti | 89.71 |
| Mo | 5.60 |

**Figure 2.** Chemical composition of Ti-6Al-6Mo by using EDS

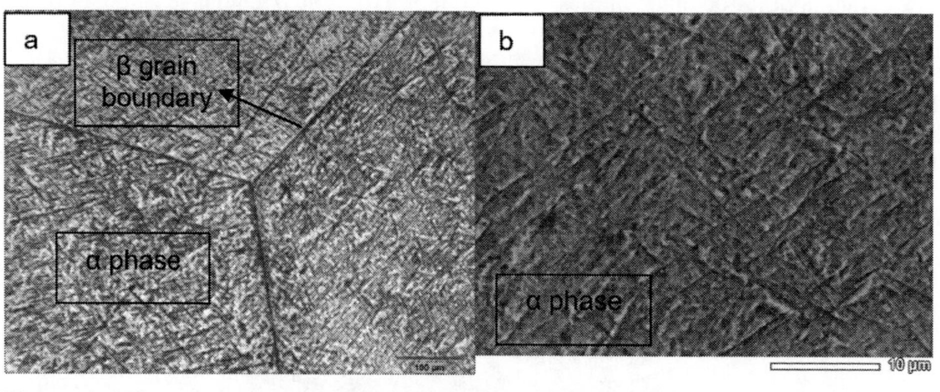


**Figure 3.** Microstructures of Ti–6Al-6Mo alloy. (a) SEM image and (b) OM image. Etched Kroll Reagent

Figure 3 shows microstructure of as-cast Ti-6Al-6Mo. The microstructure of as-cast Ti-6Al-6Mo was basketweave structure with $\alpha$ lath and $\beta$ grain was observed. The $\alpha$ lath width was approximately less than 1 μm. Thus, the microstructural results are identical to Fengying result [11]. Fengying process Ti-6Al-Mo by using Laser Solid Formed (LSF) and confirmed the expected effect of replacing vanadium by molybdenum. Figure 4 shows diffraction pattern of Ti-6Al-6Mo. The $\alpha$ and $\beta$ phases were observed. The metallographic result is shown in Figure 3.

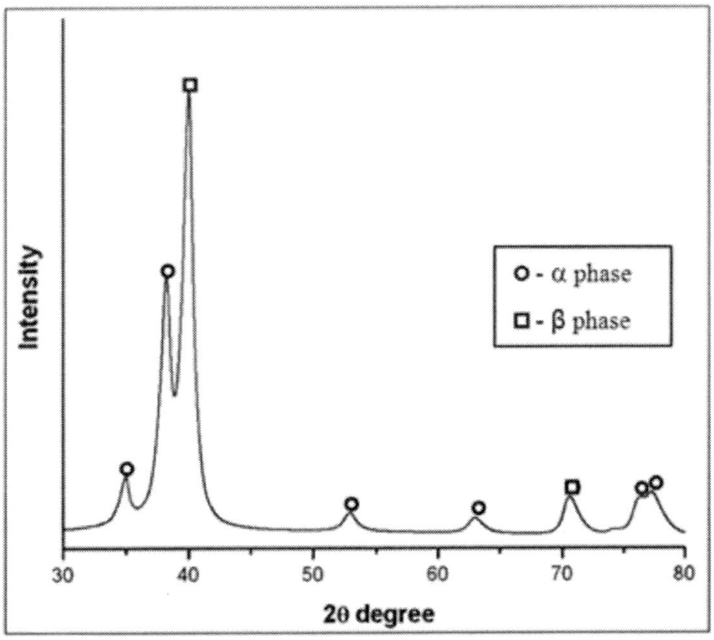


**Figure 4.** XRD Pattern of Ti-6Al-6Mo

Figure 5 shows hardness value of as-cast Ti-6Al-6Mo and as-cast another titanium alloy for the biomedical purpose. As-cast Ti-6Al 6Mo has hardness value of 398.7 HV, which is higher than pure Ti (186 HV), Ti-15Mo-5Nb (246HV), and Ti-15Mo-10NB (257HV), but lower than Ti-5Sn-5Cr (503 HV). Ti-6Al-6Mo has a higher hardness than pure titanium due to increasing of $\beta$ volume fraction. The $\beta$ volume fraction was influenced by the addition of Mo as a $\beta$ stabilizer in titanium $\alpha + \beta$ system.

In this case (Ti-6Al-6Nb alloy) has the higher hardness than pure titanium due to solid solution strengthening or crystal structure effect/phase ($\alpha$, $\beta$, $\acute{\omega}$) [6,11]. Another study by A. Kumar in Ti-6Al-4V alloy showed that the increasing of $\beta$ volume fraction could increase hardness value in $\alpha + \beta$ system [12].

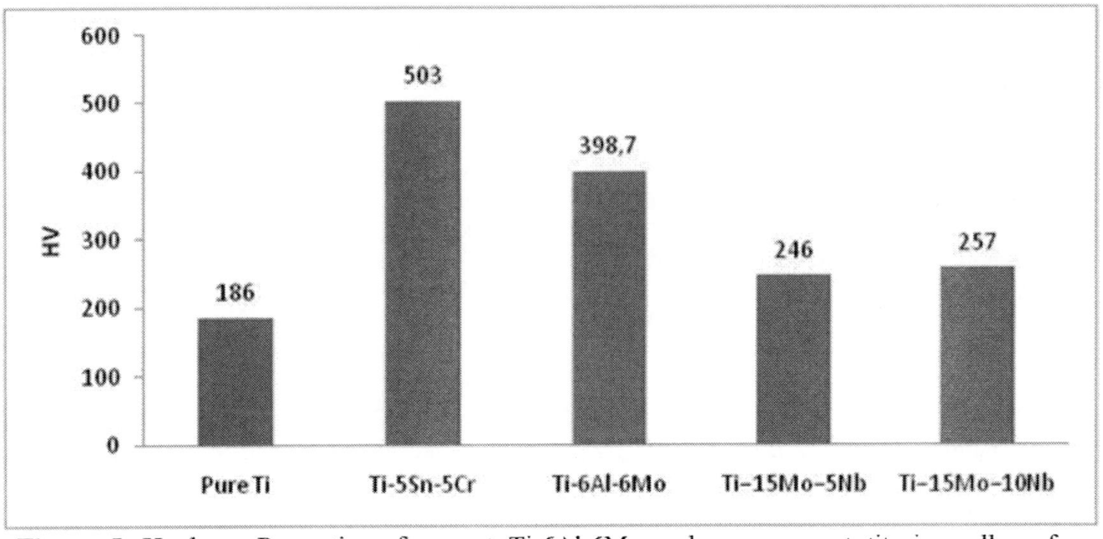


**Figure 5.** Hardness Properties of as-cast Ti-6Al-6Mo and some as-cast titanium alloys for biomedical application

Figure 6 shows tafel diagram corrosion test of as cast Ti-6Al-6Mo in simulated body fluid Hank's Solution. The corrosion rate of as-cast Ti-6Al-6Mo was 0.0016 mmpy (mm/year). Compared to another implant materials such as stainless steel 316L (0.004 mm/y), Co alloy (0.00127 mm/y), and Ti-6Al-4V (0.00018 mm/y) Ti-6Al-6Mo shows better corrosion resistant than Stainless Steel 316L and Co alloy, but lower than the common Ti6Al4V [13].

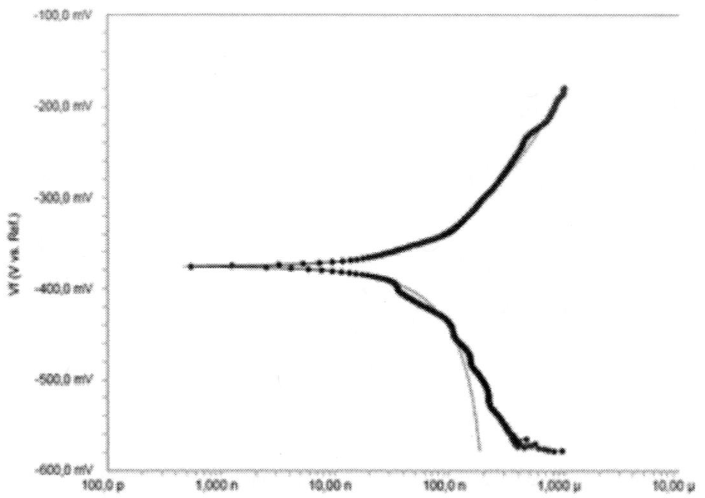


**Figure 6.** Tafel Diagrams of Ti-6Al-6Mo alloy in Hank's Solution

## 4. Conclusion

The characteristics of microstructure, mechanical properties, and biocompatibility of as-cast Ti-6Al-6Mo have been observed in this study. The as-cast Ti-6Al-6Mo has similar microstructure as the conventional implant of Ti-6Al-6V. The microstructure was lamellar, α and β phases were observed in the microstructure and diffraction pattern result. The hardness of Ti-6Al-6Mo was higher than pure Ti and lower than β type alloy Ti-5Sn-5Cr. The corrosion rate of as cast Ti-6Al-6Mo was 0.0016 mmpy.

## 5. References

[1] Niinomi M *et al* 2012 *Acta Biomaterialia* **8** 3888-3903
[2] Lautenschlager E P *et al* 1993 *International Dental Journal* **43** 245-253
[3] Kuroda D *et al* 1998 *Materials Science and Engineering: A* **243** 244-249
[4] Lutjering G *et al* 2007 *Titanium*
[5] Senopati G *et al* 2016 *AIP Conference Proceedings*
[6] Cardozo F F *et al* 2014 *Journal of The Mechanical Behavior of Biomedical Materials* **32** 31-38
[7] Ho W F *et al* 2011 *Biomaterials* **20** 2115-2122
[8] Kent D *et al* 2013 *Journal of the Mechanical Behavior of Biomedical Materials* **28** 15-25
[9] Hsu L C *et al* 2014 *Material Science and Engineering A* **606** 157-164
[10] Xu L J *et al* 2013 *Trans. Nonferrous Met. Soc. China* **23** 692-698
[11] Fengying z *et al* 2013 *Rare Metal Materials and Engineering* **42** 1332-1336
[12] Kumar A *et al* 2012 *Advanced material research* **585** 381-386
[13] Gurappa I 2002 *Materials Characterization* **49** 73-79

**Acknowledgments**
This work was financially supported by 2016 thematic program of Indonesian Institute of Sciences

IOP Conf. Series: Materials Science and Engineering **202** (2017) 012035 doi:10.1088/1757-899X/202/1/012035

# Corrosion Behavior of Magnesium Based Foam Structure in Hank's Solution

**Franciska P L, A Erryani, D Annur and I Kartika**

Research Center for Metallurgy and Materials, Indonesian Institute of Sciences (LIPI) Gedung 470, Kawasan Puspiptek Serpong Tangerang Selatan, 15314, Indonesia

Email: franciskapramuji@gmail.com

**Abstract.** Metal foam is a new class of materials with promising applications and a unique combination of physical, chemical, and mechanical properties. The purpose of biodegradable implants is to support tissue regeneration and healing in a particular application by material degradation and implant replacement through the surrounding tissue. Magnesium alloys are expected to be degraded in the body and its corrosion products not deleterious to the surrounding tissue. In the present study, the foam metal was manufactured via powder metallurgy with a different variation of sintering temperature and $TiH_2$ used as a foaming agent which are added to Mg-1Ca-3Zn alloy as much 3% wt $TiH_2$. The sintering temperatures were 500, 550 and 600 °C with a constant holding time of 5 hours. It's critical that the sintering temperature is carefully selected in consideration of their corrosion behavior. This paper reports the study of the behavior of the Mg-Ca-Zn alloy metal foam which evaluated by SEM, EDX, and electrochemical corrosion test in Hank's solutions. After exposure, the SEM result of Mg-Ca-Zn-3$TiH_2$ to Hank's solution, a volcano-like structure is formed. The streams of $H_2$ bubbles form at local sites on the Mg alloy surface where electrochemical reactions are taking place, leading to the particular structure with around shape and often with a hole in the center. The corresponding EDS result maps reveal enrichment of O, Ca, P and Mg as corrosion product. Potentiodynamic polarization experiments conducted at 37 °C and pH 7.4 indicated the increased biodegradation rates resulted from porous structure of foam samples. Corrosion rate in 500oC sintering temperature were 1.99 millimiles per year (mmpy) with corrosion current density ($I_{corr}$) 87.3.10-6 A/cm$^2$, corrosion rate in 550 °C sintering temperature were 2,16 mmpy with $I_{corr}$ 94.4.10-6 A/cm$^2$ and rate in 600 °C sintering temperature were 2.41 mmpy with $I_{corr}$ 105.10-6 A/cm$^2$. The results showed that the increasing of sintering temperature could influence the corrosion resistance of Mg alloy.

**Keyword:** magnesium, alloy, corrosion, in vitro, Hank's solution

## 1. Introduction

Research and development work on magnesium and magnesium-based alloys for bio-implant applications have increased significantly in recent years [1-3]. Magnesium (Mg) and magnesium alloys as a new kind of biomedical material have attracted increasing attention due to their unique advantages. The mechanical properties of Mg and its alloys such as Young's modulus of elasticity ($E$ = 41-45 GPa) and density (1.74-1.84 g/cm$^3$) are known to be similar that of bone ($E$ = 15-25 GPa and density = 1.8-2.1 g/cm$^3$). This is lower than other biodegradable materials such as iron-manganese (Fe-

Mn) and zinc (Zn) based alloys [4]. Corrosion of metal implants is critical due to it can adversely affect the mechanical properties and biocompatibility. Simultaneously, the materials used should not cause any adverse biological reaction in the body, and they must be stable retaining their functional properties. Mg element is the fourth most abundant in the human body, and $Mg^{2+}$ is an essential element for the human body and the daily intake of $Mg^{2+}$ for a normal adult is about of 300-400 mg. A high content of Mg is naturally found in the skeleton [5]. Excess magnesium ions are harmless and can be excreted in the urine, while toxic element releases lead to metal allergy and skin disease, such as Ni in 316L. Magnesium alloys can be biodegraded in the human body by corrosion, which can eliminate second surgery for implant removal. Some in vivo studies showed that the degradation of the magnesium alloy is harmless [3,6,7]. Additionally, it has been reported that magnesium stents can be successfully degraded and have a minimal effect on the surrounding tissue [8,9].

As bone regeneration increases, the resorption of the implant material introduces an augmented load transmission to the bone. A period of 6–18 months is desired for the remodeling process to be completed [10]. Since the stiffness of metals greatly exceeds the stiffness of bones, the cellular metals with low young's modulus have been proposed to avoid the resulting problems of stress shielding [11-13]. Cellular metals typically fall within the stiffness range of cancellous bone [14]. Metal foams enable bone cells in growth and blood vessels incorporation promoting implant stabilization [10,11,15]. In the ideal case, progressive osteointegration on the one hand and degradation of the implant, on the other hand, guarantee an optimal adaptation to the corresponding strength state at any time [10].

Mg and its alloys are now viewed as a potential alternative for making a scaffold for tissue regeneration application due to a combination of their excellent mechanical properties and degradability. Mg is necessary for calcium incorporation into bone, and so the release of Mg ions is expected to be beneficial for bone healing [16]. The cellular or foam structure magnesium materials have been developed and examined in clinical tests as implant materials in bone surgery [17] and cardiovascular surgery [18]. It was found that even though this material is highly biocompatible and features excellent osteoconductivity; it corrodes at such a high speed that the newly established bone is not yet able to carry the load necessary. Biodegradable Mg foam alloys are breaking the current mold in biomaterial science of developing only corrosion resistant metals. In biomedical applications, Mg alloy also ideal implants due to their outstanding comprehensive performance compared with other materials, such as stainless steel, polymers, and ceramic.

Various *in vitro* and *in vivo* tests have to be carried out in order to identify appropriate materials for use as surgical implants. It is desirable to keep the metal ion release to a minimum by the use of corrosion-resistant materials because corrosion is the first consideration for a material of any type that is to be employed in the body because metal ion release takes place mainly due to corrosion of surgical implants. Some individuals are sensitive to metals, and some may develop metal sensitivity later after receiving an implant. The environment of the body is extremely well buffered so that the pH is maintained at around 7.4 at 37 °C. Corrosion behavior of materials can be studied either *in vivo* or *in vitro* using artificial physiological fluids in which the oxygen content is controlled at a suitably low value at 37 °C. Hank's solution is an example of an artificial solution, which has been used over a period of corrosion testing in the laboratory [19].

This work aims at examining the electrochemical degradation behavior of Mg-Ca-Zn metal foam (3 wt% of $TiH_2$) with variation sinter temperature (500, 550, and 600 °C) using electrochemical techniques in Hanks solution. Scanning Electron Microscope (SEM), Energy Dispersive Spectrometer (EDS) and electrochemical measurement were used for corrosion, post-corrosion and fracture surface analysis.

## 2. Experimental Method

Mg–Zn–Ca alloys were prepared by powder metallurgy method. The starting materials were high purity Magnesium (Merck KGaA, 98.5% pure, D 0.06-0.3 mm), Zinc (Merck KGaA, D < 45 μm), Ca granule (Merck KGaA, ≥ 98.5% pure, D < 2.6 mm) and $TiH_2$ fine powder serving as foaming agent (98% pure) were used to prepare foamable precursors. The $TiH_2$ was heat-treated at 450 °C for 120 min in argon in order to shift the hydrogen release range to higher temperatures, which is essential for a good foam structure [20]. The compositions were designed in this work was Mg–3Zn–1Ca with 3% $TiH_2$ in weight percentages. The powders were mixed in shaker mill for 120 min. As-milled powders were subsequently compacted at room temperature in a cylindrical die of 10 mm in diameter and 10 mm in height (green compact) under a pressure of 100 MPa for 2 min then 200 MPa for 3 min by using a hydraulic press. The green compact was sintered in an alumina crucible at the sintering temperature 500, 550, and 600 °C with constant holding time for 5 hours in argon condition and with natural cooling rate to reach ambient temperature. The heating rate was programmed to be 5 °C/minute.

For microstructural examination using optical microscopy, the specimens were prepared by standard metallographic procedures and were etched in a solution containing 3.5 g picric acid, 6.5 ml acetic acid, 20 ml water, and 100 ml ethanol. A detailed study of microstructure and the secondary phase particles distributed in the alloys was done using scanning electron microscope (SEM) and energy dispersive X-ray (EDX) analyzer for chemical microanalysis technique. SEM provides detailed high-resolution images of the sample by focusing the electron beam on the surface and detecting secondary or backscattered electron signal. An EDX is also used to provide elemental identification and quantitative compositional information.

In this study, the in vitro corrosion were carried out in the electrochemical studies were conducted using Gamry Instruments G750 potentiostat, interfaced to a computer. A three-electrode cell was used for electrochemical measurements. The counter electrode was made of carbon, and the reference electrode was saturated calomel electrode (SCE). The exposed area of the working electrode was Mg alloy foam to the solution was 1 cm2. All the measurements were carried out on an electrochemical workstation at the temperature of 37 °C and pH value of 7.4 in Hank's solution [21]. The simulated body liquid electrolyte was Hank's solution, and composition of Hank's solution used was in Table 1. The freshly prepared solution was used for each experiment. A constant electrolyte temperature of 37 ± 2 °C was maintained. The potentiodynamic polarization tests were carried out from -200 mV to +200 mV at a scanning rate of 0.5 mV/s. The corrosion rate was determined using the Tafel extrapolation method.

**Table 1.** Composition of Hank's Solution [21]

| Component | Concentration (g/L) |
|---|---|
| NaCl | 8.0 |
| KCl | 4.0 |
| CaCl2 | 0.14 |
| MgSO4.7$H_2O$ | 0.1 |
| $NaHPO_4$ | 0.048 |
| $NaHCO_3$ | 0.35 |
| Glucose | 1.0 |
| $KH_2PO_4$ | 0.6 |
| $MgCl_2.6H_2O$ | 0.1 |

## 3. Results and Discussion

### *3.1. EDX and SEM before corrosion process*

The EDX spectra of the corresponding microstructures of the three sintering temperatures of Mg metal foam alloys are shown in Figure 1. Quantitative EDX data of the spot composition are given in Table 1. The increasing of the sintering temperature of Mg alloy foam have increased the pore size of magnesium alloy. The thermal pre-treatment of $TiH_2$ with argon was a useful property since it improves metal foam generation. The effect was attributed to oxide layers around the cores of TiH2 particles which build up during heat treatment and act as diffusion barriers to hydrogen [22]. Oxide layers on titanium hydride particles modify their hydrogen desorption behavior because they form an efficient diffusion barrier [23, 24].

**Table 2.** Spot element of mg alloys before electrochemical reaction.

| Temperature sinter (°C) | Element in mass% | | | |
|---|---|---|---|---|
| | Mg | Ca | Zn | Ti |
| 500 | 79.78 | 5.79 | 6.61 | 7.82 |
| 550 | 80.89 | 5.18 | 6.76 | 7.17 |
| 600 | 82.87 | 5.05 | 6.19 | 5.89 |

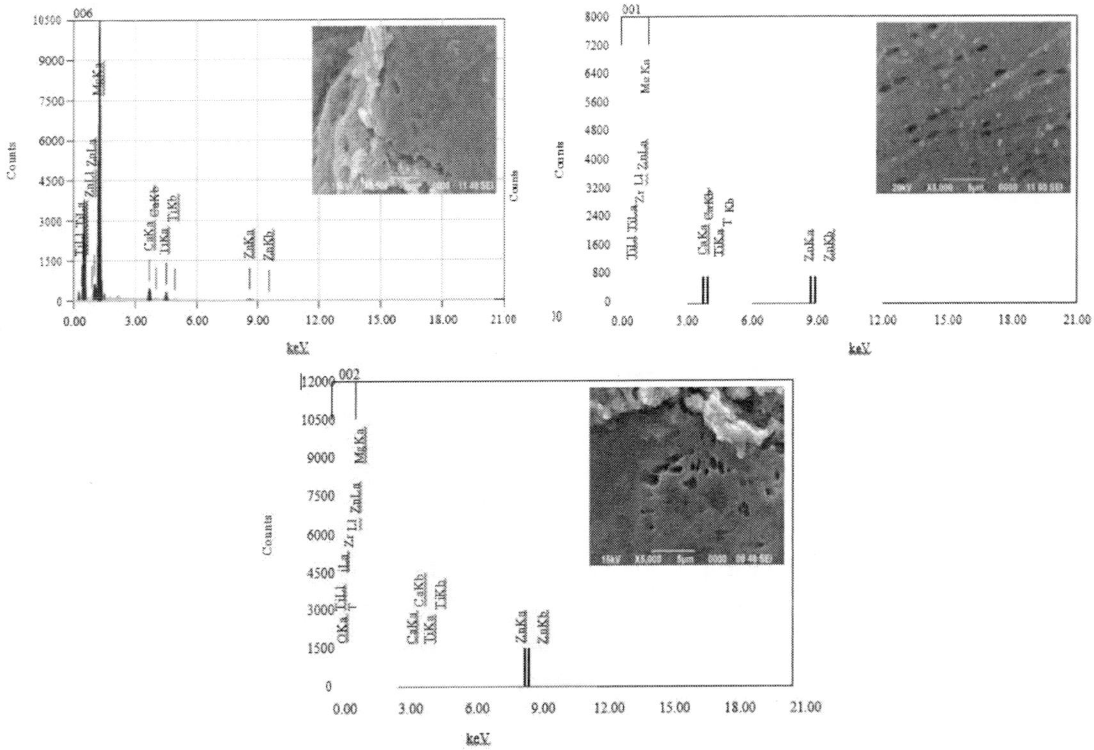


**Figure 1.** EDX spectra of Mg-Ca-Zn-$TiH_2$ at sintering temperatures of (left) 500 °C, (right) 550 °C and (center) 600 °C. The corresponding optical microstructures are presented as insets.

The oxide layer in $TiH_2$ foaming agent make the foam more stable, so that need the higher sinter temperature can make the layer break, which resulted in the release of hydrogen in Ti and formed the foam structure [25]. It correlates with the pore diameter changed significantly with an increasing in higher sinter temperature. The higher heat of reaction seems to make the viscosity of the molten or semi-solid specimen lower. Under this condition, the growth of pores (the growth of small pores or coalescence of pores) becomes easier, and the size of pores becomes larger. The EDX quantitative spot element composition analysis shows in Mg-Ca-Zn-$TiH_2$ alloy shift is not significant substantially changes with increasing the sintering temperature.

### *3.2. Electrochemical measurement*

The increasing of biodegradable or corrodible alloys for medical become higher because the degradation or decomposition the material which was corroded in the body's solution has a potential for orthopedic or cardiovascular application. Corrosion potential ($E_{corr}$) and current density ($I_{corr}$) is calculated by extrapolation intersection from anode and cathode of line Tafel. The electrochemical polarization curves of the alloy with three different sinter temperatures, as shown in Figure 2. The corresponding electrochemical data are given in Table 3.

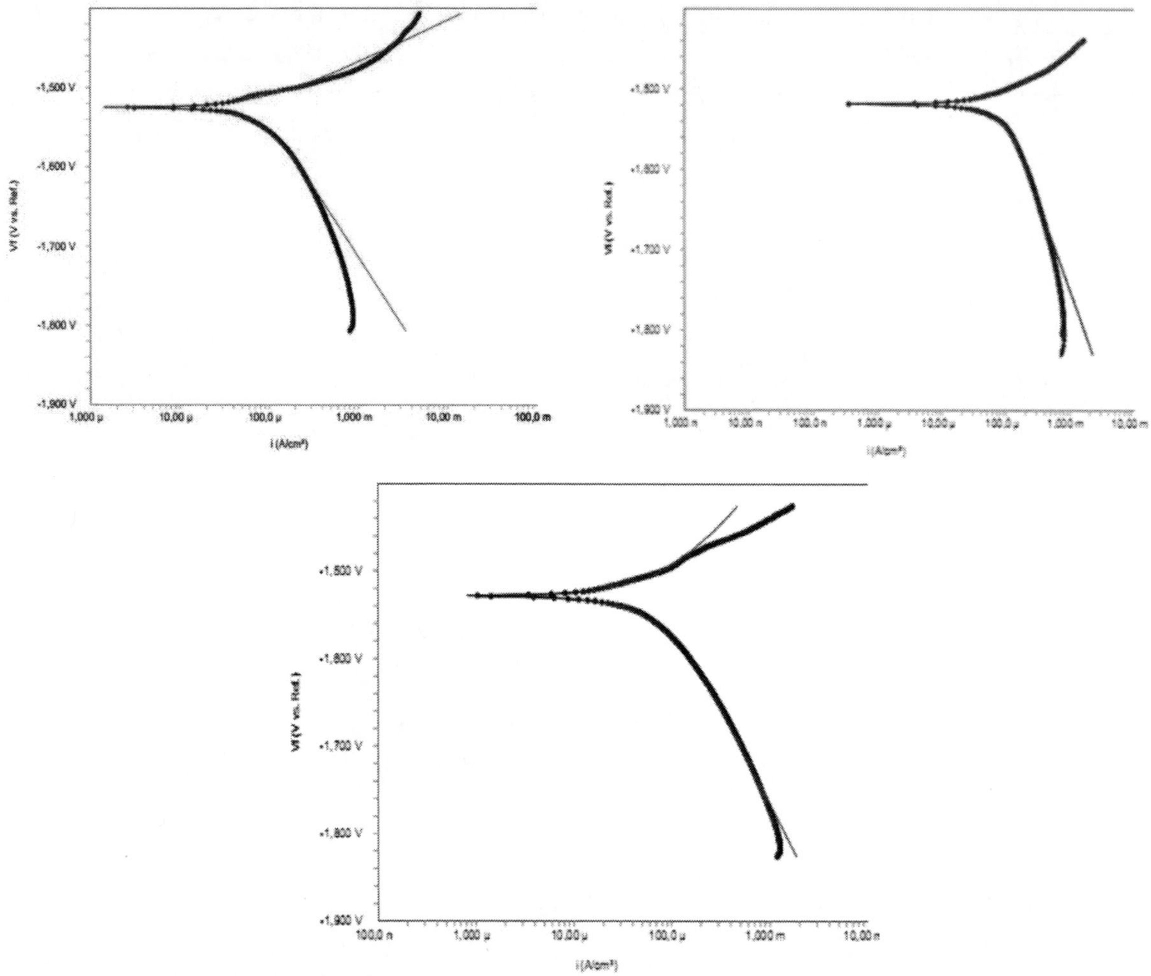


**Figure 2.** Potentiodynamic polarization curves Mg-1Ca-3Zn-3Ti$H_2$ with sintering temperatures of (left) 500 °C, (right) 550 °C Ti$H_2$, and (center) 600 °C

The corrosion rate values show that the increasing sintering temperature to magnesium alloy shifts the corrosion potential towards more corroded. The anodic polarization curves show a passivation-like region in all the alloys. However, the current density in this region is substantially lower in lower sintering temperature. All the alloys showed a similar break-down potential at 1.5 V. The increasing of sinter temperature, as generally resulted in the increasing of porosity, so that can decrease the corrosion potential. The positive line in Tafel diagram indicated any decreasing the corrosion rate. Significant value occurred at sinter temperature 500 and 600 °C. At sinter temperature 500 °C, produce lower foam in Mg alloy foam. It can be formed the passive layer in Mg material that exposed in atmosphere condition and becomes a protective layer [26]. Song etc. reported that magnesium oxide layer could hinder the direct contact with a solution [27]. Ca/P compound precipitated on hydroxide layer from hank solution as corrosion product. The increasing corrosion rate in Mg alloy was association with potential corrosion become more negative. $Cl^-$ ion can be penetrated from corrosion layer, distract the layer structure and reacted with the metal substrate.

However, pores are inherent defects that make the materials mechanically weak while electrochemically active; there is a decrease in corrosion resistance due to the larger surface area exposed to the electrolyte compared to non-porous materials. A previous study [25] has shown that the sintered porous Mg alloy with a higher porosity ratio underwent more severe corrosion than the sample with a lower porosity because of the larger real surface area. With higher porosity and larger pore size have poorer corrosion resistance.

### *3.3. SEM and EDS investigation after electrochemical process*

In Figure 3, showed the details foam material surface after corrosion. Foam material surfaces were rough and damage. Corrosion propagation was seen in Figure 3 (b) and (c) because of the highest sintering temperature. In a small pore, the solution can become acidic or oxygen deficient, resulting in an autocatalytic process by which localized corrosion accelerates. From Figure 3, an increase of the porosity, many pores exhibit irregular shape. And the inner walls of the pores are generally unsmooth. These features may also be vulnerable sites and imperfections in any passive films. When the porosity increases, the interconnectivity of larger pores is clearly increased. For the samples with a lower porosity, the majority of the pores are the middle pores and small pores, which are unlikely to have trapped appreciable volumes of solution. This causes the samples relative resistance to induce corrosion. When the porosity increases, the interconnectivity of the larger pores is markedly improved. The interconnected channels allow the free flow of the liquid and fewer sites are available to induce corrosion.

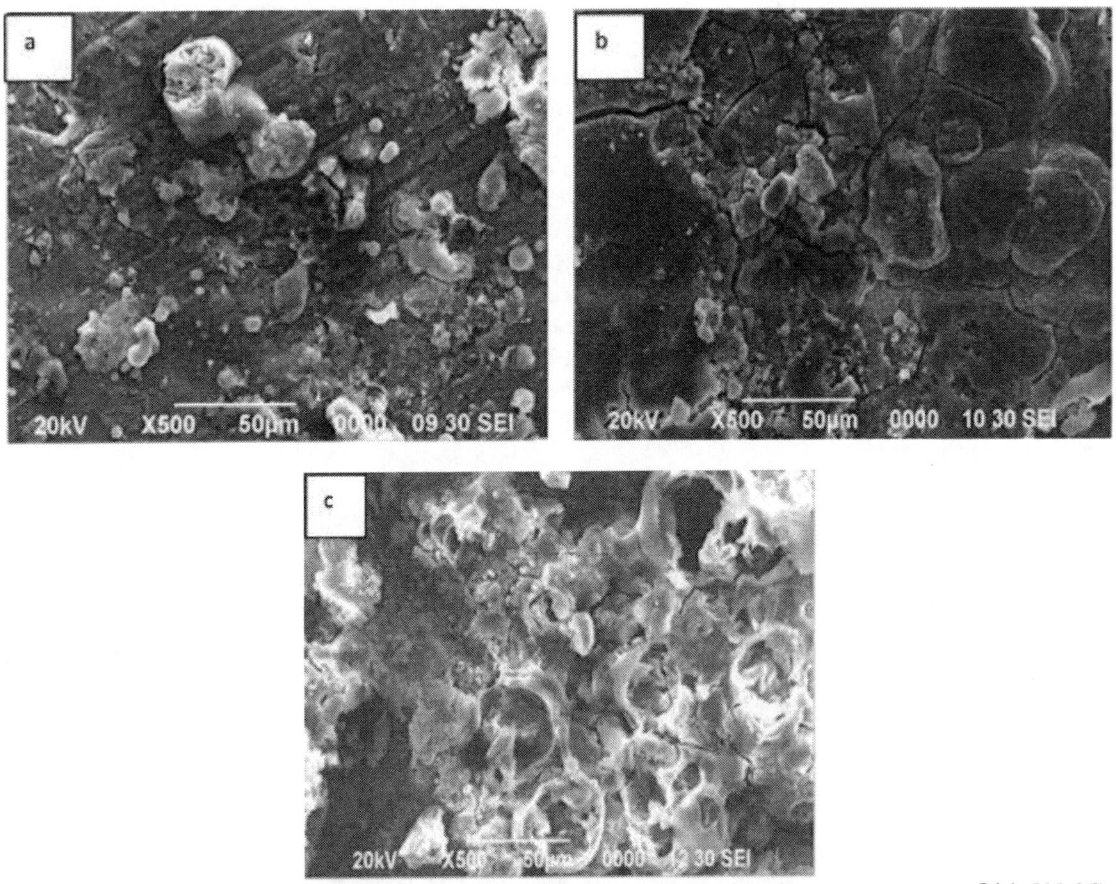


**Figure 3.** Corrosion morphology after electrochemical process at sintering temperatures of (a) 500 °C, (b) 550 °C, and (c) 600 °C.

It would be expected that Mg and Mg alloys be vulnerable to general corrosive attack because they are highly reactive. The corrosive attack on Mg is local rather than general although Mg is highly reactive. After exposure of Mg-Ca-Zn-$TiH_2$ to Hank solution, a volcano-like structure formed. It is clearly evident that the localized corrosion occurs, as shown in Figure 3. The spherical corrosion morphology suggests that H2 bubbles form at the corrosion sites. The Mg dissolution is accompanied by $H_2$ evolution. The $H_2$ evolution occurs in the form of bubbles. The steams of $H_2$ bubbles form at local sites on the Mg alloy surface where electrochemical reactions are taking place, leading to the peculiar structure with a round shape and often with a hole in the center as shown in Figure 4. $H_2$ evolution is always from cathodic sites. Thus the second phases within the spherical structure are the water reduction areas which act as cathodes [28].

Figure 4 presents the genesis of the volcano-like corrosion morphology in Mg alloy schematically, showing the micro-galvanic corrosion process suffered by Mg alloys. The Mg dissolution results in the formation of Mg oxide products ($Mg(OH)_2$) on the Mg surface, according to the reaction 5.1 to 5.3.

| | | | |
|---|---|---|---|
| Anodic | : | $Mg \rightarrow Mg^{2+} + 2e^-$ | (1) |
| Cathodic | : | $2H_2O + 2e^- \rightarrow H_2 + 2OH^-$ | (2) |
| Product | : | $Mg^{2+} + 2OH^- \rightarrow Mg(OH)_2$ | (3) |

According to the cathodic reaction (in Equation 2), the local pH increases due to local generation of $OH^-$ ions. Consequently, corrosion products (such as $Mg(OH)_2$) are easily formed at areas adjacent to the second phases. The volcanic-like structure adjacent to the cathodic is accumulated by the $Mg(OH)_2$. The formation of $Mg(OH)_2$ is stimulated by the cathodic water reduction releasing $OH^-$ and by anodic Mg corrosion releasing $Mg^{2+}$.

Besides the formation of $Mg(OH)_2$, other corrosion products are generated during exposure to the complex electrolyte (Hank's solution). The presence of $OH^-$ near the Mg surface promotes the transformation of $HPO_4^{2-}$ in Hank's solution into $PO_4^{3-}$ . It is expected that $Ca^{2+}$ and $PO_4^{3-}$ ions diffuse from Hank's solution towards the Mg surface and are incorporated in the corrosion layer $Mg(OH)_2$ as corrosion products. The solubility product constant of $Ca_5(PO_4)_3(OH)$ and $Mg_3(PO_4)_2$ is smaller than for $Mg(OH)_2$ [29]. Thermodynamically, $Ca_5(PO_4)_3(OH)$ and $Mg_3(PO_4)_2$ are more stable than $Mg(OH)_2$. Thus, the final products are $Mg(OH)_2$, $Ca_5(PO_4)_3(OH)$ and $Mg_3(PO_4)_2$ [30].

Yu Lu reported that the localized corrosion has occurred in the Mg matrix rather than in the second phase, and the elemental distribution at a localized corrosion site in Mg alloy suggesting an enrichment of O, P, Ca and Mg. From EDS result in Figure 5 after electrochemical test also produced O, Mg, Ca, P also Zn and Ti. The cluster of corrosion products is because the increase of local pH around the corroded site promotes their formation. The water reduction at the cathodic particle results in a locally increased pH.

*Volcano-like morphology*

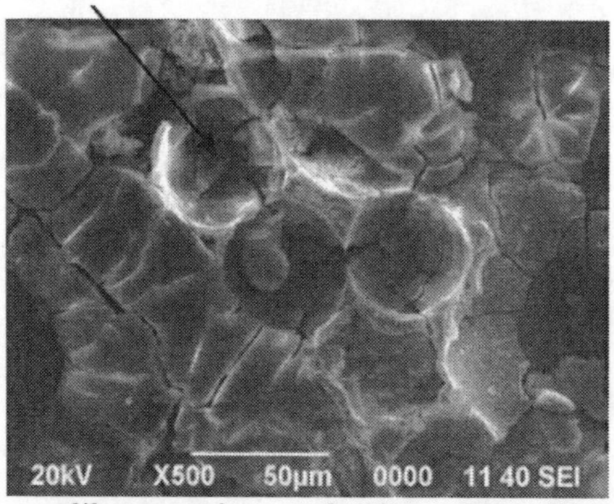


**Figure 4.** Schematic of volcano-like corrosion mechanism of Mg Alloy Mg-Ca-Zn-$TiH_2$ metal foam volcano-like morphology.

IOP Conf. Series: Materials Science and Engineering **202** (2017) 012035 doi:10.1088/1757-899X/202/1/012035

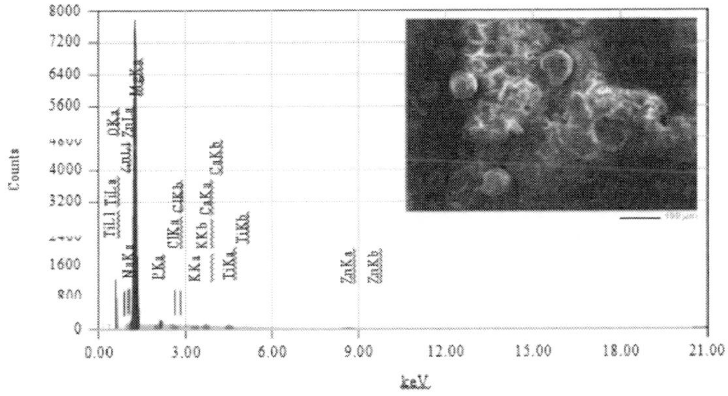


| Element | Mass% | Atom% |
|---|---|---|
| O K | 45.92 | 57.06 |
| Na K* | 2.89 | 2.50 |
| Mg K | 47.99 | 39.25 |
| P K* | | |
| Cl K* | 0.04 | 0.02 |
| K K* | 0.07 | 0.04 |
| Ca K* | 0.50 | 0.25 |
| Ti K | 0.98 | 0.41 |
| Zn K* | 1.68 | 0.51 |
| Total | 100.00 | 100.00 |

(a)

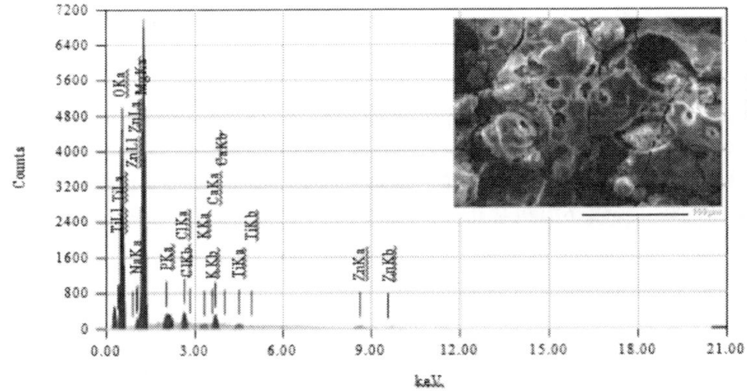


| Element | (Mass% | Atom% |
|---|---|---|
| O K | 50.61 | 62.07 |
| Na K* | 2.91 | 2.48 |
| Mg K | 40.18 | 32.43 |
| P K* | 0.59 | 0.37 |
| Cl K | 1.83 | 1.01 |
| K K* | 0.08 | 0.04 |
| Ca K | 2.05 | 1.00 |
| Ti K* | 0.59 | 0.24 |
| Zn K* | 1.16 | 0.35 |
| Total | 100.00 | 100.00 |

(b)

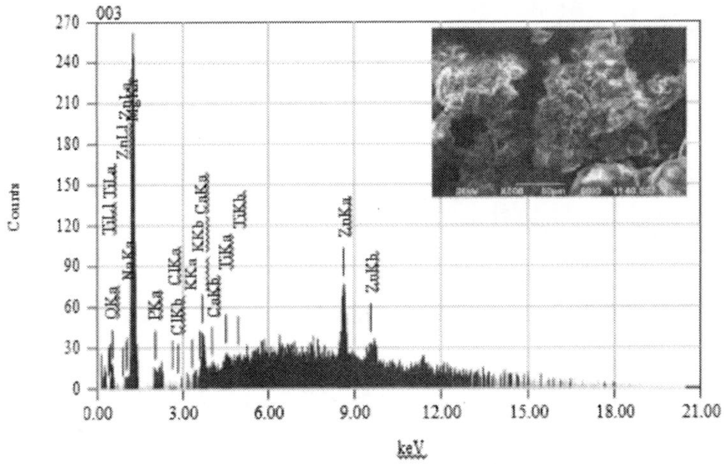


| Element | Mass% | Atom% |
|---|---|---|
| O K | 3.04 | 6.90 |
| Na K* | 1.99 | 3.15 |
| Mg K | 37.30 | 55.79 |
| P K* | 0.49 | 0.58 |
| K K* | 0.10 | 0.09 |
| Ca K | 3.84 | 3.49 |
| Ti K | 1.94 | 1.47 |
| Zn K | 51.30 | 28.53 |
| Total | 100.00 | 100.00 |

(c)

**Figure 5.** EDS Spectra on Mg-Ca-Zn-$TiH_2$ alloy foam at sintering temperatures of (a) 500 °C, (b) 550 °C, and (c) 600 °C.

## 4. Conclusion

The corrosion behavior of Mg-Ca-Zn alloy foam in hanks solution showed that the corrosion resistances magnesium alloy, in generally, decrease as the increasing the sintering temperature. Higher sintering temperatures resulted in the higher pores in metal foam and the samples with higher porosities suffer severer corrosion than the ones with lower porosities because the real surface areas of porous generally increase with the increasing porosities. The localized corrosion is observed in all Mg-Ca-Zn alloy foam in 3 different sinter temperature. There are prominent volcano-shaped corrosion pits indicated by the accumulation of corrosion products locally. The volcano-shaped structure is formed around central which act as cathodes.

## 5. References

[1] Hou L, Li Z, Pan Y, Dua L and Li X, Zheng Y, Li L 2014 In vitro and in vivo studies on biodegradable magnesium alloy *Progress in Natural Science: Materials International* **24** 466–471

[2] Witte F, Ulrich H, Rudert M and Willbold E 2007 Biodegradable magnesium scaffolds: Part I: Appropriate inflammatory response *Journal of Biomedical Materials Research Part A* **81A**(3) 748-756

[3] Witte F, Kaese V, Haferkamp H, Switzer E, Meyer-Lindenberg A and Wirth C J 2005 In vivo corrosion of four magnesium alloys and the associated bone response *Biomaterials* **26** 3557e63

[4] Staiger M P, Pietak A M, Huadmai J and Dias G 2006 Magnesium and its alloys as orthopedic biomaterials: a review *Biomaterials* **27** 1728-1734

[5] Hartwig A 2001 Role of magnesium in genomic stability *Mutation Research Fundamental and* Molecular *Mechanisms of Mutagenesis* **475**(1-2) 113-121

[6] Witte F, Ulrich H, Rudert M and Willbold E 2007 Biodegradable magnesium scaffolds: Part I: Appropriate inflammatory response *Journal of Biomedical Materials Research Part A* **81A**(3) 748-756

[7] Saris N E L, Mervaala E, Karppanen H, Khawaja J A and Lewenstam A 2000 Magnesium-An update on physiological, clinical and analytical aspects *Clinica Chimica Acta* 2000 **294**(1-2) 1-26

[8] Heublein B, Rohde R, Kaese V, Niemeyer M, Hartung W and Haverich A 2003 Biocorrosion of magnesium alloys: a new principle in cardiovascular implant technology? *Heart* **89**(6) 651-656

[9] Zartner, Buettner M, Singer H and Sigler M 2007 First biodegradable metal stent in a child with congenital heart disease: Evaluation of macro and histopathology *Catheterization and Cardiovascular Interventions* **69**(3) 443-446

[10] Wegener B, Sievers B, Utzschneider S, Müller P, Jansson V, Rößler S, Nies B, Stephani G, Kieback B and Quadbeck P 2011 *Mater. Sci. Eng. B* **176** 1789

[11] Ryan G, Pandit A and Apatsidis D P 2006 Fabrication methods of porous metals for use in orthopaedic applications *Biomaterials* **27** 2651-2670

[12] Webster T J, Schadler L S, Siegel R W and Bizios R 2001 *Bioactive glasses fundamental,* technology *and application Tissue Eng* **7** 291

[13] Robertson D M, Pierre L and Chahal R 1976 Preliminary observations of bone ingrowth into porous materials *J. Biomed. Mater. Res.* **10** 335

[14] Wadley H N G 2002 *Cellular metals manufacturing Adv. Eng. Mater.* **4** 726

[15] Gibson L J 2000 *Mechanical behavior of metallic foams Annu. Rev. Mater. Sci.* **30** 191

[16] Wang Y B, Xie X H, Li H F, Wang X L, Zhao M Z, Zhang E W, Bai Y J, Zheng Y F and Qin L 2011 *Biodegradable CaMgZn bulk metallic glass for potential skeletal application Acta Biomater.* **7** 3196-08

[17] Wen C E, Mabuchi M, Ymada Y, Shimojina K, Chino Y and Asahina T 2001 *Scr. Mater.* **45** 147

[18] Heublein B, Rohde R, Kaese V, Niemeyer M, Hartung W and Haverich A 2003 Biocorrosion of magnesium alloys: a new principle in cardiovascular implant technology? *Heart* **89** 651

[19] de Assis S L, Wolynec S and Costa I 2005 Corrosion characterization of titanium alloys by electrochemical techniques. *Electrochimica Acta* **35** 28-36

[20] Jiménez C, Garcia-Moreno F, Pfretzschner B, Klaus M, Wollgarten M M, Zizak I, Schumacher G, Tovar M and Banhart 2011 Decomposition of $TiH_2$ studied in-situ by

synchrotron X-ray and neutron diffraction *Acta Materialia* **59** 6318-30
[21] Kuwahara H, Al-Abdullat Y, Mazaki N, Tsutsumi S and Aizawa T 2001 Precipitation of magnesium apatite on pure magnesium surface during immersing in hank's solution *Mater Trans* **42** 1317–21
[22] Matijasevic B and Banhart J 2006 Decomposition behaviour of as-received and oxidized $TiH_2$ powder *Scripta Mater* **54** 503
[23] Schoenfelder C W and Swisher J H 1973 *J Vac Sci Technol* **10** 862–870
[24] Fromm E 1998 *Kinetics of gas-metal interactions at low temperatures* (Berlin: Springer)
[25] Lestari F P, Erryani A, Amal M I, Annur D, Munir B and Kartika I 2015 Mechanical and potential corrosion studies of Mg alloy metal foam with TiH2 pre-treated as foaming agent *The 2$^{nd}$ International Conference on Materials and Metallurgical Technology*
[26] Saw B A 2003 *Corrosion Resistance of Magnesium Alloys* ASM Handbook 130
[27] Song G and Atrens A 2003 *Adv. Eng. Mater.* **5** 837
[28] Lu Y 2014 *Microstructure and degradation behaviour of Mg-Zn-(Ca) alloys* University of Birmingham
[29] Lide D R 2004 *CRC Handbook of chemistry and physics*
[30] Kalb H, Rzany A and Hensel B 2012 Impact of microgalvanic corrosion on the degradation morphology of WE43 and pure magnesium under exposure to simulated body fluid *Corrosion Science* **57** 122-130

# The Effect of Substrate Temperature on Surface Modification of Polystyrene by using Nitrogen Plasma

**Novi A F, D J D H Santjojo and Masruroh**

Jurusan Fisika FMIPA Univ. Brawijaya, Jl. Veteran Malang 65145, Indonesia

Email: dsantjojo@ub.ac.id

**Abstract**. Studies on the effect of substrate temperature on the modification process of a thin layer of polystyrene samples deposited on glass by using spin coating technique have been conducted. The polystyrene thin film surface was modified by using nitrogen plasma generated in a vacuum reactor by a capacitive plasma generator with a frequency of 40 kHz. The substrate temperature was set by variation of 60, 70 and 80 °C during the modification process to determine the effect of substrate temperature on the level of roughness and hydrophobicity of the film. Observations were conducted by using optical microscopy, optical micro-profiler, and measuring the contact angle of liquid. The results showed a decrease in roughness due to modification at a temperature of 60 °C and 70 °C. On the other hand, the treatment at a temperature of 80 °C resulted in an increase of roughness. By measuring the contact angle, it was proven that hydrophobicity is not only influenced by the surface roughness of thin layers.

**Keywords**: polystyrene, nitrogen plasma, roughness, hydrophobicity.

## 1. Introduction

Quartz Crystal Microbalance (QCM) is a piezoelectric sensor that has been widely used because it can record real-time response and high sensitivity to a change in mass [1]. In the field of medical instrumentations, the QCM is equipped with a layer of a particular material so that it can function as a bio-immunosensor for a specific protein [2]. Enhancement of the performance of QCM biosensor can be done by coating the surface of the sensor with a thin layer of binding biomolecules such as polystyrene [3]. The QCM sensor surface that is generally hydrophobic as it can bind more protein through physical adsorption processes. The hydrophobicity of the surface of solids can be improved by modifying the surface morphology. According to previous studies, UV-irradiation of polystyrene resulted in the more hydrophilic surface which can be observed by a contact angle measurement [4]. Other studies observed the effect of different solvent on the polystyrene surface roughness. All of the studies implied that a rougher surface of the polystyrene thin film enhanced the immobilization of biomolecules [5].

Another technique to modify a polymer surface is by means of nitrogen plasma treatment. The plasma treatment can change microstructure and character of the polymer surface resulting in surface functionalization of materials. The process is carried out in a vacuum vessel filled with nitrogen gas. The gas is then subjected to a high electric field between two electrodes. A nitrogen plasma is produced in the form of glowing species of high energy nitrogen. Interactions between the nitrogen and the surface can result in reaction and diffusion of the nitrogen on the surface of objects [6]. Controlling the reactions on the surface of the thin film is critical to produce certain surface roughness.

One important parameter affecting the reactions is the substrate temperature. This study is intended to investigate the effect of the substrate temperature on the roughness and hydrophobicity of the polystyrene film.

## 2. Experimental Method

Samples in this study were a thin film of polystyrene on the surface of a glass substrate. The film was deposited by using a spin coating technique. First, raw polystyrene material was dissolved in a chloroform solution at a concentration of 3%. Then, the solution was dripped onto the glass surface rotating at high speed (2000 rpm). After the deposition, the sample was annealed for 1 hour at a temperature of 100 °C to remove moisture and the remaining solvent. The polystyrene surfaces were then modified by using nitrogen plasma generated by a plasma generator with a frequency of 40 kHz in a vacuum chamber. The following figure shows a schematic representation of the nitrogen plasma system used in this study.

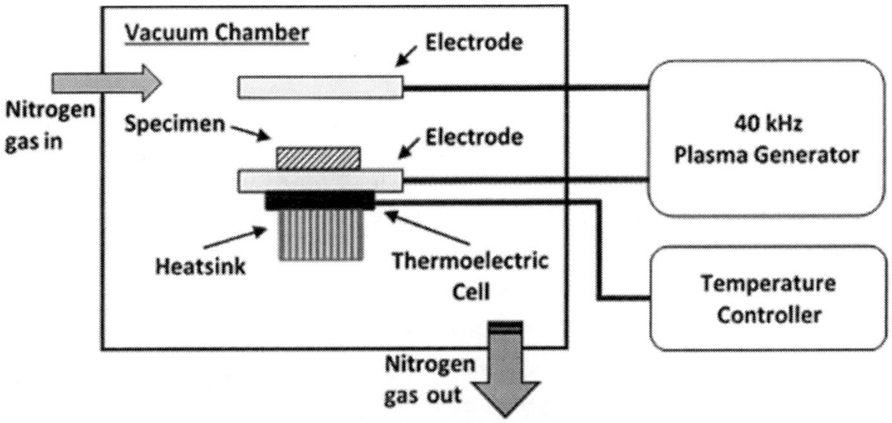


**Figure 1**. Nitrogen plasma system

A capacitive plasma was produced by a set of two parallel plate electrodes. The distance between the electrodes was 0.8 cm. A cooling system was attached at the bottom of the lower electrode consisting of thermoelectric and thermal radiators.The samples were loaded into the reactor and placed on the top of the lower electrode.

In this study, the reactor was vacuum pumped until the pressure reached 0.0188 torrs. Then, the nitrogen gas was introduced into the chamber. The flow of the nitrogen gas was set to 20 mL/min. The plasma was then ignited, and the power was kept to 40 watts. When the plasma was formed, the pressure in the reactor increased to 0.3198 torrs. The temperature of the lower electrode was controlled by adjusting the thermoelectric voltage. During the modification process, the substrate temperature was set to 60, 70 and 80 °C respectively.

Observations of the samples were carried out before and after the modification of the polystyrene. Surface morphology was first observed by using an optical microscope. An optical micro-profiler was then used to measure surface roughness of the samples. Hydrophobicity of the samples was observed by a contact angle of liquid measurement.

## 3. Results and Discussion

The spin coating technique employed in this study produced homogeneous polystyrene thin film. The homogeneity of the polystyrene thin film was examined before the plasma treatment process by using an optical microscope. This examination ensured similar conditions of the samples before the treatment. Typical microstructure of the polystyrene film deposited on a glass substrate is shown in Figure 2. From Figure 2, it can be seen that the polystyrene thin film formed a grainy pattern. The size of the grains is relatively uniform.

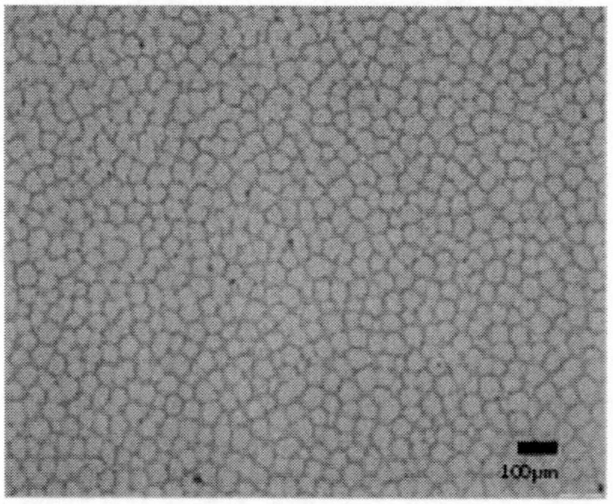


**Figure 2**. Optical microscope surface image of polystyrene thin film.

Topological structure and roughness of the samples which was measured by the optical micro-profiler are shown in Figure 3. Determination of the roughness is carried out by two methods, each of which produces a value Sz and Sa. Sz is the value obtained from roughness measurement technique which is calculated from the value of the highest peak and the deepest valley. Sa represents the average roughness value per area.

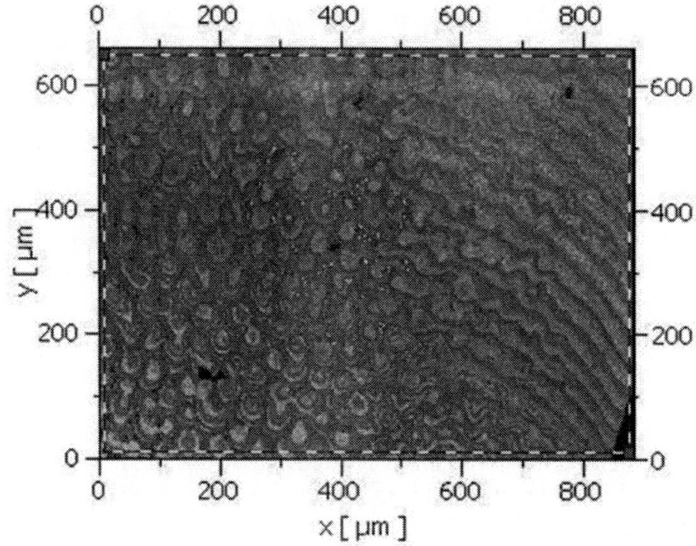


**Figure 3**. Micro profile image obtained by using TMS 2000

Roughness measurements on samples before the plasma modification show that the roughness of the polystyrene surface was 7.0 nm (Sz) and 274 nm (Sa). Surface roughness measurements performed on the plasma treated samples are shown in Figure 4. The roughness of the surfaces increased significantly at a low substrate temperature of 60 °C i.e. 7.37 nm (Sz) and 307 nm (Sa). However, the average roughness (Sa) decreased after 60 °C.

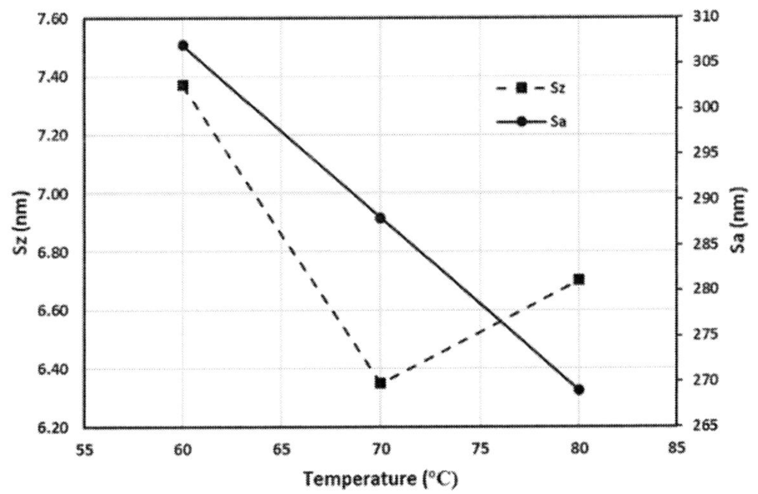


**Figure 4**. Roughness measured with Sa and Sz due to variation of substrate temperature

Sz, on the other hand, decreased from 60 °C to 70 °C then started to increase after 70 °C. The increase was due to the very active surface of the polystyrene approaching its melting point (90 °C). In general, plasma consists of high-energy and charged particles. These particles can produce collisions between ions and the substrate causing the increase of substrate temperature. This situation can affect the polystyrene surface. Besides the physical bombardment, some chemical reactions are possible during the treatment [7].

In this study, we also conducted a study on the effect of surface roughness and hence the substrate temperature on the hydrophobicity of the film. A contact angle measurements performed before and after treatment of liquid nitrogen plasma to determine the level of hydrophobicity. Nitrogen plasma treatment affects the contact angle of the polystyrene layer as shown in Figure 5.

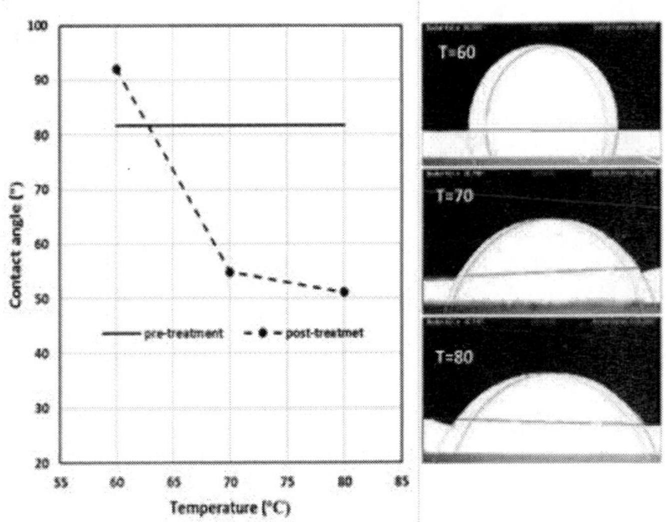


**Figure 5**. The effect of substrate temperature on contact angle

Figure 5 shows that the plasma treatment increased the hydrophobicity of the polystyrene surface significantly. The increase is strongly related to the roughness of the surface. Higher contact angle i.e.

higher hydrophobicity was obtained on the rougher surface. The results obtained in this study indicate that after plasma treatment process at the substrate temperature of 60 $^{o}$C, the roughness of the polystyrene increased by 0,37 nm (Sz) or 33 nm (Sa). The increase produced a high contact angle of about 92 degrees which is more hydrophobic than untreated sample. A similar study showed that the increase of roughness decreases the surface energy resulted in higher hydrophobicity [8].

The higher the substrate temperature, the contact angle of polystyrene-water tends to decrease. As the temperature of the substrate increased, the polystyrene microstructure becomes more dynamic. The smoother surface is a result of the reaction between the nitrogen and the more dynamic microstructures. Some damages were noticed on specimens treated at 80 $^{o}$C.

The decrease of the contact angle at higher substrate temperature is linear with the average roughness (Sa) but is not linear with the change of roughness (Sz). The surface energy becomes smaller at the smoother surface. However, the surface energy does not depend on the roughness, but also the microstructure of the materials. Some evidence of this phenomenon was also observed by another researcher [8].

## 4. Conclusion

Based on this research, it was found that the influence of the substrate temperature on the surface modification process of polystyrene thin film was significant. The highest roughness of the polystyrene after treatment was achieved at 60 °C. The roughness becomes lower after treatment at higher than 60 °C. The very active surface at the temperature close to the melting temperature could be the reason of the smoothing. However, the smoothing phenomena from 70-80 °C showed a small decrease of the hydrophobicity. From these results, it can be concluded that the surface roughness does not entirely influence the hydrophobicity. New microstructures on the surface induced by the plasma treatment may cause a lower surface energy.

## 5. References

[1] Kurosawa et.al 2006 Quartz crystal microbalance immunosensors for environmental monitoring *Biosensors and Bioelectronics* 22 pp.473–481.
[2] Marx K A 2003 Quartz Crystal Microbalance: A Useful Tool for Studying Thin Polymer Films and Complex Biomolecular Systems at the SolutionSurface Interface *Biomacromolecules* 4(5) pp 1099-1120
[3] Sakti S P et.al 2012 Improvement of Biomolecule Immobilization on Polystyrene Surface by Increasing Surface Roughness *Journal of Biosensors & Bioelectronics* **03** pp.3–7
[4] Irawati F, Sakti S P & Unggul P J 2013 Immobilisasi BSA pada Sensor QCM dengan Modifikasi Sifat Hidrofobik-Hidrofilik Permukaan Polistiren Menggunakan Radiasi Ultraviolet. *Natural-B* **2**(2) pp 117–121
[5] Masruroh et al 2014 Solvent effect on morphology of polystyrene coating and their role to improvement for biomolecule Immobilization in application of QCM based biosensor. *Applied Mechanics and Materials* **531** pp. 54-57
[6] Santjojo D J, Istiroyah and Aizawa T 2015 Dynamics of Nitrogen and Hydrogen Species in a High Rate Plasma Nitriding of Martensitic Stainless Steel. *Proceeding of The 9$^{th}$ SEATUC Symposium* pp 311-314
[7] Landgraf R et al 2009 Functionalization of Polymer Sensor Surfaces by Oxygen Plasma Treatment *Procedia Chemistry* **1**(1) pp 1015-1018
[8] Choudhury et al 2011 Investigations of the hydrophobic and scratch resistance behavior of polystyrene films deposited on bell metal using RF-PACVD process *Applied Surface Science* **257**(9) pp 4211-4218

**Acknowledgements**
The authors would like to thank the Directorate of Research and Community Service, Ministry of

Research, Technology and Higher Education who provides a research grant in accordance with Letter of Agreement number: 137/SP2H/ LT/DRPM/III/2016.

IOP Conf. Series: Materials Science and Engineering **202** (2017) 012037 doi:10.1088/1757-899X/202/1/012037

# Optimizing Heat Treatment Process of Fe-13Cr-3Mo-3Ni Martensitic Stainless of Steel

**M S Anwar, S Prifiharni and E Mabruri**

Research Center for Metallurgy and Materials – Indonesian Institute of Sciences (LIPI), Kawasan PUSPIPTEK Building 470 – South Tangerang – Banten – 15314, Indonesia

Email : moch026@lipi.go.id

**Abstract**. The Fe-13Cr-3Mo-3Ni stainless steels are modified into martensitic stainless steels for steam turbine blades application. The working temperature of steam turbine was around 600 – 700 °C. The improvement properties of turbine blade material is necessary to maintain steam turbine work. The previous research revealed that it has corrosion resistance of Fe-13Cr-3Mo-3Ni which is better than 13Cr stainless steels in the chloride environment. In this work, the effect of heat treatment on microstructure and hardness of Fe-13Cr-3Mo-3Ni stainless steels has been studied. The steel was prepared by induction melting followed by hot forging. The steels were austenitized at 1000, 1050, and 1100 °C for 1 hour and were tempered at 600, 650, and 700 °C for 1 hour. The steels were then subjected to metallographic observation and hardness test of Rockwell C. The optimal heat treatment of Fe-13Cr-3Mo-3Ni was carried out austenitized in 1050 °C and tempered in 600 – 700 °C.

**Keywords:** Fe-13Cr-3Mo-3Ni, martensitic, stainless steel, turbine blade, heat treatment.

## 1. Introduction

The development of materials for steam turbine blade is still a concern for the researcher. They need the best mechanical and physical development to achieve high-efficiency turbine work to result in electrical power. The resistance to wear, fatigue, and corrosion is a property that has to be available in the steam turbine blade materials. However, the failure phenomenon still occurs on the steam turbine blade caused by lack of their properties [1-4]. Some of the efforts to minimize failure of turbine blade materials are the alloying process followed by heat treatment, maintaining service condition in the normal stage without fluctuating loads, low corrosive environment, and temperature stabilization [5].

The martensitic stainless steel is preferred to steam turbine blade application because it is hardenable by heat treatment. Since the first patent claim in 1912 [6], the several studies for martensitic stainless steel are still conducted until present. Liu Yu-Rong et al. [7] and Bojack A et al. [8] have been studying the influence of heat treatment of 13Cr super martensitic stainless steel on the austenite formation with a nickel content of 4 – 6 wt% and molybdenum content of 2 wt%. Liu Y. et al. [9] have been studying the effect of heat treatment of AISI 431 martensitic stainless steel on the microstructure and tensile properties. Refer to these previous studies, the heat treatment process on the martensitic stainless steel can rise retained austenite that disperses in the martensitic microstructure. It can enhance the mechanical and physical properties of this steel. However, the increasing temperature of heat treatment also can decrease the hardness of blade materials [10]. In this study, the optimizing heat treatment process of 13Cr martensitic stainless steel containing 3 wt% Ni and Mo has been investigated.

IOP Conf. Series: Materials Science and Engineering **202** (2017) 012037 doi:10.1088/1757-899X/202/1/012037

## 2. Experimental Method

In the study, the material was Fe-13Cr-3Mo-3Ni martensitic stainless steel prepared in an electric induction furnace to carry out melting process. Then, the liquid metal was cast in the mold of 5 x 5 x 10 cm. Liquid metal became ingot in the solid phase in the several times. The ingot was carried out hot forged at the initial temperature of 1125 °C to achieve ingot cross section size around 3 x 3 $cm^2$. After the forging process, the specimen was cut into 1 cm in thickness for chemical composition test by using OES (optical emission spectrometer) instrument. The chemical composition of Fe-13Cr-3Mo-3Ni steel (in weight percent) is 0.1 C, 0.61 Mn, 0.24 Si, 12.73 Cr, 2.93 Ni, 2.52 Mo and balance Fe.

The specimen was conducted austenitizing process at 1000, 1050, 1100 °C for 60 min and quenching in the oil. Then, austenitizing alloy steel was tempered at 600, 650, 700 °C for 60 min and followed by air cooling to room temperature. The heat treated alloy steel was prepared to metallographic test by using optical microscopy, which Kalling reagent as an etchant solution. Hardness was measured with a Rockwell C hardness (HRC) tester with a load of 150 kgf and a diamond of 120° in angle as an indentor. HRC hardness test was measured six times and averaged.

## 3. Results and discussion

### *3.1. Microstructure analysis*

*3.1.1. Tempering at 600 °C.* Figure 1a – c shows the microstructure of Fe-13Cr-3Mo-3Ni steel austenitizing at 1000, 1050, and 1100 °C followed by tempering at 600 °C.

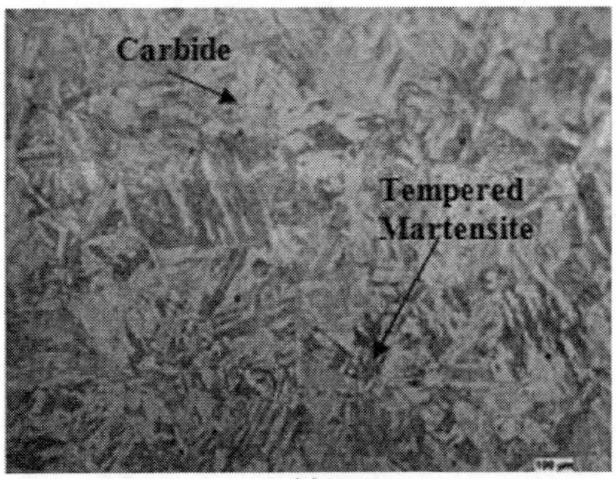


(a)

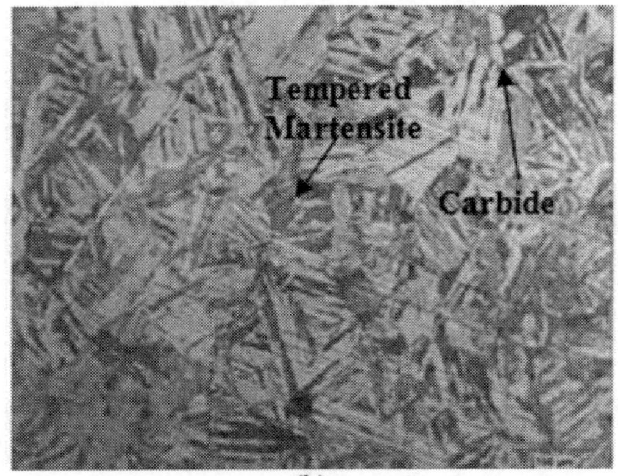


(b)

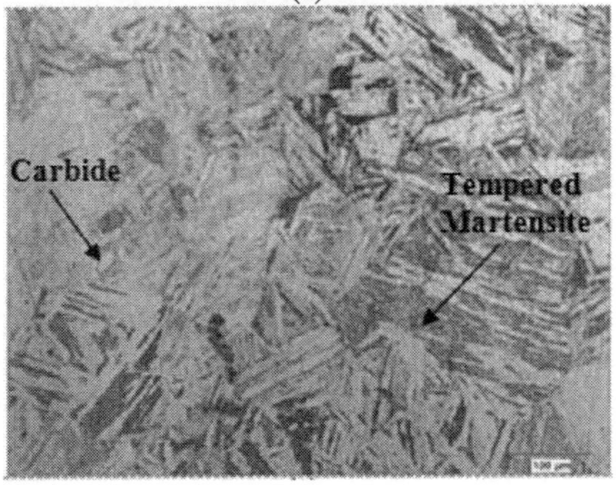


(c)

**Figure 1**. Microstructure of Fe-13Cr-3Mo-3Ni austenitizing at (a) 1000 °C, (b) 1050 °C, (c) 1100 °C followed by tempering at 600 °C, in magnification of 200X

The tempered martensite and carbide are found in the microstructure. After tempering at 600 °C, the Fe-13Cr-3Mo-3Ni steel of the austenitizing of 1100 °C has a lower growth of tempered martensite and much reversed austenite formed in the microstructure when compared with Fe-13Cr-3Mo-3Ni steel in the austenitizing of 1000 and 1050 °C. Refer to the previous study [9], the shape of carbide of Fe-13Cr-3Mo-3Ni steel shows irregular circle that precipitates in the martensitic microstructure. The tempered martensite and carbide were also found in the previous study [11].

*3.1.2. Tempering at 650 °C.* Figure 2a – c shows the microstructure of Fe-13Cr-3Mo-3Ni steel austenitizing at 1000, 1050, and 1100 °C followed by tempering at 650 °C. The tempered martensite is found almost fully in the microstructure. It means that the transformation of the reverse austenite easily becomes martensite after tempering at 650 °C in the various austenitizing temperature. The carbide also precipitates in the martensitic microstructure.

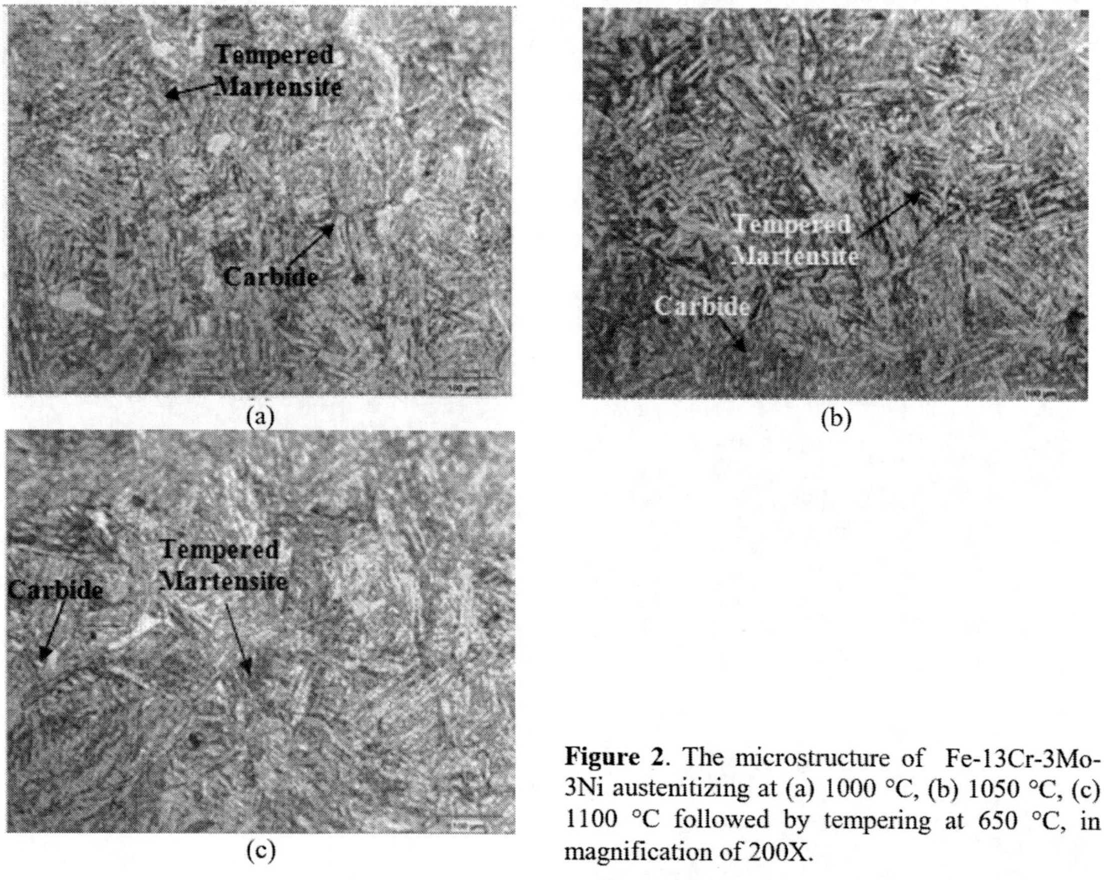


**Figure 2**. The microstructure of Fe-13Cr-3Mo-3Ni austenitizing at (a) 1000 °C, (b) 1050 °C, (c) 1100 °C followed by tempering at 650 °C, in magnification of 200X.

*3.1.3. Tempering at 700 °C.* Figure 3a– c shows the microstructure of Fe-13Cr-3Mo-3Ni steel austenitizing at 1000, 1050, and 1100 °C followed by tempering at 700 °C. After tempering at 700 °C, the Fe-13Cr-3Mo-3Ni steel in the austenitizing of 1050 °C has a higher growth of tempered martensite when compared with Fe-13Cr-3Mo-3Ni steel in the austenitizing of 1000 and 1100 °C. It means that the transformation of the reverse austenite is more easily become martensite after tempering at 700 °C followed by cooling in the austenitizing of 1050 °C. The microstructure of Fe-13Cr-3Mo-3Ni steel austenitizing at 1050 °C followed by tempering at 700 °C in the Figure 3b is different with the others. The growth of their tempered martensite is more coarse when compared with the microstructure of Fe-13Cr-3Mo-3Ni steel austenitizing at 1000, and 1100 °C. It is probably the higher temperature tempering applied in the Fe-13Cr-3Mo-3Ni steel.

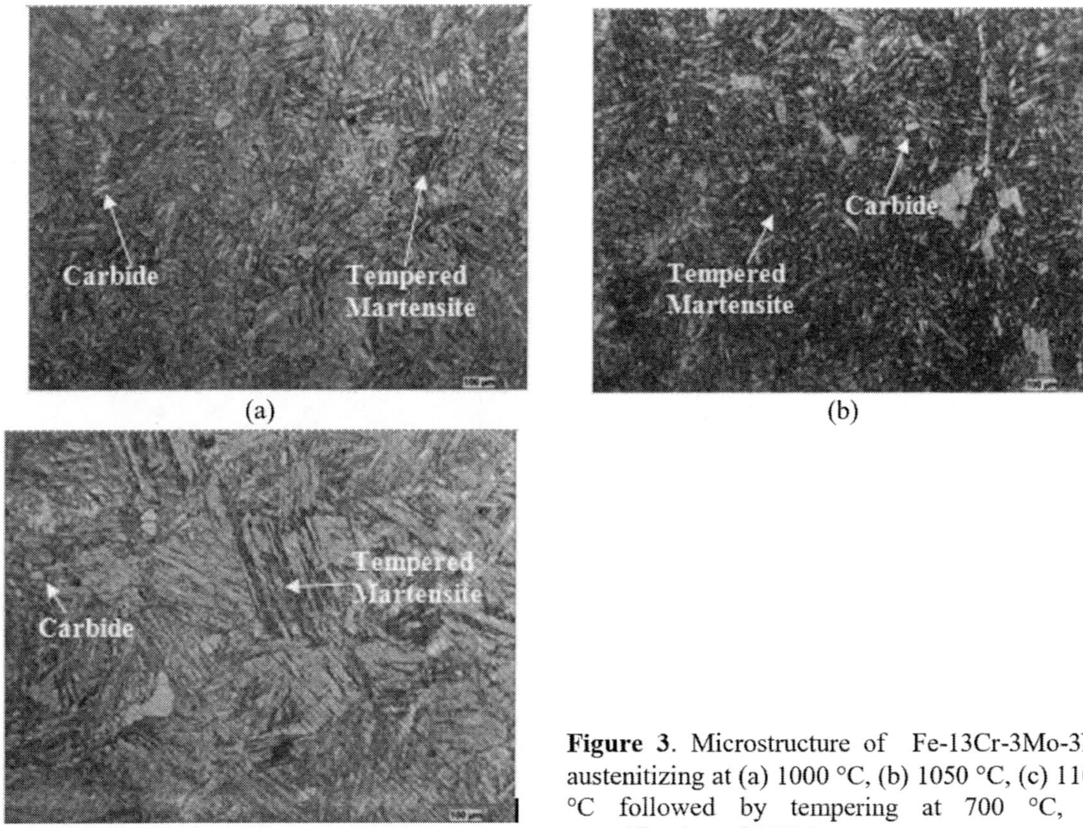


**Figure 3**. Microstructure of Fe-13Cr-3Mo-3Ni austenitizing at (a) 1000 °C, (b) 1050 °C, (c) 1100 °C followed by tempering at 700 °C, in magnification of 200X

*3.2. HRC hardness analysis*

Figure 4 shows the change of HRC hardness at various austenitized and tempering temperatures. The HRC hardness decreases at all austenitized and tempering temperatures. It notices that the higher temperature of tempering process can decrease fragile nature of Fe-13Cr-3Mo-3Ni steel, but increase their ductility and toughness [12]. This steel becomes softener because the carbide starts coarsen and martensite becomes less tetragonal [13]. It is confirmed in the Figure 1–3.

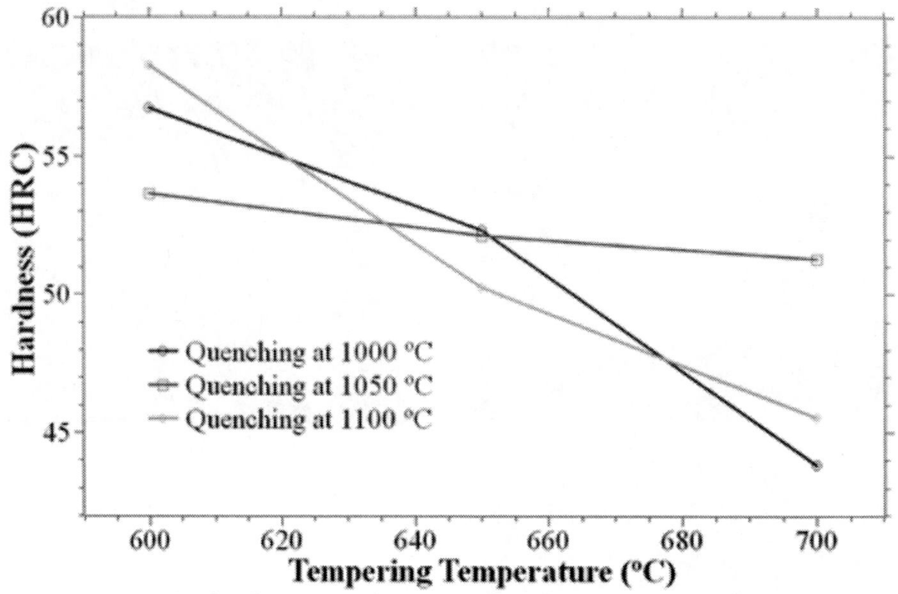


**Figure 4**. HRC hardness at various austenitized and tempering temperatures

Table 1 shows the HRC hardness loss percentage. The lowest HRC hardness loss percentage was found on the Fe-13Cr-3Mo-3Ni steel when it carried out austenitized in the temperature of 1050 °C and tempering in the temperature of 650 – 700 °C. It means that this steel can be used as turbine blade material in the steam turbine's work in the temperature of 600 – 700 °C; their hardness will be almost stable. Also, the wear and fatigue resistance properties of this steel are better when compared with Fe-13Cr-3Mo-3Ni steel austenitized in 1000 and 1100 °C and tempered in 600 – 700 °C. It implies that the optimal heat treatment of Fe-13Cr-3Mo-3Ni was carried out austenitized in 1050 °C and tempered in 600 – 700 °C.

**Table 1**. HRC hardness loss percentage (%)

| Tempering Temperature (°C) | Austenitized Temperature (°C) | | |
|---|---|---|---|
| | 1000 | 1050 | 1100 |
| 600 | 0 | 0 | 0 |
| 650 | 7.81 | 2.8 | 13.81 |
| 700 | 22.82 | 4.44 | 21.9 |

## 4. Conclusion

Refer to results and discussion, the conclusions comprise that the 13Cr martensitic stainless steel containing 3 wt% Ni and Mo ( Fe-13Cr-3Mo-3Ni) can be used as turbine blade material in the steam turbine's work in the temperature of 600 – 700 °C. The heat treatment process of Fe-13Cr-3Mo-3Ni exhibits tempered martensite and carbide in the microstructure. The increment of tempering temperature at the various austenitization temperature can effect on the coarser of tempered martensite microstructure and hardness Rockwell C (HRC) of the Fe-13Cr-3Mo-3Ni steel. The lowest HRC hardness loss percentage was found on the Fe-13Cr-3Mo-3Ni steel when austenitized temperature was 1050 °C at tempering temperature of 650 – 700 °C. The optimal heat treatment of Fe-13Cr-3Mo-3Ni was carried out austenitized in 1050 °C and tempered in 600 – 700 °C.

## 5. References

[1] Ahmad M, Schatz M and Casey M V 2013 *Wear* **303** 83-86

[2] Azevedo C R F and Sinátora A 2009 *Engineering Failure Analysis* **16** 2290-2303
[3] Stifanese R, Belsanti L, Toselli M, Letardi P and Traverso P 2016 *Engineering Failure Analysis* **64** 58-66.
[4] Ilieva Ilieva G 2016 *Engineering Failure Analysis* **70** 90-104
[5] Maleque M A and Salit M S *Material selection and design* Chapter 2: Mechanical Failure of Material Springer ISBN: 978-981-4560-37-5
[6] Setovirta M 2013 *Handbook of Stainless Steel* Outokumpu Finland
[7] Liu Yu-rong , YE Dong, YONG Qi-long, SU Jie, ZHAO Kun-yu and JIANG Wen 2011 *Journal of Iron and Steel Research, International* **18**(11) 6C-66
[8] Bojack A, Zhao L, Morris P F and Sietsma J 2016 *Metallurgical And Materials Transactions A* **47A** 1996 – 2009.
[9] Liu Y, Li A, Cheng X, Zhang S Q and Wang H M 2016 *Materials Science & Engineering A* **666** 27–33.
[10] Johnpaul C. Ezechidelu1 , Samuel O. Enibe *International Advanced Research Journal in Science, Engineering and Technology* **3 (4)** 6-12
[11] Barlow L D and Du Toit M 2012 *Journal of Material engineering and performance* **21 (7)** 1327 – 1336.
[12] Rajasekhar A 2015 *International Journal of Scientific & Engineering Research* **6 (4)** 2229-5518
[13] Salih A A, Omar M Z, Junaidi S and Sajuri Z 2011 *Australian Journal of Basic and Applied Sciences* **5(12)** 867-871

**Acknowledgments**
The authors thank Research Center for Metallurgy and Materials, Indonesian Institute of Sciences (LIPI) for funding this research in the fiscal year of 2016.

IOP Conf. Series: Materials Science and Engineering **202** (2017) 012038 doi:10.1088/1757-899X/202/1/012038

# Fe-doped ZnO Supported with Montmorillonite: Synthesis, Characterization, and Photocatalytic Activity

**M I Pratiwi[1,2], N Afifah[1,2] and R Saleh[1,2]**

[1]Department of Physics, Faculty of Mathematics and Natural Sciences, Universitas Indonesia, Kampus UI Depok, Depok 16424, Indonesia
[2]Integrated Laboratory of Energy and Environment, Faculty of Mathematics and Natural Sciences, Universitas Indonesia, 16424 Depok, Indonesia

Email: rosari.saleh@ui.ac.id

**Abstract**. In this study, Fe-doped ZnO/MMT has been prepared by using co-precipitation method with the various amount of MMT (10, 20, 30, and 40 wt%). The samples were characterized by using X-ray diffraction (XRD), Fourier Transform Infrared Spectroscopy, and (FTIR), Brunauer-Emmett-Teller (BET) surface area analysis. The crystallite structure of ZnO did not change with the additional of dopant and MMT. The presence of MMT could be confirmed by using FTIR, which showed the bending vibration and stretching vibration of Si-O-Al and Si-O-Si. The degradation of methylene blue and methyl orange were examined by using montmorillonite (MMT) modified Fe-doped ZnO catalyst in photocatalytic process under UV light irradiation. The photocatalytic results indicated that certain amount of MMT could increase photocatalytic performance in degrading methylene blue and methyl orange. Methylene blue degradation increased with the increasing of pH value while the opposite trend occurred for methyl orange degradation.

## 1. Introduction

Chemical, textile, and paint industries have played a big role in dye-contaminated water. Some methods have been used to eliminate dyes from wastewater including adsorbtion [1], activated carbon [2], and photocatalysis [3]. In the last decades, photocatalysis has been a promising way to degrade dyes in aqueous solution because it is low cost, highly efficient, and less polluting.

Zinc oxide (ZnO) nanoparticle is a popular photocatalyst beside $TiO_2$. ZnO has a direct bandgap of 3.37 eV which is in the ultra violet (UV) region. However, pure ZnO still perfoms the low photocatalytic activity. Doping ZnO with transition metal iron (Fe) and modifying it with montmorillonite (MMT) clay are good ways to enhance the photocatalytic performance[4,5]. Moreover, MMT is a good supporter in nanoparticle because of its high surface area and excellent adsorption ability[6].

In our previous work [7], we had studied the photodegradation of methylene blue (MB) and methyle orange (MO) by using Fe:ZnO/MMT nanocomposite. In expectation to obtain better dyes' photodegradation, we varied the amount of MMT in nanocomposites and studied their photocatalytic activity in this work. The effects of dosage concentration and pH in the photocatalytic activity were also examined.

## 2. Experimental Method

### *2.1. Materials*
$FeSO_4.7H_2O$ (Merck), $ZnSO_4.7H_2O$ (Merck), NaOH (Merck), and MMT clay (Nanocor) were analytical grade and used without further purification. Distilled water was used in the whole experiment.

### *2.2. Synthesis of the Fe:ZnO/MMT nanocomposites*
The Fe:ZnO/MMT nanocomposites were prepared through co-precipitation method as described in our previous work [7]. The nanocomposite was synthesized four times with the amount of montmorillonite 10-40%.

### *2.3. Characterization*
The specific surface area, optical properties, and structural properties of the samples were characterized by using Burneur-Emmett-Teller (BET; Nova Quantachrome 2000) surface area analysis, X-ray diffraction (XRD; Rigaku Miniflex 600), and fourier transform infrared spectroscopy (FTIR; FEI Spectrophotometer Shimazu IR Prestige-21).

### *2.4. Catalytic performance*
The photocatalytic performance of Fe:ZnO/MMT nanocomposites was examined by using methylene blue (MB) and methyl orange (MO) as the models of organic pollutant under a 40 W UV lamp as light irradiation source. The degradation rate of MB and MO was observed in 15 minutes intervals for two hours by using UV-Vis Hitachi UH5300 spectrophotometer. To obtain the optimal condition in the photocatalytic activity, the effect of pH was also studied.

## 3. Results and Discussion
The crystal structure of the samples were studied by using XRD spectroscopy. Table 1 summarizes the lattice parameter value, obtained by the Rietveld refinement technique by using the program of MAUD, and the crystallite size that was calculated through the Debye-Scherer formula [8]. Figure 1 shows the diffraction patterns of Fe:ZnO with various amount of MMT. The diffraction pattern of MMT and ZnO was also displayed as comparison. The diffraction patterns of Fe:ZnO/MMT nanocomposites show the existence of diffraction peaks at $2\theta = 31.54°$, 34.34°, 36.13°, 47.46°, 56.44°, 62.77°, 66.02°, 67.81°, and 68.91° which represent the areas of (100), (002), (101), (102), (110), (103), (200), (112), and (201) which are from the hexagonal wurtzite structure of ZnO [19]. They indicated that the nanocomposites were successfully synthesized without changing the structure of ZnO after doping with metal iron. Furthermore, the intensity of MMT in four samples does not increase significantly as its loading increases. There were also no impurities detected from the XRD patterns. Specific surface area of the synthesized samples were characterized by using Burneur-Emmett-Teller (BET) and summarized in Table 1. It shows that the increasing amount of MMT in Fe:ZnO could increase the surface area of the nanocomposite.

**Table 1.** Lattice Parameters, Unit Cell Volumes, Crystal Size, and Specific Surface Area of Nanocomposites

| Sample | MMT Ammount | Lattice Parameter | | Unit cell volume (Å$^3$) | Crystal Size (nm) | Surface Area (m$^2$/g) |
|---|---|---|---|---|---|---|
| | | a = b (Å) | c (Å) | | | |
| Fe:ZnO/MMT | 10% | 3.259 | 5.222 | 47.695 | 17 | 23.64 |
| | 20% | 3.261 | 5.221 | 48.097 | 18 | 35.58 |
| | 30% | 3.261 | 5.225 | 48.126 | 19 | 42.85 |
| | 40% | 3.263 | 5.226 | 48.156 | 20 | 48.31 |

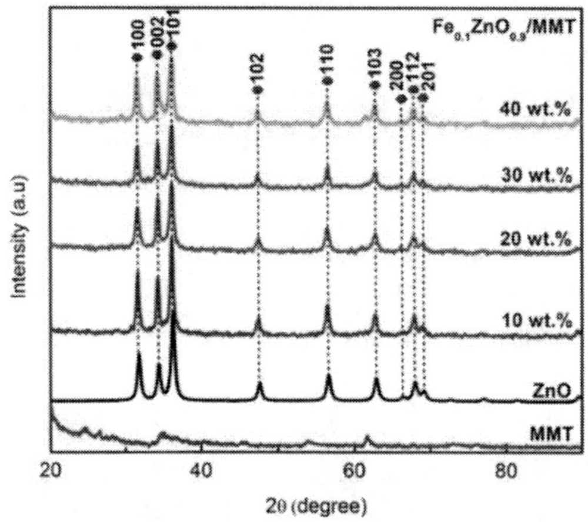


**Figure 1.** XRD Patterns of MMT, ZnO, and Fe:ZnO/MMT Nanocomposites

FTIR spectroscopy was used to confirm the existence of the MMT in the nanocomposites. Figure 2 shows the absorption spectra of montmorillonite and the nanocomposites with percentage of MMT from 10-40%. Respectively, the Si-O-Al bending vibration and Si-O-Si stretching vibration were found at wave number of 970 $cm^{-1}$ and 1090 $cm^{-1}$, confirming the existence of MMT in the sample [10,11]. There are also stretching vibrations of C-O, O-H, and C=O at the absorption peaks of wave numbers of 1049, 1384, and 1723 $cm^{-1}$ [12]. The spectroscopy also detected a vibration mode of the O-H bond from the broad peak at wave numbers of 3000-3500 $cm^{-1}$ [13]. The photodegradation of methylene blue and methyl orange were observed under UV light irradiation for 2 hours. Figure 3 shows the degradation of MB and MO using nanocomposites with various amount of MMT. It shows that increasing amount of MMT to 30% results in improved photocatalytic degradation efficiency of MB and MO. The increase of photocatalytic efficiency may be due to the increasing num of the specific surface area value. Therefore, the surface active site also increased, which yielded to a higher interfacial charge carrier transfer for photocatalysis [14]. However, when the amount of MMT further increased to 40%, the photocatalytic efficiency decreased. It may be due to the active sites get blocked by the over maximum content of MMT.

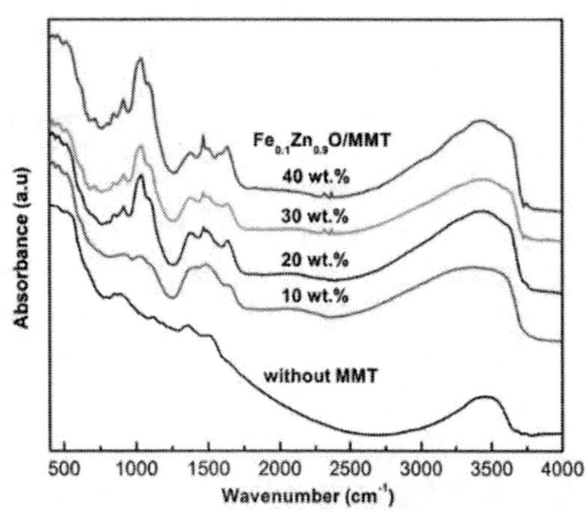


**Figure 2.** Infrared Spectra of Fe:ZnO Nanoparticle and Fe:ZnO/MMT Nano composites

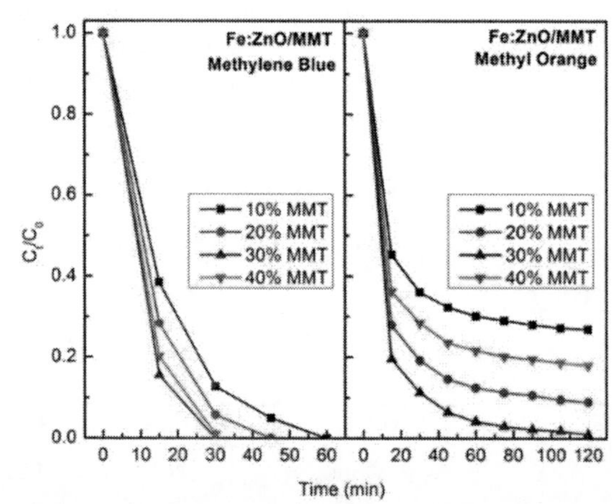


**Figure 3.** Photodegradation Percentage of MB and MO using Fe:ZnO with various amount of MMT.

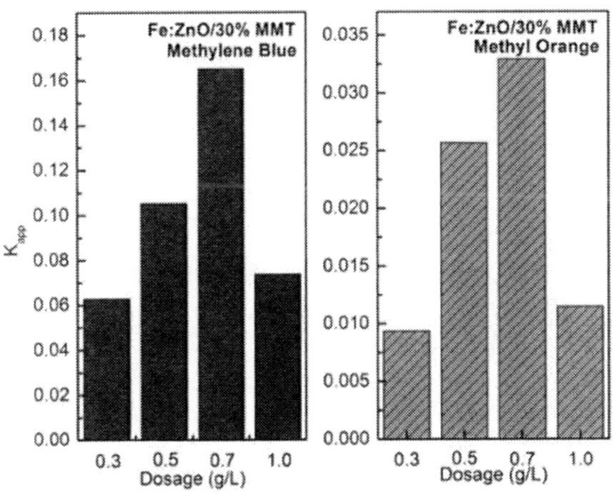

**Figure 4.** The apparent reaction rate constant ($k_{app}$) of MB and MO degradation with different catalyst

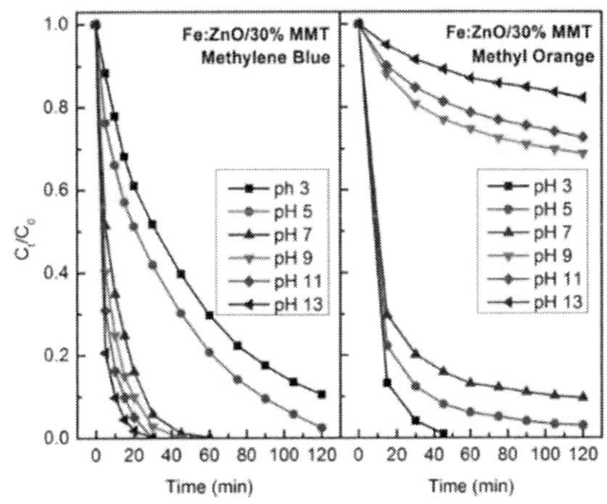

**Figure 5.** The degradation curve of MB and MO under different pH value.

Figure 4 shows the apparent rate constant ($k_{app}$) of MB and MO with the effect of catalyst dosage which varied from 0.3g/L to 1.0 g/L. Increasing catalyst dosage from 0.3 g/L to 0.7 g/L led to enhancing degradation of both dyes but decreasing degradation when the dosage was further loaded to 1.0 g/L. These results indicate that the increasing catalyst amount can add the number of active sites to interact with dye molecules. However, the excessive amount of catalyst can decrease the photocatalytic efficiency due to the opacity of the suspension increasing [5].

Figure 5 shows the photocatalytic performance of Fe:ZnO/30% MMT under acidic, neutral, and alkaline conditions. As can be seen in the figure, the MB photodegradation rate increased rapidly in alkaline conditions but decreases in acidic conditions while MO photodegradation rate exhibits the opposite trend, like had been observed in our previous works [7]. The mechanism is described as follows. ZnO was reported to have zero point charge of 9±0.3 [15] which makes it will be positively charged under acidic and neutral condition and negatively charged under alkaline condition. Since MB is a cationic dye, it is positively charged in aqueous solution. In the acidic condition, the Coulomb repulsive force occurs between positive charges of ZnO and MB, which can decrease the photocatalytic efficiency because MB cannot provide hydroxyl group to form hydroxyl radicals (•OH) formation [16]. However, in alkaline condition, the opposite charges of ZnO and MB attract each other so they can interact to decompose MB molecules from the water by providing a higher concentration of hydroxyl ions that can react with holes to form •OH [17].

The degradation of MO under different pH condition can be described similarly to the degradation of MB above. MO is an anionic dye, resulting positively charged ZnO's surface in pH value, which is acidic condition, lower than its zero point charge can enhance the photocatalytic activity in degrading MO. Meanwhile, the pH value which is higher than its zero point charge cannot decompose MO's complex molecule effectively.

## 4. Conclusion

The Fe:ZnO nanocomposites with various amount of montmorillonite have been successfully synthesized by using co-precipitation method. The results show that the amount of 30% MMT plays important role in enhancing the photocatalytic activity of Fe:ZnO. The pH value of 13 and 3 are the best condition for degrading MB and MO, respectively.

## 5. References

[1] Tian G, Wang W, Kang Y, and Wang A 2016 *J. Taiwan Inst. Chem. Eng.* **58** 417–23.
[2] Malik PK 2004 *J Hazard Mater* **113** 81-88.
[3] Djaja NF Saleh R 2014 *Spect. Acta A: Mol. Bio. Spect.* **130** 581–590.

[4] Nur Afifah, Djaja NF and Saleh R 2015 *Mat Sci For* **827** 19-24
[5] Djaja NF and Saleh R 2014 *Superlatt. Microstruct.* **74** 217-233.
[6] Praveen Kumar J, Ramacharyulu PVRK, Prasad GK and Beer Singh 2015 *Appl. Clay Sci.* 116-**117** 236-272.
[7] Nur Afifah, Siti Andriani, Djaja NF and Saleh R 2015 *Adv. Mater. Res.* **1123** 295-302.
[8] Khan EM, Alam F, Parveen A and Naqvi AH 2013 *J. Adv. Micros. Res.* **8** 45-52.
[9] Estévez-Hernández O, Santiago-Jacinto P and Reguera E 2013 *Mater. Focus* **2** 438–442.
[10] Du E, Yu S, Zuo L, Zhang J, Huang X and Wang Y 2011 *Appl. Clay Sci.* **51** 94–100
[11] Slavutsky AM, Bertuzzi MA and Armada M 2012 *Brazil J. Food Tech.* **15** 208-218.
[12] Stankovich S, Piner RD, Nguyen ST and Ruoff RS 2006 *Carbon* **44** 3342–3347.
[13] Xing S, Zhou Z, Ma Z and Wu Y 2011 *Appl. Catal. B Environ.* **107** 386–392.
[14] Wang XH, Li JG, Kamiyama H, Moriyoshi Y and Ishigaki T 2006 *J. Phys. Chem. B* **110** 6804–6809.
[15] Zhang Y, Wan J and Ke Y 2010 *J. Hazard. Mater.* **177** 750–754.
[16] Barakat MA, Schaeffer H, Hayes G and Ismat-Shah S 2004 *Appl. Catal. B: Environ.* **57** 23–30.
[17] Xiao Q, Zhou Q, Zhang J and Ouyang L 2009 *J. Alloy. Compd.* **468** L9–L12.

# Study of Sigma Phase in Duplex SAF 2507

**D M Fellicia, Sutarsis, B A Kurniawan, D Wulanari, A Purniawan, and A T Wibisono**

Department of Materials and Metallurgical Engineering, Faculty of Industrial Technology, Institut Teknologi Sepuluh Nopember (ITS), Surabaya 60111, Indonesia

Email: dian@mat-eng.its.ac.id

**Abstract.** Super duplex stainless steel is one of the stainless steel which has a combination between high strength properties and excellent corrosion resistance. However, the resistance can decrease by precipitation of sigma phase which is formed at high temperature, for example after welding processes. A series of experiments has been performed to study the effect of solution annealing to existence of sigma phase on super duplex SAF 2507. Variations of solution-annealing temperatures were 1000 °C, 1065 °C and 1125 °C with holding time of 15 and 30 minutes for each temperature. Effect of solution annealing process was characterized by using XRD, SEM, and Optical Microscopy. The result showed precipitation of sigma phase completely dissolved at 1065 °C and 1125 °C because it reformed to austenite. After it was heated at 1065 °C, chromium carbide appeared in ferrite site and grain boundary. The amount of chromium carbide increased with the increasing of solution annealing temperature.

**Keywords:** Super duplex, annealing, sigma phase, SAF 2507

## 1. Introduction

Super duplex stainless steel SAF 2507 is an austenitic-ferritic stainless steel containing 25% Cr and 7% Ni [1]. This material is used for service in highly corrosive condition and high temperature environment due to its excellent corrosion resistance as the result of solution annealing treatment at 1025 °C-1125 °C on the manufacturing process. Sandvik SAF 2507 also has good mechanical properties and impact strength with brittle-ductile transition temperature is below -50 °C (-58 °F). Sandvik SAF 2507 is a duplex stainless steel that specially designed for service in aggressive chloride-containing environments [2]. However, the excellence of mechanical properties and corrosion resistance of SAF 2507 decrease significantly due to the existence of sigma phase.

Sigma phase ($\sigma$) is an intermetallic phase with tetragonal crystal structure which is formed at high temperature (320 °C to 955 °C) after holding at a certain period. Sigma phase precipitation occurs when ferrite transforms to secondary austenite and sigma phase ($\alpha \rightarrow \sigma + \gamma_2$). Precipitation takes place at ferrite sites with rich chromium content of the ferritic stainless steel, grain boundary between austenite and ferrite of duplex stainless steel, and austenite grain boundary of austenitic stainless steel [3]. Sigma phase precipitates at a ferrite grain site because ferrite contains high concentration of chromium, where chromium is known as ferrite former and stabilizer. Nucleation of sigma phase initiates at grain boundary and grows to ferrite grain until transformation ends. Figure 1 shows a correlation between sigma phase formation and energy at the grain boundary. According to [4], when steel is heated to 800 °C-900 °C internal energy of super duplex stainless steel increased. Furthermore, energy is used by atom for transformation process. Elements of ferrite formers will form ferrite phase, while austenite formers will form austenite phase. Because chromium elements is also a phase stabilizer, then the chromium moves and remains on the center of the grain, hence grain boundary is a site with high energy. When the system search for equilibrium, chromium is attracted to grain boundary and forms sigma phase.

The purpose of this research was to recover corrosion resistance of SAF 2507 by dissolving sigma phase through solution treatment and followed by quenching. The solution annealing temperature to dissolve sigma phase was 1025 °C-1125 °C. In this research SAF 2507 was the solution annealed and followed by water quenching to dissolve sigma phase precipitation to recover the material. The solution annealing temperatures were 1000 °C, 1065 °C and 1125 °C with holding time of 15 and 30 minutes for each temperature.

## 2. Experimental Method

The test specimen is a seamless tube of super duplex SAF 2507 with a chemical composition as following in Table 1. Specimens are divided into two, namely the control specimen that had sigma phase and heat treatment specimen. The heat treatment performed is solution annealing followed by cooling water quench. Based on ASTM A923[5], the intermetallic phase is formed when the material is exposed to heat at a temperature of 320 °C – 955 °C. To dissolve sigma phase, the specimens is heated at a temperature of solution annealing, held for some time and then water quenching is performed. Temperature variations used were 1000 °C, 1065 °C and 1125 °C. Meanwhile, the variations of holding time used for each temperature were 15 minutes and 30 minutes.

To analyze the microstructure changes of super duplex SAF 2507, XRD, SEM Image-J and OM were conducted. X-Ray Diffraction (XRD) was performed to observe phases, its compound on each sample. Furthermore, XRD was used to investigate the existence of sigma phase as the result of solution annealing. The purpose of Scanning Electron Microscope (SEM) is to observe the morphology of SAF 2507 samples. The optical microscope (OM) was used for microstructure observation of SAF 2507 with different treatment.

**Table 1.** Chemical composition of super duplex SAF 2507 [2]

| C | Si | Mn | P | S | Cr | Ni | Mo | Others |
|---|---|---|---|---|---|---|---|---|
| ≤0.03 | ≤0.8 | ≤1.2 | ≤0.025 | ≤0.015 | 25 | 7 | 4 | N=0.3 |

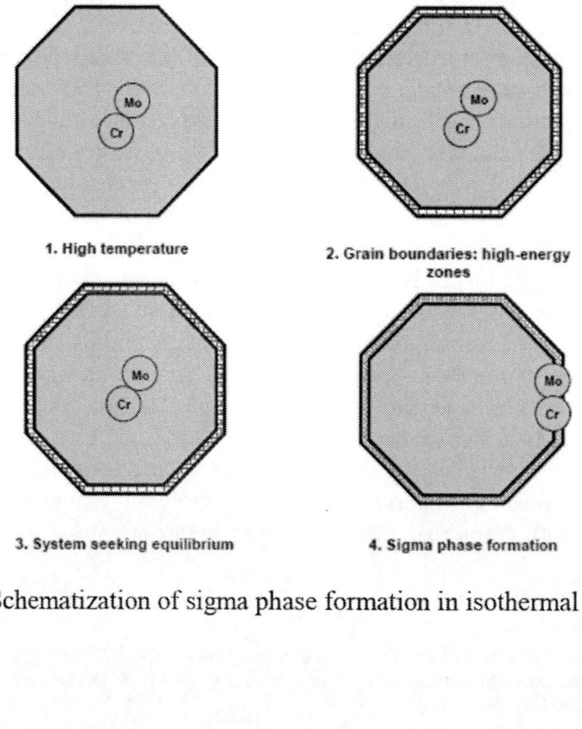


**Figure 1.** Schematization of sigma phase formation in isothermal condition [3]

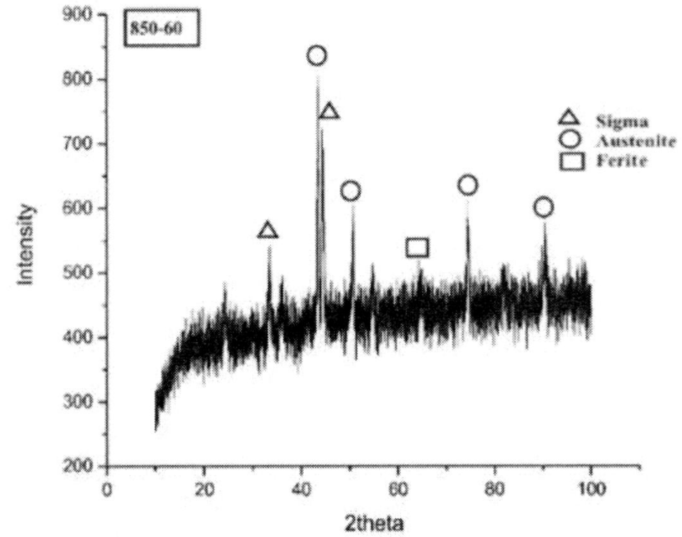


**Figure 2.** XRD Diffractogram of aging SAF 2507

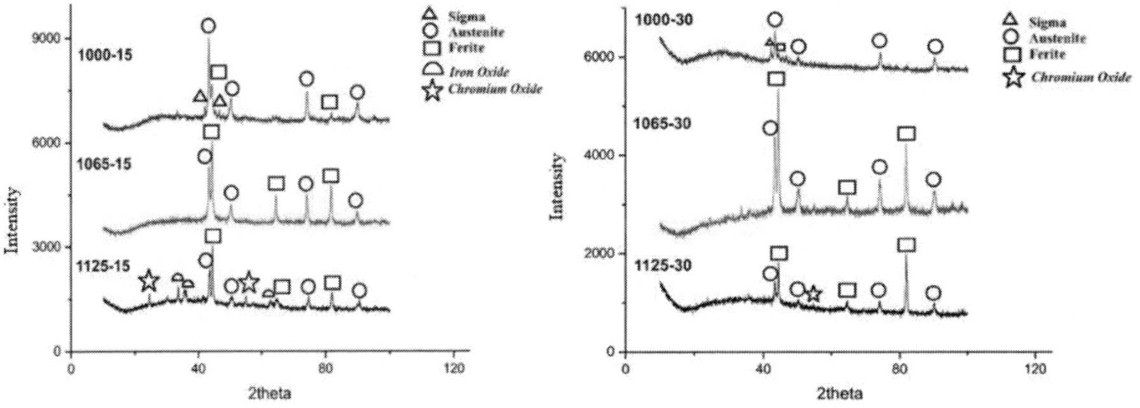


(a) (b)

**Figure 3.** XRD Diffractograms of solution annealed samples (1000, 1065 and 1125°C) with holding time of (a)15 minutes and (b) 30 minutes

## 3. Results and Discussion

### *3.1. X-Ray Diffraction Test*

X-ray diffraction has been carried out with an angle range of 10-100° and wavelength Cu-Kα of 1.54060 Å. XRD generate peaks with certain intensity, and peaks which has high intensity were matched with PDF card to obtain the phases or compounds contained in the sample.

Figure 2 shows the graphic XRD of aging SAF 2507. The result is match with PDF card 47-1417 which shows austenite, PDF card 03-065-9117 which shows sigma phase and also PDF card 85-1410 which shows ferrite phase. This result is confirmed by Ternary diagram of Fe-Cr-Ni in alloying 25%Cr-7%Ni which found sigma phase at a temperature of 1120 K.

XRD of solution annealed samples with holding time of 15 and 30 minutes are shown in Figure 3. Analysis by using PDF card yields phases formed at 1000 $^0$ C was austenite, ferrite and sigma. At 1065 °C the austenite and ferrite phases were formed. Meanwhile,at 1125 $^0$ C, austenite, ferrite, iron oxide and chromium oxide phases were formed. Sigma phase at heating temperature 1000 °C still existed because it incomplete dissolved, while iron oxide and chromium oxide were found at temperature of 1125 °C due to the high temperature oxidation. The decreasing of sigma phase is supported by metallography, SEM, and phase calculation by using image-J.

### *3.2. Metallography Test*

Metallography testing initially was performed to analyze the effect of temperature and holding time of solution annealing to the existence of sigma phase in sample aging of Super Duplex SAF 2507. Observation result is shown in figure 4. The phase formed wereγ-austenite α-ferrite, and σ-sigma. This result is in accordance with a research of [7], which stated austenite phase formed like island, ferrite phase color in black and sigma phase is light colored area surrounding austenite.

Figure 5 shows microstructure of SAF 2507 after solution annealing and holding time of 15 minutes, at 1000 °C, 1065 °C, and 1125 °C. Figure 6 shows microstructure of SAF 2507 after solution annealing and holding time of 30 minutes, at 1000°C, 1065 °C, and 1125 °C. After heating to 1000°C and holding for 15 minutes, microstructures found on SAF 2507 were γ-austenite α-ferrite, and σ-sigma. Sigma phase was still found because it had not dissolved completely. According to ASTM A-480 [6], the minimum temperature of solution annealing was 1025°C. Figure 5b and 5c show the absence of sigma phase. This case is also in accordance to ternary diagram of Fe-Cr-Ni which shows two phase only, austenite and ferrite. Agreed with [3], the metallography result showing the solubility of sigma phase is accordance with growth of ferrite area. The reduction of ferrite area was caused by sigma phase consumes Cr and Mo in ferrite. Besides, as a ferrite former and stabilizer, Cr and Mo are also recognized as a sigma phase former. Due to this characteristic, sigma phase precipitate is in area containing high Cr and Mo.

### *3.3. Scanning Electron Microscopy Test*

Figure 7 is SEM result of an aging sample of SAF 2507. This figure shows austenite phase (island form), ferrite phase (dark area), and sigma phase (boundary region). Ferrite grain size is narrowed due to sigma precipitation which consumes Cr and Mo in ferrite.

Figure 8 is SEM image of sample of the solution annealing with holding time of 15 minutes at 1000°C, 1065°C and 1125°C. While Figure 9 is SEM image of sample of the solution annealing with holding time of 30 minutes at 1000°C, 1065°C and 1125°C. According to Figure 8a, holding time at 1000°C for 15 minutes still left sigma phase due to incomplete dissolution. At temperatures of 1065°C and 1125 °C (Figure 8b and 8c), the sigma phase already vanishds. Phase formed were austenite and ferrite, and carbide found as well on the sample. Number of carbide found at 1125 °C was higher compared to sample at 1065 °C. This is in accordance to research conducted by [8], which stated that chromium carbide precipitates at high temperatures in the ferrite zone because the ferrite forming elements such as chromium is able to diffuse more than austenite. Carbide was formed right after completion of holding time process, where there was a small time interval before quenching occurred.

Table 2 is the result of the calculation of sigma phase in the SEM sample by using image-J software. Sample with aging of SAF 2507 showed number of sigma phase of 34.364%. Sample with solution annealing at temperature of 1000 °C yields number of sigma phase of 26.937% and 26.077% for holding time of 15 minutes and 30 minutes respectively. Based on the calculation phase above, sigma phase tends to decrease and dissolve after solution annealing process. At temperature of 1000 °C, sigma phase decreased with an increasing of holding time.

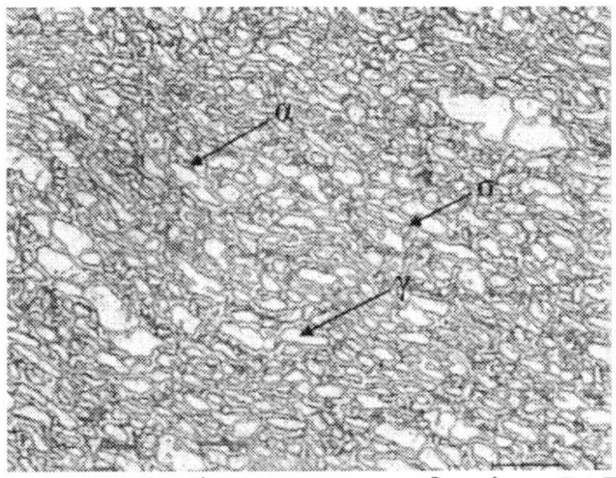

**Figure 4.** Microstructure of aging SAF 2507 with 500× magnification

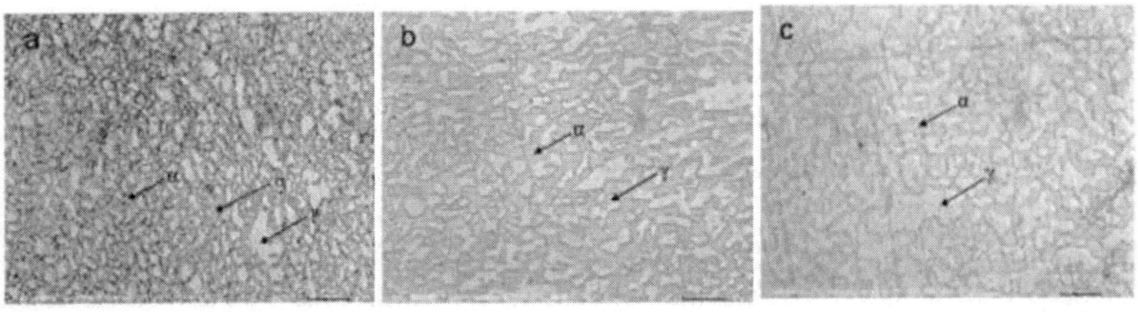


**Figure 5.** Microstructure of solution annealing SAF 2507, holding time 15 minutes at (a) 1000 °C (b) 106 5 °C and (c) 1125 °C

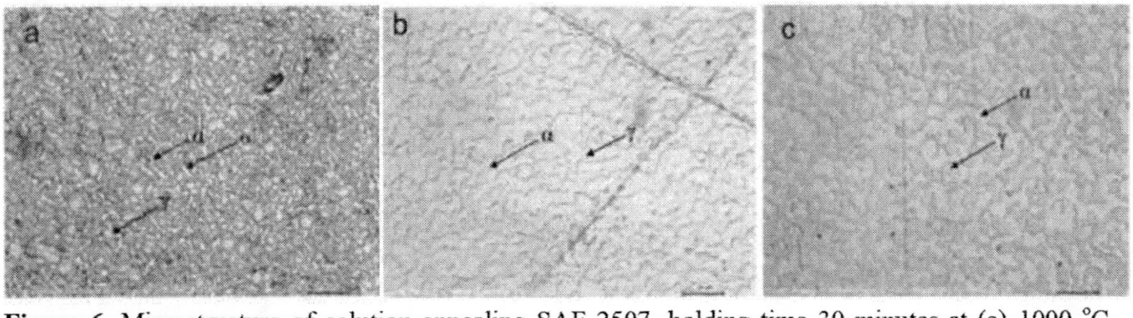


**Figure 6.** Microstructure of solution annealing SAF 2507, holding time 30 minutes at (a) 1000 °C (b) 1065 °C and (c)1125 °C

**Table 2.** Calculation of phase percentage

| Sample | Austenite (%) + Ferrite (%) | Sigma (%) |
|---|---|---|
| Aging | 65.636 | 34.364 |
| SA 1000 – 15 | 73.063 | 26.937 |
| SA 1000 – 30 | 73.923 | 26.077 |
| SA 1065 – 15 | - | - |
| SA 1065 – 30 | - | - |
| SA 1125 – 15 | 100 | - |
| SA 1125 – 30 | - | - |

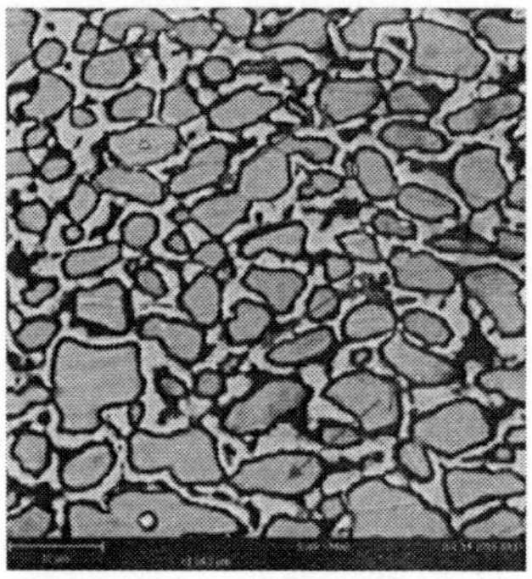

**Figure 7.** SEM of aging SAF 2507 with 5000× magnification

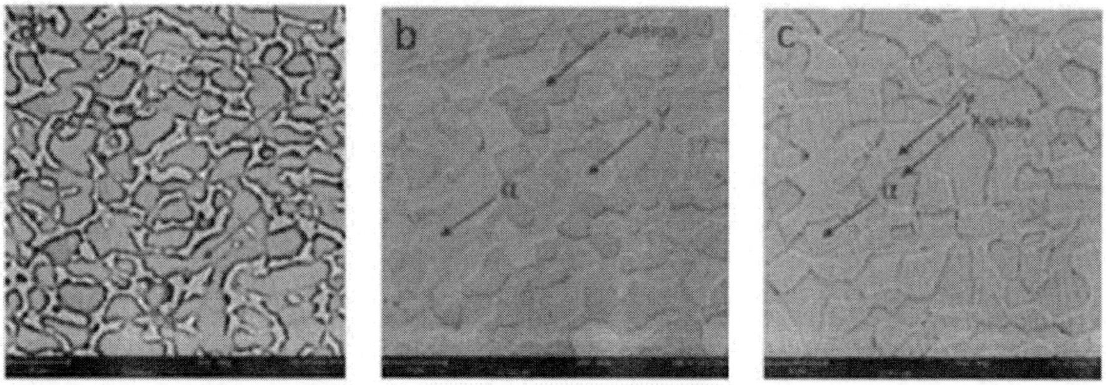

**Figure 8.** SEM with 5000× magnification of solution annealing sample with holding time of 15 minutes at (a) 1000 °C, (b) 1065 °C and (c) 1125 °C

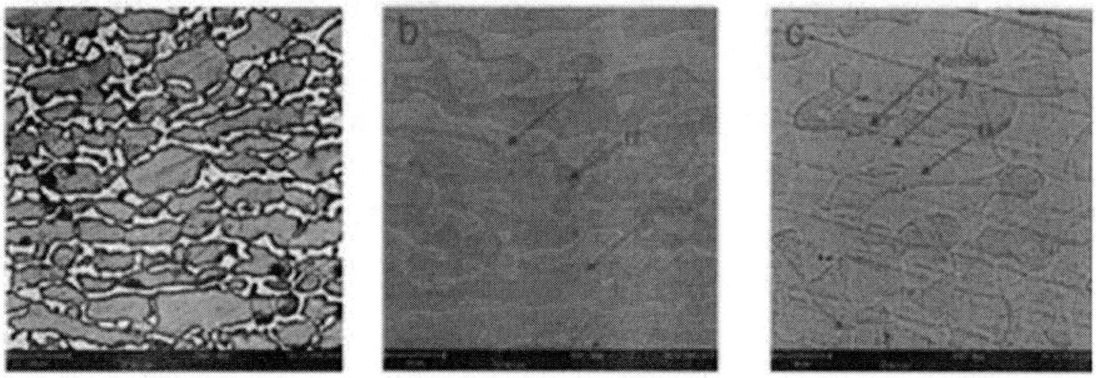

**Figure 9.** SEM with 5000× magnification of solution annealing sample with holding time of 30 minutes at (a) 1000 °C, (b) 1065 °C and (c) 1125 °C

## 4. Conclusion

The effects of solution annealed on precipitation of sigma phase were investigated. Aged SAF 2507 which contains sigma phase remained to exist after it annealed at temperature of 1000°C. While, sigma phase which existed on aged SAF 2507 vanished after solution annealing at temperature of 1065°C and 1125°C. The austenite and ferrite phases exist. Moreover, carbide also was discovered on entire samples.

## 5. References

[1] *Duplex Stainless Steel* 2013 *Wear* **303**

[2] Sandvik 2015 *SANDVIK SAF 2507 Tube and pipe seamless datasheet*

[3] *Hip fracture incidence map* International Osteoporosis Foundation (viewed 20 November 2013) http://www.iofbonehealth.org/facts-and-statistics/hipfracture-incidence-map

[4] Villalbos D, Albiter and Maldonado C 2009 *Microstructural change in SAF 2507 super duplex stainless steel produced by thermal cycle* Revista Matéria **14**(3) 1061-69

[5] ASTM A 923 2003 *Standard test method for detecting detrimental intermetallic phase in duplex austenitic/ferritic stainless steels*

[6] ASTM A-480 2003 *Standard specification for general requirements for flat-rolled stainless and heat-resisting steel sheet, and strip*

[7] Fargas G, Maestra A and Mateo A 2013 *Effect of sigma phase on the wear behavior of a super duplex stainless steel* Wear **303**(s 1–2) 584–590

[8] Kashiwar A 2012 *Effect of solution annealing temperature on precipitation in 2205 duplex stainless steel* Materials Characterization **74** 55–63

## Acknowledgements

The author would like to acknowledge *PENELITIAN PEMULA PNBP* ITS 2016 for funding.

# Study of Raman Spectra of Aluminum Powder-Substituted Barium Hexaferrite (BaM) $BaFe_{12-x}Al_xO_{19}$ as a Result of Solid State Reaction Process

**S Mustofa[1], R Rizaldy[2] and W A Adi[1]**

[1]Center for Science and Technology of Advanced Materials (PSTBM), National Nuclear Energy Agency of Indonesia (BATAN), Kawasan PUSPIPTEK Serpong, Tangerang Selatan, Banten, Indonesia 15314
[2]Dept. of Physics, Faculty of Science and Technology, UIN Syarief Hidayatullah Jakarta, Jl. Ir. H. Juanda No. 95 Ciputat, Tangerang Selatan, Banten, Indonesia 15412

Email: s_mustofa@batan.go.id

**Abstract**. Aluminum-substituted barium hexaferrite (BaM) that have the general formula of $BaFe_{12-x}Al_xO_{19}$ (x: 0, 1, 2, 3, 4, and 5) was obtained through the process of solid state reaction. The samples of Aluminum-substituted barium hexaferrite (BaM) were characterized using XRD to determine the phase structure, and by Raman Spectroscopy to know the chemical composition and structure. The single phase was obtained for the samples x = 0 - 4 from the weight percentage of the phase, and the other phase i.e. $Al_2O_3$ and BaO have been achieved for the sample x = 5. The characterization of Raman spectrum on Aluminum-substituted barium hexaferrite (BaM) has been carried out using Bruker SENTERRA instrument. The presence of sharp peaks corresponded to $A_{1g}$ mode shows the possibility of Al substitution in the barium hexaferrite system without any distortion in crystal symmetry. It is also observed that the substitution of $A_l$ in the $BaFe_{12}O_{19}$ system leads to an increase in the intensity of resonance band when compared with the parent compound, which indicates a large polarizability variation during the vibrations in the Aluminum-substituted $BaFe_{12-x}Al_xO_{19}$ compounds.

**Keywords** : Raman, Al, BaM, $BaFe_{12-x}Al_xO_{19}$, polarizability

## 1. Introduction

Since the discovery in 1930, hexagonal ferrite or hexaferrite is very interesting for developed.. Hexaferrite materials are materials that are crucial both commercially and technologically, which is marked by the global fabrication of magnetic material for many kinds of applications and uses of these materials. Barium hexaferrite is often applied as a permanent magnet for the applicationsof magnetic recording material (recorder), data storage (hard drive), and a component in electronic devices (sensor), as well as an electromagnetic wave absorbing material [1].

Hexagonal ferrites family are aunique materials that divided into six types, namely M-type ($BaFe_{12}O_{19}$ or $SrFe_{12}O_{19}$), Z-type ($Ba_3Me_2Fe_{24}O_{41}$), Y-type ($Ba_2Me_2Fe_{12}O_{22}$), W-type ($BaMe_2Fe_{16}O_{27}$), X-type ($Ba_2Me_2Fe_{28}O_{46}$), and U-type ($Ba_4Me_2Fe_{36}O_{60}$), where Me can be inserted with the elements of Sr, Co, Ni, or Zn. All hexaferrite families have ferromagnetic properties. Principally, the magnetic properties of hexaferrite family are associated with the crystal structure because hexaferrite has Magnetocrystalline Anisotropy (MCA), which means the magnetization has a tendency towards the orientation in the crystal structure [2]. One of the ferrite families that attract interest from many researchers in the last few decades is the family of M-type hexaferrite ($BaFe_{12}O_{19}$), which has the empirical equation of BaO.6 ($Fe_2O_3$). This compound is very cheap and easy to be produced because of its more simple chemical formula than those of other materials in hexaferrite family. The bafe12o$_{19}$ compound has a high electrical resistivity of about 108 Ω cm, a high uniaxial magnetic anisotropy along the c axis, and a high magnetic saturation value that can potentially be applied as a permanent magnet [3-4]. In crystallography, $BaFe_{12}O_{19}$ has a hexagonal crystal structure, with a space group of P 63/mmc (194) and lattice parameters of $a = b = 5.9291$ Å and $c = 23.4070$ Å [5].

In the progress of research developments, the $BaFe_{12}O_{19}$ compound is often modified in such a way for further purposes and different applications, such as for absorption of electromagnetic waves in a high frequency (GHz scale), as the radar absorption material, whether in the field of health or stealth technology [6]. In such applications, Barium Hexaferrite is often substituted with other ions. Several studies have been focused on the ion substitution by $Fe^{3+}$, which is the only ion that has a magnetic moment, by recombination with other ions that can give an effect to the magnetic properties of the material. Such material is required to have permeability and permittivity with ratios appropriate [7] for the applications as an absorption of the electromagnetic wave. Barium Hexaferrite has a quite high level of permeability and permittivity. Engineering material is needed by substituting the $Fe^{3+}$magnetic ion with a non-magnetic ion, such as $Al^{3+}$ [8-9] to improve the absorption ability.

In this research, we developed the $Al^{3+}$ ion-substituted barium hexaferrite to obtain the formula of barium hexaferrite that has high permeability and permittivity, and also good absorption ability for the application as an absorbing material for an electromagnetic wave in order to prepare raw materials of anti-radar paint. In accordance with this entire research, this report was based on the analysis of the X-ray diffraction pattern and the crystal structure of the material using the Rietveld method to obtain the accurate crystal structure from the sample, by replicating the crystal structure that already existed in the database so that it complied with the diffraction pattern observation using least squares contained in GSAS-EXPGUI software. Furthermore, an analysis using Raman Spectroscopy to know the chemical composition and structure was also conducted.

### *1.1. Material of Barium Hexaferrite*

The M-type of Hexagonal Ferrite, with the general formula of $BaO.6Fe_2O_3$, has a lattice constant of $a$ = 5.89 Å and $c$ = 23.19 Å in a hexagonal closed packed of crystal system, which is composed of Oxygen and Barium with Fe in octahedral (12k, 4f2 and 2a), tetrahedral (4f1), and bipyramidal trigonal / hexagonal (2b). Fe cation is a single source of the magnetic moment in this system, which have an upward spin direction in the positions of 12k, 2a, and 2b, and have a downward spin direction [10] in the position of 4f1 and 4f2, as shown in Figure 1. The general structure of the M-type Hexaferrite ($AO.6Fe_2O_3$ or $AFe_{12}O_{19}$, where A is the divalent ions, such as $Ba^{2+}$, $Sr^{2+}$, $Pb^{2+}$, and so on) was hexagonal with a space group of P 63/mmc, consisting of 4 blocks with the symbols of S, S*, R, and R* as shown in Figure 1. Oxygen atoms stacked covered with Ba and Fe atoms in an interstitial place. There are 10 layers of oxygen atoms along with the c axis and Fe atoms located at five different places in crystallography [10].

Among the twelve ions of $Fe^{3+}$ in the formula unit, Fe atom in the position of 4f1 is coordinated tetrahedrally with oxygen, while the Fe atom in the position of 2b was coordinated with five oxygen ions. There is a short distance of Fe-Fe in the structure, and the distance of Fe-Fe Å at the position of 4f2 was ±2.7. Fe ions in the position of 12k of a network with each Fe connected with four ions of another Fe in the same layer [10].

In the form of spin, one ion in the 2b layer on the R block was pointing upwards, and two ions of octahedral led down, and seven octahedral ions on the block S led down. Because each of $Fe^{3+}$ ion contributed $5\mu_B$ at absolute zero of magnetic moment, the total magnetization at zero temperature can be calculated. As known, eight $Fe^{3+}$ ions have an upward spin, and four ions have a downward spin, which resulted in four $Fe^{3+}$ ions with an upward spin. Therefore, the total magnetization per unit molecule is $(1 - 2 + 7 - 2)\ x5\mu_B = 20\mu_B$ [10].

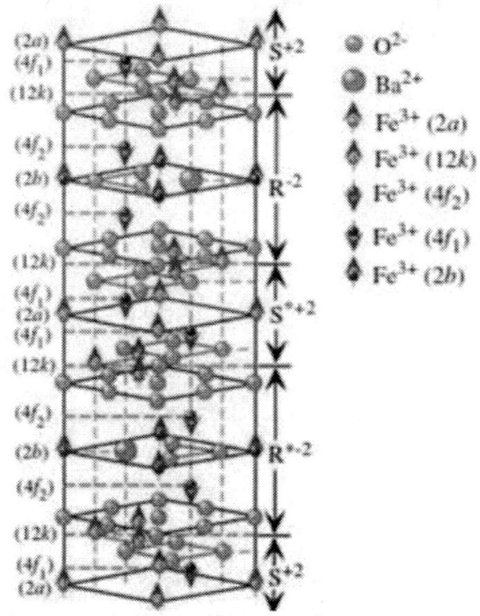


**Figure 1.** Schematic crystal structure of Barium Hexaferrite [11]

### 1.2. *Aluminum Substituted Barium Hexaferrite*

The combination of the intrinsic magnetic properties and electrical properties of such Ferrite are placing a magnetic material of Ferrite as a buffer of electromagnetic waves, especially in micro scale, including the waves with frequencies used in Radar [12]. Nevertheless, the modification by ion substitution is often carried out to increase the non-magnetic permittivity properties so that it can add higher frequency of the power absorption. For example as reported by *Priyono & Azwar* by substituting $Ti^{4+}$ and $Mn^{2+}$ with $Fe^{3+}$ inside the material of Barium Hexaferrite [12]. Evidently, the power absorption of electromagnetic waves could be increased higher than before the substitution. In addition to the substitution of $Fe^{3+}$ ion, substituting $Sr^{2+}$ ion with $Ba^{2+}$ ion was also reported and resulting in a more maximum absorption power [7].

The studies on the substitution of various cations with the crystal structure of Barium Hexaferrite have indicated that it can change the magnetic properties. The focus of this research was the substitution of $Al^{3+}$ ion. According to Dianne & Yokoviev et al. (1969), $Al^{3+}$ ion could substitute $Fe^{3+}$ ion in the system of $BaFe_{12}O_{19}$, depending on the number and positions occupied by $Al^{3+}$ ion, and therefore it could change the magnetic properties [13]. It is well known that $Fe^{3+}$ ion that has a magnetic moment of $5\mu_B$ will be replaced by $Al^{3+}$ ion that has a zero magnetic moment. The magnetic moment of $Al^{3+}$ ion ($0\mu_B$) cannot be removed if it has downward moment direction of spin from the $Fe^{3+}$ ($5\mu_B$). Therefore, the substitution of $Al^{3+}$ ion lead to the weakening superexchange interactions from $Fe_A^{3+} - O - Fe_B^{3+}$, which led to the collapse of magnet collinearity in the crystal lattice structure of Barium Hexaferrite [14].

## 2. Experimental Method

### 2.1. *Rietveld refinement method*

The basis for multiphase profile analysis of powder diffraction patterns was firstly introduced completely by Rietveld in 1969. Rietveld showed the possibility of replicating the results of diffraction pattern measurements with pattern calculation. The advantage of such pattern calculation is when there is an occurrence of an error that is caused by an intensity deviation in an imperfect model structure, it will tend to leave the remaining intensity, both negative and positive.

The basic principle of Rietveld analysis is to match (fitting) the peak profile of the calculation with the peak profile of the observation. The profile matching (fitting) is carried out by applying the procedure of a nonlinear least square calculation. Expressed as a minimum objective function as follows:

$$S_{Min} = \sum_{i=1}^{N} w_i [y_i(obs) - y_i(cacl)]^2 \quad (1)$$

$w_i$ is weighting factor, i.e. $1/[y_i(obs)]$ that equal FWHM, $y_i(obs)$ is the intensity obtained from observation, and $y_i(calc)$ is the intensity obtained from calculations [15].

The intensity value of profile diffraction pattern at certain positions of Bragg angle can be calculated after correction by entering a function of scale, profile, and background, that is:

$$y_i(calc) = \sum_h s.|F_{hkl}|_h^2 . p_h . L(\theta). M_h . pV(\Delta T)_h + y_{ib}(calc) \quad (2)$$

Where $y_i(calc)$ is the intensity taken from calculation, $s$ is scale factor, $F$ is structure factor, $p$ is the multiplicity factor, $L(\theta)$ is Lorentz-polarization factor, M is temperature factor, $pV(\Delta T)$ is geometry factor (function profile: Pseudo-Voigt), $y_{ib}(calc)$ is function of background [16].

### *2.2. Degrees of Conformity*

To determine the success of a smoothing, it needs a standard value which indicates the quality of the smoothing. In GSAS use least square smoothing that is indicated by some residual function, which is set in a particular way in each histogram.

### *2.3. Software GSAS-EXPGUI*

GSAS stands for General Structure Analysis System created by R.B. Von Dreele and A.C. Larson. This program is based on the Rietveld smoothing method. GSAS is a set of programs for processing and analyzing single crystal diffraction data and powder obtained by the characterization of x-ray diffraction. GSAS can handle data of powder diffraction pattern from phase mixture by softening the structure parameters of each phase. The collected data that are ready to be input into the EXPGUI GSAS will be processed through qualitative and quantitative analysis by Rietveld smoothing method.

### *2.4. Preparation of Powder Materials of $BaAl_xFe_{12-x}O_{19}$ (x: 0 – 5)*

The powder materials of $BaAl_xFe_{12-x}O_{19}$ (x: 0 - 5) used for characterization by XRD (SHIMADZU 700 XRD) and Raman Spectroscopy have been prepared through the process of synthesis using basic powder materials of $BaCO_3$, $Fe_2O_3$, and $Al_2O_3$. The synthesis process was performed by solid state reaction method using HEM (High Energy Milling) equipment for 7 hours, which consist of 5-hour milling and 2-hour break, and then the powder materials were burnt for 5 hours at a temperature of 1000 °C.

### *2.5. Characterization of Raman Spectra*

The samples were characterized using Raman Spectrometer (Bruker SENTERRA) for microstructural analysis. The Raman Spectroscopy system employed in this study consisted of 1 mW – 100 mW lasers with 785 nm wavelength, single grating 0.5 m spectrometer single grating, charged – coupled device (CCD) detector, and holographic optics. A spectrum range of 400 to 1800 $cm^{-1}$ at room temperature was examined in this study using a grating of 1800 grooves/mm. A 60-minute exposure time was used for recording the Raman spectrum.

## 3. Results and Discussion

### *3.1. XRD Pattern*

Figure 2 shows the XRD pattern of the samples, which is a modification of the Barium Hexaferrite system by the addition of Al element into Fe with variations of x = 0, 1, 2, 3, 4 and 5. The XRD results of Barium Hexaferrite samples showed that the suspected different phase came after the addition of Al into Fe. A further phase identification is required concerning the COD. Phase identification of raw data was carried out by using software of XRD Match! 2 that refer to COD based on the information of peak position and intensity as shown in Figure 2.

From the matching peak using software for further identification, the phase of $BaFe_{12}O_{19}$, $BaAl_{2.18}Fe_{9.82}O_{19}$, $Al_2O_3$, and BaO were obtained, which respectively refer to the research findings of Obradors et al. [5] at a ground state level (x = 0), Sandiumenge [8] for x = 1 - 5, and Wang et al. [17] and Wyckoff et al. [18] for the additional phase at x = 5.

After the information phases from peak matching using software have been obtained, then a further analysis was carried out to determine the effect of Al substitution on the crystal structure of Barium Hexaferrite. To analyze the pattern of XRD and to find out more information about crystallography from the samples, the Rietveld smoothing method with the help of software GSAS-EXPGUI was needed. Samples x = 0 is the ground state of the system Barium Hexaferrite, without any substitution of $Al^{3+}$ ion. Rietveld refining results are illustrated in the form of the red plot for data calculation and green for observation data, along with the blue curve after subtraction of the background function, as shown in Figure 3.

The results of smoothing have a matching degree that is convergent. Therefore it is believed that there was a single phase of Barium Hexaferrite at x = 0. The crystallographic information and the level of compatibility are shown in Table 1.

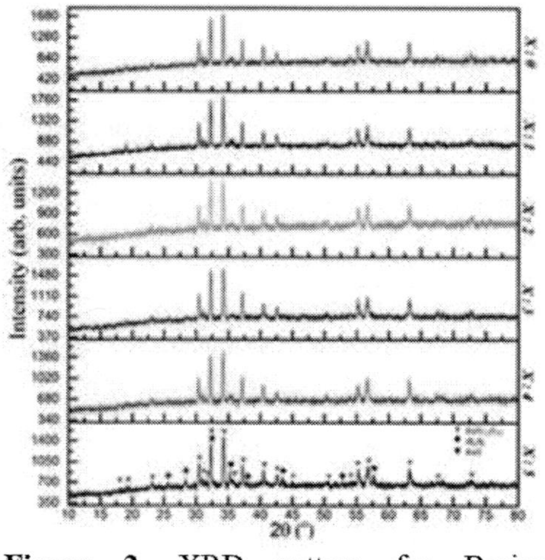


**Figure 2.** XRD pattern for Barium Hexaferrite $BaFe_{12-x}Al_xO_{19}$ with $Al^{3+}$ ion substitution (x = 0 - 5)

IOP Conf. Series: Materials Science and Engineering **202** (2017) 012040 doi:10.1088/1757-899X/202/1/012040

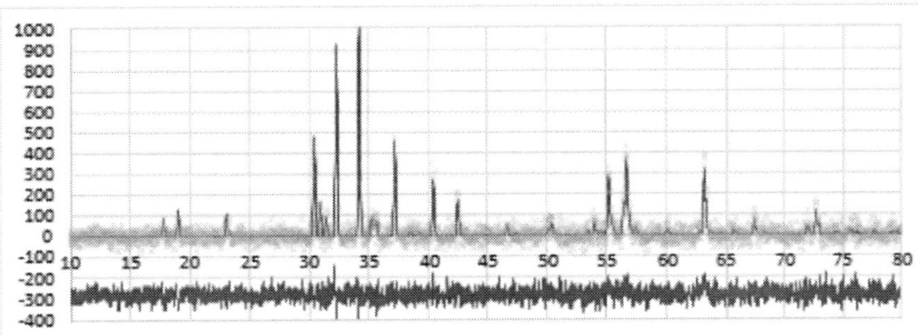


**Figure 3.** The smoothing results of XRD pattern of $BaAl_xFe_{12-x}O_{19}$ (x = 0)

**Table 1.** Parameter structure, factor R and $x^2$ of $BaAl_xFe_{12-x}O_{19}$ (x = 0)

| | | Lattice Parameters | | |
|---|---|---|---|---|
| Space Group : | P 63/m m c | | | |
| Crystal system : | Hexagonal | $a$ = 5.886965 Å | $b$ = 5.886965 Å | $c$ = 23.188545 Å |
| Volume : | 695.964 Å$^3$ | | | |
| Article I. Density (ρ) : | Article II. 5.304 g.cm$^{-3}$ | $\alpha$ = 90° | $\beta$ = 90° | $\gamma$ = 120° |
| Factor R | | $w$Rp = 3.86 | Rp = 3.06 | $x^2$ = 1.121 |

**Table 2.** Parameter structure, factor R and $x^2$ of $BaAl_xFe_{12-x}O_{19}$ (x = 1)

| | | Lattice Parameters | | |
|---|---|---|---|---|
| Space Group : | P 63/m m c | | | |
| Crystal system : | Hexagonal | $a$ = 5.88659 Å | $b$ = 5.88659 Å | $c$ = 23.18459 Å |
| Volume : | 695.757 Å$^3$ | | | |
| Density (ρ) : | 5.136 g.cm$^{-3}$ | $\alpha$ = 90° | $\beta$ = 90° | $\gamma$ = 120° |
| Factor R | | $w$Rp = 3.72 | Rp = 2.96 | $x^2$ = 1.103 |

**Table 3.** Parameter structure, factor R and $x^2$ of $BaAl_xFe_{12-x}O_{19}$ (x = 2)

| | | Lattice Parameters | | |
|---|---|---|---|---|
| Space Group : | P 63/m m c | | | |
| Crystal system : | Hexagonal | $a$ = 5.88303 Å | $b$ = 5.88303 Å | $c$ = 23.17027 Å |
| Volume : | 694.448 Å$^3$ | | | |
| Density (ρ) : | 5.016 g.cm$^{-3}$ | $\alpha$ = 90° | $\beta$ = 90° | $\gamma$ = 120° |
| Factor R | | $w$Rp = 4.00 | Rp = 3.15 | $x^2$ = 1.133 |

**Table 4.** Parameter structure, factor R and $x^2$ of $BaAl_xFe_{12-x}O_{19}$ (x = 3)

| | | Lattice Parameters | | |
|---|---|---|---|---|
| Space Group : | P 63/m m c | | | |
| Crystal system : | Hexagonal | $a$ = 5.882072 Å | $b$ = 5.882072 Å | $c$ = 23.16498 Å |
| Volume : | 694.102 Å$^3$ | | | |
| Density (ρ) : | 4.898 g.cm$^{-3}$ | $\alpha$ = 90° | $\beta$ = 90° | $\gamma$ = 120° |
| **Factor R** | | $w$Rp = 4.00 | Rp = 3.21 | $x^2$ = 1.154 |

**Table 5.** Parameter structure, factor R and $x^2$ of $BaAl_xFe_{12-x}O_{19}$ (x = 4)

| | | Lattice Parameters | | |
|---|---|---|---|---|
| Space Group : | P 63/m m c | | | |
| Crystal system : | Hexagonal | $a$ = 5.88111 Å | $b$ = 5.88111 Å | $c$ = 23.16509 Å |
| Volume : | 693.879 Å$^3$ | | | |
| Density (ρ) : | 4.787 g.cm$^{-3}$ | $\alpha$ = 90° | $\beta$ = 90° | $\gamma$ = 120° |
| Factor R | | $w$Rp = 4.12 | Rp = 3.27 | $x^2$ = 1.119 |

**Table 6.** Parameter structure, factor R and $x^2$ of $BaAl_xFe_{12-x}O_{19}$ (x = 5)

| | | Lattice Parameters | | |
|---|---|---|---|---|
| Space Group : | P 63/m m c | | | |
| Crystal system : | Hexagonal | $a$ = 5.87847 Å | $b$ = 5.87847 Å | $c$ = 23.15467 Å |
| Volume : | 692.945 Å$^3$ | | | |
| Density (ρ) : | 4.745 g.cm$^{-3}$ | $\alpha$ = 90° | $\beta$ = 90° | $\gamma$ = 120° |
| Factor R | | $wRp$ = 4.25 | Rp = 3.39 | $x^2$ = 1.177 |

As for the case of x = 1 to 4 (see Table 2 – Table 6), no significant changes occurred. It indicated that after the addition of $Al^{3+}$ ion does not form another phase and overall well-substituted with barium hexaferrite into the system. The lattice parameters decreased compared with the sample x = 0. The larger the value of x, the greater the decline in the value of the lattice parameter, and the value of $Al^{3+}$ ion population factor (Factor R) was also gradually increasing. Many ion $Al^{3+}$ substituted for $Fe^{3+}$ in position 2a, followed on 12k, 2b, 4f1, and 4f2. While the sample x = 5, very different results than x = 0 - 4 is obtained, and also other phases of $BaAl_{2.18}Fe_{9.82}O_{19}$, namely $Al_2O_3$ and BaO were gained. Also, the sample x = 5 is also not a single phase. The lattice parameter in the sample x = 5 does not decline significantly as x = 0 - 4, the increase of the value of $Al^{3+}$ ion population factor is not significant anymore, but it is not much different from the sample x = 4.

### *3.2. Raman Spectra*

All peaks in the spectra are single phases of $BaFe_{12}O_{19}$, with no presence of any additional phase or an impurity. Since the laser power is limited, 10mW, no transformation of the compound into other phases or degradation was observed.

The peaks at 713 and 684 cm$^{-1}$ can be assigned to $A_{1g}$ vibrations of Fe-O bonds at the tetrahedral 4f1 and bipyramidal 2b sites, respectively. Other peaks at 614, 512, 467, 453 and 317 cm$^{-1}$ are due to $A_{1g}$ vibrations of Fe-O bonds at the octahedral 4f2, 2a, and 12k sites, whereas peak at 409 cm-1 is due to $A_{1g}$ vibration at the octahedral 12k dominated site.

The peaks at 527, 285 and 212 cm$^{-1}$ were due to $E_{1g}$ vibrations, while the peak at 335 cm$^{-1}$ was due to $E_{2g}$ vibration. The peak at 184 and 173 cm$^{-1}$ resulted from $E_{1g}$ vibrations of the whole spinel block. The assignment of Raman spectra to the corresponding lattice symmetries are summarized in Table of Figure 4.

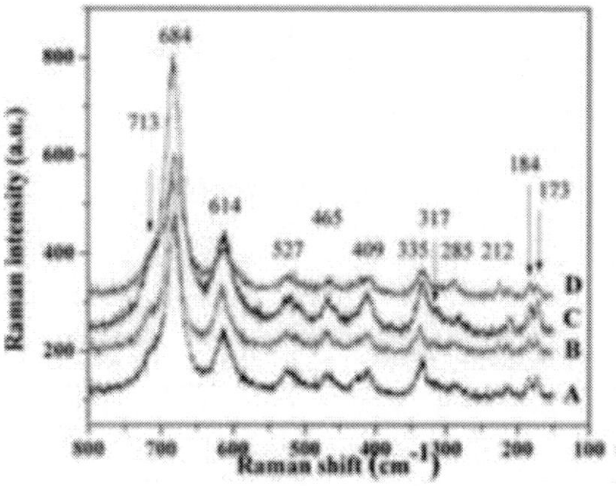


| Raman active modes | Observed frequency (± 3 cm$^{-1}$) | Assigned polyhedra |
|---|---|---|
| $A_{1g}$ | 713 | Tetrahedra ($4f_1$) |
| | 684 | Bipyramid ($2b$) |
| | 614 | Octahedra ($4f_2$) |
| | 512, 467, 453 | Octahedra ($2a$ + $12k$) |
| | 409 | Octahedra ($12k$) dominated |
| | 317 | Octahedra $2a$ |
| $E_{1g}$ | 527, 285, 212 | --- |
| | 184, 173 | Whole spinel block |
| $E_{2g}$ | 335 | Octahedra $12k$ |

**Figure 4**. Raman spectra for Barium Hexaferrite $BaFe_{12-x}Al_xO_{19}$ with $Al^{3+}$ ion substitution.

## 4. Conclusion

Substitution of $Al^{3+}$ ion by Barium Hexaferrite gave an impact on the changes in crystal structure. The factor of $Fe^{3+}$ ion population is slowly taken over by $Al^{3+}$ ion in each position. The distances between the atoms became increasingly closer. Lattice parameter values decreased, resulting in decreasing the volume of the unit cell and atomic density. The ability of $Al^{3+}$ ion substitution by an $Fe^{3+}$ ion in this

Barium Hexaferrite system only up to the limit x until 4, which form the structure of $BaFe_8Al_4O_{19}$. The addition of $Al^{3+}$ ion for x > 4 into the system of Barium Hexaferrite will form three phases, $BaFe_8Al_4O_{19}$, BaO and $Al_2O_3$, and the weight percent of BaO and $Al_2O_3$ phase will continue to increase. From the result of Raman Spectrum, the substitution of Al in the $BaFe_{12}O_{19}$ system leads to an increase in the intensity of resonance band when compared with the parent compound that indicates a significant variation of polarizability during the vibrations in the aluminum-substituted $BaFe_{12-x}Al_xO_{19}$ compounds.

## 5. References

[1] Pullar R C 2012 Hexagonal ferrites: A review of the synthesis, properties and applications of hexaferrite ceramics *Progress in Material Science* **57** 1191-1334.
[2] Arkel A E V *et al* 1936 *Rev. Trv. Chim.* **55** 331.
[3] Smit J *et al* 1959 Ferrites *Philips Technical Library* Eindhoven
[4] Nowosielski *et al* 2007 Barium ferrite powders prepared by milling and annealing *Journal of Achievements in Materials and Manufacturing Engineering* **28 (12)** 735-742.
[5] Obradors X *et al* 1985 X-ray analysis of the structural and dynamic properties of $BaFe_{12}O_{19}$ hexagonal ferrite at room temperature *Journal of Solid State Chemistry* **56** 171-181.
[6] Matsumoto M and Miyata Y 1996 A Gigahertz-range electromagnetic wave absorber with bandwidth made of hexagonal ferrite *J. Appl. Phys.* **79 (8)** 5486-5488.
[7] Adi W A and Manaf A 2012. Structural and absorption characteristics of Mn-Ti substituted Ba-Sr hexaferrite synthesized by mechanical alloying route *J. Basic. Appl. Sci. Res.* **2 (8)** 7826-7834.
[8] Sandiumenge F *et al* 1988. X-ray profile analysis of cation distribution in $SrAl_xFe_{12-x}O_{19}$ solid solution *Materials Research Bulletin* **23** 685-692.
[9] Shirtcliffe N J *et al* 2007 *Mater Res Bull* **42** 281.
[10] Umit Ozgur et al 2009 Microwave ferrites, part 1: fundamental properties J Mater. Sci: Mater. Electron. **20** 789 – 834
[11] Matsumoto M, Miyata Y 1996 A Gigahertz-range Electromagnetic Wave Absorber with Bandwidth Made of Hexagonal Ferrite. J. Appl. Phys., 79 (8): 5486-5488.
[12] Priyono and Azwar Manaf 2010 Magnetic material of M-type barium hexaferrite for material anti radar in the range of s-band frequency *Indonesian Journal of Materials Science* **11** 75 – 78.
[13] D Mishra *et al* 2003 X-ray diffraction studies on aluminium-substituted barium hexaferrite *Materials Letters* **58** 1147 – 1153.
[14] Vinod N *et al* 2011 Structural and magnetic behaviour of aluminium doped barium hexaferrite nanoparticles synthesized by solution combustion technique *Physica B* **406** 789 – 792.
[15] E J Mittemeijer and U Welzel 2013 Modern diffraction methods. *Wiley-VCH Germany* **xxvi** 519.
[16] Allen C Larson and Robert B Von Dreele 2004 GSAS : General structure analysis system *Los Alamos National Laboratory Report LAUR. California* 231.
[17] Wang X L *et al* 1994 Neutron diffraction measurements of the residual stresses in $Al_2O_3$ - $ZrO_2$ (CeO) ceramic composites *Journal of the American Ceramic Society* **77 (6)** 1569-1575.
[18] Wyckoff R W G 1963 Rocksalt structure. Crystal Structures *Second edition. Interscience Publishers, New York* **1** 85-237.

**Acknowledgments**
This work is supported in part by the DIPA of BATAN from Ministry of Finance Year 2015.

# Analysis of Distribution of Polyvinyl Alcohol Hydrogel Nanocrystalline by using SAXS Synchrotron

**Sunaryono[1,2], A Taufiq[1,2], N Mufti[1,2], N Hidayat[1,2], S Rugmai[3], S Soontaranon[3], E G R Putra[4,5] and Darminto[6]**

[1]Department of Physics, Faculty of Mathematics and Natural Sciences, State University of Malang, Jalan Semarang No 5, Malang 65145, Indonesia
[2]Center for Minerals and Advanced Materials, Faculty of Mathematics and Natural Sciences, Universitas Negeri Malang (State University of Malang) Jl. Semarang 5 Malang 65145, Indonesia
[3]Synchrotron Light Research Institute, 111 University Venue, Muang District, Nakhon Ratchasima 30000, Thailand
[4]Center for Science and Technology of Advanced Materials, National Nuclear Energy Agency of Indonesia (BATAN), Kawasan Puspiptek Serpong, Tangerang15134, Indonesia
[5]Sekolah Tinggi Teknologi Nuklir, National Nuclear Energy Agency of Indonesia, Yogyakarta 55281, Indonesia
[6]Department of Physics, Faculty of Mathematics and Natural Sciences, Sepuluh November Institute of Technology (ITS), Kampus ITS Sukolilo, Surabaya 60111, Indonesia

Email: sunaryono.fmipa@um.ac.id

**Abstract:** Polyvinyl alcohol (PVA) hydrogel has been successfully synthesized through freezing-thawing (F-T) process by using time-variation. This work is aimed to investigate the distribution of nanocrystalline from the hydrogel. Fourier Transform Infrared (FTIR) Spectroscopy, Differential Thermal Analysis/Thermogravimetric (DTA/TG), and Synchrotron Small-Angle X-ray Scattering (SAXS) were used as the instruments in characterizing the PVA hydrogel, respectively to observe the frequency of absorption, thermal degradation, and structural dimensions. The functional groups which represent the PVA polymer chains were verified on the wavenumber of 1450-1480 $cm^{-1}$ and 850-870 $cm^{-1}$ which is in accordance with the stretching of $-CH_2$ vibration mode. The absorption band of PVA polymer chains was also found on the wavenumber of 1090-1150 $cm^{-1}$ which is in accordance with the stretching of carboxyl vibration mode (CO), and this wavenumber gave a contribution towards the crystallinity of PVA polymer. Furthermore, the PVA polymer only interacted with the distilled water in the sample of PVA hydrogel without experiencing any chemical interactions between the PVA polymer and other substances. Meanwhile, the graphic of PVA hydrogel thermal degradation shows three thermal decompositions which are indicated by three areas in which there was sample weight loss. The second decomposition with sample weight loss was equivalent to 61.62%-73.04% occurred at the temperature of 282-376 $^{o}C$ which became the highest sample weight loss due to polymer chain degradation. Teubner-Strey and Beaucage models were used to analyze the characterization of structural dimension and distribution of PVA Hydrogel nanocrystalline with SAXS Synchrotron. With a high compatibility between the model data and the experiment, the average structural dimension of PVA hydrogel nanocrystalline is the equivalent of 3.96 nm, with an inter-crystalline average distance of 16.9 nm. These results indicate that PVA hydrogel is very potential to be applied as a primary material for human implants.

**Keywords:** distribution, nanocrystalline, structural dimension, PVA hydrogel, SAXS

## 1. Introduction

Polyvinyl Alcohol (PVA) Hydrogel is a flexible material consists of polymer chain network structure. The network structure can be made of a physical cross-binding agent which hereinafter shall be referred to physical gel and a chemical binding agent which hereinafter shall be referred to as chemical gel. The physical gel is formed by an interaction between ion, water and hydrophobic, hydrogen bond, and Van der Walls force, while the chemical gel is formed by covalent bond [1]. On the other hand, PVA hydrogel can be formed by using various methods, such as $\gamma$-irradiation, cross-binding process, and freezing-thawing cycle.

The freezing-thawing method is one of the most used methods by researchers in fabricating PVA hydrogel. This method was chosen since it has a simpler process and it is easy to be conducted. The potential of PVA hydrogel to be utilized and developed in medical, industrial and nanotechnology fields has been reported for many times. In medical field, PVA hydrogel can be utilized and developed, among others, for implants of human body network system [2], artificial cartilage [3-5], contact lenses [6-8], heart valves [9-10], and cornea implants for rabbits [11]. However, PVA hydrogel has some disadvantages for its application in the medical fields, such as weak hydrogel network, weak hydrogel endurance in high temperature, high level of hydrogel porosity and the incompatibility of hydrogel network with human body network.

PVA hydrogel quality improvement is necessary to overcome such problems. One effort which can be done is fixing the fabrication process and PVA hydrogel network structure. The PVA hydrogel network structure can be observed by identifying the structural and dynamic nature of PVA polymer and PVA hydrogel by using a sophisticated tool such as Small-Angle X-ray Scattering (SAXS). Studies on PVA hydrogel network structure have been reported by the authors and several previous researchers. The authors have successfully characterized the PVA hydrogel structure for various numbers of freezing-thawing cycles. The results show that the dimension of nanocrystalline average diameter is 4 nm with an average inter-crystalline distance of 18 nm [12]. These results are similar to the results reported by previous researchers, such as the fact the isotropic PVA hydrogel structure consists of 3 nm structural dimensions and is distributed with approximately 19 nm distance [13]. Furthermore, PVA hydrogel consists of polymer crystal with the radius of approximately 2.8-3.3 nm and with the average inter-crystal distance of 15-17.5 nm [14]. However, the period of freezing-thawing cycles for fabricating the PVA hydrogel and the observation results on PVA hydrogel structure have not frequently been reported.

Therefore, this research focused on the fabrication and characterization of PVA hydrogel structure with various periods of freezing-thawing cycles. The structural dimension and nanocrystalline distribution inside the PVA hydrogel were characterized by using SAXS instrument and global fitting analysis technique. The fabrication method and data analysis technique used in this research is one alternative to improve the quality and characteristic technique of PVA hydrogel structure. The frequency of absorption and thermal degradation of PVA hydrogel were also reported.

## 2. Experimental Method

### *2. 1. Material*

PVA and distilled water were used as the materials. PVA polymer used has a hydrolysis degree of $\geq$ 98%, and the weight of molecules is of approximately 60,000 g/molecule, purchased from Merck Schuchardt OHG, Germany. The distilled water is utilized as a cross-binding agent on the synthetic process of the hydrogel.

IOP Conf. Series: Materials Science and Engineering **202** (2017) 012041 doi:10.1088/1757-899X/202/1/012041

*2. 2. PVA Hydrogel Fabrication*

PVA hydrogel fabrication was initiated by dissolving the PVA polymer and distilled water with a certain composition of weight. The mixture was stirred by using a magnetic stirrer and heated on a hot plate magnetic stirrer with the temperature of 70-90 °C in order to enhance the level of solubility of the PVA polymer in the water; this step is similar to the previous research [15]. The mixture transformation from a liquid into paste indicated that the PVA polymer solution had been perfectly mixed with water. Furthermore, the mixture was poured into the mold, and the process of freezing-thawing was done by using 3 cycle process. The freezing process was performed with the temperature of – 10 °C, while the thawing process was performed with the temperature of 27 °C. The variation of time periods on the freezing process of 1, 2, 3, 4, 5, and 6 hours are the main objects of research to determine the distribution of PVA hydrogel nanocrystalline.

*2. 3. Characterization*

*2.3. 1. Spectroscopy of Fourier Transform Infrared (FTIR).* The FTIR spectroscopy was utilized to determine the chemical bond, chemical function, and organic compound structure consisted in PVA hydrogel. The FTIR spectrum had been recorded in the frequency range of 500-4000 $cm^{-1}$ by using Shimadzu spectrometer.

*2.3. 2. Thermogravimetric Analysis (TGA).* TGA was used to characterize the thermal degradation of PVA hydrogel sample. TA TGA Q500 Analyzer was used to measure the thermal degradation. The sample measuring operation was performed by dynamic mode in the nitrogen with the heating rate of 10 °C/minute on the temperature of 25 until 600 °C. Similar with the XRD characterization, the sample of PVA hydrogel was drained under vacuum on the temperature of 37 °C in 48 hours and then characterized by using TGA.

*2.3. 3. Small Angle X-ray Scattering (SAXS).* SAXS Instrument was used to analyze the distribution of PVA hydrogel nanocrystalline. This instrument is a sophisticated tool which is capable of providing information on the size, morphology, inhomogeneity and interaction among the Nanocrystalline accurately. This is made possible since the SAXS instrument is a tool which will not damage the sample and this instrument has a characteristic of being capable to properly transmit the nanocrystalline light from the sample.

The SAXS measurement was performed in Siam Photon Laboratory of Synchrotron Light Research Institute (SLRI), Thailand. This measurement was made in order to observe the structural dimension and distribution of PVA hydrogel nanocrystalline. The distribution of PVA hydrogel Nanocrystalline was observed by using CCD detector with the sample-to-detector-distance (SDD) of 1.8 m and 4.5 m with the range of $q$ momentum scattering of approximately 0.12 until 2 $nm^{-1}$. The SAXS initial data processing was initiated by the processes of normalization, calibration, and verification of the sample scattering intensity towards the background of the instrument by using SAXSIT Software [16]. The two SAXS data configurations which are resulted from two different SDD distances were combined and analyzed by using SASfit data analysis program [17].

## 3. Results and Discussion

As shown in Figure 1, FITR spectroscopy showed the configuration of absorption intensity of PVA hydrogel sample as the function of frequency. The PVA hydrogel absorption band was observed on five main peak areas which represent the mixture of PVA Polymer absorption band with the distilled water. The functional groups which represent the PVA polymer chains were verified on the frequency of $\upsilon$ = 1450-1480 $cm^{-1}$ and $\upsilon$ = 850-870 $cm^{-1}$ which are associated with the stretching of $-CH_2$ vibration mode and the absorption wavenumber of 2850-3000 $cm^{-1}$ which is associated with the carboxyl of vibration mode stretching (C-O) and this wavenumber gives a significant contribution on the PVA crystallization [18-19].

Such five peaks of the absorption band which were observed from the PVA hydrogel were the real supports of the PVA polymer and distilled water. As we know, the chemical structure of PVA polymer was delivered by ($CH_2$-CH-OR) equation, where R = H and R = $COCH_3$ respectively for fully hydrolyzed PVA polymer and partly hydrolyzed PVA polymer. Therefore, the vibration mode stretching of $-CH_2$, C-O, and CH is in accordance with the characteristics of partly-hydrolyzed PVA polymer. It is confirmed by the material used in this research, namely PVA polymer with 98% hydrolysis. Furthermore, the intensity absorption support on the peak of $\upsilon = 3200\text{-}3570\ cm^{-1}$ frequency is associated with the –OH vibration mode or hydrogen mode stretching. This result is equivalent with the cross-binding agent of the distilled water which is used in the production of PVA hydrogel. Of the most important matters to be paid attention to is that –OH stretching band is a connecting factor between the hydrogen intra-molecule and inter-molecule and the PVA polymer chain which is obtained from a high hydrophilic process [20]. But in the other side, PVA polymer only interacted with the distilled water in the PVA hydrogel sample without being continued by chemical interaction between PVA polymer and other substances.

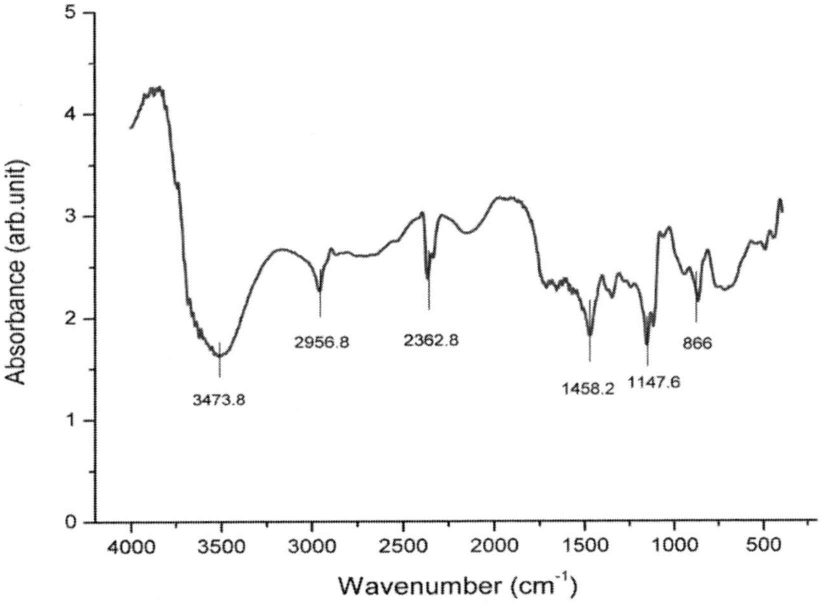

**Figure 1.** FITR Spectrum of PVA hydrogel for 3-hour freezing cycle

Thermal degradation on PVA hydrogel with various periods of the freezing-thawing cycle is shown in Figure 2. The graphic of PVA hydrogel thermal degradation shows three thermal decompositions which are indicated by three areas of sample weight-loss for all variation of cycle periods. The first decomposition with sample weight loss of 1.14-3.87% occurred in the temperature range of 181-192 °C. It can be clarified that the thermal decomposition on the first area was caused by crystal decomposition of water, condensation reaction, and evaporation of solvent trapped in the PVA hydrogel. This reaction is the process of water molecule release from the PVA hydrogel sample, where the group of hydroxyl (OH) moved from one sub-unit of PVA polymer chain and the hydrogen (H) atom was released from the other polymer chains.

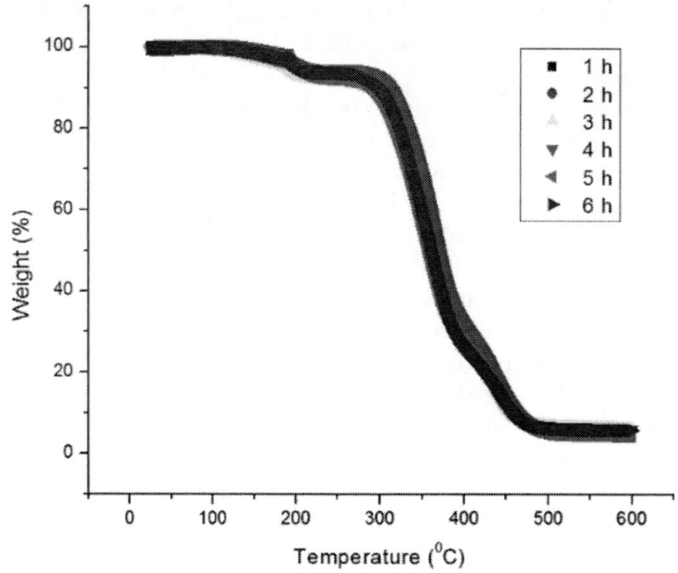


**Figure 2.** Profile of TGA thermogram of PVA hydrogel with the variation of freezing-thawing periods

For the second decomposition, sample weight loss of approximately 61.62-73.04% occurred in the temperature range of 282-376 °C. On this stage, the decomposition was resulted by the degradation of polyene structure from the polymer chain [21-22], while the third decomposition with sample weight loss of approximately 16.64-23.32% occurred in the temperature range of 390-460 °C. The third decomposition is the continuance of the polyene structure residue by producing carbon and hydrocarbon [22-24]. Besides research on sample decomposition, research on the temperature of maximum sample weight-loss level ($T_p$) was also reported. Based on the DTA/TG characterization results, the $T_p$ value of the various cycle periods did not experience significant changes. The average $T_p$ value of all samples is around 345.97 °C. However, such $T_p$ value is higher than the $T_p$ value of PVA polymer on about 230 °C [25]. It showed that the temperature of PVA hydrogel thermal degradation increased more significantly than PVA polymer. Such result indicated that the PVA hydrogel thermal stability was properly composed by the formation of hydrogen bond among the polymer [26]. The thermal stability of PVA hydrogel shows the orientation of stability of polymer chain and saturation of cross-binding agent diffusivity of the distilled water in the PVA hydrogel with a similar number of cycle. The thermal stability of the PVA hydrogel will be more stable along with the decreasing diffusivity of the distilled water cross-binding agent in PVA hydrogel by increasing the number of the freezing-thawing cycle [24-27]. Therefore, the period of the freezing-thawing cycle does not significantly affect the mobility of polymer chain and the diffusivity of the cross-binding agent.

The data of PVA hydrogel with various freezing-thawing cycle periods which are characterized by using SAXS instrument are shown in Figure 3 and Table 2. The combination of two model of data analysis, namely Teubner-Strey and Beaucage, was used in the scattering data analysis in order to observe the structure and distribution of PVA hydrogel nanocrystalline. The Teubner-Strey model provides a description on the inter-crystalline distance (d) and the length of correlation (ξ) of the PVA hydrogel sample. The scattering intensity equation is formulated as follow [28].

$$I(q) = \frac{8\pi\langle\eta^2\rangle\xi}{a^2 - 2bq^2 + q^4} \quad (1)$$

Where $\eta^2$ is contrasted scattering square with a long density contrast, $a$ is positive quantity

$$a = k^2 + \frac{1}{\xi^2} \tag{2}$$

and b can be positive or negative, depends on the amount of wave vector $k = 2\pi / d$ and ξ.

$$b = k^2 - \frac{1}{\xi^2} \tag{3}$$

Meanwhile, the Beaucage model [29] is capable of determining the sample dimension structure with the following equation.

$$I(q) = G\exp(-q^2 R_g^{\ 2} / 3) + B\left[erf(qR_g / \sqrt{6})\right]^{3P} / q^P \tag{4}$$

Where $R_g$ is the radius of gyration of the structure, $P$ is the fractal dimension characteristic, $G$ is big structured factor Guinier, $B$ is the pre-factor of the scattering specification and the equation of $\left[erf\left(qR_g / \sqrt{6}\right)\right]^{3P}$ is the transitional factor between high $q$-range and low $q$-range.

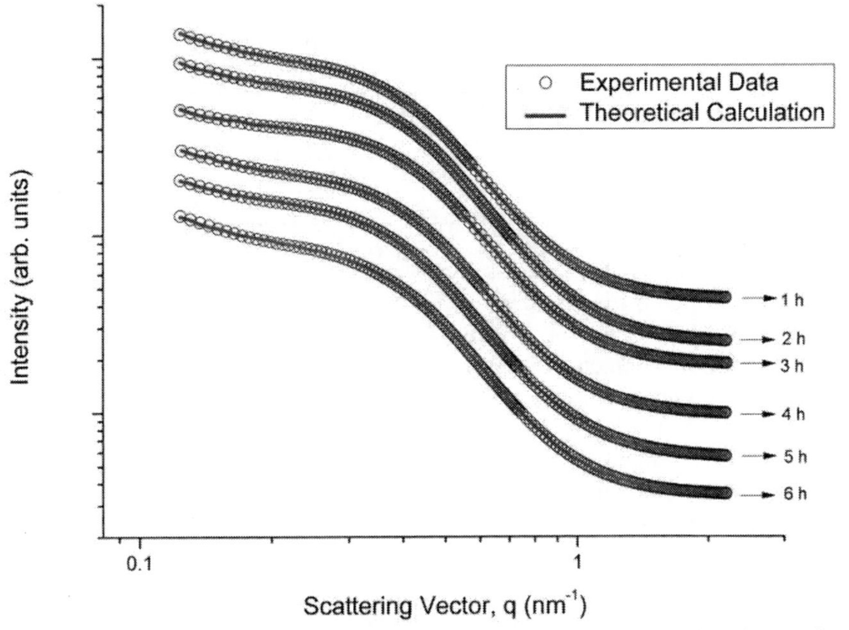


**Figure 3.** Data of PVA hydrogel with the variation of freezing-thawing cycle periods which is characterized by using the SAXS instrument

The data of PVA hydrogel has been successfully analyzed by using the Teubner-Strey and Beaucage models. The size of nanocrystalline structural dimension and distance has been properly confirmed by the data of *fitting* (Figure 3) between the data of experimental results (black circle line) from Teubner-Strey and Neaucage models which were improved by using the SASfit software. The results of SAXS data analysis show that the size of PVA hydrogel crystal experienced degradation along with the increasing period of F-T cycle from 1 hour (4.0 nm) up to 3 hours (3.8 nm) and which increased one more time on 6-hour cycle (4.1 nm) as shown in Table 2. Even though the crystal size on 1 hour cycle period was closed with the size crystal size on 6-hour cycle, the inter-crystal distance on 1-hour cycle (17.0 nm) is longer than the 6-hour cycle (16.0 nm). It can be confirmed that the sample homogeneity on 6-hour cycle has improved along with the degrading inter-crystal distance.

Therefore, the condition of PVA hydrogel structure with 6-hour cycle treatment was relatively more stable than the treatments with other cycle periods.

Such results show that the PVA hydrogel with 6 hour cycle period has an optimum crystal density which is produced from the decreasing amorphous phase in the *polymer-rich* domain. On the other hand, the PVA hydrogel with 3 hour cycle period has an optimum cross-binding power which can be seen on the smallest crystal. Such results are in accordance with the results reported by Millon *et al.* [13]. They revealed that the anisotropic mechanical nature of PVA-$D_2O$ hydrogel was affected by big-scaled pore structure of the gel and not by the PVA crystal structure which acted as a physical cross-binding agent.

Furthermore, Table 1 shows that the average structural dimension of PVA hydrogel is about 3.96 nm and the average inter-PVA hydrogel nanocrystalline distance is 16.9 nm. These results are similar to the report of the previous researchers that, among others, the isotropic PVA hydrogel structure consisted of the structural dimension of about 3 nm and was distributed with the average distance of approximately 2.8-3.3 nm and the average inter-crystal distance was approximately 15-17.5 nm [14].

**Table 1.** Parameter of structural dimension ($\xi$) and inter-crystalline distance ($d$) of PVA hydrogel

| The variation of time periods on the freezing process (hours) | $\xi$ (nm) | $d$ (nm) |
|---|---|---|
| 1 | 4.00 | 17.0 |
| 2 | 3.89 | 17.1 |
| 3 | 3.84 | 17.5 |
| 4 | 3.93 | 17.1 |
| 5 | 4.07 | 17.0 |
| 6 | 4.12 | 16.0 |

Based on the above results of the discussion, it can be confirmed that PVA hydrogel has a fairly good thermal stability which is identified with its average maximum sample weight-loss ($T_p$) of 345.97 °C which is higher than the $T_p$ value of PVA polymer of approximately 230 °C [25]. Such results were supported with the SAXS characterization which shows that the PVA hydrogel had an optimum crystal density along with the decreasing amorphous phase inside the polymer-rich domain. Therefore, the PVA hydrogel consists of a structure which consists of weak polymer chain which referred to as amorphous phase and combined with the nanocrystalline referred to as polymer-rich phase which forms the PVA hydrogel structure. With its unique natures, PVA hydrogel is very potential to be applied as a primary material of human implant [11].

## 4. Conclusion

The weight of PVA hydrogel sample experienced the most significant drop in the temperature range of 282-376 °C as a result of the PVA polymer chain degradation. The average $T_p$ value of PVA hydrogel is much higher than the $T_p$ value of PVA polymer. It showed that the temperature of PVA hydrogel thermal degradation increased more significantly than the PVA polymer. Such results were supported by the SAXS characterization which showed that the PVA hydrogel had the optimum crystal density along with the decreasing amorphous phase in the polymer-rich domain. Measured by using the Teubner-Strey and Beaucage models, the average structural dimension of PVA hydrogel nanocrystalline is in the range of about 3.96 nm with the inter-nanocrystalline distance of 16.9 nm.

## 5. References

[1] Osada Y, Ping Gong J and Tanaka Y 2004 Polymer Gels *J. Macromol. Sci. Part C* **44** 87–112

[2] Liu Y, Vrana N E, Cahill P A and McGuinness G B 2009 Physically crosslinked composite hydrogels of PVA with natural macromolecules: structure, mechanical properties, and endothelial cell compatibility. *J. Biomed. Mater. Res. B Appl. Biomater.* **90** 492–502

[3] Peppas N A 1975 Turbidimetric studies of aqueous poly(vinyl alcohol) solutions *Makromol. Chem.* **176** 3433–3440
[4] Peppas N A 1979 Characterization of Homogeneous and Pseudocomposite Homopolymers and Copolymers for Articular Cartilage Replacement *Biomater. Med. Devices Artif. Organs* **7** 421–33
[5] Stammen J A, Williams S, Ku D N and Guldberg R E 2001 Mechanical properties of a novel PVA hydrogel in shear and unconfined compression. *Biomaterials* **22** 799–806
[6] Li J K, Wang N and Wu X S 1998 Poly(vinyl alcohol) nanoparticles prepared by freezing-thawing process for protein/peptide drug delivery. *J. Control. Release Off. J. Control. Release Soc.* **56** 117–26
[7] Schmedlen R H, Masters K S and West J L 2002 Photocrosslinkable polyvinyl alcohol hydrogels that can be modified with cell adhesion peptides for use in tissue engineering. *Biomaterials* **23** 4325–32
[8] Hyon S H, Cha W I, Ikada Y, Kita M, Ogura Y and Honda Y 1994 Poly(vinyl alcohol) hydrogels as soft contact lens material. *J. Biomater. Sci. Polym. Ed.* **5** 397–406
[9] Wan W K, Campbell G, Zhang Z F, Hui A J and Boughner D R 2002 Optimizing the tensile properties of polyvinyl alcohol hydrogel for the construction of a bioprosthetic heart valve stent. *J. Biomed. Mater. Res.* **63** 854–61
[10] Jiang W, Yang H C, Yang S Y, Horng H E, Hung J C, Chen Y C and Hong C-Y 2004 Preparation and properties of superparamagnetic nanoparticles with narrow size distribution and biocompatible *J. Magn. Magn. Mater.* **283** 210–4
[11] Vijayasekaran S, Fitton J H, Hicks C R, Chirila T V, Crawford G J and Constable I J 1998 Cell viability and inflammatory response in hydrogel sponges implanted in the rabbit cornea. *Biomaterials* **19** 2255–67
[12] Sunaryono, Taufiq A, Putra E G R, Okazawa A, Watanabe I, Kojima N, Rugmai S, Soontaranon S, Zainuri M, Triwikantoro, Pratapa S and Darminto 2015 Small-Angle X-Ray Scattering Study on PVA/$Fe_3O_4$ Magnetic Hydrogels *Nano* **11** 1650027
[13] Millon L E, Nieh M-P, Hutter J L and Wan W 2007 SANS Characterization of an Anisotropic Poly(vinyl alcohol) Hydrogel with Vascular Applications *Macromolecules* **40** 3655–62
[14] Puspitasari T, Raja K M L, Pangerteni D S, Patriati A and Putra E G R 2012 Structural Organization of Poly(vinyl alcohol) Hydrogels Obtained by Freezing/Thawing and γ-Irradiation Processes: A Small-Angle Neutron Scattering (SANS) Study *Procedia Chem.* **4** 186–93
[15] Sunaryono, Taufiq A, Munaji, Indarto B, Triwikantoro, Zainuri M and Darminto 2013 Magneto-elasticity in hydrogels containing $Fe_3O_4$ nanoparticles and their potential applications *AIP Conf. Proc.* **1555** 53–6
[16] Rugmai S and Soontaranon S 2013 *Manual for SAXS/WAXS data processing using SAXSIT* (Synchrotron Light Research Institute, Thailand: Siam Photon Laboratory)
[17] Kohlbrecher J 2012 *User guide for the SASfit software package* (Paul Scherrer Institute Laboratory for Neutron Scattering (LNS) CH-5232 Villigen PSI joachim.kohlbrecher@psi.ch)
[18] Thomas J, Lowman A and Marcolongo M 2003 Novel associated hydrogels for nucleus pulposus replacement. *J. Biomed. Mater. Res. A* **67** 1329–37
[19] Hassan C and Peppas N 2000 Structure and Applications of Poly(vinyl alcohol) Hydrogels Produced by Conventional Crosslinking or by Freezing/Thawing Methods *Biopolymers · PVA Hydrogels, Anionic Polymerisation Nanocomposites* Advances in Polymer Science vol 153 (Springer Berlin Heidelberg) pp 37–65
[20] Mansur H S, Oréfice R L and Mansur A A P 2004 Characterization of poly(vinyl alcohol)/poly(ethylene glycol) hydrogels and PVA-derived hybrids by small-angle X-ray scattering and {FTIR} spectroscopy *Polymer* **45** 7193–202

[21] Alexy P, Káchová D, Kršiak M, Bakoš D and Šimková B 2002 Poly(vinyl alcohol) stabilisation in thermoplastic processing *Polym. Degrad. Stab.* **78** 413–21

[22] Zhang X, Takegoshi K and Hikichi K 1992 Phase separation and thermal degradation of poly(vinyl alcohol)/poly(methacrylic acid) and poly(vinyl alcohol)/poly(acrylic acid) systems by 13C c.p./m.a.s. n.m.r. *Polymer* **33** 718–24

[23] Aggour Y A 2001 Reaction kinetics of graft copolymerization and thermochemical studies of the degradation of poly(vinyl alcohol) graft copolymer *Polym. Int.* **50** 347–53

[24] Sivaram S and Singh R P 1991 Degradation and stabilization of ethylene-propylene copolymers and their blends: A critical review *Polymer Compositions Stabilizers/Curing* Advances in Polymer Science vol 101 (Springer Berlin Heidelberg) pp 169–215

[25] Nata I F, Wang S S-S, Wu T-M and Lee C-K 2012 Carbonaceous hydrogels based on hydrothermal carbonization of glucose with chitin nanofibers *Soft Matter* **8** 3522–5

[26] Guan Y, Bian J, Peng F, Zhang X-M and Sun R-C 2014 High strength of hemicelluloses based hydrogels by freeze/thaw technique *Carbohydr. Polym.* **101** 272–80

[27] López D, Cendoya I, Torres F, Tejada J and Mijangos C 2001 Preparation and characterization of poly(vinyl alcohol)-based magnetic nanocomposites. 1. Thermal and mechanical properties *J. Appl. Polym. Sci.* **82** 3215–3222

[28] Teubner M and Strey R 1987 Origin of the scattering peak in microemulsions *J. Chem. Phys.* **87** 3195–200

[29] Beaucage G 1995 Approximations Leading to a Unified Exponential/Power-Law Approach to Small-Angle Scattering *J. Appl. Crystallogr.* **28** 717–28

**Acknowledgements**
The authors would like to thank to DRPM the Ministry of Research, Technology, and Higher Education for the financial support from the "Penelitian Produk Terapan" 2016-2017 for S. The authors would also like to thank the SLRI, Thailand for the given opportunity to perform sample characterization with application code No. 1498.

IOP Conf. Series: Materials Science and Engineering **202** (2017) 012042 doi:10.1088/1757-899X/202/1/012042

# Crystal Structure, Optical, and Electrical Properties of SnSe and SnS Semiconductor Thin Films Prepared by Vacuum Evaporation Techniques for Solar Cell Applications

**Ariswan[1], H Sutrisno[2] and R Prasetyawati[1]**

[1]Department of Physics Education, Faculty of Mathematics and Natural Sciences, Yogyakarta State University, Kampus Karangmalang, Jl. Colombo 1, Yogyakarta, 55281, Indonesia
[2]Department of Chemistry Education, Faculty of Mathematics and Natural Sciences, Yogyakarta State University, Kampus Karangmalang, Jl. Colombo 1, Yogyakarta, 55281, Indonesia

Email: ariswan@uny.ac.id

**Abstract.** Thin films of SnSe and SnS semiconductors had been prepared by vacuum evaporation techniques. All prepared samples were characterized on their structure, optical, and electrical properties in order to know their application in technology. The crystal structure of SnSe and SnS was determined by X-Ray Diffraction (XRD) instrument. The morphology and chemical composition were obtained by Scanning Electron Microscopy (SEM) coupled with Energy Dispersive of X-Ray Analysis (EDAX). The optical property such as band gap was determined by DR-UV-Vis (Diffuse Reflectance-Ultra Violet-Visible) spectroscopy, while the electrical properties were determined by measuring the conductivity by four probes method.The characterization results indicated that both SnSe and SnS thin films were polycrystalline. SnSe crystallized in an orthorhombic crystal system with the lattice parameters of $a$ = 11.47 Å, $b$ = 4.152 Å and $c$ = 4.439 Å, while SnS had an orthorhombic crystal system with lattice parameters of $a$ = 4.317 Å, $b$ = 11.647 Å and $c$ = 3.981 Å. Band gaps ($Eg$) of SnSe and SnS were 1.63 eV and 1.35 eV, respectively. Chemical compositions of both thin films were non-stoichiometric. Molar ratio of Sn : S was close to ideal which was 1 : 0.96, while molar ratio of Sn : S was 1 : 0.84. The surface morphology described the arrangement of the grains on the surface of the thin film with sizes ranging from 0.2 to 0.5 microns. Color similarity on the surface of the SEM images proved a homogenous thin layer.

**Keywords:** Vacuum evaporation, solar cell, SnSe semiconductor, SnS semiconductor.

## 1. Introduction

The studies of semiconductor materials, i.e. tin monoselenide or monosulfide are mostly done by researchers because the materials are semiconductor materials that can be applied in several areas of technology. One of their applications is in solar cell technology, due to the band gap of both materials that falls in the photon energy range of the spectrum of solar cells [1]. The study of SnSe has been widely conducted by several researchers, for example, research related to the structure and optical properties of SnSe thin film obtained by Chemical Bath Dedosition (CBD) [2]. Other researchers obtained SnSe using Aqueous Solution Method [3], and had also been reported that SnSe thin film were prepared by photoelectrochemical techniques deposition [4]. Research on SnS thin film had also been successfully produced by spray pyrolysis method [5], electrodeposition technique [6] and of course with other techniques.

## 2. Experimental Procedure

### *2.1 Preparation of SnSe and SnS Semiconductors*

At first, the vacuum system equipment consisting of a primary pump in the form of rotary van pump capable of suppressing the pressure up to $10^{-3}$ Torr was prepared. The primary pump was connected to the secondary pump in the form of diffusion pumps which were capable of decreasing the pressure of evaporator chamber at $10^{-6}$ Torr. Evaporator was equipped with boat-shaped crucible made of molybdenum (Mo) and connected to a power supply system that can generate electrical currents of up to 300 A. Crucible was mounted above the support where the substrates can be put and equipped with electrodes that can be connected to a power supply to set the substrate temperature.

SnSe powder was placed on the crucible, and then heated by slowly flowing current to crucible until the temperature reached above 1000 °C. The powder reached the point of vapor and deposited on the substrate at a temperature of 500 °C. After the deposition,the thin layers were left for 12 hours and then the evaporator was opened and a number of thin films were obtained. An equal treatment was done for the SnS powder.

### *2.2 Characterization*

To determine the structures and parameters of the crystal lattice, the SnSe and SNs thin film were characterized by XRD. Both thin films were examined using powder X-ray diffraction (XRD). The XRD patterns were obtained on a Rigaku MiniFlex 600-Benchtop XRD instrument, operated in the Bragg configuration using Cu Kα radiation ($\lambda$ = 1.5406 Å). The XRD instrument ran at 40 kV and 15 mA. The intensities were determined in the 2θ interval ranging from 3° to 90°. The surface morphology and chemical composition of SnSe and SnS thin films were investigated byScanning Electron Microscopy (SEM) coupled with Energy Dispersive Analysis of X-Ray (EDAX). To calculate the energy gap of the prepared samples, Diffuse Reflectance UV-Vis-UV 1700 Pharmaspec Spectroscopy at wavelengths ranged from 200 nm up to 900 nm was used. To specify the electrical properties such as conductivity and resistivity of the thin films, four-point probe technique was used.

## 3. Results and Discussion

SnSe and SnS thin films have been successfully obtained by vacuum evaporation techniques through the implementation of following variables variation. Variable control include vacuum pressure, the mass of material, and the temperature source (crucible) and the distance between the crucible and substrate, while the independent variables in this study were four variations of the temperature of the substrate, namely without heating, 250, 350, 500 and 550 °C [7]. In this article, only substrate temperature at 500 °C was reported to get the best values of the physical parameters after the characterization had been done. The characterization includes the structure and parameters of the crystal, the composition of the sample, optical properties and electrical properties of the thin film prepared using vacuum evaporation techniques.

### *3.1 XRD spectra of SnSe and SnS thin films*

XRD (X-Ray Diffraction) can be used to determine the crystal lattice parameters and the structure of thin films. The data obtained from XRD analysis were in the form of a graph of intensity ($I$) and the spectrum of diffraction angle ($2\theta$). Diffractogram displays spectral peaks that appeared in the samples. XRD analysis showed the distance among the crystal planes *hkl* ($d_{hkl}$). Diffraction patterns obtained from the sample were then compared with the data of JCPDS (Join Committee on Powder of Diffraction Standards) so it can be determined the Miller index from the diffraction peaks was formed.

The calculation of parameters can also be done once the crystal structure was known. Figure 1 shows the diffractogram of SnSe and SnS samples. The lattice parameters of orthorhombic crystal system can be calculated using the following equation [8].

$$2d \sin\theta = n\lambda$$

where $d$ can be calculated by $\frac{1}{d^2} = \frac{h^2}{a^2} + \frac{k^2}{b^2} + \frac{l^2}{c^2}$, so the following relationship can be applied:

$$\sin^2\theta = \frac{\lambda^2}{4}\left(\frac{h^2}{a^2} + \frac{k^2}{b^2} + \frac{l^2}{c^2}\right)$$

where $d$ is the distance of crystal planes, $\theta$ is the diffraction angle, $\lambda$ is X-ray wavelengths, $h\ k\ l$ is Miller Index, and $a$, $b$, $c$ are the crystal lattice parameters. The calculation of each SnSe and SnS sample presented the crystal parameters of $a$= 11.47 Å, $b$= 4.152 Å and $c$ = 4.439 Å for SnSe, and $a$ = 4.317 Å, $b$ = 11.647 Å and $c$ = 3.981 Å for SnS. These results were consistent with other researchers' research findings [9,10].

*3.2 SEM and EDAX*

Scanning Electron Microscopy (SEM) is an instrument used to determine the morphology of the surface of a material. The morphology of the surface of SnSe and SnS thin films were obtained through capturing and processing the secondary electrons emitted by a thin film when subjected to electron beams. The results of SEM characterization were images that depicted the surface of a crystal. The surface morphology characterization of SnSe and SnS samples with 40,000 times magnification is shown in Figure 2.

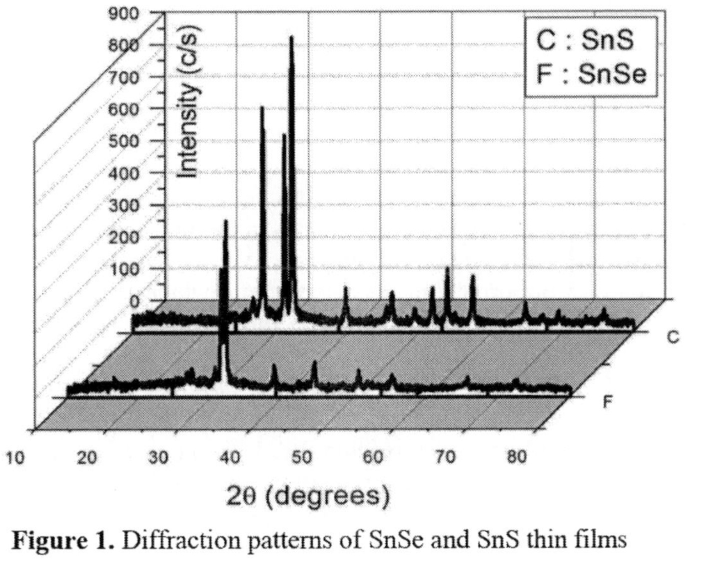


**Figure 1.** Diffraction patterns of SnSe and SnS thin films

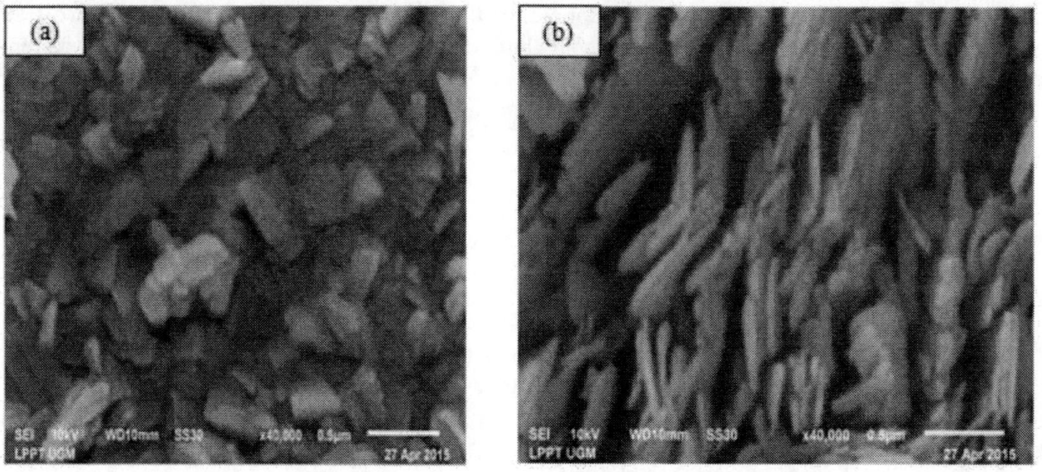


**Figure 2.** SEM images of (a) SnSe and (b) SnS thin films

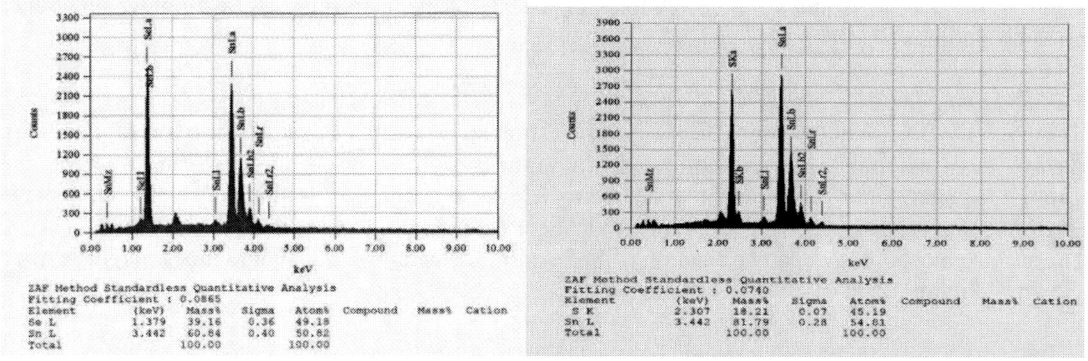


(a) (b)

**Figure 3.** EDAX on chemical compositions of (a) SnSe and (b) SnS samples

Based on the morphological images resulted from the characterization by SEM as in Figure 2, it appears that there is homogeneity of Se and S crystals in the thin film. The SEM images showed that the structures of the sample are in bead and angled beam shapes and scattered on a regular basis. The granular form according to orthorhombic crystal structure which has an axis perpendicular to each other and had one parameter longer compared with the two other lattice parameters. The chemical composition of the thin films analyzed by EDAX can be seen in Figure 3.

EDAX is integrated with SEM and it cannot work without SEM. Therefore, the EDAX was integrated with SEM device. The results of EDAX are in the form of aspectrum describing the relationship among the energy intensities resulting from firing electron beam on the sample. EDAX testing would produce two forms of analysis: qualitative and quantitative analyses. Qualitative analysis was used to calculate the type of element that was present in the sample where energy peaks appear on the curve being analyzed, whereas quantitative analysis was used to determine the composition of the constituent elements of the material being analyzed. From this quantitative analysis, the identification of elements completed with mass and atomic percentages are shown. The expected value of molar ratio of Sn and Se, also of Sn and S is 1: 1. Based on EDAX characterization results, the SnSe and SnS thin films contained Tin (Sn), Selenium (Se) and sulfur (S). But the result was slightly deviate from expectations. The molar ratio of Sn and Se in SnSewas51.03% : 48.97% or 1 : 0.96, and of Sn andS in SnS was 54.47% : 45.53% or 1:0.84.

*3.3 Optical and electrical properties*

The sample characterization for optical properties was carried out by UV-VIS spectroscopy with a wavelength ranged from 200 nm up to 900 nm. This equipment was not perfect because it could not reach up to the NIR with a wavelength of about 4000 nm. However, specifically for SnSe and SnS materials, it can be used to show the band gap width of both materials in the wavelength range of the spectrum. The absorption of wavelength can occur when radiation passes through the band gap of the material.

The phenomenon arose when inside the energy is passed so that the absorbance is very smalla band gap. When a photon of energy was equal to or greater than the band gap, the absorption will occur so that the absorbance (*A*) would go up significantly. The absorbance spectrum as a function of photon energy can be shown in Figure 4.

The calculation of the band gap of materials can be done using the formulation as follows. Absorption coefficient $\alpha$ is proportional to the absorbance $A$, so it can be stated that $\alpha = 2.303A/d$, where $d$ is the thickness of the thin films. Then the relationship between $\alpha$ and the photon energy $hv$ for direct band gap semiconductor material was given by the equation of $(\alpha, hv)^2 = B^2(hv - Eg)$, where $B$ is a constant between $10^5$ and $10^6$ $cm^{-1}eV^{-1}$ [11]. The result of this calculation produced the width of SnSe and SnS band gaps of 1.63 eV and 1.35 eV respectively. The results are consistent with the results obtained by other researchers [12,13].

The data, such as shown in Figure 4, can be processed to produce $(\alpha, hv)^2$ as a function of photon energy $hv$ as shown in Figure 5. Based on the measurement of electrical properties, it can be seen that all SnSe and SnS samples hadthe same conductivity type,which is p-type, with average specific resistivity of 4.6 $10^{-2}$ ohm-cm and 4.4 x $10^{-2}$ ohm-cm, respectively. Both samples were prepared atthe same substrate temperature which was 500 °C. This result was in accordance with the range of resistance values of semiconductor materials.

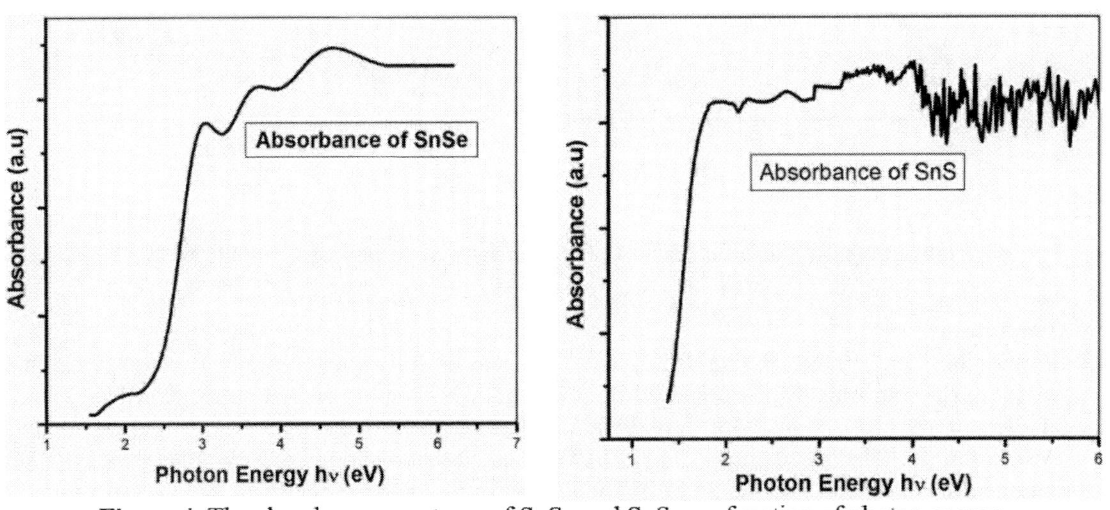


**Figure 4.** The absorbance spectrum of SnSe and SnS as a function of photon energy

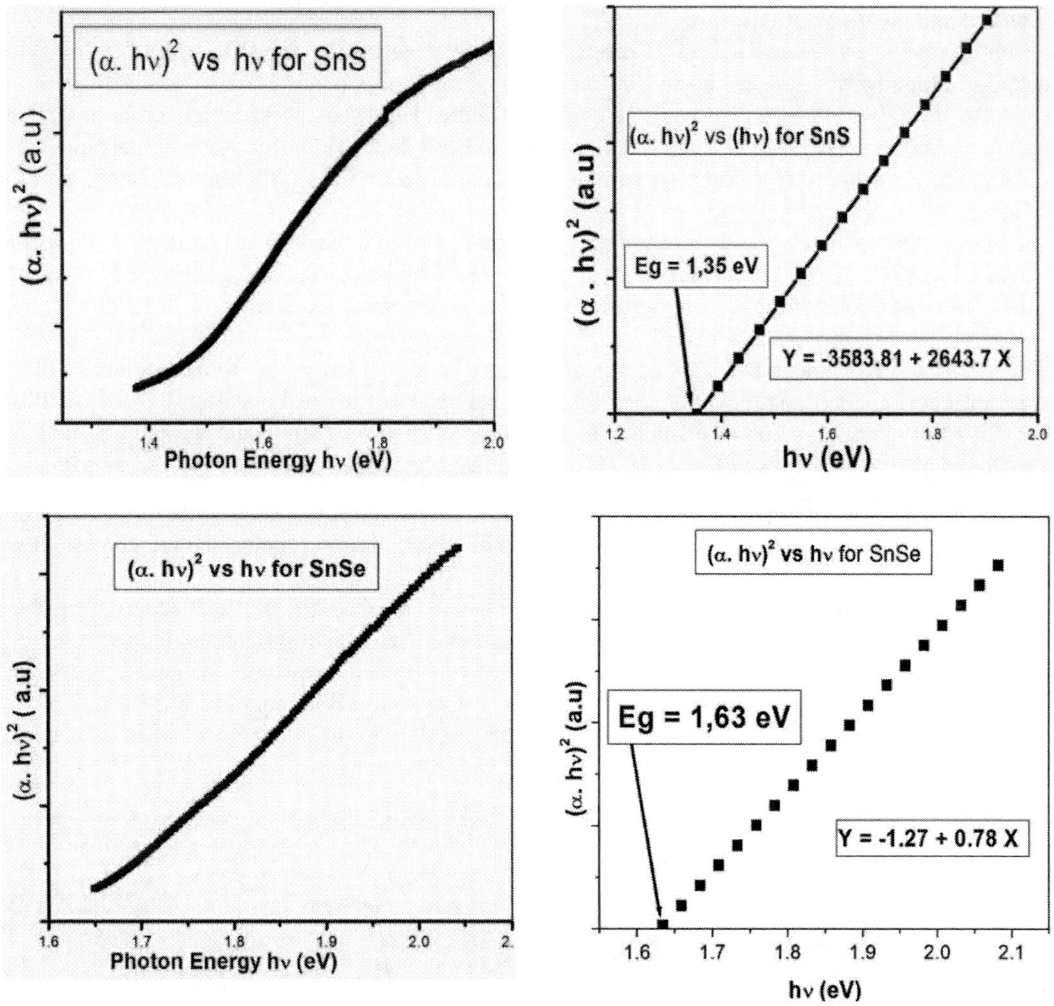


**Figure 5.** $(\alpha.h\nu)^2$ as a function of photon energy hν of SnS and SnSe thin layers

## 4. Conclusion

SnSe and SnS semiconductor thin films had been successfully prepared by vacuum evaporation techniques. The thin films were composed of atoms in orthorhombic crystal structure with the lattice parameters of $a$ = 11.47 Å; $b$= 4.152 Å and $c$ = 4.439 Å for SnSe, and $a$= 4.317 Å, $b$ = 11.647 Å and $c$ = 3.981 Å for SnS. The observations from EDAX indicated that SnSe and SnS samples were both non-stoichiometric with molar ratio of Sn:Se = 1:0.96 and Sn:S = 1:0.84. Electrical and optical properties of SnSe and SnS indicated that both materials have p-type conductivity with an average of resistivity of 4.6 $10^{-2}$ ohm-cm and 4.4 × $10^{-2}$ohm-cmand with band gaps of ($Eg$) of 1.63 eV and 1.35 eV, respectively.

## 5. References

[1] Kumar N, Sharma V, Parihar U, Sachdewa R, Padha N and Panchal C J 2011 *J. of Nano Electron. Physics* **3** 117-126
[2] Okereke N A and Ekpunobi A J 2010 *Journal of Chalcogenide Letters* **7**(9) 531
[3] Solanki G K, Gosai N N and Patel K D 2015 *Research Journal of Chemical Sciences* **5**(3) 1-5

[4] Sunyoung H A M, Choi S, Chae Y, Lee W J, Paeng K J, Kim W G and Bull N M 2010 *Korean Chem Soc* **31**(11) 3043-3046
[5] Helal M O A and Boshta 2012 *Journal of American science* **8**(8) 61-63
[6] Maripan R, Ragavendar M and Ponnuswamy V 2011 *Optica Applicata* **XLI**(4) 989-997
[7] Zouaoui A, Lachab M, Hidalgo M L, Chaffa A, Llinares C and Kesri N 1999 *Thin Solid Films* **339** 10
[8] Suryanarayana C and Norton M G 1998 *X- Ray Diffraction A Practical Approach* Plenum Press- New York and London
[9] Salgado E B, Nair M T S and Nair P K 2014 *ECS Journal of Solid State Science and Technology* **3**(8) 169- 175
[10] Yanuar, Guastavino F and Llinares C 2001 *Jurnal of Materials Sciences*
[11] Sakata H and Ogawa H 2000 *Solar Energy Materials and Solar Cells* **63** 25
[12] Kumar N, Parihar U, Kumar R, Patel K J, Panchal C J, Padha N 2012 *American Journal of Materials Science* **2**(1) 41-45
[13] Ogah O E, Zoppi G, Forbes I, Milles R W 200 *Properties of Thin Films of Tin Sulphide Produced Using The Thermal Evaporation Method* 23rd European Photovoltac Solar Energy Conference

**Acknowledgments**
Researchers would like to thank to Yogyakarta State University (UNY) that has funded this research. Also, thanks to the technical expert of Chemistry Department's Laboratory, Gadjah Mada University, for the measurement of X- Ray Diffraction and thin films' optical properties.

# Synthesis of Polyvinylpyrrolidone (PVP)-Green Tea Extract Composite Nanostructures using Electrohydrodynamic Spraying Technique

**Kamaruddin[1], D Edikresnha[1,2], I Sriyanti[1], M M Munir[1,2] and Khairurrijal[1,2]**

[1]Department of Physics, Faculty of Mathematics and Natural Sciences, Institut Teknologi Bandung, Jalan Ganesa 10, Bandung 40132, Indonesia
[2]Bioscience and Biotechnology Research Center, Institute for Research and Community Services, Institut Teknologi Bandung, Jalan Ganesa 10, Bandung 40132, Indonesia

Email: krijal@fi.itb.ac.id

**Abstract**. Green Tea Extract (GTE) as an active substance has successfully loaded to PVP nanostructures using electrohydrodynamic spraying technique. The precursor solution was the mixture of ethanolic polyvinylpyrrolidone (PVP) with a molecular weight of 1,300 kg/mol and ethanolic GTE solutions at a weight concentration of 4 wt.% and 2 wt.%, respectively, and it was estimated that the entanglement number was 2. The electrospraying was conducted at the voltage of 15 kV, the flow rate of 10 μL/min., and the distance between the collector and the tip of the nozzle of 10 cm. The SEM images showed that the PVP/GTE nanostructures had a combination of agglomerated beads (less spherical particles) and nanofibers. This occurred because if the PVP concentration is low, the PVP/GTE composite has weak core structures that cause the shell to be easily agglomerated each other. The intermolecular interaction between PVP and GTE in the PVP/GTE nanostructures occurred as confirmed by the peak at 3396 $cm^{-1}$, which is the carboxyl group, proving that the PVP/GTE nanostructures contained water, alcohols, and phenols. The peak at 1040 $cm^{-1}$, which is the stretching of C-O group in amino acid, gave another proof to the intermolecular interaction.

**Keywords:** Polyvinylpyrrolidone, Green tea extract, Nanostructures, Intermolecular interaction

## 1. Introduction

Green Tea or *Camellia sinensis* L. is a tropical plant that is widely consumed for generations as beverages. It is beneficial for human since it has high antioxidant activity [1-3]. Moreover, it has various polyphenolic compounds containing active components known as catechins [1]. These polyphenols have been reported to have higher antioxidant activity than those vitamin C and E, water soluble, non-toxic [2], is good enough to prevent prostate and colorectal cancer [3,4], and can also extend the shelf-life of various food products [1]. However, the amount of catechins that can be absorbed by the body is just less than 1-2% when consumed orally [5]. Since the molecular weight of catechins is very high, it results in a low bioavailability [1-5]. An alternative to overcome this problem is by reducing the size of the carrier system. Several methods can be used to improve the bioavailability of catechins or EGCC (epigallocatechin-3-gallate), which is the most abundant green tea constituent, such as encapsulation using oil-in-water sub-micrometer-sized emulsions, liposomes, and microparticles of protein/polyphenols [4]. However, for all these methods, the size is still in the micrometer scale, which must still be reduced down to the nanoscale. One way to obtain particle with nanoscale is through electrohydrodynamic spraying method.

Electrohydrodynamic spraying, or famously shortened as electrospray, is an easy and expensive way to produce nanoparticles. In this technique, a high voltage is being applied to induce charges on the surface of a polymer solution. The charged polymer solution will then be attracted towards the collector by the electrostatic force to form particles or fibers [6-8]. The polymer can be used as the carrier of an active content such as green tea. Therefore, by mixing the extract of green tea with a polymer, composite nanostructures (nanoparticles or nanofibers) may be created, and it is expected that the bioavailability of green tea extract will increase.

Several studies have used polyvinylpyrrolidone (PVP) as the carrier medium in various fields of applications that involved the formation of nanoparticles or nanofibers [9-11]. Noting that PVP has been recognized by FDA (Food and Drug Administration), USA, and is widely used as a drug carrier to increase the drug solubility [10], it was then selected as the carrier of green tea extract (GTE) in the present study. To the limit of our knowledge, PVP/GTE composite nanostructures produced by using electrohydrodynamic spraying have not been reported. The morphology of the PVP and PVP/GTE composite particles was characterized using a scanning electron microscope (SEM). A Fourier transform infrared spectrometer was used to detect the presence of typical functional groups of PVP and PVP/GTE composite.

## 2. Experimental Method

Polyvinylpyrrolidone (PVP) with a molecular weight of 1,300 kg/mol was purchased from Sigma-Aldrich, Indonesia. Analytical grade ethanol purchased from Merck, Indonesia, was used as the solvent. Green tea extract (GTE) powder was produced using ethanolic maceration method at Laboratory of Natural Materials, ITB. PVP and GTE were dissolved in ethanol at a weight concentration of 4 wt.% and 2 wt.%, respectively. PVP solution was made as the precursor to produce PVP nanostructures through the electrospraying technique. Another PVP solution was mixed with the GTE solution and stirred for ±2 hours at room temperature until the formation of a homogeneous solution. During electrospraying, each solution of PVP and PVP/GTE was placed in a 12-mL syringe and was connected to an applied voltage of 15 kV, a flow rate of 10 μL/min., and at a distance of 10 cm between the needle and the collector. The electrospray was conducted at a constant humidity of 55% and at room temperature. The electrospray process lasted for approximately 5-8 hours to obtain nanostructured samples.

The viscosities, the surface tensions, and the conductivities of PVP and PVP/GTE solutions were measured using Ostwald Viscometer (Fisher), du Noüy Tensiometer (Fisher), and conductometer (Mettler-Toledo), respectively. The morphology of the nanostructures was observed using a scanning electron microscope (SEM) (Hitachi, SU3500) operating at an accelerating voltage of 25 kV with the magnification between 5,000 and 60,000 times. Meanwhile, to identify the existing functional groups and the intermolecular interactions of PVP and GTE, a Fourier transform infrared (FTIR) spectrometer (Bruker, Alpha) with the wavenumber in the range of 500 to 4000 $cm^{-1}$ was used.

## 3. Results and Discussion

Syntheses of PVP nanoparticles and nanofibers were successfully conducted as reported in many previous works [8,12]. However, performing electrospray with a polymer of high molecular weight added by a bioactive substance such as green tea extract is a quite new. Electrosprayed PVP/GTE composite nanostructures were obtained using the entanglement analysis $(n_e)_{soln}$ [12]. Using ethanolic PVP and ethanolic GTE solutions at a weight concentration of 4 wt.% and 2 wt.%, respectively, it was estimated that the entanglement number $(n_e)_{soln}$ ~2, where particles will form. Figure 1 depicts the colors of PVP solution, GTE solution, and the mixture of PVP and GTE solutions. The white PVP solution mixed with the dark GTE solution resulted in the PVP/GTE solution with a light brown color. Two main parameters that influence the morphology and size on the electrospray are solution and process parameters [7]. The solution parameters include the concentration of polymer (as viscosity, surface tension, and conductivity), the characteristics of the solvent, and the types of additives. The process parameters include the voltage, the flow rate, the diameter of the needle and the distance between the needle and the collector. The viscosities, conductivities, and surface tensions of the PVP, GTE, and PVP/GTE solutions were measured, and the results are displayed in Table 1. The PVP solution is much more viscous than the GTE solution, and the viscosity of PVP/GTE solution is also higher than that of GTE solution but lower than that of PVP solution. The conductivity of GTE solution is higher than that of PVP solution, and therefore the conductivity of PVP/GTE solution is in between those of PVP and GTE solutions. The surface tensions of solutions have the same characteristics of their viscosities.

(a) (b) (c)

**Figure 1.** The colors of (a) PVP solution of 4 wt.%, (b) PVP/GTE solution with the concentration ratio of 8:2, and (c) GTE solution of 2 wt.%.

**Table 1.** The measurement of viscosity, conductivity, and surface tension at room temperature

| Solution | Viscosity (centiPoise) | Conductivity (μS/cm) | Surface Tension (dyne/cm) |
|---|---|---|---|
| PVP 4 wt.% | 7.14 | 0.08 | 29.0 |
| GTE 2 wt.% | 0.85 | 1.47 | 28.5 |
| PVP/GTE (8:2) | 3.80 | 0.51 | 28.8 |

IOP Conf. Series: Materials Science and Engineering **202** (2017) 012043 doi:10.1088/1757-899X/202/1/012043

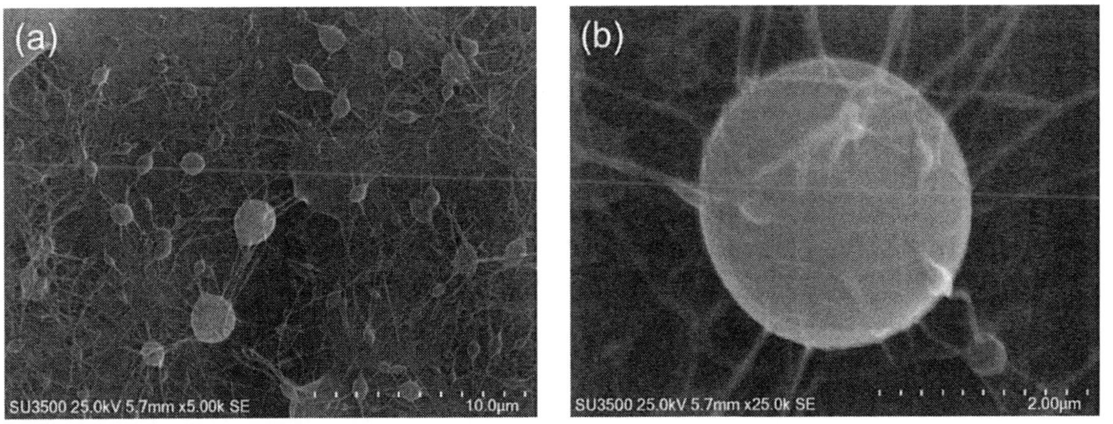


**Figure 2.** SEM images of PVP nanostructures being magnified at (a) 5,000 and (b) 25,000 times

The SEM images of the PVP nanostructures made by the electrospray technique are displayed in Figure 2. It was shown that the nanostructures had a combination of beads (less spherical particles) and nanofibers with a wide distribution of size. The diameter distribution of PVP beads was within the range of 291 to 2212 nm and the average diameter of PVP nanofibers was about 84 nm. This finding is consistent with the previous work reported that the combination of a high molecular weight and a low concentration of PVP being electrospun at high voltage and low flow rate and the estimated entanglement of about 2 would result in the formation of beaded fibers [8].

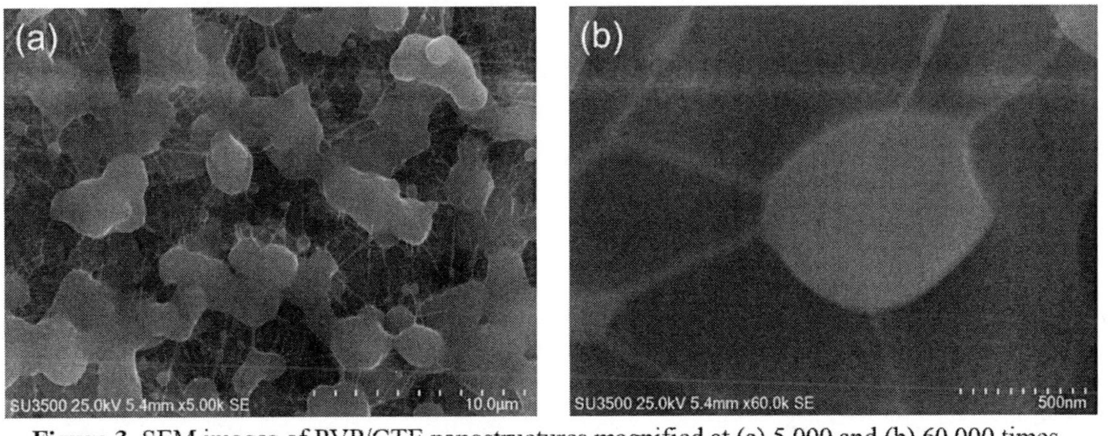


**Figure 3.** SEM images of PVP/GTE nanostructures magnified at (a) 5,000 and (b) 60,000 times

As given in Figure 3, the PVP/GTE nanostructures are the agglomerations of beads with nanofibers in between. It happened since the addition of GTE content reduced the concentration of PVP. When the PVP concentration was low, the PVP/GTE composite had weak core structures that caused the shell to be easily agglomerated each other [13]. The diameter distribution of PVP/GTE agglomerated beads was within the range of 231 to 1164 nm and the average diameter of PVP/GTE nanofibers was about 49 nm.

The FTIR spectra of PVP nanostructures, GTE solution, and PVP/GTE nanostructures are shown in Figure 4. The FTIR spectrum of GTE had a broad peak at 3319 $cm^{-1}$ which are the stretching vibration of O-H group in water, alcohols, and phenols [15]. The peaks of 2944 and 2832 $cm^{-1}$ were identified as asymmetric C-H alkanes stretch and O-H carboxyl acid stretch, respectively [15]. The group of C=C stretch in the aromatic ring and C=O stretch in polyphenols were characterized at 1655 $cm^{-1}$, and C-O stretch in amino acids was detected at 1022 $cm^{-1}$ [15]. Therefore, these findings revealed that the GTE was rich in polyphenols, carboxylic acids, and amino acids.

The FTIR spectrum of PVP had a peak at 3424 $cm^{-1}$ which indicates O-H stretching [16]. The peaks at 2950 and 1652 $cm^{-1}$ proved the existence of asymmetric stretching of $CH_2$ and stretching of C-O, respectively [16]. The C-H bending and $CH_2$ wagging were observed at 1423 $cm^{-1}$ and 1288 $cm^{-1}$, respectively [16]. The peaks at 1018 and 568 $cm^{-1}$ were identified as the $CH_2$ rock and N-C=O bending, respectively [16].

The FTIR spectrum of PVP/GTE composite resulted in the following peaks. The peak at 3396 $cm^{-1}$ was the carboxyl group hence proving that it contained water, alcohols, and phenols [15]. Moreover, this peak could also be considered as a result of intermolecular interaction between PVP and GTE, where its position and intensity were in between those of PVP and GTE. Other phenomena happened at the wavenumbers of 1018 and 1040 $cm^{-1}$. In the FTIR spectrum of PVP, there was rock bending of $CH_2$ at 1018 $cm^{-1}$ with low intensity. After being loaded by GTE, the peak became increasingly high and shifted to 1040 $cm^{-1}$, which is the stretching of C-O group in amino acid. Thus, the PVP/GTE composite had successfully been synthesized as confirmed by phenolic compounds and amino acids.

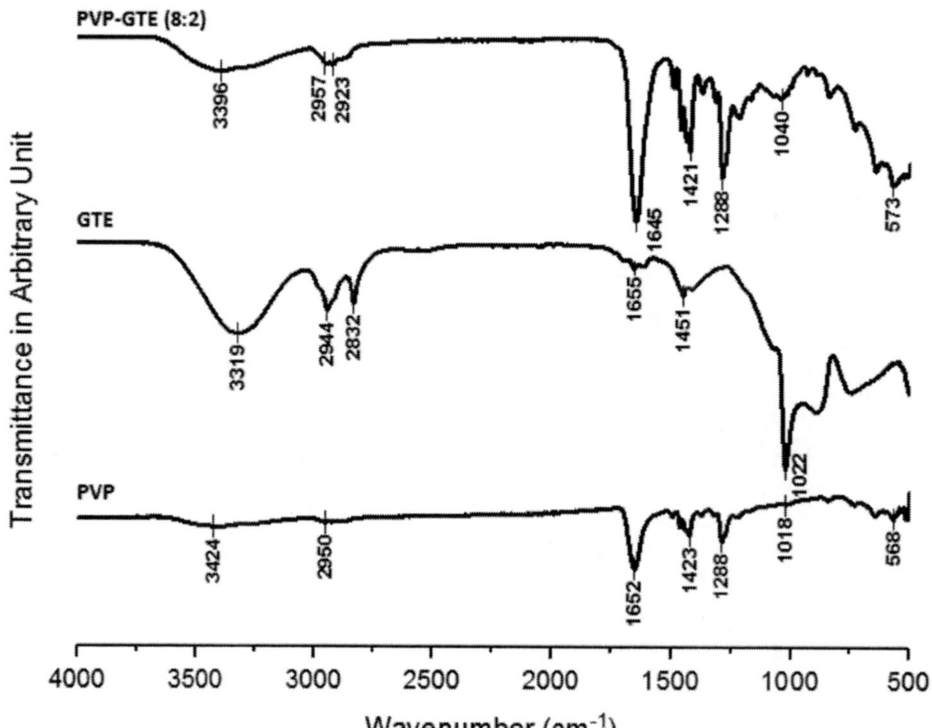


**Figure 4.** FTIR spectra of PVP nanostructures, GTE solution, and PVP/GTE nanostructures

## 4. Conclusion

The synthesis of polyvinylpyrrolidone (PVP) composite nanostructures loaded with green tea extract (GTE) as the active ingredient using electrohydrodynamic spraying technique has successfully been done. From the SEM images, it was shown that the PVP/GTE nanostructures had a combination of agglomerated beads (less spherical particles) and nanofibers. This occurred because if the PVP concentration is low, the PVP/GTE composite has weak core structures that cause the shell to be easily agglomerated each other. The intermolecular interaction between PVP and GTE in the PVP/GTE nanostructures was confirmed by the peak at 3396 $cm^{-1}$, which is the carboxyl group, proving that the

PVP/GTE nanostructures contained water, alcohols, and phenols. Another confirmation was from the peak at 1040 $cm^{-1}$, which is the stretching of C-O group in amino acid.

## 5. References

[1] Senanayake S P J N 2013 Green Tea Extract: Chemistry, Antioxidant Properties, and Food Applications-A Review *Journal of Functional Foods* **5** 1529 -1541.

[2] Kerio L C, Wachira F N, Wanyoko J K, Rotich M K 2013 Total Polyphenols, Catechin Profiles and Antioxidant Activity of Tea Products from Purple Leaf Coloured Tea Cultivars. *Food Chemistry* **136** 405-1413.

[3] Chow H H S and Hakim I A 2011 Pharmacokinetic and Chemoprevention Studies on Tea in Humans. *Pharmacological Research* **64** 105-112.

[4] Naumovski N, Blades B L, Roach P D 2015 Article: Food Inhibits the Oral Bioavailability of the Major Green Tea Antioxidant Epigallocatechin Gallate in Humans. *Antioxidants* **4** 373-393.

[5] Yashin A, Nemzer B, Yashin Y 2012 Bioavailability of Tea Components. *Journal of Food Research* **1** No. 2, 281-290.

[6] Ramakrishna S, Fujihara K., Teo W E, Lim T C, Ma Z 2005 An Introduction to Electrospinning and Nanofibers *World Scientific Publishing Co. Pte. Ltd, Singapore*

[7] Andrady A L 2008 Science and Technology of Polymer Nanofibers. *John Willey and Son, Inc.*, USA.

[8] Munir M M, Suryamas A B, Iskandar F, Okuyama K 2009 Scaling Law on Particle to Fiber Formation during Electrospinning. *Polymer* **50** 4935–4943.

[9] Dai X Y, W Nie W Y C Wang Y Shen Y Li and Gan S J 2012 Electrospun Emodin Polyvinylpyrrolidone Blended Nanofibrous Membrane: A Novel Medicated Biomaterial for Drug Delivery and Accelerated Wound Healing. *Journal of Materials Science: Materials in Medicine* **23** 2709-2716.

[10] Nuryantini, Edikresnha A Y, Munir D, Khairurrijal M M, 2015 Electrospun Polyvinylpyrrolidone as a Carrier for Leaves Extracts of Anredera Cordifolia (Ten.) Steenis *Materials Science Forum* ***827*** 91-94.

[11] Samprasit W, Akkaramongkolpon P, Ngawhirunpat T, Rojanarata T, Kaomongkolgit R, Opanasopit P 2015 Fast releasing oral electrospun PVP/CD nanofiber mats of taste-masked meloxicam *International Journal of Pharmaceutics* **487** 213–222

[12] Shenoy S L, Bates W D, Frisch H L, Wnek, G E 2005 Role of Chain Entanglements on Fiber Formation during Electrospinning of Polymer Solutions: Good Solvent, Non-Specific Polymer-Polymer Interaction Limit. *Polymer* **46** 3372-3384.

[13] Zarrabi, A. and Vossoughi, M., (2009): Electrospray: Novel Fabrication Method for Biodegradable Polymeric Nanoparticles for Further Applications in Drug Delivery Systems. *NANOCON* **20. - 22**. 10. 2009, Roznov pod Radhostem, Czech Republic, EU

[14] Morota K., Matsumoto H., Mizukoshi T., Konosu Y., Minagawa M., Tanioka A., Yamagata Y., Inoue K 2004 Poly (Ethylene Oxide) Thin Films Produced by Electrospray Deposition: Morphology Control and Additive Effects of Alcohols on Nanostructure *Journal of Colloid and Interface* Science **279** 484-492.

[15] Senthilkumar, S. R. dan Sivakumar, T. (2014): Green Tea (Camellia Sinensis) Mediated Synthesis of Zinc Oxide (ZnO) Nanoparticles and Studies on Their Antimicrobial Activities, *International Journal of Pharmacy and Pharmaceutical Sciences*, **6**, 461- 465.

[16] Rahma, A., Munir, M.M., Khairurrijal, Prasetyo A., Suendo, V., and Rachmawati, H., (2016): Intermolecular Interactions and The Release Pattern of Electrospun Curcumin-Polyvinyl(Pyrrolidone) Fiber. *Biological and Pharmaceutical Bulletin*, **39** No.2, 163-173.

**Acknowledgements**
This research was supported by Directorate of Research and Community Engagement of Ministry of Research, Technology, and Higher Education, the Republic of Indonesia under the University Excellence Research (PUPT) Grant in the fiscal year of 2015-2016. The first author (Kamaruddin) is thankful for the scholarship provided by the Endowment Fund for Education, the Republic of Indonesia.

IOP Conf. Series: Materials Science and Engineering **202** (2017) 012044 doi:10.1088/1757-899X/202/1/012044

# Photodegradation of Rhodamine B by using $ZnFe_2O_4$ Nanoparticles Synthesized through Precipitation Method

**Karnaji and I Nurhasanah**

Departement of Physics, Faculty of Mathematics and Natural Sciences, Diponegoro University, Jl.Prof. Soedartho, S.H, Semarang, Indonesia

Email: karnaji@st.fisika.undip.ac.id

**Abstrak**. $ZnFe_2O_4$ nanoparticles were synthesized through precipitation method from the mixed aqueous solution of ferric nitrate and zinc nitrate. The precipitate was calcined at the temperature of 600, 700, and 800 °C for 3 hours to obtain crystalline nature. The effects of calcination temperature on the properties of $ZnFe_2O_4$ nanoparticles were investigated. The structure, morphology, and optical properties of $ZnFe_2O_4$ nanoparticles were characterized by using X-ray diffractometer (XRD), scanning electron microscope (SEM), and UV-vis spectrophotometer, respectively. The photocatalytic activity of $ZnFe_2O_4$ nanoparticles was examined to Rhodamine B degradation in water under UV irradiation. The results show that the calcination temperature influences on $ZnFe_2O_4$ phase, crystallite size, and the energy gap of $ZnFe_2O_4$ nanoparticles. $ZnFe_2O_4$ nanoparticles calcined at 700 °C shows the best photocatalytic activity to Rhodamine B degradation due to smaller crystallite size and higher absorption.

**Keywords:** $ZnFe_2O_4$, Precipitation, Rodhamine B, Photocatalytic

## 1. Introduction

Considerable research and development in searching for a safe, abundant, and inexpensive photocatalyst have been carried out in several years. Semiconductor photocatalysts show promising photocatalytic activity for the degradation of various dye organic pollutants in water [1,2]. $TiO_2$ powder is the main photocatalyst, which is widely used and developed for dye degradation due to its high photo efficiency and stability under UV light irradiation [3,4]. However, the large band gap energy and rapid recombination of photogenerated charge carriers of that photocatalyst lead limitation in its implementation. The powder photocatalysts were difficult to remove for reuse in aqueous solution. The strategy to overcome its limitation is to investigate the photocatalytic properties of magnetic nanoparticles. $ZnFe_2O_4$ nanoparticles are one of the proposed magnetic nanoparticles that can be easily removed from aqueous solution by applying an external magnetic field [5,6]. Moreover, the energy gap of $ZnFe_2O_4$ is smaller than $TiO_2$. Thus $ZnFe_2O_4$ has attracted attention as semiconductor mediated photocatalysis with wide wavelength spectrum from UV to visible [7].

$ZnFe_2O_4$ nanoparticles have been synthesized by various methods including ball mill [6], hydrothermal [8,9], electrochemical reduction [10], sol-gel [11,12], microwave [13], solvothermal [3] and precipitation [2]. Calcination or sintering improve the properties of ZnFe2O4 nanoparticles. The calcination influencing on optical properties was found on $ZnFe_2O_4$ nanoparticles synthesized by sol-gel method [12]. Another research reports the $ZnFe_2O_4$ nanoparticles synthesized by sol-gel method with sintering have good properties for catalyst [14]. Comparison to other methods, precipitation is a promising method for producing nanoparticles because of its simplicity and low cost. Usually, the product of precipitate was then calcined to form crystalline nanoparticles. Therefore in this research, precipitation method was used for producing $ZnFe_2O_4$ nanoparticles followed by calcination to study the effect of calcination temperature on the properties of $ZnFe_2O_4$ nanoparticles. The calcination temperatures of 600, 700, and 800 °C have been selected based on previous reference [14]. The photocatalytic activity of $ZnFe_2O_4$ nanoparticles was investigated for degradation of Rodhamine B dye under UV light irradiation. Calcination temperature of 700 °C yields $ZnFe_2O_4$ nanoparticles with good photocatalytic activity.

## 2 Experimental Method

### *2.1 Synthesize of $ZnFe_2O_4$ nanoparticles*

$ZnFe_2O_4$ nanoparticles were synthesized through precipitation method by mixing of Zinc Nitrate (Merc, Germany) and Iron (III) Nitrate (Sigma-Aldrich, Germany) with a ratio mole of 1:2 in 100 ml of distilled water. The solution was stirred at 60 °C, and 3 M ammonia solution was dropwise until pH of 10 to form the precipitate. The precipitated product was filtered and washed with ethanol for several times. The product was then dried at 100 °C and further calcined at 600, 700, and 800 °C for 3 hours.

### *2.2 Characterization*

Phase and structure of $ZnFe_2O_4$ nanoparticles were identified by Phillips PW 1710 X-ray diffractometer with Cu K$\alpha$ irradiation in the scan range from 20 to 70°. The morphology and particle size average were analyzed by scanning electron microscope (JEOL/EO version 1.1, Type: JSM-6360). UV-vis absorption spectrum was recorded by using UV-vis spectrophotometer (SHIMADZU) with a wavelength in the range 200 to 800 nm.

### *2.3 Photodegradation experiment*

Photodegradation of 20 ppm Rhodamine B (RhB) was carried out to evaluate photocatalytic activity of $ZnFe_2O_4$ nanoparticles. 35 mg nanoparticles $ZnFe_2O_4$ was dispersed in 70 ml RhB solution and stirred in the dark for 30 minutes to reach absorption-desorption equilibrium. Then the solution was irradiated by UV lamp for 150 minutes. At given time intervals, 10 ml the solution was sampled and separated by centrifuge to measure the remaining concentration of RhB by using Uv-vis spectrophotometer. The characteristic absorption peak of RhB was at $\lambda = 554$ nm.

## 3 Results and Discussion

### *3.1 Structure and morphology of $ZnFe_2O_4$*

Figure 1 shows the diffraction pattern of $ZnFe_2O_4$ nanoparticles calcined at 600, 700, and 800°C. The peaks of (220), (311), (400), (511), and (440) appear for all samples, indicate the formation of spinel cubic structure (JCPDS No. 22-1012). To confirm that finding, the lattice constant was determined for diffraction peaks of (311) by using Equation 1.

$$a = \frac{\lambda\sqrt{h^2 + k^2 + l^2}}{2 \sin \theta} \tag{1}$$

$a$ is lattice constant, $\lambda$ is the wavelength of X-ray, $\theta$ is the Bragg diffraction angle and *(hkl)* is index miller [5]. The lattice constant was found to be 0.841-0,842 nm. Those results corresponded to a lattice constant of bulk cubic spinel structure of $ZnFe_2O_4$ [5]. Furthermore, increasing calcination temperature from 600 to 700°C enhanced the crystallinity of $ZnFe_2O_4$ nanoparticles. It is represented by the diffraction peak intensity as can be observed in Figure 1. The diffraction peaks intensity slightly increased at 700°C. However, the diffraction peak intensity slightly dropped back at the calcination temperature of 800°C. In addition, a peak of $\alpha$-$Fe_2O_3$ phase observed at $2\theta = 33°$ indicates that the single phase of cubic spinel structure $ZnFe_2O_4$ is not yet completely formed. The $\alpha$-$Fe_2O_3$ phase that is formed in nanoparticles can be calculated by using peak intensity ratio of phase $\alpha$-$Fe_2O_3$ with peak intensity of phase $ZnFe_2O_4$ [15]. The results show the $ZnFe_2O_4$ nanoparticles calcined at a temperature of 600, 700, and 800 °C composed by the $\alpha$-$Fe_2O_3$ phase of 12.44%, 13.48%, and 13,36% respectively. The $\alpha$-$Fe_2O_3$ phase causes crystallite size reduction of $ZnFe_2O_4$ nanoparticles. The crystallite size average of $ZnFe_2O_4$ nanoparticles was calculated by using the full width at half maximum (FWHM) of the dominant diffraction peak of (311) from the XRD patterns based on Scherrer formula (Equation 2) [2,5].

$$D = \frac{0.9\lambda}{\beta\cos\theta} \quad (2)$$

$D$ is the crystallite size, $\lambda$ is the wavelength of X-ray, $\beta$ is FWHM, and $\theta$ is the Bragg diffraction angle. These results are presented in Table 1. Generally, the calcination temperature caused an increase in the crystallite size of nanoparticles, however in this research that is not observed. This case is due to the phase of $\alpha$-$Fe_2O_3$ formation in the nanoparticles $ZnFe_2O_4$.

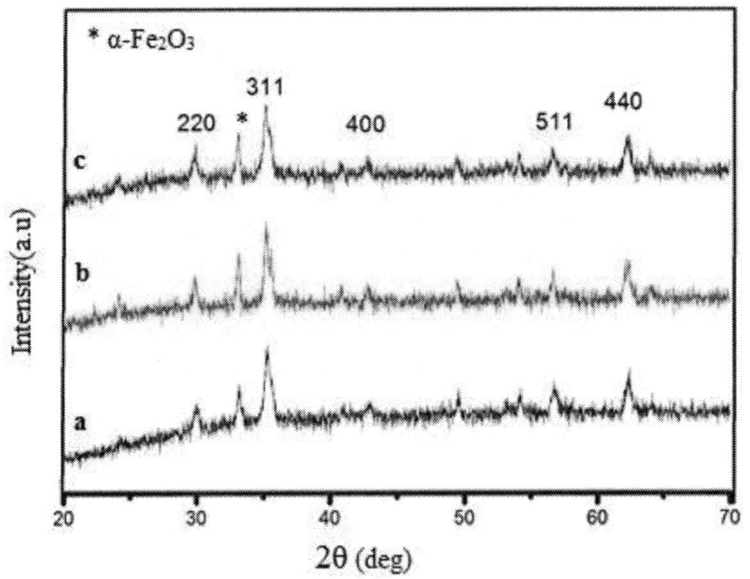


**Figure 1.** XRD patterns of $ZnFe_2O_4$ nanoparticles (a) 600°C; (b) 700°C and (c) 800°C

**Table 1.** Crystallite size, lattice constant, and particle size of $ZnFe_2O_4$ nanoparticles

| Calcination temperature (°C) | Crystallite size (nm) | Lattice constant (nm) | Particle size (nm) |
|---|---|---|---|
| 600 | 19.95 | 0.841 | 51.8 |
| 700 | 13,80 | 0.842 | 84.4 |
| 800 | 15.23 | 0.842 | 69.5 |

The morphological nature of $ZnFe_2O_4$ nanoparticles was identified by using SEM images as presented in Figure 2. The spherical-like morphology with an average size of 51.8, 84.4 and 69.6 nm observed for $ZnFe_2O_4$ nanoparticles with calcination temperature of 600, 700, and 800°C, respectively. The estimated average size of SEM images is different with the calculated average crystallite size using Scherrer formula indicating the nanoparticles composed by several crystallites as a consequence of polycrystalline nature. Moreover, high agglomeration appears for $ZnFe_2O_4$ nanoparticles calcined at

IOP Conf. Series: Materials Science and Engineering **202** (2017) 012044 doi:10.1088/1757-899X/202/1/012044

700°C, because its crystallite size was smaller than the other. This finding is similar to $ZnFe_2O_4$ nanoparticles synthesized by sol-gel method [8].

**Figure. 2.** SEM image of $ZnFe_2O_4$ nanoparticles calcined at (a) 600, (b)700, and (c) 800 °C

*3.2 Optical properties of $ZnFe_2O_4$ nanoparticles*

To verify the UV-vis light absorptivity of $ZnFe_2O_4$ nanoparticles, the absorbance spectrum of $ZnFe_2O_4$ nanoparticles was recorded and shown in Figure 3. It can be seen, that $ZnFe_2O_4$ nanoparticles absorb UV-vis light from 200 to 600 nm. The Calcination at 700°C resulted in $ZnFe_2O_4$ nanoparticles with higher absorption in the UV-vis region and its absorption edge exhibited red-shift to higher wavelength compared with other calcination temperature. Thessseee phenomena were due to the phase of $\alpha$-$Fe_2O_3$ composed in the nanoparticles was higher than other. It is expected that $ZnFe_2O_4$ nanoparticles can show higher photocatalytic activity.

The band gap energy of $ZnFe_2O_4$ nanoparticles was determined through touc' plot method based on direct transition as expressed in Equation 3 [16]:

$$(\alpha h\nu)^2 = A(h\nu - E_g) \quad (3)$$

Where, $\alpha$ is absorptivity constant, $h$ is Planck constant, $\nu$ is light frequency, $A$ is a constant and $E_g$ is band gap energy. The graph of $(\alpha h\nu)^2$ vs the photon energy $(h\nu)$ is shown in Figure 4. The band gap energy of $ZnFe_2O_4$ nanoparticles was obtained to be 2.43, 1.85, and 2.39 eV for calcination temperature of 600, 700, and 800 °C.

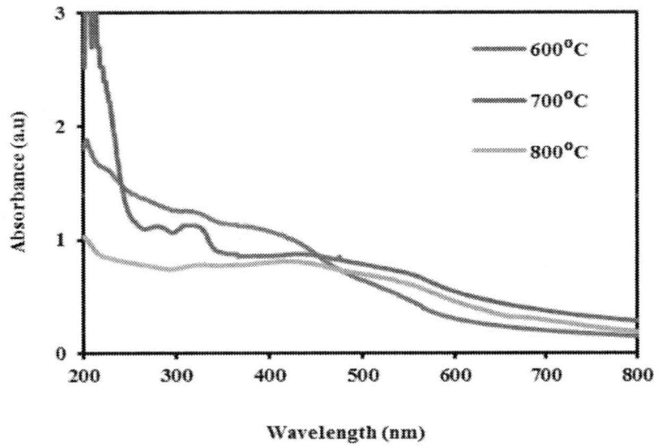


**Figure 3.** Absorbance spectrum of $ZnFe_2O_4$ nanoparticles

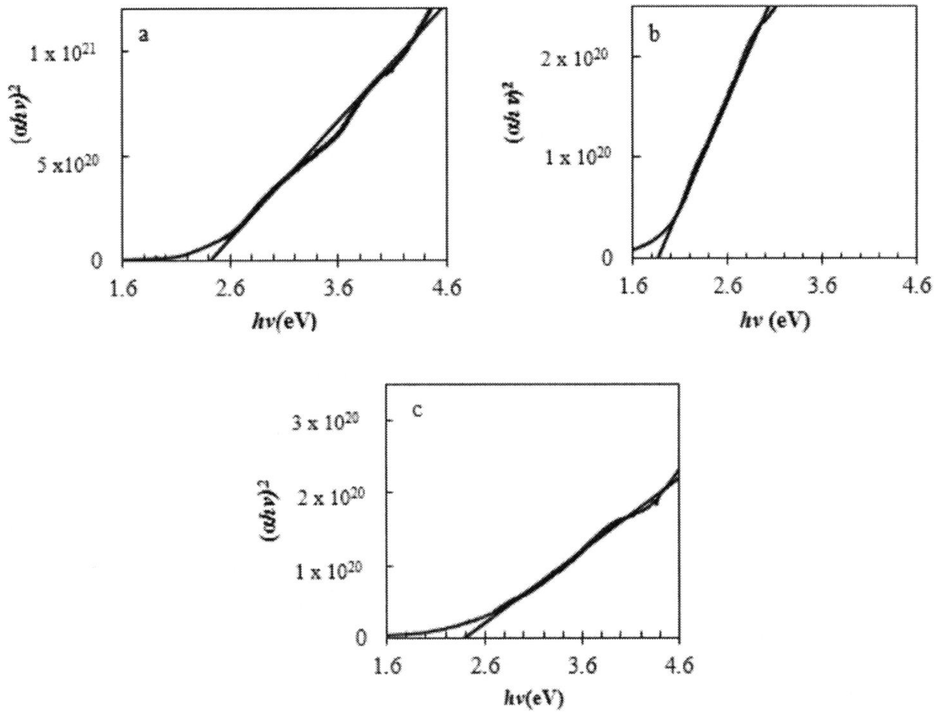


**Figure 4.** Tauc's plot of $ZnFe_2O_4$ nanoparticles calcined at (a) 600, (b) 700, and (c) 800 °C

3.3 *Photodegradation of RhB solution*

Photodegradation of RhB solution under UV light irradiation was carried out to evaluate the effect of calcination temperature on the photocatalytic activity of $ZnFe_2O_4$ nanoparticle. As depicted in Figure 5, RhB degradation without $ZnFe_2O_4$ nanoparticles almost did not appear after 120 min irradiation. Sharp degradation up to 60% occurred after 60 min irradiation, and then the degradation increased slightly. The photocatalytic activity of $ZnFe_2O_4$ nanoparticles calcined at 600 and 800°C were almost similar. Whereas, the $ZnFe_2O_4$ nanoparticles calcined at 700°C exhibited higher photocatalytic activity than the other. It could be correlated to smaller crystallite size and higher UV absorption of that $ZnFe_2O_4$ nanoparticles. Small crystallite size leads large surface area that results in more electron-hole pairs covering the surface of $ZnFe_2O_4$ nanoparticles reacting with RhB solution.

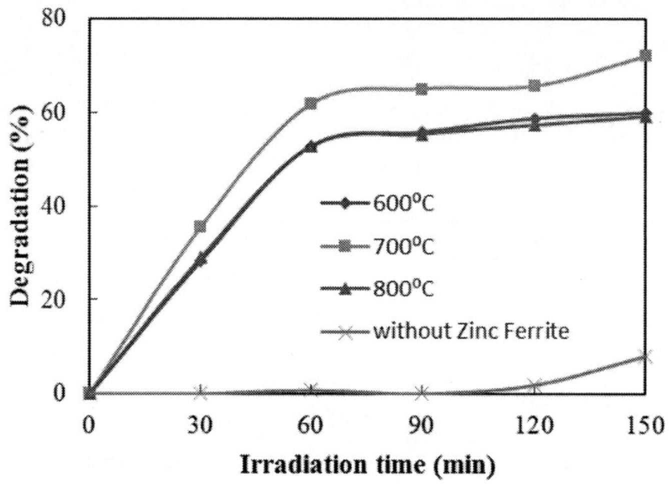


**Figure. 5**. Photodegradation activity of $ZnFe_2O_4$

Calcination temperature of 700 °C yields good photocatalytic activity of $ZnFe_2O_4$ nanoparticles. Similar phenomena was reported for sintered $ZnFe_2O_4$ nanoparticles prepared by the sol-gel method at 700°C for four h [14]. Photocatalytic of $ZnFe_2O_4$ nanoparticles as described in other research shows the degradation of 10 ppm MG solution under UV irradiation for 60 min was 50% [17], while in this research the degradation of 20 ppm RhB solution was 60%. Therefore comparison to other research, our $ZnFe_2O_4$ nanoparticles show good photocatalytic activity. The degradation of RhB will be improved by prolonging irradiation.

## 4. Conclusion

$ZnFe_2O_4$ nanoparticles were synthesized by precipitation method with calcination temperature of 600, 700, and 800°C. The band gap energy of $ZnFe_2O_4$ nanoparticles was obtained to be 2.43, 1.85, and 2.39 eV, respectively. The $ZnFe_2O_4$ nanoparticles at calcination temperature of 700°C possess high crystallinity, high UV absorption, and good photocatalytic activity.

## 5. References

[1] Shirzadi A and Alireza N 2016 *J. Mol. Catalst A: Chem* **411** 222-229
[2] Jia Z *et al* 2015 *Suplatt Micro* **82** 174-187
[3] Krasae N and Wantala K 2015 *App Surf Sc* **380** 309-317
[4] Taoda H 2008 *Res. Chem. Intermed.* **34** 417-426
[5] Patil R P *et al Reslt in Phys.* **3** 129-133
[6] Thankachan R M *et al* 2015 *J. Pwr Sourc.* **282** 462-470
[7] Chun C *et al* 2016 *App. Cats B: Envirn* **182** 456-468
[8] Haifeng D *et al Ceramint.* **42** 10,520-10,525
[9] Jie Z *et al* 2015 *Snb: Chem.* **221** 55-62
[10] Yao Z *et al* 2016 *Chin. J. Chem Eng* **00471** 5
[11] Veith M *et al* 2005 *Chem. Mater.***17** 95-101
[12] Iqbal F *et al* 2016 *Proc. Eng* **148** 787-794
[13] Zhang Y *et al* 2016 *Jmmm* **409** 6-9
[14] Sun M *et al* 2016 *Elec.chimica* **190** 186-192.
[15] Suryanaraya C and Norton M G 1998 *Plenm.Press:New York*
[16] Truffault L *et al* 2010 *Matresbu* **45** 527-535.
[17] Kaneva N V and Dushkin D C 2011, *Col. and Surf a: Phys.chem and eng. Asp* **382** 211-218

# Effect of Synthesis Parameter on Crystal Structures and Magnetic Properties of Magnesium Nickel Ferrite ($Mg_{0.5}Ni_{0.5}Fe_2O_4$) Nanoparticles

**R Maulia, R A Putra and E Suharyadi**

Department of Physics, Universitas Gadjah Mada, Sekip Utara BLS 21 Yogyakarta, Indonesia

Email: esuharyadi@ugm.ac.id

**Abstract**. $Mg_{0.5}Ni_{0.5}Fe_2O_4$ nanoparticles have been successfully synthesized by using co-precipitation method and varying the synthesis parameter, i.e. synthesis temperature and NaOH concentration. X-ray Diffraction (XRD) pattern showed that nanoparticles have cubic spinel structures with an additional phase of $\gamma$-$Fe_2O_3$ and particle size varies within the range of 4.3 – 6.7 nm. This variation is due to the effect of various synthesis parameters. Transmission Electron Microscopy (TEM) image showed that the nanoparticles exhibited agglomeration. The observed diffraction ring from selected area electron diffraction showed that the sample was polycrystalline and confirmed the peak appearing in XRD. The coercivities showed an increasing trend with an increase in particle size from 44.7 Oe to 49.6 Oe for variation of NaOH concentration, and a decreasing trend with an increase in particle size from 46.8 to 45.1 Oe for variation of synthesis temperature. The maximum magnetization showed an increasing trend with an increase in the ferrite phase from 3.7 emu/g to 5.4 emu/g possessed in the sample with variations on NaOH concentration. The maximum magnetization for the sample with variations on synthesis temperature varied from 4.4 emu/g to 5.7 emu/g due to its crystal structures.

**Keywords:** Crystal structure, Magnetic Property, $Mg_{0.5}Ni_{0.5}Fe_2O_4$

## 1. Introduction

Ferrite nanoparticles ($MFe_2O_4$, M = Co, Ni, Mn, Zn, Mg) have attracted the interest of researcher in the field of functional materials due to their high electrical properties, high chemical and thermal stabilities [1]. Ferrite nanoparticles are widely used in technologies like drug delivery, magnetic recording media, Magnetic Resonance Imaging (MRI) contrast agent and tomography imaging [1, 2]. The properties of ferrite nanoparticles strongly depend on preparation method and synthesis parameter. Among ferrite nanoparticles, $NiFe_2O_4$ and $MgFe_2O_4$ have been widely investigated. $NiFe_2O_4$ has an inverse spinel structure with high magnetization, good conductivity and bad chemical leachability [3]. On the other hand, $MgFe_2O_4$ has a partial inverse spinel structure with high coercivity, high resistivity and good chemical leachability [4]. The introduction of $NiFe_2O_4$ to $MgFe_2O_4$ will provide alloy ferrite nanoparticles such as $Mg_{0.5}Ni_{0.5}Fe_2O_4$. The $Mg_{0.5}Ni_{0.5}Fe_2O_4$ is a soft magnetic nanoparticle which better conductivity than $MgFe_2O_4$ and better chemical leachability than $NiFe_2O_4$ [5-7]. The $Mg_{0.5}Ni_{0.5}Fe_2O_4$ nanoparticles have been synthesized by using several methods such as ceramic method, a sol-gel method, a solvent evaporation method, combustion, and co-precipitation [8].

Coprecipitation methods have been widely used due to easier controlling particle size and crystal structure of nanoparticles. In co-precipitation methods, the concentration of coprecipitation as the source of $OH^-$ in synthesis, and synthesis temperature that will influence the reaction rate of synthesis, were found to affect the change of the particle size [9-10]. The change of particle size and crystal structure will also change the magnetic properties of nanoparticles [11]. To explore the magnetic properties of $Mg_{0.5}Ni_{0.5}Fe_2O_4$ nanoparticles, variations on synthesis parameter such as NaOH concentration and synthesis temperature remains to be investigated. In this paper, the effect of NaOH concentration (as co-precipitant) and temperature on crystal structures and magnetic properties of $Mg_{0.5}Ni_{0.5}Fe_2O_4$ nanoparticles are investigated.

## 2. Experimental Method

The precursors such as $FeCl_3.6H_2O$ (Merck, Germany), $MgCl_2.6H_2O$ (Merck, Germany), $NiCl_2.6H_2O$ (Merck, Germany) and NaOH (Merck, Germany) were weighed carefully in accordance with the required stoichiometric proportion of salt. The $MgCl_2.6H_2O$ salt and $NiCl_2.6H_2O$ salt were mixed and dissolved them in 25 ml distilled water. Each $FeCl_3.6H_2O$ salt and NaOH salt were dissolved in 25 ml of distilled water. Then, it was mixed with the metal solution and added with 3.37 ml of HCl 37% (Merck, Germany). The precipitation was carried out by dropping the alloys solution into 3M, 6M, 10M and 12M of NaOH solution called sample A, B, C, and D respectively, at temperature of 30, 60, 90, and 150 °C called sample E, F, G and H respectively, for 1 hour and stirred at 1000 rpm. The precipitate was then washed six times for 30 minutes and dried at 95 °C for 5 hours.

The crystal structures were analyzed by using X-ray diffractometer (XRD) Shimadzu XD with CuKα1 radiation ($\lambda$ = 1.5406 Å). The magnetic properties were analyzed by using Vibrating Sample Magnetometer (VSM) Riken Denshi Co. Ltd. with $H_{max}$= 15 kOe and $M_{max}$ = 2.5 emu at room temperature. The morphology was analyzed by using Transmission Electron Microscopy (TEM) JEOL Jem-4000. The bonding analysis was characterized by using IR spectrometer Shimadzu Prestige-21. Particle size (*D*) was calculated from XRD peak broadening by using Scherrer formula, the values of interplanar distance (*d*) were estimated by using Bragg's equation, peaks reflection (*hkl*) was used to estimate the lattice constant (*a*) [6].

## 3. Results and Discussion

### 3.1 *Structural Studies*

Structural studies were done by using XRD measurement and then compared to standard data (JCPDS no 17-0464) in the $2\theta$ range of 25° – 80°. X-ray diffraction pattern for various NaOH and temperature variations is presented in Figure 1 and in Figure 2, respectively. All the diffractograms showed the highest peak corresponding to the (311) plane which indicates the formation of cubic spinel phase with an additional peak of $\gamma$-$Fe_2O_3$ phase (JCPDS no 39-1346). The broadening of diffraction peak confirms all samples experience the high value of microstrain [9]. The X-Ray density and microstrain of all samples were also calculated as shown in Table 1.

There was a peak shift at about 0.45°-0.82° from the standard data. This phenomenon may be due to the nonuniform microstrain in the crystal samples [12]. The lattice parameter of all samples were close to the value from the bulk. Lattice parameter is increasing relatively with the increase of NaOH concentration and decreasing with the increase of synthesis temperature. These could be explained as the effect of the increase of NaOH concentration leading to the decrease of cohesive energy, so the lattice parameter is increasing. The increase of synthesis temperature leads to the increase of cohesive energy, so the lattice parameter decreases [10]. The X-ray density value showed linear relation to the lattice parameter. X-ray density values tend to be constant because the composition of Mg and Ni is fixed for all samples.

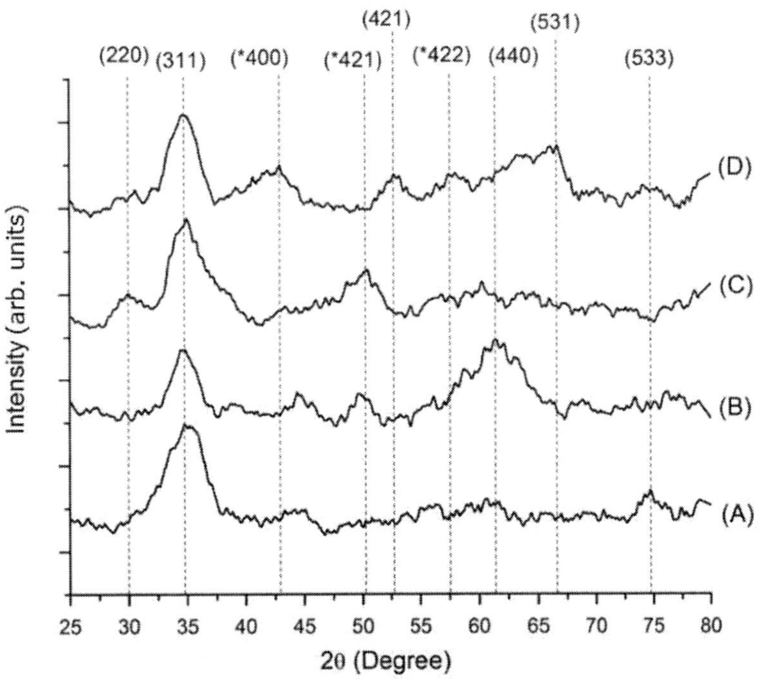


**Figure 1.** XRD pattern of $Mg_{0.5}Ni_{0.5}Fe_2O_4$ with NaOH concentration (A) 3, (B) 6, (C) 10, and (D) 12M.

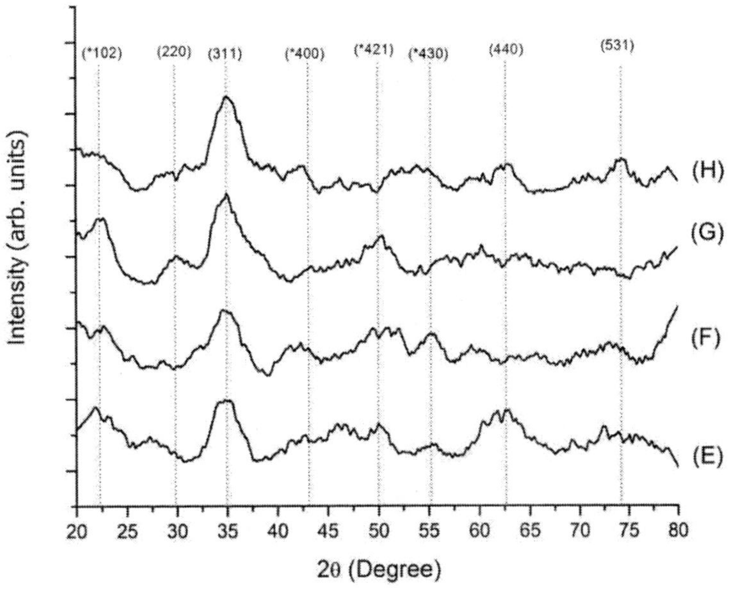


**Figure 2.** XRD pattern of $Mg_{0.5}Ni_{0.5}Fe_2O_4$ with synthesis temperature (E) 30, (F) 60, (G) 90, and (H) 150 °C

**Table 1.** Particle size, lattice constant, X-Ray density and microstrain of $Mg_{0.5}Ni_{0.5}Fe_2O_4$ nanoparticles synthesized at different NaOH concentration and synthesis temperature

| Sample | Synthesis parameter variable | Temperature (°C) | NaOH Concentration (M) | Lattice constant (Å) | X-Ray Density (g/cm$^3$) | Particle size (nm) | Microstrain |
|---|---|---|---|---|---|---|---|
| A | NaOH Concentration | 90 | 3 | 8.534 | 4.47 | 4.4 | 0.1059 |
| B | | 90 | 6 | 8.558 | 4.43 | 6.8 | 0.0684 |
| C | | 90 | 10 | 8.472 | 4.57 | 4.6 | 0.1004 |
| D | | 90 | 12 | 8.564 | 4.42 | 5.9 | 0.0786 |
| E | Synthesis Temperature | 30 | 10 | 8.541 | 4.46 | 5.3 | 0.0879 |
| F | | 60 | 10 | 8.563 | 4.42 | 4.7 | 0.0997 |
| G | | 90 | 10 | 8.472 | 4.57 | 4.6 | 0.1004 |
| H | | 150 | 10 | 8.478 | 4.56 | 5.8 | 0.0791 |

The particle size is not linear to the increase of the concentration of NaOH and the increase of synthesis temperature. This phenomenon may be due to the different stage that happens in every variation of the synthesis parameter [13]. The smaller size particles in the NaOH concentration of 3M and 10 M were nucleation stage when the structures are not crystallized better [13-15], and the values of microstrain are higher. The bigger particles in the NaOH concentration of 6M and 12M were in the crystallization stage [13-15] when the structures are crystallized better, and its microstrain values are lower. For the sample with synthesis temperature variations, at the temperature of 30° C to 90° C, the nucleation process still happens so that the structures of the samples are unstable and the microstrain values are increasing. The higher microstrain value, the more unstable particle in the nucleation process, so that the particles size decrease. The values of microstrain are not linear to the increase of synthesis parameter variations. The high value of microstrain caused by synthesis process that used low-temperature process.

*3.2 Morphological Studies*

The result of TEM characterization of the sample is depicted in Figure 3. Figure 3(a), and 3(b) shows the image of $Mg_{0.5}Ni_{0.5}Fe_2O_4$ nanoparticles synthesized at 90 °C and 10M NaOH.

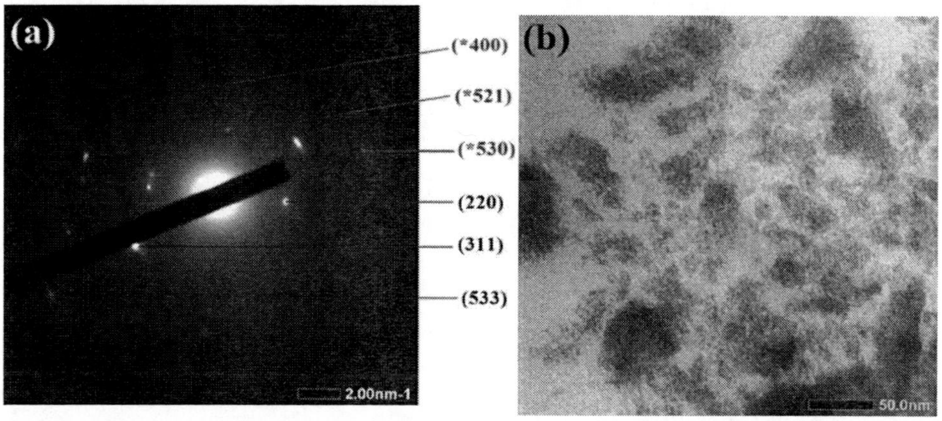


**Figure 3.**TEM Image of $Mg_{0.5}Ni_{0.5}Fe_2O_4$ (a) diffraction rings and (b) sample shape

The diffraction rings in Figure 3(a) confirms the peak as shown in XRD pattern. The diffraction spots are not clearly sharp due to come from not single crystal and small size of the nanoparticles. It coouls also possible that the satellites inside the grains. The image shows that the particles form an irregular shape and aggregate (Figure 3(b)). The distribution of particles size agrees well with the histograms obtained by XRD. The particles tend to aggregate because of the tiny size of particles. Smaller particles have higher surface potential energy, so they tend to aggregate to reduce it.

*3.3 IR Spectral Studies*

IR spectra of $Mg_{0.5}Ni_{0.5}Fe_2O_4$ nanoparticles was synthesized at 90 °C and 10M of NaOH shown in Figure 4. There are two main absorptions in the wavenuber of 362 $cm^{-1}$ and band of 609.51 $cm^{-1}$ corresponding to the stretching vibration of metal ions of the tetrahedral and octahedral site, respectively [6,8]. This case confirms that all $Mg_{0.5}Ni_{0.5}Fe_2O_4$ samples have cubic spinel structures. The absorption in 1357.89 $cm^{-1}$ and 1627.92 $cm^{-1}$ corresponds to bending vibration of O-H because of $H_2O$ molecules [6]. This phenomenon may due to the sample drying process which is not perfectly done. The sharp peaks absorption in the wavenumber of 3417.86 $cm^{-1}$ corresponds to stretching O-H and may be due to the not reacted $OH^-$ because of high concentration of NaOH.

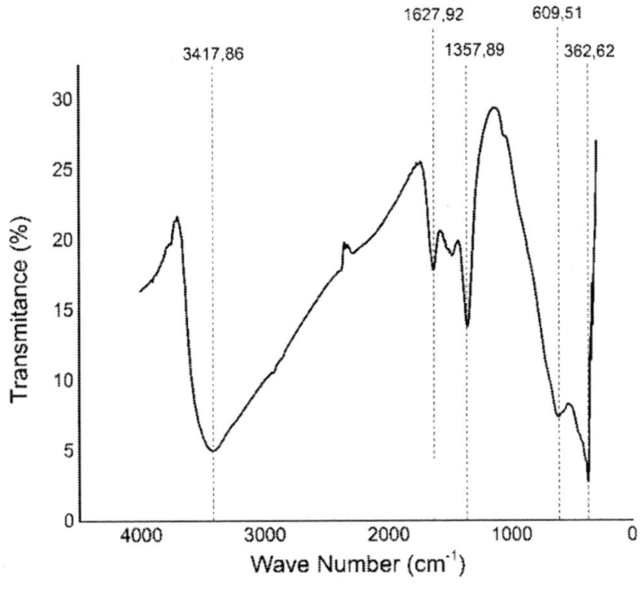


**Figure 4.** IR spectra of $Mg_{0.5}Ni_{0.5}Fe_2O_4$ systems

*3.4 Magnetic Behavior*

The magnetic properties of all samples were studied by using VSM. The results from the measurement of coercivity (*Hc*), maximum magnetization ($M_{max}$), magnetic anisotropy constant and magnetic moment [8] are shown in Table 2. Calculation of magnetic anisotropy (*K*) constant is by using Equation 1 [6].

$$K = Hc \times M_{max} \quad (1)$$

The value of coercivities increases relatively with increasing particle size for the sample with variations of NaOH concentration. This phenomenon leads to the particle with variations of NaOH concentration which has single domain [16]. For the sample with variations of synthesis temperature, the coercivities relatively decrease with increasing particle size confirming that the samples have multidomain [16]. Magnetic anisotropy constant is not linear with the coercivity. This case is due to the irregular shape of the particles [16].

**Table 2.** Particle size, ferrite phase ratio, coercivity, maximum magnetization, magnetic anisotropy constant and magnetic moment of $Mg_{0.5}Ni_{0.5}Fe_2O_4$ nanoparticles with variations of parameter synthesis

| Synthesis parameter variable | Particle size (nm) | Ferrite Phase Ratio (%) | *Hc* (Oe) | $M_{max}$ (emu/g) | *K* (erg/g) | *Magnetic moment* ($\mu_B$) |
|---|---|---|---|---|---|---|
| NaOH Concentration | 4.4 | 54.9 | 45.18 | 3.71 | 167.89 | 0.15 |
| | 4.6 | 57.1 | 46.83 | 4.70 | 220.20 | 0.18 |
| | 5.9 | 72.8 | 44.72 | 5.47 | 244.54 | 0.21 |
| | 6.8 | 66.2 | 49.56 | 3.45 | 170.96 | 0.14 |
| Synthesis Temperature | 4.6 | 57.1 | 46.83 | 4.70 | 220.20 | 0.18 |
| | 4.7 | 46.8 | 42.97 | 4.43 | 190.48 | 0.17 |
| | 5.3 | 44.5 | 46.25 | 5.30 | 245.12 | 0.21 |
| | 5.8 | 59.9 | 45.13 | 5.77 | 260.58 | 0.22 |

## 4. Conclusion

Powder of $Mg_{0.5}Ni_{0.5}Fe_2O_4$ nanoparticles has been successfully synthesized by co-precipitation method. Variations on synthesis parameter i.e. NaOH concentrations and synthesis temperature were found to influence the particle size. TEM image confirms the peaks shown in XRD. IR spectral showed vibration in tetrahedral and octahedral sites that conform spinel structures. The coercivity increases relatively by increasing particle size with variations of NaOH concentration. The coercivity for samples with variations on synthesis parameter decreases relatively by increasing particle size that leads to be varied in the magnetic properties of $Mg_{0.5}Ni_{0.5}Fe_2O_4$ nanoparticles.

## 5. References

[1] Goldman A, 2006, *Modern Ferrite Technology,* New York: Springer Science
[2] Schmided G., 2010, *Nanoparticles: From Theory to Application,* Weinhem: WILEY-VCH Verlag GmbH & Co. KGaA
[3] Sivakumar P, Ramesh R, Ramanand A, Ponnusamy S and Mutamizchelvan C, 2013, *Journal of Alloy and Compounds,* vol. **536** p 6-11
[4] Naseri M G, Ara M H and Saion E B, 2014, *Journal of Magnetism and Magnetic Materials* vol **350** p 141
[5] Naeem M, Shah N A, Gull H and Masqood A, 2009, *Journal of Alloys and Compounds* vol **487**
p 739
[6] Moradmard H, Shayesteh S F, Tohidi P, Abbas Z and Khalegi M, 2015, *Journal of Alloys andCompounds* vol **650** p 116
[7] Berchmans L J, Selvan R K, Kumar P N S and Augustin C O, 2003, *Journal of Magnetism and Magnetic Materials* vol **279** p 103
[8] Hankare P P, Jadav S P, Sankpa l U B, Chavan S S, Chougule B K and Waghmare B K, 2009, *Journal of Alloys and Compounds* vol **475** p 926
[9] Briceno S, Escamilla W B, Silva P, Delgado G E, Plaza E, Palacios J, and Canizales E, 2012, *Journal of Magnetism and Magnetic Materials* vol **324** p 2926-2931
[10] Godbole B, Badera N, ShrivastaS B, Jain D, Sharath L S, Chadra and Ganesan V, 2013, *Physics Precedia* vol **49** p 58-66
[11] Mathew D S and Juang R, 2006, *Chemichal Engineering Journal* vol **129** p 51-56
[12] Cullity B D and Graham C D, 2009, *Introduction to Magnetic Materials Second Edition,* USA:

IEEE Press
[13] WeddemanA, Ennen I, Regtmeier A, Albon C, Wolff A, Eckstädt K, Mill N, Peter M K H, Mattay J, Plattner C, Sewald N and Hütten A, 2010, *Beilstein Journal of Nanotechnology* vol **1** p 75-93
[14] Arshadi S, Moghaddam J and Eskardian M, 2014, *Korean Journal of Chemical Engineering* vol **31** p 2020-2026
[15] Callister W D, 2007, *Material Science and Engineering,* USA: John Wiley & Sons Inc
[16] Cullity B D, 1956, *Element of X-Ray Diffraction*, USA: Addison Wesley Publishing Company Inc

**Acknowledgements**
The reported work was financially supported by *Hibah Kompetensi* (HIKOM) of the Ministry of Research and Technology of High Education for the period of 2015 – 2017, and Nanofabrication Platform Consortium Project of Nagoya University, Minister of Culture, Sports, Science and Technology (MEXT) Nano-Project Platform, Japan of Period 2012 - 2016. Authors would like to thank Institute of Materials and Systems for Sustainability, Nagoya University for providing VSM instrument to magnetic measurement.

# Study on The Influence of Crystal Structure and Grain Size on Dielectric Properties of Manganese Ferrite ($MnFe_2O_4$) Nanoparticles

**Kurnia, Heriansyah and E Suharyadi**

Department of Physics, Gadjah Mada University, Yogyakarta, Indonesia

Email: esuharyadi@ugm.ac.id

**Abstract**. Manganese ferrite ($MnFe_2O_4$) has been successfully synthesized by using co-precipitation methods with a various particles size of 13.1, 13.8, 16.2, 18.1, and 18.4 nm. Analysis of crystal structure properties which has been carried out by using XRD (X-Ray Diffraction) method revealed that all samples belong to cubic spinel structure. The measurement used computerized impedance spectroscopy system in the frequency range of 10-120 Hz. The real dielectric constant ($\varepsilon'$), imaginary dielectric constant ($\varepsilon''$), and impedance (Z) were determined at room temperature as a function of frequency. The dielectric constant (real and imaginary) and impedance decreased with frequency up to 70 kHz, and then it was constant at a frequency more than 70 kHz. The dielectric properties depending on the crystal structure are also investigated, it consists of the degree of crystallinity, lattice parameter, strain, X-ray density and the impurities of the sample. In the range of grain size of 13.1-18.4 nm, the real ($\varepsilon'$) and imaginary ($\varepsilon''$) dielectric constants, and loss tangent increased with the increase of the grain size. In addition, impedance (Z) decreased with increasing grain size. The dielectric constant was found to be the highest for $MnFe_2O_4$ with a particle size of 18.1 nm because of low strain value of 0.028 and high X-ray density value of 5.11. Therefore, a sample with the highest dielectric properties is a sample with grain size of 18.1 nm with real dielectric constant of 311; imaginary of 118; and impedance of 328 kΩ at 10 kHz. The frequency variation of ($\varepsilon'$ and $\varepsilon''$) is explained by Knop phenomenological theory and hopping of electrons between $Fe^{2+}$ and $Fe^{3+}$ on octahedral sites.

**Keywords:** $MnFe_2O_4$ Nanoparticle, Complex Dielectric Constant, Grain Size, and Crystal structure

## 1. Introduction

Magnetic nanoparticles such as manganese ferrite possess excellent magnetic and dielectric properties. These spinel ferrite nanoparticles are generally used in magnetic recording media, a microwave device, and magnetic resonance imaging [1]. The major advantages of the ferrite material are that they perform higher efficiency, low cost, and appropriate dielectric loss and hence meet potential applications in microwave device and memory cores [2]. The dielectric properties of the spinel ferrites are generally affected by factors of preparation method, cation distributions, grain size, etc. [3]. Rani [4] studied the dielectric properties of spinel ferrite and reported that the value of dielectric constant increases with increasing particle size. These phenomena may be due to the fact that when the particle size increases, the number of polarizable ferrous ions on the octahedral site increases. Rezlescu and Rezlescu [5] explain the composition dependence of the dielectric constant with the assumption that the mechanism of dielectric polarization is similar to that of the conduction. They have observed that the electron exchange between $Fe^{2+}$ and $Fe^{3+}$ which results in local displacement of electrons in the direction of the electric field determines polarization. Narayanasamy [6] reports that the number of Fe ions at the B sites decreases with the grain size reduction. This phenomenon results in a reduction of the number of $Fe^{2+} \leftrightarrow Fe^{3+}$ pairs at B sites and therefore the polarization also decreases with the reduction in grain size. However, the dielectric properties of nanocrystalline manganese ferrites have not yet been investigated in detail.

In this paper, the influence of crystal structure and grain size on dielectric properties of manganese ferrite nanoparticles that were prepared through co-precipitation method is investigated. The permittivity and impedance with various frequencies at room temperature of dielectrics are analyzed.

## 2. Experimental Method

Manganese ferrite has been synthesized by using co-precipitation method with $FeCl_3.6H_2O$ (Merck, Germany) and $MnSO_4.H_2O$ (Merck, Germany). The synthesis was started by putting 3.37 mL of HCl (37%) into 0.84 gram of $MnSO_4.H_2O$ and 2.703 gram of $FeCl_3.6H_2O$ which were dissolved into 20 mL of distilled water and mixed until homogeneous. The solution was dropped to NaOH and stirred with 770 rpm velocity at a certain temperature. Furthermore, the solution was precipitated by external field (permanent magnet). The precipitation was washed in seven times by using distilled water to obtain a pure MnFe2O4. Samples were dried in a furnace at 80 °C for four hours. The crystal structure and grain size of samples were analyzed by using X-Ray Diffractometer (XRD) Shimadzu model XD-3H (Cu-K$\alpha$ =1.5406 Å), and Transmission Electron Microscope (TEM) Jeol Jem-1400. In this study, the synthesis parameter of $MnFe_2O_4$ performed with NaOH concentration and time centrifugation is shown in Table 1.

**Table 1.** Synthesis of Parameters Variation of $MnFe_2O_4$ Nanoparticles

| Sample name | Mass of $MnSO_4H_2O$ (g) | Mass of $FeCl_36H_2O$ (g) | Volume of HCl (mL) | NaOH (M) | Stirring time (minutes) | Synthesis temperature (°C) |
|---|---|---|---|---|---|---|
| N5 | 0.84 | 2.703 | 3.37 | 1.5 | 120 | 80 |
| N3 | 0.84 | 2.703 | 3.37 | 5 | 120 | 80 |
| N4 | 0.84 | 2.703 | 3.37 | 10 | 120 | 80 |
| N2 | 0.84 | 2.703 | 3.37 | 10 | 90 | 80 |
| N1 | 0.84 | 2.703 | 3.37 | 10 | 150 | 80 |

Dielectric properties were measured by using computerized impedance spectroscopy on a circuit of AC in the frequency range from 10 to120 kHz. Powder samples were firstly compacted into pellets of 1.23 cm in diameter at the compressive force of 50.000 N. The dielectric properties measured such as complex dielectric and impedance as a function frequency.

## 3. Results and Discussion

The XRD spectra of the samples are shown in Figure 1. All of the main peaks are indexed as the spinel of $MnFe_2O_4$. The average crystalline size of $MnFe_2O_4$ samples was calculated from X-Ray line broadening of reflections of (220), (311), (400), (333), and (440) by using Scherer's equation. The broadening of the reflection peaks indicates that the samples are composed of small particles [1]. Figure 1 shows a sample of N1 with a small grain size that has broadening of the reflection profile. The broadening of the reflection profiles is not only caused by the small particle or the crystallite size but also due to the lattice strain, instrumental errors, dislocation, impurities, etc. [7].

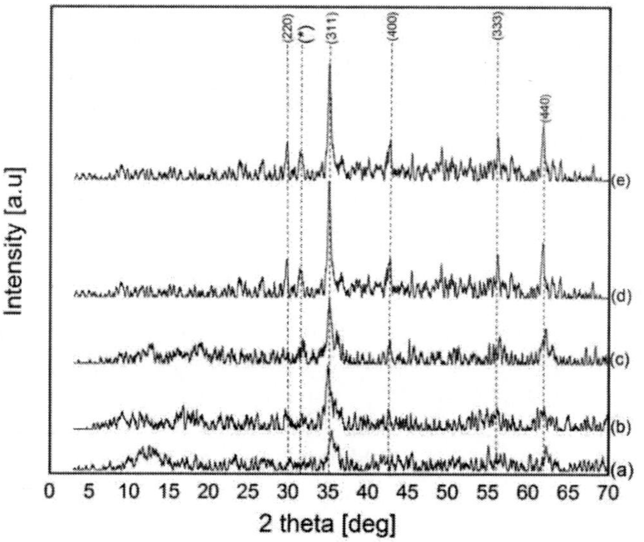


**Figure 1**. XRD pattern of $MnFe_2O_4$: (a) N1; (b) N2; (c) N3; (d) N4 and (e) N5

**Table 2**. XRD Analysis of $MnFe_2O_4$

| Sample name | Grain size (nm) | Lattice parameter (Å) | X-ray Density ($g/cm^3$) | Strain |
|---|---|---|---|---|
| **N1** | 13.1 | 8.36 | 5.26 | 0.034 |
| **N2** | 13.8 | 8.48 | 5.04 | 0.035 |
| **N3** | 16.2 | 8.44 | 5.11 | 0.029 |
| **N4** | 18.1 | 8.44 | 5.11 | 0.028 |
| **N5** | 18.4 | 8.48 | 5.04 | 0.025 |

The lattice parameter, strain, X-ray density and particle size of different parameter synthesized are compared in Table 2. the average grain size of the nanoparticle is calculated from the highest peak of X-Ray diffraction of (311) by using Scherer's equation.

$$t = \frac{k\lambda}{B\cos\theta}, \tag{1}$$

where $t$ is grain size, $\lambda$ is the wavelength of the Cu-K$\alpha$ radiation ($\lambda$ = 1.5406Å), $\theta$ is the Bragg's diffraction angle, and $B$ is the full width at half maximum (FWHM) of the peak in the X-ray diffraction pattern [2].

X-Ray density depends on the lattice constant and molecular weight of the sample which is calculated by using the relation [2].

$$d_x = \frac{8M}{Na^3} \tag{2}$$

where $M$ is the molecular weight of the ferrite sample, $N$ is the Avogadro number, $a$ is lattice constant and 8 represents the number of molecules per unit cell. The strain is considered as a crystal defect from nanoparticles $MnFe_2O_4$. The presence of this strain weakens the electrical response of the sample.

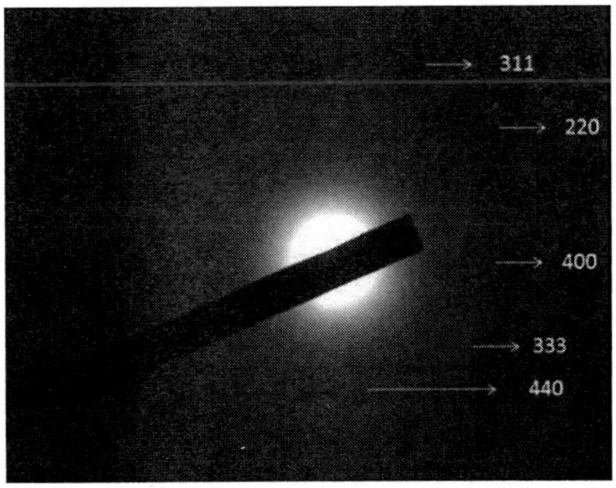


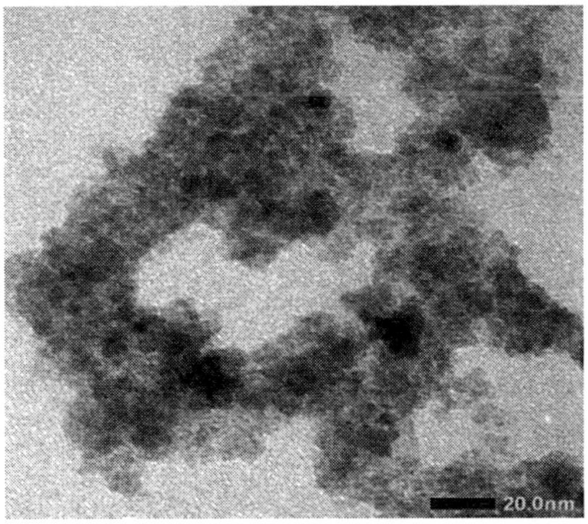


**Figure 2.** (a) Diffraction pattern and (b) Morphology of nanoparticles $MnFe_2O_4$

Ring diffraction patterns are closely related to XRD analysis as shown in Figure 2(a). which corresponds to the crystal plane (440), (333), (400), (311) and (220). Ring discontinue indicated the polycrystalline structure of nanoparticles and high value of grain size. Morphology of nanoparticles looks more dispersed and distributed relatively homogeneous as shown in Figure 2(b). The dielectric properties of $MnFe_2O_4$ as a function of frequency is shown in Figure 2(b).

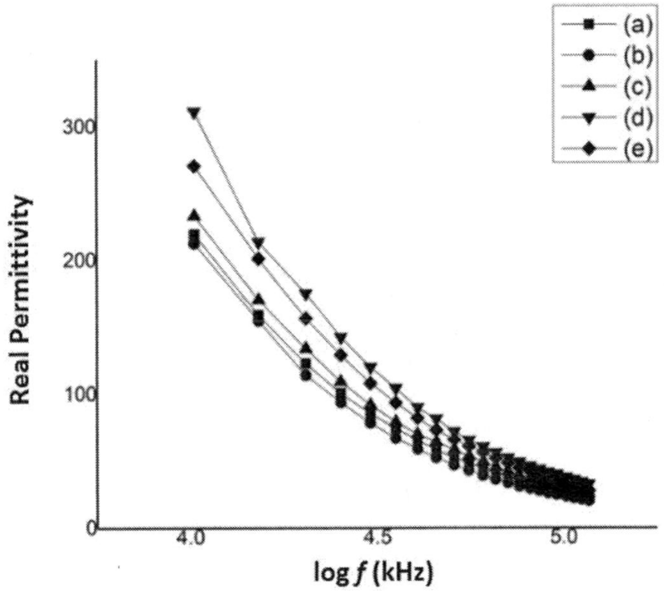


**Figure 3**. Real Permittivity of $MnFe_2O_4$ (a) N1; (b) N2; (c) N3; (d) N4 and (e) N5

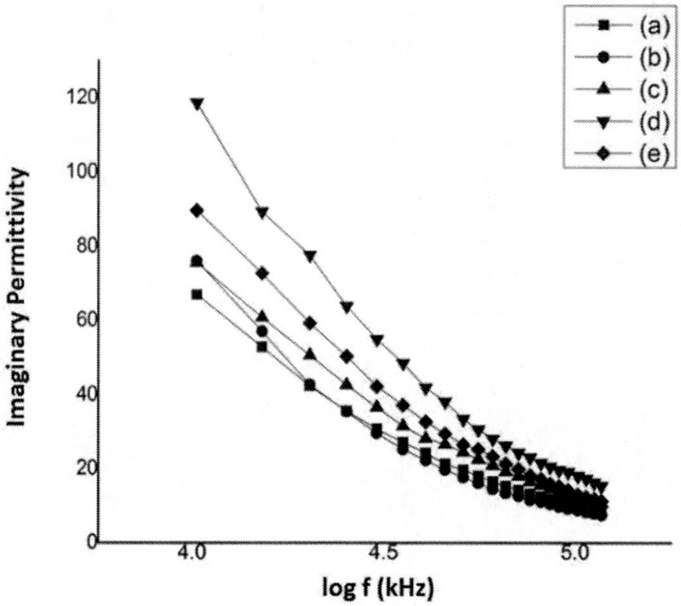


**Figure 4.** Imaginary Permittivity of $MnFe_2O_4$ (a) N1; (b) N2; (c) N3; (d) N4 and (e) N5

Figures 3 and 4, show that dielectric properties (Real and Imaginary) decrease as a function frequency. The highest value of dielectric constant at a lower frequency can be explained by the Maxwell-Wagner interfacial polarization and Koop's phenomenological theory [8]. The low-frequency dispersion may be due to the inhomogeneous nature of studied dielectric of which compound has two layers [9, 10]. The first layer is the larger ferrite grains of high conducting elements separated by the second thin layer of relatively weak conducting substance (grain boundary). The grain boundaries and grains are more effective at low and high frequencies respectively. As grain boundaries offer high resistance, the electron gets crowded there and thus produces an enhanced space charge polarization. As a result, the material has larger permittivity in the low-frequency region [7]. It is expected that the dielectric constant may decreases with the increase of the frequency until it reaches a nearly constant because beyond a certain frequency of the electric field. Further, the moment dipole on the material may not follow the alternating field, so it reaches a constant value [11]. The real and imaginary dielectric constants tend to increase as the grain size increases. According to the Iwauchi, for a different structure of large particles contains more ferrous ions than the small particles and becomes more conductive. The influence of crystal structure on dielectric properties can be explained by the structure of ferrite.

Ferrite has spinel structure into a cubic close-packed structure of oxygen ions. The ions occupy two types sites. One of them is called tetrahedral sites of which is surrounded by four oxygen ions in tetrahedral sites. The other sites are known as the octahedral of which the cation is coordinated by six oxygen ions in octahedral site. The exchange of $Fe^{2+}$ and $Fe^{3+}$ ions in spinel structure will result in local displacement that is caused by an electric field, so more and more sites in spinel structure can increase the polarization because of increasing the number of Fe ions. Increasing grain size in ferrite causes increasing the number of ion Fe in site octahedral, this phenomenon results in an increasing the number of $Fe^{2+}$ $Fe^{3+}$ pairs in octahedral sites, and therefore the polarization also increases with the increasing grain size [4]. Rezlescu and Rezlescu have studied the dielectric properties of Cu-Mn and Cu-Ni Ferrites by varying the composition, frequency, and temperature. They explained the composition dependence of the dielectric constant with the assumption that the mechanism of dielectric polarization is similar to the conduction process. They have observed that the electron exchange between $Fe^{2+}$ and $Fe^{3+}$ which results in local displacement of electrons in the directions of the electric field determines the polarization [5]. Figure 5 shows Cole-Cole plot of $MnFe_2O_4$ nanoparticles. The Cole-Cole plot is capable to show the relaxation of material and provide information on the kind of electrical processes occurring within the material such as polarization process [6]. It is also observed that relaxation time increases as a grain size increases.

Impedance as a function frequency obtained by the impedance spectroscopy is shown in Figure 6. The impedance decreases as the frequency increases. The impedance at low frequencies is high because the influence of grain boundaries are not conductive. The effect of grain boundaries reduced is due to the increase of electrical energy applied so that conductive grains are more predominant at high frequency. Figure 6 also shows the impedance increases as a grain size decreases. It may be due to the particle surface that poor conductor becomes thicker than the higher grain size [12].

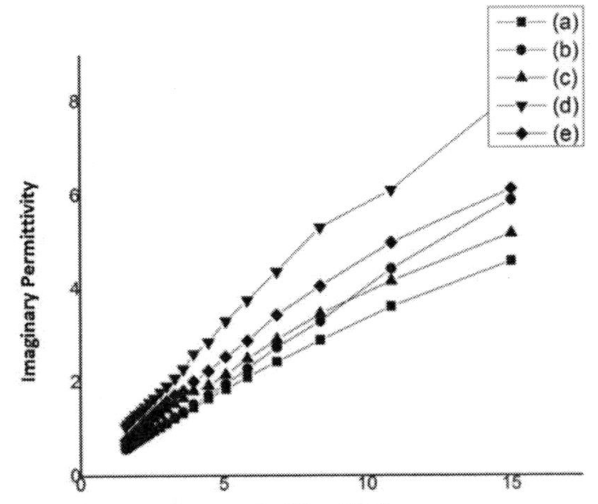


**Figure 5.** Cole-Cole plot of $MnFe_2O_4$: (a) N1; (b) N2; (c) N3; (d) N4 and (e) N5

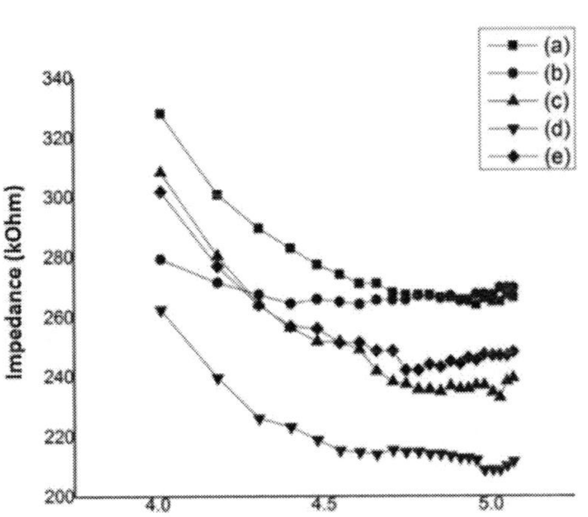


**Figure 6**. Impedance as a function of frequency of $MnFe_2O_4$ (a) N1; (b) N2; (c) N3; (d) N4 and (e) N5

IOP Conf. Series: Materials Science and Engineering **202** (2017) 012046 doi:10.1088/1757-899X/202/1/012046

## 4. Conclusion

$MnFe_2O_4$ nanoparticles were successfully prepared by coprecipitation method. The formation of spinel ferrite has been investigated by X-Ray Diffraction method. It is also obtained that dielectric properties increase as the particle size increases. It may be due to the different structure of large particles containing more ferrous ions than the small particles, and becoming more conductive. In addition, relative permittivity decreases as a function of frequency. It is because above a certain frequency of the external field, the dipole is not able to follow the frequency of alternating current.

## 5. References

[1] Kumar E R, Kamzin A S, Prakash T 2015 *J. Magn. Magn. Mater* **378** 389-396
[2] Hashim M, Alimuddin, Kumar S, Koo B H, Shirsath S E, Mohammed E M J, Shah, Kotnala R K, Choi H K, Chung H, Kumar R 2012 *J. Alloys and Compounds* **518** 11-18
[3] Samaha F S H A, Ismail M I M, 2014 *Mater Science in Semiconductor Processing* **19** 50-56
[4] Kumar E R, Aleksandr Kamzin S, Prakash T 2015 *J. Magn. Magn. Mater* **378** 389-396
[4] Ponpandian N, Narayanasamy A 2002 *Journal of Applied Physic* **92** 5
[5] Rezlescu N and Rezlescu E 1974 *Phys. Status Solid* A **23** 575
[6] Behera C, Choudhary R, Das P 2015 *Ceramics International* **41** 13042-13054
[7] Behera C, Choudhary R N P, Piyush R D 2015 *J. Mater. Sci.* **26** 2343-2356
[8] Koops C G 1951 *Phys. Rev.* **83** 121
[9] Jain H, Hsieh C H, 1994 *J.Non-Cryst. Solids* **172** 1408-1412
[10] Swapna S N M G, Malinil K A, Joy P A, Anantharaman M R, 2007 J. *Phys D: Appl Phys* **40** 1593-1602
[11] Muthafar F, Al-Hilli I 2011 *Journal of Physics* **14** 6-12
[12] Heriansyah, Mustawarma, Suharyadi E 2015 *Spectra Jurnal* **16** 3

**Acknowledgments**

The authors would like to thank the *Hibah Kompetensi* (HIKOM) of the Ministry of Research and Technology of High Education for the period of 2015-2017

# Study of Structural and Magnetic Properties of Silica and Polyethylene Glycol (PEG-4000)-Encapsulated Magnesium Nickel Ferrite ($Mg_{0.5}Ni_{0.5}Fe_2O_4$) Nanoparticles

**F Deswardani, R Maulia and E Suharyadi**

Department of Physics, Universitas Gadjah Mada, Sekip Utara BLS 21 Yogyakarta Indonesia

Email: esuharyadi@ugm.ac.id

**Abstract.** $Mg_{0.5}Ni_{0.5}Fe_2O_4$ has been successfully synthesized by using co-precipitation method. Two series of $Mg_{0.5}Ni_{0.5}Fe_2O_4$ silica encapsulated have been prepared by varying the concentration of silica and variation of PEG-4000 concentration. Analysis of X-Ray Diffraction (XRD) pattern showed that nanoparticles contained $Mg_{0.5}Ni_{0.5}Fe_2O_4$ spinel phase and $\gamma$-$Fe_2O_3$ phase with a particle size of 5.1 nm. The various of silica encapsulation give rise to produce a new phase of $SiO_2$ and increase the particle size to 16.1 nm. PEG-4000 encapsulation affected to create a new phase of $\gamma$-FeO(OH), and reduce the particle size down to 4.5 nm. Fourier Transform Infra Red (FTIR) for $Mg_{0.5}Ni_{0.5}Fe_2O_4$ showed absorption peaks around 300-600 $cm^{-1}$ which are M-O bond vibration. After silica encapsulation, there was new bond vibration typical of silica such as Si-O-Si (1049.28 $cm^{-1}$), Si-OH (779.24 $cm^{-1}$), and Si-O-Fe (570.93 $cm^{-1}$). The PEG-4000 encapsulation creates a new vibration for typical of PEG-like of C-O (1103.28 $cm^{-1}$) and C-H (925.83, 1481.33, and 2924.09 $cm^{-1}$). Both of encapsulations series have M-O bond vibration indicating the presence of $Mg_{0.5}Ni_{0.5}Fe_2O_4$. After silica encapsulation, the coercivity of $Mg_{0.5}Ni_{0.5}Fe_2O_4$ decreased from 47 Oe to 10 Oe due to the decrease of particle size. Even though, the discrepancy of particle size as the effect of PEG-4000 encapsulation, the coercivity just slightly reduced to 46 Oe. The saturation magnetization of $Mg_{0.5}Ni_{0.5}Fe_2O_4$ decreased from 4.7 emu/g to 1 emu/g after silica encapsulation because diamagnetic $SiO_2$. Otherwise, the saturation magnetization increased to 7.7 emu/g after PEG-4000 encapsulation because of domination of $Mg_{0.5}Ni_{0.5}Fe_2O_4$ phase ratio.

**Keywords:** Magnesium nickel ferrite ($Mg_{0.5}Ni_{0.5}Fe_2O_4$), coprecipitation, encapsulation, silica, PEG-4000

## 1. Introduction

Recently, research on magnetic nanoparticles (MNPs) has been extensively done. The magnetic nanoparticles have quantum size effect (the transition region between atoms and bulk), and the large surface area dramatically changes some of the magnetic properties and exhibits superparamagnetic phenomena [1]. Spinel ferrite is one of the most important magnetic nanoparticles because their structure, chemical, electrical, and magnetic properties make them ideal for many applications. Spinel ferrite has formula $MFe_2O_4$, where M is a divalent cation.

$MgFe_2O_4$ exhibits soft magnetic, high resistivity, non-cytotoxic, and magnetic anisotropy which are lower than of other spinel ferrites because $Mg^{2+}$ is carrying no magnetic moment, so the magnetic coupling is purely originated from the magnetic moment of Fe cations [2, 3]. On the other hand,

$NiFe_2O_4$ exhibits soft magnetic, high resistivity, cytotoxic [2], and magnetic anisotropy which are higher than $MgFe_2O_4$ because $Ni^{2+}$ has a magnetic moment. When $Mg^{2+}$ ions are combined with $Ni^{2+}$ ions with the same composition, $Mg_{0.5}Ni_{0.5}Fe_2O_4$ will be found.

According to properties of ferrites used for biomedical application, ferrite should be biocompatible and non-toxic. Magnetic nanoparticles tend to agglomerate because of their large surface area. Therefore ferrite has encapsulated with organic materials (like chitosan, Polyvinyl alcohol (PVA), Polyethylene glycol (PEG), etc.) or inorganic materials (like gold, silica, etc.) [4]. Silica has silanol groups which can easily react with coupling agents providing strong attachment of surface ligands on magnetic nanoparticles; encapsulated silica increases the stabilization of magnetic nanoparticles in liquid, and it also can prevent agglomeration by increasing the surface charges hence the electrostatic repulses among particles [4]. The silica can produce by using precursor like tetraethyl orthosilicate (TEOS), tetramethyl orthosilicate (TMOS), and sodium silicate ($Na_2SiO_3$). The precursors like TEOS and TMOS are environmentally unfriendly, toxic, and expensive. $Na_2SiO_3$ precursors can be used as a source of silica because they are environmentally friendly, non-toxic, and inexpensive [5, 6]. On the other hand, PEG is biocompatible, low-cost, non-toxic, and hydrophilic [7]. PEG with molecular weight under 1000 g/mol is expensive, so PEG-4000 can be used as encapsulation material because it is inexpensive.

Synthesized of $MgNiFe_2O_4$ by coprecipitation has been investigated [8, 9]. In the recent years, the research about silica encapsulated nanocomposites as $NiFe_2O_4$ [10], $KFeO_2$ [11], etc. has gained much attention. PEG also has much attention as encapsulation materials such as for $CaFe_2O_4$ [12], $Mn_{(1-x)}Zn_xFe_2O_4$ [13], etc. However, the study of the influence of silica and PEG encapsulated $Mg_{0.5}Ni_{0.5}Fe_2O_4$ on the structural and magnetic properties has not been paid much attention. In this work, $Mg_{0.5}Ni_{0.5}Fe_2O_4$ will be synthesized by using coprecipitation method and encapsulated by varying the concentration of silica and PEG-4000. The influence of silica and PEG-4000 on crystal structure and magnetic properties of $Mg_{0.5}Ni_{0.5}Fe_2O_4$ will also be investigated.

## 2. Experimental Method

Two series of silica and PEG-4000 encapsulated magnetic nanoparticle $Mg_{0.5}Ni_{0.5}Fe_2O_4$ were prepared by two steps for each encapsulation materials. Firstly, $Mg_{0.5}Ni_{0.5}Fe_2O_4$ was prepared by using the coprecipitation method. In the second step, silica encapsulated $Mg_{0.5}Ni_{0.5}Fe_2O_4$ was prepared by varying silica concentration along with the using of sodium silicate (SS) precursor. The PEG-4000 encapsulation also prepared by varying PEG concentration. XRD patterns were recorded on Shimadzu XD by using $CuK_\alpha$ radiation. FTIR transmission spectra were taken by using FTIR spectrometer Shimadzu Prestige-21. Magnetic properties of samples were studied by using Vibrating Sample Magnetometer (VSM) Riken Denshi.

### *2.1. Preparation of magnetic nanoparticle*

The magnetic nanoparticle of $Mg_{0.5}Ni_{0.5}Fe_2O_4$ was prepared by coprecipitation method, by using aqueous solutions of magnesium, nickel, and iron salts. The stoichiometric amount of 2.703 g $FeCl_3.6H_2O$ (Merck, Germany) were dissolved in 25 mL of distilled water, 0.508 g $MgCl_2.6H_2O$ (Merck, Germany), and 0.594 g $NiCl_2.6H_2O$ (Merck, Germany) were also dissolved in 25 mL of distilled water and then 3.37 mL HCl was added. The solution was then added in dropwise to 10 M NaOH (Merck, Germany) solution. The reaction was kept under constant stirring at 90°C for 1 hour. Precipitated ferrite nanopowder was washed with distilled water for six times. Finally, the sample was dried in a furnace at 95 °C for 5 hours.

### *2.2. Silica-encapsulated $Mg_{0.5}Ni_{0.5}Fe_2O_4$*

The nanoparticle of $Mg_{0.5}Ni_{0.5}Fe_2O_4$ dissolved to 30 mL silica solution with different concentration (Table 1). Silica solution was prepared by adding sodium silicate to the distilled water stirring at room temperature for 30 minutes. Then, 0.6 g powder of $Mg_{0.5}Ni_{0.5}Fe_2O_4$ was added to silica solution and stirred at room temperature for 5 hours. Encapsulated ferrite was washed by using distilled water for five times. Finally, the sample was dried at room temperature and then dried in a furnace at 50 °C for 30 minutes.

**Table 1** Sample of $Mg_{0.5}Ni_{0.5}Fe_2O_4$ with different silica concentration

| Sample | $Mg_{0.5}Ni_{0.5}Fe_2O_4$ : Silica solution |
|---|---|
| **A** | 0.6 g : 50% (15 mL SS + 15 mL aquades) |
| **B** | 0.6 g : 30% (9 mlL SS + 21 mL aquades) |
| **C** | 0.6 g : 20% (6 mL SS + 24 mL aquades) |
| **D** | 0.6 g :15% (4.5 mL SS+25.5 mL aquades) |
| **E** | 0.6 g : 10% (3 mLSS + 27 mL aquades) |
| **F** | 0.6 g : 5% (1.5 mL SS +28.5 mL aquades) |

*2.3. PEG-4000 encapsulated $Mg_{0.5}Ni_{0.5}Fe_2O_4$*

$Mg_{0.5}Ni_{0.5}Fe_2O_4$ nanoparticles were dissolved in 10 mL of PEG-4000 solution with different concentration (Table 2). The PEG -4000 solution was prepared by adding PEG -4000 to the distilled water stirred at room temperature for 30 minutes. Then, 0.6 g of powder $Mg_{0.5}Ni_{0.5}Fe_2O_4$ was added to PEG-4000 solution and stirred at room temperature for 1 hour. Encapsulated ferrite was washed by using distilled water for five times. Finally, the sample was dried at room temperature.

**Table 2** Sample of $Mg_{0.5}Ni_{0.5}Fe_2O_4$ with different PEG-4000concentration

| Sample | $Mg_{0.5}Ni_{0.5}Fe_2O_4$ : PEG-4000 | PEG concentration |
|---|---|---|
| **G** | 3:1 (0.6 g : 0.2 g) | 25% |
| **H** | 2:1 (0.6 g : 0.3 g) | 33% |
| **I** | 1:1 (0.6 g : 0.6 g) | 50% |
| **J** | 1:2 (0.6 g : 1.2 g) | 67% |
| **K** | 1:3 (0.6 g : 1.8 g) | 75% |
| **L** | 1:4 (0.6 g : 2.4 g) | 80% |

## 3. Results and Discussion

*3.1. Silica-Encapsulated $Mg_{0.5}Ni_{0.5}Fe_2O_4$ Nanoparticles*

Figure (1a) shows the XRD spectra of $Mg_{0.5}Ni_{0.5}Fe_2O_4$, and (1.b) silica encapsulated of $Mg_{0.5}Ni_{0.5}Fe_2O_4$. The peaks existed in XRD spectra are indexed as (220), (311), (222), and (511) planes. These peaks are the characteristics planes of spinel ferrite (matched with JCPDS card no 36-0398). Figure. 1.a also shows two weak peaks of $\gamma$-$Fe_2O_3$ phase (matched with JCPDS card no 39-1346). The silica encapsulation as depicted in Figure 1b, showed peaks planes of $SiO_2$ phase (matched with JCPDS card no 45-0131). The silica encapsulation shifted the peak position of (311) to a higher diffraction angle from 35° to 36° which indicates the decrease of the average lattice spacing due to the strain by the silica. The shift of the (311) peak indicates the existing of the interaction between nanoparticles and silica as the material encapsulation.

As shown in Table 3, the calculated sizes of the $Mg_{0.5}Ni_{0.5}Fe_2O_4$ particles are 5.1 nm, and the encapsulation enlarges to 16.1 nm. The encapsulation provides the nanoparticles begin to react with silica, and hence silica successfully encapsulated the $Mg_{0.5}Ni_{0.5}Fe_2O_4$ nanoparticles [11]. $Mg_{0.5}Ni_{0.5}Fe_2O_4$ nanoparticles became the core and surrounded by silica with a thickness that can manipulate the size of the particle of $Mg_{0.5}Ni_{0.5}Fe_2O_4$. An increase of X-ray density indicates atom density per unit volume at nanoparticles became higher, and the strain value decreases indicating the reduction of the average lattice spacing due to the appearance of Silica or individual Si and O atoms to nanoparticles.

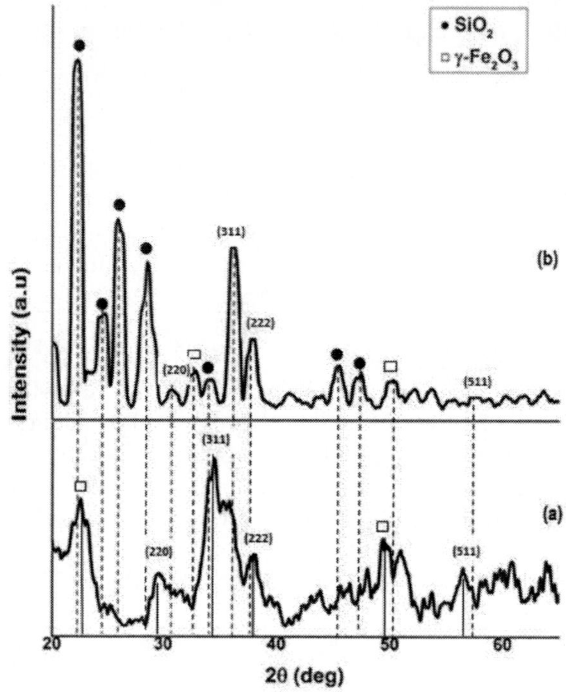


**Figure. 1** XRD pattern of (a) $Mg_{0.5}Ni_{0.5}Fe_2O_4$ and (b) silica encapsulated $Mg_{0.5}Ni_{0.5}Fe_2O_4$

**Table 3.** Data of particles size, lattice parameters, X-ray density, and strain of $Mg_{0.5}Ni_{0.5}Fe_2O_4$ and $Mg_{0.5}Ni_{0.5}Fe_2O_4$ + silica (50%)

| Sample | $2\theta$ | Lattice parameters (nm) | Particles size (nm) | X-ray density (g/cm$^3$) | Strain (ε) |
|---|---|---|---|---|---|
| **$Mg_{0.5}Ni_{0.5}Fe_2O_4$** | 35 | 0.85 | 5.1 | 4.7 | $9.2 \times 10^{-2}$ |
| **$Mg_{0.5}Ni_{0.5}Fe_2O_4$ + Silica (50%)** | 36 | 0.82 | 16.1 | 5.2 | $2.8 \times 10^{-2}$ |

The FTIR spectrum of $Mg_{0.5}Ni_{0.5}Fe_2O_4$ at Figure. 2.a shows the bands at 3417.86 cm$^{-1}$ and 1627.92 cm$^{-1}$ are assigned to the O-H stretching and bending vibration, respectively [11, 13, 14]. The bands at 362.62 cm$^{-1}$ correspond to M-O stretching mode of the octahedral group, and 609.51 cm$^{-1}$ corresponds to M-O stretching mode of the tetrahedral group [8, 11, 12]. In the FTIR spectrum of silica (Figure. 2.c) the bands exist at 3487.30 cm$^{-1}$ and 1651.07 cm$^{-1}$ which are assigned to the O-H stretching and bending vibrations. The band at 447.49 cm$^{-1}$ is O-Si-O bending [11, 15, 16], at 794.67 cm$^{-1}$ is Si-OH stretching [11], and at 1087.85 cm$^{-1}$ is Si-O-Si stretching asymmetric [6, 15]. Furthermore, in the FTIR spectrum of silica encapsulated $Mg_{0.5}Ni_{0.5}Fe_2O_4$ nanoparticles (Figure. 2.b) show the bands of nanoparticles and also silica. The bands at 3448.72 cm$^{-1}$ and 1658.78 cm$^{-1}$ are due to O-H stretching and bending vibrations, respectively. The wavenumbers of O-H stretching and bending increase after silica encapsulated $Mg_{0.5}Ni_{0.5}Fe_2O_4$, affected the silica surface is terminated with silanols that have adamant polar interaction. The silanol groups are hydrogen bonded to water and constitute at least part of strongly held water [6]. The O-Si-O bending at 455.2 cm$^{-1}$ [11, 15, 16],

Si-Fe-O at 570.93 cm$^{-1}$ [6, 17], Si-OH stretching at 779.24 cm$^{-1}$ [11], and Si-O-Si stretching asymmetric at 1049.28 cm$^{-1}$ [6, 15] clearly show the presence of silica on $Mg_{0.5}Ni_{0.5}Fe_2O_4$. The bands at 300.9 cm$^{-1}$ and 347.19 cm$^{-1}$ correspond to M-O stretching mode of the octahedral group, and the band at 686.66 cm$^{-1}$ corresponds to M-O stretching mode tetrahedral group [8, 11, 13].

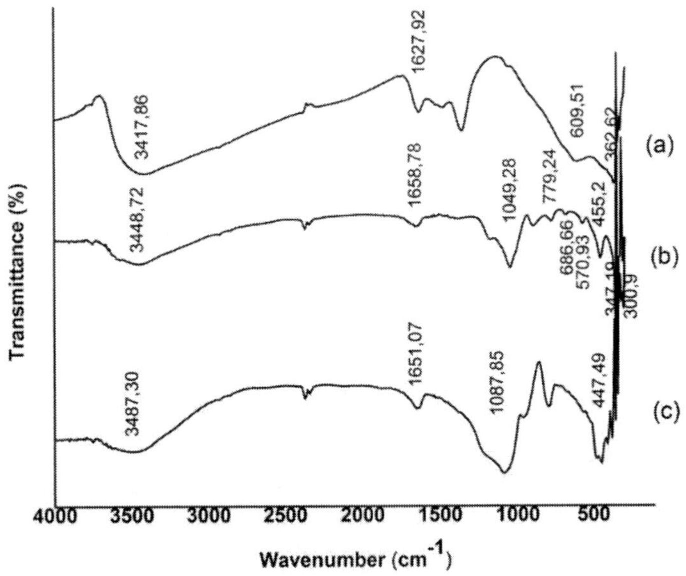


**Figure. 2** FTIR spectrum of (a) $Mg_{0.5}Ni_{0.5}Fe_2O_4$, (b) silica encapsulated $Mg_{0.5}Ni_{0.5}Fe_2O_4$, and (c) silica

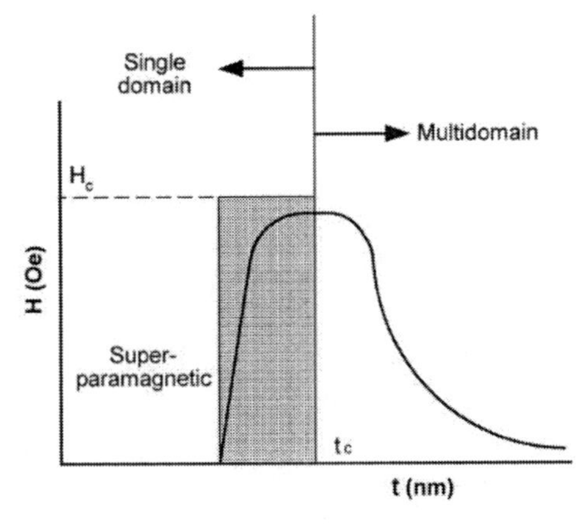


**Figure. 3** Illustration of the dependence of coercivity on particle size [1]

Table 4 shown the decrease of coercivity from 47 Oe to 10 Oe affected by increase of silica encapsulation onto $Mg_{0.5}Ni_{0.5}Fe_2O_4$. The coercivity dependence of particle size, if the coercivity is proportional to $1/t$, then the nanoparticles are multidomain. If the coercivity is proportional to $t$, then the nanoparticles are single-domain. This behavior is illustrated in Figure. 3. Therefore, all samples before and after silica encapsulation are multi-domain due to the increase of particles size after silica encapsulation. The diamagnetism of $SiO_2$ phase [1] on nanoparticles also caused coercivity value decreasing as increasing silica concentration. The saturation magnetization of $Mg_{0.5}Ni_{0.5}Fe_2O_4$ was 4.7 emu/g. The saturation magnetization decreased as increasing silica concentration, due to diamagnetism properties of $SiO_2$ at nanoparticle. $SiO_2$ caused the magnetic response of $Mg_{0.5}Ni_{0.5}Fe_2O_4$ weaker, so, by increasing silica concentration, the magnetic anisotropy became weaker than before encapsulation.

**Table 4.** Magnetic parameters at various silica encapsulated $Mg_{0.5}Ni_{0.5}Fe_2O_4$

| Sample | Silica concentration (%) | $H_c$ (Oe) | $M_s$ (emu/g) | $K$ (erg/g) |
|---|---|---|---|---|
| **A** | 50 | 10 | 1.0 | 10 |
| **B** | 30 | 12 | 1.1 | 13 |
| **C** | 20 | 23 | 1.8 | 41 |
| **D** | 15 | 22 | 1.6 | 35 |
| **E** | 10 | 39 | 2.0 | 78 |
| **F** | 5 | 40 | 3.2 | 128 |
| **$Mg_{0.5}Ni_{0.5}Fe_2O_4$** | 0 | 47 | 4.7 | 221 |

### *3.2. PEG-4000 Encapsulated $Mg_{0.5}Ni_{0.5}Fe_2O_4$ Nanoparticles*

The XRD pattern of $Mg_{0.5}Ni_{0.5}Fe_2O_4$ at Figure 4a shows that it is indexed as (220), (311), (222), and (511), whereas the XRD pattern of PEG-4000 encapsulated $Mg_{0.5}Ni_{0.5}Fe_2O_4$ at Figure. 3.b shows that it is indexed as (311) and (440). After PEG-4000 encapsulation, Figure 4b shows another phase beside $Mg_{0.5}Ni_{0.5}Fe_2O_4$ and $\gamma$-$Fe_2O_3$ phase, that is $\gamma$-FeO(OH) phase. Peak position from peak (311) after PEG-4000 encapsulation is shifted to the lower angle from 35° to 34.8° indicating the increase of the average lattice spacing. The change of the peak position was resulted from the interaction between the nanoparticle and PEG-4000 as material encapsulation. This interaction caused the substituted O atom of O-H group from PEG-4000 to nanoparticles.

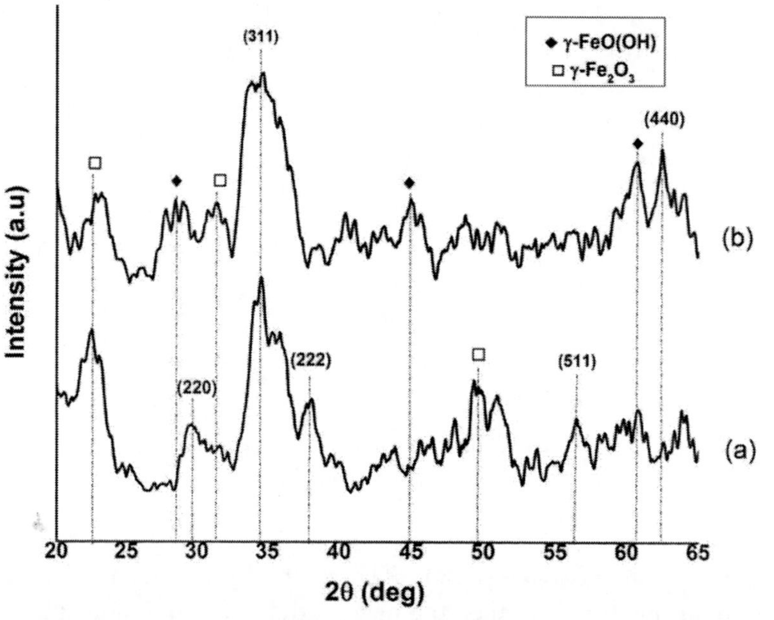


**Figure 4.** XRD pattern of (a) $Mg_{0.5}Ni_{0.5}Fe_2O_4$ and (b) PEG-4000 encapsulated $Mg_{0.5}Ni_{0.5}Fe_2O_4$

**Table 5.** Data of particles size, lattice parameters, X-ray density, and strain of $Mg_{0.5}Ni_{0.5}Fe_2O_4$ and $Mg_{0.5}Ni_{0.5}Fe_2O_4$ + PEG-4000 (50%)

| Sample | $2\theta$ | Lattice parameters (nm) | Particles size (nm) | X-ray density ($g/cm^3$) | Strain ($\varepsilon$) |
|---|---|---|---|---|---|
| **$Mg_{0.5}Ni_{0.5}Fe_2O_4$** | 35 | 0.85 | 5.1 | 4.7 | $9.2 \times 10^{-2}$ |
| **$Mg_{0.5}Ni_{0.5}Fe_2O_4$ + PEG-4000 (50%)** | 34.8 | 0.86 | 4.5 | 4.6 | $10.2 \times 10^{-2}$ |

Table 5 shows the particles size of $Mg_{0.5}Ni_{0.5}Fe_2O_4$ is 5.1 nm and PEG-4000 encapsulation became 4.5 nm. PEG-4000 has larger molecular weight and a long chain of monomer, so there are a lot of particles trapped by polymer chain and then the growth of crystal decreases and the particle size becomes smaller [18]. Also,Table 5 shows that X-ray density has no significant difference after PEG-4000 encapsulation. Otherwise, the strain became higher than before encapsulation due to the increase of the average lattice spacing because of the substituted O atoms to nanoparticles.

The FTIR spectrum of PEG-4000 (Figure 5c) has bands at 3448.72 $cm^{-1}$ and 1635.64 $cm^{-1}$ which correspond to the O-H stretching and bending vibrations. The bands at 956.69 $cm^{-1}$ is C-H bending (out of plane), at 1465.9 $cm^{-1}$ is C-H in plane bending (scissoring), at 2885 $cm^{-1}$ is C-H stretching (symmetric), and at 1064.71 $cm^{-1}$ and also 1111 $cm^{-1}$ are stretching ether C-O from $-CH_2$-OH group in PEG-4000 [12, 19]. Finally, in the FTIR spectrum of PEG-4000 encapsulated $Mg_{0.5}Ni_{0.5}Fe_2O_4$, (Figure 5b) it shows the bands of PEG-4000 and $Mg_{0.5}Ni_{0.5}Fe_2O_4$. The bands at 3410.15 $cm^{-1}$ and 1627.92 $cm^{-1}$ are assigned to O-H stretching and bending vibrations. The wavenumber of O-H stretching decreases after PEG-4000 encapsulated $Mg_{0.5}Ni_{0.5}Fe_2O_4$ because of PEG-4000 reducing water molecule in nanoparticles. The C-H bending (out of plane) at 925.83 $cm^{-1}$, C-H in plane bending (scissoring) at 1481.83 $cm^{-1}$, C-H stretching (asymmetric) at 2924.09 $cm^{-1}$, and stretching ether C-O at 1103.28 $cm^{-1}$ [12, 19] clearly show the presence of PEG-4000 on $Mg_{0.5}Ni_{0.5}Fe_2O_4$. The bands at 324.04 $cm^{-1}$ and 354.9 $cm^{-1}$ correspond to M-O stretching mode of the octahedral group, and the band at 609.51 $cm^{-1}$ correspond to M-O stretching mode of the tetrahedral group [8, 9, 10].

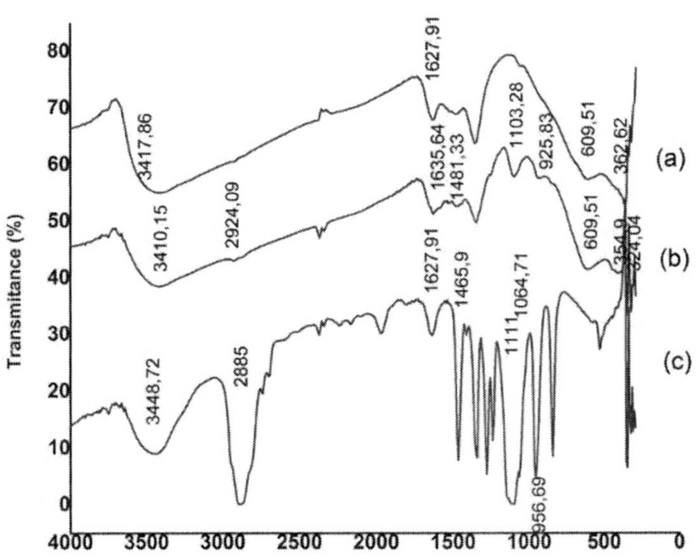


**Figure 5.** FTIR spectrum of (a) $Mg_{0.5}Ni_{0.5}Fe_2O_4$, (b) PEG-4000 encapsulated $Mg_{0.5}Ni_{0.5}Fe_2O_4$, and (c) PEG-4000

Table 6 shows that the coercivity after PEG-4000 encapsulation has no significant difference with the coercivity of $Mg_{0.5}Ni_{0.5}Fe_2O_4$, which is due to no significant difference in particle size before and after encapsulation that indicates all samples are still in multidomain. The saturation magnetization and magnetic anisotropy increase as increasing PEG-4000 concentration. Increasing of that magnetic parameters is because of the domination $Mg_{0.5}Ni_{0.5}Fe_2O_4$ phase and also because there are two other phases which are $\gamma$-$Fe_2O_3$ and $\gamma$-FeO(OH). All phases contributed magnetic moment with different magnetic portion. $Mg_{0.5}Ni_{0.5}Fe_2O_4$ and $\gamma$-$Fe_2O_3$ are ferrimagnetic, and $\gamma$-FeO(OH) is antiferromagnetic, so the nanoparticle has random moment magnetic orientation and high magnetic anisotropy.

**Table 6.** Magnetic Poperties at various PEG-4000 concentration of $Mg_{0.5}Ni_{0.5}Fe_2O_4$

| Sample | PEG-4000 concentration (%) | $H_c$ (Oe) | $M_s$ (emu/g) | $K$ (erg/g) |
|---|---|---|---|---|
| **L** | 80 | 46 | 5.8 | 267 |
| **K** | 75 | 44 | 6.1 | 268 |
| **J** | 67 | 44 | 7.6 | 334 |
| **I** | 50 | 46 | 7.7 | 354 |
| **H** | 33 | 41 | 5.7 | 234 |
| **G** | 25 | 45 | 5.7 | 256 |
| **$Mg_{0.5}Ni_{0.5}Fe_2O_4$** | 0 | 47 | 4.7 | 221 |

## 4. Conclusion

The $Mg_{0.5}Ni_{0.5}Fe_2O_4$ nanoparticles have been successfully synthesized by using coprecipitation method. Two series of $Mg_{0.5}Ni_{0.5}Fe_2O_4$ silica encapsulated were prepared at various of the concentration of silica and the concentration of PEG-4000. Silica encapsulation gives rise to produce a new phase, and also change its particle size. The increase of silica encapsulation might reduce the coercivity of $Mg_{0.5}Ni_{0.5}Fe_2O_4$ logarithmically from 47 Oe to 10 Oe which is strongly related to the decline of particle size. The saturation magnetization of $Mg_{0.5}Ni_{0.5}Fe_2O_4$ also decreases from 4.7 emu/g to 1 emu/g as silica increase. On the other hand, it was obtained that the increase of PEG-4000 encapsulation does not alter the coercivity of $Mg_{0.5}Ni_{0.5}Fe_2O_4$ which is due to no significant difference in particle size. This situation was also shown by the saturation magnetization. The saturation magnetization of $Mg_{0.5}Ni_{0.5}Fe_2O_4$ after silica encapsulation also decreases from 4.7 emu/g become 1 emu/g and after PEG-4000 encapsulation the saturation magnetization increases to 7.7 emu/g.

## 5. References

[1] Mathew D S and Juang R S 2007 *Chemical Engineering Journal* **129** pp 51-56
[2] Khot V M 2013 *Synthesis of Magnesium Merrite Nanoparticles and Studies on Their Induction Heating for Biomedical Applications* (Kolhapur: D. Y. Patil University)
[3] Tomitaka A, Jeun M, Bae S and Takemura Y 2011 *Journal of Magnetics* **16(2)** pp 164-168
[4] Umut E 2013 *Surface Modification of Nanoparticles Used in Biomedical Applications* In Tech, chapter 8 pp 185-208
[5] Setyawan H and Balgis R 2011 *Asia-Pac. J. Chem. Eng* **7** pp 448-454.
[6] Setyawan H, Fajaro F, Widiyastuti W, Winard, Sugeng I, Lenggoro, Wuled and Nandang M., 2012 *J Nanopart Res* **14** 807
[7] Covaliu C I, Jitaru I, Paraschiv G, Vasile E and Biris S S 2013 *Powder Technology* **273** pp 415-426
[8] Hankare P P, Jadhav S D, Sankpal U B, Chavan S S, Waghamare K J and Chougule B K 2009 *Journal of Alloys and Coumpounds* **475** pp 926-929
[9] Moradmard H, Shayesteh S F, Tohidi P, Abbas Z and Khaleghi M 2015 *Journal of Alloys and Coumpounds* **650** pp 116-122

[10] Nadeem K, Krenn H, Sarwar W and Mumtaz M 2014 *Applied Surface Science* **288** pp 677-681
[11] Khanna L and Verma N K 2013 *Materials Science and Engineering* ***B*** **178** pp 1230-1239
[12] Khanna L and Verma N K 2013 *Physica* ***B*** **427** pp 68-75
[13] Kareem S H, Ati A A, Shamsuddin M and Lee S L 2015 *Ceramic International* **41** pp 11702-11709
[14] Ahmadipour M, Mohzhgan H and Kalagadda V R 2012 *Advances in Nanoparticle* **1** pp 37-43
[15] Mojić B, Giannakopoulos K P, Cvehić Ž and Srdić V V 2012 *Ceramic International* **38** pp 6635-6641
[16] Gharagozlou M, Naderi R, and Baradaran Z 2016 *Progress in Organic Coating* **90** pp 407-413
[17] Zhang S, Dong D, Sui Y, Liu Z, Wang H, Qian Z and Su W 2006 *Journal of Alloys and Compounds* **415** pp 257-260
[18] Perdana F A, Baqiya M A, Mashuri, Triwikantoro and Darminto 2011 *Jurnal Material dan Energi Indonesia* **01** pp 1-6
[19] Pavia D L, Lampman G M, Kriz G S, and Vyvyan J R 2009 *Introduction to Spectroscopy Fourth Edition* (USA: Brooks/Cole Cengage Learning)

**Acknowledgements**
The reported work was financially supported by *Hibah Kompetensi* (HIKOM) from the Ministry of Research and Technology of High Education for the period of 2015 – 2017 and Nanofabrication Platform Consortium Project of Nagoya University, Minister of Culture, Sport, Science and Technology (MEXT) Nano-Project Platform, Japan for the Period of 2012-2016.

IOP Conf. Series: Materials Science and Engineering **202** (2017) 012048 doi:10.1088/1757-899X/202/1/012048

# Effect of Synthesis Temperature and NaOH Concentration on Microstructural and Magnetic Properties of $Mn_{0.5}Zn_{0.5}Fe_2O_4$ Nanoparticles

**N Siregar[1], I P T Indrayana[1], E Suharyadi[1], T Kato[2] and S Iwata[3]**

[1] Department of Physics, Universitas Gadjah Mada, Yogyakarta, Indonesia
[2] Department of Electrical Engineering and Computer Science, Nagoya University, Nagoya, Japan
[3] Institutes of Materials and Systems for Sustainability, Nagoya University, Nagoya, Japan

Email: nurhasanah.siregar@mail.ugm.ac.id

**Abstract.** $Mn_{0.5}Zn_{0.5}Fe_2O_4$ nanoparticles have been successfully synthesized through coprecipitation method by varying NaOH concentrations from 0.5 M to 6 M and synthesis temperatures from 30 to 120 °C. The X-ray diffraction (XRD) pattern indicates samples consisting of multiphase structures such as spinel of $Mn_{0.5}Zn_{0.5}Fe_2O_4$, $\alpha$-$MnO_2$, ZnO, $\lambda$-$MnO_2$, and $\gamma$-$Fe_2O_3$. The crystallite size of $Mn_{0.5}Zn_{0.5}Fe_2O_4$ is in the range of 14.1 to 26.7 nm. The Transmission electron microscope (TEM) image shows that sample was agglomerate. The hysteresis loops confirm that nanoparticles are soft magnetic materials with low coercivity ($H_c$) in the range of 45.9 to 68.5 Oe. Those values increased relatively with increasing particles size. For NaOH concentration variation, the maximum magnetization of the sample increased from 10.4 emu/g to 11.6 emu/g with increasing ferrite content. Meanwhile, the maximum magnetization increased from 7.9 to 15.7 emu/g for samples with various synthesis temperature. The highest coercivity of 68.5 Oe was attained for a sample of 6 M NaOH under 90 °C. The highest magnetization of 15.7 emu/g was achieved for a sample of 1.5 M NaOH under 120 °C caused by the maximum crystallinity of sample.

**Keywords:** magnetic property, microstructural, NaOH concentration, synthesis temperature

## 1. Introduction

Magnetic nanoparticles have attracted interest in the field of nanomaterial because of the potential applications such as catalytic, biomedicine, magnetic resonance imaging (MRI), data storage [1]. Ferrite nanoparticle has formula $MFe_2O_4$, which M is divalent metal ions i.e., $Zn^{2+}$, $Mn^{2+}$, $Fe^{2+}$, $Cu^{2+}$, $Mg^{2+}$, and $Co^{2+}$. Metal cation M can also be a combination of two different metal cations, for example, $Mn^{2+}$ and $Zn^{2+}$. The $Mn_{0.5}Zn_{0.5}Fe_2O_4$ nanoparticle is a soft magnetic material which has high initial magnetic permeability, high magnetization, high resistivity, and low power loss [2]. The nanoparticle can be prepared by using various techniques like combustion [2,3], coprecipitation [4,5], hydrothermal [6], mechanochemical processing [7], and spray pyrolysis [8]. Among of them, coprecipitation method is relatively simple, short reaction time, high reaction yield, and small particle size.

The microstructure of nanoparticle directly affects their magnetic, electrical, dielectric, and optical properties [9]. The microstructure of nanoparticle can be tuned by controlling synthesis parameters, such as metal precursors type, pH, reaction temperature and reaction rate [6]. All of the parameters directly affect the crystallite size of nanoparticle [6]. Treatment temperature sintering for $Mn_{0.5}Zn_{0.5}Fe_2O_4$ would be affected by crystal structure and magnetic properties [2]. In the case of coprecipitation synthesis, a reaction temperature of solutes and pH is very sensitive in determining the reaction yield. Therefore, the presented research would like to control the synthesis temperature and pH of solutes by varying co-precipitant agent concentrations, in this case, belong to NaOH concentration.

## 2. Experimental Method

The metal precursors such as $MnCl.H_2O$ (Merck, Germany), $ZnSO_4.7H_2O$ (Merck, Germany) and $FeCl_3.6H_2O$ (Merck, Germany) were used to prepare metal cation of $Mn^{2+}$, $Zn^{2+}$, and $Fe^{3+}$. Those hydroxides were dissolved in 40 mL of distilled water. The HCl solution of 3.37 mL than was added to the precursor solution. The mixture was stirred for 5 minutes at room temperature with constant stirring of 500 rpm. The obtained solution was dropped wisely to the co-precipitant solution of 50 mL NaOH (1.5 M) under the constant stirring rate of 1000 rpm and different synthesis temperature of room temperature (A5), 60 °C (A6), 90 °C (A2), and 120 °C (A7). In the case of NaOH concentration, samples were synthesized under a temperature of 90 °C and NaOH concentration of 0.5 M (A1), 1.5 M (A2), 3 M (A3), and 6 M (A4). The synthesis reaction was kept for 60 minutes until the formation of $Mn_{0.5}Zn_{0.5}Fe_2O_4$ nanoparticle formed which is indicated by the dark brown mixture. The mixture then was cooled naturally until thermal equilibrium occurred with the room temperature. During the cooling process, dark brown slurry of nanoparticles will separate from the solvent. The slurry was washed for six times to remove the incomplete chemical precursors during synthesis. Those processes will provide a dark brown slurry which is ready to be heated in a furnace to a constant temperature of 90 °C for 4 hours. At last, the dark brown powder of nanoparticles will be obtained. The crystal structure, microstructure, vibrational spectra, and magnetic properties of the powder were characterized by using XRD Shimadzu XD, TEM JEOL Jem-4000, IR spectrometer Shimadzu Prestige-21 and VSM Riken Denshi Co. Ltd., respectively.

## 3. Results and Discussion

### *3.1. X-Ray diffraction pattern of nanoparticles*

The diffraction pattern of samples for both NaOH concentration and synthesis temperature variation is presented in Figure 1 and 2. The diffraction pattern referred to the standard data of $Mn_{0.5}Zn_{0.5}F_2O_4$ (JCPDS no. 10-0467). The profile of diffraction pattern confirmed the formation of spinel structure and some other structures, i.e. ZnO (36-1451), $\lambda$-$MnO_2$ (44-0992), $\alpha$-$MnO_2$ (44-0141), $\gamma$-$Fe_2O_3$ (39-1346). Those phases were estimated from metal precursors which exhibited incomplete chemical reaction during synthesis. Principally, the stage of coprecipitation was initialized by the formation of hydroxide compound via coprecipitation reaction of a metal salt precursor with NaOH. The HCl solution was taken place to break the ionic bond between metal cation and chloride anions or sulfate group. Those metal cations were highly attracted by hydroxide anion $OH^-$ to form hydroxide compound. Each hydroxide compound of $Zn(OH)_2$, $Mn(OH)_2$ and $Fe(OH)_3$ has different reaction rate to form ferrite nanoparticles. The hydroxide which fails to form ferrite nanoparticles will be oxidized by oxygen in the atmosphere to form further oxidation state. The complete chemical reaction of formation the oxidation state can be presented in the following reactions,

$$Zn(OH)_2 \rightarrow ZnO + H_2O \quad (1)$$

$$2Mn(OH)_2 + O_2 \rightarrow 2MnO_2 + 2H_2O \quad (2)$$

$$2Fe(OH)_3 \rightarrow \gamma\text{-}Fe_2O_3 + 3H_2O \quad (3)$$

Regarding Figure 1, in sample A2 (3M),the presence of ZnO phase of which composition is greater than ferrite $Mn_{0.5}Zn_{0.5}Fe_2O_4$ and $\lambda$-$MnO_2$ phases are higher than $\alpha$-$MnO_2$ is due to the maximum oxidation process exhibited by $Zn(OH)_2$ and $Mn(OH)_2$. According to diffraction profile, sample A4 has the lowest degree of crystallinity due to the highest NaOH concentration that contributes to the largest number of $OH^-$ anion that presents in solution. The number of $OH^-$ is very sensitive to affect the environmental condition for the reaction. In other hand, samples A1 to A4 have a better degree of crystallinity and sample A3 of 1.5 M NaOH has the highest degree of crystallinity.

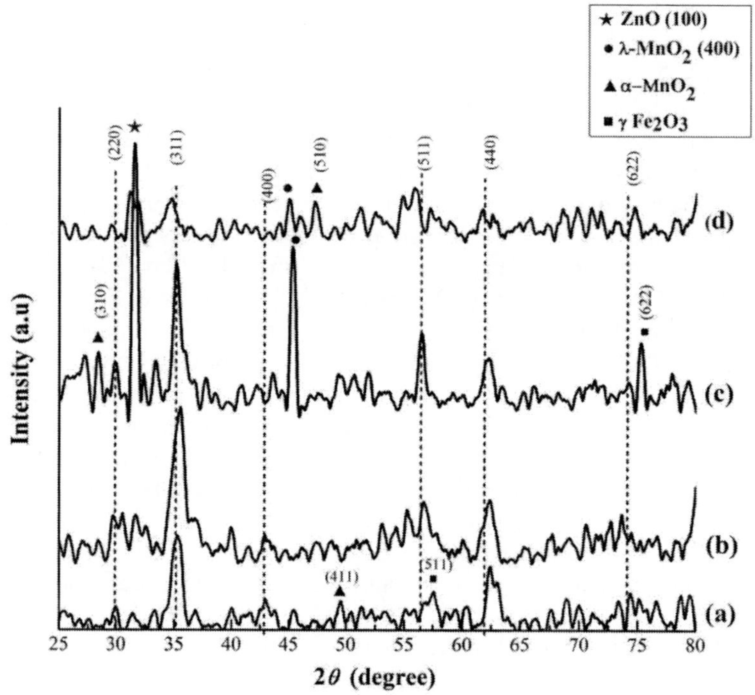


**Figure 1.** XRD pattern of $Mn_{0.5}Zn_{0.5}Fe_2O_4$ with different NaOH concentration of (a) 0.5M, (b) 1.5M, (c) 3M, and (d) 6M.

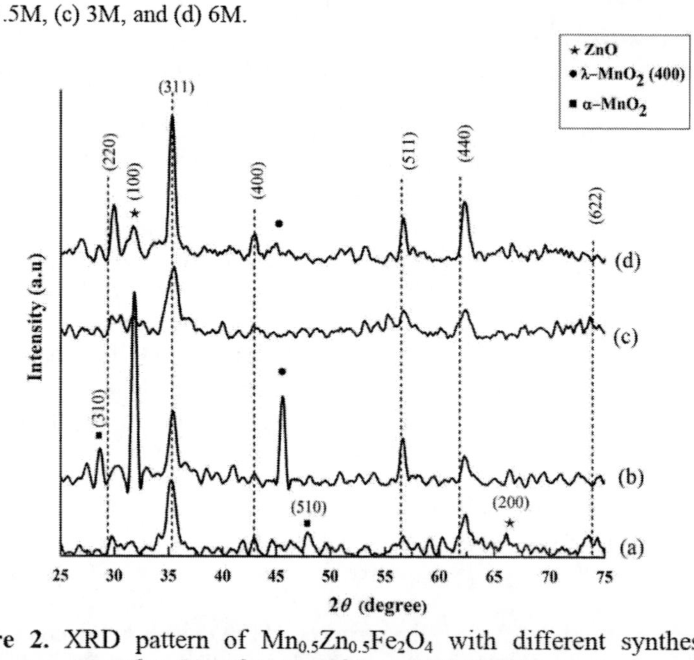


**Figure 2.** XRD pattern of $Mn_{0.5}Zn_{0.5}Fe_2O_4$ with different synthesis temperature (a) 30°C, (b) 60°C, (c) 90°C, and (d) 120°C.

In relation to synthesis temperature variation, the diffraction pattern of samples can be presented in Figure 2. In diffraction data, the appearance of ZnO, λ-$MnO_2$, and α-$MnO_2$ phases were also investigated. Sample A6 was dominated by ZnO phase compared than ferrite $Mn_{0.5}Zn_{0.5}Fe_2O_4$ phase due to the condition for which the formation stage of ZnO phases. The diffraction pattern of sample A7 shows a clear diffraction peaks of (400). The diffraction peaks possessed by all samples were very broad due to microstrain effect and dislocation density of crystal lattice. This defect might be caused by reaction temperature. The reaction temperature gives thermal energy for metal cation to grow as nanoparticles. Nucleation process is very sensitive to reaction temperature that agitates thermal energy for the reaction. If the nucleation does not work well, so defect might occur on crystal nanoparticles. The broadening of diffraction peaks as an indication of that defect. The uninformed microstrain occurring in the crystal lattice of nanoparticles makes change the site of metal cations. The crystal site of a metal cation and also anion can determine ion distribution of nanoparticles. Thus, ion distribution will contribute to lattice size and crystallite size of nanoparticles [10]. Quantitatively, the result of calculation for crystallite size (*t*) by using Scherer's formula, the values of interplanar distance (*d*), the estimated lattice size (*a*), X-Ray density, and lattice microstrain can be presented in Table 1.

**Table 1.** Interplanar distance, lattice constant, X-Ray density and microstrain of $Mn_{0.5}Zn_{0.5}Fe_2O_4$ nanoparticles synthesized with different NaOH concentration and synthesis temperature

| Synthesis parameter variable | Sample | Interplanar Distance (Å) | Lattice constant (Å) | X-Ray Density ($g/cm^3$) | Particle size (nm) | Microstrain |
|---|---|---|---|---|---|---|
| NaOH Cont. | A1 | 2.545 | 8.44 | 5.21 | 16.5 | 0.028 |
| | A2 | 2.535 | 8.41 | 5.27 | 14.1 | 0.032 |
| | A3 | 2.546 | 8.44 | 5.20 | 20.8 | 0.022 |
| | A4 | 2.587 | 8.58 | 4.96 | 14.5 | 0.032 |
| Synthesis Temperature | A5 | 2.542 | 8.43 | 5.23 | 16.7 | 0.027 |
| | A6 | 2.534 | 8.40 | 5.28 | 22.1 | 0.021 |
| | A2 | 2.535 | 8.41 | 5.27 | 14.1 | 0.032 |
| | A7 | 2.540 | 8.42 | 5.24 | 26.7 | 0.017 |

The lattice sizes of samples for variation of NaOH concentration are not different significantly compared to the bulk size. In detail, the increase of NaOH concentration tends to affect to an increase of lattice size of nanoparticles. Sample A4 has the largest values of lattice parameter which can be estimated by considering the largest broadening of diffraction peaks. This phenomenon is an indication that sample A4 exhibited the largest strain and interplanar distance as discussed. The density of nanoparticles decreased with an increase of NaOH concentration. The increase of lattice size causes volume expansion of crystal lattice. The volume expansion was not followed by an increase of total weight of crystal.

Increasing synthesis temperature does not affect significantly to a change of crystal lattice nanoparticles. Lattice parameter and the crystal density are relatively constant. Both NaOH and synthesis temperature variations do not give a particular trend on crystallite size value of nanoparticles. The smallest crystallite size was attained from sample A2 which was 14.1 nm, and the greatest was achieved from samples A7 which was 26.7 nm. Increasing crystallite size of nanoparticles was followed by the decrease of lattice microstrain.

### *3.2. Microstructure and Morphology of nanoparticles*

The TEM image of sample $Mn_{0.5}Zn_{0.5}Fe_2O_4$ for synthesis temperature of 120°C and NaOH concentration of 1.5 M can be shown in Figure 3. The following selected area electron diffraction (SAED) also can be shown.

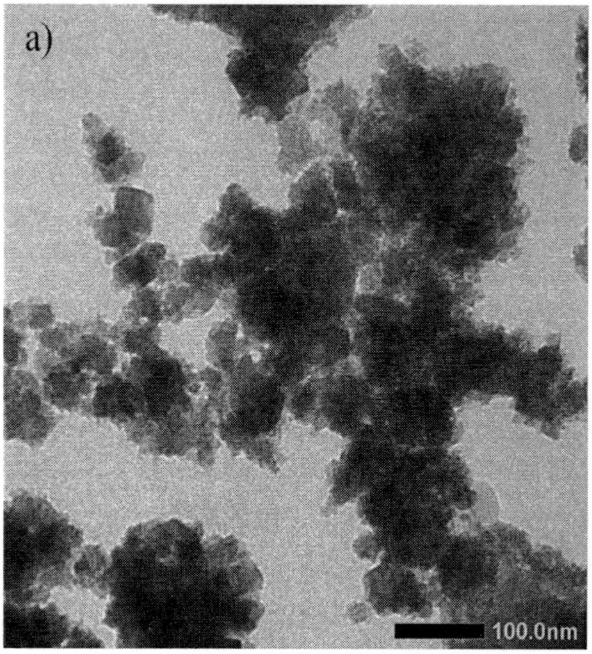

**Figure 3.** (a) TEM Image and (b) SAED image of $Mn_{0.5}Zn_{0.5}Fe_2O_4$

Figure 3 shows that nanoparticle is in the form of aggregate. The sample exhibited strongly agglomeration due to a small particle size which has higher surface energy density. The surface energy is enhanced by the largest number of charges occupying the surface of nanoparticles. Therefore, the surface tension of particle is higher than larger particles. Furthermore, the SAED image shows that the formation of diffraction rings composes the non-continuous diffraction pattern. This SAED image confirms the formation of polycrystalline $Mn_{0.5}Zn_{0.5}Fe_2O_4$ nanoparticles which possess low crystallinity compared than bulk material.

### *3.3. Functional group bonding spectra*

Functional groups bonds between ions can be shown in Figure 4 regarding IR spectra of $Mn_{0.5}Zn_{0.5}Fe_2O_4$ nanoparticles. According to Figure 4, absorption peaks in wavenumber of 455.20 $cm^{-1}$ and 578.64 $cm^{-1}$ were investigated which corresponded to stretching vibration of metal-oxygen at tetrahedral and octahedral sites [11]. These bands confirm that the formation of tetrahedral and octahedral sites are in the crystal lattice of $Mn_{0.5}Zn_{0.5}Fe_2O_4$ nanoparticles. Band absorption of 933.55 $cm^{-1}$ corresponds to bending vibration of Mn-O-H [8]. This band confirms that the existences of hydration and oxidation reaction between sample of nanoparticles and hydrogen in the atmosphere. This case confirms the formation of the nonspinel structure according to diffraction data. The absorption bands of 1342.46 $cm^{-1}$, 1627.92 $cm^{-1}$ corresponds to bending vibration of O-H [11,12] and 3410.15 corresponds to stretching vibration of O-H [11] due to KBr oxidation state during the preparation of disk sample for FTIR characterization.

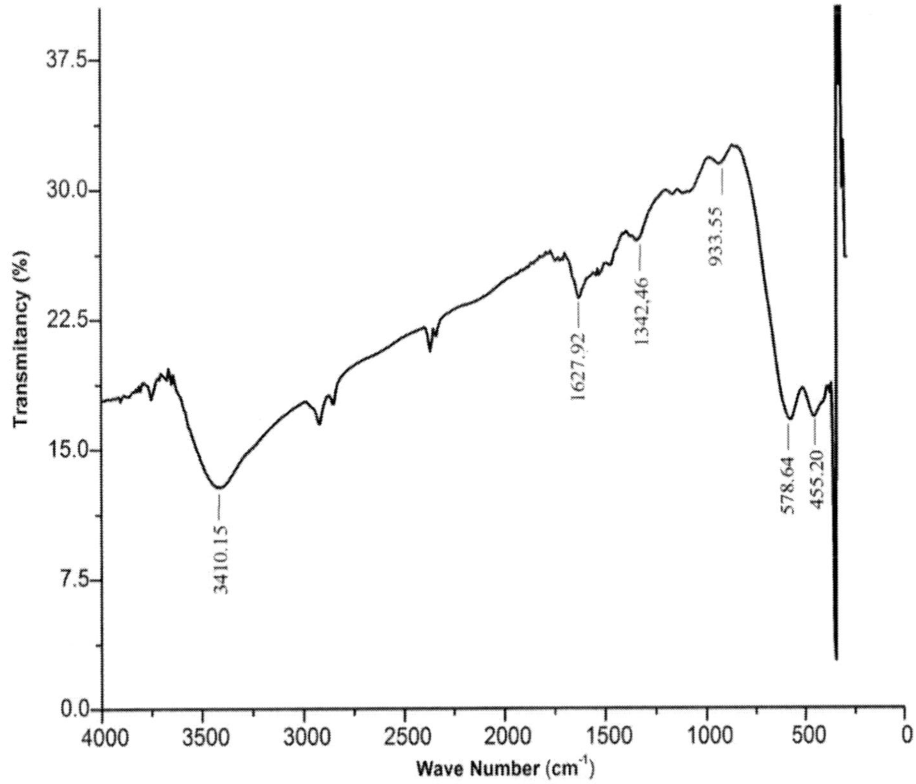


**Figure 4.** IR spectra of $Mn_{0.5}Zn_{0.5}Fe_2O_4$ systems

### *3.4. Magnetic Measurement*

The magnetic properties of samples were measured by using VSM. The examples of hysteresis loop can be shown in Figure 5. The measurement results of *Hc*, maximum magnetization of 15 kOe ($M_{max}$), and magnetic anisotropy constant (*K*) is indicated in Table 2.

**Table 2.** Ferrite phase ratio, coercivity, maximum magnetization, and magnetic anisotropy constant of $Mn_{0.5}Zn_{0.5}Fe_2O_4$ nanoparticles with variations of parameter synthesis

| Parameters | Sample | Ferrite Phase Ratio (%) | *Hc* (Oe) | $M_{max}$ (emu/g) | *K* (erg/g) |
|---|---|---|---|---|---|
| Varying NaOH | A3 | 32.4 | 48.89 | 11.32 | 553.49 |
| | A4 | 47.7 | 68.51 | 11.48 | 787.02 |
| | A1 | 83.1 | 47.18 | 10.42 | 491.70 |
| | A2 | 87.7 | 46.03 | 11.62 | 535.03 |
| Varying Temperature | A6 | 27.3 | 47.85 | 7.87 | 376.34 |
| | A5 | 80.2 | 45.95 | 7.99 | 367.15 |
| | A2 | 87.7 | 46.03 | 11.62 | 535.03 |
| | A7 | 91.9 | 47.53 | 15.69 | 741.41 |

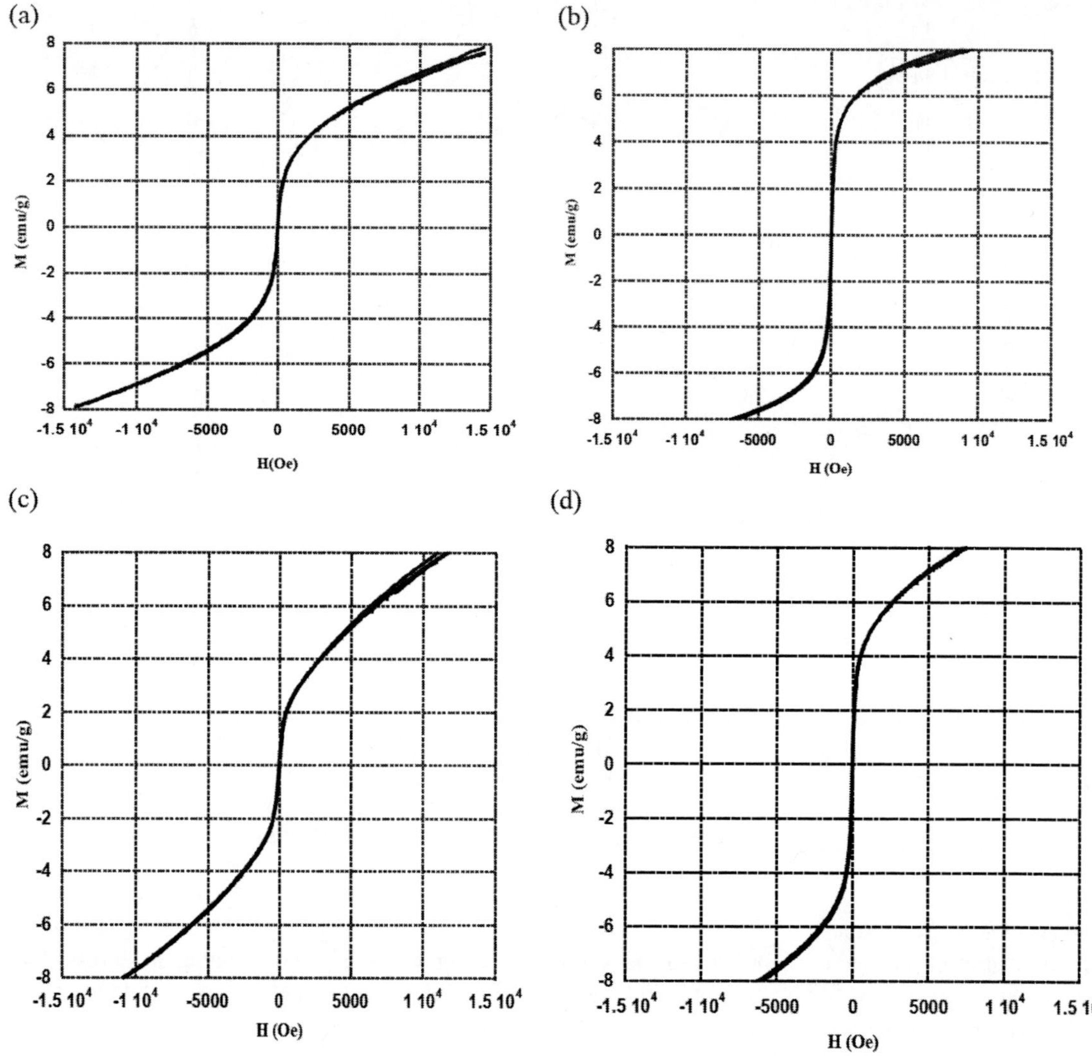


**Figure 5.** Hysteresis curve (a) sample A1, (b) sample A2, (c) sample A4, and (d) sample A7.

According to Table 2, the values of maximum magnetization at 15 kOe of samples increase relatively with an increase of concentration NaOH and synthesis temperature. The value of maximum magnetization is strongly influenced by the degree of crystallinity of samples. Sample A7 has the highest maximum magnetization due to the highest degree of crystallinity among all samples. The anisotropy increases with the increase of coercivity and maximum magnetization.

For NaOH variation, the coercivity of nanoparticles increase with an increase of crystallite size up to 14.6 nm [68.5 Oe] and decrease up to 20.8 nm [48.9 Oe]. Hence, for synthesis temperature variation, the increase of crystallite size is followed by the increase of coercivity. Graphically, the relation between crystallite size and coercivity for each variation can be shown in Figure 6. According to Figure 6, sample A4 has the highest coercivity. Some samples with coercivity have increased due to their cation distribution of $Fe^{3+}$ in a spinel structure which exhibits super-exchange interaction with cations in tetrahedral and octahedral sites [3].

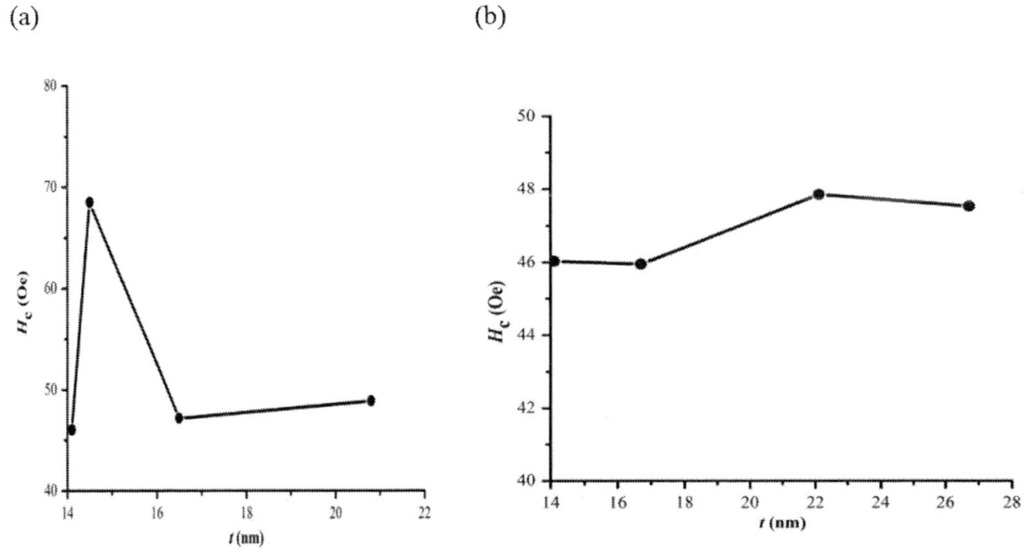


**Figure 6**. Correlation between grain size and coercivity of samples $Mn_{0.5}Zn_{0.5}Fe_2O_4$: a) different NaOH concentration and b) synthesis temperature.

## 4. Conclusion

The $Mn_{0.5}Zn_{0.5}Fe_2O_4$ nanoparticles have been successfully synthesized. The incomplete formation of $Mn_{0.5}Zn_{0.5}Fe_2O_4$ was investigated for 6M NaOH concentration and low synthesis temperature of room temperature and 60°C. The highest maximum magnetization was observed in a sample of 1.5 M NaOH and synthesis temperature 120 °C due to the highest ferrite content. The NaOH concentration and synthesis temperature strongly affected crystal structure, microstructure, crystallinity and magnetic properties of $Mn_{0.5}Zn_{0.5}Fe_2O_4$ nanoparticle.

## 5. References

[1] Lu A H, Salabas L and Schüth F 2007 Angewandte Chemic Inter. **46** 1222-1244
[2] Hu P, Yang H B, Pan D A, Wang H, Tian J J, and Zhang S G 2010 *J. Magn. Magn. Mater*. **322** 173-177
[3] Deraz N M and Alarifi A 2012 *Int. J. Electrochem. Sci.* **7** 5828-5836
[4] Mathur P, Thakur A and Singh M 2008 *J. Magn. Magn. Mater*. **320** 1364-1369
[5] Irfan S, Nabi M A, Jamil Y and Amin N 2013 *Int. Symp. Adv. Mater*. **60** 012-048
[6] Desai R, Davariya V, Parekh K and Upadhyay V R 2009 *J. Phys*. **37** 765-780
[7] Isfahani M J, Myndyk M, Menzel D, Feldhoff A, Amighian J and Sepelak V 2009 *J. Magn. Magn. Mater*. **321** 152-156
[8] Maria I, Dzmitry K, Vladimir P and Yulia F 2009 *Proc. Symp. Multifunc. Adv. Comp. Mater.* 14-18
[9] Gubin S P 2009 *Magnetic Nanoparticles* Wiley Weinheim
[10] Cullity B D 1956 *Element of X-Ray Diffraction* New Jersey Prentice Hall
[11] Kareem S H, Ooi Y K, Abdulnoor S S, Samsuddin M and Lee S L 2014 *J. Sci. Eng*. **69** 103-106
[12] Dar M A, Kotnala R K, Verma V, Shah J, Siddiqui W A and Alam M 2012 *J. Phys. Chem*. **116** 5277-5287

**Acknowledgements**
The reported work was financially supported by:

1. *Hibah Kompetensi* (HIKOM) from the Ministry of Research and Technology of High Education for the period of 2015 – 2017, and
2. Nanofabrication Platform Consortium Project of Nagoya University, Minister of Culture, Sports, Science and Technology (MEXT) Nano-Project Platform, Japan for the Period of 2012 - 2016.

Authors would like to thank to the Institute of Materials and Systems for Sustainability, Nagoya University for providing VSM instrument to magnetic measurement.

# Effect of Cu-Dopant on the Structural, Magnetic and Electrical Properties of ZnO

**D Aryanto[1], C Kurniawan[1], A Subhan[1], T Sudiro[1], P Sebayang[1], M Ginting[1], S M K Siregar[2] and Nasruddin M N[2]**

[1)]Research Center for Physics, Indonesian Institute of Sciences, Serpong 15314, Tangerang Selatan, Banten, Indonesia
[2)]Postgraduate Program, Universitas Sumatera Utara, Medan 20155, Indonesia

Email: masno.ginting@gmail.com

**Abstract.** $Zn_{1-x}Cu_xO$ ($x$ = 0, 2, 3, and 4 at.%) was synthesized by using solid-state reaction technique. The ZnO and CuO powders were mixed and then milled by using high-speed shaker mill. The influence of Cu dopants on the structure, magnetic, and electrical properties was investigated by using XRD, VSM, and *I-V* and *C-V* measurements. The XRD analysis showed that the $Zn_{1-x}Cu_xO$ had hexagonal wurtzite polycrystalline. The diffraction intensity decreased and the peak position shifted directly to a higher $2\theta$ angle with increasing the dopant concentration. Furthermore, the lattice parameters decreased when the ZnO was doped with $x$ = 0.04, which indicated that the crystal structure changed. The increase of Cu dopants was believed to affect the magnetic and electrical properties of ZnO.

**Keywords:** ZnO, structural, magnetic, electrical properties, and Cu-dopant

## 1. Introduction

ZnO is a semiconductor which has a band gap energy, Eg, of 3.37 eV. It was potentially used for some applications such as solar energy conversion, storage, spintronics, gas sensing, optoelectronics, photovoltaic, varistors, photocatalysts, piezoelectric devices, LED, UV light emitters, luminescent devices, field emission electron sources and nanoscale power generator [1-4]. ZnO doped with transition metals of Mn, Fe, Co and Ni shows a ferromagnetic behavior which makes ZnO as a strong candidate in the realization of spintronics [5,6]. The ferromagnetic properties of ZnO at room temperature due to transition metal dopants have widely been investigated [7]. Cu, which is one of the dopant materials to ZnO, is intensively studied as a potential alternative material for DMS (Dilute Magnetic Semiconductor) [8]. The study by Owens [7] showed that Cu-doped ZnO has a ferromagnetic behavior at above room temperature. Various methods were used to synthesize Cu-doped ZnO, such as solid state reaction [4,9] coprecipitation [1,10], and citrate sol-gel [11,4].

Zheng *et al.* [6] showed theoretically and experimentally that ZnO with a variation of Cu-dopant concentrations has ferromagnetic properties at room temperature. Another study by Karamat *et al.* [12] showed that ZnO thin film doped with 1-3 at.% Cu was ferromagnetic whereas when doped with 5 at.% Cu it had diamagnetic behavior. Cu-doped ZnO nanoparticles synthesized through sol-gel method and annealed at $N_2$, $O_2$, and Ar environments have wurtzite hexagonal structure and ferromagnetic behavior [4,13,14]. Iqbal *et al.* [15] and Yildirim & Durucan [16] have successfully substituted Cu ions to ZnO matrix by using the co-precipitation method so that Cu-doped ZnO became ferromagnetic.

However, only a few studies which reported the synthesis of ZnO-based DMS material by using solid state reaction method. The advantages of solid state reaction compared with the other methods such as ion implantation and thin film deposition is a relative simple, cheap, and easily scaled-up [7]. This paper reports the syntheses of Cu-doped ZnO by using solid state reaction method. The effects of varying doping composition to the structures, electrical and magnetic properties were studied by using XRD, *I-V* meter, *C-V* meter, and VSM will also be discussed.

## 2. Experimental Method

*2.1 Syntheses of $Zn_{1-x}Cu_xO$*

Syntheses of $Zn_{1-x}Cu_xO$ ($x$ = 0, 2, 3, and 4 at.%) was done through the solid state reaction. ZnO and CuO powders were utilized as starting materials. They were mixed and milled in the cylindrical stainless steel vial with stainless steel balls. The ratio of the mixed powder and balls was 1: 10. The process was done by wet milling for 3 hours with toluene as solvent. Afterward, the milled powder was dried in an oven at 100 °C for 3 hours. The dried, milled powder was then pressed into bulk with 15 mm in diameter and 2 mm in thickness by using an automatic axial hydraulic press with 1500 kgf/cm$^2$ in load. All bulk samples were finally sintered at 900 °C for 8 hours in the air with a heating rate of 10 °C/ min.

*2.2 Characterization*

The phase and crystal parameters of all samples were identified by XRD (Rigaku, SmartLab) using Cu $K_\alpha$ radiation ($\lambda$ = 1.5406 Å). *I-V* meter (Fluke, 8842A multimeter high impedance) and *C-V* meter (Keithley, 590 CV Analyzer) were used to analyze their electrical properties. The magnetic properties were measured by using Vibrating Sample Magnetometer (VSM 250, Dexing Magnet).

## 3. Results and Discussion

Figure 1(a) shows the XRD patterns of the bulk samples of $Zn_{1-x}Cu_xO$ ($x$ = 0, 2, 3 and 4 at.%) synthesized by using the solid state reaction. The lattice planes at the $2\theta$ angle of 31.77°, 32.68°, 3.42°, 36.26°, 47.54°, 56.59°, 62.85°, 66.37°, 67.95°, 69.08°, 72.55°, 76.96°, 81.37° and 89° were (100), (002), (101), (102), (110), (103), (200), (112), (201), (103), (202), (004), (202), (104) and (203), respectively. The results indicated that all samples were polycrystalline with wurtzite hexagonal structure. The Cu-doped samples showed tenorite (CuO) peaks at the $2\theta$ angle around 32.63°, 35.62° and 38.80° (see Figure 1.(b)). The XRD patterns also revealed the FWHM of (002) plane becoming wider when ZnO was doped with 2-3 at.% Cu (Figure1.(c)). However, the FWHM became narrow when ZnO was doped with 4 at.% Cu. Furthermore, the (002) plane shifted to a higher $2\theta$ angle. It is believed that $Cu^{2+}$ ions have substituted $Zn^{2+}$ on the ZnO lattice [17, 18, 19]. However, some amount of CuO did not react with ZnO which was indicated by the presence of CuO on the XRD patterns. From XRD analysis there was no indication that the secondary phases such as $Cu_5Zn_8$ or CuZn were formed [7].

The crystal parameters, such as lattice constant, lattice parameter ratio, *d*-spacing and crystal size could then be determined. The lattice parameters ($a$ and $c$) from hexagonal phase were determined from (100) and (002) planes by using the following equations.

$$a = \frac{\lambda}{\sqrt{3}\sin\theta}\sqrt{h^2 + hk + k^2} \quad (1)$$

and

$$c = \frac{\lambda}{2\sin\theta} l \quad (2)$$

where $\lambda$ is the wavelength, $\theta$ is the Bragg's angle of diffraction peak (002), and $h$ $k$ $l$ are the Miller indices. The obtained crystal parameters are given in Table 1. The values of $a$ and $c$ obtained at

IOP Conf. Series: Materials Science and Engineering **202** (2017) 012049 doi:10.1088/1757-899X/202/1/012049

present were smaller compared with the those of theory (JCPDS No. 00-005-0664) where the point $a$ is 3.249Å and $c$ is 5.205 Å. These results indicated that, partly, $Cu^{2+}$ ions substituted $Zn^{2+}$ ions. The ionic radius of $Zn^{2+}$ (0.60 Å) which was greater than that of $Cu^{2+}$ (0.57 Å) was believed to contribute to the changes in the lattice parameters. The ratio $c/a$ of $Zn_{1-x}Cu_xO$ samples was constant, which means that the dopants have mostly done the substitution [20].
The crystallite size was calculated by using Equation (3).

$$D = \frac{0.9\,\lambda}{\beta \cos\theta} \quad (3)$$

where $\beta$ is the FWHM of the diffraction peak of (002). The crystal sizes of all samples were in the range of 49.9 to 50.9 nm.

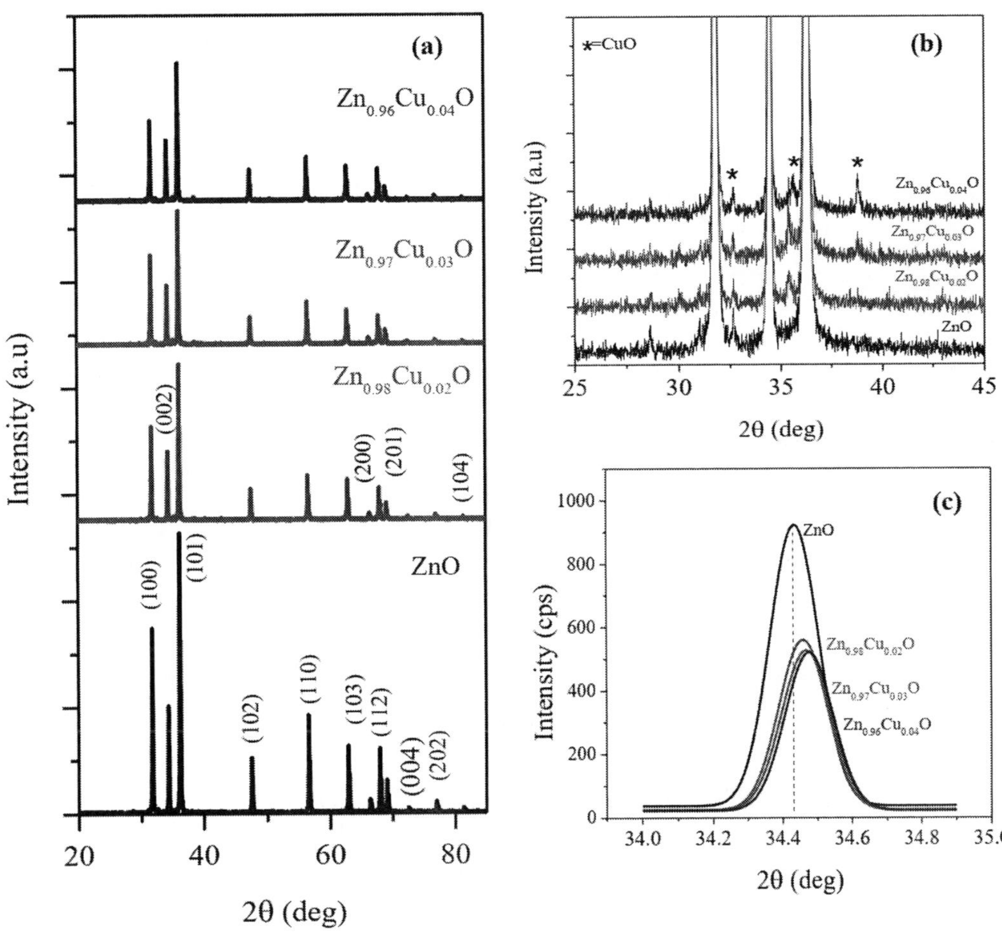


**Figure 1.** (a) The XRD patterns of $Zn_{1-x}Cu_xO$ ($x$ = 0, 2, 3 and 4 at%), (b) Detail observations on $2\theta$ of 25° to 45° and (c) plane peak of (002)

Figure 2 shows the current ($I$)-voltage ($V$) properties of $Zn_{1-x}Cu_xO$ ($x$= 0, 2, 3 and 4 at.%) measured at the atmospheric environment and room temperature. The measurement results showed the almost linear relationship between current and voltage for all samples. The calculated resistivity ($\rho$) and conductivity ($\sigma$) are given in Table 2. Based on the calculated results, the conductivity increased with increasing Cu dopant. This phenomenon can be explained by noting that some amount of $Cu^{2+}$ ions having substituted $Zn^{2+}$ ions on the ZnO lattice increased with the increase in dopant concentration. Cu has greater conductivity than Zn, where the Cu conductivity is 58.5x10$^6$ *S/m* [21] and Zn is 16.6 x 10$^6$ *S/m* [22]. The measured conductivities of all samples were around of 2.159 x 10$^{-8}$ S/cm to 2.624 x 10$^{-8}$ S/cm. Table 2 also shows the measured dielectric constant ($\varepsilon_r$). The dielectric constant changed due to the addition of dopant concentration. It was believed that the changes in dielectric constant value were due to the crystal defects on the bulk of $Zn_{1-x}Cu_xO$ as the effect of $Cu^{2+}$ ions substituting ions at the ZnO crystal. The strain happening due to dopant effect caused dislocation defects on grain boundaries [23]. Besides, the presence of CuO phase in $Zn_{1-x}Cu_xO$ bulk was also believed to cause the presence of crystal defect. This case was confirmed by XRD patterns in Figure 1.

**Table 1.** The calculated lattice parameters *a, c, c/a* and the crystallite size (D) of $Zn_{1-x}Cu_xO$ ($x$= 0, 2, 3 and 4 at.%) samples on the (002) plane

| $Zn_{1-x}Cu_xO$ (at.%) Sample | *a* (Å) | *c* (Å) | *c/a* | FWHM (*deg*) | *D* (nm) |
|---|---|---|---|---|---|
| $x = 0$ | 3.249 | 5.205 | 1.602 | 0.154 | 50.6 |
| $x = 2$ | 3.248 | 5.201 | 1.601 | 0.151 | 50.5 |
| $x = 3$ | 3.248 | 5.200 | 1.601 | 0.154 | 49.9 |
| $x = 4$ | 3.247 | 5.199 | 1.601 | 0.150 | 50.9 |

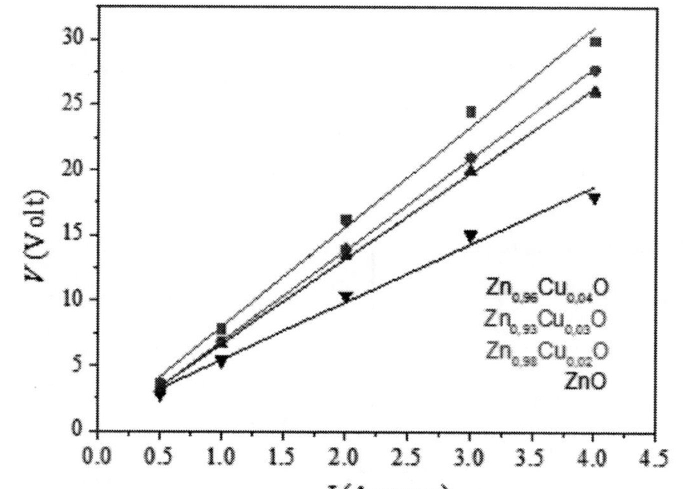


**Figure 2.** *I-V* curve of $Zn_{1-x}Cu_xO$

**Table 2.** The electrical properties of $Zn_{1-x}Cu_xO$ samples

| Sample $Zn_{1-x}Cu_xO$ (at%) | $\rho$ ($\Omega$cm) | $\sigma$ (*S*/cm) | $\varepsilon_r$ |
|---|---|---|---|
| $x = 0$ | 3.259 x 10$^7$ | 3.068 x 10$^{-8}$ | 12.52 |
| $x = 2$ | 4.631 x 10$^7$ | 2.159 x 10$^{-8}$ | 14.87 |
| $x = 3$ | 4.279 x 10$^7$ | 2.337 x 10$^{-8}$ | 14.14 |
| $x = 4$ | 3.812 x 10$^7$ | 2.624 x 10$^{-8}$ | 15.92 |

Figure 3 shows the hysteresis curve of $Zn_{1-x}Cu_xO$ O ($x$ = 0, 2, 3 and 4 at%) samples at room temperature below 20 kOe. It can be seen that the ZnO powder without heat treatment behaved as a diamagnetic (the black line in Figure 3.(a)). This phenomenon means that undoped ZnO powder used in this research is diamagnetic material [24]. ZnO became paramagnetic if the number of dopants was increased (Figure 3.(a)). The detailed observation on hysteresis curve for ZnO doped with 2 at.% Cu showed the minor effect of diamagnetic behavior up to 20 kOe (Figure 3.(b)). Increasing the concentration of Cu to be 3 at.% caused the minor diamagnetic effect observed up to 14 kOe (Figure 3.(c)). Meanwhile, the addition of 4 at.% Cu dopant caused the magnetic property changed from diamagnetic to paramagnetic (Figure 3.(d)). The change in magnetic property was influenced by CuO phase for Cu-doped ZnO (as shown in the XRD patterns). In this research, the source of doping was obtained from CuO. However, the presence of CuO in the sample did not cause ferromagnetism because it behaves as paramagnetic above 250 K [7]. However, the present result supported the previous report of Hsu *et al.* [17], where the addition of 1.93 and 3.01 at.% Cu on ZnO produced the paramagnetic material. Likewise, Vachhani *et al.* [25] showed that, in general, Cu-doped ZnO sintered at high temperature becomes paramagnetic. This case could be due to the CuO phase formed at the higher sintering temperature.

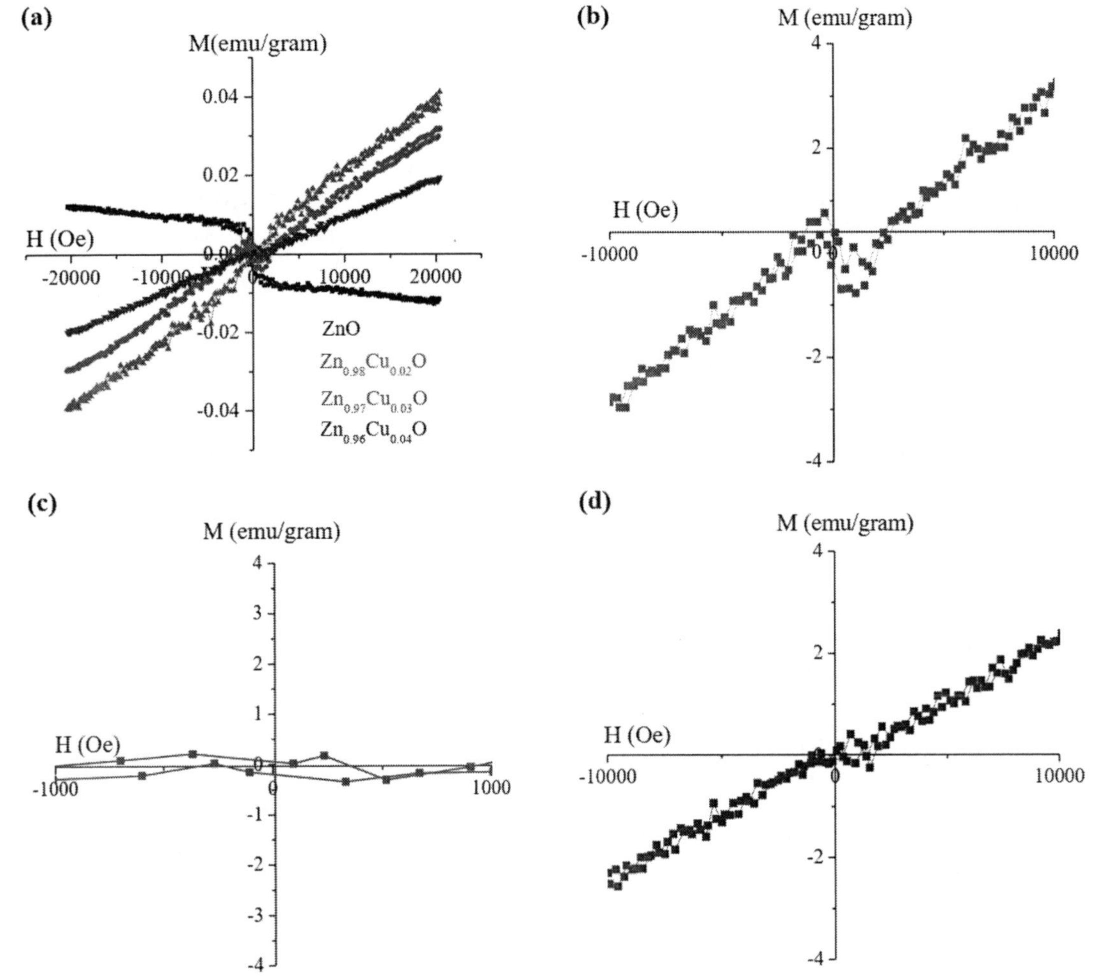


**Figure 3**. (a) The magnetic hysteresis loops of $Zn_{1-x}Cu_xO$ ( $x = 0, 2, 3$ and 4 at%) samples at room temperature below 20 kOe, (b) The magnetic hysteresis loop of $Zn_{0.98}Cu_{0.02}O$, (c) The magnetic hysteresis loop of $Zn_{0.97}Cu_{0.03}O$ and (d) The magnetic hysteresis loop of $Zn_{0.96}Cu_{0.04}O$

## 4. Conclusion

$Zn_{1-x}Cu_xO$ ($x$ = 0, 2, 3, and 4 at%) has been synthesized by using solid-state reaction, and their structures, electrical and magnetic properties were studied. Their XRD patterns showed that $Zn_{1-x}Cu_xO$ was polycrystalline with wurtzite hexagonal structure. The detailed XRD analysis indicated that some parts of Cu dopants substituted Zn on the $Zn_{1-x}Cu_xO$ system and some still in CuO form. The conductivity of $Zn_{1-x}Cu_xO$ was increased by increasing the dopant concentration. Crystal defects formed during the synthesis process of $Zn_{1-x}Cu_xO$ affected the capacitance and dielectric constant. $Zn_{1-x}Cu_xO$ became paramagnetic by increasing Cu dopants.

## 5. References


[1] Ashokkumar M and Muthukumaran S 2014(a) *Journal of Magnetism and Magnetic Materials* **374** 61-66

[2] Vojisavljević K, Žunić M, Branković G, Srećković T 2006 *Science of Sintering* **38** 131-138

[3] Wang Z L 2004 *Journal of Physics Condensed Matter* **16** 829-858

[4] Balamurugan S and Melba K 2014 *Journal of Nanoscience and Nanotechnology* **14** 1-9

[5] Potzger K and Zhou S 2009 *Phys. Status Solidi B* **246** 1147-1167

[6] Zheng J H, Song J L, Li X J, Jiang Q, and Lian J S 2011 *Cryst. Res. Technol* **46** 1143-1148

[7] Owens F J. 2009 *Journal of Magnetism and Magnetic Materials* **321** 3734-3737

[8] Wang Q, Wang J, Zhong X, Tan Q, Zhou Y 2012 *Solid State Communications* **152** 50-52.

[9] Tseng Y W and Huang C L 2015 *Journal of Alloys and Compounds* **638** 29-33.

[10] Ashokkumar M and Muthukumaran S 2014(b) *Optical Materials* **37** 671–678

[11] Liu H, Yang J, Hua Z, Zhang Y, Yang L, Xiao L, Xie Z 2010 *Applied Surface Science* **256** 4162-4165

[12] Karamat S, Rawat R S, Tan T L, Lee P, Springham S V, Rehman A, Chen R, Sun HD 2013 *J Supercond Nov Magn* **26** 187-195

[13] Liu W, Tang X, Tang Z, Chu F, Zeng T, Tang N 2014 *Journal of Alloys and Compounds* **615** 740-744.

[14] Zhao S, Bai Y, Chen J, Bai A and GaoW 2014 *Journal of Magnetics* **19** 68-71

[15] Iqbal J, Safdar N, Jan T, Ismail M, Hussain S S, Mahmood A, Shahzad S, Mansoor Q 2014 *Journal of Materials Science & Technology* **437** 00215-1

[16] Yıldırim Ö A and Durucan C 2016 *Ceramics International* **42** 3229-3238

[17] Hsu C L, Gao Y D, Chen Y S and Hsueh T J 2014 *Appl Mater Interfaces* **6** 4277-4285

[18] Wan W, Huang J, Zhu L, Hu L, Wen Z, Sun L and Ye Z 2013 *Cryst Eng Comm* **15** 7887-7894.

[19] Hu L, Zhu L, He H, Guo Y, Pan G, Jiang J, Jin Y, Sun L and Ye Z 2013 *Nanoscale* **5** 9577-9581

[20] Fabbiyola S, Kennedy L J, Dakhel A A, Bououdina M, Vijaya J J, Ratnaji T 2016 *Journal of Molecular Structure* **1109** 89-96..

[21] Giancoli D 2009 *Prentice Hall.* **658.** ISBN 0-13-149508-9.

[22] Physical constants. Retrieved on 2011-12-17.

[23] Kiesel P, Schmidt O, Geis A and Johnson N 2014 *Pramana-J. Phys* **X** 1-15

[24] Mala N, Ravichandran K, Pandiarajan S, Srinivasan N, Ravikumar B, Pushpa K C S, Swaminathan K, Arun T 2016 *Ceramics International* **42** 7336-7346

[25] Vachhani P S, Dalba G, Ramamoorthy R K, Rocca F, Sipr O and Bhatnagar A K *Journal Of Physiscs Condensed Matter* **24** 506001(1-7).



**Acknowledgements**

The authors are grateful to Laboratory of Advanced Magnetic Materials, Research Center for Physics-LIPI, Indonesia.

# Effect of Growth Time on the Characteristics of ZnO Nanorods

**R Idiawati[1], N Mufti[1,2], A Taufiq[1,2], H Wisodo[1,2], I K R Laila[1], A Fuad[1,2], Sunaryono[1,2]**

[1]Department of Physics, Faculty of Mathematics and Natural Sciences, Universitas Negeri Malang (State University of Malang), Jl. Semarang 5, Malang, 65145, Indonesia
[2]Central Laboratory for Mineral and Advance Materials, Faculty of Mathematics and Natural Sciences, Universitas Negeri Malang (State University of Malang), Jl. Semarang 5, Malang, 65145, Indonesia

Email: nandang.mufti.fmipa@um.ac.id

**Abstract**. ZnO nanorods were successfully synthesized via two stages, that is deposition of ZnO seed layers on ITO substrate and growth of ZnO nanorods via solvothermal method. A characterization by using XRD was employed to investigate the structure and size of the crystals. SEM characterization was used to study the size and morphology of the particles. UV-Vis and FTIR characterizations were used to investigate the bandgap and functional group of the samples. The data analysis presented that the ZnO particles had a crystal structure of hexagonal with rod morphology. The diameter size of the ZnO nanorods growth on the substrate surface for the growth time of 4, 6, and 8 hours were respectively 138, 230, and 236 nm with the length of the rods of 570, 934, and 1280 nm, respectively. The bandgaps of the ZnO nanorods for growth times of 4, 6, and 8 hours were respectively 3.14, 3.12, and 3.05 eV, while the ZnO seed layers had 3.22 eV band gap. FTIR spectrum showed that the main peaks were 400-515, 870, and 1250 $cm^{-1}$ showing the functional groups of ZnO and H-C-N.

**Keywords**: ZnO, nanorod, solvothermal, optical properties, band gap.

## 1. Introduction

The synthesis of ZnO semiconductor has become a topic that catches researchers' attention in the few decades. Several unique characteristics offered by ZnO, such as its wide band gap energy of 3.37 eV, large exciton binding energy of 60 meV, and high electron mobility, make ZnO is a multifunction material [1]. Based on such characteristics, the ZnO semiconductor has been widely applied as, among others, the main component of numerous optical or electronic devices such as fluid sensor [2], gas sensor [3], UV emission devices [4], light emitting diode (LED) [5], a photocatalyst [6], and solar cells [7].

Along with nanoscience and nanotechnology development, the synthesis of nano-scaled ZnO has been a focus. It is due to the better electronic and optical characteristics offered by nano-scaled materials compared with the bulk materials [8]. Some researchers have performed synthesis of nano-size ZnO in varied types of morphology for numerous applications. Some of such morphology include nanorod [9], nanowire [10], nanotube [11], nanoflower [12], nano comb-like [13], nanosheet [14], nanoflake [15], nanocastle [16], and nanoforest [17]. The formation of vertically oriented ZnO in rod morphology, among other types of morphology, on a substrate has been intensively studied until recently.

Vertically oriented ZnO nanorods have been successfully synthesized by several methods, either the vapor or liquid phase method. Several vapor-phase methods such as vapor-liquid-solid [18], metal-organic chemical vapor deposition [19], and chemical vapor transport [6] are proven to be effective in forming vertically oriented ZnO nanorods on a substrate. However, some of such vapor-phase synthesis methods have several drawbacks, i.e. their demand for a relatively high temperature, expensive operational cost, large energy consumption, and a relatively long synthesis time [18]. Compared to the vapor-phase synthesis method, liquid phase synthesis method is more economical and easier to be applied in a wider scale.

Some liquid-phase synthesis methods, such as sol-gel, hydrothermal, electrodeposition, and chemical bath deposition, are developed more intensively by the researchers [19]. By all of the methods, a hydrothermal method is chosen by many researchers due to its superiorities. This method is mostly used for directing the formations of various metal oxides since it requires a simple process, a low temperature, and it can be applied on a wide scale [20]. The performance of ZnO semiconductor could be improved by optimizing the rod length and diameter that will influence the surface area. A high surface area of the semiconductor applied as an indicator electrode in the solar cell allows more absorption of dye- sensitizers so that the solar cell energy conversion can improve [21]. The length and diameter of the rod can be optimized by varying the hydrothermal time during the synthesis process. It was reported that the synthesis of ZnO in nanoflower morphology was conducted using hydrothermal method. The hydrothermal time that was increased to a certain limit could improve the length of the rod that formed the nanoflower so that it would have a larger surface area [22].

In relation to the above fact, another work reported that among the variety of nanostructures, the formation of ZnO nanorods that is vertically oriented on the substrate has been an extensively studied topic until recently [21]. It is also crucial to increase the electron transport that has an impact on the improvement of indicator electrode photosynthesis. Such rod-like 1-dimension nanostructure has been greatly studied and it is known that the morphology of such nanostructure could increase the diffusion of an electron on the photoelectrode film by directly providing the conduction pathway. Such direct pathway along the nanostructure is expected to be able to reduce the charge recombination that is likely to occur on randomly-oriented conventional polycrystalline nanoparticles [23].

According to the above descriptions, the development of synthesis methods to produce ZnO nanorods in a variety of shapes and sizes in order to boost the performance becomes crucial. One way that could be taken is varying the growth time of the ZnO synthesis on the substrate. The particle structure, morphology and size, functional group, and band gap were certainly included in this research.

## 2. Experimental Methods

The primary materials of Zinc Acetate Dihydrate ($Zn(CH_3COO_2).2H_2O$) PA, Ethanol $C_2H_5OH$ (EtOH) PA, and Monoethanolamine $C_2H_7NO$ (MEA) were used in the synthesis of ZnO nanoparticles as seed layers. The materials used for the ZnO nanorods growth were Zinc Nitrate Tetrahydrate ($Zn(NO_3).4H2O$), Hexamethylenetetramine $C_6H_{12}N_4$ (HMT), Deionized Water, and Indium Tin Oxide (ITO) as a substrate. The following process was the formation of ZnO solution by dissolving $Zn(CH_3COO_2).2H_2O$ in ethanol. For the $Zn(CH_3COO_2).2H_2O$ to be completely dissolved, a hot plate magnetic stirrer was utilized at 70 °C for 45 minutes. Such process was followed by an addition of MEA into the solution while stirred at a constant temperature of 70 °C for 2 hours. The molar ratio concentration of $Zn(CH_3COO_2).2H_2O$: MEA was 1:1. Then, the solution was stored at room temperature for 24 hours until a homogeneous and transparent solution was formed. Next, the ITO substrate was put on the spin coater during the formation process of seed layer ZnO thin films. After that, the solution was dropped on the ITO substrate to be then spun at 2500 rpm for 25 seconds, continued by the preheating process at 150 °C for 10 minutes. After the third repetition of such process, an annealing process was performed at 550 °C for 2 hours. Then, after the coating of seed layer ZnO thin films was grown on the ITO substrate, the ZnO nanorods were grown by using the hydrothermal method. The solution growth was performed using Zinc Nitrate Tetrahydrate ($Zn(NO_3)_2.4H_2O$), HMT, and Deionized Water. The utilized precursor concentration was 75 mM, with a molar comparison between $Zn(NO_3)_2.4H_2O$ and HMT of 1:1. The solution was then stirred by a magnetic stirrer for 45 minutes at room temperature. Then, the seed layer ZnO film was dipped into the prepared solution. During the growth process, the solution was heated at a temperature of 90 °C with the growth-time variation of 4 hours, 6 hours, and 9 hours for each concentration. The sample was rinsed by Deionized Water to remove the impurity from the surface area to be then dried at room temperature. Subsequently, the ZnO nanorods on the ITO substrate were annealed at a temperature of 550 °C for 2 hours.

The ZnO nanorods were characterized using X'Pert Pro XRD with Cu-Kα 1.540 Å to identify the structure of the crystal, while INSPECT-S50 SEM-EDAX Type FEI characterized the morphology, elemental composition, and size of the samples. FTIR, with Shimadzu-IR Prestige 21 brand was utilized to determine the functional group. Further, a test by using UV-Vis Spectrophotometer was conducted to characterize the band gap.

## 3. Results and Discussion

The diffraction patterns of X-ray films of ZnO nanorods/ITO with a precursor concentration of 75 mM for a growth time variation of 4, 6, and 8 hours using the solvothermal method are shown in Figure 1. The ZnO nanorods phase analysis informed that the formed diffraction patterns were in in good agreement with standard JCPDS card No. 36-1451 with zincite phase, which had a space group of P 63 m c, hexagonal crystal systems, unit cell of a = 3,2533 Å and c = 5,2073 Å. The phase that was formed in accordance with the research findings reported by Rai et al. (2014) stating that ZnO nanorods were formed in the zincite phase [24]. Figure 1 shows the diffraction peaks located at the (100), (002), and (101) plans. The peak at (002) plane is the highest peak intensities and increases by increasing growth times. This phenomenon indicates that ZnO nanorods growth along c-axis with excellent crystallinity. It is in agreement with the research result reported by Selvan et al. (2014) [25].

Furthermore, the results of crystal size measurement based on the XRD data are presented in Table 1. The crystal size was estimated using Scherrer equation along (002) plane. Table 1 presented that there was an increase in the crystal size along with the improved growth time of nanorods from 44.2 nm to 51.3 nm for growth times of 4 hours and 8 hours, respectively. However, the growth speed rate of ZnO nanorods at growth time 4 hours and 8 hours decreases respectively from 11.05 nm/h to 6.41 nm/h.

**Table 1.** Crystallite sizes of the ZnO nanorods.

| Sample | FWHM (rad) | Crystal Size (nm) |
|---|---|---|
| ZnO nanorods 4 hours | 0.003282 | 44.2 |
| ZnO nanorods 6 hours | 0.002973 | 48.8 |
| ZnO nanorods 8 hours | 0.002827 | 51.3 |

**Table 2.** Dimensions of the ZnO nanorods.

| Type of Sample | Length (nm) | Diameter (nm) |
|---|---|---|
| ZnO nanorods 4 hours | 570 | 138 |
| ZnO nanorods 6 hours | 934 | 230 |
| ZnO nanorods 8 hours | 1280 | 236 |

IOP Conf. Series: Materials Science and Engineering **202** (2017) 012050 doi:10.1088/1757-899X/202/1/012050

**Figure 1.** SEM images of the ZnO nanorods in variation of the growth time of (a) 4 hours, (b) 6 hours, and (c) 8 hours.

Figure 1 shows the cross-section and top-section morphology of ZnO nanorods films with a variation of growth time of 4, 6, and 8 hours. The results of SEM characterization identified the formation of a hexagonal rod structure on the entire surface of the substrate. The SEM characterization results were analyzed and tabulated in Table 2.

It can be seen in Table 2 that the rod diameters on the variation of growth time of 4, 5, and 8 hours were 138 nm, 230 nm, and 236 nm, respectively, while the rod lengths were 570 nm, 934 nm, and 1280 nm, respectively. A longer the ZnO nanorods growth time produced a larger and a higher length of rod diameter. The nanorods diameter was controlled by the precursor concentration. A higher concentration would create a larger nanorods diameter [26], while the length of the nanorods was controlled through by the length of the growth time. A longer growth time would result in a longer rod [27]. The varied diameter sizes were due to the combination of the adjacent rods forming new rods with larger diameters. It was possible to happen when the distance between one rod and another became shorter along with the increased solvothermal period. Besides, it was also enabled by the influence of the different substrate position during the growth process. The ZnO films with the up-facing substrate would have longer rods compared with the ZnO films with the down-facing substrate. The morphology of nanorods with a higher density can be seen in the ZnO nanorods synthesized for a growth time of 8 hours. The longer growth time by hydrothermal method affects the increased density of the Nanorods, which was formed along with the growing length of the rods.

Such results were in accordance with the statement pointed out by Baruah and Dutta (2009) stating that nanorods growth is initiated by the initial nucleation process that took quite long time to be then followed by the growth process at a relatively faster rate until it reached the maximum rod length. Finally, there was not any growth in the rate because the content of $Zn^{2+}$ ion in the solution phase was reduced. The further radial growth did not cause the increase on the diameter size, yet by the combinations of several single nanorods forming new nanorods resulted from the increasing density of rods along with the addition of the hydrothermal period. In the following step, there was an excessive deposition when the reactant was still available in the solution phase while, on the other hand, the substrate could not provide a place for the initial nucleation process of the ZnO nanorods [28]. The increased density of rods had positive impacts on the orientation and direction of rods formed on the substrate. Nanorods with a more compact structure could result in a vertical orientation. It was possible since the rods with a more compact structure would support each other so that the possibilities of them to tilt or fall could be minimized.

A characterization by UV-Vis spectrometer on the wavelength ranged from 350-800 nm was performed to identify the transmittance spectrum and the optical properties of ZnO nanorods shown in Figure 2. Figure 2 indicates that all ZnO nanorods with 75 mM precursor concentration with a variation of growth time of 4, 6, and 8 hours showed the highest excitonic absorption peak at the wavelength of ~376 nm. The results of this study were closely in agreement with the results obtained by Foo et al. (2014) with the highest peak of excitonic absorption at the wavelength of 378 nm [29]. Such a peak indicated that ZnO nanorods have a good optical quality and a strong exciton binding energy. The strongest peak of ZnO nanoparticle film was found at the wavelength of ~360 nm, such wavelength was the sharp peak from the free exciton. Exciton is a combination of neutral bond electron and hole that causes a rapid increase on the peak absorption. The absorption coefficient ($\alpha$) of the ZnO nanorods transitional directions is formulated through the Equation 1.

$$\alpha = \frac{\ln(1/T)}{d} \tag{1}$$

Where $T$ represents the transmission of ZnO film and $d$ is the thickness of the film [29].

The characterization of the energy bandgap ($E_g$) was done based on the Kubelka Munk equation for direct band gap semiconductor, which is determined from a Tauc plot between $(ahv)^2$ vs. $hv$, as can be seen in Figure 3. Based on the interpretation of Figure 3, the intersection between the line extrapolation and x-axis shows a thin energy band gap in each film. the band gap values of ZnO nanorods were obtained to be 3.14 eV , 3,12 eV, and 3,05 eV. In general, the bandgap of ZnO nanorods are lower rather than ZnO bulk (3,37 eV) due to optical confinement effect of the formation of ZnO nanorods [30]. The bandgap decreases with increasing growth time due to reduction of surface area [31].

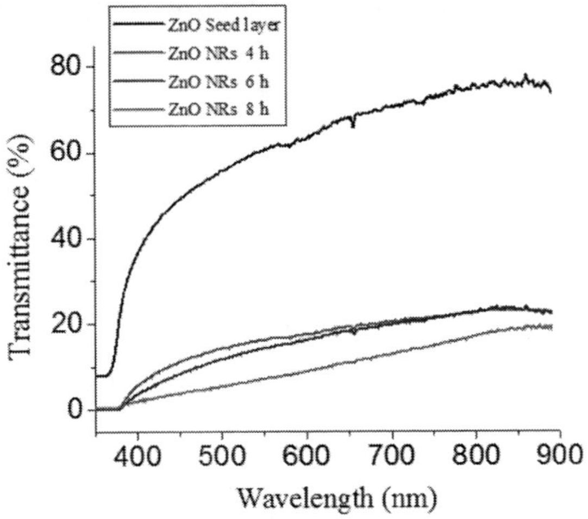


**Figure 2.** UV-Vis spectra of ZnO nanorods with variation of growth time.

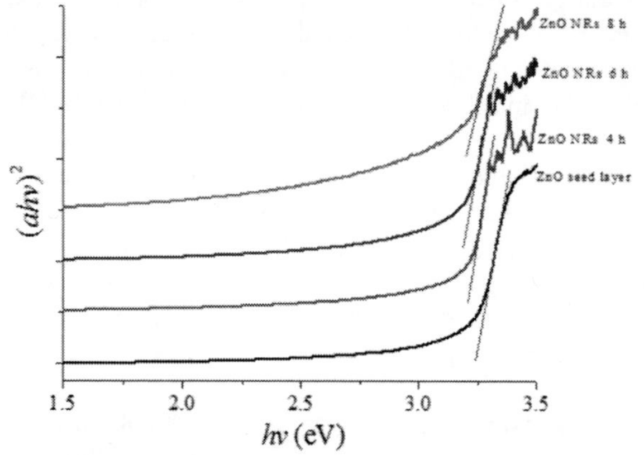


**Figure 3.** Touch graph of ZnO nanorods with variation of growth time.

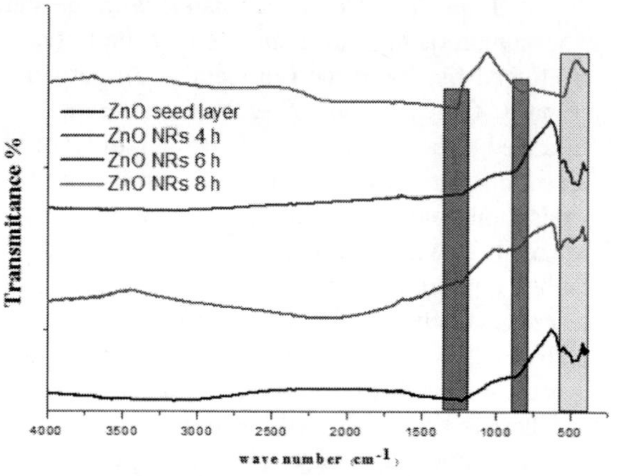


**Figure 4.** FTIR Spectrum of ZnO nanorods with variation of growth time.

Figure 4 shows the FTIR Spectrum test results on ZnO seed layer (A) and ZnO nanorods/ZnO seed layer/ITO with a precursor concentration of 75 mM with a variation of hydrothermal growth periods of 4 hours (B), 6 hours (C), and 8 hours (D) in the wavelengths ranged from 400-500 $cm^{-1}$ at room temperature. Figure 8 identified three main peaks on the wavelengths ranged from 400-515 $cm^{-1}$, 870 $cm^{-1}$, and 1250 $cm^{-1}$. Generally, ZnO absorptions were found in the wavelengths ranged from 500-515 $cm^{-1}$, in line with a report made by Gayen et al. (2010) [32], while the H-C-N functional groups formed at the wavelengths of 870 $cm^{-1}$ and 1250 $cm^{-1}$ from HMT material with a chemical formula of $C_6H_{12}N_4$, as reported by Cottin et al. (2002) [33] and Garcia et al. (2010) [34].

## 4. Conclusion

ZnO nanorods have been successfully synthesized through ZnO seed layer deposition on the ITO substrate surface by solvothermal method. The crystal structure formed in ZnO nanorods is hexagonal zincite. The crystal sizes formed by employing 75 mM precursor concentration and with a variation of growth time of 4, 6, and 8 hours were 44, 48, and 51 nm, respectively. Also, the respective band gap values were identified to be 3.14, 3.12, and 3.05 eV. The SEM results showed that the rod diameter of each sample was 138, 230, and 236 nm, while the respective rod lengths were 570, 934, and 1280 nm. There were three primary peak absorptions on the wavelengths ranged from 400-515 $cm^{-1}$, 870 $cm^{-1}$, and 1250 $cm^{-1}$ showing that the ZnO absorption was identified on the wavelengths ranged from 400-515 $cm^{-1}$, while the absorption of H-C-N functional group was identified at the wavelengths of 870 $cm^{-1}$ and 1250 $cm^{-1}$.

## 5. References

[1] Solís-Pomar F, Martínez E, Meléndrez M F and Pérez-Tijerina E 2011 Growth of vertically aligned ZnO nanorods using textured ZnO films *Nanoscale Res. Lett.* **6**(1) 524

[2] Water W, Chen S E, Meen T H and Ji L W 2012 ZnO thin film with nanorod arrays applied to fluid sensor *Ultrasonics* **52**(6) 747–752

[3] S L Patil, S G Pawar, A T Mane, M A Chougule and V B Patil 2010 Nanocrystalline ZnO thin films: optoelectronic and gas sensing properties *J. Mater. Sci. Mater. Electron.* **21**(12) 1332–36

[4] Zhang L *et al* 2010 Controllable synthesis and shape-dependent photocatalytic activity of ZnO nanorods with a cone and different aspect ratios and of short-and-fat ZnO microrods by varying the reaction temperature and time *Appl. Phys. A,* **100**(4) 1061–67

[5] Park S H, Kim S H and Han S W 2007 Growth of homoepitaxial ZnO film on ZnO nanorods and light emitting diode applications *Nanotechnology* **18**(5) 055608

[6] Kuo T J, Lin C N, Kuo C L and Huang M H 2007 Growth of ultralong ZnO nanowires on silicon substrates by vapor transport and their use as recyclable photocatalysts *Chem. Mater.* **19**(21) 5143–47
[7] Wang C *et al* 2010 Synthesis of nanostructural ZnO using hydrothermal method for dye-sensitized solar cells *Sci. China Technol. Sci.* **53**(4) 1146–49
[8] Zhao Q *et al* 2007 Size-and orientation-dependent photovoltaic properties of ZnO nanorods *J. Phys. Chem. C* **111**(45) 17136–45
[9] Li H, Liu H, Li Y and Liu Q 2017 Controllable Growth of the ZnO Nanorod Arrays on the Al Substrate and Their Reversible Wettability Transition *J. Nanotechnol.* **2017** 1–4
[10] Kjeldstad T, Thøgersen A, Nilsen O, Monakhov E and Galeckas A 2017 Controllable template approach for ZnO nanowire growth: Controllable template approach for ZnO nanowire growth *Phys. Status Solidi A* **214**(2) 1600480
[11] Xu H, Zhang R Q, Zhang X, Rosa A L and Frauenheim Th 2007 Structural and electronic properties of ZnO nanotubes from density functional calculations *Nanotechnology* **18**(48) 485713
[12] Zhang S, Chen H S, Matras-Postolek K and Yang P 2015 ZnO nanoflowers with single crystal structure towards enhanced gas sensing and photocatalysis *Phys Chem Chem Phys* **17**(45) 30300–06
[13] Li W *et al* 2016 Hydrothermal synthesis of a 3D double-sided comb-like ZnO nanostructure and its growth mechanism analysis *Chem Commun* **52**(53) 8231–34
[14] Demel J, Pleštil J, Bezdička P, Janda P, Klementová M and Lang K 2011 Few-layer ZnO nanosheets: preparation, properties, and films with exposed {001} facets *J. Phys. Chem. C* **115**(50) 24702–06
[15] Kaneti Y V, Yue J, Jiang X and Yu A 2013 Controllable synthesis of ZnO nanoflakes with exposed $(10\bar{1}0)$ for enhanced gas sensing performance *J. Phys. Chem. C.* **117**(25) 13153–62
[16] Wang X, Song J and Wang Z L 2006 Single-crystal nanocastles of ZnO *Chem. Phys. Lett.* **424**(1–3) 86–90
[17] Ko S H *et al* 2011 Nanoforest of Hydrothermally grown hierarchical ZnO nanowires for a high efficiency dye-sensitized solar cell *Nano Lett.* **11**(2) 666–671
[18] Yuan Z, Yu J and Jiang Y 2011 Growth of diameter-controlled ZnO nanorod arrays by hydrothermal technique for polymer solar cell application *Energy Procedia* **12** 502–507
[19] Kumar R, Kumar G, Al-Dossary O and Umar A 2015 ZnO nanostructured thin films: depositions, properties and applications—a review *Mater. Express* **5**(1) 3–23
[20] Polsongkram D *et al* 2008 Effect of synthesis conditions on the growth of ZnO nanorods via hydrothermal method *Phys. B Condens. Matter.* **403**(19) 3713–17
[21] Charoensirithavorn P and Yoshikawa S 2006 Dye-sensitized solar cell based on ZnO nanorod arrays *in The 2nd Joint International Conference on Sustainable Energy and Environment* 21–23.
[22] Chen C W *et al* 2006 Anomalous blueshift in emission spectra of ZnO nanorods with sizes beyond quantum confinement regime *Appl. Phys. Lett.* vol. 88, no. 24, p. 241905, 2006.
[23] Wanit M *et al* 2012 ZnO nano-tree growth study for high efficiency solar cell *Energy Procedia* **14** 1093–98
[24] Rai P, Jeon S H, Lee C H, Lee J H and Yu Y T 2014 Functionalization of ZnO nanorods by CuO nanospikes for gas sensor applications *RSC Adv.* **4**(45) 23604
[25] selvan P T, Venkatachalam, Saroja M, Gowthaman P, Ravikumar S and Shankar S 2014 Influence of growth time on zinc oxide nano rods prepared by dip coating method **3**(9)
[26] Babu E S, Hong S K, Vo T S, Jeong J R and Cho H K 2015 Photoelectrochemical water splitting properties of hydrothermally-grown ZnO nanorods with controlled diameters *Electron. Mater. Lett.* **11**(1) 65–72
[27] Shabannia R 2105 Vertically aligned ZnO nanorods on porous silicon substrates: Effect of growth time *Prog. Nat. Sci. Mater. Int.* **25**(2) 95–100

[28] Baruah S and Dutta J 2009 pH-dependent growth of zinc oxide nanorods *J. Cryst. Growth* **311**(8) 2549–54

[29] Foo K L, Hashim U, Muhammad K and Voon C H 2014 Sol–gel synthesized zinc oxide nanorods and their structural and optical investigation for optoelectronic application *Nanoscale Res. Lett.* **9**(1) 429

[30] Foo K L, Hashim U, Muhammad K and Voon C H 2014 Sol–gel synthesized zinc oxide nanorods and their structural and optical investigation for optoelectronic application *Nanoscale Res. Lett.* **9**(1) 429

[31] Lin H, Huang C, Li W, Ni C, Shah S and Tseng Y 2006 Size dependency of nanocrystalline $TiO_2$ on its optical property and photocatalytic reactivity exemplified by 2-chlorophenol *Appl. Catal. B Environ.* **68**(1–2) 1–11

[32] Khun K, Ibupoto Z, AlSalhi M, Atif M, Ansari A and Willander M 2013 Fabrication of well-aligned ZnO nanorods using a composite seed layer of ZnO nanoparticles and chitosan polymer *Materials* **6**(10) 4361–74

[33] Cottin H, Bachir S, Raulin F and Gazeau M C 2002 Photodegradation of hexamethylenetetramine by VUV and its relevance for CN and HCN extended sources in comets *Adv. Space Res* **30**(6) 1481–88

[34] García B B *et al* 2010 Hexamethylenetetramine multiple catalysis as a porosity and pore size modifier in carbon cryogels *J. Non-Cryst. Solids* **356**(33) 1620–25

## Aknowledgments

This research was financially supported by The Ministry of Research Technology and Higher Education (RISTEKDIKTI) under research grant 2016-2017.

# Synthesis and Characterization of Magnetic Elastomer based PEG-Coated $Fe_3O_4$ from Natural Iron Sand

**C Kurniawan[1], Eko A S[1], Ayu Y S[1], P T A Sihite[2], M Ginting[1], P Simamora[2] and P Sebayang[1]**

[1] Research Center for Physics – LIPI, South Tangerang, Banten, Indonesia
[2]Department of Physics, Universitas Negeri Medan, Medan, North Sumatera, Indonesia

Email : pard003@lipi.go.id

**Abstract.** Magnetic elastomer nanocomposite based PEG-coated $Fe_3O_4$ with silicone rubber binder have been prepared from natural iron-sand by using coprecipitation method. The samples were characterized by using X-ray Diffractometer, X-ray Fluorescence, Fourier Transform Infra-Red, tensile strength test, and Vibrating Sample Magnetometer to analyze the physical and magnetic properties. We observed that all samples were formed by single phase cubic spinel magnetite ($Fe_3O_4$) crystalline structure. The atomic bonding analysis by FTIR showed that the C-O-C and C-H ordering were understood as the PEG – $Fe_3O_4$ bonding characteristics. We have observed that the Young modulus of elastomer based PEG-coated $Fe_3O_4$ slightly decreased compared to the natural iron-sand based elastomer. The magnetic properties of PEG-coated $Fe_3O_4$ were known to be magnetically softer with the lowest coercivity without losing its magnetization saturation value. We propose that the PEG-coated $Fe_3O_4$ is a promising candidate to be applied as magnetorheological elastomer due to a good mechanical and magnetic characteristic and also promising as microwave absorbing materials.

**Keywords:** Magnetic elastomer, nanomagnetite, PEG-6000, coprecipitation.

## 1. Introduction

Natural iron sand is one of the popular minerals which has a ferromagnetic behavior. This case is because the iron sand is mainly composed of hematite ($Fe_2O_3$) and magnetite ($Fe_3O_4$) phases. In nanometer size, the magnetite has an interesting property called as superparamagnetic which has a broad range of applications such as biosensor, magnetic hyperthermia, and drug delivery system[1–5]. The superparamagnetic characteristic is achieved if the magnetite particle size approaches its critical value that produces single domain magnetic and zero coercivity field [2].

The surface coating such as PEGylation is one of the standard methods to make the magnetic nanoparticles stable from aggregation. Recent publications stated that number and type of functional groups of PEG as the coated polymer on nanoparticle magnetic surface was affected to the stabilization of the colloidal system especially for biomedical applications [6-7]. Therefore, understanding any route to highly effective synthesis the PEG-coated nano-magnetite became emerged to produce high quality stable magnetic nanoparticles.

However, very little reports have been known for investigating the nanoparticle stabilization effect in the fabrication of magnetic elastomers. The variation of loaded magnetic nanoparticles in a silicone based elastomer has increased the compressive modulus for about 300% [8]. Another research reported that in the latex based magnetic elastomer, the composition of magnetite was affected by the magnetic properties and also its mechanical properties [9].

In this paper, we have investigated the effect of PEG-coated magnetite nanoparticles by using co-precipitation method from natural iron-sand for magnetic elastomer application. Characterizations performed were XRD, XRF, FTIR, and tensile strength tests to analyze the mechanical and microstructural characteristics. Magnetic properties were studied by hysteresis curve built by vibrating sample magnetometer.

## 2. Experimental Method

Natural iron-sand from Bingei River, North Sumatera was employed in this experiment. Iron-sand was milled by using planetary ball mill for 15 hours to get the pure powder until passes #200 standard-meshes sieve (≤ 76 μm). From the XRD analysis (Figure 1), we observed that the milled natural iron sand was composed by single phase $Fe_3O_4$ (magnetite). The co-precipitation method was performed by mixing 16 g of magnetite (from milled iron sand) with HCl (37%, 12M) until 100 mL. The solution was stirred for 90 minutes at a temperature of 70 °C. PEG-6000 crystal from Merck was melted and mixed with magnetite solution (with a ratio of 1:3 v/v) and stirred for 60 minutes at temperature of 70 °C. Ammonia ($NH_4OH$) was added to the solution with a ratio of 1:5 v/v and stirred again for 90 minutes (70 °C) until the precipitate is settled down driven by a permanent magnet. The sediment of magnetite-PEG was washed with distilled water until reaching the normal pH (~ 7). The sludge was then dried at a temperature of 80 °C for 24 hours. The co-precipitation process reaction is compiled as follow [10,11]:

$$3Fe_3O_{4(s)} + 8HCl_{(l)} \rightarrow 2FeCl_{3(l)} + FeCl_{2(l)} + 3Fe_2O_{3(s)} + 3H_2O_{(l)} + H_{2(g)}$$
$$2FeCl_{3(l)} + FeCl_{2(l)} + H_2O_{(l)} + 8NH_4OH_{(l)} \rightarrow \text{nano-}Fe_3O_{4(s)} + 8NH_4Cl_{(l)} + 5H_2O_{(s)}$$

To make an elastomer, PEG-coated magnetite as a filler blended with silicone rubber RTV-683 with fixed binder composition of 20 wt.%. The elastomer was cast with a thickness of 8.5 mm. For comparison, iron-sand and uncoated $Fe_3O_4$ based elastomers were also produced by using the same composition of silicone rubber. The natural iron-sand, uncoated and PEG-coated $Fe_3O_4$ powders were characterized by using Rigaku Smartlab x-ray diffractometer (XRD), Torontech x-ray fluorescence (XRF), Shimadzu Fourier transform infrared spectroscopy (FTIR), and Dexing VSM250 vibrating sample magnetometer (VSM). The tensile strength of elastomer of PEG-coated $Fe_3O_4$ with silicone rubber binder was characterized by using the universal testing machine.

## 3. Results and Discussion

The X-ray diffraction pattern of milled iron-sand, uncoated and PEG-coated $Fe_3O_4$ is displayed in Figure 1. By using the semi-quantitative analysis from Match by Crystal Impact and using COD database, we have analyzed that all samples are composed by single-phase magnetite ($Fe_3O_4$) with the cubic spinel crystalline structure. The lattice constant of $a = b = c = 8.419$ Å is depicted by (220), (311), (400), (442), (511), and (440) planes. It means that the encapsulation of $Fe_3O_4$ by PEG using co-precipitation process does not affect the crystalline structure.

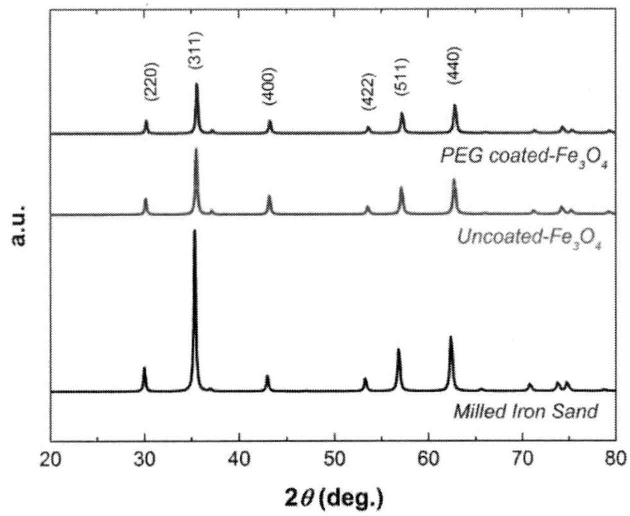


**Figure 1**. X-ray Diffraction pattern (XRD) of milled iron-sand, uncoated and PEG-coated $Fe_3O_4$ samples.

Comparison of elemental composition from raw material to uncoated and PEG-coated magnetite nanoparticles is shown in Table 1 which was performed by XRF analysis. The table shows that the co-precipitation process was affected by elemental composition especially for uncoated and PEG-coated nanoparticles. It shows that the Fe composition was slightly increased after the chemical process. On the other side, some impurities such as Ti, Si and Al were decreased. It means that the chemical process was slightly purified and thrown out other elements during co-precipitation process.

**Table 1**. XRF analysis of milled iron-sand, uncoated $Fe_3O_4$ and PEG-coated $Fe_3O_4$ samples.

| Element | Iron-sand (at.%) | Nano-$Fe_3O_4$ (at.%) | PEG-$Fe_3O_4$ (at. %) |
|---|---|---|---|
| Fe | 91.82 | 93.35 | 93.58 |
| Ti | 4.11 | 3.83 | 3.91 |
| Si | 1.83 | 0.61 | 0.49 |
| Al | 1.01 | 0.95 | 0.81 |
| Mn | 0.78 | 0.79 | 0.82 |
| V | 0.34 | 0.36 | 0.33 |
| Others | 0.11 | 0.11 | 0.06 |
| Total | 100.00 | 100.00 | 100.00 |

The Fourier transform infrared (FTIR) analysis was performed in this experiments to analyze atomic bonding from magnetic nanoparticles and PEG as the coated materials. FTIR curve as shown in Figure 2 shows that the transmittance peak for natural iron sand and uncoated-$Fe_3O_4$ from co-precipitation did not change for all wavenumber. On the other hand, the PEG-coated $Fe_3O_4$ transmittance curve shows new peak addition at 1095.5 and 2337.7 $cm^{-1}$. We believed that this peak contributed from PEG bonding as shown in Table 2. We found that there was a shifting peak at Fe-O bonding region, from 578.6 $cm^{-1}$ to 570.9 $cm^{-1}$ for natural iron-sand and PEG-coated (and uncoated) $Fe_3O_4$, respectively. This Fe-O shifting peak was suggested that the particle size of $Fe_3O_4$ had already reduced during the chemical process and this also predicted that the PEG had already coated the $Fe_3O_4$ particles [12-13]. Another sign that PEG already coated the magnetic nanoparticle is the emerging of C-O-C and C-H bonding with stretch vibration at 1095.5 and 2337.7 $cm^{-1}$.

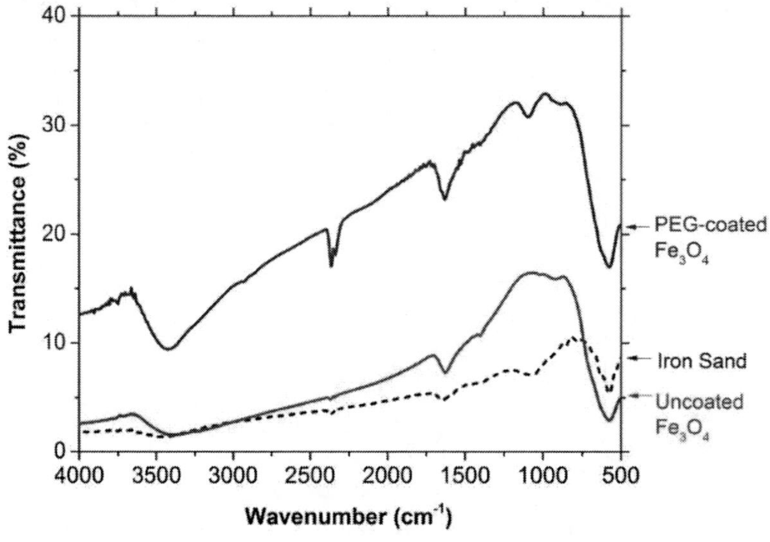


**Figure 2.** FTIR analysis of the samples: milled iron-sand, uncoated-$Fe_3O_4$, and PEG-coated $Fe_3O_4$.

**Table 2.** FTIR bond analysis of the materials.

| Bond | Wave number, $cm^{-1}$ | | | Vibration |
|---|---|---|---|---|
| | Milled iron sand | Uncoated $Fe_3O_4$ | PEG-coated $Fe_3O_4$ | |
| (Fe-O) | 401.19 | 401.19 | 401.19 | Stretch |
| (Fe-O) | 578.64 | 570.93 | 570.93 | Stretch |
| (C-O-C) | - | - | 1095.57 | Stretch |
| (O-H) | 1635.64 | 1627.92 | 1627.92 | Bend |
| (C-H) | - | - | 2337.72 | Stretch |
| (O-H) | 3425.58 | 3387.00 | 3425.58 | Stretch |

All powder samples were embedded in a silicone rubber (20 wt.%) to make magnetorheological elastomers. To analyze the mechanical properties of the elastomers, we have performed a tensile test for all samples according to ASTM D3039. The tensile test result is displayed in Figure 3. The Young modulus was calculated from the stress-strain graph which produced the values of 0.47, 0.53, and 0.32 MPa for natural iron-sand, uncoated nano-$Fe_3O_4$, and PEG-coated $Fe_3O_4$ composites, respectively. We observed that the chemical process which produced a refined particle of $Fe_3O_4$ did not change the mechanical properties of the elastomers significantly. As shown in Figure 3, the Young modulus of iron sand and uncoated $Fe_3O_4$ elastomers did not relatively change. On the other hand, the PEG-coating to $Fe_3O_4$ particles was affected to slightly decrease the Young modulus. This effect is expected to be caused by the densification of the PEG-coated magnetic nanoparticles. The decrease of particle size was known to reduce the pore and strengthen the bonding between filler and matrix in nanocomposite materials [14]. In this case, the increase of bonding strength is related to the increase of the elastomer stiffness.

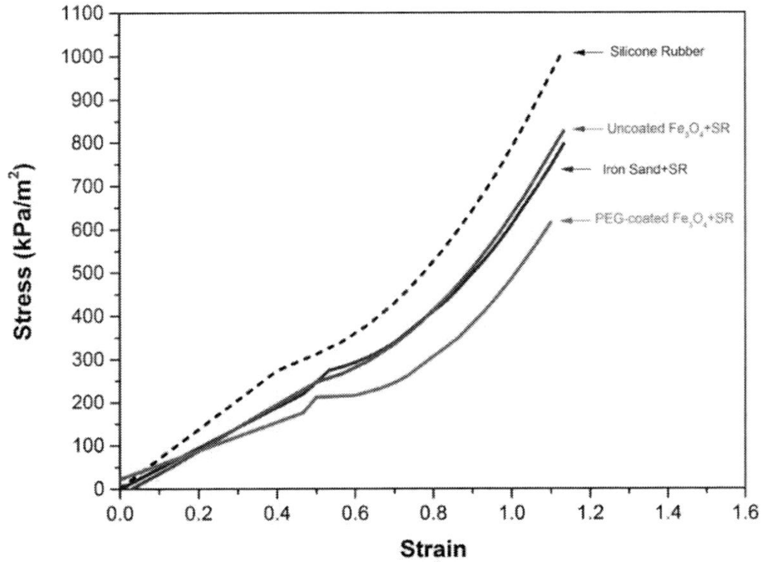


**Figure 3.** Stress-Strain analysis of elastomer material made of 100% silicone rubber (SR), 20% SR + uncoated $Fe_3O_4$, 20% SR + PEG-coated $Fe_3O_4$, and 20% SR + Milled iron-sand

To understand the magnetic characteristics in the form of powders and as magnetic elastomer of PEG-coated $Fe_3O_4$, we have performed hysteresis analysis by vibrating sample magnetometer as shown in Figure 4. We observed that the co-precipitation products which are uncoated and PEG-coated $Fe_3O_4$ showed a decrease of magnetic remanence and coercivity field compared to the milled iron-sand. This effect was closely looked to the superparamagnetic phenomena. PEG-coated $Fe_3O_4$ sample depicted the lowest coercivity both in the form of powder and a magnetic elastomer with the value of 85.9 and 78.4 Oe, respectively. The magnetic characteristics summary of milled iron-sand, uncoated and PEG-coated $Fe_3O_4$ is displayed in Table 3. It shows that the PEG-coated $Fe_3O_4$ behaves differently compared to the natural iron-sand and uncoated $Fe_3O_4$ in the form of powder and elastomer. The coercivity of PEG-coated $Fe_3O_4$ decreases while increasing trend is observed for milled iron-sand and uncoated $Fe_3O_4$. It means that the polymeric coated nanoparticle was stabilized by the nanoparticle distribution in the form of elastomer materials and prevents it to be aggregated during elastomer fabrication. Therefore PEG-coated $Fe_3O_4$ elastomer is a promising candidate to be a good magnetorheological elastomer due to the good mechanical and magnetic characteristics.

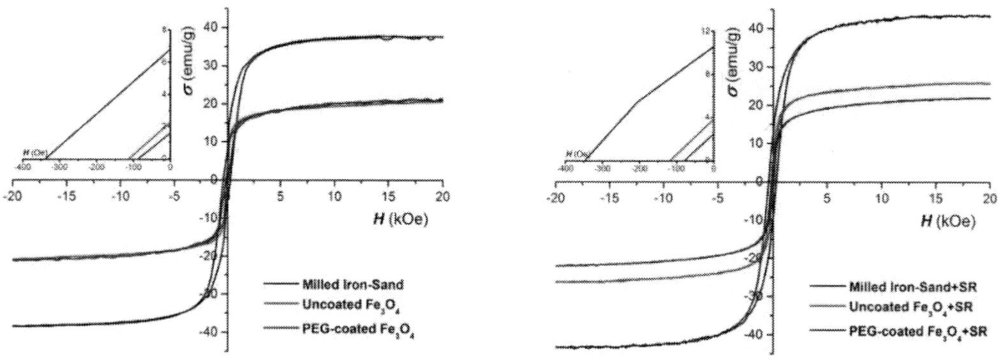


**Figure 4.** Hysteresis Curve of milled iron-sand, nano-$Fe_3O_4$ and PEG-coated $Fe_3O_4$ at powder condition (a) and as a magnetic elastomer (b).

**Table 3.** Magnetic characteristics of the milled iron-sand, uncoated and PEG-coated $Fe_3O_4$ materials in the form of powder and magnetic elastomer.

| Sample Form | Sample Type | $\sigma_s$ (emu/g) | $\sigma_r$ (emu/g) | $H_{ci}$ (Oe) |
|---|---|---|---|---|
| | Milled Iron-Sand | 37.5 | 6.8 | 340.7 |
| Pow□er | Uncoated $Fe_3O_4$ | 20.6 | 2.2 | 113.1 |
| | PEG-Coated $Fe_3O_4$ | 21.2 | 1.6 | 85.9 |
| | Milled Iron-Sand + SR | 43.2 | 13.4 | 353.6 |
| Elastomer | Uncoated $Fe_3O_4$ + SR | 25.7 | 3.8 | 119.3 |
| | PEG-Coated $Fe_3O_4$ + SR | 21.8 | 2.4 | 78.4 |

## 4. Conclusion

Magnetic elastomer based on PEG-coated $Fe_3O_4$ has been prepared by co-precipitation method from natural iron sand. The milled natural iron-sand and uncoated nano-$Fe_3O_4$ were also prepared as the compared materials. The XRD analysis showed that all samples were composed by single phase magnetite, $Fe_3O_4$. The co-precipitation process contributed to the reduction of particle size to the nanometer scale (< 100 nm) and vanishing impurity from the natural iron-sand. The addition of PEG to the chemical process was to prevent aggregation of the nanoparticles by coating the particle surfaces. The atomic bonding analysis by FTIR showed that the C-O-C and C-H ordering in the PEG-coated which is known as the PEG – $Fe_3O_4$ bonding characteristics. We have observed that the elastomer based PEG-coated $Fe_3O_4$ has decreased the Young modulus compared to the natural iron-sand based elastomer. The magnetic characteristic of PEG-coated $Fe_3O_4$ was known to be magnetically softer with the lowest coercivity without losing its magnetization saturation value. We propose that the PEG-coated $Fe_3O_4$ elastomer is a promising candidate to be a good magnetorheological elastomer due to the good mechanical and magnetic characteristics and also promising to be microwave absorber materials.

## 5. References

[1] Kodama R . 1999 Magnetic nanoparticles *J. Magn. Magn. Mater.* **200** 359–72
[2] Gubin S P, Koksharov Y A, Khomutov G B and Yurkov G Y 2005 Magnetic nanoparticles: preparation, structure and properties *Russ. Chem. Rev.* **74** 489–520
[3] Neuberger T, Schöpf B, Hofmann H, Hofmann M and von Rechenberg B 2005 Superparamagnetic nanoparticles for biomedical applications: Possibilities and limitations of a new drug delivery system *J. Magn. Magn. Mater.* **293** 483–96

[4] Lu A-H, Salabas E L and Schüth F 2007 Magnetic Nanoparticles: Synthesis, Protection, Functionalization, and Application *Angew. Chem. Int. Ed.* **46** 1222–44

[5] Wu A, Ou P and Zeng L 2010 Biomedical Applications Of Magnetic Nanoparticles *Nano* **5** 245–70

[6] Cao X, Zhang B, Zhao F and Feng L 2012 Synthesis and Properties of MPEG-Coated Superparamagnetic Magnetite Nanoparticles *J. Nanomater.* **2012** 1–6

[7] Illés E, Tombácz E, Szekeres M, Tóth I Y, Szabó Á and Iván B 2015 Novel carboxylated PEG-coating on magnetite nanoparticles designed for biomedical applications *J. Magn. Magn. Mater.* **380** 132–9

[8] Evans B A, Fiser B L, Prins W J, Rapp D J, Shields A R, Glass D R and Superfine R 2012 A highly tunable silicone-based magnetic elastomer with nanoscale homogeneity *J. Magn. Magn. Mater.* **324** 501–7

[9] Hou C, Gao L, Yu H, Sun Y, Yao J, Zhao G and Liu Y 2016 Preparation of magnetic rubber with high mechanical properties by latex compounding method *J. Magn. Magn. Mater.* **407** 252–61

[10] Safari J, Zarnegar Z and Hekmatara H 2016 Green Synthesis of $Fe_3O_4$ Nanoparticles and Survey their Magnetic Properties *Synth. React. Inorg. Met.-Org. Nano-Met. Chem.* **46** 1047–52

[11] Mahdavi M, Ahmad M, Haron M, Namvar F, Nadi B, Rahman M and Amin J 2013 Synthesis, Surface Modification and Characterisation of Biocompatible Magnetic Iron Oxide Nanoparticles for Biomedical Applications *Molecules* **18** 7533–48

[12] Waldron R D 1955 Infrared Spectra of Ferrites *Phys. Rev.* **99** 1727–35

[13] Zhao D-L, Teng P, Xu Y, Xia Q-S and Tang J-T 2010 Magnetic and inductive heating properties of Fe3O4/polyethylene glycol composite nanoparticles with core–shell structure *J. Alloys Compd.* **502** 392–5

[14] Friedrich K, Fakirov S and Zhang Z 2005 *Polymer composites: from nano-to-macro-scale* (New York: Springer)

# Synthesis, Investigation on Structural and Magnetic Behaviors of Spinel M-Ferrite [M = Fe; Zn; Mn] Nanoparticles from Iron Sand

**S Bahtiar[1], A Taufiq[1,2], Sunaryono[1,2], A Hidayat[1,2], N Hidayat[1,2], M Diantoro[1,2], N Mufti[1,2] and Mujamilah[3]**

[1]Department of Physics, Faculty of Mathematics and Natural Sciences, Universitas Negeri Malang (State University of Malang), Jl. Semarang 5, Malang 65145, Indonesia
[2]Center for Minerals and Advanced Materials, Faculty of Mathematics and Natural Sciences, Universitas Negeri Malang (State University of Malang), Jl. Semarang 5, Malang 65145, Indonesia
[3]Center for Technology of Nuclear Industry Materials, National Nuclear Energy Agency (BATAN), Kawasan Puspiptek Serpong, Tangerang 15314, Indonesia

Email: ahmad.taufiq.fmipa@um.ac.id

**Abstract.** Spinel M-ferrite [M = Fe; Zn; Mn] nanoparticles were prepared from iron sand using a coprecipitation-sonochemical approach. The purified $Fe_3O_4$ from iron sand, $ZnCl_2$ and $MnCl_2.4H_2O$, HCl, and $NH_4OH$ were used as raw materials. X-Ray Diffractometer (XRD), Fourier Transform Infra-Red (FTIR) spectroscopy, Transmission Electron Microscopy (TEM), and Vibration Sample Magnetometer (VSM) were employed to characterize the crystal structure, functional groups, particle size, morphology, and magnetic behavior of the prepared samples, respectively. From the XRD data analysis, M-ferrite particles exhibited a single phase in spinel structure. Furthermore, the M-ferrite particle increased their lattice parameter and crystal volume tracking the metallic-ionic radii of M. The particle size of the M-ferrites particles varied with M, whereas the biggest and lowest were for Zn and Mn, respectively. Based on the magnetization curve, the M-ferrite nanoparticles tended to perform a superparamagnetic behavior and their saturation magnetization as a function of their M ion and particle size.

**Keywords:** M-ferrite [M = Fe; Zn; Mn], nanoparticle, coprecipitation, sonochemical, structure, magnetization

## 1. Introduction

In the past few decades, the interest of spinel ferrite particles preparation especially in nanometric size has been reinforced by many nanoscience and nanotechnology experts due to the numerous potential applications of the nanostructured ferrites [1]. M-ferrite or $MFe_2O_4$ particles in the spinel structure, where M is the metallic ions such as Fe, Zn, Mn, Ni, Co, and others, become a group of functional materials that have created interest due to their physical and chemical behaviors. Therefore, various preparation approaches have been intensively performed by many researchers to achieve the spinel M-ferrite nanoparticles in different sizes and morphologies. Moreover, in order to fabricate the spinel M-ferrite nanoparticles in a mass production, the preparation method has also been developed by finding alternative precursors as raw materials from natural resources that are abundantly available in the world. One of the natural resources that can potentially be utilized as raw materials is iron sand.

Several papers reported that the M-spinel ferrite nanoparticles in the forms of nanopowders, films, magnetic fluids, and magnetic gels were prepared from iron sand through a simple coprecipitation method by varying several physical and chemical parameter processes [2–8]. The coprecipitation method was successfully developed for preparing the M-spinel ferrite nanoparticles in several sizes and morphologies in relatively simple, quick, and inexpensive ways. However, the phenomenon of the magnetic nanoparticle agglomeration has not been overcome yet. Therefore, it is necessary to find and or develop a method for overcoming the problem. One way that can be conducted is by combining the coprecipitation method with the sonochemical route. The main reason of such combination is due to the performance of the acoustic cavitation in producing high temperature and pressure in the liquid [9].

In this paper, we report the successful combination method of coprecipitation and sonochemical approaches in the fabrication of M-ferrite [M = Fe; Zn; and Mn] nanoparticles from the natural resource of Indonesia. The M-ferrite nanoparticles for M = Fe, Zn, and Mn were selected regarding their performances that can be potentially applied in several fields such as for drug delivery vehicles and MRI contrast agents [10], for the detection of pathogenic bacteria [11], as anode materials for lithium-ion batteries [12], and so forth. Furthermore, the structural, morphological, and magnetic behaviors of the prepared M-ferrite nanoparticles are also discussed.

## 2. Experimental Method

### *2.1 Materials*

In this experiment, the materials used in preparing M-ferrite nanoparticles were iron sand from East Java Indonesia, zinc chloride ($ZnCl_2$), manganese (II) chloride tetrahydrate ($MnCl_2.4H_2O$), ammonium hydroxide ($NH_4OH$), hydrochloric acid (HCl), and distilled water.

### *2.2 Preparation*

M-ferrite [M = Fe; Mn; Zn] nanoparticles were prepared by a combination of coprecipitation-sonochemical methods. The iron sand was utilized to obtain magnetite powder. The magnetite was then dissolved in the hydrochloric acid for 20 minutes to form a solution containing $Fe^{2+}$ and $Fe^{3+}$ salts. The formation mechanism of the iron salts is presented in the following equation [13].

$$Fe_3O_4 + 8HCl \rightarrow 2FeCl_3 + FeCl_2 + 4H_2O \quad (1)$$

The product of reaction (1) was then dropped wisely with the ammonium hydroxide to form magnetite nanoparticles. Meanwhile, the product was mixed with $ZnCl_2$ and $MnCl_2.4H_2O$ before adding the ammonium hydroxide to form Zn-ferrite and Mn-ferrite, respectively. These processes were combined with the sonochemical method using an ultrasonic bath at a frequency of 40 kHz. All of the operations were kept at the ambient temperature. In order to obtain the M-ferrite particles in nanopowders, the process was continued by the washing process for several times and followed by drying process at 100 °C.

### *2.3 Characterizations*

All prepared samples were characterized by using XRD (X-ray Diffractometer) to investigate their phase purity, crystallite size, and crystal structure. Meanwhile, the FTIR (Fourier Transform Infra-Red) spectroscopy was employed to study the functional groups of the samples. Furthermore, the TEM (Transmission Electron Microscopy) was also used to determine the morphology and particle size of the samples. Finally, the magnetic behavior of the samples was investigated by means of VSM (Vibration Sample Magnetometer). All characterization processes were kept at ambient temperature.

## 3. Results and Discussion

The chemical formation of the prepared spinel M-ferrite particles can be expressed by the following equations.

$$2FeCl_3 + FeCl_2 + 4H_2O + 8NH_4OH \rightarrow Fe_3O_4 + 8NH_4Cl + 8H_2O \tag{2}$$

$$2(2FeCl_3 + FeCl_2 + 4H_2O) + 3ZnCl_2 + 22NH_4OH + \frac{1}{2}O_2 \rightarrow 3ZnFe_2O_4 + 22NH_4Cl + 19H_2O \tag{3}$$

$$2(2FeCl_3 + FeCl_2 + 4H_2O) + 3MnCl_2.4H_2O + 22NH_4OH + \frac{1}{2}O_2 \rightarrow 3MnFe_2O_4 + 22NH_4Cl + 31H_2O \tag{4}$$

The iron salts containing $Fe^{2+}$ and $Fe^{3+}$ ions resulted from one stage reaction of the purified iron sand with HCl were directly wisely dropped by $NH_4OH$ to form $Fe_3O_4$ particle according to reaction (2). The $ZnCl_2$ and $MnCl_2.4H_2O$ were also mixed with the iron salts and then followed by dropping $NH_4OH$ carefully to form $ZnFe_2O_4$ and $MnFe_2O_4$ particles. The $NH_4Cl$ and $H_2O$ as unwanted products were cleaned or removed through washing and drying processes to obtain the spinel M-ferrite particles in a single phase. In order to investigate the success of the synthesis process, the details of the phase purity, crystal structure, and crystallite size were discussed in the following section.

Figure 1 shows the X-ray diffraction patterns of the M-ferrite particles prepared by using a combined coprecipitation-sonochemical route. Based on the picture, it is clear that all samples have similar diffraction patterns which correspond to the similar phase. All diffraction patterns have Miller indexes (hkl) on the planes (220), (311), (400), (422), and (511) representing the diffraction patterns of ferrites. The qualitative analysis of the XRD data showed that all samples construct a single phase without any impurities. The Zn-ferrite particle has the sharpest peak indicating that it has the largest particle size, while the Mn-ferrite particle has the opposite one. Furthermore, the quantitative analysis of the XRD data pointed out that the refinement model represented by the solid line fits well the data represented by a circle. The detailed results of the fitness are tabulated in Table 1. From the table, the lattice parameter values are 0.837, 0.839, and 0.842 nm for iron, zinc, and manganese ferrites, respectively. In accordance with the results of this work, Syue *et al.* reported that the manganese ferrite nanoparticle prepared from commercial precursors by combustion method has the largest lattice parameters than that of zinc ferrite nanoparticle [14]. Logically, the increasing lattice parameters are originated from the contribution of the ionic radii of the M-ions, where the $Mn^{2+}$ has larger ionic radii (~0.82 Å) than that of $Zn^{2+}$ (~0.74 Å), and $Fe^{2+}$ (~0.78 Å). These metallic ions were distributed randomly in the spinel crystal structure, both in the octahedral and tetrahedral positions.

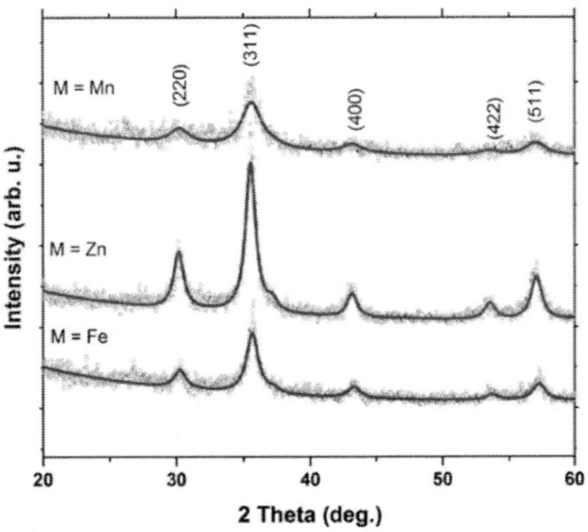


**Figure 1.** XRD patterns of the M-ferrites [M = Fe; Zn; Mn] from iron sand

The crystallite size ($D$) of the prepared samples was determined by using Debye-Scherrer's formula (5).

$$D = \frac{K\lambda}{B_o \cos\theta} \tag{5}$$

Where $K$ is constant, $B_0$ is the full width at half maximum, and $\theta$ is the diffraction angle. From the calculation, as shown in Table 1, the samples have the crystallite in nanometric size ranging from about 5.1 to 8.7 nm. Remarkably, the crystallite size of the M-ferrite particles in this work was smaller than that of M-ferrites produced by a combustion method [1]. Therefore, it can be concluded that the combined method of sonochemical and coprecipitation approaches in this work was able to prepare the M-ferrite particles from natural iron sand in a single phase in a nanometric size.

**Table 1.** Results of Rietveld refinement of the M-ferrites [M = Fe; Zn; Mn] XRD patterns

| Sample | M = Fe | M = Zn | M = Mn |
|---|---|---|---|
| Lattice parameter (nm) | 0.837 | 0.839 | 0.842 |
| *GoF* | 0.6 | 0.6 | 0.6 |
| $R_p$ | 14.0 | 14.7 | 14.3 |
| $R_{wp}$ | 19.8 | 19.1 | 19.7 |
| $D$ (nm) | 8.7 | 10.3 | 5.1 |

The particle size of the prepared samples was also confirmed by the TEM characterizations as presented in Figure 2. Based on the figure, it can be seen that both of $FeFe_2O_4$ and $ZnFe_2O_4$ have an average particle size below 100 nm. The TEM images show that the magnetic nanoparticles tended to form spherical-particle agglomerations. Quantitatively, the $ZnFe_2O_4$ average particle size corresponded to the XRD data analysis more than that of $FeFe_2O_4$. Further analysis in studying the functional groups of the prepared samples will be discussed in the following section.

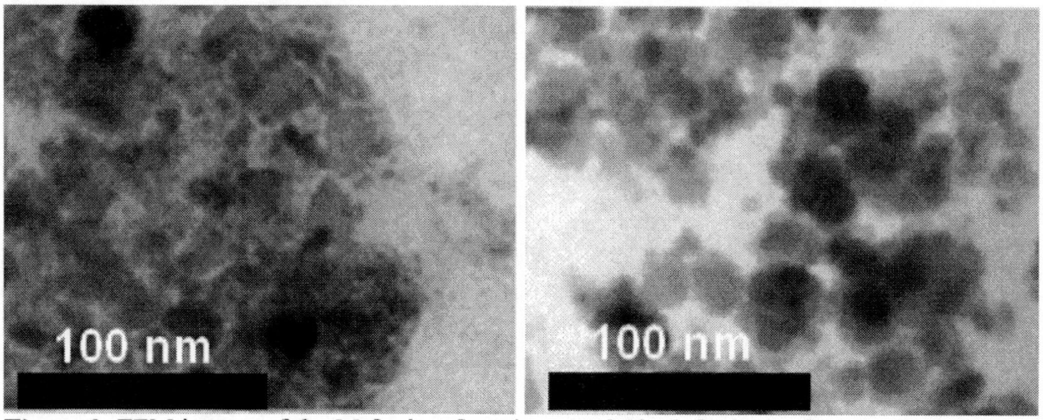


**Figure 2.** TEM images of the M-ferrites from iron sand (the left picture for M = Fe and the right one for M = Zn)

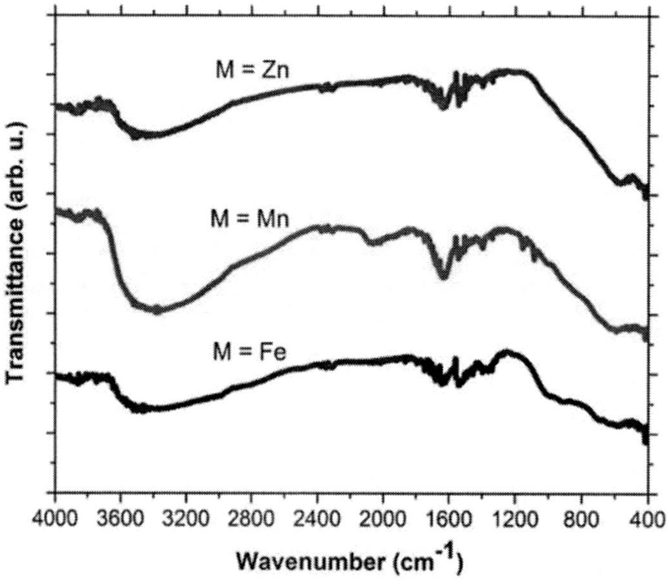


**Figure 3.** FTIR spectra of the M-ferrites [M = Fe; Zn; Mn] from iron sand

Figure 3 presents the FTIR spectra of the prepared M-ferrite nanoparticles in the electromagnetic waves ranging from 4000 to 400 cm$^{-1}$. Based on the quantitative analysis, there are several peaks of about 420-590, 1640, and 3420 cm$^{-1}$. The OH group of waters absorbed in the samples can be represented at the wavenumber of around 3420 cm$^{-1}$. Furthermore, the vibration of Metal-O bonds both in the octahedral and tetrahedral sites are described by the peaks in the range of 420 – 590 cm$^{-1}$. These results are in good agreement with the FTIR data of the ferrite systems that have a general spectrum at the frequency of about 400-600 cm$^{-1}$ [15]. Additionally, another group pointed out that the vibration at around 1640 cm$^{-1}$ was contributed by the O-H bending [16]. Therefore, these spectra became alternative evidence that the M-ferrite nanoparticles were successfully prepared from iron sand using a coprecipitation-sonochemical approach.

The magnetic properties of the M-ferrite [M = Fe; Zn; Mn] nanoparticles are presented in Figure 4. The figure represents the magnetization (*M-H*) curve by varying the external magnetic field from -1 to 1 T. Qualitatively, all of the samples have an *S*-shaped hysteresis curve indicating that the prepared M-ferrite nanoparticles performed a superparamagnetic behavior. However, in order to ensure that the prepared samples in this work present a superparamagnetic character, it is still necessary to conduct further research by FC-ZFC experiment. Further quantitative analysis showed that the saturation magnetization (*Ms*) values of the M-ferrite nanoparticles for M = e, Zn, and Mn are of about 42.1, 29.8, and 17.4 emu/g, respectively. Based on the theoretical calculation, because $Mn^{2+}$ has a higher magnetic moment (5 μB) than that of $Fe^{2+}$ (4 μB) and $Zn^{2+}$ (0 μB), the Mn-ferrite particle should have the largest net magnetic moment among the others [17]. However, based on the experimental results, the Mn-ferrite particle has the smallest saturation magnetization among the others. These phenomena can be explained by the effect of the particle size of the M-ferrite nanoparticles where the Mn-ferrite particle has the smallest size according to the XRD data analysis. Another work reported that the magnetization saturation of ferrite nanoparticles decreased along with the reducing particle size [18]. Furthermore, Taufiq *et al.* reported that the magnetic properties of the ferrite nanoparticles, as well as their magnetization saturation and blocking temperature, are contributed not only by the metallic ions compositions, but also by the primary and secondary particles, cluster, and fractal dimension [3]. Therefore, the magnetization saturation of the M-ferrite particles in this experiment is contributed not only by the type of M ions but also by the particle size. For final remark, after successfully fabricating the M-ferrite nanoparticles from iron sand via the combined methods, it is also essential to conduct a further experiment by utilizing these ferrites for advanced applications as well as for sensors, for antibacterial systems, and as radar absorbing materials.

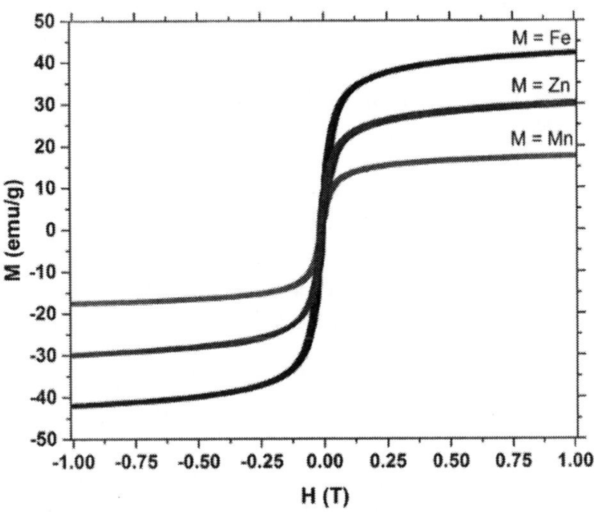


**Figure 4.** Magnetization curves of the M-Ferrites [M = Fe; Zn; Mn] from iron sand

## 4. Conclusion

The iron sand was utilized as a raw precursor to prepare the spinel M-ferrite [M = Fe; Zn; Mn] nanoparticles via a combined coprecipitation-sonochemical approach. In this work, all prepared samples formed a single phase with the spinel cubic structure. The M-ferrite particles formed in nanometric size and tended to agglomerate. The lattice parameter of the M-ferrite nanoparticles varied with M types with the largest value for M = Mn originated from the $Mn^{2+}$ ionic radii. The saturation magnetization value of the M-ferrite nanoparticles was contributed by the M ion and particle size.

## 5. References

[1] Murugesan C and Chandrasekaran G 2016 Structural and Magnetic Properties of $Mn_{1-x}Zn_xFe_2O_4$ Ferrite Nanoparticles *J. Supercond. Nov. Magn.* **29** 2887–97

[2] Triwikantoro, Baqiya M A, Heriyanto T, Mashuri and Darminto 2017 Nano-coating of Aluminum Surface Using Fe3O4-based Magnetic Fluids *J. Supercond. Nov. Magn.* **30** 555–60

[3] Taufiq A, Sunaryono, Rachman Putra E G, Okazawa A, Watanabe I, Kojima N, Pratapa S and Darminto 2015 Nanoscale Clustering and Magnetic Properties of $Mn_xFe_{3-x}O_4$ Particles Prepared from Natural Magnetite *J. Supercond. Nov. Magn.* **28** 2855–63

[4] Sunaryono, Taufiq A, Putra E G R, Okazawa A, Watanabe I, Kojima N, Rugmai S, Soontaranon S, Zainuri M, Triwikantoro, Pratapa S and Darminto 2015 Small-Angle X-Ray Scattering Study on PVA/$Fe_3O_4$ Magnetic Hydrogels *Nano* **11** 1650027

[5] Taufiq A, Sunaryono, Rachman Putra E G, Pratapa S and Darminto 2015 Nano-Structural Studies on $Fe_3O_4$ Particles Dispersing in a Magnetic Fluid Using X-Ray Diffractometry and Small-Angle Neutron Scattering *Mater. Sci. Forum* **827** 213–8

[6] Sunaryono, Taufiq A, Mashuri, Pratapa S, Zainuri M, Triwikantoro and Darminto 2015 Various Magnetic Properties of Magnetite Nanoparticles Synthesized from Iron-Sands by Coprecipitation Method at Room Temperature *Mater. Sci. Forum* **827** 229–34

[7] Sunaryono, Taufiq A, Munaji, Indarto B, Triwikantoro, Zainuri M and Darminto 2013 Magneto-elasticity in hydrogels containing $Fe_3O_4$ nanoparticles and their potential applications *AIP Conf. Proc.* **1555** 53–6

[8] Setiadi E A, Sebayang P, Ginting M, Sari A Y, Kurniawan C, Saragih C S and Simamora P 2016 The synthesization of $Fe_3O_4$ magnetic nanoparticles based on natural iron sand by co-precipitation method for the used of the adsorption of Cu and Pb ions *J. Phys. Conf. Ser.* **776** 12020

[9] Mirzaie A, Musabeygi T and Afzalinia A 2016 Sonochemical synthesis of magnetic responsive $Fe_3O_4$@TMU-17-NH2 composite as sorbent for highly efficient ultrasonic-assisted denitrogenation of fossil fuel *Ultrason. Sonochem.*

[10] Beg M S, Mohapatra J, Pradhan L, Patkar D and Bahadur D 2017 Porous $Fe_3O_4$-$SiO_2$ core-shell nanorods as high-performance MRI contrast agent and drug delivery vehicle *J. Magn. Magn. Mater.* **428** 340–7

[11] Pal M, Lee S, Kwon D, Hwang J, Lee H, Hwang S and Jeon S 2017 Direct immobilization of antibodies on Zn-doped $Fe_3O_4$ nanoclusters for detection of pathogenic bacteria *Anal. Chim. Acta* **952** 81–7

[12] Tang X, Hou X, Yao L, Hu S, Liu X and Xiang L 2014 Mn-doped $ZnFe_2O_4$ nanoparticles with enhanced performances as anode materials for lithium ion batteries *Mater. Res. Bull.* **57** 127–34

[13] Baqiya M A, Taufiq A, Sunaryono, Ayun K, Zainuri M, Pratapa S, Triwikantoro and Darminto 2017 Spinel-Structured Nanoparticles for Magnetic and Mechanical Applications *Magnetic Spinels: Synthesis, Properties and Applications* (InTechOpen) pp 253–72

[14] Syue M-R, Wei F-J, Chou C-S and Fu C-M 2011 Magnetic and electrical properties of Mn–Zn ferrites synthesized by combustion method without subsequent heat treatments *J. Appl. Phys.* **109**07A324

[15] Wiriya N, Bootchanont A, Maensiri S and Swatsitang E 2014 Magnetic properties of $Zn_{1-x}Mn_xFe_2O_4$ nanoparticles prepared by hydrothermal method *Microelectron. Eng.* **126** 1–8

[16] Suharyadi E, Setiadi E A, Shabrina N, Kato T and Iwata S 2014 Magnetic Properties and Microstructures of Polyethylene Glycol (PEG)-Coated Cobalt Ferrite

(CoFe<sub>2</sub>O<sub>4</sub>) Nanoparticles Synthesized by Coprecipitation Method *Adv. Mater. Res.* **896** 126–33

[17] Cullity B D and Graham C D 2008 *Introduction to Magnetic Materials* (Hoboken, N.J: Wiley-IEEE Press)

[18] Feng J, Mao J, Wen X and Tu M 2011 Ultrasonic-assisted in situ synthesis and characterization of superparamagnetic $Fe_3O_4$ nanoparticles *J. Alloys Compd.* **509** 9093–7

**Acknowledgements**

The authors appreciate Kemenristekdikti, Republic of Indonesia, in supporting "research grand" for AT 2016-2017.

IOP Conf. Series: Materials Science and Engineering **202** (2017) 012053 doi:10.1088/1757-899X/202/1/012053

# Variation of Carbon Coating on $Li_2Na_2Ti_6O_{14}$ as Anode Material of Lithium Battery

**B Prihandoko[1], S Priyono[1], A Subhan[1] and A Mulya[2]**

[1]Research Center for Physics, Indonesian Institute of Sciences, Tangerang Selatan Indonesia
[2]University of North Sumatera, Medan Indonesia

Email: bamb012@lipi.go.id

**Abstract**. $Li_2Na_2Ti_6O_{14}$ was developed from $Li_5Ti_4O_{12}$ as a good active material for anode of the lithium battery. $Li_2Na_2Ti_6O_{14}$ was prepared by a preliminary formation of $Li_2Na_2Ti_6O_{14}$ through solid state reaction with calcination temperature at 700 °C, and sintering temperature at 800 °C for 8 hours. By analysis of XRD patterns, $Li_2Na_2Ti_6O_{14}$ shows spinel structure which similar as $Li_5Ti_4O_{12}$. Carbon coating on $Li_2Na_2Ti_6O_{14}$ was done by using pyrolysis method at a temperature of 700 °C for 2 hours. The process of carbon coating was done under variation of comparison between $Li_2Na_2Ti_6O_{14}$ and tapioca powder as a carbon source, i.e. 8:1, 10:1 and 12:1. The working voltage of $Li_2Na_2Ti_6O_{14}$/C product was 1,2 volt. The other analysis were conductivity, cyclic voltammeter, and charge – discharge.

## 1. Introduction

Lithium ion batteries have been widely applied as power sources for electronic devices such as cameras, mobile phones, computers and other related equipment. The Recent development of lithium ion batteries as electric sources for electric and hybrid electric vehicles (EVs and HEVs) also widely investigated[1]. In fact, high power and high energy density of lithium ion batteries is the major key to application in EVs and HEVs [2]. A search of new electrode materials exhibiting high charge/discharge current rates are urgently required. The spinel lithium titanate ($Li_4Ti_5O_{12}$) has been demonstrated as a potential candidate for the anode electrode material in high power Li-ion batteries as well as hybrid supercapacitors because it has some unique characteristics as compared with carbon-based anode materials [3].

Spinel $Li_4Ti_5O_{12}$ is a promising anode material for high safety due to the absence of solid electrolyte interphase (SEI) film [4, 5,6]. Furthermore, its discharge platform at about 1.55 V, which is higher than the reduction potential of most organic electrolytes, can restrain the origination of SEI film and avoid the formation of metallic lithium dendrite [7]. However, bare $Li_4Ti_5O_{12}$ suffering from low electronic conductivity ($10^{-13}$ S $cm^{-1}$) shows poor lithium storage properties at high rates, which have a huge negative effect on its commercial applications [8,9].

In this problem, the most improvements have been focused on improving the electronic conductivity and electrochemical properties of $Li_4Ti_5O_{12}$ by developing nanostructures of material, by using different doping of element and coating techniques [10, 11]. $Li_4Ti_5O_{12}$ was successfully doped with Na. This doping changed to new anode active material of $Li_2Na_2Ti_6O_{12}$ and made the lower working potential of 1.3 V. $Li_2Na_2Ti_6O_{12}$ has been regarded as a novel promising anode material to replace $Li_4Ti_5O_{12}$ in recent years [7]. Unfortunately, it also suffers from low electronic conductivity which hinders its further research. In this work, the problem will be an improvement with a carbon coating. The coating technique was used for increasing of electronic conductivity. Under carbon coating, that carbon can give an effect on the electrochemical characterization of $Li_2Na_2Ti_6O_{14}$. Carbon coating with local cassava powder as carbon sources was done as well.

## 2. Experimental Method

Preparation of $Li_2Na_2Ti_6O_{12}$ has been carried out with the technical grade raw materials of $TiO_2$ (Japan), $Na_2CO_3$ (Merck), and $LiOH.H_2O$ (Germany) using powder metallurgy method. After the stoichiometric weighing and mixing, the mixture was calcined at 700°C for 1 hour. After milling, calcined powder material was sintered at 800 °C for 8 hours. Carbon coating on $Li_2Na_2Ti_6O_{14}$ was done with pyrolysis method at a temperature of 700 °C for 2 hours [12]. The process of carbon coating was done under variation of comparison between $Li_2Na_2Ti_6O_{14}$ and tapioca powder as a carbon source, i.e. 8:1, 10:1 and 12:1. The crystal structure of $Li_2Na_2Ti_6O_{14}/C$ powders was observed by using X-Ray. For the electrochemical test, the working electrode was prepared by dispersing a mixture of as-prepared active material, carbon black, and polyvinyl difluoride binder with a weight ratio of 8.5:0.5:1 in DMAC solvent to form a homogeneous slurry and then passing the slurry on copper foil to form a thin film. This film was dried at 80 °C. In the coin-type half cells, metallic lithium foils were used as the counter and reference electrodes; the electrolyte was 1 mol L_1 solution of $LiPF_6$. The method of EIS (Electrochemical Impedance Spectroscopy) from HIOKI equipment was used primarily to test the conductivity of the material. Cyclic Voltammeter of sample was done by using WBCS3000 automatic battery cycler.

## 3. Results and Discussion

The XRD results crystal structure in Figure 1 formed a phase of $NaLiTi_3O_7/C$ or $Li_2Na_2Ti_6O_{14}/C$ that was orthorombic (space group = 69: Fmmm) with lattice parameters of a = 16.485 Å, b = 5.7382 Å, c = 11.221 Å, with the same angle of $\alpha = \beta = \gamma = 90°$. The phase of $TiO_2$ was seen in the results of XRD pattern. $TiO_2$ was tetragonal (space group = 136: $P4_2/mnm$) with a lattice parameter of a = 4.590 Å, b = 4.590 Å, c = 2.962 Å, with the same angle of $\alpha = \beta = \gamma = 90°$. $Li_2Na_2Ti_6O_{14}$ phase is shown according to standard ICDD (International Center for Diffraction Data) with the number of 00-052-0690 and TiO2 phase is shown according to standard ICDD (International Center for Diffraction Data) with the number of 04-004-4337.

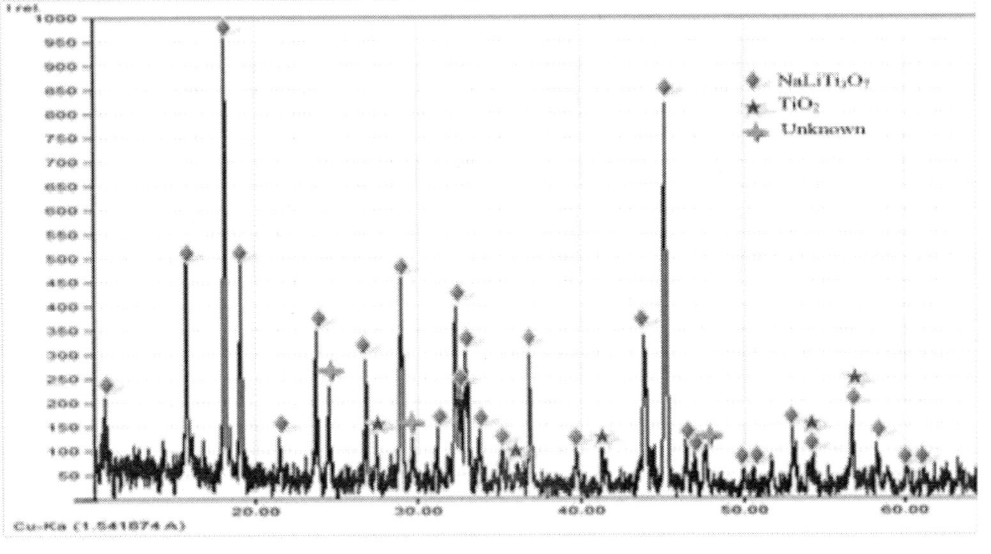


**Figure 1**. The X-ray pattern of carbon coated $Li_2Na_2Ti_6O_{14}$.

Detail data of crystal structure analysis can be seen in Table 1. In generated data by quantitative analysis, the purity of sodium lithium titanium oxide $Na_2Li_2Ti_6O_{14}$ phase is 96%, while the remaining phase (4%) is rutile, syn $TiO_2$.

**Table 1.** The crystal structure (lattice parameter) of $NaLiTi_3O_7$ and $TiO_2$

| Phase name | Chemical composition | *a* (Å) | *b* (Å) | *c* (Å) | $\alpha=\beta=\gamma$ | Density (g/cm$^3$) |
|---|---|---|---|---|---|---|
| Lithium Sodium Titanium Oxide | $NaLiTi_3O_7$ | 16.485 | 5.7382 | 11.221 | 90° | 3.577 |
| Rutile,syn | $TiO_2$ | 4.590 | 4.590 | 2.962 | 90° | 4.28 |

The test results for the four samples of CV can be seen from the graph of cyclic voltammograms in Figure 2. The working electrode of samples was prepared by dispersing a mixture of as-prepared active material, carbon black and polyvinyl difluoride as a binder. In the coin-type half cells, the samples were tested with a scan rate of 160 mV/s.

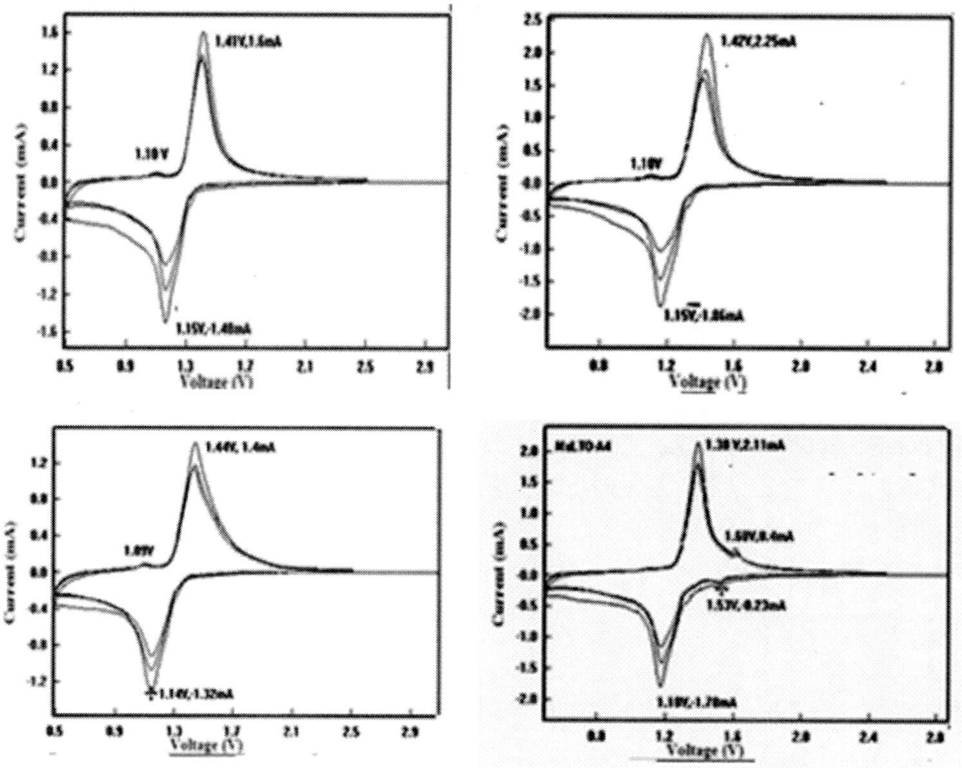

**Figure 2**. Results of cyclic voltammograms

From Figure 3 we can see the working voltages. Table 2 gives the detail data for working voltage of samples. From Table 2, it is obtained that each sample with a different ratio of carbon coating produced a relatively similar working voltage between 1.28 volt and 1.29 volt. The electrochemical reaction occurs nearly equally in the three samples, but uncoated samples A4 showed other signal. The working voltage of lithium titanate appeared in sample A4. In the tree samples, a coated carbon could erase an influence of the other small phase.

**Table 2.** The results of working voltage analysis by using cyclic voltammogram

| Sample Code | Carbon coating composition | Working voltage [V] |
|---|---|---|
| A1 | 12:1 | 1.28 |
| A2 | 10:1 | 1.285 |
| A3 | 8:1 | 1.29 |
| A4 | uncoated | 1.28 and 1.56 |

Figure 3 shows the charge-discharge results of the tests used to determine the capacity of the material $Na_2Li_2Ti_6O_{14}$ in the same coin-type half cells. The rate of 0.1 C was used in this experiment. One cycle is equal to one charge process (the oxidation reaction occurs) and one discharge (reduction reaction occurs).

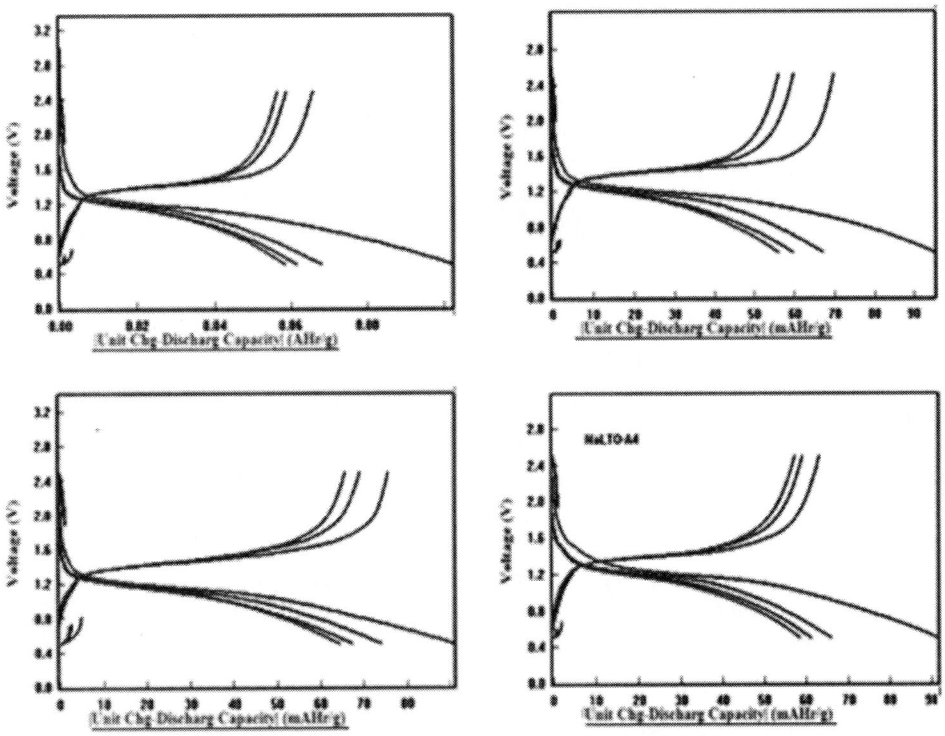

**Figure 3**. Results of charge–discharge test

Figure 3 shows the analysis of charge-discharge retention capacity gives a performance information to determine the materials that well interact in this process.

**Table 3.** The results of the analysis of charge-discharge retention capacity

| Sample | Active mass [mg] | $-C_i$ [mAh/g] | $-C_f$ [mAh/g] | $-C_i$ / $-C_f$ [%] |
|---|---|---|---|---|
| A1 | 10 | 101.7 | 60.9 | 61.193 |
| A2 | 12 | 94.0 | 59.1 | 55.554 |
| A3 | 9.6 | 90.2 | 67.1 | 60.,52 |
| A4 | 12 | 90.0 | 60.7 | 50.,63 |

A detail numerical analysis results are displayed in Table 3. The sample A1 showed the biggest value of charge-discharge retention capacity while sample A4 showed the lowest one. Sample with the smallest coated carbon had the best process of charge–discharge. The uncoated carbon sample was the lowest capacity.

Electrochemical Impedance Spectroscopy (EIS) was being done to identify conductivity of the samples. In Figure 4, we can see the results showing single semicircle phase. The results of wide semicircle was shown by sample A4, while smaller semicircle was demonstrated by the sampleA1. The smallest coated carbon of sample A1 got the highest conductivity.

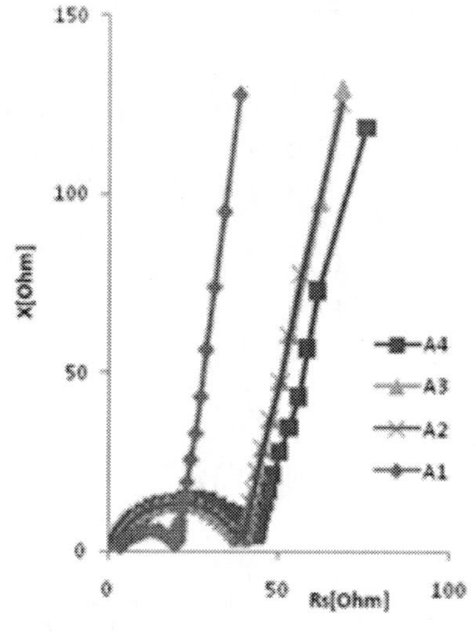


**Figure 4.** Cole-cole plot of the samples

## 4. Conclusion

The fabrication of $Na_2Li_2Ti_6O_{14}/C$ composite was successfully done with 96% purity obtained from crystal structure analysis. The working voltage of $Na_2Li_2Ti_6O_{14}/C$ was around 1.29 V. The coated carbon on anode material of $Na_2Li_2Ti_6O_{14}$ gave a positive effect on the electrochemical properties. We also obtained that the sample with the smallest ratio exhibited the best performance in charge-discharge process. The conductivity of $Na_2Li_2Ti_6O_{14}/C$ was achieved by the smallest amount of coated carbon with a ratio of 12:1.

## 5. References

[1] Nelson R F 2000 *J. Power Sources* **91** 2
[2] C.Jiang *et al* 2006 Nanotoday **1** 28
[3] Xing Li *et al* 2010 Preparation and electrochemical performance of Li4Ti5O12/carbon/carbon nano-tubes for lithium ion battery *Electrochimica Acta* **55** 2978–2982
[4] Yi T F *et al* 2012 *J.Power Sources* **214** 220-226
[5] Yi T F *et al* 2012 *J. Power Sources* **215** 258-265
[6] Zhu Y R *et al* 2013 *J. Alloy. Compd.* **547** 107-112
[7] Wu K *et al* 2014, Copper/carbon coated lithium sodium titanate as advanced anode material for lithium-ion batteries *Journal of Power Sources* **259** (2014) 177-182
[8] Chen J S *et al* 2010 *J. Am. Chem. Soc.* **132** 6124-6130
[9] Doherty C M 2009 *Chem. Mater.* **21** 2895-2903
[10] Wang W 2014 *J. Power Sources* **245** 624-629
[11] Guo X F *et al* 2012 *J. Power Sources* **214** 107-112

IOP Conf. Series: Materials Science and Engineering **202** (2017) 012053 doi:10.1088/1757-899X/202/1/012053

[12] Prihandoko B *et al* 2013 Electrochemical Behavior of Li4Ti5O12 under In Situ Process of Sintering and Surface Coating with Cassava Powder *Advanced Materials Research* **789** 21-27

**Acknowledgments**
This work was financially supported by Research Program of LIPI and INSINAS of Ministry of Research and Technology and Higher Education Republic of Indonesia in the research consortium of secondary lithium battery for an electrical vehicle.

# Optical Properties of $Fe_3O_4$ Magnetic Fluid from Iron Sand

**A Puspitaningrum[1], A Taufiq[1,2], A Hidayat[1,2], Sunaryono[1,2], N Hidayat[1,2] and Samian[3]**

[1]Department of Physics, Faculty of Mathematics and Natural Sciences, Universitas Negeri Malang (State University of Malang), Jalan Semarang No 5, Malang 65145, Indonesia
[2]Center for Minerals and Advanced Materials, Faculty of Mathematics and Natural Sciences, Universitas Negeri Malang (State University of Malang), Jl. Semarang 5, Malang 65145, Indonesia
[3]Department of Physics, Faculty of Science and Technology, Universitas Airlangga, Jalan Sutorejo, Surabaya 60115, Indonesia

Email: arif.hidayat.fmipa@um.ac.id

**Abstract:** Nowadays, a high sensitive sensor for the magnetic field has become an essential tool that vastly desired in several fields, especially in biomedical application. Therefore, the development of preparing material for the magnetic sensor becomes crucial to be conducted. In this experimnet, we propose the use of $Fe_3O_4$ magnetic fluid prepared from a local iron sand in Indonesia as a material for a magnetic sensor. In this work, optical activities of the $Fe_3O_4$ magnetic fluid as the effect of magneto-optics were performed under varying external magnetic field. The polarization direction change of the laser was detected as a function of the external magnetic field with the exponential function. Moreover, the intensity collected by a photodetector exhibited a linear correlation with the external magnetic field. These phenomena become strong evidence that the prepared $Fe_3O_4$ magnetic fluid opens potential to be applicated further as sensors, especially as a high sensitive optics-based sensor for the magnetic field.

**Keywords:** Magnetic fluid, $Fe_3O_4$, iron sand, polarization, magneto-optic.

## 1. Introduction

Magnetic fluid or often called as ferrofluid is a stable colloidal suspension consisting of magnetic nanoparticles with single-domain character and dispersing in a suitable liquid carrier [1]. The surfactant-based magnetic fluid has good stability broadening its applications in many fields. The applications cover such as for controlling lubricant migration [2], for antibacterial and anticancer [3], as a contrast agent for magnetic resonance imaging [4], for magnetic field sensor [1], and so forth. Furthermore, for application in a big amount, the development of synthesis method of local natural substance based magnetic fluid which is inexpensive and abundant is significant by using natural iron sand, for example.

Magnetic nanoparticle in the ferrofluid forms a colloid magnetic system caused by its permanent magnetic moment [5]. The distributed particles act as single-domain magnetic particles in the position of random Brownian motion without the influence of external magnetic field. However, if it is exposed to an external magnetic field, the particles motions will be regular since they undergo a tensile strength along the field direction. It means that the dipolar interaction of the magnetic domains tends to increase. When the dipolar interaction becomes strong enough, the particles undergo the structure change [6-7]. First, the head-tail of the magnetic particles oriented randomly knits each other as long as the field direction by forming a number of chains. This chain-like structure is the result of the competition between the magnetic dipolar interaction and thermal interaction [8]. After that, these chains experience aggregation again through lateral smelting forming a bundle of the chain [9]. This phenomenon is known as zippering effect as well. Two effects happening simultaniously are interaction induced by thermal fluctuation and inhibitor of the local lateral field caused by the topology defect in the dipolar chain correlated to the type of this structure formation. As a consequence, there are many significant changes in the fluid characteristics observed. Regarding this case, specifically, the characteristics of magneto-optic and magneto-viscus are essential things and have attracted many experts' attention in the last years. Therefore, in this work, a study focused on the optical characteristic of $Fe_3O_4$ magnetic fluid detected by using magneto-optic effect is significant.

The magneto-optic effect is a symptom related to the light wave interaction with the external magnetic field applied in a material. The light polarization describes the vector orientation of electrical field from the light wave of a certain point in one period of vibration. In the light polarized circularly, the direction of propagation and the orientation of electrical field form either right and left round [10]. When the light wave propagates to penetrate the active optical material, the right and left polarization vector of the light wave moves with different velocity [11]. This difference is shown as the different magneto-optic effect. The magneto-optic effect causes changes of optical parameters of materials [12]. The optical activity of material changes with the application of magnetic field. The magneto-optic effect can give information on the change of optical characteristics of the material that can be measured through the material ability in changing the polarization angle of light wave penetrating the material. The light polarization orientation in a material caused by the external magnetic field can be learned by using Faraday rotation effect. If the fluid is used as an active material in the Faraday rotation effect, the weak external magnetic field is enough to see the rotation of polarization angle and the light intensity change. This phenomenon is interesting in developing the magnetic field sensor especially for detecting the weak magnetic field like the magnetic field within a body.

Based on the explanation above, it is paramount to study the optical characteristic of $Fe_3O_4$ magnetic fluid made from iron sand. In this work, specifically, the study is focused on the magneto-optic effect happening on the $Fe_3O_4$ magnetic fluid.

## 2. Experimental Method

The $Fe_3O_4$ magnetic fluid was prepared by using a coprecipitation- sonochemical method. The purified $Fe_3O_4$ powders as a main precursor was selected from natural iron sand following our previous works [13–17]. However, in this work, the synthesis method was developed by combining the coprecipitation as reported in our previous works with the sonochemical route using ultrasonic bath at a frequency of 40 KHz at room temperature for 1 hour. Furthermore, the preparation of the magnetic fluid was tracked by the previous work [18] but modified by adding more water to enhance the homogeneity of the magnetic particles. The chemical reaction of the magnetite particle formation in this work is presented in the equation (1).

$$Fe^{2+} + 2Fe^{3+} + 8OH^{-} \rightarrow Fe_3O_4 + 4H_2O \qquad (1)$$

The phase purity, crystal structure, and particle size of the sample were characterized by means of X-Ray Diffractometer (XRD) at ambient temperature. Furthermore, the optical properties of the magnetic fluid as a function of the external magnetic field was investigated by using a set magneto-optical experiment as presented in the following figure.

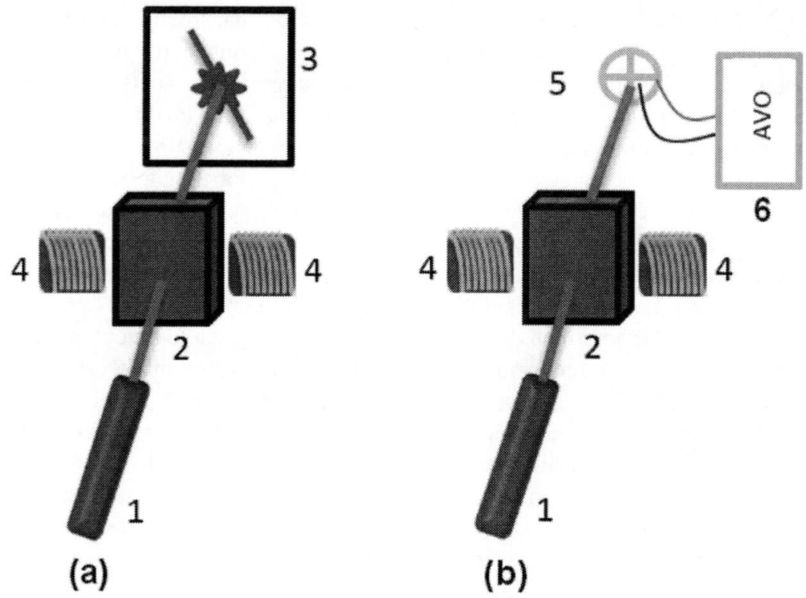


**Figure 1.** Experimental set-up of the optical properties of the $Fe_3O_4$ magnetic fluid: (a) The optical properties collected by a screen and (b) collected by optical detector

In Figure 1, number 1 represents the laser He-Ne "CVI Melles Griot" input: 115/230 VAC, max = 39.5 Hz: 50 - 400, number 2 represents the glass with the size of 2 cm × 1.5 cm × 0.3 cm with the glass thickness of 1 mm, number 3 represents the screen, number 4 represents the external magnetic field, number 5 represents the optical detector "THORLABS" Si Amplified Detector, PDA 100 A – EC, number 6 represents the digital AVO-meter FLUKE 179 with the accuracy of 0.001 mV. Practically, the experiment was conducted by placing the magnetic fluid in the glass. The laser was the directed to the magnetic fluid and detected by the screen and or the optical detector. The external magnetic field was varied during the experiment and measured by using a digital Teslameter MG – 3002 with the resolution of 0.01 mT. The data were collected both from the set of (a) and (b) experiments regarding Figure 1.

## 3. Results and Discussion

X-ray diffraction pattern of the $Fe_3O_4$ magnetic fluid is shown in Figure 2. Based on the quantitative analysis by using Rietica program [19], the $Fe_3O_4$ particle was structured in the inverse spinel with the space group of *F d -3 m Z*. The statistic parameters of the quantitative analysis were of $GoF = 0.34$, $R_{wp} = 15.18$, $R_p = 10.78$. It means that the refinement model was highly acceptable in fitting the experimental data. The lattice parameters and crystal volume were respectively $a = b = c = 8.368$ Å and $V = a \times b \times c = 585.97$ Å$^3$. These results are similar to the crystal parameters of the $Fe_3O_4$ nanoparticles preparing from iron sand by other synthesis method [13]. Furthermore, the particle size of the $Fe_3O_4$ particle was approximately 9 nm after determined by using a Debye Scherer's formula. Therefore, this particle size becomes a substantial evidence that the $Fe_3O_4$ particle prepared in this work is suitably dispersed in a liquid carrier to form a stable magnetic fluid.

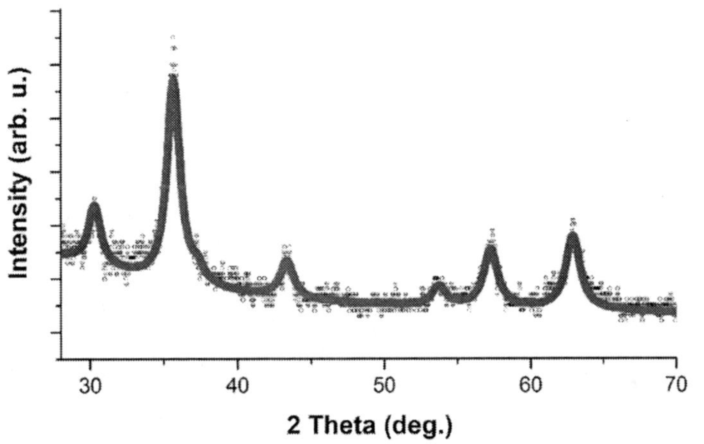


**Figure 2.** XRD pattern of the dried $Fe_3O_4$ fluid at room temperature

Figure 3 presents the effect of magnetic field on the optical characteristic of the $Fe_3O_4$ magnetic fluid indicated by the shift of laser beam on the screen. The amount of beam shift is shown visually in Figure 3. In Figure 3 (a), we can see that the laser light spreads over the screen. This condition happens when the laser light holds the screen directly without passing a glass beam containing ferrofluid and external magnetic field. When the glass beam containing ferrofluid is placed between laser and screen (laser shot to the glass beam containing ferrofluid), a beam arises directing to the angle of 50° as visualized in Figure 3 (b). After giving external magnetic field like in Figure (c) until Figure (n), we can see that the laser light beam seen on the screen gets widening and moving far from the x-axis.

Qualitatively, the amount of polarization angle shift influenced by the external magnetic field is shown in Figure 4. This figure shows the rotation graph of polarization angle due to the symptom of magneto-optic in the $Fe_3O_4$ magnetic fluid. In the range of 0 mT up to 123.8 mT, the beam shift happens in the amount of 50° to 70°. The polarization angle rotation tends to increase linearly in the magnetic field range of 0 mT to 30 mT; while above 30 mT, the polarization angle rotation increases gradually or exponentially. Generally, when a light beam penetrates an active optical object, it will transform into two light beams polarized circularly [20]. The phenomenon of polarization angle rotation happens through Faraday effect. Generally, Faraday effect is expressed as $\theta = V \times H \times I$, where $V$ is constant of Verdet material, $H$ is the given magnetic field, and $I$ is the length of the optical path or the thickness of $Fe_3O_4$ magnetic fluid [21].

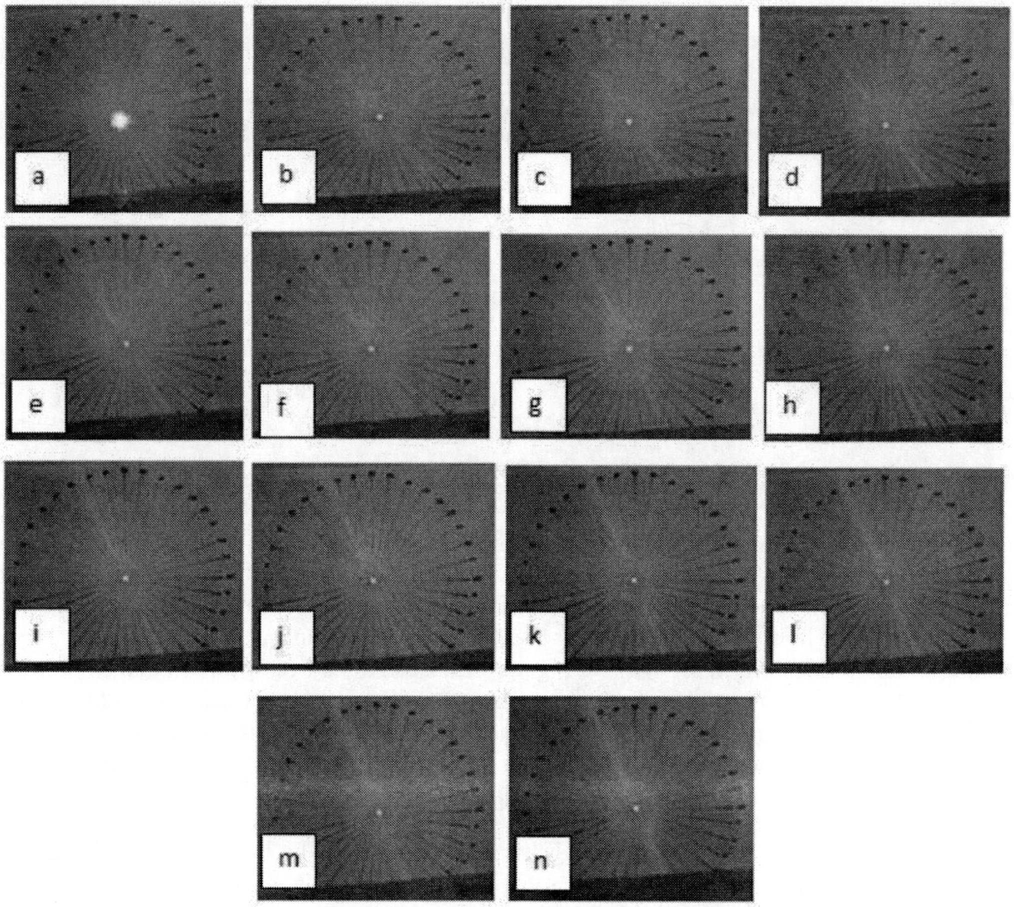


**Figure 3.** The polarization angle of the laser (a) for direct beam of the laser without $Fe_3O_4$ magnetic fluid and in the absence of external magnetic field; for the $Fe_3O_4$ magnetic fluid under magnetic field of (b) 0 mT (c) 0.7 mT (d) 2.2 mT (e) 3.6 mT (f) 5.1 mT (g) 21.8 mT (h) 36.4 mT (i) 50.9 mT (j) 65.5 mT (k) 80.1 mT (l) 94.7 mT (m) 109.2 mT, and (n) 123.8 mT

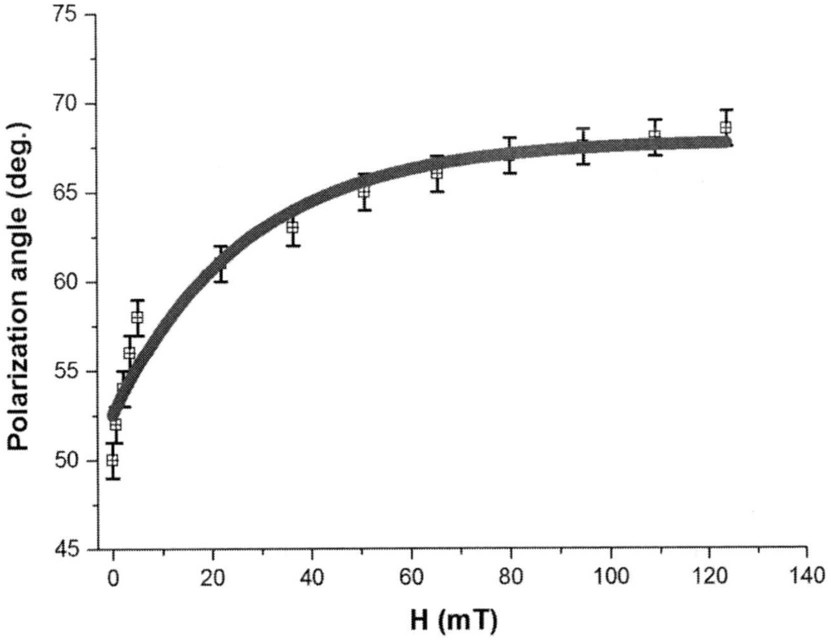


**Figure 4.** Polarization angle vs. external magnetic field (H) of the $Fe_3O_4$ magnetic fluid

Physically, the polarization angle rotation causes the increase of light intensity since the magnetic field interaction of the light wave with the external static magnetic field given to $Fe_3O_4$ magnetic field as a magneto-optic effect. The graph of intensity change as a function of the magnetic field is shown in Figure 5. One of the interesting phenomena in this work is the appealing of polarization although the external magnetic field was not given to the $Fe_3O_4$ magnetic fluid. In physics, this phenomenon occurs because laser brings the magnetic field causing the magnetic moment of $Fe_3O_4$ in fluid undergoes orientation. The magnetic moment orientation of $Fe_3O_4$ induces its particle orientation as well. Since $Fe_3O_4$ magnetic fluid is super-paramagnetic, how small the external field is given, it can induce the magnetic moment of the particle. Meanwhile, the flexibility of fluid compared to film, powder or bulk contribute as well to the easiness of magnetic particle orienting in the magnetic fluid. This case is different from the laser light intensity increasing linearly as a function of external magnetic field. The linear relationship between intensity and external magnetic field has an ideal potential to optical based magnetic field sensor. This experiment result is in line with the experiment result conducted by Nair et al. [14].

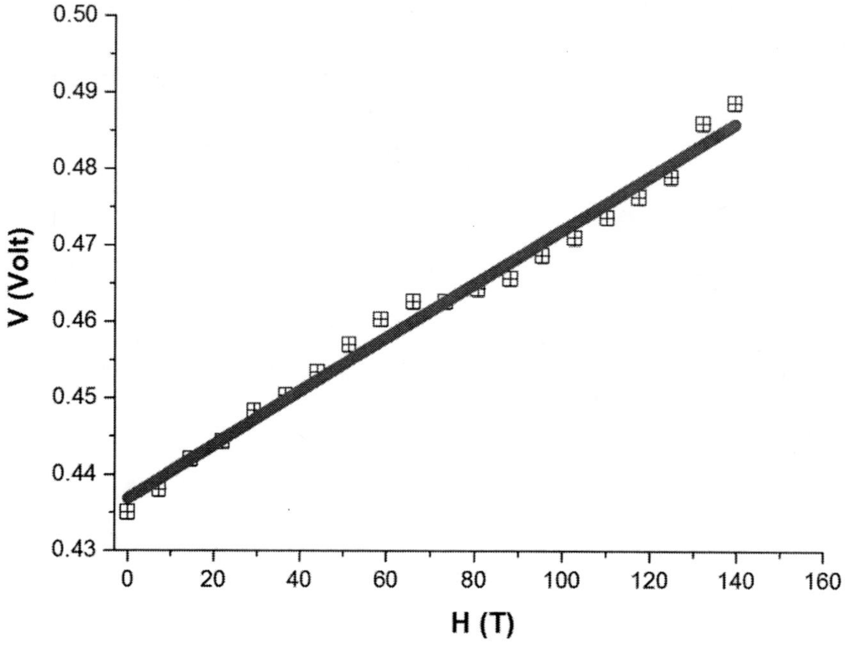


**Figure 5.** Voltage vs. external magnetic field of the magnetic fluids. V represents the intensity measured by using photodetector (in Volt unit)

Some reports in the literature show the occurrence of magneto-optic effect in the magnetic fluid [22,23]. The most important magneto-optic effect shown by the magnetic fluid is Faraday rotation, Faraday Ellipticity, Kerr Effect, linear dichroism, birefringence, and so forth [24]. Furthermore, it has been shown as well that Faraday rotation can happen in a railroad like-way. This quantization is correlated to the magnetic moment tunneling of resonance in the case of smaller size-quantum particles [25]. The theory of no-reciprocity demonstrates that the rotation direction of the polarized light depends on only the magnetic field direction [26]. Furthermore, the reciprocity of Faraday effect shown by the magnetic fluid and magneto-optic glass explains that although the magnetic field direction is reversed, the rotation of the polarized light is still similar to the first case [27]. Regarding this case, there are many tools developed based on the magnetic fluid such as a sensor, isolator, and modulator designed efficiently by controlling the magneto-optic effect [28,29]. Thereby, based on the result of this study, the $Fe_3O_4$ magnetic fluid as the preparation result of iron sand has an ideal potential to be developed further as a high sensitive optics-based magnetic sensor.

## 4. Conclusion

We have successfully prepared the $Fe_3O_4$ magnetic fluid from natural iron sand. The $Fe_3O_4$ magnetic fluid exhibited a magneto-optical phenomenon. The $Fe_3O_4$ magnetic fluid induced the change of polarization angle and its intensity originated from the external magnetic field. The change of polarization angle tended to increase exponentially as a function of external magnetic field. Furthermore, the intensity of the laser increases linearly with increasing external magnetic field. These phenomena become an essential data that the prepared $Fe_3O_4$ magnetic fluid opening the high potential for developing sensor, especially as a high sensitive optics-based magnetic sensor.

## 5. References

[1] Zheng Y, Dong X, Chan C C, Shum P P and Su H 2015 Optical fiber magnetic field sensor based on magnetic fluid and microfiber mode interferometer *Opt. Commun.* **336** 5–8

[2] Ke H, Huang W and Wang X 2016 Controlling lubricant migration using ferrofluids *Tribol. Int.* **93, Part A** 318–23

[3] Sheikh L, Vohra R, Verma A K and Nayar S 2015 Biomimetically Synthesized Aqueous Ferrofluids Having Antibacterial and Anticancer Properties *Mater. Sci. Appl.* **6** 242–50

[4] Casula M F, Corrias A, Arosio P, Lascialfari A, Sen T, Floris P and Bruce I J 2011 Design of water-based ferrofluids as contrast agents for magnetic resonance imaging *J. Colloid Interface Sci.* **357** 50–5

[5] Rosensweig R E 2014 *Ferrohydrodynamics* (Mineola, New York: Dover Publications, Inc)

[6] Mendelev V S and Ivanov A O 2004 Ferrofluid aggregation in chains under the influence of a magnetic field *Phys. Rev. E* **70**

[7] Zubarev A Y and Iskakova L Y 2006 Direct and inverse domain structures in ferrofluids *Phys. Stat. Mech. Its Appl.* **367** 55–68

[8] Laskar J M, Philip J and Raj B 2009 Experimental evidence for reversible zippering of chains in magnetic nanofluids under external magnetic fields *Phys. Rev. E* **80**

[9] Born M, Wolf E and Hecht E 2000 *Principles of Optics: Electromagnetic Theory of Propagation, Interference and Diffraction of Light Phys. Today* **53** 77–8

[10] Hopster H and Oepen H P 2005 *Magnetic microscopy of nanostructures* (Berlin; New York: Springer)

[11] Iizuka K 2002 *Elements of photonics* (New York, N.Y: Wiley-Interscience)

[12] Marand Z R, Farimani M H R and Shahtahmasebi N 2014 Study of magnetic and structural and optical properties of Zn doped $Fe_3O_4$ nanoparticles synthesized by co-precipitation method for biomedical application *Akush. Ginekol. (Sofiia)* **15** 238–47

[13] Taufiq A, Sunaryono, Rachman Putra E G, Okazawa A, Watanabe I, Kojima N, Pratapa S and Darminto 2015 Nanoscale Clustering and Magnetic Properties of $Mn_xFe_{3-x}O_4$ Particles Prepared from Natural Magnetite *J. Supercond. Nov. Magn.* **28** 2855–63

[14] Pratapa S, Susanti L, Insany Y A S, Alfiati Z, Hartono B, Mashuri, Taufiq A, Fuad A, Triwikantoro, Baqiya M A, Purwaningsih S, Yahya E and Darminto 2010 XRD line-broadening characteristics of M-oxides (M = Mg, Mg-Al, Y, Fe) nanoparticles produced by coprecipitation method *AIP Conf. Proc.* **1284** 125–8

[15] Sunaryono, Taufiq A, Munaji, Indarto B, Triwikantoro, Zainuri M and Darminto 2013 Magneto-elasticity in hydrogels containing $Fe_3O_4$ nanoparticles and their potential applications *AIP Conf. Proc.* **1555** 53–6

[16] Sunaryono, Taufiq A, Putra E G R, Okazawa A, Watanabe I, Kojima N, Rugmai S, Soontaranon S, Zainuri M, Triwikantoro, Pratapa S and Darminto 2015 Small-Angle X-Ray Scattering Study on PVA/Fe3O4 Magnetic Hydrogels *Nhano* **11** 1650027

[17] Sunaryono, Taufiq A, Mashuri, Pratapa S, Zainuri M, Triwikantoro and Darminto 2015 Various Magnetic Properties of Magnetite Nanoparticles Syntesized from Iron-Sands by Coprecipitation Method at Room Temperature *Mater. Sci. Forum* **827** 229–34

[18] Taufiq A, Sunaryono, Rachman Putra E G, Pratapa S and Darminto 2015 Nano-Structural Studies on $Fe_3O_4$ Particles Dispersing in a Magnetic Fluid Using X-Ray Diffractometry and Small-Angle Neutron Scattering *Mater. Sci. Forum* **827** 213–8

[19] Hunter B A 1998 Rietica - A visual Rietveld program *Newsl. Int. Union Crystallogr. Comm. Powder Difraction*

[20] Khandanlou R, Ahmad M, Shameli K and Kalantari K 2013 Synthesis and Characterization of Rice Straw/$Fe_3O_4$ Nanocomposites by a Quick Precipitation Method *Molecules* **18** 6597–607

[21] Nair S S, Rajesh S, Abraham V S and Anantharaman M R 2010 Ferrofluid Thin Films as Optical Gaussmeters Proposed for Field and Magnetic Moment Sensing *Indian Acad. Sci.* **34 no 2** 245–9

[22] Lacoste D, Donatini F, Neveu S, Serughetti J A and Van Tiggelen B A 2000 Photonic Hall effect in ferrofluids: Theory and experiments *Phys. Rev. E* **62** 3934–43
[23] Davies H W and Llewellyn J P 1980 Magneto-optic effects in ferrofluids *J. Phys. Appl. Phys.* **13** 2327–36
[24] Kooij E S, Gâlcă A C and Poelsema B 2006 Versatile transmission ellipsometry to study linear ferrofluid magneto-optics *J. Colloid Interface Sci.* **304** 261–70
[25] Jamon D, Donatini F, Siblini A, Royer F, Perzynski R, Cabuil V and Neveu S 2009 Experimental investigation on the magneto-optic effects of ferrofluids via dynamic measurements *J. Magn. Magn. Mater.* **321** 1148–54
[26] Donatini F, Jamon D, Monin J and Neveu S 1999 Experimental investigation of longitudinal magneto-optic effects in four ferrite ferrofluids in visible-near infrared spectrum *IEEE Trans. Magn.* **35** 4311–7
[27] Deb P, Gogoi M and Karmakar P K 2012 Anomalous magneto-optic Faraday rotation behavior due to resonant tunneling of magnetic moment *J. Opt.* **41** 41–7
[28] Massard C, Taverdet J L, Brouillet S and Donnet C 2007 Hybrid Sol–gel thin films for magneto-optical applications: Chemical, optical and tribological study *Surf. Coat. Technol.* **202** 1067–72
[29] Wang S, Sun C, Du L, Yao C and Yang Y 2012 Reciprocity of Faraday effect in ferrofluid: Comparison with magneto-optical glass *Opt. - Int. J. Light Electron Opt.* **123** 553–8

**Acknowledgements**
The authors (AT and AH) very appreciate *DRPM-KEMENRISTEKDIKTI*, Republic of Indonesia, for supporting "*HIBAH PENELITIAN*" 2016-2017.

# High-Performance Silver Nanowire Film on Flexible Substrate Prepared by Meyer-rod Coating

**Junaidi[1], K Triyana[2], Harsojo[2] and E Suharyadi[2]**

[1]Departement of Physics, Universitas Lampung, Bandar Lampung, 35145 Indonesia
[2]Departement of Physics, Universitas Gadjah Mada, Yogyakarta, 55281 Indonesia

Email: triyana@ugm.ac.id

**Abstract.** Flexible and transparent conducting (FTC) films based on silver nanowires (AgNWs) were fabricated onto a polycarbonate (PC) substrate by a simple Meyer-rod coating process. AgNWs were synthesized using polyvinyl alcohol by polyol process with a diameter and length of 500 nm and 10 μm, respectively. The FTC film was fabricated by varying the number of layers. The FTC film exhibited a sheet resistance of 12.1 $\Omega.sq^{-1}$ with a transparency of 89.5%. The optical conductivity of FTC films was obtained about 4.7 - 13.1 × $10^5$ $S.m^{-1}$ with a refractive index about 1.2 to 1.7. The number of layers is an important to influence on the sheet resistance and transmittance of the FTC films. The sheet resistance and transmittance decreases with the number of layers.

**Keywords:** conductive film, flexible substrate, silver nanowire, transparent electrode.

## 1. Introduction

Transparent conductive electrode (TCE) has emerged rapidly and attracted much attention due to its fascinating features and applications in recent years. For the reasons, many TCE materials, such as indium tin oxides (ITO), graphenes, carbon nanotubes (CNTs), and silver nanowires (AgNWs). ITO has excellent transparency throughout the visible spectrum and low sheet resistance. However, ITO has certain limitations in application for next-generation devices: (i) indium is a rare material, so it has a high price, (ii) the process of making ITO is also expensive because it requires high temperature and vacuum to control its thickness [1], (iii) ITO is brittle and can easily wear out or crack when used in the applications which involve bending, and (iv) ITO has a high refractive index. Such disadvantages make ITO unsuitable for display applications because it will reflect light and reduce the brightness of the screens [2].

The deposition techniques are performed on transparent substrates such as glass, quartz, and poly (ethylene terephthalate) or PET [3–5]. The main focus of the TCE manufacture is getting low sheet resistance ($R_s$) and high optical transmittance ($T$). TCE based on AgNWs can simultaneously realize outstanding flexibility, conductivity, and transparency compared with other electrode materials, and its low-cost deposition techniques [6]. The spin coating method for producing AgNWs films was demonstrated by Chen et al. and produced sheet resistance and transmittance of about 10 Ω/sq and $T \approx$ 80%, respectively [7]. Margulis et al. reported solution-process of AgNWs films by spray coating method with $R_s \approx$ 18 Ω/sq and $T \approx$ 86% [8]. Sachse et al. reported a dip coating method for AgNWs film fabrication with $R_s \approx 15$ Ω/sq and $T \approx 85\%$ [9]. Fabrication of AgNWs film with a meyer-rod coating has been done by Kumar et al. with $R_s \approx 300$ Ω/sq and $T \approx 90\%$ [10].

In this paper, we performed the synthesis of AgNWs using polyvinyl alcohol (PVA) as a capping agent and stabilizer. Furthermore, the resulted AgNWs created a thin layer on polycarbonate (PC) substrates using Meyer-rod coating method. The number of AgNWs layers was varied to produce optimum optical and electrical properties. The produced AgNWs films can be used as TCE for various optoelectronic applications, such as touch screens, organic light-emitting diodes (OLEDs), solar cells, and transparent heaters.

## 2. Experimental Method

### *2.1 Materials*

The materials used included: silver nitrate ($AgNO_3$, 99%, Merck), polyvinyl alcohol (PVA, Mw. 31000-50000 g/mol, Sigma-Aldrich), copper (II) chloride dihydrate ($CuCl_2.2H_2O$, Sigma-Aldrich), ethylene glycol (EG, 99%, Merck), and ethanol (EtOH, 98%, Merck).

### *2.2 Synthesis of AgNWs*

The AgNWs were synthesized by polyol method with several modifications [11]. Firstly, 0.5 g of PVA was mixed with 10 mL of ethylene glycol (EG) and 20 µL of $CuCl_2.2H_2O$ 0.1 M using an Erlenmeyer flask at 150 °C for 1 h. After that, 5 mL of $AgNO_3$ 0.5 M in EG was added dropwise using a syringe for about 10 min followed by stirring at 350 rpm for 1 h. Furthermore, the solution containing the produced AgNWs then cooled naturally to room temperature. It was followed by several times of centrifugation using ethanol at a speed of 6000 rpm.

### *2.3 Fabrication of AgNWs Films*

Firstly, the PC substrates were ultrasonicated in ethanol at room temperature for 30 min and rinsed thoroughly with deionized water. The PC substrates had been dried at room temperature under ambient conditions. The AgNWs with a diameter ($\phi$) and length ($l$) of ($\phi$ = 500 nm, $l$ = 10 µm) as shown in SEM images in Figure 1 were used in this study. The samples of AgNWs were dried to determine their mass and concentration by evaporating ethanol in a heating chamber at a temperature of 60 °C. The as-prepared AgNWs was dispersed for one more time in an ethanol solution with a concentration of 10 wt. %. The dispersion solution of the AgNWs was coated onto 25 mm × 25 mm PC substrates by varying the number of layers manually using a bar-coater (RDS-20). After coating the AgNWs, the samples were dried on a hot plate at 90 °C for 15 min to remove the solvent.

### *2.4 Characterization*

The crystal structure of AgNWs suspension was analyzed via X-ray diffraction (Shimadzu R6000) by CuKα ($\lambda$ = 1.54184 Å) with a $2\theta$ scanning in the temperature range of 20 to 90°. Furthermore, the electrical and optical properties of the flexible AgNWs films were recorded using a four-point-probe (Keithley) and UV-vis spectroscopy (Shimadzu, UV-1700) in the wavelength range of 300 to 800 nm. The morphology, size, and thickness of AgNWs films were observed using a transmission electron microscope (JEOL, JEM-1400) and scanning electron microscope (JEOL, JSM-6510).

## 3. Results and Discussion

Figure 1(a) shows the UV-vis spectra of AgNWs in an ethanol solution. From the UV-vis spectra, the absorption peaks of AgNWs solution formed at a wavelength of 360 nm and 378 nm when synthesized using PVA as a capping agent and stabilizer [12]. The characteristics and the optical phenomenon of metal nanomaterials, especially Ag nanostructures regarding the absorption spectrum depend on the geometry and size. The absorbance peak that occurs at a specific wavelength results from the interaction of light and matter (sample).

The XRD pattern of the AgNWs is depicted in Figure 1(b). The crystal structure of AgNWs produced four diffraction peaks. All peaks can be indexed to Ag cubic phase of which each angle of diffraction 2θ was at 38.21° (111), 44.49° (200), 65.59° (220), 77.41° (311), and 81.59° (222). The number in the parentheses indicates the corresponding crystal plane. According to standard JCPDS card of 04-0783 from ASTM, the XRD pattern indicates the crystallization of the AgNWs. The crystalline of the AgNWs can be identified as a face-centered cubic (fcc).

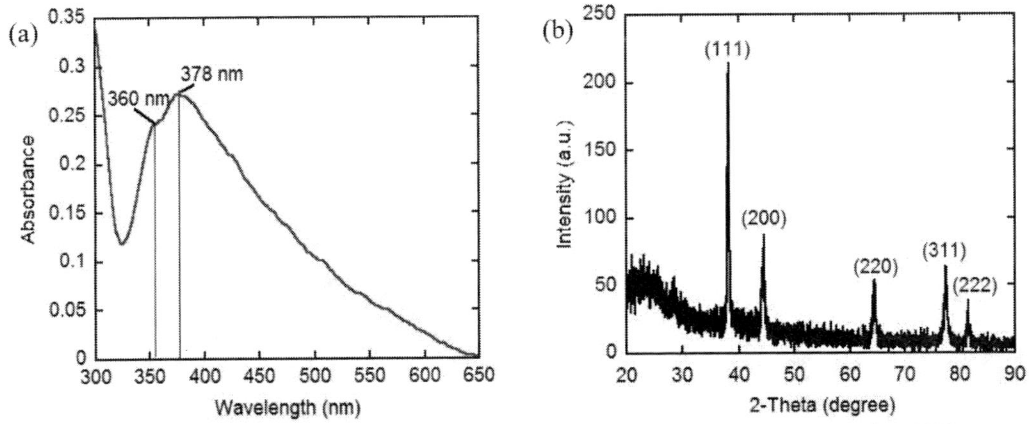


**Figure 1**. (a) UV-vis Spectra of AgNWs solution and (b) XRD pattern of AgNWs.

The AgNWs synthesized in this study also showed a high aspect ratio. This case showed the diffraction signal peak of XRD at (111) was larger than that at (200). The calculated lattice constant according to the spacing distance ($d_{hkl}$) of the {111} planes was 4.080 Å. Such lattice constant was very close to the literature value of 4.085 Å [12]. The SEM images of AgNWs solution and AgNWs film are shown in Figure 2.

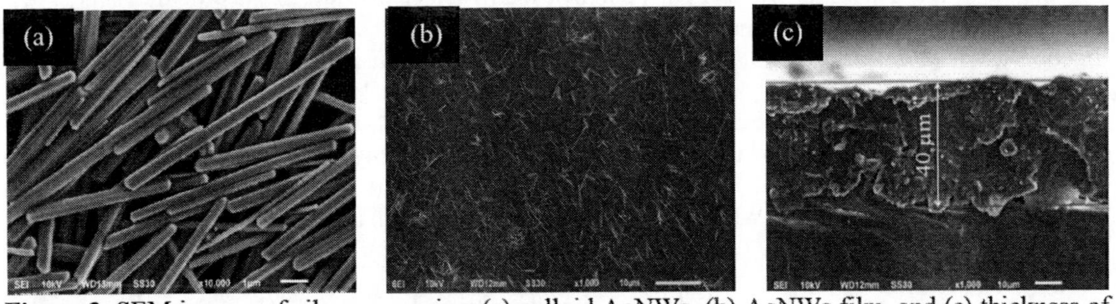


**Figure 2**. SEM images of silver nanowires (a) colloid AgNWs, (b) AgNWs film, and (c) thickness of AgNWs film.

Figure 2(a) shows that the diameter and length of AgNWs were (500 ± 100) nm and (10 ± 5) µm. The diameter and length of different AgNWs will have an impact on changes in optical and electrical properties when used as TCEs. The SEM image of the AgNWs film with one layer was shown in Figure 2(b). The thickness of AgNWs film was analyzed using cross-sectional techniques. The thickness of AgNWs film obtained was about of 40 µm (Figure 2(c)).

PVA is an effective polymer capping agent and stabilizer in the synthesis of AgNWs. PVA can control the process of forming the Ag nanocrystal seeds with various sizes. The presence of PVA can change the free surface energy that influences anisotropic formation mechanism of AgNWs. Anisotropic growth is retained by the PVA on (100) facets. The mechanical strength of PVA caused a single crystal and MTP seeds grow into AgNWs. The bonding capabilities of PVA depend on the shape of the Ag nanocrystal. The Ag atoms will be attached to the (111) facets that cause MTPs seeds grew and formed pentagonal NWs [13,14]. The NWs structure can grow to form Ag-O bond on the (111) facets. PVA can effectively inhibit the growth on the (100) so as to control the growth on the

(111) with various Ag nanocrystals [15,16]. The transmittance and band gap energy of AgNWs film are illustrated in Figure 3.

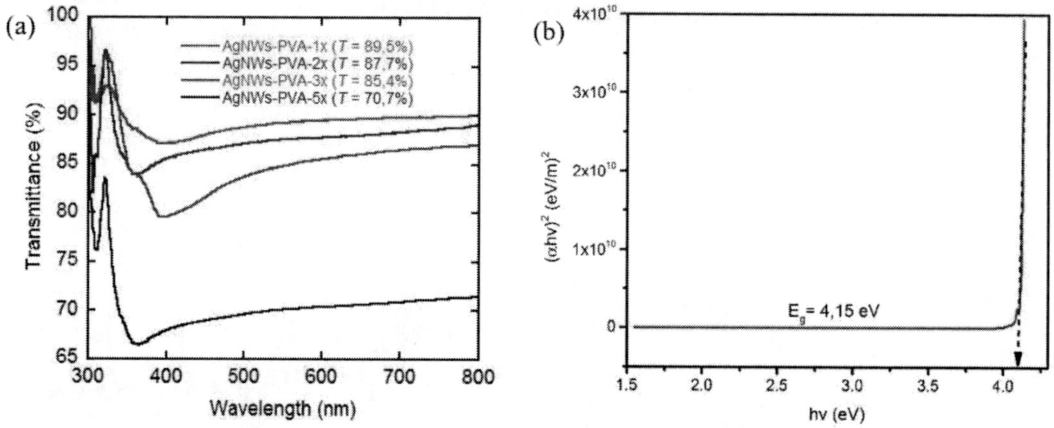


**Figure 3.** The curve of (a) transmittance and (b) band gap energy of AgNWs film.

Figure 3(a) shows that the transmittance of AgNWs film was approximately 70.7 to 89.5%. The AgNWs film transmittance decreased along with the increasing number of layers. The measurement of absorbance as a function of wavelength and thickness of thin films conducted to determine the value of the band gap energy ($E_g$) of the thin films. The energy gap of AgNWs film was about 4.15 eV as shown in Figure 3(b).

The extension coefficient ($k$) is a measure of the fraction of lost light due to scattering and absorption of light per distance units of the medium or thickness. The extension coefficient decreased together with the increasing wavelength of light which subjected to the surface of the sample. The decline in the extension coefficient resulted in the increasing of the wavelength shows that the light fraction has been reduced or lost due to the decreasing light intensity scattered and absorbed by the sample [17].

The refractive index was calculated using the reflectance values, and the extension coefficient. The resulting refractive index is used to calculate the optical conductivity of the AgNWs film. The optical conductivity is the ability of material to conduct photon energy. The curves of the refractive index and optical conductivity of the AgNWs film as a function of wavelength are presented in Figure 3.

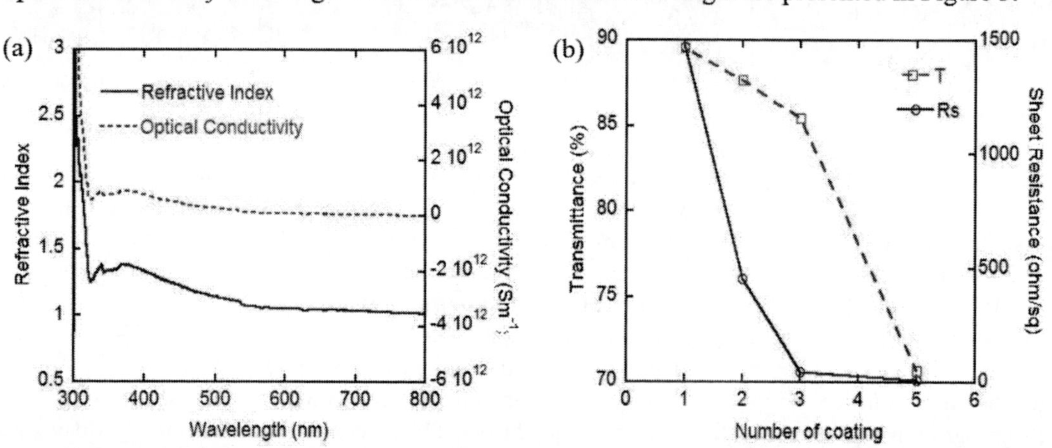


**Figure 4.** (a) Refractive index and optical conductivity and (b) transmittance and a sheet resistance of AgNWs film.

Figure 4(a) shows that the refraction index decreased in line with the increasing wavelength. The refractive index experienced saturation at a wavelength of about 550 nm, i.e. about 1.2 to 1.7 for the AgNWs film [18]. The obtained optical conductivity onto the PC substrate for AgNWs film was about $4.7 \times 10^5$ $Sm^{-1}$. The optical conductivity of AgNWs film is lower than the optical conductivity of metallic silver, i.e. $6.3 \times 10^7$ $Sm^{-1}$ [19,20].

I-V curves were used to determine the sheet resistance ($R_s$) of AgNWs film that measured using a four-point probe on a direct current from 0 to 0.5 mA as shown in Figure 4(b). Figure 4(b) shows that the sheet resistance of AgNWs film in one layer was 1470.5 $\Omega sq^{-1}$. A sheet resistance continues to decline consecutively about 454.2, 43.9, and 12.1 $\Omega sq^{-1}$ when the coating is done as much as 2x, 3x and 5x. Decrease of sheet resistance indicates the increasing in electrical conductivity ($\sigma_{dc}$) of AgNWs film. The number of layers can increase the electrical conductivity and decrease the optical transmittance of AgNWs film as seen in Figure 4(a) [21].

## 4. Conclusion

The synthesis AgNWs were conducted by polyol process using polyvinyl alcohol as a capping agent and stabilizer. The absorption peaks of AgNWs solution formed at the wavelengths of 360 nm and 378 nm with the face-centered-cubic (fcc) crystal structures. The Diameter and length of AgNWs solution are (500 ± 100) nm and (10 ± 5) μm. The AgNWs were coated onto the PC substrate by various numbers of coating. The thickness of AgNWs film obtained about of 40 μm prepared by Meyer-rod coating. The transmittance of AgNWs films was approximately 70.7 to 89.5% and the sheet resistance of AgNWs film was 12.1 $\Omega sq^{-1}$. The obtained optical conductivity for AgNWs film was about $4.7 \times 10^5$ $Sm^{-1}$.

## 5. References

[1] Hecht D S, Hu L and Irvin G 2011 Emerging transparent electrodes based on thin films of carbon nanotubes, graphene, and metallic nanostructures *Adv. Mater.* **23** 1482–513

[2] Wang J, Jiu J, Araki T, Nogi M, Sugahara T, Nagao S and Koga H 2015 Silver nanowire electrodes : conductivity improvement without post-treatment and application in capacitive pressure sensors *Nano-Micro Lett.* **7** 51–8

[3] Lim J W, Cho D Y, Jihoon-Kim, Na S I and Kim H K 2012 Simple brush-painting of flexible and transparent Ag nanowire network electrodes as an alternative ITO anode for cost-efficient flexible organic solar cells *Sol. Energy Mater. Sol. Cells* **107** 348–54

[4] Naito K, Yoshinaga N, Tsutsumi E and Akasaka Y 2013 Transparent conducting film composed of graphene and silver nanowire stacked layers *Synth. Met.* **175** 42–6

[5] Tokuno T, Nogi M, Jiu J and Suganuma K 2012 Hybrid transparent electrodes of silver nanowires and carbon nanotubes: a low-temperature solution process *Nanoscale Res. Lett.* **7** 281

[6] Kim D, Zhu L, Jeong D J, Chun K, Bang Y Y, Kim S R, Kim J H and Oh S K 2013 Transparent flexible heater based on hybrid of carbon nanotubes and silver nanowires *Carbon N. Y.* **63** 530–6

[7] Chen T-G, Huang B-Y, Liu H-W, Huang Y-Y, Pan H-T, Meng H-F and Yu P 2012 Flexible silver nanowire meshes for high-efficiency microtextured organic-silicon hybrid photovoltaics. *ACS Appl. Mater. Interfaces* **4** 6857–64

[8] Margulis G Y, Christoforo M G, Lam D, Beiley Z M, Bowring A R, Bailie C D, Salleo A and McGehee M D 2013 Spray deposition of silver nanowire electrodes for semitransparent solid-state dye-sensitized solar cells *Adv. Energy Mater.* **3** 1657–63

[9] Sachse C, Müller-Meskamp L, Bormann L, Kim Y H, Lehnert F, Philipp A, Beyer B and Leo K 2013 Transparent, dip-coated silver nanowire electrodes for small molecule organic solar cells *Org. Electron. physics, Mater. Appl.* **14** 143–8

[10] Kiran Kumar A B V, Wan Bae C, Piao L and Kim S H 2013 Silver nanowire based flexible electrodes with improved properties: High conductivity, transparency, adhesion and low

haze *Mater. Res. Bull.* **48** 2944–9

[11] Junaidi, Triyana K, Sosiati H, Suharyadi E and Harsojo 2015 Effect of temperature on silver nanorods synthesized by polyol method *Adv. Mater. Res.* **1123** 256–9

[12] Junaidi, Triyana K, Harsojo and Suharyadi E 2016 Chloride ion addition for controlling shapes and properties of silver nanorods capped by polyvinyl alcohol synthesized by polyol method **020092** 020092

[13] Sun Y, Mayers B, Herricks T and Xia Y 2003 Polyol synthesis of uniform silver nanowires: a plausible growth mechanism and the supporting evidence *Nano Lett.* **3** 955–60

[14] Zhang Q and Wang Q 2013 Particle size distribution of silver nanoparticles in different water conditions *J. Mater. Environ. Sci.* **4** 617–20

[15] Junaidi, Yunus M, Harsojo, Suharyadi E and Triyana K 2016 Effect of stirring rate on the synthesis silver nanowires using polyvinyl alcohol as a capping agent by polyol process *Int. J. Adv. Sci. Eng. Inf. Technol.* **6** 365–9

[16] Junaidi, Triyana K, Hui H, Wu L Y L, Suharyadi E and Harsojo 2016 The silver nanowires synthesized using different molecule weight of polyvinyl pyrrolidone for controlling diameter and length by one-pot polyol method **020014** 020014

[17] Bakr N A, Funde A M, Waman V S, Kamble M M, Hawaldar R R, Amalnerkar D P, Gosavi S W and Jadkar S R 2011 Determination of the optical parameters of a-Si:H thin films deposited by hot wire-chemical vapour deposition technique using transmission spectrum only *Pramana - J. Phys.* **76** 519–31

[18] Khanarian G, Joo J, Liu X, Eastman P, Werner D, Connell K O, Trefonas P, Khanarian G, Joo J, Liu X, Eastman P, Werner D and Connell K O 2016 The optical and electrical properties of silver nanowire mesh films The optical and electrical properties of silver nanowire mesh films **024302**

[19] De S, Higgins T M, Lyons P E, Doherty E M, Nirmalraj P N, Blau W J, Boland J J and Coleman J N 2009 Silver nanowire networks as flexible, transparent, conducting films: extremely high DC to optical conductivity ratios *ACS Nano* **3** 1767–74

[20] Mi H Y, Li Z, Turng L S, Sun Y and Gong S 2014 Silver nanowire/thermoplastic polyurethane elastomer nanocomposites: Thermal, mechanical, and dielectric properties *Mater. Des.* **56** 398–404

[21] Dinh D A, Hui K N, Hui K S, Kumar P and Singh J 2013 Silver nanowires: a promising transparent conducting electrode material for optoelectronic and electronic applications *Rev. Adv. Sci. Eng.* **2** 1–22

## Acknowledgments

This work was supported by a research grant of “International Research Collaboration and Scientific Publication through Contract No. LPPM-UGM/998/LIT/2015 with the Directorate General of Higher Education (DIKTI), Ministry of Education and Culture, Republic of Indonesia.

IOP Conf. Series: Materials Science and Engineering **202** (2017) 012056 doi:10.1088/1757-899X/202/1/012056

# Effect of NiO and Light Intensity on Dielectric Constant of $SiO_2$-$B_2O_3$-$Bi_2O_3$-$Na_2CO_3$ Glass Based on Silica Gel of Natural Sands

**M Diantoro[1,2], Z Muniroh[1], B Zaini[1], A A Mustikasari[1], Nasikhudin[1,2,3], A Hidayat[1,2], A Taufiq[1,2], Sunaryono[1,2], and N.Mufti[1,2]**

[1]Department of Physics, Faculty of Mathematics and Natural Sciences, Universitas Negeri Malang (State University of Malang), Jl. Semarang 5, Malang 65145, Indonesia
[2]Center for Minerals and Advanced Materials, Faculty of Mathematics and Natural Sciences, Universitas Negeri Malang (State University of Malang), Jl. Semarang 5, Malang 65145, Indonesia
[3]Departmet of Physics, Faculty of Mathematics and Natural Sciences, Universitas Gadjah Mada

Email: markus.diantoro.fmipa@um.ac.id

**Abstract.** The use of silica in various fields is significantly increasing. One common application is silica based functional glass which has naturally show specific dielectric, optical, and magnetic properties. Many studies have been performing to explore the influence of dopant, composition, and other processing parameters as well as employing various characterization. In the previous work, we report the use of silica from silica sands. To reduce the melting temperature, we used silica sol-gel beside the utilization of some oxides such as $B_2O_3$, $Na_2CO_3$, and $Bi_3O_3$. We also used NiO as dopant explore the glass properties. We have prepared a series of sample with the composition of $50SiO_2$-$25B_2O_3$-(6.5-x) $Bi_3O_{3\text{-}18.5}$ $Na_2CO_3$-xNiO (x = 0, 1, 2, 3 and 4 wt%). After weighting process, the composition was blended, then heated to 450 °C for 120 minutes and then raised at 950 °C for 60 minutes in the crucible. Then samples of glass separated from the crucible and in the characterization of the structure using the DTA, XRD, SEM-EDAX and FTIR and measuring dielectric constant using a capacitance meter. The increase of NiO dopant resulted in increasing the dielectric constant of glass. On the other hand, the dielectric constant gradually decreases with the increase of light intensity. One can be noted that the applied intensity give rise to the step-like decrease of the dielectric constant. Whereas, the increasing magnetic field indicate the increase of dielectric constant.

**Keywords:** NiO, light intensity, dielectric constant, borosilicate glass, natural sand

## 1. Introduction

One of the abundant natural resources of Indonesia is silica. Silica can be obtained from many resources, such as silica sands, pyrophyllite, bauxite, and from many parts of plants. The content of silica of natural sand of Tuban district east java is approximately 50.1% [1]. The huge amount of silica source needs to be developed for many kinds of application. The use of silica for daily life is becoming diverse. The common use of silica is for producing architectural glass. Recently, glass can be fabricated to more functional materials. Some examples of modern silica application are smart glass, solar glass, electro-optic devices, optic-fibers, absorbing x-rays [2] and even absorbing gamma rays which have a very high energy [3].Silica glass can fabricate by solely use of silica without any material. But it is not practical due to very high temperatures, i.e., 1760-1870 °C for synthesis. One way to lower the melting point of silica is the modifying to a smaller size and feed the raw silica as an amorphous form. To fit the desired silica, the sol-gel method can be implemented. In the other hand, the addition of another oxide can also give the lower melting point of silica.

The addition of metal on the structure of the glass will affect the optical, electrical and magnetic properties [4]. Silica glass dipped in a solution of $Ni(NO_2)_3.6H_2O$ (0.5 M) with a relatively high heat treatment between 760-850 K produces a rise the field polarization of the glass up to of 500 V/cm [5]. NiO can also be said to have good optical properties, in the form of the film has an energy of optical band gap of 3.58 eV [6]. Ni is expected to be a good candidate fo absorbing of electromagnetic waves properties reaching at 12-18 GHz [7]. Nickel also has potential to provide optical properties, electrical, and magnetic of glass materials.

So far in Indonesia has not been found the study of glass with multi-functional properties, namely optical glass that has the properties, dielectric and magnetic, comprehensively. Hence in the research of silica glass added NiO show to provide high optical, dielectric and magnetic properties. The addition oxide such as B2O3 [8]. add optical properties of glass, $B_2O_3$ [9], and $Na_2CO_3$ are mostly to reduce the melting point as well as to improve the optical or dielectric properties of the glass.

## 2. Experimental Methods

### *2.1 Materials and Instrumentation*
The instrument is the balance of Digital HF-3000 (0.019), Furnace Brother XD-1700M, Hot Plate, Beaker, Burret, Mortar, and Pestle. The materials of synthesis $SiO_2$ are HCl 32%, 5M HCl, and 5M NaOH. The raw material of glass based on the natural sand beach of $SiO_2$ Bancar, $B_2O_3$ Merck 100169, B2O3 Merck 101862, and $Na_2CO_3$ NiO Merck 106723.

### *2.2 Preparation of $SiO_2$*
Silica sand from the beach of Tuban Bancar washed six times using aquades, then crushed up to 100 mesh. It already soaked with 32% HCl for 12 hours and rinsed using aquades up to a pH 7 hence dried.

### *2.3 Sol-Gel Method*
Sixty grams of silica sand and 1920 mL NaOH [5 M] was stirred by magnetic stirrer using 600 rpm at 140 °C until no the moisture content. Added water as much as 1920 mL stirring for 3 hours then keep under ambient temperature. The filtrate was separated by decantation, then followed by stirring using a magnetic stirrer at 600 rpm while titrated up to 5M HCl that is showing pH 7. The gel was formed then dried for 24 hours at 125 °C. The obtained sample then was characterized by using XRF to obtain the purity of silica.

### *2.4 Synthesis of Glass $50SiO_2$-$25B_2O_3$-(6.5-x)B2O3-$18.5Na_2CO_3$-xNiO*
The glass of $50SiO_2$-$25B_2O_3$-(6.5-x) $B_2O_3$-18.5 $Na_2CO_{3\text{-}x}$NiO with x = 0, 1, 2, 3, and 4 wt% were mixed until homogeneous and followed by heated using Furnace Brother XD-1700M at 450 °C for 2 hours and rinsed again up to 950 °C for 1 hour in a ceramics crucible. Then cooled to room temperature. The sample was separated from the crucible and characterized by DTA, XRD, SEM-EDX and capacitance meter.

## 3. Result and Discussion

### *3.1 Silica Sol-Gel*
The elemental composition of silica powder was characterized by XRF ass shown in Table 1. It is showed the composition of silica sand before and after via sol-gel process. The purity of silica increase from 88.1% to 95.5%. The results of the sol-gel method are able to reduce some of the impurities contained in the silica sand, such as K, Ca, Ti and Fe. These elements are remain in the sol-gel because the element or in the form of its compounds are insoluble completely with NaOH and HCl.

**Table 1.** Elemental composition of silica

| Element | before Sol-gel (%) | after Sol-gel (%) |
|---|---|---|
| Si | 88.1 | 95.5 |
| K | 4.36 | 0.00 |
| Ca | 2.76 | 1.2 |
| Ti | 1.37 | 0.44 |
| Fe | 2.97 | 1.92 |

### *3.2 The Influence NiO on Structure of $SiO_2$-$B_2O_3$-B2O3-$Na_2CO_3$ Glass*
In this study, the structure of the $50SiO_2$-$25B_2O_3$-(6.5-x) $Bi_3O_3$-18.5 $Na_2CO_{3\text{-}x}$NiO glass have been showed by DTA, and XRD diffraction pattern, and FTIR. Previously the samples of glass have been characterized by SEM-EDX to obtain the composition contained therein. Table 2 presents the composition of the prepared glass from EDX analysis.

*3.2.1 DTA Pattern of $50SiO_2$- $25B_2O_3$- (6.5-x)$Bi_3O_3$- $18.5Na_2CO_3$- xNiO Glass.* The typical DTA pattern of the $50SiO_2$ -$25B_2O_3$-(6.5-x) B2O3-18.5 $Na_2CO_{3\text{-}x}$NiO glass is shown in Figure 1. The pattern of DTA can be seen from the range of temperature between 0-1000 °C. Changing patterns of heat flow and baseline. This alteration is showed by the transition temperature of glass [2].

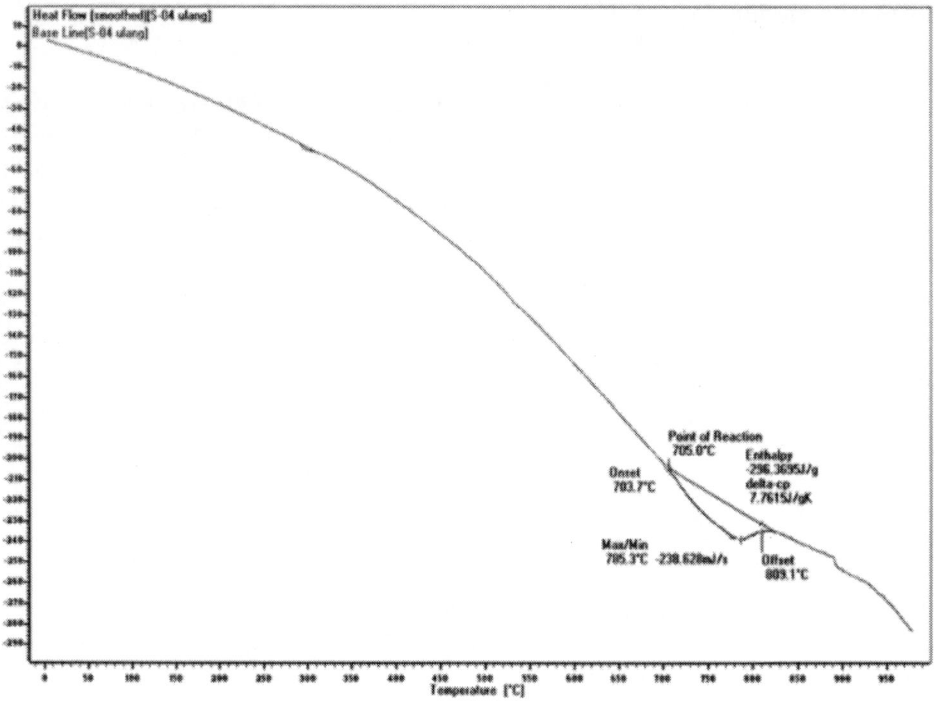


**Figure 1.** The Pattern DTA of $50SiO_2$- $25B_2O_3$- (6.5-*x*)$Bi_3O_3$- $18.5Na_2CO_3$- *x*NiO x=3 wt% Glass

It is obtained that occurrence of an endothermic process is at 705-809.1 °C. This temperature is also indicated the onset of melting $B_2O_3$ and $Na_2CO_3$ with a maximum at 785.3 °C.

*3.2.2 Difraction Patterns of Glass.* The Diffraction patterns of $50SiO_2$-$25B_2O_3$-(6.5-x) $Bi_3O_3$-18.5 $Na_2CO_{3-x}$NiO Glass with x = 0, 1, 2, 3, 4 wt% which has normalized shown in Figure 2.

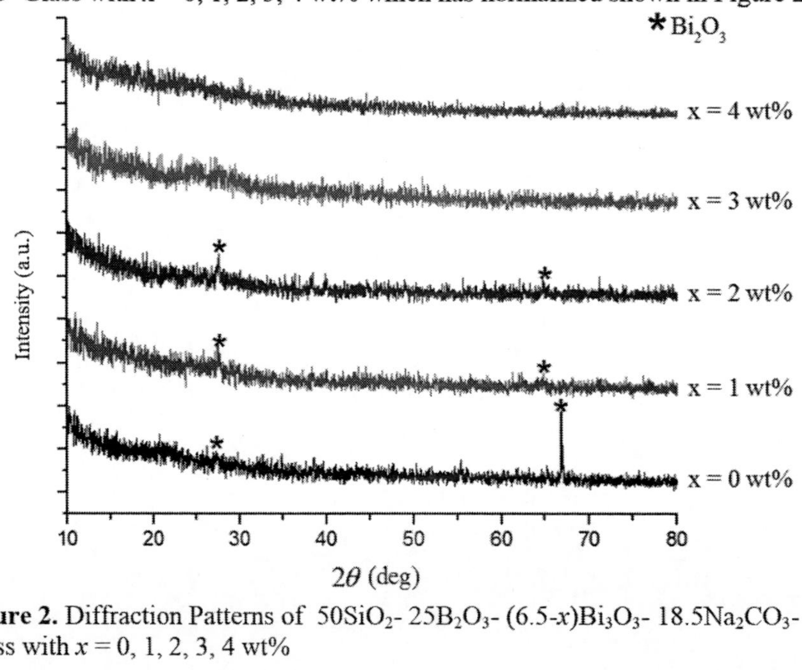


**Figure 2.** Diffraction Patterns of $50SiO_2$- $25B_2O_3$- (6.5-*x*)$Bi_3O_3$- $18.5Na_2CO_3$- *x*NiO Glass with *x* = 0, 1, 2, 3, 4 wt%

It looks that the samples are amorphous. For sample with x = 0 wt%, x = 1 wt%, and x = 2 wt%, there is a weak peak which indicates a few crystalline of $B_2O_3$phase. The higher of NiO induced in the sample, the glassy phase tend to higher purity. Samples with higher content of NiO provide the homogeneity of the compositional diffusion to maintain in forming an amorphous phase. It looks that increase of NiO reduce the crystallinity of $50SiO_2$-$25B_2O_3$-(6.5-x) $Bi_3O_3$-6, $5Na_2CO_{3\text{-}x}NiO$ glass. These conform with the SEM results.

**Table 2.** Composition of $50SiO_2$- $25B_2O_3$- (6.5-*x*)$Bi_3O_3$- $18.5Na_2CO_3$- *x*NiO Glass

| **Element** | Composition (at%) | | | | | Element | EDX Result (at%) | | | | |
|---|---|---|---|---|---|---|---|---|---|---|---|
| | **S0** | **S1** | **S2** | **S3** | **S4** | | **S0** | **S1** | **S2** | **S3** | **S4** |
| **$SiO_2$** | 60.34 | 60.34 | 60.34 | 60.34 | 60.34 | **Si** | 33.61 | 21.20 | 24.13 | 28.34 | 25.25 |
| **$B_2O_3$** | 26.01 | 26.01 | 26.01 | 26.01 | 26.01 | **B** | - | - | - | - | - |
| **B2O3** | 1.01 | 0.85 | 0.69 | 0.53 | 0.38 | **Bi** | 01.15 | 00.31 | 00.29 | 00.26 | 00.14 |
| **$Na_2CO_3$** | 12.64 | 12.64 | 12.64 | 12.64 | 12.64 | **Na** | 14.27 | 16.32 | 16.40 | 15.02 | 15.13 |
| **NiO** | 0.00 | 0.96 | 1.91 | 2.84 | 3.76 | **Ni** | 00.00 | 00.37 | 00.80 | 1.30 | 01.41 |
| | | | | | | **O** | 56.92 | 61.79 | 58.38 | 55.09 | 58.07 |

*3.2.3 Morphology of $50SiO_2$-$25B_2O_3$-(6.5-x)B2O3-18.5$Na_2CO_{3-x}NiO$ Glass by SEM.* Based on morphology of $50SiO_2$-$25B_2O_3$-(6.5-x)$B_2O_3$-18.5$Na_2CO_{3-x}NiO$ glass which can be seen in Figure 3.

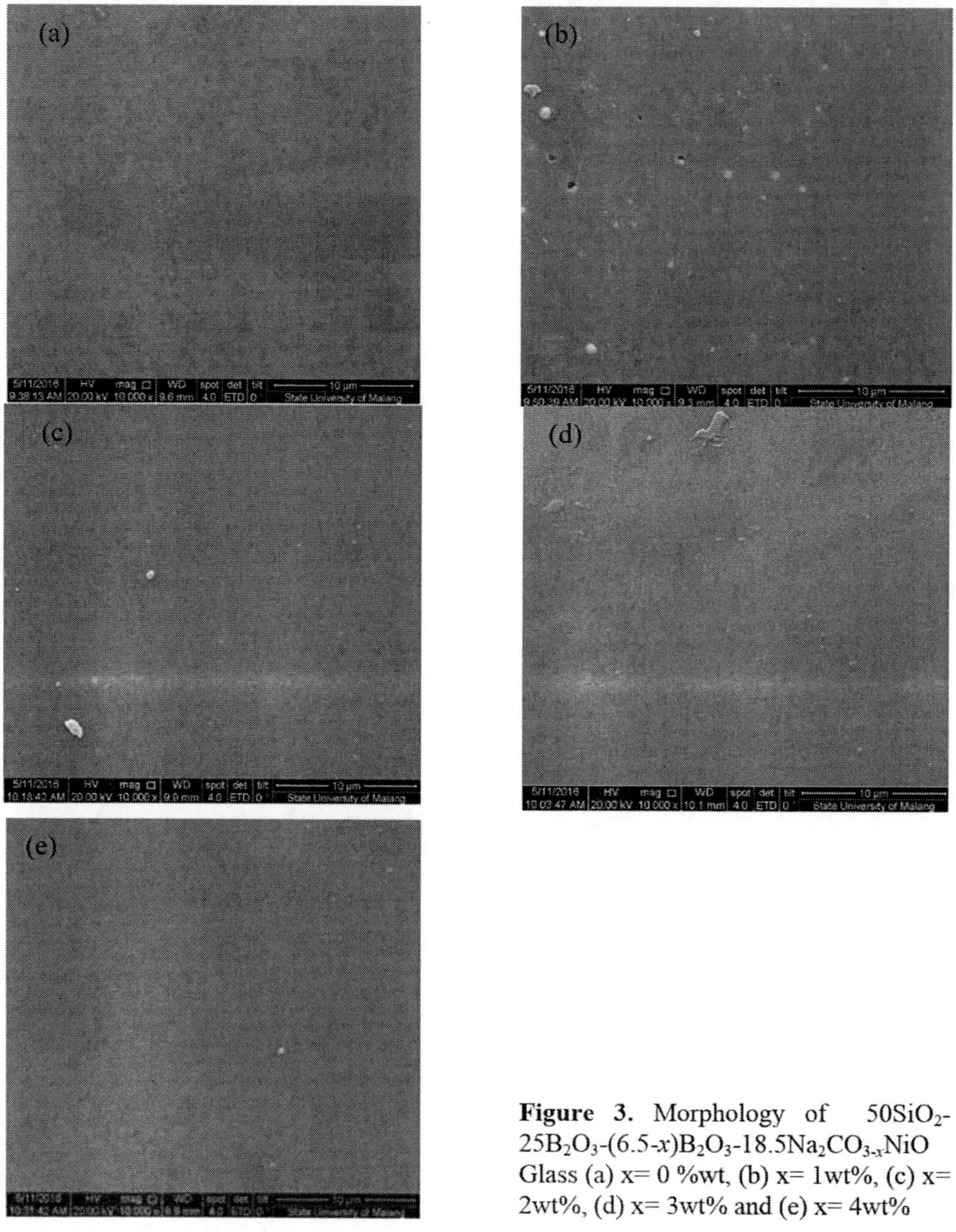


**Figure 3.** Morphology of $50SiO_2$-$25B_2O_3$-(6.5-x)$B_2O_3$-18.5$Na_2CO_{3-x}NiO$ Glass (a) x= 0 %wt, (b) x= 1wt%, (c) x= 2wt%, (d) x= 3wt% and (e) x= 4wt%

Figure 3 represents the morphology of $50SiO_2$-$25B_2O_3$-(6.5-x) -18, 5$Na_2CO_3$-$B_2O_3$ xNiO glass.

The small white spots indicate a presence of $B_2O_3$ which is also revealed the existing of $B_2O_3$ phase from XRD histogram. It looks that the surface of the sample is relatively homogeneous, with increasing NiO doping give rise to reduce the white spots and porosity.

IOP Conf. Series: Materials Science and Engineering **202** (2017) 012056 doi:10.1088/1757-899X/202/1/012056

*3.2.4 The Binding Vibration of $50SiO_2$-$25B_2O_3$-(6.5-x)$B_2O_3$-18.5$Na_2CO_3$-xNiO Glass by FTIR.* Bond vibration on $50SiO_2$-$25B_2O_3$-(6.5-x) $B_2O_3$-18.5 $Na_2CO_3$-xNiO glass at sample 0-4 can be seen in Figure 4. Bonding of silica group looks at wavenumbers between 950 – 1200 $cm^{-1}$ [10][11]. On the wave-number 950-1000 $cm^{-1}$ shows the variation of the antisymmetric stretching vibration on bonds from the bonds of O-Si-O tetrahedral $SiO_4$ units. It also appears of the wave-numbers of 1000-1200 $cm^{-1}$ showing te asymmetry strain of Si-O. With the rise in the concentration of NiO there is some change of FTIR patterns. The appearance of the Valley in the 1200-900 $cm^{-1}$ from a sample of ten 0-4 takes part at a wavelength of 1116, 1139, 1141, 1161, and 1172 $cm^{-1}$. This shows that the addition of NiO, changing the structure of the network, marked with the location of the peaks are formed differently. These results is in contrast with previous research which shows that the addition of NiO did not change the structure of the functional groups of glass [12]. However, this research has been showed that the phase of NiO tended to form as crystalline, and the NiO is also bound with other oxides.

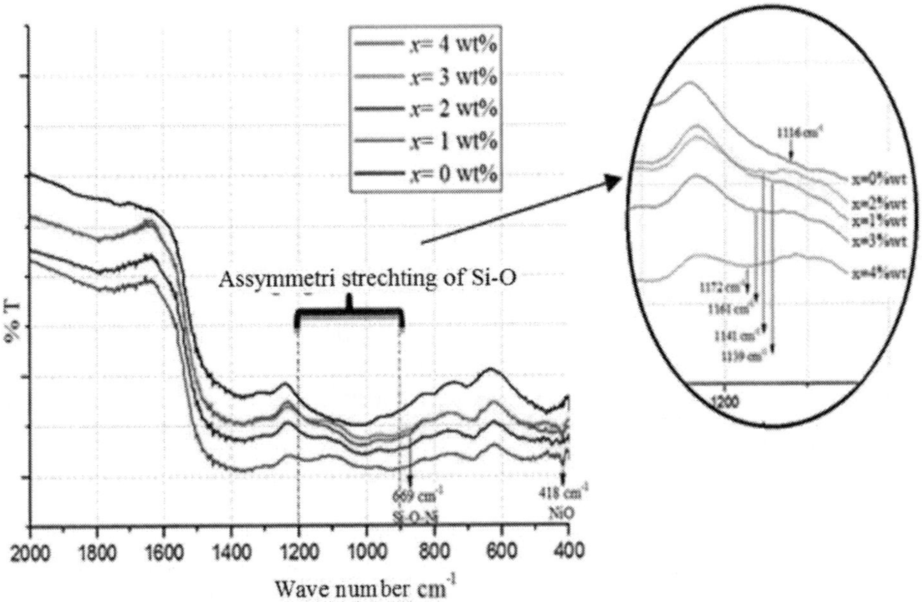


**Figure 4.** Vibration Peak of $50SiO_2$-$25B_2O_3$-(6.5-$x$)$B_2O_3$-18.5$Na_2CO_3$-$x$NiO Glass

Vibrational NiO bonds appear at 412, 418, 445, 449 $cm^{-1}$ and at 669 $cm^{-1}$ represent the bonds of Si-O-Ni. Based on Figure 4, some peaks have been showed the presence of Ni-O bond vibration on the wavenumber of 418 $cm^{-1}$. NiO is linked bonds crossed with silica [13].

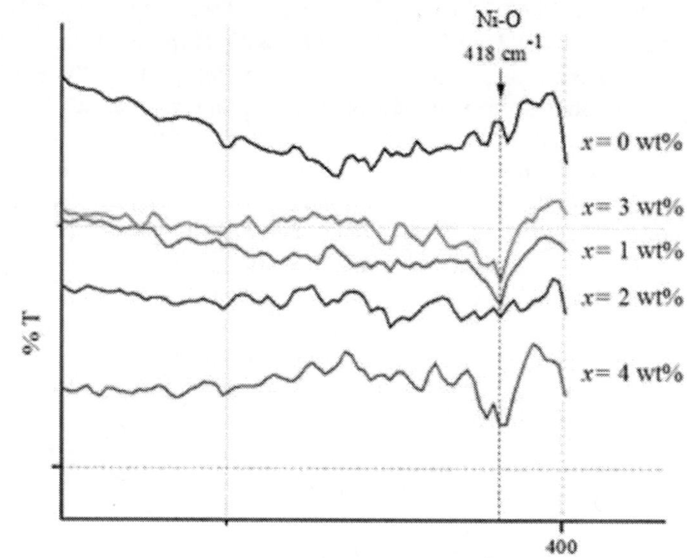


**Figure 5.** Binding Vibration of Bi-O-Bi on $50SiO_2$-$25B_2O_3$-$(6.5$-$x)B_2O_3$-$18.5Na_2CO_3$-$x$NiO Glass.

*3.3 The Influence of NiO to the Dielectric of $SiO_2$-$B_2O_3$-$B_2O_3$-$Na_2CO_3$ Glass*

The value of these measurements by capacitance meter is the capacitance. By using the Equation 1. knowable large dielectric constants whose results have been presented in the form of graphs in Fig 6. it shows the influence of the addition of NiO at the constant dielectric glass in dark circumstances or the intensity of zero. NiO with dielectric constants, look also increasing. The improvement of dielectric constant (ε) reached up to $12.68 \times 10^3$ which was calculated using the Equation (1).

$$\varepsilon_r = \frac{C.d}{\varepsilon_0.A} \quad (1)$$

The data were obtained from the measurement of electric capacitance ($C$) with respect to samples' surface area ($A$) and thickness ($d$). The results from the calculation for all samples show that the higher the addition of NiO, dielectric constants are also increased as depicted in Figure 6. This indicates that the $Ni^{2+}$ ion has been modifying the glass sample. This modification cause defects in the glass chain which resulted in rising space charge polarization leading to a high dielectric constant.

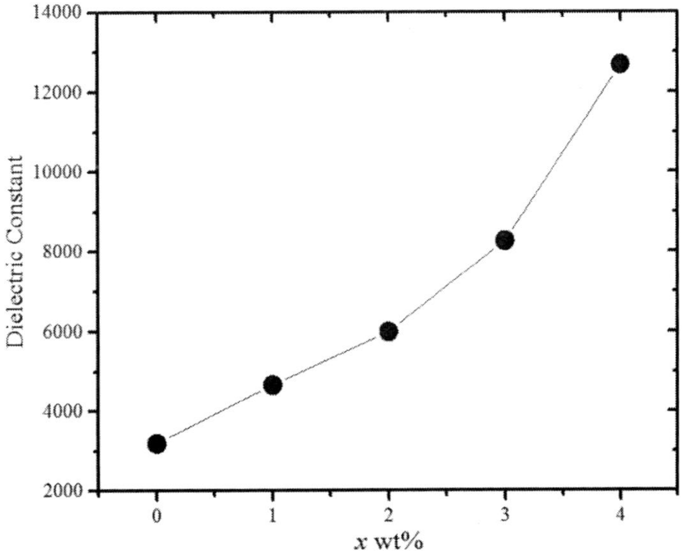


**Figure 6.** The influence of NiO on Dielectric Constant in the absence of Light Intensity

*3.4 The Influencing intensity of light to dielectric of 50$SiO_2$- 25$B_2O_3$- (6.5-x)$B_2O_3$-18.5$Na_2CO_3$- xNiO Glass.*

The influence of light intensity on the dielectric constant of 50$SiO_2$-25$B_2O_3$-(6.5-x) $B_2O_3$-18.5 $Na_2CO_3$-xNiO glass has also been explored using a capacitance meter. Equipped with a lux meter to obtain the intensity of a light source of 100 W Philips 5d. The range of the Intensity used was from 0-2500 lux. Using Equation 1 the dielectric constant can also be obtained. The results of the dielectric constant calculation are presented in Figure 7. It can be seen from Figure 7 that for all samples are obviously showing that employing light intensity on the samples reduce the dielectric constant. This suggests the existence of an interaction between light and glass samples. The intensity of light on the sample may drive to the excitation of electrons within atoms. With the greater radius of the atom, the electron binding energy becomes weak, allowing for the electrons jump up to the excited state. The mechanism resulted in reducing the dielectric constant.

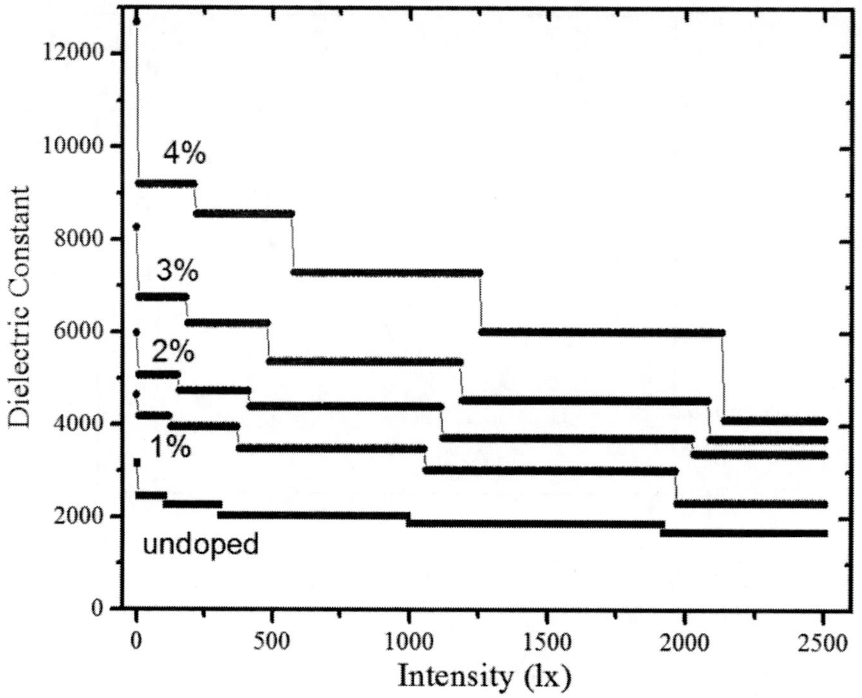


**Figure 7.** The influence of intensity on dielectric Constant

When we look closer in Figure 7, it is an interesting phenomenon observed. The reducing dielectric constant by the intensity of light is not a continuum. A step-like curve of the dielectric constant exists in all sample at a particular value of intensity. Following the description of excitation, the intensity of light represents the energy of photons of $E = nh\upsilon$ which is absorbed by the electrons to excite to the next outer energy levels. When this phenomenon related to the electronics polarization, the energy from the intensity of light is absorbed. The energy used for electrons to excite to occupy a higher energy level also quantized. This step-like features also shown by alumina doped orosilicate glass [14]

*3.5 The Influencing magnetic field to dielectric $50SiO_2$- $25B_2O_3$- $(6.5\text{-}x)B_2O_3\text{-}18.5Na_2CO_3$- xNiO Glass*

Influence of magnetic field against dielectric constant of glass $50SiO_2$-$25B_2O_3$-(6.5-x) $B_2O_3$-18.5 $Na_2CO_3$-xNiO was measured using a capacitance meter. Using the same Equation 1 the dielectric constant of the glass can be obtained. Figure 8 is a graph of the results of calculation of dielectric constants. It can be seen that along with the rising of the magnetic field applied; the dielectric constant is also rising. This shows that the NiO was able to increase the magnetic field response of sample glass. On the experimental results also brings about the relationship between the external magnetic field is applied to the dielectric constant of the material with antiferromagnetic is linear [14].

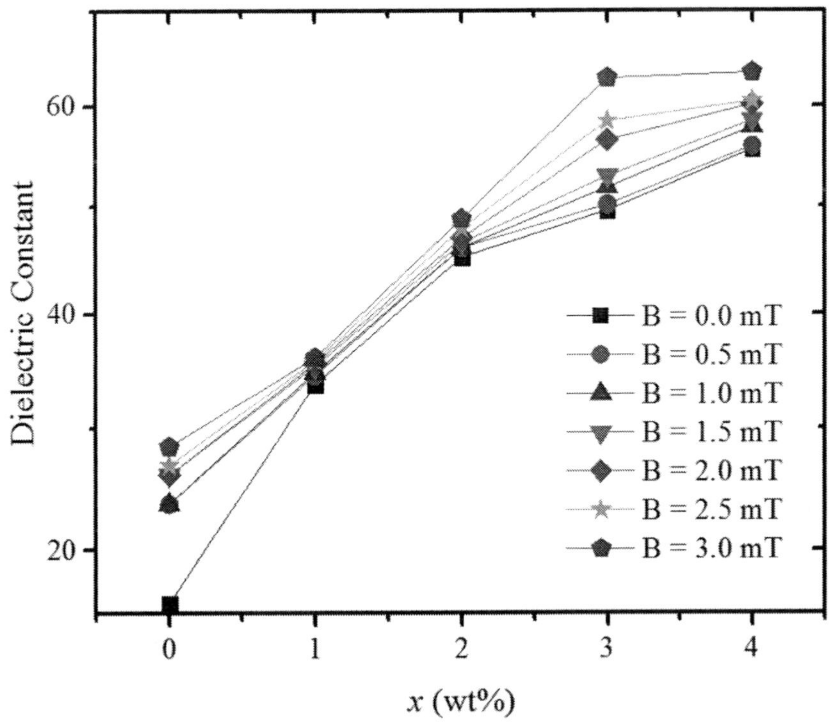


**Figure 8.** The influence of magnetic field on dielectric Glass

## 4. Conclusion

It can be concluded that the addition of NiO give rise to maintain the glassy structure of $50SiO_2$-$25B_2O_3$-(6.5-x) $B_2O_3$-18.5 $Na_2CO_3$-xNiO, and modify the functional bonds. The increase of NiO in the glass increase the dielectric constant. However, in the presence of light intensity onto the samples, the dielectric constants decrease with a distinct step-like at a certain amount. On the other hand, the applied of a magnetic field was able to increase the dielectric constant of the glass.

## 5. References

[1] Diantoro M, Yuwita P E, Olenka D and Nasikhudin 2014 Fabrication of $CuAl_{1-x}M_xO_2$ (M = Fe, Cr)/Ni film delafossite compounds using spin coating and their microstructure and dielectric constant *AIP Conf. Proc.* **1617** 170–4

[2] C Gautam, Kumar D, Parkash, O and Singh, P 2013 Synthesis, IR, crystallization and dielectric study of (Pb, Sr)$TiO_3$ borosilicate glass–ceramics **36** 461–9

[3] R K, S S and Pandey O.P 2012 FTIR structural investigation of gamma irradiated BaO–$Na_2O$–$B_2O_3$–$SiO_2$ glasses. *Phys B Condens MatterPhys B Condens Matter* **407** 4765–4769

[4] Moustafa F A, Fayad A M, Ezz-Eldin F M and El-Kashif I 2013 Effect of gamma radiation on ultraviolet, visible and infrared studies of NiO, $Cr_2O_3$ and $Fe_2O_3$-doped alkali borate glasses *J. Non-Cryst. Solids* **376** 18–25

[5] Macalik B, Suszynska M, Rysiakiewicz-Pasek E, Krawczyk L, Morawska-Kowal T and Okal J 2005 Spectroscopic and dielectric characteristics of nickel-doped porous silica glasses *Opt. Appl.* **35** 761

[6] Patil P S and Kadam L D 2002 Preparation and characterization of spray pyrolyzed nickel oxide (NiO) thin films *Appl. Surf. Sci.* **199** 211–221

[7] Yan S J, Xu C Y, Jiang J T, Liu D B, Wang Z Y, Tang J and Zhen L 2014 Strong dual-frequency electromagnetic absorption in Ku-band of C@ FeNi 3 core/shell structured microchains with negative permeability *J. Magn. Magn. Mater.* **349** 159–164

[8] He F, He Z, Xie J, Li Y and others 2014 IR and Raman Spectra Properties of $Bi_2O_3$-ZnO-$B_2O_3$-BaO Quaternary Glass System *Am. J. Anal. Chem.* **5** 1142

[9] Singh L P, Agarwal S K, Bhattacharyya S K, Sharma U and Ahalawat S 2011 Preparation of silica nanoparticles and its beneficial role in cementitious materials *Nanomater. Nanotechnol.* **1** 9

[10] Suresh B, Reddy M S, Reddy A S S, Gandhi Y, Kumar V R and Veeraiah N 2015 Spectroscopic features of $Ni^{2+}$ ion in PbO–$Bi_2O_3$–$SiO_2$ glass system *Spectrochim. Acta. A. Mol. Biomol. Spectrosc.* **141** 263–271

[11] X G, H M, M L, J Y and Li, B 2012 Facile synthesis route to NiO–$SiO_2$ intercalated clay with ordered porous structure: Intra gallery interfacially controlled functionalization using nickel–ammonia complex for deep desulfurization **148** 25–33

[12] Yao R, Liao S, Chen X, Tang G, Wang G and Zheng F 2016 Effects of ZnO and NiO on material properties of microwave absorptive glass-ceramic tile derived from iron ore tailings *Ceram. Int.* **42** 8179–8189

[13] Twej W A 2013 Spectral analysis of electronic transitions for low doping concentration of nickel ions in silica matrices at different phases *J. Non-Cryst. Solids* **382** 45–51

[14] Lu Y P, Ji C X, Sun Y L, Fang Y, Zhang L, Han Z D, Qian B, Jiang X F and Zhou W P 2016 Magnetic, dielectric and magnetoelectric properties of polycrystalline $Nb_2Co_4O_9$ *J. Alloys Compd.* **679** 213–217

# Composites of $Fe_3O_4/SiO_2$ from Natural Material Synthesized by Co-Precipitation Method

**Munasir[1], A S Dewanto[1], A Yulianingsih[1], I K F Saadah[1], Z A I Supardi[1], A Mufid[2] and A Taufiq[3]**

[1]Department of Physics, Faculty of Mathematics and Sciences, Universitas Negeri Surabaya (UNESA), Jl. Ketintang, Surabaya 60231, Indonesia.
[2]Department of Physics, Faculty of Mathematics and Sciences, Institut Teknologi Sepuluh Nopember (ITS), Jl. Arif Rahman Hakim, Surabaya 60111, Indonesia.
[3]Department of Physics, Faculty of Mathematics and Natural Sciences, Universitas Negeri Malang (UM), Jl. Semarang 5, Malang 65145, Indonesia.

Email: munasir_physics@unesa.ac.id

**Abstract**. Synthesis of $Fe_3O_4$/a-$SiO_2$ nanoparticle has been conducted based on natural materials by using the co-precipitation method. The used silica was a-$SiO_2$ collected from silica sand from Bancar Tuban, Indonesia. The $Fe_3O_4$ collected from iron sand's Lumajang, Indonesia. The samples was set for $Fe_3O_4$ : a-$SiO_2$: PEG composition of 1 : 2 : 1. The process was maintained by melting the PEG 4000 and following by adding a-$SiO_2$ and $Fe_3O_4$. The samples were characterized by using X-Ray Diffractometry (XRD), Transmission Electron Microscopy (TEM), and Fourier Transform Infra-Red (FTIR) spectroscopy. The analysis of the XRD data showed that additional silica into $Fe_3O_4$ tended to change the diffraction pattern that represents the existent of both $Fe_3O_4$ and silica. The SEM images presented that the samples formed in nanometric size with core-shell structure. Furthermore, the FTIR spectra showed that Si-O-Fe bound appeared at 555-568 $cm^{-1}$. Thus, this work was able to produce the $Fe_3O_4$/a-$SiO_2$ nanocomposites by using co-precipitation method in a core-shell structure.

**Keyword**: Composite, nanoparticle, $Fe_3O_4$, $SiO_2$, coprecipitation

## 1. Introduction

Indonesia is a country with abundant natural resources. Some of the natural resources extracted from minerals are quartz sand and iron sand. Most of the content of quartz sand is silica ($SiO_2$) with the purity of 70%. Quartz sand has a characteristic of its yellowish white color (Bancar area, Tuban-East Java, Indonesia) [1], its density of 2.50–2.70 g/cm$^3$, and having melting point of around 1715 °C [2]. Based on the previous research, the iron sand from Lumajang had a noticeably high level of Ferrous that could easily produce magnetic nanoparticles with high purity [3,4]. The latest reports show that Indonesia has a large number of iron minerals that consist of the main iron ore (17%), iron ore (8%) and laterite iron ore (75%) [4].

There are many applications of nanocomposites in industrial and health sectors especially the use of this material for small particles from micro to nanometric size. Thus, we need to advanced and the right technology to process it. It needs a special treatment in the process of obtaining the nanometric materials. A lot of methods were used to get the nanocomposites with the best result including the co-precipitation method, sol-gel, etc. [2,5–8]. Nanocomposite material is a combination of two or more composing materials with a different characteristic material (matrix and filler). In this research, the magnetic particle ($Fe_3O_4$) was combined with silica because it has multiple advantages including the ability of its pure materials to synthesize in large scale and the large capacity of cation exchange. Silica comes from a natural or synthetic material and usually is in the form of crystalline or amorphous solid [1,6,9]. Silica is a metal oxide compound that is widely available in nature, but its existence is not in the free state. It is bounded to another compound physically and chemically. This material is used in many industries and medical field because of its unique nature and morphology such as the large surface area and pore volume, and the ability to absorb various substances like water, oil, and radioactive material. In general, Silica has hydrophobic or hydrophilic properties based on its structure and morphology [10].

In composite materials, the component did not undergo a chemical change. $Fe_3O_4/SiO_2$ nanocomposite is a composite material because its constituent materials have nanometric size and exist without any new compound formed. Another work reported that the $Fe_3O_4/SiO_2$ nanocomposites were succesfully prepared by using the co-precipitation method with $FeCl_3$ and silica solution as the basic material [11]. In this work, we were interested to develop synthesis method of $Fe_3O_4/SiO_2$ nanocomposites by exploring the abundant natural material via a co-precipitation method. This method can be utilized to prepare an inorganic compound based on the precipitation process of substances. The advantage of this method is having a relatively low temperature in a short process [12,13]. Interestingly, the method can prepare the $Fe_3O_4/SiO_2$ nanoparticle that was illustrated to form a composite formation, where the $Fe_3O_4$ ferromagnetic particle is covered inside the hollow of $SiO_2$ particle [11,14].

## 2. Experimental Method

### *2.1. Materials*

The equipment used in this research were a beaker glass, measuring glass/graduated cylinder, magnetic stirrer, filter paper, digital scales, spatula, pH meter, pipette, spatula, funnel, crucible, mortal and pestle permanent magnet, cardboard and a 60-watt lamp. The material used in this study were Bancar-Tuban sand, Lumajang iron sand, NaOH, HCl 2M, $NH_4OH$, PEG 4000, and distilled water (DI). The manipulation variables were the a-$SiO_2$mass and the PEG mass. The Control Variable was the mass of $Fe_3O_4$. The response variable was the phase of $Fe_3O_4$/a-$SiO_2$ nanocomposites.

### *2.2. Experiment*

The step taken to synthesize the nanocomposites of $Fe_3O_4$/a-$SiO_2$ was melting the PEG via the stirrer at a temperature of 60 °C. It was then added with a-$SiO_2$ powder while it was being stirred continuously until evenly combined, after that it was added with $Fe_3O_4$ powder and it was stirred for another 3 minutes, so that it formed precipitate. Those steps were repeated for 3 times with different compositions. The precipitate was then dried and thus formed $Fe_3O_4$/a-$SiO_2$ powder. This product was then characterized by using XRD to determine the occurrence of the composite on both $Fe_3O_4$and a-$SiO_2$. Furthermore, it was also characterized by using FTIR to determine the functional group possessed by the sample of $Fe_3O_4$/a-$SiO_2$. The size and distribution of the particles were characterized by using SEM/TEM.

## 3. Results and Discussion

### *3.1. XRD Characterization of Composite of $Fe_3O_4$/a-$SiO_2$*

XRD Diffractogram result showed the diffraction pattern of amorphous silica and $Fe_3O_4$ shown in Figure 1. The curve indicates that the peaks in sample #1, and the base materials which were $Fe_3O_4$ and $SiO_2$. The XRD results proved the existence of peak particle combination of $Fe_3O_4$ and a-$SiO_2$

particles. The diffraction peak occured at $2\theta = 30.07°, 35.47°, 43.17°, 53.33°, 56.90°$ which showed the cubic structure of the $Fe_3O_4$ particle, with each of crystal planes (220), (311), (400), (422), and (511). Therefore, undergoing a composite process did not change the structure of $Fe_3O_4$ particle . In sample #1, the wide peak on $2\theta \approx 24°$ showed the existence of amorphous peak silica form. This peak intensity will increase along with the addition of silica, where the position of polymer form is very influential in the composite [15]. The higher the variation of silicate addition into $Fe_3O_4$ the more declined the diffraction peak pattern of $Fe_3O_4$ [11,16,17].

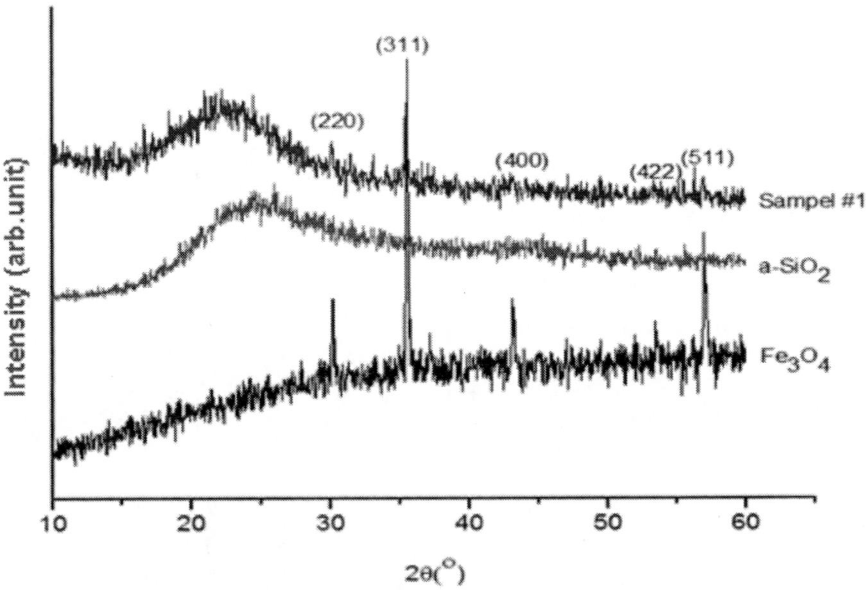


**Figure 1.** XRD patterns of $Fe_3O_4$/a-$SiO_2$

*3.2. Data of FTIR Characterization of $Fe_3O_4$/a-$SiO_2$ Composites*

After a test by using XRD, the $Fe_3O_4$/a-$SiO_2$ nanoparticles was characterized by using FTIR to determine the functional groups of the $Fe_3O_4$/a-$SiO_2$ sample. The functional groups corresponding to $Fe_3O_4$@$SiO_2$ (Sample #1) therein are depicted in Table 1.

**Table 1.** Functional group of the synthesized $Fe_3O_4$/a-$SiO_2$

| Peak | Wavenumber ($cm^{-1}$) | Functional Group | Reference |
|---|---|---|---|
| | $Fe_3O_4$@$SiO_2$ (Sample #1) | | |
| 1 | 468 | Si-O-Si / O-Si-O | [1,7] |
| 2 | 555 | Si-O-Fe | [12] |
| 3 | 796 | Si-O-Si | [1,7,14] |
| 4 | 950 | Si-O | [12] |
| 5 | 1093 | Si-O-Si | [1,14] |
| 6 | 1647 | H-O-H | [12] |
| 7 | 2914 | H-O-H | [12] |
| 8 | 3443 | H-O-H | [1,14] |

Based on the data, compared to the previous research, our FTIR result of $Fe_3O_4$/a-$SiO_2$ synthesis had compatibility with the previous and appropriate range of wavenumbers. In the FTIR spectra, the wavenumber of 1641-3464 $cm^{-1}$ was a stretching vibration and bending of H-O-H bond. In 1097, the wavenumber was an antisymmetric vibration of the Si-O-Si bond which was hooked by oxygen movement. There was a bending vibration of Si-O-Si or O-Si-O on the 468 $cm^{-1}$ wavenumber. On the 796 $cm^{-1}$ wavenumber, there was a symmetric stretch vibration of Si-O-Si in the $SiO_2$ [1,6,7]. In 950 $cm^{-1}$ wavenumber, there was a symmetric functional group of Si-O and lastly, the 555-568 $cm^{-1}$ wavenumber showed the existence of Si-O-Fe bond. Thus, the existence of $Fe_3O_4$ nanoparticles has been identified.

### *3.3. Photo SEM of $Fe_3O_4$/a-$SiO_2$ Composites*

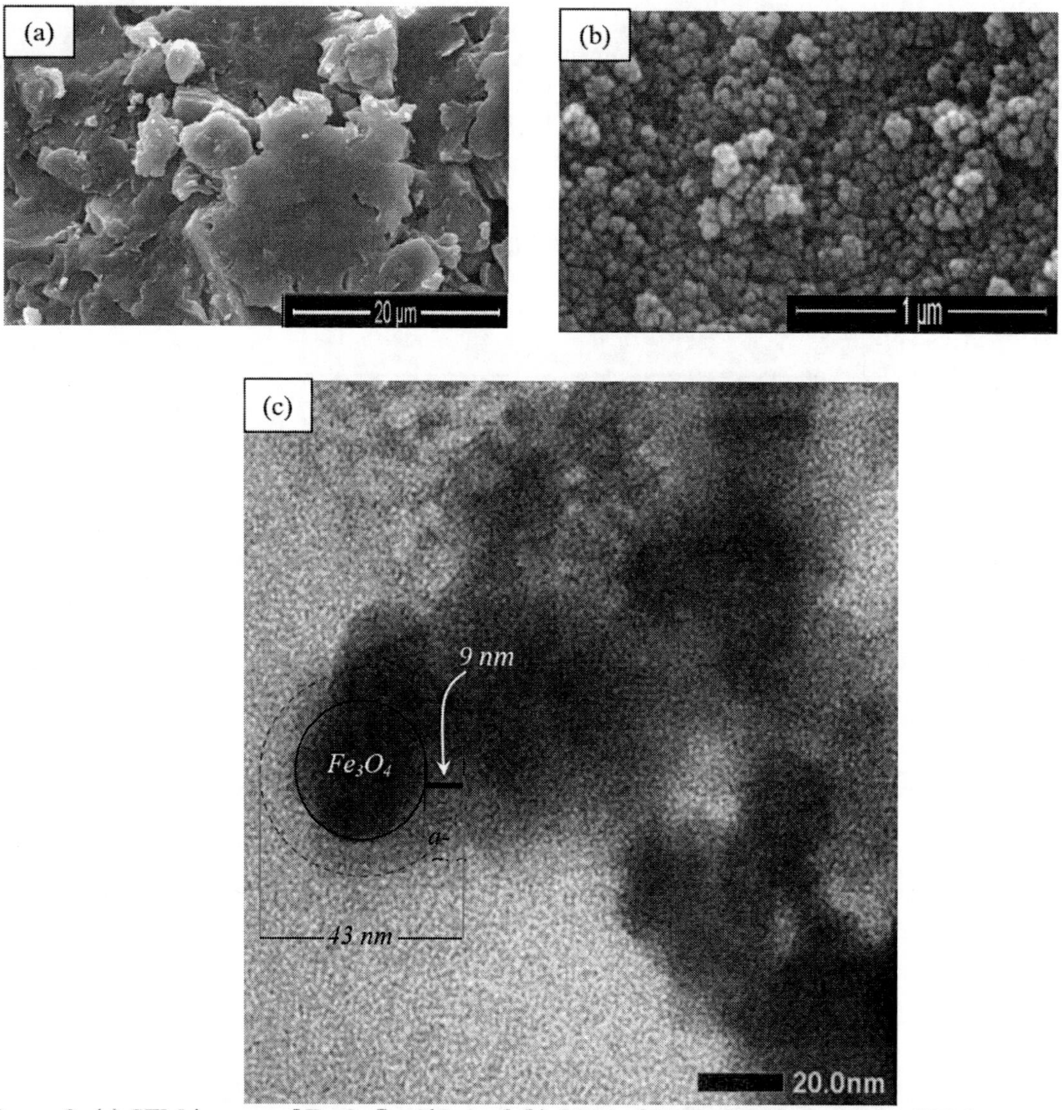


**Figure 2**. (a) SEM images of $Fe_3O_4$@a-$SiO_2$ and (b) SEM of $Fe_3O_4$@c-$SiO_2$, and (c) TEM test result of $Fe_3O_4$@$SiO_2$ (Sample #1)

IOP Conf. Series: Materials Science and Engineering **202** (2017) 012057 doi:10.1088/1757-899X/202/1/012057

TEM characterization aimed for knowing the morphological structure of $Fe_3O_4$/a-$SiO_2$ nanocomposites that formed as a $Fe_3O_4$@a-$SiO_2$ core shell. Here is the result of TEM test of the $Fe_3O_4$@a-$SiO_2$ core shell (Figure 2c). It appears that the $Fe_3O_4$ particles are black-colored and coated with $SiO_2$ particle shown by a lighter cloud around it. $Fe_3O_4$@a-$SiO_2$ particle has a round-shaped property and agglomerated with other particles. The size of the $Fe_3O_4$@*a*-SiO2 particle was about ~25 nm and the thickness is ~9 nm. Thus, the $Fe_3O_4$/a-$SiO_2$ nanocomposites have successfully formed a $Fe_3O_4$@a-$SiO_2$ core shell [14,18]. There were significant different shape and size of the composite particles of $Fe_3O_4$@ a-$SiO_2$ amorphous (#sample 1) and $Fe_3O_4$@ c-SiO2 (crystal), as shown in Figure 2 (a) and Figure 2 (b); composite particle size of Fe3O4/a-SiO 2 was greater than the particle size of the $Fe_3O_4$@c-$SiO_2$ composite.

## 4. Conclusion

Based on the data analysis, we can conclude that $Fe_3O_4$/a-$SiO_2$ nanocompositeS can be fabricated by using PEG as the binder of Fe3O4 and a-SiO2 particleS, where $Fe_3O_4$, a-$SiO_2$ can be easily synthesized by using the co-precipitation method. The Characteristic of $Fe_3O_4$/a-$SiO_2$ based on XRD crystal structure and functional group of FTIR analysis shows that the $Fe_3O_4$/a-$SiO_2$ nanocomposites have formed a sharp peak of diffraction pattern and a distinct of functional group of Si-O-Fe on wavenumbers of 555-568 $cm^{-1}$.

## 5. References

[1] Munasir, Triwikantoro, Zainuri M and Darminto 2015 Synthesis of $SiO_2$ nanopowders containing quartz and cristobalite phases from silica sands *Mater. Sci.-Pol.* **33** 47–55

[2] Munasir, Sulton A, Triwikantoro, Zainuri M and Darminto 2013 Synthesis of silica nanopowder produced from Indonesian natural sand via alkalifussion route *AIP Conference Proceedings* International Conference On Theoretical And Applied Physics (LCTAP 2012) vol 1555(AIP Publishing)pp 28–31

[3] Taufiq A, Sunaryono, Rachman Putra E G, Pratapa S and Darminto 2015 Nano-Structural Studies on Fe<sub>3</sub>O<sub>4</sub> Particles Dispersing in a Magnetic Fluid Using X-Ray Diffractometry and Small-Angle Neutron Scattering *Mater. Sci. Forum* **827** 213–8

[4] Rochani S, Pramusanto, Sariman and Anugrah R 2008 The Current Status of Iron Minerals in Indonesia *Indones. Minning J.* **11** 1–17

[5] Trabelsi W, Benzina M and Bouaziz S 2009 Physico-chemical characterisation of the Douiret sand (Southern Tunisia): Valorisation for the production of Silica Gel *Phys. Procedia* **2** 1461–7

[6] Kalapathy U, Proctor A and Shultz J 2000 A simple method for production of pure silica from rice hull ash *Bioresour. Technol.* **73** 257–62

[7] Jal P K, Sudarshan M, Saha A, Patel S and Mishra B K 2004 Synthesis and characterization of nanosilica prepared by precipitation method *Colloids Surf. Physicochem. Eng. Asp.* **240** 173–8

[8] Affandi S, Setyawan H, Winardi S, Purwanto A and Balgis R 2009 A facile method for production of high-purity silica xerogels from bagasse ash *Adv. Powder Technol.* **20** 468–72

[9] Mori H 2003 Extraction of silicon dioxide from waste colored glasses by alkali fusion using potassium hydroxide *J. Mater. Sci.* **38** 3461–8

[10] Koch K and Barthlott W 2009 Superhydrophobic and superhydrophilic plant surfaces: an inspiration for biomimetic materials *Philos. Trans. R. Soc. Math. Phys. Eng. Sci.* **367** 1487–509

[11] Suryani T and Suharyadi E 2015 Sythesis of Magnetite Nanoparticle ($Fe_3O_4$) with Silica (SiO2) Template and Characterization of Magnetic Properties *Indones. J. Appl. Phys.* **5** 23–30

[12] Ahangaran F, Hassanzadeh A and Nouri S 2013 Surface modification of $Fe_3O_4$@$SiO_2$ microsphere by silane coupling agent *Int. Nano Lett.* **3** 23

[13] Alizadeh A, Khodaei M M, Beygzadeh M, Kordestani D and Feyzi M 2012 Biguanide-Functionalized $Fe_3O_4$ /$SiO_2$ Magnetic Nanoparticles: An Efficient Heterogeneous Organosuperbase Catalyst for Various Organic Transformations in Aqueous Media *Bull. Korean Chem. Soc.* **33** 2546–52

[14] Chae H S, Kim S D, Piao S H and Choi H J 2016 Core-shell structured $Fe_3O_4$@$SiO_2$ nanoparticles fabricated by sol–gel method and their magnetorheology *Colloid Polym. Sci.* **294** 647–55

[15] Zamani F and Izadi E 2014 Polyvinyl amine coated $Fe_3O_4$@$SiO_2$ magnetic microspheres for Knoevenagel condensation *Chin. J. Catal.* **35** 21–7

[16] Music' S, N. F-V and L S 2011 Precipitation of Amorphous $SiO_2$ Particles and Their Properties *Braz. J. Chem. Eng.* **28** 89–94

[17] Zawrah F H, A.A. E-K and Abd-Elaal H M 2009 Facile and Economic Synthesis Of Silica Nanoparticles *J. Ovonic Res.* **5** 129–33

[18] Hui C, Shen C, Tian J, Bao L, Ding H, Li C, Tian Y, Shi X and Gao H-J 2011 Core-shell $Fe_3O_4$ @$SiO_2$ nanoparticles synthesized with well-dispersed hydrophilic $Fe_3O_4$ seeds *Nanoscale* **3** 701–5

# Fabrication of $TiO_2$/Carbon Photocatalyst using Submerged DC Arc Discharged in Ethanol/Acetic Acid Medium

**T E Saraswati, A O Nandika, I F Andhika, Patiha, C Purnawan, S Wahyuningsih and S B Rahardjo**

Department of Chemistry, Mathematic and Natural Sciences Faculty, Sebelas Maret University, Jl. Ir. Sutami 36A Kentingan, Surakarta 57126 Indonesia

Email: teguh@mipa.uns.ac.id

**Abstract**. This study aimed to fabricate a modified photocatalyst of $TiO_2$/C to enhance its performance. The fabrication was achieved using the submerged direct current (DC) arc-discharge method employing two graphite electrodes, one of which was filled with a mixture of carbon powder, $TiO_2$, and binder, in ethanol with acetic acid added in various concentrations. The arc-discharge method was conducted by flowing a current of 10–20 A (~20 V). X-ray diffraction (XRD) patterns showed significant placements of the main peak characteristics of $TiO_2$, C graphite, and titanium carbide. The surface analysis using Fourier transform infrared spectroscopy (FTIR) revealed that fabricated $TiO_2$/C nanoparticles had stretching vibrations of Ti—O, C—H, C═O, C-O, O-H and C═C in the regions of 450–550 $cm^{-1}$, 2900–2880 $cm^{-1}$, 1690–1760 $cm^{-1}$, 1050–1300 $cm^{-1}$, 3400–3700 $cm^{-1}$ and ~1600 $cm^{-1}$, respectively. In addition, the study investigated the photocatalysts of unmodified and modified $TiO_2$/C for photodegradation of methylene blue (MB) dye solution under mercury lamp irradiation. The effectiveness of the degradation was defined by the decrease in 60-minute absorbance under a UV-Vis spectrophotometer. Modified $TiO_2$/C proved to be significantly more efficient in reducing dye concentrations, reaching ~70%. It indicated that the oxygen-containing functional groups have been successfully attached to the surface of the nanoparticles and played a role in enhancing photocatalytic activity.

## 1. Introduction

Titanium dioxide is an excellent photocatalyst because of the high oxidizing power of photogenerated holes and its chemical inertness and non-toxicity [1]. However, due to its low adsorption, titanium dioxide must be modified by carbon to increase its photocatalytic activity. It is known that carbon-based nanoparticles can be produced using the arc-discharge method. Sano et al. (2002) fabricated carbon onions using the arc-discharge method in water [2]. It has been reported that nanoparticles produced via arc-discharge in a liquid medium have been of higher purity, larger quantity, and lower cost than those produced by other methods.

Several fabrication routes for carbon-modified $TiO_2$ have been reported, including oxidative annealing, hydrolysis followed by carbonization [3,4], solvothermism [5], electrospinning [6,7], chemical vapor deposition (CVD) [8], hydrothermism [9] and sol-gel [10]. To prevent carbon from being deposited directly into bulk $TiO_2$, an appropriate technique is needed to allow the co-assembly of carbon and $TiO_2$ into carbon-modified $TiO_2$, which has a nanoarchitecture structure. Torimoto et al. (1996) reported another simple method of synthesizing a composite of $TiO_2$/carbon: sonication of the raw materials of $TiO_2$ and activated carbon [11]. The material produced using this approach had enhanced adsorption and photodegradation abilities. Saraswati *et al.* confirmed that arc discharge could produce $TiO_2$ which is encapsulated within graphite layers in liquid media, including titanium dioxide suspension in ethanol [12]. However, problems remained, including the dispersion of nanoparticles in water. The commonly used carbon product was hydrophobic, so the interaction of the nanoparticle surfaces with a water-based environment could not optimally be achieved.

Oxygen-containing functional groups, including hydroxyl (-OH), play major roles in degrading and mineralizing colored dyes. Surface modification using oxygen-containing functional groups can be performed by plasma [13, 14], chemical [15, 16] and electrochemical [17] routes. Introducing an oxidation agent containing functional groups into the liquid medium used in the submerged arc-discharge method has been considered a useful way to modify the condensed carbon material formed during the arc discharge. Supplying suitable oxidizing agents to the pores of the carbon matrix has been believed to be the key to successfully introduce oxygen-containing functional groups to the surface of the carbon material [18]. A successful surface modification enhances the photodegradation activity of titanium dioxide.

The present study investigated the synthesis of carbon modified $TiO_2$ ($TiO_2$/C) functionalized by oxygen-containing functional groups using an arc-discharge method performed in an ethanol/acetic acid liquid medium. To the best of our knowledge, this has not previously been reported. The fabrication process was followed by photocatalytic experiments using methylene blue (MB) dye. The activity of the $TiO_2$/C photocatalyst was observed by measuring the efficiency of photocatalytic degradation of aqueous MB in 60 minutes under mercury lamp irradiation in batch mode.

## 2. Experimental Method

### *2.1. Fabrication of $TiO_2$/Carbon*

Nanoparticles were fabricated using an arc-discharge method in an ethanol/acetic acid medium employing graphite taken from a dry-cell battery for the electrodes. The anode was filled with carbon powder, titanium dioxide, and silica glue as a binder at a ratio of 3:1:1 w/w and was formed precisely to facilitate the electrical breakdown. Ethanol (35.7%) and acetic acid (5.0%, 12.6%, 25.3%, 37.9% and 50%) formed the liquid medium.

The anode and cathode electrodes were submerged in the liquid medium with a 1–2-mm gap between them. The discharge current was ~10–20 A. When the arc-discharge process began, there was a bright area between the two electrodes, indicating the arc-plasma region. This arc caused the carbon graphite in the cathode to evaporate. Furthermore, the interaction between carbon graphite and $TiO_2$ produced the $TiO_2$/C nanoparticles. This fabrication included two types of powder: floating and sedimentary. The floating material had a hydrophobic surface, and the sedimentary material had a hydrophilic one. Therefore, the sediment was analyzed further using several instruments.

### *2.2. Photodegradation of methylene blue (MB)*

A photocatalyst nanoparticle of $TiO_2$/C (0.01 g) was added into dye sample in a vial containing 3-ppm MB solution. The sample was sonicated for ~30 min in a dark room and then irradiated under a 125-W mercury lamp for ~60 min. The solution was centrifuged to separate the nanoparticle from the MB solution. The absorbance of the MB was measured every 15 min. Each sample was tested to identify its photodegradation efficiency.

## 3. Results and Discussion

The arc-discharge method used two electrodes, i.e. anode, and cathode, which were momentarily brought into contact, striking an arc. Then, the distance between the two electrodes was reduced and a constant gap maintained between them by adjusting the position of the anode. The high temperature sublimated the carbon in the anode during this process. During cooling in the cool zone, thesublimated carbon became carbon nanomaterials, including carbon onions, fullerenes and other carbon allotropes with $sp^2$ hybridization carbon structures. In addition, the fabrication process produced $TiO_2/C$ resulted from the interaction between the carbon graphite and $TiO_2$ were found as floating and sedimentary nanoparticles. The floating nanoparticles were typically hydrophobic while the sedimentary nanoparticle was hydrophilic, meaning that the latter had better dispersion in the liquid medium. The sedimentary powder from the surface of the liquid medium was characterized using several instruments.

Figure 1 shows the X-ray diffraction (XRD) patterns of the original materials, $TiO_2$ anatase, and carbon graphite. The patterns showed high crystallinity of $TiO_2$ at some peaks, including hkl (101) $2\theta = 25.33°$; hkl (200) $2\theta = 48.06°$ and hkl (105) $2\theta = 53.95°$, compared to the standard for $TiO_2$ anatase (JCPDS No. 86-1157) in hkl planes: (101) $2\theta = 25.32°$; hkl (200) $2\theta = 48.06°$ and hkl (105) $2\theta = 53.98°$. The XRD pattern of the graphite showed peaks at hkl (002) $2\theta = 26.60°$, which matched the standard for graphite (JCPDS No. 41-1487) at hkl (002) $2\theta = 26.62°$. Futhermore, the peak at $2\theta = 20$ to $25°$ was indicated to $C_{60}$ (fullerene), carbon onions and other carbon allotropes with $sp^2$ hybridization carbon structures.

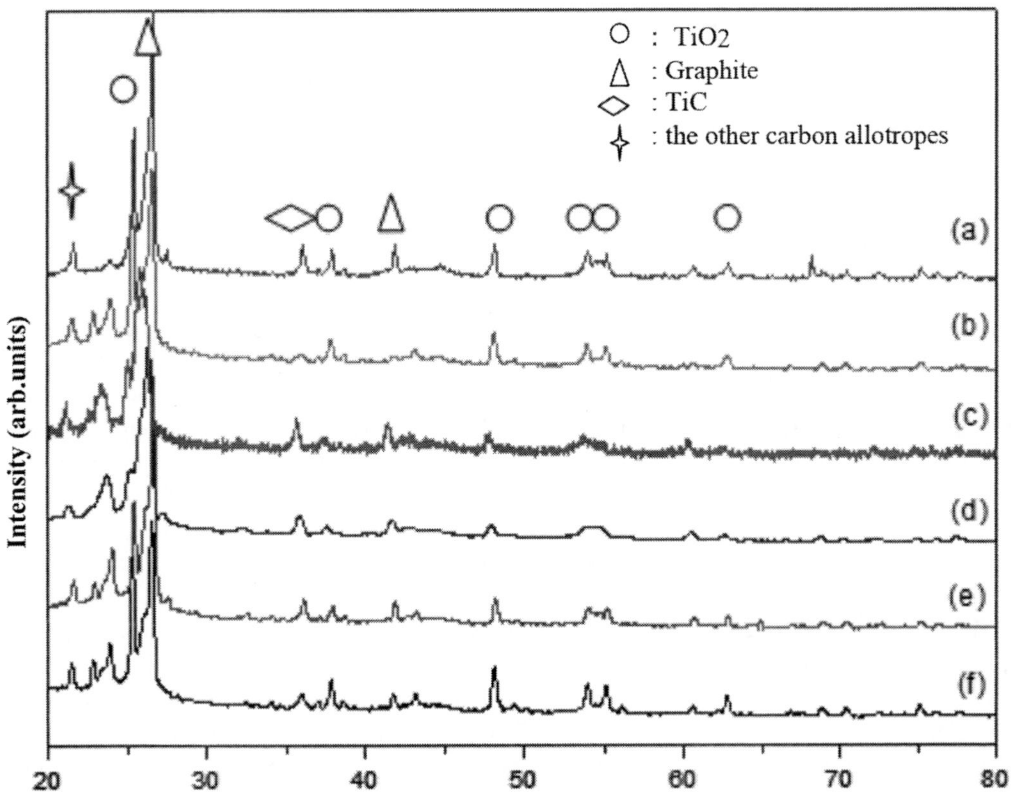


**Figure 1.** XRD patterns of the nanoparticles of (a) $TiO_2/C$ 0, (b) $TiO_2/C$ 5, (c) $TiO_2/C$ 12, (d) $TiO_2/C$ 25, (e) $TiO_2/C$ 37 and (f) $TiO_2/C$ 50.

Figure 1 also reveals the XRD patterns of the $TiO_2/C$ nanoparticles, which had peaks similar to those of the original materials. However, Figure 1 also shows new peaks at hkl (202) $2\theta = 36.02°$ (JCPDS TiC No. 72-2496) and hkl (222) $2\theta = 20.75°$ (JCPDS $C_{60}$ No. 47-0787) that refer to TiC and $C_{60}$. It indicated that the nanoparticles were composites in which the main peak characteristic of the original material still appeared, along with newly generated peaks. The formation of titanium carbide has been shown in the reaction of $TiO_{2(s)} + 3C_{(s)} \rightarrow 2CO_{(g)} + TiC_{(s)}$ [19].

Fourier transform infrared spectroscopy (FTIR) was used to characterize the functional groups on the surface of the carbon graphite by comparing transmittance of the peak lines. The FTIR technique was mainly used qualitatively to assess the surface character of the carbon layer. The original material, graphite carbon, absorbed almost all radiation in the visible spectra, and the peaks obtained were usually a sum of the interaction of the various types generated by the functional groups.

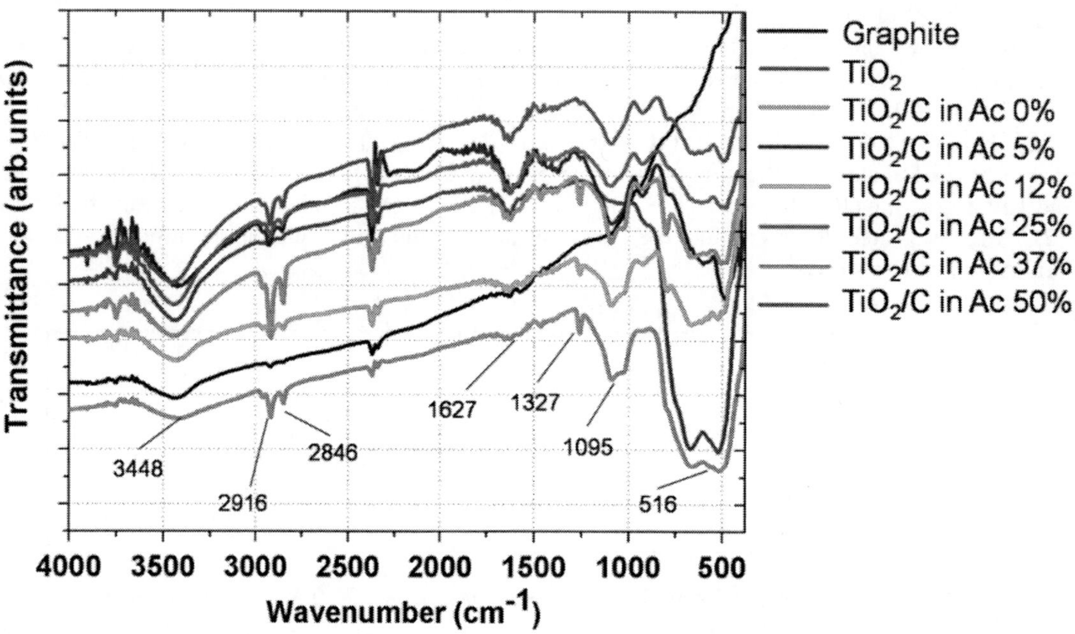


**Figure 2.** FTIR spectrum of carbon graphite, $TiO_2$ anatase, $TiO_2/C$ 0, $TiO_2/C$ 5, $TiO_2/C$ 12, $TiO_2/C$ 25, $TiO_2/C$ 37 and $TiO_2/C$ 50.

Figure 2 shows the spectra of the carbon graphite and the $TiO_2$ anatase and compares them with those of the $TiO_2/C$ nanoparticles treated by adding acetic acid. This comparison indicated that the organic acid interacted with the surface of the graphite through the carboxylic group, as expected. The region in the range of 450–550 $cm^{-1}$ corresponded to the stretching vibration of the Ti-O. A broadband at 3300–2500 $cm^{-1}$ corresponded to the stretching of the –OH group of alcohol or phenol on the adsorbent surface. The peaks in the region of 2850–3000 $cm^{-1}$ corresponded to the stretching of C-H. A broad peak at about 1600 $cm^{-1}$ signified the presence of C=C bonds in aromatic rings. There was a C=O bond at about 1700 $cm^{-1}$, although it is not prominent. Similarly, the presence of C-O of alcohol, ether, carboxylic acid, and ester was signaled in the range from 1000–1320 $cm^{-1}$. As can be seen in Figure 2, the FTIR spectrum of $TiO_2/C$ 12 nanoparticles had a different pattern, and the area of -OH vibration was larger than the others. The vibration areas of the samples $TiO_2/C$ 0, $TiO_2/C$ 5, $TiO_2/C$ 12, $TiO_2/C$ 25, $TiO_2/C$ 37 and $TiO_2/C$ 50 were 745.6, 132.3, 911.2, 452.8, 439.8 and 576.4, respectively. These results indicated that $TiO_2/C$ 12 nanoparticles had better dispersion in the liquid medium because there were interactions by which hydrogen bonding formed.

Also, the samples were investigated using a scanning electron microscope (SEM) to identify the morphology and structures of the constituents in the $TiO_2$ anatase and $TiO_2/C$ 12. Figure 3 shows the results of morphological examination of $TiO_2$ anatase and $TiO_2/C$ nanoparticles. The SEM image with 50,000 x magnification showed that the nanoparticles were in spherical structures, with $TiO_2/C$ 12 clusters varying in size from 140–180 nm.

A dispersion test was applied to confirm that $TiO_2/C$ 12 nanoparticles had better dispersion than $TiO_2/C$ 0 (without an addition of acetic acid). In this case, distilled water was used as the liquid medium and ethanol was employed as the comparison. Figure 4 shows the dispersion test results of each nanoparticle sample.

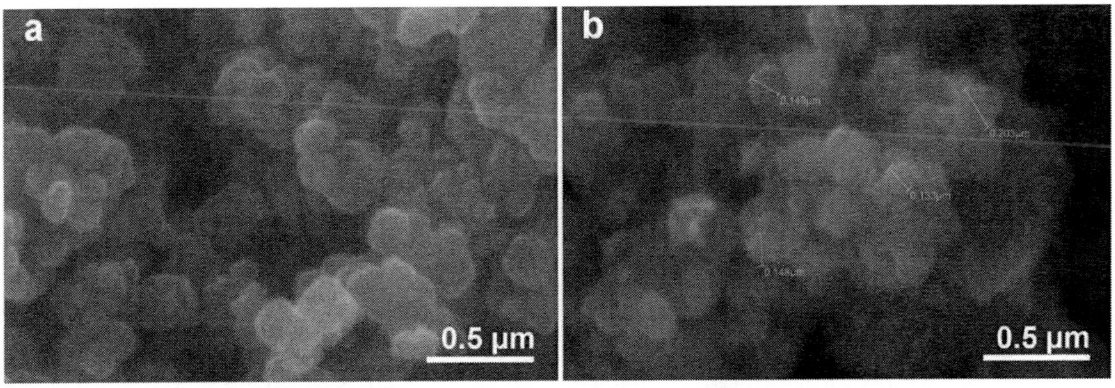


**Figure 3.** SEM images of (a) $TiO_2$ anatase and (b) $TiO_2/C$ 12.

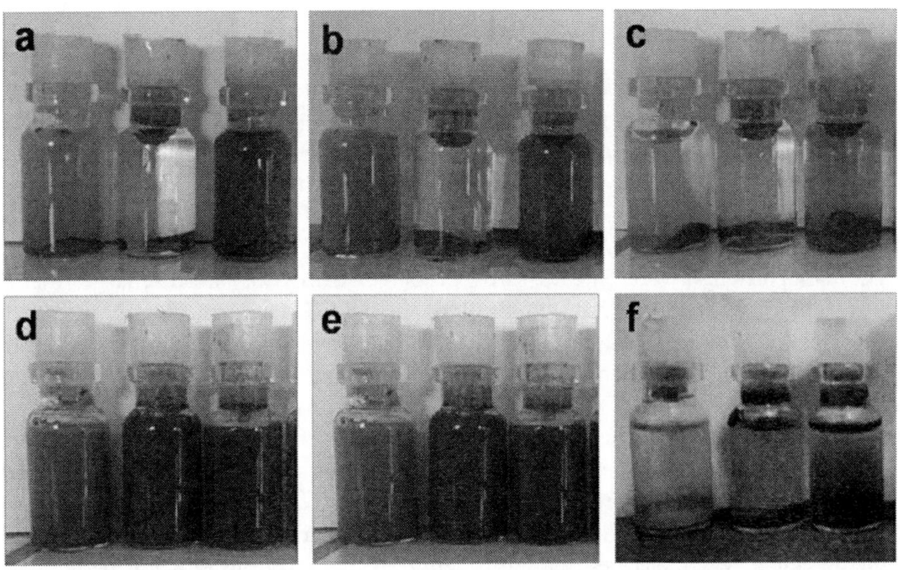


**Figure 4.** Dispersion of $TiO_2/C$ nanoparticles (left to the right) $TiO_2/C$ 0, $TiO_2/C$ 5, and $TiO_2/C$ 12, in (a) distilled water and (b) ethanol.

As presented in Figure 4, the $TiO_2/C$ 12 nanoparticles had better dispersion than the other samples in both distilled water and ethanol. It indicated that oxygen-containing functional groups, including –OH, C-O and C=O, successfully attached to the surface of the nanoparticles. An experiment was performed using an MB dye solution as a model substance to evaluate the photocatalytic activity of the samples ($TiO_2$ anatase, $TiO_2/C$ 0 and $TiO_2/C$ 12). The initial MB concentration was 3 ppm. The direct photolysis used a 125-W mercury lamp. The adsorption of MB was measured every 15 minutes by a Uv-vis spectrophotometer. Figure 5 shows the percentages of MB removed.

As shown in figure 5, photodegradation of MB was optimally reached when $TiO_2$-C 12 fabricated using submerged arc discharge in ethanol/acetic acid medium were applied as a photocatalyst. This best result was achieved due to the effects of both adsorption and photodegradation. The presence of carbon graphite layers and also $C_{60}$ (fullerene), carbon onions and other carbon allotropes with $sp^2$ hybridization carbon structures played a role in adsorbing the dye molecules from the solution. The adsorbed molecules on the surface of the graphite came in contact with the $TiO_2$, causing photodegradation. Therefore, both effects of adsorption and photodegradation increased the efficiency of the photodegradation.

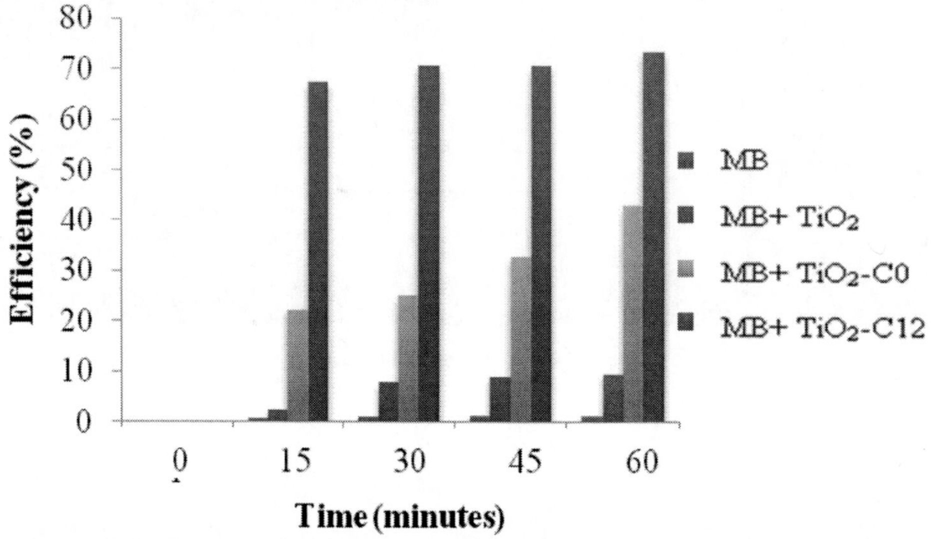


**Figure 5**. Percentage of MB removal (without photocatalyst), MB removal using photocatalyst of $TiO_2$ anatase powder ($MB+TiO_2$), MB removal using $MB+TiO_2/C$ 0 photocatalyst, and MB removal using $TiO_2/C$ 12 photocatalyst.

## 4. Conclusion

The $TiO_2/C$ nanoparticles were successfully prepared using an arc-discharge method in an ethanol/acetic acid liquid medium. XRD analysis of the $TiO_2/C$ nanoparticles showed that the main peaks of the original materials ($TiO_2$ and carbon) had high crystallinity, indicating that the materials were composites. The FTIR spectra of the $TiO_2/C$ showed the presence of oxygen-containing functional groups, including –OH, C=O and C-O. Therefore, this material had better dispersion. SEM studies showed that the nanoparticles product had spherical structures with clusters that varied in size from 140–180 nm. The better dispersion exhibited by $TiO_2/C$ 12 nanoparticles indicated that this material was more capable of increasing the efficiency of photodegradation. MB removal using $TiO_2/C$ 12 photocatalyst had a higher efficiency of photodegradation (73.60%) than MB removal using $TiO_2/C$ 0 which was fabricated by submerging arc discharge in ethanol medium without the addition of acetic acid (43.01%).

## 5. References

[1] Fu D G, Zhang Y, Wang X, Liu J-z and Lu Z-h 2002 *Chem. Lett.* **30**
[2] Sano N, Wang H, Alexandrou I, Chhowalla M, Teo K B K, Amaratunga G A J and Iimura K 2002 *J. Appl. Phys.* **92** 2783-8
[3] Morawski A W, Janus M, Tryba B, Inagaki M and Kałucki K 2006 *C. R. Chim.* **9** 800-5
[4] Sakthivel S and Kisch H 2003 *Angew. Chem. Int. Ed.* **42** 4908-11
[5] Zhao L, Chen X, Wang X, Zhang Y, Wei W, Sun Y, Antonietti M and Titirici M-M 2010 *Adv.Mater.* **22** 3317-21
[6] Kim C H, Kim B-H and Yang K S 2012 *Carbon* **50** 2472-81
[7] Peining Z, Nair A S, Shengjie P, Shengyuan Y and Ramakrishna S 2012 *ACS Appl. Mater. Interfaces* **4** 581-5
[8] Wojtoniszak M, Dolat D, Morawski A and Mijowska E 2012 *Nanoscale Res. Lett.* **7** 1-6

[9] Zhang H, Lv X, Li Y, Wang Y and Li J 2009 *ACS Nano* **4** 380-6
[10] Morawski A, Janus M, Tryba B, Toyoda M, Tsumura T and Inagaki M 2009 Carbon modified $TiO_2$ photocatalysts for water purification. In: *Pol. J. Chem. Technol.*, p 46
[11] Torimoto T, Ito S, Kuwabata S and Yoneyama H 1996 *Environmental Science & Technology* **30** 1275-81
[12] Saraswati T E, Andika I F, Anwar M, Purnawan C, Wahyuningsih S and Patiha 2015 *Advanced Research Materials* **1123** 285-8
[13] Boudou J P, Paredes J I, Cuesta A, Martínez-Alonso A and Tascón J M D 2003 *Carbon* **41** 41-56
[14] Pittman C U, Jiang W, He G R and Gardner S D 1998 *Carbon* **36** 25-37
[15] Pittman C U, He G R, Wu B and Gardner S D 1997 *Carbon* **35** 317-31
[16] Xie F, Phillips J, Silva I F, Palma M C and Menéndez J A 2000 *Carbon* **38** 691-700
[17] Hu C-C and Wang C-C 2004 *J. Power Sources* **125** 299-308
[18] Shen W, Li Z and Liu Y 2008 *Recent Patents on Chemical Engineering* **1** 27-40
[19] Swift G A and Koc R 1999 *Journal of Materials Science* **34** 3083-93

**Acknowledgments**
This work was supported in part by research grants provided by the Minister of Research, Technology and Higher Education, Republic of Indonesia, under project No. 458/UN27.21/LT/2016 and 632/UN27.21/LT/2016.

# Synthesis and Characterization of Microwave Absorber $SiO_2$ by Sol-Gel Methode

**S Wardiyati, W A Adi and Deswita**

Center for Science and Technolgy of Advance Materials,
National Nuclear Energy Agency. Kawasan Puspiptek Serpong 15314,
Tangerang Selatan

Email: siti-war@batan.go.id

**Abstract.** In recent years, information technology is growing rapidly, such as communication devices. However, there are still many shortcomings, for example, confidential information leaks caused by the leakage of electromagnetic waves used. A coating of electromagnetic materials or formation composite of electromagnetic material with other materials such as $SiO_2$ is needed to overcome these problems. For such needs, it is necessary to study the manufacture of $SiO_2$ which is simple, cheap, and effective. In this research, manufacture of $SiO_2$ by sol-gel method used a solution of sodium silicate ($Na_2SiO_3$) as precursors and $H_2SO_4$ as a catalyst. The parameters tested in this experiment is the effect of sintering temperature on the properties of the resulting $SiO_2$. The purpose of this study was to obtain an amorphous $SiO_2$ powder, which is in nano-sized and has a high surface area. The characterization of prepared samples were performed by using an X-ray Diffraction (XRD), Fourier Transmission Infra Red (FTIR), Scanning electron microscopy-energy dispersive spectrometer (SEM-EDS), Transmission Electron Microscopy (TEM), and Surface Area Analyzer (SAA). Based on the experimental results, the $SiO_2$ amorphous structure was obtained with a particle size of 15-20 nm, the surface area of 298 $m^2/g$, and sintering temperature of 100 °C.

**Keywords:** Microwave absorber, nanomaterial, $SiO_2$, sol-gel

## 1. Introduction

In the recent years, the development of information technology gives rise to rocketing growth in the development of microwave electronic systems and telecommunications in high frequency. However, many problems have appeared along with it. It could be originated from the misoperation of precise electronic equipment, and leaks of secret information occurred by leakage of electric waves [1]. To solve above problems, the electromagnetic wave absorbers with wider absorption band and more effective absorption properties constantly become crucial. Electromagnetic wave absorbing materials are being investigated as anti-electromagnetic interference coatings in self-concealing technology, and properties such as dielectric loss, conductivity loss, and magnetic loss. This material includes metal oxides, alloy-epoxy composite, and especially metallic magnetic material [2]. The magnetic composites have attracted intense attention for performance in the various application, such as magnetic resonance imaging, optoelectronic devices, and microwave absorbing [3]. One of promising of composite magnetic materials is silica ($SiO_2$) and iron (Fe). The use of silica because silica has several characteristics such as biocompatibility in a biological system, long suspension in various solvents, surface conductivity modification and improved chemical stability of magnetic composite [3]. The silica for their application, especially for microwave absorbing must be in nano size so that the permeability and permittivity of nano composing ferromagnetic metals (Fe) formed increasing and high electrical resistance ceramics ($SiO_2$) [4]. Therefore, an experiment study of making nanoscale $SiO_2$, which next will be composited with Fe to form the electromagnetic wave absorbers Fe-$SiO_2$ should be conducted.

Over the last decade, there has been rapid progress in research in the field of nanoscience and nanotechnology. Nanoparticles are defined as particles with a diameter of less than 100 nm. Nanomaterials have specific properties, such as small size, large surface area, form a special structure, because of the nature of the unique shape and morphology that the nanomaterial can be manipulated physically and chemically, and is widely used in industrial and biomedical processes [5]. With the development of nanotechnology may increase the economic value of a product.

The nano-sized material has a larger surface area, compared with the bulk material. These factors cause this material has a reaction rate, optical properties, magnetic and catalytic increased [6]. Silicon dioxide ($SiO_2$) nanoparticles can be used as a support material that is ideal for magnetic nanoparticles because it is very easy to experience functionalization; prevent anisotropic dipolar magnetic attraction when given an external magnetic field, and improve the corrosion resistance of the magnetic nanoparticles [7]. Amorphous and dielectric properties of $SiO_2$ are suitable to be used as microwave absorber electric for anti-radar, which in this year is being developed in PSTBM BATAN Indonesia.

Several methods can be employed for the manufacture of $SiO_2$, including precipitation [8-10], and sol-gel [11-13]. The sol-gel method is widely used because the process takes place at low temperature, the simple and the resulting product has a high purity and nanoscale. This method is one of the wet method because the process involves the solution as a medium. In the sol-gel method, the change in the phase of the solution becomes sol (colloid has suspended solids in the solution) and then it turns into a gel (colloidal but has a solid fraction larger than the sol). Basically, the synthesis of materials by the sol-gel method is through 4 (four) stages of the process, i.e. (1) hydrolysis, (2) condensation, (3) maturation (aging), and (4) drying [7].

In general, the synthesis of $SiO_2$ by sol-gel method is by using a precursor containing alkoxy ligands, such as TEOS (Triethyl orthosilicate) and TMOS (Tetraethoxysilane) [14]. On the use of precursors containing ligand alkoxy, in the hydrolysis reaction happen of replacement ligand alkoxy (-OR) with a hydroxyl group (- OH). Whereas, at this stage of the hydroxyl groups condensation will react with ligands alkoxy and the hydroxyl group to another, forming Si-O-Si as shown in Equations of reaction 1,2 and 3 [13].

$$\equiv Si-OR+H_2O \rightarrow \equiv Si-OH+ROH \qquad \text{Hydrolisis} \qquad (1)$$

$$\equiv Si-OH+\equiv Si+OR \rightarrow \equiv Si-O-Si\equiv +ROH \qquad (2)$$

$$\equiv Si-OH+\equiv Si-OH \rightarrow \equiv Si-O-Si\equiv +H_2O \qquad \text{Condensation} \qquad (3)$$

In the process of maturation, the gel network formation is more rigid and stronger; while in the drying process the evaporation of unwanted fluid occurs to obtain $SiO_2$ having a high surface area. Alkoxide precursor materials are relatively expensive and dangerous. This material can be replaced by other cheaper precursor and soluble in water such as $Na_2SiO_3$ (sodium silicate or water glass). In the process of hydrolysis, the sodium silicate reacts with water to produce silicic acid (see Equation of reactionof reactionof reaction 4) and then this acid polymerizes to form silica gel as shown in Equation of reactionof reaction of reaction 5 [7].

$$Na_2SiO_3 + H_2O + HCl \rightarrow Si\,(OH)_4 + 2NaCl \qquad (4)$$

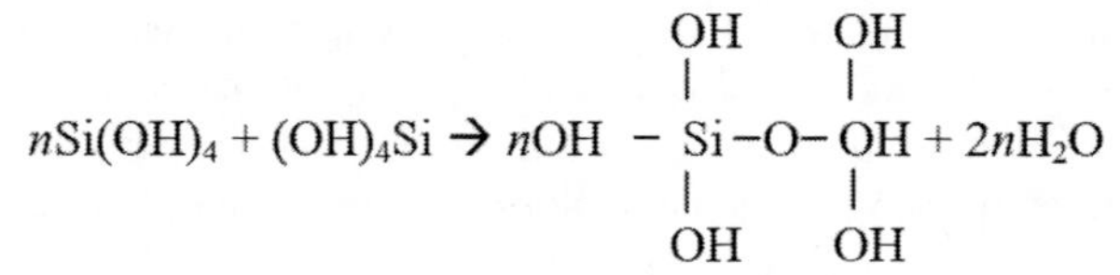


(5)

In this study, the synthesis of $SiO_2$ by sol-gel method using sodium silicate as a precursor and $H_2SO_4$ as a catalyst according to following reaction was conducted (Equation of reaction 6) [11].

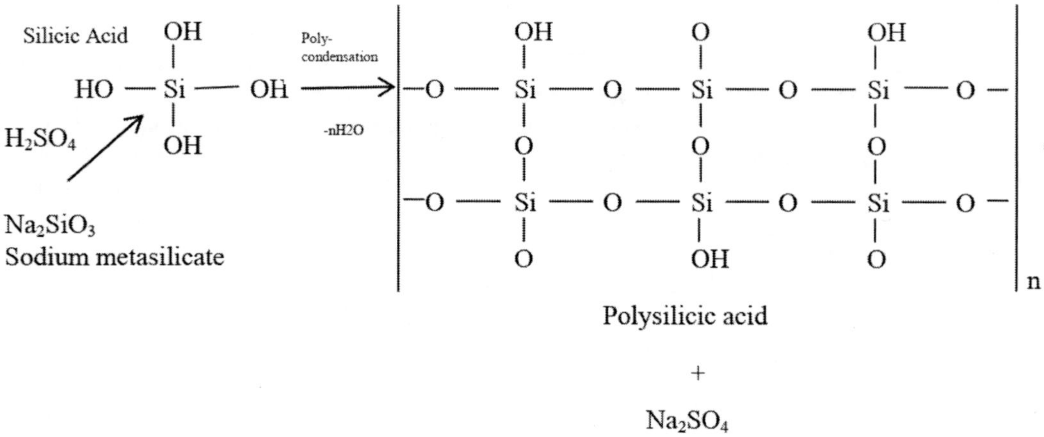


(6)

The purpose of this study is to obtain an amorphous $SiO_2$ powder which is in nano-sized, has a high surface area, and is ready to be 'composited with Fe to form Fe-$SiO_2$ microwave absorber. To achieve these objectives, in this experiment was carried out the variation of sintering temperature was carried out, which has not been done in the previous studies. The interest in sintering temperature variations is to determine the effect of heating to the physical structure of $SiO_2$ that occurs, such as surface area and particle size.

## 2. Experimental Method

### *2.1 Materials and equipment*

The chemicals used were sodium silicate ($Na_2SiO_3$ from Sigma-Aldrich), $H_2SO_4$ (Merck), NaCl (Merck), $BaCl_2$ (Sigma-Aldrich), demineralized water. All materials were used in the pure analysis grade.

For the characterization we used X-ray diffractometer (XRD) to determine the structure, FTIR (Fourier Transformation Infra Red) to determine the functional bonds, scanning electron microscopy - energy dispersive spectrometer (SEM-EDS) to analyze microstructural and determine the chemical composition of the microstructural features that are formed, transmission electron microscopy (TEM) to determine the morphology and particle size, and surface area analyzer (SAA) to determine of surface area of the particles.

### *2.2 Preparation of Silica oxide from Sodium silicate*

250 mL of sodium silicate solution ($Na_2SiO_3$) 25% was heated while stirring with a magnetic stirrer until the temperature of 80-90 °C, then a solution of 10% $H_2SO_4$ was added dropwise by using a peristaltic pump until a pH of 9.5 - 10. After that, the solution was cooled to room temperature and neutralized with $H_2SO_4$ 10% followed by stirring for 30 minutes [8]. If the reaction has been completed, a solution of NaCl was added as much as 1-5 g/L and stirred for 30 minutes to bind the sulfate ions. After that, the precipitates formed were washed with water until free from $SO_4^{-2}$ ion. $SO_4$

ion test was done by adding $BaCl_2$ into the washing water. When white precipitate ($BaSO_4$) raised, there was still $SO_4^{-2}$ ion. Furthermore, $SiO_2$ formed was dried in an oven with different heating temperature of 100, 600, and 700 °C for 6 hours, and then the $SiO_2$ product was ready to be characterized.

## 3. Results and Discussion

### *3.1 X-ray diffraction*
X-Ray diffractometer (XRD) analysis was carried out in the $2\theta$ range of 10°-70° by using PANanalytical diffractometer employing CuKα radiation (0.154060 nm) source operated at 40 kV and 40 mA. The result of X-Ray diffraction pattern of $SiO_2$ is shown in Figure 1.

Figure 1 shows the XRD pattern of $SiO_2$ synthesized by sol-gel method at 100 °C sintering temperature. This pattern is the same with the XRD pattern of $SiO_2$ produced by Asmaa Mourhly, et al. where the hump was widened and a strong peak at Bragg angle 22°-23°. This result indicated that $SiO_2$ product is an amorphous phase. The diffraction pattern in Figure 1 showed only one peak which possibly indicated that $SiO_2$ product has a high degree of purity.

### *3.2 Fourier transform infrared spectroscopy*
Observation of Fourier transforms infrared spectroscopy (FTIR) study was carried out by using Brucker, Germany with KBr as a reference in the wave number range of 500 – 4000 $cm^{-1}$. This observation needs to be done for determining the type bonding of compound formed. Spectrogram of $SiO_2$ observed by FTIR is shown in Figure 2.

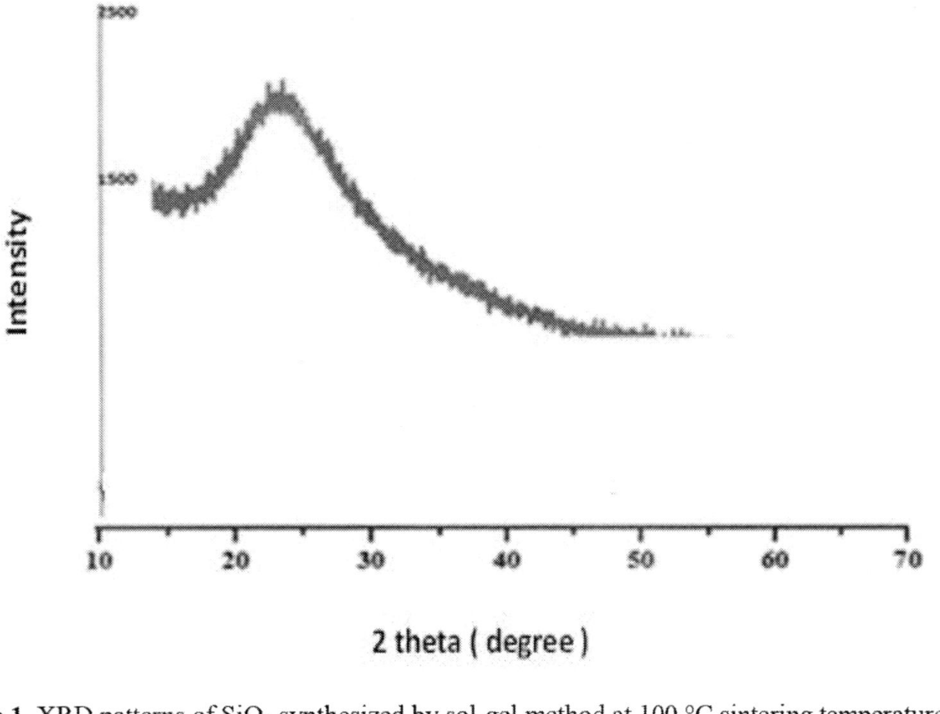


**Figure 1.** XRD patterns of $SiO_2$ synthesized by sol-gel method at 100 °C sintering temperature.

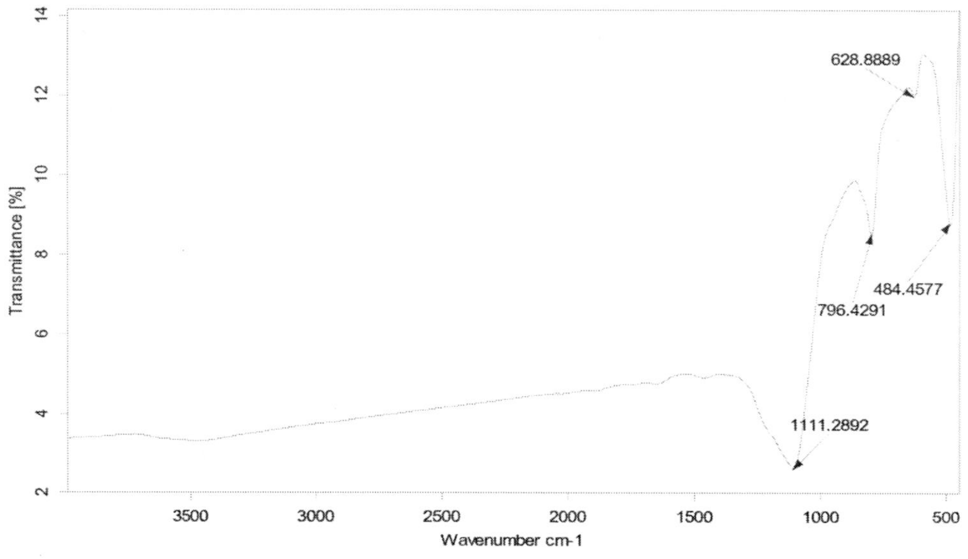


**Figure 2.** Infra Red Spectrum of $SiO_2$ by sol-gel method by using $Na_2SiO_3$ precursor.

**Figure 3.** SEM image of $SiO_2$ synthesized by sol-gel method.

In Figure 2, the three IR spectra were found, i.e. spectrum at a wavenumber of 484.4577 $cm^{-1}$ which is bending vibrations of O-Si-O; IR spectrum at 800 $cm^{-1}$ which is symmetric stretching vibration of Si-O-Si and IR spectrum 1111 which is the asymmetric stretching vibration of Si-O-Si. If these spectrums are compared to experiment of Music *et al.* [8] on the synthesis of $SiO_2$ by by using precipitation method, there are some differences. The $SiO_2$ results of the experiment conducted by Music showed the presence of stretching vibration of $H_2O$ molecules in the IR band ± 3485 $cm^{-1}$ and vibrational bending $H_2O$ molecules in the IR band ± 1632 $cm^{-1}$. Meanwhile, in this experiment, they were not found. This phenomenon shows that the sol-gel method is better to be used for synthesizing $SiO_2$ because the water content is lower, so it is more stable.

IOP Conf. Series: Materials Science and Engineering **202** (2017) 012059 doi:10.1088/1757-899X/202/1/012059

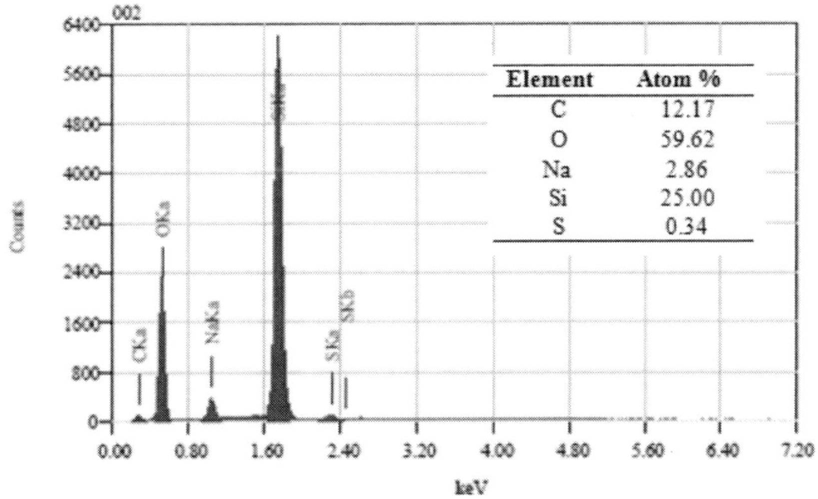


**Figure 4.** EDS spectrum and chemical composition ( % atom) of $SiO_2$ synthesized by sol-gel method.

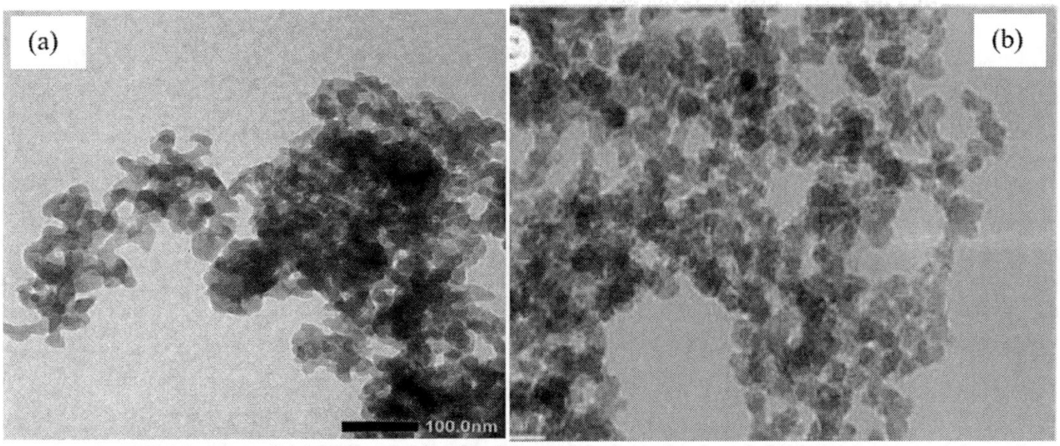


**Figure 5.** TEM image of (a) $SiO_2$ synthesized by sol-gel method and (b) $SO_2$ commercial product

*3.3 Scanning electron microscopy - energy dispersive spectrometer (SEM-EDS)*

Scanning electron microscopy - energy dispersive spectrometer (SEM-EDS) used in this experiment was JEOL from Japan. The SEM-EDS was used for observation of microstructural and analysis of chemical composition of ions formed. Figure 3 shows the microstructural of $SiO_2$ observed by SEM and analysis result of the chemical composition of $SiO_2$ by using EDS presented in Figure 4. Amorphous $SiO_2$ from irregular shape and aggregate of $SiO_2$ particle is detected in Figure 3.

Figure 4 shows the presence of Si atoms with relatively high counts ± 6200 at 1.6 keV and O ions at an energy value of 0.525 keV with counts of ± 2850. It shows that Si and O atoms exist. Besides Si (25%) and O (59.62%) atoms, the presence of Na, S and C atoms are also shown although they are very small. Na atom was derived from reactants of $Na_2SiO_3$ with a value of 2.86%; S atom came from $H_2SO_4$ used to neutralize the silicate of 0.34%, and the C atom came from the tape used for preparation sample on measurement by SEM-EDS with a value of 12.17%. The existence of Na and S ions on $SiO_2$ was possible because the washing process was less than perfect.

IOP Conf. Series: Materials Science and Engineering **202** (2017) 012059 doi:10.1088/1757-899X/202/1/012059

*3.4 Transmission electron microscope (TEM)*

The electron microscopy (TEM) instrument used in this experiment was JEOL from Japan. TEM was used to analyze the microstructural, and determine the size particle of $SiO_2$ formed. The observation results of microstructural and size particle of $SiO_2$ by TEM are presented in Figure 5.

Figure 5(a) shows the image of $SiO_2$ structure synthesized by sol-gel method with precursor of $Na_2SiO_3$, $H_2SO_4$ catalyst, and sintering temperature of 100 °C. Meanwhile, Figure 5(b) is TEM image of commercial $SiO_2$ from Nano Nanjing High Technology Co. Ltd. product. Both Figures have the similarities, namely the shape and size of the particles, the shape is amorphous, and particle size is about 20 nm. These result proved that the $SiO_2$ product of this experiment was good because the products have the same shape and size with the commercial $SiO_2$.

*3.5 Surface Area Analyzer (SAA)*

Determination of surface area of $SO_2$ product was carried out by using Surface area analyzer (SAA). From the observation, the surface area of $SiO_2$ was 298.3 $m^2/g$. These results were lower than the surface area of $SiO_2$ generated from Pumice Rock by Asmaa Mourhly et al. which reached 422 $m^2/g$ [15]. This difference is possibly due to the differences in the raw materials used. Pumice Rock was used as raw materials in the formation of $SiO_2$ having high porosity values so that the resulted $SiO_2$ was higher as well.

*3.6 The Effect of sintering temperature on the characteristic of $SiO_2$ synthesized*

The sintering temperature is one of the factors that affects the properties of the resulting $SiO_2$ characteristics, especially the surface area and particle size. The Changes of surface area can be observed by using SAA. Table 1 shows the surface area value of $SiO_2$ product observed by using SAA with the variations of sintering temperature.

Table 1. shows that the surface area of the resulting $SiO_2$ was getting smaller with the increase of the sintering temperature. This phenomenon is similar to the statement Wenbo Liu et.al. [16] stating that the higher the sintering temperature, the possibility of joining crystal lattice with another crystal is getting bigger so that the crystal size becomes large and the surface area decreases. Besides surface area, sintering temperature differences also affect against the diffraction pattern of obtained $SiO_2$. Effect of sintering temperature $SiO_2$ at temperatures of 100, 600, and 700 °C are shown in Figure 6.

**Table 1.** Surface area of SiO2 on different sintering temperatures

| Sintering temperature [°C] | Surface area of $SiO_2$ [$m^2/g$] |
|---|---|
| 100 | 298.31 |
| 600 | 236.99 |
| 700 | 85.60 |

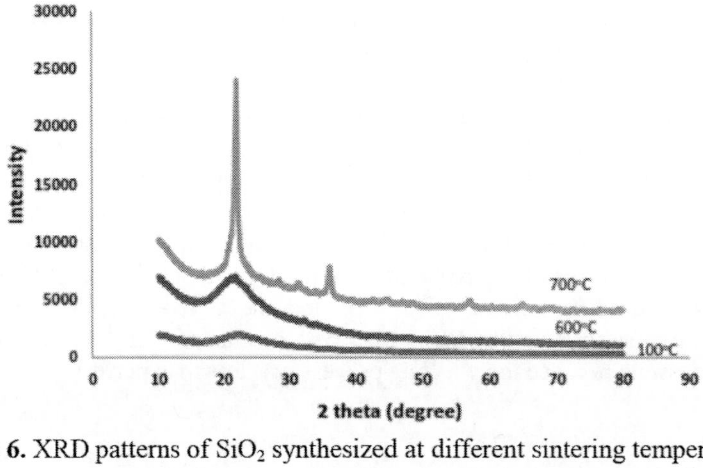


**Figure 6.** XRD patterns of $SiO_2$ synthesized at different sintering temperatures.

**Table 2.** Size crystalline of $SiO_2$ on different sintering temperatures.

| Compound | Temperature of calcination (°C) | *B* observed (FWHM) | *B* standard (FWHM) | Crystallite Size (nm) |
|---|---|---|---|---|
| $SiO_2$ | 100 | 4.9108 | 0.0923 | 16.19 |
| $SiO_2$ | 600 | 4.0216 | 0.0923 | 20.23 |
| $SiO_2$ | 700 | 0.5777 | 0.0923 | 167.16 |

Figure 6 shows that the sintering temperature affects the shape of the $SiO_2$ peaks. The shape peak of $SiO_2$ was initially widened. The peak became more cone as the increase of the sintering temperature. The change in peak form this will affect the size of $SiO_2$ particles produced, where the peak is getting cone, size particle increasing, this can be proved by using Debey-Scherrer Equation 7 [17].

$$D = \frac{0.89\lambda}{B\cos\theta} \tag{7}$$

where D is the size of the crystals, a factor of 0.89 is characteristic of round object, $\lambda$ is the wavelength of X-ray, *B* is the FWHM (full width at half maximum), and θ is the diffraction angles. Equation of reactionof reaction1 shows the relationship between the crystal diameter (*D*) with FWHM, where *D* is inversely proportional to the FWHM. The smaller the FWHM, the diameter (*D*) particles formed increase. Table 2 shows the size crystalline of $SiO_2$ on the different sintering temperatures.

Based on the results of $SiO_2$ characterization by using SAA and XRD, we can see that at a high-temperature sintering, the particle size became large and another phase appeared. However, from Tables 1 and 2, we can see that the temperature sintered up to ≤ 600 °C, the particle size was still in the nanoscale, i.e., about 20 nm and a surface area of over 237 $m^2/g$, this phenomenon indicated that $SiO_2$ produced was quite stable.

## 4. Conclusion

The potential of microwave absorber of $SiO_2$ amorphous materials with nanoscale was successfully synthesized by Sol-gel method using a precursor of $Na_2SiO_3$ and catalyst of $H_2SO_4$ at a sintering temperature of ≤ 600 °C. It was obtained that the size of the particle was around 20 nm. Meanwhile, the surface area was 298.31 $m^2/g$. Spectroscopic studies showed that functional bond of O-Si-O exists at wavenumbers of 484.4577 $cm^{-1}$, Si-O-Si at 800 $cm^{-1}$, and 1111 $cm^{-1}$ which indicated the existing of $SiO_2$. The vibrations bending and stretching of $H_2O$ from observation by FTIR were not visible. It means that the synthesis of $SiO_2$ by using sol-gel method at a sintering temperature ≤ 600 °C produced $SiO_2$ with low water contents, so it was more stable as compared with the previous research.

## 5. References

[1] Chen L *et al.* 2010 Influence of $SiO_2$ fillers on microwave absorption of carbonyl iron/carbon black double-layer coatings *Material & Design* **08** 021

[2] Jin F *et al.* 2006 Structure and microwave absorbing properties of Fe-particle containing alumina prepared by micro-arc discharge oxidation *Surface & Coating Technology* **201** 292-295

[3] Yang Y 2013 Synthesis and microwave absorbing properties of Fe/$SiO_2$ particles with core/shell structure

[4] Fujieda F 2002 Microwave absorption properties of Fe-$SiO_2$ nanocomposite powder *Journal of the Japan Institute of Metals* **66**(3)135-138

[5] Fibroblasts L, Qiang Z X, Hong Y L, Meng T and Pu P Y 2011 ZnO, $TiO_2$, $SiO_2$, and $Al_2O_3$ nanoparticles-induced toxic effects on human fetal *Biomed Environ Sci* **24**(6) 661-669

[6] Xi-gui W, Xia Q I, Su-ling B O, Mi-la N A 2011 Preparation and luminescence properties of nanomaterials $TiO_2$-$SiO_2$ doped with $eu^{3+}$, spectroscopy and spectral analysis **31**(5) 1193-96

[7] Yamahata C et al. 2012 Influence of the precursor concentration on the characteristics of silica powder obtained from $Na_2SiO_3$ by a facile low-temperature synthesis process *Journal of Materials Science and Engineering B* **2**(8) 429-436

[8] Musić S, Filipović-Vincेković N and Sekovanić L 2011 Precipitation of amorphous $SiO_2$ particles and their properties *Brazilian Journal of Chemical Engineering* **28**(01) 89 - 94

[9] Mohanraj K et al. 2012 Preparation and characterization of nano $SiO_2$ from corn cob ash by precipitation method *Optoelectronics and Advanced Materials-Rapid Communications* **6**(3-4) 394-397

[10] Cui J, Sun H, Luo Z, Sun J and Wen Z 2015 Preparation of low surface area $SiO_2$ microsphere from wheat husk ash with a facile precipitation process *Materials Today Proceedings* **2**(4–5) 3575–3579

[11] Essien E R, Olaniyi O A, Adams L A and Shaibu R O 2012 Sol-gel-derived porous silica: economic synthesis and characterization *Journal of Minerals and Materials Characterization and Engineeringc* **11** 976-981

[12] Dubey R S, Rajesh Y B R D and More M A 2015 Synthesis and characterization of $SiO_2$ nanoparticles via sol-gel method for industrial applications *Materials Today Proceedings* **2** 3575-79

[13] Rahman I A and Padavettan V 2012 Synthesis of silica nanoparticles by sol-gel: size-dependent properties, surface modification, and application in silica-polymer nanocomposites-a review *Journal of Nanomaterial* **2012** article ID 132424

[14] Zhang H, Hong C and Qiao Y 2011 Synthesis, structural and thermal properties of nano-porous $SiO_2$-based aerogel, advances in nanocomposites-synthesis, characterization and industrial aplications, Dr. Boreddy Reddy (ed), ISBN: 978-953-307-165-7

[15] Mourhly A, Khachani M, El Hamidi A, Kacimi M, Halim M and Arsalane S 2015 The synthesis and characterization of low-cost mesoporous silica $SiO_2$ from Local Pumice Rock *Nanotechnology and Nanomaterial* ISSN 1847-9804

[16] Liu W, JiY, Tan P, Zang H, He C, Yun D, Zhang C and Yang Z 2016 Irradiation induced microstructure evolution *Nanostructured Materials* **9** 105

[17] Monshi A, Foroughi M R and Monsh M R 2012 Modified Scherrer Equation of reactionof reactionto estimate more accurately nano-crystallite size using XRD *World Journal of Nano Science and Engineering* **2** 154-160

**Acknowledgements**

The authors would like to thank all those who have helped the experiment. This research was funded from DIPA funds in 2014 and 2015 with the title of The Optimization of waste treatment with photocatalytic method and advanced smart magnetic materials and magnetic oxide, Center for Science and Technolgy of Advance Materials, National Nuclear Energy Agency.

IOP Conf. Series: Materials Science and Engineering **202** (2017) 012060 doi:10.1088/1757-899X/202/1/012060

# Natural Silica Sand/Alumina Ceramic Composites: Promising Candidates for Fuel-Cell Sealants

**N Hidayat[1], Istiqomah[2], M Y H Widianto[3], A Taufiq[1,4], Sunaryono[1,4], Triwikantoro[3], M Zainuri[3], M A Baqiya[3], G Aristia[5] and S Pratapa[3]**

[1]Department of Physics, Faculty of Mathematics and Natural Sciences, Universitas Negeri Malang (State University of Malang), Jl. Semarang 5, Malang 65145, Indonesia
[2]National Standardization Agency of Indonesia, Jl. M. H. Thamrin No 8, Kebon Sirih, Jakarta Pusat 10340, Indonesia
[3]Department of Physics, Faculty of Mathematics and Natural Sciences, Institute of Technology Sepuluh Nopember, Jl. Arif Rachman Hakim, Surabaya 60111, Indonesia
[4]Center for Minerals and Advanced Materials, Faculty of Mathematics and Natural Sciences, Universitas Negeri Malang (State University of Malang), Jl. Semarang 5, Malang 65145, Indonesia
[5]Department of Physical and Theoretical Chemistry, Institute of Chemistry and Biochemistry, Freie Universität Berlin, Kaiserswerther Str. 16-18, 14195, Germany

Email: nurul.hidayat.fmipa@um.ac.id

**Abstract.** An attempt has been developed to establish the prospect of the useful application of Indonesian natural silica sand, instead of commercially expensive materials, as a future fuel-cell sealant. The sand was initially washed and ball-milled at 150 rpm for 60 minutes and then heated at 1000 °C for the same duration. The resulting powder was then mixed with alumina powder at various amounts and shaped into discs before sintering at 1150 °C and 1250 °C to produce compact ceramics. The diameter shrinkage, porosity, and density of the ceramics were evaluated by Archimedes method. Their crystalline phase composition was quantified by Rietveld refinement analysis on the X-ray diffraction (XRD) data and the phase weight fraction was then used for coefficient of thermal expansion (CTE) evaluation. It was observed that the bulk density increased while the porosity decreased with alumina addition. The XRD data analysis revealed that the prepared silica sand contains a very high purity of quartz-$SiO_2$, i.e. 97.8(18)%. The sintering temperatures of 1150 °C and 1250 °C transformed some quartz-$SiO_2$ to crystobalite-$SiO_2$. All the calcite-$CaCO_3$ exhibited reaction sintering with $SiO_2$ forming wollastonite-$CaSiO_3$. Therefore, the ceramic composites contained $SiO_2/Al_2O_3/CaSiO_3$. Regarding CTE, all of the composites meet the criteria for fuel-cell sealants, in the range of 9-12 ppm/°C.

**Keywords:** Indonesian silica sand, alumina, phase quantification, coefficient of thermal expansion, fuel-cell sealant.

## 1. Introduction

In light of finding future clean energy technologies, solid oxide fuel cells (SOFCs) are recognized as one of the most environmentally friendly and feasible energy conversion devices [1,2]. It is due to their very long life expectancy [3], high efficiency [4], and fuel flexibility properties [5]. The SOFCs, devices that convert the chemical energy of fuels into electricity, have been extensively studied to reduce $CO_2$ emission and fossil fuel consumption [6]. In spite of this, it is confronted with a challengeable task on preventing leakage when an SOFC operates at high temperature (800-1000 °C) [7]. Consequently, a proper gas-tight sealing material plays a decisive role to obtain the best operation and performance of SOFCs [8]. The very basic criteria for fuel cell sealing material are the CTE, which must be in the range of 9-12 ppm/°C [9], and thermo-mechanically and chemically stable [10]. The matching of CTE of the fuel-cell sealing materials to other cell components is essential to minimize the thermal stresses [11].

Recently, ceramic-based composites and glass-ceramics have become highly recommended sealing materials for various SOFCs attributable to their superior characteristics compared with those of metallic sealants, mainly to resist both reducing and oxidizing environments [12,13]. Additionally, they also perform excellent gas tightness and provide CTE similar to those of the other stack components [14]. Numerous investigations have been conducted in search of great fuel cell sealants, e.g. $30CaO–40SiO_2–20B_2O_3–10A_2O_3$ (A = Y, La) [7], $Al_2O_3$-glass composite [13], and $(1-x)[SrO–SiO_2–B_2O_3]–xAl_2O_3$ [15]. $Al_2O_3/SiO_2$-based sealants show an excellent performance in maximizing the SOFC operation [16]. However, rarely has a research exclusively explored the use of natural silica-based materials as the $SiO_2$ source for fuel cell sealants production.

In this present study, we introduce the prospect of using Indonesian silica sand as the fuel cell sealing material. This breakthrough is crucial not only to minimize the fabrication cost of alumina/silica-based fuel cell sealants but also to elevate the added-value of the abundant natural silica sand resources. As one of the most abundant naturally-occurring materials, $SiO_2$ can be predominantly found in silica sand [17]. Indonesia becomes a country that serves a significant amount of natural silica sand. It can be revealed from the facts that (1) Indonesia is the largest archipelagic nation with roughly 67% sea and (2) it is one of the world's top five countries with the longest coastlines imply many advantages regarding its mineral resources [18]. Also, our previous preliminary studies [19,20] reported that Indonesian silica sand from coastal district of Tuban comprised of very high quartz-$SiO_2$ content. The results lead to a must-solved research question of establishing the potential of natural silica sand/alumina ceramic composites for fuel-cell sealants that is described in this paper.

## 2. Experimental Methods

High purity quartz-$SiO_2$ was extracted from natural silica sand from Indonesian Beach, particularly located in Tuban area. The selected sands were simply washed by using water and dried overnight. The dried silica sands were then mechanically sieved to obtain fine powders and magnetically extracted to eliminate the magnetic constituents. The typical silica sands were ball-milled at 150 rpm for 60 minutes followed by sintering at 1000 °C for other 60 minutes. The elemental contents of the prepared $SiO_2$ were reported elsewhere [19,20]. Alumina (PA, Sigma-Aldrich) with various concentrations were ground with the prepared silica before 4-hour sintering at 1150 °C and 1250 °C.

The prepared powder was then uniaxially pressed with a force of 600 N to produce a cylindrical-shaped sample with a diameter of 1.3(1) cm for density-porosity measurement. X-ray diffraction (XRD) using Cu-K$\alpha$ radiation was applied to evaluate the phase profiling in the sample. Density-porosity was measured by means of Archimedes method following the Australian standard of 1774.5. Rietveld refinement approach was executed to analyze the relative weight fractions of each phase [21] via Rietica software [22] as the crystalline phase identification was completely acomplished by means of X'Pert High Score Plus software [23]. The relative weight fractions were then converted to volume fraction for CTE composite evaluations. Additionally, the nomenclatures for sample identifications were given by using the following naming convention. "NSS" refers to natural silica sand, "A" refers to alumina, and "1150" (or "1250") relates to the sintering temperature. Therefore, "NSS1A" means a composite of natural silica sand with the addition of 10% alumina. "NSS3A1250" means a composite of natural silica sand with 30% of alumina addition and 1250 °C sintering temperature.

## 3. Results and Discussion

The physical data for all composites in terms of diameter shrinkage, porosity, and bulk density (*note*: a terminology of "density" in this paper always refers to "bulk density" unless there is another detail) are depicted in Table 1. Based on Table 1, it can be seen that the highest diameter shrinkage is -0.19(1) mm for NSS1250. Such shrinkage is believed due to the external pressure and sintering treatment [24]. Other factors that affect shrinkage are inhomogeneous particle shape [25] and porous shape [26].

In general, an addition of $Al_2O_3$ increases the composite's density and reduces its porosity. At a sintering temperature of 1250 °C, the porosity of NSS1250 decreases 1.22% from the porosity of NSS1150. The porosities of NSS1A1250, NSS2A1250, and NSS2A1250 are, respectively, 29.61(4)%, 29.07(4)%, and 28.98(4)%. The highest density was obtained when the sample has 30% $Al_2O_3$ content for both sintering temperatures. The addition of 30% $Al_2O_3$ to the $SiO_2$ increases the density of as much as 0.19 g/cm$^3$ and 0.31 g/cm$^3$, respectively at 1150 °C and 1250 °C sintering temperatures. Additionally, those increments of the composites density result from the presence of crystalline phases formed in the samples.

Phase identification and quantification are necessary to evaluate the coefficient of thermal expansion (CTE) of the composites. The results are also essential to justify chemical reactions between silica and alumina. The XRD profiles of the composites after sintering are depicted in Figure 1. The phase compositions of natural silica sand/alumina composites are given in Table 3. The phase identification verifies that all the XRD profiles contain quartz-$SiO_2$ (PDF reference code: 00-046-1045), corundum-$Al_2O_3$ (01-075-1862), wollastonite-$CaSiO_3$ (01-075-1396), and cristobalite-$SiO_2$ (01-077-1317). An example of the phase identification analysis result (sample NSS1A1150) is listed in Table 2. Full pattern Rietveld refinement gives more precise phase quantification data for the representative samples as depicted in Table 3.

Table 3 tells that calcite does not further exist. Wollastonite occurred as a consequence of reaction between $SiO_2$ and CaO during the sintering process. CaO could come from $CaCO_3$ decomposition within the natural silica sand at a temperature of, at least, 895 °C [27]. $CaCO_3$ was detected in the as prepared silica sand. The processed silica sand, based on Rietveld refinement on XRD data, contained quartz-$SiO_2$ and calcite-$CaCO_3$ with weight fractions of 97.8(18)% and 2.2(2)%, respectively (the XRD data is not shown). The contribution of calcium oxide in wollastonite formation is not merely influenced by CaO, or $CaCO_3$ decomposition. It also comes from the alumina itself (from XRF data of alumina, it contains 1.1(1)% calcium), though the contribution is subtle. The wollastonite formation is given in Equation 1 [28,29]:

$$CaO_{(s)} + SiO_{2(s)} \rightarrow CaSiO_{3(s)} \quad (1)$$

**Table 1.** Physical properties of natural silica sand/alumina ceramic composites (the values in parentheses are the estimated standard deviation to the left).

| Sample | $\Delta d$ (mm) | $P$ (%) | $\rho$ (g/cm$^3$) |
|---|---|---|---|
| NSS1150 | -0.09(1) | 36.37(9) | 2.65(1) |
| NSS1A1150 | -0.09(1) | 30.21(4) | 2.64(1) |
| NSS2A1150 | -0.07(1) | 29.94(4) | 2.71(1) |
| NSS3A1150 | -0.07(1) | 29.80(4) | 2.84(1) |
| NSS1250 | -0.19(1) | 35.15(9) | 2.46(1) |
| NSS1A1250 | -0.05(1) | 29.61(4) | 2.57(1) |
| NSS2A1250 | -0.03(1) | 29.07(4) | 2.62(1) |
| NSS3A1250 | -0.02(1) | 28.98(4) | 2.77(1) |

Note: $\Delta d$ = diameter shrinkage, $P$ = porosity, and $\rho$ = density.

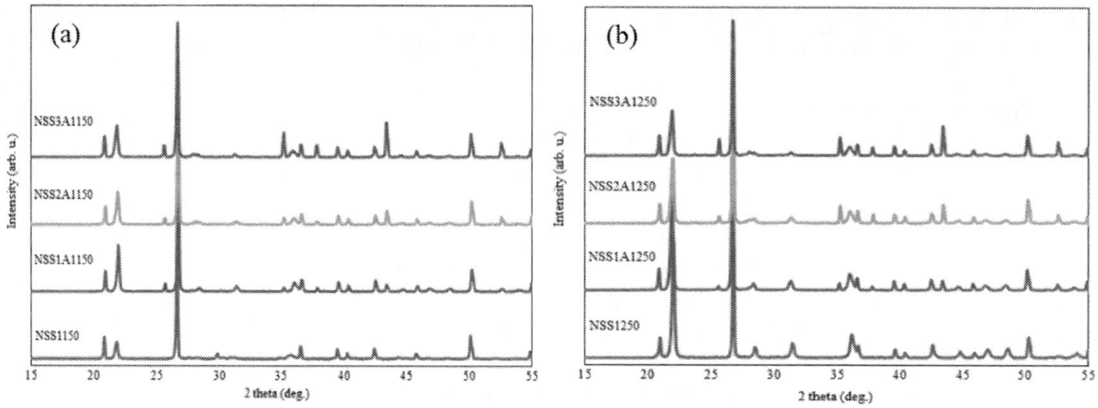


**Figure 1.** XRD profiles (Cu-Kα radiation) of the processed silica sand and the composites; (a) after sintered at 1150 °C and (b) after sintered at 1250 °C.

**Table 2.** Phase identification of NSS1A1150 by means of X'Pert High Score Plus.

| Position (2$\theta$ deg.) | Relative Intensity (%) | Identified Crystalline Phase |
|---|---|---|
| 20.96 | 15.10 | Quartz-$SiO_2$ |
| 21.99 | 32.68 | Cristoblite-$SiO_2$ |
| 25.71 | 6.25 | Corundum-$Al_2O_3$ |
| 26.75 | 100.00 | Quartz-$SiO_2$ |
| 28.43 | 2.46 | Cristoblite-$SiO_2$ |
| 31.40 | 3.36 | Wollastonite-$CaSiO_3$ |
| 35.24 | 2.94 | Corundum-$Al_2O_3$ |
| 36.06 | 6.14 | Cristoblite-$SiO_2$ |
| 36.66 | 7.92 | Quartz-$SiO_2$ |
| 37.88 | 2.44 | Quartz-$SiO_2$ |
| 39.58 | 6.80 | Quartz-$SiO_2$ |
| 40.40 | 3.93 | Quartz-$SiO_2$ |
| 42.53 | 7.99 | Quartz-$SiO_2$ |
| 43.41 | 4.83 | Corundum-$Al_2O_3$ |
| 44.73 | 1.39 | Cristoblite-$SiO_2$ |
| 45.88 | 4.03 | Quartz-$SiO_2$ |
| 46.88 | 1.99 | Cristoblite-$SiO_2$ |
| 48.60 | 1.61 | Cristoblite-$SiO_2$ |
| 50.22 | 17.52 | Quartz-$SiO_2$ |
| 52.58 | 2.16 | Corundum-$Al_2O_3$ |
| 54.02 | 1.11 | Cristoblite-$SiO_2$ |

Meanwhile, the cristobalite within the samples is caused by phase transformation of the quartz-$SiO_2$ due to thermal treatment since quartz transformed to be cristobalite at a temperature above 1000 °C [30]. Such fact is in line with the result found in this experiment; cristoballite occurred at the temperature of 1150 °C. The addition of $Al_2O_3$ decreases the concentration of $SiO_2$. Therefore, the reduction of cristobalite as the addition of $Al_2O_3$ undoubtedly happens.

**Table 3.** Rietveld relative weight fraction of crystalline phases in natural silica sand/alumina ceramic composites (the values in parentheses are the estimated standard deviation).

| Sample | Relative Weight Fraction (%) | | | |
|---|---|---|---|---|
| | Quartz ($SiO_2$) | Corundum ($Al_2O_3$) | Wollastonite ($CaSiO_3$) | Cristobalite ($SiO_2$) |
| NSS1150 | 61.3(19) | - | 7.0(7) | 31.7(10) |
| NSS1A1150 | 55.4(21) | 10.4(28) | 1.8(1) | 33.2(27) |
| NSS2A1150 | 49.7(13) | 19.1(15) | 1.4(3) | 29.8(10) |
| NSS3A1150 | 41.4(20) | 30.2(17) | 1.3(1) | 27.1(38) |
| NSS1250 | 26.5(19) | - | 6.1(8) | 67.4(22) |
| NSS1A1250 | 40.3(16) | 10.2(8) | 1.0(2) | 48.6(31) |
| NSS2A1250 | 37.8(10) | 20.4(10) | 0.9(2) | 40.8(13) |
| NSS3A1250 | 37.6(18) | 30.6(34) | 1.0(2) | 30.8(19) |

Once $SiO_2$-$Al_2O_3$ phase diagram is taken into consideration to explain the phenomena observed in this experimental results, a question should arise as follows "why does not mullite, $3Al_2O_3{\cdot}2SiO_2$ exist?". According to the $SiO_2$-$Al_2O_3$ phase diagram [31], the solidus of mullite shifts as the alumina increases above the eutectic point. Nucleation and growth of mullite start to occur above the sintering temperature of 1587 °C [31] since quartz completely transforms to cristobalite at 1570 °C [30]. Hence, it makes sense that mullite was not observed by XRD yet.

An understanding of the composite performance, based on the constituent materials, is important not only for practical purposes like predicting the composite properties, e.g. CTE but also as fundamental knowledge to develop new materials [32]. After converting the weight fractions in Table 3 to volume fractions, CTE values prediction can be made by applying Halpin-Tsai equation [33,34] as given in Equation 2. The CTE characteristics of the samples are shown in Figure 3.

$$\alpha_c = \alpha_m \frac{1+2sqV_f}{1-qV_f} \quad (2)$$

Where $q=(\alpha_r-1)/(\alpha_r+2s)$ and $\alpha_r=\alpha_f/\alpha_m$, $\alpha$ represents CTE, $V$ represents volume, and $s$ represents the dimensional ratio between reinforcement and matrix; for isotropic composite, $s$ = 1. Subscripts $c$, $f$, and $m$, respectively refer to composite, filler, and matrix.

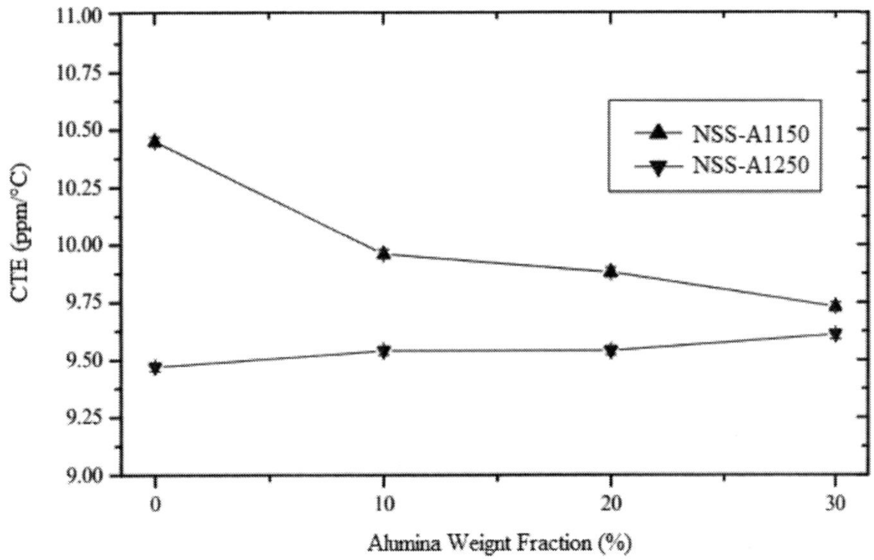


**Figure 2.** CTE prediction values for $SiO_2/Al_2O_3/CaSiO_3$ after XRD phase compositional Rietveld analysis.

As noticeably pictured in Figure 2, an addition of $Al_2O_3$ for the sintering temperature of 1150 °C reduces the CTE of the composites. It is because of the smaller CTE of corundum compared with quartz as the predominant phase in the composites. This finding is in good agreement with another literature [35]. Moreover, at the same time, an addition of $Al_2O_3$ for the sintering temperature of 1250 °C conversely elevates the CTE of the composites. It is because of the increasing amount of cristobalite (see Table 3) and volume fraction of corundum which is no longer able to decrease the CTE of the composites. On top of that, all $SiO_2/Al_2O_3/CaSiO_3$ composites show excellent CTE matching to those of fuel-cell components, with CTE values in the range of 9-12 ppm/°C.

## 4. Conclusion

This present study concludes that the natural silica sand from Tuban area is a promising candidate as a fuel cell sealing material with the incorporation of alumina. The 4-hour sintering at temperatures of 1150 °C and 1250 °C has transformed some quartz-$SiO_2$ to crystobalite-$SiO_2$. Calcite-$CaCO_3$ within the samples completely reacted with $SiO_2$ forming wollastonite-$CaSiO_3$. Thus, the ceramic composites encompassed $SiO_2/Al_2O_3/CaSiO_3$. Finally, regarding CTE, all of the composites meet the criteria for fuel cell sealants.

## 5. References

[1] Sabato A G *et al* 2016 Glass-ceramic sealant for solid oxide fuel cells application: Characterization and performance in dual atmosphere *J. Power Sources* **328** 262–270

[2] Sharifzadeh M, Meghdari M and Rashtchian D 2017 Multi-objective design and operation of solid oxide fuel cell (SOFC) triple combined-cycle power generation systems: Integrating energy efficiency and operational safety *Appl. Energy* **185**(1) 345–361

[3] Choudhury A, Chandra H and Arora A Application of solid oxide fuel cell technology for power generation-A review *Renew. Sustain. Energy Rev.* **20** 430–442

[4] Irvine J T S, Neagu D, Verbraeken M C, Chatzichristodoulou C, Graves C and Mogensen M B 2016 Evolution of the electrochemical interface in high-temperature fuel cells and electrolysers *Nat. Energy* **1** 15014

[5] Yang L *et al* 2009 Enhanced sulfur and coking tolerance of a mixed ion conductor for SOFCs: $BaZr_{0.1}Ce_{0.7}Y_{0.2-x}Yb_xO_{3-\delta}$ *Science* **326**(5949) 126–129

[6] Sandhu N K *et al* 2016 Electrochemical performance of a short tubular solid oxide fuel cell stack at intermediate temperatures *Appl. Energy* **183** 358–368

[7] Kumar V, Rupali, Pandey O P and Singh K Thermal and crystallization kinetics of yttrium and lanthanum calcium silicate glass sealants for solid oxide fuel cells *Int. J. Hydrog. Energy* **36**(22) 14971–76

[8] Reddy A A *et al* 2013 Diopside–Ba disilicate glass–ceramic sealants for SOFCs: Enhanced adhesion and thermal stability by Sr for Ca substitution *Int. J. Hydrog. Energy* **38**(7) 3073-86

[9] Mahapatra M K and Lu K 2010 Seal glass for solid oxide fuel cells *J. Power Sources* **195**(21) 7129–39

[10] Malzbender J, Wakui T and Steinbrech R W 2006 Curvature of planar solid oxide fuel cells during sealing and cooling of stacks *Fuel Cells* **6**(2) 123–129

[11] Ghosh S, Kundu P, Sharma A D, Basu R N and Maiti H S 2008 Microstructure and property evaluation of barium aluminosilicate glass–ceramic sealant for anode-supported solid oxide fuel cell *J. Eur. Ceram. Soc.* **28**(1) 69–76

[12] Bause T, Malzbender J, Pausch M, Beck T and Singheiser L 2013 Damage and failure of silver based ceramic/metal joints for SOFC stacks *Fuel Cells* **13**(4) 578–583

[13] Zhang W *et al* 2016 Development of flexible ceramic-glass seals for intermediate temperature planar solid oxide fuel cell *Int. J. Hydrog. Energy* **41**(14) 6036–44

[14] Nguyen X V *et al* 2016 Study of sealants for SOFC *Int. J. Hydrog. Energy* **41**(46) 21812–19

[15] Kim E A, Choi H W and Yang Y S 2015 Effects of $Al_2O_3$ on $(1-x)[SrO–SiO_2–B_2O_3]–xAl_2O_3$ glass sealant for intermediate temperature solid oxide fuel cell *Ceram. Int.* **41**(10) Part B 14621–26

[16] Dai Z, Pu J, Yan D, Chi B and Jian L 2011 Thermal cycle stability of $Al_2O_3$-based compressive seals for planar intermediate temperature solid oxide fuel cells *Int. J. Hydrog. Energy* **36**(4) 3131–37

[17] Nesse W 2011 *Introduction to Mineralogy* 2$^{nd}$ edition (New York: Oxford University Press)

[18] Brotosusilo A, Apriana I W A, Satria A A and Jokopitoyo T 2016 Littoral and coastal management in supporting maritime security for realizing indonesia as world maritime axis *IOP Conf. Ser. Earth Environ. Sci.*, **30**(1) 012016

[19] Aristia G A G, Istiqomah, Hidayat N, Triwikantoro, Baqiya M A and Pratapa S 2015 Phase analysis of natural silica-sand-based composites as potential fuel-cell seal material *Adv. Mater. Res.*.**1112** 294–298

[20] Hidayat N, Triwikantoro, Baqiya M A and Pratapa S 2013 Thermal expansion coefficient prediction of fuel-cell seal materials from silica sand *AIP Conf. Proc.* **1555** 99–101

[21] Rietveld H M 1969 A profile refinement method for nuclear and magnetic structures *J. Appl. Crystallogr.* **2**(2) 65–71

[22] Hunter B A 1998 Rietica-A visual Rietveld program *Newsl. Int. Union Crystallogr. Comm. Powder Difraction*

[23] Degen T, Sadki M, Bron E, König U and Nénert G 2014 The HighScore suite *Powder Diffr.* **29**(S2) S13–S18

[24] Olevsky E and Skorohod V 1993 Deformation aspects of anisotropic-porous bodies sintering *J. Phys. IV* **03**(C7) 739-742

[25] Zavaliangos A, Missiaen J M and Bouvard D 2006 Anisotropy in shrinkage during sintering *Sci. Sinter.* **38**(1) 13–25

[26] Ch'ng H N and Pan J 2007 Sintering of particles of different sizes *Acta Mater.* **55**(3) 813–824

[27] L'vov B V, L&rsquo, Vov B V, Polzik L K and Ugolkov V L 2002 Decomposition kinetics of calcite: a new approach to the old problem *Thermochim. Acta* **390**(1–2) 5–19

[28] Deer W A, Howie R A and Zussman J 1997 *Single-Chain Silicates* Geological Society

[29] Tavangarian F and Emadi R 2010 Synthesis of nanocrystalline forsterite ($Mg_2SiO_4$) powder by combined mechanical activation and thermal treatment *Mater. Res. Bull.* **45**(4) 388–391

[30] Day A L and Shepherd E S 1906 The lime-silica series of minerals *J. Am. Chem. Soc.* **28**(9) 1089–114

[31] Tse J S and Klug D D 1991 The structure and dynamics of silica polymorphs using a two-body effective potential model *J. Chem. Phys.* **95**(12) 9176–85

[32] Karadeniz Z H and Kumlutas D 2007 A numerical study on the coefficients of thermal expansion of fiber reinforced composite materials *Compos. Struct.***78**(1) 1–10

[33] Affdl J C H and Kardos J L 1976 The Halpin-Tsai equations: A review *Polym. Eng. Sci.* **16**(5) 344–352

[34] Pascual M J, Guillet A and Durán A 2007 Optimization of glass–ceramic sealant compositions in the system $MgO–BaO–SiO_2$ for solid oxide fuel cells (SOFC) *J. Power Sources* **169**(1) 40–46

[35] Wang X *et al* 2013 Optimization of $Al_2O_3$–glass composite seals for planar intermediate-temperature solid oxide fuel cells *J. Power Sources* **226** 127–133

# $Ag/Fe_3O_4/ZrO_2$ Composite: Ternary Magnetically Separable UV-light-driven Photocatalyst for Removal Methylene Blue Dyes

**Y Kristianto[1,2], A Taufik[1,2] and R Saleh[1,2]**

[1] Department of Physics, Faculty of Mathematics and Sciences, Universitas Indonesia, Kampus UI Depok, Depok 16424, Indonesia
[2] Integrated Laboratory of Energy and Environment, Fakultas MIPA, Universitas Indonesia, Kampus UI Depok, Depok 16424, Indonesia

Email: rosari.saleh@gmail.com

**Abstract.** In this research, we demonstrate the preparation of ternary $Ag/Fe_3O_4/ZrO_2$ composites as a novel magnetically separable UV-light-driven photocatalyst. The $Fe_3O_4/ZrO_2$ composites were successfully synthesized by using two-step methods, that is, sol-gel followed by ultrasonic-assisted method. Furthermore, the $Fe_3O_4/ZrO_2$ composites were coupled with silver through ultrasound-assisted method. The prepared samples were fairly characterized by using X-Ray Diffraction and UV-Vis DRS spectroscopy. The photocatalytic activity of this composite was tested to degrade methylene blue dyes under UV-light irradiation. The influence of catalyst dosage, initial concentration, and scavengers of active species were investigated and the results were discussed. Compared with $Fe_3O_4/ZrO_2$ sample, the $Ag/Fe_3O_4/ZrO_2$ composite with three weight ratio (5%, 15%, and 25%) of Ag to $Fe_3O_4/ZrO_2$ display higher photocatalytic activity. Among three variations of the weight ratio of Ag to $Fe_3O_4/ZrO_2$, 15% weight ratio of Ag in $Ag/Fe_3O_4/ZrO_2$ exhibits the superior photocatalytic performance followed by weight ratio of 25% and 5%. Moreover, the composite was found to be a reusable catalyst.

**Keyword :** $Ag/Fe_3O_4/ZrO_2$, photocatalyst, UV driven, reusable

## 1. Introduction

In recent years, wastewater treatment from industrial organic dyes has attracted particular attention because of an environmental hygiene-related. Persistent organic pollutant, such as methylene blue (MB) has a complex aromatic structure that causes it is not degraded easily by biological decomposition [1]. Moreover, the presence of organic dyes in the water system can cause serious damage to human health.

To overcome this issue, several methods, such as membrane filtration, coagulation, electrochemical oxidation, adsorption and advanced oxidation process (AOPs) have been established for wastewater treatment. Among these methods, AOPs has received much attention due to the excellent performance for mineralization of many non-biodegradable organic pollutants [2,3].

Many metal oxide semiconductors (MOS), such as $TiO_2$, $SnO_2$, $ZrO_2$, have been frequently employed as a supportive catalyst. Zirconium dioxide ($ZrO_2$) as a photocatalyst has attracted considerable attention due to its several advantages such as its low cost, non-toxicity, environment-friendly, and high stability [4,5]. However, the main drawback of $ZrO_2$ is the low catalytic performance because of fast recombination of electron-hole pairs.

Combining MOS with noble metal catalyst, such as Ag, has been carried out as an effort to improve the photocatalytic performance by diminishing the recombination of the photoexcitation electron-hole pairs by trapping electron in conduction band [6].

Much effort has been focused on the incorporation of magnetic components, like $Fe_3O_4$, into MOS photocatalyst, which can be easily separated from the treated solution by an external magnetic field. This technique is very useful in reusability for practical applications. Herein, the as prepared ternary $Ag/Fe_3O_4/ZrO_2$ composites were employed as a highly UV light-driven photocatalyst towards the degradation of MB. The physical and optical characteristics of the samples were examined by using XRD and UV-Vis spectroscopy. We also report that the prepared samples stable and recyclable for the degradation of MB.

## 2. Experimental Method

### *2.1 Materials*

All the chemical reagents were of an analytical grade and were used without further purification. Iron(II) sulfate heptahydrate ($FeSO_4.7H_2O$, 99%), silver nitrate ($AgNO_3$), sodium dodecyl sulfate (SDS), sodium carbonate ($Na_2CO_3$), ethylene glycol (EG), zirconium(IV) chloride ($ZrCl_4$) and sodium hydroxide (NaOH) were purchased from Merck. NGP and graphene were purchased from Angstron Materials.

### *2.2 Preparation of $Fe_3O_4$ nanoparticle*

The $Fe_3O_4$ nanoparticles were synthesized by using the same method as described previously [7].

### *2.3 Preparation of $ZrO_2$ nanoparticle*

The $ZrO_2$ nanoparticles were synthesized by using the sol-gel method. Briefly, $ZrCl_4$ and NaOH were dissolved separately in distilled water through magnetic stirring. Then, the $ZrCl_4$ solution was added to the NaOH solution slowly, followed by magnetic stirring for 3 h, which allowed a homogenous solution to be achieved. The temperature of solution was maintained at 80 °C during the stirring process. The mixed solution was then centrifuged to separate the products that were precipitated from the solution. The precipitates were rinsed with distilled water and ethanol several times to remove the impurities. The product was heated at 120 °C for 12 h. Then, it was calcined at the temperature of 550 °C. Calcination took place in the air for 5 h, at a heating rate of 10 °C $min^{-1}$.

### *2.4 Preparation of Ag nanoparticle*

The Ag nanoparticles were synthesized by using a microwave-assisted method. $AgNO_3$ and SDS were separately dissolved in EG solution through magnetic stirring for 30 min. Meanwhile, $Na_2CO_3$ was dissolved in distilled water. Then, $Na_2CO_3$ solution was added into the SDS solution followed by adding $AgNO_3$ solution into $Na_2CO_3$ and SDS mixture solution. The solution was kept stirred for 10 min. Then it was placed in a microwave oven (800 W) for 1 min followed by magnetic stirring for 30 min. The mixed solution was centrifuged to separate the products that were precipitated from the solution. The precipitates were rinsed with distilled water and ethanol several times to remove the impurities. The product was heated at 120 °C for 5 h to obtain Ag nanoparticles.

### *2.5 Preparation of $Fe_3O_4/ZrO_2$ composite*

The $Fe_3O_4/ZrO_2$ composites were synthesized by using ultrasonic-assisted method. Briefly, $ZrO_2$ nanoparticles were mixed with $Fe_3O_4$ in aqueous and ethanol mixture. Then, the mixed solution was given ultrasonic for 2 h followed by centrifugation to obtain the product on its precipitation. The obtained product was dried in vacuum condition at 80 °C for 12 h to obtain $Fe_3O_4/ZrO_2$ composites which were labeled as FZ.

*2.6 Preparation of Ag/$Fe_3O_4$/$ZrO_2$ composite*
The ternary Ag/FZ with various Ag weight percentage (wt%) were synthesized by using the same method and procedure as synthesizing FZ composites.

*2.7 Characterization*
X-ray Diffraction (XRD) performed on a Rigaku Miniflex 600 diffractometer with Cu K-α radiation ($\lambda$ =1.5406 Å) was used to obtain the crystal structure of samples. Study on the optical properties of samples was carried out by using UV-Vis spectroscopy (UV-Vis).

*2.8 Photocatalytic activity*
The photocatalytic activity of samples was investigated via degradation of an MB aqueous solution as a model of organic pollutant at room temperature. In the process, the photocatalyst was dissolved into MB solution. Then, the solution was continuously stirred in the dark for 30 min to ensure that an equilibrium was reached between the adsorption and desorption MB molecules on the catalyst surface. Subsequently, the solution was exposed to UV light irradiation. A small amount of the water sample was collected through centrifugation in every 15 min to measure the MB concentration with UV-Vis spectrophotometer. As a source of irradiation light, two lamps of 40 W UV-C were used.

*2.9 Scavenger experiment*:
To determine the species involved in photocatalytic activities, the different radical scavengers were added to the solutions of MB. Di-ammonium oxalate, sodium sulfate and tertbutyl alcohol (TBA) were used as scavengers for holes, electrons, and hydroxyl radical, respectively. The same measurements as described above were used.

## 3. Result and Discussion
As we can see in Figure 1, XRD patterns of Ag/FZ composites with various weight percent of Ag exhibit the series peaks that can be characterized as tetragonal phase of $ZrO_2$ ($2\theta$ = 30.18°, 35.03°, 50.36°, and 59.94°, which correspond to the [111], [200], [220], and [311] planes, respectively), cubic spinel phase of $Fe_3O_4$ ($2\theta$ = 30.14°, 35.49°, 43.28°, 53,76°, 57.20° and 62.83°, which correspond to the [220], [311], [400], [422], [511] and [440] planes, respectively) and cubic phase of Ag ($2\theta$ = 38.14° and 44.26°, which correspond to the [111] and [200] planes, respectively). However, no typical diffraction peaks of Ag were observed within the 5wt% Ag/FZ composites, which probably due to the low content of Ag in sample. Furthermore, there were no others peaks from impurities, suggesting that all prepared samples have good purities. The lattice parameter and grain size of all prepared samples were calculated by using Debye-Scherrer's formula and Rietveld refinement. The result are tabulated in Table 1.

**Table 1.** Lattice constant and grain size of FZ, 5 wt% Ag/FZ, 15 wt% Ag/FZ and 25 wt% Ag/FZ.

| Samples | Lattice constant (Å) | | | | <D> (nm) | | |
|---|---|---|---|---|---|---|---|
| | Ag | $Fe_3O_4$ | $ZrO_2$ | | | | |
| | a=b=c | a=b=c | a=b | c | Ag | $Fe_3O_4$ | $ZrO_2$ |
| FZ | - | 8.374 | 3.612 | 5.153 | - | 42 | 13 |
| 5wt% Ag/FZ | - | 8.368 | 3.612 | 5.153 | - | 42 | 13 |
| 15wt% Ag/FZ | 4.085 | 8.361 | 3.609 | 5.152 | 13 | 41 | 13 |
| 25wt% Ag/FZ | 4.083 | 8.357 | 3.608 | 5.151 | 13 | 40 | 13 |

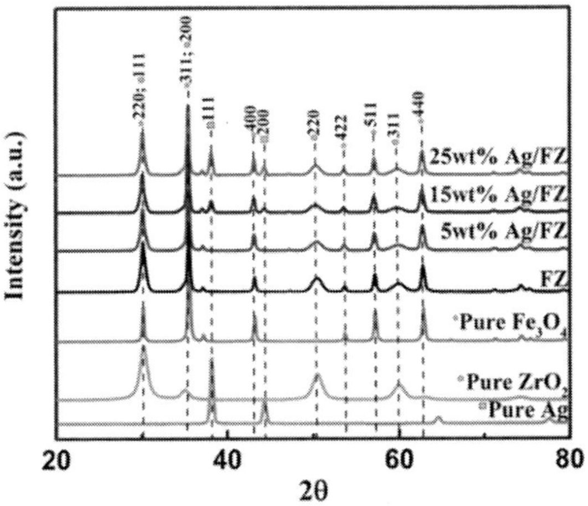


**Figure 1.** XRD patterns of Ag, $ZrO_2$, $Fe_3O_4$, FZ, 5 wt% Ag/FZ, 15 wt% Ag/FZ and 25 wt% Ag/FZ.

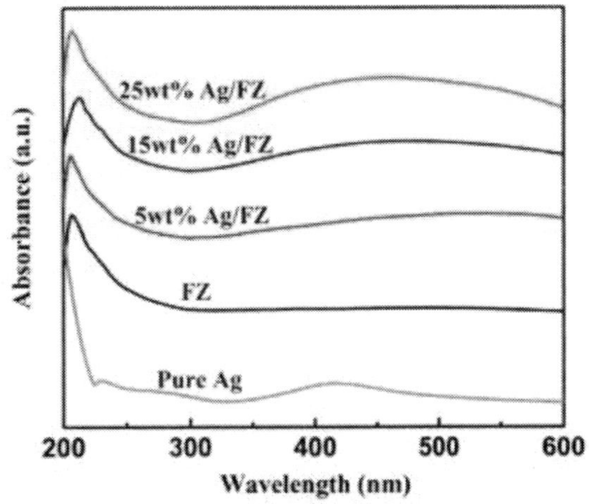


**Figure 2.** UV-Vis absorbance spectra of all prepared samples

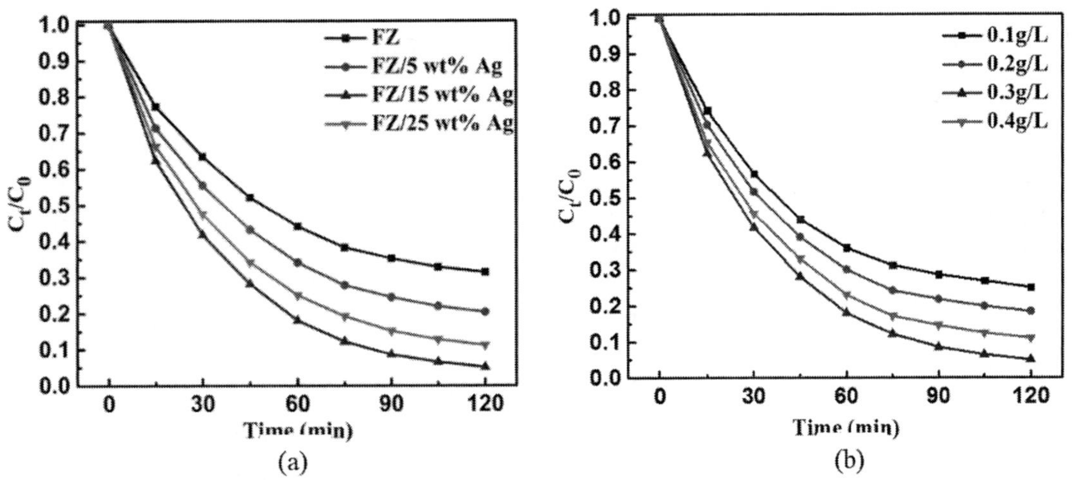


**Figure 3.** (a) The degradation of MB in the presence of different photocatalyst. (b) Effect of catalyst dosage on the degradation of MB

Figure 2 shows the UV-Vis absorbance spectra of Ag, FZ, 5wt% Ag/FZ, 15wt% Ag/FZ and 25wt% Ag/FZ. As can be seen, there are broad peaks around 400 - 450 nm in Ag spectra which indicate the surface plasmon resonance (SPR) as a characteristic of Ag [8]. These broad peaks also can be seen at the samples of 5wt% Ag/FZ, 15wt% Ag/FZ and 25wt% Ag/FZ, indicating the presence of Ag on the prepared samples.

The photocatalytic performance was investigated by using pseudo-first-order kinetic model, which expressed:

$$\frac{C_t}{C_0} = e^{-kt}$$

where $C_0$ is the initial concentration, $C_t$ is the concentration at $t$, $t$ is the irradiation time, and $k$ is the pseudo first-order kinetic rate constant. The results of photocatalytic performance are shown in Figure 3 (a). As shown in Figure 3 (a), after 120 min UV light irradiation, the degradation efficiency of MB is reached about 69%, 80%, 95% and 89% for sample of FZ, 5wt% Ag/FZ, 15wt% Ag/FZ and 25wt%

Ag/FZ, respectively, at the condition of 0.3 g/L of catalyst dosage and 20 mg/L of initial concentration. These results indicate that the presence of Ag in samples can increase the photocatalytic performance. The reason is that Ag can form Schottky junction in composite, which can inhibit the rate of recombination of electron-hole pairs [6,9]. The degradation efficiency increased as increasing wt% of Ag until 15wt%, but decreased for 25wt% of Ag. This probably due to the decreased number of active sites brought by $ZrO_2$ as increasing wt% of Ag.

The effect of catalyst dosage on the degradation of the MB was performed. Catalyst dosages were varied from 0.1 g/L until 0.4 g/L for sample of 15 wt% Ag/FZ with 20 mg/L of MB initial concentration and the result is shown in Figure 4 (b). As can be seen, the degradation efficiency increased as the increase of catalyst dosage from 0.1 g/L until 0.3 g/L. Increasing catalyst dosage means increasing the number of active sites, which consequently increases the population of hydroxyl radicals as a reactive agent to degrade MB. However, when it overdoses, agglomeration of catalyst may occur with increasing turbidity, which turns to a decrease of light penetration and an increase of light scattering [10].

The effects of initial concentration MB were also carried out by varying it from 20 mg/L until 60 mg/L for sample of 15 wt% Ag/FZ with optimum catalyst dosage of 0.3 g/L. The results are shown in Figure 4 (a). This figure shows that the catalytic activity tends to decrease with the increase of MB initial concentration. The reason is that the solution transmittance decreased with increasing initial concentration, resulting in reducing the number of photons that reach the photocatalyst surface [11].

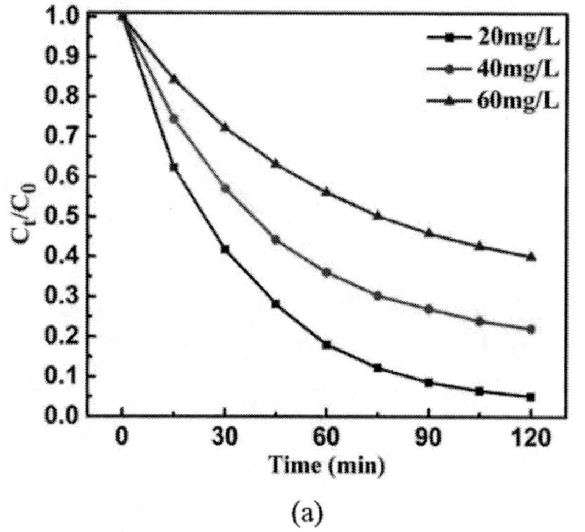


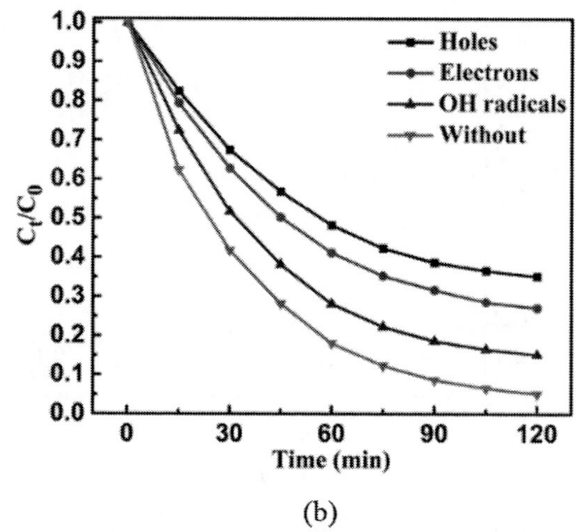


**Figure 4.** (a) Effect of initial concentration on the degradation of MB. (b) Effect of various scavenger on the degradation of MB

To elucidate the mechanism of photocatalytic, the effect of various active species which participate in photocatalytic activity were studied at optimum condition (0.3 g/L of catalyst dosage and 20 mg/L of initial concentration). The roles of the main active species were investigated by hole, electron and hydroxyl radical trapping experiments and the results obtained are shown in Figure 4 (b). The degradation efficiency of MB furthest decreased upon the addition of hole scavenger. This result demonstrated that the photocatalytic activity was mainly governed by holes, followed by electrons and hydroxyl radicals.

The photocatalytic stability of the 15 wt% Ag/FZ was studied by repeating the photocatalytic process four times at optimum condition. After each reaction finished, the photocatalyst was separated from the solution by an external magnet to be reused with newly MB solution. In Figure 5, we can see that after four-times cycles, the photocatalytic activity didnot significantly decrease. The acceptable stability in repeated cycles demonstrated that the 15wt% Ag/FZ photocatalyst composite exhibited good activity for the repeated degradation of MB.

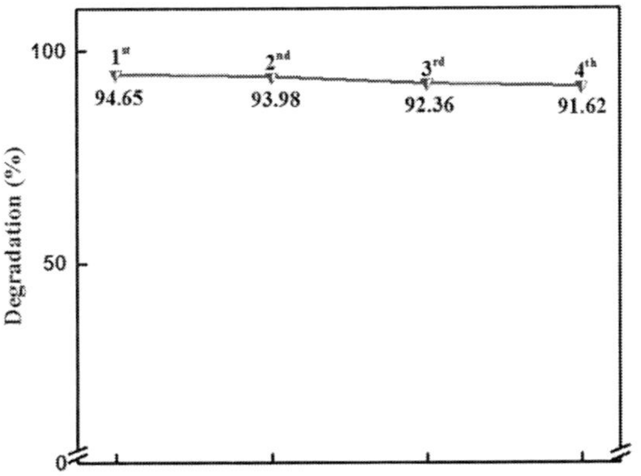


**Figure 5.** Reusability of 15wt% Ag/FZ on degradation MB

## 4. Conclusion

In this study, the photocatalyst ternary $Ag/Fe_3O_4/ZrO_2$ with various wt% of Ag were successfully prepared through ultrasonic-assisted method confirmed by XRD and UV-Vis spectroscopy. The best photocatalytic performance is attained with a 15wt% Ag/FZ composite. To understand the mechanism of the reaction, the Di-ammonium oxalate, sodium sulfate, and tertbutyl alcohol (TBA) were used as scavengers for holes, electrons, and hydroxyl radical, respectively. The result revealed that the hole is the main active species in the photocatalytic process, followed by electron and hydroxyl radical. The catalyst can be reused with no significant loss in photocatalytic activity for the four-times cycles.

## 5. References

[1] Lee S M and Ong S T 2014 *APCBEE Proc.* **9** 165-169.
[2] Wang P, Wang L, Sun Q, Qiu S, Liu Y, Zhang X, Liu X, and Zheng L 2016 *Materials Letters* **183** 61-64.
[3] Yu L, J. Chen, Z. Liang, W. Xu, L. Chen, D. Ye 2016 *Separation and Purification Technology* **171** 80-87.
[4] Guo G Y and Chen Y l 2005 *Journal of Solid State Chemistry* **178** 1675-82
[5] Tahmasebpour, Babaluo A A and Razavi M K 2008 *Journal of the European Ceramic Society* **28** 773-778
[6] Tian Y and Tatsuma T 2005 *Journal of the American Chemical Society* **127** 7623-37
[7] Taufik A, Susanto I K and Saleh R 2015 *Materials Science Forum* **827** 37-42
[8] Patil S S, Mali M G, Tamboli M G, Patil D R, Kulkarni M V, Yoon H, Kim H, Al-Deyab S S, Yoon S S, Kolekar S S and Kale B B 2016 *Catalysis Today* **260** 126-134
[9] Zhang X, Wang L, Liu C, Ding Y, Zhang S, Zeng Y, Liu Y and Luo S 2016 *Journal of Hazardous Materials* **313** 244-252
[10] Tao H, Liang X, Zhang Q and Chang C 2015 *Applied Surface Science* **324** 258-264
[11] Wang R F, Li B, Dong L, Zhang F, Fan M and Zhou L 2014 *Materials Letters* **135** 99-102

# Effect of $Fe_3O_4$ on the Electro-Optic and Magneto-Electric Characteristics of (PANI/$Fe_3O_4$)-Ag Film

**M Diantoro[1,2], D Pradhana[1], A A Mustikasari[1], A D Kusumawati[1], A Taufiq[1, 2], Sunaryono[1, 2], N Mufti[1, 2] and H Nur[3]**

[1]Department of Physics, Faculty of Mathematics and Natural Sciences, Universitas Negeri Malang (State University of Malang), Jl. Semarang No 5, Malang 65145, Indoensia
[2]Central Laboratory of Minerals, and Advanced Materials, Faculty of Mathematics and Natural Sciences, Universitas Negeri Malang (State University of Malang), Jl. Semarang No 5, Malang 65145, Indoensia
[3]Centre for Sustainable Nanomaterials, Ibnu Sina Institute for Scientific and Industrial Research Universiti Teknologi Malaysia, 81310 UTM Skudai, Johor, Malaysia

Email: markus.diantoro.fmipa@um.ac.id

**Abstract.** (PANI/$Fe_3O_4$)-Ag films have been successfully synthesized on a glass substrate. In this study, PANI functioned as a matrix, Ag as a catalyst enhancing the conductivity of PANI, and $Fe_3O_4$ as the influence of magnetic properties. (PANI/$Fe_3O_4$)-Ag was synthesized by a combination of chemical polymerization method to introduce Ag doping into PANI, and coprecipitation method of $Fe_3O_4$. The samples were characterized by using FTIR, X-RD, SEM, EDAX, 4-probe electrical conductivity under the employing of magnetic fields, and under the influence of light intensity. It was shown that there was a $Fe_2O_3$ phase as the consequence of $Fe_3O_4$ oxidized as $AgNO_3$ protonation on the ultrasonication treatment.
introducing $Fe_3O_4$ into (PANI/$Fe_3O_4$)-Ag film, increase the surface density, but reduce the electrical conductivity. The increasing of magnetic field might also reduce the conductivity of (PANI/Fe3O4)-Ag. Further, We also obtained that the increasing of light intensity employed to the samples enhances the electrical conductivity. The electrical conductivity of (PANI/Fe3O4)-Ag can be tuned by controlling the Fe3O4 , magnetic field, as well as the light intensity

**Keywords**: PANI, Ag, $Fe_3O_4$, microstructure, magneto-electric, electro-optic

## 1. Introduction

A research of Polyaniline (PANI) conductive polymer has become a fascinating topic recently due to its promising application potential along with a search of another natural polymer for electronics [1-2]. Many PANI related research have been focused on the application such as for gas sensor [1], the raw matrix material composition of lithium battery [2], and super-capacitor [3]. The polymer is basically as an insulator. The polymer can be modified until raising conductive characteristic through synthesis development. The synthesis method of PANI with polymerization of a chemical reaction is because it is mostly used since the synthesis is relatively simple, easy, inexpensive and efficient [4].

The effort to increase the magneto-electric on PANI composite by using PMME has been reported resulting in the highest conductivity which was 73 $Scm^{-1}$ [5]. PANI/$Fe_3O_4$ produced conductivity of

$10^{-9}$ $Scm^{-1}$ [6], while the study on an electro-optical characteristic of PANI/chlorophyll could enhance the conductivity from 1 S/cm-8 S/cm [7]. Magneto-electric is an interesting study because of the application potential on the electronic and sensor [8,9], but in the synthesis conducted, it is still challenging to get a significant and excellent material for optimizing the magneto-electric characteristic to [10]. Besides the idea of magneto-electric characteristic, the electro-optical characteristic was considered as well [11].

So far, it is rare to find the study of Ag, and $Fe_3O_4$ doped PANI films. It is also hard to find the reports of the films primarily effect of magnetic fields and light intensity on the electrical conductivity of the films. In this work, we report on the fabrication of PANI/Fe3O4/Ag films and its microstructure, electrical conductivity as a function of magnetic fields, and light exposures for various $Fe_3O_4$ dopant. It is necessary to know the correlation of electrical and magnetic characteristic of PANI. The electrical characteristic of PANI/Ag is a function of the magnetic characteristic of $Fe_3O_4$. Those parameters are presumably interrelated. This works also would like to infer fundamental knowledge of how to tune the electrical properties of PANI/Fe3O4/Ag by controlling dopant, magnetic fields, and light intensity.

## 2. Research Method

### *2.1. Instrument and Material*

The instruments used in this work were Digital Scale of HF-3000 (0.019), Ultrasonic Bath LUC-405, Spin coater TC100, Furnace Brother XD-1700M, Heater and magnetic stirrer, beaker glass and Pyrex vessel (250 mL) and petri dish. The synthesis of $Fe_3O_4$, coprecipitation and chemistry reaction polymerization of PANI used several raw basic materials such as iron sand, aniline (PA 99.9% Merck), HCl (PA 99.9%), $NH_4OH$ (PA 99.9%), deionized water, APS, CSA, acetone (PA 99.9%), and chloroform (PA 99.9%).

### *2.2. Procedure*

*2.2.1 $Fe_3O_4$ Coprecipitation.* Firstly, the iron sand was extracted and purified by using permanent magnetic tube to get iron sand powder. The sand powder was reacted with HCl by using magnetic stirrer at a temperature of 140 °C for 30 minutes. After that, the solution was filtered by using filter paper. The solution penetrating the filter paper was taken since $Fe_3O_4$ dissolved in HCl. The solution then was dropped with $NH_4OH$ gradually while being stirred for around 30 minutes until formed a black precipitate. The formed precipitated was then washed using distilled water up to pH 7 (normal pH) to get the real sample which was free from impurity. The obtained precipitate was dried on a hotplate at a temperature of 70 °C.

2.2.2 *PANI/$Fe_3O_4$ Polymerization.* PANI polymerization was conducted through chemical reaction method following the method in the literature [1]. The first solution was made from aniline monomer (1.82 mL) reacted with HCl (50 mL; 0.2 M). Another solution was prepared by mixing $(NH_4)_2S_2O_8$ (5.71 g) with 50 mL distilled water. Both solutions were mixed and then stirred mechanically by using magnetic stirrer for 30 minutes. After stirring, the mixed solution was let precipitating for 24 hours until a green deposit was obtained. The precipitated materials were then washed with deionized water (DI water) many times until it looked like a crystal. PANI obtained from the first polymerization was PANI-ES 1 and it was then added in dropwise with $NH_4OH$ (0.5 M; 10 mL) + $Fe_3O_4$ of every mass variation (0.242; 0.464; 0.696; 0.928; 1.16 gram) while it was being stirred for 4 hours. These five samples are designated as 1 M, 2 M, 3 M, 4 M, and 5 M of $Fe_3O_4$. This deprotonation process was done at a room temperature, and it was let untreated for 24 hours to get reaction and precipitation process occurred. The obtained deposit was blue showing an emeraldine base (EB) polyaniline + $Fe_3O_4$ (PANI-ES-2). The samples subsequently were annealed at 80 °C for five hours for FTIR characterization.

*2.2.3 PANI/Ag/$Fe_3O4$ Protonation.* The fabrication of PANI/Ag/$Fe_3O4$ was basically followed the previous work [3] with several modifications. Firstly, introducing Ag doping was done by grinding 0.5 g of ready PANI/$Fe_3O_4$ mixed with 0.5 g CSA. The mixture was ground for 15 minutes. Along with this, an $AgNO_3$ was dissolved in acetone (5 mL; 0.5 M). All prepared materials were mixed and dissolved in 10 mL chloroform for chemical reaction breaking of the atoms bonds. The sonification process was done for 45 minutes for one hour. The last step was precipitating the solution for 24 hours for protonation process. The precipitated was then filtered to obtained PANI/Ag/$Fe_3O_4$ paste which will be used for XRD and FTIR characterization. The PANI/$Fe_3O_4$ were similarly treated and designated as PAF1M, PAF2M, PAF3M, PAF4M, and PAF5M.

## 3. Results and Discussion

### *3.1. Functional Groups of (PANI/$Fe_3O_4$)-Ag*

Spectrum of FTIR for PANI, PANI/$Fe_3O_4$, and PANI/Ag/$Fe_3O_4$ is shown in Figure 1.

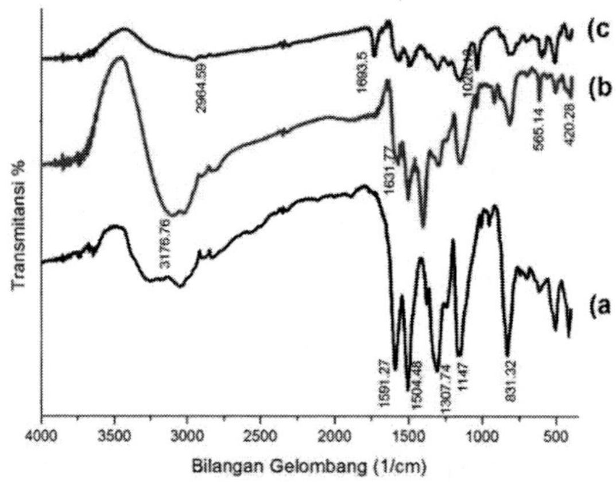


**Figure 1.** The FTIR Result of (a) PANI-EB, (b) PANI/$Fe_3O_4$; 5 M, (c)PANI/Ag/ $Fe_3O_4$.

Synthesizing PANI-EB into PANI (ES II)/$Fe_3O_4$ was performed by adding the mass variation of $Fe_3O_4$ composite. The addition of $Fe_3O_4$ caused the matrix on PANI bound and covered $Fe_3O_4$ during the process of polymerization occurring. The result of FT-IR showed a functional group of PANI-EB at the wavenumber of 1591,27$cm^{-1}$, 1504,48 $cm^{-1}$, 1307,74 $cm^{-1}$, 1147,65 $cm^{-1}$, and 831,32 $cm^{-1}$. This result is closely similar to our previous work [1]. The peak at 1591,27$cm^{-1}$ occurred due to the stretch vibration of C-N bond on quinoid ring; the peak at 1504,48 $cm^{-1}$ occurred due to benzoid C-N bond; the peak at 1307,74 $cm^{-1}$ was a stretch vibration of bipolar structure of C-N bond, the wavenumber of 1147,65 $cm^{-1}$ was a bending section result of aromatic C-H bond, and the wavenumber at the range of 831,32 $cm^{-1}$ was the consequence of aromatic C-H bond. The absorption difference of IR $Fe_3O_4$ infused when PANI-ES polymerization occurred toward PANI-EB peak is shown in Figure 1 (b). From the figure, we can see that the transmittance plain was at the wave number of 3176,76 $cm^{-1}$, 1631,77 $cm^{-1}$, 565,14 $cm^{-1}$, and 420,28 $cm^{-1}$. The absorption of IR $Fe_3O_4$ at 3176,76 $cm^{-1}$ identified the function cluster of -OH- occurring in the range of 3000-4000 $cm^{-1}$, the wavenumber of 1631,77 $cm^{-1}$ identified function cluster of $H_2O$, the wavenumber of 565.14 $cm^{-1}$ happened as a consequence of the presence of ($Fe^{3+}$-$O^{2+}$) and for the wavenumber of 420.28 $cm^{-1}$, it was a functional group of (Fe-O). From Figure 1 (b), it was identified that $Fe_3O_4$ was bound with PANI matrix during the polymerization process. When Ag added to PANI/$Fe_3O_4$ it might change the origin bonds of PANI and were replaced by $Ag^+$ ion from $AgNO_3$ of which bond was broken. The identification of FTIR for Ag ion at the spectrum of 2965.15 $cm^{-1}$ occurred due to the presence of C-Ag bond; the wavenumber at 1693,5 $cm^{-1}$

showed a functional group of –N-H- enabling –H- atom attached with–N- was replaced by Ag into –N-Ag-. While the wave number of 1026,13 $cm^{-1}$ represent a function of -OH-.

### *3.2. Crystal Structure of (PANI/$Fe_3O_4$)-Ag*

The XRD diffraction patterns is provided to study the crystallinity and phase formed from PANI/Ag/$Fe_3O_4$ film. The un-doped and $Fe_3O_4$ doped (PANI/$Fe_3O_4$)-Ag is displayed in Figure 2.

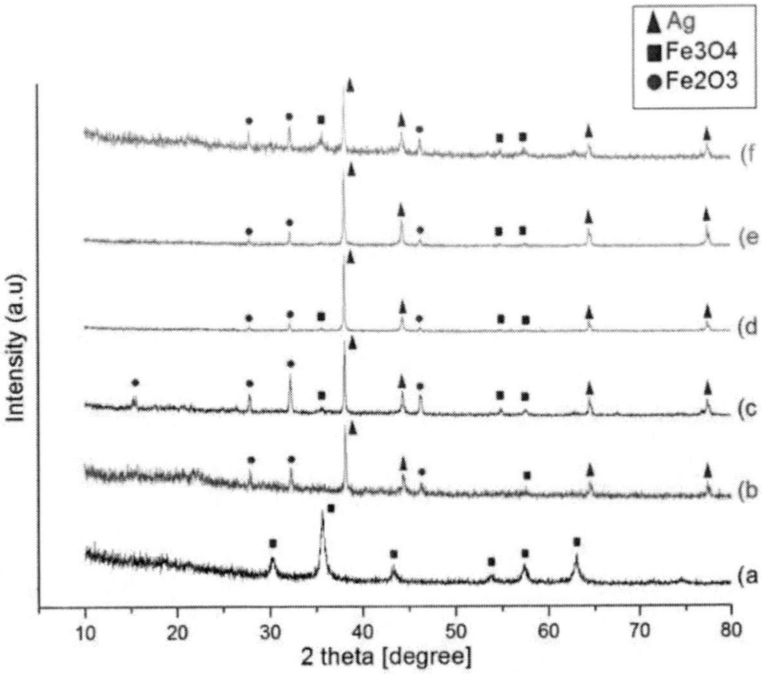


**Figure 2.** XRD Patterns of (a). PANI/$Fe_3O_4$; 5M, (b). PAF1M, (c). PAF2M, (d). PAF3M, (e). PAF4M, and (f). PAF5M.

The undoped PANI/ $Fe_3O_4$ sample as depicted in Figure 3.3 (a) we detect only $Fe_3O_4$ peaks are appear. Introducing silver ion in the sample PAF1M give rise to the appearance of other phases of $Fe_2O_3$, and Ag. It is observed that the Bragg peak of PAF1M sample indicated at the position of $2\theta = 57°$ for $Fe_3O_4$ phase. It is also detected a $Fe_2O_3$ phase showing by appearing peaks position at $2\theta$ of 28°, 32.25°, and 46.11°. The remaining are the peak of Ag $2\theta$ = 38,2°, 44,3°, 64.5° and 77.4°. The peaks of For PAF2M indicating the appearance of $Fe_3O_4$ which were observed at 30, 53°, and 57°. A weak peak at $2\theta = 46.11°$ assigned of $Fe_2O_3$ phase. Similar to the other samples, the PAF3M, PAF4M, and PAF5M also showing the appearance of $Fe_3O_4$, $Fe_2O_3$, and Ag phases.

A few parts of $Fe_3O_4$ particle transformed into $Fe_2O_3$ during the process introducing Ag dopant. Although the initial dopant was $Fe_3O_4$ phase, from the XRD patterns, we could easily see that $Fe_2O_3$ also appeared in all doped samples. It means that the $Fe_2O_3$ found act as dopant phase. The partial transformed of $Fe_3O_4$ into $Fe_2O_3$ was caused by several factors such as the process of incorporating of Ag into PANI/$Fe_3O_4$, and also possible came from sonochemical process. In sonification step, breaking $AgNO_3$ bond may occurred and transformed $Ag^+$ and $NO_3^-$. The $Ag^+$ separated from $NO_3^-$ might enter to PANI –C=H- bond cluster and possibly transformed into–C=Ag- and –N-H- bond cluster changed into –N-Ag-. The process of bond separation in sonification made some $Fe_3O_4$ within PANI/$Fe_3O_4$ were broken into $Fe^+$ and $O^-$, so some $Fe_2O_3$ phase appeared in (PANI/$Fe_3O_4$)-Ag phase. The other factor of the appearance of $Fe_2O_3$ phase in PANI/Ag/$Fe_3O_4$ is because of Ag+ ion is too reactive to $O_2$. Therefore, when $Ag^+$ is separated from $NO_3^-$, it can be reactive to the $Fe_3O_4$ particle until it is oxidized becomes $Fe_2O_3$. The third factor is $Fe_3O_4$ particle becomes reactive in nano-size, enabling such condition may happen in the process of giving Ag doping by some of $Fe_3O_4$ oxidized becomes $Fe_2O_3$ phase.

### *3.3. Microstructure of (PANI/$Fe_3O_4$)-Ag*

The effect of $Fe_3O_4$ to the (PANI/$Fe_3O_4$)-Ag film on its microstructure could be seen based on the topography surface by SEM characterization as depicted in Figure 3.

**Figure 3.** The Microstructure of (a1) PAF1M, (b1) PAF2M, (c1) PAF3M, (d1) PAF4M, (e1) PAF5M

These SEM images do not contain the same features with the metallic doped of PANI [1]. It is seen that the morphology of (PANI/$Fe_3O_4$)-Ag samples are plate-like. It might be due to the synthesis of PANI was by using chemical reaction polymerization. In Figure 3 (A1), the elemental content of $Fe_3O_4$ was fewer than the other samples, so the granular microstructure of (PANI/$Fe_3O_4$)-Ag surface remained plate-like and smooth. The size of PANI formed was around 0.3-0.5 μm. In Figure 3. (B1), Fe increased to represent the existing of $Fe_3O_4$ or $Fe_2O_3$. The influence of $Fe_3O_4$ affected the (PANI/$Fe_3O_4$)-Ag microstructure tended to rougher and covered with the cavity of granular PANI. The $Fe_3O_4$ particle formed in (PANI/$Fe_3O_4$)-Ag looked cubic spinel, even the $Fe_2O_3$ formed from $Fe_3O_4$ oxidation had cubic spinel matched with AMCSD 0020515. Figure 3 (C1, D1, and E1) show that the influence of $Fe_3O_4$ on PANI microstructure made the rough surface became rougher as the amount of $Fe_3O_4$ increased.

### *3.4. Magneto-Electric of (PANI/$Fe_3O_4$)-Ag by Using 4 Probes*

The magneto-electric characteristic of (PANI/$Fe_3O_4$)-Ag film is an interesting complement study in which the $Fe_3O_4$ has a magnetic characteristic as the PANI of which conductivity is increased by giving Ag doping.

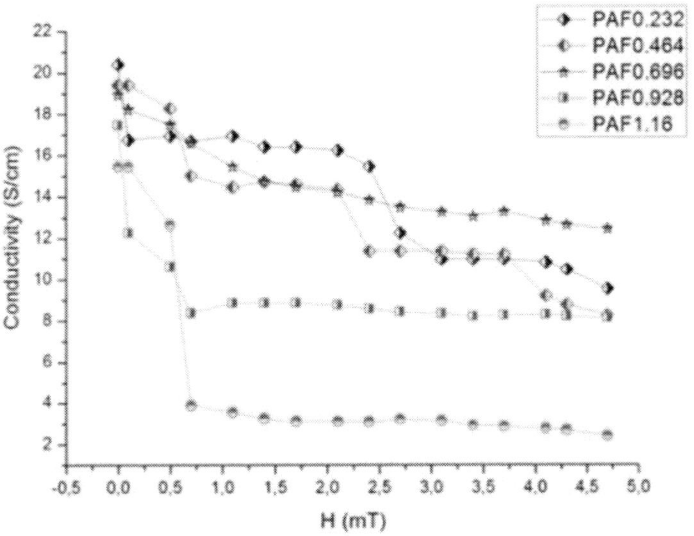


**Figure 4.** Conductivity as a function of Magnetic Field of ($PANI/Fe_3O_4$)-Ag Samples

It is observed in Figure 4, that employing magnetic field resulting to reduce electrical conductivity for all ranges of $Fe_3O_4$ doping. This feature is similar to the pure extracted PANI as reported by other work [8]. Another group, reported that at low magnetic field, electrical conductivity is obviously increased by increase of magnetic field. [16]. When the more $Fe_3O_4$ compound was added to $PANI/Ag/Fe_3O_4$, the conductivity is getting reduced. In general also observed that the magnetic field affected to decrease the electrical conductivity. Interesting phenomena achieved in this figure, both increased of $Fe_3O_4$ dopant as well as increase of magnetic intensity give rise to reduced the electrical conductivity. It means that, we can control electrical conductivity of $PANI/Ag/Fe_3O_4$ films by controling $Fe_3O_4$ dopant and or employed external magnetic field. The magnetic field applied to the ($PANI/Fe_3O_4$)-Ag material made the magnetic dipoles of inactive $Fe_3O_4$ became magnetized by external magnetic field (H) applied. The magnetic dipole influenced by magnetic field made the electron dissipation brought the charge on the PANI bond chain or from ferrites compounds. The dissipation disturbed the electron path in the PANI bond chain, so the electron became harder to move to deliver the electric charge since it was quantized by the magnet dipole of $Fe_3O_4$.

### *3.5. Electro-Optic Characteristics of ($PANI/Fe_3O_4$)-Ag Film by Using 4 Probes*

The electric-optic characteristic of ($PANI/Fe_3O_4$)-Ag sample is an influence of light intensity on the conductivity. Inducing doping of $Fe_3O_4$ as well as the light intensity significantly change the electrical conductivity of ($PANI/Fe_3O_4$)-Ag. The light intensity applied on ($PANI/Fe_3O_4$)-Ag made the electrical conductivity of material increased. It is presumably that the intensity caused thermal energy causing electron energy increased in the PANI bond chain within the valence band. Such condition made the electron easily moved to the conduction band. The band energy from the sample could be derived by giving light intensity in the range in which quantum energy $h\nu$ should exceed the band energy. The light intensity applied on ($PANI/Fe_3O_4$)-Ag sample was about 550 to 2770 Lux. This feature may be similar to the thermal applied to PANI which also increase as the temperature increase [16]. It is revealed that the effect of light intensity below 550 Lux did not affected to the ($PANI/Fe_3O_4$)-Ag conductivity. There is no step-like patterns of intensity induced of electrical conductivity as in borosilicate glass [17].

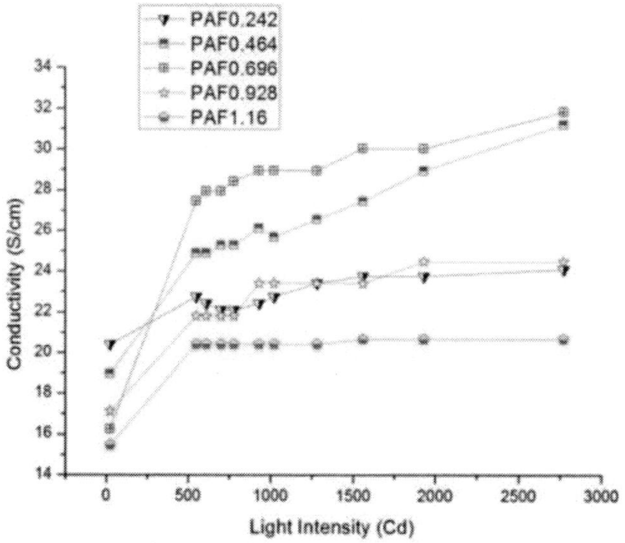


**Figure 5.** The Graph of Correlation Between Light Intensity and Conductivity of PAF1M; PAF2M;PAF3M; PAF4M; and PAA5M

## 4. Conclusion

Based on the characterization of microstructure by using SEM, the bigger the addition of $Fe_3O_4$, the density of (PANI/$Fe_3O_4$)-Ag film increases as well. The increase of surface density also induces to the reduce of electrical conductivity. This implies that introducing $Fe_3O_4$ into (PANI/$Fe_3O_4$)-Ag film, is not in line with the increase in surface density. The response of increasing of magnetic field might also reduce the conductivity of (PANI/$Fe_3O_4$)-Ag film. Further, we also obtained that the increasing of light intensity employed to the samples enhances the electrical conductivity. The electrical conductivity of (PANI/$Fe_3O_4$)-Ag can be tuned by controlling the $Fe_3O_4$, magnetic field, as well as the light intensity.

## 5. References

[1] Fratoddi I, Venditti I, Cametti C and Russo M V 2015 Chemiresistive polyaniline-based gas sensors: A mini review *Sens. Actuators B Chem.* **220** 534–48

[2] Wang D, Wang X, Yang X, Yu R, Ge L and Shu H 2015 Polyaniline modification and performance enhancement of lithium-rich cathode material based on layered-spinel hybrid structure *J. Power Sources* **293** 89–94

[3] Kim J, Ju H, Inamdar A I, Jo Y, Han J, Kim H and Im H 2014 Synthesis and enhanced electrochemical supercapacitor properties of Ag–MnO2–polyaniline nanocomposite electrodes *Energy* **70** 473–7

[4] Ghaffari-Moghaddam M and Eslahi H 2014 Synthesis, characterization and antibacterial properties of a novel nanocomposite based on polyaniline/polyvinyl alcohol/Ag *Arab. J. Chem.* **7** 846–55

[5] Angappane S, Srinivasan D, Rangarajan G, Prasad V, Subramayam S . and Wessling B 2000 Transport and magneto-transport study on some conducting polyanilines *Phys. B* 284–8

[6] Beata Z, Patrycja A, Petr J, Petrovsky E, Pavel B and Danniel H 2016 Magnetoconductive maghemite core/polyaniline shell nanoparticles: Physico-chemical and biological assessment *Colloids Surf. B Biointerfaces*

[7] Manna J S, Basu S, Mitra M K, Mukherjee S and Das G C 2012 Evolution of chlorophyll/polyaniline nanorod network with enhanced electro-optic property *Mater. Lett.* **76** 147–50

[8] Saha J, Sharma G, Kaushik S D, Rani V, Sudesh, Siruguri V and Patnaik S 2016 Magneto-electric coupling in Ca3CoMnO6 thin films *Proc. 20th Int. Conf. Magn. Barc. 5-10 July 2015* **400** 282–5

[9] Tripathy S N, Mishra B G, Shirolkar M M, Sen S, Das S R, Janes D B and Pradhan D K 2013 Structural, microstructural and magneto-electric properties of single-phase BiFeO3 nanoceramics prepared by auto-combustion method *Mater. Chem. Phys.* **141** 423–31

[10] Zhou L, Fu Q, Zhou D, Xue F, Tian Y and Hao L 2015 Magneto-electric coupling study in multiferroic La0.7Ba0.3MnO3–BaTiO3 composite ceramic at room temperature *Ceram. Int.* **41** 2367–72

[11] Wu J, Wang W, Wang L, Liu J, Chen K and Bo S 2016 Introduction of fluorine to change the dielectric environment of nonlinear optical chromophores for improved electro-optic activities *Mater. Lett.* **164** 636–9

# Effect of Crystallite Structure and Graphene Incorporation on Photocatalytic Performance of $LaFeO_3$

**N Afifah[1,2] and R Saleh[1,2]**

[1] (Department of Physics, Faculty of Mathematics and Sciences) Departemen Fisika, Fakultas MIPA, Universitas Indonesia, Kampus UI Depok, 16424 Depok, Indonesia
[2] Integrated Laboratory of Energy and Environment, Fakultas MIPA, Universitas Indonesia, Kampus UI Depok, 16424 Depok, Indonesia

Email: rosari.saleh@gmail.com, rosari.saleh@ui.ac.id

**Abstract**. In this study, $LaFeO_3$ photocatalysts were synthesized by using co-precipitation method at different calcination temperatures of 600 °C, 700 °C, and 800 °C. Graphene was also introduced into $LaFeO_3$ to improve the photocatalytic performance. The photocatalysts were characterized by using X-Ray Diffraction (XRD), Brunauer-Emmet-Teller (BET) surface area analysis, and Fourier Transform Infrared (FTIR) spectroscopy. The prepared photocatalysts were also evaluated to determine their photocatalytic degradation activities against methylene blue (MB) under visible light irradiation. $LaFeO_3$ with the calcination temperature of 800 °C with graphene supported showed orthorhombic structure and the highest photocatalytic performance. The different structure might affect the photocatalytic performance of samples, while the role of graphene on the nanocomposite could act as an electron transport layer and inhibited the recombination of photogenerated electron-hole pairs. The stability of the catalyst on the photocatalytic process was also studied.

**Keywords:** Graphene, photocatalytic, $LaFeO_3$.

## 1. Introduction

$LaFeO_3$ with the multifunctional properties has received considerable attention due to its promising application in advanced technology such as fuel cells, gas sensor, solid electrolytes, superconductors, and catalyst [1-2]. As a catalyst, $LaFeO_3$ exhibits strong catalytic abilities due to its unique optoelectronic properties and its narrow optical gap which can be exploited as a highly effective catalyst in the visible light region [3]. The photocatalytic activity of $LaFeO_3$ nanoparticles has been reported in several studies [3-4]. However, there are few reports about the effect of calcination temperature on the crystal structure and photocatalytic activity of $LaFeO_3$ nanoparticles.

The photocatalytic activity of $LaFeO_3$ nanoparticles calcined at 800 °C has been reported in our previous work [5]. However, $LaFeO_3$ nanoparticles can not completely degrade MB pollutant due to its high electron and hole recombination. Therefore, to enhance the photocatalytic performance, graphene with electron transport layer capacity is introduced to $LaFeO_3$ to inhibit the electron and hole recombination and enhance the photocatalytic performance. Graphene with its large surface area also can increase the photocatalytic performance by increasing the number of the active site involves in the degradation process [6-7]. The aim of this work is to study the effect of crystal structure and graphene incorporation on the photocatalytic performance of $LaFeO_3$ nanoparticles. The stability of the catalyst on the photocatalytic experiment is also checked for four-times cycling processes.

## 2. Experimental Method

### *2.1 Catalyst preparation*

All chemicals are analytical grade and were used without further purification. $LaCl_3 \cdot 7H_2O$, $FeCl_2 \cdot 4H_2O$, and sodium hydroxide (NaOH) were used to synthesize $LaFeO_3$ nanoparticles based on our previous work [5]. The $LaFeO_3$ nanoparticles were calcined at a different temperature of 600, 700, and 800 °C for 2 h under vacuum condition. The obtained $LaFeO_3$ (untreated, 600, 700, and 800 °C) were then mixed with graphene, purchased from Angstron Materials, to obtain the $LaFeO_3$/graphene composites. In the typical procedure, graphene was dissolved into distilled water and ethanol through ultrasonic treatment for 2 h. Then, $LaFeO_3$ nanoparticles were added to the graphene solution, followed by stirring for 2 h. The suspension was then centrifuged and dried at 70 °C for 12 h. The weight ratio of graphene to $LaFeO_3$ was held constant at 5 wt%.

### *2.2 Characterization*

The samples were characterized by using X-Ray Diffraction (XRD), Brunauer-Emmet-Teller (BET) surface area analysis, and Fourier Transform Infrared (FT-IR) spectroscopy.

### *2.3 Photocatalytic*

The photocatalytic activities of prepared samples were measured by observing the degradation of MB under visible light irradiation. First, the catalyst was dissolved into MB solution. Then, the mixture was stirred in the dark for 30 min to allow it to reach absorption–desorption equilibrium. After that, the suspension was irradiated under visible light irradiation for 2 h. The degradation of MB was monitored by using a Hitachi UH5300 UV-visible spectrophotometer at regular intervals. The percentage degradation of MB was recorded as $C_t/C_0$, where $C_t$ is the concentration of MB at each measurement interval during irradiation and $C_0$ is the initial concentration of MB after reaching adsorption–desorption equilibrium. The stability of the catalyst was also checked in this research by reusing the catalyst for four-times cycling processes.

## 3. Results and Discussion

Figure 1a shows the XRD patterns of $LaFeO_3$ for different calcination temperatures. For comparison, the XRD patterns of untreated $LaFeO_3$ is also shown in the figure. The result showed that the XRD pattern for untreated $LaFeO_3$ exhibited mixed phases of $La_2O_3$, $Fe_2O_3$, and $Fe_3O_4$, respectively [8-9]. After being calcined at 800 °C, the XRD pattern of the sample exhibited a single perovskite phase with orthorhombic structure [4, 10]. However, the secondary phase of $La_2O_3$ was still detected at 600 °C. The effect of graphene incorporation in the untreated and treated $LaFeO_3$ XRD pattern is also shown in Figure 1b. The result showed that the incorporation of graphene in the composite did not change the untreated and treated $LaFeO_3$ structure. The lattice parameter, crystallite size <*D*>, and specific surface area values of the samples are also summarized in Table 1. The lattice parameter values were obtained by MAUD Rietveld refinement method, while the crystallite size <*D*> and specific surface area values were calculated by using Scherrer's equation [11], and BET method, respectively.

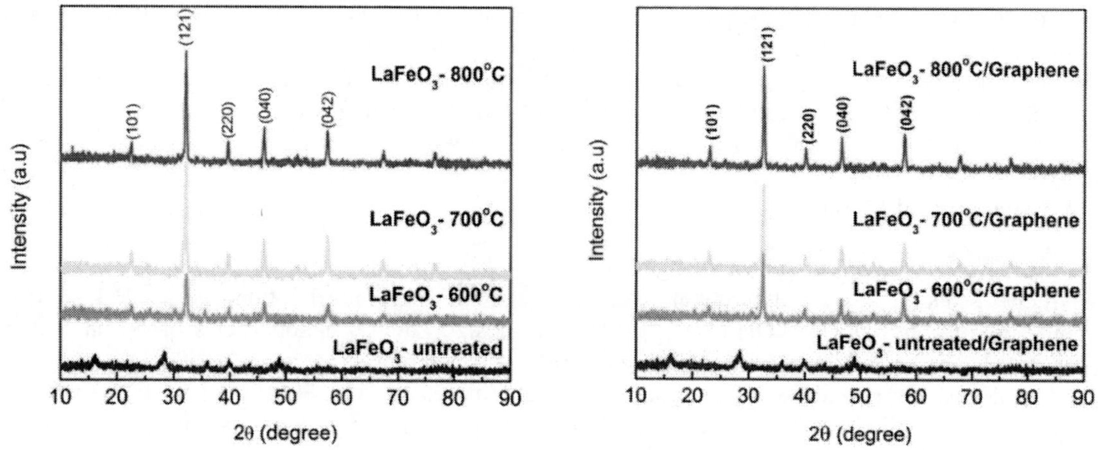


(a) (b)

**Figure 1**. The XRD pattens of LaFeO3 with different calcination temperature without (a), and with additional graphene (b).

**Table 1.** The lattice parameter, grain size <*D*>, and surface area of $LaFeO_3$ with different calcination temperature without and with additional graphene.

| **Sample** | **Lattice Parameter** | | | **<D> (nm)** | **Surface Area ($m^2/g$)** |
|---|---|---|---|---|---|
| | ***a*** **(Å)** | ***b*** **(Å)** | ***c*** **(Å)** | | |
| $LaFeO_3$- 600°C | 5.5647 | 7.8551 | 5.5560 | 22 | 47 |
| $LaFeO_3$- 700°C | 5.5355 | 7.8513 | 5.5553 | 24 | 38 |
| $LaFeO_3$- 800°C | 5.5436 | 7.8456 | 5.5528 | 58 | 31 |
| $LaFeO_3$- 600°C/graphene | 5.4351 | 7.8144 | 5.5586 | 20 | 137 |
| $LaFeO_3$- 700°C/graphene | 5.4592 | 7.8340 | 5.5556 | 24 | 129 |
| $LaFeO_3$- 800°C/graphene | 5.4934 | 7.8403 | 5.5590 | 26 | 123 |

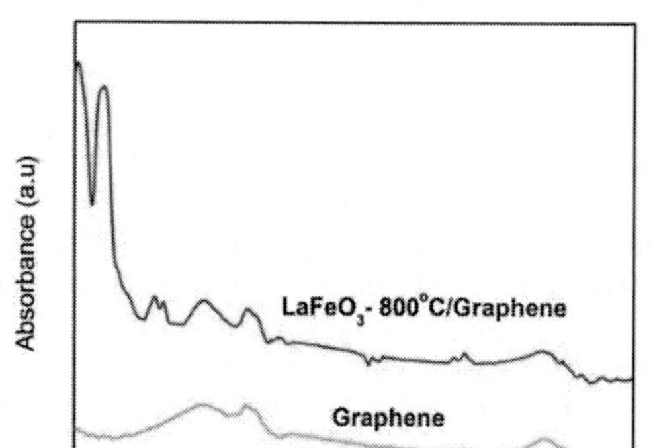


**Figure 2.** FTIR spectra of pure graphene and $LaFeO_3$- 800 °C/graphene composite

IOP Conf. Series: Materials Science and Engineering **202** (2017) 012063 doi:10.1088/1757-899X/202/1/012063

The incorporation of graphene in the composite could be confirmed by using FT-IR measurements. Figure 2 shows the FT-IR spectra of pure graphene and the $LaFeO_3$- 800 °C/graphene composite. The FT-IR spectra of graphene and its composites show a strong absorption at wavenumber 3440 $cm^{-1}$, which corresponds to the O–H stretching vibration [12]. The C=O and C-OH stretching vibrations of graphene are also observed at 1460–1750 $cm^{-1}$ and 1236 $cm^{-1}$, respectively, indicate the formation of graphene in the composite [12-13]. In addition, the presence of $LaFeO_3$ also could be detected with the existence of the Fe–O stretching vibration at wavenumber of 555 $cm^{-1}$ [14]. The 555 $cm^{-1}$ band was due to the Fe–O stretching vibration of octahedral $FeO_6$, which is the characteristic of the perovskite $LaFeO_3$ [14-15].

Figure 3a shows the photocatalytic activity of $LaFeO_3$ with different calcination temperature. In comparison, the photocatalytic activity of untreated $LaFeO_3$ is also shown in the figure. The results showed that the photocatalytic performance of $LaFeO_3$ increased with the increasing calcined temperature. It indicates that the higher crystallinity of the samples could improve the photocatalytic performance of $LaFeO_3$. It was reported that the higher crystallinity could induce the higher carrier mobility, thus providing more exposed polar faces for hydroxyl radicals formation. These radicals were responsible for the degradation of organic pollutant which could increase the photocatalytic performance [16]. The effect of graphene incorporation on the photocatalytic activities of the untreated and treated $LaFeO_3$/graphene composites are also shown in the Figure 3b. The result showed that the incorporation of graphene could further increase the degradation efficiency. Graphene with the large surface area could increase the photocatalytic performance by increasing the surface active site that plays an important role in the degradation process. Moreover, its electron transport layer capacity could inhibit the photo-generated of electron-hole pair recombination. Thus it increased the photocatalytic performance [6-7]. The degradation of MB using the untreated and treated with $LaFeO_3$ nanoparticles and $LaFeO_3$/graphene composites followed the pseudo-first-order kinetic rate equation [17]. The obtained results presented in the Figure 4a that shows the higher crystallinity by increasing calcined temperature could increase the photocatalytic performance. In addition, the incorporation of graphene in the composite exhibited better photocatalytic performance. The stability of the catalyst was also evaluated for four-times repetitions. The results are shown in Figure 4d is indicating that the catalyst has good stability after four cycling processes in the photocatalytic experiment.

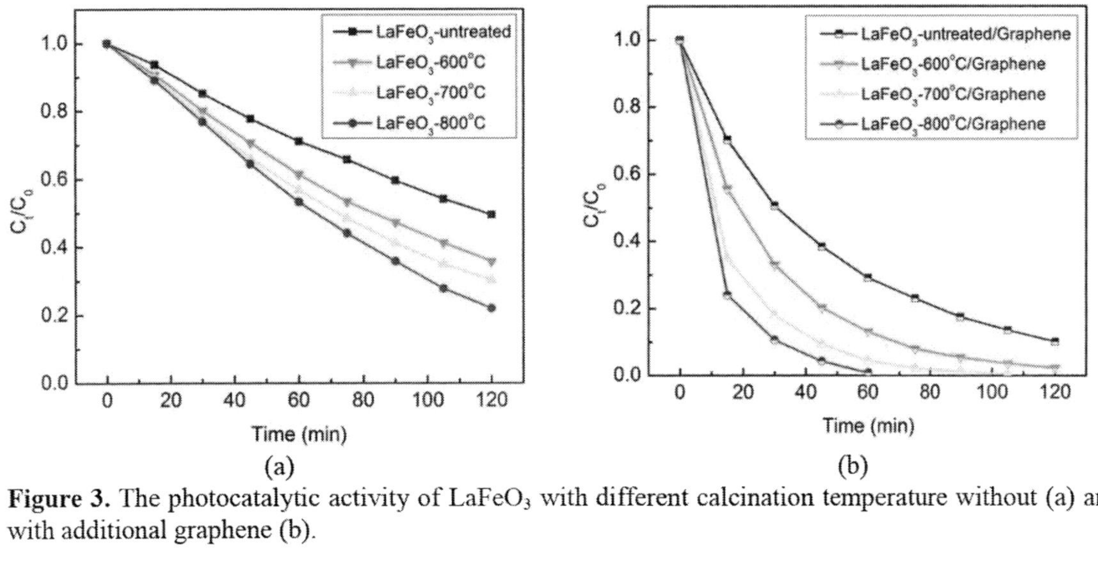


**Figure 3.** The photocatalytic activity of $LaFeO_3$ with different calcination temperature without (a) and with additional graphene (b).

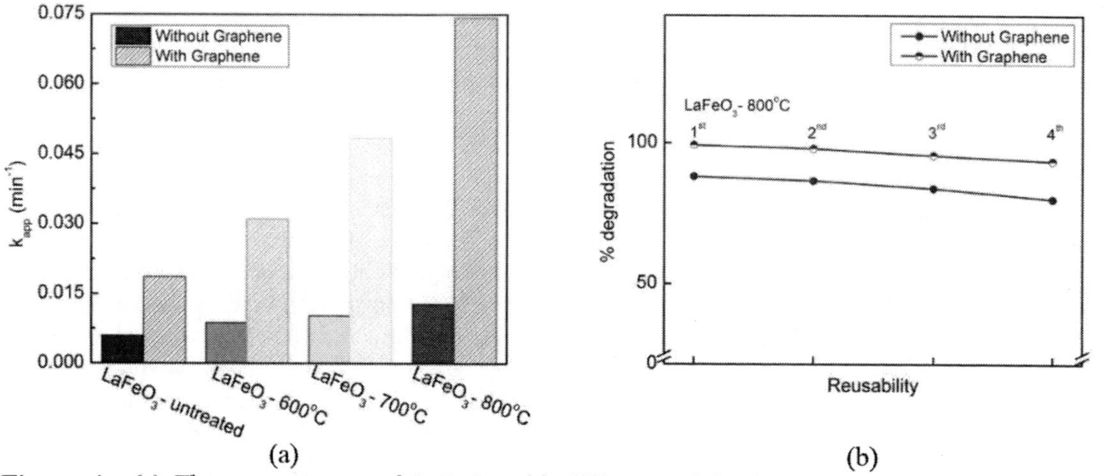


**Figure 4.** (a) The rate constant of $LaFeO_3$ with different calcination temperature without and with additional graphene; and (b) The stability of the catalyst in photocatalytic process

## 4. Conclusion

The photocatalytic activities of the untreated and treated with $LaFeO_3$ nanoparticles and $LaFeO_3$/graphene composites have been investigated under visible light irradiation. The effect of crystallinity of the catalyst improve the photocatalytic performance of $LaFeO_3$. We also obtained that the incorporation of graphene in the composite exhibited better photocatalytic performance. The prepared catalyst also has good stability in the photocatalytic experiment.

## References

[1] Wang Y, Zhu J, Zhang L, Yang X, Lu L, and Wang X 2006 Mater. Lett. **60** 1767
[2] Zhang Q and Saito F 2001 J. Mater. Sci. **36** 5643
[3] Tang P, Tong Y, Chen H, Cao F and Pan G 2013 Curr. Appl. Phys. **13** 340
[4] Abazari R and Sanati S 2013 Superlattice. Microst. **64** 148
[5] Afifah N and Saleh R 2016 J. Phys. Conf. Ser. **710** 012030
[6] Hu J, Men J, Ma J and Huang H 2014 J. Rare Earth. **32** 1126
[7] Hu J, Ma J, Wang L and Huang H 2014 J. Alloy. Compd. **583** 539
[8] Parida K M, Reddy K H, Martha S, Das D P, and Biswal N 2010 Int. J. Hydrogen Energ. **35** 12161
[9] Kato H and Kudo A 2003 Catal. Today **78** 561
[10] Phokha S, Pinitsoontorn S, Maensiri S and Rujirawat S 2014 J. Sol-Gel Sci.Techn. **71** 333
[11] Köferstein R, Jäger L and Ebbinghaus S G 2013 Solid State Ionics **249–250** 1
[12] Fujii T, Matsusue I, Nakatsuka D, Nakanishi M and Takada M 2015 Mater. Chem. Phys. **129** 805
[13] Tang L, Nguyen V H, Lee Y R, Kim J and Shim J J 2015 Synthetic Met. **201**54
[14] Noroozifar M, Khorasani-Motlagh M, Ekrami-Kakhki M-S and Khaleghian-Moghadam R 2014 J. Power Sources **248** 130
[15] Li J, Pan X, Xu Y, Jia L, Yi X and Fang W 2015 Int. J. Hydrogen Energ. **40** 13918
[16] Pudukudy M, Hatieqa A, and Yaakob Z 2014 Appl. Surf. Sci. **319** 221
[17] Saleh R and Djaja N F 2014 Superlattice. Microst. **74** 217

# Factors Influence the Structural and Magnetic Properties of Ag-$Fe_3O_4$ Nanocomposites Synthesized by Reduction Method

**F Fajaroh and Nazriati**

Department of Chemistry, Faculty of Mathematics and Natural Sciences, Universitas Negeri Malang (State University of Malang), Jl. Semarang 5, Malang 65145, Indonesia

Email: fauziatul.fajaroh.fmipa@um.ac.id

**Abstract**. Silver nanoparticles integrated with a magnetic matrix such as $Fe_3O_4$ to form Ag-$Fe_3O_4$ nanocomposites show some advantages in their applications as an antibacterial agent and heterogeneous catalyst. This material can be synthesized by a combination of electrochemical and reduction methods supported by ultrasonic route. The purposes of this study are to synthesize the nanocomposites by reduction method and to study the effect of synthesis parameters on the structural and magnetic properties of the generated nanocomposites. This research consisted of three main stages. The first step was the electrochemical synthesis of magnetite nanoparticles using electro-oxidation of iron in water. Second, adsorption of $Ag^+$ on the surface of $Fe_3O_4$ carried out by sonication of a mixture of $AgNO_3$ solution and powder of $Fe_3O_4$ nanoparticles. The third stage was the reduction of $Ag^+$ to $Ag^0$ with glucose as reductant and NaOH as an accelerator, where it was conducted under a variation of $AgNO_3$ mole ratios to glucose and NaOH concentration. The characterizations of Ag-$Fe_3O_4$ nanocomposites were performed using XRD and VSM. The results of the characterizations showed that Ag-$Fe_3O_4$ nanocomposites have been successfully synthesized. There was an optimum concentration of NaOH at pH (11) and mole ratio of $AgNO_3$ to glucose (1 : 8) in the synthesis process producing the composite with the highest crystallinity.

**Keywords:** Ag, $Fe_3O_4$, nanocomposite, crystallinity, reduction method

## 1. Introduction

Composite is a combination of two or more materials to improve certain specific properties. The nanocomposite is a composite in nanometer dimensions. Nanocomposite has physical, chemical and mechanical properties which are better than the similar materials in macro scale [1]. The states of the nanocomposites in their applications can be colloids, powders, fibers, membranes or films. One of the nanocomposites is a composite between silver and magnetite, Ag-$Fe_3O_4$, which has many benefits, e.g. as a catalyst and an antibacterial [2-3]. The applications of Ag-$Fe_3O_4$ nanocomposite as an antibacterial agent related to the activity of silver that affect the metabolisms of cells and inhibit cell's growth. Silver penetrates into the cell membrane and potent to prevent the synthesis of protein to further and ultimately lead the cell to die [4]. Due to the mentioned properties, the nanocomposite can be applied as an antibacterial in water treatment [5].

The methods for Ag-$Fe_3O_4$ nanocomposite synthesis include reduction method, sonochemical, solvothermal, sol-gel, and co-precipitation methods [6]. The synthesis of the composite in this study based on reduction method, started from the synthesis of $Fe_3O_4$ nanoparticles by electro-oxidation of iron in water [7], and then followed by synthesis of the Ag-$Fe_3O_4$ composite using a combination of adsorption and reduction processes. The adsorption of $Ag^+$ ion on the surface of magnetite was assisted by ultrasonic. Ultrasonic waves will make the agglomeration of magnetite rupture so that the size of the magnetite decreases and it creates greater surface area. A larger surface area of magnetite enhances its adsorption capacity. After the adsorption process, a reduction using sugar as a reducing agent and sodium hydroxide as an accelerator was conducted. Factors suspected to affect the characters of the resulting nanocomposite are the pH of the system and the mole ratio of $AgNO_3$ to glucose. The purposes of this study was to synthesize the nanocomposite by using reduction method and study the effect of $AgNO_3$ mole ratio to glucose and the concentration of NaOH or the pH system to the character of Ag-$Fe_3O_4$ nanocomposite.

## 2. Experimental method

This research consisted of three main stages. The first step was the electrochemical synthesis of magnetite nanoparticles i.e. electro-oxidation of iron in water under DC 70V. The iron resulted via electroplating process using $FeSO_4.7H_2O$ (Merck) as the electrolyte. Second, adsorption of $Ag^+$ on the surface of $Fe_3O_4$ was carried out by sonication of a mixture of a solution of $AgNO_3$ (Merck) and powder of $Fe_3O_4$ nanoparticles under the frequency of 42 kHz. The third stage was the reduction of $Ag^+$ to $Ag^0$ with glucose as a reductant and NaOH (Merck) as an accelerator. The last step was conducted with a variation of $AgNO_3$ mole ratios to glucose and the concentration of NaOH. Characterization of the Ag-$Fe_3O_4$ nanocomposite was done using XRD (Panalytical Xpert Pro) and VSM (Type 1.2H VSM, Oxford). The XRD spectra of the composite were then confirmed by the XRD standard pattern of magnetite (JCPDS Card No. 19-629) and silver (AMCSD Code No. 0011135).

## 3. Results and Discussion

The synthesis of Ag-$Fe_3O_4$ nanocomposite assisted by ultrasonic was conducted by dissolution of silver nitrate in demineralized water, so $AgNO_3$ dissociates into $Ag^+$ and $NO_3^-$, and then they were mixed with $Fe_3O_4$ and subsequently sonicated at room temperature. Glucose was used as a reducing agent to change silver (1) to silver (0). NaOH was added to the system for being the accelerator in the reduction process. The reaction equation of such process is described in the following equations.

$$Ag^+ + Fe_3O_4 \rightarrow [\,Ag(Fe_3O_4)]^+ \quad (1)$$
$$2\,[\,Ag(Fe_3O_4)]^+ + 2\,OH^- + C_5H_{11}O_5\text{-}CHO \rightarrow 2\,Ag + 2\,Fe_3O_4 + H_2O + C_5H_{11}O_5\text{-}COOH \quad (2)$$

Theoretically, as shown in equation (2), the concentration of NaOH and the mole ratio between $AgNO_3$ and $Fe_3O_4$ will affect the process. The statement is confirmed by the results of the product's characterization.

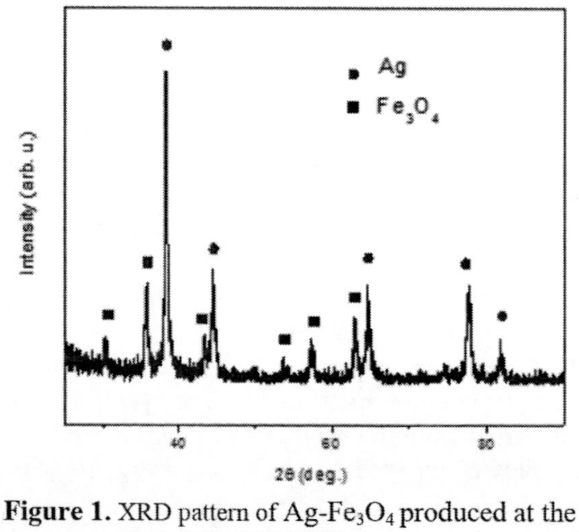


**Figure 1.** XRD pattern of Ag-$Fe_3O_4$ produced at the concentration of 0.04 NaOH M

Figure 1 presents the result of XRD analysis of Ag-$Fe_3O_4$ produced in the concentration of 0.04 M NaOH. As shown in Figure. 1, such spectrum presents two kinds of peaks belong to $Fe_3O_4$ and Ag, respectively, i.e. at 2θ 30,2°; 35,6°;53,6°;57,1°; 62,6°, and 38,1°, 44,3°, 64,4°, 77,5°, 81,5°. This diffraction data proved that a high purity of Ag-$Fe_3O_4$ has been successfully prepared. This conclusion was also supported by the corresponding magnetization curve that was produced by VSM analysis as shown in Figure. 2. As described in Figure 2, the non-magnetic material of silver that was coated at magnetite's surface reduced the magnetization saturation ($M_s$) of the particles. The same result was found by Jiang [2]. The SEM image of the particles is presented in Figure 3. After quantitative analysis, it can be seen that the average size of the particles is about 42.1 nm.

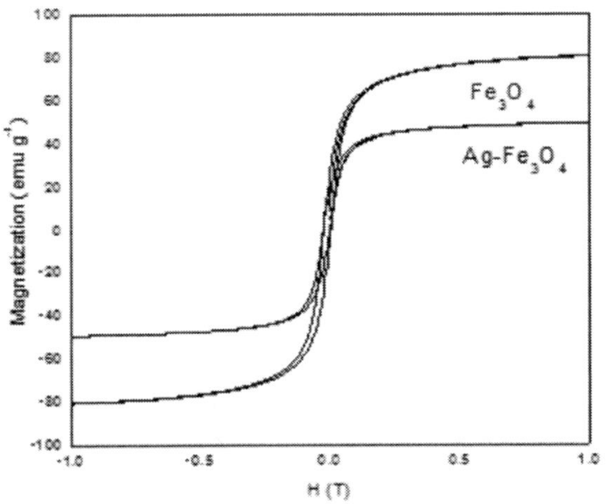


**Figure 2.** Magnetization curve of Ag-$Fe_3O_4$ produced at the concentration of 0.04 NaOH M

**Figure 3.** SEM image of Ag-$Fe_3O_4$ produced at the concentration of 0.04 NaOH M

The presence of silver on the surface of the magnetite particles improved its crystallinity. The XRD patterns presented in Figure. 4 shows the highest intensity of the samples at 0.04 M concentration of NaOH that corresponds to pH = 11. Such pH is the optimal pH of the synthesis process. The crystallinity initially increased along with the increasing concentration of NaOH, until the optimum condition was reached, but an addition of a higher concentration of NaOH decreased the crystallinity. At a higher concentration of NaOH, the excess $OH^-$ prevented the reduction of silver ions.

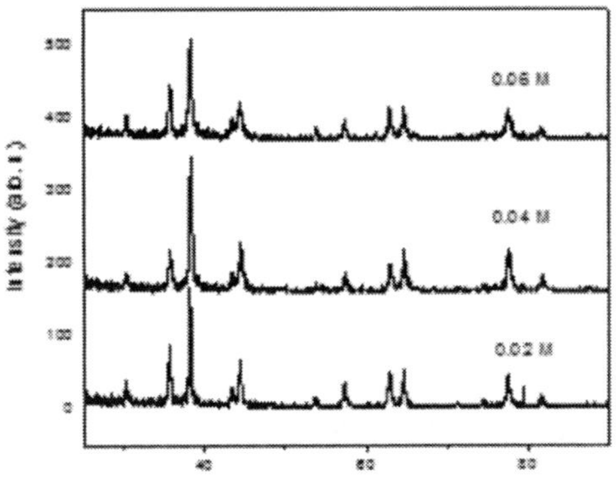


**Figure 4.** XRD patterns of Ag-$Fe_3O_4$ under a variation of NaOH concentration

As shown in Figure 5, in the synthesis of Ag-$Fe_3O_4$ with **a** variation of mole ratio of $AgNO_3$ to glucose, the highest intensity was obtained at the mole ratio of 1: 8 as shown in Figure 5. However, NaOH had the more reductant inhibition performance as an accelerator.

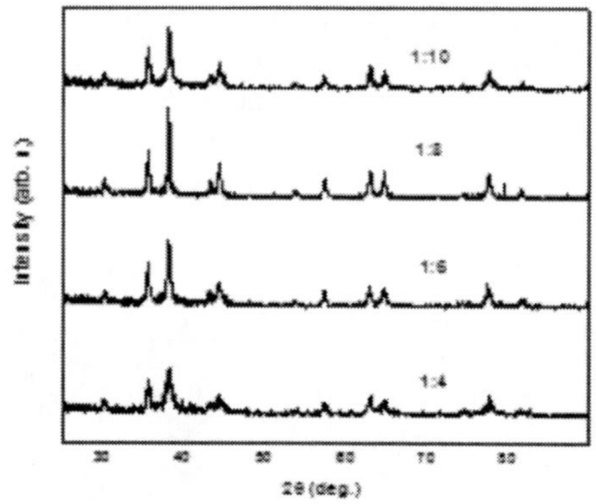


**Figure 5.** XRD pattern of the particles generated under various $AgNO_3$ mole ratios to glucose

## 4. Conclusion

Ag-$Fe_3O_4$ nanocomposite has been successfully synthesized by a combination of electrochemical and reduction methods. The pH of the synthesis system and the mole ratio of reactants influence the character of the particles. There is an optimum pH = 11 and the mole ratio 1 : 8 of reactants in the formation of Ag-$Fe_3O_4$.

## 5. References

[1] Rao T K, Viswanath I V K, Murthy Y L N 2016 A comparative study on antibacterial activity of silver-coated cobalt ferrit nanocomposite and nanosilver particles. *International Journal of Advanced Technology in Engineering and Science* **4** Special issue 01

[2] Wanquan Jiang, Yufeng Zhou,Yanli Zhang, Shouhu Xuanb and Xinglong Gong 2012 Superparamagnetic Ag@$Fe_3O_4$ core–shell nanospheres: fabrication, characterization and application as reusable nanocatalysts. *Dalton Trans* **41** 4594.

[3] Dong-Hui Zhang, Guo-Dong Li, Ji-Xue Li and Jie-Sheng Chen 2008 A magnetically recyclable and efficient catalyst for epoxidation of styrenew *Chem. Commun* **2008** *3414–3416.*

[4] Montazer M, Hajimirzabab H, Rahimi M K Alibakhshi, S 2012 Durable Anti-bacterial Nylon Carpet Using Colloidal Nano Silver. *FIBRES & TEXTILES in Eastern Europe* **4**:96-101.

[5] Amarjargal A, Leonard D, Tijing & Cheol S K 2014 Simultaneous preparation of Ag/$Fe_3O_4$ core-shell nanocomposites with enhanced magnetic moment and strong antibacterial and catalytic properties *Chemical Engineering Journal* **4**:54.

[6] Xueping Zhang, Wanquan Jiang, Xinglong Gong, Zhong Zhang 2010 Sonochemical synthesis and characterization of magnetic separable $Fe_3O_4$ / Ag composite and its catalytic properties. *Journal of alloys and compound* **508** 400-405.

[7] Fajaroh F, Sutrisno, Nazriati N, Wonorahardjo S 2014 To enhance the purity and crystallinity of magnetite nanoparticles prepared by surfactant-free electrochemical method by imposing higher voltage. *AIP proceeding* **1586** 179.

IOP Conf. Series: Materials Science and Engineering **202** (2017) 012065 doi:10.1088/1757-899X/202/1/012065

# Sonocatalytic Degradation of Methylene Blue with $LaMnO_3$ Supported by Different Surface Area of Graphene

**N Afifah[1,2] and R Saleh[1,2]**

[1] Department of Physics, Faculty of Mathematics and Natural Sciences, Universitas Indonesia, Kampus UI Depok, 16424 Depok, Indonesia
[2]Integrated Laboratory of Energy and Environment, Fakultas MIPA, Universitas Indonesia, Kampus UI Depok, 16424 Depok, Indonesia

Email: rosari.saleh@ui.ac.id

**Abstract**. The present study compared the sonocatalytic degradation ability of $LaMnO_3$ supported by different surface area of graphene on methylene blue (MB) as a model of the organic pollutant under ultrasonic irradiation. Both sonocatalysts were synthesized by using co-precipitation method and characterized by using X-Ray Diffraction (XRD), Fourier Transform Infrared, and Brunauer-Emmet-Teller (BET) surface area analysis. The orthorhombic structure of $LaMnO_3$ was detected on the composites, while the existence of graphene was also confirmed. The present of graphene in the samples could improve the magnetic properties and surface area of composites as well as the sonocatalytic performance. $LaMnO_3$ supported by different surface area of graphene (SA-1 and SA-2) showed the significant improvement of sonocatalytic activity compared to $LaMnO_3$. $LaMnO_3$/SA-2 composites with the largest surface area showed the highest degradation efficiency. To detect the possible reactive species involved in the degradation of MB, control experiments with introducing scavenger into the solution of MB were carried out. The results indicated that photo-generated hole played an important role in the degradation of MB.

## 1. Introduction

Photocatalytic degradation of organic pollutant using $LaMnO_3$/graphene composite as a catalyst is promising for water purification. $LaMnO_3$ has been chosen due to its unique physical and chemical properties, its low cost, its high catalytic activity and environmental friendliness. In addition, $LaMnO_3$ also shows good respond as a visible light catalyst which makes it a promising candidate to maximize the utilization of solar energy in the catalytic field [1]. The graphene with a large specific surface area, excellent adsorption capacity, and high chemical stability also can be compounded with $LaMnO_3$ to enhance the catalytic performance[2-3]. It was reported that the catalyst with the large surface area could increase the catalytic activity by increasing the number of active sites which could further degrade the organic pollutant. In order hand, graphene also exhibits the electron transport layer capacity which can inhibit the electron-hole recombination and enhance the catalytic performance [1, 4].

The photocatalytic activity can degrade organic pollutant in wastewater as well as the sonocatalytic activity. Recently, the ultrasonic irradiation is also used for degrading of organic contaminants based on the hot spot and sonoluminescence mechanism [5]. The sonoluminescence involves a broad range of wavelength such as UV light, which makes the catalyst behaves like photocatalyst during sonication

[6-7]. The photocatalytic activity of $LaMnO_3$/graphene composite has been reported by several studies [1, 4]. However, very few reports are available on the sonocatalytic activity using $LaMnO_3$/graphene composite as a catalyst. Therefore, in this work $LaMnO_3$ composite with a different surface area of graphene ($LaMnO_3$/SA) has been synthesized and its sonocatalytic activity has been evaluated by using methylene blue (MB) as a model organic pollutant under ultrasonic irradiation.

## 2. Experimental details

*2.1 Catalyst preparation*

$LaMnO_3$ nanoparticles were prepared by using the co-precipitation method based on our previous study [8]. Lanthanum (III) chloride heptahydrate ($LaCl_3.7H_2O$), manganese (II) chloride tetrahydrate ($MnCl_2.4H_2O$), and sodium hydroxide (NaOH), purchased from Merck, were used as the reagents. Nanographene platelets and graphene with different surface areas of 30 $m^2/g$ (SA-1) and 400 $m^2/g$ (SA-2), purchased from Angstron Materials, were used to synthesize the $LaMnO_3$/SA composites, respectively. The $LaMnO_3$/SA composites were prepared by following these steps: Briefly, graphene with different surface areas were dissolved into a solution of water and ethanol through ultrasonic treatment for 2 hours, respectively. After that, $LaMnO_3$ was added to each solution then stirred for 2 hours. The mixture was then heated at 120 °C for 3 h to influence the deposition of $LaMnO_3$ on the graphene sheets. Then, centrifugation was used to separate the precipitate and the solution. At last, the $LaMnO_3$/SA composite powders were obtained by drying the precipitate at 70 °C for 12 hours. The content of graphene in $LaMnO_3$ was constantly held and estimated to be 5 wt.%.

*2.2 Characterization*

The samples were characterized by using X-Ray Diffraction (XRD), Fourier Transform Infrared, and Brunauer-Emmet-Teller (BET) surface area analysis spectroscopy.

*2.3 Sonocatalytic test*

The sonocatalytic activities of prepared samples were observed by measuring the degradation of MB under ultrasonic irradiation. The detailed procedure has been described in our previous work [8]. As an irradiation sources, the commercial ultrasonic bath which operated at fixed frequency and power of 40 kHz and 150 W, respectively, was used. The degradation of MB was monitored by using a Hitachi UH5300 UV-visible spectrophotometer. The degradation of MB was recorded as $C_t/C_0$, where $C_t$ is the concentration of MB at each measurement interval during irradiation and $C_0$ is the initial concentration of MB after reaching adsorption–desorption equilibrium. The maximum dosage concentration on the sonocatalytic process was checked by varying the amount of catalyst from 0.05/L to 0.40 g/L. The main active species involved in the sonocatalytic degradation of MB were studied by adding several substances such as sodium sulfate, di-ammonium oxalate, and tert-butyl alcohol, which are recognized as scavengers for electron, hole, and hydroxyl radical species, respectively.

## 3. Results and Discussion

The XRD patterns of $LaMnO_3$ composites with the different surface area of graphene ($LaMnO_3$/SA) are presented in Figure 1a. The XRD patterns of $LaMnO_3$ nanoparticles, nanographene platelets (SA-1), and graphene (SA-2) are also shown in this figure. The orthorhombic structure is detected on both the $LaMnO_3$ nanoparticle and its composites. For $LaMnO_3$/SA-1 composite, the diffraction peaks revealed the presence of graphite structure which confirmed from the observed 2θ values at 26.4° (002) for nano graphene platelets (NGP). However, the present of graphene on $LaMnO_3$/SA-2 composite did not change the main diffraction peaks of $LaMnO_3$. This may be due to the low content of graphene in the composite and the broad XRD peak of graphene. The absence of other phases of $LaMnO_3$ nanoparticles, NGP and graphene diffraction peaks in the spectra confirmed that the synthesized samples were the desired materials, $LaMnO_3$/SA composites. The lattice parameter and crystallite size <D> values were obtained by using the Rietveld refinement method using MAUD program and were calculated by using Scherrer's formula [9] and also summarized in Table 1.

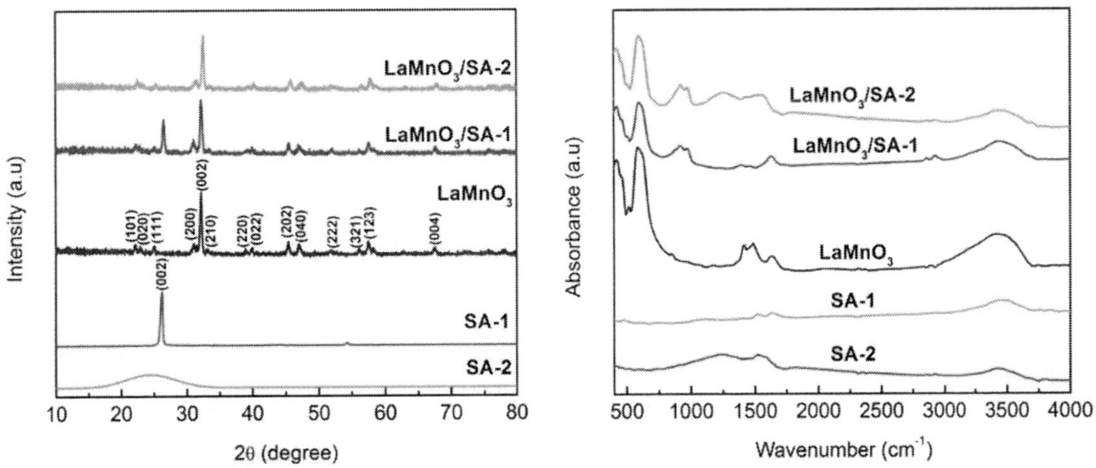


(a) (b)

**Figure 1.** (a) XRD pattern; and (b) FT-IR spectra of $LaMnO_3$ composites with different surface area of graphene.

**Table 1.** The lattice parameter, grain size <D>, surface area and rate constant of $LaMnO_3$ composites with different surface area of graphene.

| Sample | Lattice Parameter | | | <D> (nm) | Surface Area ($m^2/g$) | Rate Constant ($min^{-1}$) |
|---|---|---|---|---|---|---|
| | a (Å) | b (Å) | c (Å) | | | |
| $LaMnO_3$ | 5.7419 | 7.7215 | 5.5403 | 44 | 17 | 0.0156 |
| $LaMnO_3$/SA-1 | 5.7586 | 7.7090 | 5.5464 | 23 | 34 | 0.0232 |
| $LaMnO_3$/SA-2 | 5.6859 | 7.7207 | 5.4925 | 21 | 87 | 0.0689 |

Figure 1b displays the FT-IR spectra of the $LaMnO_3$/SA composites with a different surface area of graphene. As a comparison, the FT-IR spectra of NGP (SA-1) and graphene (SA-2) are also shown. As seen in the figure, the spectra of both composites shows a strong absorption peak at 593 $cm^{-1}$, which could be ascribed to the stretching vibration of metal-oxygen bond in $LaMnO_3$ [10]. The incorporation of NGP and graphene could also be confirmed by the presence of absorption peaks at a wavenumber of 1460–1750 $cm^{-1}$ and 3440 $cm^{-1}$, which are attributed to the C=O and O-H stretching vibration, respectively [11-13]. The C-OH stretching vibration at a wavenumber of 1236 $cm^{-1}$ was also found; it indicated the formation of graphene in the $LaMnO_3$/SA-2 composite [13].

Figure 2a shows the sonocatalytic degradation rate of MB in the presence of $LaMnO_3$ nanoparticles and $LaMnO_3$/SA composites with a different surface of graphene. As clearly seen in the figure, the incorporation of NGP and graphene in the composites could increase the sonocatalytic performance of $LaMnO_3$ nanoparticles. The rate constant ($k_{app}$) of the samples was calculated by using the pseudo-first-order kinetic rate equation [4]. The result (Table 1) shows that the incorporation of graphene with a larger surface area in the composites exhibits the best sonocatalytic performance. It has been suggested that the surface area of the catalyst is one of the important factors that plays the roles in degradation efficiency. The catalyst with the large surface area could improve the degradation efficiency by increasing the contact probability between light and its active site. Therefore, the rate at which the oxidation process happens may further increase [1, 4]. The surface area of $LaMnO_3$/SA composites obtained by using BET method is also presented in Figure 2b. The composite BET surface area increases with the increase of graphene surface area. This result indicates that $LaMnO_3$/SA composite with the largest surface area also shows the highest sonocatalytic performance.

To obtain the maximum dosage concentration for degrading MB under ultrasonic irradiation, the amount of $LaMnO_3$/SA-2 composite with the largest surface area was varied from 0.05 g/L to 0.40 g/L. The results that are shown in Figure 3a suggested that 0.1 g/L of $LaMnO_3$/SA-2 composite shows maximum dosage concentration for degrading MB in the sonocatalytic experiment.

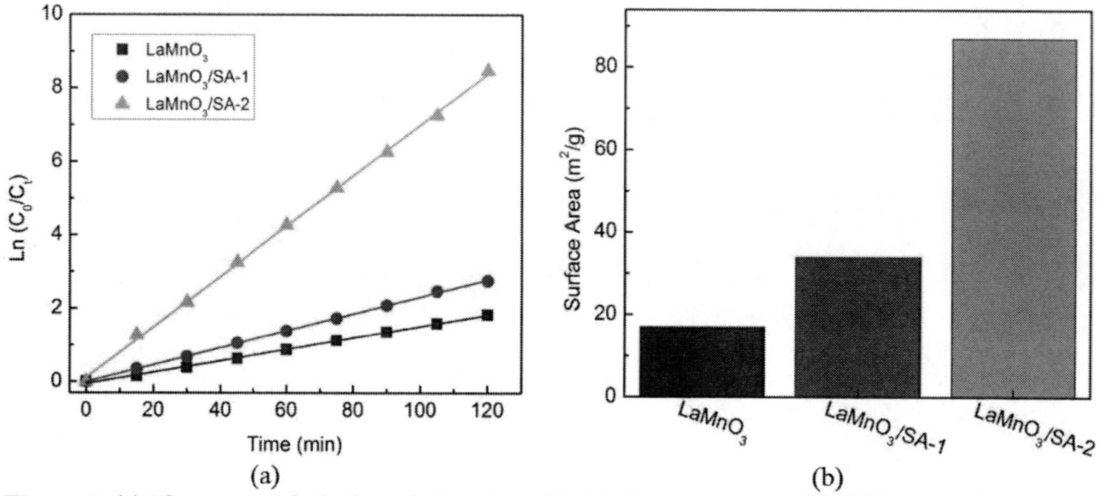


**Figure 2.** (a) The sonocatalytic degradation rate of $LaMnO_3$ composites with different surface area of graphene; (b) The BET surface area of the samples

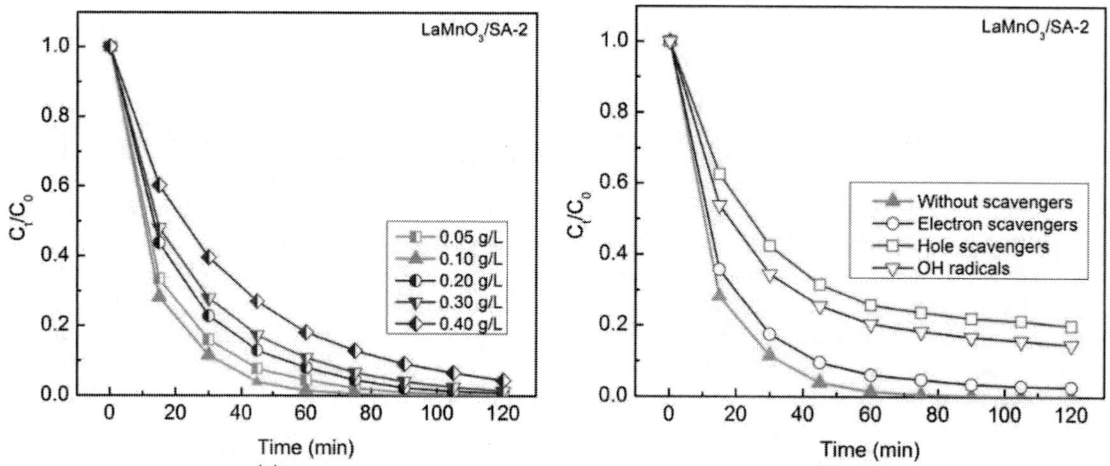


**Figure 3.** (a) The effect of catalyst dosage; and (b) scavengers on photocatalytic degradation of MB by using $LaMnO_3$/SA-2 composite;

After getting the maximum dosage concentration of the catalyst, several active species such as an electron, hole, and hydroxyl radical scavengers were added into MB solutions to study the main active species in the sonocatalytic experiment. $Na_2S_2O_8$, diammonium oxalate, and tertbutyl alcohol were used as an electron, hole, and hydroxyl radical scavengers. The results (shown in Figure 3b) revealed that the presence of scavenger could decrease the degradation rate of MB. The largest decrease was obtained with the present of di-ammonium oxalate as a hole scavenger. It indicates that hole plays an important role in the sonocatalytic experiment. The large surface area of graphene could increase the sonocatalytic performance as well as its electrons transport layer capacity. During the irradiation time, graphene collected the photo-generated electron in the conduction band of $LaMnO_3$ into its layers to interact with oxygen ($O_2$) and form superoxide radicals ($\bullet O_2^-$), thus inhibiting the recombination of photogenerated electron-hole pairs [1, 4]. At the same time, the hole left in the valence band of $LaMnO_3$ could move freely to be able to react with water molecules, resulting in the total hydroxyl radicals (•OH) obtained in the sonocatalytic process increase [15-16]. Hence, this result confirmed that hole plays important role in the sonocatalytic experiment.

## 4. Conclusion

The sonocatalytic activity of $LaMnO_3$/SA composites with a different surface area of graphene has been evaluated for degrading MB under ultrasonic irradiation. The presence of graphene in $LaMnO_3$/SA composites could increase the sonocatalytic performance of $LaMnO_3$ nanoparticles. $LaMnO_3$/SA-2 with the largest surface area exhibited the best sonocatalytic performance. The maximum dosage concentration of the catalyst was 0.1 g/L and the hole played a major role in the degradation of MB.

**References**

[1] Hu J, Ma J, Wang L and Huang H 2014 *J. Alloy. Compd.* **583** 539
[2] Kim T W, Park M, Kim H Y and Park S J 2016 *J. Solid State Chem.* **239** 91
[3] He G L, Zhong Y H, Chen M J, Li Z, Fang Y P and Xu Y X 2016 *J. Mol. Catal. A: Chem.* **423** 70
[4] Hu J, Men J, Ma J and Huang H 2014 *J. Rare Earth.* **32** 1126
[5] Sathishkumar P, Mangalaraja R V and Anandan S 2016 *Renew. Sust. Energ. Rev.* **55**, 426
[6] Hu B, Wu C, Zhang Z and Wang L 2014 *Ceram. Int.* **40** 7015
[7] Bremner D, Burgess A and Chand R 2011 *Curr. Org. Chem.* **15** 68
[8] Afifah N and Saleh R 2016 *Mater. Sci. Forum* **864** 99
[9] Köferstein R, Jäger L and Ebbinghaus S G 2013 *Solid State Ionics* **249–250** 1
[10] Wenwei W, Jinchao C, Xuehang W, Sen L, Kaituo W, and Lin T 2013 *Adv. Powder Technol.* **24** 154
[11] Taufik A and Saleh R 2016 *AIP Conf. Proc.* **1725** 020089
[12] Fujii T, Matsusue I, Nakatsuka D, Nakanishi M and Takada M 2015 *Mater. Chem. Phys.* **129** 805
[13] Tang L, Nguyen V H, Lee Y R, Kim J and Shim J J 2015 *Synthetic Met.* **201**54
[14] Saleh R and Djaja N F 2014 *Superlattice. Microst.* **74** 217
[15] Reddy R D, Dinesh G K, Anandan S and Sivasankar T 2016 *Chem. Eng. Process.* **99** 10
[16] Gogate P and Pandit A 2004 *AIChE J.* **50** 1051

# Nanoneedles of Lanthanum Oxide ($La_2O_3$): A Novel Functional Material for Microwave Absorber Material

**W A Adi, S Wardiyati and S H Dewi**

[1]Centre for Science and Technology of Advanced Materials, BATAN, Kawasan Puspiptek Serpong Tangerang Selatan Banten, Indonesia

Email: dwisnuaa@gmail.com

**Abstract.** Synthesis and characterization of $La_2O_3$ nanoneedles have been successfully carried out by using hydrolysis reaction with $NH_4OH$ as a solvent. $La_2O_3$ was prepared from $LaCl_3$ solution. $LaCl_3$ was preheated on a hotplate and stirred by using a magnetic stirrer, then added solution of $NH_4OH$ 1 M with the variation speed drops of 3 ml, 6 ml and 9 ml to obtain white precipitation $La(OH)_3$ until a pH of 9. Stirring was continued until 1 hour. A white precipitate formed is separated from the effluent by centrifugation method with 3000 rpm and neutral pH. Then, the precipitate was dried in an oven at a temperature of 70 °C and sintered at 1000 °C to get $La_2O_3$ nanoneedles. The identification of sample phase was carried out by using X-ray diffractometer (XRD). The refinement results of X-ray diffraction pattern shows that the samples have a single phase of $La_2O_3$. The morphology of the nanoneedles was observed by using the transmission electron microscope (TEM). Meanwhile, the reflection and transmission of electromagnetic wave were measured by using the vector network analyzer (VNA) at X-band frequency.

**Keywords**: Lanthanum oxide, nanoneedle, rare earth, microwave absorber.

## 1. Introduction

Rare earth metals consist of 15 elements in the lanthanide group from the periodic table, however, scandium (Sc) and yttrium (Y) can be categorized as rare earth metals the similarities in nature and also the existence in nature tend to be along with elements of the lanthanide. Rare earth metals in Indonesia are found in the mineral monazite which is associated minerals incorporated with lead and radioactive element [1-2]. Because as a by-product of tin processing results then only sold at very cheap prices, even at some points thrown away, whereas in the mineral contained rare earth elements having high sales value and much needed in some industries. The dominant elements of rare earth mineral contained in monazite are lanthanum (La), cerium (Ce), neodymium (Nd) and praseodymium (Pr) [3]. The economic potential of monazite previously into tin mining by-products can be increased as added value which is in line with their needs as large as the raw material of advanced industry, in addition to the China policy about export restrictions for the rare earth metals.

Recently, utilization of the rare earth metals is very spacious, especially in the development of high technology. Europium (Eu) is used for computer monitors, erbium (Er) for a cable of fibre optic, neodymium (Nd), samarium (Sm), gadolinium (Gd), dysprosium (Dy), and praseodymium (Pr) for permanent magnets, while lanthanum (La) has been applied widely as functional magnetic materials. There is much more utilization of LTJ which has not been replaced by another material yet [4-6].

An interesting case to be studied is the use of nanoparticles of lanthanum oxide as absorbing microwaves [7-10]. Lanthanum (La) is the lightest element in the lanthanide group, and has been widely studied in the form of oxides, hydroxides, and phosphates. Various chemical methods have been used to make nanostructures of lanthanum oxide. Synthesizing $La(OH)_3$ nanotubes has been succeeded through hydrothermal method. $La(OH)_3$ nanorods have been made through chemical methods. In addition, and a simple method to make $La(OH)_3$ nanowires has been reported[11-16].

This research will perform the synthesis of lanthanum oxide which focuses on the process of forming nanoneedles. The method used is a hydrolysis method. In addition, this section also discusses the factors that influence during the synthesis process namely pH value and temperature. So, it is expected to find the factors that play role important in controlling the morphology of these nanoneedles. Besides, this paper will present several challenges of scientific with respect to further investigation about microwave absorption capability at this material.

## 2. Experimental Method

Powders of $La_2O_3$ were synthesized by the hydrolysis method. The raw materials used were a solution of $LaCl_3$ and ammonia. The first stage of the process conducted for this synthesis was the solution of $LaCl_3$ was heated by using a hotplate at a temperature of 80 °C and stirred by using a magnetic stirrer with a speed of 70 rpm, then added with a solution of ammonia of 1 M by using a peristaltic pump with variations of drops speed of 3 ml/min, 6 ml/min and 9 ml/min for the deposition process up to reach a pH of 9. Stirring was continued for 1 hour. The third of white precipitate formed is separated from the effluent through the method of centrifugation at 3000 rpm until the pH back to neutral. Then the precipitate was dried in an oven at a temperature of 70 °C. After that, the precipitate furthermore was sintered in the electric chamber furnace at 1000 °C for 5 hours to obtain crystalline powder of $La_2O_3$ materials.

The reactions that occur during the hydrolysis process are as follows:

$$LaCl_3 + 3H_2O + 3NH_3 \rightarrow La(OH)_3 + 3NH_4Cl$$

$$2La(OH)_3 \rightarrow La_2O_3 + 3H_2O\uparrow$$

The particle morphology of $La_2O_3$ powder resulting from differences in drops speed of 3 ml/min, 6 ml/min and 9 ml/min was observed by using a transmission electron microscope (TEM) JEOL brand. Meanwhile, the phase identification of the three samples was characterized by using X-ray diffraction (XRD) Analytical Pan brand with CuKα radiation ($\lambda$ = 1.5406 Å). The crystal structure analysis of the sample is calculated based on analysis by using the Rietveld method namely GSAS programs (general structure analysis system) to determine the crystal structure parameters of each sample and the presence of crystal growth in the hydrolysis process results [17]. The calculating electromagnetic wave absorption in this material was by using a vector network analyzer (VNA) with a frequency range of 8-12 GHz (X-band).

## 3. Results and Discussion

IOP Conf. Series: Materials Science and Engineering **202** (2017) 012066 doi:10.1088/1757-899X/202/1/012066

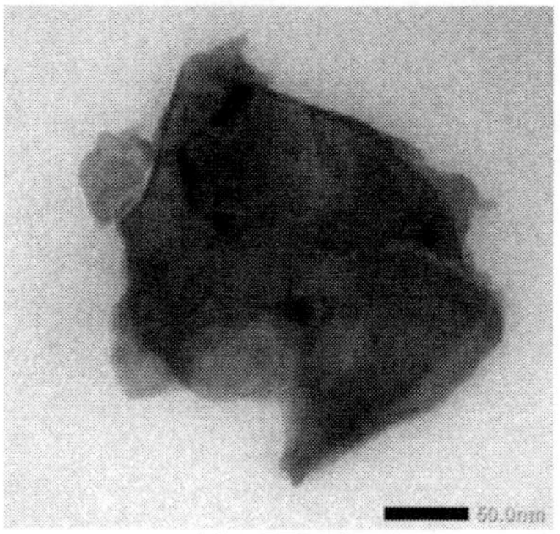

(a) TEM image

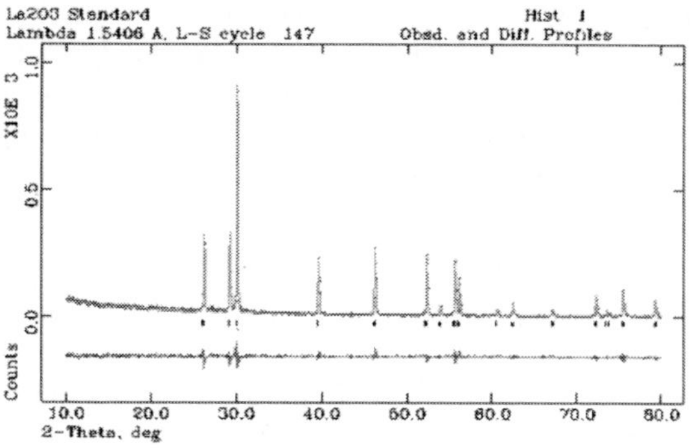


(b) Refinement result of X-ray difraction pattern

**Figure 1**. Morphology and phases analysis of $La_2O_3$ nanoparticle

At the beginning of this study, the the nanoparticles morphology of $La_2O_3$ commercial product (standard) was observed to get a general idea about particle morphology of $La_2O_3$. This morphological observation using a transmission electron microscope (TEM) with a magnification scale of 50 nm is shown in Figure 1(a).

Figure 1(a) shows nanoparticles morphology of $La_2O_3$ standard in which polygonal solid shaped with particle sizes was around 200 nm. The interesting thing about the image that the polygonal shape is a basic form of nanoparticles $La_2O_3$ corresponding to the shape of this crystal structure. The measurement results of X-ray diffraction pattern of the standard $La_2O_3$ nanoparticles is shown in Figure 1(b).

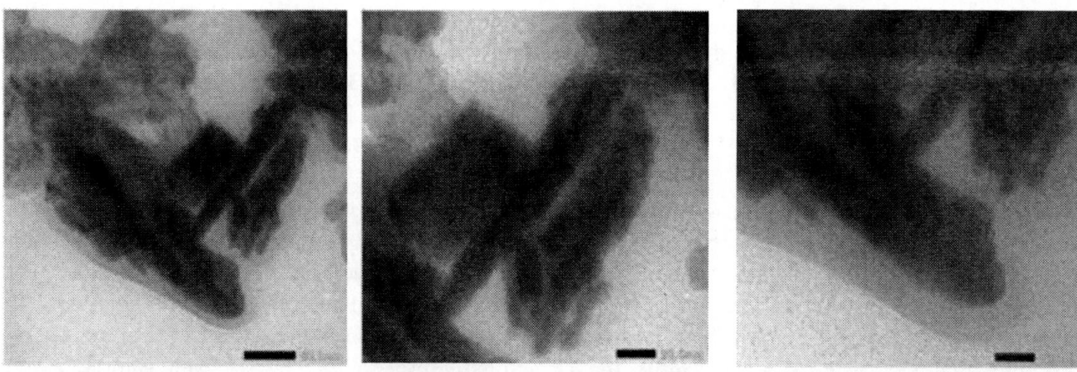

(a) Drops speed of 3 ml/min

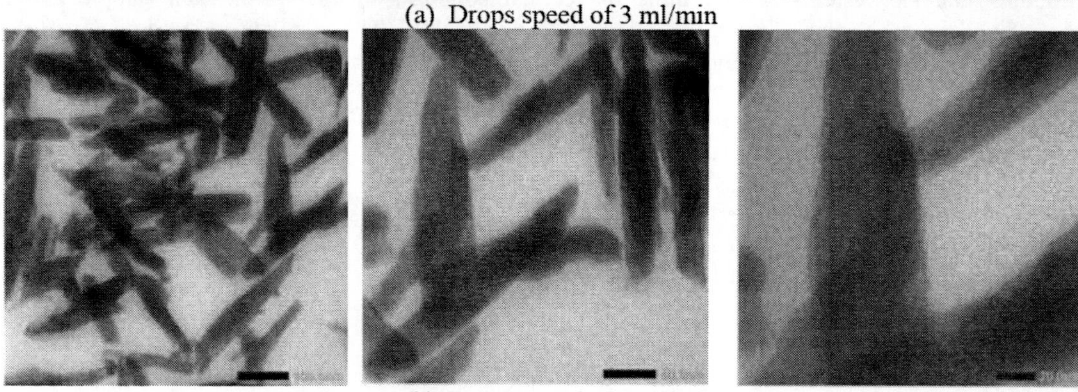

(b) Drops speed of 6 ml/min

(c) Drops speed of 9 ml/min

**Figure 2**. Nanoparticles morphology of $La_2O_3$ nanoneedles

The refinement of XRD profile for $La_2O_3$ standard as shown in Figure 1(b) appears that the fitting X-ray diffraction pattern has a very good fitting quality based on the criteria of fit (Rwp) and goodness of fit ($\chi^2$) in accordance with the agreement [17]. Rwp is the weight ratio of the difference between the XRD pattern of observation and calculation (ideal value of Rwp < 10%). Meanwhile, χ2 (chi-squared) is the ratio of the XRD pattern of observation results comparable with expectations (ideal value of $1 < \chi^2 < 1.3$). The refinement results of x-ray diffraction pattern confirmed that the sample is a single phase of $La_2O_3$ with a hexagonal structure, space group of P 63/m m c, meanwhile, the unit cell parameters for the $La_2O_3$ phase is summarized as the following: $a = b = 3.9346(1)$ Å and $c = 6.1265(1)$ Å, $\alpha = \beta = 90°$ and $\gamma = 120°$, V = 82.13(1) Å$^3$, $\rho$ = 5.030 gr.cm$^{-3}$,*w*Rp = 6.7% and $\chi^2$ (chi-squared) = 1.14. The results of nanoparticle morphology is similar to the results of research of Mazloumi [18] who has managed to make nanostructure lanthanum hydroxide powders by using chemical process with KOH and NaOH as the solvents.

However, the process result of this hydrolysis has different morphological forms. The TEM image observations have obtained the morphology of $La_2O_3$ nanoparticles shaped nanoneedles. Figure 2 shows the morphology of $La_2O_3$ nanoneedles according to a drops speed of 3 ml/min, 6 ml/min and 9 ml/min ammonia. The looks growth occurs on the specific field orientation in the $La_2O_3$ crystal.

Figure 2(a) is a TEM image of $La_2O_3$ nanoneedles with the drops speed of 3 ml/min. It appears that the growth of $La_2O_3$ nanoneedles has been successfully formed although there are still several different particle shapes. This phenomenon needs to be confirmed based on the analysis phases result of the X-ray diffraction pattern. $La_2O_3$ nanoneedles formed has a diameter of 30 nm by variation of the length in ranges from 50-150 nm. Figure 2(b) is a TEM image of $La_2O_3$ nanoneedles with the drops speed of 6 ml/min of which growth results shape was much better compared with the results of the drops speed of 3 ml/min.

At drops speed of 3 ml/min, $La_2O_3$ nanoneedles formed has a shape and size that are relatively uniform in which particle diameter is 20 nm and an average length around 250 nm. However, these particles seem still wrapped with the form of other particles so they need also confirmation with the results of phase analysis by using XRD. Figure 2(c) is a TEM image of $La_2O_3$ nanoneedles synthesized with the drops speed of 9 ml/min. At this speed, the results of nanoneedles obtained relatively looks better compared to the drops speed of 3 and 6 ml/min. Nanoneedles morphological forms appear more clearly and relatively homogeneous. Average diameter of particle is about 10 nm with an average length of about 250 nm. The interesting case in this study is the changing shape of the particles from polygonal shape into nanoneedles. This case means that the preferred orientation (PO) of crystal is thought to occur in the direction of a particular field so that an understanding of the crystal structure and phases analysis needs to be studied further.

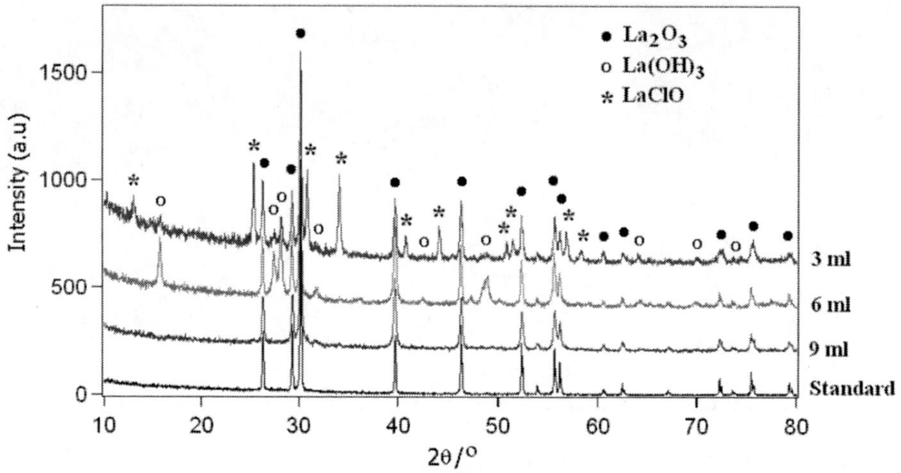


**Figure 3**. X-ray diffraction pattern of the $La_2O_3$ nanoneedle with variations in drops speed of (3 ml/min, 6 ml/min, 9 ml/min, and standard)

The X-ray diffraction profiles of $La_2O_3$ nanoneedle are shown in Figure 3 in which respective profiles are 3 ml/min, 6 ml/min, 9 ml/min, and the standard compared. It is shown that the three diffraction patterns exhibit a pattern different between ones each other.

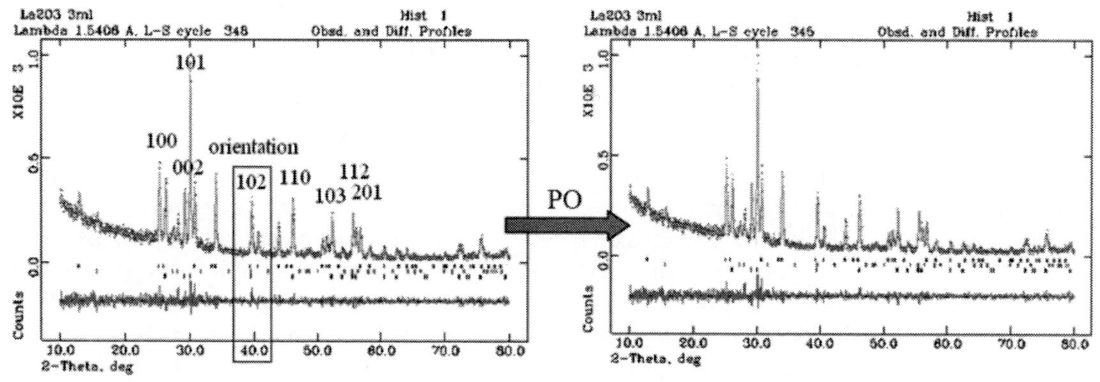


(a) 3 Drops speed of 3 ml/min

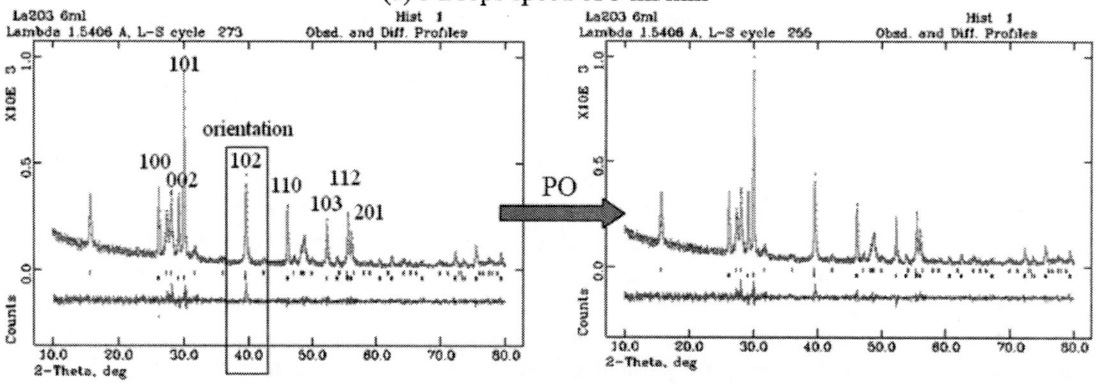


(b) Drops speed of 6 ml/min

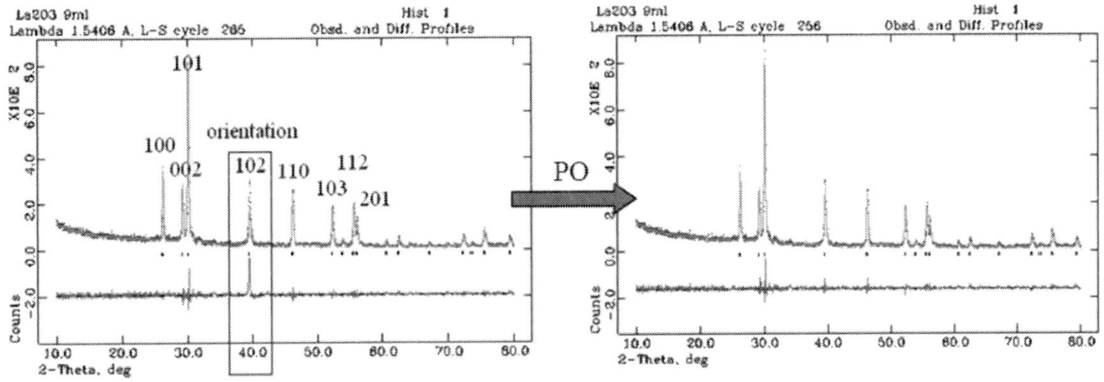

(c) Drops speed of 9 ml/min

**Figure 4**. The refinement results of XRD pattern of the $La_2O_3$ nanoneedles.

Figure 3 shows the results of measurements of X-ray diffraction pattern of the nanoneedle $La_2O_3$ with variations in drops speed of (3 ml/min, 6 ml/min, 9 ml/min, and standard). Based on the results of the phase identification, the reaction has successfully formed a single phase $La_2O_3$ on the drops speed of 9 ml/min, while for the drops speed of 3 ml/min and 6 ml/min, the sample can not react perfectly so the sample consists of multi-phases. The results of the phase identification for drops speed of 3 ml/min consist of three phases namely $La_2O_3$ (lanthanum oxide), $La(OH)_3$ (lanthanum hydroxide) and LaClO (lanthanum oxychloride), meanwhile for drops speed of 6 ml/min there are two phases namely $La_2O_3$ and $La(OH)_3$. Thus, the results require further analysis to determine the changes of the crystal structure parameters, the amount of mass fraction formed, and fitting quality as shown in Figure 4.

Figure 4 shows the results of refinement X-ray diffraction pattern of the nanoneedle $La_2O_3$ with the variations of drops speed (3 ml/min, 6 ml/min, and 9 ml/min). Qualitative and quantitative analysis refers to the Crystallography Open Database with the card number (COD: 2002286), (COD: 4031381) and (COD: 2101549) respective for phases of $La_2O_3$, $La(OH)_3$ and LaClO.

The complete summary of the results of X-ray diffraction pattern of the nanoneedle $La_2O_3$ with the variations of drops speed (3 ml/min, 6 ml/min, and 9 ml/min) for all of samples are shown in Table 1.

**Table 1.** The value of structure parameters, criteria of fit ($R_{wp}$), goodness of fit ($\chi^2$) and the mass fraction of phase formed in the $La_2O_3$ nanoneedle with variations of drops speed (3 ml/min, 6 ml/min, and 9 ml/min).

| Sample | Phase | Lattice parameter (Å) | | | V | P | Fraction | $R_{wp}$ | $\chi^2$ |
|---|---|---|---|---|---|---|---|---|---|
| | | *a = b* | *C* | *PO* | (Å$^3$) | (g/cm$^3$) | wt% | (%) | |
| 3 ml/min | $La_2O_3$ | 3.9341(2) | 6.1275(4) | [102] (0.94) | 82.13(1) | 5.989 | 22.97 | 7.81 | 1.26 |
| | $La(OH)_3$ | 6.5274(3) | 3.8494(2) | - | 142.0(1) | 4.371 | 15.92 | | |
| | LaClO | 4.1196(3) | 6.8743(8) | - | 116.6(2) | 5.419 | 61.11 | | |
| 6 ml/min | $La_2O_3$ | 3.9347(1) | 6.1235(3) | [102] (0.82) | 82.10(8) | 5.691 | 73.28 | 7.57 | 1.25 |
| | $La(OH)_3$ | 6.5243(8) | 3.8544(6) | - | 142.1(1) | 4.368 | 26.72 | | |
| 9 ml/min | $La_2O_3$ | 3.9344(2) | 6.1241(4) | [102] (0.76) | 82.10(1) | 5.371 | 100.00 | 7.31 | 1.15 |

Figure 4 and Table 1 show that the refinement results of X-ray diffraction pattern has also a good fitting quality. Table 1 shows that based on the refinement of XRD pattern, especially for phases of $La_2O_3$, it is concluded for all of the samples, that the lattice parameters, volume of unit cell, and atomic density do not change. It means that the crystal structure of $La_2O_3$ phase is not deformed due to the effect of drops speed. However, the drops speed affect the total mass fraction of $La_2O_3$ crystal growth and preferred orientation of $La_2O_3$ crystal as shown in Figure 5.

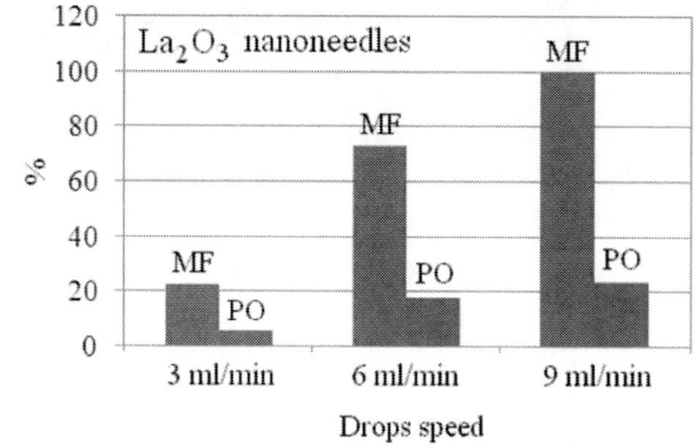


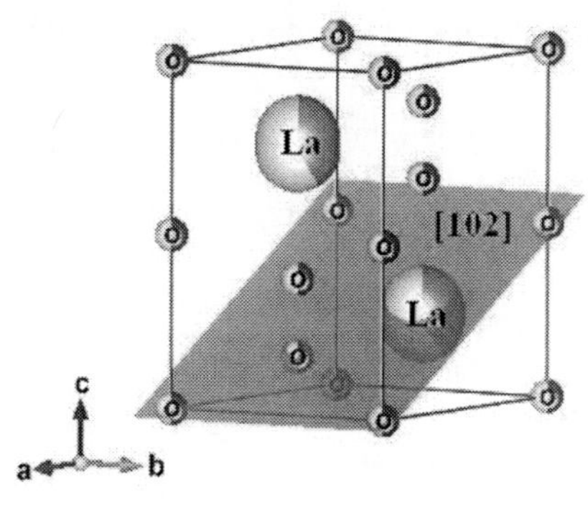


(a) Mass fraction (MF) and preferred orientation (PO) (b) Crystal orientation

**Figure 5**. Effect of the drops speed on the mass fraction and preferred orientation of $La_2O_3$ crystal

In Figure 5, it appears that the $La_2O_3$ crystal growth and degree of preferred orientation [102] increased in line with the increase of the drops speed. This phenomenon is due to the drops speed of ammonia can affect the amount of $La_2O_3$ crystal growth. The $La_2O_3$ crystal has grown naturally on the direction of the field [102]. The present orientation crystal is generally caused by change of particle shape from polygonal to nanoneedles shape.

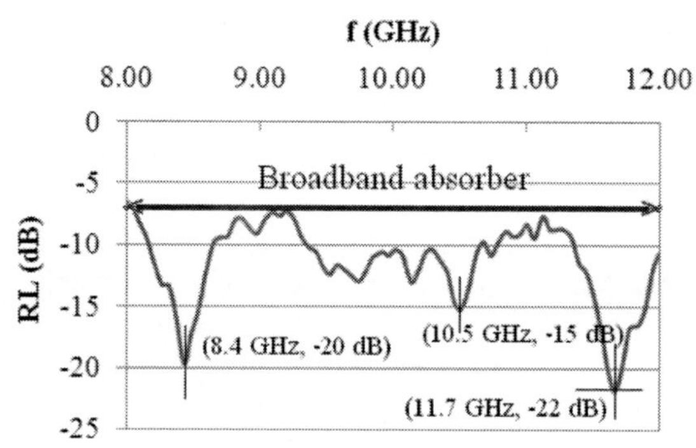


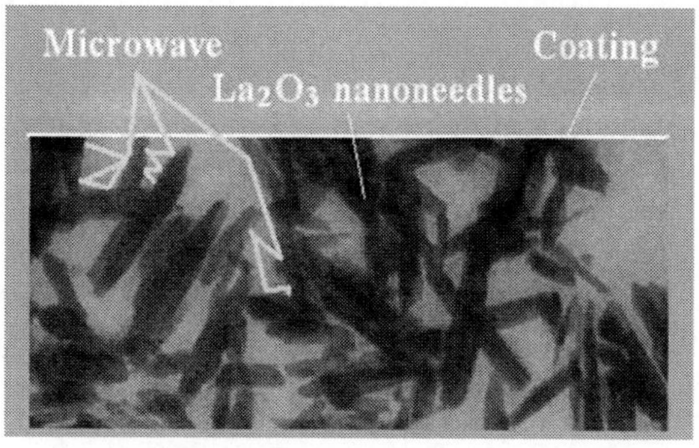


(a) Reflection loss (RL) (b) Scattering mechanism of microwave

**Figure 6**. Characteristics of microwave absorber on the $La_2O_3$ nanoneedles

The intrinsic and extrinsic characteristics of microwave absorber material make a unique phenomenon to be studied. It is because the main requirement is needed as a material absorbing microwave is the presence of intrinsic characteristics (properties of magnetic loss and dielectric loss) on the material in addition to the extrinsic characteristics (factors of the particle geometry). It means that the factor of particle geometry also determines the microwave absorption capability.

Figure 6(a) shows the relation between the reflection loss (RL) of $La_2O_3$ nanoneedles and the microwave at frequency in range of 8-12 GHz when the thickness of sample is 2 mm. There are at least three absorption peaks observed within the frequency X-band, of which value of the absorption peaks are around -20 dB at 8.4 GHz; -15.0 dB at 10.5 GHz and -22 dB at 11.7 GHz. Meanwhile, the effect of particle geometry on the microwave absorber is shown in Figure 6(b). Figure 6(b) shows that the mechanism of microwave scatters on the particle morphology so that the microwave will not be reflected back to the surface.

## 4. Conclusion

The synthesis of $La_2O_3$ nanoparticles has successfully made by using hydrolysis method. Based on the TEM image observations, the morphology of $La_2O_3$ nanoparticles shaped nanoneedles. The result of morphological of $La_2O_3$ nanoneedles indicate that better sample is obtained when the sample is synthesized with the drops speed of 9 ml/min. Nanoneedles morphological forms appear more clearly and relatively homogeneous. The average diameter of the particle is about 10 nm with an average length of about 250 nm. The changing shape of the particles from polygonal shape into nanoneedles is affected by the presence of preferred orientation (PO) of crystal on the direction of the field [102]. The factor of the particle geometry also determines the microwave absorber capability. The microwave characteristic of the $La_2O_3$ nanoneedles has indicated certain microwave absorption properties in the frequency range of 8-12 GHz when the thickness of sample is 2 mm. There are at least three absorption peaks observed within the frequency X-band, of which value of the absorption peaks are around -20 dB at 8.4 GHz; -15.0 dB at 10.5 GHz and -22 dB at 11.7 GHz.

## 5. References

[1]. Sumarni, Riesna Prassanti, Kurnia Trinopiawan, Sumiarti and Hafni Lissa N 2011 Eksplorium **32** No. 2 115 – 124
[2]. Riesna Prassanti 2012 Proceedings of the Seminar on Nuclear Geology and Natural Resources Mines (Indonesia: BATAN) 305-325.
[3]. Kurnia Setiawan Widana, Bambang Priadi and Yustina Tri Handayani Eksplorium 2014 **35** No. 1 1 - 12
[4]. Mao G, Zhang H, Li H, Jin J and Niu, S., J. Electrochem. Soc., 2012, **159**, p. J48.
[5]. Das T, Mahata C, Mallik S, Varma S, Sutradhar G,Bose P K and Maiti C K 2012 J. Electrochem. Soc. **159** p. H323.
[6]. Rocha K O, Santos J B O, Meira D, Pizani P S, Marques C M P, Zanchet D and Bueno J M C 2012 Appl. Catal. A: General **431** p. 79.
[7]. Cong-Ju Li, Bin Wanga, Jiao-Na Wang 2012 Journal of Magnetism and Magnetic Materials **324** 1305–1311
[8]. Shuyuan Zhang, Quanxi Cao 2012 Materials Science and Engineering B **177** 678– 684.
[9]. Wisnu Ari Adi and Azwar Manaf 2013 Journal of Advanced Materials Research **789** pp 97-100
[10]. Nurul Taufiqu Rochman and Wisnu Ari Adi 2014 Journal of Advanced Materials Research Vol. **896** pp 423-427
[11]. Zheng D, Shi J, Lu X and Wang C 2010 Cryst. Engin. Commun. **10** p. 4066
[12]. Wu Y, Chen Y and Zhou J 2013 Mater. Lett. **95** p. 5.
[13]. Khosrow pour F, Aghazadeh M, Dalvand S and Sabour B 2013 Mater. Lett **104** pp. 61–63.
[14]. Aghazadeh M, Malek Barmi A A, Mohammad Shiri H and Sedaghat S 2013 Ceram. Int. **39** p. 1045.
[15]. Jamali Sheini F and Yousefi R 2013 Ceram. Int. **39** p. 3715.
[16]. Cong-Ju Li, Bin Wanga, Jiao-Na Wang, 2012 Journal of Magnetism and Magnetic Materials **324** 1305–1311
[17]. Toby B H 2001 EXPGUI A Graphical User Interface for GSAS J. Appl. Crystallogr. **34** 210-221
[18]. Mazloumi M, Zanganeh S, Kajbafvala A, Shayegh M R and Sadrnezhaad S K 2008 IJE Transactions B: Applications **21** No. 2 169-176.

**Acknowledgements**
The authors gratefully acknowledge the support of the Centre for Science and Technology of Advanced Materials, BATAN for the research facilities.

IOP Conf. Series: Materials Science and Engineering **202** (2017) 012067 doi:10.1088/1757-899X/202/1/012067

# Effect of Thermal Processes on the Electrical and Optical Properties of $Fe_2TiO_5$ Ceramics

**R Fajarin[1], Widyastuti[1], M A Baqiya[2] and I Y S Putri[1]**

[1]Department of Materials and Metallurgical Engineering, Institut Teknologi Sepuluh Nopember (ITS) Surabaya, Kampus ITS Surabaya 60111, Indonesia.
[2]Department of Physics, Institut Teknologi Sepuluh Nopember (ITS) Surabaya, Kampus ITS Surabaya 60111, Indonesia

Email: fajar@mat-eng.its.ac.id

**Abstract**. Pseudobrookite ($Fe_2TiO_5$) is one of the Fe-Ti oxides that have been commonly studied. It is the most stable phase among the Fe-titanates. The multiferroic properties of $Fe_2TiO_5$ make the material can be used as a potential candidate for new applications due to the combination of semiconducting, magnetic, dielectric, and optical properties. In this research, $Fe_2TiO_5$ ceramics were synthesized using mechanical milling method for 7 h with various temperatures of 1100 °C, 1200 °C, and 1300 °C. Scanning electron microscopy (SEM) observation and x-ray diffraction (XRD) measurements were performed to analyze the microstructures and crystal structures of the $Fe_2TiO_5$ ceramics. In order to investigate the band gap of the $Fe_2TiO_5$, the UV-Vis Diffuse Reflectance measurements were conducted. It has been found that the $Fe_2TiO_5$ ceramic can be applied as a promising candidate for semiconducting devices in which the electrical conductivity and the band gap of the $Fe_2TiO_5$ ceramic were 1.73 $\times 10^{-7}\ \Omega^{-1}.cm^{-1}$ and 1.71 eV, respectively.

## 1. Introduction

Iron titanate (Fe-Ti) oxides have four stable phases, namely ilmenite ($FeTiO_3$), pseudobrookite ($Fe_2TiO_5$), pseudorutile ($Fe_2Ti_3O_9$), and ulvospinel ($Fe_2TiO_4$). Among those Fe-Ti oxides, $Fe_2TiO_5$ is the most stable phase and has been intensively studied due to the multi-functional behaviors in one material. $Fe_2TiO_5$ particles have been successfully synthesized by solid state reaction [1], sonochemical [2], sol-gel [3, 4], precipitation [5], and mechanical alloying methods [6]. $Fe_2TiO_5$ phase also can be prepared from the mixture of $Fe_3O_4$/$Fe_2O_3$ and $TiO_2$ at a high temperature of 1423 K [7]. Recently, the synthesis of porous $Fe_2TiO_5$ microparticulate through a hydrothermal route might improve the electrochemical properties [8] and as a candidate for Li-ion battery anode material [9].

$Fe_2TiO_5$ nanopowders have been found to behave as ferrimagnetic and paramagnetic from the vibrating sample magnetometer (VSM) observation at room temperature [10]. The electronic structures and the optical properties, have been theoretically studied based on the LSDA and LSDA+U calculations. As a result, it has been found that the band gap of $Fe_2TiO_5$ was about 1.81 eV [11]. This result showed that $Fe_2TiO_5$ is an n-type semiconducting oxide.

The purpose of this work was to investigate the effect of the thermal process on the electrical and optical properties of the synthesized $Fe_2TiO_5$ ceramics through the combination of coprecipitation method to obtain $Fe_3O_4$ as one of the raw materials and mechanical alloying method to get the $Fe_2TiO_5$

ceramics. The temperature variation of heat treatment was conducted to understand the phase formation of $Fe_2TiO_5$ affecting the electrical and optical properties of the samples.

## 2. Experimental Method

$FeCl_2.4H_2O$ was used as a raw material to synthesize $Fe_3O_4$ by coprecipitation method. $FeCl_2.4H_2O$ was dissolved in distilled water, then mixed and stirred with HCl solution at room temperature. The $NH_4OH$ solution was added into the dissolved $FeCl_2$ solution at a temperature of 70 °C to obtain the black precipitate of $Fe_3O_4$. This precipitate was then washed with distilled water until reaching the pH value of 7.0 and then dried to obtain $Fe_3O_4$ powders. In order to obtain the $Fe_2O_3$ powders, the $Fe_3O_4$ powders were heated at 900 °C for 5 h.

$Fe_2TiO_5$ ceramics were prepared by the mechanical alloying technique with $Fe_2O_3$ and $TiO_2$ as the starting materials. The composition ratio of $Fe_2O_3 : TiO_2 = 3 : 2$ was used according to the phase diagram in Ref. [12]. In this technique, the milling process was conducted for 7 h at the rate of 300 rpm and BPR of 15:1. The milled powders were then heated at various temperatures of 1100 °C, 1200 °C, and 1300 °C for 2 h to obtain $Fe_2TiO_5$ ceramic powders.

The prepared $Fe_2TiO_5$ powders were first characterized by X-ray diffractometry (XRD) to identify the phase formation in the samples. Scanning Electron Microscope (SEM) measurement with FEI Inspect S50 was performed to observe the particle morphology. The electrical conductivity measurement was conducted by IV meter connected to LabView development system to study the electrical properties of the samples. The Ultraviolet-Visible Spectrophotometer (UV-Vis) measurement using Shimadzu 2450 with double beam system and diffuse reflectance spectroscopic (DRS) was performed to investigate the optical properties of the $Fe_2TiO_5$ ceramics.

## 3. Results and Discussion

Figure 1 shows the XRD patterns for $Fe_2TiO_5$ powders after milling process and various heating temperatures of 1100 °C, 1200 °C, and 1300 °C for 2 h. The XRD data show that there is not any impurity detected in all samples after the heating process, suggesting a single phase of $Fe_2TiO_5$ has been formed by only performing heat treatment at 1100 °C. This temperature is within the temperature range in which the interdiffusion took place [13]. It can also be seen that the degree of crystallinity of $Fe_2TiO_5$ powders increased along with the increasing heating temperature, indicating the increase in particle size as well as the crystalline size of $Fe_2TiO_5$ powders. The particle morphology after milling process and various heat treatments can be seen in Figure 2. It is confirmed that the higher temperature results in the larger particle size due to a higher diffusion rate among atoms in the crystals which are comparable with the results of research conducted by Wiendartun *et al.* [14].

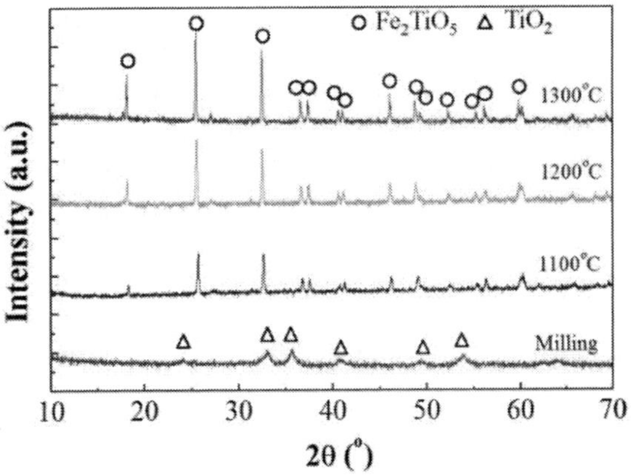


**Figure 1.** XRD patterns of the milled $Fe_2TiO_5$ powders and the milled powders after heat treatment with various temperatures of 1100 °C, 1200 °C, and 1300 °C for 2 h

**Figure 2.** Particle morphologies observed by SEM for the milled $Fe_2TiO_5$ powders after heat treatment at the temperatures of (a) 1100 °C, (b) 1200 °C, and (c) 1300 °C for 2 h.

Table 1 shows the average value of electrical conductivity of all prepared samples at room temperature. It can be seen that there was a slight increase of conductivity value, which was in the order of $10^{-7}$ together with the increasing temperature of heat treatment. This was probably due to the effect of a large particle size implying a less number of grain boundaries. It is well known that grain boundaries may act as the second resistance besides the first resistance from the grain itself in the ceramic materials. Higher conductivity value can be achieved by decreasing the number of grain boundaries in the ceramics.

**Table 1.** The electrical conductivity of $Fe_2TiO_5$ powders with different temperatures of heat treatment.

| Temperature (°C) | Resistivity (ρ) (x$10^{-6}$ Ω.cm) | Conductivity (σ) (x$10^{-7}$ $Ω^{-1}$.$cm^{-1}$) |
|---|---|---|
| 1100 | 6.84 | 1.46 |
| 1200 | 4.63 | 2.16 |
| 1300 | 4.18 | 2.39 |

UV-Vis measurements can be used to obtain the value of gap energy of a semiconducting material. In this report, each of $Fe_2TiO_5$ ceramic prepared by mechanical alloying technique shows semiconducting behavior since the conductivity value is in the order of $10^{-7}$. Figure 3 shows the UV-Vis absorption spectra for the milled $Fe_2TiO_5$ before and after heat treatment at the temperatures of 1100 °C, 1200 °C, and 1300 °C for 2 h. By analyzing these data using $E = hc/\lambda$, the band gap energies of these samples can be obtained and summarized in Table 2. It is found that the band gap energy of the prepared $Fe_2TiO_5$ samples has a smaller value compared to the previous result [11]. It is probably due to the formation of localized bands near conduction band. A smaller band gap is important in the photocatalytic reaction. With the small value of band gap, the sample would be more responsive to visible light [15]. It has been indicated that the photocatalytic activity of the addition of electronic stated within the gap [16]. From Table 2, it seems that the change of the band gap value was relatively small along with the increasing temperature. Accordingly, the band gap value has been found to be almost independent of the variation of heat treatment process.

**Table 2.** Band gap values of $Fe_2TiO_5$ powders with different temperatures of heat treatment.

| Sample | Band Gap (eV) | λ (cut-off wavelength) |
|---|---|---|
| 1100 °C | 1.71 | 723 |
| 1200 °C | 1.73 | 710 |
| 1300 °C | 1.77 | 701 |
| $Fe_2TiO_5$ (*) | 1.81 | - |

The (*) is the result of the former paper [11] written together as a comparison with the present data.

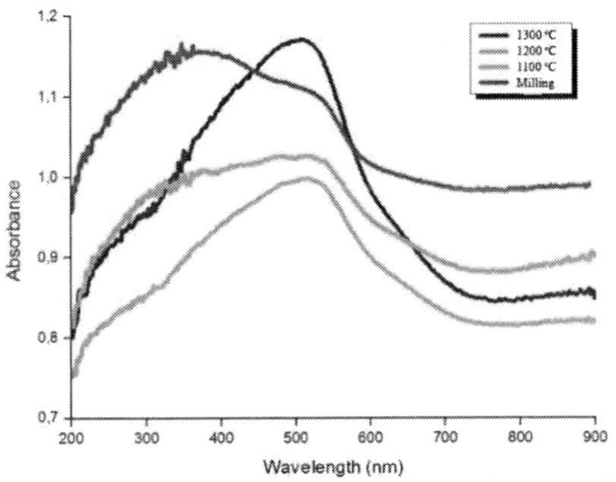


**Figure 3.** The UV-Vis spectra of $Fe_2TiO_5$ powders with different temperatures of heat treatment.

## 4. Conclusion

The effect of the thermal process on the electrical and optical properties of $Fe_2TiO_5$ ceramics has been investigated. For this purpose, the $Fe_2TiO_5$ ceramics have been successfully synthesized by a combination of coprecipitation and mechanical alloying methods continued by the temperature variation of heat treatment at 1100 °C, 1200 °C, and 1300 °C. It has been found that the electrical conductivity of the $Fe_2TiO_5$ ceramics slightly increased with the increasing temperature due to the increasing crystallinity and decreasing number of grain boundaries. Based on the analysis of UV-Vis spectra, it has been revealed that the prepared $Fe_2TiO_5$ ceramics have a semiconducting behavior with the band gap energy of around 1.71 eV. It has been found that the thermal process played a minor role in the change of bandgap energy in $Fe_2TiO_5$ ceramics.

## 5. References

[1] Hirota K., Bradt R.C 1991 *Anal. Sci.* **7** 1275-1278.
[2] Huang W.P., Tang X.H., Felner I., Koltypin Y., Gedanken A 2002 *Mater. Res. Bull.* **37** 1721-1735.
[3] Phani A.R., Ruggieri F., Passacantando M., Santucci S 2008 *Ceram. Int.* **34** 205-211.
[4] Ramezani M., Davoodi A., Malekizad A., Hosseinpour-Mashkani S.M. 2015 *J. Mater. Sci. - Mater. Electron.* **26** 3957-3962.
[5] Gao X.M., Li M.W., Hou Y.L., Wang C.Y. 2015 *Mater. Res. Innovations* **19** 1-6.
[6] Fajarin R., Purwaningsih H., Widyastuti, Susanti D., Helmy R.K 2014 *AIP Conf. Proc.* **1617** 63-66.
[7] Ye F.X., Tsumura T., Nakata K., Ohmori A 2008 *Mater. Sci. Eng. B* **148** 154-161.
[8] Guo S., Wang S., Wu N., Liu J., Ni Y., Liu W 2015 *RSC Advances* **5** 103767-103775.
[9] Min K.-M., Park K.-S., Lim A.-H., Kim J.-C., Kim D.-W 2012 *Ceram. Int.* **38** 6009-6013.
[10] Enhessari M., Razi M.K., Etemad L., Parviz A., Sakhaei M 2014 *J. Exp. Nanosci.* **9** 167-176.
[11] Jin L., Zhou C. 2013 *Mater. Int.* **23** 413-419.
[12] Kimura S., Muan A. 1971 *Am. Mineral.* **56** 1332-1346.
[13] Ren Z.-s., Hu X.-j., Li S.-y., Xue X.-x., Chou K.-c. 2013 *Int. J. Miner. Metall. Mater.* **20** 273-278.
[14] Wiendartun, Risdiana, Fitrilawati, Siregar R.E. (2016) *J. Phys: Conf. Ser.* **739** 012140.
[15] Bard A.J., Fox M.A 1995 *Acc. Chem. Res.* **28** 141-145.
[16] Yalçın Y., Kılıç M., Çınar Z 2010 *Appl. Catal. B* **99** 469-477.

# Mechanosynthesis of A Ferritic ODS (Oxide Dispersion Strengthened) Steel Containing 14% Chromium and Its Characterization

**A K Rivai, A Dimyati, and W A Adi**

Center for Science and Technology of Advanced Materials, BATAN, Kawasan PUSPIPTEK Gedung 71, Tangerang Selatan, Banten 15314

Email: rivai.abukhalid@batan.go.id

**Abstract**. One of the advanced materials for application at high temperatures which is aggressively developed in the world is ODS (Oxide Dispersion strengthened) steel. ODS ferritic steels are one of the candidate materials for future nuclear reactors in the world (Generation IV reactors) because it is able to be used in the reactor above 600 °C. ODS ferritic steels have also been developed for the interconnect material of SOFC (Solid Oxide Fuel Cell) which will be exposed to about 800 °C of temperature. The steel is strengthened by dispersing homogeneously of oxide particles (ceramic) in nano-meter sized in the matrix of the steel. Synthesis of a ferritic ODS steel by dispersion of nano-particles of yttrium oxide (yttria: $Y_2O_3$) as the dispersion particles, and containing high-chromium i.e. 14% has been conducted. Synthesis of the ODS steels was done mechanically (mechanosynthesis) using HEM (High Energy ball Milling) technique for 40 and 100 hours. The resulted samples were characterized using SEM-EDS (Scanning Electron Microscope-Energy Dispersive Spectroscope), and XRD (X-ray diffraction) to analyze the microstructure characteristics. The results showed that the crystal grains of the sample with 100 hours milling time was much smaller than the sample with 40 hours milling time, and some amount of alloy was formed during the milling process even for 40 hours milling time. Furthermore, the structure analysis revealed that some amount of iron atom substituted by a slight amount of chromium atom as a solid solution. The quantitative analysis showed that the phase mostly consisted of FeCr solid-solution with the structure was BCC (body-centered cubic).

**Keywords:** Mechanosynthesis, Ferritic, ODS, Chromium, HEM

## 1. Introduction

High chromium steels are one type of steel-based materials for high-temperature service operation of many power plants such as nuclear reactors [1-2]. However, the conventional high chromium steels are limited to about 600 °C of service temperature. Therefore, development of advanced steel-based materials for higher temperature service operation is needed. Oxide dispersion strengthened (ODS) steels are one type of steel-based high-temperature material candidates which are mostly developed for advanced nuclear fission reactors (so-called Generation IV reactors) and nuclear fusion reactors because of the promising materials to be utilized at high temperature under severe irradiation exposure environment [2-11]. Development of ODS steels with the addition of several significant elementals has been done by several researchers. Development of Al added high-Cr ODS steels had been reported in order to improve the corrosion resistance at elevated temperature [9-10]. Furthermore, it has been reported that the Al addition to ODS ferritic steel considerably improves the compatibility between U–Zr fuel and the ODS steel [11].

ODS steels have been developed by powder metallurgy method using mechanical alloying technique so-called mechanosynthesis to obtain a dispersion of oxide particles in a steel matrix [2-11]. Mechanical alloying is a technique to acquire homogeneous materials from elemental powder materials. Welding, breaking, and re-welding processes of the powder particles were performed in a milling machine during mechanical alloying. Finally, fine dispersion of oxide particles (mostly $Y_2O_3$ nano-particles) inside the matrix of the steel will be obtained. Therefore, the optimum amount of $Y_2O_3$ in the steel is significantly important. It has been reported that the tensile strength and hardness of ODS steel increases as the content of $Y_2O_3$ increases, but it should be noted that the content of $Y_2O_3$ needs to be controlled within a range [12]. A novel route for synthesizing of ODS steels has also been developed using a sol–gel method combining with hydrogen reduction [13]. Furthermore, it has been reported about mechanically alloyed of ODS ferritic steels in different atmospheres which give an effect of the mechanical and microstructural properties [14].

In this paper, the study of mechanosynthesis of a ferritic ODS steel containing 14% chromium is presented. The amount of nano-sized ceramic $Y_2O_3$ powder was 0.25 wt%. The ferritic ODS steel was synthesized using HEM (*High Energy ball Milling*).

## 2. Experimental Method

Iron (Fe), chromium (Cr), titanium (Ti), molybdenum (Mo) elemental powders were mixed with nano-sized $Y_2O_3$ powder for the mechanosynthesis of a ferritic ODS steel. The powders of Fe, Cr, Ti, Mo, and $Y_2O_3$ were bought from a chemical company without any further treatment. The ratio of Fe:Cr:Ti:Mo:$Y_2O_3$ was 84.45:14:1:0.3:0.25 as the composition in weight percentage.

The ratio of BPR (ball-to-powder mass ratio) of the samples was 5:1. In order to avoid or to minimize oxidation and impurities during milling, a toluene solution was added as process control agent in the vial which the vial has already inserted the powders and the balls. The mixed powders were milled using HEM (High Energy ball Milling) for 40 and 100 hours. The speed of the vial during milling was 1,000 rpm (rotation per minute). The characterizations were done using SEM (Scanning Electron Microscope) - EDS (Energy Dispersive X-Ray Spectroscope) and XRD (X-Ray Diffraction) for microstructure analyses.

## 3. Results

The SEM-EDS analysis of ODS powders after milling for 40 and 100 hours are shown in Figures 1-4. Figure 1 of SEM micrograph with 500 times magnification of the sample after milling for 40 hours shows the various morphology i.e. small, medium and big chunks. However, the powder was dominated by small chunks. During mechanical alloying, welding, breaking, re-welding, and re-breaking processes of the powder particles occurred in a vial in the milling machine. Therefore, some of the powders were still in welding or re-welding stage, and some of the powders were in breaking or re-breaking stage. Figures 2 show SEM-EDS analysis with 2,000 times magnification of the sample after milling for 40 hours, (a) SEM micrograph, (b) EDS analysis at position-1, (c) EDS analysis at position-2, (d) EDS analysis at position-3. The EDS analysis was presented in quantitative of elemental percentages as shown in Table 1. The results showed that the chromium content slightly decreases from the original composition. It might occur because some percentage of impurity from the vial and toluene such as carbon was unexpected added during the milling process.

**Figure 1.** SEM micrograph with 500 times magnification of the sample after milling for 40 hours

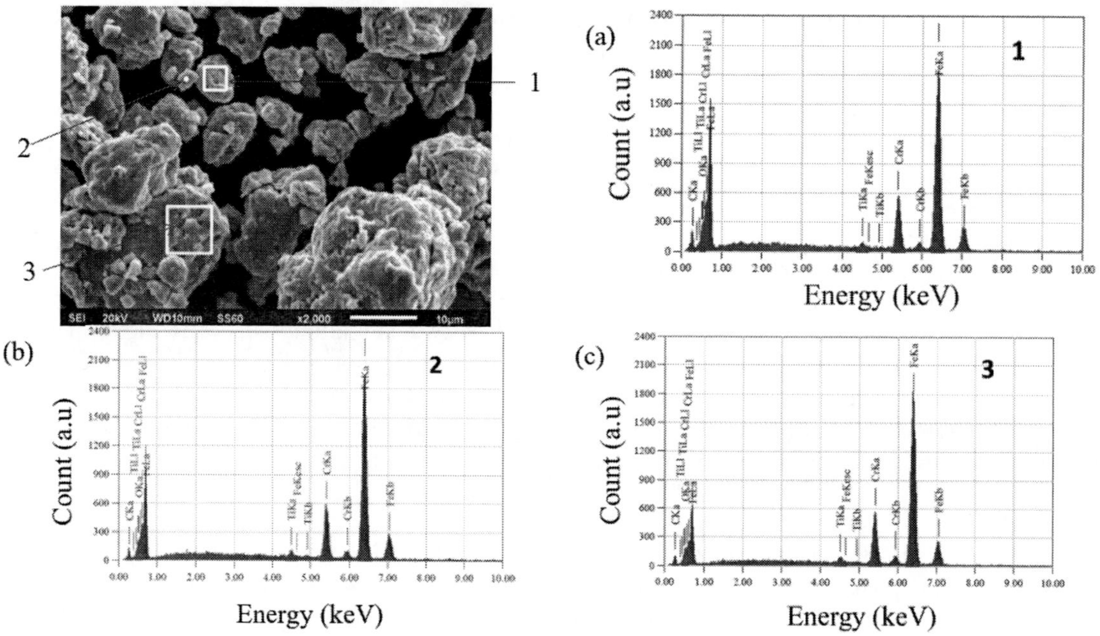


**Figure 2.** SEM-EDS analysis with 2,000 times magnification of the sample after milling for 40 hours (a) SEM micrograph (b) EDS analysis at position-1 (c) EDS analysis at position-2 (d) EDS analysis at position-3

**Table 1.** Chemical compositions of the sample after milling for 40 hours

| Element/area | Original (mass%) | Area-1 (mass%) | Area-2 (mass%) | Area-3 (mass%) |
|---|---|---|---|---|
| Cr | 14 | 12.63 | 13.32 | 13.66 |
| Ti | 1 | 0.94 | 1.3 | 0.96 |

Figure 3 of SEM micrograph with 500 times magnification of the sample after milling for 100 hours shows that the powder was dominated by small chunks. In general, from SEM micrograph of morphology, it is able to be analyzed that the sample after milling for 100 hours was smaller than the sample after milling for 40 hours. Therefore, most of the powders were in breaking or re-breaking stage and a small amount of the powders welding or re-welding stage. Figures 4 show SEM-EDS analysis with 2,000 times magnification of the sample after milling for 10 hours, (a) SEM micrograph,

(b) EDS analysis at position-1, (c) EDS analysis at position-2, (d) EDS analysis at position-3. The EDS analysis was presented in quantitative of elemental percentages as shown in Table 2. The results showed that the chromium content slightly decreases from the original composition. It is able to be analyzed that the content of chromium in the sample after milling for 100 hours was more decrease than the sample after milling for 40 hours. It might occur because some percentage of impurity from the vial and toluene such as carbon was unexpected added during the milling process. Therefore, more milling time more impurity increases.

**Figure 3.** SEM micrograph with 500 times magnification of the sample after milling for 100 hours

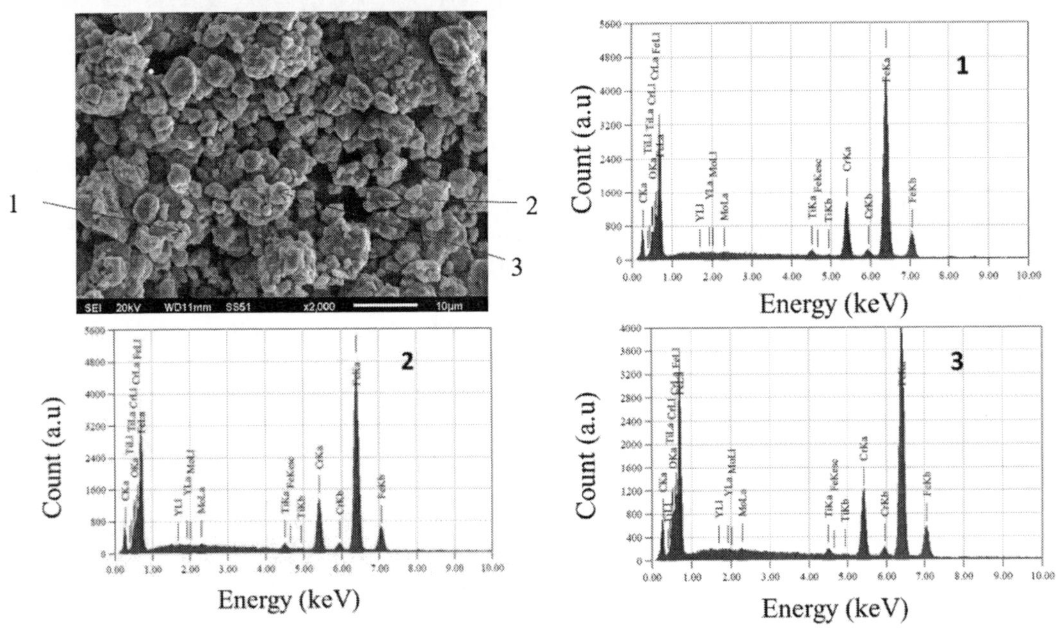


**Figure 4.** SEM-EDS analysis with 2,000 times magnification of the sample after milling for 100 hours : (a) SEM micrograph, (b) EDS analysis at position-1, (c) EDS analysis at position-2, (d) EDS analysis at position-3

**Table 2.** Chemical Compositions of the sample after milling for 100 hours

| Element/area | Original (mass%) | Area-1 (mass%) | Area-2 (mass%) | Area-3 (mass%) |
|---|---|---|---|---|
| Cr | 14 | 11.93 | 12.02 | 12.63 |
| Ti | 1 | 0.86 | 0.97 | 0.94 |

Figure 5 showed XRD pattern of the samples after milling for 40 and 100 hours. For the sample after milling 40 hours, the peaks of 2θ diffraction pattern are at 44.51°, 65.13°, 82.14°, 98.57° which is FeCr phase as in the JCPDS 34-0396. For the sample, after milling for 100 hours, the XRD pattern tend to grow as an amorphous structure. The phenomenon occurred because of a milling time. The results also showed that the XRD pattern of the sample after milling 100 hours was broader than the sample after milling 40 hours. It happened because the longer the milling time make a reduction of crystal particle size. Therefore, it is found that the grain size was decreased with the increasing of the milling time which is shown by the decreasing and broadening of the spectrum.

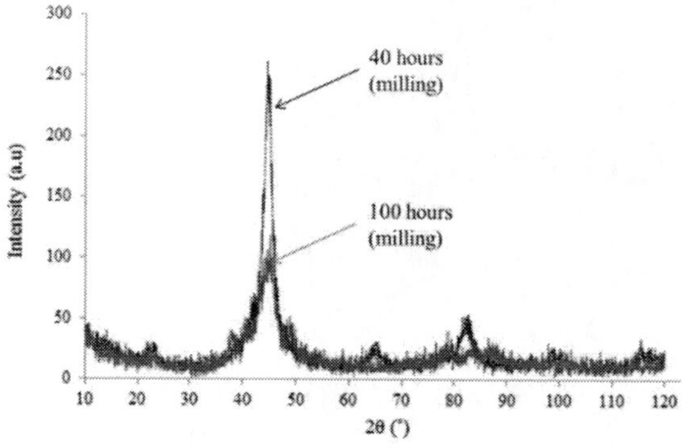


**Figure 5.** XRD patterns of the samples after milling for 40 and 100 hours

Figure 6 shows XRD pattern analysis of the sample after milling for 40 hours in more detail with refinement. The results showed that the XRD pattern consists of 3 phases i.e. FeCr alloy, FeCr solid solution, and $FeO_2$. The quantitative analysis showed that content of the phases were 1,3%; 92,2%; 6,5% for FeCr alloy, FeCr solid-solution and $FeO_2$ phases, respectively. It is able to investigate that the structure of the sample was BCC (body-centered cubic) as shown in Figure 6. The structure analysis revealed that some amount of iron atom substituted by a slight amount of chromium atom as a solid solution. Therefore, the formation of the solid solution might occur during milling process beside of homogeneous composition process of the powder. This process of solid solution reaction might occur due to diffusion through interfaces and grain/subgrain boundaries of the metals during the milling process.

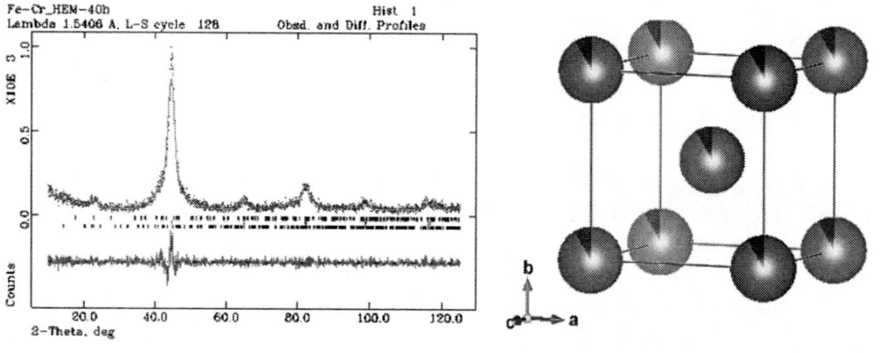


**Figure 6.** XRD pattern analysis with refinement of the sample after milling for 40 hours

It showed that some of the oxygen from environment reacted with iron to form $FeO_2$ during the milling process. Furthermore, the phase of the sample after milling for 40 hours was dominated by FeCr solid solution phase. The phenomenon of this kind of process has also been reported [16]. However, some amount of alloy was formed even for the 40 hours milling process.

## 4. Conclusion

Mechanosynthesis of a ferritic ODS steel containing 14% chromium using high energy ball milling for 40 and 100 hours and its characterization have been done. The results showed that the grain size was decreased with the increasing of the milling time. The structure of the sample was BCC (body-centered cubic) with some amount of iron atom was substituted by a slight amount of chromium atom as a solid solution. Some of the oxygen from environment reacted with iron to form $FeO_2$ during the milling process. Nevertheless, some amount of alloy was formed during the milling process. It is able to analyze that to improve the quality of the synthesis process at least two factors should be done i.e. first, increasing the BPR (ball to powder mass ratio) to increase the mechanical impact that will increase the alloy phase and second, more protect the vial and machine environment to avoid or to minimize the impurities.

## 5. References

[1] Rivai A K and Takahashi M 2010 *Journal of Nuclear Materials* **398** 146-153
[2] Yvon P and Carré F 2009 *Journal of Nuclear Materials* **385** 217–222
[3] Hirata A T, Fujita Y R, Wen *et al*, 2011 *Nature Materials* **10** 922-926
[4] Oksiutaa Z, Lewandowska M, Unifantowicz P *et al* 2011 *Fusion Engineering and Design* 86 2417–2420
[5] Unifantowicz P, Oksiuta Z, Olier P *et al* 2011 *Fusion Engineering and Design* **86** 2413–2416
[6] Rivai A K, Anggarini D, Dimyati A *et al* 2012 *Proceeding of the 3rd International Conference on Physics 2012* ISBN: 979-95620-2-3, 235-238
[7] Rivai A K, Anggarini D, Sebleku P *et al* 2012 *Proceeding of National Seminar of Materials and Metallurgy 2012* - LIPI, ISBN: 979-95620-2-3 235-238
[8] Tresnaputri F, Rivai A K and Fajarin R 2013 *Journal of Undergradute ITS* 1-4
[9] Kimura A, Kasada R, Iwata N *et al* 2011 *Journal of Nuclear Materials* **417** 176–179
[10] El-Dasher B, Farmer J, Ferreira J *et al* 2011 *Journal of Nuclear Materials* **419** 15–23
[11] Ohtsuka S, Kaito T, Ukai S *et al* 2013 *Journal of Nuclear Materials* **441** 286–292
[12] Yangl F, Guo Z, Guo L *et al* 2013 *Journal of Nanomedice and Nanotechology* **4:1** 1-3
[13] Sun Q X, Zhang T, Wang X P *et al* 2012 *Journal of Nuclear Materials* **424** 279–284
[14] Iwata N Y, Liu T, Dou P *et al* 2011 *Journal of Nuclear Materials* **417** 162–165
[15] Toualbi L, Ratti M, André G *et al* 2011 *Journal of Nuclear Materials* **417** 225–228
[16] Pasebania S, Dutt A K, Burns J *et al* 2015 *Materials Science and Engineering A* **630** 155–169

## Acknowledgements

Authors would like to express the gratitude to Fitriani Tresnaputri and Drs. Bambang Sugeng for their help in preparation of samples and characterization, and Head of Center For Science and Technology of Advanced Materials for the support and encouragement.

IOP Conf. Series: Materials Science and Engineering **202** (2017) 012069 doi:10.1088/1757-899X/202/1/012069

# Methyl Ester Production via Heterogeneous Acid-Catalyzed Simultaneous Transesterification and Esterification Reactions

**S Indrayanah[1], Erwin[2], I N Marsih[3], Suprapto[4] and I K Murwani[1]**

[1] Department of Chemistry, Institut Teknologi Sepuluh Nopember, Surabaya 60111, Indonesia
[2] Department of Chemistry, Universitas Mulawarman, Samarinda 75119, Indonesia
[3] Department of Chemistry, Institut Teknologi Bandung, Bandung 40132, Indonesia
[4] Department of Chemical Engineering, Institut Teknologi Sepuluh Nopember, Surabaya 60111, Indonesia

Email: irmina@chem.its.ac.id

**Abstract.** The heterogeneous acid catalysts ($MgF_2$ and $ZnF_2$) have been used to catalyze the simultaneous transesterification and esterification reactions of crude palm oil (CPO) with methanol. Catalysts were synthesized by sol-gel method (combination of fluorolysis and hydrolysis). The physicochemical, structural, textural, thermal stability of the prepared catalysts was investigated by $N_2$ adsorption-desorption, XRD, FT-IR, SEM and TG/DTG. Both $MgF_2$ and $ZnF_2$ have rutile structures with a different phase. The surface area of $ZnF_2$ is smaller than that of $MgF_2$, but the pore size and volume of $ZnF_2$ are larger than those of $MgF_2$. However, these materials are thermally stable. The performance of the catalysts is determined from the yield of catalysts toward the formation of methyl ester determined based on the product of methyl ester obtained from the reaction. The catalytic activity of $ZnF_2$ is higher than $MgF_2$ amounted to 85.21% and 26.82% with the optimum condition. The high activity of $ZnF_2$ could be attributed to its pore diameter and pore volume but was not correlated with its surface area. The yield of methyl ester decreased along with the increase in molar ratio of methanol/CPO from 85.21 to 80.99 for $ZnF_2$, respectively.

**Keyword.** Acid catalyst, crude palm oil, methyl ester, simultaneous transesterification and esterification reactions

## 1. Introduction

The use of methyl ester based on renewable vegetable oils as a starting material or intermediate for several oleochemical industries such as in the cosmetics, surfactants (MES), plastics, biolubricants, pharmacy, foodstuff, and energy production is considered and proposed as a new alternative in the development of Green-Sustainable chemistry [1-6]. Methyl ester can be obtained from the processing of vegetable oils such as rapeseed oil, jatropha oil, soybean oil, karanja oil, rubber seed oil, safflower oil, and crude palm oil [7-13]. In Indonesia, a source of vegetable oil that is used the most widely is derived from crude palm because it is abundant, cheap, eco-friendly and renewable [14-16].

Methyl ester can be obtained through the transesterification of palm oil and alcohol with a base catalyst such as KOH or NaOH [17]. However, the catalyst is very sensitive to triglyceride which contains high free fatty acid (FFA). FFA reacts readily with a base catalyst to form soap so that the transesterification reaction with the base catalyst cannot be conducted directly. Also, methyl ester can also be obtained by simultaneous transesterification and esterification reactions using a homogeneous acid catalyst such as $H_2SO_4$, HF, $H_3PO_4$, HCl, toluensulfonate acid [18], $SnCl_2$ [19], $AlCl_3$ and $ZnCl_2$ [20]. However, the reaction in homogeneous phase is very dangerous because the acid used as the catalyst is very corrosive. Therefore, the use of the solid phase heterogeneous acid catalyst in simultaneous transesterification and esterification reactions is considered as an alternative to minimize environmental damage [21].

Some solid acids have been tested as catalysts for simultaneous transesterification and esterification reactions. Zirconium oxide is used the most frequently because it has the acidity [22]. Samart et al. reported that the number of acids is one criterion for the selection of catalysts for simultaneous transesterification and esterification reactions [23]. In addition, $V_2O_5$ phosphate can also be used as a catalyst for simultaneous transesterification and esterification reactions. The ability of $V_2O_5$ phosphate as a catalyst in this reaction related with the acid side [24]. Synthesis of methyl ester has also been successfully carried out with the $MoO_3$ supported by $SiO_2$ as a catalyst [25] and $Al_2O_3$ [26]. However, the use of catalysts in the above still have disadvantages such as high price e.g. for zirconium, easily deactivated e.g. $V_2O_5$ phosphate, and the problem of *leaching* to become inactive e.g for $MoO_3$.

Recently, the fluoride-based materials are one of the solids that are known to have acid side and are often used as catalysts in reactions that require acid side such as in the synthesis of vitamin E [27], vitamin K [28], benzyl benzene [29], and acetylation of glycerol [30]. In addition to having an acid side, the catalytic performances of this catalyst are related to its chemical structure, high surface area, and mesoporous pore diameter [31]. Therefore, we would use fluoride-based materials such as $MgF_2$ and $ZnF_2$ as catalysts in the simultaneous transesterification and esterification reactions. The use of $MgF_2$ and $ZnF_2$ as catalysts in the reaction of crude palm oil (CPO) with methanol to methyl ester production was due to the similarity between Mg and Zn. In this experiments were demonstrated the synthesis, characterization and catalytic activity of $MgF_2$ and $ZnF_2$ as catalysts to methyl ester production by the simultaneous transesterification and esterification reactions.

## 2. Experimental Method

### *2.1. Synthesis of catalysts*

The $MgF_2$ material was synthesized by sol-gel method (combination of fluorolysis and hydrolysis). Firstly, Mg turning was dissolved in dry methanol at room temperature for 12 hours and heated under reflux condition for 3 hours. Furthermore, the HF (48%) solution was added and stirred for 3 hours. After the aging process at room temperature, the resulting gel was dried under vacuum at a temperature of 70 °C for 5 hours. Zn acetate was used as a precursor to prepare the $ZnF_2$ material. The precursor was dissolved in dry methanol at the room temperature without heated at reflux for 3 hours. Then, the steps were repeated for the synthesis of $MgF_2$.

### *2.2. Characterization of Catalyst*

All samples were characterized by XRD (X-ray diffractometer) Philips X-Pert with Cu-$K_\alpha$ radiation source ($\lambda$ = 1.54056) at 30 kV, and the speed of 1°/ min at $2\theta$ = 20-80°. The obtaining XRD patterns were compared by diffractogram of the JCPDS-PDF database. The infrared spectrums were recorded using KBR pellets method in the range of 4000-400 $cm^{-1}$ with an FT-IR Shimadzu spectrophotometer. TG/DTG was performed on METTLER TOLEDO through the sample-holder system (Pt/PtRh10 thermocouple). The pore size and surface area were measured by adsorption-desorption of nitrogen ($N_2$), while the morphology and topography of samples were observed by Scanning Electron Microscope (SEM). The methyl esters, obtained from the catalytic test, were conducted with HP6890 GC chromatography equipped with FID detector, capillary column, HP 19095N-123-HP INNOWax Polyethylene Glycol, l = 30 m, d = 0.53 mm.

### *2.3. Catalytic tests*

The resulting materials were tested in simultaneous transesterification and esterification reactions to methyl ester production. The crude palm oil (CPO) was obtained from PT Wilmar, Indonesia. Furthermore, the simultaneous transesterification and esterification reactions of CPO with methanol at molar ratios of methanol to CPO of 10:1, 20:1, 30:1 and 40:1 at a temperature of 150 °C for 5 hours with a stirring speed of 600 rpm and 5 wt.% concentration of CPO as the catalyst were done. Then, the mixtures were centrifuged for 15 minutes, and the residue of methanol was evaporated with a rotary evaporator. The methyl esters were analyzed by GC. The yield of catalysts toward the formation of methyl ester was determined based on the product of methyl ester obtained from the reactions. The yield of methyl ester was calculated using Eq. (1) [32]:

$$\text{Yield}(\%) = \frac{\text{weight of methylesters(g)}}{\text{weight of CPO(g)}} \times 100\% \quad (1)$$

## 3. Results and Discussion

X-ray diffraction techniques were employed to obtain the information about the structure, crystallinity, and composition of the prepared material. Figure 1 shows the XRD patterns of $MgF_2$ and $ZnF_2$. As shown in Figure 1, the $MgF_2$ material has three main peaks at $2\theta$ = 27.29, 40.56 and 53.47, which indexed to the rutile structure. However, the $MgF_2$ sample has a structural disorder and very broad as presented in Figure 1(a). Figure 1 (b) shows the diffractogram pattern of $ZnF_2$ sample. As displayed in Figure 1(b), the $ZnF_2$ catalyst had the main peaks at $2\theta$ = 26.69°, 34.39° and 52.98°, which matched to rutile structure. However, the $ZnF_2$ has a higher crystallinity structure than $MgF_2$ with the same structure of rutile.

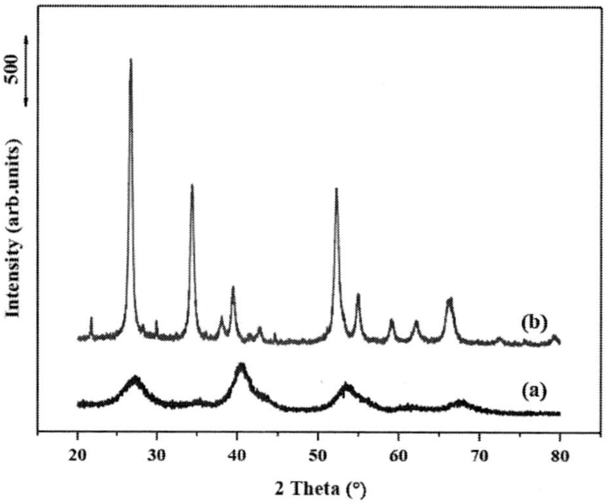


**Figure 1.** X-ray diffraction patterns of $MgF_2$ (a) and $ZnF_2$ (b)

The $MgF_2$ and $ZnF_2$ catalysts were also characterized by infrared spectrophotometer (FT-IR). The use of this characterization aimed to determine the chemical bond of $MgF_2$ and $ZnF_2$. Figure 2 shows the spectra of $MgF_2$ and $ZnF_2$ materials. As illustrated in Figure 2, both $MgF_2$ and $ZnF_2$ have the spectral bands at 3000-3700 $cm^{-1}$ (♣), which indicated the presence of the adsorbed $H_2O$ molecules and of bridged –OH group, while the bending vibrations of the H-O-H and OH group at 1640 $cm^{-1}$ (♠) and 792 $cm^{-1}$ (●). The bands at 418, and 490 $cm^{-1}$ (♦) were attributed to Zn-F and Mg-F bending. As presented in Figure 2 (a)-(b), the spectral bands at 1563, 1453, 702, and 615 $cm^{-1}$ referred to acetate group were not seen. Moreover, the bands appearing at 1026 and 945 $cm^{-1}$ in all prepared samples, which were correlated with the existence of C-C and C-H vibrations, also did not appear. These results indicated that these samples were free of acetate group and methanol as precursors and solvents.

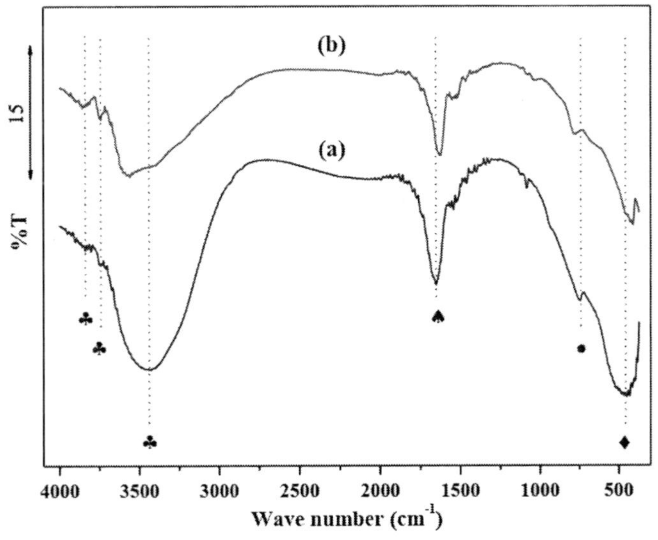


**Figure 2.** Fourier Transform Infrared (FT-IR) Spectra of $MgF_2$ (a) and $ZnF_2$ (b)

The information about the thermal stability of the $MgF_2$ and $ZnF_2$ samples was gathered using TG/DTG. Both TG and DTG profiles in Figure 3 and 4 showed that all samples had similar behaviors. The TG/DTG profiles of $MgF_2$ and $ZnF_2$ are presented in Figure 3 and 4. As shown in Figure 3 and 4, these samples only had one single destruction phase (150-300 °C) that indicated the loss of the -OH group on the surface of solids. As shown in Figure 3 and 4, it was not appeared the removal of HF, which was usually observed at a higher temperature, i.e. 700 °C (both TG and DTG curves). This result indicated that the $MgF_2$ and $ZnF_2$ materials were thermally stable.

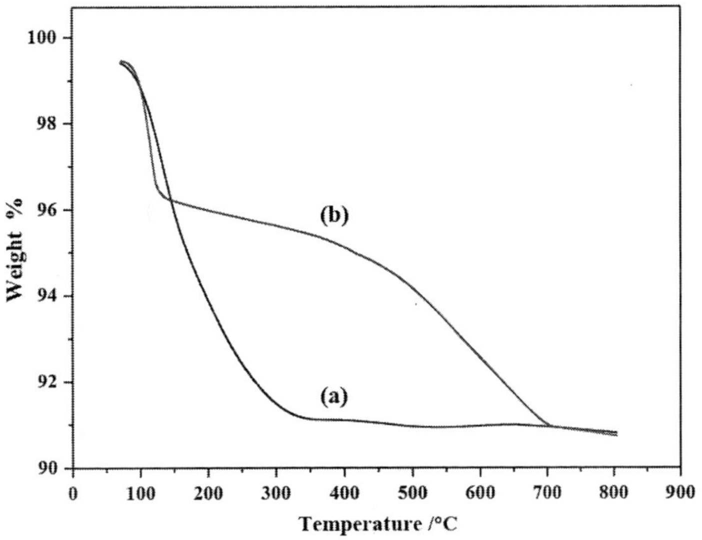


**Figure 3**. TG profiles of $MgF_2$ (a) and $ZnF_2$ (b)

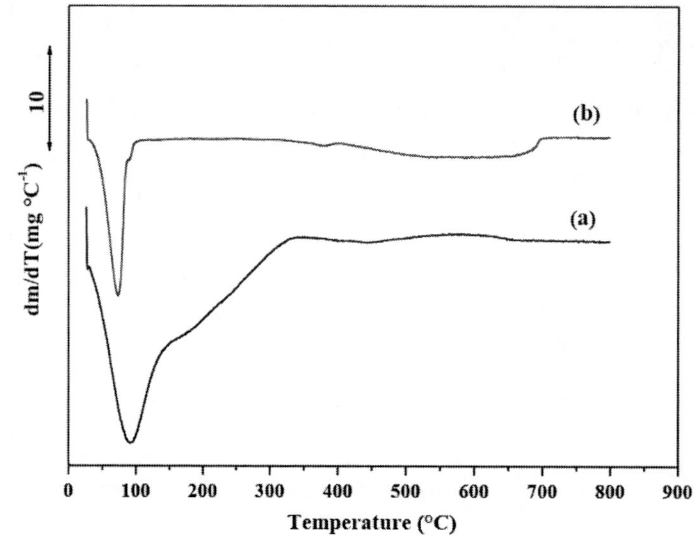


**Figure 4.** DTG profiles of $MgF_2$ (a) and $ZnF_2$ (b)

The porous features of catalysts were measured through the $N_2$ adsorption-desorption measurements. The $N_2$ adsorption-desorption isotherms of materials are shown in Figure 5. Based on the IUPAC classification, the $MgF_2$ has typical IV with H1 hysteresis loop, which showed a specific pore. While the $ZnF_2$ sample had typical II with H3 hysteresis loop, which referred to pores with non-uniform size. Furthermore, from pore size distribution analysis as shown in Table 1, the pore sizes of $MgF_2$ and $ZnF_2$ were 3.63 and 50.16 nm. These results showed that both of $MgF_2$ and $ZnF_2$ represent a mesoporous structures. The surface area from BET measurement of $MgF_2$ and $ZnF_2$ were 110.973 and 17 $m^2/g$, respectively.

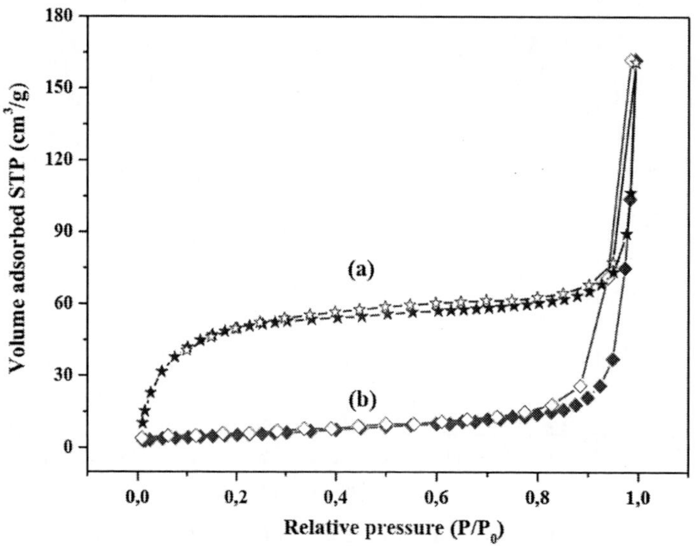


**Figure 5.** $N_2$ Adsorption-desorption isotherms of $MgF_2$ and $ZnF_2$

**Table 1.** Physicochemical properties of $MgF_2$ and $ZnF_2$

| Catalysts | $S_{BET}$ ($m^2$/g) | Pore diameter (nm) | Pore volume ($cm^3$/g) |
|---|---|---|---|
| $MgF_2$ | 110.973 | 3.6330 | 0.1008 |
| $ZnF_2$ | 17 | 50.1568 | 0.2515 |

The morphology and topography of the catalysts were observed by Scanning Electron Microscope (SEM). The SEM image of $MgF_2$ and $ZnF_2$ is shown in Figure 6. From SEM images, it can be obtained the information that the particle size of $MgF_2$ and $ZnF_2$ was around 0.2-3.0 μm, while the morphology of these samples was similar as shown in Figure 7.

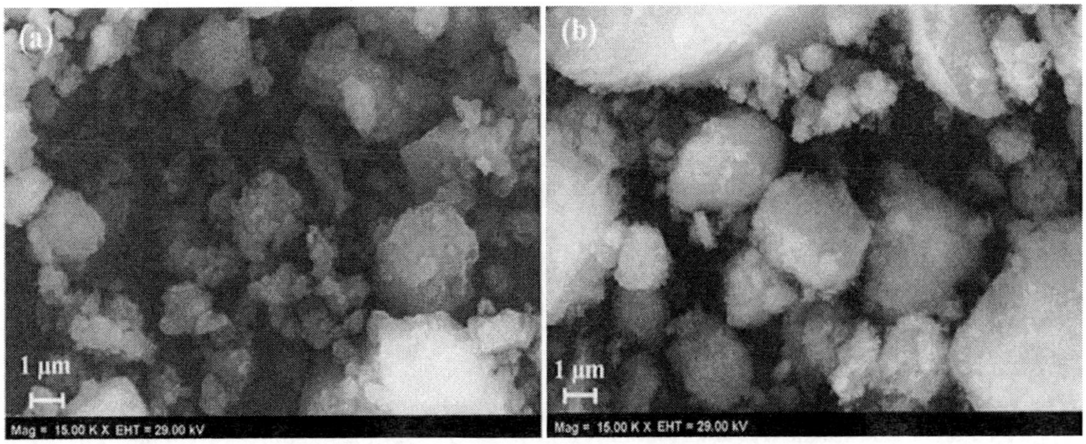


**Figure 6.** SEM images of $MgF_2$ (a) and $ZnF_2$ (b)

The catalytic evaluation of these catalysts in the simultaneous transesterification and esterification reactions was summarized in Table 2. As presented in Table 2, the $ZnF_2$ catalyst showed a higher methyl ester yield than $MgF_2$ of 85.21 and 26.82%, respectively. This result indicated that the crystallinity of material had advantages in the reaction with synthesis methyl ester than disorder phase. In addition, the higher activities could also be attributed to the pore diameter and pore volume of $ZnF_2$. It was clarified that the large pore size of the catalyst was able to adsorb more reactants and resulted in a higher catalytic activity. On the other hand, although the surface area of $MgF_2$ was greater than that of $ZnF_2$, it cannot help to improve the contact between the reactant and active site of the catalyst. It was clear that catalytic activity of these catalysts did not correlate with the surface area. Moreover, as shown in Table 2, the reaction had also been carried out with different molar ratio and showed different results. The yield of methyl ester increased together with the increase in molar ratio of CPO to methanol from 13.97 to 85.21% ($ZnF_2$), respectively. According to the previous research, the molar ratio of CPO to methanol is one of the important variables that influence the yield of methyl ester. Stoichiometrically, this reaction required three moles of methanol to react with one mole of triglyceride to obtain three moles of methyl ester (transesterification) and one mole of methanol to react one mol of free fatty acids to produce one mole of methyl ester (esterification). Therefore, it needed excess methanol for product formation. However, the excess of methanol could increase the polarity of reaction mixtures, and then decrease the yield of methyl ester from 85.21 to 80.99% ($ZnF_2$), respectively.

**Table 2.** The catalytic evaluation of the catalysts

| Catalysts | Molar ratio | Temperature (°C) | Time (h) | Yield (%) |
|---|---|---|---|---|
| $MgF_2$ | 1:10 | 150 | 5 | 2.16 |
| $MgF_2$ | 1:20 | 150 | 5 | 10.59 |
| $MgF_2$ | 1:30 | 150 | 5 | 26.82 |
| $MgF_2$ | 1:40 | 150 | 5 | 20.27 |
| $ZnF_2$ | 1:10 | 150 | 5 | 13.97 |
| $ZnF_2$ | 1:20 | 150 | 5 | 54.76 |
| $ZnF_2$ | 1:30 | 150 | 5 | 85.21 |
| $ZnF_2$ | 1:40 | 150 | 5 | 80.99 |

## 4. Conclusion

The synthesis of $MgF_2$ and $ZnF_2$ has been successfully prepared by sol-gel process. The resulting samples performed as heterogeneous catalysts in the simultaneous transesterification and esterification reactions for the production of methyl ester. These samples have rutile structure with a different phase (amorphous and crystal). The surface area of $ZnF_2$ is lower than $MgF_2$, but the pore size and volume of $ZnF_2$ are higher than that of $MgF_2$. However, the catalytic activity of $ZnF_2$ is greater than $MgF_2$ of 85.21% and 26.82% with the same reaction condition. In addition, the optimum conditions of reactions for finding the highest yield of methyl ester are obtained at a molar ratio of oil/methanol of 1:30, a reaction temperature of 150 °C for 5 hours with a stirring speed of 600 rpm and 5 wt.% of CPO concentration as a catalyst. The increasing methanol decreased the yield of methyl ester from 85.21 to 80.99% for $ZnF_2$, respectively.

## 5. References

[1] Rathore V, Newalkar B L and Badoni R P 2016 *Energy for Sustainable Development* **31** 24-49
[2] Rashed M M, Kalam M A, Masjuki H H, Mofijur M, Rasul M G and Zulkifli N W M 2016 *Industrial Crops and Products* **79** 70-76
[3] Zhu L, Cheung C S and Huang Z 2016 *Applied Thermal Engineering* **94** 249-255
[4] Danov K D, Stanimirova R D, Kralchevsky P A, Basheva E S, Ivanova V I and Petkov J T 2015 *Journal of Colloid and Interface Science* **457** 307-318
[5] Issariyakul T and Dalai A K 2014 *Renewable Sustainable Energy Review* **31** 446-471
[6] Aransiola E F, Ojumu T V, Oyekola O O, Madzimbamuto T F and Ikhu-Omoregbe D I O 2014 *Biomass-Bioenergy* **61** 276-297
[7] Thunyaratchatanon C, Luengnaruemitchai A, Chollacoop N and Yoshimura Y 2016 *Fuel* **163** 8-16
[8] Kouzu M, Kajita A and Fujimori A 2016 *Fuel* **182** 220-226
[9] Syamsuddin Y, Murat M N and Hameed B H 2016 *Bioresource Technology* **214** 248-252
[10] Onoji S E, Iyuke S E, Igbafe A I and Nkazi D B 2016 *Energy Conversion and Management* **110** 125–134
[11] Helwani Z, Aziz N, Kim J and Othman M R 2016 *Renewable Energy* **86** 68-74
[12] Eryilmaz T and Yesilyurt M K 2016 *Renewable Energy* **95** 233-247
[13] Amani H, Asif M and Hameed B H 2015 *Journal of the Taiwan Institute of Chemical Engineers* **000** 1-9
[14] Poosumas J, Ngaosuwan K, Quitain A T and Assabumrungrat S 2016 *Energy Conversion and Management* **120** 62–70
[15] Li H, Niu S, Lu C and Li J 2016 *Fuel* **176** 63–71
[16] Ye W, Gao Y, Ding H, Liu M, Liu S, Han X and Qi J 2016 *Fuel* **180** 574-579
[17] Shahbazi M Z, Khoshandam B, Nasiri M and Ghazvini M 2012 *Journal of the Taiwan Institute of Chemical Engineers* **43** 504-510
[18] Carmo Jr, A C, De Souza L K C, Costa C E F, Longo E, Zamian J R and Rocha Filho G N 2009 *Fuel* **88** 461-468

[19] Cardoso A L, Neves S C G and Da Silva M J 2009 *Energy & Fuels* **23** 1718-1722
[20] Soriano N R, Venditti R Jr and Argyropoulos D S *Fuel* **88** 560-565
[21] Shao G N, Sheikh R, Hilonga A, Lee J E, Park Y and Kim H K 2013 *Chemical Engineering Journal* **215-216** 600-607.
[22] Iglesias J, Melero J A, Bautista L F, Morales G, Sánchez-Vázquez R and Andreola M T 2011 *Catalysis Today* **167** 46-55
[23] Samart C, Chaiya C and Reubroycharoen P 2010 *Energy Conversion and Management* **51** 1428-31
[24] Domingues C, Correia M J N, Carvalho R, Henriques C, Bordado J and Dias A P S 2013 *Journal of Biotechnology* **164** 433-440
[25] Bail A, dos Santos V C, de Freitas M R, Ramos L P, Schreiner W H, Ricci G P, Ciuffi K J and Nakagaki S 2013 *Applied Catalysis B: Environmental* **130-131** 314-324
[26] Sankaranarayanan T M, Pandurangan A, Banu M and Sivasanker S 2011 *Applied Catalysis A: General* **409- 410** 239-247
[27] Wuttke S, Coman S M, Krhnert J, Jentoft F C and Kemnitz E 2010 *Catalysis Today* **152** 2-10
[28] Coman S M, Parvulescu V I, Wuttke S and Kemnitz E 2010 *WILEY-VCH Verlag GmbH & Co. KGaA Weinheim* **2** 92-97
[29] Candu N, Wuttke S, Kemnitz E, Coman S M and Parvulescu V I 2011 *Applied Catalysis A: General* **391** 169-174
[30] Troncea S B, Wuttke S, Kemnitz E, Coman S M and Parvulescu V A 2011 *Applied Catalysis B: Environmental* **107** 260-267
[31] Wuttke S, Coman S M, Schulz G, Kirmse H, Vimont A, Daturi M, Schroeder S L M and Kemnitz E 2008 *WILEY-VCH Verlag GmbH & Co. KGaA, Weinheim* **14** 11488-11499
[32] Rodríguez D A T, Romero-Ibarra I C, Ibarra I A and Pfeiffer H 2016 *Renewable Energy* **93** 323-331

**Acknowledgements**
The authors thank Sepuluh Nopember Institute of Technology (ITS) for the given research funding, Laboratory of Materials and Energy, Department of Chemistry of ITS, as well as all those who helped.

# Synthesis of Zinc Oxide Nanoparticles using Anthocyanin as a Capping Agent

**N L W Septiani[1], B Yuliarto[1,2], M Iqbal[1] and Nugraha[1,2]**

[1]Department of Engineering Physics, Institut Teknologi Bandung, Indonesia
[2]Research Center for Nanosciences and Nanotechnology, Institut Teknologi Bandung, Bandung, Indonesia

Email: brian@tf.itb.ac.id

**Abstract:** Zinc Oxide nanoparticles have been successfully synthesized by utilizing anthocyanin as a capping agent by thermal decomposition of precursor route. The influence of the high and low concentrations of the anthocyanin to the shape and size of ZnO was investigated in this work. The anthocyanin was obtained from Indonesia black rice extract with methanol as a solvent. The crystallinity and morphology properties were characterized by X-Ray Diffractometer (XRD), and Scanning Electron Microscope (SEM), respectively. XRD result showed that ZnO was formed with good crystallinity without any second phase and had a hexagonal wurtzite crystal structure. SEM result revealed that ZnO with a low concentration of anthocyanin has a spherical shape with a uniform size of about 16 nm while ZnO with a high concentration of anthocyanin has a rod-like shape. The size of spherical ZnO in this work is smaller than ZnO from the same method of synthesis without anthocyanin (~30 nm).

**Keywords:** Zinc oxide, nanoparticle, anthocyanin, capping agent.

## 1. Introduction

Zinc oxide (ZnO) is one of n-type semiconductors that has a wide band gap of 3.37 eV and high exciton binding energy of 60 meV at room temperature [1-3]. Moreover, ZnO has some characteristics such as antimicrobial activity, photocatalytic activity, UV absorbance, and piezoelectric that make ZnO can be applied to various applications such as for solar cell, gas sensor, photodiode, photodetector, and anti-bacterial agent [4]. Besides, this material is used in biomedical field because it is non-toxic, biocompatible, and flexible [5].

ZnO has several advantages, i.e. relatively low-temperature synthesis, controllable morphology, and the particular properties that depend on its size [6]. Nano-scaled ZnO has significantly different properties with their bulk. Until this time, there are two kinds of routes utilized to synthesize this material, namely dry process, and wet process. However, the wet processes like hydrothermal, solvothermal and sol-gel are preferable because it is easy to control the parameters and they have relatively low costs. In this route, the activity of controlling the particle size is generally done by controlling the calcination temperature [7]. However, because of their high surface energy, agglomeration frequently occurred [7, 8]. Therefore, it must be restricted by a capping agent. Some works reported the utilization of a chemical-based capping agent to inhibit the growth of nanoparticles

[6]. However, a chemical-based capping agent yields high toxic residual. Therefore, researchers are attracted to use an organic capping agent to replace chemical-based capping agent [9-12] This work reports the use of anthocyanin as a capping agent to fabricate ZnO nanoparticles. The influence of anthocyanin on the morphology of ZnO was also investigated in this study.

## 2. Experimental Method

The first step of this experiment was the extraction of black rice. The black rice was obtained from Cirebon, West Java, Indonesia. In a typical procedure, black rice was smashed until the soft powder was formed. 100 g black rice was then diluted in 150 ml methanol and stirred for 24 h. The solution was centrifuged to obtain the black rice extracted liquid. Next, 23.7 mmol of $Zn(NO_3)_2$ $4H_2O$ was diluted in water and stirred for 30 minutes. 1 ml black rice extract was dropped wisely and stirred at 60 °C until the precipitation was formed. The precipitation was then rinsed with ethanol and deionized water for several times. The result was subsequently calcined at 400 °C for 3h to get white powder. The synthesis process of ZnO without anthocyanin was performed in the same way without adding the black rice extract. The crystallinity and morphology of ZnO were characterized by X-Ray Diffraction (XRD) and Scanning Electron Microscope (SEM), respectively.

## 3. Results and discussion

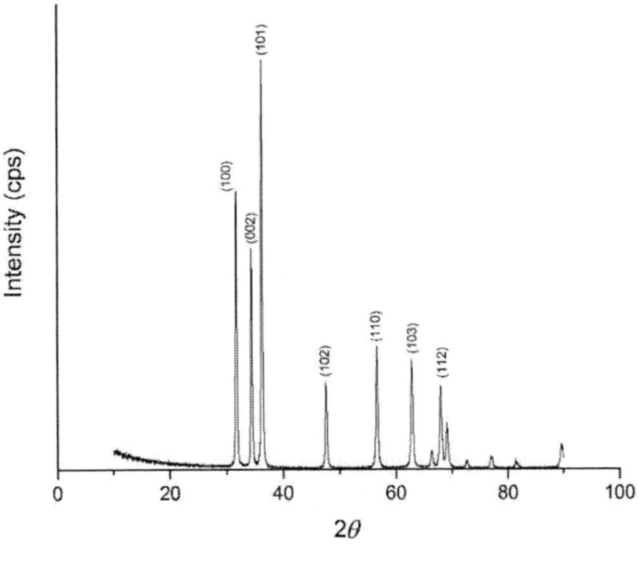


**Figure. 1** Diffraction patterns of anthocyanin-capped Zinc Oxide

The diffraction patterns of anthocyanin-capped ZnO are shown in Figure 1. From the patterns, it can be seen that there was only one phase of the ZnO sample. The sharp peak of the diffraction patterns indicated that the ZnO had excellent crystallinity. In this case, ZnO had hexagonal wurtzite crystal structures with 2θ of 31.775, 34.75, 36.294, 47.498, 56.598, 62.912, 66.379, 67.988, and 69.164. The highest peaks came from the planes of (100), (002), (101), (102), (110), (103), and (113). The data matched the JCPDS card no. 790206. Moreover, there was more than one peak in the patterns indicating that ZnO was polycrystal with more than one orientation. The addition of anthocyanin did not affect the crystallinity of ZnO at all.

Figure 2 (a & b) shows the scanning electron microscopic result of ZnO and anthocyanin-capped ZnO. From the figure, both ZnO and low anthocyanin-capped ZnO had spherical shapes, and the sizes were relatively uniform. The average size of ZnO was measured by Scherrer formula in equation 1, where λ is a wavelength of Cu Kα radiation 0.154 nm, D is the size of the crystal, β is FWHM, and K is Scherrer constant, in this case, was 0.9. From this formula, the average size of ZnO without anthocyanin was found to be about 30 nm, and the present of anthocyanin could decrease the average size until about 16 nm. Anthocyanin is one of flavonoid substances that can act as a capping agent. Anthocyanin is anthocyanidin with the sugar group mainly consisting of glucoside. Five hydroxyl groups from glucoside and some from anthocyanidin could interact with the surface of ZnO and made the surface passive. They could suppress the growing of ZnO particles during the synthesis and calcination processes. They also could prevent the agglomeration process. Moreover, Figure 2c indicated that a high concentration of anthocyanin could change the shape of ZnO from spherical to rode-like. This phenomenon indicated that the concentration of anthocyanin affected the shapes of ZnO. In high concentration, anthocyanin did not act as a capping agent but as a template for ZnO. The chain of sugar group could serve as a template and make rod-like shaped ZnO. This phenomenon was also observed by other results where glucose or sucrose was used as a capping agent [13,14]. This result is similar to a recent study where the high concentration of glucoside could make rod-like shaped CdO [15].

$$D = \frac{K\lambda}{\beta cos\theta} \quad (1)$$

**Figure. 2** Morphology of ZnO with a spherical shape (a), and low anthocyanin-caped ZnO with a spherical shape (b), and high anthocyanin-caped ZnO with a rod-like shape.

## 4. Conclusion
The black rice extract or anthocyanin is successfully fabricated and utilized as a capping agent in the synthesis of ZnO nanoparticles. The low concentration of anthocyanin capping agent could decrease the size of ZnO nanoparticles from 30 nm to 16 nm. The high concentration of anthocyanin could

change the shape of ZnO into rod-like. Moreover, the presence of anthocyanin did not affect the crystallinity of ZnO.

## 5. References

[1] Yuliarto B, Nulhakim L, Ramadhani M F, Iqbal M, Nuruddin A,2015 Improved Performances of Ethanol Sensor Fabricated on Al-Doped ZnO Nanosheet Thin Films, **15** 4114–4120.

[2] Yuliarto B, Julia S, Iqbal N L W S M, Ramadhani F 2015 The Effect of Tin Addition to ZnO Nanosheet Thin Films for Ethanol and Isopropyl Alcohol Sensor Applications **47** 76–91. doi:10.5614/j.eng.technol.sci.2015.47.1.6.

[5] Sim M 2013 Applied Surface Science Anionic 11-mercaptoundecanoic acid capped ZnO nanoparticles **282** 342–347. doi:10.1016/j.apsusc.2013.05.130

[7] Sangeetha G, Rajeshwari S, Venckatesh R 2011 Green synthesis of zinc oxide nanoparticles by aloe barbadensis miller leaf extract : Structure and optical properties *Mater. Res. Bull.* **46** 2560–2566. doi:10.1016/j.materresbull.2011.07.046

[8] Sangeetha G, Rajeshwari S, Venckatesh R 2011 Green synthesis of zinc oxide nanoparticles by aloe barbadensis miller leaf extract : Structure and optical properties *Mater. Res. Bull.* **46** 2560–2566. doi:10.1016/j.materresbull.2011.07.046.

[9] Maryanti E, Damayanti D, Gustian S Y *S* I *2014* Synthesis of ZnO nanoparticles by hydrothermal method in aqueous rinds extracts of Sapindus rarak DC, *Mater. Lett.* **118** 96–98. doi:10.1016/j.matlet.2013.12.044.

[10] Javed R, Usman M, Tabassum S, Zia M 2016 Applied Surface Science Effect of capping agents : Structural , optical and biological properties of ZnO nanoparticles *Appl. Surf. Sci.* **386** 319–326. doi:10.1016/j.apsusc.2016.06.042.

[11] Gunalan S, Sivaraj R, Venckatesh R 2012 Spectrochimica Acta Part A : Molecular and Biomolecular Spectroscopy Aloe barbadensis Miller mediated green synthesis of mono-disperse copper oxide nanoparticles : Optical properties *Spectrochim. ACTA PART A Mol. Biomol. Spectrosc.* **97** 1140–1144. doi:10.1016/j.saa.2012.07.096.

[12] Diallo A, Ngom B D, Park E, Maaza M 2015 Green synthesis of ZnO nanoparticles by Aspalathus linearis : Structural & optical properties *J. Alloys Compd.* **646** 425–430. doi:10.1016/j.jallcom.2015.05.242.

[13] Jeyarani W J, Tenkyong T, Bachan N, Kumar D A, Shyla J M 2016 An investigation on the tuning effect of glucose-capping on the size and bandgap of CuO nanoparticles *Advanced Powder Technology* **27** 338-346. doi: 10.1016/j.apt.2016.01.006

[14] Arani M M, Niasari M S 2016 Effect of carbohydrate sugars as a capping agent on the size and morphology of pure $Zn_2SnO_4$ nanostructures and their optical properties *Materials Letters* **174** (2016) 71-74. doi: 10.1016/j.matlet.2016.03.084.

[15] Faleni N, Moloto M J, Africa S 2013 Effect of Glucose as Stabilizer of ZnO and CdO *Nanoparticles on the morphology and optical* **14** 127–135.

## Acknowledgments

This research was partially supported by Advance Functional Materials Laboratory (AFM), Bandung Institute of Technology (ITB).

# Microstructure Transformation of Mg-1.6Gd during Hot Rolled at High Deformation Ratio

**O Susanti, M A Mochtar and S Harjanto**

Department of Metallurgical and Materials Engineering, Faculty of Engineering Universitas Indonesia, Kampus UI Depok16424, Indonesia

Email : harjanto@metal.ui.ac.id

**Abstract.** Mg-1.6Gd alloy was prepared by hot rolling to produce a potentially degradable implant materials. Microstructural transformation during hot rolling was observed and correlated to mechanical properties. The objective of this study was to examine the constituents of the microstructure such as grain size, phases which are contributed to mechanical properties. Hot rolled Mg-1.6Gd alloy was performed at a temperature range of recrystallization (400-550 °C) with a speed of 10 mm/min and reduction of 95%. The hot rolled was employed by two methods: unidirectional rolling (UR) and cross rolling (CR). The results show that both of the hot rolling techniques reveal refine microstructure with the grain size range of 50 -130 μm. The number of lamellar in every grain were shown on the UR, while for the CR resulted in many precipitates, especially at the highest temperature hot rolling. The hot rolled sample of 95% reduction had the relatively smaller of the grain size than the hot rolled sample of 30% reduction (200 μm).

**Keywords:** Mg-rare earth element, microstructure, lamellar, precipitate, biodegradable, prospective implant materials

## 1. Introduction

Binary Mg-Gd alloys have been received much attention as degradable implant materials [1, 2]. The alloys are well known for their low density, high strength, creep resistance, and inherent biocompatibility. The elastic modulus is about 45 GPa, which is close to that of human bone properties. It is suitable and is used as the requirements of implant materials[3-6]. However, magnesium alloys have been limited in application due to their low ductility[7, 8]. Many efforts have been performed to improve mechanical properties and/or degradable characteristic of the alloy, such as the addition of other elements and/another mechanical process. In terms of mechanical properties, that can be attributed to the microstructure control that is responsible for the properties relevant to applications, such as strength, hardening, ductility, and elastic modulus[9-18]. Therefore, Mg-Gd alloys have a great potential as candidates for the future design of degradable material.

Recently, Peng *et al.* reported that the melt-spun Mg-20Gd alloy are found in a few metastable (amorphous) phases, and they transform into a stable phase ($Mg_5Gd$) after at the high temperature (350- 400 °C)[10]. The tensile strength of the melt spun is higher than that of as-cast alloy, while the elongation of the two samples is less than 2%. The fracture of the alloy sample is a combination of brittle fracture and ductile fracture.Other study showed that hot extrusion is one of technique to refine the microstructure morphology and to improve the mechanical properties of Mg-Gd alloys. Mg-1.6 Gd (wt%) alloys exhibit the tensile strength of 232 MPa, the yield strength of 142 MPa and elongation to failure of 72%. These alloys have been fabricated by direct extrusion at the temperature of recrystallization range of 400-550 °C [12]. Rolling process is also an effective way to refine the microstructure which will affect the mechanical properties. Cizek *et al.* studied the properties of cold rolled Mg-15Gd alloy with a reduction ratio of 8%, 21%, and 29%. Precipitation was found from the rolled samples which cause strong hardening[11]. Luo *et al.* also reported that the microstructure and the mechanical properties could be improved by the addition of Mg-xGd (x = 0.8, 1.5, and 2.5) [16]. Further study is still needed to understand the microstructure transformation of Mg-1.6Gd alloys that performs the thermo-mechanical process. In this paper, severe hot rolled processing, in terms of unidirectional (only 0° rolling direction) and cross rolling (simultaneously rolling in 0° and 90° direction) with a total reduction of 95% is conducted to Mg-1.6Gd (wt%) alloys. Microstructure morphology and their transformation of the alloy sheet will be observed and discussed.

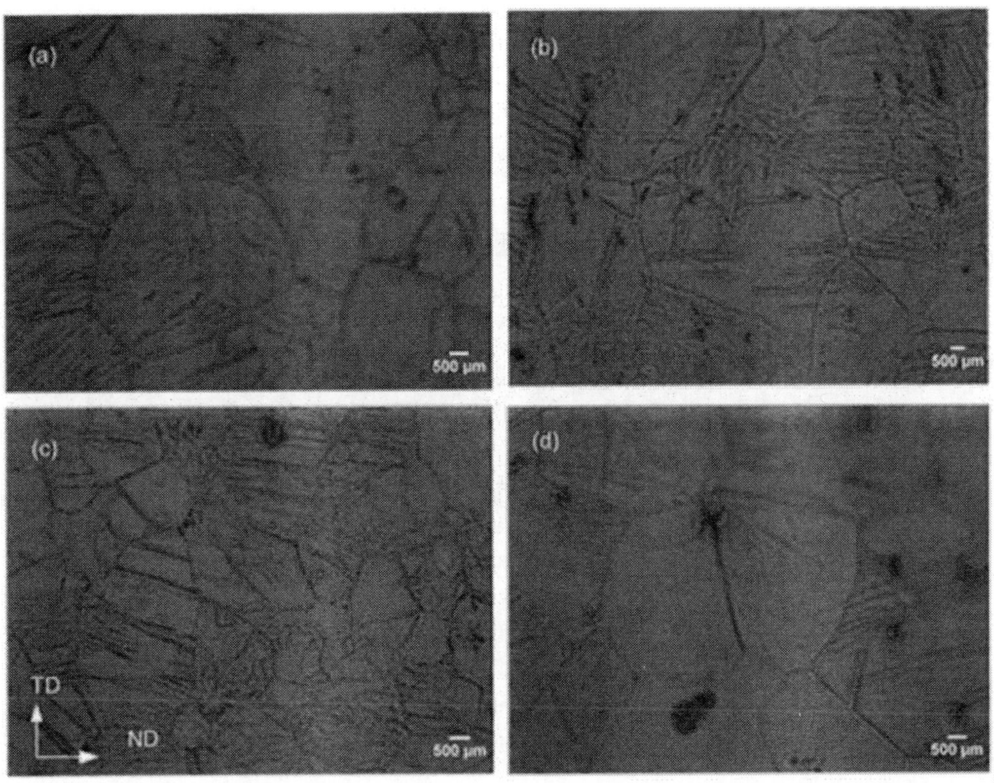


**Figure 1**. Microstructures of hot rolled Mg-1.6Gd alloy at various temperatures with reduction of 95%: (a) 400 °C, (b) 452 °C, (c) 491 °C, and (d) 560 °C unidirectional rolling (UR)

## 2. Experimental Method

Mg-1.6 Gd (wt%) alloy ingot were machined for hot rolling specimens to the initial size of 50 mm in length, 25 mm in width and 10 mm in thickness. The alloy was homogenized at 500 °C for 5 hours and then followed by normalizing at room temperature (about 37 °C). The process was employed by two methods: unidirectional rolling (UR) and cross rolling (CR) at a total reduction of 95%. The UR has employed one direction each pass, while the CR was employed by 0° rolling direction first and followed by changing the rolling direction to 90° each pass (for four times). The hot rolling, both UR and CR was carried out at a temperature range of recrystallization ( 400-550 °C). The specimens were rolled by multi-pass to the final thickness (h = 3 mm) with speed of 10 mm/min. Each the rolling samples were reheated to 400 - 550 °C for 15 min before per pass rolling. The sheets rolling were examined by standard metallography procedures. Samples were ground by using up to 2000 grade silicon carbide paper and then polished with alumina paste. An etching solution was prepared containing 20 mL acetic acid +1 mL nitric acid +60 mL ethylene glycol +20 mL aquadest for about 15 seconds. Then the microstructure of the specimen was observed using optical microstructure after cooled to room temperature after hot rolling.

## 3. Results and Discussion

The microstructures of Mg-1.6Gd alloys reveal nearly similar grain morphology with tendency fine equiaxed for each temperature for both of UR and CR as shown in Figure 1 and 2. Figure 1 shows the hot rolled of UR for different temperatures. Generally, the grain size of UR is smaller than CR. The average of grain size for UR is about 70 μm. At a temperature of 400 °C the grain size is about 69 μm. The grain size increases as temperature increases where the grain size growth occurs at a temperature of 452 and 560 °C are 72 and 89 μm, respectively. The grain size drops significantly to 49 μm at a temperature of 491 °C. In this temperature, the hot rolled process can influence clearly to the grain size refinement. All of the hot rolled samples show lamellar phases inside the grain which grows in the different orientation. A sharp form of Mg-rich phase (α-Mg) is also observed in the grain of the samples which hot rolled unidirectional at a temperature of 491 °C.

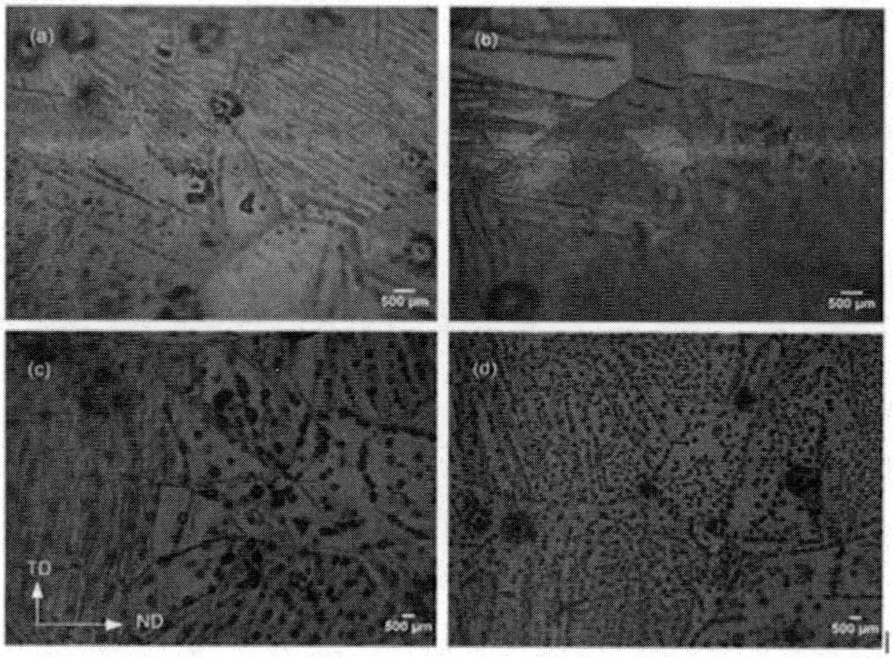


**Figure 2**. Microstructures of hot rolled Mg-1.6Gd alloy at various temperatures with reduction of 95%: (a) 410 °C, (b) 455 °C, (c) 521 °C, and (d) 561 °C of Cross rolling (CR)

Figure 2 shows the microstructure of the hot rolled for CR which has larger grain size compared to the UR which is visually more pretending homogenized. At a temperature of 410 °C has the biggest grain size about 127 μm, and the grain size goes down to 64 μm at the highest temperature of 561 °C. The smallest grain size is found at a temperature of 521 °C with the size of 56 μm. The microstructure also exhibited the number of lamellar in every the grain. The lots of lamellar are found in the hot rolled sample of 410 °C. The formation of precipitates occurs at higher temperature. It can be predicted that these precipitates particles in grain are $Mg_5Gd$ as explained by Hort *et al.* in phase diagram [9]. The Mg-1.6Gd alloy in the different conditions at a temperature below 500 °C are composed mainly of α-Mg solid solution and second phase or precipitates of $Mg_5Gd$. However, the precipitation during the hot rolled is hardly detected because of the limited particles size at a temperature range of 400-550 °C. Moreover, the grain size and shape of the as-rolled plays an important role in the high strength of the Mg-1.6Gd alloy. It should be noted that the formation of precipitates only occurs in cross rolled samples. The formation of precipitation increases at higher temperature.

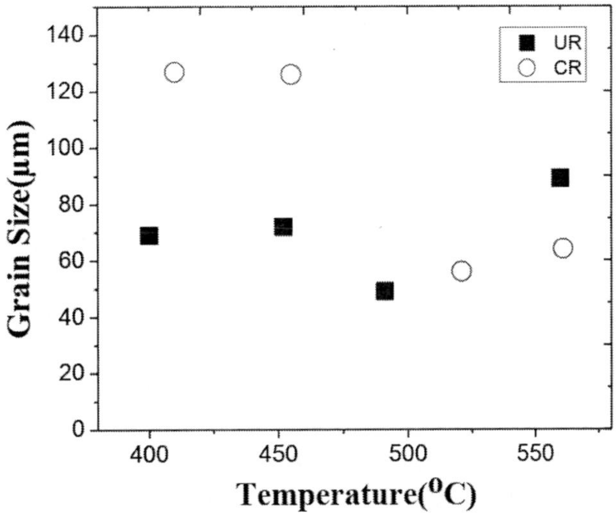


**Figure 3**. The grain size of Mg-1.6Gd alloys after unidirectional hot rolling (UR) and cross rolling (CR) at various temperatures.

The temperature effect on the grain size of both rolling methods, *i.e.* UR and CR is shown in Figure 3. The grain size of UR samples is relatively similar at the range of rolling temperature. However, for CR samples, at a temperature below 500 °C give higher grain size than that of UR samples at the same range of temperature. The grain size refinement of CR samples occurs clearly at a temperature higher than 500 °C.

**4. Conclusion**

In this study, the microstructure observation of the hot rolled Mg-1.6Gd had been conducted by two methods i.e. UR and CR at a temperature range of 400-550 °C. It is obtained that the lamellar phase which is a Mg-rich phase (α-Mg), are found in both of UR and CR samples at all range of the temperatures. It is also found that CR samples which were rolled at a temperature higher than 500 °C, there is precipitation of $Mg_5Gd$ that spread in the grain. The microstructure of the samples of UR has smaller grain size compared to CR samples at a temperature below 500 °C. However, refinement of grain size of CR samples occurs at rolling temperature higher than 500 °C.

IOP Conf. Series: Materials Science and Engineering **202** (2017) 012071 doi:10.1088/1757-899X/202/1/012071

## 5. References


[1] Hort N, Huang Y, Fechner D, Störmer M, Blawert C, Witte F *et al.* 2010 Magnesium alloys as implant materials – Principles of property design for Mg–RE alloys *Acta Biomaterialia* **6** pp. 1714-1725

[2] Grillo C A, Alvarez F and de Mele M A F L 2014 Cellular response to rare earth mixtures (La and Gd) as components of degradable Mg alloys for medical applications *Colloids and Surfaces B: Biointerfaces* **117** pp. 312-321

[3] Zheng W-w, Li X-g, Zhang K, Li Y-j, Ma M-l and Shi G-12015 Reactive diffusion in Mg–Gd binary system at 773 K *Transactions of Nonferrous Metals Society of China* **25** pp. 3904-3908

[4] Fronzi M, Kimizuka H and Ogata S 2015 Atomistic investigation of vacancy assisted diffusion mechanism in Mg ternary (Mg–RE–M) alloys *Computational Materials Science* **98** pp. 76-82

[5] Qiuzu L, Xiaofeng D, Yanping L and Xiaojia W 2017 Analysis on micro-structure and mechanical properties of Mg-Gd-Y-Nd-Zr alloy and its reinforcement mechanism *Journal of Alloys and Compounds* **690** pp. 961-965

[6] Liu X, Zhang Z, Le Q, and Bao L 2016 Effects of Nd/Gd value on the microstructures and mechanical properties of Mg–Gd–Y–Nd–Zr alloys *Journal of Magnesium and Alloys* **4** pp. 214-219

[7] Drynda A, Deinet N, Braun N and Peuster M 2009 Rare earth metals used in biodegradable magnesium-based stents do not interfere with proliferation of smooth muscle cells but do induce the upregulation of inflammatory genes *Journal of Biomedical Materials Research Part A* **91** pp. 360-369

[8] Mueller W D, Fernández Lorenzo de Mele M, Nascimento M L and Zeddies M 2009 Degradation of magnesium and its alloys: dependence on the composition of the synthetic biological media *Journal of biomedical materials research Part A* **90** pp. 487-495

[9] Hort N, Huang Y, Fechner D, Störmer M, Blawert C, Witte F *et al.* 2010 Magnesium alloys as implant materials–principles of property design for Mg–RE alloys *Acta biomaterialia* **6** pp. 1714-1725

[10] Peng Q, Wu Y, Fang D, Meng J and Wang L 2006 Microstructures and properties of melt-spun and as-cast Mg-20Gd binary alloy *Journal of Rare Earths* **24** pp. 466-470

[11] Čížek J, Prochazka I, Smola B, Stulíková I and Očenášek V Influence of deformation on precipitation process in Mg–15wt.% Gd alloy *Journal of alloys and compounds* **430** pp. 92-96

[12] Susanti O, Harjanto S and Mochtar M 2015 Microstructure and Mechanical Properties of Extruded Mg-1.6Gd as Prospective Degradable Implant Materials *Advanced Material Research* **1112** p. 592

[13] Kulyasova O, Islamgaliev R, Kil'mametov A and Valiev R Superplastic behavior of magnesium-based Mg-10 wt% Gd alloy after severe plastic deformation by torsion *The Physics of Metals and Metallography* **101** pp. 585-590

[14] Nie J, Gao X and Zhu S 2005 Enhanced age hardening response and creep resistance of Mg–Gd alloys containing Zn *Scripta Materialia* **53** pp. 1049-1053

[15] Peng Q, Wu Y, Fang D, Meng J and Wang L 2007 Microstructures and properties of Mg–7Gd alloy containing Y *Journal of Alloys and Compounds* **430** pp. 252-256

[16] Luo S, Yang G, Liu S, Wang J, Li J and Jie W 2016 Microstructure evolution and mechanical properties of directionally solidified Mg-xGd (x = 0.8, 1.5, and 2.5) alloys *Materials Science and Engineering: A,* **662** pp. 241-250

[17] Cao G, Zhang D, Zhang W and Qiu C 2015 Microstructure evolution and mechanical properties of Mg–Nd–Y alloy in different friction stir processing conditions *Journal of Alloys and Compounds* **636** pp. 12-19

[18] Li X, Qi W, Zheng K and Zhou N 2013 Enhanced strength and ductility of Mg–Gd–Y–Zr alloys by secondary extrusion *Journal of Magnesium and Alloys* **1** pp. 54-63, 3// 2013.

**Acknowledgements**
This study is supported by the Directorate of Research and Community Service Universitas Indonesia through University Doctoral Research 2016 (Hibah Publikasi International TerIndek Tugas Akhir Mahasiswa-PITTA). The author would like to thank Prof. Dr. Ir. Gunawarman for supporting his laboratory facility (SEM) from Univ. Andalas Padang and Dr. Eng Jaka Fajar Fatriansyah for the discussion from Universitas Indonesia.

IOP Conf. Series: Materials Science and Engineering **202** (2017) 012072 doi:10.1088/1757-899X/202/1/012072

# The Effect of Molar Ratio on Crystal Structure and Morphology of $Nd_{1+X}FeO_3$ (X=0.1, 0.2, and 0.3) Oxide Alloy Material Synthesized by Solid State Reaction Method

**V Zharvan, Y N I Kamaruddin, S Samnur and E H Sujiono**

Laboratory of Material Physics. Department of Physics, *Universitas Negeri Makassar,* Makassar 90224, Indonesia.

Email: e.h.sujiono@unm.ac.id

**Abstract**. Perovskite is an oxide alloy which has a structure of $ABO_3$ (A= La, Nd, Sm, Gd; B= Fe, CO, Ni) and has an excellent catalytic activity and gas-sensitive properties. $NdFeO_3$ and its derivatives are important candidates for gas sensors. In this study, the effect of molar ratio (x=0.1, 0.2 and 0.3) on crystal structure and morphology of $Nd_{1+x}FeO_3$ synthesized by solid state reaction method has been studied. $Nd_{1+x}FeO_3$ samples were prepared using $Nd_2O_3$ (99.99 %) and $Fe_2O_3$ (99.99 %) as precursors. All of the samples were characterized using XRD to identify the phase and using SEM to identify the morphology. The synthesized $Nd_{1+x}FeO_3$ samples showed that molar ratio strongly influences the intensity, FWHM, and crystalline size. The samples of $Nd_{1+x}FeO_3$ have homogenous morphology and have three major phases, i.e. $NdFeO_3$, $Nd(OH)_3$ and $Nd_2O_3$ with crystalline sizes of $NdFeO_3$ of 137.0±0.1 nm, 152.2±0.1 nm and 137.0±0.1 nm for $Nd_{1.1}FeO_3$, $Nd_{1.2}FeO_3$, and $Nd_{1.3}FeO_3$, respectively. These results indicated that the sample of $Nd_{1.2}FeO_3$ was a good candidate for a gas sensor material.

**Keywords**: Molar ratio; crystal structure; morphology; $NdFeO_3$ oxide alloy; solid state method.

## 1. Introduction

The rare-earth orthoferrite, $ReFeO_3$ (Re: rare-earth element), is a well-known crystalline in the orthorhombic structure derived from a perovskite structure belonging to Pbnm space group and has exhibited physical and chemical properties due to their ionic and electronic defects [1]. Further, it is known that nano-crystalline perovskite oxides, $ReBO_3$ (Re = rare-earth elements, B = 3d transition metals and O = oxygen) on gas sensors have been studied [2]. Among these perovskite oxides, $NdFeO_3$ was reported to have excellent gas-sensing properties [3]. The other result showed that $NdFeO_3$ has high catalytic activities and high sensitivity with CO and HCs [4] comparing with other rare-earth elements.

There are various methods for the synthesis of $NdFeO_3$ alloy oxide such as sol-gel method [5], gel combustion method [1], hydrothermal method [6], solid state reaction method [7], and so on. There are advantages of using solid state reaction because this method is cheap and easy to complete. Also, the product of the reaction has a high purity and excellent crystallinity. The expected results from using this approach are to obtain a larger size, better crystallinity, and a higher purity than other methods. However, in solid-state reaction method, there are many properties that influence the crystal structure and morphology of $NdFeO_3$ such as time and calcined temperature [1], and also milling process [8] but there is not a report that explains the effect of molar ratio on the synthesis of $NdFeO_3$ by using this method employing $Nd_2O_3$ and $Fe_2O_3$ precursors.

In this paper, we present the influence of molar ratio (x = 0.1, 0.2 and 0.3) on crystal structure and morphology of $Nd_{1+x}FeO_3$which is synthesized by solid state reaction method from $Nd_2O_3$ and $Fe_2O_3$ precursors.

## 2. Experimental method

The raw materials of $Nd_2O_3$ (Strem Chemicals, 99.99%) and $Fe_2O_3$ (Sigma-Aldrich, 99.99%), were mixed based on the stoichiometric calculation to get an oxide alloy of $Nd_{1.1}FeO_3$ and were ground using mortar for 3 hours to maximize the reaction and to increase the homogeneity. The mixed materials were then calcined using furnace at 950 °C for 30 hours. After the calcination process was complete, the sample was then prepared into a pellet with 1.25 cm diameter under a 16-ton pressure. The pellet was then sintered at a temperature of 950 °C for 20 hours and slowly cooled to a temperature of 450 °C to be subsequently sintered again for 20 hours. The same procedure was repeated for the variation of the molar ratio at x = 0.2 and x = 0.3.All of the pellets of $Nd_{1+x}FeO_3$ (x = 0.1, 0.2 and 0.3) were then characterized by XRD (Xpert Pro) using $CuK_{\alpha}$ radiation (λ= 1.54060 Å) to obtain phase and by SEM (Inspect S50) to obtain the morphology contained in the pellets.

## 3. Results and discussion

Figure 1 shows the X-ray diffraction patterns of the $Nd_{1+x}FeO_3$ samples as the variations of molar ratio were x = 0.1, 0.2 and 0.3, respectively.

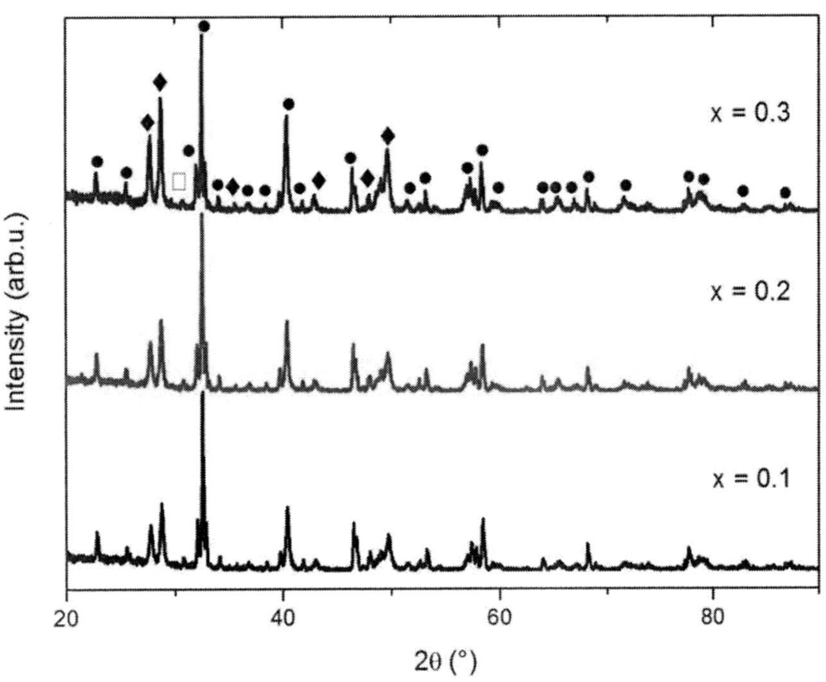


**Figure 1**. XRD pattern of $Nd_{1+x}FeO_3$ as variations of molar ratio of x =0.1, 0.2 and 0.3, respectively (● = $NdFeO_3$, □ = $Nd_2O_3$, ♦ = $Nd(OH)_3$).

The phases analysis using *X'Pert High Score Plus* software revealed that there were three phases in each sample, i.e. $NdFeO_3$ (PDF reference code 01-074-1473), $Nd(OH)_3$ (PDF reference code 01-070-0215) and $Nd_2O_3$ (PDF reference code 01-074-1147) with the highest intensity for $NdFeO_3$. It indicated that the samples were polycrystalline materials with a dominant phase of $NdFeO_3$ with an orthorhombic structure. Lou Xiangdong et al has also done synthesis of $NdFeO_3$ with a temperature of 800 °C [9] and Yabin Wang et al. with a temperature of 1000 °C [10], and they get similar results that the $NdFeO_3$ phase existed at $2\Theta = 32.56°$ with a hkl value of 121. Based on Figure 1, there were two minor $Nd(OH)_3$ phases due to the absorption of hydrogen and oxygen by neodymium [8] and $Nd_2O_3$ that existed due to the low temperature during the processes of calcination.

**Table 1.** Position (2θ), intensity and FWHM value of $NdFeO_3$ phase as a variation of molar ratio

| Molar ratio | 2 θ ( °) | Intensity (counts) | FWHM ( °) |
|---|---|---|---|
| 0.1 | 32.5649±0.001 | 757.8600±0.001 | 0.1181±0.001 |
| 0.2 | 32.5352±0.001 | 795.8900±0.001 | 0.0984±0.001 |
| 0.3 | 32.5063±0.001 | 625.2100±0.001 | 0.1181±0.001 |

Table 1 shows the influence of molar ratio on the position of 2Θ, intensity, and the FWHM value on the phase of $NdFeO_3$ for each sample. As illustrated, the increasing molar ratio of Nd content gradually shifted the XRD peaks towards a lower angle or higher *d* value due to larger ionic radii of Nd than Fe [11]. Peak position value $2\Theta = 32.5°$ had the highest intensity compare with other peaks and this peak was related to the $NdFeO_3$ phase with 121 hkl value based on the analysis using *Match!* software. Relating to the intensity, the molar ratio has influenced the intensity of the sample. Based on Table 1, it can be seen that the sample with a molar ratio of x = 0.2 had the narrowest and highest intensity which indicated the growth of $NdFeO_3$ crystallite and also gave information that the $Nd_{1.2}FeO_3$ sample has an excellent crystallinity and conductivity. The sample with a molar ratio of x = 0.2 has lowest FWHM value among others. This result indicates that the sample with a molar ratio of x = 0.2 has a similar crystal structure [12,13]. The determination of the crystalline size can be calculated using *Debye-Scherrer* equation [2]:

$$D = \frac{0.89\,\lambda}{\beta \cos\theta} \tag{1}$$

where λ is the wavelength of X-ray (0.15406 nm), β is FWHM value at 121, and θ is diffraction angle. Based on the calculation, the value of the crystalline size of $Nd_{1+x}FeO_3$ which was synthesized with the variations of molar ratio of x = 0.1, 0.2 and 0.3 were 137.0 nm, 152.2 nm, and 137.0 nm, respectively. Table 2 shows the lattice parameters of $Nd_{1+x}FeO_3$ sample obtained by Rietveld refinement using *Rietica* software.

**Table 2**. Lattice parameters of $Nd_{1+x}FeO_3$ obtained by Rietveld refinement

| Sample | a (Å) | b (Å) | c (Å) | Rp | Rwp | Re | $\chi^2$ |
|---|---|---|---|---|---|---|---|
| $Nd_{1.1}FeO_3$ | 5.5837±0.0004 | 7.7620±0.0006 | 5.4519±0.0004 | 19.58 | 26.06 | 15.83 | 2.71 |
| $Nd_{1.2}FeO_3$ | 5.5844±0.0005 | 7.7612±0.0006 | 5.4510±0.0004 | 20.10 | 27.36 | 15.84 | 2.98 |
| $Nd_{1.3}FeO_3$ | 5.5852±0.0005 | 7.7623±0.0007 | 5.4520±0.0005 | 20.63 | 27.56 | 15.55 | 3.14 |
| A1 | 5.579100 | 7.763700 | 5.443000 | - | - | - | - |

It can be seen from Table 2 that the increasing molar ratio gradually increased the lattice parameter due to differences in ionic radii between Nd and Fe that caused the *d*-spacing became larger and

IOP Conf. Series: Materials Science and Engineering **202** (2017) 012072 doi:10.1088/1757-899X/202/1/012072

shifted the x-ray peaks to a lower degree. The A1 sample is a result of the $NdFeO_3$ synthesis performed by Pedro et al. as a reference value [8].

SEM investigated the morphology, structure and particle size of the $Nd_{1+x}FeO_3$ samples with variations of molar ratio. Figure 2 shows the SEM image of $Nd_{1+x}FeO_3$ samples.

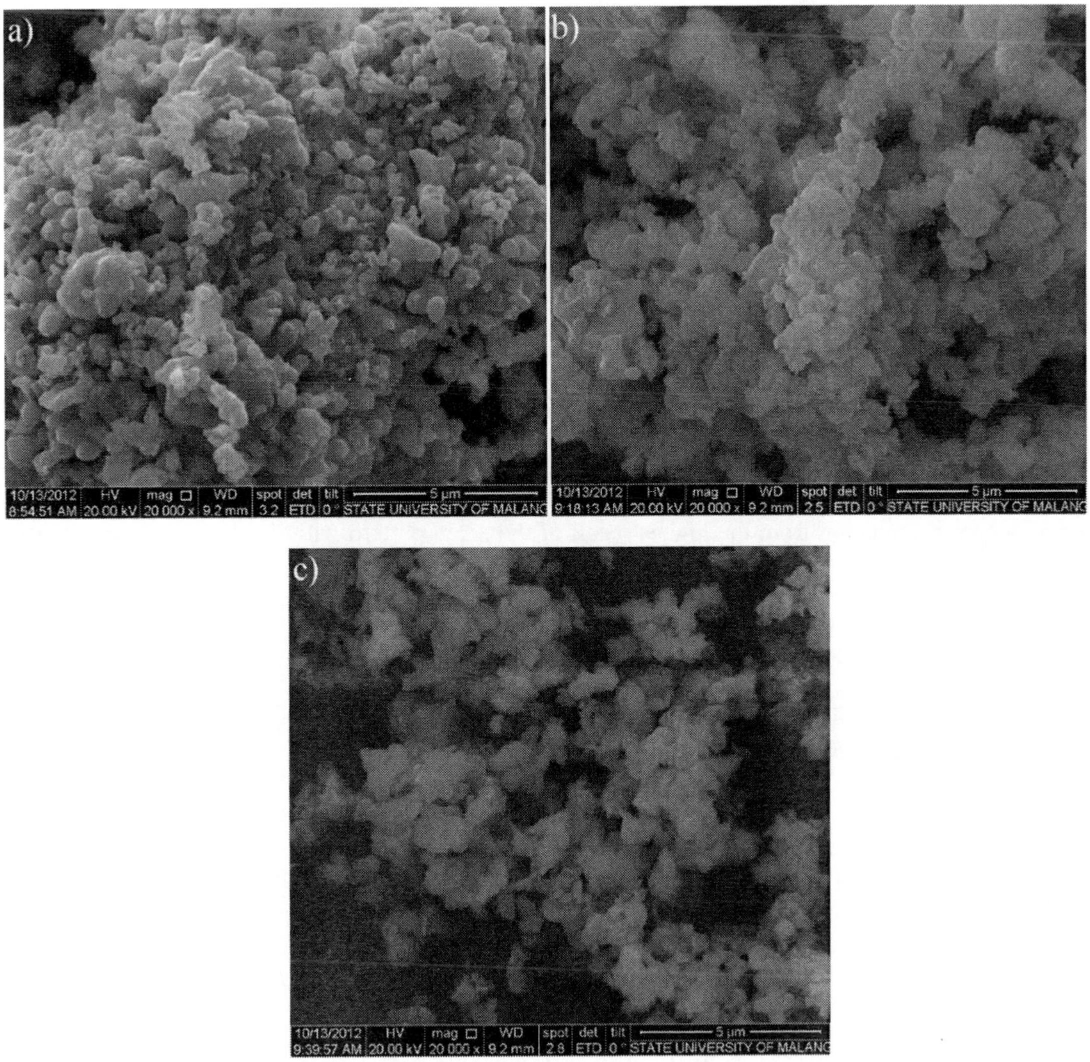


**Figure 2**. $Nd_{1+x}FeO_3$ sample morphology with the variations of molar ratio of a) 0.1, b) 0.2 and c) 0.3, respectively.

Based on Figure 2, it can be seen that all of the samples had non-homogeneous morphology. There were still many agglomerates made due to the temperature and mechanical treatment. The particle size of the samples was about 1 μm. These powders have high porosity, and this was an advantage for the gas sensor materials. A higher magnification of $Nd_{1+x}FeO_3$ at $x = 0.2$ can be seen in figure 3.

**Figure 3**. Morphology of $Nd_{1.2}FeO_3$ sample at a higher magnification

A higher magnification for the $Nd_{1.2}FeO_3$ sample showed that the grains were composed of fine particles with different shapes. The crystalline size of the $Nd_{1+x}FeO_3$ sample was determined using Debye-Scherrer's equation from the highest peak of the x-ray pattern with the crystalline sizes varied from 137.0 nm to 152.2 nm. The similar results were obtained by Khorasani-Motlagh M. et al. that synthesized $NdFeO_3$ nanocrystals with a good porosity and with the crystalline size ranging from 28.82 nm to 200.71 nm [14].

## 4. Conclusion

The $Nd_{1+x}FeO_3$ powders with the variations of molar ratio (x = 0.1, 0.2 and 0.3) were successfully synthesized using solid-state reaction method. The variation of molar ratio has a significant effect on the diffraction intensity, FWHM, and the crystalline size of the samples. Based on the calculation, it can be identified that the crystalline size of the samples was ranging from 137.0 – 152.2 nm.

All of the samples have non-homogeneous morphology with many different shapes and high porosity. The sample with the molar ratio of x = 0.2 has the highest intensity, largest crystalline size, and highest porosity which indicated that the $Nd_{1.2}FeO_3$ sample is a good candidate for a gas sensor material.

## 5. References

[1] Anhua W *et al* 2009 *Asia-Pac. J. Che. Eng.* **4** 518-521
[2] Ho T G *et al* 2011 *Sensors and Actuators B.* **158** 246-251
[3] Ru Z *et al* 2010 *Journal of Rare Earths.* **28** 591-595
[4] Truong G H *et al* 2011 *Adv. Nat. Sci: Nanosci. Nantotechnol.* **2** 015012 4pp
[5] Zhan-lei W *et al* 2010 *International Journal of Minerals, Metallurgy, and Materials.* **2** 141
[6] You W *et al* 2014 *CrystEngComm.* **16** 858
[7] Shujuan Y *et al* 2011 *Journal of Applied Physics.* **109** 07E141
[8] Pedro V S *et al* 2016 *Materials Research.* **19(2)** 389-393
[9] Lou X *et al* 2005 *Journal of Rare Earths.* **23** 328
[10] Yabin W *et al* 2011 *Journal of Crystal Growth.* **318** 927-931
[11] Prabeer B *et al* 2010 *Journal of Material Chemistry.* 1-8
[12] Husain H *et al* 2015 *Jurnal Sainsmat.* **4(1)** 1-6
[13] Wickenden A E *et al* 1994 *Journal of Applied Physics.* **75** 5367
[14] Khorasani-Motlagh M *et al* 2013 *Int. J. Nanosci. Nanotechnol.* **9(1)** 7-14

**Acknowledgement**
This research was funded by the Directorate of Research and Community Services, Directorate General of Research and Development, Ministry of Research, Technology, and Higher Education, Republic of Indonesia, under the research scheme of *Hibah Kompetensi* 2016

# Modification of Pseudobrookite $Fe_{2-x}Mn_xTiO_5$ with Solid State Reaction Method using a Mechanical Milling

**Y Sarwanto and W A Adi**

Centre for Science and Technology of Advanced Materials, BATAN, Kawasan Puspiptek Serpong Tangerang Selatan Banten, Indonesia

Email: yosef@batan.go.id

**Abstract.** Modification of pseudobrookite $Fe_{2-x}Mn_xTiO_5$ with solid state reaction method using a mechanical milling has been synthesized. Raw materials used to prepare these samples were $Fe_2O_3$, $MnCO_3$, and $TiO_2$. $Fe_2O_3$ and $TiO_2$ powders (ratio of 1:1) were mixed with $MnCO_3$ powder at various composition of x = 0; 0.1; 0.2; 0.3; 0.4; 0.5; and 1, which each composition was added with 50 ml ethanol and then milled for 5 hours through high energy milling, after that sintered at 1000 °C for 5 hours by using box furnace. The phases of $Fe_{2-x}Mn_xTiO_5$ were measured by using X-ray diffraction (XRD) and then identified by using Match program. The crystal structure was analyzed by using the program of General Structure Analysis System (GSAS). Quality fitting of Rwp and $\chi^2$ (chi-squared) are relatively good because based on the curve of normalized error distribution looks just left background and its normal probability plot shows the value of comparable between observation and expectation. The refinement analyses of X-ray diffraction patterns showed that the samples formed single phase for $x \leq 0.3$. However, the samples of $x > 0.3$ were multi-phases. The single phase of sample had composition of pseudobrookite $Fe_2TiO_5$ with orthorhombic structure, space group of C m c m (63), the lattice parameters of a = 3.7390 Å, b = 9.7790 Å, and c = 9.9780 Å, $\alpha = \beta = \gamma = 90°$, V = 364.83 Å$^3$, and $\rho = 4.360$ g.cm$^{-3}$. Meanwhile, the other phase analysis for the composition of $x > 0.3$ is bixbyite ($FeMnO_3$). The bixbyite has a cubic structure, under the space group of I a - 3 (206), the lattice parameters of a = b = c = 9.40 Å, $\alpha = \beta = \gamma = 90°$, V = 830.58 Å$^3$, and $\rho$ = 5.078 g.cm$^{-3}$.

**Keywords:** Pseudobrookite, $Fe_{2-x}Mn_xTiO_5$, Milling, Modification, Crystal Structure.

## 1. Introduction

Absorber materials of electromagnetic waves is now becoming one of the interesting topics to be studied and understood for electronic application [1,2]. The main requirement needed as a material absorbing electromagnetic waves is the presence of intrinsic characteristics, namely magnetic loss and dielectric loss on the materials. At first, the absorber materials developed until now are a carbon-based material that has a high dielectric loss [3]. However, recently it is found that many studies of electromagnetic wave absorber materials began to lead to the use of magnetic materials [4-5]. This study has done structure engineering on the paramagnetic material of pseudobrookite $Fe_2TiO_5$-based. The main reason for choosing this system is the magnetic material has a stable structure up to high temperatures. Pseudobrookite $Fe_2TiO_5$ may also be obtained from the phase transformation of ilmenite ($FeTiO_3$) [6-7], while ilmenite itself can be obtained from the iron sand where their reserves is very large in Indonesia. Therefore, this material becomes an excellent product at low price as well as the raw materials are easily obtained.

Pseudobrookite $Fe_2TiO_5$ previously can be used for multiple applications such as microelectronics materials, gas sensors, non-linear optics, magnetic applications, filter optics, photocatalyst, photoelectrode, anode batteries, pigments, and a membrane at high temperatures for fuel cell applications [8-10]. Thus the application of Pseudobrookite $Fe_2TiO_5$ material is very wide and makes this material is one of the multi-functional materials, so the further understanding of this material is very interesting to study. Previous research has carried out preliminary studies about the effect of composition on the phase and parameter of crystal structure pseudobrookite $Fe_{2+x}Ti_{1+x}O_5$ [11]. This research conducted modification through structural engineering on the pseudobrookite $Fe_{2-x}Mn_xTiO_5$ materials by solid-state reaction using a wet milling process. The addition of nickel (Mn) atoms is expected able to replace most of the position of the iron (Fe) atoms so that the interaction between the magnetic spin $Fe^{3+}$ ions with $Mn^{3+}$ occur it and can affect the magnetic properties of this material. Thus, the aim of this study is to synthesize and characterize pseudobrookite $Fe_{2-x}Mn_xTiO_5$ material with variations in composition (x = 0.1, 0.2, 0.3, 0.4, 0.5 and 1). The discussion will be focused on the analysis of the crystal structure (phase and parameter structures), the morphology of particles and primary content of the pseudobrookite $Fe_{2-x}Mn_xTiO_5$ due to the addition of Mn atoms in the system.

## 2. Experimental Method

Pseudobrookite $Fe_{2-x}Mn_xTiO_5$ was synthesized by solid state reaction method with variations in composition (x = 0.1, 0.2, 0.3, 0.4, 0.5, and 1). The raw materials used were $Fe_2O_3$, $TiO_2$ (anatase) and $MnCO_3$. The three materials were weighed according to the calculation of stoichiometric composition, then mixed and placed in a media made of stainless steel. Once it was added with ethanol and balls also made of stainless steel with a mass ratio between the balls and the material are 1: 2. The mixture was then milled for 5 hours by using high energy milling equipment of Spex8000. The mixture of milling result was then dried and compacted with a pressure of 5000 psi. Furthermore, the samples were sintered at 1000 °C for 5 hours in a furnace.

Crystalline phase identification was measured by X-ray diffractometer (XRD), Pan Analytical. Measurement of the diffraction pattern by using X-ray tube with a wavelength of $\lambda$ = 1.5406 Å, mode: continuous scan, step size: 0.02°, and time per step: 0.5 seconds and qualitative-quantitative phases analysis formed in the sample was used by GSAS (General Structure Analysis System) software. Meanwhile, the particle morphology was observed by using transmission electron microscope (TEM) of JEOL brand.

## 3. Results and Discussion

Figure 1 shows the results of measurements of X-ray diffraction pattern of the sample pseudobrookite $Fe_{2-x}Mn_xTiO_5$ with variations in composition (x = 0.1, 0.2, 0.3, 0.4, 0.5, and 1). Based on the results of phase identification appears that the reaction successfully formed a single phase $Fe_{2-x}Mn_xTiO_5$ on the composition of x = 0.1 – 0.3, while for the composition of x > 0.3, the sample could not react perfectly so the sample consisted of two phases, namely phases of $Fe_{2-x}Mn_xTiO_5$ and $FeMnO_3$. Results of phase identification are fascinating to understand because for x = 0.1 – 0.3 show that atoms of manganese have succeeded in substituting partially of the atoms Fe in the structure $Fe_{2-x}Mn_xTiO_5$, and for x > 0.3 is thought to occur reaction imbalance when the amount of Fe content was reduced, the Mn content increased. In Figure 1 we can see that the increasing Mn content (x > 0.3) in the sample caused the phases of $FeMnO_3$ increased as well. Thus, it required further analysis to determine the changes in the crystal structure parameters, the amount of mass fraction formed, and cationic distribution on the results of substitution Mn into the Fe atom as shown in Figure 2.

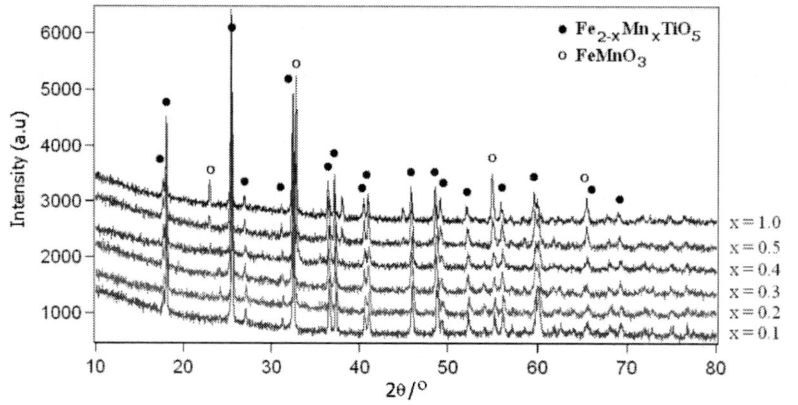


**Figure 1.** X-ray diffraction pattern of the sample of pseudobrookite $Fe_{2-x}Mn_xTiO_5$ with variations in composition (x = 0.1, 0.2, 0.3, 0.4, 0.5, and 1)

Figure 2 shows the results of refinement X-ray diffraction pattern of the sample pseudobrookite $Fe_{2-x}Mn_xTiO_5$ with the variations in composition (x = 0.1, 0.2, 0.3, 0.4, 0.5, and 1).

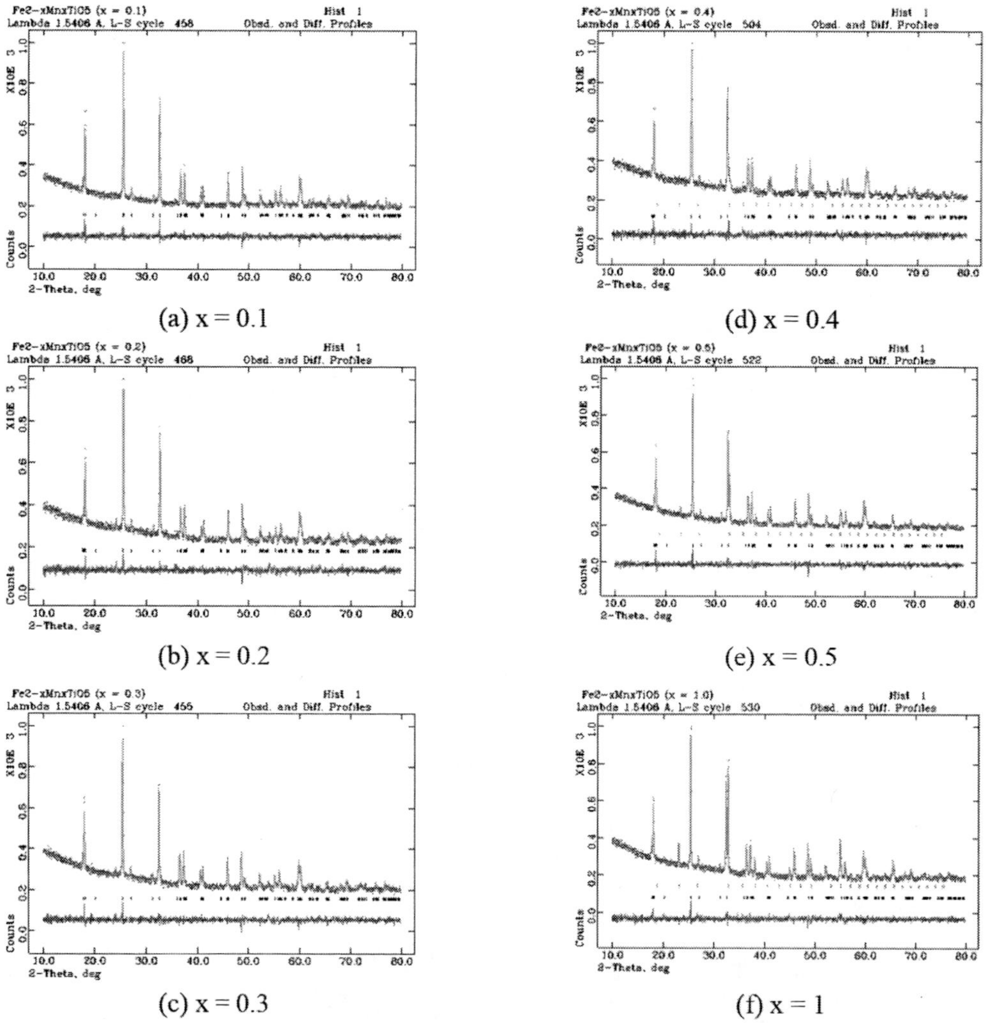


**Figure 2.** The refinement results of XRD pattern of the sample pseudobrookite $Fe_{2-x}Ni_xTiO_5$

Figure 2(a – c) show the results of refinement of the XRD patterns for x = 0.1 – 0.3 that formed Bragg diffraction peaks with a single phase following the $Fe_2TiO_5$ structure. Figure 2(d – f) show the results of refinement of the XRD patterns for x = 0.4 – 1.0 that formed Bragg diffraction peaks with a multi-phases, which follows the structure of $Fe_2TiO_5$ and $FeMnO_3$. Qualitative and quantitative analysis refers to the Crystallography Open Database with the card number (COD: 1011342) and (COD: 9008068) respective for phases of $Fe_2TiO_5$ and $FeMnO_3$.

Refinement complete summary of the X-ray diffraction pattern results of the pseudobrookite $Fe_{2-x}Mn_xTiO_5$ sample with variations in composition (x = 0.1, 0.2, 0.3, 0.4, 0.5, and 1) for all of samples is shown in Table 1.

**Table 1.** The value of structure parameters, criteria of fit ($R_{wp}$), the goodness of fit (χ2) and the mass fraction of phase formed in the pseudobrookite $Fe_{2-x}Mn_xTiO_5$ sample with the variations in composition (x = 0.1, 0.2, 0.3, 0.4, 0.5, and 1).

| **Mn** | **Phase** | **Lattice parameter (Å)** | | | **V** | **ρ** | **Fraction** | $R_{wp}$ | $\chi^2$ |
|---|---|---|---|---|---|---|---|---|---|
| **(x)** | | *a* | *b* | *c* | **(Å$^3$)** | **(g/cm$^3$)** | **wt%** | **(%)** | |
| 0 [11] | $Fe_2TiO_5$ | 3.7188(4) | 9.7569(1) | 9.9466(1) | 360.90(2) | 4.360 | 100.00 | - | - |
| 0.1 | $Fe_2TiO_5$ | 3.7254(1) | 9.7766(4) | 9.9636(4) | 362.89(3) | 4.297 | 100.00 | 3.29 | 1.25 |
| 0.2 | $Fe_2TiO_5$ | 3.7287(2) | 9.7790(6) | 9.9634(6) | 363.30(5) | 4.292 | 100.00 | 3.42 | 1.29 |
| 0.3 | $Fe_2TiO_5$ | 3.7292(2) | 9.7791(6) | 9.9636(5) | 363.36(4) | 4.291 | 100.00 | 3.45 | 1.31 |
| 0.4 | $Fe_2TiO_5$ | 3.7304(2) | 9.7795(6) | 9.9647(6) | 363.52(5) | 4.288 | 95.04 | 3.45 | 1.32 |
| | $FeMnO_3$ | 9.4310(1) | 9.4310(1) | 9.4310(1) | 838.84(1) | 5.030 | 4.96 | | |
| 0.5 | $Fe_2TiO_5$ | 3.7311(1) | 9.7805(5) | 9.9649(5) | 363.63(4) | 4.283 | 87.64 | 3.54 | 1.30 |
| | $FeMnO_3$ | 9.4372(1) | 9.4372(1) | 9.4372(1) | 840.48(4) | 5.035 | 12.36 | | |
| 1 | $Fe_2TiO_5$ | 3.7318(9) | 9.7811(5) | 9.9654(5) | 363.74(4) | 4.281 | 72.54 | 3.65 | 1.31 |
| | $FeMnO_3$ | 9.4379(7) | 9.4379(7) | 9.4379(7) | 840.69(2) | 5.018 | 27.46 | | |

Figure 2 and Table 1 show that the refinement results of X-ray diffraction pattern have a very good fitting quality based on the criteria of fit ($R_{wp}$) and the goodness of fit (χ2) in accordance with the agreement [12]. Table 2 shows that based on the refinement pattern of X-ray diffraction pseudobrookite $Fe_{2-x}Mn_xTiO_5$ samples with variations in composition (x = 0.1, 0.2, 0.3, 0.4, 0.5, and 1), the samples having a single phase are in the composition of x = 0.1 – 0.3. Data supporting the other is from the analysis of changes in the volume of the unit cell as shown in Figure 3.

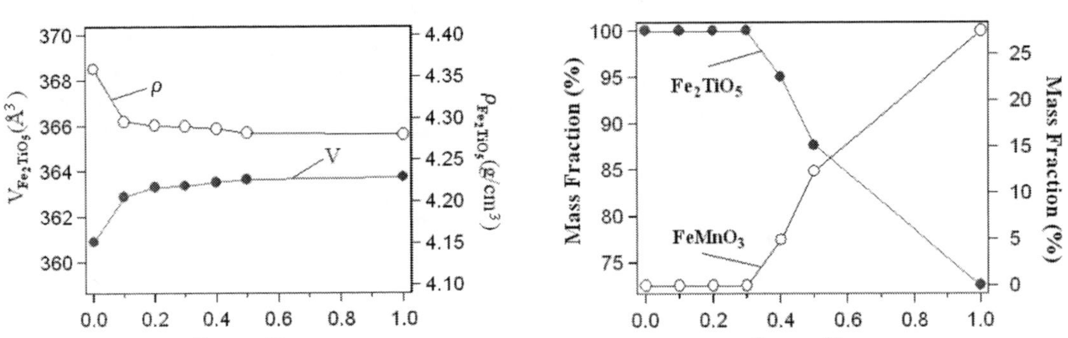


(a) Volume and density of unit cell (b) Mass fraction

**Figure 3.** The volume of unit cells as a function of composition

In Figure 3(a) it appears that when composition x increased, the expansion of unit cell volume occurred up to x = 0.3. After that, the composition (x > 0.3) under net saturation. This phenomenon was due to the atomic radius of Mn (r = 1.79 Å) was longer than the atomic radius of Fe (r = 1.72 Å). Thus it resulted in lattice parameter which also increased for the third axis. Then, for the composition x > 0.3, the unit cell volume gradually saturated caused by the lattice distortion and by the presence of another phase formed. The addition of Mn excess (x > 0.3) resulted in an imbalance of the reaction so that Mn preferred to bind Fe to form Ferromanganite since the composition of these compounds was relatively stable compared with $Fe_{2-x}Mn_xTiO_5$. Meanwhile, another Fe will bind to Mn forming $FeMnO_3$. Figure 3 (b) shows that on the composition of x > 0.3, there was no change in the unit cell volume on the phases of $FeMnO_3$. It means that Mn had reacted with Fe to form $FeMnO_3$. This phenomenon can be explained accordance to the results of the analysis of mass fractions as shown in Figure 4.

These refinement results are also supported by observations of the surface morphology of particles for both single phases using transmission electron microscope (TEM) as shown in Figure 3. Figure 3 shows that the particle morphology of the composition of x = 0.3 had a good particle homogeneity and uniform in across the sample surface with a polygonal particle shape and the particle size varied from 20 – 100 nm. In general, a single phase characteristic of polycrystalline samples by observation of TEM image is homogeneity and uniformity particle shape in across the sample surface.

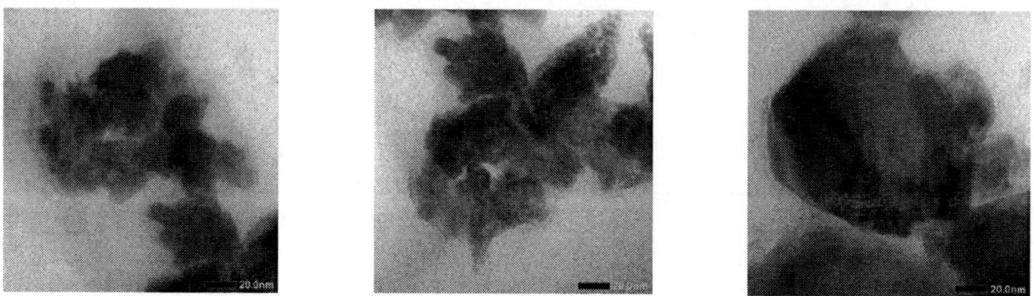


(a) (b) (c)

**Figure 4.** The particle morphology of the pseudobrookite $Fe_{2-x}Mn_xTiO_5$ (x = 0.3) by TEM

The presence of this Mn can be proven based on the result of EDS (Energy Dispersive Spectroscopy) data as shown in Figure 5 and Table 2.

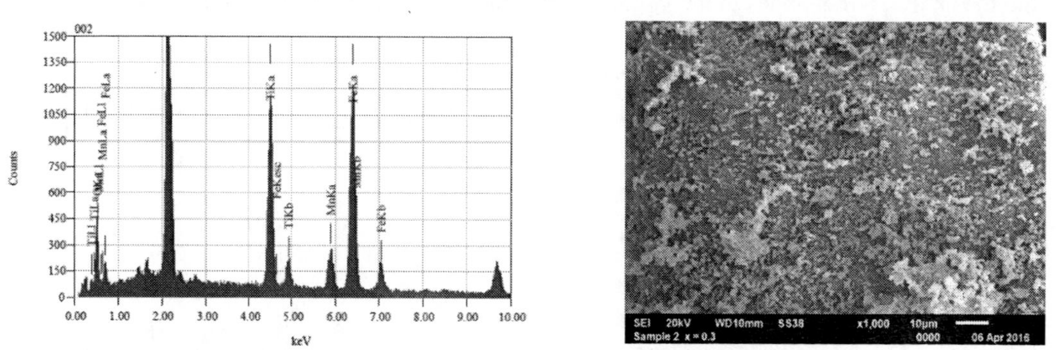


**Figure 5.** EDS spectrum of the pseudobrookite $Fe_{2-x}Mn_xTiO_5$ (x = 0.3)

The interesting thing is both materials have a single phase with the same structure, but their compositions are different. On the composition of x = 0.3, there was a 0.3 % Mn atoms succeeded substituting partially Fe atoms. Thus, the substitution of Mn into Fe on pseudobrookite $Fe_2TiO_5$ system was only capable of 0.3 at.% without changing the crystal structure of the material.

**Table 2.** Element distribution on the $Fe_{2-x}Mn_xTiO_5$ (x = 0.3) measured by EDS.

| Element | Energy (keV) | Mass fraction (%) | |
|---|---|---|---|
| | | wt.% | at.% |
| O | 0.525 | 17.63 | 41.61 |
| Ti | 4.508 | 23.35 | 18.40 |
| Mn | 5.894 | 7.61 | 5.23 |
| Fe | 6.398 | 52.41 | 34.75 |

## 4. Conclusion

Synthesis of pseudobrookite $Fe_{2-x}Mn_xTiO_5$ with variations in composition (x = 0.1, 0.2, 0.3, 0.4, 0.5, and 1) has been successfully carried out. The refinement pattern of X-ray diffraction samples of pseudobrookite $Fe_{2-x}Mn_xTiO_5$ shows that the sample has a single phase in the composition of x = 0.1 – 0.3. The particle morphology of the composition of x = 0.1 – 0.3 has a good particle homogeneity and is uniform in across the sample surface with a polygonal particle shape, and the particle size varies. Thus, substitution of Mn into Fe on the pseudobrookite $Fe_{2-x}Mn_xTiO_5$ is only capable of 0.3 at.% without changing the crystal structure of the material.

## 5. References

[1]. Eswaraiah V, Sankaranarayanan V and Ramaprabhu S 2011 Functionalized Graphene–PVDF Foam Composites for EMI Shielding *Macromol. Mater. Eng.* **296** 1-5.

[2]. Wu Z and Li M 2011 Electromagnetic interference shielding of carbon nanotube macrofilms *Scripta Mater* **64** 809-12

[3]. Wang X 2011 Investigation of Electromagnetic Shielding Effectiveness of Nanostructural Carbon Black/ABS Composites Journal of Electromagnetic Analysis and Applications **3** 160-164.

[4]. Zhang S, Cao Q 2012 Electromagnetic and microwave absorption performance of some transition metal doped $La_{0.7}Sr_{0.3}Mn_{1-x}TM_xO_{3\pm1}$ (TM = Fe, Co or Ni) *Materials Science and Engineering B* **177** 678– 684.

[5]. Adi W A and Manaf A 2012 Structural and Absorption Characteristics of Mn-Ti Substituted Ba-Sr Hexaferrite Synthesized by Mechanical Alloying Route *Journal of Basic and Applied Scientific Research* **2** (8) 7826-7834.

[6]. Xiao W, Lu X, Zou X, Wei X and Ding W 2013 Phase Transitions, Micro-Morphology and its Oxidation Mechanism in Oxidation of Ilmenite ($FeTiO_3$) Powder., *Trans. Nonferrous Met. Soc. China* **23** 2439–2445.

[7]. Rochman N T and Adi W A 2013 Preliminary study of the development of absorber electromagnetic wave materials by modifying iron sand. *International Journal of Academic Research Part A* **5**(4) 91-96.

[8]. Jin L and Zhou 2013 Electronic structures and optic properties of $Fe_2TiO_5$ using LSDA+U approach Progress in Natural Science Materials International 23 (4) 413–419.

[9]. Khosravi E, Razi M K and Enhessari M 2012 Photocatalytic and Magnetic Behavior of $Fe_2TiO_5$-5A Zeolite Nanocomposites Int. J. Bio-Inorg. Hybd. Nanomat Vol. 1 No. 2 115-122.

[10]. Houshyari K, Javanbakht M, Naji L and Enhessari M 2012 Physical and Electrochemical Study of Nafion/$Fe_2TiO_5$ Nanocomposite Membranes for High Temperature PEM Fuel Cells *Proceedings of the 4th International Conference on Nanostructures* (ICNS4) 12-14.

[11]. Yosef S and Wisnu A A 2015 Effect of Pseudobrookite $Fe_{2+x}Ti_{1+x}O_5$ Composition on the Changes in Phase and Parameter of Crystal Structure Proc. National Seminar of X-ray and Neutron Scattering 97-101

[12]. Toby B H 2001 EXPGUI A Graphical User Interface for GSAS *J. Appl. Crystallogr.* **34** 210-221

**Acknowledgments**
This work is supported by the program for research and development of smart magnetic material (DIPA 2015), Center for Science and Technology of Advanced Materials, National Nuclear Energy Agency. The authors are thankful to Drs. Bambang Sugeng, for his kind help to characterize using XRD.

# Effect of Precursor Concentration Ratio on The Crystal Structure, Morphology, and Band Gap of ZnO Nanorods

**A Fuad[1,2], A A Fibriyanti[1], Subakti[1,2], N Mufti[1,2] and A Taufiq[1,2]**

[1]Department of Physics, Faculty of Mathematics and Natural Sciences, State University of Malang (Universitas Negeri Malang), Jl. Semarang No 5 Malang, Indonesia 65145
[2]Research Center for Minerals and Advanced Materials, State University of Malang (Universitas Negeri Malang), Jl. Semarang No 5 Malang, Indonesia 65145

Email: abdulloh.fuad.fmipa@um.ac.id

**Abstract**: In this present study, the ZnO nanorods were synthesized by using the hydrothermal method with various precursor concentrations. The ZnO nanorods were characterized by X-Ray Diffractometer (XRD) to identify the crystal structure, Scanning Electron Microscopy (SEM) to determine the morphology, and UV-Vis Spectrophotometer to determine the band gap. It was shown that the ZnO nanorods have hexagonal wurtzite crystal structure with the c-axis orientation at (002) Bragg's plane. The morphology of the ZnO particles was in a rod structure growing over the surface of the substrate. The diameter of the ZnO nanorods resulted ifrom the concentrations of 30, 35 mM, and 45 mM were 94.2, 156.3, and 76.7 nm, respectively. The data analysis of the UV-Vis data showed that the band gap of ZnO nanorods (~3.18 eV) was smaller than that of ZnO nanoparticles (~3.46 eV).

**Keywords:** ZnO, nanorod, crystal structure, band gap, solar cell.

## 1. Introduction

Electrical energy has a major role in supporting the national development. In term of the total use of electrical power, its growth rate was approximately 8.3 % annually from 222 TWh in 2015 becomes 305 TWh in 2019 [1]. Thereby, the availability of fossil as the primary source of energy in the world gradually reduces, so the development of the alternative electrical power including renewable energy should be conducted continually. In the last five years, solar cells with active material combined with halide organic/inorganic material and perovskite structure have caught many researchers' attention due to their ability in developing great power photovoltaic tools with high efficiency of up to 19.3 % [2].

Efficiency is influenced by the design of a solar cell involving the optimization of the layer structure, the characteristic of every layer like semiconductor layer, and the equipment parameter comprising a lifetime of charge carrier, the diffusion coefficient of minority charge carrier, and the surface recombination rate [3]. The level of exciton-dissociation efficiency and the charge carrier rate are definitely influenced by the surface morphology. The surface morphology of the active layer is one of the important factors affecting the efficiency of the hybrid solar cell.

ZnO is a semiconductor with the type-n of II-VI class with the bandgap of about 3.37 eV and excitonic bound energy of about 60 meV at room temperature [4]. On the other side, ZnO is categorized as a transparent conducting oxide material which is low-cost, looks promising, and it has been applied in many commercial applications such as light-emitting diode [5], UV sensor [6], nanolaser [7], gas sensor [8], and solar cell [9].

In semiconductor nanostructure material, one-dimension (1D) ZnO nanostructure can be modified into nanowires [10], nanobelts [11], and nanorods [12]. Each structure modification has a different physical characteristic. Nanorods structure, for example, can influence particular material to enhance its light absorption ability regarding its higher surface area [13]. In general, the fabrication of ZnO nanostructures can be done by using thin film techniques such as spray pyrolysis, sputtering, MOCVD, and hydrothermal method [14]. The hydrothermal method has been widely used in producing the transparent oxide by using a simple, safe and low-cost technique [15].

In this research, the ZnO particles were modified to form nanorods initiated through the process of making the ZnO nanoparticle layered on an Indium Tin Oxide (ITO) substrate. In this work, the effect of precursor concentration on the crystal structure, morphology, and band gap of the ZnO nanorods is discussed.

## 2. Experimental Method

ITO was dipped into acetone in the ultrasonic cleaner for ten minutes and followed by the cleaning process using water before being dried by using a blower. Zinc acetate dehydrate (ZnAc) was dissolved in the ethanol and stirred by using a magnetic stirrer for 45 minutes. The PH of the solution was varied from 6 to 7. Furthermore, the ZnAc was mixed with MEA with the composition of 1:1 molar. The clean ITO substrate was placed on the spin coater holder by dropping the solution on the glass surface at the rotational speed of 3000 rpm. Preheating process was then conducted at 150 °C for 10 minutes followed by the annealing process at 550 °C for 2 hours. The solution was prepared by using the precursors of $Zn(NO_3)_2.4H_2O$ and hexamethylenetetramine/HMT in the water. Precursor concentrations used were 30, 35, and 45 mM with the molar comparison between $Zn\ (NO_3)_2.4H_2O$ and HMT of 1:1. The solution was then stirred at room temperature. Furthermore, the ZnO nanoparticles were dipped into the prepared solution. During the growing process, the solution was heated at a temperature of 90 °C for 6 hours. Water was used to clean the remaining salt from the sample surface. The obtained sample was then dried by using a mini electric blower. Lastly, the ZnO nanorods on the ITO substrate were annealed at a temperature of 550 °C for 2 hours. The samples were characterized by using X-Ray Diffractometer (XRD), Scanning Electron Microscopy (SEM), and UV-Vis spectrophotometer.

## 3. Results and Discussion

The X-ray diffraction patterns of the ZnO nanoparticles are shown in Figure 1. The diffraction patterns showed the highest peak indicating the different crystal orientation. The ZnO nanoparticle with pH 6 had the highest peak intensity at the plane (101), while the ZnO nanoparticles with pH 7 were at the plane of 002. The ZnO nanoparticle with pH 6 is in line with the result reported by Zhang *et al.* [16]. Meanwhile, the ZnO nanoparticle with pH 7 is in line with the study reported by Polsongkram *et al.*[17].

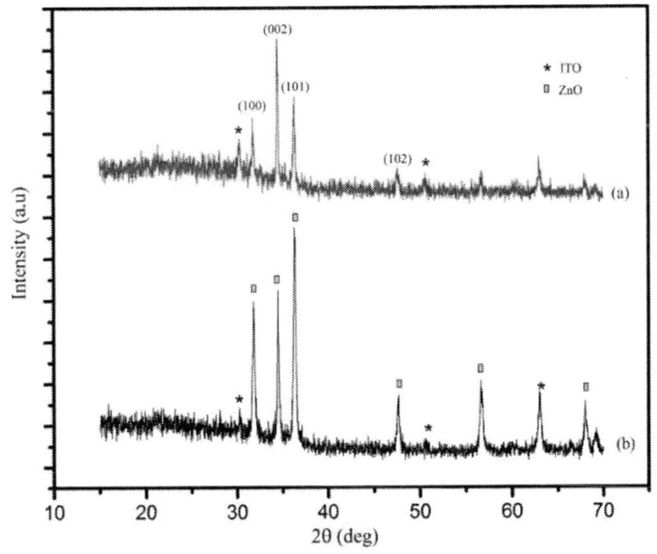


**Figure 1.** X-Ray Diffraction Patterns of The Zno Nanoparticles with Ph (a) 7 and (b) 6

In this work, the ZnO nanorods were synthesized by using the hydrothermal method with ZnO nanoparticles as a buffering layer in the concentrations of 30, 35, and 45 mM. The X-ray diffraction patterns are presented in Figure 2.

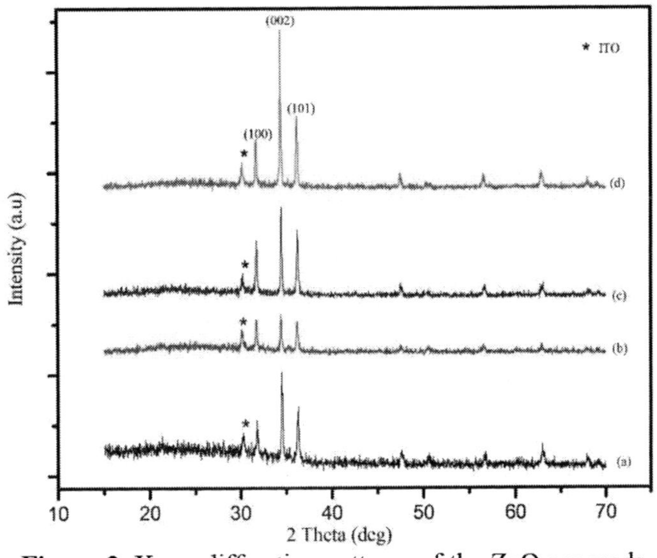


**Figure 2.** X-ray diffraction patterns of the ZnO nanorods with the precursor concentration on the buffering layer.

Figure 2 shows the X-ray diffraction pattern of ZnO nanoparticles and ZnO nanorods in various precursor concentrations. The three highest peaks were (100), (002), and (101). The crystal preferred orientation of the samples can be indicated by the highest peak intensity shown by Bragg plane of (002). It implied that the growth orientation was dominated by grains which were directed to (002) or perpendicular to the substrate. It is also obtained that the increase of precursor concentration caused the growth of the diffraction peak intensity. The highest peak intensity was obtained with the concentration of 45 mM. It could be inferred that the greater precursor concentration lead to be prefer of the crystal orientation of the ZnO nanorods to the c-axis at the (002) plane. Based on the Rietveld refinement method, the analysis results of the diffraction data are presented in Table 1.

**Table 1.** Refinement results of the samples

| Parameter | ZnO Model | ZnO nanoparticles | ZnO nanorods 30 Mm | ZnO nanorods 35 mM | ZnO nanorods 45 mM |
|---|---|---|---|---|---|
| $a = b$ (Å) | 3.2494 | 3.258 | 3.263 | 3.244 | 3.251 |
| $c$ (Å) | 5.2038 | 5.213 | 5.224 | 5.193 | 5.206 |
| $c/a$ | 1.6 | 1.6 | 1.6 | 1.6 | 1.6 |
| $V$ (Å$^3$) | - | 47.5 | 48.17 | 47.35 | 47.65 |
| *GoF (%)* | - | 1.25 | 1.30 | 1.24 | 1.25 |
| *Rwp* | - | 19.29 | 33.89 | 31.01 | 32.31 |
| *Rp* | - | 15.11 | 25.81 | 23.52 | 23.76 |

Table 1 shows that the lattice parameter for all samples has a $c/a$ ratio of about 1.60. It indicated that the structure of ZnO nanorods was formed in hexagonal wurtzite. Furthermore, the X-ray diffraction data was also analyzed by using Scherrer's equation to determine the particle size of ZnO nanorods. The calculation results are presented in Table 2.

**Table 2.** Particle size of the ZnO nanorods

| Precursor Concentration (mM) | FWHM (rad) | Particle Size (nm) |
|---|---|---|
| 30 | 0.20311 | 44.58 |
| 35 | 0.15385 | 58.87 |
| 45 | 0.1699 | 53.16 |

The particle size of the ZnO nanorods in the concentrations of 30, 35, and 45 mM were 44.58, 58.87, and 53.16 nm, respectively. We can see that the size increased from the concentration of 30 to 35 mM, and decreased from 35 to 45 mM. It was probably due to the effective reaction condition also provided by the sample with 35 mM. The temperature change at the period of growing can probably change the FWHM value and the average size of the crystal grain.

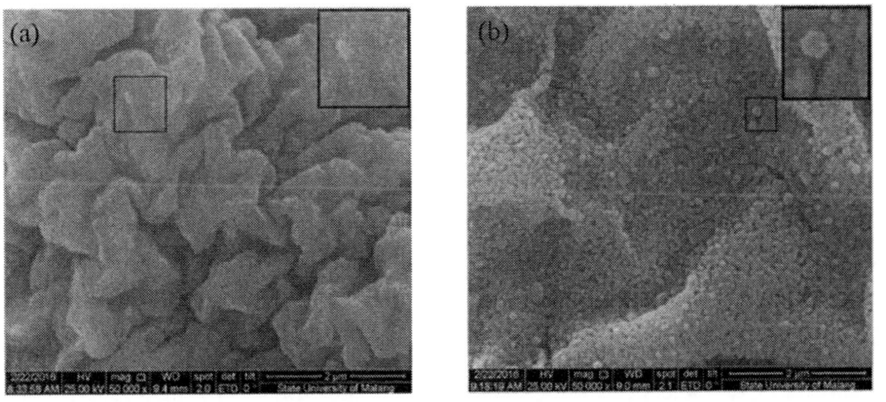

**Figure 3.** SEM images of ZnO films (a) pH 6; (b) pH 7

Based on Figure 3, we can see that the surface morphology showed a typical shape and size of ZnO. The figure also demonstrates that the sample obtained for pH 7 has a different morphology from pH 6. Furthermore, the SEM images of ZnO nanorods with the variation of precursor concentrations are displayed in Figure 4.

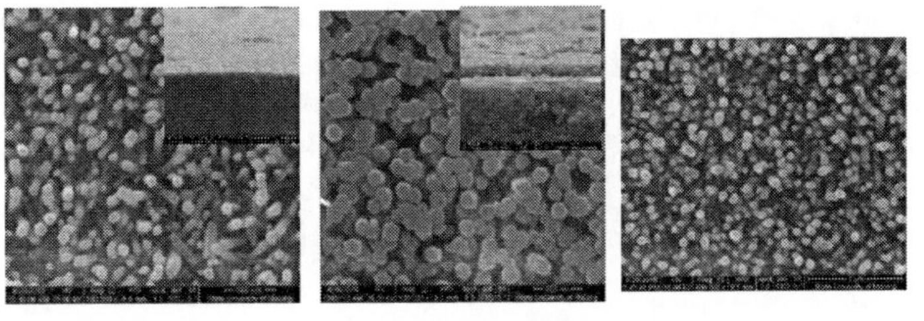

(a) (b) (c)

**Figure 4.** SEM images of the ZnO Nanorods with precursor concentrations of (a) 30 mM; (b) 35 mM; (c) 45 mM

Figure 4 showed that the results of synthesis presented that the ZnO nanorods has rod morphology with a hexagonal structure on the substrate surface. The particle size was analyzed and tabulated in Table 3. Meanwhile, further analysis results of average thickness of the films are listed in Table 4.

**Table 3.** Average diameter of the ZnO nanorods

| Concentration (mM) | Diameter (nm) |
|---|---|
| **30** | 94.21 |
| **35** | 156.3 |
| **45** | 76.68 |

**Table 4.** Average thickness of the ZnO Nanorods Films

| Thickness | ZnO/ITO (μM) | ZnO Nanorods /ITO (μM) | | |
|---|---|---|---|---|
| | | 30 mM | 35 mM | 45 mM |
| 1 | 0.943 | 1.049 | 2.39 | 1.925 |
| 2 | 0.765 | 1.253 | 2.419 | 1.807 |
| 3 | 0.884 | 1.312 | 2.331 | 1.925 |
| 4 | 0.706 | 1.02 | 2.709 | 1.78 |
| 5 | 1.002 | 1.079 | 2.361 | 1.681 |
| **Average** | **0.86** | **1.1426** | **2.442** | **1.8236** |

Table 3 shows the diameters of the samples for the concentrations of 30, 35, and 45 mM that were 94.21, 156.3, and 76.68 nm, respectively. Theoretically, the nanorod is one of the nanostructures with the dimension length range of about 1-100 nm (Orhan, 2012). Therefore, in this work, the ZnO nanorods films that had the smallest size were the ones in the concentration of 35 mM.

The characterization of UV-Vis spectrophotometer was conducted to identify the absorbance and film transmittance of ZnO nanoparticles and nanorods. The wavelength used was in the range of 300-800 nm. The measurement results shown in Figure 5 presented that the absorbance value of the samples reduced along with the increasing wavelength. The ZnO nanorods absorption for all samples with different precursor concentrations was relatively similar. The figures show that the nanorods structure can enhance the light absorption. The absorbance value is in inverse with the transmittance, so the light passed along by the ZnO nanorods is less than the light passed along by the ZnO nanoparticles. The strongest excitonic absorption peak of all ZnO nanorods samples was located along the wavelength of ~375 nm, while the ZnO nanoparticles were located along the wavelength of ~350 nm. The wavelength was an emission peak of UV from free exciton.

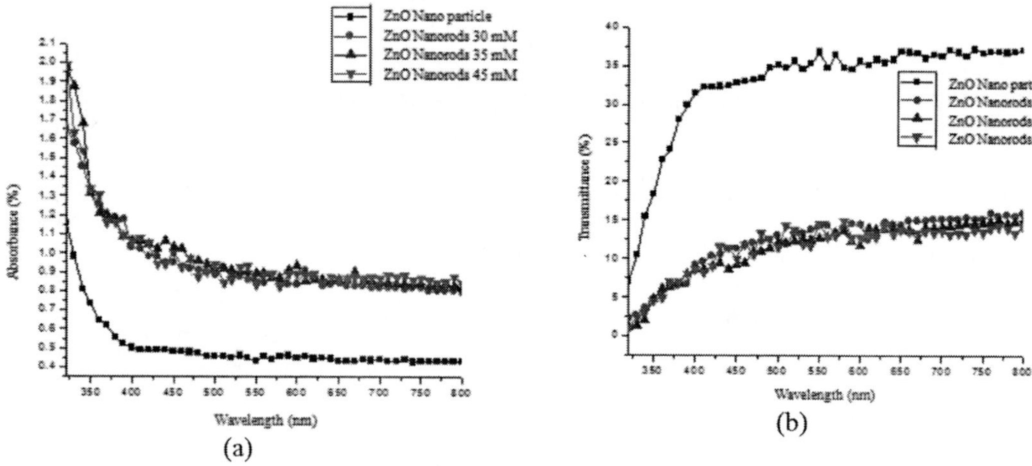


(a) (b)

**Figure 5.** Result of UV-Vis Test on (a) Absorbance; (b) Transmittance

A band gap can be determined through the maximum and minimum transmittance values, film layer thickness, as well as the coefficient of thin layer absorption. The orientation of band gap is addressed through a position of graphic Tauc $(\alpha h\nu)^2$ versus. hv. This method was used to measure the difference between the valence band and the conduction band. The interceptions between the tangent to the linear portion of the curve and hv-axis in the ZnO nanoparticles film and the ZnO nanorods with the variation of precursor concentrations are presented in Figure 6. The band gap of the samples is shown in Table 5.

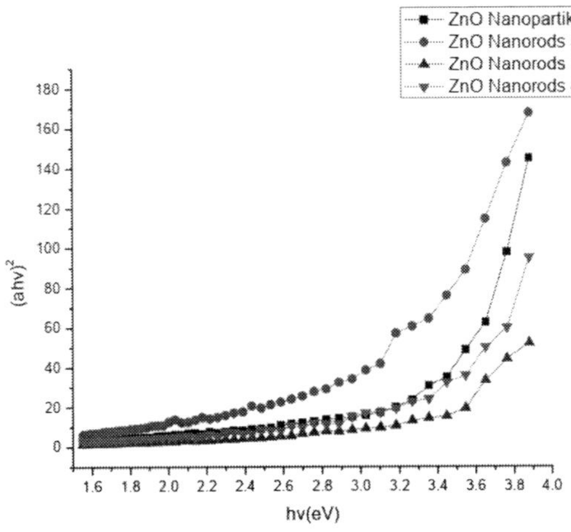


**Figure 6.** Tauc plot results of the ZnO nanoparticles and ZnO nanorods

**Table 5.** Band gaps of ZNO nanoparticles, and ZnO nanorods Films.

| Sample | Band Gap (eV) |
|---|---|
| Film ZnO Nanoparticle/ITO | 3.46 ± 0.99 |
| Film ZnO Nanorods/ ITO 30 mM | 3.14±0.99 |
| Film ZnO Nanorods/ ITO 35 mM | 3.16±0.98 |
| Film ZnO Nanorods/ ITO 45 mM | 3.18±0.96 |

From Table 5, it can be seen that the band gap of the ZnO nanoparticles film is ~3.46 eV. This result is in the range of the band gap of ZnO nanoparticles film which is 3.2-3.5 eV (Kasuma, 2012). The ZnO nanorods with the different precursor concentrations had a lower band gap than ZnO nanoparticles film. This phenomenon was originated from the ZnO nanorods that have a larger surface than that of ZnO nanoparticles. The ZnO nanorods can relatively increase the light absorption and lower the band gap of the material. The band gap of the ZnO nanorods for all samples with different precursor concentrations is relatively similar.

## 4. Conclusion

The crystal structure of the ZnO nanorods was hexagonal. The greater precursor concentration lead to grow the crystal orientation of the ZnO nanorods parallel to c-axis at the plane of (002). The particle size of the ZnO nanorods with the concentrations of 30, 35, and 45 mM, were 44.58 nm, 58.87 nm, and 53.16 nm, respectively. The greater the precursor concentration, the bigger rod diameter will be. The ZnO nanorods with the concentrations of 30, 35, and 45 mM had a relatively similar band gap, that was ~3.18 eV, while the band gap of the ZnO nanoparticles was ~3.46 eV.

## 5. References

[1] Sugiyono A *et al* 2014 *Outlook Energi Indonesia 2014* Jakarta: Pusat Teknologi Pengembangan Sumberdaya Energi (PTPSE).

[2] Jung H S and Nam G P 2015 Perovskite Solar Cells: From Materials to Devices **11**(1) 10-25

[3] Prakoso A B *et al* 2012 Optimasi Pembentukan Lapisan Boron P+ pada Permukaan Belakang Sel Surya untuk Peningkatan Efisiensi Sel Surya berbasis Wafer Silikon Monokristal. *TEKNOFISIKA*.**1**(1)

[4] Ozgur U *et al* 2005 A Comprehensive Review of ZnO Materials and Devices *J.Appl. Phys.* **98**(4) 1-103.

[5] Tsai K J *et al* 2012 n-ZnO nanorods/p$^+$-Si (111) heterojunction light emitting diodes *Nanoscale Research Letters* **7** (664) 1-6

[6] Chai G Y *et al* 2011 Fabrication and characterization of an individual ZnO microwire-based UV photodetector *Solid State Sciences* **13** 1205-1210

[7] Visimberga G E E *et al* 2010 Nanolaser from ZnO nanorods as natural resonance cavities *Phys. Status Solidi C7* **7**(6): 1668-1671

[8] Pawar A *et al* 2015 Growth of ZnO Nanorods using Hydrothermal for Gas Sensing Application *Research Journal of Physical Science* **3**(2): 1-3

[9] Loh L *et al* 2013 Perovskite Enhanced Solid State ZnO Solar Cells *Journal of Physics* **476**: 1-6

[10] Dhara S and Giri P K 2011 Enhanced UV photosensitivity from rapid thermal annealed vertically aligned ZnO nanowires *Nanoscale Research Letters* **6**:504

[11] Ding Y and Zhong L W 2009 Structures of planar defects in ZnO nanobelts and nanowires." Micron **40** (3): 335–42

[12] Foo K L *et al* 2014 Sol-Gel Synthesized Zinc Oxide Nanorods And Their Structural And Optical Investigation For Optoelectronic Application Malaysia *Nanoscale Research Letters*

[13] Mehrabian M and Aslyousefzadeh S 2015 Highly Efficient Hybrid Solar Cell Using ZnO Nanorods and Assessment of Changes in Cell Performance by Varying the Growth Period *Journal of The Korean Physical Society* **66** (10): 1527-1531

[14] Ibupoto Z H *et al* 2013 Hydrothermal Growth of Vertically Aligned ZnO Nanorods Using a Biocomposit Seed Layer of ZnO Nanoparticles *Materials* **6**: 3584-3597

[15] Kathalingam A, Park HC, Kim S D, Kim H S, Velumani S & Mahalingam T 2015 *Synthesis of ZnO nanorods using different precursor solution and their two terminal devices characterization* **26**:5724-5734

[16] Zhang X *et al* 2014 Effect of aspect ratio and surface defects on the photocatalytic activity of ZnO nanorods. *Scientific Reports*. **4**: 4596

[17] Polsongkram D *et al* 2008 Efeect Of Synthesis Conditions on the growth of ZnO nanorods via hydrothermal method *Physica B* **403**: 3713-3717

**Acknowledgements**

The authors would like to thank the KEMENRISTEKDIKTI RI, Mineral and Advanced Material Laboratory of Faculty of Mathematics and Natural Sciences, State University of Malang for the use of laboratory facilities.

# Physical and Magnetic Properties of $La_{0.5}Ca_{0.5}Mn_{0.9}Cu_{0.1}O_3$ at Temperature in the Range of 10-100 K

**Y E Gunanto[1], W A Adi[2], E Steven[3], B Kurniawan[4], T Ono[5] and H Tanaka[6]**

[1] Department of Physics Education, Faculty of Education, University of Pelita Harapan, Karawaci, Tangerang, Indonesia 15811
[2] Center for Science and Technology of Advanced Materials, BATAN, Tangerang Selatan 15314, Indonesia
[3] Department of Physics, National High Magnetic Field Lab., Tallahase, Florida, USA
[4] Department of Physics, University of Indonesia, Depok 16424, Indonesia
[5] Department of Physics, Osaka Prefecture University, Sakai, Osaka 599-8531, Japan
[6] Department of Physics, Tokyo Institute of Technology, Meguro-ku, Tokyo 152-8551, Japan

Email: yohanes.gunanto@uph.edu

**Abstract**. We have successfully measured physical and magnetic properties of the $La_{0.5}Ca_{0.5}Mn_{0.9}Cu_{0.1}O_3$ at temperature in the range of 10-100 K. Physical properties have covered electrical resistivity and specific heat. The properties, which serve as a function of temperature, were investigated by using Physics Properties Measurement System (PPMS) with and without an external magnetic field up to 8.5 Tesla. The magnetic structure was characterized using High-Resolution Powder Diffraction (HRPD) at room and low temperatures. Furthermore, the magnetization was measured by using Magnetic Properties Measurement System (MPMS) with the external magnetic field of 0-7 Tesla. In the temperature interval of 10-53 K, the resistivity of the sample increased with the increasing of external magnetic field. At temperatures $T < 53$ K, the samples were crystalline metal. This was revealed by fitting the graph Ln $R$ as a function of $1/T$. At temperatures $T > 53$ K, the sample was an insulator according to fittings Ln $R$ as a function of $1/T^{0.25}$. The result was also shown by the decreasing value of the resistivity of the sample. The transition temperature of the metal-insulator ($T_{MI}$) decreased with the increasing of external magnetic field, i.e. from 60 K without an external magnetic field to 53 K with external magnetic field 8.5 Tesla. A coefficient electron specific heat and specific heat of spin wave increased, and phonon specific heat was relatively unchanged in the presence of an external magnetic field. Results of HRPD experiment were analyzed by using Fullprof program. At room temperature, the sample had the paramagnetic phase, while at low temperatures of 20 K it had antiferromagnetic phase. Temperature phase transition from the paramagnetic to antiferromagnetic phase or Neel temperature was around 225 K. However, the saturation magnetization of the sample was not observed.

**Keywords:** metal-insulator; transition temperature; specific heat; magnetic structure; paramagnetic-antiferromagnetic

IOP Conf. Series: Materials Science and Engineering **202** (2017) 012075 doi:10.1088/1757-899X/202/1/012075

## 1. Introduction

The compounds of perovskite manganite with doping Cu, $R_{1-x}A_xMn_{1-y}Cu_yO_3$ are still attractive to some researchers at last decade [1-7]. Study on these materials has not reached any consensus yet, such as resistivity, the nature magnetoresistance, and magnetization. The previous study of the Cu-doped at $La_{0.47}Ca_{0.53}Mn_{1-x}Cu_x$ with Cu = 0.13 is still not clear. This sample can be in the metal-insulator state according to the model Arrhenius and Mott-model. Separating in one model is difficult. The fitting $LnR \approx \frac{1}{T}$ has the same result with the fit $LnR \approx \frac{1}{T^{0.25}}$ [1]. All previous studies, however, show that the Cu-doped at Mn site can change the resistivity.

The Cu substitution resulting decline insulator-metal transition temperature is related to the primary mechanism of double exchange which can disrupt network $Mn^{3+}$-O-$Mn^{4+}$ [3]. The Cu-doped can increase a number of ions $Mn^{4+}$, so it will strengthen the interaction of super-exchange antiferromagnetic between Mn ions.

This paper discusses the influence of an external magnetic field in the range of 0-9 T on the resistivity, specific heat, and magnetic transition temperature of the $La_{0.5}Ca_{0.5}Mn_{0.9}Cu_{0.1}O_3$.

## 2. Experimental Method

A sample of $La_{0.5}Ca_{0.5}Mn_{0.9}Cu_{0.1}O_3$ was prepared by using the conventional solid-state reaction method [1] from materials $La_2O_3$, $CaCO_3$, $MnO_2$ and CuO with purity above 99%. High energy milling (HEM) was used to mix these materials. The sample was heated at a temperature of 1350$^0$C for 6 hours, after which it was milled again. Finally, it was kept at 1100$^0$C for 24 hours. XRD was used to characterize the crystal structure. While, the homogeneity of the sample was carried out by using HRPD ($\lambda$ = 1.8223) in the Laboratory for Neutron Scattering BATAN, Serpong, Tangerang.

Resistance measurements at various temperatures and magnetic fields were characterized by using a Physical Properties Measurement System (PPMS) in four probe configurations. Polycrystalline powders were compressed into pellets of equal thickness for each sample types. The sample was then cut to ~ 7 mm (length) by 3 mm (width) where 4-probe carbon paste contacts were applied at an equally spaced distance from each other to minimize the resistance variation due to the geometry of the electrodes. Magnetization measurements were investigated by using a SQUID Magnetic Properties Measurement Systems (MPMS).

## 3. Results and Discussion

Plot a graph of resistivity as a function of temperature R(T), both without an external magnetic field and the external magnetic field up to 8.5 T can be seen in Figure 1. The sample without an external magnetic field, at a temperature of $T < 53$ K, shows that the increasing of the external magnetic field, the resistivity of the sample is also increasing following the crystals model of Arrhenius, where the sample is more fulfilling plot graph Ln $R \sim 1/T$. But for the temperature $T > 53$ K, it shows the increasing of the external magnetic field that affects it decreases and is more in line with the model of Mott insulator Ln $R$ ~ 1/(T^0.25) (see Figure 2 a-b). The results indicate that the metal-insulator transition temperature $T_{MI}$ occurs at 53 K. Furthermore, there are changes in metal-insulator transition temperature $T_{MI}$ for some value of the external magnetic field. Those values are 53 K, 55 K, 58 K and 60 K for the value of an external magnetic field of 8.5 T, 6 T, 3 T and 0 T, respectively.

A resistivity peak corresponding to the magnetic transition is present. There is a clear field-induced shift of maximum resistivity for all samples, that is from 58 K for magnetic field 3 T into 53 K for 8.5 T. In the double exchange mechanism, the mobility of the charge carrier $e_g$ electrons improves if the localized spins are polarized. The applied field aligns the canted electron spin which should reduce the scattering of itinerant electrons with spins and thus the resistivity is reduced [8]. With the increasing value of the external magnetic field, metal-insulator transition temperature $T_{MI}$ becomes lower. This can be explained that the magnetic field is able to rotate electron spins which are not parallel to become more aligned, so it will reduce the value resistivity of the sample. While at low temperatures, the spins of electrons are relatively parallel. Due to the external magnetic field, it will have an impact on the activation energy $E_a$ of samples. The amount of activation energy is calculated by equation (1).

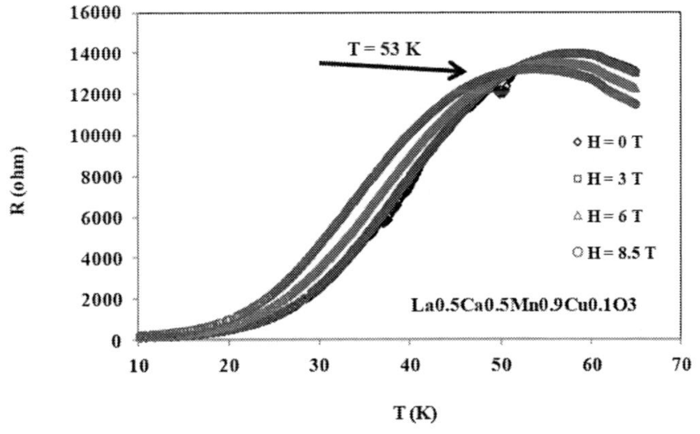


**Figure 1**. Resistivity vs temperature in apllied magnetic field $H = 0,3,6,8.5$ T.

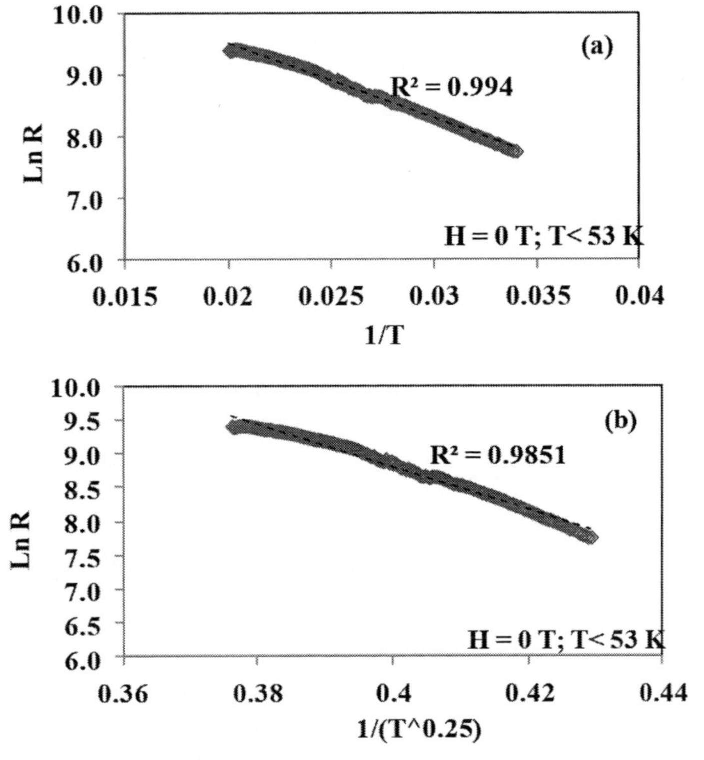


**Figure 2**. Plot graph (a) Ln $R$~$1/T$ dan (b) Ln $R$~$1/T$^0.25 for $H$ = 0 T at the temperature < 53 K.

The amount of activation energy due to the influence of an external magnetic field is 1.64 x $10^{-6}$, 1.53 x $10^{-6}$, 1.21 x $10^{-6}$, and 1.12 x $10^{-6}$ eV to an external magnetic field of 0, 3, 6, and 8.5 T, respectively.

$$R = C\ exp(E_a/kT) \qquad (1)$$

Where $R$ is the electrical resistance, $C$ = factor pre-exponential containing among others the concentration of charge-carrier, $T$ = absolute temperature, $k$ = Boltzmann's constant, and $E_a$ = activation energy.

The results characterization using high-resolution powder diffraction (HRPD) can be seen in Figure 3. The X-ray diffraction data were successfully analyzed by using FullProf program. The results show that the sample classified as a single phase. At a temperature of 300 K, the sample categorized as paramagnetic, while at the temperature of 20 K is antiferromagnetic. The orientation of the magnetic spin of the AFM to the direction of the axis b and c can be seen in the Figure 4.

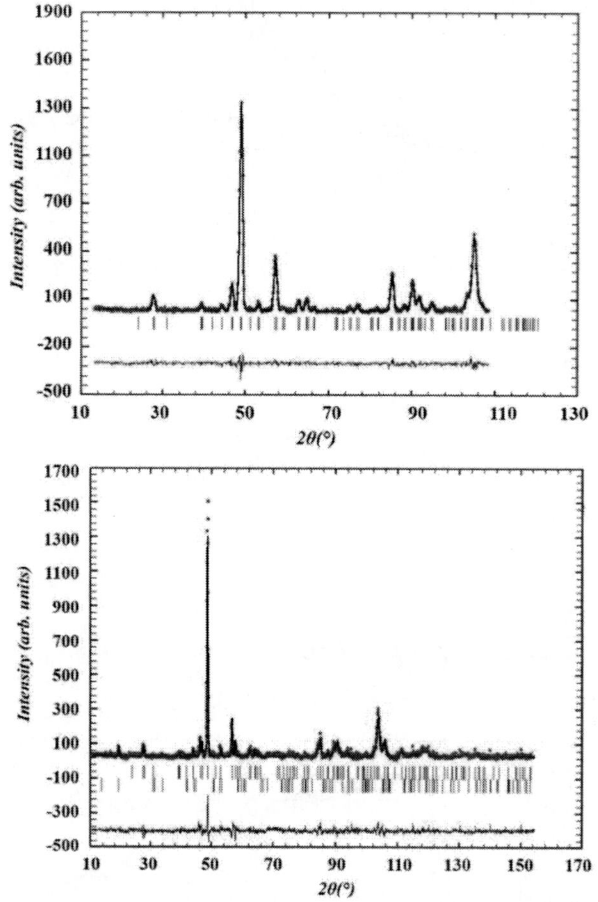


**Figure 3.** Diffraction patterns from neutron scattering (a) at room temperature and (b) at low temperature (20 K)

The lattice constants, Mn(Cu)-O bond length and distortion parameters at room temperature and low temperature (20K) can be shown in Table 1. The volume and the Mn(Cu)-O bond length decrease with the lowering of the temperature. This will lead to increase distortion at the position B, which literally lead to increase the concentration of $Mn^{3+}$ or $Cu^{2+}$ associated with the Jahn-Teller effect. The variation of the deformation of the octahedral ($Mn(Cu)O_6$) related to B places is quite complex [9]. The lattice parameter of Cu is very small at low temperatures. The parameter of Cu is not only in a state of $Cu^{2+}$ but also $Cu^{3+}$ where the radius of $Cu^{3+}$ is 0.54 Å, smaller than the radius of $Mn^{3+}$ (0.645 Å), but larger than the radius $Mn^{4+}$ (0.53 Å).

IOP Conf. Series: Materials Science and Engineering **202** (2017) 012075 doi:10.1088/1757-899X/202/1/012075

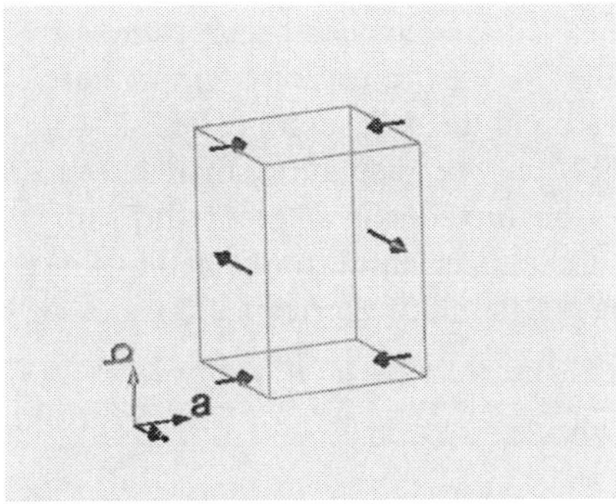


**Figure 4.** The orientation of the magnetic spins at a temperature of 20 K.

**Table 1.** The lattice constants (*a*, *b*, *c*), The Mn(Cu)-O distance, and Jahn-Teller distortion at ($S_d$) 300 K and 20 K

| *T* (K) | *a* (Å) | *b* (Å) | *c* (Å) | *V* (Å$^3$) | Mn-O$_1$ (Å) | Mn-O$_2$ (Å) | Mn-O$_3$ (Å) | <d> | $S_d$ |
|---|---|---|---|---|---|---|---|---|---|
| 300 | 5.4205 | 7.6334 | 5.4387 | 225.0374 | 1.9365 | 1.9408 | 1.9488 | 1.9420 | $6.502.10^{-3}$ |
| 20 | 5.4341 | 7.5394 | 5.4602 | 223.7022 | 1.736 | 1.927 | 2.1466 | 1.9365 | $2.122.10^{-1}$ |

Generally, the specific heat can be calculated by equation

$$C(T) = \gamma T + \beta_3 T^3 + \beta_5 T^5 + \delta T^n + \alpha T^{-2} \tag{2}$$

Where $\gamma T$ is the electronic specific heat, $\beta_3 T^3 + \beta_5 T^5$ is the phonon specific heat, $\delta T^n$ is the spin wave specific heat ($n = 3/2$ for ferromagnetic spin wave, n = 2 for antiferromagnetic spin wave), and $\alpha T^{-2}$ is the specific heat of the hyperfine structure [10]. Plot a graph of specific heat as a function of temperature C(T), both without an external magnetic field and the external magnetic field 9 T can be seen in Figure 5. From the analysis of the results of neutron scattering, it was found that the magnetic structure of the sample is antiferromagnetic, so n = 2. From our experiments, the results of fitting of the data, without the magnetic field and the magnetic field 9 T can be seen in Table 2.

Our experiments showed that the coefficient of electronic specific heat γ without applied magnetic field is 11.6 mJ.mol$^{-1}$.K$^{-2}$, whereas the application of a magnetic field 9 T is resulting $\gamma$ value by 78 mJ.mol$^{-1}$.K$^{-2}$. The resulting value of γ without the magnetic field, not much different from that produced by Z.Han *et al.* at 15 mJ.mol$^{-1}$.K$^{-2}$ [10]. The value of γ is related to electronic energy state density around the Fermi energy $\gamma = (\pi^2/3)k_B^2 N(E_F)$. The electronic energy state density around the Fermi energy is $N(E_F)$ = 2.97 x 10$^{24}$ eV$^{-1}$.cm$^{-3}$ without magnetic field and 1.99 x 10$^{25}$ eV$^{-1}$.cm$^{-3}$ with external magnetic field 9 T. Of the value of β, we can determine the Debye temperature $\theta_D$ by using equation

$$\theta_D = (12\pi^4 pR/5\beta)^{1/3} \tag{3}$$

Where $R$ = Ideal gas constant, $p$ = 5 is the number of atoms per formula unit. We found the Debye temperature is 460 K.

The sample magnetization of $La_{0.5}Ca_{0.5}Mn_{0.9}Cu_{0.1}O_3$ that measured in zero-field cooling (ZFC), field-cooled cooling (FCC), and field-cooled warming (FCW) at 2 mT can be seen in Figure 6. It appears that the curve FCC and ZFC are irreversible. It is as a consequence of the magnetic anisotropy of the compound. The antiferromagnetic transition temperature ($T_N$) and the charge-ordered antiferromagnetic transition temperature ($T_{CO}$) measured on cooling were obtained from the maximum of the low-field 2 mT. In Tthe ZFC curve, the peak occurs at temperatures around 225 K. This indicate the Neel temperature $T_N$, the phase transition temperature between the paramagnetic to antiferromagnetic phase occurs around this temperature. The $T_{CO}$ transition temperature is around 115 K. The heating and cooling M(T) curves display a small hysteresis close to the $T_N$. Probably this hysteresis is a consequence of the mixed character of the sample. We suspect there is a cluster ferromagnetic (FM) trapped in the cluster antiferromagnetic (AFM). At lowest temperature, a small anomaly in the M(T) curves marks the antiferromagnetic ordering, in complete agreement with the scan of neutron diffraction data.

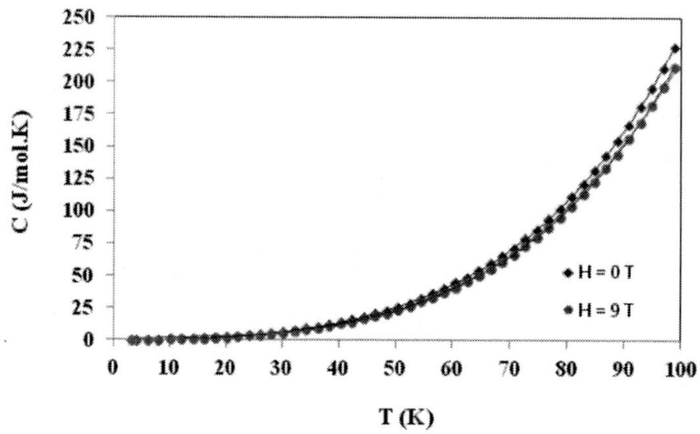


**Figure 5**. The specific heat versus temperature curve of $La_{0.5}Ca_{0.5}Mn_{0.9}Cu_{0.1}O_3$ under different magnetic field 0 T and 9 T at a temperature in the range of 10 K-100 K.

**Table 2**. Parameters under different magnetic fields fitting

| H (T) | $\gamma$ (J.mol$^{-1}$.K$^{-2}$) | $\beta_3$ (J.mol$^{-1}$.K$^{-4}$) | $\beta_5$ (J.mol$^{-1}$.K$^{-6}$) | $\delta$ (J.mol$^{-1}$.K$^{-3}$) | $\alpha$ (J.mol$^{-1}$.K$^{-1}$) |
|---|---|---|---|---|---|
| 0 | $1.16 \times 10^{-2}$ | $10^{-4}$ | $10 \times 10^{-9}$ | $3.4 \times 10^{-3}$ | $1.4 \times 10^{-3}$ |
| 9 | $7.8 \times 10^{-2}$ | $10^{-4}$ | $9 \times 10^{-9}$ | $3.5 \times 10^{-3}$ | $1.4 \times 10^{-3}$ |

Figure 7 shows the magnetization data M(H) of the sample measured at 120 and 140 K. It can be revealed that the isothermal Magnetization the data at 120 and 140 K is matched very well. The results of these two measures, both not saturated even though the external magnetic field up to 7 T. At temperature 120 K and 140 K, the value of $M$ rises slowly up to the value of the magnetic field 5 T, but to a temperature of 120 K, it began to appear hysteresis loop, indicating an increase in the value of the magnetic moment [1].

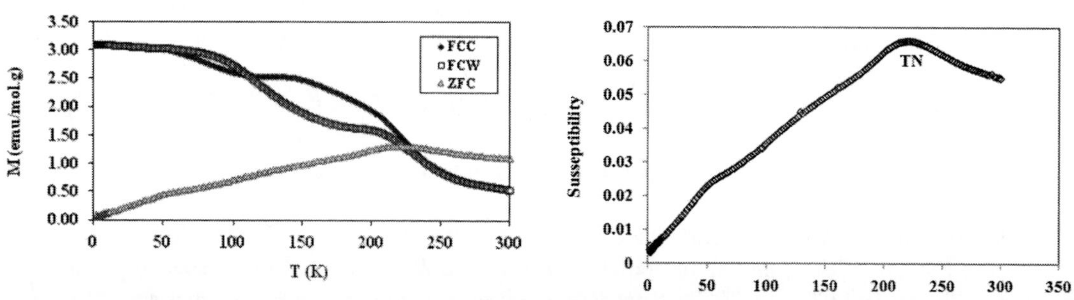


**Figure 6**. (a). ZFC, FCC, and FCW magnetization of $La_{0.5}Ca_{0.5}Mn_{0.9}Cu_{0.1}O_3$ as a function of temperature. (b) The curve of susceptibility as a function temperature.

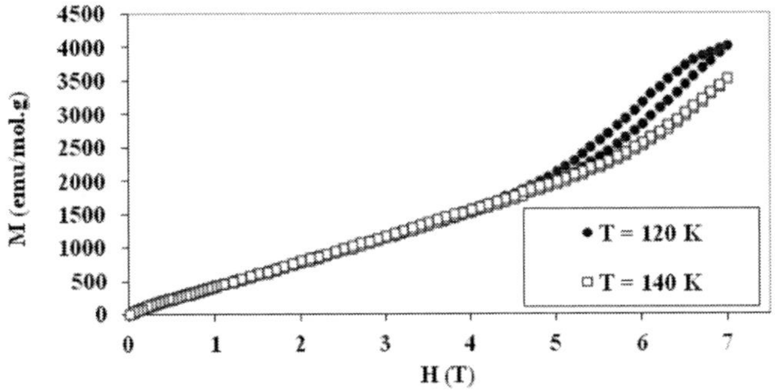


**Figure 7.** Magnetization as a function of field of $La_{0.5}Ca_{0.5}Mn_{0.9}Cu_{0.1}O_3$ sample at temperature 120 K and 140 K.

## 4. Conclusions

The effect of external magnetic field on physical and magnetic properties of $La_{0.5}Ca_{0.5}Mn_{0.9}Cu_{0.1}O_3$ has been studied. The transition temperature of the metal-insulator $T_{MI}$ decreased with the increasing external magnetic field. While at a temperature of less than 53 K, the samples followed the model of Arrhenius. However, at the temperatures higher than 53 K, the samples followed the model of Mott-insulator. A coefficient of electronic specific heat $\gamma$ increased quite dramatically in the presence of an external magnetic field 8.5 T. At room temperature the samples were paramagnetic, while at low temperatures were antiferromagnetic. The transition temperature between the paramagnetic phase to the antiferromagnetic phase, referred to the Neel temperature $T_N$ of about 225 K.

## 5. References

[1] Gunanto Y E, Sinaga K, Kurniawan B, Poertadji S, Tanaka H, Ono T, Steven E, and Brooks 2015 Advanced Materials Research **1123** 73-77

[2] Gunanto Y E, Purwanto A, Andika F, Mugirahardjo H, Wisnu A A, Kurniawan B, and Poertadji S 2012 *Indonesian Journal of Materials Science* **14** No. 1 8-12

[3] Kumar P and Dwivedi R K 2011 Journal of the Korean Physical Society **58** No.1 58-63

[4] Kim M S, Yang J B, Medvedeva J, Yelon W B, Paris P E, and James W J 2008 *J.Phys.:Condens.Matter* **20** 255228

[5] El-Hagary M, Shoker Y A, Mohammad S, Moustafa A M, El-Aal A A, Michor H, Reissner M, Hilscher G, and Rama A A 2008 *J.Alloys Compd.* doi:10.1016/j.jallcom.2008.01.048

[6] Yunoki S, Moreo A, Dagotto E, Okamoto S, Kancharla S S, and Fujimori A 2007 *Phys.Rev. B* **76** 064532

[7] Yang J, Li Y, and Lee Y P 2007 *Physica Status Solidi* (b), **244** 12 4546-4549

[8] Kim M S, J.B.Yang, P.E. Parris, Q.Cai, X.D.Zhou, W.J.James, W.B.Yelon, D.Buddhikot, and Mali S K 2005 *J.Appl.Phys.* **97** 10H714

[9] Craus M L and Lozovan M 2006 *Journal of Optolectronic and Advance Materials* **8** No.3 1008-1011

[10] Han Z and Jing Z 2014 *Advances in Condensed Matter Physics* **2014** Article ID 394296. http://dx.doi.org/10.1155/2014/394296

# Effect of Heat Treatment on The Crystal Structur, Electrical Conductivity and Surface of $Ba_{1.5}Sr_{0.5}Fe_2O_5$ Composite

**P Purwanto, WA Adi and Yunasfi**
Center for Science and Technology of Advanced Materials, National Nuclear Energy Agency of Indonesia (BATAN), Kawasan Puspiptek Serpong, Tangerang Selatan 15314, Indonesia

e-mail: patricius@batan.go.id

**Abstract**. The Composite of $Ba_{1,5}Sr_{0,5}Fe_2O_5$ has been synthesized by using powder metallurgy technique. The $Ba_{1.5}Sr_{0.5}Fe_2O_5$ were prepared from $BaCO_3$, $SrCO_3$ and $Fe_2O_3$ raw materials with a specific weight ratio. The three materials were synthesized by powder metallurgy under heat treatment at 800 °C, 900 °C, and 1000 °C for 5 hours. All the three samples were characterized by using X-ray Diffraction (XRD) to determine the crystal structure and crystal size, LCR meter to determine the conductivity, and Scanning Electron Microscope (SEM) to observe the morphological of the composites. The phase analysis result showed that the composite consists of several minor phases such as $BaO_2$, $SrO_2$, and $Fe_2O_3$. The Crystal size of composite $Ba_{1.5}Sr_{0.5}Fe_2O_5$ decreased while increases the strain of crystal with increasing of sintering temperature. The crystal size of the $Ba_{1.5}Sr_{0.5}Fe_2O_5$ composite is 3.55 nm to 7.23 nm and value of strain is 8.47% until 3.90%. Based on the conductivity measurement, it was obtained that the conductivity of the $Ba_{1.5}Sr_{0.5}Fe_2O_5$ composite decreased with increasing sintering temperature. It was also noticed that the conductivity increased with increasing of frequency. The conductivity ranged from $6.619\times10^{-7}$ S/cm to $65.659\times10^{-7}$ S/cm. The energy dispersive spectroscopy (EDS) analysis showed that several dominant elements were a good agreement with the phase analysis.

**Keywords:** Composite, $Ba_{1.5}Sr_{0.5}Fe_2O_5$, X-ray diffraction, conductivity, surface.

## 1. Introduction

The magnetic properties to determine with characteristic include field strength magnet, the magnetic moment is saturated and has a stable chemical properties [1-3]. Impurity phases within the magnet depending on the synthesis process and are associated with homogeneity and heat treatment [4,5]. Ferrite magnetic material contained in the iron in the form of iron oxide ($Fe_2O_3$) called hematite and magnetite ($Fe_3O_4$). Hematite is used as a raw material while the hard magnetic magnetite as soft magnetic materials [6]. Effect of temperature on the magnetic properties and crystal structure and electrical properties have been studied [7]. Properties of ionic conductivity in solids as well as the theoretical approach of the ionic conductivity [8,9].

The synthesis of the $Ba_{1.5}Sr_{0.5}Fe_2O_5$ magnetic composite so far has been proposed to study the materials for an electromagnetic absorber, electrical conductivity, crystal size and surface properties. These properties are very useful for the designing for an absorber of electromagnetic especially for of $Ba_{1.5}Sr_{0.5}Fe_2O_5$ magnetic composite. So far, many reports of $Ba_{1.5}Sr_{0.5}Fe_2O_5$ related compound can be found in the literature [6-9]. Unfortunately, a comprehensive study of this compound as a function of weight ratio, sintering temperature, porosity, and electrical conductivity, as well as the absorbtion response of electromagnetic wave have rarely been informed. Therefore, we proposed to synthesize a composite of $Ba_{1.5}Sr_{0.5}Fe_2O_5$. The preparation of the composite was also proposed by utilizing of powder metallurgy method from preparing from several starting materials i.e. $Fe_2O_3$, $BaCO_3$, and $SrCO_3$. Furthermore, in this work, we discuss the structure and electrical properties of the composite.

## 2. Experimental Method

The synthesis of composite $Ba_{1.5}Sr_{0.5}Fe_2O_5$ was conducted by using powder metallurgy method. The materials used were $BaCO_3$ (Aldrich 99.9% purity), $SrCO_3$ (Aldrich 99.9% purity), and $Fe_2O_3$ (Aldrich 99.9% purity). The weight compositions of the starting materials were $BaCO_3$ = 6.300, $SrCO_3$ = 0.845, and $Fe_2O_3$ = 2.855 g. The chemical equation for the composite is:

$$(2\text{-}x)BaCO_3 + (x)SrCO_3 + Fe_2O_3 \rightarrow Ba_{(2-x)}Sr_{(x)}Fe_2O_5 + 2CO_2$$

The starting materials consist of $BaCO_3$, $SrCO_3$, and $Fe_2O_3$ were mixed into a container vial and milled for 2 hours. After milling process, the materials were heated for 5 hours at a temperature of 800, 900, and 1000 °C. The samples were characterized using X-ray diffractometry to investigate the phase and crystallite size. Using LCR-meter and SEM, the conductivity and morphology of the samples were respectively characterized. Finally, the electrical properties of the samples were investigated by using LCR-meter.

## 3. Results and Discussion

### *3.1. X- Ray Diffraction of $Ba_{1.5}Sr_{0.5}Fe_2O_5$*

Figure 1 is the X-ray diffraction peak of the $Ba_{1.5}Sr_{0.5}Fe_2O_5$ composite with variation sintering temperature. The identification results in Figure 1 looks that the sample consists of other small phases of BaO, SrO, and $Fe_2O_5$. The diffraction peaks of $Ba_{1.5}Sr_{0.5}Fe_2O_5$ composite then analyzed by using Igor Lorentzian program. The result of the analysis was the Brag's position (2 θ) and full width half maximum ($\beta$) as depicted in Table 1.

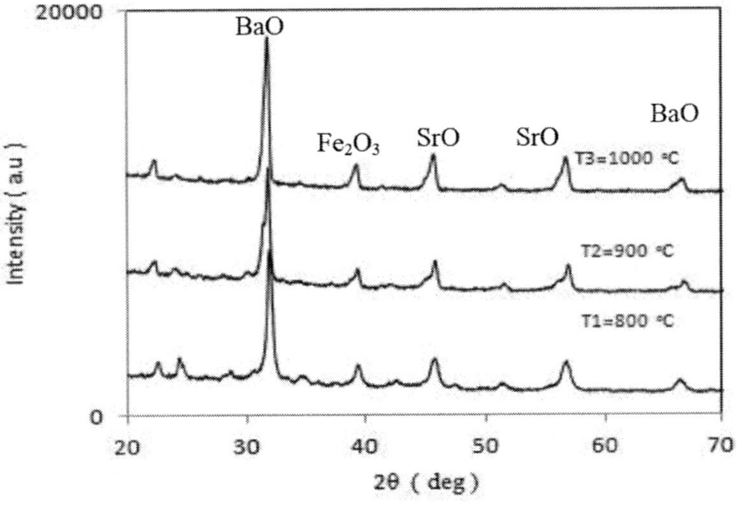


**Figure 1.** Phases identification of $Ba_{1.5}Sr_{0.5}Fe_2O_5$ X-ray diffraction peak.

**Table 1.** The calculated of $2\theta$ and full width at half maximum (FWHM) on the composite of $Ba_{1,5}Sr_{0,5}Fe_2O_5$.

| $T_{sinter}$ (°C) | $2\theta$ (°) | $\beta$ (rad) | sin $\theta$ | $\beta$ cos $\theta$ |
|---|---|---|---|---|
| 800 | 22.270 | 0.083 | 0.193 | 0.082 |
| | 31.761 | 0.106 | 0.274 | 0.102 |
| | 39.159 | 0.168 | 0.335 | 0.159 |
| | 45.509 | 0.320 | 0.387 | 0.295 |
| | 56.560 | 0.149 | 0.474 | 0.132 |
| | 66.310 | 0.256 | 0.547 | 0.215 |
| 900 | 22.255 | 0.026 | 0.193 | 0.026 |
| | 31.818 | 0.078 | 0.274 | 0.075 |
| | 39.300 | 0.048 | 0.336 | 0.046 |
| | 45.731 | 0.078 | 0.389 | 0.072 |
| | 56.887 | 0.047 | 0.476 | 0.042 |
| | 66.754 | 0.195 | 0.550 | 0.163 |
| 1000 | 22.320 | 0.038 | 0.194 | 0.038 |
| | 31.794 | 0.058 | 0.274 | 0.055 |
| | 39.199 | 0.095 | 0.335 | 0.089 |
| | 45.607 | 0.089 | 0.388 | 0.082 |
| | 56.688 | 0.133 | 0.475 | 0.117 |
| | 66.479 | 0.094 | 0.548 | 0.079 |

**Table 2.** Crystal size (D) and lattice strain ($\eta$) of $Ba_{1.5}Sr_{0.5}Fe_2O_5$.

| $T_{sinter}$ (°C) | $D$ ( nm ) | $\eta$ (%) |
|---|---|---|
| 800 | 3.55 | 8.47 |
| 900 | 5.66 | 6.42 |
| 1000 | 7.23 | 3.90 |

From Table 1, it can be transformed curves between ($\beta$ cos $\theta$) against sin $\theta$ as shown in Figure 2. By using a model of Williamson-Hall [10] as written in Equation (1), it can be calculated a value of crystal size and strain.

$$B(\cos\theta) = 0.9\lambda / D + 4\eta \sin\theta \tag{1}$$

Where $D$ is the crystal size, and $\eta$ is the strain of crystal. From the curve, we calculated the crystal size and lattice strain. The results are shown in Table 2.

By using Equation 1, the crystal size size and lattice strain can be obtained by a straight line which is the slope of the line is a constant strain crystal and crystal size. It is seen that the crystal size increases with the increase of sintering temperature.On the other hand, the lattice strain decreases with the increase of sintering temperature. The increase of the crystal size effected by sintering temperatures indicated by the growth of granularity. We can expect that the conductivity may increase the grain size also increase. This phenomenon may depict as lowering the pore size. The lattice strain reduced as an increase of sintering temperature affected to the space between the grains get smaller. This is also driven to better electrical conductivity. Empirically, it is difficult to obtain information about the relation between the size and strain to the electrical conductivity of the compound.

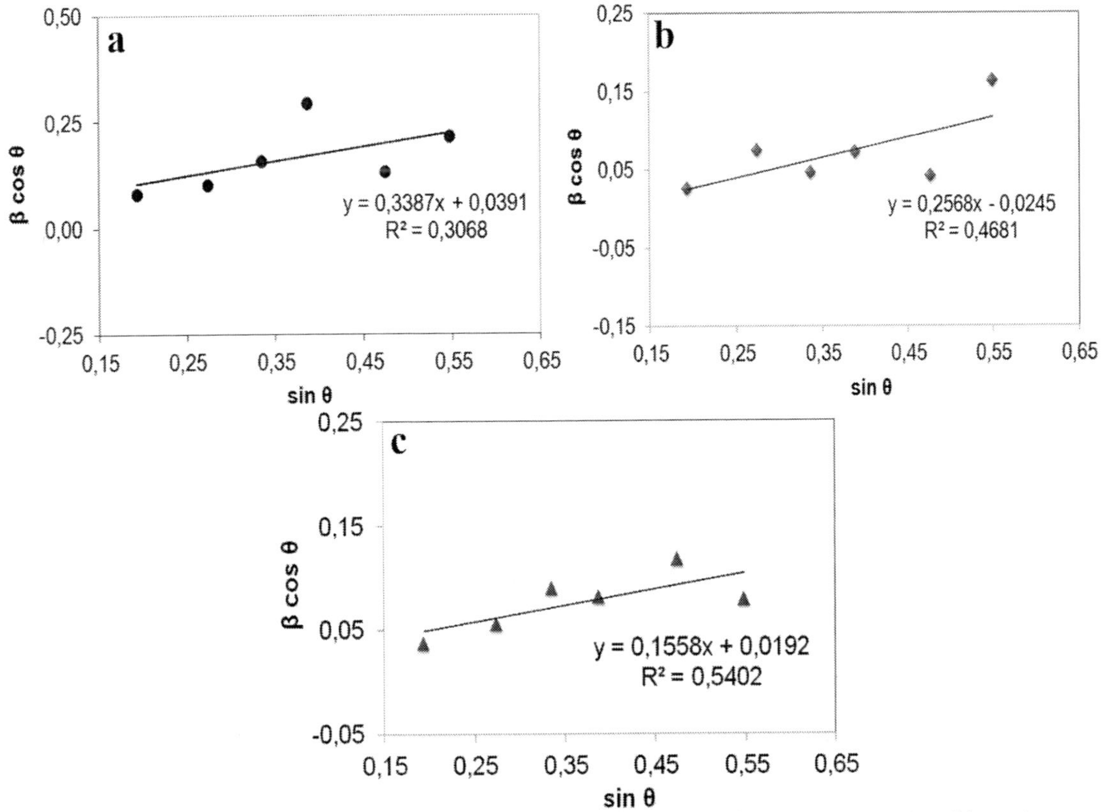


**Figure 2.** Relation curve of ($\beta$ cos $\theta$ versus sin $\theta$ on composite of $Ba_{(1.5)}Sr_{(0.5)}Fe_2O_5$ with various sintering temperature (a) $T_1 = 800$ °C, (b) $T_2 = 900$ °C, and (c) $T_3 = 1000$ °C.

### 3.2. *Conductivity of $Ba_{1.5}Sr_{0.5}Fe_2O_5$.*

Curves conductivity on the composite of $Ba_{1.5}Sr_{0.5}Fe_2O_5$ is shown in Figure 3. From Figure 3, then calculating value conductivity of $Ba_{(1.5)}Sr_{(0.5)}Fe_2O_5$ by using a model [9], with a equation:

$$\sigma = \sigma_0 f^s \tag{2}$$

where $\sigma$ is a conductivity (S/cm), $\sigma_o$ is conductivity that does not depend on the frequency and $s$ is an exponential factor. From Equation (2), it can be converted to the logarithm form, as in Equation (3)

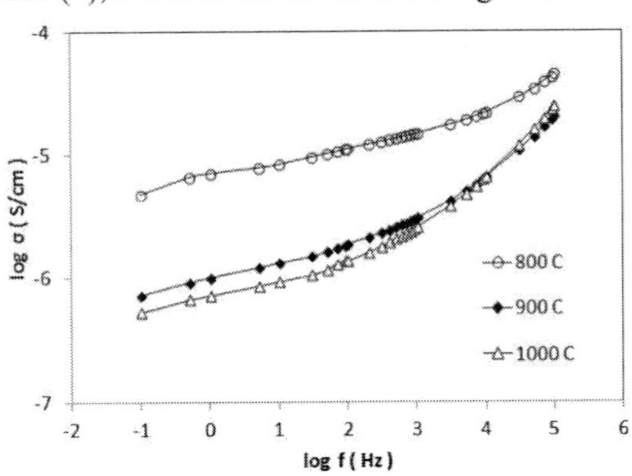


**Figure 3.** Conductivity value of $Ba_{1.5}Sr_{0.5}Fe_2O_5$.

**Table 3.** Electricl conductivity of $Ba_{1.5}Sr_{0.5}Fe_2O_5$.

| $T_{sintering}$ (°C) | $\sigma_0^a$ ( S/cm ) |
|---|---|
| 800 | $65.659 \times 10^{-7}$ |
| 900 | $9.399 \times 10^{-7}$ |
| 1000 | $6.619 \times 10^{-7}$ |

$\sigma_0^a$ = conductivity at the frequency of 0 Hz

$$\log \sigma = \log \sigma_o + s \log f \quad (3)$$

By using Equation (3), it can be calculated the electrical conductivity of $Ba_{1.5}Sr_{0.5}Fe_2O_5$. The result is shown in Table 3.

It is observed from Table 3 that the electrical conductivity of the composite is $6.619 \times 10^{-7}$ S/cm until $65.659 \times 10^{-7}$ S/cm. The conductivity decrease with increasing of sintering temperature. It is indicated in Figure 3 that the conductivity of the composite $Ba_{1.5}Sr_{0.5}Fe_2O_5$ increase with increasing frequency. Conductivity at an interval to begin frequency of 0.1 Hz to 1000 Hz, this shows the conductivity of $Ba_{1.5}Sr_{0.5}Fe_2O_5$ is dependent on the frequency. In the sintering process of thermal energy to be absorbed by the atoms, so that the vibration occurs between atoms. The thermal energy influences the arrangement of the grains, so that it can affect the diffusivity arrangement and homogeneity of the material during heat treatment process.

### *3.3. The surface of* $Ba_{1.5}Sr_{0.5}Fe_2O_5$.

The surface morphology of $Ba_{1.5}Sr_{0.5}Fe_2O_5$ with magnification at 10.000×, shown in Figure 4.

**Figure 4.** Surface of $Ba_{1.5}Sr_{0.5}Fe_2O_5$ and Spectrum EDS with various sintering temperatures (a) $T_1$ = 800 °C, (b) $T_2$ = 900 °C, and (c) $T_3$ = 1000 °C.

**Table 4**. Analysis element EDS of $Ba_{1.5}Sr_{0.5}Fe_2O_5$ (% weight).

| $T_{sinter}$ (°C) | Fe | Ba | Sr | C | Cu |
|---|---|---|---|---|---|
| 800 | 19.07 | 54.34 | 9.51 | 3.56 | 3.90 |
| 900 | 20.46 | 52.49 | 11.31 | 4.36 | 1.89 |
| 1000 | 21.94 | 56.87 | 9.38 | 2.56 | 1.96 |

The surface morphology of composite $Ba_{1.5}Sr_{0.5}Fe_2O_5$ is shown in Figure 4. The morphology of granules of the samples transforms to more ordered as the increase of sintering temperature. These SEM images showing the crystallinity of the samples also increase as the increase of the sintering temperature. This data clearly support the spectra of XRD patterns as depicted in Figure 1. As we cannot measure the shape of the grains quantitatively, so it is insignificant to relate of the conductivity to the surface morphology. The electrical conductivity is better described by the size and the strain of the crystal than to its morphology.

The result of EDS spectrum of all sintering temperature samples is presented in percent weight. Analysis of EDS spectrum of the samples can be obtained from the original data, and the results are shown in Table 4 excluded oxygen content. It is observed from Table 4 that increasing of sintering temperature increase Fe element but changes irregularly for other elements.

## 4. Conclusion

The composite of $Ba_{1.5}Sr_{0.5}Fe_2O_5$ has been synthesized by using powder metallurgy technique. The analyses of XRD spectra showed that the are other phases formed e.g. $BaO_2$, $SrO_2$, and $Ba_2Fe_2O_5$. The $Ba_{1.5}Sr_{0.5}Fe_2O_5$ composites revealed that the crystal size increase while the strain decrease with increasing of sintering temperature. The observation of the electrical measurement by LCR meter indicated that the conductivity decrease with increasing sintering temperature. On the other hand, the conductivity increase with increasing frequency. The changes of surface morphology of the $Ba_{1.5}Sr_{0.5}Fe_2O_5$ composite which is mainly originated from it grains is not significantly affected by sintering temperature. From these results, it can be concluded that heat treatment (sintering temperature) on the composite of $Ba_{1.5}Sr_{0.5}Fe_2O_5$ can affect the crystal size, strain, electrical conductivity, and surface morphology.

## 5. References

[1] Sebayang P and Mulyadi 2011 *Study of Journal Science and Technology* **29** 55-62
[2] Katoch A, Singh T and Sandhu B S 2013 *Int.J. of Res. in Adv.Tech.* **1** 456-46
[3] Srikrishna Ramya S I and Mahadevan C K 2014. *Int.J.of Res Eng and Tech.* **3** 570-573
[4] Abdel Latif I A 2012 *J.of Phys.* **1** 15-31
[5] Molaei M J, Ataie A, Raygan S, Picken S J, Mendes E and Tichelaar F D 2012 *Powder.Tech* **211** 292-295
[6] Singh, A, Narang S B, Sing Hul K, Pandey O P, Andey and Kotmala R K 2010 *J. of Cer. Proces. Res.* **11** 241-249
[7] Novizal, Manaf A, Hikam M 2014 *Int. J. of Eng and Tech.* **14** 88-92
[8] Kumar P P and Yashonath S 2006 *J.Chem. of Sci.* **118** 135-154
[9] Sunandana C S and Kumar P S 2004 *Bull. Mat. Sci.* **27** 1-17
[10] Mote V D, Purushotham Y and Dale B N 2012 *J. of Theoretic. App. Phys* **6** 1-8

**Acknowledgements**

The authors would like to thank colleagues in BSBM-PSTBM who have help in this study for the use of XRD instrument, LCR and SEM. The authors also thank the support fund DIPA-2015 and my colleague Mr Ade Mulyawan.

# Monitoring Microalgae Population Growth by using $Fe_3O_4$ Nanoparticles-based Surface Plasmon Resonance (SPR) Biosensor

**D T Nurrohman[1], M Oktivina[1], E Suharyadi[1], E A Suyono[2] and K Abraha[1]**

[1]Department of Physics, Universitas Gadjah Mada, Indonesia
[2]Faculty of Biology, Universitas Gadjah Mada, Indonesia

Email : esuharyadi@ugm.ac.id

**Abstract.** The population growth of microalgae has been successfully monitored by using the surface plasmon resonance (SPR)-based biosensor activated by using PEG-4000 functionalized $Fe_3O_4$ magnetic nanoparticles. The SPR-based biosensor in this work used Kretchmann configuration and Ha-Ne laser beam ($\lambda = 633$nm). The population of microalgae were estimated by using calibration curve. The shift of SPR angle happens due to the change of dielectric medium from prism/Au/air system to prism/Au/$Fe_3O_4$+PEG+microalgae/cover glass system. The change of dielectric medium occurs due to deposition of $Fe_3O_4$+PEG+microalgae samples that were monitored for seven days. The population of microalgae on the first day was estimated of 1188146 cell/ml and SPR angle had shifted from $42.8^0$ to $69.0^0$. The SPR angle shifted as much as $26.2^0$. The population of microalga grew significantly on the third day as much as 1649106 cell/ml and the SPR angle had shifted $43.8^0$ to $71.1^0$. The SPR angle shifted as much as $27.3^0$. On fourth to seventh day, the population of microalgae decreased. The population of microalgae on the seventh day was estimated of 1206961 cell/ml and the SPR angle had shifted from $44.0^0$ to $68.8^0$. The SPR angle shifted as much as 24.8 degrees. The shift of SPR angle was influenced by the population of microalgae so that the SPR-based biosensor has the potential for use in monitoring the microalgae population.

**Keywords:** Surface Plasmon Resonance (SPR), Magnetic $Fe_3O_4$ nanoparticle, Attenuated total reflection (ATR), Microalgae

## 1. Introduction

Surface plasmon resonance (SPR) is an optical sensing technique based on the electromagnetic phenomenon in which surface plasmon polaritons (SPP) waves are used to probe the interactions between target molecules and receptor molecules [1]. The interaction between target molecules and receptor molecules will produce a local change in the refractive index (RI) at the metal-dielectric interfaces. The RI change will then further affect the propagation constant of the SPPs which can be measured by a phenomenon termed as attenuated total reflection (ATR).

Surface plasmon resonance (SPR) appears as a sharp minimum (plasmon dip) in the internal reflectance of a metal film deposited on the face of a prism [2]. The position of plasmon dip can be monitored in term of incident angle versus reflectance interrogation. These approaches are capable of recognizing biological events on their sensing surfaces [3]. Today, SPR-based biosensors are increasingly not only in gas sensing [4], but also in many other important applications such as detection of low levels of *Escherichia coli* [5], *Salmonella* [6], and sensing of living cells [7]. Another potential of SPR-based biosensor that may be applied is for monitoring the growth of microalgae.

During this time, determiningthe growth of microalgae is carried out by a direct calculation by using the hemocytometer. This method requires a long time and will be difficult when there is a lot of cells in the hemocytometer. Another device that can be used is UV-Vis spectrophotometer. However, this device is an additional tool only for avoiding repeated calculations by using the hemocytometer.

In this paper, an SPR-based biosensor has been developed for growth monitoring of microalgae population based on the shift of the plasmon dip at reflectivity graph. $Fe_3O_4$ magnetic nanoparticles have been used in this study because of some of its unique properties (superparamagnetic properties, perfectly reacts with biomolecules and larger surface area)., $Fe_3O_4$ nanoparticles must be functionalized by encapsulation using polyethylene glycol 4000(PEG-4000) to be able for binding microalgae. In this study, growth monitoring of microalgae population obtained from the shift of the plasmon dip at reflectivity graph is compared with the data obtained by using UV-Vis spectrophotometer.

## 2. Experimental Method

### *2.1 Deposition of Au thin layer on the prism*

Pure gold (Au) with mass 15,4 mg was deposited on the surface of prism by evaporating it at pressure point of $3.10^{-3}$ Pa and an electric current of 50 A for 7 minutes. The prism-coated was then called prism/Au/air system.

### *2.2 Device set-up*

SPR device set up consists of the He-Ne laser (OSK Ogawa Seiki Co Ltd, $\lambda = 632.8$ nm), a polarizer, a beam splitter, two detectors (laser power meter) and a BK7 prism (Figure 1).

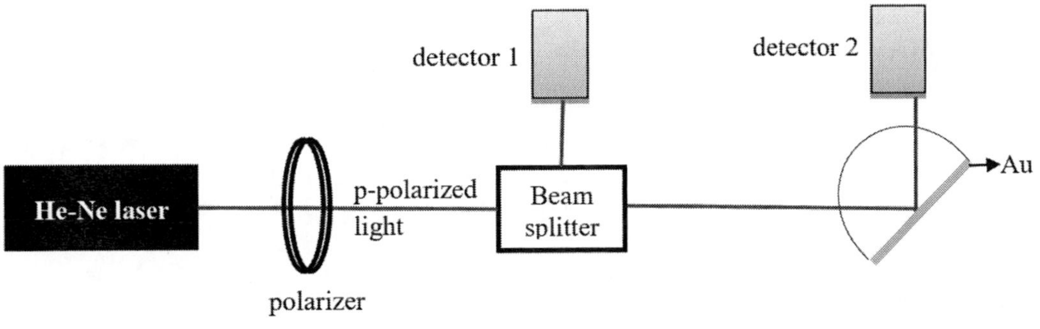


**Figure 1.** SPR set-up device in Kretschmann configuration.

Reflectivity graph of SPR was observed by varying the light incident angle ($\theta$) within the angle range of $30^0$ - $80^0$ and recording the reflectivity ($R$). Reflectivity in this experiment is defined by the power ratio recorded by detector 2 and detector 1. These values were plotted in a $R$ vs $\theta$ reflectivity graph.

### *2.3 $Fe_3O_4$ synthesis and functionality*

$Fe_3O_4$ magnetic nanoparticles were synthesized by using co-precipitation method. Iron II sulfate heptahydrate with iron III chloride hexahydrate were dissolved in distilled water, and then the solution formed was titrated by 60 mL amonia (with the amonia concentration of 10%). The solution was stirred at a temperature of $30^0$C with the angular speed of 450 rpm for 90 min. After that, it was placed on the permanent magnet and washed with distilled water. The precipitated solution was dried at $80^0$C for 2 hours by using a furnace. Finally, it was ground to powder and kept in an air-tight container.

The functionality of $Fe_3O_4$ nanoparticles was done with PEG applying the mass ratio of 1:1 for $Fe_3O_4$ and PEG. $Fe_3O_4$ and PEG were respectively dissolved in distilled water for 30 minutes, and then both solutions were mixed in a steady state in the stirrer for 1 hour. The solution that has been formed was dried to obtain $Fe_3O_4$ nanoparticles that have been functionalized with PEG ($Fe_3O_4$ + PEG).

*2.4 Microalgae culture*
Glagah microalga isolates were used in this study. This microalgae was isolated from the coast Glagah, Kulon Progo, Yogyakarta. The culture of microalgae is by Bold Bassal Medium method (BBM). The culture medium is made in a laminar air flow (LAF) and sterilized by using an autoclave for 2 hours. Microalgae is inserted into a medium which has been sterilized and its growth was then observed. The growth of microalgae was monitored by using UV-Vis spectrophotometer (Thermo scientific Genesis 10uv Scanning) and SPR-based biosensors for seven days, each of them was in the intervals of 24 hours. The population of microalgae with UV-Vis spectrophotometer was determined by using a calibration curve.

*2.5 Calibration curve*
Calibration curve gives the relationship between concentration and absorbance value. The number of cells for five samples of microalgae are calculated by using Haemocytometer Nuebauer of 1 mm to obtain the concentration value. The concentration of the number of cells from each sample was then determined by using Equation 2 as follows :

$$Concentration\left(\frac{cell}{ml}\right) = number\ of\ cell\ \times 5 \times 10^4 \tag{2}$$

Absorbance of each sample was then measured by using a UV-Vis spectrophotometer. After that, concentration and absorbance values were plotted to obtain a linear equation expressing their relation.

*2.6 Microalgae Deposition*
$Fe_3O_4$ + PEG were dissolved in distilled water for 30 minutes. The dispersed solution was then mixed with samples of microalgae that was being monitored. The sample is called as $Fe_3O_4$ + PEG + microalgae sample. After that, this sample was deposited on prism/Au/air system by using a micropipette so that the prism system changed into prism/Au/$Fe_3O_4$ + PEG + microalgae/cover glass system. The FTIR characterization was carried out on $Fe_3O_4$, PEG, $Fe_3O_4$ + PEG and $Fe_3O_4$ + PEG + microalgae samples to study the interaction between the the $Fe_3O_4$ + PEG and microalgae.

## 3. Results and Discussion

*3.1 Interaction of $Fe_3O_4$ + PEG with microalgae*
The FTIR spectra of the samples are presented in Figure 2. The type of molecular vibration that occurs in the sample is shown in Table 1. The peaks located in 408.91 and 586.36 $cm^{-1}$ (in Figure 2a) are due to the stretching vibrations of Fe – O bonds in octahedral and tetrahedral sites [8]. The peak at 3433.29 $cm^{-1}$ is attributed to the stretching vibrations of –OH group [9]. In the $Fe_3O_4$+PEG sample (in Figure 2c), the peak of Fe-O bond has shifted from 408.91 to 401.19 $cm^{-1}$. Besides that, the peak located in 2893.92 and 3448.72 $cm^{-1}$ are –OH group which indicates the presence of PEG in the $Fe_3O_4$ nanoparticles. From this result, $Fe_3O_4$+PEG sample is strong evidence for the formation of PEG-coated nanoparticles.

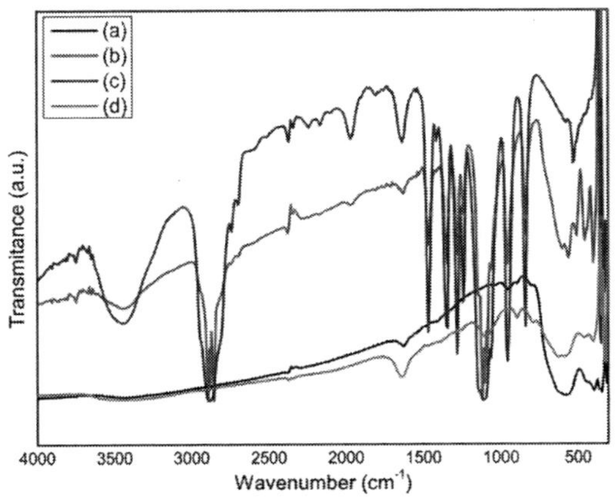


**Figure 2.** FTIR Spectra (a) $Fe_3O_4$ (b) PEG (c) $Fe_3O_4$+PEG and (d) $Fe_3O_4$+PEG+microalgae.

In $Fe_3O_4$+PEG+microalgae samples, the peak at Fe – O in tetrahedral sites has the same peak with $Fe_3O_4$+PEG samples. This phenomenon indicates that $Fe_3O_4$ nanoparticles have been successfully encapsulated by PEG and so there is no direct interaction between F – O in $Fe_3O_4$ nanoparticles with microalgae. Besides that, the shift of absorption peak has occurred in C – O – C, C – O and $CH_2$. The shift of absorption peak in C – O – C and C – O bonds from the sample of $Fe_3O_4$ + PEG and $Fe_3O_4$ + PEG + microalgae has occurred because of the C – O – C bond with carbohydrate of polysaccarides and carboxylic acid. The shift of absorption peak in H – C – H in wavenumber of 1465.90 cm$^{-1}$ and 2885.51 cm$^{-1}$ from $Fe_3O_4$+PEG samples has occurred due to the interaction between this sample and lipid as well as protein in microalgae.

**Table 1.** Assignment of FTIR spectra of $Fe_3O_4$, PEG, $Fe_3O_4$+PEG and $Fe_3O_4$+PEG+microalgae shown in Figure 2

| Band | $Fe_3O_4$ | PEG | $Fe_3O_4$ + PEG | $Fe_3O_4$+PEG + microalgae |
|---|---|---|---|---|
| Fe –O octahedral | 408.91 | - | 401.19 | - |
| Fe–O tetrahedral | 586.36 | - | 586.36 | 586.36 |
| C – C | - | 840.96 | 840.96 | - |
| C – C | - | 956.69 | 956.69 | - |
| C – C | - | 1056.99 | 1056.99 | - |
| C – O – C | - | 1103.28 | 1111.0 | 1095.57 |
| C – O | - | 1242.16 | 1242.16 | - |
| C – O | - | 1280.73 | 1280.73 | - |
| C – O | - | 1342.46 | 1342.46 | 1381.03 |
| H – C – H | - | 1465.90 | 1465.90 | 1473.62 |
| H – C – H | - | 1635.64 | - | 1635.64 |
| H – C – H | - | 2877.79 | 2885.51 | 2939.52 |
| -OH | - | 2893.22 | 2893.22 | - |
| -OH | 3433.29 | 3448.72 | 3448.72 | 3441.01 |

*3.2 Calibration Curve*

Figure 3 shows the relationship between concentration and absorbance for sample calibration curve. $x$-axis shows concentration and $y$-axis shows absorbance value. Based on the fitting result obtained from IGOR Pro software, linear equation is obtained as expressed in the form of $y = bx + \text{a}$. Error bar is obtained from standart deviation $\Delta b$ and $\Delta\text{a}$ that can be written as $\Delta y = \Delta bx + \Delta\text{a}$. From Figure 3, the relationship between the absorbance and concentration can be expressed by the following equation:

$$\text{Absorbance} = 3.19 \times 10^{-7}(\text{concentration}) - 0.14 \quad (2)$$

To determine the growth of microalgae, Equation 2 can be re-expressed as :

$$\text{Concentration} = \frac{\text{absorbance} + 0.14}{3.19 \times 10^{-7}} \quad (3)$$

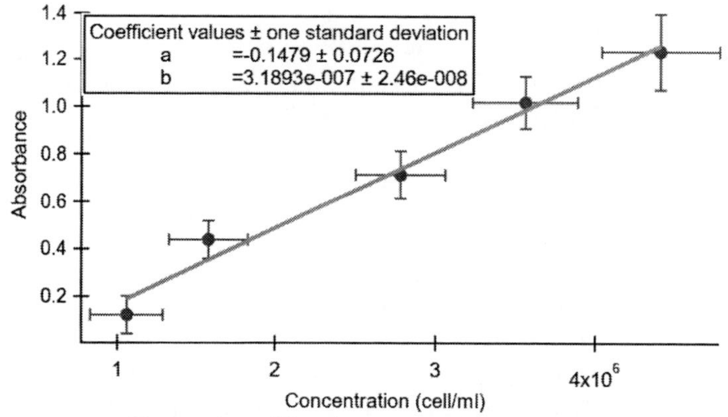


**Figure 3.** Calibration curve of microalgae

*3.3 Microalgae population growth by UV-Vis spectrophotometer*

Table 2 illustrates the microalgae population that is observed by using UV-Vis spectrophotometer for 7 days. The concentration of microalgae can be determined by Equation 3. The microalgae population decreases on the second day because microalgae is still adapting to the medium. The growth of microalgae increases significantly on the third day. The growth of microalgae declines in the fourth to the seventh day and slowly begins to decline from day 4 to day 7.

**Table 2.** The growth of microalgae for 7 day observation.

| Day | Absorbance | Concentration (cell/ml), $10^6$ |
|---|---|---|
| 1 | $0.23 \pm 0.08$ | $(1.19 \pm 0.23)$ |
| 2 | $0.20 \pm 0.08$ | $(1.12 \pm 0.23)$ |
| 3 | $0.38 \pm 0.08$ | $(1.65 \pm 0.24)$ |
| 4 | $0.28 \pm 0.08$ | $(1.34 \pm 0.24)$ |
| 5 | $0.27 \pm 0.08$ | $(1.30 \pm 0.24)$ |
| 6 | $0.24 \pm 0.08$ | $(1.22 \pm 0.23)$ |
| 7 | $0.24 \pm 0.08$ | $(1.20 \pm 0.23)$ |

*3.4 Monitoring microalgae population growth by using surface plasmon resonance-based biosensor.*

*3.4.1 The influence of destilled water and $Fe_3O_4$+PEG to the shift of SPR angle.* In this study, microalgae are detected in the form of liquid sample. Therefore, the observation of SPR phenomenon begins with observations to determine the effect of distilled water as medium for the culture of microalgae and $Fe_3O_4$ + PEG samples as active nanoparticles to bind microalgae. The shift of SPR angle has been occurred due to distilled water and $Fe_3O_4$+PEG deposition in the prism system. The red curve shows prism/Au/air system and the blue curve shows prism/Au/destilled water/cover glass (Figure 4a) and prism/Au/$Fe_3O_4$+PEG/cover glass (Figure 4b). The shift of SPR angle due to influence of distilled water and $Fe_3O_4$ + PEG are 23.5 and 24.1 degrees, respectively. The $Fe_3O_4$+PEG has showed larger shifting of SPR angle because refractive index of $Fe_3O_4$+PEG is bigger than destilled water. Refractive index of $Fe_3O_4$+PEG is 1.63 [10] and refractive index of distilled water is 1.33[10].

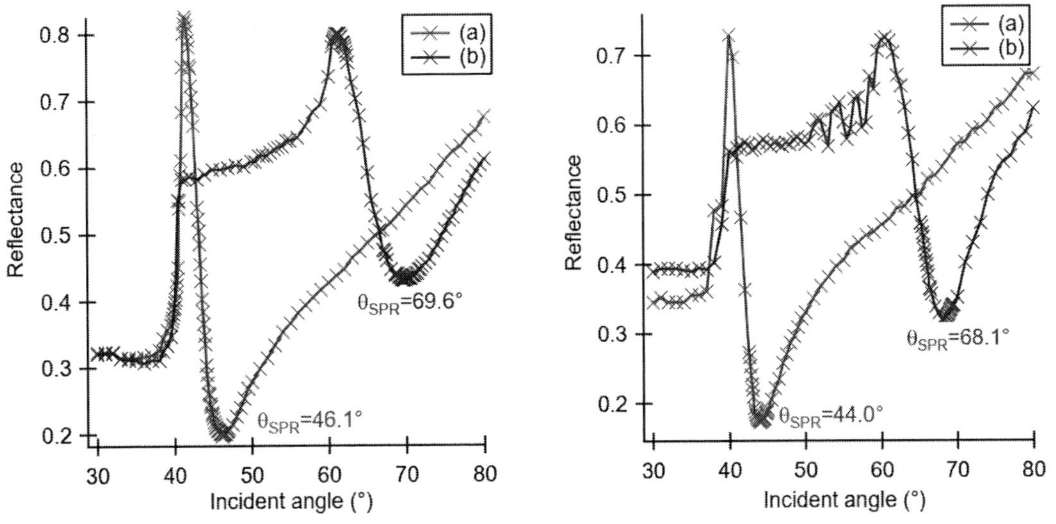


(a) (b)

**Figure 4.** (a) The shift of SPR angle due to distilled water (b) The shift of SPR angle due to $Fe_3O_4$+PEG.

*3.4.2 Monitoring microalgae population growth by SPR-based biosensor.* Table 3 shows the amount of shift of SPR angle due to deposition of a mixture $Fe_3O_4$ + PEG+microalgae. Microalgae that mixed are microalgae samples taken during the 7 days of observation using UV-Vis spectrophotometer (Table 2). Table 3 shows the response of SPR-based biosensor that is expressed by the shift of SPR angle. Prism/Au/air is initial condition (before microalgae deposition) and prism/Au/$Fe_3O_4$+PEG+ microalgae/cover glass is SPR system after deposition $Fe_3O_4$+PEG+ microalgae. Based on Table 3, the shift of SPR angle is different from each other. The concentration of microalgae influences the shift of SPR angle.

**Table 3.** The shift of SPR angle for 7 days observation of microalgae.

| Time (day) | Prism/Au/air | Prism/Au/$Fe_3O_4$ +PEG+microalgae/cover glass | The shift of SPR angle ($\Delta\theta_{SPR}$) |
|---|---|---|---|
| 1 | 42.8 ± 0.1 | 69.0 ± 0.1 | 26.2 ± 0.1 |
| 2 | 45.8 ± 0.1 | 70.0 ± 0.1 | 24.2 ± 0.1 |
| 3 | 43.8 ± 0.1 | 71.1 ± 0.1 | 27.3 ± 0.1 |
| 4 | 45.7 ± 0.1 | 72.8± 0.1 | 27.1 ± 0.1 |
| 5 | 43.6 ± 0.1 | 70.4 ± 0.1 | 26.8 ± 0.1 |
| 6 | 45.0 ± 0.1 | 71.3 ± 0.1 | 26.3 ± 0.1 |
| 7 | 44.0 ± 0.1 | 68.8 ± 0.1 | 24.8 ± 0.1 |

Figure 5 shows the results of monitoring the growth of microalgae with SPR based biosensor and UV-Vis spectrophotometer. The blue line shows the respon of SPR based biosensor based on the shift in SPR angle, while the red line shows the microalgae growth charts derived from UV-Vis spectrophotometer. The growth of microalgae with SPR-based biosensor shows the same pattern graphics as indicated by UV-Vis spectrophotometer. Therefore, SPR-based biosensors have the potential to be further developed for monitoring the growth of microalgae so that the process of monitoring the growth of microalgae is more effective and has the potential to be monitored in real time.

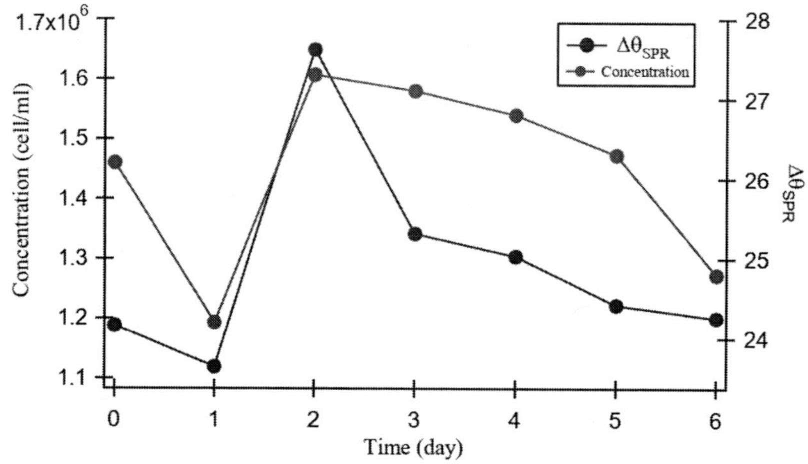


**Figure 5.** Growth monitoring of microalgae by using SPR based biosensor and UV-Vis spectrophotometer.

## 4. Conclusions

$Fe_3O_4$ nanoparticles have been successfully functionalized by PEG-4000 for monitoring of microalgae. It is proven by the discovery of interaction between $Fe_3O_4$ + PEG-4000 with components of microalgae which consists of carbohydrates, protein, carboxyl group and lipid in the FTIR spectra. The microalgae population growth has been successfully monitored by using a SPR-based biosensor. Finally, the shift of SPR angle can be used for monitoring the microalgae population growth.

## 5. References

[1] Raether P D H 1988 Surface plasmons on smooth surfaces *Surface Plasmons on Smooth and Rough Surfaces and on Gratings* Springer Tracts in Modern Physics (Springer Berlin Heidelberg) pp 4–39

[2] Islam M S and Kouzani A Z 2013 Simulation and Analysis of a Sub-Wavelength Grating Based Multilayer Surface Plasmon Resonance Biosensor *J. Light. Technol.* **31** 1388–98

[3] Yang D, Lu H-H, Chen B and Lin C-W 2010 Surface Plasmon Resonance of SnO2/Au Bi-layer Films for Gas Sensing Applications *Sens. Actuators B Chem.* **145** 832–8

[4] Linman M J, Sugerman K and Cheng Q 2010 Detection of low levels of Escherichia coli in fresh spinach by surface plasmon resonance spectroscopy with a TMB-based enzymatic signal enhancement method *Sens. Actuators B Chem.* **145** 613–619

[5] Mazumdar S D, Barlen B, Kämpfer P and Keusgen M 2010 Surface plasmon resonance (SPR) as a rapid tool for serotyping of Salmonella *Biosens. Bioelectron.* **25** 967–971

[6] Chabot V, Cuerrier C M, Escher E, Aimez V, Grandbois M and Charette P G 2009 Biosensing based on surface plasmon resonance and living cells *Biosens. Bioelectron.* **24** 1667–1673

[7] Jafari A, Shayesteh S F, Salouti M and Boustani K 2015 Effect of annealing temperature on magnetic phase transition in Fe 3 O 4 nanoparticles *J. Magn. Magn. Mater.* **379** 305–312

[8] Sun J, Zhou S, Hou P, Yang Y, Weng J, Li X and Li M 2007 Synthesis and characterization of biocompatible Fe3O4 nanoparticles *J. Biomed. Mater. Res. A* **80** 333–341

IOP Conf. Series: Materials Science and Engineering **202** (2017) 012077 doi:10.1088/1757-899X/202/1/012077

[9] Duygu D Y, Udoh A U, Ozer T B, Akbulut A, Erkaya I A, Yildiz K and Guler D 2012 Fourier transform infrared (FTIR) spectroscopy for identification of Chlorella vulgaris Beijerinck 1890 and Scenedesmus obliquus (Turpin) Kützing 1833 *Afr. J. Biotechnol.* **11** 3817–3824
[10] Hale G M and Querry M R 1973 Optical constants of water in the 200-nm to 200-μm wavelength region *Appl. Opt.* **12** 555–563

**Acknowledgments**
The reported paper was financially supported by *Hibah Kompetensi* (HIKOM) from the Ministry of research and Technology of High Education for the period of 2015- 2017 and thesis scholarship of *Lembaga Pengolahan Dana Pendidikan* (LPDP) batch 3 for the period of 2016.

# The Addition of Graphene and Magnetite Materials in $TiO_2$/CuO Catalyst for Enhancing Photosonocatalytic Performance and Reusability

**A Taufik[1,2], A Muzaki[1,2] and R Saleh[1,2]**

[1] Integrated Laboratory of Energy and Environment, Fakultas MIPA-Universitas Indonesia, Kampus UI Depok, Depok 16424, Indonesia
[2] Departemen Fisika, Fakultas MIPA-Universitas Indonesia, Kampus UI Depok, Depok 16424, Indonesia

Email: rosari.saleh@ui.ac.id; rosari.saleh@gmail.com

**Abstract.** In the present study, $TiO_2$/CuO with the addition of graphene and magnetite materials were investigated as a catalyst for waste water removal through the photosonocatalytic process. $TiO_2$/CuO/graphene and $Fe_3O_4$/$TiO_2$/CuO/graphene were synthesized by using the sol-gel method. Their physical properties were characterized by using X-ray Diffraction, Fourier Transform Infra-red Spectroscopy, Thermogravimetric Analysis, Transmission Electron Microscope, and Breneur Emmet Teller (BET) surface area analyzer. The Photosonocatalytic activity was examined under visible light irradiation and 40 kHz ultrasound frequency with methyl orange (MO) as a model of organic pollutant. Compared to photocatalytic and sonocatalytic process, photosonocatalytic activity can improve the degradation process due to a synergistic effect between photocatalytic and sonocatalytic process at the same time. The addition of graphene materials can improve the charge carrier separation and prevent recombination electron and holes. Therefore, the photosonocatalytic performance increases significantly with the addition of graphene materials. The addition of magnetite materials is not only can enhance the photosonocatalytic performance but also can improve the separation process for reuse the catalyst. The photosonocatalytic mechanism was correlated with their physical properties and visible as well as ultrasound irradiation.

**Keywords:** Graphene, magnetite, TiO/CuO, photosonocatalytic.

## 1. Introduction

Semiconductor materials have been considered as promising photocatalyst materials for removal of organic dye from the environment. Among many others materials, the pioneer photocatalyst materials ($TiO_2$) still attract much attention due to its outstanding performance in the photocatalytic activity [1-2]. $TiO_2$ has excellent properties such as low cost, eco-friendly, and high stability [3]. However, $TiO_2$ has a limitation due to the high rate recombination electron and holes. Moreover, $TiO_2$ with direct band gap 3.2 eV can only be activated by UV light irradiation which takes 4% from the sunlight irradiation [4]. For practical application, it is crucial to modify or create photocatalyst that can be activated not only by UV light irradiation but also under visible light irradiation. Therefore, it is important to modify the $TiO_2$ semiconductor photocatalyst to extend and improve its photocatalytic performance.

To modify $TiO_2$ properties, they can be coupled with other materials. Many reports have proven that coupled $TiO_2$ with another semiconductor can enhance the photocatalytic performance and also extend the activation energy to visible light irradiation [5]. Among other semiconductors, CuO has been chosen as suitably coupled materials with $TiO_2$ due to its proper energy band position that allowed each charge carrier to transfer from one semiconductor to another [6-7]. It will impact the prolonged charge carrier recombination, and the photocatalytic activity can be enhanced. Recently, the carbon materials have been known as promising materials to support the photocatalytic performance due to its unique properties such as (1) high charge carrier mobility, (2) high thermal conductivity and (3) high specific surface area. Those properties are vital in improving photocatalytic performance [8-9].

Separation photocatalyst is another challenge for enhancing the efficiency of the photocatalyst; there are many methods of the separation process of a catalyst such as centrifugation and filtration [10]. However, those methods are expensive and complex process. The use of magnetic separation process offers high efficiency with low cost and simple operation. Magnetic separation process needs magnetic external and also magnetic catalyst inside the solution to attract the catalyst by magnetic separation. Therefore, the incorporation of magnetic materials in the photocatalyst is very promising in separation efficiency. Magnetite ($Fe_3O_4$) has been considered as supported photocatalyst due to high adsorption capacity, unique magnetic properties, and can prolong charge carrier recombination.

Many researchers have reported that the photocatalytic activity still has a limitation due to the low light penetration ability and agglomeration of particles [11]. The addition of ultrasonic wave in the photocatalytic experiment has been considered as the best solution to overcome the limitation. Ultrasonic wave has better penetration ability compared to light wave in any water medium. Even the existence of ultrasonic wave can promote dispersion of catalyst, and the agglomeration of particles can be prevented [11].

Based on the explanation above, in this work $TiO_2$/CuO modified by graphene and magnetite material is used as a catalyst for removing methyl orange as a model of organic pollutant. The physical properties of the samples were characterized by using X-ray diffraction (XRD), Fourier transforms infrared spectroscopy (FT-IR), transmission electron microscope (TEM), Thermogravimetric analysis (TGA), and brenneur-Emmet-Teller (BET) surface area analysis). The removal ability of the sample was carried out by using the photosonocatalytic method. The reusability and separation of catalyst were also investigated.

## 2. Experimental Method

All reagents we used were copper sulfate pentahydrate ($Cu(SO)_4.5H_2O$), Iron sulfate heptahydrate $Fe(SO)_4.7H_2O$, Titanium dioxide ($TiO_2$), Sodium hydroxide (NaOH), Ethanol, ethylene glycol were purchased from Merck, while graphene was purchased from Angstron materials. All reagents were used without further purifications.

$Fe_3O_4$ and CuO were synthesized based on our previous reports [12]. $TiO_2$/CuO was synthesized by using the sol-gel method. In the typical process, $TiO_2$ were dissolved in water, and then NaOH solution was added slowly into the $TiO_2$ solution. CuO nanoparticles were dispersed in water and poured into the above solution. The reaction condition was kept at 80 $^oC$ for 2h, and then the precipitates were collected through centrifugation process. The final product was heated at 100 $^oC$ for 1 hours. The addition of $Fe_3O_4$ was done by pouring $Fe_3O_4$ dispersion in the above solution.

The graphene addition was carried out by using co-precipitation method. Typically graphene was dispersed in water and ethanol mixture through ultrasonic treatment. Each $TiO_2$/CuO and $Fe_3O_4$/$TiO_2$/CuO nanocomposites were immersed in the graphene dispersion and stirred by using magnetic stirrer. After stirring process, the mixture was heated at 120 $^oC$ for 3 hours. Finally, the product was collected by using centrifugation process and dried under vacuum condition.

The all prepared samples were characterized by some measurements such as X-Ray diffractions (XRD) using Rigaku Minflex 600 operated at a voltage and current of 30 kV and 150 mA, respectively. Fourier Transform infra-red spectroscopy was investigated by using Shimadzu prestige with Potassium Dichromate (KBr) as a reference, Thermogravimetric analysis (TGA) measurement was performed by using Rigaku Thermo Plus EVO2 TG8121 under argon atmosphere, and BET surface area was analyzed by using Nova Quantachrome 2000 and transmission electron microscope were characterized by using TEM FEI Tecnai G2 Super Twin. Photosonocatalytic activity was tested on the degradation of methyl orange from the aqueous solution. An amount of 0.03 g catalyst were immersed in 100 mL beaker glass containing 20 mg/L methyl orange concentration. Prior the irradiation, the adsorption process was performed for 4 hours to ensure equilibrium stage of adsorption-desorption. After the adsorption process, the samples were irradiated by using 40 kHz ultrasound and 40 W visible light irradiation simultaneously. Every 15 minutes the methyl orange solution was carried, and its absorbance was measured by using UV-Vis absorbance spectroscopy.

## 3. Results and Discussion

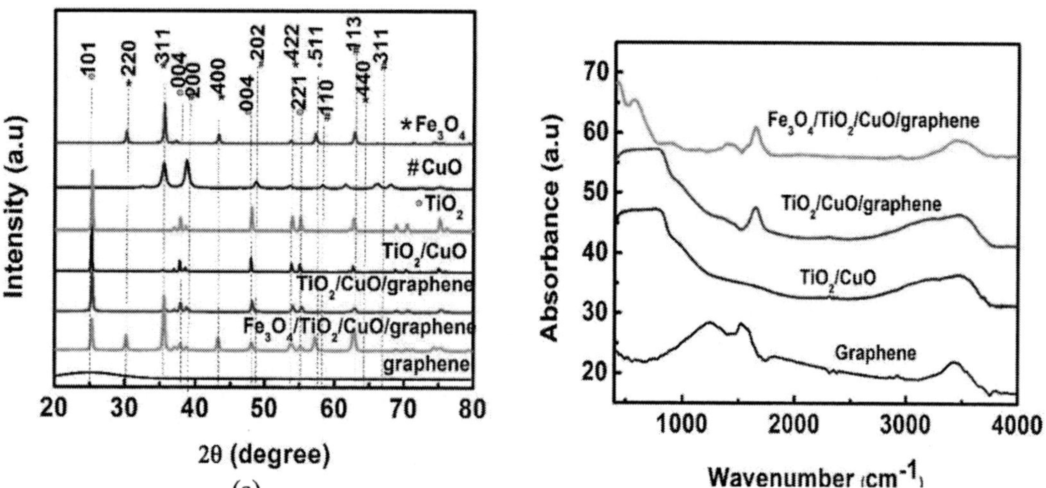


**Figure 1.** (a) XRD spectra of $Fe_3O_4$, CuO, $TiO_2$ nanoparticles, graphene, $TiO_2$/CuO, $TiO_2$/CuO/graphene, and $Fe_3O_4$/$TiO_2$/CuO/graphene. (b) FT-IR spectra of $TiO_2$/CuO, $TiO_2$/CuO/graphene, and $Fe_3O_4$/$TiO_2$/CuO/graphene

**Table 1.** Lattice parameter and grain size of $Fe_3O_4$, CuO, $TiO_2$ nanoparticles, graphene, $TiO_2$/CuO, $TiO_2$/CuO/graphene, and $Fe_3O_4$/$TiO_2$/CuO/graphene.

| Sample | Lattice parameter (Å) | | | | | | Grain Size (nm) | | |
|---|---|---|---|---|---|---|---|---|---|
| | $Fe_3O_4$ | $TiO_2$ | | CuO | | | | | |
| | a=b=c | a=b | c | a | B | c | $Fe_3O_4$ | $TiO_2$ | CuO |
| $Fe_3O_4$ | 8.363 | - | - | - | - | - | 45 | - | - |
| CuO | - | - | - | 4.691 | 3.425 | 5.138 | - | - | 15 |
| $TiO_2$ | - | 3.795 | 9.525 | - | - | - | - | 67 | - |
| $TiO_2$/CuO | - | 3.792 | 9.521 | 4.683 | 3.425 | 5.134 | - | 59 | 16 |
| $TiO_2$/CuO/graphene | - | 3.786 | 9.516 | 4.612 | 3.428 | 5.167 | - | 59 | 16 |
| $Fe_3O_4$/$TiO_2$/CuO/graphene | 8.361 | 3.782 | 9.510 | 4.602 | 3.437 | 5.223 | 45 | 58 | 15 |

Figure 1a shows the XRD spectra of $TiO_2$/CuO with the addition of magnetite and graphene. The XRD spectra of $Fe_3O_4$, CuO, $TiO_2$, graphene and CuO/$TiO_2$ were plotted as a comparison. From the XRD spectra of $TiO_2$/CuO with the addition graphene, several peaks could be found that could correspond to the existence of monoclinic structures from CuO at $2\theta \approx 38^o$ (220), and anatase structures from $TiO_2$ nanoparticles at $2\theta \approx 26^o$ (101) and $48^o$ (004). Moreover, with the incorporation of magnetite material in the nanocomposites, several peaks at $2\theta \approx 30$ (220), $36^o$ (311), and $44^o$ (400) which correspond to the existence of cubic spinel structure of $Fe_3O_4$. The presence of graphene could not be detected by using XRD measurement due to the small amount of graphene and also the broad XRD peak of graphene. The lattice parameter and grain size of the samples were calculated by using Rietveld refinement method and debye Scherrer equation. The results are listed in Table 1. The result indicates that there are no significant differences between the lattice parameters of each nanocomposites component compared to its pure nanoparticles.

The molecular vibration of the sample was analyzed by using FT-IR measurement, and the results were plotted in Figure 1b. As a comparison, the FT-IR spectra of graphene, as well as CuO/TiO2, are also plotted in the figures. From the FT-IR measurement, the existence of graphene could be detected at wavenumber of 1050 and 1621 $cm^{-1}$ which could be attributed to the molecular vibrations from C-O and C=C vibrational mode, respectively [13]. As can be seen from the figure the vibrational mode of $TiO_2$/CuO does not change with the incorporation of graphene. The vibrational mode around 400-700 $cm^{-1}$ could be attributed to the existence of Cu-O and Ti-O-Ti vibration [14-15]. The incorporation of magnetite material on the nanocomposite resulted in the sharp peak around 500-700 $cm^{-1}$ that could be attributed to the existence of Fe-O vibration [16]. The broad peak around 3000-3400 $cm^{-1}$ express the O-H vibrations from water molecules [16].

Thermal degradation of samples was measured to analyze the stability of the samples. Thermal stability was recorded at room temperature until 1000 $^oC$. Figure 2 shows the TGA measurement of $Fe_3O_4/TiO_2$/CuO/graphene. As a comparison, the TGA measurement of $TiO_2$/CuO and $TiO_2$/CuO/graphene composites were also performed. From the figures, we can find that the $TiO_2$/CuO performed good stability until 1000 $^oC$. However, the incorporation of graphene the thermal stability of samples slightly decreased. At the region of 25 $^oC$ until 100 $^oC$ the thermal degradation could be due to the desorption of physically adsorbed water and residual solvent [16]. The thermal degradation from 400 $^oC$–850 $^oC$ could be attributed to the combustion of carbon materials, and at 850 $^oC$ to 1000 $^oC$ the thermal degradation could be due to the phase transformation between CuO to $Cu_2O$.

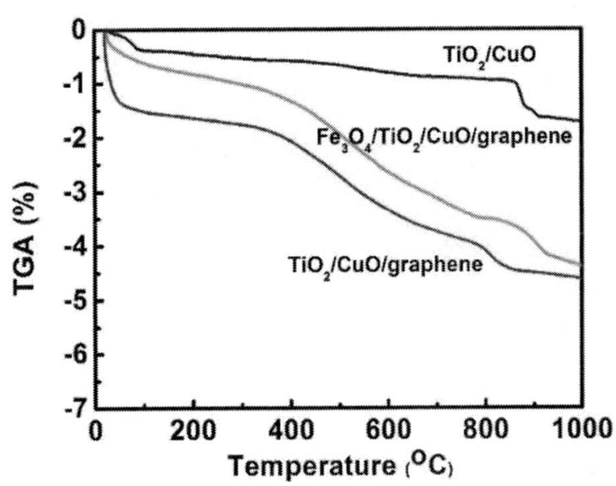


**Figure 2.** A typical TGA of $TiO_2$/CuO, $TiO_2$/CuO/graphene, and $Fe_3O_4/TiO_2$/CuO/graphene.

The surface area of the samples was measured by using BET measurement. The results were listed in Table 2. The specific surface area of CuO/$TiO_2$ is 5.78 $m^2/g$. The incorporation of graphene enhances the specific surface area of the samples until 45.43 $m^2/g$, however with the inclusion of magnetite $Fe_3O_4$ the specific surface area of the sample decreased to 34.32 $m^2/g$. The specific surface

IOP Conf. Series: Materials Science and Engineering **202** (2017) 012078 doi:10.1088/1757-899X/202/1/012078

area is important in catalytic performance. The presence of surface areas is favorable to absorb pollutants, harvest light and separate the photogenerated electron-hole pairs, which are factors for the promotion of catalytic activity.

**Table 2.** A surface area analysis of $TiO_2$/CuO, $TiO_2$/CuO/graphene, and $Fe_3O_4$/$TiO_2$/CuO/graphene.

| Sample | Surface area ($m^2$/g) |
|---|---|
| $TiO_2$/CuO | 5.78 |
| $TiO_2$/CuO/graphene | 45.43 |
| $Fe_3O_4$/$TiO_2$/CuO/graphene | 34.32 |

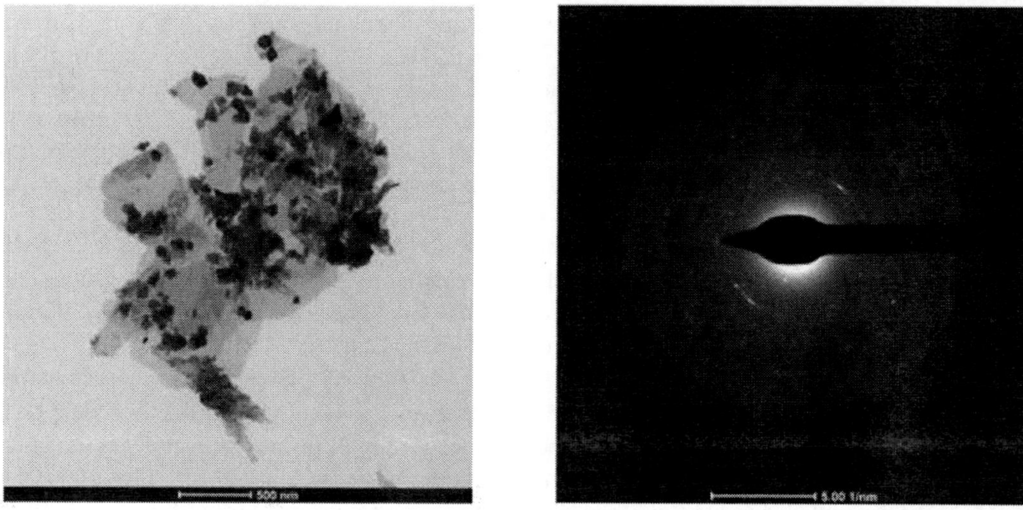

(a) (a)

**Figure 3.** (a)TEM and (b) SAED pattern $Fe_3O_4$/$TiO_2$/CuO/graphene.

The morphology of the samples were investigated by using TEM measurement. Figure 3 represents the TEM and SAED pattern of $Fe_3O_4$/$TiO_2$/CuO/graphene. As can be seen from the figure that the material $Fe_3O_4$, $TiO_2$, and CuO could be dispersed in the graphene sheets. The SAED pattern indicates that the samples pose good crystallinity.

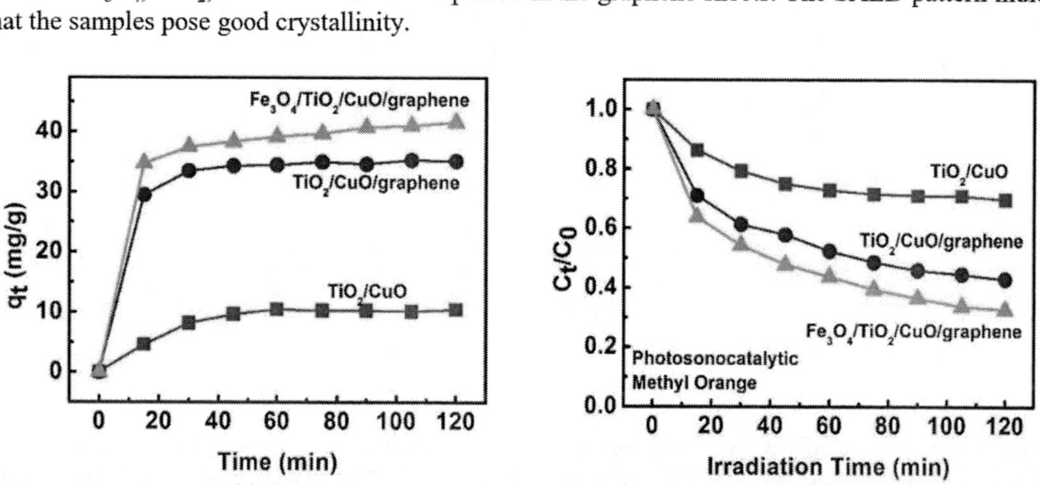


(a) (b)

**Figure 4.** (a) Adsorption capacity and (b) photosonocatalytic activity of $TiO_2$/CuO, $TiO_2$/CuO/graphene and $Fe_3O_4$/$TiO_2$/CuO/graphene

Before investigating the photosonocatalytic performance, we evaluated the adsorption ability of the samples. The adsorption experiment was represented with adsorption capacity vs. time as can be seen in Figure 4a. The amount of dye adsorbed at different time intervals was calculated by using the following equation:

$$q_t = \frac{(C_0 - C_t)V}{m} \tag{1}$$

where $q_t$ is the amount of dye adsorbed per unit mass of adsorbent (mg/g) at time t and $C_t$ is concentration of dye (mg/L) at time t, $m$ is amount of adsorbent (g), and $V$ is amount of dye solution (L). The all prepared samples were significantly adsorbed for early 30 min. After 30 min the adsorption ability achieved the equilibrium stage. The adsorption capacity gradually increased with the addition of graphene and $Fe_3O_4$. The adsorption capacity was found increased from 10 mg/g to 41 mg/g for $CuO/TiO_2$ and $Fe_3O_4/TiO_2/CuO$/graphene respectively. The increase of adsorption ability from $Fe_3O_4/TiO_2/CuO$/graphene is probably due to the higher specific surface area and the higher adsorption capacity of $Fe_3O_4$.

After investigating the adsorption ability of the samples, now we investigated the photosonocatalytic activity for all samples. The result is plotted in Figure 4b. The induction of $Fe_3O_4$ and graphene significantly increased the photosonocatalytic activity for removing organic dye from the aqueous solution. $TiO_2/CuO$ has degradation efficiency about 31%; with the addition of graphene the degradation efficiency increased to 58%. The degradation efficiency further increased with the incorporation of $Fe_3O_4$ with the maximum degradation efficiency around 68%. The higher catalytic performance was probably due to the retarding of recombination electron and holes and the enhancement of the specific surface area. The retarding of recombination electron and holes could promote the interaction between holes and electron with water and oxygen molecules, respectively to produce free radicals for attacking the organic dye.

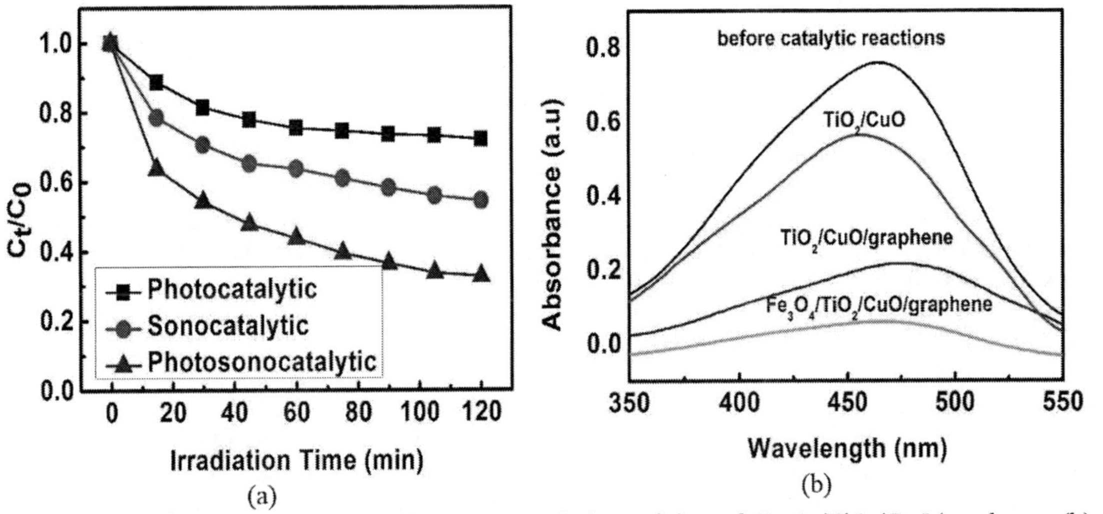


**Figure 5.** (a) photo-, sono-, and photosono-catalytic activity of $Fe_3O_4/TiO_2/CuO$/graphene. (b) Methyl orange absorbance spectra after photosonocatalytic reaction by using $TiO_2/CuO$, $TiO_2/CuO$/graphene and $Fe_3O_4/TiO_2/CuO$/graphene.

Figure 5a represents the comparison between photo-, sono-, and photosono- catalytic performance of $Fe_3O_4/TiO_2/CuO$/graphene. From the figure, we can found that the photocatalytic activity exhibits lowest degradation ability than sonocatalytic and photosonocatalytic process. It could be due to the lowest light penetration under photocatalytic process [11]. The using ultrasound irradiation has the advantage to penetrate any aqueous medium. Therefore, the degradation ability increased [11].

Moreover, after the combination between both photocatalytic and sonocatalytic process, the degradation ability increased significantly. It indicates that the synergy between photo- and sono-catalytic is very potential for waste water removal.

Figure 5b shows the absorbance spectra vs. wavelength of methyl orange at the different catalyst. According to absorbance spectrum of methyl orange, maximum absorbance is at 465 nm. It can be seen from the figure that the adsorption spectra gradually decreased after photosonocatalytic reaction. It indicates that the methyl orange is degraded by using photosonocatalytic process. The methyl orange absorbance from $Fe_3O_4/TiO_2/CuO$/graphene exhibits the lowest absorbance spectra indicating the highest degradation ability for removing methyl orange.

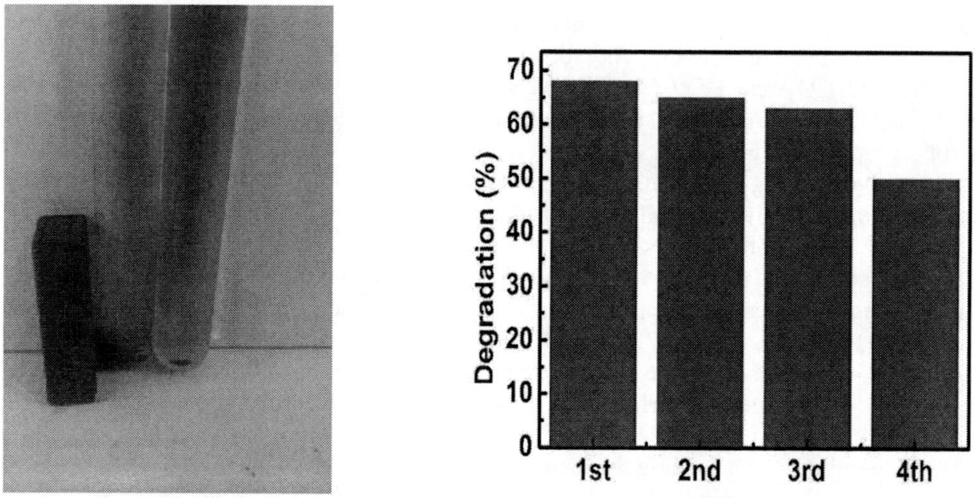


(a) (b)

**Figure 6.** (a) Magnetic separation process. (b) Reusability of catalyst

The reusability is also an important parameter for practical application. The reusability experiment was carried out by using the external magnetic bar to collect the magnetic catalyst. The magnetic separation process is depicted in Figure 6a. As can be seen from the figure, the samples could easily be attracted by the presence of an external magnetic bar. It indicates that the magnetic separation process is an effective method for separating catalyst from the solution. After the separation process, the new MB were poured into the beaker glass containing a catalyst. After that, dark process and photosonocatalytic process were repeated for four-times used. The reusability result is plotted in Figure 6b. From the figure, we can see that the samples possess good stability after four-times cycling process.

## 4. Conclusion

The $Fe_3O_4/TiO_2/CuO$/graphene was successfully synthesized by using the hydrothermal method. The addition of graphene materials could enhance the adsorption and photosonocatalytic performance. The addition of magnetite materials could increase the photosonocatalytic performance as well as the stability of the samples. The enhancement of surface area and retarding or recombination electron and holes are probably the main important aspect of the increasing photosonocatalytic performance. The higher degradation of photosonocatalytic performance compared to photocatalytic and sonocatalytic is due to the synergistic effect of ultrasound and visible light irradiation.

## 5. References

[1] Yao Y C, Dai X R, Hu X Y, Huang S Z, Jin Z 2016 *Appl. Surf. Sci.* **387** 469–476
[2] Hu S Z, Li F Y, Fan Z P, Chang C C 2011 *Appl. Surf. Sci.* **258** 182–188
[3] Wei N, Cui H, Song Q, Zhang L, Song X, Wang K, Zhang Y, Li Y, Wen J, Tian J 2016 *Appl.*

*Catal., B* **198** 83–90
[4] Ghassemi H, Harlow W, Mashtalir O, Beidaghi M, Lukatskaya M R, Gogotsi Y, Taher M L 2014 *J. Mater. Chem. A* **2** 14339–14343
[5] Wang Z, Liu B, Xie Z, Li Y, Shen Z Y 2016 *Ceramics International* **42** 13664–13669
[6] Luna A L, Valenzuela M A, Justin C C, Vázquez P, Rodriguez J L, Avendano J R, Alfaro S, Tirado S, Garduno A, la Rosa J M D 2016 *Appl. Catal., A* **521** 140–148
[7] Yu Y H, Chen Y P, Cheng Z 2015 *Int. J. Hydrogen Energy* **40** 15994-16000
[8] Yang Y, Xu L, Wang H, Wang W, Zhang L 2016 *Mater. Des.* **108** 632–639
[9] He G L, Zhong Y H, Chen M J, Li X, Fang Y P, Xu Y H, 2016 J. *Mol. Catal. A: Chem.* **423** 70–76
[10] Jia X, Dai R, Lian D, Han S, Wu X, Song H 2017 *Appl. Surf. Sci.* **392** 268–276
[11] Zhou M, Yang H, Xian T, Yang Y, Zhang Y 2015 Chinese Journal of Catalysis 36 1987–1994
[12] Taufik A, Saleh R 2016 Mater. Sci. Forum **864** 106-111
[13] Shang Y S, Li T, Li H, Dang A, Zhang L, Yin Y, Xiong C, Zhao T 2016 Composites Part B **99** 106-111
[14] Li Z Q, Wang H L, Yun Zi L, Zhang J J, Zhang Y S 2015 *Ceram. Int.* **41** 10634–10643
[15] Sonia S, Jayram N D, Kumar P S, Mangalaraj D, Ponpandian N, Viswanathan C 2014 *Superlattices and Microstructures* **66** 1–9
[16] Lin C C, Ho J M 2014 *Ceram. Int.* **40** 10275–10282
[17] Song J, Fan Q, Zhu Q, Wang R, Dong Z 2016 *Mater. Lett.* **165** 14–18

IOP Conf. Series: Materials Science and Engineering **202** (2017) 012079 doi:10.1088/1757-899X/202/1/012079

# Synthesis and Characterization of Acrylic-Based Photopolymer as a Candidate for Denture Base Material

**S T Wicaksono, Rasyida and H Ardhyananta**

Materials and Metallurgical Engineering, Faculty of Industrial Technology, Sepuluh Nopember Institute of Technology
Jl. Arief Rahman Hakim, Surabaya 60111 Indonesia

Email: sigit@mat-eng.its.ac.id

**Abstract:** Denture base is a denture part that rests on the soft tissue covering the jawbone and becomes an anchor of a denture. The material that commonly used for this purpose is poly (methyl methacrylate). However, it lacks in mechanical properties due to high water absorption. The aim of this research was to improve the physical and mechanical properties of poly (methyl methacrylate) by making a copolymer with styrene via photopolymerization process. In this method was used the addition of styrene monomer at 10, 20, 30, 40, and 50 wt% into the acrylic resin to form copolymer materials via photopolymerization process. The amount of 1.5 wt% Irgacure 784's photoinitiator was added as a photoinitiator. The results showed that the addition of 40% by weight of styrene copolymer is the best performance compare to the addition styrene of 10, 20, 30, and 50%. The samples with an addition styrene of 40 wt% showed excellent properties such as high water absorption value of 2.405 μg/mm$^3$, the solubility of 0.434 μg/mm$^3$, the flexural strength of 69.336 MPa, a flexural modulus of 1.236 GPa, and a hardness value of 82.583 HD. *Poly (methyl methacrylate-co-styrene)* copolymer with the addition of styrene 40 wt% has the closest value to the requirements for a denture base material.

**Keywords:** denture base, poly(methyl methacrylate-co-styrene), photopolymerization, Irgacure784.

## 1. Introduction

Many cases of tooth loss tend to increase along with the increasing growth rate of a country's population. It indicates that the use of dentures is growing. According to the Health Research conducted by the Ministry of Health of the Republic of Indonesia in 2007, the percentage of denture prostheses use in Indonesia reached 4.5% of the population, and 14.5% of the denture use was by the elderly over the age of 65 years [1]. The increasing use of denture prostheses also increased the demand for denture base [2].

Denture base is used as an anchor or base of the denture in soft tissue. Denture base must have biocompatible properties with the environmental bodies and have good mechanical properties so that it can withstand the forces due to the process of chewing and wear resistant to oral conditions [3]. In the fabrication of denture base, one type of polymer material that is frequently used is poly (methyl methacrylate), or PMMA. PMMA has excellent mechanical properties, but it has some drawbacks, such as poor water resistance and low mechanical strength due to water absorption [4].

Several attempts were made to improve the mechanical properties of PMMA, such as modifying the chemical structure and changing its polymerization techniques. The modification techniques commonly used a chemical structure of copolymer. In PMMA, the addition of copolymers does not only modify the physical properties of the conventional PMMA but also serves to improve toughness and impact resistance and prevent crack propagation [5]. Nearly 95% of denture base material uses conventional polymerization method by heat or heat polymerization [6]. This approach has several disadvantages such as the requirement of energy (heat), low resistance to cracking, and the requirement of a complex mold design (mold). Photopolymerization Method is one novel method that has been used in dentistry to minimize those adverse effects resulted in such conventional polymerization methods [7]. When compared with the use of thermal polymerization, the polymerization process by using light has several advantages including small space requirement, less energy requirement, low temperature, waste-free (do not use solvent), and a simple work system [8]. The aim of this research was to improve the physical and mechanical properties of PMMA to form a copolymer of *poly (methyl methacrylate-co-styrene)* via photopolymerization process as the candidate material for denture base.

## 2. Experimental Method

The two processes were done in this study include the powder synthesis of poly (methyl methacrylate) as a powder component of acrylic resin and the copolymer synthesis of *poly (methyl methacrylate-co-styrene)* as a candidate material for denture base.

### *2.1 Synthesis of Poly-(methyl methacrylate) Powder*

The synthesis process of poly (methyl methacrylate) powder was done for the purpose of producing the powder component in the acrylic resin and increasing the viscosity of the base material on the photopolymerization process. At this stage, methyl methacrylate monomer and heat-initiator benzoyl peroxide were used. The synthesis process began with the mix of methyl methacrylate monomer and 0.5% benzoyl peroxide by weight for 10 minutes at room temperature. Subsequently, the mixture was polymerized at a temperature of 85-90 $^{\circ}$C for 1 hour. Then, the formed solid poly (methyl methacrylate) was crushed into fine powders. The poly (methyl methacrylate) was then characterized using the instrument of FTIR test and viscosity measurement (to determine the value of the resulted polymer molecular weight).

### *2.2 Synthesis of Poly-(methyl methacrylate-co-styrene)*

The acrylic resins used was a mixture of the powder component and the liquid ratio with the comparison value of 1:3. The component of powder used was a powder poly (methyl methacrylate) which had been synthesized. Additionally, the liquid component consisted of a mixture of methyl methacrylate monomer and styrene with the addition of styrene varied to 10, 20, 30, 40, and 50 wt%. The acrylic resin that had been formed was then added an initiator of Irgacure 784, 1.5 wt%. Then the mixture was stirred for 10 seconds until it became homogeneous. The technique used was a fluid resin technique. The following photopolymerization process was carried out using the help of LED light curing units (LED-LCUs) via continuous curing methods (900 mW/cm$^2$ power density). This process was done gradually (layer by layer). The layering process was done by pouring the acrylic resins into the mold until it reached a thickness of 1-1.5 mm and then given up to a half polymerized blue light (half-cured). The allowed thickness of the acrylic resins was about 2-2.5 mm so that the light can get into the bottom layer. The light intensity should be recorded, while the LED LCUs located as close as possible to the surface of acrylic resin into the mold of ± 1 cm in order to harden the entire layer [9]. Furthermore, on the surface of the polymerized resin, the acrylic resin was poured back by using the same procedure. It was repeatedly done until the thickness of the polymer was synthesized to meet the allowed thickness of the mold. The white light exposure process was performed until the acrylic resin had been polymerized completely in the last layer. The copolymer of *poly (methyl methacrylate-co-styrene)* was then characterized using the instruments of FTIR, SEM, water absorption, flexural, and hardness test.

## 3. Results And Discussion

### *3.1 Characterization of Poly (methyl methacrylate) Powder*

The comparison of infrared spectra of monomers methyl methacrylate (MMA) with poly (methyl methacrylate) (PMMA) FTIR test results is shown in Figure 1.

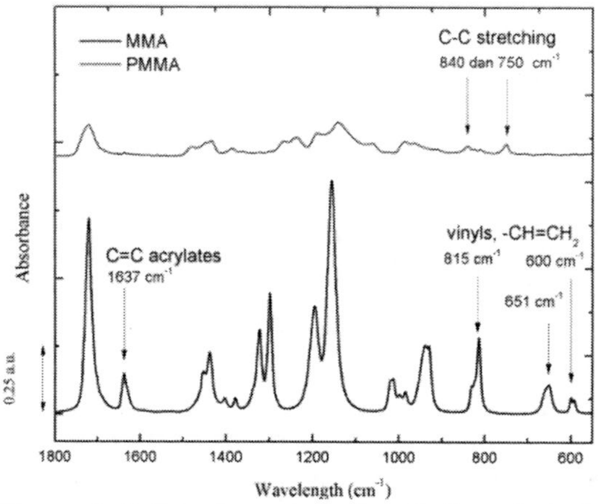


**Figure 1.** Infrared spectra of methyl methacrylate and poly (methyl methacrylate)

Based on the infrared spectra presented in Figure 1, there was a change in the absorption peak of methyl methacrylate in the forms of monomers and polymers. The shift was a shift in the intensity of the absorption in the wavelength of ~1637 $cm^{-1}$ (group C=C acrylates), ~ 600, 651, and 815 $cm^{-1}$ (-CH=CH2 vinyl group), and ~750 and 840 $cm^{-1}$ (C-C stretching group) [10]. At the wavelength of ~1637 $cm^{-1}$, the intensity of the absorption decreased in the form of a polymer. In the form of monomers, there appeared the wavelength absorption peak at ~600, 651, and 815 $cm^{-1}$. In addition to the polymer, the wavelength of ~750 and 840 $cm^{-1}$ appears on the wave absorption peak which did not appear in the infrared spectrum monomer. It indicated the termination of the C=C bond to -C-C- bond into the polymerization process. The severance of the bond indicated an addition to polymerization process performed to turn the methyl methacrylate into poly (methyl methacrylate) [11].

The results of the molecular weight calculation using the viscosity approach are given in Table 1.

**Table 1.** Molecular weight value of poly (methyl methacrylate) powder

| Intrinsic Viscosity | Viscosity Parameter | | Molecular weight ($M_w$) |
|---|---|---|---|
| | K ($10^{-3}$ mL/g) | α | |
| 36.76 | 7.7 | 0.70 | 180.115 |

The table showed the value of molecular weight poly (methyl methacrylate) powder, where K and α is a constant viscosity parameter of poly (methyl methacrylate) when dissolved in acetone using the method of viscosity [12]. The molecular weight poly (methyl methacrylate) which acted as the powder component in an acrylic resin might affect the mechanical properties of the denture base. The result of the molecular weight synthesized poly (methyl methacrylate) calculation in the research was 180.115 g/mol. It will tend to form long chain polymers with excellent physical and mechanical properties [13] with a molecular weight of above 100.000.

*3.2. Characterization of Copolymer Poly (methyl methacrylate-co-styrene)*

The infrared spectrum FTIR test results with a variety of additional styrene monomer in the copolymer poly (methyl methacrylate-co-styrene) were showed in Figure 2. In this figure, it appeared that the changes in FTIR spectra indicated the forming of a styrene functional group in the polymer chain. There were some changes to wave absorption intensity at a particular wavelength along with the addition of styrene [14] on the spectrum. Such changes occured to the certain wavelengths of 700 and 780 cm$^{-1}$ which indicate the group mono-substituted benzene in styrene, 1000-1300 cm$^{-1}$ indicating the C-O group in the area of methyl methacrylate, 1600 cm$^{-1}$ which inicating the C=C aromatic ring area of benzene in styrene, and at a wavelength of 1720 cm$^{-1}$ indicating C=O groups on the carboxyl methyl methacrylate.

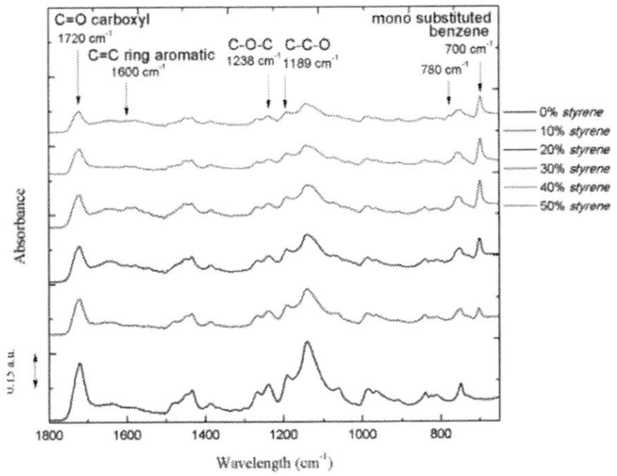


**Figure 2.** Infrared spectra of *poly (methyl methacrylate-co-styrene)* with a variation of styrene addition

Mono-substituted benzene group has two strong wavelengths of wave absorption, i.e. the wavelengths of 780 cm$^{-1}$ and 700 cm$^{-1}$. Usually, only one wavelength alone raises the intensity of absorption in the spectrum [15]. Such fact indicated that the rate of absorption at the wavelength of 700 cm$^{-1}$ was low and could only be identified by a high level monomer styrene addition (50%) to the copolymer, while the addition of low-level monomer styrene (10%) at the wavelength of 780 cm$^{-1}$ to copolymers could already bring the peak in the spectrum of FTIR results. The presence of styrene which is included in the chain methyl methacrylate also caused dilation of peaks indicating C-O bond in methyl methacrylate groups [14]. It can be seen in the FTIR results presented in Figure 2 where the area from 1100 to 1300 cm$^{-1}$ indicated the C-O bond dilation peak along with the addition of styrene. The peak was widening mainly at the wavelength of 1189 cm$^{-1}$, which indicated the C-C-O stretching peak, and at the wavelength of 1238 cm$^{-1}$, which indicated the C-O-C bending. The wavelength of ~1600 cm$^{-1}$ indicated an area that showed the C=C stretching vibration in the aromatic ring appeared on styrene, while the wavelength of ~1720 cm$^{-1}$ represented the area that indicated the carboxyl group C=O on the methyl methacrylate. The intensity of wave absorption at ~1720 cm$^{-1}$ decreased along with the increasing addition of styrene concentrations. Along with the addition of styrene, there appeared a wavelength peak at ~1600 cm$^{-1}$ indicating the benzene aromatic ring groups, but the value was small. The phenyl bond at 1600 cm$^{-1}$ was weaker when compared with the carbonyl bond at 1720 cm$^{-1}$ in the spectrum [14]. Figure 3 showed porosities that appeared in the cross-section copolymers with 1,000 times of magnification. The pictures show the porosity, which was indicated by a red circle, either on the polymer poly (methyl methacrylate) shown in Figure 3a or *poly (methyl methacrylate-co-styrene)* copolymer presented in Figure 3b until 3f. The addition of styrene monomer to 30 wt% (Figure 3d) caused the appearance of more cavity than other variations of additions. Then, the addition of 50 wt% of styrene (Figure 3f) found that the cavity had a larger size than that in the addition of 40 wt% of styrene, and there were also small-sized cavities around the large cavity.

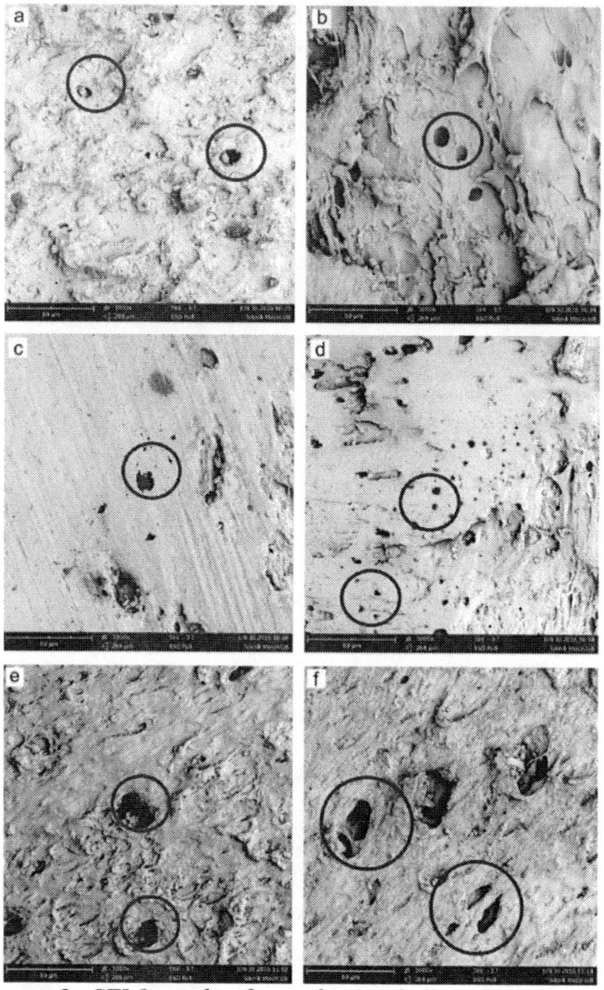


**Figure 3.** SEM results for *poly (methyl methacrylate-co-styrene)* (a) without the addition of styrene, (b) addition of 10 wt%, (c) the addition of 20 wt%, (d) addition of 30 wt%, (e) the addition of 40 wt%, and (e) the addition of 50 wt% of styrene

Porosity emerged as a result of the rudmentary mixing process between powder and liquid components. It resulted in some areas that contain more resin monomer than others. During the polymerization process, such areas shrank more than the other regions and tend to produce shrinkage cavities [6]. The use of the fluid resin technique with the layer-by-layer system also can cause the synthesis of the copolymer. It was because of the air trapped during the gradual process of the resin casting which also caused the porosity that can arise. The presence of the porosity or cavities on the synthesized copolymers can affect the mechanical properties of the copolymer. The increasing number of cavities that appear led to the decreasing mechanical properties of the polymer [11]. Table 2 and Figure 4 show that the addition of styrene groups to the main chain of the copolymer was able to lower its water absorption performance. The amount of water absorption of denture base depends on the degree of hydrophobicity and porosity [16]. Methyl methacrylate contains carboxyl groups that can

IOP Conf. Series: Materials Science and Engineering **202** (2017) 012079 doi:10.1088/1757-899X/202/1/012079

form hydrogen bonds with water molecules. With the addition of styrene, copolymer groups can bring hydrophobic interactions with aromatic rings [17]. The water absorption capacity of a denture base not only depends on the polarity of the constituent materials, but diffusion mechanisms also play a role in the process of water molecule entry into the polymer chain [11]. According to the copolymer SEM results shown in Figure 3, the increasing value of water absorption in copolymers with the addition of styrene monomer content of 30 wt% and 50 wt% also caused the cavities that appeared on both additions. The presence of the cavity made it easier for the water molecules to extend into the copolymer which had been synthesized so as to increase the capacity of its water absorption.

**Table 2.** The effect of the styrene addition on water absorption capacity of the copolymer sample

| The styrene addition into the copolymer sample ( wt% ) | The average water absorption value (mg/mm$^3$) |
|---|---|
| 0 | 6.54 |
| 10 | 4.25 |
| 20 | 3.50 |
| 30 | 4.10 |
| 40 | 2.40 |
| 50 | 2.95 |

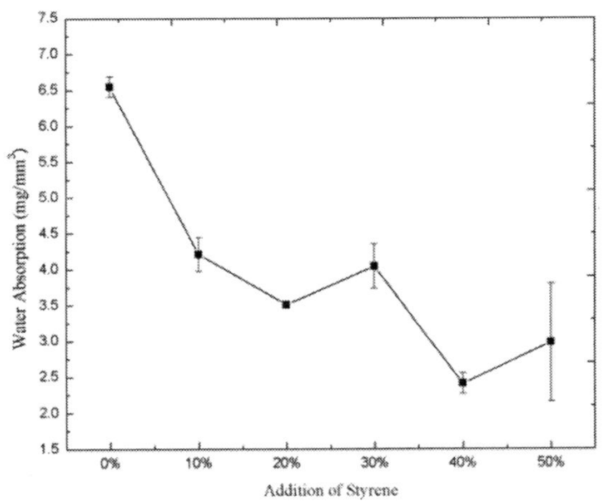


**Figure 4.** The effect of styrene addition on water absorption of the copolymer sample

**Table 3.** The effect of the styrene addition on the copolymer sample solubility

| The styrene addition into the copolymer sample ( wt% ) | The average water absorption value (mg/mm$^3$) |
|---|---|
| 0 | 1.87 |
| 10 | 0.99 |
| 20 | 0.9 |
| 30 | 1.25 |
| 40 | 0.42 |
| 50 | 1.52 |

Table 3 and Figure 5 demonstrated that the addition of styrene to the main chain of the copolymer was able to lower the value of its solubility. The increased value of solubility in the addition of 30 wt% and 50 wt% of styrene indicated a higher level residual monomer when compared to copolymers with the addition of 10, 20, and 40 wt%. The value solubility in the denture base had a correlation with the possible emergence of residual monomer.

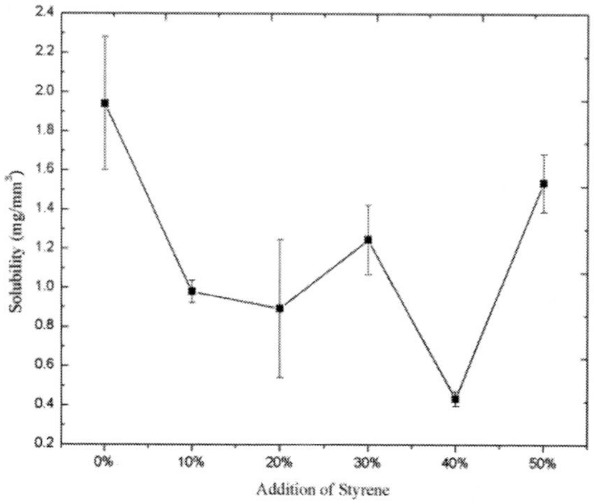


**Figure 5.** The effect of the styrene addition on the solubility of the copolymer sample

There is a correlation between residual monomer and water absorption [16] which would increase the value of absorption and solubility of the residual monomer. Such indication would also be correlated with the results of the mechanical testing. A relation between residual monomer and mechanical properties was also obtained. The residual monomer acted as a plasticizer tended to degrade the mechanical properties of denture base materials [17]. Flexural testing was done because it was considered as most suitable for comparing the denture base materials. The test can reflect the complex pressures that occur during the process of chewing and provides an indication of the strength of the material. Figure 6 shows the effect of the styrene monomer addition on the value of flexural strength and flexural modulus copolymer of *poly (methyl methacrylate-co-styrene)* was successfully synthesized.

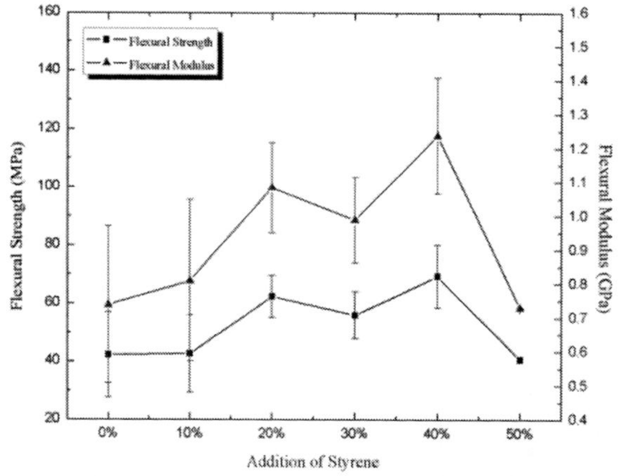


**Figure 6.** The effect of the styrene addition to copolymer *poly (methyl methacrylate-co-styrene)* flexural strength and flexural modulus

Figure 6 indicates that the obtained value of flexural strength and flexural modulus copolymer of *poly (methyl methacrylate-co-styrene)* increased together with the addition of styrene. However, there was a decline in the values of flexural strength and flexural modulus when given the addition of styrene 30 wt% and 50 wt%. Several factors influence the flexural properties of the acrylic resins, such as residual monomer, water absorption, and porosity [18]. It was consistent with the results of water absorption test shown in Figure 4 and the solubility shown in Figure 5. If the water absorption test result was related to its solubility, an indication of residual monomer affects the flexural strength of a denture base because of its plasticizing properties [19]. The plasticizing parts of such plasticizing properties caused the decreased mechanical properties of the polymer. An indication of residual monomer also caused porosity that can affect the mechanical properties [20]. It was consistent with the presence of cavities which were visible on the SEM test results shown in Figure 3.

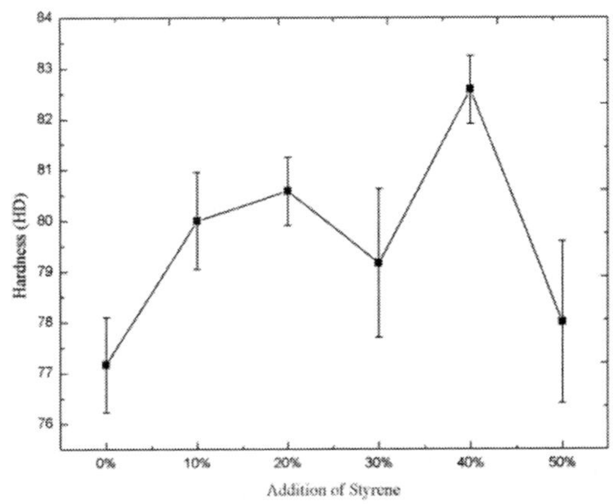


**Figure 7.** The effect of the styrene addition on the copolymer *poly (methyl methacrylate-co-styrene)* hardness

**Table 4.** The effect of the styrene addition on the hardness of the copolymer sample

| The styrene addition to the copolymer sample ( wt% ) | The average hardness value (HD) |
|---|---|
| 0 | 77.3 |
| 10 | 80.0 |
| 20 | 80.5 |
| 30 | 79.2 |
| 40 | 82.5 |
| 50 | 78.0 |

Their benzene rigid group was added into the polymer main chain, along with the addition of styrene monomer, and then the hardness of the modified acrylic compound would show an increasing trend [21]. The hardness values were also associated with the residual monomer in the acrylic denture base resin as shown in Table 4 and Figure 7. The hardness values also decreased along with the increasing value of water absorption. The lessened hardness in the addition of 30 wt% and 50 wt% had a common trend with the values of water absorption, solubility, flexural strength, and flexural modulus.

The results of the test were then compared with the standard *ANSI/ADA Specification No. 12 (ISO 1567) for Denture Base Resin* to determine whether or not the synthesized copolymer fulfills the requisites for denture base material. In Table 5, a comparison of the physical and mechanical properties of copolymers that had been synthesized with the physical and mechanical properties that are necessary to become a denture base material.

Based on Table 5, it was found that the addition of styrene copolymer with a composition of 40 wt% was the closest composition to the necessary requirements determined by *ANSI/ADA Specification No. 12 (ISO 1567) for Denture Base Resin*.

**Table 5.** Mechanical properties of the copolymer and denture base material specifications

| Properties | ISO 1567 | Addition of the styrene monomer | | | | | |
|---|---|---|---|---|---|---|---|
| | | 0% | 10% | 20% | 30% | 40% | 50% |
| Water Absorption (μg/mm$^3$) maximum | 32 | 6.54 | 4.20 | 3.50 | 4.04 | 2.40 | 2.97 |
| Solubility (μg/mm$^3$) maximum | 1.6 | 1.94 | 0.97 | 0.89 | 1.24 | 0.43 | 1.53 |
| Flexural Strength (MPa) minimum | 65 | 42.22 | 42.61 | 62.29 | 55.88 | 69.33 | 40.73 |
| Flexural Modulus (GPa) Minimum | 2.0 | 0.73 | 0.80 | 1.08 | 0.98 | 1.23 | 0.72 |
| Hardness (HD) | - | 77.16 | 80 | 80.58 | 79.16 | 82.58 | 78 |

## 4. Conclusion

The addition of styrene copolymer *poly (methyl methacrylate-co-styrene)* led to a decrease in the value of water absorption and solubility, an increasing value of flexural strength and flexural modulus, and an increasing value of its hardness. The copolymers with the addition styrene of 40 wt% has physical and mechanical properties that are close to the requirements of denture base materials.

## 5. References

[1] Indonesia Health Ministry 2007 Riset Kesehatan Dasar Nasional Tahun 2007 Jakarta
[2] Ningsih, Agustini R, Noor H and Nanang K 2015 Perbedaan Kekuatan Tekan Basis Gigi Tiruan Berbahan Termoplastik Nilon pada Beberapa Ketebalan Jurnal Ilmiah FKG Universitas Muhammadiyah Surakarta
[3] Karthick R, Sirishiha P and Ravi Sankar M 2014 Mechanical and Tribological Properties of PMMA-Seashell Based Biocomposite for Dental Application 3rd International Conference on Material Processing and Charcterization doi: 10.1016/j.mspro.2014.07.234
[4] Kurata, Shigeaki, Morishita K, Shimoyama K and Uemoto K 2008 Basic Study on the Application of Novel Functional Monomer to a Denture Resin. Dental Materials Jurnal **27**(2) 273-277
[5] Hashem, Mohamed, Samah O, Alsaleem, Assery M K, Abdeslam E B, Vellappally S, dan Sukumaran Anil 2014 A Comparative Study of the Mechanical Properties of the Light-cure and Conventional Denture Base Resins OHDM **13** No 2
[6] Manappallil, and John J 2003 Basic Dental Material New Delhi : Jaypee Brothers Medical Publisher (P) Ltd.
[7] Chiu S H, Wicaksono S T, Chen K T and Pong S H 2014 Morphology and properties of a photopolymer/clay nanocomposite prepared by a rapid prototyping system *Sci Eng Compos Mater* **21**(2): 205–210
[8] Wicaksono S T, Chiu S, Chen K and Pong S H 2016 Effect of nano-$BaTiO_3$ on thermal, mechanical and electrical properties of HDDA/TPA prepared by digital light processor RP machine, *Sci Eng Compos Mater* **23**(4) (a head of print)

[9] Susanto, Annette A 2005 Pengaruh Ketebalan Bahan dan Lamanya Waktu Penyinaran terhadap Kekerasan Permukaan Resin Komposit Sinar. Majalah Kedokteran Gigi **38** No 1 (2005): 32-35

[10] Stuart B 2004 InfraRed Spectroscopy : Fundamental and Applications. USA : John Wiley & Sons

[11] Anusavice, Kenneth J, Chiayi S dan Ralph R S H 2013 Phillips' Science of Dental Materials, 12th Ed. USA : Elsevier Science

[12] Wagner H L 1986The Mark-Houwink-Sakurada Relation for Poly(methyl methacrylate) *J. Phys. Chem. Ref.* Data **16** 2

[13] Kawaguchi T, Lippo V J L, Ai T, Yukata T and Pekka K V 2011 Influence of Molecular Weight of Poly(methyl methacrylate) Beads on the Properties and Structure of Cross-linked Denture Base Polymer *Journal of The Mechanical Behavior of Biomedical materials* **4** 1846-1851

[14] Mori S 1989 Compositional Analysis and Infrared Spectra of Styrene- Methyl Methacrylate Random Copolymers *Journal of Applied Polymer Science* **38** 547-555

[15] Socrates G2002 Infrared and Raman Characteristic Group Frequencies 3rd Ed. USA : John Wiley & Sons, Ltd

[16] Shah J, Nilesh B, Shilpa K, Riddhi S, dan Dilip K 2014 Comparative Evaluation of Sorption, Solubility and Microdardness of Heat Cure Polymethylmethacrylate Denture Base Resin & Flexible Denture Base Resin *Journal of Clinical and Diagnostic Research* **8(8);** ZF01-ZF04; doi : 10.7860/JCDR/2014/8707.4770

[17] Pfeiffer P and Ernst-Ulrich R 2004 Residual Methyl Methacrylate Monomer, Water Sorption, and Water Solubility of Hypoallergenic Denture Base Materials *Journal of Prosthetic* Dentistry **92** no 1; doi: 10.1016/j.prosdent.2004.04.003

[18] Saen-isara, Tassanaporn, Sirasa Y, Surachai D, Niwat A, Toemsak S, Siriporn T, Theeralaksna S and Wassana W 2013 Comparative Study of the Flexural Strength and Flexural Modulus of Local Made Orthodontic Resin : A Pilot Study *Journal of Advanced Materials Research* **746** pp 303-307; doi : 10.4028/www.scientific.net/AMR.746.303

[19] Gharechahi, Jafar, Nafiseh Asadzadeh, Foad Shahabian, dan Maryam Gharechahi 2014 Flexural Strength of Acrylic Resin Denture Bases Processed by Two Different *Methods Journal of Dental Research Dental Clinics Dental Properties* **8**(3): 148-152; doi: 10.5681/joddd.2014.027

[20] Dogan A, Bek B, Cevik N N and Usanmaz A 1995 The Effect of Preparation Condition of Acrylic Denture Base Materials on the Level of Residual Monomer Mechanical Properties and Water Absorption *Journal of Dentistry* **23**(5); 313-318

[21] Zhang, Yanhong, Yan Wang, Xianchao Kong, dan Dongmei Z 2014 The Polarity of Poly(methyl methacrylate) Copolymers. *Journal of Applied Mechanics and Material* Vols **687-691** pp 4411-4414 doi: 10.4028/www.scientific.net/AMM.687-691.4411

# Composite Based Chitosan/Zinc-Doped HA as a Candidate Material for Bone Substitute Applications

**S T Wicaksono, A Rasyida, A Purnomo, N N Pradita, H Ardhyananta and M I P Hidayat**

Department of Materials and Metallurgical Engineering, Faculty of Industrial Technology, Institut Teknologi Sepuluh Nopember (ITS) Surabaya, Indonesia, 60111

Email: amaliyarasyida@gmail.com

**Abstract**. The composite based Zinc-doped in Chitosan/Hydroxyapatite was successfully prepared by wet mixing method through the addition of 10, 15, and 20wt% of chitosan. The addition of Chitosan increased the compressive strength and the modulus elasticity. However, it decreased the density and the surface hardness of HA-Zn. Mechanical characterization revealed that these composites are suitable as a candidate of a cancellous bone substitute. Composite with 10% chitosan has compressive strength and modulus elasticity of 57.03 MPa and 0.15 GPa, respectively. Hence, it has the physical and mechanical properties that meet the standards as a cancellous bone substitute material. Also, in vitro biocompatibility test against BHK-21 cells exhibited non-toxic materials.

**Keywords**: composite, chitosan, zinc, HA, bone

## 1. Introduction

Tissue engineering (TE) is a multidisciplinary field focused on the development and application of knowledge in chemistry, physics, engineering, life and clinical sciences to the solution of medical problems, such as tissue loss and organ failure. It involves the fundamental understanding of structure and function relationships in normal and pathological tissues as well as the development of biological substitutes that restore, maintain or improve tissue function [1].

Different efforts have been addressed to the development of research in the field of biomaterials namely by using various modifications in the manufacture of synthetic biomaterials. Composites comprising calcium phosphates and natural biopolymers are widely used as biomaterials for bone tissue repair and engineering [2]. Hydroxyapatite, $Ca_{10}(PO_4)_6(OH)_2$, has been used as a principal inorganic component of synthetic materials for orthopedic application for many years. However, this bioceramics are fragile, and the structures are unstable when mixed with body fluids or blood of patients [3,4]. Meanwhile, the addition of ions such as Zn, Mg, and Si has been performed to improve the mechanical properties of HA. Zn ion addition to the HA bioceramics can effectively improve the mechanical properties as compared to Mg and Si ions [5].

Moreover, natural polymer such as chitosan has been extensively studied and used in many tissue engineering applications as well, such as for bone substitution [1,5]. This biopolymer is widely used to modify the mechanical properties of Hydroxyapatite (HA)-based composite [5-7]. In addition, chitosan is bioresorbable, biocompatible, non-toxic, non-antigenic, biofunctional and osteoconductive which can accelerate the growth of osteoblasts.

The aim of this research was to improve the mechanical properties of HA by doping Zn ion, followed by adding different amounts of chitosan, making materials potentially interesting for tissue engineering applications.

## 2. Experimental Method

Chitosan derived from crab with a degree of deacetylation of 84% obtained from CV. Biochitosan Indramayu, West Java. Zn-doped HA was prepared by sol-gel method in accordance with the reference [8]. The synthesis process of composite Chitosan / HA-Zn is performed using a wet mixing method. Chitosan was first dissolved in 100 mL of 3% acetic acid solution and then stirred at a constant temperature of 700 °C for one hour. HA-Zn powder was then added and stirred for 1 hour at a temperature of 700 °C. The composites prepared with weight percent ratio of chitosan / HA-Zn WERE 10/90; 15/85; 20/80.

The functional group characterization of the composite was performed using the Thermo Scientific Nicolet iS 10 (Thermo Fisher Scientific Inc., MA, USA) with a wavelength of 4000-400 $cm^{-1}$. X-Ray Diffraction (XRD) were recorded using X'Pert PRO PANanalytical at room temperature by using filtered Cu K$\alpha$ radiation ($\lambda$=1.54 Å). Scanning electron microscopy (SEM) was carried out using PhenomProX (Phenom-World, Eindhoven, NL) on the fracture surfaces of the specimens, previously coated by sputtering with gold and by using a SEM apparatus equipped with energy dispersive spectroscopy (EDS).

The measurement of bulk density was performed according to DIN 2738 for testing in determining the density of the porous material.Compression mechanical properties of samples were evaluated using a traction-compressive device. Compression molded cylindrical samples were used and positioned between parallel plates. Hardness testing was conducted in accordance with ASTM C 1327 – 08, using Vickers indenter with a diameter of 2.5 mm, a load of 30 kg and indentation time for 10 seconds.

Biocompatibility of composite chitosan/HA-Zn was evaluated through an in vitro cytotoxicity test in accordance with reference [1] using BHK-21 Cells. The samples were incubated at a temperature of 37 °C for 24 hours.

## 3. Results and discussion

### *3.1. X-Ray Diffraction*

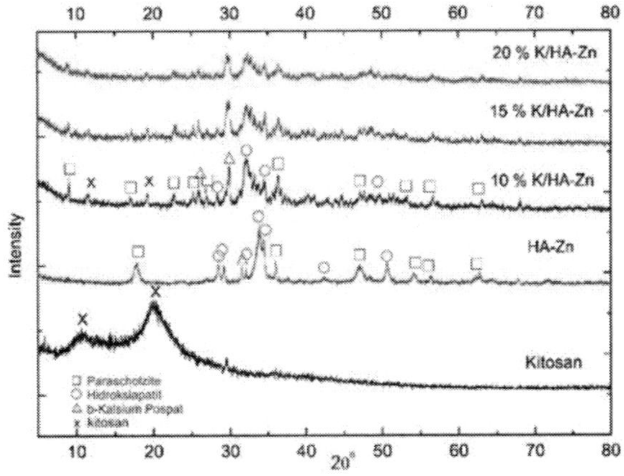


**Figure 1.** XRD Pattern of Chitosan, HA-Zn, and its composite

Figure 1 shows the XRD pattern of chitosan and HA-Zn and its composites, it can be identified that there were peaks belong to HA-Zn and Chitosan. Chitosan is characterized by a typical peak at 2θ 9-

IOP Conf. Series: Materials Science and Engineering **202** (2017) 012080 doi:10.1088/1757-899X/202/1/012080

10° and 19-20° [9]. It can be evaluated that its composite has a shifted peak and lower intensity. Likewise, the HA-Zn peaks show more ramps along with the increased addition of chitosan added. It indicates the occurrence of hydroxyapatite and chitosan molecular interactions [5,10].

### *3.2. Fourier transform infrared spectroscopy (FTIR)*

The functional group characterizations are shown in Figure 2. It can be seen that HA-Zn and Chitosan have the same spectral pattern with their composite. The uptake in the waves that indicated a chitosan identified by OH peak at around 3400 $cm^{-1}$ and NH peak at around 1600 $cm^{-1}$. In addition to chitosan, phosphates, hydrogen phosphate, and carbonate, which indicated HA appear in the range of 1000, 900, and 1400 $cm^{-1}$, respectively. Moreover, the OH peak of the composite in the range of 3400 $cm^{-1}$ increased intensively along with the increasing addition of chitosan. In addition, the amide peak at around 1600 $cm^{-1}$ showed the same behavior. These results are in agreement with previous XRD data proved that the synthesis of K/HA-Zn composite via sol-gel method was successful without any formation of a new compound. The decreasing intensity and broadening of the peak indicated the interactions between the constituent elements of the composite, such as hydrogen bonding between chitosan and hydroxyapatite [11]

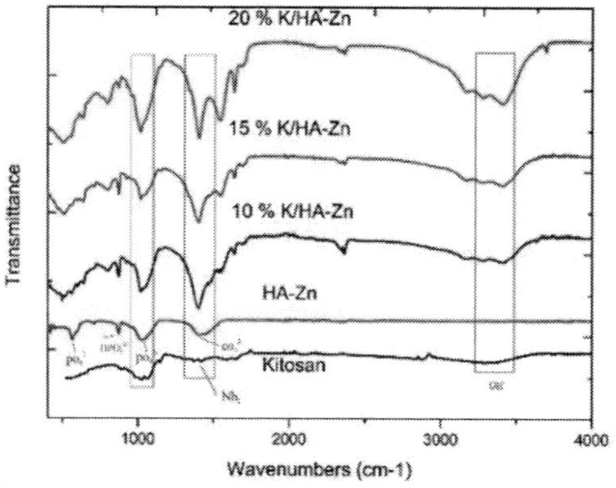


**Figure 2.** FTIR patterns of chitosan, and HA-Zn and its composite

### *3.3. Scanning Electron Microscopy (SEM)*

The Scanning Electron Microscopy of Figure 3a revealed that the HA-Zn particles have morphology as a rock. The results of the elemental analysis performed on the HA-Zn elements showed the presence of the elements of Ca, P, O and Zn as ions doped. The Ca / P ratio of HA-Zn was calculated to be 1.81, which was greater than the Ca / P pure hydroxyapatite which was 1.67. It was likely to occur because the carbonate groups replaced phosphate groups in the structure of hydroxyapatite. Although there were not materials having carbonate groups in the process of synthesis, it can form inside the apatite [12].

Figure 3b shows that there is a chitosan compound characterized by the presence of fibers connecting the HA-Zn particles to one another. In Chitosan-HA composites, a HA particle will act as a filler which is dispersed evenly when there is a high amount of chitosan in the composites. However, along with the decrease of chitosan, it may not cover the HA particles and tend to act as a binder [12].

Figure 4 show that chitosan was able to bind HA-Zn particles. Figure 4B illustrates the presence of particles that agglomerated HA-Zn and spread evenly in a relatively small form. This form seemed bigger in Figure 4C and 4D when the amount of chitosan addition increased to 15% and 20%. From these, it can be assumed that the greater the amount of Chitosan added, the higher its ability to bind HA-Zn particles.

IOP Conf. Series: Materials Science and Engineering **202** (2017) 012080 doi:10.1088/1757-899X/202/1/012080

**Figure 3.** SEM patterns of HA-Zn (a) and 20% K/HA-Zn (b)

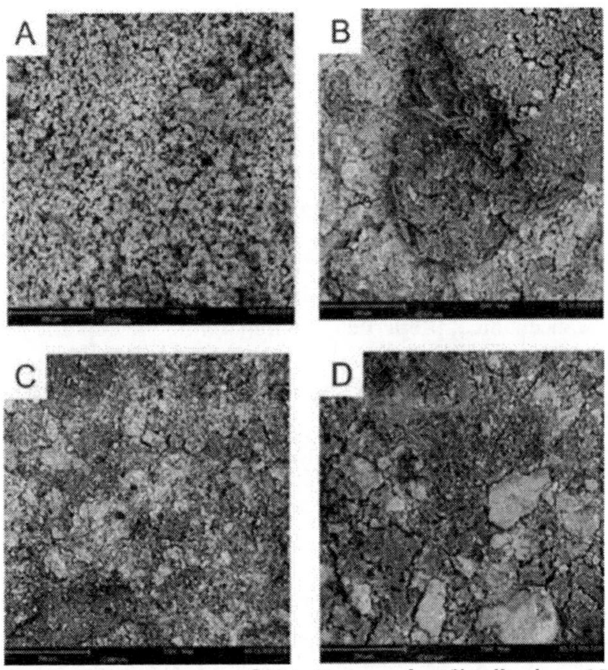


**Figure 4.** SEM patterns of HA-Zn powder distribution, 10% K / HA-Zn (B), 15% K / HA-Zn (C) and 20% K / HA-Zn(D) composites

Table 2 shows that the addition of chitosan against HA-Zn will reduce the density of the composites since chitosan is a polymer which has a specific gravity that is lower than the HA-Zn which is a ceramic material. This result is in agreement with the previous results [13]. Moreover, bone is a composite material consisting of calcium phosphate and biopolymers. Therefore, the bone has a lower density equivalent to 1.8 g / $cm^3$ [14]. Based on this theory, it is known that the densities of all composites were in the range of bone density values.

**Table 1.** The effect of chitosan addition on the densities of composites

| Materials | Density (g/$cm^3$) |
|---|---|
| HA-Zn | 2.74 |
| Composite 10% K / HA-Zn | 1.94 |
| Composite 15% K / HA-Zn | 1.86 |
| Composite 20% K / HA-Zn | 1.85 |

### 3.4. Compression test

The addition of chitosan increased the compressive strength value of the composite (see Table 2). The compressive force of the cortical and cancellous bone was 137.8 MPa and 41.4 MPa, respectively. Thus, the values of the compressive strength of the composite are in the range of cancellous bone compressive strength. Moreover, the modulus of elasticity of the cortical bone was 7-30 GPa, while the elastic modulus of the cancellous bone was approximately 10 - 1500 MPa [14]. Thus, we may say that a composite with 10% chitosan meets the standard value of cancellous bone mechanical properties.

**Table 2.** The effect of chitosan addition on the modulus and compressive strength of composites

| Material | Modulus Elasticity (GPa) | Compressive strength (MPa) |
|---|---|---|
| HA-Zn | 0.29 | 14.05 |
| Composite 10% K/HA-Zn | 0.15 | 57.03 |
| Composite 15% K/HA-Zn | 0.18 | 61.09 |
| Composite 20% K/HA-Zn | 2.25 | 66.93 |

### 3.5. Hardness test

The highest hardness value was obtained from the HA-Zn composite of 29.2 VHN. 20% K / HA-Zn composite had the lowest hardness value of 20.6 VHN. It can be concluded that the composite hardness values decreased together with the increasing amount of chitosan. It was due to the addition of chitosan that lowered the surface hardness of the composite material since chitosan is a polymeric material which has lower mechanical properties as compared to the ceramic material.

**Table 3.** The effect of chitosan addition on hardness of composite

| Material | Hardness (VHN) |
|---|---|
| HA-Zn | 29.2 |
| Composite 10% K/HA-Zn | 25.3 |
| Composite 15% K/HA-Zn | 21..3 |
| Composite 20% K/HA-Zn | 20.6 |

### 3.6. In vitro biocompatibility test

The most important requirement for a material to be used in medical applications is its compatibility, not only regarding physical and mechanical properties but also in those that define its behavior at the time it is in contact with the body. MTT assays were conducted to the composites to test their cytotoxicity. BHK-21 fibroblast cell lines were selected in the MTT assay. The viability of cells that were exposed to composites is summarized in Table 4. MTT results reveal no visible reduction in viability after incubation for 24 hours, presenting that these materials had good biological safety and displayed as almost non-cytotoxic. Also, the cells incubated in the composite suspension can be clearly observed using an inverted microscope (Fig. 5). It is possible to observe that the cell viability is in a similar rate against time of incubation, which was in agreement with the MTT assay. In accordance with the references [5], these composites can be categorized as non-toxic materials.

**Table 4.** Cell viability of HA-Zn and its composites

| Samples | Cell Viability (%) |
|---|---|
| HA-Zn | 33.8 |
| 10% K/HA-Zn | 48.7 |
| 15% K/HA-Zn | 46.2 |
| 20% K/HA-Zn | 48.4 |

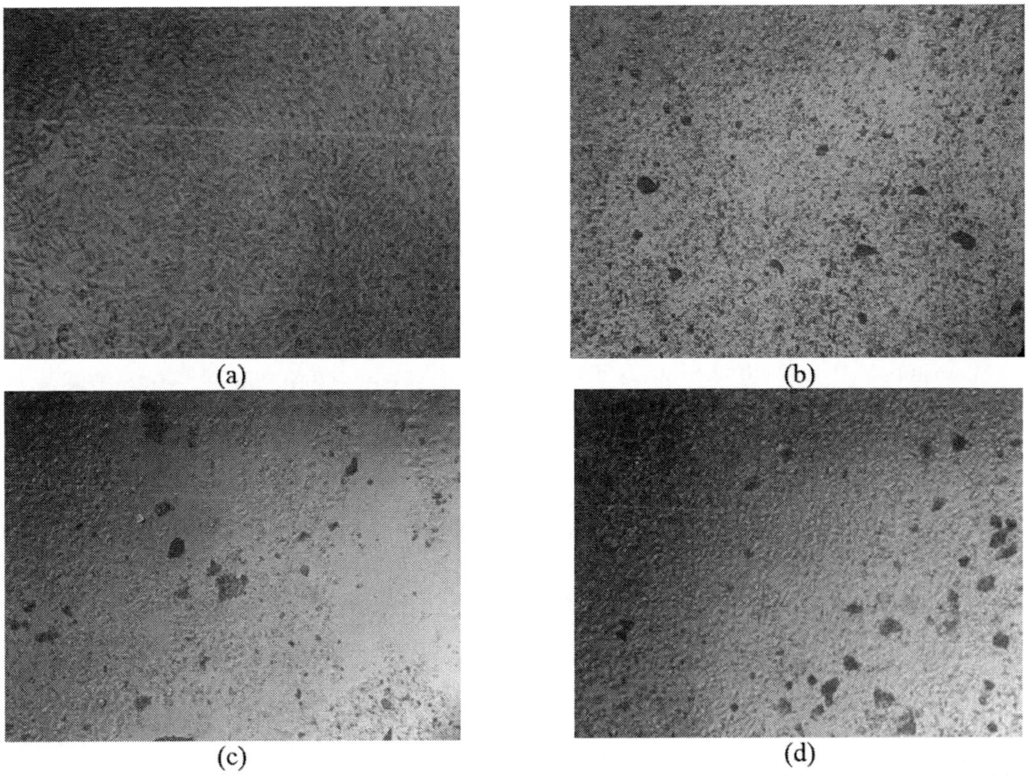


**Figure 5**. Cell phase contrast micrographs, control (a) 10% K/HA-Zn (b) 15% K/HA-Zn (c) 20% K/HA-Zn (d)

## 4. Conclusion

The addition of chitosan resulted in a more evenly distributed bond between chitosan and HA-Zn particles. The density and hardness values of chitosan/HA-Zn composites decreased together with the increasing compressive strength. The composite with 10% Chitosan has physical and mechanical properties that meet the standards of material as a candidate for substitute material for cancellous bone. In addition, biocompatibility test showed that these composites are non-toxic materials.

## 5. References

[1] Fukushima K, Wu M H, S Bocchini, Rasyida A, Yang M C 2012 PBAT based nanocomposites for medical and industrial applications *Materials Science and Engineering C* **32** pp 1331–1351

[2] Danilchenko S N 2009 Chitosan–hydroxyapatite composite biomaterials made by a one step co-precipitation method: preparation, characterization and in vivo tests*Journal of Biological Physics and Chemistry* **9** No.3 2009 pp.119-126

[3] Darmawan W Y D 2008 Sintesis dan karakterisasikomposithidroksiapatit (HA) sebagai graft tulangsintetik*A Scientific Journal For the Application of Isotopes and Radiation*, pp. 143-153.

[4] Samik G 2012 Densification and mechanical characterization of pure and doped hydroxyapatite *National Institute of Technology*.

[5] Istifarah, Aminatun P W 2012 Sintesis dan Karakterisasi Komposit Hidroksiapatit dari Tulang Sotong (Sepia sp.)-Kitosan untuk Kandidat Aplikasi Bone Filler. Surabaya: Universitas Airlangga.

[6] Kong L, Yuan G, Cao, Wenling., Gong, Yandao., Zhou, Naming., Xang, Ziufang 2005

Preparation and characterization of nano-hydroxyapatite/chitosan composite scaffolds *Wiley Inter Science* (Wiley Inter Science) 275-282.

[7] Zhang L 2005 Preparation and in vitro investigation of chitosan/nano-hydroxyapatite composite used as bone substitute materials *Journal of materials science: materials in medicine* 213-219.

[8] Norhidayu D 2008 Development of Zinc Doped Hydroxyapatite for Bone Implant Application *ICCBT - F* – (**24**) pp 257-270.

[9] Arafat A 2015 Preparation and characterization of chitosan from shrimp shell waste*international journal of science and engineering research* **6** 5 pp 538-542, 2015.

[10] Nikhpour M R 2012 Synthesis and characterization of hydroxyapatite/chitosan nanocomposite materials for medical engineering applications*Elsevier. Composites: part B* **43** pp. 1881-1886.

[11] Warastuti Y B E D 2012 Sintesis dan karakterisasi membran komposit Hidroksiapatit Tulang Sapi-Kitosan-poli(vinil alkohol) untuk aplikasi biomaterial *Jurnal sain materi Indonesia*, pp. 1411-1098

[12] Kong L 2005 Preparation and characterization of nano-hydroxyapatite/chitosan composite scaffolds *Wiley Inter Science*, pp. 275-282.

[13] Yildrim 2004 Preparation and Characterization of Chitosan/Calcium Phosphate Based Composite Biomaterials*Izmir Institute of Technology*. Izmir, Turki

[14] Yuehuei H 2000 Mechanical Testing of Bone and the Bone-Implant Interface *New York: CRC PRESS.*

[15] Shi D 2004 Bimaterials and Tissue Engineering-Biological and Medical Physics-Biomedical Engineeirng *New York: Springer*.

**Acknowledgements**
The author would like to acknowledge the PENELITIAN PEMULA PNBP ITS 2016 for funding.

# Investigation on the Mechanical Properties of A356 Alloy Reinforced AlTiB/SiC$_p$ Composite by Semi-Solid Stir Casting Method

**E I Bhiftime[1] and N F D S Gueterres[2]**

[1]Department of Mechanical Engineering, University of Cenderawasih, Kamwolker Street, Jayapura 99351, Papua - Indonesia
[2]Department of Mechanical Engineering, Dili Institute of Technology, Ai-Mutin Street, Dili 55718, East Timor

Email: irianto.bhiftime@yahoo.com

**Abstract.** Employing the semi-solid stir casting method to strengthen MMCs by SiC$_p$ particle was the simplest way in the casting process. The strength and toughness of a composite material can be obtained through a combination of the A356 / Al7Si composites and SiC$_p$ particle. The purpose of this study was to investigate the difference in the mechanical properties of $Al_7Si_1Mg/SiC_p$ and $Al_7Si_1Mg_{1.5}TiB/SiC_p$ composites, as well as the changing effects on the extra level of SiC$_p$ particle. $Al_7Si$ was used as matrix strengthened by SiC$_p$ with the percentage variations of 10, 15, 20 wt%. The additional level of 1.5 wt% AlTiB aimed for increasing the matrix grain refinement. The semi-solid stir casting method was performed to spread the SiC$_p$ particles evenly in the liquid matrix. The results of the study were the tensile strength value of as much as 143 MPa or the increase of 22.16%. The biggest yield strength on the $Al_7Si_1Mg_{1.5}TiB/SiC_p$ composite was averagely as much as 106 MPa. The highest impact toughness was averagely amounted to 4.74 J/mm$^2$. The lowest porosity value was averagely 2.10%. The morphology of the composite between the reinforcing particle and the matrix was able to unite and to be dispersed evenly. The present study was conducted through density test, tensile test, impact test, microstructure test, and SEM.

**Keywords:** Mechanical properties, A356 Alloy, Semi-solid stir casting

## 1. Introduction

Recently, the development of Metal Matrix Composites (MMCs) in science and technology is considered a critical need. The Al-Si composite has a major role in the field of aluminum composite casting. The material has been widely used in the production of automotive, planes, outer space materials and industrial construction because of its excellent resistance. The composite material can be welded easily, and it is resistant towards high corrosion, and it also has other superior properties [1-2].

The production of MMCs combines the aluminum composite of A356 / $Al_7Si$ as matrix and the particle of ceramics silicon carbide (SiC$_p$) as the reinforcement. The addition of the composite of Titanium Diboride (TiB) as the grain refiner is performed because its property could increase the metal strength and toughness [3]. SiC$_p$ is employed as the reinforcement because it has an excellent mechanical property with the density of 3.2 g/cm$^3$, the yield strength of 600 MPa, the hardness of 2480 Knop (2170 HB), and the modulus of elasticity of 400 GPa, and it does not cause oxidation on metal

[4-5]. The A356 composite has several strengths such as having light property (density of 2.7 g/cm$^3$), the yield strength of 172 MPa, and the corrosion resistance. Nevertheless, it has low level of hardness of as much as 60 HB [5]. Moreover, Mg (1 wt%) is added to improve the wettability of $Al_7Si$ matrix towards $SiC_p$ particle. The function of improving the wettability is to wet the entire surface of the reinforcing particle that may affect an increase on the strength of matrix bond with the $SiC_p$ particle [6-7].

The results of the study about $AlSi_7Mg_2$ matrix reinforced by SiC (5-15 wt.%) indicated the attained maximum tensile strength on SiC (10 wt.%) was as much as 280MPa. The hardness value increased on SiC (15 wt.%), i.e. that was 98HB, was increased of as much as 48% [8]. The employment of A356 matrix, the reinforcing SiC particle and the graphite with the SiC (0-9 wt.%) could cause the maximum hardness value on SiC (9 wt.%) of as much as 144VH. However, in this case, the hardness increased, but the elongation decreased [9]. In the case of Al6062 Matrix with the reinforcing particles (5, 10, 15, 20 wt.%), the results of the study showed that the highest hardness value was attained on SiC (20 wt.%), that was 83 HRB [10]. A study about the employment of 98.41 % Matrix with the SiC reinforcing particles (5, 10, 15, 20, 25, 30 wt.%) indicated that the best hardness value was obtained on the SiC percentage of 25%; that was 45.5 BHN [11]. A study investigating AlTiC and AlTiB with variations of 0.2%, 0.5%, and 1% showed that the addition of 1% TiB caused the grain size to become smaller [12]. The Al7178 composite with the variations of TiB (1, 2, 3, 4 wt.%) indicated that, with the addition of TiB between 1-4 wt.%, the grains underwent a significant decrease with the addition of 1 wt.% TiB; that was as much as 140 μm, on 4 wt.% to be 55 μm [13]. A study employing the AlTiB master alloy with the variation from 0.03% to 0.15% indicated that the smallest grain size was 50 μm, attained on the variation of 0.13 % TiB [14].

The mixing process of the $SiC_p$ ceramic particle in the liquid matrix of aluminium composite has two main weaknesses. First, the $SiC_p$ particle of which surface is not usually wetted by liquid matrix. Second, the $SiC_p$ particle tends to settle or float on the liquid matrix surface. It may cause an uneven distribution of $SiC_p$ particles on the matrix surface [6]. The study conducted by employing the semi solid stir casting was aimed to ease the mixing process of $SiC_p$ particle into the liquid matrix [15]. The purpose of this study was to investigate the differences in the mechanical properties of $Al_7Si_1Mg/SiC_p$ composite and $Al_7Si_1Mg_{1.5}TiB/SiC_p$ composite as well as the changing effects toward the addition of $SiC_p$ particle. This present study employed the semi solid stir casting method.

## 2. Experimental Method

The experiment of composite $Al_7Si_1Mg/SiC_p$ and $Al_7Si_1Mg_{1,5}TiB/SiC_p$ composites was done by employing the semi solid stir casting method. In this case, the A356 / $Al_7Si$ alloy played a role as matrix. Besides, 1 wt.% Mg was used to incerase the wettability. Then, 1.5 wt.% of TiB was added to refine the grains on the matrix. Next, The $SiC_p$ played a role as the reinforcement on the $Al_7Si$ matrix. The variations of $SiC_p$ employed are 10, 15, 20 wt.%. Table 1 below shows the compositions of chemical materials used.

**Table 1.** Chemical composition of materials

| Materials | Chemical composition (%) | | | | | | | | |
|---|---|---|---|---|---|---|---|---|---|
| | Al | Si | $SiC_p$ | Fe | Ti | B | Mg | Mn | Other |
| $Al_7Si$ / A356 (ingot) | 92.39 | 7.26 | - | 0.147 | - | - | 0.07 | 0.008 | 0.125 |
| Mg (ingot) | 0.022 | 0.013 | - | 0.003 | - | - | 99.93 | 0.012 | 0.02 |
| AlTiB (ingot) | 93 | 0.16 | - | 0.16 | 5.00 | 0.98 | - | - | 0.05 |
| $SiC_p$ (powder) | 0.03 | - | 98.6 | 0.1 | - | - | 0.03 | - | 1.24 |

There were several steps conducted during the casting process. First, preparing the materials to be casted, including $Al_7Si$, Mg, and AlTiB. Then, those materials were cut by machine according to the variations of the casting compositions. After that, the materials were set aside to get the wt.%

composition set. The variations of the casting compositions could be seen on Table 2. Then, the $Al_7Si$, Mg, and Al-TiB were added to the electric furnace stir casting shown in Figure 1a.

**Table 2.** Variations in the composition of foundry

| Composition (wt.%) | $Al_7Si$ (gram) | Mg (gram) | AlTiB (gram) | $SiC_p$ (gram) | Total (gram) |
|---|---|---|---|---|---|
| $Al_7Si_1Mg/SiC_p$ 0% | 990 | 10 | - | - | 1000 |
| $Al_7Si_1Mg/SiC_p$ 10% | 890 | 10 | - | 100 | 1000 |
| $Al_7Si_1Mg/SiC_p$ 15% | 840 | 10 | - | 150 | 1000 |
| $Al_7Si_1Mg/SiC_p$ 20% | 790 | 10 | - | 200 | 1000 |
| $Al_7Si_1Mg_{1.5}TiB/SiC_p$ 0% | 975 | 10 | 15 | - | 1000 |
| $Al_7Si_1Mg_{1.5}TiB/SiC_p$ 10% | 875 | 10 | 15 | 100 | 1000 |
| $Al_7Si_1Mg_{1.5}TiB/SiC_p$ 15% | 825 | 10 | 15 | 150 | 1000 |
| $Al_7Si_1Mg_{1.5}TiB/SiC_p$ 20% | 775 | 10 | 15 | 200 | 1000 |

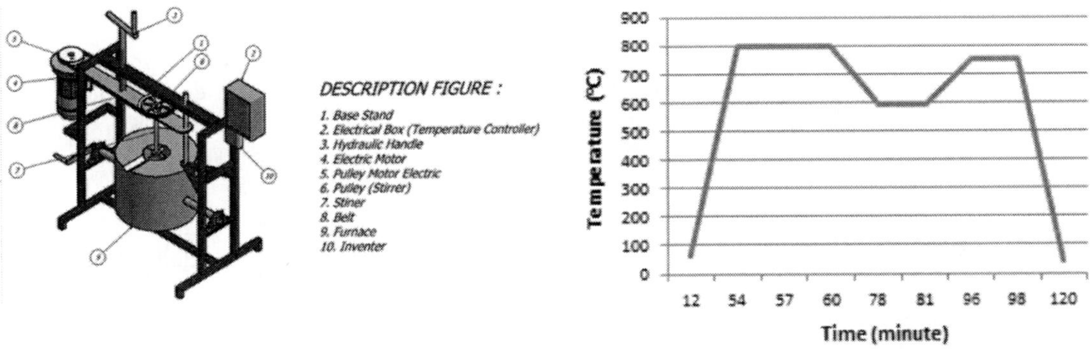


**Figure 1.** (a) Electric furnace stir casting, (b) Temperature vs. time semi solid stir casting process

Then, they were heated to the temperature of 800 °C to get a liquid state. On the other hand, the temperature was lowered to 590 °C to get a semi solid state. Meanwhile, the $SiC_p$ particle was heated separately place inside an electric oven at a temperature of 400 °C for 120 seconds. After that, the $SiC_p$ was added to the electric furnace stir casting and stirred thoroughly by using a mechanical mixer. The mixing speed of the mixer was 500 rpm for 180 seconds. The metal mold was also heated to the temperature of 300 °C. After all materials were mixed and stirred in semi solid state, the temperature was increased to 750 °C to get a casting condition. The casting product was refrigerated at room temperature of 36 °C which could be seen in Figure 1b, the graphic of semi solid temperature vs. time. After that, the composite was cut according to the test specimen. The composite specimen was tested by employing density test, tensile test, impact test, and microstructure photo and SEM. Figure 2 below illustrates the specimen of casting product.

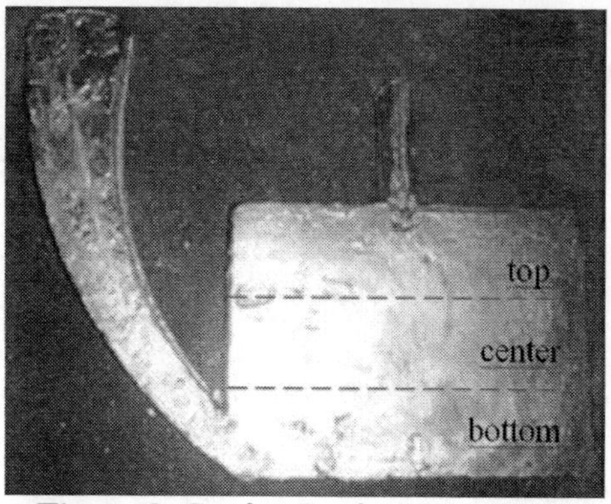


**Figure 2.** Casting Product Composite

## 3. Results and Discussion

### *3.1. Density and Porosity*

The specimens for the density test had the size of 2 x 2 x 2 cm by using a VIBRA digital balance. The density test results and its measurement on the porosity of $Al_7Si_1Mg/SiC_p$ and $Al_7Si_1Mg_{1.5}TiB/SiC_p$ composites are shown on Table 3.

**Table 3.** Density test results and porosity calculations

| Composites (wt.%) | $\rho_{actual}$ (g/cm$^3$) | $\rho_{theoretic}$ (g/cm$^3$) | Porosity |
|---|---|---|---|
| $Al_7Si_1Mg/SiC_p$ 0% | 2.63 | 2.66 | 1.50 |
| $Al_7Si_1Mg/SiC_p$ 10% | 2.65 | 2.71 | 2.58 |
| $Al_7Si_1Mg/SiC_p$ 15% | 2.66 | 2.72 | 2.57 |
| $Al_7Si_1Mg/SiC_p$ 20% | 2.69 | 2.76 | 2.89 |
| $Al_7Si_1Mg_{1.5}TiB/SiC_p$ 0% | 2.64 | 2.68 | 1.49 |
| $Al_7Si_1Mg_{1.5}TiB/SiC_p$ 10% | 2.65 | 2.73 | 2.19 |
| $Al_7Si_1Mg_{1.5}TiB/SiC_p$ 15% | 2.67 | 2.76 | 2.22 |
| $Al_7Si_1Mg_{1.5}TiB/SiC_p$ 20% | 2.70 | 2.78 | 2.51 |

The graphics of the density and porosity of the $Al_7Si_1Mg/SiC_p$ and $Al_7Si_1Mg_{1.5}TiB/SiC_p$ composites are shown on Figure 3.

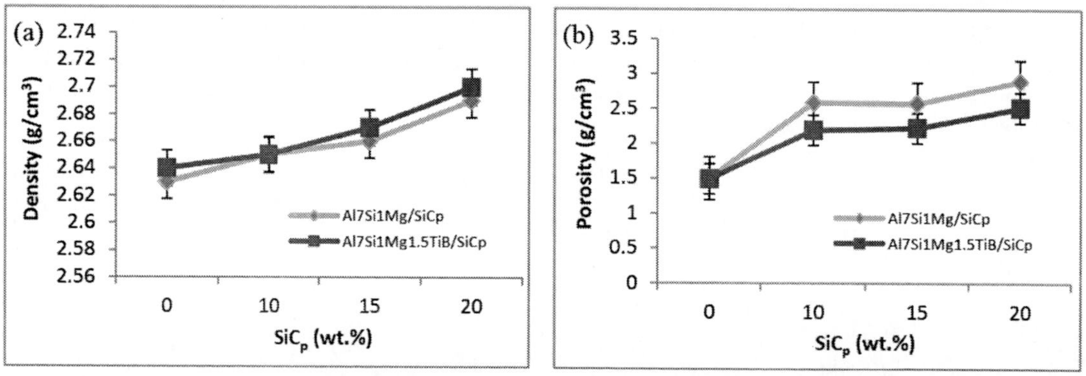


**Figure 3.** Graph (a) composite density, (b) composite porosity

The density graphic indicated that the $Al_7Si_1Mg/SiC_p$ composite underwent an average increase of 1.2%. Whereas, the $Al_7Si_1Mg_{1.5}TiB/SiC_p$ composite went through a decrease on the variation of $SiC_p$

of 15 wt.%, and increased of as much as 1.4% on the variation of $SiC_p$ of 20 wt.%. Figure 3b showed that the porosity of the $Al_7Si_1Mg/SiC$ composite increased averagely of as much as 28 %. Whereas, the $Al_7Si_1Mg_{1.5}TiB/SiC_p$ composite decreased on the variation of $SiC_p$ (15 wt.%), and then, increased of as much as 10.9% on the variation of $SiC_p$ (20 wt.%). On the process of stir casting, the porosity might potentially occur. This occurence may be caused by the chemical reaction between matrix and reinforcing particle, so that it may result in the trapped gas during the mixing process. Consequently, the more the number of $SiC_p$, the more the gas reaction occurs leading to the porosity tendency to increase. These results, according to the results of previous research, indicated that porosity may potentially occur in the process of stir casting because of the chemical reaction between the reinforcing particle and matrix which caused gas, as well as gas which was trapped during the stirring process [6]. The addition of 1% Mg could reduce the porosity. Therefore, Mg would reduce the layer of $SiO_2$ on the surface of $SiC_p$, and form $MgAl_2O_4$ with high wettability on the interface [16]. The porosity formed on the interface of matrix and reinforcing particles would disap+pear or get reduced [17].

### *3.2. Tensile Test*

The tensile strength was acquired from the measurement of maximum load divided by cross-sectional area of tensile test specimen. The tensile test was conducted according to the testing standard of ASTM E 8M-04 [18]. The following Table 4 presents the result of the tensile test.

**Table 4.** Test results of tensil strength measurement

| Composites (wt.%) | $\sigma_u$ (MPa) | $\sigma_y$ (MPa) | %EL |
|---|---|---|---|
| $Al7Si1Mg/SiC_p$ 0% | 77 | 64 | 1.01 |
| $Al7Si1Mg/SiC_p$ 10% | 106 | 94 | 1.21 |
| $Al7Si1Mg/SiC_p$ 15% | 116 | 102 | 1.64 |
| $Al7Si1Mg/SiC_p$ 20% | 125 | 107 | 2.41 |
| $Al7Si1Mg1.5TiB/SiC_p$ 0% | 93 | 82 | 1.16 |
| $Al7Si1Mg1.5TiB/SiC_p$ 10% | 120 | 94 | 1.86 |
| $Al7Si1Mg1.5TiB/SiC_p$ 15% | 134 | 116 | 2.46 |
| $Al7Si1Mg1.5TiB/SiC_p$ 20% | 143 | 132 | 2.98 |

The effect of $SiC_p$ percentage to the tensile stength of $Al_7Si_1Mg/SiC_p$ and $Al_7Si_1Mg_{1.5}TiB/SiC_p$ composites is shown in Figure 4.

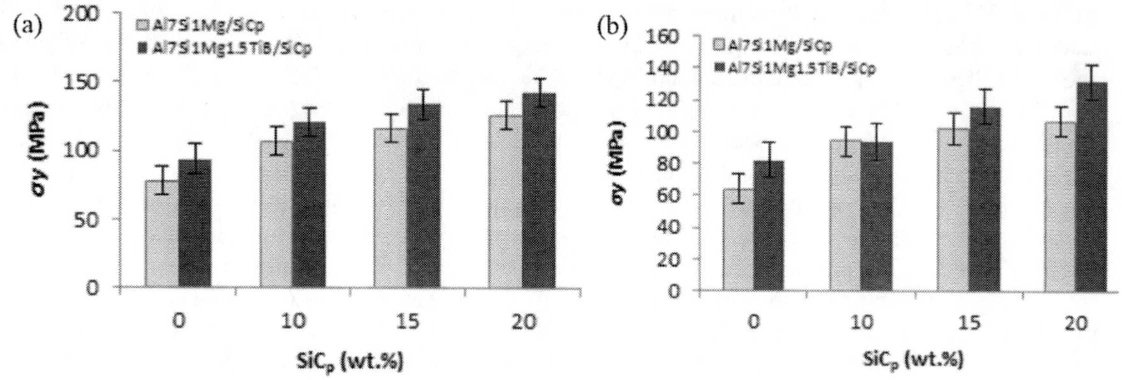


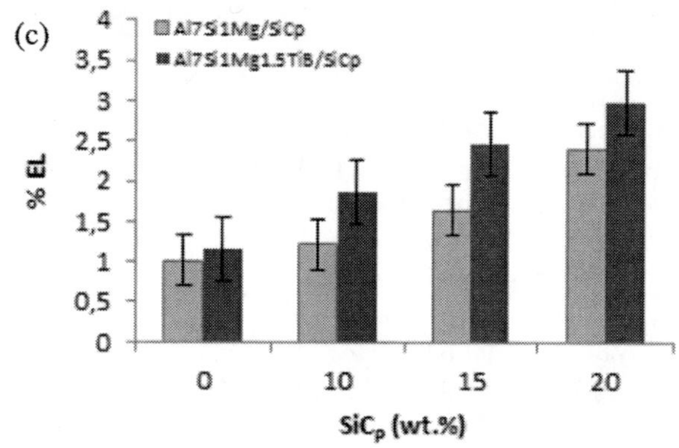


**Figure 4.** Graph (a) Tensile strength, (b) Yield strength, (c) Elongation of Al7Si1Mg/SiC$_p$ and $Al_7Si_1Mg_{1.5}TiB/SiC_p$ Composites

Figure 4a The graphic of tensile strength of $Al_7Si_1Mg/SiC_p$ composite was bigger than $Al_7Si_1Mg_{1.5}TiB/SiC_p$ composite with the difference of average of 16.18 %. The strength of $Al_7Si_1Mg_{1.5}TiB/SiC_p$ composite increased on the variation of $SiC_p$ of 10 wt.% of as much as 120 MPa and on variation of $SiC_p$ of 20 wt.% became 143 MPa, or improved of as much as 22.16 %. For the $Al_7SiMg/SiC_p$ composite, the tensile strength was increasing on the variation of $SiC_p$ of 10 wt.% of as much as 106.84 MPa, and on the variation of $SiC_p$ of 20 wt.% became 125 MPa, or as much as 19.01%. The addition of 1.5% TiB could increase the tensile strength. Figure 4b indicates that the biggest yield strength was on the $Al_7Si_1Mg_{1.5}TiB/SiC_p$ composite, which was as much as 106 MPa. The difference on the yield strength of the $Al_7Si_1Mg/SiC_p$ and $Al_7Si_1Mg_{1.5}TiB/SiC_p$ composites was of as much as 14 MPa. The yield strength of the $Al_7Si_1Mg_{1.5}TiB/SiC_p$ composite was bigger than that of the composite with $Al_7Si_1Mg/SiC_p$ matrix. It could be concluded from Figure 4c that the strain between the $Al_7Si_1Mg/SiC_p$ and $Al_7Si_1Mg_{1.5}TiB/SiC_p$ composites increased with the difference of as much as 0.55%EL. The strain on the $Al_7Si_1Mg/SiC_p$ composite was averagely 1.56%EL. whereas the strain of the $Al_7Si_1Mg_{1.5}TiB/SiC_p$ composite was averagely 2.11%EL. The force on the $Al_7Si_1Mg_{1.5}TiB/SiC_p$ composite was bigger than that on the $Al_7Si_1Mg/SiC_p$ composite. It was based on the discovery of the previous researchers that the extra level of TiB with certain percentage would repair the grain structure of the composites [10,19].

### *3.3. Impact Test*

The impact test was conducted according the standard of ASTM E-23 02a [20]. The test results could be seen on Table 5.

**Table 5.** Results of impact test

| Composites (wt.%) | Load of hitter (Kg) | Impact value ($J/mm^2$) |
|---|---|---|
| $Al_7Si_1Mg/SiC_p$ 0% | 1 | 4.43 |
| $Al_7Si_1Mg/SiC_p$ 10% | 1 | 3.93 |
| $Al_7Si_1Mg/SiC_p$ 15% | 1 | 4.93 |
| $Al_7Si_1Mg/SiC_p$ 20% | 1 | 5.46 |
| $Al_7Si_1Mg_{1.5}TiB/SiC_p$ 0% | 1 | 4.56 |
| $Al_7Si_1Mg_{1.5}TiB/SiC_p$ 10% | 1 | 6.06 |
| $Al_7Si_1Mg_{1.5}TiB/SiC_p$ 15% | 1 | 4.73 |
| $Al_7Si_1Mg_{1.5}TiB/SiC_p$ 20% | 1 | 5.63 |

The graphic of relationship between the impact value and the percentage variation of $SiC_p$ is shown in Figure 5.

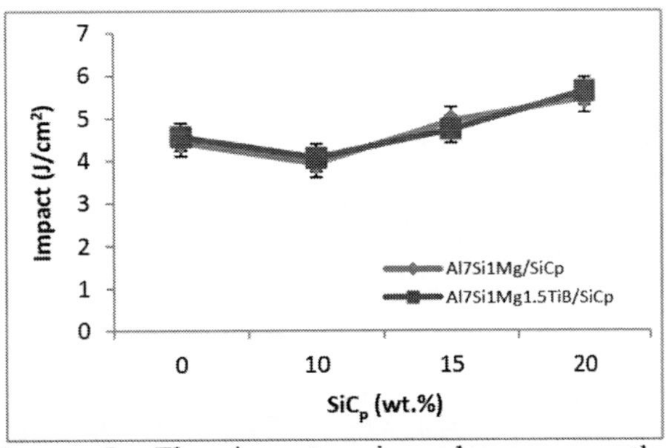


**Figure 5.** The impact price changes on the $Al_7Si_1Mg/SiC_p$ and $Al_7Si_1Mg_{1.5}TiB/SiC_p$ composites

The impact value of $Al_7Si_1Mg_{1.5}TiB/SiC_p$ composite was higher than that of $Al_7Si_1Mg/SiC_p$ composite. On the $Al_7Si_1Mg_{1.5}TiB/SiC_p$ composite, the impact value for the variations of $SiC_p$ of 10, 15, and 20 wt.% increased averagely of 4.74 $J/cm^2$. On the $Al_7Si_1Mg/SiC_p$ composite with 10, 15, 20 wt.% variations of $SiC_p$, the impact price increased averagely of 4.68 $J/cm^2$. The difference between $Al_7Si_1Mg/SiC_p$ and $Al_7Si_1Mg_{1.5}TiB/SiC_p$ composites was of as much as 0.06 $J/cm^2$. The highest impact value was on found on the $Al_7Si_1Mg_{1.5}TiB/SiC_p$ composite. The extra level of TiB could bear a bigger impact load.

### *3.4. Micro Structure*

The micro structure test was conducted by employing an Olympus optical microscope. The results of micro-photos of $Al_7Si_1Mg/SiC_p$ and $Al_7Si_1Mg_{1.5}TiB/SiC_p$ composites are shown on Figure 6.

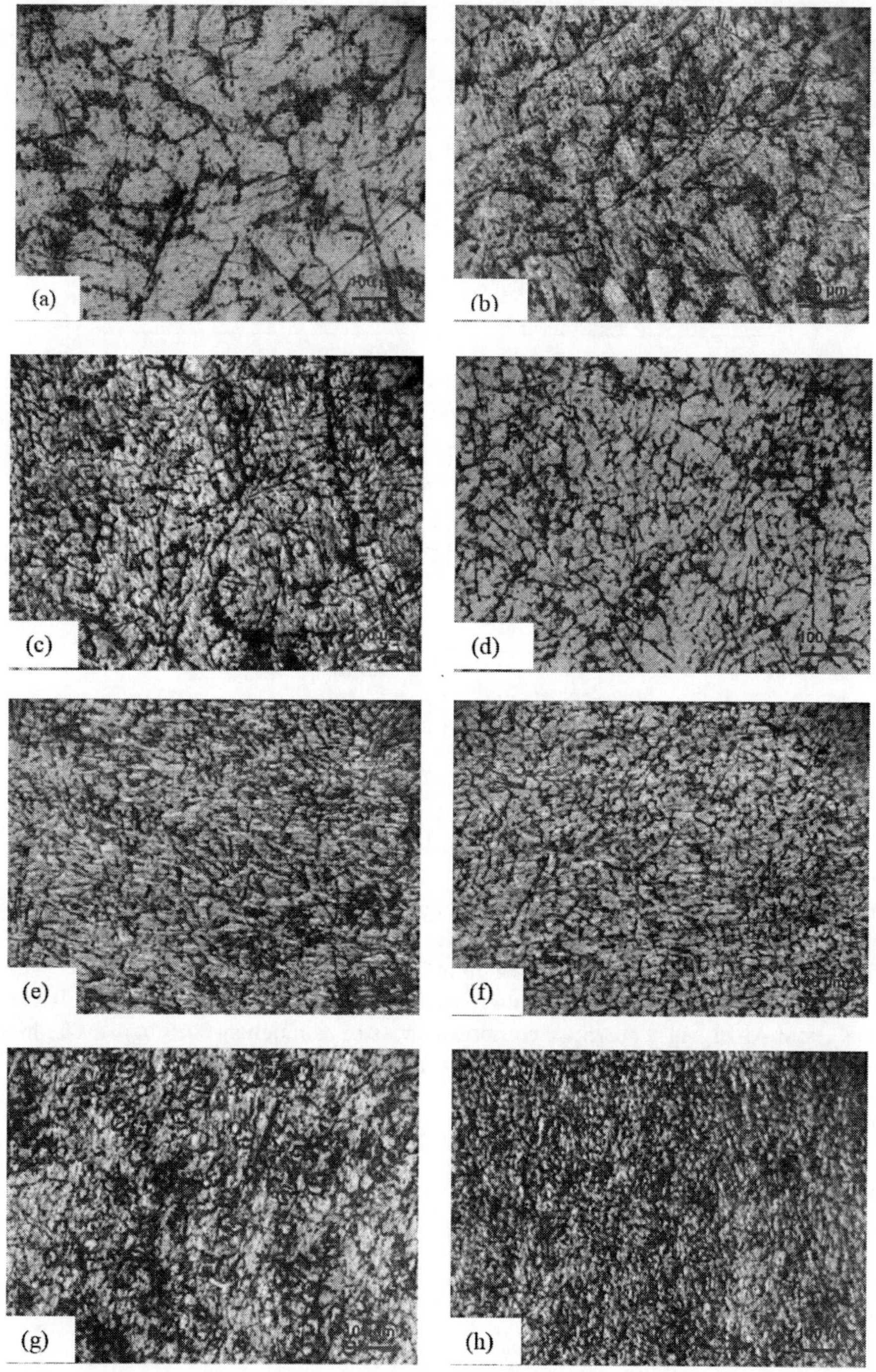

**Figure 6.** Micro photos of: (a) 0% $Al_7Si_1Mg/SiC_p$, (b) 0% $Al_7Si_1Mg_{1.5}TiB/SiC_p$, (c)10% $Al_7Si_1Mg/SiC_p$ (d)10% $Al_7Si_1Mg_{1.5}TiB/SiC_p$ (e)15% $Al_7Si_1Mg/SiC_p$ (f) 15% $Al_7Si_1Mg_{1.5}TiB/SiC_p$, (g) 20% $Al_7Si_1Mg/SiC_p$ (h) 20% $Al_7Si_1Mg_{1.5}TiB/SiC_p$ composites

It could be seen in Figure 6 that the higher the level of $SiC_p$ in the composite, the more $SiC_p$ was bound in the matrix. On the $Al_7Si_1Mg/SiC_p$ composite, the Si was the grey substance with irregular round grains. Figure 8 b, d, f, and h show the $AlSi_1Mg_{1.5}TiB/SiC_p$ composite in the form of black-colored fine grain boundaries. Finer grains on that composite were also noticed. The $SiC_p$ particle was dispersed randomly and evenly in the $Al_7Si_1Mg/SiC_p$ matrix as well as $Al_7Si_1Mg_{1.5}TiB/SiC_p$ matrix. The more $SiC_p$ was dispersed, the harder the composite could be. It was based on the previous study by Hashim et al. (2001) and Lin et al. (2010) that indicated that the employment of semi solid stir casting may help increasing the wettability and dispersing the reinforcing particle homogenously [7].

### *3.5. Morphology*

The test of Scanning Electron Microscopy (SEM) is known to be able to measure the spread of $SiC_p$ particle in the Al7Si matrix. The SEM observation result was on the form of morphology with 10.000x magnification in which it produced SEM photos of $AlSi_1Mg/SiC_p$ and $AlSi_1Mg_{1.5}TiB/SiC_p$ which are shown on Figure 7 below.

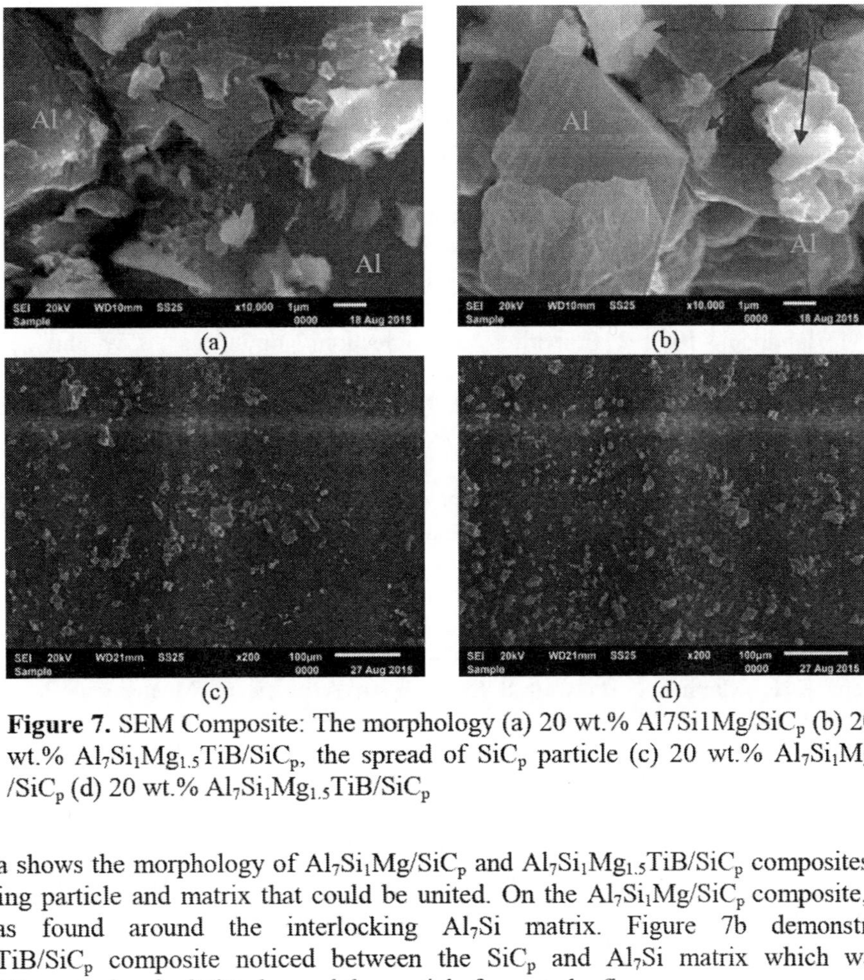


**Figure 7.** SEM Composite: The morphology (a) 20 wt.% $Al7Si1Mg/SiC_p$ (b) 20 wt.% $Al_7Si_1Mg_{1.5}TiB/SiC_p$, the spread of $SiC_p$ particle (c) 20 wt.% $Al_7Si_1Mg$ $/SiC_p$ (d) 20 wt.% $Al_7Si_1Mg_{1.5}TiB/SiC_p$

Figure 7a shows the morphology of $Al_7Si_1Mg/SiC_p$ and $Al_7Si_1Mg_{1.5}TiB/SiC_p$ composites between the reinforcing particle and matrix that could be united. On the $Al_7Si_1Mg/SiC_p$ composite, the $SiC_p$ particle was found around the interlocking $Al_7Si$ matrix. Figure 7b demonstrates the $Al_7Si_1Mg_{1.5}TiB/SiC_p$ composite noticed between the $SiC_p$ and $Al_7Si$ matrix which was mixed thoroughly. The extra level of TiB changed the particle forms to be finer.

## 4. Conclusion

In the present study, the researchers derived several conclusions as follow: The morphology of the $Al_7Si_1Mg/SiC_p$ and $Al_7Si_1Mg_{1.5}TiB/SiC_p$ composites between the reinforcing particle and matrix could be united and dispersed evenly. The effect of $SiC_p$ wt.% was noticed from the value of tensile strength from two kinds of composites. The $Al_7Si_1Mg_{1.5}TiB/SiC_p$ composite went through an increase on the variation of $SiC_p$ of 20 wt.% became 143 MPa or increased of as much as 22.16 %. The biggest yield strength was found on the $Al_7Si_1Mg_{1.5}TiB/SiC_p$ composite that was averagely amounted to 106 MPa. The difference in the yield strength between those two composites was 14 MPa.%EL in which the highest one was found on the $Al_7Si_1Mg_{1.5}TiB/SiC_p$ composite with the average value of 2.11%EL. The $Al_7Si_1Mg_{1.5}TiB/SiC_p$ composite had a higher impact value compared to the $Al_7Si_1Mg/SiC_p$ composite. The highest impact toughness was averagely amounted to 4.74 $J/mm^2$. The porosity value of the $Al_7Si_1Mg/SiC_p$ composite was averagely 2.38%. On the $Al_7Si_1Mg_{1.5}TiB/SiC_p$ composite, the porosity value was averagely 2.10%. The smallest porosity was found on the $Al_7Si_1Mg_{1.5}TiB/SiC_p$ composite. The difference in the increase of porosity percentage on those two composites was 0.13 %. The extra level of 1.5 wt.% TiB could make the grain refinement to be smoother. The extra level of TiB could result in an increase on the tensile strength value and impact value, and from the different morphology from micro structure and SEM.

## 5. References

[1] Zhu M, Jian Z Y, Yang G C, Zhou Y H 2012 Effects of T6 heat treatment on the microstructure tensile properties, and fracture behavior of the modified A356 alloys. *Mater* 36:243–9

[2] Sajjadi S A, Ezatpour H R, Torabi Parizi M 2012 Comparison of microstructure and mechanical properties of A356 aluminum alloy/$Al_2O_3$ composites fabricated by stir and compo-casting processes. Mater Des 2012;34:106–11

[3] Saheb D A 2011 Aluminum Silicon Carbide and Aluminum Graphite Particulate Composites *Journal of Engineering and Applied Science* **6** pp. 41-46.

[4] Sharma P, Chauhan G, Sharma N 2011 Production of AMC by Stir Casting An Overview *International Journal of Contemporary Practices* **2** pp. 23-26

[5] ASM Handbook 1990 **2** Properties and Selection Nonferrous Alloy and Special-Purpose Materials *ASM International*

[6] Hashim J, Looney L, Hashmi, MSJ., 1999 Metal Matrix Composites: Production by The Stir Casting Method *Journal of Material Processing Technology* **92-93** pp. 1-7

[7] Hashim J, Looney L, Hashmi MSJ 2001 The Wettability of SiC Particle in Cast Aluminium Matrix Composites *Journal of Material Processing* Technology **119** pp. 329-335.

[8] Ozben T et al 2008 Investigation Mechanical and Machinability Properties of SiC Particle Reinforced Al-MMC *Journal of Materials Processing Technology* 1 **98** 220-225e.

[9] Viswanatha B M, Kumar P M, Basavarajappa S 2013 Mechanical Properties Evaluation of A356/SiCp/Gr Metal Matrix Composites *Journal of Engineering Science and Technology* **8** pp. 754-763.

[10] Meena K L, Manna A, Banwait S S 2013 An Analysis of Mechanical Properties of The Developed Al/SiC-MMC. *American Journal of Mechanical Engineering* **1** pp. 14-19.

[11] Das S, 2004 Development of Aluminium Alloy Composites for Engineering Aplication *Trans. Indian Inst. Met* **57** No. 4, pp. 325-334

[12] LI Jian-guo, YE Wei, LI Li, HUANG Min 2004 Preparation of AlTiC Master Alloy Under Ultrasonic Vibration. CN 200410103904.3 ( in Chinese )

[13] Mondal D.P, Nidhi JHA, Anshul Badkul, DAS 2012 Effect of Al-TiB master alloy addition on microstructure, wear and compressive deformation behavior of aluminium alloys *Trans. Nonferrous Met. Soc. China* **22**

[14] Wang T., Fu W., Chen Z., Xu Jun, Zhu Jing, Cao Fei, Li Tingju 2012 A Novel Fading Resistant Al-3Ti-3B Grain Refiner for Al-Si Alloy *Journal of Alloys and Compounds* **511** pp. 45– 49

[15] Soe Y H, Kang C G 1995 The Effect of Aplied Pressure on Particle Dispersion Characteristic and Mechanical Properties in Melt Stiring Squeeze Cast SiC/Al Composites *J. Mater Process, Technol* **55** pp.370-379.

[16] Lin G, Hongwei Z, Hoaze Li, Lina G 2010 Effect of Mg Content on Microstructure and Mechanical Properties of SiCp/Al-Mg Composites Fabricated by Semi Solid Stirring Technique *Trans Nonferrous et. Soc.* **20** pp. 1851-1885.

[17] Wang X J, Nie K B, Sa X J, Hu X S, Wu K, Zheng M Y 2012 Microstructure and Mechanical Properties of SiCp/Mg Zn Ca Composites Fabricated by Stir Casting *Materials Science and Engineering* **A 534** 60-67.
[18] ASTM E 8M-04 2004 Standard Test Methods for Tension Testing of Metallic Materials
[19] Zhang D, Sugio K, Sakai K, Fukushima H 2008 Effect of Volume Fraction on The Flow Behaviour of Al-SiC Composites Considering Spatial Distribution of Delaminated Particles *Materials Transactions* **49** No. 3 pp. 661-670
[20] ASTM E-23 2003 Standard Test Methods for Notched Bar Impact Testing of Metallic Materials

# Poly (1,8 Octanediol-co-Citrate) Hydroxyapatite Composite as Antibacterial Biodegradable Bone Screw

**P Widiyanti[1,2], I Sholikhah[1], A Isfandiary[1], NAF Hasbiyani[1], M B Lazuardi[1] and R D Laksana[1]**

[1]Biomedical Engineering, Faculty of Science and Technology, Universitas Airlangga, C Campus, Jalan Mulyorejo, Surabaya 60115, Indonesia
[2]Institute of Tropical Disease, Universitas Airlangga, C Campus, Jalan Mulyorejo, Surabaya 60115, Indonesia

Email: pwidiyanti@fst.unair.ac.id

**Abstract.** The high bone fracture rates reaching up to 300-400 cases per month have been treated with surgical procedure of internal fixation. Nevertheless, the commonly used metal screw has shown several weaknesses. Therefore, it is required bone screw of which primary characteristics include being biocompatible, bio-functional, biodegradable, and anticorrosive. The study aimed to synthesize Antibacterial Poly 1,8-Octanediol-co-Citrate (POC) and investigated the effect of chitosan on the antibacterial and compatibility characteristics of POC-HA composite as antibacterial biodegradable bone screw. The characterization were conducted on POC-HA composite to assess its functional cluster, antibacterial activity, cytotoxicity, degradation capacity, and morphology. Pre-polymer POC was composited with 62% nano-HA, followed by post-polymerization treatment. The sample then coated by chitosan with composition variations of 1%, 3%, and 5%. The nano-HA marked with the appearance of phosphate cluster on the wavenumber of 872.17 $cm^{-1}$and 559.51 $cm^{-1}$, while the chitosan marked with C=O stretch cluster of esther at 1729 $cm^{-1}$ from Fourier Transform Infra-Red (FTIR) measurement. The best result was obtained with 3% chitosan coating. The POC-HA composites showed bacterial inhibiting ability of 16.92 mm with non-toxic characteristics. These results indicated that chitosan coating Poly 1,8-Octanediol-co-Citrate (POC)-Nano Hydroxyapatite composite is a potential candidate for an antibacterial biodegradable bone screw.

**Keywords:** Poly-1,8-octanediol-co-citrate (POC), hydroxyapatite, chitosan, nanomaterial, condensation polymerization, antibacterial biodegradable bone screw.

## 1. Introduction

Fracture is cracking in a bone structure resulting from excessive force. The overall estimated annual incidence of fractures was 100 cases per 10,000 populations for males and 81 cases per 10,000 populations for females. The hip fracture incidence map of the International Osteoporosis Foundation indicates that the hip fracture rate in Indonesia is 119/100,000 per year both for men and women [1]. The orthopedic surgeon repositions the bone fragments with external and internal fixations using a bone screw. The surgical procedures for bone fractures incidents have implemented internal fixation by using bone screw [2]. Bone screw comprises special screw for bones as a method for treating fractures [3]. Sri Rahayu (2012) [4] invented a simpler design for

IOP Conf. Series: Materials Science and Engineering **202** (2017) 012082 doi:10.1088/1757-899X/202/1/012082

bone fracture fixation using screw on the bone cortex made of a mixture of titanium and stainless steel adaptive to the body. Also, Respati (2010) [5] claimed that stainless steel is mechanically adaptive for bones, yet less compatible with the body tissues. Stainless steel cannot be used for long-term application as it may cause local corrosion in body fluid environment. This disadvantage calls for an innovation of material that is not only mechanically adaptive but also compatible with body tissues, such as a combination of biocompatible polymers or composites. A commonly used polymer for medical biomaterial is polyethylene. Some studies have reported that *poly (1,8-Octanediol-co-Citrate) (POC)* biomaterial has exhibited the characteristics of being biocompatible, non-toxic, easily available and inexpensive [6].

POC was chosen for its flexible degradation time. *Hydroxyapatite (HA)* is ceramic material belonging the apatite mineral group with biocompatible and bioactive chemical formula of $Ca_{10}(PO_4)_6(OH)_2$ of which mineral content chemically and physically resembles bones [5]. Nano-size hydroxyapatite may offer a significant advantage in its function as filler for it is expected to improve the mechanics and compatibility of cells in bone screw application and to accelerate the growth of osteoblast. Moreover, chitosan is a polysaccharide from crustaceans group with biocompatible and antibacterial characteristics [7, 8]. There are reported cases [9] where a metal screw had corrosion in bone tissue, and it caused neurological spasm. Besides that, metal screw needs to be taken out from the body (re-operation). Metal stainless steel bone screw is good for the bone, but less suitable for the tissue [10]. Therefore, it is necessary to study on the synthesis of *chitosan coating poly (1,8-Octanediol-co-Citrate) (POC) nano-hydroxyapatite* as a candidate for antibacterial biodegradable bone screw that would results in a biocompatible, biodegradable, antibacterial material with excellent mechanical characteristics and capability of activating osteoblast growth.

## 2. Experimental Method

### *2.1. Materials*

Materials used for the study comprised of 1,8-*octanediol* (98%) which was obtained from Sigma-Aldrich (St. Louis, MO: USA), citrate acid (99.5%) [Mw: 210.14] from SAP, Hydroxyapatite (HA) [Mw: 502.32, assay > 90%; particle size < 100nm from PAIR-BATAN, chitosan from Biochitosan, *Ethanol for analysis* (99.99%), and *Phosphate Buffer Saline* (PBS), with pH 7.4, at 37 °C..

### *2.2. Methods*

Bone screw making was initiated with a synthesis of Poly (1,8-octanediol-co-citrate), mixing citric acid concentration and 1,8-octanediol by adding it to 250 ml three neck flask round-bottom equipped with inlet and outlet adapters. The mixture melts at 160-165 °C under nitrogen gas flow while stirred. The temperature of the system was later reduced to 140-145°C for 1 hour with stirring to make pre-polymers. The synthesis of this polymer pre-POC dissolves into dioxane (50% b/v).

The nano-hydroxyapatite and poly (1,8-octanediol-co-citrate) (POC) Composites (POC/HA) were synthesized with the composition of HA is 62%. Then 10% glycerol was added to the composite as crosslinker. The composite post-polymerization occurred after vacuumed at 120 °C for 1 day, 3 days, and 6 days. The synthesized composite poured and centered in a square shaped iron mold with a size of 15 cm x 15 cm and a thickness of about 0.5 mm. The purpose of the composite was so that the produced form is perfect and to keep the film thickness all across the surface. Chitosan will coat the composite. The chitosan powder dissolved in 2% (b/b) acetic acid was homogenized. The composite was then immersed into a homogeneous mixture after the coating, and the sample was subsequently dried at room temperature for 12 hours. The process was continued with drying in a vacuum oven at 50 °C.

## 3. Results and Discussion

The results of this study were in the form of the development of antibacterial bone screw from composite hydroxyapatite and *poly (1,8-Octanediol-co-Citrate) (POC)* that is biodegradable, biocompatible, and antibacterial.

### *3.1. Functional Cluster Analysis by Fourier Transform Infra-Red*

The *Fourier Transform Infrared* (FTIR) data showed that a cross-combination occurred on C=O stretch cluster which was an esther cluster on 1729 $cm^{-1}$ as a result of the following chemical reaction. The peaks at 1729 $cm^{-1}$ indicated successful polymerization and formation of polyester groups [6,11].

Figure 2 shows the POC-HA composite absorption spectrum after polymerization. The POC molecule formation is marked with the four-wave peaks showing the functional clusters: $CH_3$ (wave number 2928.51 $cm^{-1}$), C-H (wave number 2854.9 $cm^{-1}$), C=O stretch (esther) (wave number 1929.55 $cm^{-1}$), and C-H wag ($-CH_2X$) (alkyl halide) (wave number 1184.68 $cm^{-1}$). Hydroxyapatite molecule formation was marked with the five wave peaks showing the functional clusters of $CO_3^{2-}$ (wave number 1406.26 $cm^{-1}$), $HPO_4^{2-}$ (wave number 872.17 $cm^{-1}$), and $PO_4^{3-}$ (wave number 101.41, 599.74, and 559.51 $cm^{-1}$). Those peaks were also present in FTIR spectra recorded for POC–HA composites and related to the phosphate groups of HA [12]. Functional cluster analysis on the POC-HA composite IR spectrum did not show a new cluster formation. This is because the composite comprising of the POC polymer as the matrix and the hydroxyapatite as the filler is not tied chemically but only physically.

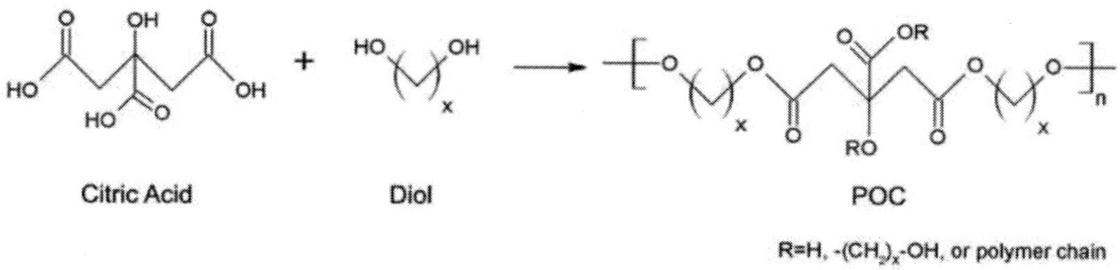


**Figure 1.** Chemical reaction of POC synthesis

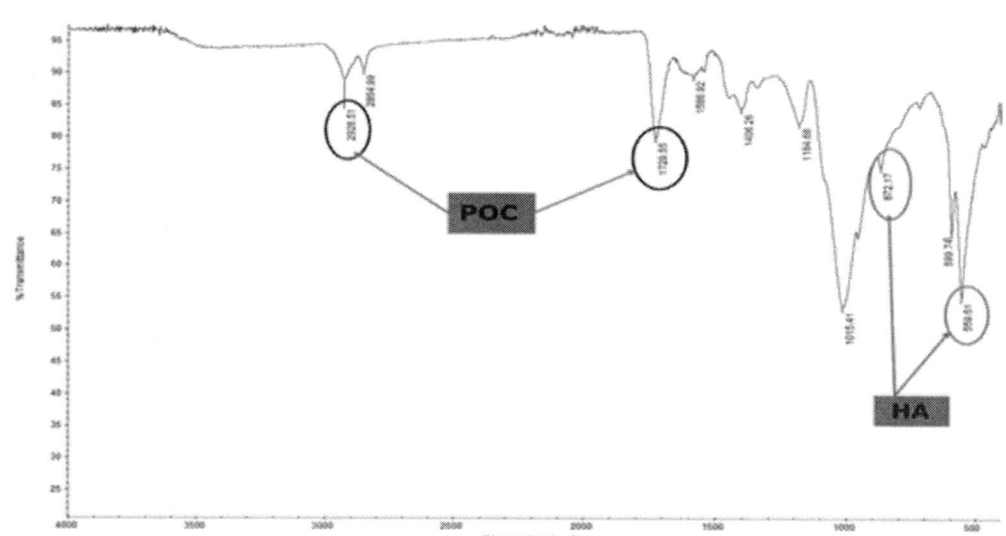


**Figure 2.** POC-HA composite IR spectrum

IOP Conf. Series: Materials Science and Engineering **202** (2017) 012082 doi:10.1088/1757-899X/202/1/012082

### *3.2. Antibacterial Assay*

The antibacterial characteristic of the sample was tested by *Staphylococcus aureus*, which is a pathogenic agent of several infectious diseases commonly detected in musculoskeletal infection [13]. The incubated samples are shown in the following Figure 3.

The observation results revealed that the 3% chitosan coating is the best concentration to inhibit *S-aureus* bacteria growth, shown by its greatest clarity zone diameter of 16.32 mm among other variations.

In 5% chitosan concentration, there has not shown the bacterial inhibition activity. It was suspected to consider that the high chitosan concentration will produce a viscous solution. A viscous solution will be difficult to diffuse. As a result, the inhibition diameter data showed a decline in the inhibitory activity. Chitosan solution which applied in a well could not diffuse. Chitosan molecules could not penetrate the bacterial cell wall. The inhibition zone diameter declined together with the increasing chitosan concentration [14].

The antibacterial effectiveness was based on Greenwood's classification of bacterial growth inhibition responses [15]. The 3% chitosan coating sample showed a medium inhibition capacity as it occurred on the bacteria inhibition zone of > 16-20 mm signifying that the sample has an antibacterial characteristic. Chitosan had a tendency to act as bacteriostatic rather than bactericidal [16], although the exact mechanism was not fully understood and several other factors might contribute to the antibacterial action [17].

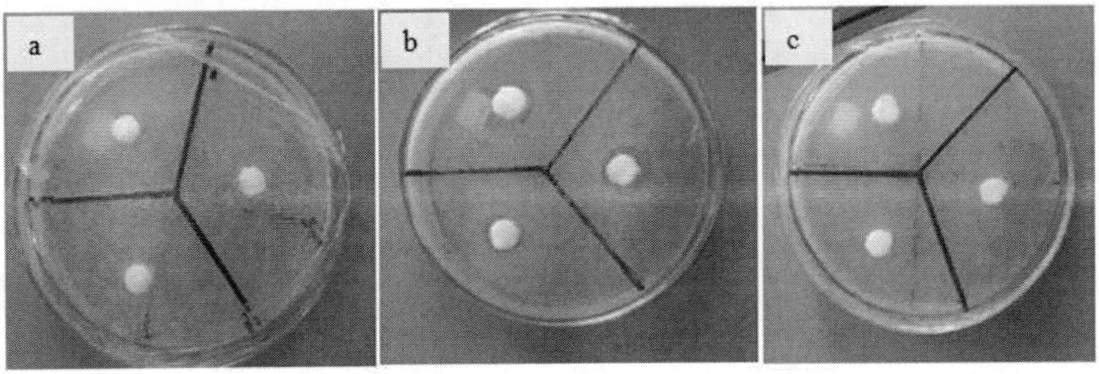


**Figure 3.** (a) 1% Chitosan sample, (b) 3% Chitosan sample, (c) 5% Chitosan sample

**Table 1.** Antibacterial test results

| Sample | Clarity zone diameter (mm) | Average clarity zone diameter (mm) |
|---|---|---|
| 1% Chitosan | 8.9<br>10.6<br>0 | 9.9 |
| 3% Chitosan | 15.93<br>16.13<br>16.92 | 16.32 |
| 5% Chitosan | 0<br>0<br>0 | 0 |

**Table 2.** Classification of bacterial growth inhibition capacity [15]

| Inhibition Zone Diameter | Growth Inhibition Capacity |
|---|---|
| > 20 mm | strong |
| 16 - 20 mm | medium |
| 10 - 15 mm | weak |
| < 10 mm | none |

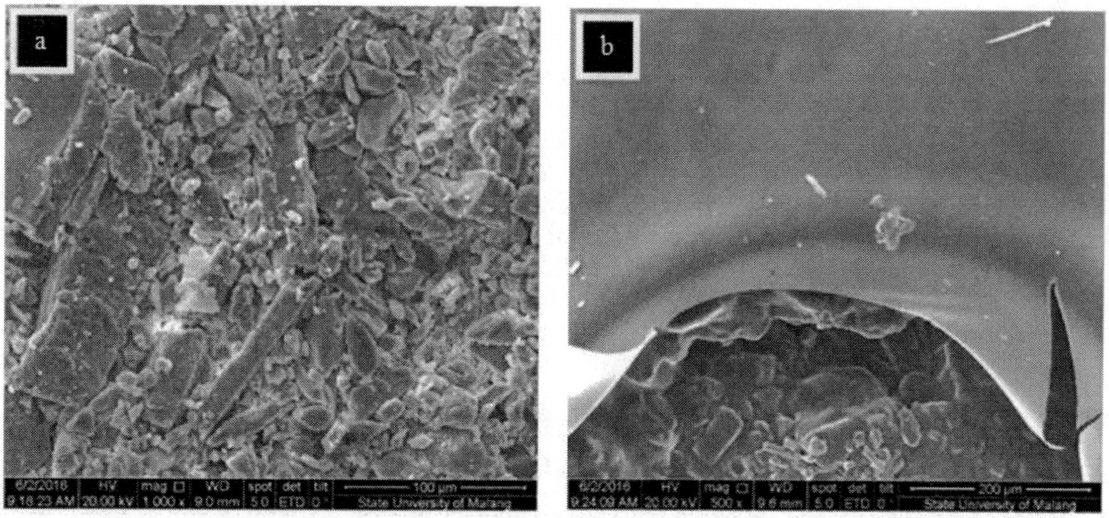


**Figure 4.** (a) Pre-coating sample morphology, (b) Post-coating sample morphology

**Table 3.** Elemental content of the pre-coating sample

| *Element* | *Wt%* | *At%* |
|---|---|---|
| *OK* | 41.22 | 61.58 |
| *PK* | 19.27 | 14.87 |
| *CaK* | 39.51 | 23.56 |
| *Matrix* | Correction | ZAF |

**Table 3.** Elemental content of the post-coating sample

| *Element* | *Wt%* | *At%* |
|---|---|---|
| *CK* | 39.54 | 49.36 |
| *NK* | 09.59 | 10.26 |
| *OK* | 36.70 | 34.39 |
| *PK* | 06.24 | 03.02 |
| *CaK* | 07.93 | 02.96 |
| *Matrix* | Correction | ZAF |

Notes: *Wt%* = weight percentage, *At%* = Atom percentage, *CK* = Carbon, *OK* = Oxygen, *NaK* = Natrium, *PK* = Phosphor, *CaK* = Calcium, and NK = Nitrogen.

The EDX spectra of HA/POC are illustrated in Figure 8 and 9. The Figures indicated the element content differences of the samples before and after coating, and the emergence of a new element, namely N, which is the chitosan amino cluster. Hence, this proves that the coating process occurred in the samples. The presence of HA in the HA/POC composite was confirmed by an appearance of calcium, phosphorous, and oxygen characteristics in the peaks, which are the main components of hydroxyapatite [18].

### *3.3. Degradation Assay*

Degradability is a capability of being slowly broken down into simple parts or chemically degraded. The actual mass loss of the biodegradable bone screw occurs due to release of soluble

IOP Conf. Series: Materials Science and Engineering **202** (2017) 012082 doi:10.1088/1757-899X/202/1/012082

degradation products, phagocytosis by macrophages and intracellular degradation [19]. The comparison of coating variation effects on some samples for degradation speed is illustrated in a graphic presented in Figure 5.

The degradation process was conducted by incubating the samples in *Phosphate Buffer Saline* on pH 7 liquid and static temperature for four weeks. The total degradation calculation with chitosan coating samples shows a total degradation of a month. The length of total degradation on this sample is in accordance with Appley's maximum bone recovery process of 21 months [20].

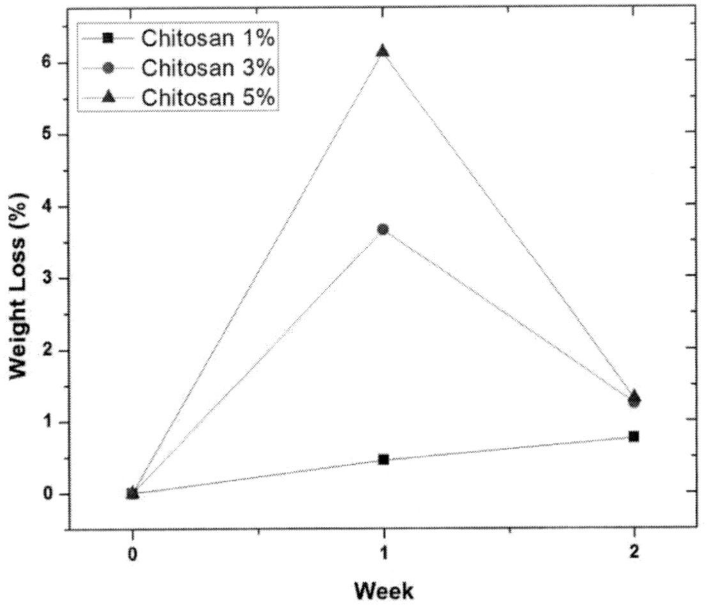


**Figure 5.** The result of degradation test

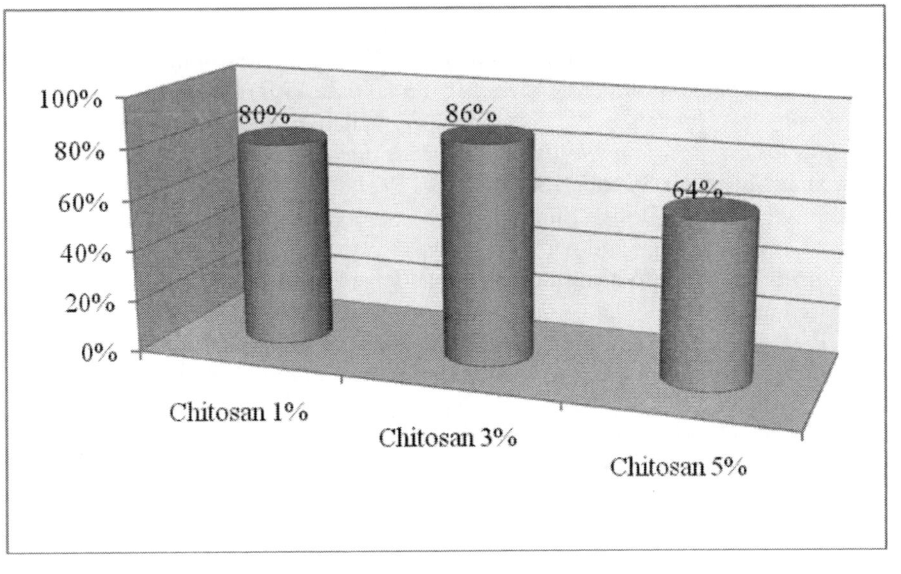


**Figure 6.** Cytotoxicity test graphic

### *3.4. Cytotoxicity Assay*

The body will recognize the implanted antibacterial Biodegradable Bone Screw as a foreign matter. Therefore, the material should be compatible with the body. Accordingly, a cytotoxicity test or MTT was conducted. The MTT principle is the reduction of tetrazolium MTT (3-[4,5-dimetiltiazol-2-yl]-2,5-difenil tetrazolium bromide) yellow salt by reduction system, then its absorption capacity was measured with ELISA *reader* [21] using the Hepatocyte cell. The cytotoxicity test results can be seen in Figure 6.

Figure 6 illustrates the percentage of living cells in each sample. The best result was presented by chitosan coating sample of 86%. All composition variations were within the non-toxic range due to the cell viability which was higher than 50% [22].

## 4. Conclusion

The POC-HA composite was successfully synthesized. The FTIR Result showed that The POC marked by C=O stretch esther cluster chain at 1729 $cm^{-1}$ , while HA marked with phosphate cluster at 872.17 $cm^{-1}$ and 559,51 $cm^{-1}$. The highest degradation speed in antibacterial activity was obtained in the TOC-HA Composite with 3% chitosan. The POC-HA composites showed antibacterial and non-toxic characteristics. The chitosan coating poly(1,8-Octanediol-co-Citrate) (POC) nanohydroxyapatite composite was a potential candidate for the antibacterial biodegradable bone screw. These results indicated that chitosan coating Poly 1,8-Octanediol-co-Citrate (POC)-Nano Hydroxyapatite composite is a potential candidate for an antibacterial biodegradable bone screw.

## 5. References

[1] *Hip fracture incidence map* International Osteoporosis Foundation (viewed 20 November 2013) http://www.iofbonehealth.org/facts-and-statistics/hipfracture-incidence-map
[2] Asrizal and Aditya R 2014 *Closed fracture 1/3 middle femur dextra Medula* **2**(3)
[3] Riyadini W and Suhardi P M 2009 *Pola dan determinan sosiodemografi cedera akibat kecelakaan lalu lintas di Indonesia* Majalah Kedokteran Indonesia **59**(10)
[4] Sri R 2012 *Titanium Bone-Screw: Alternatif fiksasi intermaksilar pada fraktur mandibula sederhana* Majalah Kedokteran FK UKI **XXVIII**(2)
[5] Respati S M B 2010 *Bahan biomaterial stainless steel dan keramik* Momentum **6**(1) 5-8
[6] Yang J, Antonio R, Webb, Samuel J P, Hageman G and Guillermo A A 2006 *Synthesis and evaluation of poly(diol citrate) biodegradable elastomers* Biomedical Engineering Department, Northwestern University IL 60208 USA
[7] Rejane C G, de Britto Douglas and Odilio B G A 2009 *A Review of the Antimicrobial Activity of Chitosan* Polímeros: Ciência e Tecnologia **19**(3) 241-247
[8] Bill G, Zhang X, Damian E M, Gordon G W, Milan B and Peter F M C 2014 *Bioactive coatings for orthopedic implants-recent trends development of implant coatings* International Journal of Molecular Science **1S**(7) 11878-11921
[9] Fastenal Engineering & Design Support, www.engineer@fastenal.com (accessed in 2015)
[10] American Orthopedic Foot and Ankle Society, http://www.aofas.org/footcaremd/treatments/Pages/Hardware-Used-in-Surgery.aspx (accessed in 2015)
[11] Moutos F T, Freed L Ea and Guilak F 2007 *A biomimetic three-dimensional woven composite scaffold for functional tissue engineering of cartilage* Nat Mater **6** 162–167.
[12] Wang S, Kempen D H R, Yaszemski M J and Lu L 2009 *The roles of matrix polymer crystallinity and hydroxyapatite nanoparticles in modulating material properties of photo-crosslinked composites and bone marrow stromal cell responses* Biomaterials **30** 3359–70
[13] Huang D, Zuo Y, Zou Q, Zhang L, Li J, Cheng L, Shen J and Li Y 2010 *Antibacterial chitosan coating on cano-hyrdoxyapatite/polyamide 66 porous bone scaffold for drug delivery* Journal of Biomaterials Science **22**(2011) 931-944

[14] Nurainy et al. 2008 *The Effect of Chitosan Concentrations on the Antibacterial Activiry with Gel Diffusion/Well Method* Jurnal Teknologi Industri dan Hasil Pertanian **13**(2)

[15] Greenwood and David 2001 *Antibacterial agents: susceptibility In Vitro and In Vivo*. In: ELS. Chichester: John Wiley & Sons Ltd

[16] Coma V, Martial-Gros A, Garreau S, Copinet A, Salin F & Deschamps A 2002 J. Food Sci. **67** 162-1169

[17] Raafat D, von Bargen K, Haas A and Sahl H G 2008 Appl. Environ. Microbiol. **74** 764-3773

[18] Adnan H, Kailash C G and Inn-Kyu K 2014 *Morphological Effects of HA on the Cell Compatibility of Electrospun HA/PLGA Composite Nanofiber Scaffolds* BioMed Research International **2014** http://dx.doi.org/10.1155/2014/308306

[19] Ambrose C G and Clanton T O 2004 *Bioabsorbable implants: review of clinical experience in orthopedic surgery* Ann Biomed Eng **32** 171-177

[20] Warwick D J *et al* 2001 *Appley's System of Orthopaedics and Fractures 8Ed* Francis: Taylor

[21] Meiyanto, Edi J, Sendy H, Adam I and Muthi 2008 *Prosedur tetap uji sitotoksisitas metode MTT: Cancer Chemoprevention Research Center* Fakultas Farmasi, Universitas Gadjah Mada

[22] Spielmann H, Hoffmann S, Botham P, Roguet R and Jones P 2007 *The ECVAM International Validation Study on In VitroTests for Acute Skin Irritation: Report on the Validity ofthe EPISKIN and EpiDerm Assays and on the Skin IntegrityFunction Test* Germany: ATLA

**Acknowledgements**
The authors would like to express their great gratitude to the Ministry of Research, Technology, and Higher Education of the Republic of Indonesia for the research grant.

# The Effects of the Addition of Silica Mol Fraction (x = 1.5; 2; 2.5) as a Solid Electrolyte on Ion Conductivity of NASICON ($Na_{1-x}Zr_2Si_xP_{3-x}O_{12}$) Using Solid-State Method

**V M Pratiwi[1], H Purwaningsih[1], Widyastuti[1], R Fajarin[1] and H Setyawan[2]**

[1]Department of Materials and Metallurgical Engineering, Faculty of Industrial Technology, SepuluhNopember Institute of Technology (ITS), Surabaya 60111 Indonesia
[2]Department of Chemical Engineering, Faculty of Industrial Technology, SepuluhNopember Institute of Technology (ITS), Surabaya 60111 Indonesia

Email: vaniamitha@mat-eng.its.ac.id

**Abstract**. Energy is a very important in modern life and need innovations to develop it. One innovation is the application of energyfor storage devices, such as batteries, capacitors, fuel cells, etc. For 30 years, the application of the NASICON ($Na_{1+x}Zr_2Si_xP_{3-x}O_{12}$) into the NASICON gas sensor material was successfully prepared by using solid-state method. The raw materials such as $SiO_2$, $Na_2CO_3$, $ZrO_2$, and $NaH_2PO_4$ with a little methanol were mixed in Ballmill equipment. The silica powder was made by the extraction of bagasse ash by using sol-gel method. The x-ray diffraction patternshowedthat the result of silica extraction was amorphous and the NASICON structure wassynthesizedto bemonoclinic. The scanning electron microscopy results indicated that silica had non-uniform surface morphology and the NASICON had good surface morphology only on the form of $Na_3Zr_2Si_2PO_{12}$. The ionic conductivty of NASICON wasshown on LCR Nyquist plot of the three compositions. The highest NASICON conductivity was found inthe composition of x = 2.0, i.e. $1.142 \times 10^{-8}$ S/m.

**Keywords :** silica, solid electrolyte, ion conductivity, NASICON

## 1. Introduction

NASICON is an acronym for sodium (Na) Super ionic Conductor, which usually refers to a family of solids with the chemical formula of $Na_{1+x}Zr_2Si_xP_{3-x}O_{12}$, $0 < x < 3$. In a broader sense, it is also used for similar compounds where Na, Zr, and/or Si are replaced by isovalent elements. NASICON compounds have high ionic conductivities, on the order of $10^{-3}$ S/cm, which is a rival of those liquid electrolytes. They are caused by hopping of Na ions among interstitial sites of the NASICON crystal lattice [1].

The main application that envisaged for NASICON materials is as a solid electrolyte in a sodium-ion battery. Some NASICON exhibits a low thermal expansion coefficient ($< 10^{-6}$ $K^{-1}$), which is useful for precision instruments and household ovenware. NASICONs can be doped with rare-earth elements, such as Eu, and used as phosphors. Their electrical conductivity is sensitive to molecules in the ambient atmosphere, a phenomenon that can be used to detect $CO_2$, $SO_2$, NO, $NO_2$, $NH_3$, and $H_2S$ gases. Other NASICON applications include as catalysis, immobilization of radioactive waste, and sodium removal from water [1].

IOP Conf. Series: Materials Science and Engineering **202** (2017) 012083 doi:10.1088/1757-899X/202/1/012083

Besides, the production of baggase ash from the waste of sugar cane industries was run over. On the other hand, the resulting ash was rich in silica of around 59%. It was as a raw material ofsilica gel and silica powder production. Generally, the silica gel was produced from the acidification of sodium silicate. Fabrication of sodium silicate was used commercially by reacting materialswhich contain silica, for example, quartz sand, and soda ash in furnaces at temperatures of over 1300 °C [2]. It was clear that this technique consumes large amounts of energy, which can prevent the sugar industry from changing the ash into silica gel. Therefore, it would be advantageous to develop a simple, inexpensive, and low energy consuming method for the production of silica gel from the produced ash so that it can support the use of waste for the economic value-added products. Reported alkali extraction method can be used to increase the concentration of silica from rice husk ash using a low-temperature reaction[3].

This research concerned to analyze the effect of silica ($SiO_2$) addition from an extraction of bagasse ash by sol-gel/alkali method on the formation of NASICON.

## 2. Experimental Method

### *2.1. Chemicals*

The chemical materials used in the experiment were Chloride Acid (99%, Merck), Sodium Oxide (99%, Merck), Sodium Dihydrogen Phosphate (99%, Sigma Aldrich), Zirconium Oxide (99%, Merck), Sodium Carbonate (99%, Sigma aldrich), and Methanol (99%, Merck). All chemical materials were used without further purification.

### *2.2. Preparation of Materials*

*2.2.1. Synthesis of Silica Powder.* In this paper, two-step synthesis method was applied to synthesize the silica powder. The first step of the method was extraction. 10 gram of bagasse ash was added into 60 mL NaOH 2 M and then stirred and heated under boiling temperature for 1 hour. Then the extraction result was filtered to separate the extract and ash residue. Secondly, the extract was titrated by HCL for gelation. Next, the silica gel was aged for 18 hours and washed by demineralized water until the salt resulted from the titration procces was not present. The pure silica gel was subsequently heated to degradate the water in an oven for 24 hours [4].

*2.2.2. Preparation of NASICON ($Na_{1+x}Zr_2Si_xP_{3-x}O_{12}$).* The NASICON material was produced by preparing $SiO_2$, $ZrO_2$, $NaH_2PO_4$, and $Na_2CO_3$ with proper compositions which were then mixed on a Ballmill with methanol sufficienly. The mol silica was varied to x=1.5, 2, and 2.5 to make the *NASICON* ($Na_{1+x}Zr_2Si_xP_{3-x}O_{12}$). The ball mill was rotated at a speed of 800 rpm for 4 hours. Then the result of pre-NASICON was calcined on horizontal furnace for 4 hours at 850 °C and was sintered by the same tool for 5 hours at 1000 °C [5].

### *2.3 General Characterization.*

The samples were characterized by X-ray Diffraction (X'Pert PRO PANalytical) to know the phase and purity of the composing materials and NASICON. Scanning Electron Microscophy (FEI Inspect S50) was employed to know the morphology of the particle surface of silica and surface pellet of NASICON, and LCR meter was used to know the ionic conductivity of the super ionic conductor of NASICON.

IOP Conf. Series: Materials Science and Engineering **202** (2017) 012083 doi:10.1088/1757-899X/202/1/012083

## 3. Result and Discussion

### *3.1. Phase and Purity of Composing Materials*

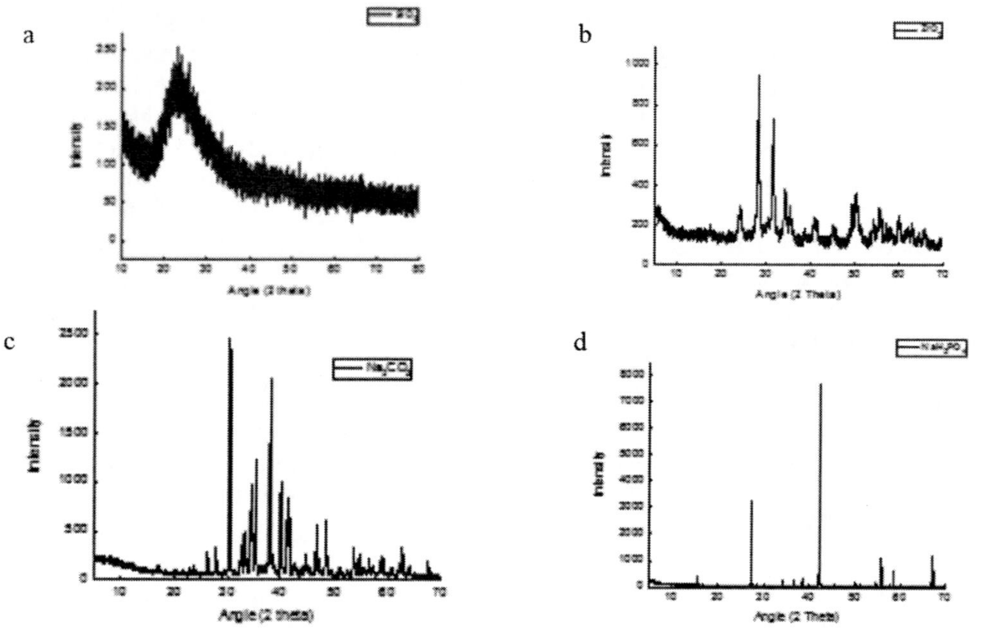


**Figure 1.** X-Ray diffraction patterns of raw materials of NASICON such as a) $SiO_2$, b) $ZrO_2$, c) $Na_2CO_3$, d) $NaH_2PO_4$

The results of x-ray diffraction analysis were shown in Figure 1. The phase of silica powder was amorf and 100% pure without impurities (Figure 1a) and the zirconia powder, sodium carbonat, and sodium dihydrogen phosphate phases were monoclic with suitable properties of conducting materials.

### *3.2. X-Ray Diffraction of NASICON*

The results of x-ray diffraction analysis for NASICON which has been formed after the milling process for 4 hours in a ball mill with a rotation rate of 800 rpm are shown in Figure 2.a. The black line shows the diffraction pattern of the sample with the variable of silica mole fraction of x = 1.5. the formed NASICON has a chemical formula of $Na_{3.12}Zr_2Si_{2.12}P_{0.88}O_{12}$ and its crystal structure is monoclinic according to JCPDS 01-084-1317. While the red line shows the diffraction pattern of the sample with the variable of silica mole fraction of x = 2. The formed NASICON has a chemical formula of $Na_{3.1}Zr_{1.78}Si_{1.24}P_{1.76}O_{12}$ and its crystal structure is rhombohedral in accorandce with JCPDS 01-077-1266 but this NASICON was not fully formed because an unreacting zirconia was still found. The blue line shows the diffraction pattern of the sample with the variable of silica mole fraction of x = 2.5. The formed NASICON has a chemical formula of $Na_{5.27}Zr_{0.5}Si_{0.5}P_{2.5}O_{12}$ and its crystal structure is rhombohedral according to JCPDF 01-087-0617.

The results of x-ray diffraction test for NASICON has done calcination to renew or change the shape of the crystal structure of the system that has been in the milling NASICON shown in Figure 2. The black line shows the diffraction pattern of the sample with variable silica mole fraction x = 1.5. NASICON formed has a chemical formula $Na_{3.12}Zr_2Si_{2.12}P_{0.88}O_{12}$ and its crystal structure is monoclinic according to JCPDS 01-084-1317. These results are similar to the diffraction test results before calcination. While the red line shows the diffraction pattern of the sample with variable silica mole fraction x = 2, NASICON formed has a chemical formula $Na_{3.252}Zr_{1.944}Si_2PO_{12}$ and its crystal structure is rhombohedral according to JCPDS 01-086-0987. After calcination, sample mole fraction x = 2 was change. Such changes occur at concentrations of zircon and silica that had not yet reacted becomes react back when done calcination. On the blue line shows the diffraction pattern of the sample with variable silica mole fraction x = 2.5, NASICON formed has a chemical formula $Na_{3.4}Zr_2Si_{2.4}P_{0.6}O_{12}$ and its crystal structure is rhombohedral according dengna JCPDF 00-036-0351. In addition to the mole fraction x = 2, variable mole fraction x = 2.5 also changes the chemical formula of sodium and phosphorus concentrations decline and the increase in the concentration of zirconia and silica. But the system did not change the crystal structure of monoclinic structure. In this variable there were silica and zirconia which have reacted yet.

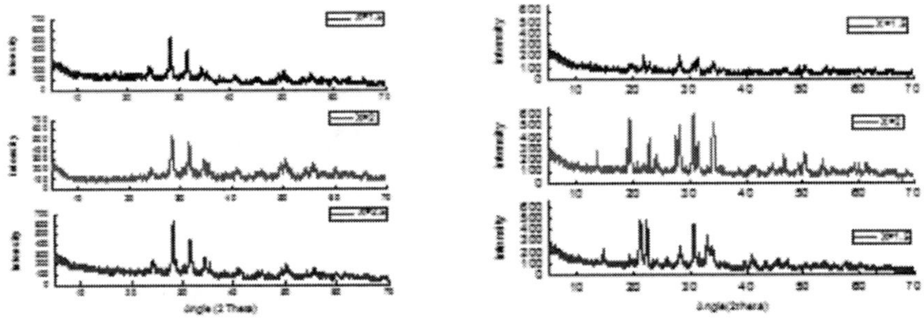

**Figure 2.** X-ray diffraction pattern of NASICON with a fraction x = 1.5, 2 and 2.5 before (a) and after (b) calcination

### *3.3. Morphology of NASICON*

Scanning electron microscopy (SEM) analysis was conducted on the powder NASICON that has heated and compacted to form pellets. The Figure 3.shown that the surface of the pellet did not have many changes in morphologies of before and after calcinating at 5000x magnification. However, the pellet surface becomes smoother and clearer morphology.

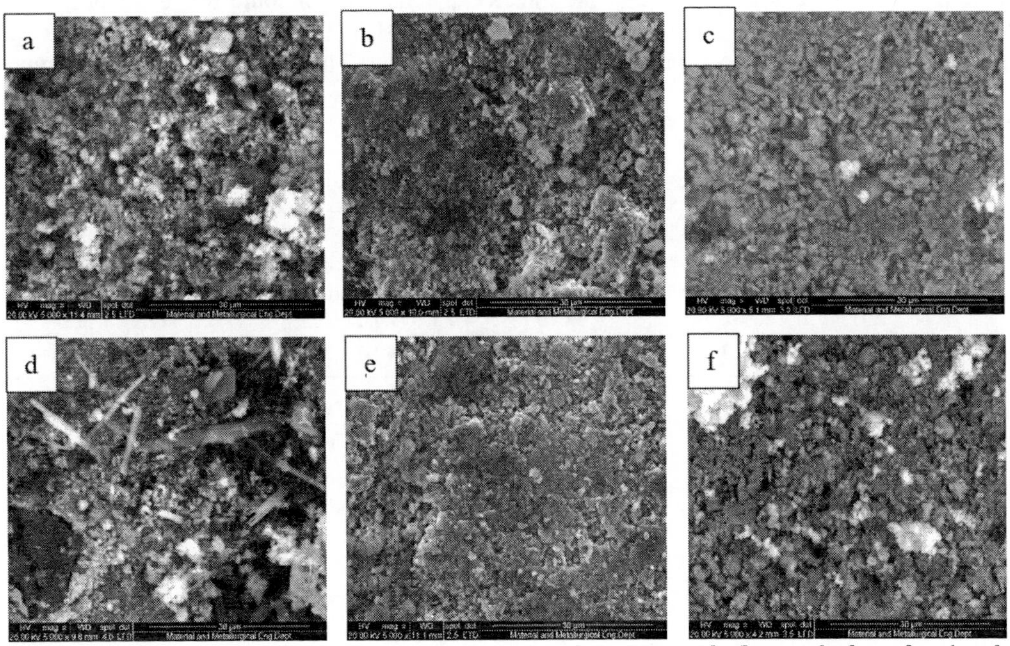


**Figure 3.** Result of Scanning Electron Microscopy of NASICON before and after of variated mol fraction of silica a&d) x=1.5, b&e) x=2.0, and c&f) x=2.5 at 5000x (30 μm)

IOP Conf. Series: Materials Science and Engineering **202** (2017) 012083 doi:10.1088/1757-899X/202/1/012083

On silica mole fraction x = 1.5, NASICON particles were unevenly distributed, not uniform and it was bigger than a fraction x = 2.0 and x = 2.5 which is smaller in size and uniform across the surface of the pellet. The mole fraction x = 1.5 is still appeared a raw material that did not react perfectly to the shape of particles obtained NASICON was no difference, irregular shape and are elongated fiber. Figure 4 shown the shape of NASICON surface with x = 2.0 was impresive. The shape like a small sheet with diameter 500 nm. There were difference between before and after calcinating, from unregular shape to be sheet shape [7]. The figure was taken on magnification 75.000x.

**Figure 4.** Result of Scanning Electron Microscopy of NASICON before and after of variated mol silica x = 2 with magnification of 75.000x (2 μm)

### *3.4. Ionic Conductivity of NASICON*

To proof of NASICON that was made is Superconductor, we must characterized with LCR meter. The result of this analysis shown in Table 1. The table shown that the higher conductivity of NASICON which is obtained on fraction of mol silica x = 2.0 as big as $1.142 \times 10^{-8}$ S/m[8]. From the surface in this variable (Figure 4.) we could know that the transfer of ions sodium could move smoothly in charge and discharge. Figure 5 shown that the nyquist plot of NASICON with three variable. The optimum charge transfer was took placed in x = 2.0 with 2 semicircle. There were two phenomenom in this figure, is chrage transfer on semicircle graph and diffusion on the body of NASICON on line straight graph.

**Table 1.** The ionic conductivity of NASICON

| Variable x | F (Hz) | R (Ohm) | G (S) | Conductivity (S/m) |
|---|---|---|---|---|
| 1.5 | 9541 | $4.97 \times 10^{5}$ | $2.94 \times 10^{-7}$ | $1.083 \times 10^{-9}$ |
| 2.0 | 7543 | $2.8 \times 10^{6}$ | $3.37 \times 10^{-7}$ | $1.142 \times 10^{-8}$ |
| 2.5 | 565 | $1.91 \times 10^{7}$ | $3.88 \times 10^{-8}$ | $1.207 \times 10^{-9}$ |

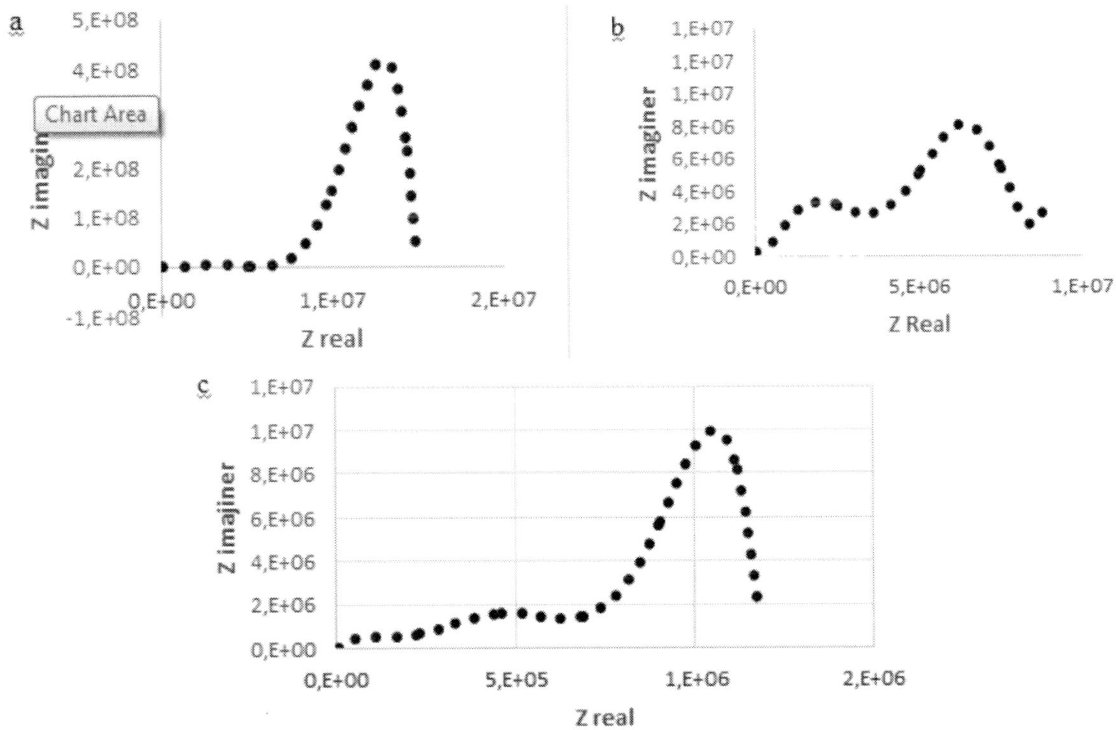


**Figure 5.** Nyquist plot form LCR meter NASICON with mol fraction of silica a) x = 1.5, b) x = 2.0, and c) x = 2.5

## 4. Conclusion

The addition of silica on the structure of NASICON can be impact on improving ion conductivity and uniformity of the surface structure. Silica was well prepared by sol-gel method had amorphous structure and the phase of NASICON were monoclicnic structure. They were the result of x-ray diffraction analysis. The greatest value of ionic conductivity was $1.142 \times 10^{-8}$ S/m from x=2.0 from LCR analysis. The stable form from the effect of silica addition is $Na_3Zr_2Si_2PO_{12}$ by x=2.0.

## 5. References

[1] Anantharamulu N, Koteswara R K, Rambabu G, Vijaya K B, Radha V, Vithal M 2011 A wide-ranging review on Nasicon type materials *Journal of Materials Science* **46** (9): 2821 Another reference

[2] Affandi S, Setyawan H, Winardi S, Purwanto A, Balgis R 2009 A Facile Method for Production of High Purity Silica Xerogel from Bagasse Ash *Advanced Powder Technology*

[3] Kalapathy U, Poctor A, Shultz J 2000 Production and properties of flexible sodium silicate films from rice hull ash silicon, *Bioresour Technol,* **72** 99 – 106

[4] Muljani S, et al 2014 A facile method for the production of higg-surface-area mesoporous silica gels from geothermal sludge. *Advanced Powder Technology* **25** 1593-1599

[5] Honma T, et al 2015 Electrical conductivity of $Na_2O$-$Nb_2O_5$-$P_2O_5$ glass and fabrication of glass-ceramics composites with NASICON type $Na_3Zr_2Si_2PO_{12}$. *Solid State Ionics* **269** 19-23

[6] Lalere F, et al 2014 An all-solid state NASICON sodium battery operating at 200°C. *Journal of Power Sources* **247** 975-980

[7] Liu Y, Junyan H, Gang L and Hongtao L 2014 Ionic Liquid-gelled polyvinylidene fluoride/polyvinyl acetate polymer electrolyte for solid supercapasitor *Chemical Engineering Journal* **258** 320-326

[8] Masashi K and Masaki K 2015 Sol-gel synthesis of $Li_{1.5}Al_{0.5}Ge_{1.5}(PO_4)_3$ *Ceramics International*

**41** 8562-8567

**Acknowledgements**
The author would like to thank to PENELITIAN PEMULA PNBP ITS 2016 for funding.

IOP Conf. Series: Materials Science and Engineering **202** (2017) 012084 doi:10.1088/1757-899X/202/1/012084

# Effects of Austenitizing and Forging on Mechanical Properties of MIL A-12560/AISI 4340 Steel

**S Herbirowo, B Adjiantoro and T B Romijarso**

Research Center for Metallurgy and Materials - Indonesian Institute of Science (P2MM-LIPI)
Kawasan Puspiptek Gedung 470, Tangerang Selatan, Indonesia

Email: satr006@lipi.go.id

**Abstract**. Laterite steels might be used for alternative armored steels. Their properties can be improved in various ways, such as by heat treatment. This paper reports the influences of tempering temperature on the hardness and microstructure of the modified MIL A-12560/ AISI 4340 steels. Samples were austenitized at 1200, 1000, and 800°C and forged at 100, 75, and 50 tons. Mechanical properties consisted of Hardness measurement was conducted by Brinell indentation and metallographic observation was conducted by scanning electron microscopy (SEM). The results showed that increasing forging force until 100 tons can decrease hardness. The formation of the microstructure consists of tempered martensite containing ferrite and dual phase perlite. The presence of void and porous can also decrease hardness. Decreasing austenizing temperature from 1200 °C to 800°C can increase material strength and hardness.

**Keyword:** Laterite steel, tempering, hardness, auztenitizing

## 1. Introduction

Armor materials having a specified goal of protection have recently found a wide use for land, air, marine and aerospace applications for civil and military purposes and also as a surface material for constructions. Recently, the heaviest threat on civil and military platforms are bullets with kinetic energy shot from modern firearms. Armor steel mentioned in this study exhibits a high level of ballistic performance against projectiles as a result of its properties, e.g. high strength, adequate toughness and good weldability [1].

Series of MIL (USA Armor Quality) are very important for many cases due to their optimal mechanical properties under kinetic and also dynamic impact to armor. On the other hand, XH129 (German Armor Quality) and Armox (Swedish Armor Quality) are known as other armor materials for protection. The mechanical properties of armor steel differ greatly in cast and heat treated forms. Quenched and tempered steels are widely used in automotive, defence, marine industries. Military steels have a martensitic/bainitic structure after being rapidly quenched. The cooling rate has to be so fast that the austenite should mainly transform within martensite and bainite ranges. In industrial applications, steels are not tough enough after a rapid quenching. So, the usable hardness -toughness balance is obtained during the tempering process by the selection of time and temperature. Parameters such as alloy design and the proper heat treatment (quenching and also tempering) are very important to determine the final properties of the product [2]. In this case, the optimum mechanical properties, for example high level of strength, hardness and also toughness, can be obtained by performing a tempering heat treatment within a certain temperature and time range [3,4]. Their properties can be improved in various ways, such as by heat treatment. This paper reports the influences of tempering temperature on hardness and microstructure of the modified MIL A-12560/ AISI 4340 steels.

## 2. Experimental Method

The modified laterite steel ingot with MIL-A-12560/ AISI 4340 standard was prepared in an electric induction furnace. The ingots of 5x5x10 cm in dimension were heat forged at various temperatures of 1200, 1000, and 800°C and forged at 100, 75, and 50 tons until the dimension of the ingots decreased to about 12 x 6 x 2 $cm^3$. The square specimens of 1 cm in thickness were cut from forged alloys for chemical composition test using Optical Emission Spectrometer (OES). The chemical composition of the alloys is presented in Table 1. The specimens were austenitized at a temperature of 1050 °C for 1 hour followed by oil quench to obtain martensite structure. Then, the specimens were tempered for 1 hour. Rockwell hardness measured all heat treated samples using standard ASTM E10 and ISO 6506 for hardness Brinell scale. The samples were polished and etched using Kalling's reagent for metallographic test, and was then examined using scanning electron microscope (SEM). Energy dispersive X-ray spectroscopy (EDS) analysis was utilized to identify the carbides during tempering.

**Table 1.** Chemical composition in wt% of the modified Laterite steel in this study

| Steel Type | C | S | P | Mn | Si | Cr | Mo | Ni | Fe |
|---|---|---|---|---|---|---|---|---|---|
| 13Cr3Mo3Ni | 0.204 | 0.0075 | 0.01 | 0.309 | 0.24 | 0.452 | 0.001 | 1.564 | Bal. |

## 3. Results and Discussion

The effect of austenizing temperature of the modified MIL A-12560 steel on hardness value is shown in Figure 1. As can be seen, the hardness decreased from the highest austenizing temperature at 1200 until 800 °C. The initial drop in hardness was due to the precipitation of M3C which causes the softening of the martensite because of carbon depletion [5]. It can be observed from Figure 1 that the highest hardness of modified MIL-A-12560 steel in temper condition was 205.03 HB at 800 °C. The decreasing hardness in the range of 50 to 100 tons can be attributed to the secondary metal forming phenomenon. It can also be seen in Fig.1 that the hardness decreased along with the increase at austenizing temperatures ranged from 800 to 1200 °C. It could occur due to the $M_7C_3$ carbides started to coarsen and partially transform to $M_{23}C_6$ carbides [6]. The reason for variations in properties can be attributed to the combined effects of (i) hardening: due to a transformation of the retained austenite and formation of fine precipitates; and (ii) softening: due to a reduction in internal stresses, decrease in dislocation density within martensite lath, and formation of reversed austenite as described in other report [7]. While tempering at higher temperatures, quenched martensite formed while the annealing solution starts softening which leads to the elimination of internal stresses, decrease in dislocation density, and an occurrence of the retained austenite [8].

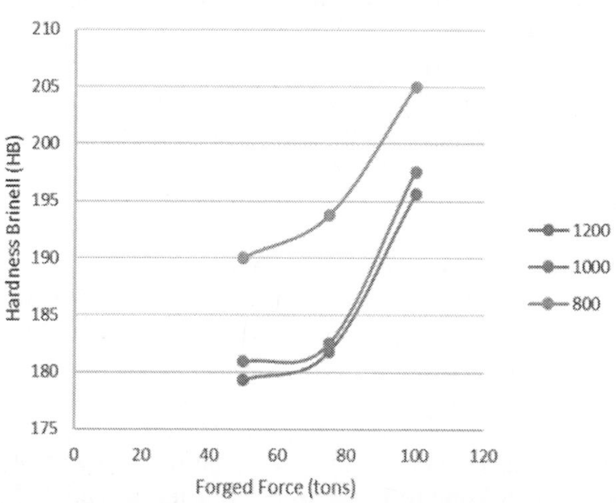


**Figure 1**. The hardness rate of the modified steel at various austenizing temperatures.

Figure 2 shows SEM microstructure of various austenizing temperatures at 1200, 1000, and 800 °C in a forging force of 75 tons. It can be observed that the microstructure of tempered laterite steel generally contain martensite lath, and carbide. The carbide formed after the tempering process was not so clear to see. These types of precipitates were mainly located in the matrix and martensitic lath structure. These thermally stable fine precipitates enhanced the long term creep resistance by impeding movement of mobile dislocations, prior austenite grain boundaries, martensite lath and subgrain boundaries, and restrict fine grain structure from recrystallization. For a constant normalizing time of 2 h, prior austenite grain size and martensite lath size increased along with the increasing austenizing temperatures of 800 °C –1200 °C, as shown in Figure 2a–c. The as-austentized (1200 °C for 2 h) microstructure had hard martensitic lath structures with high dislocation density, as observed in Figure 2b and c. Similarly, for constant austentizing temperature of 1000 and 800 °C, the martensite lath size and prior austenite grain size increased along with the increasing forged force of 100, 75 and 50 tons.

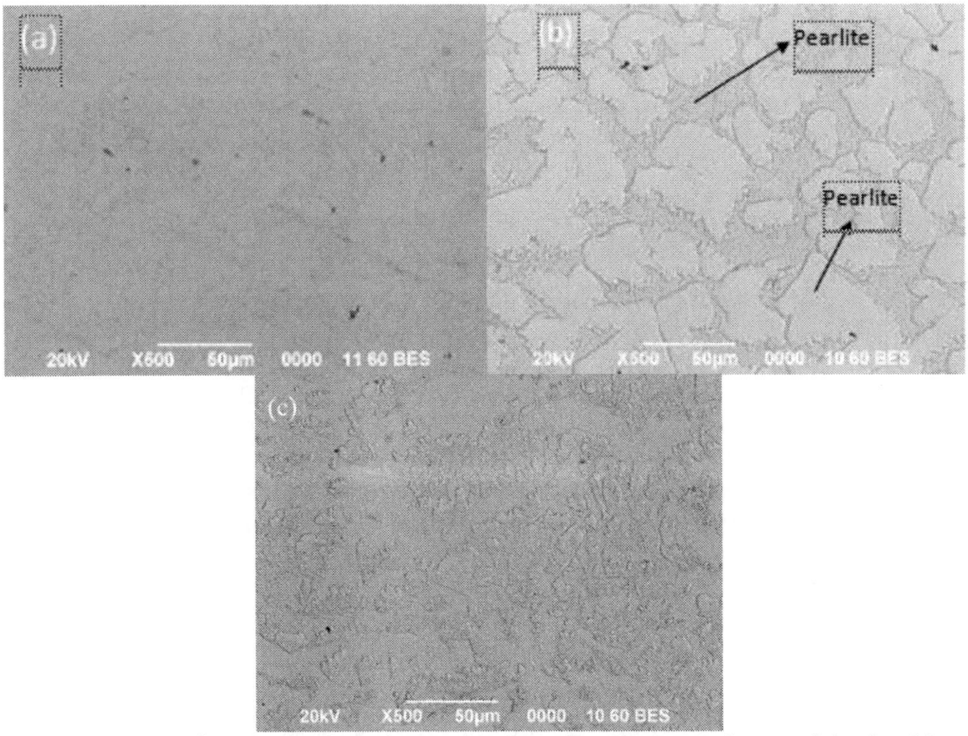


**Figure 2.** SEM micrograph of specimens that forged at 75 tons and austentisized at (a) 1200, (b) 1000, and (c) 800 °C.

During this kind of precipitation, the hardness of steel decreased as a result of the transformation from martensite to ferrite structure. However various carbide precipitates such as MC, M 2C, M7C3 can be formed with an addition of some alloying elements that have a strong affinity to carbon at high tempering temperatures. As a result, an increase in hardness can be determined and called as secondary hardness. Such increase in hardness characteristic was obtained by a complex precipitation hardening. The precipitation characteristic of some carbides formed by various alloying elements during tempering is shown in Figure 2. As seen from the figure, the formation temperatures of formed carbides increase along with the increase of the carbide stability [9-11]. Secondary hardness precipitates form preferentially upon dislocations and lath boundaries in lath-type martensitic matrix. The maximum value of secondary hardness can be obtained by precipitation of several semi-coherent nanometer sizes, while the density of M 2C is two times of MC carbides at the initial stage of over-tempering and it provides the basis of matrix hardness. This kind of precipitate grows up rapidly and its number decreased depending on its low thermo-dynamical stability during tempering.

Finally the relationship between the amount of impact strength and their forged force has been represented in Figure 3. As it can be seen from the figure, the impact strength decreased and was almost linear for the increasing forged force relationship existed for several steel based materials [12,

13]. That is why it was expected that carbides dissolve in the matrix at high austenitisation temperatures and precipitate in a fine form as secondary carbides at lower temperatures during tempering. To obtain such a microstructure, the steel must be austenitized so that the carbides of the weak and semi-strong carbide forming elements dissolved. At this heat treatment, the material was hold at the temperatures of 800-1200 °C for a certain period of time. After this, a tempering heat treatment followed at a temperature of up to 600 °C whereas the carbon and carbide forming elements dissolved in the matrix form fine dispersed secondary carbides in the solid solution and cause an enhancement of hardness and strength. To obtain a high hardness rate during tempering, it was desired that high amounts of elements like Mo and Cr should dissolve. [14].

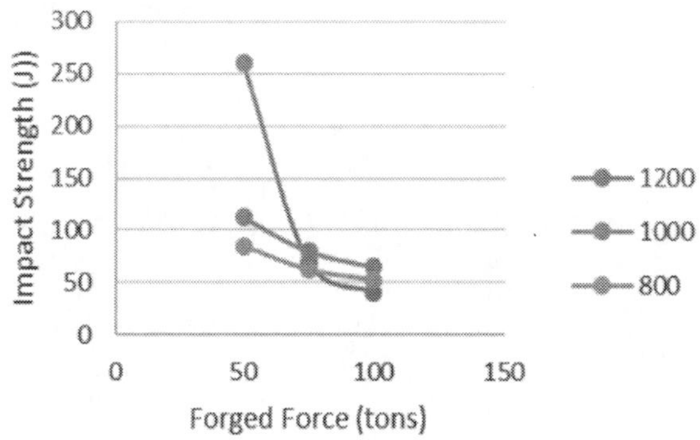


**Figure 3.** The impact strength with various forged

## 4. Conclusion

The austentizing temperature could strongly affect the hardness and microstructure of the modified MIL A-12560/ AISI 4340 steel. The increasing forged force can affect the impact strength of the modified MIL A-12560/ AISI 4340 steel. The highest hardness was obtained at 800 °C. The formation of the microstructure consists of tempered martensite containing carbides $M_{23}C_6$. Increasing forged force from 50 to 100 tons can decrease hardness and impact strength properties.

## 5. References

[1] Karagöz Ş, Yılmaz A, Atapek H 2007 Development of Quenched & Tempered Steel as Armor Materials Through Alloying and Heat Treatment *Military Science & Inteligence Magazine* p.42-45

[2] Atapek H 2006 Development of an Armor Steel along the Principles of Physical Metallurgy & Its Characterization', Master Thesis, Kocaeli University, Institue of Applied Science, Turkey

[3] Smoljan B 2007 Prediction of Mechanical Properties and Microstructure Distribution of Quenched and Tempered Steel Shaft *Journal of Materials Processing Technology* **175** p. 393 -397

[4] Matsubara H, Osuka T, Kozasu I, Tsukada K 1972 Optimization of Metallurgical Factors for Production of High Strength, High Toughness Steel Plate by Controlled Rolling *Transactions ISIJ* **12** 435 -443

[5] Balan K P, Reddy A V and Sarma D S 1998 Austenite Prepcipitaion During Tempering In 16Cr-2Ni Martensitic Stainless Steels *Scr. Mater.* **39** 7 pp. 901–905

[6] Isfahany A N, Saghafian H and Borhani G The effect of heat treatment on mechanical properties and corrosion behavior of AISI420 martensitic stainless steel *J. Alloys Compd.*, **509** 9 pp. 3931–3936, 2011.

[7] Krishna C, Kumar N, Jha A K, Pant B and George K M 2015 Microstructure and Properties of 15Cr-5Ni-1Mo-1W *Steel Reasearch* **86** 1 pp. 51–57

[8] Dening Z, Ying H, Wei Z and Xudong F 2010 Influence of Tempering Process on Mechanical Properties of 00Cr13Ni$_4$Mo Supermartensitic Stainless Steel *J. Iron Steel Researvh* **17** 8 pp. 50–54

[9] Honeycombe R W K 1981 Steels – Microstructure and Properties *Metallurgy and Materials*

*Science Series, UK*

[10] Karagoz Ş, Yılmaz A, Atapek H 2006 Development of Quenched and Tempered Sheet Steel Products', *13th International Metallurgy & Materials Congress* Istanbul-Turkey

[11] Karagoz Ş, Andren H –O 1992 Secondary Hardening in High Speed Steels Z. f. *Metallkunde* **83**, 386-394

[12] Fischmeister H, Karagoz Ş, Andren H –O 1988 An Atome Probe Study of Secondary Hardening in High Speed Steels *Acta Metall* **36** 817-825

[13] Karagoz Ş, Fischmeister H F, Andren H-O, 1992 Cai-Guang Jun Microstructural Changes During Overtempering of High Speed Tool Steels *Metall. Trans.* **23A**, 1631-1640

[14] Karagoz Ş, Atapek H, Yılmaz A 2007 Development of QT Steels as Armour Materials *4th Iron-Steel Congress* Zonguldak-Turkey

**Acknowledgements**
This paper is part of development of high strength material for laterite steels. The authors would like to be obliged by the Research Center for Metallurgy and Material of Indonesian Institute of Sciences for providing laboratory facility and funding this research.

IOP Conf. Series: Materials Science and Engineering **202** (2017) 012085 doi:10.1088/1757-899X/202/1/012085

# Influence of Austenitizing Heat Treatment on the Properties of the Tempered Type 410-1Mo Stainless Steel

**E Mabruri[1], Z A Syahlan[2], Sahlan[2], S Prifiharni[1], M S Anwar[1], S A Chandra[1], T B Romijarso[1] and B Adjiantoro[1]**

[1]Research Center for Metallurgy and Materials, Indonesian Institute of Sciences (LIPI), Kawasan Puspiptek Gd. 470 Serpong, Tangerang Selatan, Indonesia
[2]Higher Technical School of PLN, Menara PLN, Jl. Lingkar Luar Barat, Duri Kosambi, Cengkareng, Jakarta, Indonesia

Email: effe004@lipi.go.id

**Abstract.** The modified 410-1Mo stainless steel has been developed with higher tensile strength and elongation compared to the standard 410 stainless steel. This paper reports the influence of austenitizing temperature on the microstructure, hardness, impact resistance and corrosion resistance of the modified 410-1Mo steel. The steel samples were prepared by a process sequence of induction melting, hot forging, annealing, hardening, and tempering. The microstructure of the tempered steels revealed additional phase of delta ferrite at pre-austenitizing temperatures of 950 to 1050 °C and disappeared at a temperature of 1100 °C. The steels which underwent pre-austenitizing at 1100 °C showed the largest sized lath martensite and the largest amount of retained austenite. The tempered steels maintained hardness at austenitizing temperatures of 950 °C to 1000 °C and showed an increasing hardness at austenitizing temperatures from 1000 to 1100 °C. At a range of austenitizing temperatures, it was investigated that the steels exhibited higher impact resistance at 1050 °C. The tempered steels that were pre-austenitized at 950 °C and 1100 °C showed the lowest pitting potential due to the existence of carbides and coarse-high carbon martensite, respectively.

**Keywords:** Austenitizing, tempered Type 410-1Mo, impact resistance.

## 1. Introduction

The martensitic stainless steels have been widely used for many applications such as power generation, building structure, vehicle construction and mining machine due to their high strength, high toughness and moderate corrosion resistance [1-2]. However, the martensitic stainless steels have been continuously modified to improve the mechanical and corrosion properties in order to accommodate the variably severe working conditions of the services, especially for power generation application [3-6]. The improvement of the mechanical and corrosion properties of the martensitic stainless steels can be optimized by chemical composition modification and heat treatments [7-8]. The 410-1Mo martensitic stainless steel is a modification of 410 type steel recently developed by incorporation about 1 wt.% Mo which exhibits higher tensile strength, higher elongation at breaks and higher pitting resistance compared to the standard 410 type steel for steam turbine blade application [9]. The heat treatment response properties of this steel is being investigated and parts of the results have been published elsewhere [10-11]. Although Mo are containing 4xx type stainless steel already available such as 425Mod and 440 A, they contain higher carbon (> 0.5 wt.%) which are not suitable for turbine blade application. Thus, the modified 410-1Mo stainless steel with lower carbon content (<0.15 wt.%) has been developed aiming for better properties of the steam turbine blades.

It has already known that the heat treatment parameters of the martensitic stainless steels depend on their chemical composition. Once the chemical composition has been modified, the heat treatment of the martensitic stainless steels should be optimized to meet the technical requirements of the specific service. The heat treatment of the martensitic stainless steels for hardening is performed by quenching from high-temperature austenite phase providing solid martensite structure and subsequently tempering the as-quenched steel at suitable temperature and time to improve ductility and toughness. The austenitizing temperature determines the extent of carbide dissolution, dissolved alloying elements and grain growth influencing the final microstructure and properties of the tempered steels. Determining proper austenitizing temperature is required to ensure the desired microstructure and optimal properties of the tempered martensitic stainless steels. This paper presents the influence of the austenitizing temperature on the microstructure, hardness and impact energy of the tempered 410-1Mo martensitic stainless steel. The tempering of the as-quenched steels was performed at high temperature of 600 °C to facilitate the use of this martensitic stainless steel for turbine blades application operated up to such a high temperature.

## 2. Experimental Method

The induction melting furnace was used for fabricating the ingots of the investigated martensitic stainless steel of type 410-1Mo with a dimension of 5 x 5 x 10 cm. Hot forging of the steel ingots was carried out at a temperature of 900-1125 °C until the dimension of the ingots changed to about 3 x 3 x 27 cm. The square specimens with 1 cm in thickness were cut from forged steels for chemical composition testing using Optical Emission Spectrometer (OES). The chemical composition of the steels is shown in table 1. The other parts of the forged steels were annealed at a temperature of 800 °C for 20 h for softening to facilitate manufacturing the samples for testing. The steel samples were subjected to austenitizing at 1050 °C for 1 h followed by quenching in oil. The austenitized samples were then tempered at a temperature of 600 °C for 1 h followed by air cooling. The heat treated samples were surface ground to remove scales and processed through a machine to prepare samples for microstructure, impact test, hardness tests and cyclic polarization.

The samples for microstructural observation were prepared by standard metallographic operations and etching was performed by using Villela's reagent. The microstructural images were taken by an optical microscopy and Scanning Electron Microscope (SEM). Rockwell C diamond indentor measured the hardness of the steel samples at a temperature of 120 °C with indentation load and time of 15 Kgf and 3s, respectively. The impact test was conducted at room temperature by Charpy impact tester using V- notch samples with a dimension of 10 × 10 × 55 mm according to ASTM E 23. For cyclic polarization measurement, the specimens with a surface area of 2.5 cm$^2$ were polished with SiC paper from grit 120 to grit 800. The measurement was conducted in the 3.5% Sodium Chloride solution at room temperature using a Gamry G750 system.

**Table 1.** Chemical composition (wt. %) of the 410-1Mo martensitic stainless steels prepared in this work

| Steel | C | S | P | Mn | Si | Cr | Mo | Ni | Fe |
|---|---|---|---|---|---|---|---|---|---|
| 13Cr-1Mo | 0.13 | 0.003 | 0.02 | 0.94 | 0.31 | 13.13 | 0.80 | 0.29 | Bal. |

## 3. Results and Discussion

The optical microscopy images of the microstructure of the 410-1Mo martensitic stainless steel after quenched at 950, 1000, 1050, and 1100 °C and tempered at 600 °C was presented in figure 1. In general, the tempered microstructure of the martensitic stainless steels consists of tempered martensite, carbides and retained austenite. However, in figure 1 (a)-(c), the tempered microstructure of 410-1Mo martensitic steel reveals an additional feature of phase called delta ferrite, the bright island-like phase seen in the images, which was formed due to a non-equilibrium solidification, particularly found in the steels containing high ferrite-stabilizing elements such as Cr and Mo [12]. It is worth to note that at the higher austenitizing temperature of 1100 °C, the delta ferrite disappeared in the tempered microstructure as shown in figure 1(d). It may be due to the increased amount of the alloying elements including Cr and Mo going to the solution at the higher austenitizing temperature. The formation of delta ferrite in the martensitic stainless steels was also reported in literature [12-13]. Meanwhile, the brighter phase appeared in the microstructure of figure 1(d) was not delta ferrite, but it was retained austenite. The austenite phase remained unchanged to martensite during quenching. The Increasing amount of retained austenite in the as-quenched microstructure was attributed to the increasing amount of carbon and carbide-forming elements in the solution of austenite phase which depressed the martensite transformation and lowering the $M_s$ and $M_f$ temperature (Martensite start and Martensite finish, respectively) at or even below room temperature [14].

The martensite and carbides appeared in the microstructure of the tempered martensitic stainless steels were also influenced by the austenitizing temperature. It can be observed from figure 1 that the martensite appeared in different size and density of lath, the higher the austenitizing temperature, the larger lath size. However, the density of the martensite lath seemed to be the most important in the steel austenitized at 1150 °C as displayed in figure 1(c). With respect to carbides, it is observed to appear in the tempered steels with pre-austenitizing at 1000 °C and lower (figure 1 (a) and 1 (b)) and to be absence in pre-austenitizing process at 1050 °C and higher. The carbides density in the tempered steel with pre-austenitizing at 950 °C is qualitatively larger compared to that of the one with pre-austenitizing at 1000 °C. From these findings, the carbides appeared in figure 1 (a) and (b) were from the non-dissolved carbides during the austenitizing heat treatment and the carbides dissolved completely in the solution of the austenite at the temperature of 1050 °C and higher. The existence of carbides in the steels with pre-austenitizing at 950 and 1000 °C was evidenced in the images taken from SEM as presented in figure 2 (a) and 2(b), respectively, where it is not observed in the SEM images in figure 2(c) and 2(d). Small white globular precipitates indicated the carbides in the respective images.

The mechanical properties of the tempered 410-1Mo with various pre-austenitizing temperature are presented in figure 3 and figure 4. It is seen in figure 2 that the tempered steels maintained hardness at pre-austenitizing temperatures of 950 °C to 1000 °C and showed increasing hardness at a pre-austenitizing temperature from 1000 to 1100 °C. It is known that the hardness of the martensitic stainless steels strongly depended on the martensite phase involved in the microstructure. As all the steels in this work were tempered at the same temperature, the difference in hardness of these steels was attributed to the nature of the martensite resulted from different austenitizing temperatures. The lowered hardness of the steel with pre-austenitizing at 950 and 1000 °C corresponded to the tempered martensite with lower lath density and size. Whereas, the higher value of hardness in the steels with pre-austenitizing at 1100 °C corresponded to the higher carbon content of the martensite. As the austenitizing temperature increased, the amount of the carbon dissolved in the solution of austenite also increased resulting in the higher carbon content of the martensite after quenching. The higher the carbon content of the martensite, the higher the hardness of the martensite.

The correlation of pre-austenitizing temperature with an impact energy of the tempered 410-1Mo martensitic steel is shown in figure 4. The impact energy of the steels increased with the increasing pre-austenitizing temperature up to 1050 °C and decreased at the temperature of 1100 °C. The increasing impact energy as pre-austenitizing temperature increased may be due to the increasing retained austenite. Additionally, the decrease in the impact energy at the temperature 1100 °C might correspond to the brittleness of the hard high content carbon martensite and growth of the pre-austenite grains due to high-temperature austenitization [15-16].

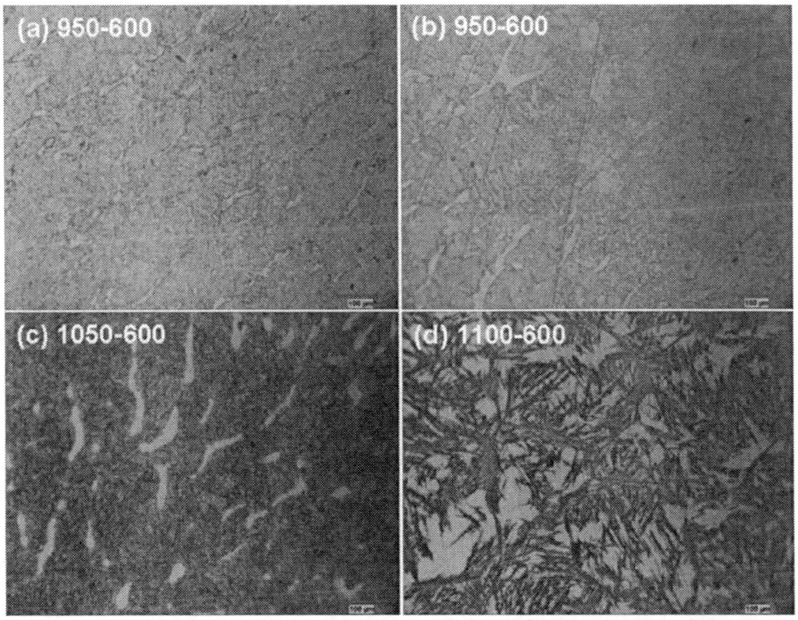


**Figure 1.** Optical microscopy images of the microstructure of the 410-1Mo martensitic stainless steels austenitized at various temperature of (a) 950, (b) 1000, (c) 1050, and (d) 1100 °C and subsequently tempered at 600 °C.

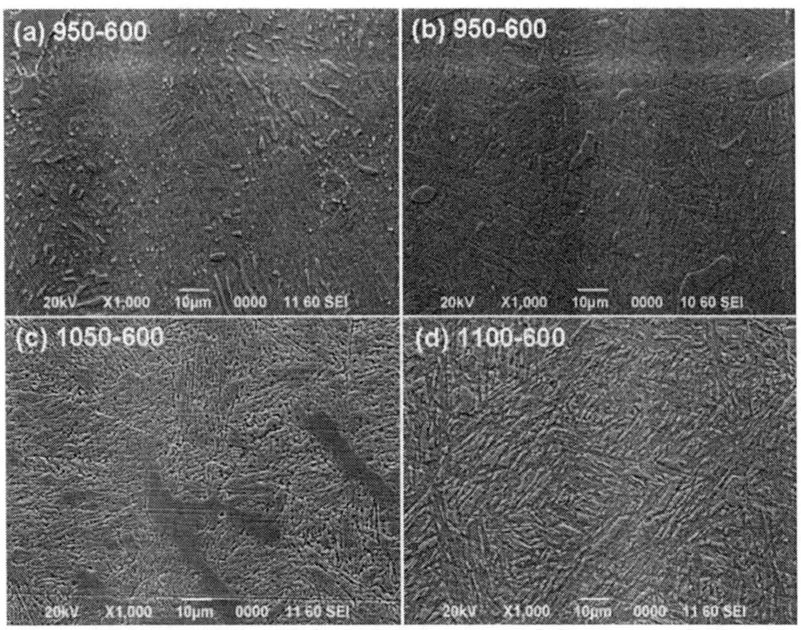


**Figure 2.** SEM images of the microstructure of the tempered 410-1Mo martensitic stainless steel austenitized at various temperatures of (a) 950, (b) 1000, (c) 1050, and (d) 1100 °C and subsequently tempered at 600 °C.

**Table 2.** Corrosion parameters of 410-1Mo martensitic stainless steel austenitized at different temperatures and tempered at 600 °C

| Heat Treatment (°C) | $E_{cor}$ (mV) | $I_{cor}$ ($\times 10^{-6}$ A/cm$^2$) | $E_{pit}$ (mV) |
|---|---|---|---|
| 950-600 | -622 | 74 | -540.4 |
| 1000-600 | -436 | 12.9 | -134.5 |
| 1050-600 | -506 | 19.5 | -291.5 |
| 1100-600 | -899 | 23.2 | -542.6 |

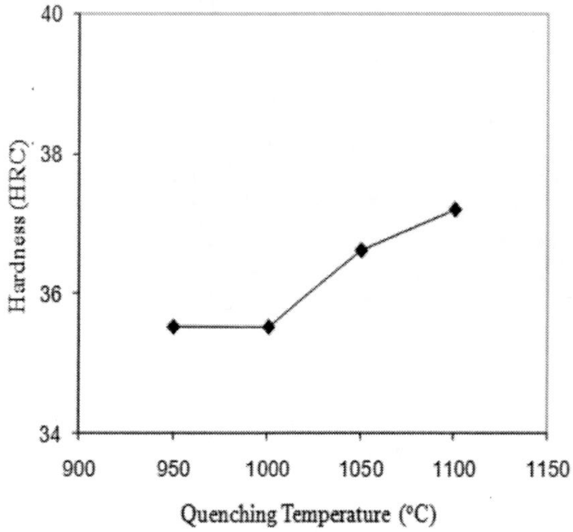


**Figure 3.** The hardness of the 410-1Mo martensitic stainless steels austenitized at various temperatures and subsequently tempered at 600 °C

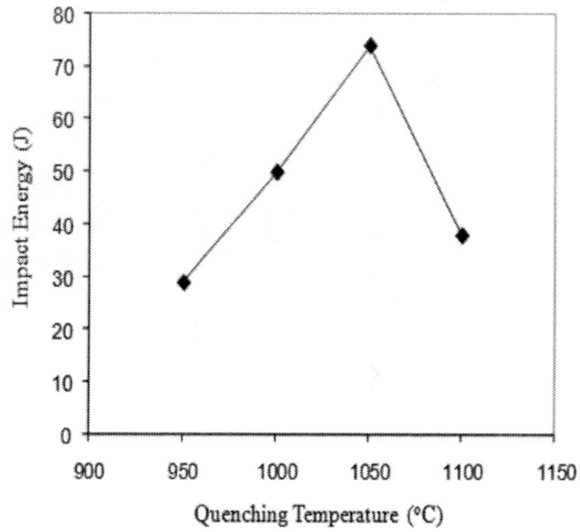


**Figure 4.** The impact energy of the 410-1Mo martensitic stainless steels austenitized at various temperatures and subsequently tempered at 600 °C

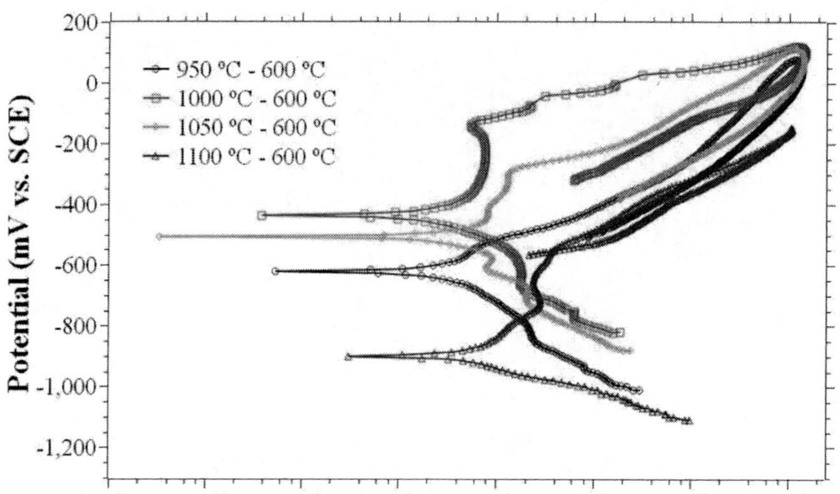


**Figure 5.** Cyclic polarization curves of the 410-1Mo martensitic stainless steel austenitized at various temperatures and subsequently tempered at 600 °C

Figure 5 shows the cyclic polarization curves of the 410-1Mo martensitic stainless steels austenitized at various temperatures and subsequently tempered at 600 °C. The curves are typical for stainless steels, displaying passivation region and breakdown potential (pitting) clearly. It was observed that the different austenitizing temperatures applied to the steels resulted in various positions of the polarization curves in the E-I plot, affecting different corrosion parameters of the tempered steels. The corrosion parameters extracted from the curves are shown in table 2. The corrosion potential and pitting potential are observed to follow the similar trends, they decreased on the increasing austenitizing temperature from 950 °C to 1000 °C, and they increased again from 1000 °C to 1100 °C, whereas the corrosion current density followed the opposite trend. Both the tempered steels that were pre-austenitized at the temperatures of 950 °C and 1100 °C exhibited high susceptibility to pitting corrosion with almost similar low potential pitting values of -540.4 and -542.6 mV, respectively. These might be due to the higher content of carbides in the tempered steels that were pre-austenitized at 950 °C and higher carbon content of the martensite in the tempered steels that were pre-austenitized at 1100 °C. However, the steels that were pre-austenitized at 1100 °C exhibited the repassivation potential on the reverse polarization scan. The repassivation process occurred in this steel might be attributed to a higher content of retained austenite with high content of carbide-forming Cr and Mo elements which were dissolved in the solid solution during a high austenitizing temperature of 1100 °C.

## 4. Conclusion

The influence of austenitizing temperature on microstructure, hardness and impact resistance of the modified 410-1Mo stainless steel has been investigated. The microstructure of the tempered steels revealed additional phase of delta ferrite at pre-austenitizing temperatures of 950-1050 °C and disappeared at 1100 °C. The steels with pre-austenitizing at 1100 °C showed the largest sized lath martensite and largest amount of retained austenite. The tempered steels maintained hardness at pre-austenitizing temperatures of 950 °C to 1000 °C and showed increasing hardness at the austenitizing temperatures from 1000 to 1100 °C. At a range of austenitizing temperature, it was investigated that the steels exhibited higher impact resistance at 1050 °C. The tempered steels that were pre-austenitized at 950 °C and 1100 °C showed the lowest pitting potential due to the existence of carbides and coarse-high carbon martensite, respectively.

## 5. References

[1] Mesa D H, Toro A and Sinatora A 2003 *Wear* **255(1)** 139
[2] Zhu X, Dai O and Zhou O 2009 *Transactions of Materials and Heat Treatment* **30(I)** 11
[3] Krishna S C, Gangwar N K, Jha A K, Pant B and George KM 2015 *Steel. Res. Int.* **86 (1)** 51-57
[4] Fan R, Gao M, Ma Y, Zha X, Hao X and Liu K 2012 *J. Mater. Sci. Technol.* **28 (11)** 1059–66
[5] Kumara B R, Sharmaa S, Mundaa P and Minz R K 2013 *Mater. Des.* **50** 392–398
[6] Cuevas, Rodriguez J A, Clemente C M, Rodríguez J M and Mariaca Y 2014 *Am. J. Mech. Eng.* **2(6)**164-168
[7] Lua S Y, Yaoa K F, Chenb Y B, Wang M H, Liua X and Ge X 2015 *Electrochim. Acta.* **165** 45–55
[8] Singh S and Nanda T 2013 *Int. J. Eng. Technol. Sci. Res.* **1(1)**
[9] Mabruri E, Anwar M S, Prifiharni S, Romijarso T B and Adjiantoro B 2016 *AIP Con. Proc.* **1725** 020039
[10] Prifiharni S, Anwar M S and Mabruri E 2016 *Widyariset* **2(1)** 9-16
[11] Mabruri E, Syahlan Z A, Sahlan, Anwar M S, Romijarso T B and Adjiantoro B 2017 *Int. J. of Eng. and Tech.* (in review)
[12] Wang P, Lu S P, Li D Z, Kang X H and Li Y Y 2008 *Acta Metall. Sin.* **44** 681–685
[13] Wang P, Lu S P, Xiao N M, Li D Z, Li Y Y 2010 *Mat. Sci. and Eng.* A. **527** 3210–16
[14] Barlow L D and Toit M D 2012 *J. of Mat. Eng. and Perf.* **21(7)** 1327.
[15] Balan K P, Venugopal R A and Sarma D S 1998 *Scripta Mater.* **39** 901–905
[16] Yu R, Ye D, Yong Q, Su J, Zhao K and Jiang W 2011 *J. Iron and Steel Res. Int.* **18(11)** 60-66

**Acknowledgments**
This work is a part of the research on the development of high-temperature materials for power plants. The authors thank Research Center for Metallurgy and Materials, Indonesian Institute of Sciences (LIPI) for supporting this research.

# Fluorescence Sensing of Nitrite Ions on Polyvinylpyrrolidone/Zinc Oxide Composites Prepared by Impregnation Method

**L Yuliati[1,2], S Z M So'ad[3], N S Alim[3] and H O Lintang[1,2]**

[1] Ma Chung Research Center for Photosynthetic Pigments, Universitas Ma Chung, Malang 65151, East Java, Indonesia
[2] Centre for Sustainable Nanomaterials, Ibnu Sina Institute for Scientific and Industrial Research, Universiti Teknologi Malaysia, 81310 Johor Bahru, Johor, Malaysia
[3] Department of Chemistry, Faculty of Science, Universiti Teknologi Malaysia, 81310 Johor Bahru, Johor, Malaysia

Email: leny.yuliati@machung.ac.id

**Abstract**. A series of polyvinylpyrrolidone/zinc oxide (PVP/ZnO) composites with different loading amounts of PVP was prepared by an impregnation method. Successful formation of the composites was analyzed by the Fourier transform infrared (FTIR), diffuse reflectance ultraviolet-visible (DR UV-Vis) and fluorescence spectroscopies. Prepared composites were then further tested as fluorescence sensors by conducting quenching studies in the presence of nitrite ions ($NO_2^-$). Among the prepared composites, the ZnO with 1% PVP exhibited the highest sensing performance for the $NO_2^-$detection ($K_{sv}$ = 0.07 μ$M^{-1}$). The efficiency of the composite was *ca.* 1.7 and 1.4 times higher than the bare ZnO and the best composite prepared by the physical mixing method, respectively. These results suggested that impregnation method is a suitable method to prepare the PVP/ZnO composites as fluorescence sensor for the $NO_2^-$detection.

**Keywords**: polyvinylpyrrolidone/zinc oxide, composite, impregnation, fluorescence sensor, nitrite ion

## 1. Introduction

Nitrite ($NO_2^-$) is commonly used as a preservative and fertilizing agent for food as well as an inorganic fertilizer. Since $NO_2^-$ has ahigh level of solubility in water, these ions also have high mobility, which has led to constant threats whenever agriculture processes are involved [1]. The emergence of $NO_2^-$as one of the hazardous contaminants has become a grave concern since $NO_2^-$can react with secondary or tertiary amines to form nitrosamines, which are carcinogenic compounds [2-4]. Ingestion of $NO_2^-$might cause endogenous nitrosation that is probably carcinogenic to human [5] and associated with colorectal and stomach cancers [6]. Furthermore, $NO_2^-$is also known to result in respiratory deficiencies as $NO_2^-$ would react with hemoglobin forming methemoglobin, which disturbs the transport of oxygen [7, 8]. Therefore, it is a must to develop tools or methods that are simple, non-toxic, selective and sensitive to detect the presence of $NO_2^-$.

For nitrite/nitrate determination, one of the most generally reported methods is spectroscopy method owing to its great limit of detection and easy protocols [1]. Particularly, fluorescence sensors have attracted many attentions since they are simple, their signal can be amplified, they can be fabricated easily into devices, and they can be combined with various outputs [9]. One of the best candidates for a simple, inexpensive, and non-toxic solid material in the detection of $NO_2^-$ is a semiconductor-based sensor. Particularly, zinc oxide (ZnO) has been widely used as a sensor for thedetection of various gasses, pollutants, and biomolecules [10-13].

As semiconductor gas sensor, ZnO has been extensively investigated for the detection of gas leaks and environmental monitoring of gaseous pollutants. For instance, ZnO showed excellent sensitivity to detect some flammable gasses as well as corrosive vapors with fast response-recovery characteristic, great selectivity, and stability [10]. ZnO was also reported serving as a sensor for sensing liquefied petroleum gas (LPG) and ethanol [11]. It was observed that ZnO also has good sensitivity at low temperature and it also showed better response and shorter recovery time. The application of ZnO as a sensor for methanal and xylene has also been investigated [12]. In addition, ZnO was also pointed out to improve signal in DNA sensor arrays [13].

In order to enhance the performance of ZnO as a sensor, ZnO can be modified with polymers [14-17]. It is recognized that grafting polymers onto the surface of ZnO was an effective way to improve its dispersion in a polymer matrix and the composites showed improved properties [14]. One of the potential polymers is polyvinylpyrrolidone (PVP), which has been reported to be a potential fluorescence sensor for the detection of nitrate ($NO_3^-$) [18]. Recently, the addition of PVP to ZnO by a physical mixing method has been reported to be able to increase the sensing efficiency of ZnO towards the $NO_2^-$detection [15]. In this study, we newly reported a higher enhancement of the fluorescence sensing capability for $NO_2^-$ detection when the PVP/ZnO composites were prepared by an impregnation method.

## 2. Experimental Method

### *2.1. Materials*

The materials used were zinc acetate dehydrate 99.5% ($Zn(CH_3COO)_2.2H_2O$, QRëc), sodium hydroxide pellet 99 % (NaOH, QRëc), cetyltrimethylammonium bromide ($C_{19}H_{42}BrN$, CTAB, Fischer Scientific), polyvinylpyrrolidone 45% in $H_2O$ ($(C_6H_9NO)_n$, PVP, Sigma-Aldrich), sodium nitrite 99.5% ($NaNO_2$, Merck), and distilled water.

### *2.2. Methods*

*2.2.1. Preparation of ZnO.* Zinc acetate dihydrate, $Zn(CH_3COO)_2.2H_2O$ was used as a precursor to synthesize the ZnO. The $Zn(CH_3COO)_2.2H_2O$ (4.5 g) was dissolved in distilled water (100 ml) as solution A. For the preparation of solution B, NaOH (6.4 g) was dissolved in distilled water and followed by the addition of CTAB (7.28 g) into the mixture and stirred for 1 hour to obtain a homogeneous solution. After 1 hour, solution A was added into solution B and heated at 70 °C for 1 hour. The remaining solid was subsequently filtered and washed with distilled water. The solid powder was dried at room temperature and then calcined at 500 °C at a rate of 1.0 °C $min^{-1}$ and further tempered for another 1 h at the same temperature.

*2.2.2. Preparation of PVP/ZnO composites.* Various PVP/ZnO composites were prepared by an impregnation method, in which the concentration ratios were varied to 0.1, 1 and 3 % (v/w). The sampleswere labeled as PVP(x)/ZnO, with x that shows the ratio of PVP to the ZnO. For a typical synthesis of PVP (1)/ZnO composite, the ZnO (1 g) was dispersed in distilled water (30 mL). The PVP (0.2020 mL) from 5 % PVP solution was then added dropwise to the mixture containing ZnO and distilled water, followed by stirring and drying at 80 °C on a hot plate.

*2.2.3. Characterizations.* As for characterizations of the prepared ZnO and PVP/ZnO composites, FTIR spectroscopy (Thermo Scientific, Nicolet iS50) was applied to investigate the functional groups in the prepared ZnO and PVP/ZnO composites. For the measurement, potassium bromide (KBr) pellet technique was used. The FTIR spectra were measured in the range of 400 to 4000 $cm^{-1}$ with 32 scan number. Nitrogen gas was purged into the system before analysis to exclude moisture for better background correction. DR UV-Vis spectroscopy (UV-2600 with an integrating sphere, Shimadzu) was used to measure the absorption spectra of the samples. The absorption spectra were recorded in the wavelengths range from 250 to 500 nm. Scan speed was at the rate of 300 nm $min^{-1}$ and the slit width was set at 1 nm. As a reference, barium sulfate ($BaSO_4$) was used in the measurement. The fluorescence spectra were recorded on the fluorescence spectrophotometer (FP-8500, JASCO) at room temperature. The emission spectra were measured when the excitation wavelength was 365 nm.

### *2.3. Sensor tests*
Sensor tests for the detection of $NO_2^-$ were carried out based on the quenching study, by monitoring the decreases in the emission intensity of the sensor when various concentrations of the $NO_2^-$ were added into the sensor. Typically, the sensor sample (0.3 g) was transferred into a fluorescence sample holder and placed at the fluorescence instrument. Initial readings for excitation and emission peaks were firstly recorded in the absence of the $NO_2^-$. As for the quenching test, the selected concentration of $NO_2^-$ (2–10 μM) with a volume of 10 μL was added slowly on top of the sample. The emission spectra were then recorded at an excitation wavelength of 365 nm for ZnO and all the PVP/ZnO composites. The efficiency for the detection of the $NO_2^-$ was further determined by Stern-Volmer plot on each of the prepared sample.

## 3. Results and Discussion

### *3.1. Functional groups*
The functional groups of PVP, ZnO, and PVP/ZnO composites were examined by using the FTIR spectroscopy. FTIR spectroscopy gives qualitative information on the type of functional groups that present in the samples. As shown in Figure 1(a), PVP gave OH stretching vibration from a water molecule at around 3447 $cm^{-1}$ due to the hydrophilic nature of the PVP, while the peaks observed around 2800-3000 $cm^{-1}$ were assigned to C-H stretching. The characteristic peaks of PVP could be seen at 1655 and 1461 $cm^{-1}$, which were corresponding to C=O and C–N stretching, respectively. The presence of C–N stretching was also shown at around 1370 $cm^{-1}$ and the peaks below 700 $cm^{-1}$ gave the evidence of C–C chains and N–C=O bending [19].

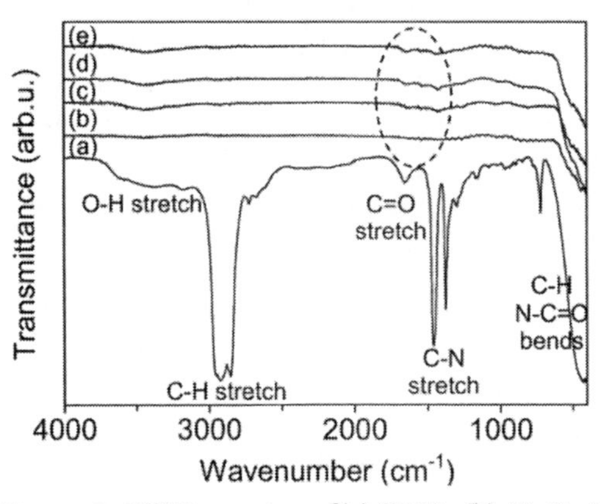


**Figure 1.** FTIR spectra of(a) PVP, (b) ZnO, (c) PVP(0.1)/ZnO, (d) PVP(1)/ZnO and (e) PVP(3)/ZnO samples

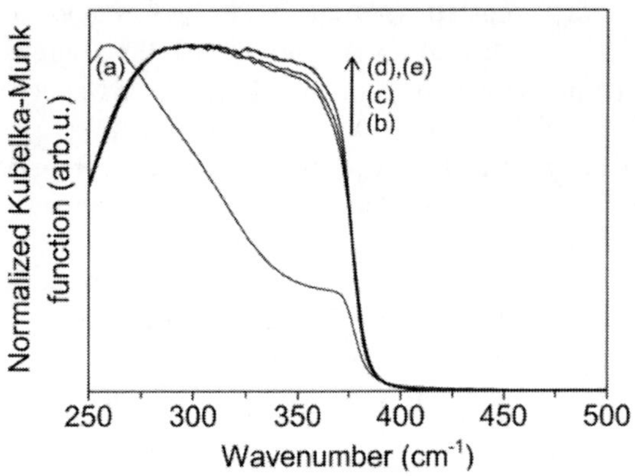


**Figure 2.** DR UV-Vis spectra of(a) PVP, (b) ZnO, (c) PVP(0.1)/ZnO, (d) PVP(1)/ZnO, and (e) PVP(3)/ZnO samples

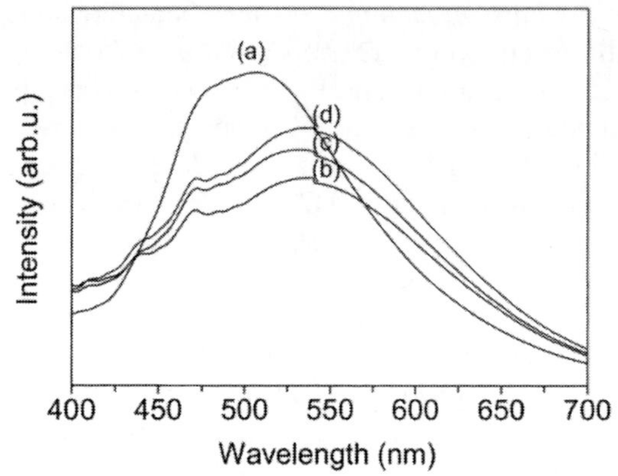


**Figure 3.** Emission spectra of(a) ZnO, (b) PVP(0.1)/ZnO, (c) PVP(1)/ZnO, and (d) in PVP(3)/ZnO samples

Displayed in figure 1(b) is the FTIR spectrum of ZnO. The ZnO showed a sharp peak at around 420 cm$^{-1}$ that could be assignedas the stretching mode of Zn−O [20, 21]. Figures 1(c)−(e) show the FTIR spectra for the PVP/ZnO composites. All the composite samples gave the presence of Zn−O stretching as well as two peaks appeared at 1647 and 1434 cm$^{-1}$ that could be related to C=O and C−N stretching, respectively. Both of the latter peaks came from the PVP. It has been suggested that the peak shift can be related to the hydrogen bond formation [19, 20]. Therefore, the shifting observed on the PVP/ZnO samples might be the outcome of the hydrogen bonding interaction between the surface hydroxyl of ZnO and the functional group in the PVP. Another responsibility that can be proposed was the existence of a strong coulombic interaction between the ZnO and the polymeric matrix [22]. When the amount of PVP in the PVP/ZnO composite increased, the intensity of C=O stretching and C−N stretching also increased owing to the increment amount of PVP contained in the PVP/ZnO composites. The FTIR spectra suggested the successful preparation of the PVP/ZnO composites.

### 3.2. *Optical properties*
The optical properties of the PVP, ZnO, and PVP/ZnO composites were investigated using the DR UV-Vis spectroscopy and are displayed in Figure 2. As shown in Figure 2(a), PVP exhibited intense narrow bands centered at 265 and 370 nm, which might be due to the presence of C=O and N−C groups, respectively [19]. On the other hand, Figure 2(b) shows a strong absorption peak of ZnO in the UV region, which was centered at 330 nm due to the electron charge transfer in the Zn−O linkage [23, 24]. The DR UV-Vis spectra of the PVP/ZnO composites are shown in Figures 2 (c)−(e). All the composite samples showed similar absorption peak to each other, where the main absorption peak of ZnO was clearly observed and the loading amount of PVP also slightly increased the absorption band of PVP. It can be observed that the addition of PVP did not give considerable influence to the band structure of the ZnO.

### 3.3. *Fluorescence properties*
Fluorescence spectroscopy is a powerful non-destructive technique to explore the fluorescence properties of thematerials. Figure 3 shows the emission spectra of the prepared ZnO and PVP/ZnO composites. Shown in Figure 3(a) is the emission spectrum for ZnO when the excitation wavelength at365 nm was used. ZnO gave one broad emission peak at 517 nm, which has been recognized as the green emission of ZnO, reflecting the presence of oxygen defect on the surface and deep level of the ZnO [22]. As can be seen from Figure 3(b)−(d), all PVP/ZnO composites have a lower emission intensity as compared to the ZnO. In addition to the reduced fluorescence intensity, the addition of PVP also shifted the emission peak to a longer wavelength. The decrease in fluorescence intensity and

the strong peak shifting indicated that there were interactions existed between the PVP and the ZnO. The most possible interactions between PVP and ZnO would occur via hydrogen bonding and/or coulombic interaction.

### *3.4. Sensing performance*

The sensing performance of the ZnO and the PVP/ZnO composites towards the $NO_2^-$was evaluated via quenching tests by monitoring the decreases in the emission intensity after the additions of various concentrations of the $NO_2^-$(0−10 $\mu$M) to the ZnO and the PVP/ZnO composites. All samples showed similar results, where the emission intensity was reduced gradually with the increase in the concentration of the $NO_2^-$. As arepresentative, the changes of the emission spectra in the absence and presence of $NO_2^-$for the PVP(1)-ZnO composite are illustrated in Figure 4(a).

The relative emission intensity can bestrongly related to the concentration of $NO_2$that investigated by a Stern-Volmer plott. In the plot, the relative emission intensity is expressed as a function of the $NO_2^-$concentrations, following the Stern-Volmer plot as shown in Equation (1).

$$I_o/I = K_{SV}[Q] + 1 \tag{1}$$

Where $I_o$ and $I$ show the fluorescence intensity without and with $NO_2^-$, respectively, $Q$shows the concentrations of $NO_2^-$, and $K_{SV}$ is the Stern-Volmer quenching constant. The calculated $K_{SV}$ indicated the efficiency of the sensor. Higher $K_{SV}$ value suggested a larger efficiency of the sensor. All the ZnO and PVP/ZnO composites gave almost linear curves, suggesting that the ZnO and the PVP/ZnO composites can be potential fluorescence sensors in the detection of the $NO_2^-$. As representative of the samples, Figure 4(b) illustrates the linear Stern-Volmer plots of the ZnO and the PVP(1)/ZnO composite in the presence of $NO_2^-$.

$K_{SV}$values of the ZnO and the PVP/ZnO composites are given in Figure 5. The $K_{SV}$ value of the ZnO was determined to be 0.04 $\mu M^{-1}$. It was evident that the addition of PVP from 0.1 to 1% improved the ZnOsensing performance, where the $K_{SV}$ value increased from 0.04 to 0.05 and 0.07 $\mu M^{-1}$, respectively. Unfortunately, the sensing performance decreased to 0.026 $\mu M^{-1}$ when the loading amount of PVP was too high (3%). It was shown that the addition of PVP reduced the ZnO intensity, thus, when the loading of PVP was too high, the reduced emission intensity would decrease the sensitivity of the sensor. This study showed that the PVP(1)/ZnO was the best sample showing the highest efficiency towards the detection of $NO_2^-$.

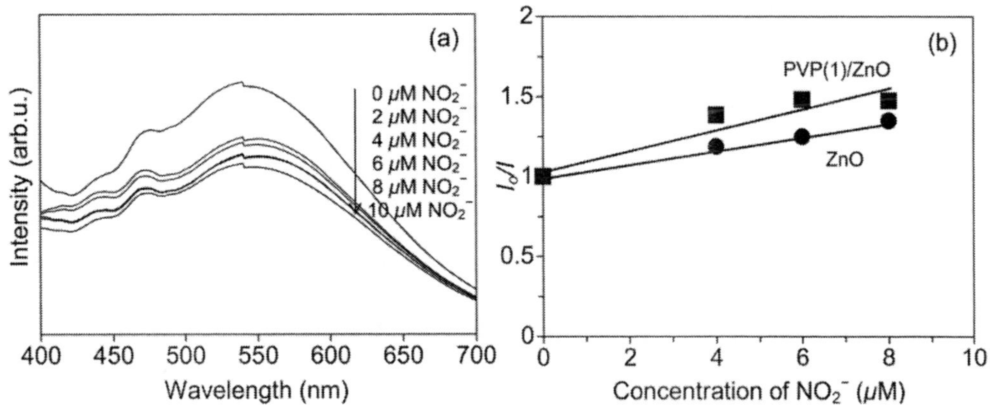


**Figure 4.** (a)Emission spectra of PVP(1)/ZnO in the absence and presence of different concentrations of $NO_2^-$and (b) Stern-Volmer plots of PVP(1)/ZnO and ZnO samples

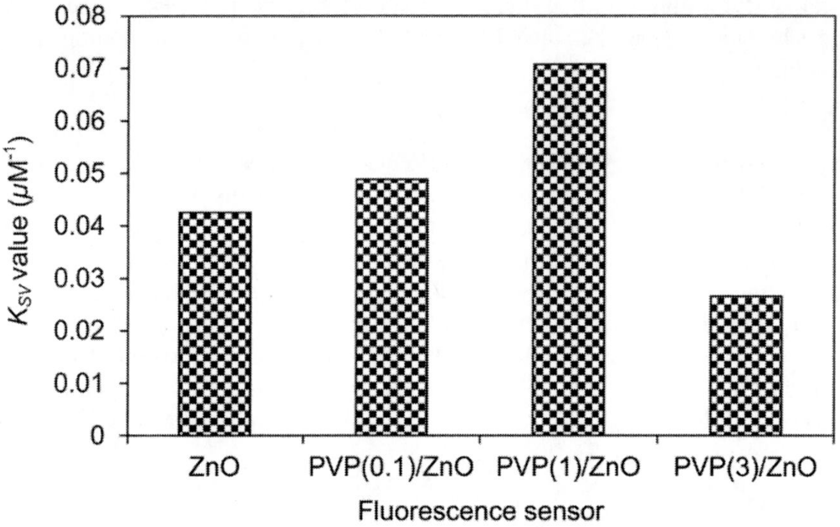


**Figure 5.** Comparison of $K_{SV}$ values of ZnO and PVP-ZnO composites

It was reported that the best PVP/ZnO composite prepared by physical mixing, the PVP(0.1)/ZnO, gave $K_{SV}$ value of 0.05 $\mu M^{-1}$ [15]. Comparison to the best PVP/ZnO composite prepared by physical mixing showed that the best PVP/ZnO composite developed by the impregnation method gave a higher $K_{SV}$ value of 0.07 $\mu M^{-1}$. It was proposed that better interaction between the PVP and the ZnO prepared by the impregnation method would be the crucial factor to get a higher efficiency. Since the impregnation method employed a sonication step, the PVP would be dispersed better on the ZnO than that prepared by physical mixing. This study demonstrated that the impregnation method was better than the physical mixing to prepare the PVP/ZnO composites as fluorescence sensor of the $NO_2^-$.

## 4. Conclusion

The PVP/ZnO composites were successfully prepared by an impregnation method as evidenced by FTIR, DR UV-Vis, and fluorescence spectroscopies. The prepared ZnO and composites gave good linear Stern-Volmer plots with various concentrations of $NO_2^-$, suggesting the potential capability of these samples as fluorescence sensors for the $NO_2^-$detection. The best composite, which was PVP(1)-ZnO, gave $K_{SV}$ value of 0.07 $\mu M^{-1}$, which was 1.7 times higher than that of the ZnO.

## 5. References

[1] Moorcroft M J, Davis J and Compton R G 2001 *Talanta* **54** 785
[2] Lijinsky W and Epstein S S 1970 *Nature* **225** 21
[3] Cassens R G 1997 *Food Technol.* **51** 53
[4] Tricker A R and Preussmann R 1991 *Mutat. Res.***259** 277
[5] Grosse Y, Baan R, Straif K, Secretan B, El Ghissassi F, Cogliano V and WHO International Agency for Research on Cancer Monograph Working Group 2006*Lancet Oncol.* **7** 628
[6] Bouvard V, Loomis D, Guyton K Z, Grosse Y, El Ghissassi F, Benbrahim-Tallaa L, Guha N, Mattock H, Straif K, and International Agency for Research on Cancer Monograph Working Group 2015*Lancet Oncol.* **16** 1599
[7] Fan A M and Steinberg V E 1996 *Regul. Toxicol. Pharmacol.* **23** 35
[8] Niţuc E, Năstase V, Mihăilescu G and Chioveanu D 2010 Rev. Med. Chir. Soc. Med. Nat. Iasi. **114** 580
[9] Kim H N, Guo Z, Zhu W, Yoon J and Tian H 2011 *Chem. Soc. Rev.* **40** 79
[10] Ma S, Li R, Lv C, Xu W and Gou X 2011 *J. Hazard. Mater.* **192** 730
[11] Baruwati B, Kumar D K and Manorama S V 2006 *Sensor Actuator B-Chem.* **119** 676

[12] Cao Y, Hu P, Pan W, Huang Y, and Jia D 2008*Sensor Actuat. B-Chem.* **134** 462
[13] Kumar N, Dorfman A and Hahm J I 2006 *Nanotechnology* **17** 2875
[14] Tshabalala M A, Dejene B F and Swart H C 2012 *Physica B* **407** 1668
[15] Tang I H, Mohd So'ad S Z, Lintang H O and Yuliati L 2015 *Malay. J. Fund. Appl. Sci.* **11** 139
[16] Tang H, Yan M, Ma X, Zhang H, Wang M and Yang D. 2006 *Sensor Actuat B-Chem.* **113** 324
[17] He J H, Ho C H and Chen C Y 2008 *Nanotechnology* **20** 1
[18] Tang I H, Sundari R, Lintang H O and Yuliati L 2016 *Malay. J. Anal. Sci.* **20** 288
[19] BashaMA F 2010 *Polym. J.* **42** 728
[20] Liufu S, Xiao H and Li Y 2004 *Powder Technol.* **145** 20
[21] Lepot N, Van Bael M K, Van den Rul H, D'Haen J, Peeters R, Franco D and Mullens J 2007 *Mater. Lett.* **61** 2624
[22] Feng W, Tao H, Liu Y and Liu Y 2006 *J. Mater. Sci. Technol.* **22** 230
[23] Hussin F, Lintang H O and Yuliati L 2015*Malay. J. Fund. Appl. Sci.***11** 134
[24] Hussin F, Lintang H O and Yuliati L 2016*Malay. J. Anal. Sci.* **20** 102

**Acknowledgements**
The work was financially supported by the Ministry of Higher Education (MOHE, Malaysia) and the Universiti Teknologi Malaysia (UTM, Malaysia) through a Tier-1 Research University Grant (cost center code: Q.J130000.2509.06H66).

# Blending of Low-Density Polyethylene and Poly-Lactic Acid with Maleic Anhydride as A Compatibilizer for Better Environmentally Food-Packaging Material

**A H Setiawan[1,2] and F Aulia[1,2]**

[1]Research Center for Chemistry, Indonesian Institute of ScienceKawasan PUSPIPTEK, Serpong 15314, Tangerang Selatan, Banten, Indonesia
[2] Technology Faculty-Pamulang University, Tangerang Selatan, Banten, Indonesia

Email: achmad.hanafi@gmail.com

**Abstract**. The common conventional food packaging materialsare using a thin layer plastic or film, which is made of a synthetic polymer, such as Low-Density Poly Ethylene (LDPE). However, the use of these polymers hasan adverse impact on the environment, because the synthetic polymersare difficult to degrade naturally. Poly-Lactic Acid (PLA) is a biodegradable polymer that can be substituted to synthetic polymers. Since LDPE and PLA have a difference in polarity, therefore the first step of research is to graft them with maleic anhydride (MAH) for increasing the properties of its miscibility. The interaction between them is confirmed by FTIR; whereas the environment issueis characterized by the water adsorption and biodegradability. The FTIR spectra indicated that there had been an interaction between LDPE and MAH and LDPE/LDPE-g-MAH/PLA blend.Increasing PLA content in the blend affected to the increasing in their water absorption and biodegradable. Poly-blend with 20% PLA content was the optimum composition for environmentally food packaging.

**Keywords**: LDPE, PLA, maleic anhydride, compatibilizer, environment; food packaging.

## 1. Introduction

The common packaging plastic material is made from synthetic polymer. However, the use of this kind of polymer hasa negative impact on the environment, due to its difficulty to degrade naturallywhether by biotic components such as micro-organisms decomposers orby abiotic components such as the sunshine. Therefore, it is necessary to develope technologyinincreasing its degradability degree or even to look for new degradable materials.Utilization of natural polymers is an alternative packaging material, and it needs to be developed.In the packaging of foodstuffs, there are two different types of packaging functions, namely the primary packaging or containers are directly attached to the foodstuffs and the secondary packaging that are not directly attached to the food [1]. Edible coating such as bio-cellulose, carrageenan, chitosan, and beeswax are the natural polymer,but they can only be used for the food primer packaging. PLA is another natural polymer which is predicted suitable for a secondary foodpackaging. However, the price ishigherthan the synthetic polymer. Therefore, blending this natural polymer with the synthetic polymer is an alternative to making a packaging product more degradable. Since all type of food packaging should be non-toxic and inert so that no chemical reaction occurs that can cause changes in color, flavor and other changes; besides also must meet special requirements such as protecting food from contamination, protecting the water content and fat, preventing the entry of odors and gases, protect food from sunlight, and resistant to pressure or impact [1]. This product can fulfill the requirements as a good packaging material.

PE is one of the elastomer synthetic material which has white and transparent properties and its melting point between 110-137 °C. In general, a property of PE is resistance to chemical materials. At room temperature, PE does not dissolve in an organic and inorganic solvent. There are some derivatives of PE (Figure 1), include *High-Density Polyethylene* (HDPE); Low-density Polyethylene (LDPE), Linear Low-Density Polyethylene (LLDPE), etc. Some PE such as LDPE has a high structure with a short and long branched. However, HDPE has a straight branched. LLDPE has a straight branched with short branched [2]. HDPE polymer is harder but workable so widely used as a kitchen utensil bucket, for example, pan, also for wire and cable coatings. LDPE is always used for thin layer food packaging, plastics bag, and raincoat.

PLA is a natural polymer has astructure as the Figure 2, derived from renewable sources with an esterification process lactic acid obtained by fermentation by bacteria using substrates starches or simple sugars [3]. PLA has no reactive group at all on the side chains [4-5]. PLA can find in many plants such as rice, cassava, maize, and sago. Therefore, they can be renewable, and able to give nature biodegradable (compostable). Another advantage of the use of PLA including; able to improve the physical and mechanical properties, during the production process using a carbon dioxide gas, providing significant energy supply, and help improve the stability of the economic and the agricultural sector [6]. PLA has a heat-resistant property, strong, and an elastic polymer and capable of providing biodegradable nature of the PE.

Development of polymer blending technology has grown rapidly, both in science and in use commercially. The advantage of this technology is able to improve the properties or performance of the low-cost material. However, themostpolyblend is an immiscible blend, where the mix tends to very low entropy, resulting in the formation of polyblend with low mechanical properties both during the process and use.Mixed Polymer which has a mixed compatibility between the constituent components of polyblends can be made using reactive compatibility or different additives. Blending compatibility may affect the morphology and properties of polyblends. Mixing compatibility strategies focusing on applications such as plastic post-consumer recycling without sorting, production of multi-layer films, packaging products, as well as the development of new materials based on plastic/plastic elastomer/elastomer mixture [7,8]. A mixture of LDPE with PLA is an immiscible blend because there are differences in their polarity. Therefore, it is necessary to add a compatibilizer such as polyethylene-grafted Maleic anhydride (PE-g-MAH) in order to increase the properties of its miscibility blend [9,10]. This addition is conducted using grafting techniques. The techniques have been utilized in various fields of application, among others to change the properties of the parent polymer with a purpose, such as increasing the adhesive strength of the polymer [11], the biodegradation of polymers [12]. Based on the description above, this research will be conducted studies on the effects of adding PLA to LDPE against water absorption, functional group change, and biodegradability of polyblend.

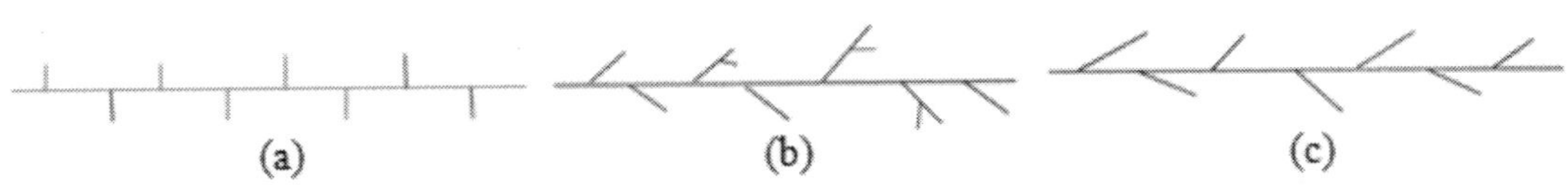


**Figure 1.** Polyethylene chain structure of (a) HDPE, (b) LDPE, and (c) LLDPE [2]

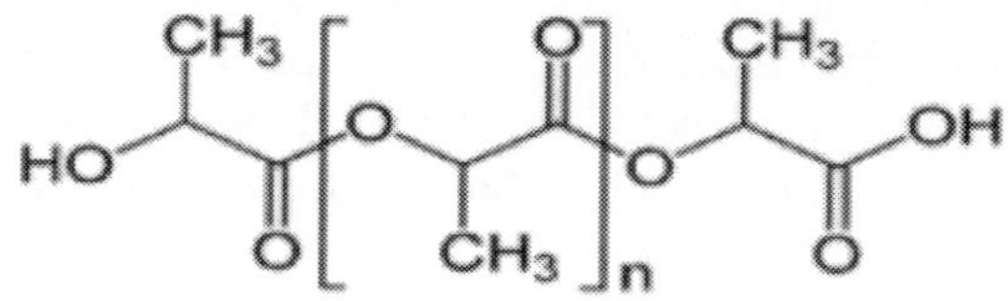

**Figure 2.** Polylactic acid structure [7]

## 2. Experimental Method

### *2.1. Materials*
All starting compounds were purchased from E.Merck and used as such without any further purification, such as Maleic Anhydride (MAH) and Chlorobenzene;Low-DensityPolypropylene (LDPE) was made of Chandra Asri Petrochemical Ltd., Poly Lactic Acid (PLA) was donated by RC Biomaterials-LIPI Cibinong.

### *2.2. Instuments*
FTIR (Shimadzu FTIR- IRPrestige-21) with a potassium bromide (KBr) disks or films was conducted to determine the functional groups of compounds.

### *2.3. Preparation of LDPE-g-MAH*
50 mL of Chlorobenzene was heated at three-neck round bottom flask until the temperature reaches 150°C. Then, 4.5 g of LDPE and 0.5 g MAH were added to the flask and stirred for 45 minutes. The mixture solution was poured into a petri dish and cooled for 24 hours in a fume hood.

### *2.4. Preparation of LDPE / LDPE-g-MAH / PLA*
50 mL of Chlorobenzene was heated at three-neck round bottom flask and heated until the temperature reaches 150°C.Then,0.15 mg LDPE-g-MAH and certain ratio of LDPE/PLA were addedso that thetotal w/w ratio were (100:0); (90:10); (80:20); (70:30); (60:40) and (50:50). Each compositionwasstirred for 45 minutes. Further, the hot solution was poured and cooled for 24 hours in a fume hood.The samples were homogenized and pelleted with KBr. Then the pellet was examined by infrared light on the instrument of Prestige-21 Shimadzu IR at a region frequency between 3500 $cm^{-1}$ to 400 $cm^{-1}$. The obtained IR spectrum was analyzed to their appearance of their functional groups.

### *2.5. Test of Water Absorption Index (WAI) ASTM-570*
The specimen was dried beforehand in an oven for 1 hour at 110°C and then weighed as the initial weight ($W_0$) and then soaked in distilled water for 24 hours. Picked up the sample and the water containing on the sample surface is removed with tissue paper, then weighted. Repeatedly the sample put back into distilled water for 10 seconds, and picked out, dried and weighted. The Immersion and weighing procedure were repeated until reached the constant weight ($W_t$). Water Absorption Index then was calculated by equation (1).

$$\%WAI = \frac{(W_t - W_0)}{W_0} \; x \; 100\% \tag{1}$$

where % $WAI$, $W_0$, and $W_t$ respectively represent water absorption index (%), initial mass (g), and final mass (g).

### *2.6. Biodegradability Test*
The sample was cut tosize about 5cm x 1cm, then dried in a desiccator to obtain a stable weight ($W_0$). Samples were then buried in the ground for six days, and then cleaned by water soaking and then

redried in the desiccator. Finally, the samples were weighed ($W_t$). The percentage of mass loss was then calculated by Equation (2).

$$\%W_{loss} = \frac{(W_0 - W_t)}{W_0} \times 100\% \tag{2}$$

where % $W_{loss}$, $W_0$, and $W_t$ respectively represent weight loss (%), initial mass (g), and final mass (g).

## 3. Results and Discussion

### *3.1. Functional groups of LDPE and LDPE-g-MAH*

MAH grafting process aims to increase the compatibility between matrixes to be reacted, particularly for a matrix which has a different polarity. In this study LDPE-g-MAH used to hold two compounds that have reactive groups with different polarity, i.e. LDPE which have reactive groups in the form of a double bond is nonpolar, and PLA has reactive groups such as hydroxyl group (OH-) which is polar.A sheet transparent and solid white has resultedafter cooling for 24 hours in a fume hood. Each of the starting material used for the synthesis of bio-composite was spectroscopicallycharacterized by FTIR. Figure 3 is a spectrum of original LDPE and LDPE-g-MAH and Figure 4 is a pure polylactic acid spectra.

As depicted in Figure 3, the wavenumber of 2848.86 $cm^{-1}$, there wasa strong peak of $CH_2$ groups, while at the wave number of 1463.97 $cm^{-1}$represent the absorption of C-H group. The functional groups H-C-Hoccurred in a wave number of 723.31 $cm^{-1}$. Those absorption bands wasa typical absorption for polyethylene [2]. The wavelength of 2916.37 $cm^{-1}$indicates the absorption of C-H stretching from MAH, and the absorption peak in the region of 1707 $cm^{-1}$shows the absorption of a carbonyl group (C = O) of MAH. In addition, the wave number of 1217.08 - 1263.37 $cm^{-1}$represents an absorption of a C-O group of MAH. These results indicated that an interaction between LDPE and MAHoccurred. Figure 5 shows the expectedreaction takes place in this work.

The characteristics of functional groups on pure PLA (Figure 4) is 3571 $cm^{-1}$ as the -OH stretch (free); 2997 $cm^{-1}$ (asym), 2946 $cm^{-1}$ (sym), 2881 $cm^{-1}$ stretch -CH-; 1757 $cm^{-1}$ -C O carbonyl stretch; 1456 $cm^{-1}$ -$CH_3$ bend. The wavenumber of 1382 $cm^{-1}$, and 1365 $cm^{-1}$indicate a -CH- deformation including symmetric and asymmetric bend; the wave number of 1225 $cm^{-1}$ -C O bend; the peak at 1194 $cm^{-1}$, 1130 $cm^{-1}$, 1093 $cm^{-1}$represent the -C-O- stretch; 1047 $cm^{-1}$ -OH bend; 956 $cm^{-1}$, 921 $cm^{-1}$ -$CH_3$ rocking; 926 $cm^{-1}$, 868 $cm^{-1}$ stretch -C-C- [6].

Figure 6 is a typical FTIR spectra of polyblend LDPE/LDPE-g-MAH/PLA. It can be seen from the figure that the methine group of the LDPE was disappeared on the polyblend spectra. The carbonyl group in MAH spectra at 1710 $cm^{-1}$ was a shift to carbonyl ester group at MAH spectra showed at 1755-1761 $cm^{-1}$. The shift position of the carbonyl group is caused by the esterification between LDPE-g-MAH and the carboxylic group of PLA. Figure 7 is a reaction mechanism proposal for the reaction amongLDPE, LDPE-g-MAH, and PLA. Figure 8, were a series of LDPE/LDPE-g-MAH/PLA spectra with the ratio of LDPE: PLA were (100:0); (90:10); (80:20); (70:30) ;( 60:40) and ( 50:50).

IOP Conf. Series: Materials Science and Engineering **202** (2017) 012087 doi:10.1088/1757-899X/202/1/012087

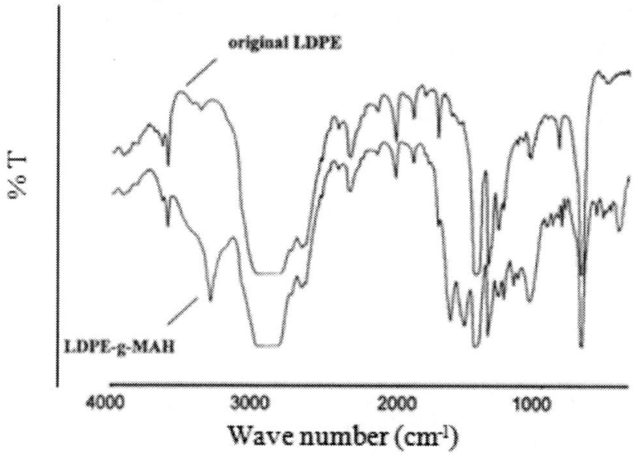


**Figure 3.** FTIR-spectra of original LDPE and LDPE-g-MAH

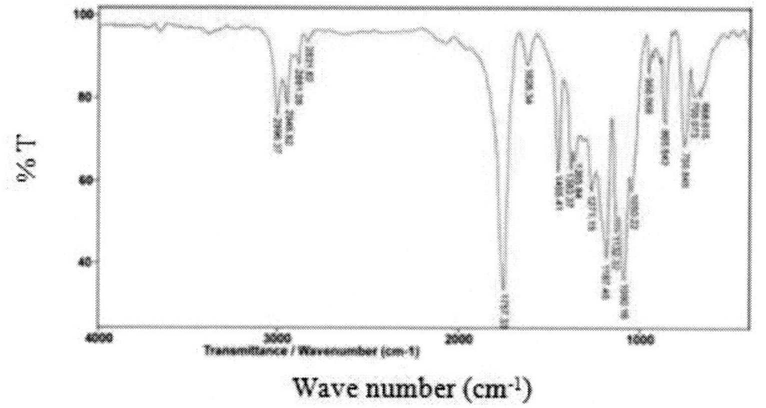


**Figure 4.** Pure Polylactic Acid Spectra

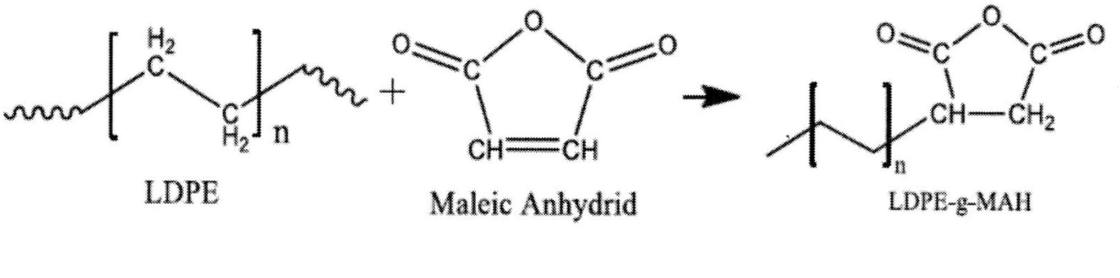


**Figure 5.** Reaction of LDPE and MAH

IOP Conf. Series: Materials Science and Engineering **202** (2017) 012087 doi:10.1088/1757-899X/202/1/012087

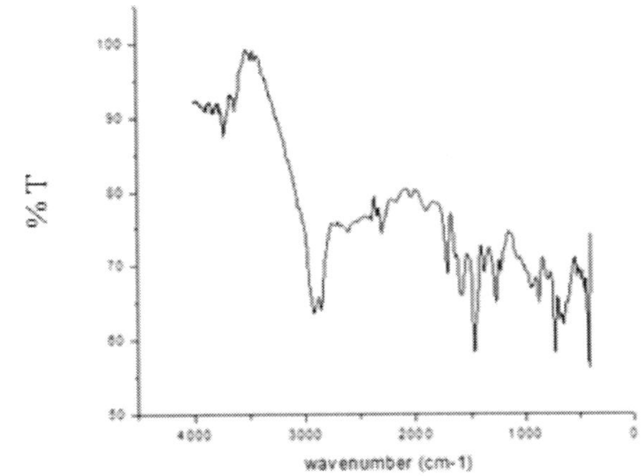


**Figure 6.** LDPE/LDPE-g-MAH/PLA FTIR spectra

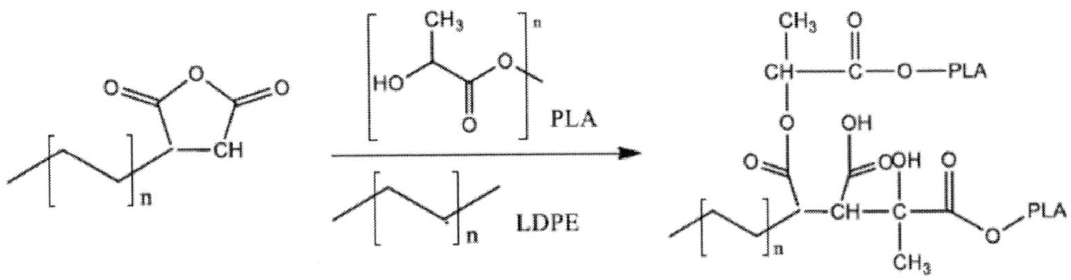


**Figure 7.** Reaction of LDPE, LDPE-g-MAH, and PLA

### *3.2. Determination of water adsorption*

Analysis of Water Adsorption Index (Absorption of Water) is necessary to know the properties of bioplastics on their water resistant. The first step of this test is an evaporation of moisture on the polyblend by using the electric furnace for 10 minutes at 100°C to remove the water that remains in the sample and then weighed. Test of Water Absorption Index (WAI) ASTM-570. Calculated Water Adsorption withthe formulawas described onthis ASTM. The water adsorption graftof all ratios of polyblends can be seen in figure 9. According to this figure, it can be concluded that generally, the more PLA added to the formulationof LDPE/LDPE-g-MAH/PLA will increase the percentage of water absorption. This can be explained since PLA has a goodbiodegradable property which able to broken down by the water, thus affecting to their degradation properties.

### *3.3. Determination of biodegradability*

The percentage degradation of polyblend by soil environmentwas carried out by weighing the mass of polyblend before and after an experiment. Figure 10 represents the percentage of biodegradability of the polyblends. From this figure, it can be estimated that there has been a reduction in the mass of polyblend after buried in the soil for 6 days. It may there is the biotic components such as micro-organisms decomposers in the soil which able to degradable the polyblend. Based on the plot of the graph, it can be concluded that the existence of PLA on the polyblend of LDPE / LDPE-g-MAH / PLA affected to their better degradation. These results are consistent with studies conducted by [13] in which the PLA to be mineralized into $CO_2$, water, and other simple biomass perfectly after 4-6 weeks.

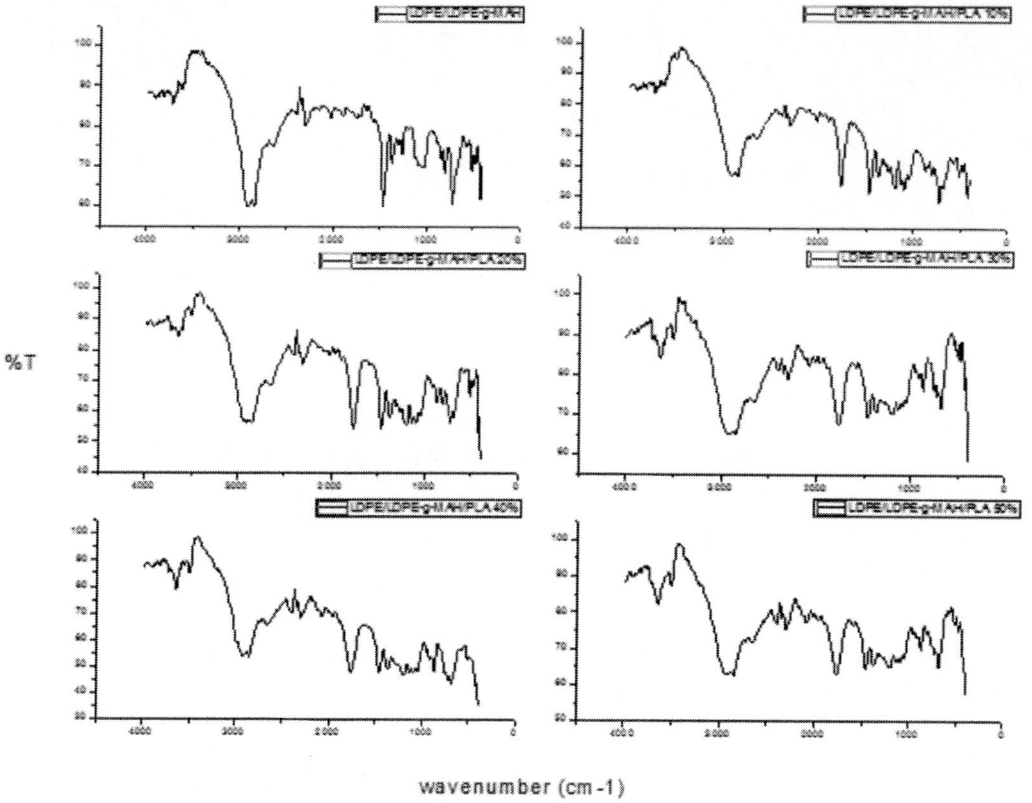


**Figure 8.** Spectra of LDPE/LDPE-g-MAH/PLA with the ratio LDPE: PLA (100:0); (90:10); (80:20); (70:30) ;( 60:40) ;( 50:50)

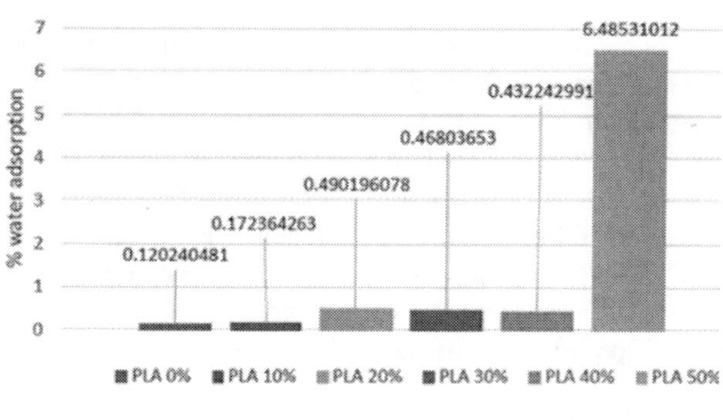


**Figure 9.** Water adsorption graft of polyblends

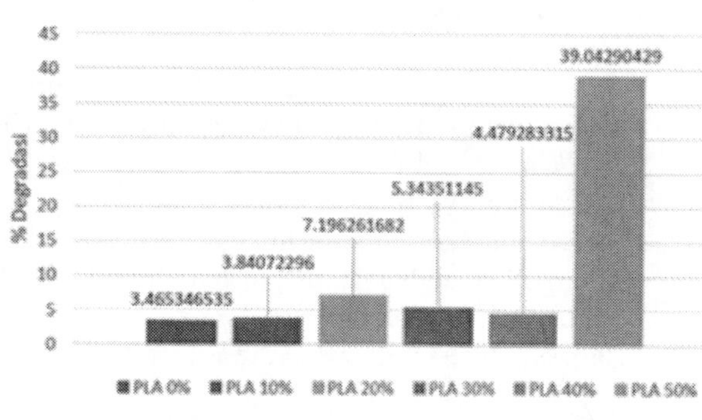


**Figure10.** Percentage of biodegradability of the polyblends

## 4. Conclusion

We concluded that the synthesis of green material polyblend LDPE/PLA grafting LDPE-g-MAH had interacted chemically. We further obtain thatthe increase of PLA percentage on the polyblend will increase the water absorption properties and their biodegradability. Poly-blend with 20% PLA content is the optimum composition for environmentally food packaging.

## 5. References

[1] Julianti E and Nurminah M 2012 Properties of plastic and paper packaging and their influence to the foods *Agriculture Technology Department* North Sumatra University (Thesis)

[2] Peacook A J 2014 *Hand Book of Polyethylene Structures, Properties, and Applications* Marcel Dekker Inc.

[3] Shogren R L, Doane W M, Garlotta D, Lawton J W and Willett J L 2004 *Biodegradation of starch/polylactic acid/poly (hydroxy ester ether) composites bars in soil* Poly. Deg. and Stab. **79**(3) 405-411

[4] Rasal R M and Hirt D E 2009 Improving the properties of the polylactic acid *Plastics Research online* Clemson, SC

[5] Wang J H and Schertz D M 2010 *Synthesis of grafted polylactic acid and polyhydroxy-alkanoate by a green reactive extrusion process* In ACS symposium series **1043** 439-453

[6] Auras R, Harte B and Selke S 2006 *An overview of polylactic acid as packaging materials* Mac. Biosci. **4**(9) 835-864

[7] Ploypetchara N, Suppakul P, Atong D and Pechyen C 2014 *A blend of polypropylene/poly(lactic acid) for medical packaging application: physicochemical, thermal, mechanical, and barrier properties* Energy Procedia **56** 201-210

[8] Marsyahyo E, Soekrisno, Rochardjo H S B and Jamasri 2005 *Investigation of chemical surface treatment of ramie fiber (buehmerianuvea) on surface topography, tensile strength, and single fiber fracture modes* Proceeding of International Seminar on Product Design and Development. Material Development, Gajah Mada University

[9] Parameswaranpillai, J, Joseph G, Jose S and Hameed N 2015 *Phase morphology, thermo mechanical, and Crystallization behavior of compatibilized and PP-g-MAH compatibilized polypropylene/polystyrene blends*. J. of App. Poly. Sci. **132**(24)

[10] Xin L, June Z, Lin Z, Hongxia J and Ruofei J 2014 *Study on mechanical properties and thermal stability behaviors of PP / PP-g-MAH / ABS-g-MAH blends*. Plas. Sci. and Tech.**9** 12

[11] Song Y W, Do H S, Joo H S, Lim D H, Kim S and Kim H J 2006 *Effect of grafting of acrylic acid onto a PET film surfaces by UV Irradiation on the adhesion of PSAs* J. Adhesion Sci. Technol. **20**(12) 1357-1365

[12] Hendri J, Ginting I and Laila A 2008 *Gamma radiation studies on the binding of acrylic acid grafted chitosan polyethylene* Proceeding of 2$^{nd}$ National Science and technology Seminar, Lampung University

[13] Hakkarainen M 2002 *Aliphatic polyesters: abiotic and biotic degradation and degradation products* In Degradable aliphatic polyesters 113-138 Springer Berlin Heidelberg

## Acknowledgements

The authors would like to express great gratitude to"Riset Unggulan 2016" of Research Center of Chemistry, Indonesian Institute of Sciences which is funding this work research.

# Mechanical and Thermal Properties of Unsaturated Polyester/Vinyl Ester Blends Cured at Room Temperature

**H Ardhyananta[1], F D Puspadewa[1], S T Wicaksono[1], Widyastuti[1], A T Wibisono[1], B A Kurniawan[1], H Ismail[2] and A V Salsac[3]**

[1]Department of Materials and Metallurgical Engineering, Institut Teknologi Sepuluh Nopember, Sukolilo, 60111 Surabaya, Indonesia
[2]School of Materials and Mineral Resources Engineering, UniversitySains Malaysia, NibongTebal, 14300 Penang, Malaysia
[3]Biomechanic and Bioengineering Laboratory (UMR CNRS 7338), Universite de Technologie de Compiegne, Compiegne, France

Email: hostaa@mat-eng.its.ac.id

**Abstract**. Unsaturated polyester (UP) resin containing aromatic ring was blended with vinyl ester (VE) at wide range composition (10, 20, 30, 40,and 80 wt.%) using mechanical blending method. The blends were cured at room temperature using methyl ethyl ketone peroxide (MEKP) (4 wt.%) as catalyst initiator without the presence of catalystaccelerator. The effect of vinyl ester composition on theenhancement of mechanical and thermal properties of unsaturated polyester/vinyl ester blends was investigated.The polymer blends were characterized by Fourier Transform Infra Red (FTIR)spectroscopy, tensile testing, hardness testing, scanning electron microscopy (SEM) and thermogravimetric analysis (TGA). IR spectra showed UP and VE peaks. The curing copolymerization reactionoccurred at vinyl (C=C) bonds. The addition of vinyl esters enhanced mechanical and thermal properties. The UP/VE blends showed homogeneous morphology, transparent and copolymer thermoset blend.

**Keywords**: unsaturated polyester, vinyl ester, polymer blend, thermal properties

## 1. Introduction

Unsaturated polyester is a widely used thermosetting resin. Unsaturated polyester has a reactive vinyl (carbon-carbon double) bond. Styrene is added to unsaturated polyester as asolvent and acrosslinking agent for curing reaction. The organic peroxide is used to initiatethe reaction. Organic salt isadded to accelerate the reaction. The polymer is achieved as a result of random condensation copolymerization reaction between styrene and unsaturated vinyl double bonds by free radical mechanismforming crosslinked network structure [1,2]. Unsaturated polyester is a resin (prepolymer) with low price, light weight, easy processing and water resistance. Unsaturated polyester commonly used in laminating resin, composites, coating, adhesive, microelectronic, aerospace, building, automotive, and shipbuilding. However, unsaturated polyester shows limited mechanical and thermal properties. It is necessary to enhance the properties of unsaturated polyester.The properties of unsaturated polyester have been enhanced by theadditionof carbon fiber , fly ash, carbon nanotube, and polyamide [3-6].

Vinyl ester is an attractivehigh-performance thermoset polymer. Vinylester has reactive double bonds with rigid epoxy structure. Vinyl ester is synthesized by the reaction of epoxy with methacrylic acid. Styrene is added as a copolymer agent (crosslinker)diluent for thepolymerization reaction. Styrene has a low price and high reactivity [2,7]. Vinyl ester has a high mechanical, thermal properties and easy processing. Vinyl ester commonly used in industrial applications such as laminating resin, composites, coating and adhesive. Vinyl ester has a potential for enhancing mechanical and thermal properties of unsaturated polyester [8,9].

This study evaluates the addition of vinyl ester (VE) into unsaturated polyester (UP)containing aromatic benzene ring forming thermoset polymer blend. The mechanical and thermal properties of the polymer blends are investigated. The unsaturated polyester/vinyl ester blends were prepared by mechanical blending method and cured at room temperature. The aim of this study is enhancing unsaturated polyesterproperties byblending with vinyl ester resin by keeping the advantages of both the resin such as easy to use and fast in processing (cured at room temperature). The polymer blend of unsaturated polyester/vinyl ester is cured using methyl ethyl ketone peroxide (MEKP) asa catalystinitiator reaction at room temperature without catalystaccelerator.

## 2. Experimental Methodology

### *2.1. Materials*

YUKALAC BQTN 157-EX unsaturated polyester (UP) resin and RIPOXY R-802 EX-1 vinyl ester (VE) resinwereobtained from Justus Sakti Raya Inc, Indonesia. The UP has specific gravity 1.1 g/cm3 at 25 °C and viscosity 4.5-5.0 poise at 25 °C. The VE has specific gravity 1.05 g/cm3 at 25 °C, acid value 7-13 KOH mg/g, viscosity 2-6 poise at 25 °C and styrene composition 46%. Methyl ethyl ketone peroxide (MEKP) was obtained from Kawaguchi Kimia Inc, Indonesia,as an initiator.

### *2.2. Preparation of the unsaturated polyester/vinyl ester blends*

The thermoset blends were achieved by adding vinyl ester (VE) into unsaturated polyester (UP) at 10, 20, 30, 40,50 and 80 wt.% by physical mechanical blending method with styrene as crosslinker diluent. Methyl ethyl ketone peroxide (MEKP) is added ascatalyst initiator at 4 wt.% without catalystaccelerator. Unsaturated polyester,vinyl ester,and MEKP were blended by mechanical mixer stirring at room temperature for 5 min affording atransparentcolorlesssolution.The mixture was poured into an aluminum mold and cured at room temperature for 30 min until 240 min by increasing vinyl ester composition to achieve cured yellowish transparent solid hard materials.

### *2.3. Characterizationof the unsaturated polyester/vinyl ester blends*

The unsaturated polyester/vinyl ester blends were examined. Fourier-transform infra-red(FTIR) spectroscopic analysis was done using Thermoscientific Nicolet I S10 spectrophotometer equipped with a smart orbit diamond crystal plate sampling technique at resolution 4 $cm^{-1}$ in the range 400-4000 $cm^{-1}$. The tensile test was done using a universal testing machine Autograph10TE at a crosshead speed of 1 mm/min using dumb-bell shaped samples according to ASTM D638 standard.Hardness test was done using durometer Shore D. Scanning electron microscopy (SEM) was done using FEI Inspect S50 at voltage 20 kV on thetensile fracture surface. The specimen was gold coated. Thermogravimetric analysis (TGA) was done using Mettler Toledo TGA/DSCat a heating rate of 10 °C/min under nitrogen atmosphere.

## 3. Results and Discussion

### *3.1. Preparation of unsaturated polyester/vinyl ester*

Vinyl ester (VE) was mixed with unsaturated polyester (UP)containing aromatic benzene ring to form UP/VE blends. The thermoset polymer blends weremixed by mechanical blending method and room temperature curing. Table 1 shows curing characteristics of UP/VE blends. UP was cured at curing time 20 min affording yellowish transparent solid hard rigid material. Figure 1 shows the appearance of cured resin blends. UP/VE blends were cured at various curing timeaffording yellowish transparent solid hard rigid material. UP/VE blends showed homogeneous appearance without phase separation corresponding to high compatibility of UP and VE. The curing time increased with increasing vinyl ester composition at 30 to 240 min. VE was not cured after 4320 min affording yellowish transparent liquid material. It suggests that VE needs catalystaccelerator such as organic salt for curing process the resin.

UP/VE blends were investigated by FTIR. Figure 2 shows the IR spectra of UP and VE virgin resin. Figure 1(a) shows IR spectra of UP virgin resin. The peak at 3482cm$^{-1}$ is O-H hydroxyl bond. The peak at 3077 cm$^{-1}$is C-H aromatic benzene ring group. The peak at 2978 cm$^{-1}$is C-H aliphatic group. The peak at 1717 cm$^{-1}$is C=O carbonyl bond. The peak at 1493 cm$^{-1}$ is C-H aromatic benzene ring. The peak at 1248 cm$^{-1}$is CO-O-C ester linkage. The peak at 1111 cm$^{-1}$is C-O-C ester linkage. The peak at 979 cm$^{-1}$is aC=C unsaturated vinyl double bond. The peak at 697 cm$^{-1}$is C-H monosubstituted aromatic ring group. Figure 1(b) shows IR spectra of VE virgin resin. The peak at 3415, 3059, 2963, 1715, 1507, 1232, 1180, 989, 696 is O-H, C-H aromatic benzene ring, C-H aliphatic, C=O, C-H aromatic benzene ring, C-O ester, C-O ether, C=C unsaturated vinyl, C-H monosubstitutedaromatic benzene ring, respectively [3, 5, 6]. The peak at 826 cm$^{-1}$ is C-H para-substituted aromatic ring group. VE shows high aromatic benzene ring composition.

**Table 1.** Curing characteristics of UP/VE blends cured at room temperature

| Material | Vinyl ester composition (wt.%) | MEKP composition (wt.%) | Curing time (min) | Appearance |
|---|---|---|---|---|
| UP | 0 | 4 | 19 | Solid, yellowish transparent |
| UP/VE(10) | 10 | 4 | 36 | Solid, yellowish transparent |
| UP/VE(20) | 20 | 4 | 64 | Solid, yellowish transparent |
| UP/VE(30) | 30 | 4 | 101 | Solid, yellowish transparent |
| UP/VE(40) | 40 | 4 | 164 | Solid, yellowish transparent |
| UP/VE(80) | 80 | 4 | 236 | Solid, yellowish transparent |
| VE | 100 | 4 | 4320 / 3 days | Liquid, yellowish transparent, Not cured |

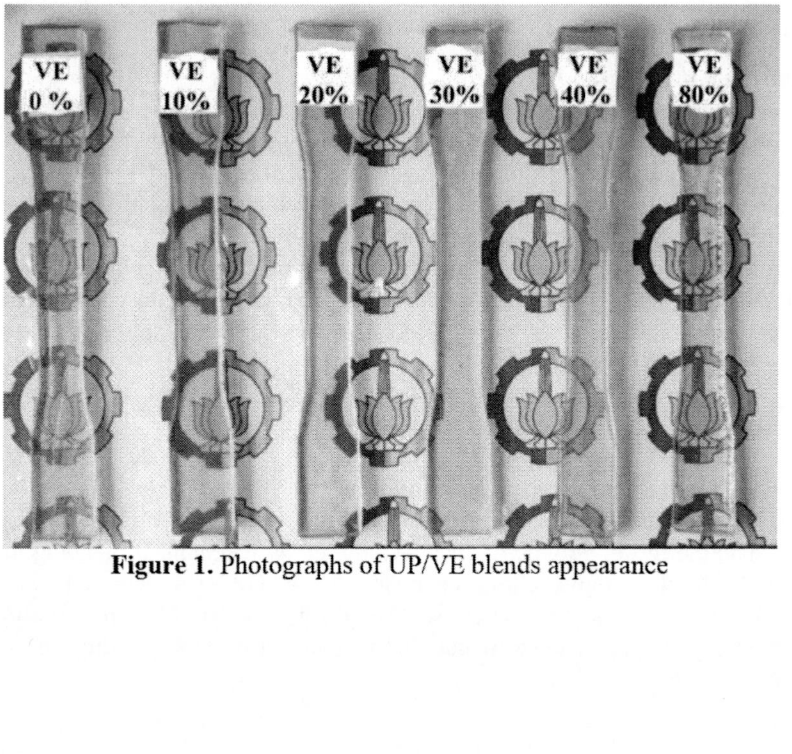


**Figure 1.** Photographs of UP/VE blends appearance

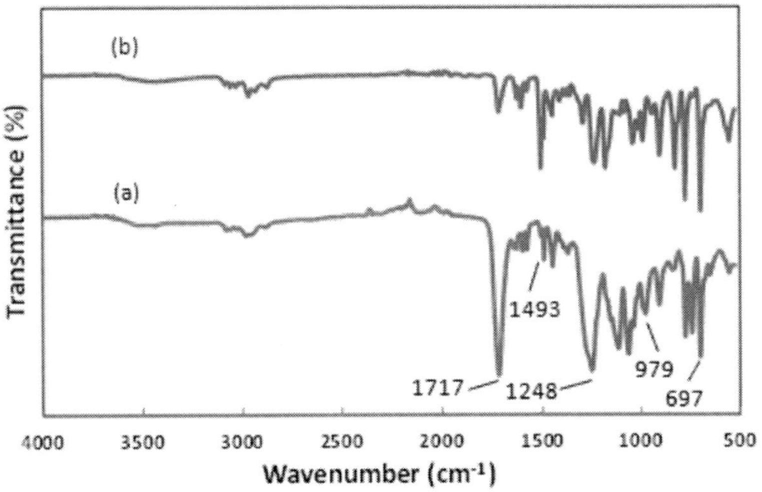


**Figure 2.** FTIR spectra of (a) UP and (b) VE

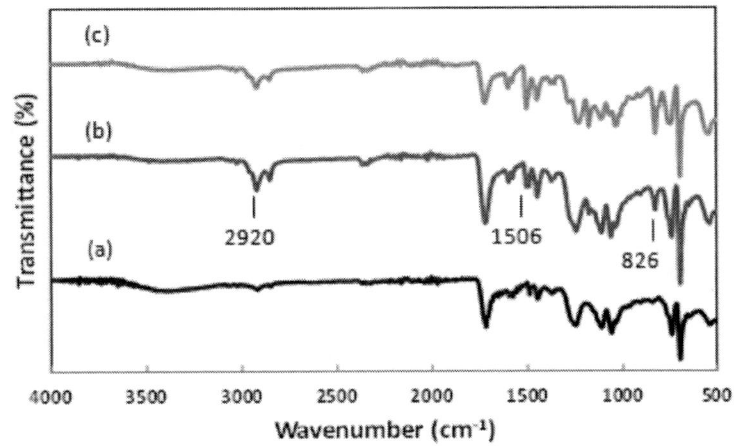


**Figure 3.** FTIR spectra of (a) UP, (b) UP/VE at 30 wt.% VE and (c) UP/VE at 80 wt.% VE cured at room temperature.

The curing process of UP/VE blends were investigated by FTIR. Figure 3 shows the IR spectra of UP/VE blends at various VE composition cured at room temperature. It is observed from Figure 3(a) and 3(b), the peak at 1506 and 826 $cm^{-1}$corresponds to C-H para-substitutedaromatic benzene ring of vinyl ester. The peak at 2920 $cm^{-1}$increased corresponding to the formation of long chain molecule structure. The peak at 979 $cm^{-1}$decreased corresponding to the vinyl functional group reaction. Figure 4 shows curing reaction and the formation of UP/VE cross-linked chemical structure network. It is suggested that UP/VE proceeded free radical chain crosslinking polymerization. It begins at the thermal decomposition of catalyst initiator MEKP producing free radical in the system. The free radical continued reaction to thevinylgroup. Afterward, it is linked to theadjacent vinyl group and styrene crosslinker achieving the formation of crosslinking molecular network structure solidUP/VE polymer. The UP/VE chemical structure suggests the formation of thermoset copolymer. Figure 3(c) shows UP/VE at high VE composition (80 wt.%). The peak at 697$cm^{-1}$corresponds to C-H monosubstituted aromatic benzene ring shows high intensity. It suggests remaining benzene ring and unreacted molecule. UP/VE at high composition VE is more difficult to cure and not easy to process.

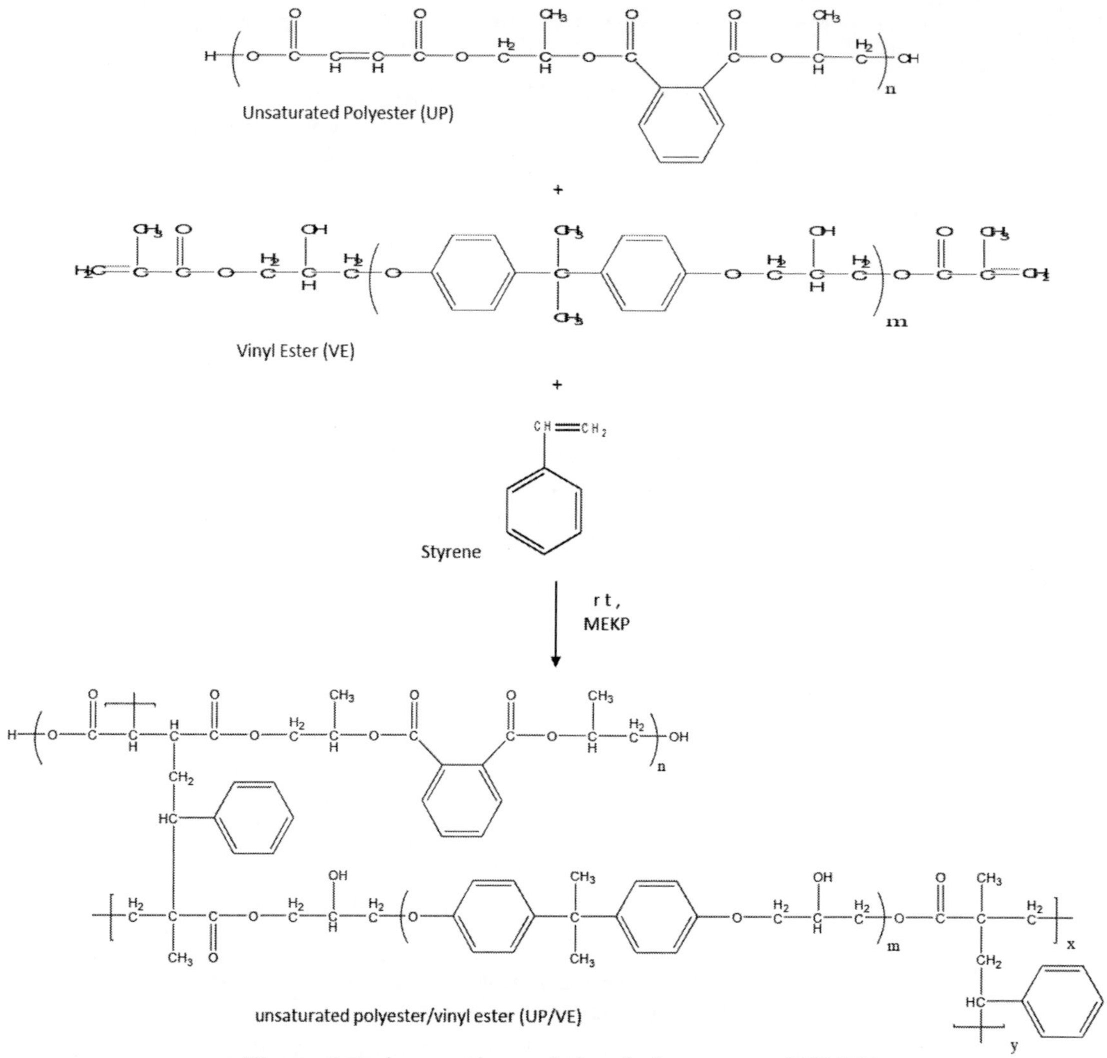


**Figure 4.**Curing reaction and chemical structure of UP/VE

### *3.2. Mechanical properties of unsaturated polyester/vinyl ester*

Mechanical properties of UP/VE blends were examined. Figure 5 shows atensile strength of UP/VE blends. Table 2 shows the results of mechanical properties of UP/VE blends. Cured UPshowed high tensile strength which corresponds to cross-linked chemical structure network [3, 9]. The addition of vinyl ester composition increased the tensile strength of UP/VE blends up to 40 wt.% of VE. The increasein tensile strength is due toincrease of rigid structure composition of aromatic benzene ring vinyl ester. At high VE composition, the tensile strength of UP/VE blends decreased due tounreacted bonds and difficultyof themoleculeto be cured.

Figure 6 shows elongation at break of UP/VE blends. Cured UP showed high elongation at break. The addition of vinyl ester composition increased the elongation at break of UP/VE even at high VE composition (80 wt.%). The increasing of elongation at break is due to thehigh long chain chemical structure of the polymer and easier movement of themolecule chain. The UP/VE shows high elongation at break and flexibility.

Figure 7 shows hardness value of UP/VE blends. Cured UP showed high hardness value. The addition of vinyl ester composition increased the hardness value of UP/VE blends up to 20 wt.% VE composition. At high VE composition, thehardness of UP/VE decreased due to theunreacted molecule.

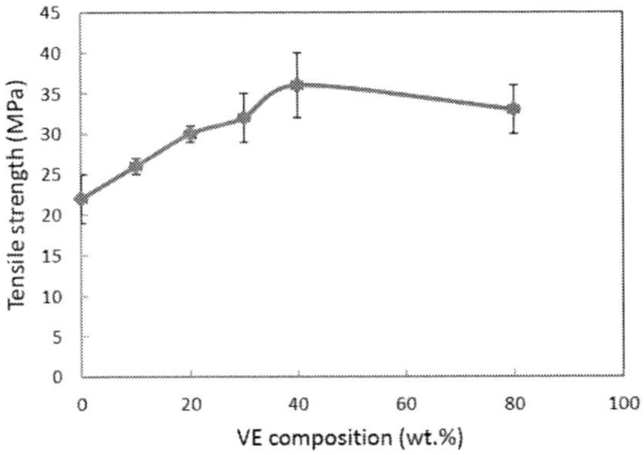


**Figure 5.** Tensile strength of UP/VE blends

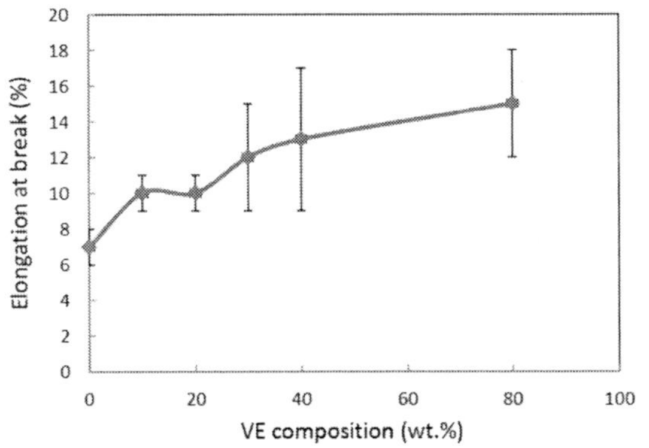


**Figure 6.** Elongation at break of UP/VE blend

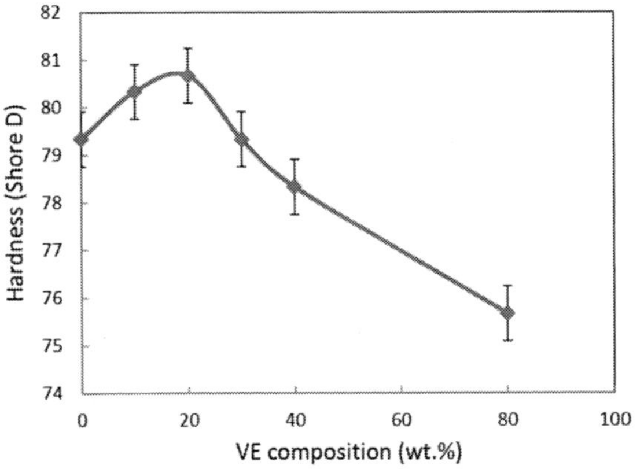


**Figure 7.** Hardness of UP/VE blends

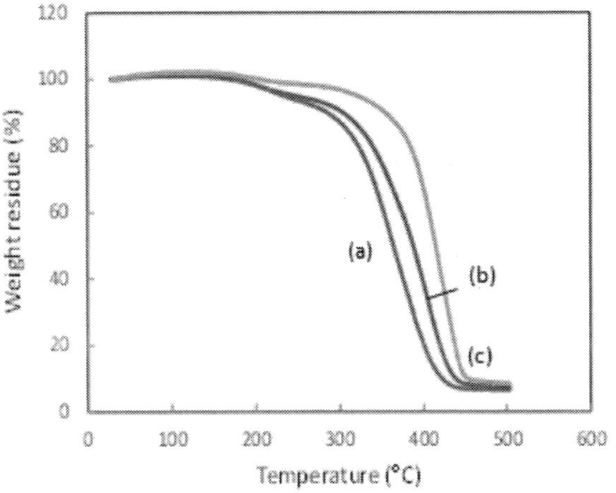


**Figure 8.** Thermogravimetric curves of (a) UP, (b) UP/VE at 30 wt.% and (c) VE and UP/VE at 80 wt.% VE

**Table 2.** Mechanical properties of UP/VE blends cured at room temperature

| Material | Vinyl ester composition (wt.%) | Tensile strength (MPa) | Elongation at break (%) | Hardness (Shore D) |
|---|---|---|---|---|
| UP | 0 | 22 ± 3 | 7 ± 1 | 79 ± 1 |
| UP/VE(10) | 10 | 26 ± 1 | 10 ± 1 | 80 ± 1 |
| UP/VE(20) | 20 | 30 ± 1 | 10 ± 1 | 81 ± 1 |
| UP/VE(30) | 30 | 32 ± 3 | 12 ± 3 | 79 ± 1 |
| UP/VE(40) | 40 | 36 ± 4 | 13 ± 4 | 78 ± 1 |
| UP/VE(80) | 80 | 33 ± 3 | 15 ± 3 | 76 ± 1 |
| VE | 100 | Not cured | Not cured | Not cured |

IOP Conf. Series: Materials Science and Engineering **202** (2017) 012088 doi:10.1088/1757-899X/202/1/012088

**Table 3.** Thermal properties of UP/VE blends cured at room temperature

| Material | Vinyl ester composition (wt.%) | Degradation Temp. at 5 wt.% (°C) | Degradation Temp. at 10 wt.% (°C) | Weight residue at 500 °C (wt.%) |
|---|---|---|---|---|
| UP | 0 | 234 | 285 | 6 |
| UP/VE(30) | 30 | 247 | 304 | 7 |
| UP/VE(80) | 80 | 322 | 355 | 8 |
| VE | 100 | Not cured | Not cured | Not cured |

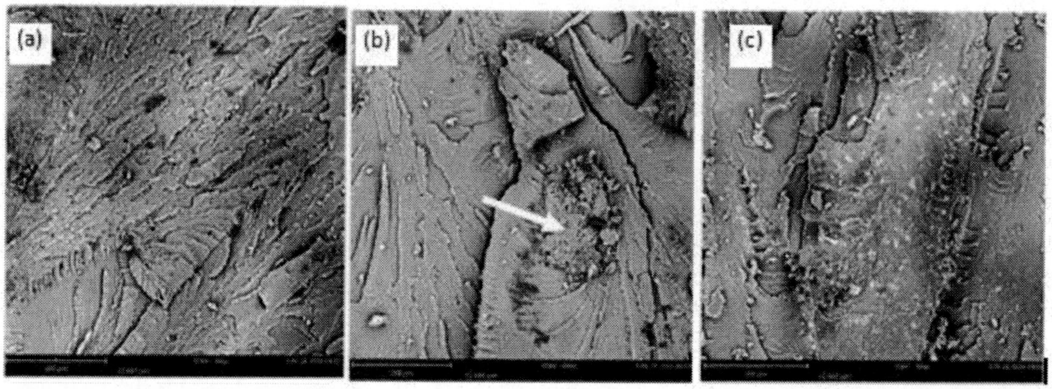


**Figure 9.** SEM tensile fracture surface of (a) UP, (b) UP/VE at 30 wt.% and (c) VE and UP/VE at 80 wt.% VE

The morphology of UP/VE blends were examined. Figure 8 shows SEM of thetensile fracture surface of UP/VE blends. Figure 9(a) showed Cured UP morphology. It showed void and homogeneousonephase which corresponds to transparent appearance. Figure 9(b) showed cured UP/VE blends. It showed homogeneoustwo-phase which corresponds to transparent appearance [3, 9]. The UP containing aromatic benzene ring shows high compatibility with VE containing aromatic benzene ring which corresponds to high compatibility and transparency.The UP/VE shows copolymer blend thermoset polymer. The surface morphology shows rough surface and plastic deformation which corresponds to high elongation at break.

### *3.3. Thermal properties of unsaturated polyester/vinyl ester*

Thermal stability of UP/VE blends was examined using TGA. Figure 8 shows TG curves of UP/VE blends. Table 3 shows the results of thermal properties of UP/VE blends. Cured UP showed high thermal degradation and weight residue which corresponds to cross-linked chemical structure network. The addition of vinyl ester composition increased the thermal stability of UP/VE blends. The thermal degradation was due to volatilization of theunreacted molecule (diluent, moisture), fragmentation of chemical structure and gasification process [5, 7, 8]. The increasing of thermal degradation temperature and weight residue is due high composition of therigid structure of aromatic benzene ring vinyl ester.

## 4. Conclusion

The unsaturated polyester (UP)/vinyl ester blend was studied. UP and VE contains anaromatic ring. UP/VE was prepared by mechanical blending method and cured at room temperature using styrene as crosslinker and MEKP as catalyst initiator. UP/VE showed solid and transparent material. VE could not be cured by this reaction condition. UP/VE reacted at thevinyl bond and formed copolymer crosslinking network chemical structure. The addition of vinyl ester composition increased the mechanical properties of UP/VE blends. UP/VE showed homogeneoustwo-phase blend morphology corresponding to high compatibility. The addition of vinyl ester composition increased the thermal properties of UP/VE blends.

IOP Conf. Series: Materials Science and Engineering **202** (2017) 012088 doi:10.1088/1757-899X/202/1/012088

**5. References**

[1] Juan S A, Emmanuel R, Bruno F and Fidele N 2016 *Polym.Degrad. Stab.* **129** 142-155
[2] Sang B L, Rocket T J and Hoffman R D *Polymer* **33** 3691-3697
[3] Trino A S M, Costa C S M F, Fonseca A C, Barata I, Julio E, Serra A C and Coelho J F J 2016 *React. Funct. Polym.* **106** 24-31
[4] Dawei J, Lixin X, Li L, Shaofan S, Qingbo Z, Zijian W, Xingru Y, Jiang G, Yudong H and Zhanhu G 2015 *Compos. Sci. Technol.* **117** 168-175
[5] Minakshi S G and Deepak S 2014 *Prog. Org. Coat.* **77** 1208-1220
[6] Yolanda R M, Thais L N, Quintela M A L and Massimo L 2014 *Polym. Degrad. Stab.* **107** 277-284
[7] Can E, Kinaci E and Palmese G R 2015 *Eur. Polym J.* **72** 129-147
[8] Umair J, Zaffar M K, Khan M B, Bassyouni M, Hamid S M S A, Aziz M H A and Syed W U H 2016 *Compos. Part B* **91** 257-266
[9] Seyhan A T, Metin T and Karl S 2009 *Mater. Sci. Eng. A.* **523** 85-92

IOP Conf. Series: Materials Science and Engineering **202** (2017) 012089 doi:10.1088/1757-899X/202/1/012089

# Characteristics of Al-Si-Mg Reinforced SiC Composites Produced by Stir Casting Route

**A Zulfia, T Zhakiah, D Dhaneswara and Sutopo**

Department of Metallurgy and Mateials, Faculty of Engineering, Universitas Indonesia, Kampus Baru UI, Depok 16424, Indonesia

Email: anne@metal.ui.ac.id

**Abstract**. Al-Si-Mg alloy that is strengthened by silicon carbide particles has the potential to have excellent mechanical properties with light weight. In this study, metal matrix composites reinforced silicon carbide from 2 vf-% to 15 vf-% and magnesium amounted to 10 wt-% as an external dopant were fabricated by stir casting route. The magnesium was added to promote the wetting between Al matrix and reinforced SiC. The process involved SiC blended inside the molten Al by a stirrer with a rotational speed of 500 rpm at 800 °C for 2 minutes and degassed with Ar gas for 4 minutes to remove all of the gas content in the molten Al. The molten composite was then cast into the plate and tensile test sample molds. The effect of SiC addition on the mechanical properties and microstructure of the composites was investigated. The result showed that the optimum tensile strength was reached at 8 vf-% SiC with the value of 175 MPa, while the elongation was 9.1%. The maximum hardness and wear rate were achieved at 10 vf-% SiC with the values of 57 HRB and 0.0022 $mm^3/m$, respectively. Such increase was related to the microstructures dominated by the presence of Chinese script, primary and eutectic $Mg_2Si$ which were contributed to the mechanical properties of the composites.

**Keywords**: Al-Si-Mg alloy, SiC, composites, stir casting, mechanical properties

## 1. Introduction

The recent development of technology requires material properties that cannot be found in metal, polymer, and ceramic. Such properties cover high specific strength, high-temperature creep resistance, and better fatigue strength. Therefore, a composite is made [1,2]. This material will be used for applications that need a high-performance material such as automotive, military, aerospace, and electric industry [3]. To satisfy this need, the demand for Al composites for automotive application has increased from time to time. Therefore, particulate reinforced composites are one of the most common materials used for many industrial applications at this moment. Aluminum is the most favorable materials selected as a matrix to produce aluminum matrix composites (AMCs) in the last two decades [4]. Many applications of AMCs have been established and one of the largest AMCs applications is in the ground transportation sector, such as automotive engine, brake shoe, train brake-lining, etc. Instead of having high wear resistance, it must be lightweight, has a high ultimate strength, hardness, high toughness and low thermal conductivity. Aluminum is the most popular metal that used as matrix due to its low density (2.7 $g/cm^3$) – almost 0.3% of the density of steel (7.83 $g/cm^3$), cuprum (8.93 $g/cm^3$) or bronze (8.53 $g/cm^3$) [5]. Pure aluminum has a low strength with a high ductility, so aluminum must be alloyed with another element such as Cu, Zn, Si, Mg or Mn. Among the aluminum based casting alloys, Al-Si-Mg alloys are most widely used in automotive industries because of their excellent casting properties accompanied by their superior mechanical properties [6]. Therefore, reinforcing aluminum alloy with SiC particles could considerably enhance their hardness and the ultimate tensile strength but decrease their ductility [7,8]. The strength of composites is also expected to be influenced by the dislocation density, dislocation-to-dislocation interaction and constraint of plastic flow due to the resistance offered by the reinforced particles [9].

Stir casting method is one of the most economical methods to cast Al-SiC composites utilizing Al-SiC liquid stirred by a straight-blade mechanical stirrer [10]. However, there are some disadvantages of stir casting method, such as segregation due to particle settling during casting. This matter could occur because of a different density between the particle and matrix. It is also hard to make homogenous dispersion due to its poor wettability, and the materials tend to agglomerate and result in high porosity content in the final product [11]. Degasification with argon and stirring with correct speed and time will minimize the possibility of a porous content.

The objective of this study was to characterize the microstructure, and mechanical properties of Al-Si-Mg aluminum reinforced SiC composites produced by stir casting method and to observe the effect of various of Vf-% SiC on microstructure and the mechanical properties of the composites was also studied.

## 2. Experimental Method

*2.1 Making of Al-Si-Mg/SiC composite*

Aluminum alloy with a chemical composition in Table 1 was used as a matrix, and SiC particles with different sizes from 25-98 μm were used as particles reinforced with various amounts from 2 to 15vf-% and commercially pure magnesium as a wetting agent with the content of 10 wt-%.

**Figure 1.** The morphology of reinforced SiC particles.

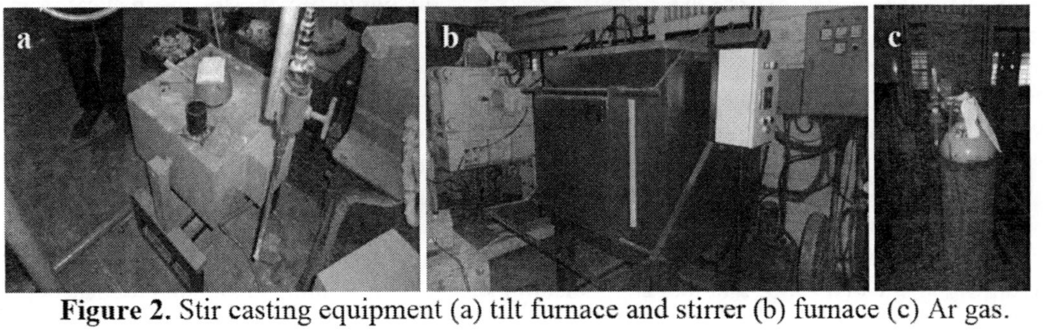


**Figure 2.** Stir casting equipment (a) tilt furnace and stirrer (b) furnace (c) Ar gas.

**Table 1.** Chemical composition of Al alloy.

| Si | Mg | Fe | Cu | Mn | Zn | Ti | Al |
|---|---|---|---|---|---|---|---|
| 6.5-7.5 | 0.25-0.45 | 0.2 | 0.2 | 0.10 | 0.10 | 0.20 | Bal. |

Aluminum alloy and magnesium were melted at 800 °C in an electric furnace while the SiC particles were preheated at 1000 °C for 1 hour in a muffle furnace. After the aluminum was fully melted, a de-slagging process was conducted to remove all slag on the surface layer of molten Al. After molten Al was cleaned, then 10 wt-% Mg was added and continue to be stirred for 10 minutes and degassed for 2 minutes to remove oxygen in the molten Al by flushing argon into the furnace. After matrix phase has prepared, it was followed by pouring the preheated SiC powder into the furnace then continue to be stirred at 500 rpm and degassed to remove the possible oxygen in the molten composites (see Figure 2). Afterward, the molten composite was casted into the tensile test mold.

*2.2 Characteristic of Al-Si-Mg/SiC composite*

The mechanical properties of composites were measured, including the tensile strength, hardness, and wear resistance. Tensile testing was carried out using GOTECH AI-7000 LA 10 in accordance with ASTM E8M-09. The tensile test was carried out using three specimens for each variable. Hardness testing was carried out using a Rocky machine with Rockwell B method in accordance with ASTM E18-11. For each sample, five hardness readings were taken on the randomly selected regions in order to eliminate the segregation effects and to get a representative value of the matrix material hardness. Wear testing was carried out using Ogoshi method with the following parameter: 12.6 kg load ($P_0$), 400 m sliding distance ($l_0$), and 1.97 m/s sliding speed. The samples were then ground using SiC paper #80 before testing. The thickness of disk employed in this testing was 3 mm, with a radius (r) of 10 mm. The metallographic preparation was conducted by grinding using emery paper started from #80, #150, #240, #400, #600, #800, #1000, #1200 to #1500, and was then polished using $TiO_2$. Keller's reagent (2 ml HF (conc.), 5 ml $HNO_3$ (conc.), 3 ml HCL and 190 ml distilled water was used as an etching agent. All sample preparations were observed using OLYMPUS BX41M-LED optical microscope and were further analyzed using Field Emission Scanning Electron Microscope (FESEM) linked to Energy Dispersive Spectrum (EDS) by INSPECT F50 FE-SEM, and phase identification was examined by X-ray diffraction (XRD).

## 3. Results and Discussion

*3.1 Microstructural Observation of Al-Si-Mg reinforced SiC composites*

The microstructures of composites with the different volume fraction of SiC are shown in Figure 3. The microstructures were mainly composed of α-Al, eutectic silicon and randomly distributed SiC particles. The dendritic structure in the Al matrix cannot clearly be seen. The SiC particles in the matrix could act as an obstacle for the dislocation movement that will improve the mechanical properties of the composites. When a higher SiC in matrix generated more interface layer in the composite that will transmit the load from the matrix to the reinforcement, the mechanical properties were improved. The second phase particles in the aluminum matrix were also found during the solidification, and according to EDS, the phases which were present in the composites were $Mg_2Si$ and $MgO.SiO_2$ (Figure 4 and Table 2). The presence of the phases in the composites was also confirmed by XRD results illustrated in Figure 6. The porosity formation in microstructure occurred when the molten composite was stirred for a longer stirring time which can increase the number of air bubbles entrapped into the molten metal and SiC particles which were prone to associate with these bubbles and remaining pores (Figure 5). Porosity or void also occurred at the interface between Al matrix and SiC particles since the wetting system was poor, while the shrinkage porosity took place during the solidification due to the different melting points of the element content in the aluminum matrix [12].

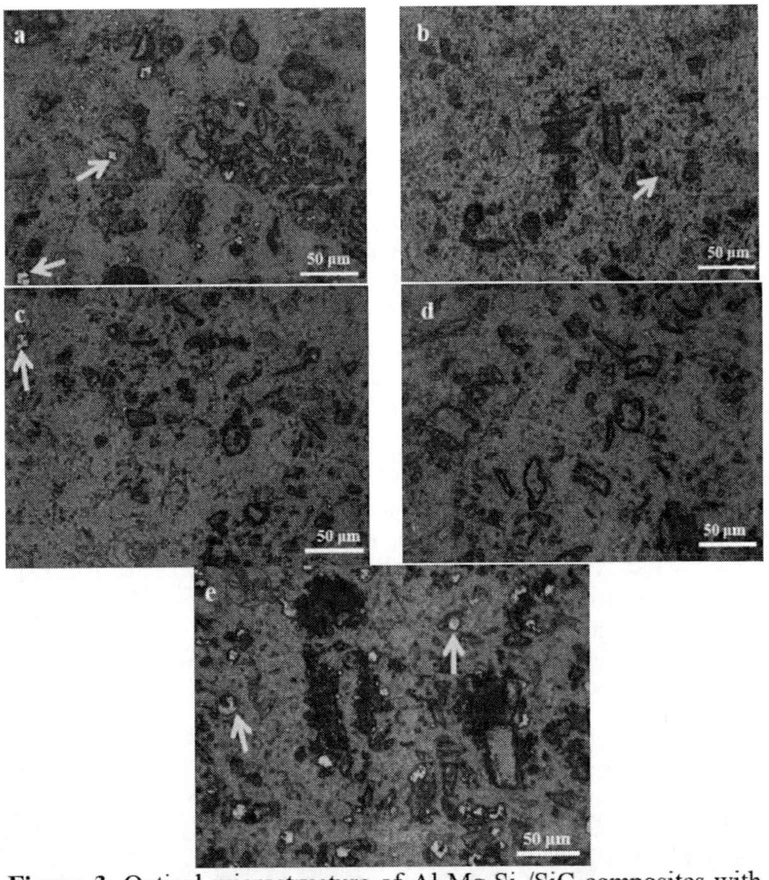


**Figure 3.** Optical microstructure of Al-Mg-Si /SiC composites with : (a) 2% SiC; (b) 5% SiC; (c) 8% SiC; (d) 10% SiC; and (e) 15% SiC

**Table 2.** Element content from points in Figure 4 analyzed by EDS

| Point | Elements content (wt%) | | | | | | | | Possible phase |
|---|---|---|---|---|---|---|---|---|---|
| | Al | Si | Mg | C | O | Sn | N | Fe | |
| 1 | 90.59 | - | 1.97 | 4.69 | 2.04 | 0.49 | - | 0.23 | α-Al |
| 2 | 48.99 | 31.93 | 2.3 | 8.25 | 6.23 | 1.23 | 1.07 | - | $MgO.SiO_2$ |
| 3 | 35.74 | 19.31 | 26.88 | 5.25 | 11.4 | 0.42 | 0.99 | - | $Mg_2Si$ |

According to SEM-EDS analysis results as demonstrated in Figure 4 and Table 2 which were confirmed by XRD patterns especially for composite with 8 wt-% SiC, there was obtained $Mg_2Si$ phase in composites, besides Al and SiC which were the base materials of sample. In the previous experiment conducted by Pawlik *et al.* [13], they have shown that many $Mg_2Si$ particles were created around the matrix and SiC. This $Mg_2Si$ precipitated from silicon content in alloy itself reacted with Mg and could also, $Mg_2Si$ precipitate produced by free Si from the reaction product of $SiO_2$ reacted with an excess of Mg in the Al alloy [14]. During the preoxidized process of the SiC particles, the $SiO_2$ layer formed on the surface of SiC particles could react with the molten Al-Mg-Si alloy to produce $MgAl_2O_4$ which was frequently surrounded by the solidification product. Other important compounds that were reported in this alloy were $MgAl_2O_4$ and $MgO.SiO_2$ as interface layer products. MgO associated with SiO and $MgAl_2O_4$ were also formed in Mg-containing alloy, but MgO occurred at a higher Mg concentration and more stable at a higher temperature than that of $MgAl_2O_4$ [15]. It was also found that other precipitates such as $Al_2Cu$, $MgZn_2$, and $Al_4C_3$ were also influencing the mechanical properties of the composites.

### *3.2 Mechanical properties of Al-Si-Mg reinforced SiC composites*

The effect of Vf% SiC on tensile strength and elongation is shown in Figure 7. The addition of SiC into Al matrix could increase the tensile strength of Al-Si-Mg matrix compared to the matrix without SiC. It could be noted that there were an enhancement strength and an elongation with the addition of SiC particles of up to 8 Vf-% SiC, then gradually decreased with the addition of SiC. Dwidevi *et al.* [7] and Amirkhanlou *et al.* [16] reported that an addition of SiC particles would increase the tensile strength continuously until it reached the optimum value and decreased due to the porosity created in the final product. Porosities in Figure 5 generated stress concentration in that spot in a matrix which could act as a source of cracking nucleation and decrease the total loading area [12]. On the other hand, the highest ultimate strength was found at 8 Vf-% SiC due to the uniform particle dispersion that can impede dislocation movement [8]. The uniform distribution of SiC particles in the aluminum matrix will also have a good effect on its mechanical properties.

However, the ductility of the composites which deteriorated significantly with a high SiC particle concentration displayed on Figure 7b, but the elongation increased with a higher SiC content of up to 8 Vf-%, this value was quite odd since the optimum strength generated an optimum elongation in this composites, where both properties should have the opposite correlation. Current research showed an in-line correlation between tensile strength and elongation, see Figure 7a and Figure 7b. It was due to ceramic solid phase addition in a matrix that would resist the dislocation movement. This experiment might be different from a study reported by V. S. Aigboidon *et al.* [8]. They said that low elongation occurred because of much porosity in a final product that generated a low tensile strength. This matter indicated that a crack at composites could be the main factor which would weaken the mechanical behavior.

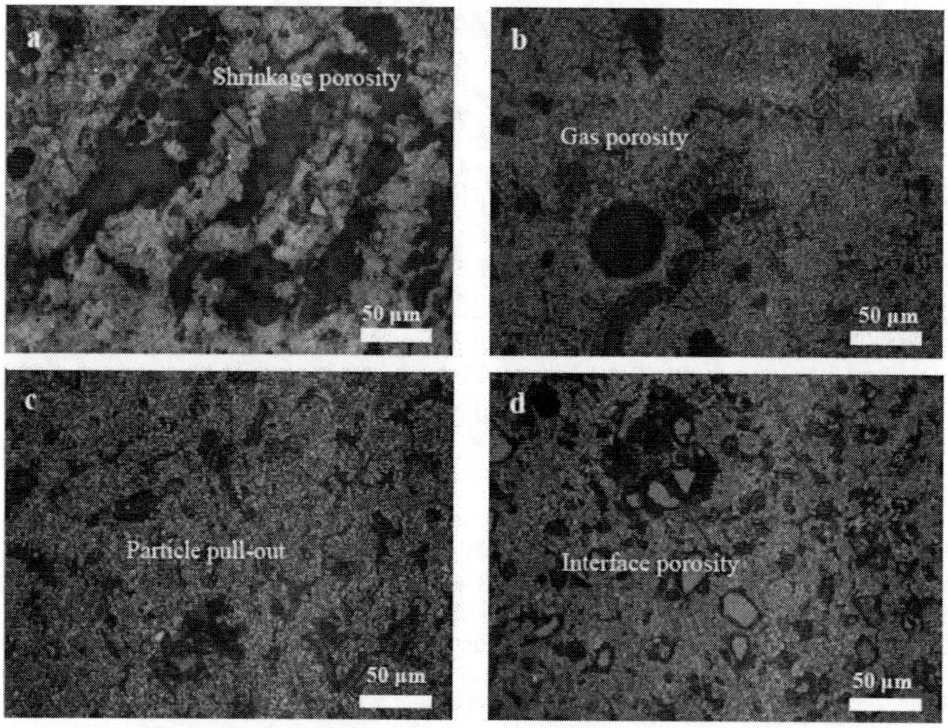


**Figure 5.** Porosity formation in the Al-Mg-Si / SiC composites.

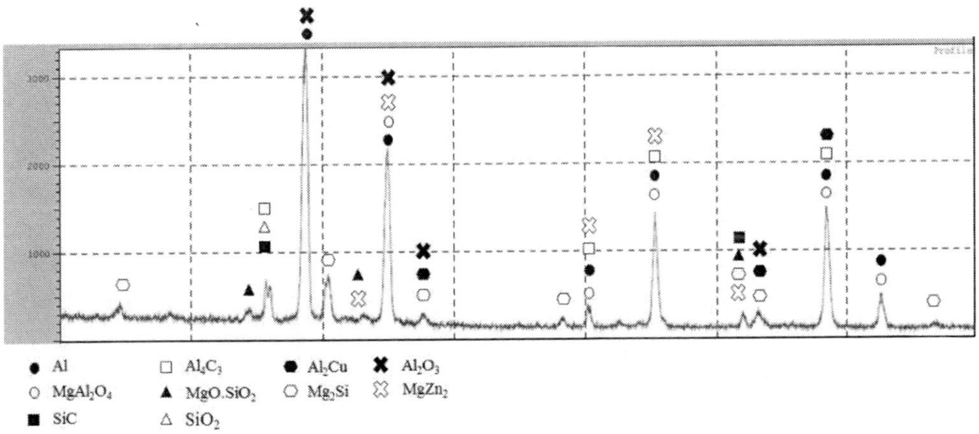


**Figure 6.** XRD pattern of Al-Mg-Si composites with 8 Vf-% SiC.

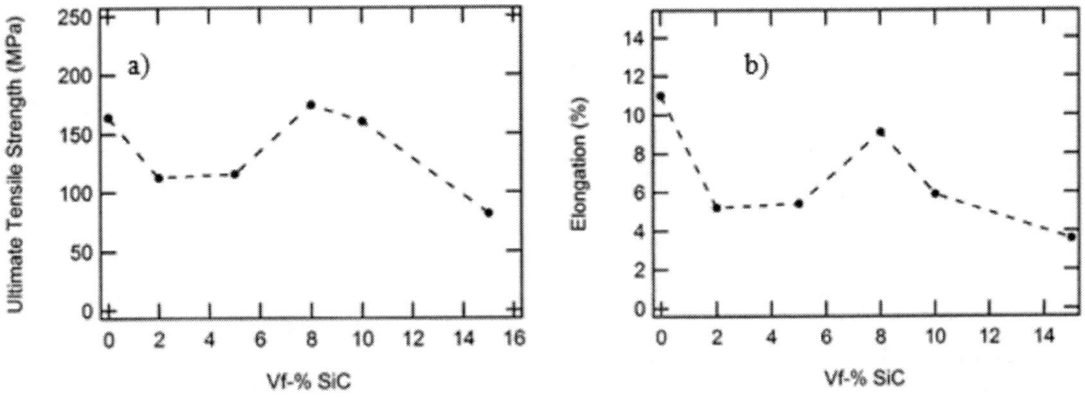


**Figure 7.** Effect of SiC on ultimate tensile strength (a) and elongation (b) of Al-Mg-Si/SiC composite.

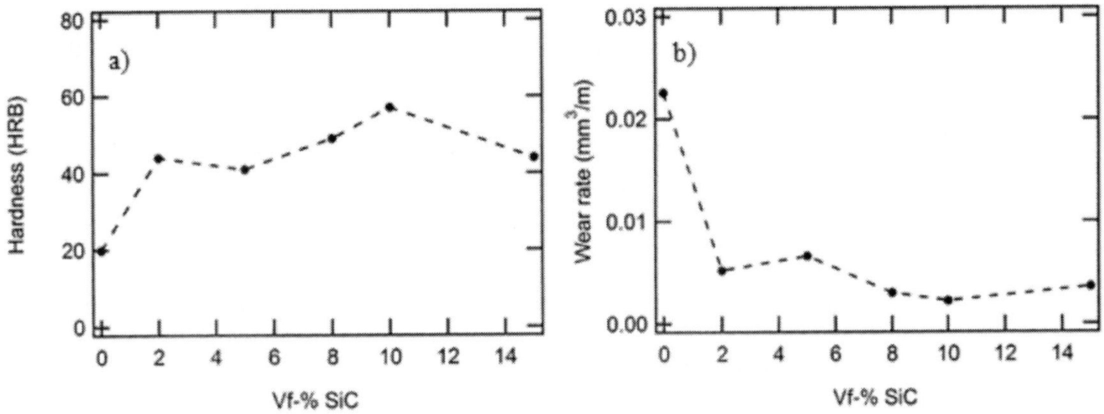


**Figure 8.** Effect of SiC on hardness (a) and wear rate (b) of Al-Mg-Si/SiC composite.

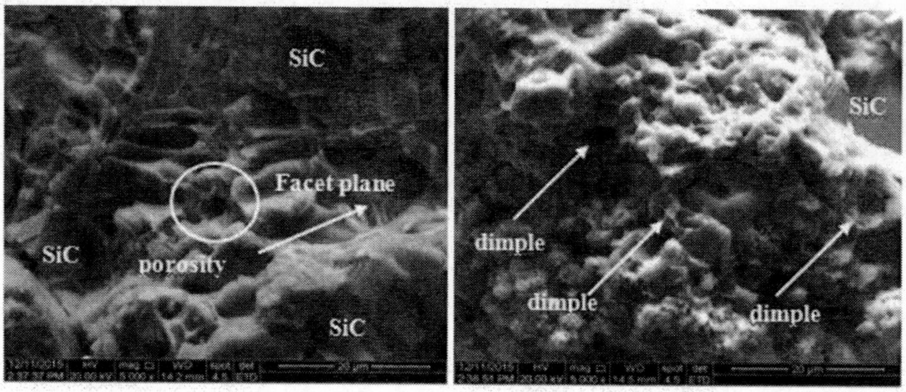


**Figure 9.** SEM fractography of tensile fracture surface of A356-10%Mg + 8% SiC composite at (a) the center and (b) the edge.

The hardness increased as the percentage of SiC particles addition in the alloy was increasing, as can be seen in Figure 8a. It was due to the increasing percentage of a hard and brittle phase of ceramic in the matrix [8]. SiC particles acted as an obstacle to dislocation movement so the dislocation density at the interface layer would increase. The difference of CTE (SiC 3 x $10^{-6}$ °$C^{-1}$ and Al 23 x $10^{-6}$ °$C^{-1}$[15]) as somehow the source of the step-up of dislocation density result in an elastic and plastic incompatibility between the matrix and the reinforcement. Hence, there was an improvement in the mechanical properties. The weakening factor might be contributed to a decreasing hardness at composites with 15 Vf-% SiC due to the formation of porosity. In the casting process, some SiC particles which were not mixed and wetted completely would stay at the bottom of the crucible. It showed that the shearing action of the stirrer blades to the liquid was weak and the inclining blade slightly generated the rising movement of SiC particles, and the centrifugal movement of SiC particles was tiny, respectively [17].

Figure 8b shows the wear rate of Al-SiC composites with different volume fractions of SiC. It was obtained that the addition of SiC into matrix generated an improvement in wear resistance compared to the unreinforced alloy. Since the interface layer is the most important role in a composite, the reinforcement should be well bonded to the matrix in order to enhance the mechanical behavior [18]. It is known that the wear rate is inversely proportional to the hardness of alloys. In the case of unreinforced Al alloy, the depth of penetration is governed by the hardness of specimen surface and applied load. However, in the case of Al composite, the depth of penetration by the harder asperities of hardened steel disk is primarily governed by protruded hard ceramic reinforcement [19]. In another word, in unreinforced Al alloy, the wear resistance is influenced by stick-formed silicon particles that were created as a result of eutectic reaction during the alloy solidification [20]. A higher wear resistance in the composite was controlled by SiC particles which act to reduce the wear rate by lowering the real contact area [21]. High wear resistance is also influenced by $MgOSiO_2$ [22] and $MgAl_2O_4$ [23] spinel at the interface layer. A good spinel layer can minimize particle pull-out that may occur during the preparation and the wear test [24].

### *3.3 Fractography*

The fracture surface of the composites was analysed by scanning electron microscope to investigate whether the material remains ductile since the reinforced particles can change the properties of composites become less ductile. Following the results in Figure 7b, the elongation or ductility or materials increased up to 8 Vf-% SiC as well as the ultimate tensile strength. It is quite strange results where tensile strength and elongation have a reverse correlation. In general, the failure in composites is due to three different sources, namely the matrix/reinforcement interfacial decohesion, reinforcement fracture and matrix failure [25].

The typical SEM micrograph of fracture surface for composites with 8Vf-% SiC is shown in Figure 9. It indicated a combination mode of fracture, between the dimples on the matrix and SiC particles decohesion. In the center of the sample, it was clearly shown facets which simply indicated a brittle fracture. Furthermore, there were cracked particles that would act as the microcrack initiators during the deformation[26]. It could occur because the interfacial cohesion between the SiC particles and matrix was strong enough. Therefore, it was no wonder for the composite with 8 Vf-% SiC to have the highest tensile strength and good ductility. Strong interfacial cohesion can improve both strength and plasticity of the composite since weak interface will nucleate microcracks at a rather low external applied stress [25]. However, on the edge of the sample, fractography showed dimples which indicated plastic deformation. These dimples may be a result of void nucleation and subsequent coalescence by strong shear deformation and fracture process on the shear plane during a tensile test.

## 4. Conclusion

It can be concluded that the addition of the reinforced SiC particles improves the mechanical properties of the composites such as strength, elongation, hardness, and wear resistance of up to 8 Vf-%. Furthermore, the ductility of the composites deteriorated significantly with the increase of SiC particles concentration. It was due to the presence of phases in composites contained of $MgO.SiO_2$ and $Mg_2Si$ in the form of Chinese script and primary $Mg_2Si$. Meanwhile, the wettability of composite was good enough. It was confirmed by the fractography image that showed the bonding cohesive between particle and matrix were good as the presence of the $MgAl_2O_4$ or spinel phase in the interface.

## 5. References

[1] Akio K, Atsushi O, Toshiro K and Hiroyuki T 1999 Fabrication, process of metal matrix composite with nano-size SiC particle produced by vortex method *J Japan Institute Light Metals* **49** 149-154

[2] Mussert K M, Vellinga W P and Bakker S 2002 A nano-indentation study on the mechanical behaviour of the matrix material in an AA6061–$Al_2O_3$ MMC *Journal of Materials Science* **37** 789-794

[3] Sajjadi S A, Ezatpour H R and Beygi H 2010 Microstructure and mechanical properties of Al-$Al_2O_3$ micro and nano composites fabricated by stir casting *In: Proceedings of 14th national conference on Materials Science and Engineering* Tehran, Iran 325-332

[4] Mazahery A and Shabani M O 2012 Nano-sized silicon carbide reinforced commercial casting aluminium alloy matrix; experimental and novel modeling evaluation *Powder technology* **217** 558-565

[5] Callister J W D 2007 *Materials Science and Engineering: an Introduction* (New York: John Wiley & Sons, Inc)

[6] Kumari S S S, Pillai R M and Pai B C 2008 Structure and properties of calcium and strontium treated Al-7Si-0.3Mg alloy *Journal of Alloys and Compounds* **460** 472-477

[7] Dwivedi S P 2014 Microstructure and mechanical properties of A356/SiC composites fabricated by electromagnetic stir casting *Procedia Materials Science* **6** 1524-532

[8] Aigbodion V and Hassan S B 2007 Effect of silicon carbide reinforcement on microstructure and properties of cast Al-Si-Fe/SiC particulate composites *Materials Science and Engineering A* **447** 3555-360

[9] Mazahery A and Shabani M O 2012 Characterization of cast A356 alloy reinforced with nano sic composites *Transaction of Nonferrous Metals Society of China* **22** 275-280

[10] Bindumadhavan P N, Chia T K, Chandrasekaran M, Wah H K and Phabakar O 2001 Experiment and simulation study on particulate distribution in A356/SiCp suction casting *Mater. Sci. Eng. A* **217** 315

[11] Hashim J, Looney L and Hashmi M S J 1999 Metal matrix composites: production by the stir casting method *Journal of Materials Processing Technology* **92-93** 1-7

[12] DONG P Y 2013 Microstructure and properties of A356-10%SiC particle composite castings at different solidification pressures *Transactios of Nonferrous Metals Society of China* **23** 2222-28
[13] Pawlik K, Gracz B, Zak P L, Szucki M, Lelito J and Krajewski W K 2006 Identification of $Mg_2Si$ phase in Az91/SiC comppposites *Poland: AGH University of Science and Technology.*
[14] Hallstedt B, Liu Z K and Agren J 1990 Reactions in $Al_2O_3$-Mg metal matrix composites *Mater. Sci. Eng. A* **129** 135-145
[15] Gu M, Jin Y, Mei Z, Wu Z and Wu R 1998 Effects on Reinforcement oxidation on the mechanical properties of SiC particulate reinforced aluminum composites *Materials Science and Engineering A* **252** 188-198
[16] Amirkhanlou S and Niroumand B 2011 Effects on reinforcement distribution on low and high temperature of low and high tensile properties of Al A356/SiCp cast composites produced by a novel reinforcement dispersion technique *Journal Materials Science and Engineering A* **528** 7186-95
[17] Du Y, Zhang P, Zhang J and Yao S 2012 Radial distribution of SiC partocles in mechanical stirring of A356-SiCp liquid *J. Material Science Technology* **28**(10) 951-955
[18] Abdulhqq A, Hamid P K, Ghosh P K, Jain S C and Subrata R 2008 The influence of porosity and particle content on dry sliding wear of cas in situ Al(Ti)-$Al_2O_3$($TiO_2$) composite *Wear* **265** 14-26
[19] Mazahery A and Shabani M O 2013 Microstructural and abrasive wear properties of SiC reinforced aluminum-based composite produced by compocasting *Transactions of Nonferrous Metals Society of China* **23** 1905-14
[20] Vencl A, Bobic I, Arostegui S, Bobic B, Marinkovic A and Babic M 2010 Structural, mechanical ad tribological properties of A356 aluminium alloy reinforced with $Al_2O_3$, SiC, and SiC + graphite particles *Journals of Alloys and Compounds* **506** 631-639
[21] Single M, Singh L and Chawla V 2009 Study of wear properties of Al-SiC composites *Journal of Minerals and Materials Characterization and Engineering* **8** 813-819
[22] Kumar T S and Kumar B D 2001 The influence of interfacial characteristics between sicp and Mg/Al metal matrix on wear, coefficient of friction and microhardness *Wear* **247** 191-201
[23] Musert K M, Vellinga W P, Bakker A and Van der Zwaag S 2002 A Nano-indentation study on the mechanical behaviour of the matrix material in an A6061-Al2O3 MMC *Journal of Material Science* **37** 789-794
[24] Devaraju A, Kumar A, Kumaraswamy A and Kotiveerachari B 2013 Influence of reinforcement (SiC and Al2O3) and rotational speed on wear and mechanical properties of aluminium alloy 6061-T6 based surface hybrid composites produced via friction stir processing *Materials and Design* **51** 331-341
[25] Mazahery A and Shabani M O 2013 Plasticity and microstrcture of A356 matrix nano composites *Journal of King Saud University - Engineering Science* **25** 41-48
[26] Wang Z, Song M, Sun C and He Y 2011 Effects of particle size and distribution on the mechanical properties of SiC reinforced Al-Cu alloy composites *Materials Science and Engineering A* **528** 1131-37

**Acknowledgements**
The authors would like to thank the Higher Education, Ministry of Research and Technology and Higher Education, Republic of Indonesia, for financial support to do this research under PUPT funding with a contract No. of 1143/UN2.R12/HKP.05.00/2016.

# The Structure, Magnetic and Absorption Properties of Zn-Ti Substituted Barium-Strontium Hexaferrite Prepared by Mechanochemical Process

**D S Winataputra[1], T L Ujiyanti[2] and W A Adi[1]**

[1]Centre for Science and Technology of Advanced Materials, National Nuclear Energy Agency
Kawasan Puspiptek Serpong, Tangerang Selatan – Indonesia 15313
[2]Study Program of Physics, Faculty of Science and Technique, Purwokerto University, Central Java – Indonesia

Email:didinsw@batan.go.id

**Abstract.** Synthesis and characterization of $(Ba,Sr)_{0.5}Fe_{12-x}Zn_xTi_xO_{19}$ M-type hexaferrite was conducted by using mechanochemical process. The composition of Zn-Ti ion in $(Ba,Sr)_{0.5}Fe_{12-x}Zn_xTi_xO_{19}$ M-type hexaferrite were varied as x = 0.0, 0.2, 0.6, and 1.0. The sample was characterized by using X-rays diffractometer (XRD), scanning electron microscope (SEM), vibrating sample magnetometer (VSM), and vector network analyzer (VNA). The XRD patterns were further analyzed by Rietveld analysis program. The Rietveld analysis result indicated that the substitution of Zn-Ti ions led to the expansion of the hexagonal lattice parameter and unit cell volume ($V_{cell}$), while the atomic density decreased with increasing of the Zn-Ti ion. The VSM measurement exhibited that the substitution of the Zn-Ti ions caused the change in magnetic properties behavior such as intrinsic magnetic coercivity ($H_{ci}$), remanence ($M_r$) and saturation ($M_s$) magnetization. The value of $H_c$ dropped significantly with the substitution of Zn-Ti ions, and then continued to decrease with the addition of Zn-Ti ions content for x = 0.6, and x = 1.0. SEM observation revealed that all the particles showed nearly hexagonal platelet shape, with the particle size distribution at most in the range between 200 – 600 nm. The variation of the reflection loss versus frequency was measured by using VNA in $(Ba,Sr)_{0.5}Fe_{12-x}(ZnTi)_xO_{19}$ hexaferrite for x= 0.0 – 1.0. The optimum reflection loss (RL) was found to be –48 dB at 14 GHz in $(Ba,Sr)_{0.5}Fe_{11.6}(ZnTi)_{0.2}O_{19}$ hexaferrite for x = 0.2.

**Keywords:** $(Ba,Sr)_{0.5}Fe_{12-x}Zn_xTi_xO_{19}$, M-type hexaferrite, Zn-Ti ions substitution, magnetic property, reflection loss.

## 1. Introduction

Barium hexaferrite is one of M-type hexaferrites. Barium hexaferrite has a large saturation magnetization, high coercivity, high Curie temperature, large uniaxial magnetic crystalline anisotropy, and excellent chemical stability. Therefore, Barium hexaferrite is promising candidates for the development of microwave absorbing materials [1–3]. Barium hexaferrite has significantly large crystalline anisotropy due to their low crystal symmetry, as it is compared with the cubic symmetry of spinel or garnet ferrites. Therefore, the resonance frequency can reach as high as GHz [4–6].

Furthermore, the location of resonance can be modified over a wide frequency range by the substitution of ions in Barium hexaferrite.

Many reports are available for the synthesis and characterization of the barium hexaferrite substituted with various cations like Sr, Pb, La for Ba and Mn-Ti, Co-Ti, Ni-Cr for Fe [7–11]. These studies revealed that the structural, magnetic, and microwave absorbing properties of barium hexaferrite are strongly influenced by the substitution of multivalent cations. Therefore, this research focused on the effect of the ions substitution on the structure, magnetic and absorbing behavior in $(Ba,Sr)_{0.5}Fe_{(12-2x)}(ZnTi)_xO_{19}$ (for x = 0.0, 0.2, 0.6, 1.0) hexaferrite. The main objectives are to fabricate $(Ba,Sr)_{0.5}Fe_{(12-2x)}(ZnTi)_xO_{19}$ (for x = 0.0, 0.2, 0.6, 1.0) using the mechanochemical method and investigate the influence of Zn-Ti ions doping concentration in composites on the structural, magnetic and microwave absorbing performance.

The synthesis was done through a mechanochemical process. The obtained materials then characterized by using X-ray diffractometer (XRD), scanning electron microscope (SEM), and vibrating sample magnetometer (VSM) to study the samples' microstructure, and magnetic properties. The characterization of reflection and transmission of microwave were carried out by using the Vector Network Analyzer (VNA, ADVANTEST R3770) with a frequency ranged from 7 to 15 GHz.

## 2. Experimental Method

As starting materials of titanium dioxide ($TiO_2$), zinc oxide ($ZnO_2$), hematite ($\alpha$-$Fe_2O_3$), barium carbonate ($BaCO_3$), and strontium carbonate ($SrCO_3$) were used. A series of $(Ba,Sr)_{0.5}Fe_{12-x}Zn_xTi_xO_{19}$ samples with x = 0.0, 0.2, 0.6, and 1.0 were synthesized with a mechanochemical method using HEM, Certi-prep 8000 mixer [12,13]. The prepared for doped samples were mixed together in a stainless steel vial. The ground mixing of each sample was performed for 5 hours. The ball and sample weight ratio was 10 : 1. Furthermore, the sample was dried in an oven at 80 °C, and followed by sintering at 1000 °C for 5 hours.

The crystalline structure was determined by using X-ray diffractometer (XRD, Philips Pan-analytical Empyrean) using Cu-K$\alpha$ radiation ($\lambda$ = 15.405 nm) in the range of 20 –70° of 2$\theta$. The XRD data was analyzed by using general structure analysis system (GSAS) program [14]. The morphology of the sample was characterized by using a scanning electron microscope (SEM, JEOL-JSM 6510). The magnetic measurements were carried out with a vibrating sample magnetometer (VSM, Oxford 1.2T) at a room temperature. The reflection loss of the absorber was measured by using microwave Network Analyzer (ADVANTEST R3770) in the frequency range of 7 – 15 GHz.

## 3. Result and Discussion

### *3.1 Structure Properties*

Figure 1 shows the XRD patterns of $Fe_2O_3$ and $(Ba,Sr)_{0.5}Fe_{12-x}Zn_xTi_xO_{19}$ (x = 0.0, 0.2, 0.6, and 1.0). Based on Figure 1(a) and 1(b), $(Ba,Sr)_{0.5}Fe_{12-x}Zn_xTi_xO_{19}$ without and with Zn and Ti doping for x = 0.2 were identified to be single phase. The diffraction peaks of M-type $BaFe_{12}O_{19}$ hexaferrite phase in Figure 1 are marked by "+". However, with the addition of Zn-Ti ions doping for x = 0.6 and 1.0, the figure shows the broadened diffraction peaks corresponding to $Fe_2O_3$ hematite phase (signed by " • "), see Figure 1(c) and 1(d).

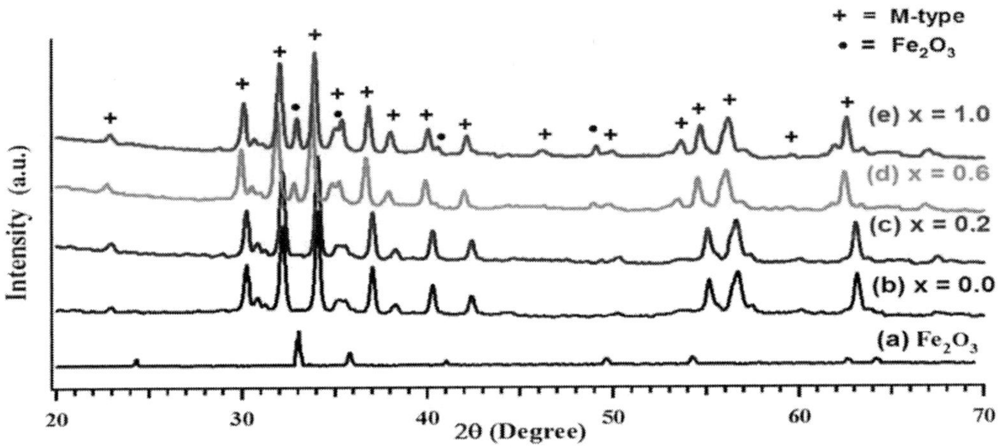


**Figure 1.** XRD pattern of $Fe_2O_3$ (a), $(Ba,Sr)_{0.5}Fe_{12-x}Zn_xTi_xO_{19}$ powder for (b) x = 0.0, (c) x = 0.2, (d) x = 0.6, and (e) x = 1.0 after heat treated at 1000 °C for 5 h.

Figure 2 shows SEM pictures of $(Ba,Sr)_{0.5}Fe_{12-x}Zn_xTi_xO_{19}$ powder (x = 0.0, 0.2, 0.6, and 1.0) prepared by mechanochemical method after sintering at 1000 °C for 5 h in the air atmosphere. From Figures 2 for all SEM images, it can be seen that the particle inside the sample was grown with relatively uniform shape. In Figure 2(a), the particles were nearly spherical-like shape. The particle size distribution was almost evenly and smaller than 600 nm. In Figure 2(b), 2(c), and 2(d), it is seen that the particle showed nearly platelet shape. However, it is clear that the particle size was smaller below 1000 nm.

**Figure 2**. The morphology of $(Ba,Sr)_{0.5}Fe_{12-x}Zn_xTi_xO_{19}$ powder for (a) x = 0.0, (b) 0.2, (c) 0.6, and (d) 1.0.

Figure 3 shows the refinement result of XRD pattern of $(Ba,Sr)_{0.5}Fe_{12-x}Zn_xTi_xO_{19}$ samples without and with Zn-Ti doping for x = 0.0, 0.2, 0.6, and 1.0. The refinement result produced a very good quality of fitting with *R* factor and goodness of fit value $\chi^2$ (chi-squared) was less than 1.2, as shown in Table 1. From Figure 3, the plus symbol (+) of red line represents the observation, while the green line is the calculation. Miller index is signed by bar symbol (I) and the difference between observation and calculation is represented by purple line symbol (-). Based on Figure 3(a), and 3(b), it can be seen that $(Ba,Sr)_{0.5}Fe_{12-x}Zn_xTi_xO_{19}$ (for x = 0.0, and x = 0.2) were identified to be single phase hexagonal structure, indicating the doping elements had been substituted into the structure successfully. The diffraction pattern produced in accordance with JCPDS # 84-1531 and the space group was $P6_3/mmc$.

A little change has been found in Figure 2(b) and 2(c) which show the peak broadening and corresponding to the impurities phase, which may be due to the difference in the molarities of the precursor. The content of impurities phase was obtained from the analysis by using GSAS program is listed in Table 1. Based on the Table 1, about 70% and 60% phase purities were achieved with substitution of Zn-Ti ions for x = 0.6 and x = 1.0 respectively. Earlier studies reported that the impurities such as $Fe_2O_3$ phase begin to dissolve at the temperature greater than 1000 °C. It helps aggregation of the particles and grains growth significantly during sintering process [1,12].

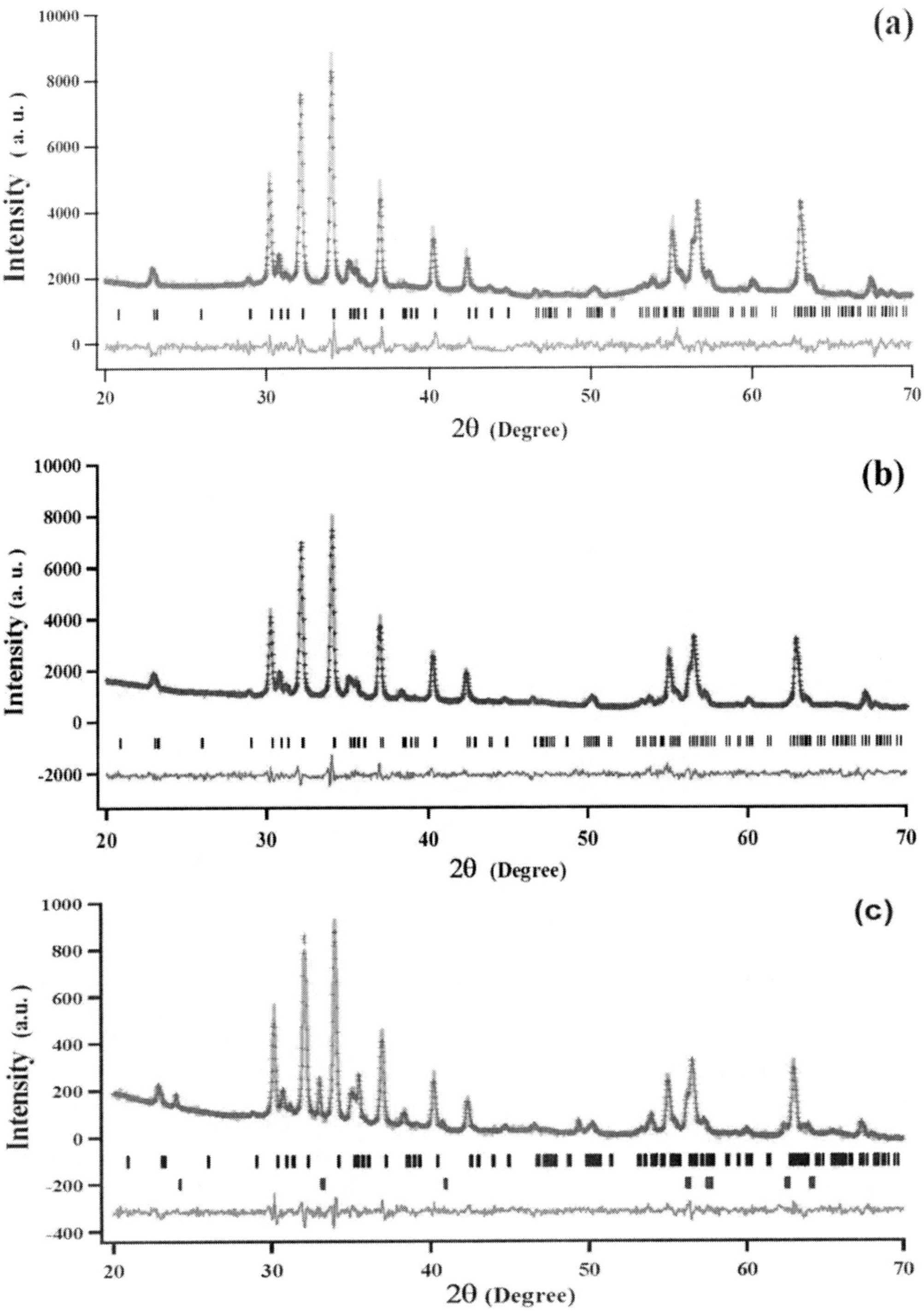
(a)
Intensity ( a. u. )
2θ (Degree)
(b)
Intensity (a. u. )
2θ (Degree)
(c)
Intensity (a.u.)
2θ (Degree)

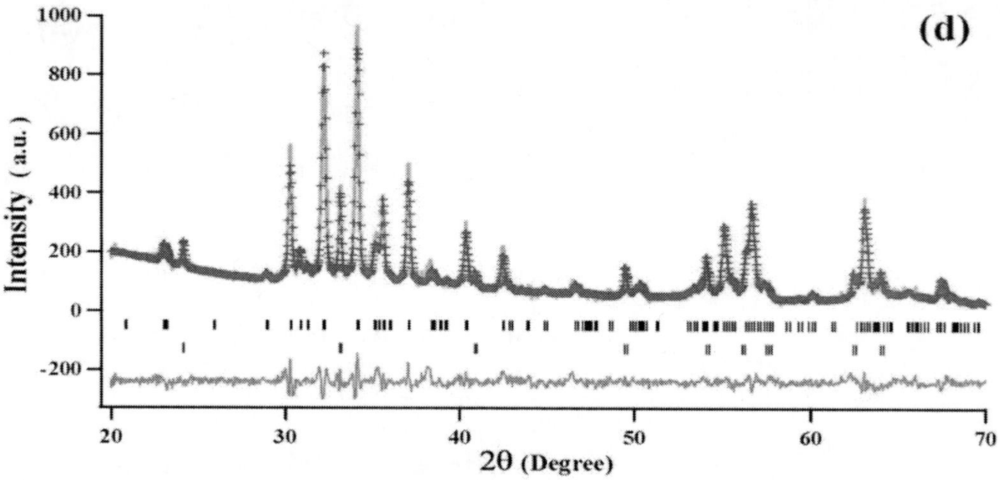


**Figure 3.** The refinement XRD pattern of $(Ba,Sr)_{0.5}Fe_{12-x}Zn_xTi_xO_{19}$ powder for (a) x = 0.0, (b) x = 0.2, (c) x = 0.6, and (d) x = 1.0.

**Table 1.** Phase quantification after Rietveld analysis on the XRD patterns of $(Ba,Sr)_{0.5}Fe_{12-x}Zn_xTi_xO_{19}$

| Comp. "x" | $(Ba,Sr)_{0.5}Fe_{(12-2x)}(ZnTi)_xO_{19}$ Hexaferrite | Phase (%) | |
|---|---|---|---|
| | | $(Ba,Sr)_{0.5}Fe_{12}O_{19}$ | $Fe_2O_3$ |
| 0.0 | $(Ba,Sr)_{0.5}Fe_{12}O_{19}$ | 100.00 | 0.00 |
| 0.2 | $(Ba,Sr)_{0.5}Fe_{11.6}(ZnTi)_{0.2}O_{19}$ | 100.00 | 0.00 |
| 0.6 | $(Ba,Sr)_{0.5}Fe_{10.8}(ZnTi)_{0.6}O_{19}$ | 77.00 | 23.00 |
| 1.0 | $(Ba,Sr)_{0.5}Fe_{10.0}(ZnTi)_{1.0}O_{19}$ | 67.00 | 33.00 |

Variation of lattice parameters (*a, c*), volume cell and atomic density as a function of Zn-Ti content is shown in Figure 4. It appears that the value of *a* and *c* are slightly increased with increasing the Zn-Ti contents as indicated by Figure 4a. The effect of Zn-Ti ions doping also led to increase the unit cell volume ($V_{cell}$) is quite significant, especially for x = 1.0, while the atomic density decreases as depicted in Figure 4(b). It indicated that the regular structure became slightly elongated along *a*-axis and *c*-axis. As a result, the dopant ions changed both of the lattice parameters, cell volume, and atomic density.

The atomic substitution of $Fe^{2+}/Fe^{3+}$ by $Zn^{2+}/Ti^{4+}$ has a significant impact on the lattice parameters. Energy perturbation and lattice distortion were expected when $Fe^{2+}/Fe^{3+}$ were replaced by $Zn^{2+}/Ti^{4+}$ ions due to the difference in their atomic radii and their orbitals. This condition caused the anisotropy energy to change, and thus the magnetic behavior changed particularly coercivity which decreased with Zn/Ti doping, as shown in Figure 5. The observed variation depended on the difference in the ionic radii of the metal ions Zn = 0.74 Å, Ti = 0.68 Å compare to Fe = 0.645 Å.

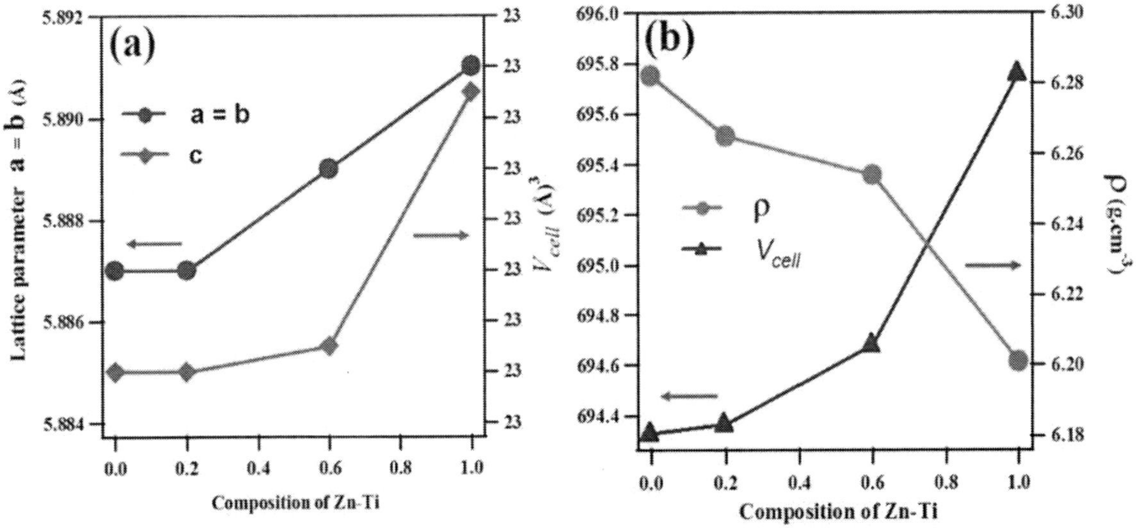


**Figure 4**. The plot of (a) lattice parameter (*a,c*), and (b) $V_{cell}$ and atomic density as a function of composition of Zn-Ti.

### *3.2 Magnetic Properties*

Figure 5 represents the sequences hysteresis loop behavior of $(Ba,Sr)_{0.5}Fe_{12-x}(ZnTi)_xO_{19}$ powder with variation of Zn-Ti ions doping for x = 0.0, 0.2, 0.6, and 1.0. The characteristic magnetic parameters are listed in Table 2. The variations of $M_s$, $M_r$, and $H_c$ values of $(Ba,Sr)_{0.5}Fe_{12-x}(ZnTi)_xO_{19}$ are shown in Figure 5(b), 5(c), and 5(d). It is seen that the substitution of the Zn-Ti ions caused decrease significantly in intrinsic coercivity, $H_{ci}$ and little affected to saturation magnetization, Ms at low substitution. Based on Table 2, $H_c$ was observed to be 4.0 kOe in the unsubstituted sample. A small amount of Zn-Ti ions doping (for x = 0.2) did not significantly affect the structure but had a great influence on magnetic properties. The $M_s$ value remained almost constant, but $H_c$ rapidly decreased from 4.00 kOe to 2.35 kOe. As x continued to increase to x = 1.0, the value of $M_s$ and Hc decreased to 74 emu/g, and $H_c$ to 1.00 kOe, respectively. Magnetic remanent, $M_r$ also decreased with the increase of ions substitution. The decrease in $M_r$ was due to the net alignment of grain magnetization caused by weak inter-grain exchange interactions. The high value of $H_c$ in undoped sample was due to uniaxial magnetocrystalline anisotropy along *c*-axis. C-axis is very sensitive to magnetic properties and anisotropy of Ba-hexaferrite and is responsible for coercivity. Thus, the increase in *c*-axis with the addition of "x" alters the distance between magnet ions, so that the exchange interaction varies which in turn changes the magnetic properties.

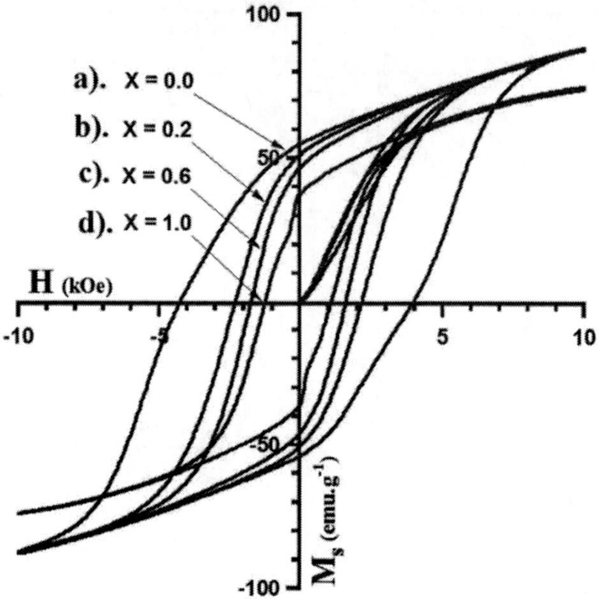


**Figure 5.** Room temperature hysteresis loop of $(Ba,Sr)_{0.5}Fe_{(12-2x)}$ $(ZnTi)_xO_{19}$ powder sample for (a) x = 0.0, (b) x = 0.2, (c) x= 0.6, and (d) x=1.0.

**Table 2** Magnetic properties of $(Ba,Sr)_{0.5}Fe_{12-x}Zn_xTi_xO_{19}$ as a function of Zn-Ti substitution level x.

| Comp. "x" | $(Ba,Sr)_{0.5}Fe_{(12-2x)}(ZnTi)_xO_{19}$ Hexaferrite | $H_{ci}$ (kOe) | emu.g$^{-1}$ | |
|---|---|---|---|---|
| | | | $M_s$ | $M_r$ |
| 0.0 | $(Ba,Sr)_{0.5}Fe_{12}O_{19}$ | 4.00 | 88.20 | 53.50 |
| 0.2 | $(Ba,Sr)_{0.5}Fe_{11.6}(ZnTi)_{0.2}O_{19}$ | 2.35 | 88.20 | 50.10 |
| 0.6 | $(Ba,Sr)_{0.5}Fe_{10.8}(ZnTi)_{0.6}O_{19}$ | 1.60 | 87.60 | 45.30 |
| 1.0 | $(Ba,Sr)_{0.5}Fe_{10.0}(ZnTi)_{1.0}O_{19}$ | 1.10 | 74.70 | 34.90 |

### *3.3 Absorption Characteristics*

Figure 6 (a-d) shows the variation of the reflection loss versus frequency, observed in $(Ba,Sr)_{0.5}Fe_{12-x}(ZnTi)_xO_{19}$ hexaferrite (for x = 0.0, 0.2, 0.6 and 1.0) at a thickness of 2 mm and in the frequency range of 7–15 GHz. The minimum value of the reflection loss exhibited a peak at approximately −10 dB at a frequency of 7 GHz for x = 0.0 (undoped) sample. For x = 0.2 (Figure 6b), the reflection loss exhibited two peaks at approximately −44 dB at a frequency of 8 GHz and -48 dB at 14 GHz, with a bandwidth of 0.5 GHz. For x = 0.6 (Figure 6c), the reflection loss also exhibited two peaks at approximately −22 dB and -26 dB at 8 GHz and 14.5 GHz. With the addition of Zn-Ti substitutions for x = 1.0 (Fig. 6d), the reflection loss exhibited three peaks approximately at -32 dB at 8 GHz, -34 dB at 11 GHz and -44 dB at 14 GHz.

The optimum reflection loss was found to be –48 dB at a frequency of 14 GHz in $(Ba,Sr)_{0.5}Fe_{11.6}(ZnTi)_{0.2}O_{19}$ hexaferrite for x = 0.2 (Figure 6b). The Zn-Ti substitution to Fe site affected to domain wall pinning and the change of magnetic anisotropy, so that the damping of domain turning and domain wall displacement increases, thus leading to the resonance enhancement and resonance frequency-position shift [6,9,15]. However, an excessive amount of ions doping may result in weak microwave absorption, due to the increase in electric resistivity. As a result, a suitable amount of Zn-Ti doping is required to improve the microwave absorption of M-type barium hexaferrite.

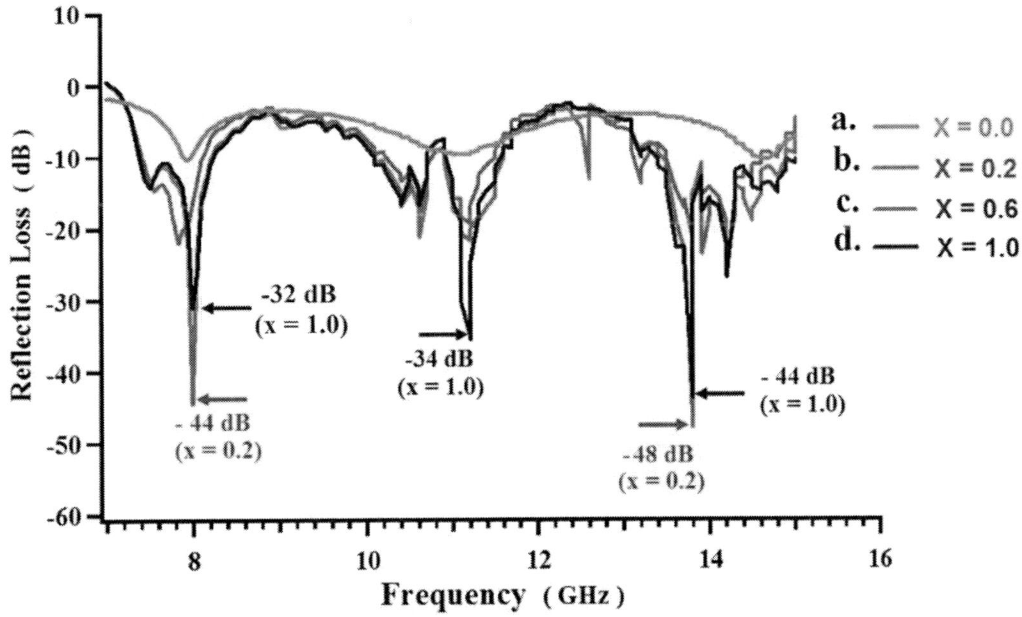


**Figure 6.** The absorption characteristics of the $(Ba,Sr)_{0.5}Fe_{12-x}(ZnTi)_xO_{19}$ hexaferrite for x at (a) 0.0, (b) 0.2, (c) 0.6, and (d) x =1.0.

## 4. Conclusion

M-type hexaferrite of $(Ba,Sr)_{0.5}Fe_{12-x}Zn_xTi_xO_{19}$ (for x = 0.0, 0.2, 0.4, 0.6, 1.0) has been synthesized through mechanochemical process. The structural, magnetic and absorption properties of the synthesized composition were studied by using XRD, SEM, VSM and VNA. The refinement of XRD pattern results showed that the M-type hexaferrite for x = 0.0 and x = 0.2 are single phase. The phase purity decreased with the increase of Zn-Ti concentration for x = 0.6 and x = 1.0. The effect of the Zn-Ti ions substitution led changes in the structural and magnetic properties. The lattice parameter increased became elongated along *a* and *c*-axis with the doping of Zn-Ti, and also led to increase the unit cell volume ($V_{cell}$). The magnetic properties behaviors such as magnetic coercivity, remanent and saturation magnetization led to decrease, as the addition of Zn-Ti substitution. The Effect doping of Zn-Ti could enhance the impedance matching and microwave attenuation, which the optimum reflection loss was found to be –48 dB at 14 GHz in $(Ba,Sr)_{0.5}Fe_{11.6}(ZnTi)_{0.2}O_{19}$ for x = 0.2.

## 5. References

[1] L Jie, Huai-Wu Z, Yuan-Xun L, Ying-Li L and M. Yan-Bing. 2012 The Structural and Magnetic Properties of Barium Ferrite Powders Prepared by Sol-gel Method *Chin Phys B* **21** 17501-017501–4

[2] Jesús F S 2014 Mechanosynthesis , crystal structure and magnetic characterization of M-type *Ceram. Int.* **40** 4033–8

[3] Trukhanov A V, Panina L V, Trukhanov S V, Turchenko V O, Kazakevich I S and Salem M M 2015 Features of Crystal Structure and Magnetic Properties of M-type Ba-hexaferrites with Diamagnetic Substitution *Int. J. Mater. Chem. Phys.* **1** 286–94

[4] Kaur R and Aul G D 2012 Effect of Reflection Property on Microwave Absorbing Materials - A Review *Int. J. Sci. Res. IJSR*

[5] Li C-J, Wang B and Wang J-N 2012 Magnetic and Microwave Absorbing Properties of Electrospun Ba(1−x)LaxFe12O19 Nanofibers *J. Magn. Magn. Mater.* **324** 1305–11

[6] Qin X, Cheng Y, Zhou K, Huang S and Hui X 2013 Microwave Absorbing Properties of W-Type Hexaferrite Ba(MnZn)xCo2(1−x)Fe16O27 *J. Mater. Sci. Chem. Eng.* **1** 8–13

[7] Haq A and Anis-ur-Rehman M 2012 Effect of Pb on Structural and Magnetic Properties of Ba-Hexaferrite *Pysica B* **407** 822–6

[8] Li W, Qiao X, Liu M and Peng H 2013 La and Co Substitutwd M-tpe Barium Ferrites Processed by Sol-gel Combustion Synthesis *Mater. Res. Bull.* **48** 4449–53

[9] Kanagesan, S, Mansor H and Sinnappan J 2014 Microwave Sintering of Zn-Nb Doped Barium Hexaferrite Synthesized via Sol-Gel Method *Mater. Sci. Appl.* **5** 171–6

[10] Ali-Sharbati, Saeed-Choopani, Ali-Ghasemi, Al-amri A, Cerqueira C F and JR. A P 2011 Synthesize and Magnetic Propertiesof of Nanocrystalline Ba3 Co2 (0.8- x) Mn0.4 Ni2x Fe24 O41 Prepared by Citrate Sol-gel Method *Digesr J. Nanomater. Biostructures* **6** 187–98

[11] González-Angeles A, Lipka J, Grusková A, Sláma J, Jančárik V and Slugeň V 2010 Magnetic comparison of BaCa and BaSr substituted hexaferrite powders *J. Phys. Conf. Ser.* **217** 12137

[12] Adi W A and Manaf A 2012 Structural and Absorption Characteristics of Mn-Ti Substituted Ba-Sr Hexaferrite Synthesized by Mechanical Alloying Route *J Basic Appl Sci Res* **2** 7826–34

[13] Ujiyanti T. L. 2014 Synthesis and Characterization of (Ba,Sr)0.5Fe12-x(ZnTi)xO19, (x = 0.2, 0.4, 0.6, and 1.0 wt %) as Electromagnetic Wave Absorber *Bachelor Thesis* 2014

[14] Larson A C and Von Dreele R B 2004 *GENERAL STRUCTURE ANALYSIS SYSTEM (GSAS), Los Alamos National Laboratory Report LAUR (2004), pp. 86-748* .

[15] Vinayasree S, Soloman M A, Sunny V, Mohanan P, Kurian P, Joy P A and Anantharaman M R 2014 Flexible microwave absorbers based on barium hexaferrite, carbon black, and nitrile rubber for 2-12GHz applications *J. Appl. Phys.* **116**

**Acknowledgements**
This research was supported by "Research and Development of Smart magnetic and Magnetic Oxide Programme for DIPA Grants 2016" at Centre for Science and Technology of Advanced Materials – National Nuclear Energy Agency, Republic of Indonesia.

# Influence of External Magnetic Field on Potential Differences of Long Josephson Junction

**A Hidayat, Y P Hardianto, E Latifah, H Wisodo and A Taufiq**

Department of Physics, Faculty of Mathematics and Natural Sciences, Universitas Negeri Malang (State University of Malang), Jl. Semarang 5, Malang 65145, Indonesia

Email: arif.hidayat.fmipa@um.ac.id

**Abstract.** The soliton phenomenon in a long Josephson junction was studied by using the particular solution of Sine-Gordon equation. Using analytical method, sine-Gordon equation was normalized and two-soliton solutions correlated to phase, and potential differences of long Josephson junction were obtained. Based on the two-soliton solutions, the influence of external magnetics fields on the dynamics of phase and potential differences of long Josephson junction were numerically studied.

**Keyword** : Josephson Effect, Long Josephson Junction, Sine-Gordon Equation, Soliton.

## 1. Introduction

Long Josephson junction is a device that consists of superconductor-insulator/metal-superconductor (SIS) junction, which in this junction there is a Josephson effect. Because of that effect, phase and potential differences of long Josephson junction are occurred [1]. Based on the theory of electrodynamics and quantum mechanics, it is revealed that the physical equation of the long Josephson junction corresponds to sine-Gordon equation. Kink-antikink soliton solutions of sine-Gordon equation in a long Josephson junction is obtained from [2-3]. Some researchers had been done to study Josephson junction in technology, for example, thermal transport, SQUID, quantum phase and solitons [4-8]

Dynamics phase of long Josephson junction is expressed in equation (1)

$$\frac{1}{v_{ph}^2}\frac{\partial^2\phi}{\partial t^2}-\frac{\partial^2\phi}{\partial z^2}=\frac{1}{\lambda_J^2}\sin(\phi) \tag{1}$$

If it is defined new variables $t'=\frac{v_{ph}}{\lambda_J}$ and $z'=\frac{1}{\lambda_J}z'$ ,thus the above equation become sine-Gordonequation,

$$\frac{\partial^2\phi}{\partial t'^2}-\frac{\partial^2\phi}{\partial z'^2}=\sin(\phi) \tag{2}$$

Using kink-antikink solutions; equation (3),

$$\phi(z',t') = 4\tan^{-1}(\frac{u(z')}{v(t')}) \tag{3}$$

as a soliton solution of equation (2). Thus, we obtained

$$\frac{(u(z'))^2}{v(t')}\frac{\partial^2 v(t')}{\partial t'^2}+\frac{(v(t'))^2}{u(z')}\frac{\partial^2 u(z')}{\partial z'^2}= \left[(v(t'))^2 - v(t')\frac{\partial^2 v(t')}{\partial t'^2}+2\left(\frac{\partial v(t')}{\partial t'}\right)^2\right]-\left[(u(z'))^2 + u(z')\frac{\partial^2 u(z')}{\partial z'^2}-2\left(\frac{\partial u(z')}{\partial z'}\right)^2\right] \tag{4}$$

equation (4) applies for all space and time. Therefore, left and right-hand side of equation (4) is permitted to derive partially with respect to $z'$ and $t'$. So it gives,

$$\frac{1}{v(t')\,{}^{\partial v(t')}/_{\partial t'}}\frac{\partial}{\partial t'}\left(\frac{{}^{\partial^2 v(t')}/_{\partial t'^2}}{v(t')}\right) = -\frac{1}{u(z')\,{}^{\partial u(z')}/_{\partial z'}}\frac{\partial}{\partial z'}\left(\frac{{}^{\partial^2 u(z')}/_{\partial z'^2}}{u(z')}\right) \tag{5}$$

Because left and right-hand side of equation (5) are differential equations having different variables, so it should be a constant.

$$\frac{1}{v(t')\,{}^{\partial v(t')}/_{\partial t'}}\frac{\partial}{\partial t'}\left(\frac{{}^{\partial^2 v(t')}/_{\partial t'^2}}{v(t')}\right) = -\frac{1}{u(z')\,{}^{\partial u(z')}/_{\partial z'}}\frac{\partial}{\partial z'}\left(\frac{{}^{\partial^2 u(z')}/_{\partial z'^2}}{u(z')}\right) = -6k^2 \tag{6}$$

Using ordinary differential method, from equation (6), we obtain two following equations:

$$\frac{\partial^2 v(t')}{\partial t'^2} = -3k^2 v^3(t') + Av(t') \tag{7}$$

and

$$\frac{\partial^2 u(z')}{\partial z'^2} = 3k^2 u^3(z') + Bu(z') \tag{8}$$

where $A$ and $B$ are constant.

The two last equations can be changed in two new equations:

$$\left(\frac{\partial v(t')}{\partial t'}\right)^2 = 2C - \frac{3}{2}k^2 v^4(t') + Av^2(t') \tag{9}$$

and

$$\left(\frac{\partial u(z')}{\partial z'}\right)^2 = 2D + \frac{3}{2}k^2u^4(z') + Bu^2(z') \tag{10}$$

The goals of this work are to investigate numerically the dynamic of phase and potential differences in long Josephson junction under the influence of external magnetic field parameter.

## 2. Numerical Method

The physical system considered in this research is long Josephson junction under external magnetic field. The external magnetic field is uniformly applied along of long Josephson junction. Heun method [Mathews, 2004] is used to determine inverse tan function. Long Josephson Junction phase difference equation that is similar to the sine-Gordon equation is converted into dimensionless equations by means of normalizing each variable.

## 3. Results and Discussion

Base on Equation (10) and (11), dynamics of phase and potential differences in long Josephson junction under the influence of external magnetic field were investigated. Three conditions were considered. The graphs under the three conditions are obtained by taking penetration depth of magnetic field ($\lambda_J$) is 1760 angstrom, insulator thickness ($d$) is 2 nm, supercurrent density ($J_C$) is 5.240.000 A/m$^2$ and integral coefficients $\alpha$ and $\gamma_{t'}$ are equal to 0,75 and $\gamma_{t'}$=0,133994, respectively.

### 3.1 *Influence of external magnetic field on the phase difference of long Josephson junction for k = 0 while A = B =α and C = D = β.*

For this condition, long Josephson junction phase difference function is written as

$$\phi(z',t') = 4\tan^{-1}\left(\frac{\left(e^{\sqrt{\alpha}(z'-\gamma_{z'})} - e^{-\sqrt{\alpha}(z'-\gamma_{z'})}\right)}{\left(e^{\sqrt{\alpha}(t'-\gamma_{t'})} - e^{-\sqrt{\alpha}(t'-\gamma_{t'})}\right)}\right) \tag{11}$$

and potential differences of the long Josephson junction is written as

$$V(z',t') = \frac{1}{2\pi}\frac{4\phi_0\sqrt{\alpha}\left(-e^{\sqrt{\alpha}(z'-\gamma_{z'})} + e^{-\sqrt{\alpha}(z'-\gamma_{z'})}\right)}{1+\left(\frac{\left(e^{\sqrt{\alpha}(z'-\gamma_{z'})} - e^{-\sqrt{\alpha}(z'-\gamma_{z'})}\right)}{\left(e^{\sqrt{\alpha}(t'-\gamma_{t'})} - e^{-\sqrt{\alpha}(t'-\gamma_{t'})}\right)}\right)^2}\frac{\left(e^{\sqrt{\alpha}(t'-\gamma_{t'})} + e^{-\sqrt{\alpha}(t'-\gamma_{t'})}\right)}{\left(e^{\sqrt{\alpha}(t'-\gamma_{t'})} - e^{-\sqrt{\alpha}(t'-\gamma_{t'})}\right)^2} \tag{12}$$

Figure 1 shows graphs of potential difference and external magnetic fields of the long Josephson junction. Based on Figure 1, we conclude that solutions are not solitons, because from $t = 0$ s to $t = 12 \times 10^{-12}$ s, graphs of the magnetic field and potential differences of long Josephson junction always change. Physically, it means that external magnet fields applied to long Josephson junction are not large enough to generate soliton solutions [10].

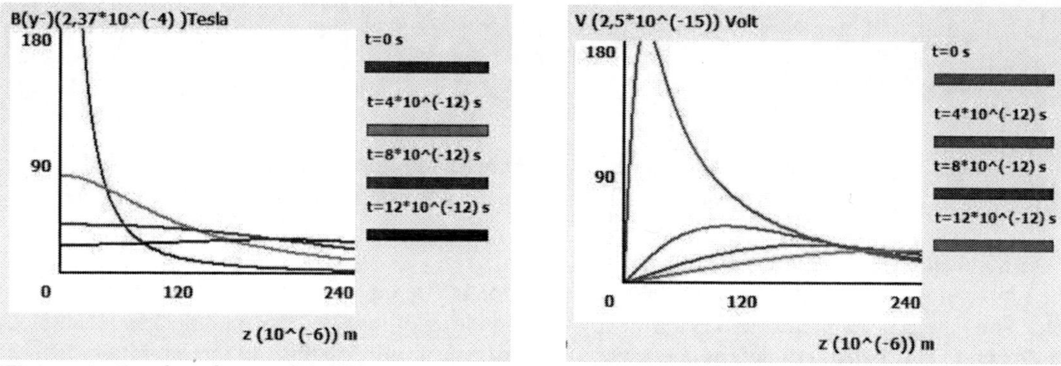


**Figure 1.** Graphs of potential difference and external magnetic fields of long Josephson junction for as a function of time; $t = 0$; $5 \times 10^{-12}$; $8 \times 10^{-12}$; $12 \times 10^{-12}$ seconds, for $k = 0$ while $A=B=\alpha$ and $C = D = \beta$

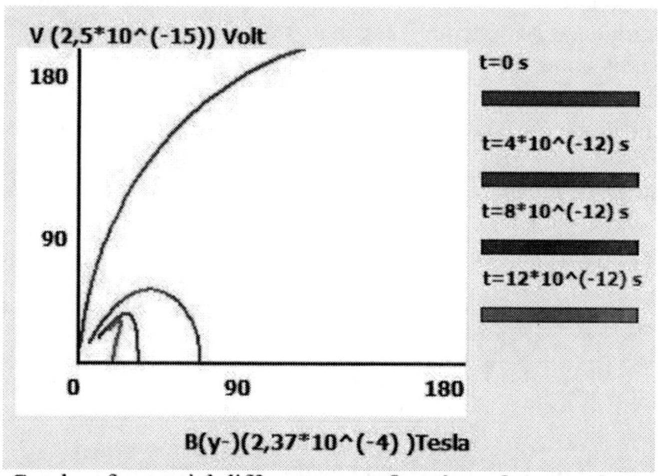


**Figure 2.** Graphs of potential difference as a function of external magnetic fields of long Josephson junction for a given times: $t = 0$; $5 \times 10^{-12}$; $8 \times 10^{-12}$; $12 \times 10^{-12}$ seconds, for $k = 0$ while $A=B=\alpha$ and $C = D = \beta$.

Because of the evolution of potential differences as a function of a magnetic field change in for different given time, the potential differences are non-solitary.

3.2 *Influence of external magnetic field on phase difference of long Josepshon junction for $C = D = k = 0$ while $A=B \neq 0$*

For this condition, long Josephson junction phase difference function is written as

$$\phi(z',t') = 4\tan^{-1}(\exp\left(\frac{z'-\beta t'}{\sqrt{1-\beta^2}}\right)) \tag{13}$$

and potential differences of the long Josephson junction is written as

$$V(z',t') = -\frac{\beta}{\sqrt{1-\beta^2}}\frac{1}{2\pi}\frac{4\phi_0}{1+\left(\exp\left(\frac{z'-\beta t'}{\sqrt{1-\beta^2}}\right)\right)^2}\exp\left(\frac{z'-\beta t'}{\sqrt{1-\beta^2}}\right) \tag{14}$$

Figure 3 shows graphs of phase and potential differences of the long Josephson junction. From the graphs shown in Figure 2, we conclude that solutions are solitons, because from $t = 0$ s to $t = 12 \times 10^{-12}$ seconds. Graphs of phase and a potential difference of long Josephson junction do not change in. Physically, it means that external magnet fields applied to long Josephson junction are enough to form soliton solutions.

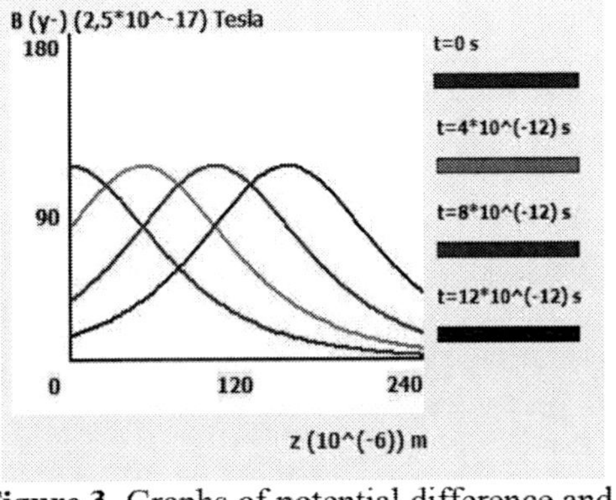


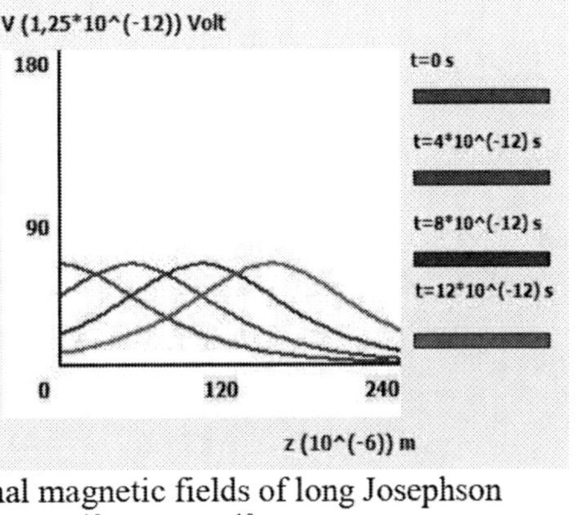


**Figure 3.** Graphs of potential difference and external magnetic fields of long Josephson junction for as a function of time; $t = 0$; $5 \times 10^{-12}$; $8 \times 10^{-12}$; $12 \times 10^{-12}$ seconds, for $\boldsymbol{C = D = k = 0}$ **while** $\boldsymbol{A=B \neq 0}$.

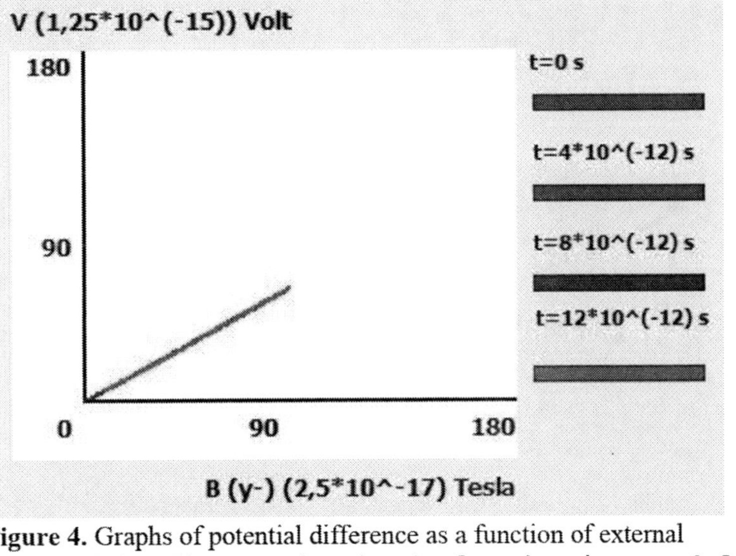


**Figure 4.** Graphs of potential difference as a function of external magnetic fields of long Josephson junction for a given times: $t = 0$; $5 \times 10^{-12}$; $8 \times 10^{-12}$; $12 \times 10^{-12}$ seconds, for $C = D = k = 0$ while $A=B \neq 0$

The potential difference of Long Josephson Junction is solitary; consequently, the external magnetic fields allow a solitary shaped pattern. So with a solitary condition, the curve of potential difference Long Josephson Junction as a function of the external magnetic fields forms linear patterns for different times. This condition is called anti-kink soliton

### 3.3 *Influence of external magnetic field on phase difference of long Josepshon junction for $C = D = 0$ while $A = B = 3/2k^2$*

For this condition, long Josephson junction phase difference function is written as

$$\phi(z',t') = 4\tan^{-1}\left(\frac{A_2}{A_1} e^{-\sqrt{\frac{3}{2}}k(z'-t')} \frac{\left(1+A_1{}^2 e^{-kt'\sqrt{6}}\right)}{\left(A_2{}^2 e^{-kz'\sqrt{6}} - 1\right)}\right) \tag{15}$$

and potential differences of the long Josephson junction is written as

$$V(z',t') = \frac{1}{2\pi}\sqrt{\frac{3}{2}}k\frac{A_2}{A_1} \frac{4\phi_0}{1+\left(\frac{A_2}{A_1} e^{-\sqrt{\frac{3}{2}}k(z'-t')} \frac{\left(1+A_1{}^2 e^{-kt'\sqrt{6}}\right)}{\left(A_2{}^2 e^{-kz'\sqrt{6}} - 1\right)}\right)^2} \frac{\left(e^{\sqrt{\frac{3}{2}}kt'} - A_1{}^2 e^{-\sqrt{\frac{3}{2}}kt'}\right)}{\left(A_2{}^2 e^{-\sqrt{\frac{3}{2}}kz'} - e^{\sqrt{\frac{3}{2}}kz'}\right)} \tag{16}$$

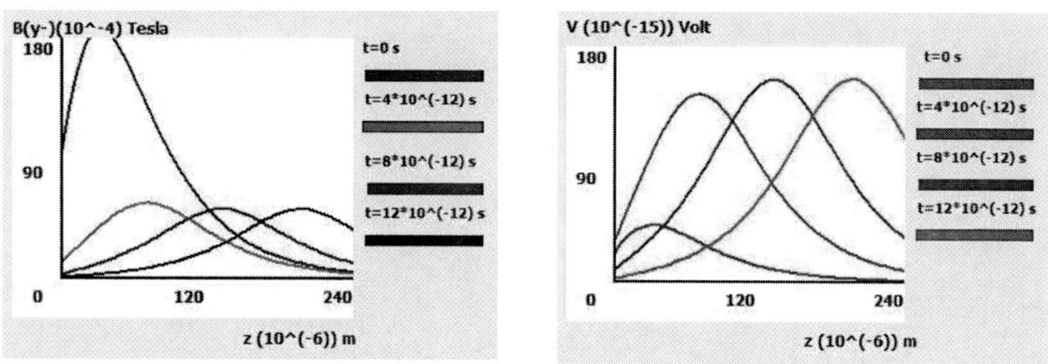


**Figure 5.** Curves of potential difference and external magnetic fields of long Josephson junction for as a function of time: $t = 0$; $5 \times 10^{-12}$; $8 \times 10^{-12}$; $12 \times 10^{-12}$ seconds, for $C = D = 0$ while $A=B = 3/2k^2$

Figure 5 shows graphs of potential differences and external magnetic fields of the long Josephson junction. From the graphs shown in Figure 5, we conclude that the solutions form the soliton solutions, because from $t = 0$ s to $t = 12 \times 10^{-12}$ seconds, the graphs of phase and a potential difference of long Josephson junction do not change in. Physically, it means that external magnet fields applied to long Josephson junction are enough to form soliton solutions. From the graphs above, the external magnetic field and the potential difference of Long Josephson Junction evolve along of the junction and attain the saturation peak. In this condition, the graphs produce the solitary solution in the form of semi fluxon [11].

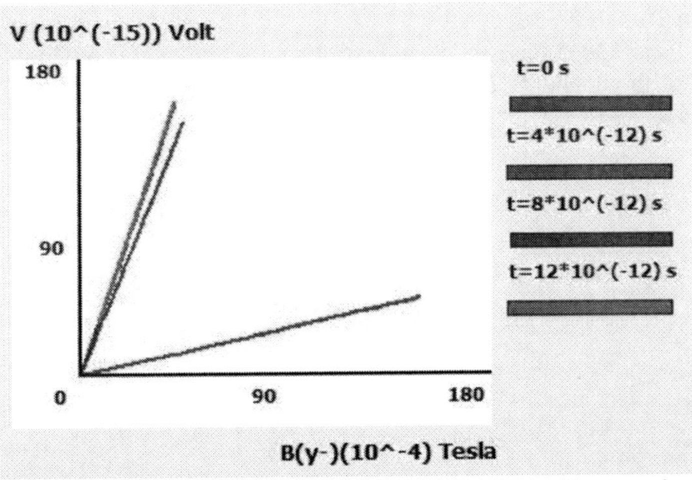


**Figure 6.** Curves of potential difference as a function of external magnetic fields of long Josephson junction for a given times: $t = 0$; $5 \times 10^{-12}$; $8 \times 10^{-12}$; $12 \times 10^{-12}$ seconds, for $C = D = 0$ while $A = B = 3/2k^2$

The influence of the external magnetic fields on the generation of a potential differences of Long Josephson junction are linear from time to time, but they attain saturation at a certain time.

## 4. Conclusion

The information on the variation of phase difference and the potential difference with the external magnetic field in a long Josephson junction is very important for the correct interpretation of the experimental results. In this paper, we investigated the dependence of phase difference and potential difference on the external magnetic field. We found that under certain external magnetic field values applied in a long Josephson junction, the dynamic of potential differences is in the form of solitons. This result gives information the change in the potential difference of Josephson Junction under the influence of external magnetic field change which is useful in SQUID application.

## References

[1] Lachennmann S G *et al* 1993 Soliton Dynamics in Two-Dimensional Josephson Tunnel Junctions *Physical Review B* **48** 22
[2] Campbell David K 1986 Kink-Antikink Interactions In The Double Sine-Gordon Equation *Physica D: Nonlinear Phenomena* **19** Issue 2 165-205
[3] Mónica A *et al* 2012 Formation of a two-kink soliton pair in perturbed sine-Gordon models due to kink–internal-mode instabilities *Phys. Rev. E* **86** 066602.
[4] Bechstein S *et al* 2017 Investigation Of Nanosquid Designs For Practical Applications IOP Publishing Ltd: *Superconductor Science and Technology* **30** 3
[5] Fornieri A *et al* 2016 Nanoscale phase engineering of thermal transport with a Josephson heat modulator *Nature Nanotechnology* **11** 258-262
[6] Lin Shizeng and Hu Xiao 2008 Possible Dynamic States in Inductively Coupled Intrinsic Josephson Junctions of Layered High-Tc Superconductors. *Phys. Rev. Lett.* **100** 247006
[7] Gallemi A *et al* 2016 Coherent quantum phase slip in two-component bosonic atomtronic circuits *New Journal of Physics* **18**
[8] Valenti D *et al* 2014 Switching times in long-overlap Josephson junctions subject to thermal fluctuations and non-Gaussian noise sources *Phys. Rev.* **B 89** 214-510

IOP Conf. Series: Materials Science and Engineering **202** (2017) 012091 doi:10.1088/1757-899X/202/1/012091

[9] Mathews J H and Fink K K 2004 Numerical Methods Using Matlab, 4th Edition 2004 474-478 Prentice-Hall Inc New Jersey USA

[10] Mazo J J and Orlando T P 2003 Discrete Breathers in Josepshon Arrays American Institute of Physics (AIP): OJPS **13** 2. 2

[11] Atanasova P P *et al* 2012 Numerical study of fluxon solutions if sine-Gordon equation under the influence of the boundary condition *MMCP 2011 LNCS* **7125** 201 – 206 Springer- Verlag Berlin Heidelberg.

# The Effect of Temperature in Induction Surface Hardening on the Distortion of Gear

**N F D S Guterres[1], Rusnaldy[2] and A Widodo[2]**

[1]Department of Mechanical Engineering, Dili Institute of Technology, Aimutin Street, Dili, Timor Leste
[2]Department of Mechanical Engineering, University of Diponegoro, Prof. Soedharto Street, Tembalang, Semarang 50275, Indonesia

Email: natalinofonseca81@gmail.com

**Abstract**. One way to extend the lifetime of gear is minimizing the distortion during the manufacturing process. One of the most important processes in gear manufacturing is induction surface hardening. Induction hardening causes a geometric distortion in gears. This study aimed to analyze the distortion of the gear after induction surface hardening. The material of gear that used in this research is S45C, and it was designed with the a module (m) of 1.75, and 29 teeth (z). In the experiment, the gear was heat-treated using induction hardening at variations of temperatures of 820, 880, and 920 °C with holding time was 35, 45, and 55 s, respectively. Then the gear quenched in water at room temperature. The microstructure analysis was conducted to explore the phase change in gear teeth. Hardness test was also performed on the gear using Rockwell test, and finally, the gear dimension was evaluated. The results showed that the hardness of gear at the teeth surface was 64.9 HRC at the case depth of 5 mm is proper to be used. The microstructure of gear at the surface is martensitic, and the deeper area is pearlite. Gear distortion of all teeth with average thickness was reduced and the highest value was 1.593% at the austenite temperature of 820 °C. While the highest reduction was produced at the temperature of 920 °C and generated the highest shrinkage. The induction surface hardening of the gear was not high, and the highest distortion was found in the outside diameter.

**Keywords:** Distortion, gear, surface hardening, S45C.

## 1. Introduction

Gearing is perhaps one of the most critical components of any machinery. Their satisfactory performance depends on the durability of their tooth surface. Pitting and spalling are amongst the common surface fatigue failures for gear teeth. The variability of the gear ratio on the teeth wear is conditioned by the variability of the angular velocity; this is the additional factor of the excitation of vibrations in gearing, especially when there is a considerable distortion of the outline of teeth because of their wear [1]. Normal wear is usually understood as a process of gradual change in size by friction, and characterized by the removal of particles of tooth material from the contact surface. Based on the quality of manufacture, the geometric characteristics and the operationalconditions, the gears as machi nes work classified in the boundary lubrication regime [2]. High-precision gears are finished by honing or grinding machine after hardening. The finishing process after hardening is used to achieve higher quality grades on pitch deviation, profile deviation, and lead deviation. One of the most important processes in gear manufacturing is induction surface hardening. Heat treatment causes a geometric distortion in gears. Minimizing distortion and enhancing compressive residual stress are the important focuses for fabricating high-quality gears [3]. The distortions can be caused by combined effects of various phenomena. The variations in the predicted distortions are typically started in the order of 2 microns. Such amount may be neglected, but in industry, only a tolerance of 10 microns is allowed [4]. The bending tendency after heat treatment to hypoid gear and pinion gear respectively indicates that the axial displacement experiences an expansion [5]. Induction hardening is a fast and environmentally friendly process to treat the surface of highly loaded steel parts such as cams, shafts, splines, and gears [6]. This study focused on changes in the dimensions of certain parts of the gears based on some reasons and consequences associated with distortions that arise during the induction surface hardening. The experiment and gear distortion analysis would be discussed in this study.

## 2. Experimental Method

The study began with the preparation of materials included the three spur gears (self-manufactured) as shown in Figure 1.

(a) (b) (c)

**Figure 1.** Spur gear types. (a) gear (A) at the temperature induction of 820 °C, (b) gear (B) at the temperature induction of 880 °C, (c) gear (C) at the temperature induction of 920 °C

The modules of the gear were similar, and the gear modules were (m) = 1.75 and (z) = 29. The geometrical parameters of gears are shown in Table 1.

**Table 1.** Geometrical parameters of gears

| Item | **Gear A** at temperature of 820 °C | **Gear B** at temperature of 880 °C | **Gear C** at temperature of 920 °C |
|---|---|---|---|
| Number of Teeth | 29 | 29 | 29 |
| Module | 1.75 | 1.75 | 1.75 |
| Pitch diameter (mm) | 52.551 | 52.552 | 52.529 |
| Face width (mm) | 8 | 8 | 8 |
| Bore diameter (mm) | 20 | 20 | 20 |
| Tooth Depth (mm) | 3. 852 | 3.828 | 3.836 |
| Pressure Angle (°) | 20 | 20 | 20 |
| Outside diameter (mm) | 56.363 | 56.377 | 56.397 |
| Tooth Thickness (mm) | 2.730 | 2.731 | 2.737 |

The gears were made from S45C medium carbon steel. Dimensional measurement, microstructure analysis, hardness Rockwell test and chemical composition analysis were conducted before the induction surface hardening of gears. Furthermore, the gears for induction surface hardening were set for different parameters of temperature. Gear A was set at a temperature of 820 °C, with a frequency of 48 kHz and holding time of 35 s. Gear B was set at a temperature of 880 °C, with a frequency of 49 kHz and holding time of 45 s, and then gear C was set at a temperature of 920 °C with a frequency of 50 kHz and holding time of 55 s. The gears were rotating at 70 Rpm when they were heated by induction hardening. The process of induction hardening is shown in Figure 2.

**Figure 2.** Induction surface hardening gears

After the heating process, the gear was quenched immediately with water at room temperature. The following step was the second measurement of the gears to identify the dimensional change, hardness Rockwell test, macro test, and microstructure analysis results. Such identification aimed to compare the number of changes in the gear characteristics and the number of changes in dimension of the gears before and after the induction surface hardening. The distortion occurred on the gears was measured based on the tooth thickness and their outside diameter as shown in Figure 3. The tooth-by-tooth measurement was performed to all gears.

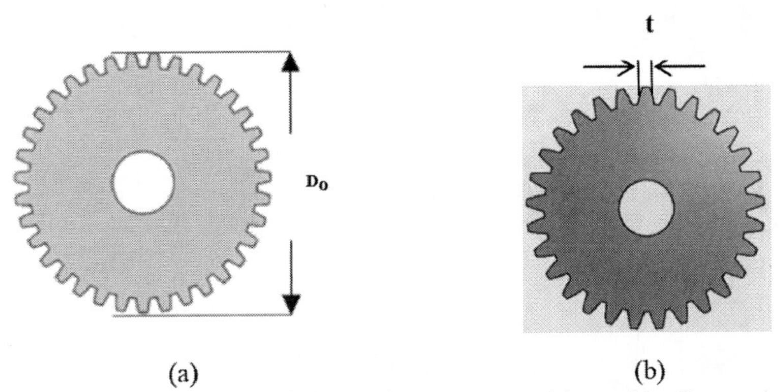


**Figure 3.** Measurement locations on the spur gear. (a) outside diameter, and (b) tooth thickness

Where $D_O$ is the outside diameter and t is the tooth thickness. The tooth thickness and outside diameter were measured by Coordinate Measurement Machine (CMM) with the type of Bright STRATO 707, Japan; travel: X = 605, Y = 605, and Z = 705. The data from Coordinate Measurement Machine would be transferred to Autodesk Inventor to get the value of the dimensional measurement. The Microstructure analysis was performed by OLYMPUS BX41M optical microscope.

The hardness test of Rockwell C (HRC) is based on the standard of ASTME18-11 with the hardening penetration position as shown in Figure 4. All points of penetration rate and indenter diameter results showed average values.

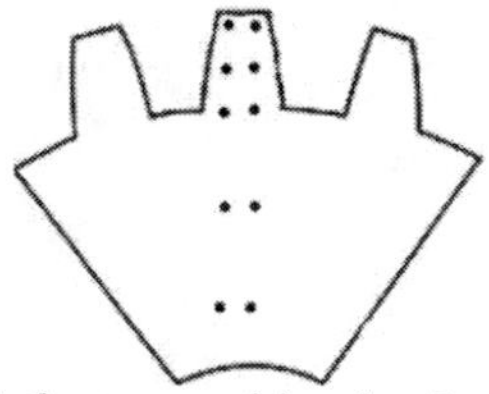

**Figure 4** Indenter position by Rockwell testers.

## 3. Results and Discussion

The chemical composition analysis showed that the gear was medium carbon steels. The Cr and Mo on the gears had lower chemical composition results as shown in Table 2.

**Table 2.** Chemical composition of the alloys

| Alloy | Fe | C | Si | Mn | P | Cr | Ni | Mo | Cu | Al | V | W |
|---|---|---|---|---|---|---|---|---|---|---|---|---|
| S45C | 98.15 | 0.41 | 0.18 | 0.63 | 0.10 | 0.32 | 0.01 | 0.01 | 0.03 | 0.01 | 0.01 | 0.10 |

### *3.1 Microstructure and macro analysis*

**Table 3.** Results of hardening layer and microstructure phase.

| Tested Gear | Macro analysis | Microstructure analysis | Types of phase | Case depth (mm) |
|---|---|---|---|---|
| Gear (A) at the temperature of 820 °C | 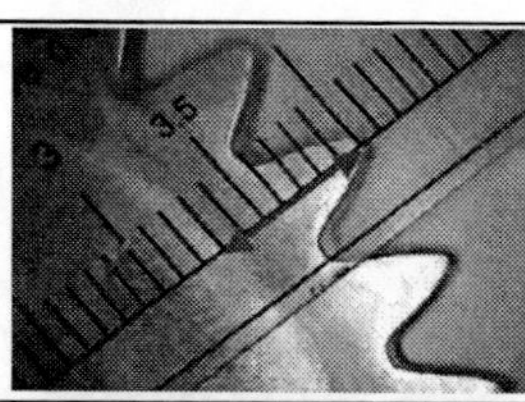 | 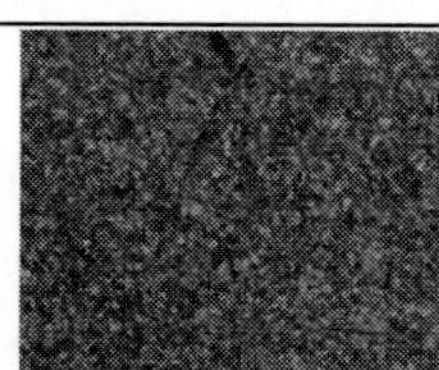 | Full martensitic | 7 mm |
| Gear (B) at the temperature of 880 °C | 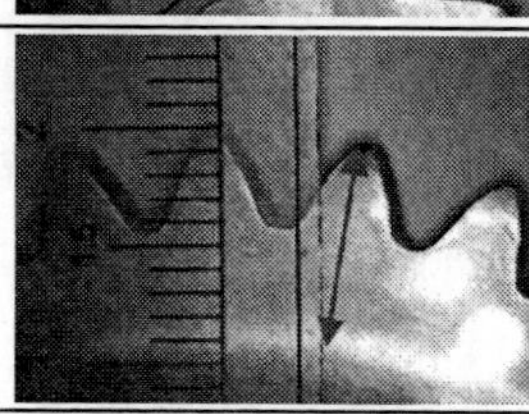 | 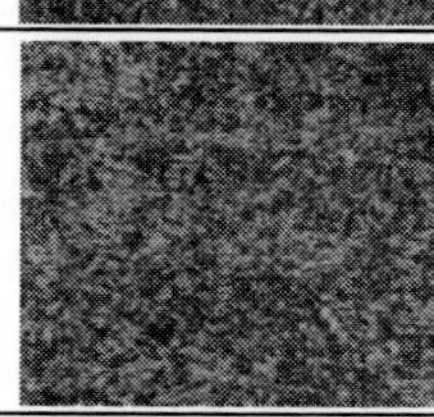 | martensitic + perlite | 9 mm |
| Gear (C) at the temperature of 920 °C | 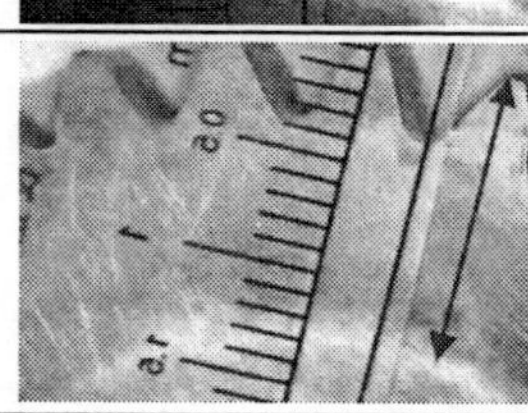 | 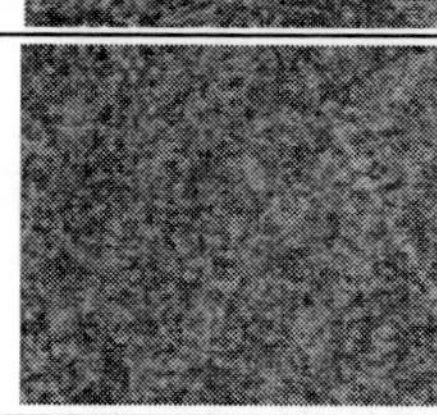 | martensitic + perlite | 13 mm |

Table 3 shows the difference in the depth of hardening at various temperatures for each gear after induction surface hardening. The tested gears were varied at the temperatures of 820 °C, 880 °C, 920 °C and the holding time were 35, 45, and 55 seconds, respectively. Then the hardness values of the case depth were respectively 7, 9, and 13 mm. The results of the hardening process was a hardened profile of all gear teeth, which was due to the limited means of induction hardening (led to insufficient capacity of frequency), and an influence of the types of coils used i.e. circular winding coil surrounding the tooth profile that allowed all teeth to be subjected to heat during the heating process. Such coil type was used in this research because of its popular pattern of induction hardening, especially for small-sized gears and sprockets. However, the heating process should be done very carefully because the hardening process to all teeth can cause cracks (Rudnev et al. 2003). From the three gears used in the experiment, it could be concluded that a higher temperature and a longer heating time resulted in a thicker layer and therefore a greater level of hardness. It happened because the higher temperature and the longer heating time result in the propagation of the austenite phase heating. After the rapid cooling process, the austenite phase was then transformed into martensitic phase.

*3.2 Hardness*

The hardness distribution in gear A, gear B and C are shown in Figure 5. In gear A (820 °C, 35 seconds at the frequency of 48 kHz), the hardness values measured from the outer diameter to the root diameter were 64.9, 62.5, 62.0, 61.8, and 61.8 HRC. While in gear B (880 °C, 45 seconds at the frequency of 49 kHz), the hardness values measured from the outer diameter of teeth to the root diameter were 65.0, 64.4, 63.7, 62.0, and 61.5 HRC. In gear C (920 °C, 55 seconds at the frequency of 50 kHz), the hardness values measured from the outer diameter of teeth to the root diameter were 65.9, 65.8, 65.6, 63.1, and 62.2 HRC.

**Table 4.** The hardness distribution of the Specimen Gears.

| Hardness Case Depth (mm) | The hardness of Gear A at the Temp. (820 °C) HRC | The hardness of Gear B at the Temp. (880 °C) HRC | The hardness of Gear C at the Temp. (920 °C) HRC |
|---|---|---|---|
| 2 | 64.9 | 65.0 | 65.9 |
| 5 | 62.5 | 64.4 | 65.8 |
| 8 | 62.0 | 63.7 | 65.6 |
| 11 | 61.8 | 62.0 | 63.1 |
| 14 | 61.8 | 61.5 | 62.2 |

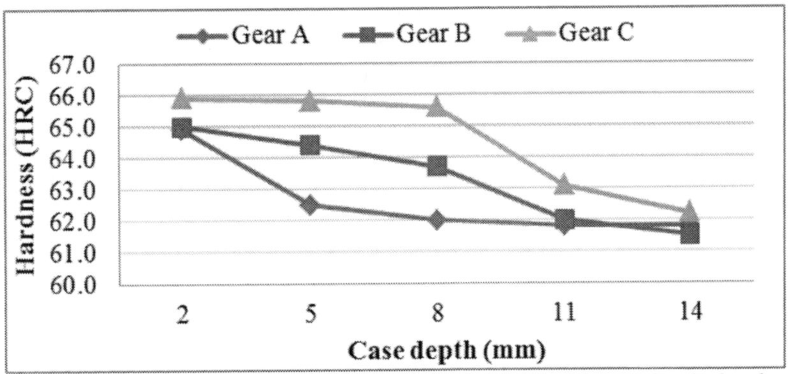

**Figure 5.** The hardness distribution of the Specimen gears after induction surface hardening for the end from the outer diameter of teeth to the root diameter

It happened because the higher temperature and the longer heating time can even result in the propagation of heating in the austenite phase. After the rapid cooling process, the austenite phase was transformed into a martensitic phase which was very hard and brittle. It caused the hardness value of the gear to increase, and the comparison of these results meet the acceptable standards of AGMA.

### *3.3 Distortion arose at the Tooth Thickness*

Figure 6 shows the results of gear distortion of all teeth, such as in the form of average values of the tooth thickness to be reduced or expanded. The highest value of reduction was 1.593%, and the maximum value of expansion was 1.396% at an austenite temperature of 820 °C as shown in Figure 6 (a). While the distortion was reduced at the temperature of 880 °C, and the highest percentage or reduction was 4.129% as shown in Figure 6 (b). Subsequently, the reduction was reduced and expanded at the temperature of 920 °C. The highest value of reduction was 7.187%, and the highest value of expansion was 2.404% as shown in Figure 6 (c).

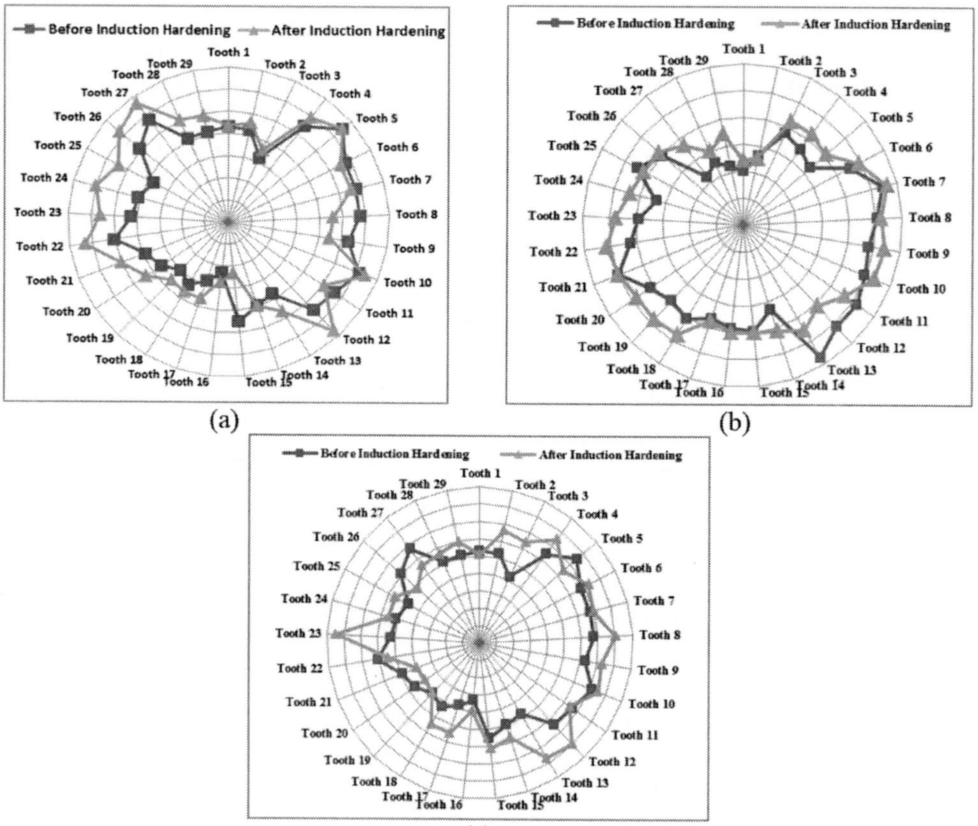


(c)

**Figure 6.** Dimensional changes in tooth thickness of the three gears; (a) gear A (b) gear B, and (c) gear C.

### *3.4 Distortion at the outside diameter*

Figure 7 shows the results of gear distortion in the form of the average values of the reduced or expanded diameter of all teeth outside diameter. The highest value of the shrinkage was 0.228%, and the maximum value of the expansion was 0.021% at the austenite temperature of 820 °C such as shown in Figure 7 (a). There were shrinkage and an expansion of distortion at the temperature of 880 °C. The highest value of shrinkage was 0.147%. Moreover, the highest value of expansion was 0.059% as shown in Figure 7 (b). Then, at the temperature of 920 °C shrinkage and an expansion of distortion were also found. The highest value of shrinkage was 0.099%, and the maximum value of expansion was 0.284% as shown in Figure 7 (c).

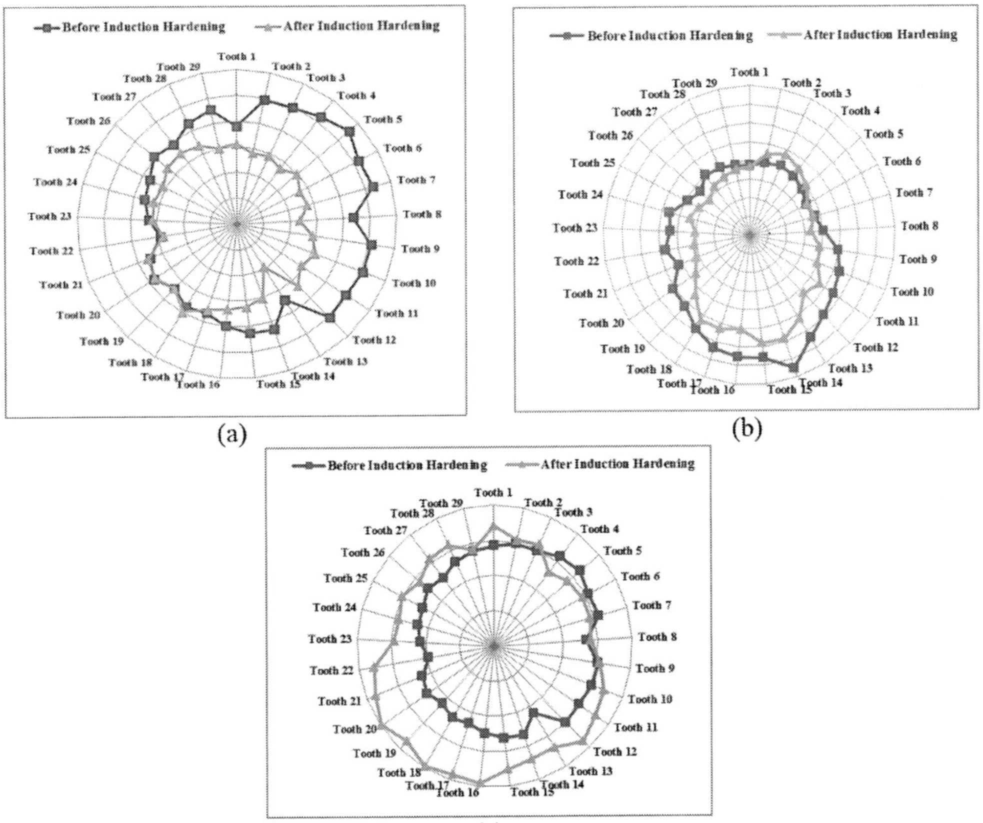


(a) (b) (c)

**Figure 7.** Dimensional changes in outside diameter of the three gears; (a) gear A (b) gear B, (c) gear C

## 4. Conclusion

Based on the data obtained from the results and the previous discussion of gears after induction surface hardening, the following summary can be drawn as follows. Induction surface hardening of gear has martensitic type and pearlite phase. The highest hardness value of induction surface was achieved at a case depth of 2-5 mm from the upper tooth profile. A higher austenitic temperature would increase the gear hardness, and more the type of martensitic would be formed. The highest type of gear distortion was a drastic shrinkage of the teeth. Distortion of induction hardening gear was found to be very high with a value of 7.187% at the temperature of 920 °C. The highest value of distortion would precisely influence inappropriate tooth contact. The highest shrinkage of the tooth thickness and outside diameter would affect the contact angle and tooth clearance because the size of gear tolerance was not standardized [7].

## 5. References

[1] Wojnarowski J and Onishchenko V 2003 Tooth wear effects on spur gear dynamics Pergamon; *Mechanism and Machine Theory* **38** pp. 161–178

[2] Onishchenko V 2015 Investigation of tooth wears from scuffing of heavy duty machine spur gear *Mechanism and Machine Theory* **83** pp. 38–55

[3] Sugianto A, Narazaki M, Kogawara M, Kim S Y and Kubota S 2007 Distortion Mechanism During Carburising-Quenching of SCr420H Helical Gear Proc. 16th IFHTSE Congress, (Brisbane) *Materials Australia* pp. 07-257

[4] Vanoverberghe L, Garcia-Aranda L, Ryckelynck D and chastel Y 2007 Anticipation of gears distortions caused by heat treatments American Institute of Physics 978-0-7354-0414-4/07
[5] Rakhit A K 2000 Heat Treatment of Gears ASM International, Materials Park, OH, p 91–100
[6] Coupard D, Palin-luc T, Bristiel P, Vicent J, Dumas C 2008 Residual stresses in surface induction hardening of steels: Comparison between experiment and simulation *Mater. Sci. Eng. A.* **487** (1-2), 328-339
[7] Sighley J E 1996 Standard handbook of machine design Third edit **17** p. 3. United States of America

**Acknowledgments**
The authors would like to thank to Dili Institute of Technology, Timor Leste, for supporting the research funding. Also, we would like to thank to Diponegoro University and PT ATMI IGI CENTER Surakarta in Indonesia for supporting a laboratory for testing.

IOP Conf. Series: Materials Science and Engineering **202** (2017) 012093 doi:10.1088/1757-899X/202/1/012093

# Effect of Temperature on the Conduction Properties of Oligo-phenyl-phenylene (OPP) Molecule

**S Setianto, H S Aswad and Y M Jajat**

Department of Physics, Universitas Padjadjaran Bandung Jl. Raya Jatinangor KM. 21, Sumedang 45363

Email: setianto@phys.unpad.ac.id

**Abstract.** In order to design molecular electronic devices, we used conjugated molecule in the form of oligo-phenyl-phenylene (OPP) system to explore and understand the molecular conductance. Although the experimentally technique has obtained information with high accuracy, the process of adsorption on the surface can be investigated in detail on the electronic structure with a theoretical method based on density functional theory (DFT). In this study, we investigated the effect of temperature on the conductivity of OPP molecule using density functional tightbinding (DFTB) and the non-equilibrium Green's function (NEGF) by implementing the code of DFTB (gDFTB) for the calculation of transport to get an electric current of this model. As the results of this simulation, the total energy of the OPP molecule increased proportionally with the temperature increase of about 1 eV. It was caused by conformation changes in the molecular structure of the OPP when getting disruption at higher temperatures.

Keywords : temperature, conduction, OPP, molecule

## 1. Introduction

A single molecule is attracting more attentions both experimental and theoretical studies for its potential application of miniaturization from sub-micrometer conventional inorganic devices to tiny active component with the size of only a few nanometers. The development of electronic devices and technology, especially in the field of single molecules, has been explored in various aspects. The most interesting aspect of the topic of this study was to explore the electronic transport properties of a material having a conjugated system such as phenyl, graphene, carbon nanotubes, fullerenes, etc. In the field of molecular electronic devices, the basic concept of electric current flow at the molecular level is an important requirement for understanding the electronic transport from single molecules to an electrode contact. Specifically, it depends on the interaction between the molecules and the metal contacts, the molecular orientation, the energy gap (HOMO-LUMO) and the length of the molecule [1]. However, the transport mechanism at the single molecule level is still opened for scientific study on how the conductance of molecular junctions can be tuned by chemically modifying the molecules. In order to design the molecular electronic devices, we used conjugated molecule in the form of oligo-phenyl-phenylene (OPP) system to explore and understand about the molecular conductance. It has been studied by several scientists by experiments [1,3,9] and theory[2,4,6]. A force liaison molecule (linker) affects the efficiency of tunnelling system-molecule-metal layer of metal using atomic microscopy (CP-AFM) [3]. Experiments [4-5] have shown that terminal groups that bind a molecule to the metal electrode can control the value of the conductance. Theoretically, it has been shown that the conductance of amine-anchored molecules has better conductance value than thiolated molecules through being connected to gold contacts [6-8]. The anchoring groups affect the energy level line up relatively to the metal Fermi energy and consequently play important role on the conductance plateaus. Although the technique has experimentally obtained information with high accuracy in the process of adsorption of the surface, it was also very important to investigate the electronic structure in detail by a theoretical method based on the density functional theory (DFT). Many theoretical studies based on DFT for isolated systems have been done [10]. Some of the results of the studies showed a good trend in the experimental data and successfully predicted the properties of a complex system such as geometry [6-8]. In this study, we used terphenyl molecules connected to copper leads. We presented a systematic study on the effect of temperature on the conductivity of OPP by using density functional tight-binding (DFTB) extended to the non-equilibrium Green's function (NEGF) approach for computation of the charge density and electronic transport.

## 2. Computational Method

*2.1The Green's Function Formalism*

The problem of quantum transport through a single molecule bridged to two contacting leads can be set by starting from the scattering states which propagate from one lead to the other. The waves scattered at the contact-molecule interface and partially reflected and transmitted across the molecule. Consider the Green's functions of the α contact:

$$g_\alpha^r = \left[ES_\alpha - H_\alpha - i\delta\right]^{-1} \tag{1}$$

where *Hα* is Hamiltonian of the contact, the overlap matrix, *S* was represented on a non-orthogonal basis set. Assume that the contacts did not have any interaction among them, the Hamiltonian of the whole system can be written as follow:

$$H = \begin{bmatrix} H_D & V_{D1} & V_{D2} \\ V_{D1}^\dagger & H_1 & 0 \\ V_{D2}^\dagger & 0 & H_2 \end{bmatrix} \tag{2}$$

where VD1(2) is the Hamiltonian of contact-molecule coupling. The complete Green's function of the system:

$$G_D^r = \left[ES_D - H_D - \Sigma^r\right]^{-1} \tag{3}$$

where $\Sigma^r$ is the total self-energy of the two contacts,

$$\Sigma_\alpha^r = \left(ES_{D\alpha} - V_{D\alpha}\right) g_\alpha^r \left(ES_{D\alpha} - V_{D\alpha}\right)^\dagger \tag{4}$$

The tunneling current flowing through the molecule that naturally arises in coherent transport is given by the Landauer formula [11] as follow:

$$I(V) = \frac{q}{h}\int dE\, T(E,V)\left[f(E,\mu_1) - f(E,\mu_2)\right] \tag{5}$$

where $T(E, V) = Tr\,[\Gamma_L G^r \Gamma_R G^a]$ is transmission function and $\Gamma_{L/R}$ is the scattering rates related to the probability of injecting an electron into the molecule from left and right electrodes. These rates also describe the width that the molecular level acquires in virtue of the coupling to the metal leads and they can be used to measure the strength of the metal-molecule coupling. Then $f(E, \mu)$ is the

Fermi-Dirac function. In equilibrium and at zero temperature ($T = 0$ K) case, the conductance can be expressed as:

$$G = \frac{2q^2}{h} T(E_F) \tag{6}$$

The advantages in using Green's function approach was that the incoherent scattering and relaxations can be included within self-energies. In this study, we observed the conductance of a single molecule by a ring system oligo-phenil-phenylene (OPP) with a contact copper lead (Cu [111]). The function of non-equilibrium Green (NEGF) implemented in the method of Density Functional Tight-Binding (gDFTB) was used to investigate the transport properties of the system. The temperature variation was given to characterize the molecular dynamics and interactions between OPP ring molecules on a short time scale of about a nano-second (ns). These molecular dynamics behavior will cause the conductance of OPP molecules fluctuate depending on the temperature and conformation.

*2.2Molecular Dynamics Simulation*

The molecular dynamics technique simulates a classical system of molecules in the most natural way: it integrates the governing equations of motion through time. The generated sequence of configurations could be run as a movie and looked very realistic. The preparation files of molecular dynamics simulations performed with the geometry optimization and energy minimized three-dimensional structure of OPP molecules using DFTB+ software [6] that runs on a multi-core computer. The parameters were used in accordance with standard molecular dynamics, namely NVT ensemble (N: the number of atoms; V: volume; T: temperature) with the velocity of Verlet algorithm. The advantage of the Verlet velocity treatment was the availability of the momenta at the same time as the positions. Further analysis of complex dynamic was based on the calculation results of the molecular dynamics simulations. The determination of molecular conformation stability of OPP molecule conducted for 100 pico seconds (ps) at the initialization phase. Simulations were performed for 700 pico seconds at a temperature of 150, 300, 450, 600, and 750 K to determine the effect of temperature on OPP molecular geometry and electronic properties. OPP three-dimensional molecular structure obtained from the use of optimization software DFTB+ then compared the results with the data from the official site database. Visualization was done by creating a script in the Jmol program which then changed into the form of movie files (mpeg files). Complex interactions among the OPP rings during molecular dynamics simulation can be seen through the output database viewer in the position format (xyz format). The visualization of the shift position of molecular conformation OPP and dynamic movements of the molecular dynamics simulation results were done by selecting the molecular conformation at certain periods until the end of the simulation time.

## 3. Results and Discussion

*3.1The geometry optimization of OPP molecules*

Three-dimensional molecular structure of OPP was obtained by minimizing the total OPP molecule energy by using the software and hardware of DFTB+ multi-core computer (parallel computing). The optimal molecular geometry of OPP is shown in Figure 1. The OPP molecule has a planar structure without ring rotation in one direction.

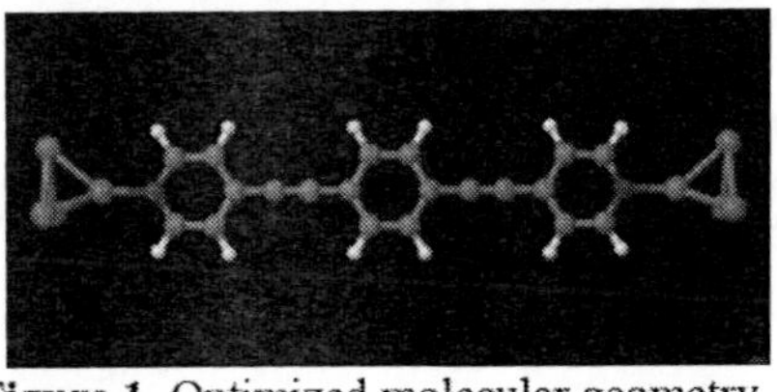

**Figure 1**. Optimized molecular geometry of OPP molecules

### *3.2Effect of temperature on molecular ring rotation*

The geometry of the OPP molecule would change over time when a certain temperature was given to the molecule. The total energy of the OPP molecule increased proportionally with the temperature increase of about 1 eV which is given in Table 1.

**Table 1**. Averages of Total Energy

| Temperature (K) | $E_{min}$ (eV) | $E_{max}$ (eV) | $E_{average}$ (eV) | $E_{max} - E_{min}$ (eV) |
|---|---|---|---|---|
| 150 | -1654.5770 | -1653.6739 | -1654.1693 | 0.9031 |
| 300 | -1653.8794 | -1652.5212 | -1653.0796 | 1.3582 |
| 450 | -1653.2590 | -1650.4037 | -1651.9914 | 2.8553 |
| 600 | -1652.6385 | -1648.9218 | -1650.8511 | 3.7167 |
| 750 | -1652.0181 | -1647.2228 | -1649.7697 | 4.7953 |

For the given temperature of 450 K, the maximum value of the rotational angle of the molecules' OPP ring produced about 60° from the initial position (flat) at simulation time of around 1750 and 3800 (a.u.) can be seen in Figure 2.

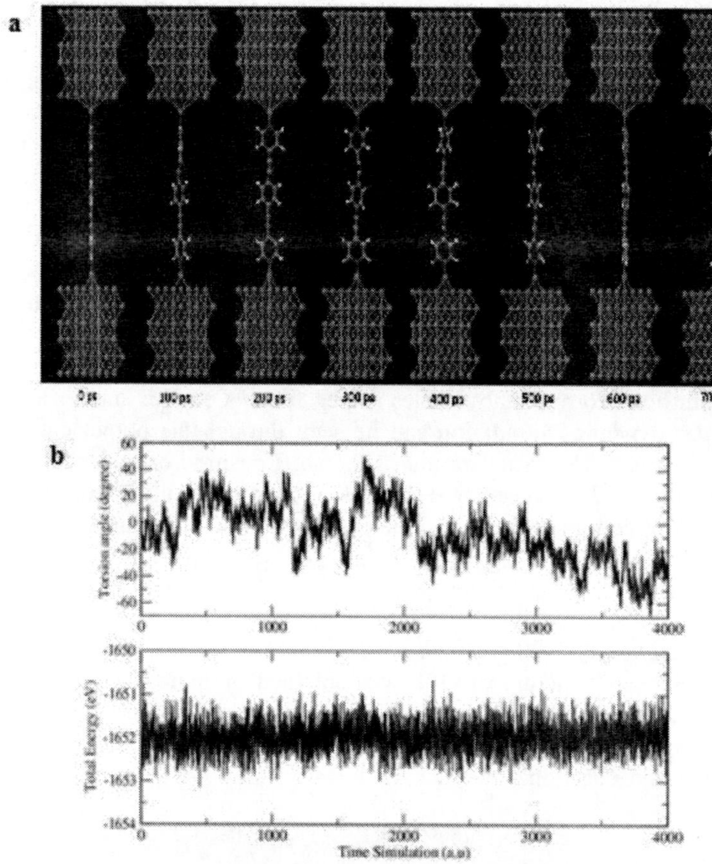


**Figure 2.** Effects of temperature on the molecular ring rotation OPP (T = 450 K): a). visualization of the OPP ring rotation b). total energy and torsion angle of the central ring in OPP molecule

The results of the calculation of the value of transmission (quantum conductivity) and the function of the local density of states (LDOS) in conditions without and with applied voltage (V = 0) were 0.5 and 1 volt, as shown in Figure 3. For the initial condition, namely OPP molecules without rotation (flat), the transmission function was higher than that rotate by 60°. This was due to a decreasing local density of OPP molecule state at around 1 to 2 eV. The voltage changes only affect the shifting LDOS (Figure. 3) to the positive axis direction (right). The increasing energy levels were caused by the provision of a positive voltage.

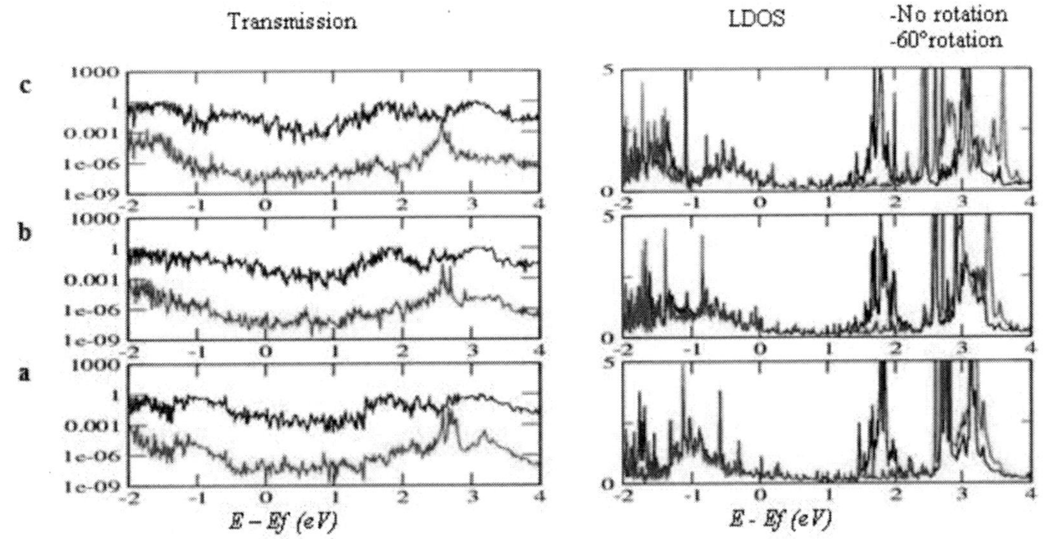


**Figure 3.** Transmission and LDOS of OPP molecular rotation with: a. applied bias 0 volt, b. applied bias 0.5, c. Volt applied bias 1 volt

## 4. Conclusion

In the first phase of research, geometrical structure of molecules OPP optimization and molecular dynamics simulation have been conducted at various temperatures of 150, 300, 450, 600, and 750 K. The results of these simulations showed that total energy of the OPP molecule increased proportionally with the increase in temperature about of 1 eV. It was caused by the conformation changes in the molecular structure of the OPP when getting disruptions at higher temperatures. At a temperature of 450 K, the maximum value of the rotational angle of the ring OPP molecules was twisted of about 60° from the initial position (flat) at simulation time of around 1750 and 3800 (a.u. of time). In OPP molecules without rotation (flat), the value of the transmission was higher than that rotated by 60°. This was due to a decrease in the local density stateof OPP molecule around 1 to 2 eV.

## 5. References

[1] Venkataraman L, Klare J E, Nuckolls C, Hybertsen M S, Steigerwalt M L, *Nature 2006* 442 904
[2] Pecchia A and Di Carlo A 2004 *Rep. Prog. Phys.* 67 1497
[3] Queck S Y, Venkataraman L, Choi H J, Louie S G, Hybertsen M S, Neataon J B 2007 *Nano Lett.* 7 3477
[4] Perdew J P, Burke K and Ernzerhof M 1996 *Phys.Rev.Lett.* 77, 3865
[5] Datta S 1995 *Electronic Transport in Mesoscopic System,* Cambridge University Press, Cambridge, (Chapter 3,8).
[6] Frauenheim Th, Seifert g, Eltsner M, Hajnal Z, Jungnickel G, Porezag D, Suhai S, Scholz R 2000 *Phys. Stat. Sol (B)* 271 41
[7] Kristensen I S, Mowbray D J, Thygesen K S, Jacobsen K W *J. Phys.: Condens. Matter 2008* 20 374101

[8] Garcia-Gil et al 2009 *Phys. Rev. B.* 79 754441
[9] Rawlett A M, Reed A M, Chen J, Prince D W, Tour J M 2001 *Appl. Phys. Lett.* 78 3735
[10] Erbeet A al *Small* 2010 6 No. 14 1529-1535

**Acknowledgements**
We would like to thank to Dr. Alessandro Pecchia for a short course of the DFTB code.

# Risk Analysis of Central Java Gas Transmission Pipeline by Risk-Based Inspection Method

**Mediansyah[1], G D Haryadi[1], R Ismail[1] and S J Kim[2]**

[1]Department of Mechanical Engineering, Diponegoro University, Jl. Prof. Soedharto, Tembalang, Semarang 50275, Indonesia
[2]Department of Mechanical and Automotive Engineering, Pukyong National University, Busan, 608-739, Korea

Email: mediansyah@rocketmail.com

**Abstract.** During the operational period of gas transmission pipeline was found a potential hazard that could result in pipeline failure. As a consequence, the problem of the pipeline failure happening more and more. Economic and environmental factors, as well as human life, be considered to involve the current challenges as structural integrity and safety standards. Therefore, the reliability of structural integrity and security of gas pipelines under various conditions, including the existence of defects should be carefully evaluated. The results of this study were the steps for setting a Risk Level on any instrument using the Risk-Based Inspection API 581 standard and the subsequent results are recommended as an effective inspection planning by Risk Level and Remaining Life Time.

**Keywords:** Pipe, risk, risk level, risk-based inspection, remaining life time.

## 1. Introduction

Oil and gas sector is an activity that has the potential hazards of high risks, capital-intensive, high-tech, as well as requiring human resources with competence and specific qualifications. The activities in this sector should constantly be monitored so as not to cause significant harm, either to workers, the public, property or the environment [1].

Natural gas is a clean fuel that has been used by many countries as fuel for power generation. This trend makes the supply of natural gas will be increasingly in demand. One crucial thing is the distribution of natural gas transportation system. Duct piping is one method that is most practical and affordable that has been applied in oil and gas transportation system since 1950. The pipe has been used as one method that is most practical and needs the lowest cost for oil and gas transportation. Installations of pipes for oil and gas transmissions increased dramatically in the past three decades [2].

Pipe is a technology that is used to drain fluid such as oil, gas or water in large amounts and long distances by sea and specific regions. The pipeline is a silent mean of transport that serves to distribute the fluid in the form of liquid or gas [3] that is widely used to transmit fluids in oil and gas industry. Its use is quite diverse, among others, used to channel the fluid from the wells to the treatment plant or offshore building (offshore facility) or of offshore platforms built directly on the land (onshore facility) [4].

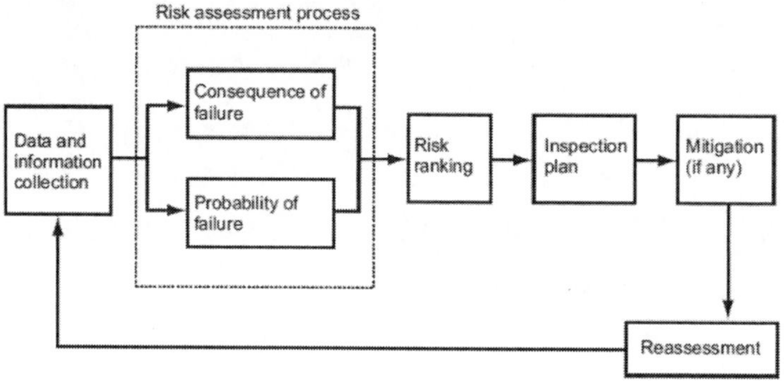


**Figure 1.** Risk-based inspection [5]

The Risk Based Inspection (RBI) is a practical way to assess the possible application of the inspection process and the impact that can occur on an instrument failure, evaluate the level of risk and recommend the type of action taken for the development of preventive measures and risk management. Risk Based Inspection (RBI) using a risk to plan and assist in the assessment of the results of the inspection, testing and monitoring [3]. The advantages of such RBI method are the increasing operational time and the work of a processing facility, which at the same time there is an increase or absence of treatment at the same risk level. The Concept of Risk Based Inspection can be seen in figure 1.

The results of this study are expected to be useful for damage prevention measures, especially in terms of materials and the development of material sciences in addition to the complete simulation data checks based on risk known as Risk Based Inspection (RBI).

## 2. Experimental Method

*2.1 Corrosion*

The piping system is a part of the most sensitive structural element of the power plant. Therefore, this system analysis and quantification of their vulnerability regarding failure probability are essential [4]. Corrosion is defined as damage to the material that results from their chemical reaction with the surrounding environment of the material. In the event of corrosion, metal would undergo oxidation, while the oxygen (air) is reduced. Metal rust is usually in the form of oxides or carbonates. The chemical formula of iron rust is $Fe_2O_3.nH_2O$, a brown-colored solid red. Corrosion is an electrochemical process. In the iron corrosion, certain sections of the metal act as the anode, where the metal undergoes oxidation.

$$Fe_{(s)} <=> Fe^{2+}_{(aq)} + 2e \quad (1)$$

Electrons are released in the anode flow to other parts of the iron which act as the cathode, where oxygen is reduced [6].

$$O_2(g) + 4H^+_{(aq)} + 4e <=> 2H_2O_{(l)} \quad (2)$$
$$O_{2(g)} + 2H_2O_{(l)} + 4e <=> 4OH^-_{(aq)} \quad (3)$$

The corrosion rate is the propagation velocity or speed of decline in the quality of materials on time.

$$CR = \frac{d - d_0}{T - T_0} \quad (4)$$

where $CR$, $T$, $d_0$, $d$, and $T_0$ respectively represent corrosion rate, year of first examination, initial thickness, examination thickness, and recent year of examination.

**Table 1.** The numerical values of probability and consequence categories

| Category of probability | | Category of consequence | |
|---|---|---|---|
| Categories | *Range* | Categories | *Range* ($ft^2$) |
| 1 | $Pf(t) \leq 2$ | A | $CA \leq 100$ |
| 2 | $2 < Pf(t) \leq 20$ | B | $100 < CA \leq 1000$ |
| 3 | $20 < Pf(t) \leq 100$ | C | $1000 < CA \leq 3000$ |
| 4 | $100 < Pf(t) \leq 1000$ | D | $3000 < CA \leq 10000$ |
| 5 | $Pf(t) > 1000$ | E | $CA > 10000$ |

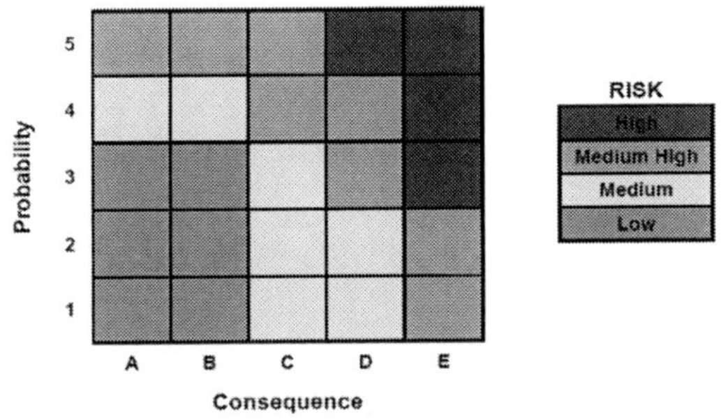


**Figure 2.** Risk matrix [5]

*2.2 Risk*

Risk is a possibility of an unwanted event that will affect an activity or object [7]. Mathematically, the definition of risk is as follows:

$$Risk = Probability \times Consequence \tag{5}$$

The reference in figure 2 can be used for determining the risk level. While the numerical values of probability and consequence categories can be shown in table 1.

*2.3 Probability of failure*

The probability of failure is a possibility of equipment or component to fail (API, 2008). Analysis of the occurrence of a failure in the component can be performed if they are currently in working conditions. The equation of the probability of failure in API RBI is given by the following equation [8]

$$P_f(t) = g_{ff} \times D_f(t) \tag{6}$$

where $P_f(t)$, $g_{ff}$, and $D_f(t)$ respectively represent probability of failure, generic failure frequency, and damaging factor.

*2.4 Consequences of failure*

The consequences of failure are the impacts that can be caused by an equipment failure. The semi-quantitative RBI API calculation involves only consequences of flammability and toxicity [9].

*2.5. Remaining lifetime*

Remaining lifetime can be defined as equipment tolerance to the type of damage. Remaining lifetime will determine the next inspection interval of time [10].

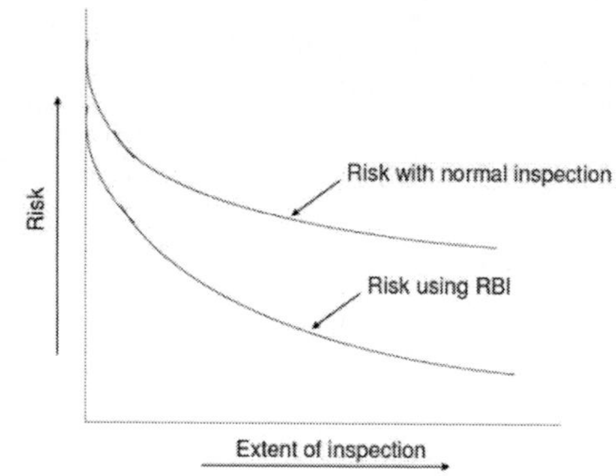


**Figure 3.** Management of *Risk Based Inspection* [*10*]

$$\text{Remaining Life Time} = \frac{d_0 - d_R}{CR} \tag{7}$$

where $d_0$, $CR$, and $d_R$ respectively represent current actual thickness, corrosion rate, and required thickness.

*2.6 Risk-Based Inspection*

Risk Based Inspection (RBI) is a relatively new method of conducting an inspection. This method is based on a risk analysis which includes the analysis of the magnitude of the possibility of a failure and the magnitude of the effects of emerging risks [5]. In general, the RBI management can be seen in figure 3.

## 3. Results and Discussion

*3.1 Data*

The data of inspected equipment are shown in Table 2.

**Table 2.** Data of equipment

| Component | Process Data | | | |
|---|---|---|---|---|
| | P. Design (psig) | P. Operation (psig) | T. Design (°F) | T.operation (°F) |
| Line pipe (inlet KP.20 Ø20") | 900 | 450 | 650 | 250 |
| Tee (inlet KP.20 Ø20") | 900 | 450 | 650 | 250 |
| Line pipe (outlet KP.20 Ø20") | 900 | 450 | 650 | 250 |

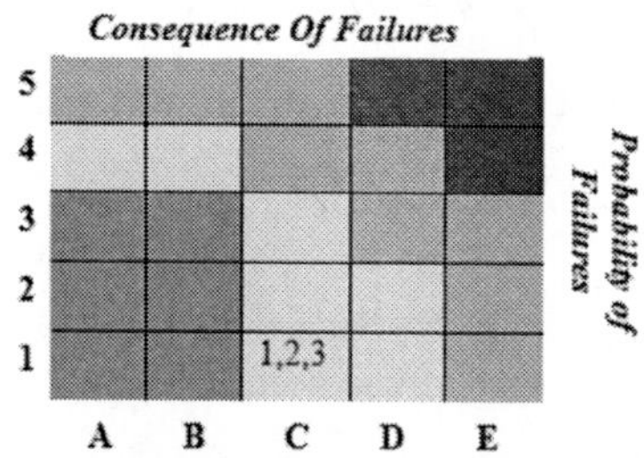


**Figure 4.** Matrix risk

Note: (1) *Inlet Automatic Shutdown valve* KP-20 Ø20", (2) *Tee* KP-20 Ø20", and (3) *Outlet Automatic Shutdown Valve* KP-20 Ø20".

*3.2 Risk level*

The risk level in three components can be seen in the Risk Matrix as presented in figure 4. From the risk matrix, we can know that the level of risk of all the three components is medium with the overall value of 1819.654 $ft^2$. The consequences of failure and the value of the probability of failure were 1.

*3.3 Remaining lifetime*

The data of the remaining lifetime of each component can be seen on the graph to make it easier to set priorities for the next inspection as illustrated in Figures 5, 6, and 7. From the graph of Remaining LifeTime Gas Pipe Straight (Automatic Shutdown Valve Inlet KP-20 Ø20"), we can know that the pipeline would fail in the years between 2025 to 2026 which was marked by an actual pipe with thickness values that approached the minimum thickness of the pipeline design.

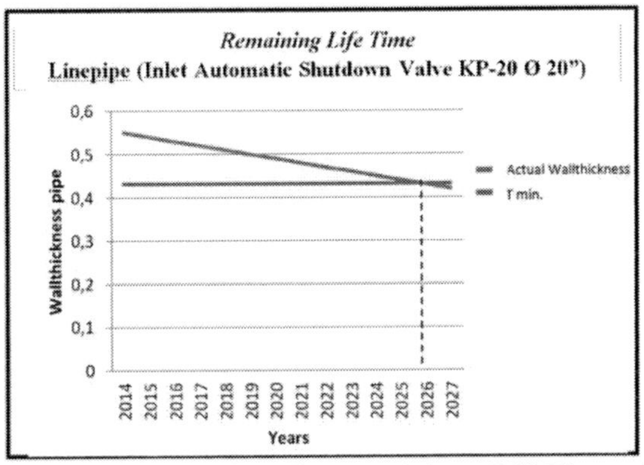


**Figure 5.** Graph of Remaining LifeTime in pipeline (Inlet Automatic Shutdown Valve type KP-20 Ø20")

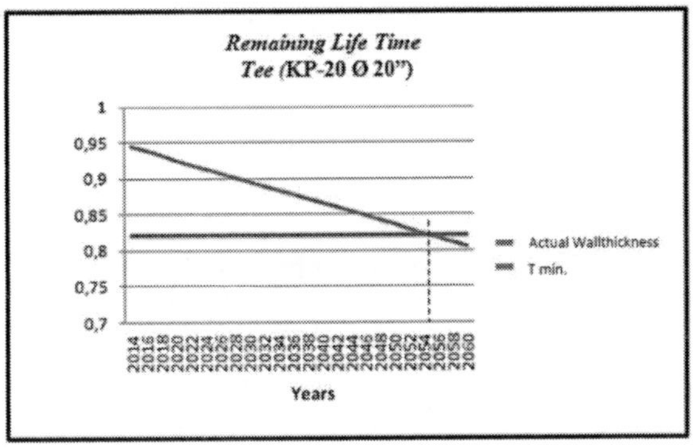


**Figure 6.** Graph of Remaining LifeTime in Tee (KP-20 Ø20")

From the graph of remaining lifetime connection Tee (KP-20 Ø20"), we can find that the pipeline would fail in the years between 2054 and 2055 was, which was marked by actual pipe thickness values that approached the minimum thickness of the pipeline design.

In the graph of Remaining LifeTime Gas Pipe Straight (Outlet Automatic Shutdown Valve) KP-20 Ø20", we can know that the pipeline would fail in the years between 2030 and 2031, which were marked by the actual pipe thickness values that approached the minimum thickness of the pipeline design.

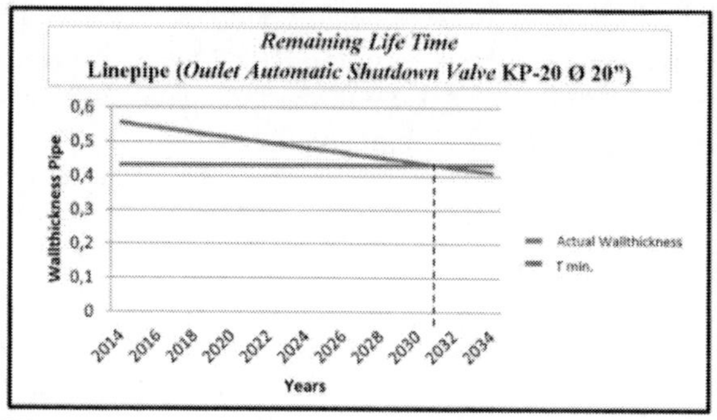


**Figure 7.** Graph Remaining LifeTime in pipeline (Outlet Automatic Shutdown Valve) KP-20 Ø20."

*3.4 Risk-Based Inspection*
Inspection planning recommendations for each component based on Table 3 and Table 4, namely:

**Table 3.** Static Equipment Integrity Inspection of RBI

| Risk Level | Type of Inspection |
|---|---|
| High | Internal Entry |
| | External NDT |
| Medium | Internal Entry |
| | External NDT |
| | Limited Internal Inspections |
| Low | Internal Entry |
| | External NDT |
| | Limited Internal Inspections |

**Table 4.** Test Results for the Type of Failure in Critical Levels

| Probability of Failure | Consequences of Failure | Inspection Method | Inspection Frequency (Month) | Wide Inspection Area |
|---|---|---|---|---|
| High | High | Ultrasonic Testing | 12 | Full |
| High | Medium | Ultrasonic Testing | 12 | Partial |
| High | Low | Ultrasonic Testing | 12 | Small |
| Medium | High | Ultrasonic Testing | 12 | Full |
| Medium | Medium | Ultrasonic Testing | 12 | Partial |
| Medium | Low | Ultrasonic Testing | 12 | Small |
| Low | High | Ultrasonic Testing | 12 | Full |
| Low | Medium | Ultrasonic Testing | 12 | Partial |
| Low | Low | Ultrasonic Testing | 12 | Small |

From Tables 3 and 4, it can be recommended:

| | Pipeline (Inlet SDV KP-20 Ø20") | Tee (KP-20 Ø20") | Pipeline (Outlet SDV KP-20 Ø20") |
|---|---|---|---|
| Frequency | every 36 months | every 36 months | every 36 months |
| Inspection Method | External NDT | External NDT | External NDT |
| Next Inspection | before the year 2020 | before the year 2035 | before the year 2023 |

## 4. Conclusion

Based on the Level Risk analysis, it can be seen that all components have a medium risk level (1C) with an overall value of 1819.654 $ft^2$. The consequences of failure and the value of the probability of failure was 1. The most critical component is the Straight Gas Pipeline (Automatic Shutdown Valve Inlet KP-20 Ø20") with the Remaining LifeTime of 9.8 years, as a result of the magnitude of corrosion rate on these components. The prioritizations of API 581 Risk-Based Inspection are the Gas Pipe Straight (Inlet Automatic Shutdown Valve KP-20 Ø20") that will be inspected in 2020 by the NDT inspection method, Gas Pipe Straight (Outlet Automatic Shutdown Valve KP-20 Ø20") that will be inspected in 2023 by NDT inspection methods and lastly Connection Tee (KP-20 Ø20") that will be inspected in 2035 via NDT inspection methods. It is concluded that RBI is an efficient method to assess risk and target inspection resources on critical items. When correctly implemented, the RBI approach is a powerful mean for increasing the safety and reducing the inspection and maintenance costs of process plant equipment.

## 5. References

[1] Tim Independen Pengendalian Keselamatan Migas 2015 *Atlas Keselamatan Migas*
[2] Pluvinage G and Elwany M H 2013 *Safety, Reliability and Risk Associated with Water, Oil and Gas Pipelines Vol. 53* Springer 1689-99
[3] Wibowo F, 2015 *Kajian* Resiko Pipa Gas Transmisi PT Pertamina Studi Kasus Simpang Km32-Palembang **3(1)** 726–33
[4] Yun F W 2008 Correlated sampling techniques used in Monte Carlo simulation for risk assessment *Int.J.of Pres.Ves. and Pip.* **85**(9) 662–669
[5] Kamsu F B 2016 Information structuring and risk-based inspection for the marine oil pipelines *Phys Procedia* **56** 132–42
[6] Vinod G, Bidhar S K, Kushwaha H S, Verma A K and Srividya A A 2003 Comprehensive framework for evaluation of piping reliability due to erosion–corrosion for risk-informed inservice inspection **82** 187–193
[7] API 2008 Risk-based inspection technology
[8] Prayogo G S, Haryadi G D, Ismail R, Kim S J 2015 Risk analysis of heat recovery steam generator with semi quantitative risk based inspection *API* **581**
[9] Noori S A Price J W H 2006 A risk approach to the management of boiler tube thinning *Nucl Eng Des.* **236(4)** 405–414.
[10] Perumal K E 2014 Corrosion risk analysis , risk based inspection and a case study concerning a condensate pipeline *Procedia Eng.* **86** 597–605
[11] Shiaw W T, Wen T H and Chih H T 2007 Study of a risk-based piping inspection guideline system *ISA Transaction* **46** 119-126
[12] API 2000 Risk-based inspection based resource document

## Acknowledgments

The authors gratefully wish to acknowledge the excellent support provided by Electrical Analysis and Engineering Measurement for Reliability Analysis Laboratory at CORES Diponegoro University, Indonesia. The authors extend their sincere thanks to Strength of Material and Reliability Evaluation Laboratory, Pukyong National University, South Korea, for the support and the constructive feedback on the work.

# Finite Element Study of the Effect of UHMWPE Liner Thickness on the Contact Area and Stress Distribution in a Bipolar Hip Joint

**J Jamari[1], E Saputra[1,2], I B Anwar[2, 3], R Ismail[1] and E V D Heide[2]**

[1]Laboratory for Engineering Design and Tribology, Department of Mechanical Engineering, University of Diponegoro, Jl. Prof. Soedharto, Tembalang, Semarang 50275, Indonesia
[2]Laboratory for Surface Technology and Tribology, Faculty of Engineering Technology, University of Twente Drienerloolaan 5, Postbox 217, 7500 AE, Enschede, The Netherlands
[3]Orthopedic and Traumatology Department, Prof. Dr. R. Soeharso Orthopaedic Hospital, Jl. A. Yani Pabelan, Surakarta 57162, Indonesia

Email: j.jamari@gmail.com

**Abstract.** Contact area and stress distribution of the polyethylene liner (PE liner) have a major influence on the wear process. The main factor that affects the contact area and stress on the PE liner is thickness. The International Standards Organization (ISO) recommends a minimum PE liner thickness of 6 mm. However, the thickness of PE liner in a bipolar hip prosthesis has a limited range of motion compared to the unipolar one due to the addition of the outer liner component. Therefore, the study of the effect of PE liner thickness on the contact area and stress distribution in the bipolar model is interesting. Theaim of this research is to investigate the effect of the PE liner thickness to the contact area and stress distribution on the surface of contact between head and PE liner and the contact between outer liner and cup in the bipolar model. This research was carried out by finite element analysis. The results showed that the highest contact stress on the liner occurred at the lowest liner thickness. The maximum contact radius on the liner surface took place at the highest liner thickness. The bipolar model with the liner thickness of 4.5 mm in this research provided the lowest contact stress.

**Keywords:** Finite element, UHMWPE, Bipolar Hip Joint

## 1. Introduction

Islam is the largest religion in Indonesia based on the central bureau of statistics in Indonesia [1]. In Islam, its adherent is called as a Muslim. One of the religious activities for Muslim is *Salat* activity. Salat activity consists of standing, bowing (*ruku'*), straightening up (*i'tidal*), the transition of standing towards prostration, prostration (*sujud*), sitting between the two prostrations, and sitting [2]. Salat is an essential activity for Muslim, including for Muslim patients who use hip prostheses. In another word, Salat activity and the general activities will affect the motion of hip prosthesis [3-5]. One of the differences between Salat activity and daily activities is the range of motion (RoM) in the hip prosthesis movement. Based on the RoM data of hip prosthesis from Soeharso Orthopedic Hospital,

Solo, some movement with a large angle of the range of motion was found. The effects of the large angle would accelerate the process of impingement which is potential to dislocation. Therefore, the design of hip prosthesis for Muslim patients must be tailored to accommodate the movement of Salat. Especially for Indonesian people, the dimension of the hip prosthesis must also be adapted to the diameter of the average hip size of Indonesian people.

In general, there are two types of the hip prosthesis in total hip replacement, i.e. unipolar model, and bipolar model. The unipolar model consists of a cup, liner, head and stem, whereas the bipolar model consists of the cup, outer liner, inner liner, head, and stem. A calculation of the RoM of Salat for the unipolar and bipolar models had been performed [6]. Based on this study, it was found that the bipolar model can accommodate Salat movement better than the unipolar model with the same head diameter. The consequence of using the bipolar model to PE liner is that the liner thickness becomes definite. By assuming the outer diameters of the cup and head of both models are the same, the liner thickness of the unipolar model is thicker than that of the bipolar model. Based on the International Organization for Standardization (ISO), the recommendation of liner thickness is equal to 6 mm [7]. This recommendation can be applied in the unipolar model. However, it is difficult to be applied in the bipolar model. It is due to the limitation of cup dimension and the addition of outer liner. In fact, the liner thickness will affect the contact stress distribution and contact area due to the contact interaction between the head and liner. The contact stress and contact area on the liner will affect the wear of the liner surface.

Several researchers investigated the effect of the liner thickness to the contact stress. In 1985, Bartel *et al.* [8] reported the effect of conformity and plastic thickness on the contact stresses in Metal-Backed plastic implants. They found that the contact stresses on the liner were very sensitive to clearance with a minimum thickness of 4-6 mm. Furthermore, in 1986, Bartel *et al.* [9] continued their research concerning the effect of conformity, thickness and material on stresses in the ultra-high molecular weight polyethylene (UHMWPE) components for total hip replacement. Their results found that the liner thickness should be maximized to reduce the contact stress that led to surface damage. In 2010, Shen *et al.* [10] investigated wear versus the thickness and other features of 5-Mrad cross-linked UHMWPE liners. They found that the reduction of the liner thickness from 6 mm to 3 mm increased the contact stress by 46%.However, the wear rate decreased by 19%. Recently, Jamari *et al.* [11] investigated the effect of the wall thickness on the wear of PE liner. Their results showed that higher contact stress wouldbe achievedby decreasing the liner thickness. The general conclusion from literature related to the liner thickness is that the liner thickness which is less than 6 mm causes a high contact stress on the liner surface that will also increase the wear process. Unfortunately, most of the contact stress investigations were only performed for the unipolar model. Therefore, the investigation of the contact stress distribution and contact radius in the bipolar model was performed.

The objective of this research is to study the contact stress distribution and contact area on the liner surface due to the contact interaction. Moreover, the contact stress distribution on the cup surface due to interaction contact between outer liner and the cup is also studied. The finite element analysis is employed for calculating the contact area and stress distribution.

## 2. Experimental Method

### *2.1. Material Model*

In general, hip prostheses are made from metal and plastic materials. In this research, the femoral head, the outer liner, and the cup components were made from the SS316L material, while the liner and the inner liner are made from UHMWPE material. For validation process, the cobalt chromium molybdenum (CoCrMo) was employed. These materials are considered as elastic materials, following Archard's wear equation [12]. The properties of UHMWPE, CoCrMo, and SS316L materials can be seen in Table 1.

IOP Conf. Series: Materials Science and Engineering **202** (2017) 012095 doi:10.1088/1757-899X/202/1/012095

**Table 1.** Material properties of UHMWPE and SS316L

| Material | Modulus Elasticity [GPa] | Poisson's ratio |
|---|---|---|
| UHMWPE [13] | 0.5-1 | 0.4 |
| CoCrMo [13] | 230 | 0.3 |
| SS316L [14] | 193 | 0.3 |

**Table 2.** Variation of thickness of the inner liner and cup

| Thickness | Inner liner | Cup |
|---|---|---|
| Thickness 1 | 3 mm | 1.9 mm |
| Thickness 2 | 3.5 mm | 1.65 mm |
| Thickness 3 | 4 mm | 1.4 mm |
| Thickness 4 | 4.5 mm | 1.15 mm |

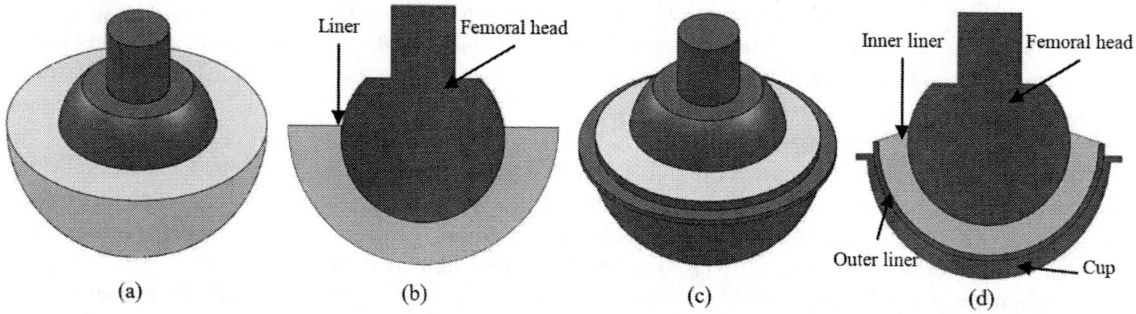


**Figure 1.** Unipolar model (UM) and bipolar model (BM) of the hip prostheses [6]: (a) UM isometric view, b) UM cross-section view, c) BM isometric view and d) BM cross-section view. (Note: dark gray color represents metal material, and light gray color represents plastic material)

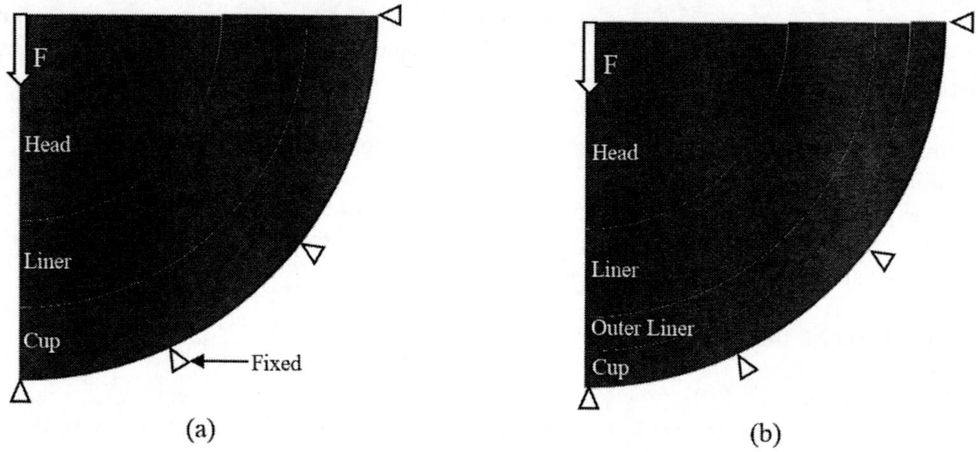


**Figure 2.** (a) Unipolar model and (b) bipolar model

*2.2. Geometrical Model*

There were two models employed in this research, i.e. unipolar and bipolar models (see Figure 1a-d for more detail). The unipolar model was used for validation process, whereas the bipolar model was used for analyzing the contact stress distribution and contact area on the liner and cup surfaces (Figure 2). The dimension of the unipolar model was adopted from the model of Puccio and Mattei [13]. The dimension of the bipolar model was designed based on the femoral head dimension average of Indonesian people. The data from the Indonesian orthopedic hospital show that the average diameter of Indonesian femoral head is 42 mm. The data were then used as a parameter to determine the dimension of outer liner and inner liner of the bipolar model. The head diameter of the bipolar model was 28 mm. To investigate the effects of the inner liner thickness on the contact stress and contact area on the inner liner, the thickness of inner liner was varied. The variations of the inner liner thickness are 3 mm, 3.5 mm, 4 mm and 4.5 mm. Further, the different thickness of the head and the inner liner was divided into two thicknesses to get the thicknesses of the outer liner and cup. Therefore, when the thickness of the inner liner increased, the thickness of cup was decreased. See Table 2 for more details. The clearances between the head-inner liner and outer liner-cup were 0.1 mm.

### *2.3. Finite Element Simulation*
The commercial finite element of ABAQUS software was employed for analysis. The axisymmetric contact model was consideredto simplify and minimize the cost of simulation. The element type of axisymmetric quadrilateral of 4-node bilinear with reduced integration andhourglass control (CAX4R) was employed, while the average number of the element was approximately 5000. The load of 3000 N was appliedat the center point of the femoral head for all models. The outer surface of the inner liner followedthe Hertzian model, and the cups in both unipolar and bipolar modelswere fixed for all directions. In order to obtain the detailed information, the area around the contact interaction was partitionedathigher mesh.

## 3. Results and Discussion

### *3.1. Validation*
In the previous research [6], the applied load was 5 N which was relatively small. Thiswas because the result was compared with Hertzian contact theory. Bartel *et al.* [8] showed that the Hertzian contact theory was not recommended for the conformal contact case. The Hertzian contact theory cannot calculate accurately for the high contact load. Therefore, this research was validated by the theory of Bartel *et al.* Bartel *et al.* theory is a set ofreferences, and all results were compared based on them. The head diameter and liner thicknesses were 28 mm and 6 mm, respectively. The material of head and liner are cobalt chrome and UHMWPE material, respectively.

Figure 3a presents the maximum contact stress on the liner surface as a function of load. The value of contact stress was takenfrom the center point of the contact surface. Figure 3a shows the results of Bartel *et al.*, Hertz, FE Puccio and Mattei, and FE present theories. Based on Figure 3a, the FE present showed a good agreement with Bartel *et al.* theory and FE Puccio and Mattei theory with an error amounted to 5 % and 0.5 %, respectively. However, the Hertz theory showedsignificantlydifferent results compared to the others.

Figure 3b shows the progress of the contact radius as a function of the load. Based on Figure 3b, the FE present also showed a good agreement with Bartel *et al.* theory and FE Puccio and Mattei with the error of 5 % and 1 %, respectively. However, the Hertz theory was significantlydifferent with Bartel *et al.*, FE Puccio and Mattei and FE present theories.

### *3.2. Maximum Contact Stress*
Figure 4a shows the maximum contact stress on the inner liner surface with a variation of liner thickness as a function of force, whereas Figure 4b shows the maximum contact stress on the cup surface. The maximum contact stress was taken in the center of the contact interaction. Based on Figure 4a, the contact stress on the inner liner increasedappropriately with the increase of the force. The highest contact stress occurredwhen the inner liner thickness was equal to 3 mm, while the lowest contact stress took place when the inner liner was equal to 4.5 mm.

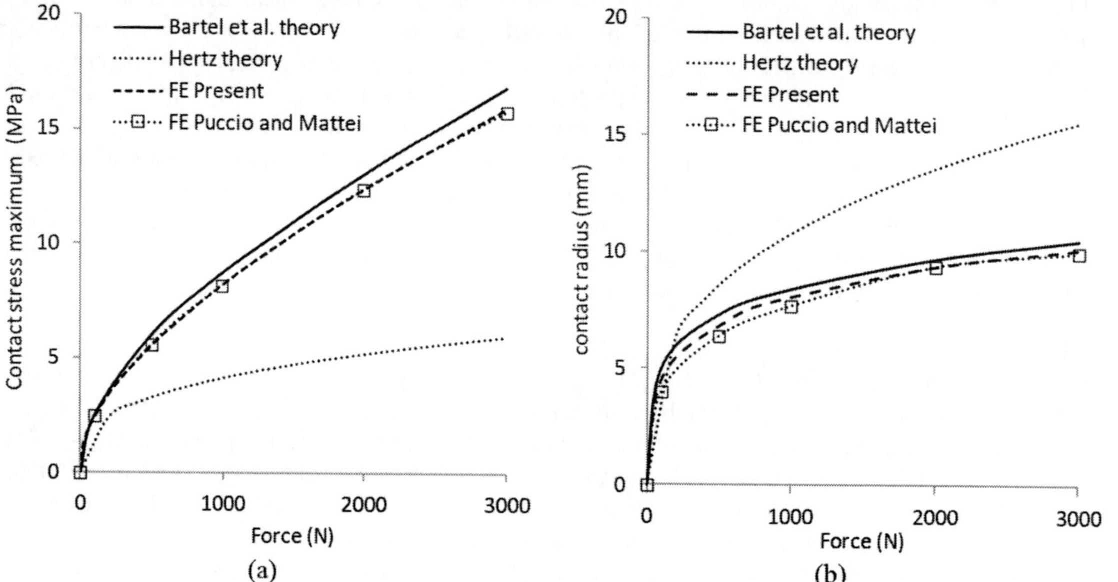


**Figure 3.** (a) The comparison between contact pressure distribution as a function of contact radius for Hertz theory and FE, and (b) the contour plot of contact pressure distribution for Hertz contact using FE simulation.

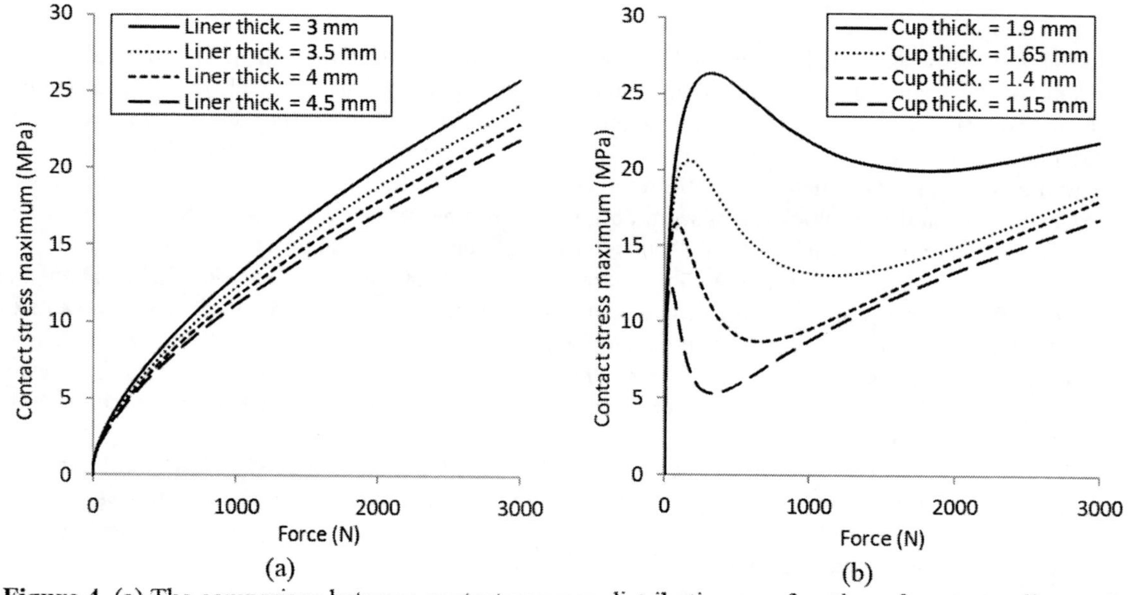


**Figure 4.** (a) The comparison between contact pressure distribution as a function of contact radius on the inner liner surface and (b) on the outer liner surface for Hertz, unipolar and bipolar condition.

Figure 4b shows the contact stress on the cup surfaces as a function of load. Based on figure 4b, the contact stress phenomenon on the cup surface was different with the contact stress phenomenon on the inner liner surfaces. The increasing contact stress on the cup surface was fluctuation. As in the cup thickness that was equal to 1.9 mm, the initial contact stress increaseduntil 27 MPa at the force of 400 N, then the contact stress was fluctuation until it reached the force of 3000 N. Such phenomenon occurred at all of the cup thickness.

### *3.3. Contact Stress Distribution and Contact Radius*

Figures 5a and 5b show the contact stress distribution and contact radius on the liner and cup surfaces as a function of the contact radius. These values were takenat the force equal to 3000 N. Figure 5a presents the contact stress distribution on the liner surface. The maximum contact stress was getting started at the center point of contact then decreased to reach the last contact radius, see Figure 6a. The highest contact stress occurredwhen the liner thickness was equal to 3 mm, while the lowest contact stress took place when the liner thickness was equal to 4.5 mm. However, the contact radius showedthecontrary phenomena. The highest contact radius occurred when the liner thickness was equal to 4.5mm, whereas the lowest contact stress took place when the liner thickness was equal to 3 mm.

Figure 5b presents the contact stress distribution on the cup surface as a function of the contact radius. The location of the highest contact stress did not appear at the center point of contact. However, it shifted to the last contact radius, see Figure 6b. The highest contact stress occurred when the cup thickness was equal to 1.9 mm, while the lowest contact stress took place when the cup thickness was equal to 1.15 mm. However, the contact radius showed different phenomena. The highest contact radius occurred when the cup thickness was equal to 1.15 mm, while the lowest contact radius happened when the cup thickness was equal to 1.9 mm.

The contact stress and contact radius on the cup and liner surface showed different phenomena. The highest contact stress on the liner surface to the liner variation occurred at the lowest liner thickness, while the highest contact stress on the cup surface to the cup variation occurred at the highest liner thickness. The maximum contact radius to the PE liner variation took place at the highest liner thickness, while the highest contact radius to the cup variation occurred at the lower cup thickness.

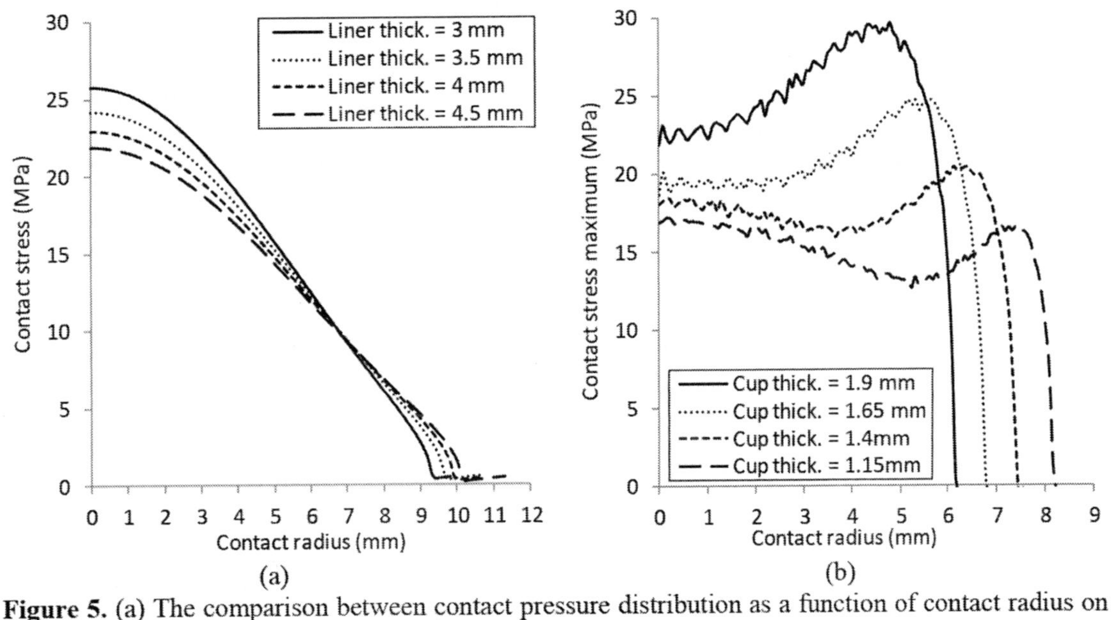


**Figure 5.** (a) The comparison between contact pressure distribution as a function of contact radius on the inner liner surfaceand (b) on the outer liner surface for Hertz, unipolar and bipolar condition.

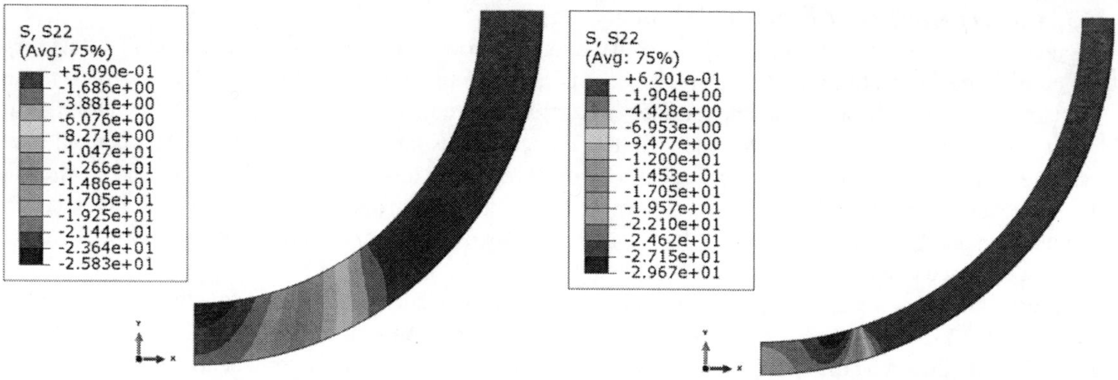


(a) (b)

**Figure 6.** (a) The contour plot of contact pressure distribution on the surface of inner acetabular liner for theunipolar model and (b) the contour plot of contact pressure distribution on the surface of outer liner for the bipolar model.

### *3.4. Von Mises Stress*

Figure 7 presents the von Mises stress on the liner and the cup. The thickness of the liner and cup was represented with the symbols of thickness 1 up to 4, see Table 2. The purpose of Figure 7 is to show the failure boundary of the liner and cup due to the contact load. Based on this figure, the von Mises stress to the liner thickness was relatively similar, while the von Mises stress to the cup thickness showed different results at each of the cup thickness. On the variation of cup thickness, the highest von Mises stress occurred at the highest cup thickness, while the lowest von Mises stress occurred at the lowest cup thickness. The contour of von Mises stress of the liner and cup can be seen in Figure 8.

Based on reference [15], the yield stress of the UHMWPE material was 23.56 MPa. Therefore, it can be concluded that the liner thickness in this bipolar model was safe because the highest von Mises stress of the liner was under the yield stress. It was also shown that the applied load was allowedin this bipolar model.

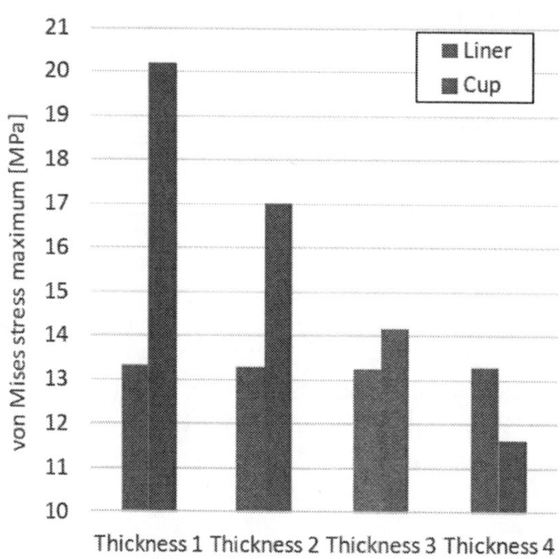


**Figure 7.** The von Mises stress of liner and cup

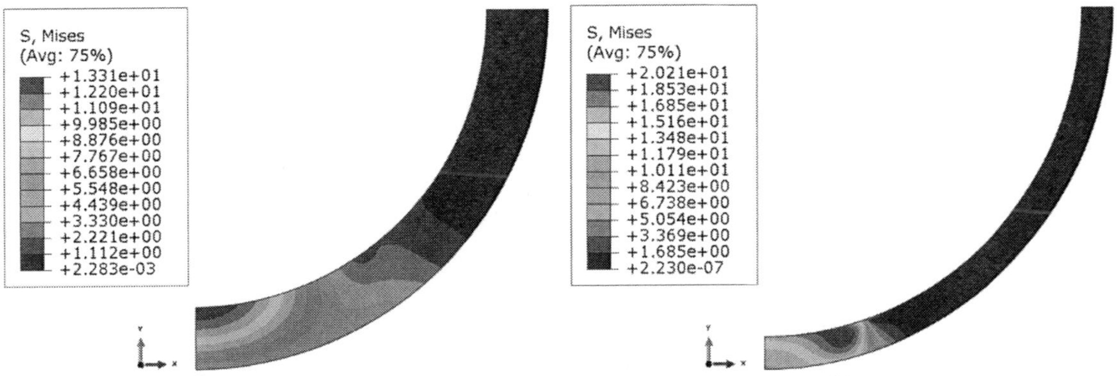


(a) (b)

**Figure 8.** (a) The contour plot of contact pressure distribution on the surface of inner liner for theunipolar model and (b) the contour plot of contact pressure distribution on the surface of outer liner for the bipolar model

## 4. Conclusion

A study of the effect of the PE liner thickness onthe contact area and stress distribution on the surface of the contact between head and PE liner and the contact between outer liner and cup in the bipolar model has been performed. This study was carried out using finite element analysis simulation. The results of the study showed that the highest contact stress on the liner occurredat the lowest liner thickness. The highest contact radius on the liner surface took place at the highest liner thickness. The bipolar model with the liner thickness of 4.5 mm in this research gavethe lowest contact stress.

## 5. References

[1] http://sp2010.bps.go.id
[2] Association of Islamic Chartiable Projects in North American. The basics of Muslim's prayer. Darulmashari 2008; 3rd Edition: 1429. http://www.aicp.org
[3] Saputra E, Anwar I B, Jamari J and van der Heide E 2013 *Procedia Eng.* **68** 102
[4] Saputra E, Anwar I B, Ismail R, Jamari J and van der Heide E 2014 *J. Teknol.* (*Sci. and Eng.*) **66** 53
[5] Ismail R, Saputra E, Tauviqirrahman M, Legowo A B, Anwar I B and Jamari J 2014 *App. Mech. and Mat.* **493** 426
[6] Saputra E, Anwar I B, Jamari J and van der Heide E 2015 *Adv. Mat. Res.* **1123** 164
[7] International Standard Organization. BS EN 12563: 1999. Non-Active Surgical Implant. Joint Replacement Implants. Specific Requirements for Hip Joint Replacement Implants.
[8] Bartel D L, Burstein A H, Toda M D and Edwards D L 1985 *J. Biomech. Eng.* **107** 193
[9] Bartel D L, Bicknell V L and Wright T M 1986 *The J. Bone and Joint Surg.* **68**-A 1041
[10] Shen F W, Lu Z and McKellop H A 2010 *Clin. Orthop. Relat. Res.* **469** 395
[11] Jamari J, Hidayat A, Saputra E, Anwar I B, Ismail R and van der Heide E 2015 *Adv. Mat. Res.* **1123** 196
[12] Archard J F 1953 *J. Appl. Phys.* **24** 981
[13] Puccio F D and Mattei L 2015 *World J. Orthop.* **6**(1) 77
[14] Yildiz F, Yetim A F, Alsaran A, Celik A and Kaymaz I 2011 *Tribol. Int.* **44** 1979
[15] Kluess D, Martin H, Mittelmeier W, Schmitz K P and Bader R 2007 *Med. Eng. & Phys.* **29** 465

# Influence of Glass Fiber Nonlinearity and Dispersion on Light propagation in Double Core Optical Fiber

**A Hidayat, A Listanti, E Latifah, H Wisodo, Nugroho A P and A Taufiq**

Department of Physics, Faculty of Mathematics and Natural Sciences, Universitas Negeri Malang (State University of Malang), Jalan Semarang No 5, Malang 65145, Indonesia

Email: arif.hidayat.fmipa@um.ac.id

**Abstract**. Influence of coupling and intermodal dispersion coefficient on pulse splitting in double core optical fibre was investigated by using solutions of normalized coupled nonlinear Schrödinger equations. It was found that if coupling coefficient and intermodal dispersion coefficient was small, and then nonlinearity cannot balance intermodal dispersion effect. Consequently, pulse was distorted. Furthermore, if intermodal dispersion coefficient was large enough, then pulse splitting occurred. Increasing coupling coefficient avoids pulse splitting and the pulse was stable.

**Keywords:** double core fiber, coupled nonlinear Schrodinger equation, intermodal dispersion.

## 1. Introduction

The communication system can transmit information from one place to another that separated by a great distance. In the optical communication system, information is carried by high carrier frequency (~ 100 THz) in the infrared region. The optical communication system is a system that uses single mode fiber optics to transmit information.

The transmission of signals through optical fiber has advantages when compared to electric wave transmission systems and microwaves. Among others, its size is very small and lightweight so easy in handling and their installation. Light waves are unaffected by electrical interference and magnetic fields. Large capacity optical fiber transmission, so that it can transmit large amounts of information and quick. One of the tools developed is a fiber optic double-core (directional coupler) that function as components of optical switching, wavelength division multiplexing device, and breaking power or power divider [1].

Double-core optical fiber (DCF) is often used in modern optical communication systems [2]. DCF fiber is a fiber that consists of two identical optical fibers, each of which has a bimodal structure that supports the mode symmetry and asymmetry mode [3]. When power is launched into one of the cores, then a power is transferred from the first core to the second core. In general, these two modes have different group velocity, so that input pulse will broaden and be split after propagating a certain distance. Effect that causes the pulse became distorted called intermodal dispersion (IMD). IMD can create distorted pulses and pulse splitting occurs. Furthermore, Tan et al showed that DCF is effective to be applied in highly power CW fiber laser construction [4].

Power transfer mechanism in two-core optical fiber is characterized by the parameter coupling coefficient. In general, the coupling coefficient depends on the wavelength. This leads to the intermodal dispersion. Intermodal dispersion can limit the bandwidth of the two-core optical fiber, as well as the pulse splitting, can occur. It was found that the pulse splitting can be suppressed by making the dispersion intermodal is quite small, the coupling coefficient is quite large, and large input power is injected into. The investigation of the propagation of light in optical fibers has been done. However, the investigation on the effect of the interaction of intermodal dispersion and nonlinearity presented by coupling coefficient to minimize the pulse distortion and even avoid the pulse splitting in two-core optical fiber in detail yet done [5-9].

## 2. Propagation of optical pulse in double-core optical fiber (DCF)

DCF is a fiber that consists of two single-mode core. The physical mechanism of transfer of power between two-core DCF can be explained using couple mode theory, as shown in Figure 1. Based on couple mode theory, when the pulse is launched in one of the cores, then part of the pulses is coupled to the other core as it passes through the area along the $L_C$ coupling interaction. Beam pulses that are not coupled to the output of cores will come out first.

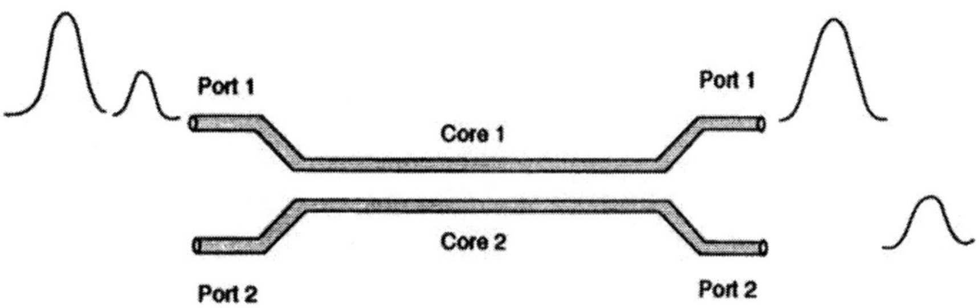


**Figure 1.** The mechanism of propagation of light in DCF [3].

DCF has a bimodal structure, which consists of even mode (symmetry) and the odd mode (anti-symmetry), respectively. The propagation of supermode into double-core optical fiber is described by the following set of normalized coupled nonlinear Schrödinger equations [3, 9] For a symmetric coupler with two identical cores, the coupled-mode equations for symmetric couplers are:

$$\frac{\partial A_1}{\partial z}+\frac{1}{v_g}\frac{\partial A_1}{\partial t}+\frac{i\beta_2}{2}\frac{\partial^2 A_1}{\partial t^2}=i\kappa A_2+i\gamma\left(|A_1|^2+\sigma|A_2|^2\right)A_1 \quad (1)$$

$$\frac{\partial A_2}{\partial z}+\frac{1}{v_g}\frac{\partial A_2}{\partial t}+\frac{i\beta_2}{2}\frac{\partial^2 A_2}{\partial t^2}=i\kappa A_1+i\gamma\left(|A_2|^2+\sigma|A_1|^2\right)A_{21} \quad (2)$$

For a symmetric coupler, the general solution of Equation (1) and Equation (2) can be written in a matrix form as

$$\begin{pmatrix} A_1(L) \\ A_2(L) \end{pmatrix}=\begin{pmatrix} \cos(Kl) & i\sin(\kappa L) \\ i\sin(\kappa L) & \cos(Kl) \end{pmatrix}\begin{pmatrix} A_1(0) \\ A_2(0) \end{pmatrix} \quad (3)$$

The determinant of the 2×2 transfer matrix on the right side is unity, as it should be for a lossless coupler. Typically, only one beam is injected at the input end. The output powers, $P_1(L)=|A_1|^2$ and $P_2(L)=|A_2|^2$, are then obtained from Equation (3) by setting $A_2(0) = 0$ and are given by

$$P_1(L) = P_0 \cos^2(\kappa L), \tag{4}$$
$$P_2(L) = P_0 \sin^2(\kappa L) \tag{5}$$

where $P_0 \equiv A_0^2$ is the incident power at the first input port. The double-core optical fiber acts as a beam splitter, and the splitting ratio depends on the parameter $\kappa L$.

$$\frac{\partial A_1}{\partial z} + \kappa_1 \frac{\partial A_2}{\partial T} + \frac{i\beta_2}{2}\frac{\partial^2 A_1}{\partial T^2} + \frac{i\kappa_2}{2}\frac{\partial^2 A_2}{\partial T^2} = i\kappa_0 A_2 \tag{6}$$

$$\frac{\partial A_2}{\partial z} + \kappa_1 \frac{\partial A_1}{\partial T} + \frac{i\beta_2}{2}\frac{\partial^2 A_2}{\partial T^2} + \frac{i\kappa_2}{2}\frac{\partial^2 A_1}{\partial T^2} = i\kappa_0 A_1 \tag{7}$$

In case of pulse propagation in a linear DCF which the nonlinear terms in Equation (1) and Equation (2) were ignored, Li .et all showed that the strong effects of intermodal dispersion will split pulses propagating in linear DCF. The results of Li [10] show clearly that group velocity dispersion ($R$) and coupling coefficient ($R_1$) exert different effects on pulse propagation, both of which will destroy the integrity of the pulse. Now, the question is whether the nonlinearity can balance the both effects of group velocity dispersion ($R$) and coupling coefficient ($R_1$) to keep the integrity of the pulse over a long distance. This questions will be numerically investigated in this article.

## 3. Research Method

The physical model of pulse propagation in dcf is derived from a set of generalized coupled nonlinear schrödinger equations, Equation (6) and Equation (7). by varying the parameters of group velocity dispersion ($r$), coupling coefficient ($r_1$) and power input, it can be analyzed the effect of variation those parameters on the pulse propagating in DCF. Pulses injected into one of the cores in double-core optical fiber are expressed by Equation (8) and Equation (9).

$$P_1(L) = P_0 \cos^2(\kappa L), \tag{8}$$
$$P_2(L) = P_0 \sin^2(\kappa L) \tag{9}$$

where $P_0$ is the input power. In terms of normalization, $P_0 = 1$ actually corresponds to one soliton power [3], the power required for soliton formation in single-core fibers. Equation (8) and Equation (9) are numerically solved with the initial condition of Equation (6) by employing pseudo-spectral method in the time domain and fourth-order Runge-Kutta method in the spatial domain with the adaptive step-size control [9- 12]. Effect of intermodal dispersion ($R_1$) and the coupling coefficient ($R$) were studied by varying $R_1$ and $R$ for two different input power values. Thus, the influence of the input power to the pulse propagation in the double core optical fiber can also be known.

## 4. Results and Discussion

Table 1 shows pulse propagation for various intermodal dispersion coefficient. Coupling coefficients are kept constant, $R = 1$. For a relatively weak power $P_0 = 1$, significant pulse distortion appears even at a very weak intermodal dispersion value as illustrated in Table 1(a) for $R_1 = -0.15$. As intermodal dispersion becomes strong, distinct pulse splitting is observed as shown by Figure 1(c) for $R_1 = -1$ and Table 1(e) for $R_1 = -2$. It is seen that pulse splitting occurs at strong IMD values.

The power $P_0 = 1$ is sufficient to balance the effects of intermodal dispersion. For a strong input power $P_0 = 4$, pulse propagations are shown by Table 1(b), Table 1(d), and Table 1(f). When input power $P_0$ is stronger; $P_0 = 4$, pulse distortion occurs (Table1(b)). As intermodal dispersion becomes strong, distinct pulse splitting is observed as shown by Table 1(d) for $= -1$ and Table 1(f) for $R_1 = -2$. It can be concluded that for small $R_1$ and $R$, pulse splitting still occurs even if the input power is stronger.

IOP Conf. Series: Materials Science and Engineering **202** (2017) 012096 doi:10.1088/1757-899X/202/1/012096

**Table 1.** Comparison of pulse propagation in a nonlinear two-core fiber between at $P_0 = 1$ and at $P_0 = 4$ with $R = 1$ for (a) and (b) $R1 = -0.15$; (c) and (d) $R1 = -1.0$; (e) and (f) $R1 = -2.0$.

| $P_0$ | R | $R_1$ | Normalized slowly varying amplitude envelope of electric field in core of DCF |
|---|---|---|---|
| 1 | 1 | −0.15 | 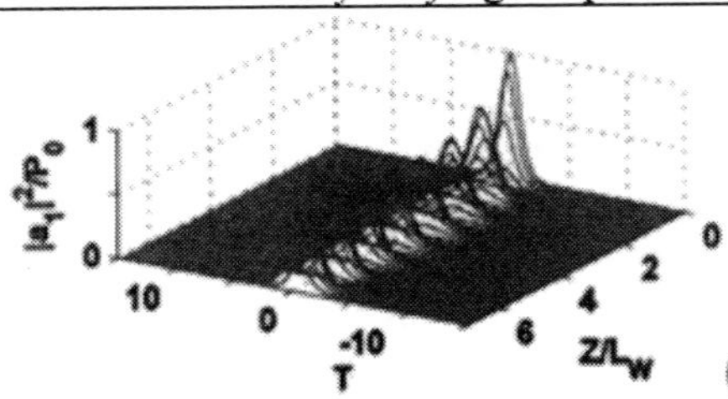 (a) 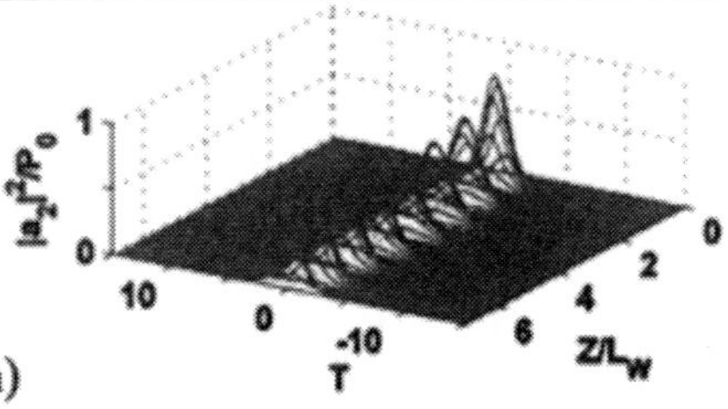 |
| 4 | 1 | −0.15 | 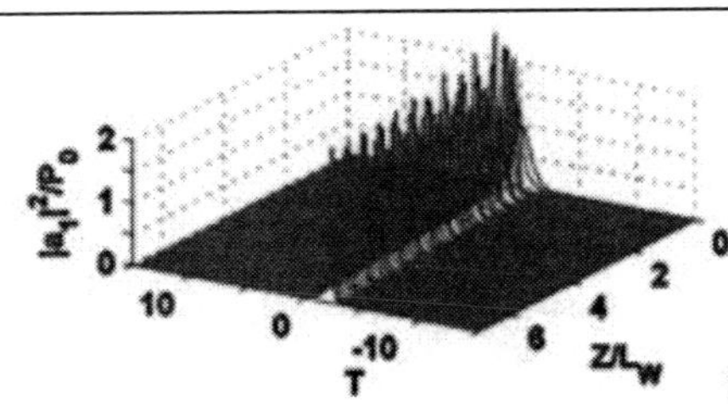 (b) 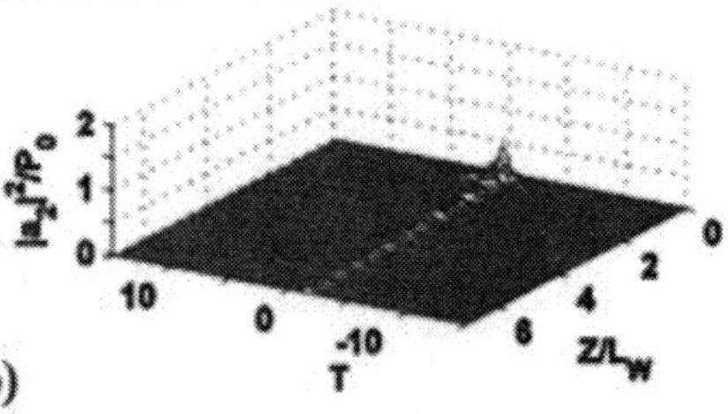 |
| 1 | 1 | −1.0 | 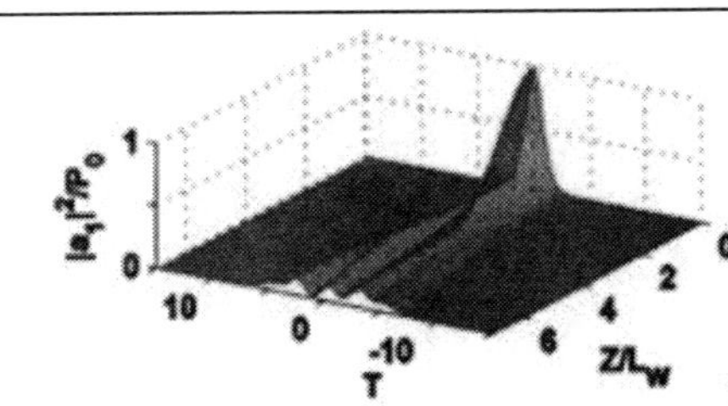 (c) 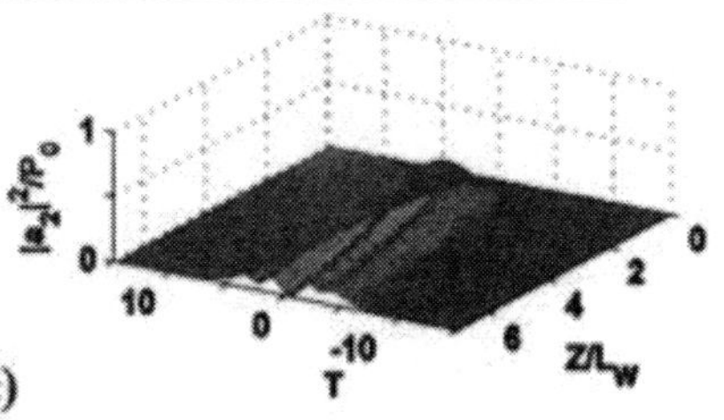 |
| 4 | 1 | −1.0 | 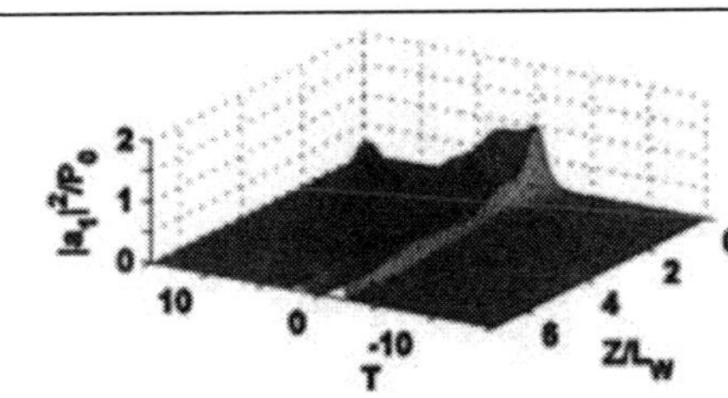 (d) 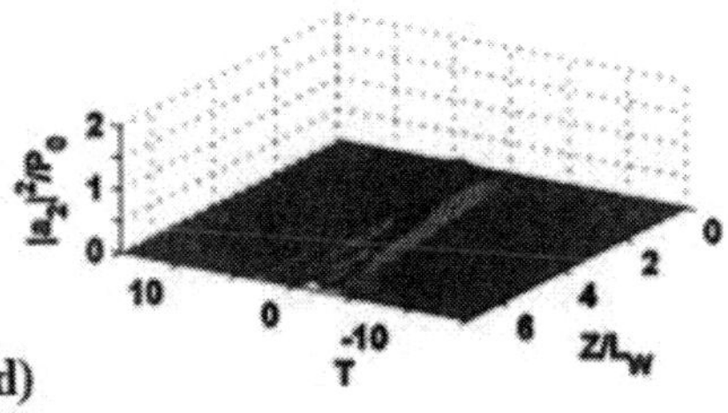 |
| 1 | 1 | −2.0 | 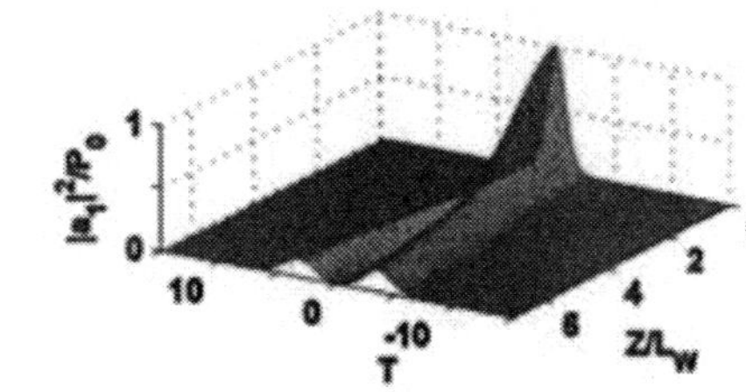 (e) 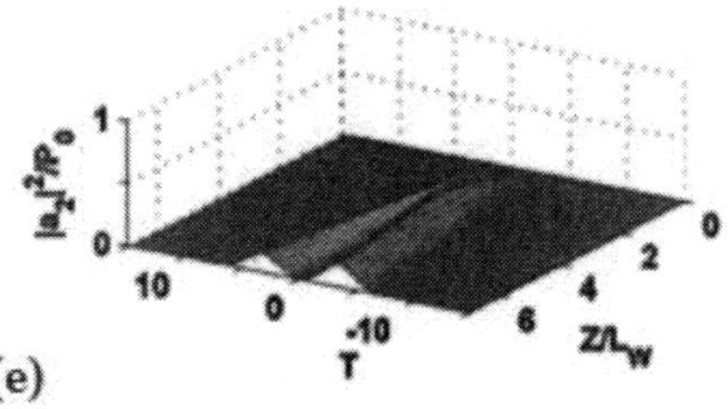 |

| | | | |
|---|---|---|---|
| 4 | 1 | −2.0 | 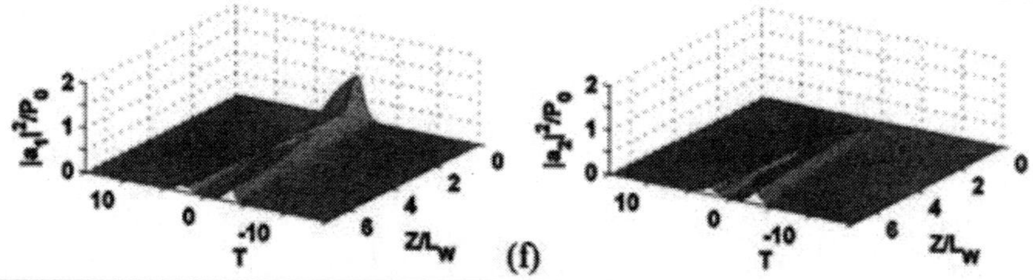 (f) |

**Table 2**. Comparison of pulse propagation in a nonlinear two-core fiber between at $P_0 = 1$ and at $P_0 = 4$ with $R_1 = -0.15$ for (a) and (b) $R = 5$; (c) and (d) $R = 10$; (e) and (f) R = 20.

| $P_0$ | R | $R_1$ | Normalized slowly varying amplitude envelope of electric field in core of DCF |
|---|---|---|---|
| 1 | 5 | −0.15 | 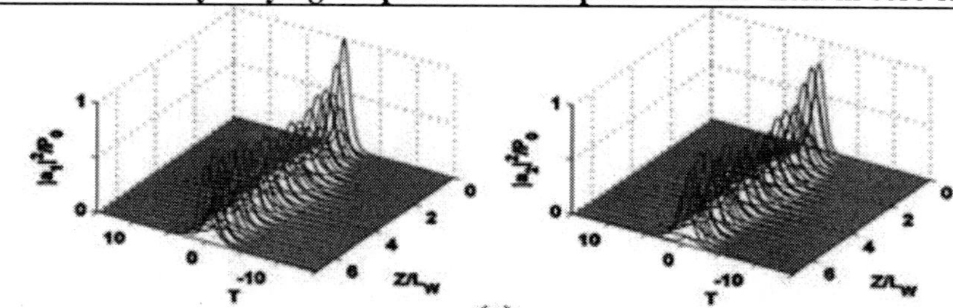 (a) |
| 4 | 5 | −0.15 | 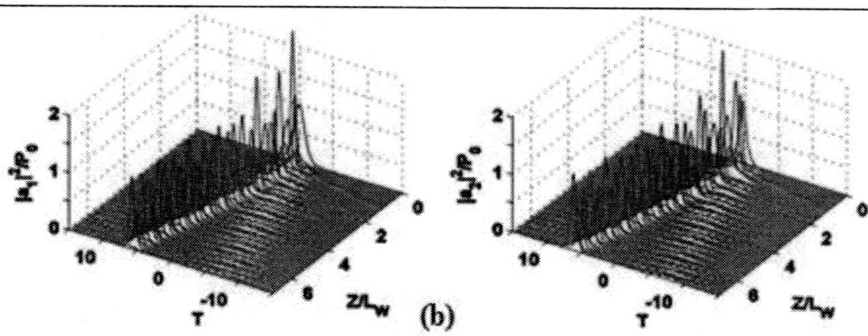 (b) |
| 1 | 10 | −0.15 | 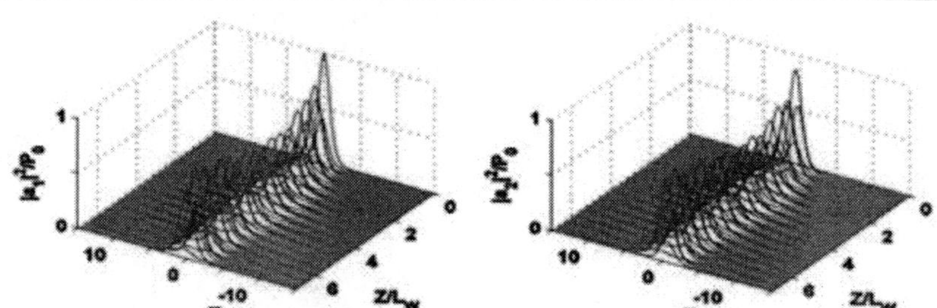 (c) |

IOP Conf. Series: Materials Science and Engineering **202** (2017) 012096 doi:10.1088/1757-899X/202/1/012096

| | | | |
|---|---|---|---|
| 4 | 1<br>0 | −0.1<br>5 | 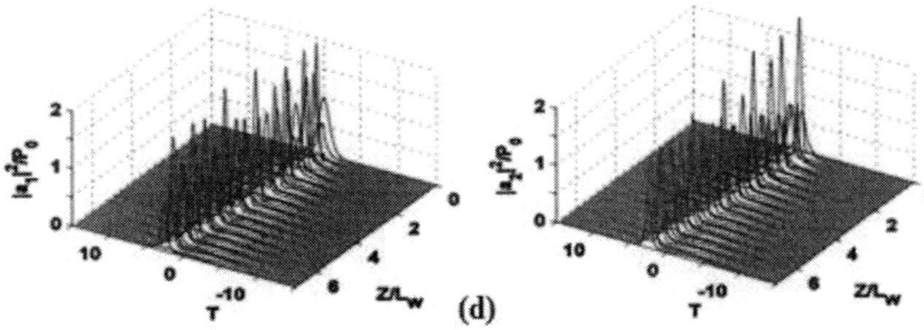 (d) |
| 1 | 2<br>0 | −0.1<br>5 | 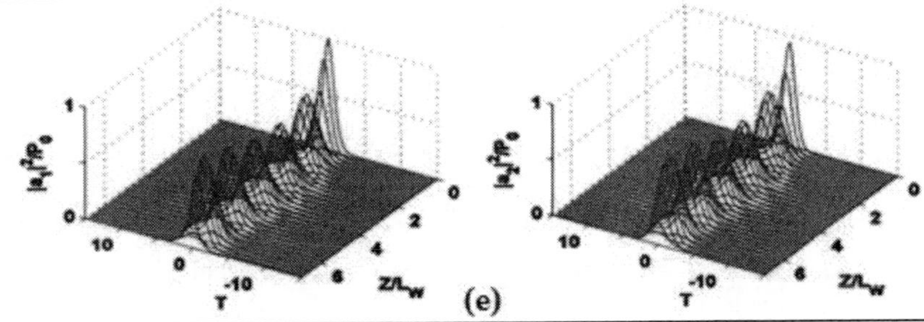 (e) |
| 4 | 2<br>0 | −0.1<br>5 | 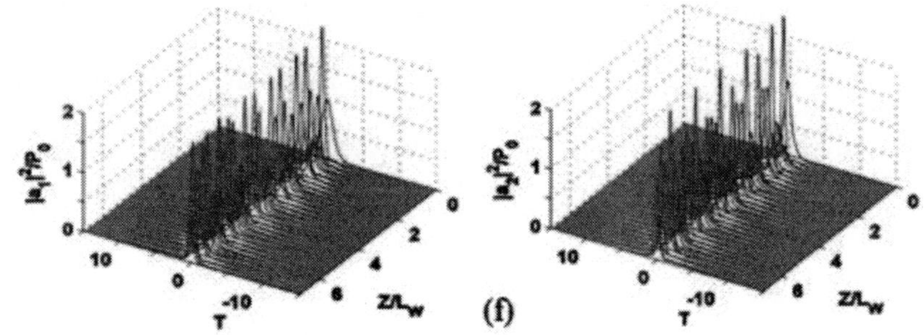 (f) |

Table 2 shows pulse propagation for various coupling coefficient. Intermodal dispersion coefficients are kept constant, $R_1$ = −0.15. For a relatively weak power $P_0$ = 1 and $R$ is small, significant pulse distortion appears as illustrated in Table 2(a) and Table 2(b) for $R_1$ = − 0.15. As coupling coefficient increases, the pulse splitting is suppressed, as shown in Table 2(c) and Table 2(d) for $R$ = 10 and Table 2(e) and Table 2(f) for $R$ = 20. When $P_0$ and $R$ are large enough, the pulse splitting vanishes as shown in Table 1(f).

## 5. Conclusion

Pulse splitting and pulse distortion can be effectively avoided by increasing the Kerr nonlinearity of the fiber. The presence of nonlinearity of optical fiber will balance the effect of dispersion. Nonlinearity of optical fiber can be increased by increasing the power $P_0$. The power of pulse required to balance the two effects must be several times of one soliton power. Intermodal dispersion coefficient $R_1$ required to balance must be smaller than 1 and the coupling coefficient $R$ must also be very strong, i.e. $R >> 1$.

## 6. References

[1]. Wayne T 2001 *Advanced Electronic Communication Systems* Prentice Hall International Inc New Jersey

[2]. Boumaza N et all 2009 Numerical simulation of nonlinear pulse propagation in a nonlinear optical directional coupler **4** 9 pp 505–513 ISSN 1992-1950 *Academic Journals*

[3]. Govind P A 2009 Applications Of Nonlinear Fiber Optics Third ed *Academic Press* New York

[4]. Tan Q, Ge T, Zhang X and Wang Z 2015 Cascaded Combiners For A High Power CW Fiber Laser *Laser Physics* **26** 2

[5]. Dror N and Malomed B A 2015 Solitons And Vortices In Nonlinear Potential Wells, *Journal of Optics* **18** 1 Special issue on singular optics and topological photonics Nov

[6]. Hasegawa A 1980 Self-confinement of multimode optical pulse in a glass fiber *Opt. Lett.* **5** 416–418

[7]. Crosignani B and Di P P 1981 Soliton propagation in multimode optical fibers *Opt. Lett.* **6** 329 – 331

[8]. Crosignani B, Cutolo A and Di P P 1982 Coupled-mode theory of nonlinear propagation in multimode and single-mode fibers: envelope and self-confinement *J. Opt. Soc. Am.* **72** 1136–1141

[9]. Chiang K S *et al* 1999 Propagation of short optical pulses in directional couplers with Kerr nonlinearity *J. Opt. Am.* B **14** 1437 – 1443

[10]. Li J H *et al* 2013 Suppression of pulse splitting in two-core optical fibers with Kerr nonlinearity **90** 2

[11]. Trefethen L N 2000 Spectral methods in MATLAB (Philadelphia, Pa.: Society for industrial and applied mathematics)

[12]. Carr L D, Kutz J N and Reinhardt W P 2001 Stability of stationary states in the cubic nonlinear Schrödinger equations: applications to the Bose-Einstein condensate *Phys. Rev. E* **63** 066604

# A Study of Chip Formation Feedrates of Various Steels in Low-Speed Milling Process

**L Prasetyo, M Tauviqirrahman and Rusnaldy**

Mechanical Engineering Department, Engineering Faculty, Diponegoro University, Jl. Prof. Soedharto, Tembalang, Semarang 50275, Indonesia, Phone: 024-7460059

Email: lisprasetyo3@gmail.com

**Abstract.** Milling is a process of metal removal by feeding the workpiece a rotating multitoothed cutter. The objective of the study was to investigate the chip characteristics (chip length, width, and thickness) during the milling process by varying the feedrates and the types of materials used based on an experimental approach. The chosen materials were AISI 1020, AISI 1045, AISI 1090, AISI D2, and AISI 4340 with a high-speed steel (HSS) as a cutter. In this work, the feedrates were varied of 5, 10, and 15 mm/minutes with the depth of cut of 0.5 mm and a low spindle speed of 70 rpm. The results show that, in general, increasing the feedrate will lead to the growth of chip length, width, and thickness for all types of materials used. Also, related to the chip shape, AISI 1020 produces the discontinuous chip which can be related to its hardness value.

**Keywords:** Chip formation, low speed, steel, milling, hardness

## 1. Introduction

One of the popular kinds of metal used in industry is steel. The use of steel is mostly for constructions and many tools. The progress of technology has produced the commercial steel. By determining the chip form from a conventional milling machine, it is believed that its mechanical properties can be determined. The research of chip characteristics has attracted many researchers from all over the world. Most of them used the cutting parameter with a high spindle speed. It is proven that the characteristic of chip varying the depth of cut and feedrates for different steels can affect the differences in chip dimension (width, length, and thickness) besides the differences in color (brown, blue, and green) [1].

Analyzing the character of chip formation can be done to control the quality of the material used. Therefore, the mechanical properties of a material can be approached. To analyze the chip, the variations of feedrates regarding low spindle speed and the depth of cut should be investigated. Then, the chips from each type of steel with a change of feedrates were observed by the perception of its formation and dimension.

In one of the previous works [2], the characteristic of the chip with face milling has ever been explored by using WCB (Wrought Carbon with grade B) varying the cutting parameters of feedrate and the depth of cut. In addition, the chip formation was also influenced by the cutting direction both in down and up milling processes [3]. Also, the formation of the chip could also be affected by the differences of spindle speed with a large range of the speed [4]. Later, in [5] the chip characterization of titanium material of Ti-6Al was also investigated with the perception method of chip in a three-dimensional model and the micro-hardness measurement regarding the length and width of the chip [5].

By extending the published results with the main frame of the chip formation, this work aimed to investigate the feedrate effect on the chip formation in different types of steel materials experimentally. Also, the resulted chip shape was also investigated related to the prediction of the type of the material used.

## 2. Experimental Method

The experiment was performed by a conventional milling machine as shown in Figure 1. Various steels used in this study were AISI 1020, AISI 1045, AISI 1090, AISI D2 and AISI 4340 which have different hardness levels. The dimension of the workpiece was Ø50.8 x 60 mm. The cutting tool with end mill cutter was 10 mm in diameter.

In this work, the operating conditions of the feedrates were varied. A digital microscope was used to analyze the chip formation. The following are the details of the experimental setup:

(1) Spindle speed : 70 rpm
(2) Depth of cut : 0.5 mm
(3) Cooling : dry processes
(4) Feedrate : feedrate 1 (5 mm/min);
feedrate 2 (10 mm/min);
feedrate 3 (15 mm/min)

**Figure 1.** Milling process of various kinds of steel using a conventional milling machine

The use of the conventional milling machine with low spindle speed and without cooling in the process was purposed for getting chips since it offered an easy chip characterization. As seen in Figure 1, the cutting edge cut the workpiece, and small chips were produced over a very short period. Mostly, sliding frictions occurred between the tool and the workpiece. Chips were quickly generated on the rake face of the cutting edge to be then separated. Finally, the bending chips flowed out along the rake face, but there were still some small chips adhering to the cutting edge.

## 3. Results and Discussion

The chips from the machining test were collected, mounted, and measured. Figure 2 shows the chip formation with different steels and feedrates. It can be seen that the chip formation for all steel types and feedrates was in the form of discontinuous chips. The chip geometry was relatively small, especially for the chip length (1.5 to 3 mm). From Figure 2, it is important to note that the chips for all feedrates and commercial steel types are in the form of Auto Reset Chips (ARC). There was an interesting result that AISI 1020, which has the lowest carbon content compared to others, tended to generate discontinuous chip for all feedrates considered in this work. It indicated that the AISI 1020 from the chip formation and in connection with the physical properties, showed a high hardness value. Note that the morphological size of the saw-tooth chip as a base of the chip formation measurement conducted in this work was described in detail [1].

In addition, from Figure 2, it can be observed that the serrations of the chip surface seemed to be neat for the feedrate values of 5 and 10 mm/min. While for a high feedrate, that is, 15 mm/min, the serrations tended to be irregular chips. Also, the chip resulted from high feedrate milling appeared to get small cracks (or called as chip breaks). Figures 3 to 5 show the effect of the feedrate on the chip length, width, and thickness, respectively, with varied types of steel materials. Based on Figure 3, it can be seen that the increasing the feedrate would increase the chip length. However, for AISI 1090 and AISI D2, there was a decrease in the chip length when the feedrate was 10 mm/min. Figure 4 demonstrates that the chip width was increased along with the improved feedrate. AISI 1020 had the largest chip width. A possible explanation was that AISI 1020 had the lowest carbon content which might resist to the applied force during milling. From Figure 4, it can be understood that the feedrate had a strong effect on the chip thickness, especially for AISI 4340. It was because AISI 4340 had the highest hardness level among others. One issue that was encountered during the experiment was that the feedrate had a strong correlation with the chip characteristics (chip length, width, and thickness) depending on the operating conditions.

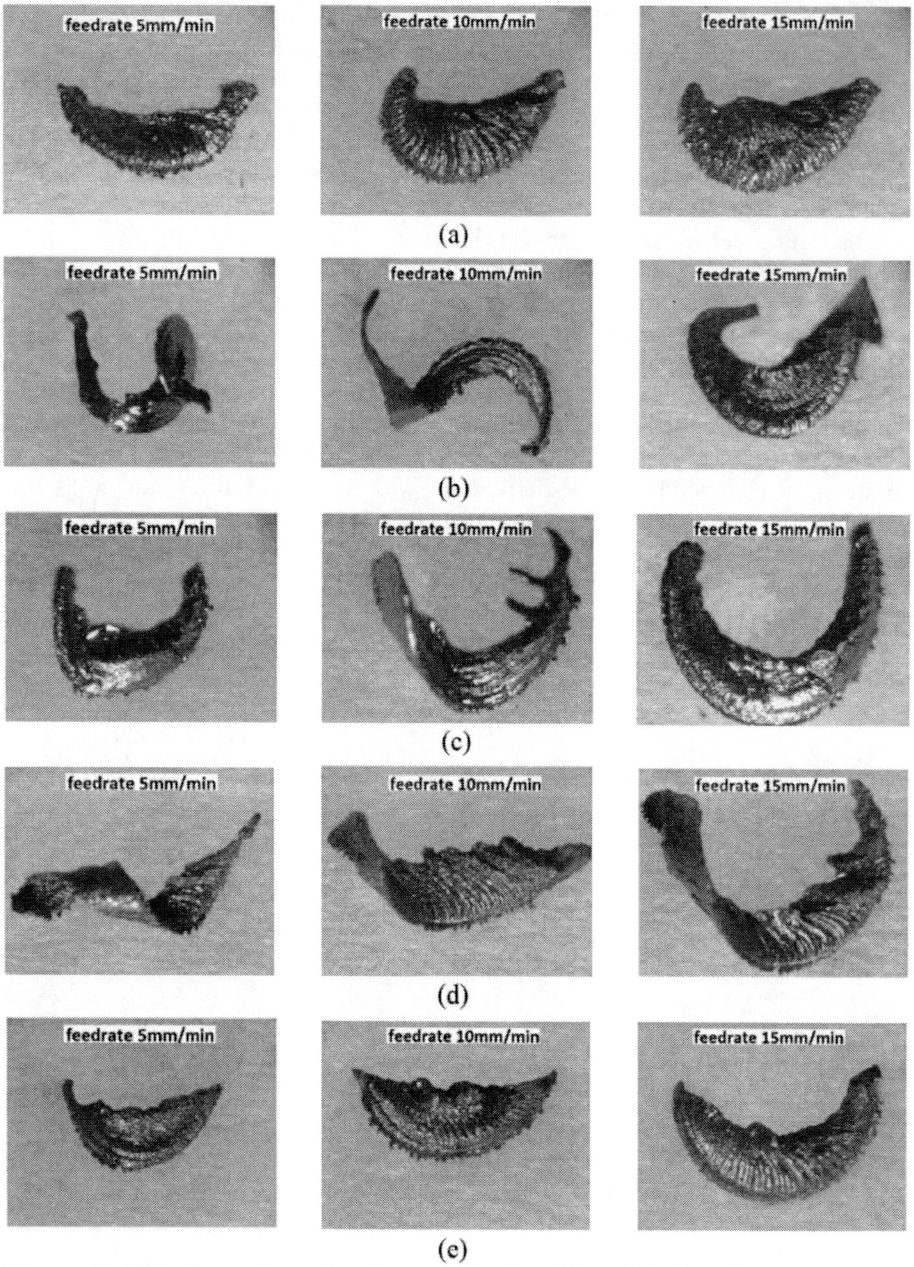


**Figure 2.** Chip formation of various steels: (a)AISI 1020, (b)AISI 1045, (c)AISI 1090, (d)AISI D2, and (e)AISI 4340

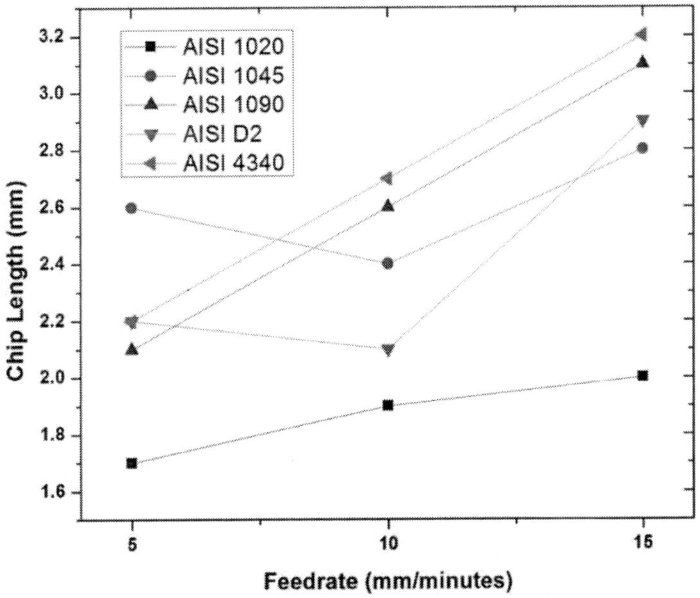


**Figure 3.** The effect of feedrate on the chip length under different steels

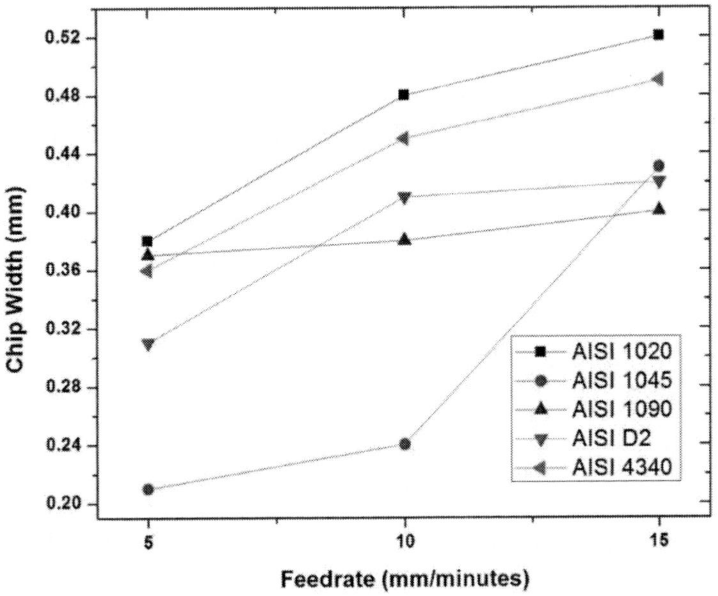


**Figure 4.** The effect of feedrate on the chip width under different steels

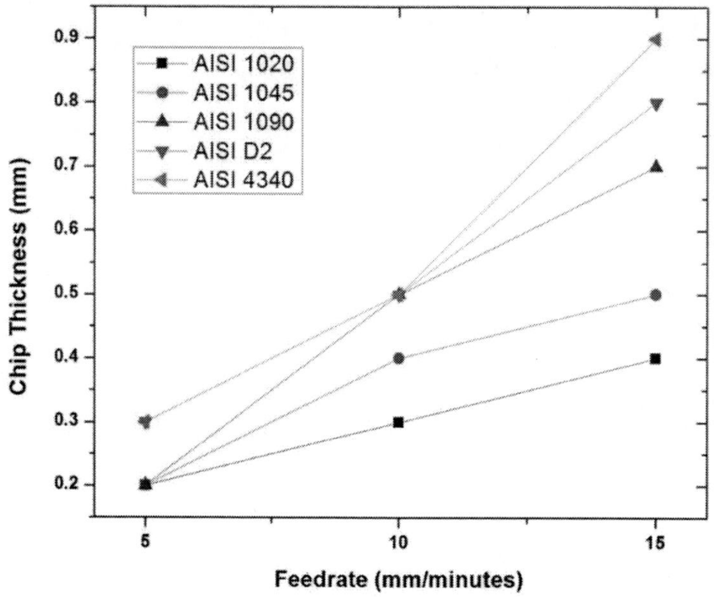


**Figure 5.** The effect of feedrate on the chip thickness under different steels of PVP/ZnO

## 4. Conclusion

The effect of feedrates on a chip characteristics varying the steel materials (AISI 1020, AISI 1045, AISI 1090, AISI D2, and AISI 4340) has been successfully investigated. Based on the discussion, the following conclusion could be drawn. Increasing the feedrate will increase the chip length, width, and thickness for all types of materials used. The significant result of this study was that the chip formation could be considered as the guide to determine the hardness level of the materials.

## 5. References

[1] Wang C, Xie Y, Zheng L, Qin Z, Tang D and Song Y Research on the Chip Formation Mechanism during the high-speed milling of hardened steel *Int. J. Mach. Tools Manuf* **79** pp. 31–48, 2014.

[2] Sheth S and George P M Experimental Investigation and Prediction of Flatness and Surface Roughness during Face Milling Operation of WCB Material," vol. 23, pp. 344–351, 2016.

[3] Wertheim R, Satran A and Ber A 1994 Modifications of the Cutting Edge Geometry and Chip Formation in Milling *CIRP Ann. - Manuf. Technol.* **43** no. 1 pp. 63–68

[4] Zhanqiang L and Guosheng S 2012 Characteristics of chip evolution with elevating cutting speed from low to very high," *Int. J. Mach. Tools Manuf.*, vol. 54–55, pp. 82–85

[5] Sun and Guo Y B A new multi-view approach to characterize 3D chip morphology and properties in end milling titanium Ti-6Al-4V *Int. J. Mach. Tools Manuf* **48** no. 12–13, pp. 1486–1494, 2008.

**Acknowledgements**

The authors would like to express their gratitude to the Laboratory of Physical Metallurgy, Diponegoro University, for the experiment. The authors would also like to appreciate the useful comments and suggestions from reviewers.

# Investigation of the Influence of Shapes-Texture on Surface Deformation of UHMWPE as a Bearing Material in Static Normal Load and Rolling Contact

**W D Lestari, R Ismail, J Jamari and A P Bayuseno**

Department of Mechanical Engineering, Faculty of Engineering, Diponegoro University, Jl. Prof. Soedarto, Tembalang, Semarang 50275, Indonesia, Phone: 024-7460059

Email: wahyu_dwilestari@ymail.com

**Abstract.** Surface texture is a common method for improving wear properties of a tribo-pair of soft and hard bearing material. The reduction of wear rates on the contacting surface material is becoming important issues. In the present study, analysis of the contact pressure on the flat surface of UHMWPE (Ultra High Molecular Weight Polyethylene) under the static- and rolling motion with the surface of steel ball used the 3D finite element method (FEM) (the ABAQUS software version 6.12). Five shaped-texture models (square, circle, ellipse, triangle, and chevron) were presented on the flat surface for analysis. The normal load of 17, 30 and 50 N was deliberately set-up for static and rolling contact analysis. The contact pressure was determined to predict the wear behavior of the shaped-texture on the flat surface of UHMWPE. The results have shown that the static normal load yielded the lowest von-Mises stress distribution on the shaped-texture of the ellipse for all values applied a load, while the square shape experienced the highest stress distribution. Under rolling contact, however, the increasing load yielded the increasing von Mises stress distribution for the texture with a triangle shape. Moreover, the texture shapes for circle, ellipse, and chevron respectively, may undergo the lowest stress distribution for all load. The wear calculation provided that the circle and square shape may undergo the highest wear rates. Obviously, the surface texture of circle, ellipse, and chevron may experience the lowest wear rates and is potential for use in the surface engineering of bearing materials.

**Keyword:** Surface texturing, texture shapes, wear, contact pressure, rolling contact, finite element method.

## 1. Introduction

Ultra-high molecular weight polyethylene (UHMWPE) has emerged as a common bearing material for biomedical application because of having excellent properties such as self-lubrication ability, and chemical inertness. It has been widely found in many joint replacement devices in medical applications, including artificial joints and face surgery. However, wear and plastic deformation are two critical mechanisms that may lead to failure of this material during contacting to harder material both in sliding and rolling contacts. The implementation of the surface technology has become a great concern in the development of wear-resistant materials. In this case, the performance and lifetime of the bearing material rely on the changes in the surface engineering level. Recently, the surface texture developed on the soft material, namely UHMWPE has become a promising method to improve tribological property [1]. Surface texture can be considered as an artificial micro-geometry which can be made by micro-groove, micro-dimples, or micro-convex scale on the surface contact.

Surface texturing can be made as an array of cavities fabricated on a surface, with specified geometry (size, shape, and distribution) and has achieved as an important part of surface engineering in the critical surface area of tribological applications such as mechanical seals, bushings, and cylinder liners [2-5]. Creating surfaces with controlled microgeometry may be adopted as an approach to achieve the best performance of tribological interfaces. A textured surface with various depth, diameter, and distribution of cavities may operate as a lubricant reservoir that causes the load carrying-pressure to diminish on the surfaces of the counterparts and hence reducing the surface friction. Surface texturing in the range of hydrodynamic and elastohydrodynamic lubrication regimes may perform according to a Stribeck curve [6-9].

The great interest of the surface texturing reflected by the previous paper is for the surface technology of bearing materials. The most of the research was mainly focus on the surface texture of hard material as ceramics and metals. The researchers have conducted study intensively on the development of surface texture for soft polymeric materials (UHMWPE). Zang B et al. [10], proposed the micro-imprint lithography technology to construct micro-scaled dimples on the UHMWPE surface. The result indicated that the surface textured on UHMWPE has an outstanding effect on the friction reduction compared to the tough materials such as metals and ceramics. Zang B et al. [11] manufactured dimple patterns on the surface of the 316 stainless steel and UHMWPE in order to compare the effect of surface texture on different materials. The texture was generated using the micro-imprint lithography and the combination of photolithography and electrolytic etching. The textured design on the UHMWPE surface provided a better wear resistance than the designed texture of the steel surface.

Further, Young et al. [12] developed a surface texture of UHMWPE by a CNC (computerized numerical controlled) milling machine. The surface texture generated yielding a substantial reduction (42%) in the friction coefficient using the bovine serum lubrication. Kustandi et al. [13] created nanoscale grooves on the surface of UHMWPE using nanoimprint lithography. The result showed that under a dry sliding condition for loads ranging from 60 to 200 mN, the surface texture of UHMWPE presented a friction reduction, whereas wear depth and width generated were lower than those of non-textured surfaces. Lopez-Cervantes *et al.* [14] studied the effect of texturing cavities on the stress distribution of the UHMWPE surface by numerical analysis using 2D and 3D finite element method. The numerical simulation was focused on the static contact under dry and wet condition. It was shown that under dry conditions, the textured surface with empty cavities has the highest von Mises stress. However, the contact pressure decreased as the liquid filled cavities. The subsequent research by Lopez-Cervantes et al. [15] reported the experimental design that resulted from tribological tests worked on UHMWPE flat disks, plain and textured, applying for a ball on disk construction to evaluate the performance of UHMWPE textured versus plain surface, under the different combinations of sliding and rolling motion. It was proposed that a lower traction coefficient will develop the performance of UHMWPE as a plastic spacer in prosthesis applications where a combination of sliding/rolling motion takes place.

Fang et al. [16-18] applied surface texturing techniques to practice attentively distributed UHMWPE particles with measured sizes and shapes. The basis of surface texture design has been established to accomplish wear particle morphology controlling in water. This accomplishment allows the bioactivity estimation of different sizes and shapes of UHMWPE wear particles. Furthermore, Fang et al. [19] examined the influences of biological lubricants on the morphology of UHMWPE wear particles produced with microfabricated surface textures. The result showed that a restricted circulation of the particles has been achieved by employing microfabrication of the surface textures. It was also shown the ability to restrain the particle size and shape by the dimensions of the surface textures. The length of the particle produced in serum is smaller than in water. Zhang et al. [20] utilized the numerical study to explore the ability usage of microstructures with different patterns to advance the tribological properties of UHMWPE. The simulation results demonstrated that the rectangular surface texture displayed interesting features over the other geometries and the result that investigated can be excellent work to enhance hydrodynamic effects.

Jamari et al. [21] examined the effect of surface texturing on the stress distribution of UHMWPE under rolling contact. As a result, the maximum von Mises stress was concentrated in the direction of rolling. Our previous study [22], the research has investigated the influence of normal load under static contact by 3D finite element method. Different diameter and separation of the dimple were introduced in this numerical study. The increasing diameter cavities and separation resulted in the reduction von Mises stress distribution. Because of the importance of surface technology for the development of wear resistant material, a numerical model of surface texture should be developed on a smaller scale or more precisely to the level of surface micro-geometry. As a result, the level of surface technology change may relate to wearing rate of the material.

In the present study, a tribo pair of hard material and soft material, that are steel and UHMWPE, was examined. Surface texture with the different geometry was created on the surface of UHMWPE. A numerical contact model using FEM was developed. The effect of different texture shape on the contacting surface was investigated at different load conditions for normal load and rolling motion. This model of the elastic and elastic-plastic behavior of the textured surfaces was subsequently run using FEM software. Deformations and elastic-plastic sub-surface stresses were determined using the data that obtained from FEM analysis. Wear prediction of the different textured-shapes on the plate surface of UHMWPE was also investigated in this study using contact pressure value that was resulted from the simulation.

## 2. Eksperimental Method

This present work was to continue our previous research [21], which was mainly focused on analyzing the surface of the circle shaped-texture under static normal load. In the present study, contact simulations by FEM performed on the various shapes of texture under a normal load and rolling motion. In order to understand the effect of surface texture during contact, the FEM software ABAQUS version 6.12 was used to analyze the contact condition on the dimple region during static and rolling motion. Figure 1 shows the geometric contact of two bodies observed during the study. In this research, the flat surface body was a presumable elastic body which has a very low modulus of elasticity compared to the high modulus elasticity of the sphere with radius R. Eventually a normal load F subjected to the flat surface yielded a contact pressure.

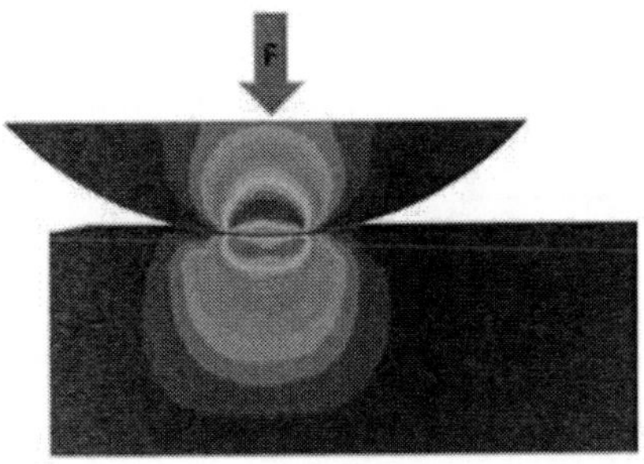


**Figure 1.** The geometric contact between spherical steel ball and flat UHMWPE.

Five different shaped-texture was generated on the surface material of UHMWPE (circle, ellipse, triangle, square, and chevron), which was represented as a flat bottom profile. The geometry of each texture pattern is depicted in Figure 2.

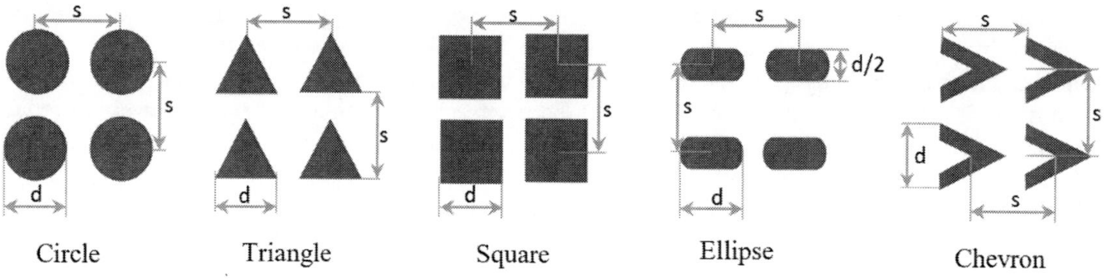


**Figure 2.** The geometrical model of textured surface with different shapes.

Material properties of steel and UHMWPE were selected to simulate the surface contact of between hard and soft material. Detail properties of contact surface are presented as follows:

Friction coefficient μ = 0.1
Young's modulus of steel $E_1$ = 210.000 MPa
Young's modulus of UHMWPE $E_2$ = 945 MPa
Poisson's ratio of steel ($v_1$) = 0.29
Poisson's ratio of UHMWPE ($v_2$) = 0.46

The steel ball as the indenter has a diameter of 10 mm, while the flat UHMWPE as the slave surface has the square dimensions of 6 x 6 mm. All shapes have the depth of grooves and separation, and it was set to 0.5 and 1 mm, respectively. The normal load applied on the indenter varied from 17, 30, and 50 N.

## 3. Results and Discussion

FEM analysis of the rolling contact between a rigid, smooth steel ball and a deformable the surface textures of UHMPWE provided the resultant stress applied on the surface. The von Mises stress distribution in the subsurface of the surface textures was determined under static and rolling contact. Plots of the resultant stress results are given in Figures 3 and 4, respectively. The texturing effect for stress analysis was made by the variation of texture shapes. In the presence of static contact under all applied load, it shows that the ellipse shape texture experienced the lowest von Mises stress distribution. However, the highest von Mises stress may be experienced by the surface texture with a square shape at the load of 30 and 50 N, triangle shape at the load of 17 N.

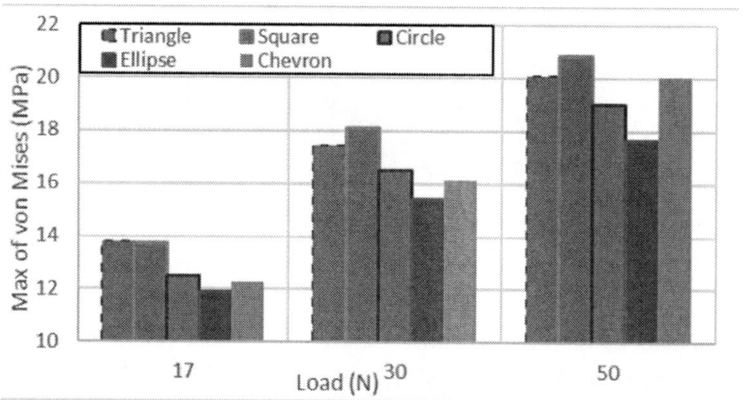


**Figure 3.** The von Mises stress distribution on the contact surface as a function of different static load

Differences in results of stress distribution occurred in the case of rolling contact was noted. The simulation procedures performed on the rolling contact of the textured surface under a normal load, shifted motion, rotation of the steel ball. The rolling contact simulation resulted in deformation and stress on the textured surface where deformation occurred in the contact area, while the stress distributed on the cavities of the texture. In the rolling contact, von Mises stress values are undergoing in the different shaped-textures (Figure 4). It shows that the surface texture of square shape experienced the highest value of von Mises stress. Moreover, for all normal load, the lowest value of von Mises stress distribution may be subject to the surface texture with the geometry of circle, chevron, and ellipse. Under rolling contact at 17 N load, the lowest von Mises stress goes to the circle shape texture, while chevron shape has the lowest von Mises stress at the load of 30 N. The increasing load to 50 N lead the texture of the ellipse to experience the lowest von Mises stress distribution.

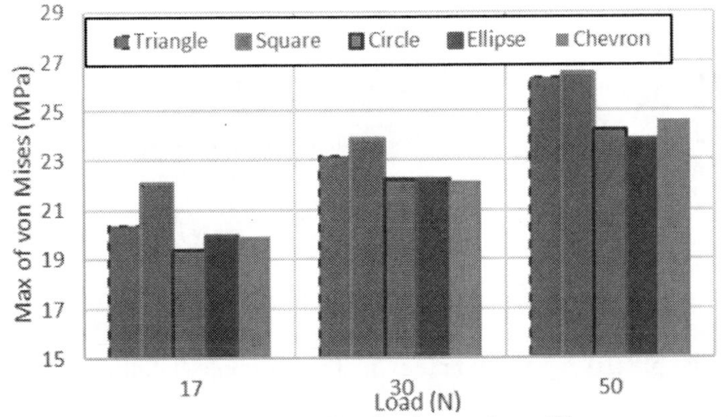


**Figure 4.** The von Mises stress distribution at the rolling contact surface as a function of different load.

The plot of the 3D contour of von Mises stress for each of surface textures in the contact rolling is shown in Figure 5. It can be seen that the concentrated high-stress fields near the surface are in the middle of the flat disk and decreased in magnitude and expand along the edge. The red color on the contours indicates the stress concentration images, which show the highest value of von Mises stress distribution. Accordingly, the stress concentration occurs when the ball contacts on the texture areas.

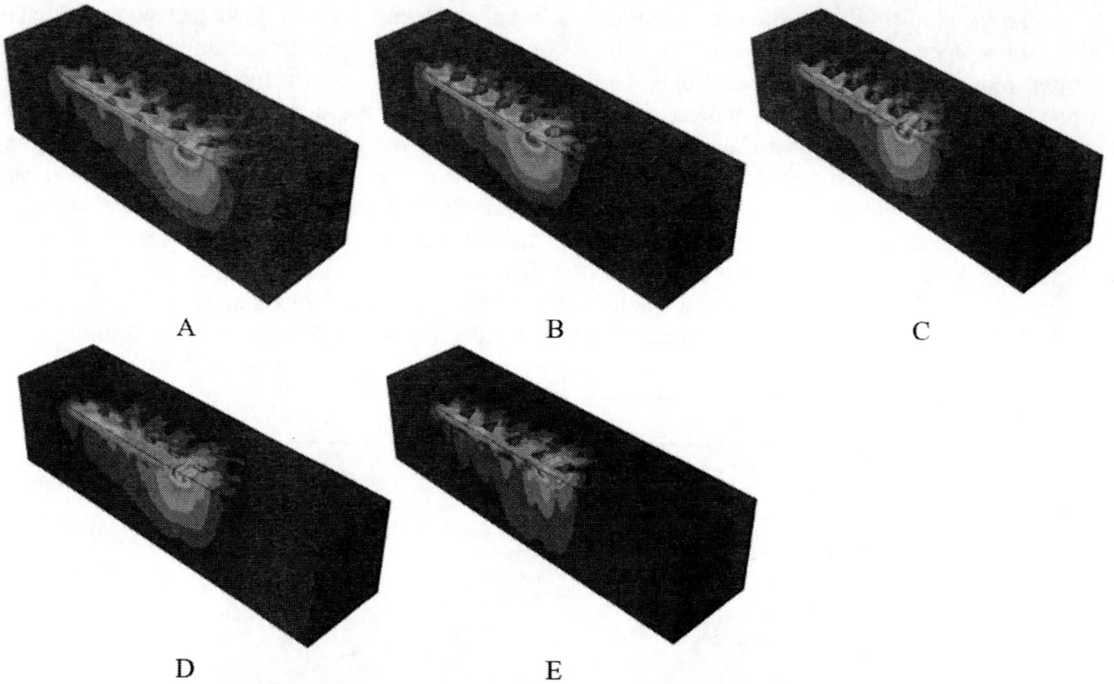


**Figure 5.** The three- dimensional von Mises stress distribution of the rolling contact for chevron dimple (A), circle dimple (B), square dimple (C), ellipse dimple (D), triangle dimple (E) at the load of 17 N

The results of FEM modeling can be used to analyze the influence of contact pressure on the shaped-texture under the static normal load of 17, 30, and 50 N (Figure 6). It shows that all shapes of texture have the similar trend of the contact pressure for all applied loads. Trend charts show straight line at the time of initial penetration pressure area, which cannot be distributed to the cavities. The chart shows increasingly contact pressure for distributed load and form a wider contact area with increasing load. There are no significant differences in the contact pressure on the shape of the texture, but the loading 17 N made the texture shape of a triangle having the highest value of the contact pressure. Moreover, in the imposition of 30 and 50N, the highest contact pressure goes to the square texture shape.

The contact pressure can be used as a prediction of premature wear, where the higher the value of contact pressure, the higher wear rates occur. The contact pressure resulting from the FEM simulation in this study are then used as input to calculate the wear. Here the average of contact pressure was obtained by adding the contact pressure on each node and then divided by the number of nodes in contact. The average value of the contact pressure obtained from simulations providing almost the same for all textures. The average value of the contact pressure at the highest loading of 17N is undergone by the circle shape (16.737 MPa), while the respective loading of 30 and 50 N in the square shaped-texture yielded the contact pressure of 19.846 and 22.989 MPa, respectively. The texture shape that has the largest contact pressure value allows for the wear is higher than in any other shape.

Based on simulation results, the different shaped-texture have insignificant values of von Mises stress and contact pressure for all operating conditions. This is due to the absence of differences in diameter and separation between cavities. The previous study [22] showed that the greater diameter and separation provided the lower von Mises stress in the texture of circle. Lopez work [14] also demonstrated that the von Mises stress and the contact pressure value are lower when the surface of the textured UHMWPE is under wet condition. Moreover, the texture of rectangular under the condition of optimum area density, radius, and depth shown capable of providing the maximum load-carrying capacity as proposed by Zhang *et al.* [20].

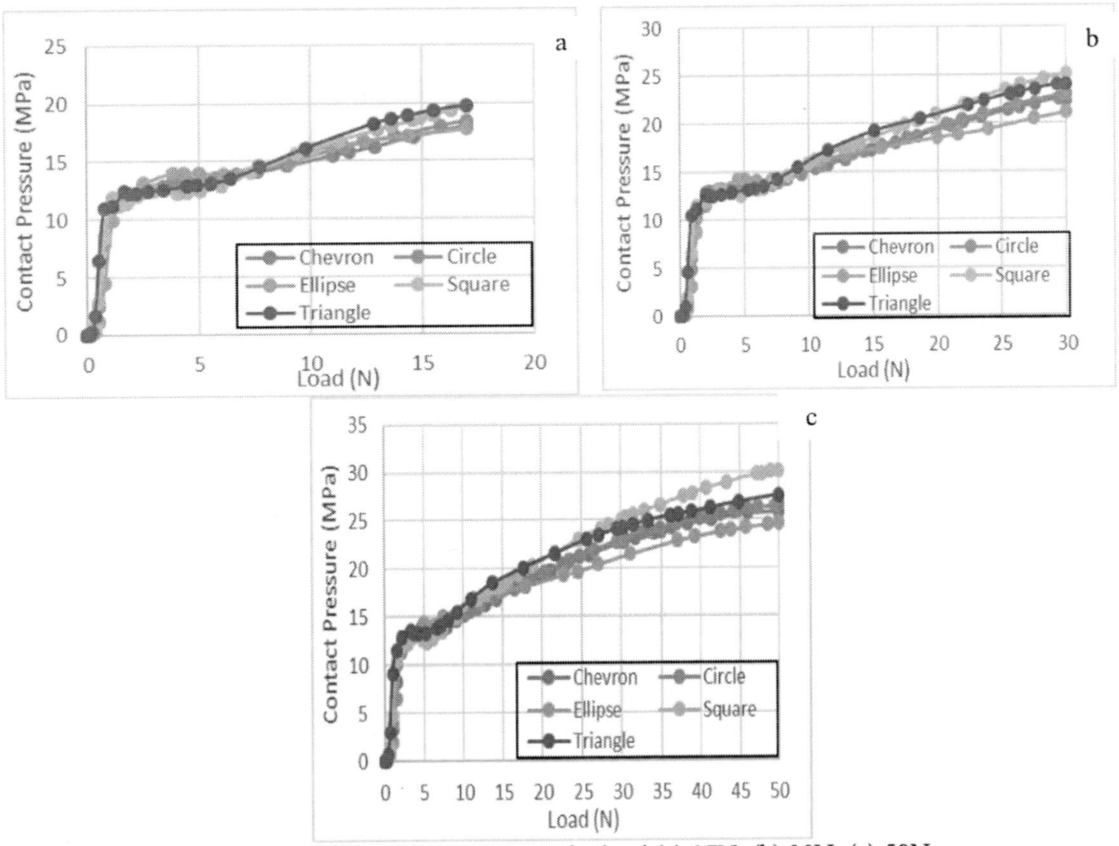


**Figure 6.** The contact pressure at the load (a) 17N, (b) 30N, (c) 50N

## 4. Conclusion

Operating conditions greatly affect the von Mises stress distribution on the surface texture of UHMWPE. In the static contact loading, the ellipse shape texture may reduce the stress distribution to a minimum value compared to the other forms. Different values of stress in the rolling contact condition are noted. The lowest value of von Mises stress distribution may reach through the texture of circle shape at the load of 17 N, chevron at the load of 30 N, and the shape of the ellipse at the load of 50N. This indicated that the surface texture of the ellipse is potential for improving wear properties. Based on the wear calculating using the average of contact pressure value with FEM, texture with shape circle and square may also be used for the surface engineering of bearing materials.

## 5. References

[1] Baena J C, Wu J and Peng Z 2015 *Lubricants* 3, 413 – 436.
[2] Antoszewski B 2009 *Hidraulica* 23, 10 – 15.
[3] Samanek O, Zimmerman M, Svoboda P, Krupka I, Vrbka M 2010 *J Eng. Mech* 17 (1), 27 – 36.
[4] Han J, Fang L, Sun J, Shirong G 2010 *Tribol T* 53 (6), 860 – 870.
[5] Ripoll M R, Podgornik B, Vizintin J 2011 *Wear* 271, 952 – 959.
[6] Burnstein L, Ingman D 1999 *J Tribol* 121, 927 – 932.
[7] Pascovici M D, Marian V, Gaman D 2004 *Friction and Biotribology. In : Proceedings of The International Nanotribology Conference Nano Sikkim II. Peeling, Sikkim, India.*
[8] Etsion I 2005 *J Tribol* 127 (1), 248 – 253.
[9] Kligerman Y, Etsion I, Shinkarenko A 2005 *J Tribol* 127, 632 – 638.
[10] Zhang B, Huang W and Wang X 2012 *Proc Inst Mech Eng Part J: J Eng Tribol* 226, 705 – 713.
[11] Zhang B, Huang W, Wang J, Wang X 2013 *Tribol. Int.* 65, 138 – 145.

IOP Conf. Series: Materials Science and Engineering **202** (2017) 012098 doi:10.1088/1757-899X/202/1/012098

[12] Young S K, Lotito M A, Keller T S 1998 *Wear* 222, 29 -37.
[13] Kustandi T S, Choo J H, Low H Y and Sinha S K 2010 *J. Phys D: Appl. Phys.* 43, 015301 (6pp).
[14] Lopez-cervantes A, Dominguez-lopez I, Barceinas-sanchez J D O, Lesso-arroyo R, Garcia-miranda J S, Garcia-garcia A L 2012 *XVIII Congr. Int. Anu. la SOMIM* 305 – 312.
[15] Lopez-cervantes A, Dominguez-lopez I, Barceinas-sanchez J D O, Garcia-garcia A L 2013 *Journal of The Mechanical Behavior of Biomedical Materials* 20, 45 – 53.
[16] Fang H W, Hsu S M, Sengers J V 2003 *Mater. Sci. Eng. Technol* 34, 976 – 988.
[17] Fang H W, Hsu S M, Sengers J V 2003 *J. Biomed. Mater. Res. B Appl. Biomater* 67B, 741 – 749.
[18] Fang H W, Hsu S M, Sengers J V 2003 *National Institute of Standars and Technology Special Publication* vol. 1002.
[19] Fang H W, Su Y C, Huang C H, Yang C B 2006 *Mater. Chem. Phys.* 95, 280 – 288.
[20] Zhang Y L, Zhang X G, Matsoukas G 2015 *Biosurface and Biotribology* 1, 270 – 277.
[21] Jamari J, Ismail R, Anwar I B, Saputra E, Tauviqirrahman M, and Heide E V D 2016 *AIP Conf. Proc. 1725*, vol. 020030, p. 020030.
[22] Lestari W D, Priyoga F H, Jamari J, Bayuseno A P 2016 *Mater. Sci. Forum* 864, 8 – 12.

**Acknowledgements**
The authors would like to acknowledge the Directorate of Higher Education, Research, and Technology, (DIKTI) Indonesia for the financial supports through PMDSU scholarship.

# Risk Analysis using Corrosion Rate Parameter on Gas Transmission Pipeline

**B Sasikirono[1], G D Haryadi[1], A Huda[1] and S J Kim[2]**

[1]Department of Mechanical Engineering, Diponegoro University, Indonesia

[2]Department of Mechanical & Automotive Engineering of Pukyong National University, South Korea

Email : bambang.sasikirono@gmail.com

**Abstract.** In the oil and gas industry, the pipeline is a major component in the transmission and distribution process of oil and gas. Oil and gas distribution process sometimes performed past the pipeline across the various types of environmental conditions. Therefore, in the transmission and distribution process of oil and gas, a pipeline should operate safely so that it does not harm the surrounding environment. Corrosion is still a major cause of failure in some components of the equipment in a production facility. In pipeline systems, corrosion can cause failures in the wall and damage to the pipeline. Therefore it takes care and periodic inspections or checks on the pipeline system. Every production facility in an industry has a level of risk for damage which is a result of the opportunities and consequences of damage caused. The purpose of this research is to analyze the level of risk of 20-inch Natural Gas Transmission Pipeline using Risk-based inspection semi-quantitative based on API 581 associated with the likelihood of failure and the consequences of the failure of a component of the equipment. Then the result is used to determine the next inspection plans. Nine pipeline components were observed, such as a straight pipes inlet, connection tee, and straight pipes outlet. The risk assessment level of the nine pipeline's components is presented in a risk matrix. The risk level of components is examined at medium risk levels. The failure mechanism that is used in this research is the mechanism of thinning. Based on the results of corrosion rate calculation, remaining pipeline components age can be obtained, so the remaining lifetime of pipeline components are known. The calculation of remaining lifetime obtained and the results vary for each component. Next step is planning the inspection of pipeline components by NDT external methods.

**Keywords :** Risk analysis, corrosion, gas transmission, pipeline

## 1. Introduction

Pipe is a major component in the transmission and distribution process of oil and gas [1-2]. Pipeline system that used to distribute oil and gas is a series of pipes that used to transport the fluids with a certain distance. Generally, the fluids have hazardous properties such as flammable and toxic. This condition would lead to increment of the risk if pipeline system experienced failure or leakage [3]. The interaction between walls of pipe in the form of metal with its environment will cause the occurrence of corrosion, the corrosion which can be one of the causes of leakage on pipes. The installation process and the different environmental conditions of the different piping network also affect the occurrence of the corrosion. The corrosion process occurs naturally and can not be prevented entirely, often take

IOP Conf. Series: Materials Science and Engineering **202** (2017) 012099 doi:10.1088/1757-899X/202/1/012099

place suddenly and outside the prediction that has been planned. The presence of corrosion in oil and gas industry led to high impacts from various aspects, such as environmental pollution, disruption of production processes due to the process of the replacement of the affected pipeline corrosion, and increment of the operating costs.

Therefore periodic inspections or checks on the pipeline system are needed. Every production facility in an industry has a level of risk for damage which is a result of probabilities and consequences of damage. The level of risk in the production facility should be at low levels, because when a production facility has a high level of risk, it will be very dangerous and may affect the safety, environment, and operational costs as well. A method that can be used for inspection of pipeline systems is to run a risk assessment [2]. Risk assessment is a systematic method to determine the level of risk. This method is defined as a whole series of identification process and damage estimation of risks such as likelihood, exposure, consequences, and safety level assessment. Risk assessment process was undertaken to identify the adverse likelihood that could possibly endanger human health, environment, production process, as well as equipment due to human activities and technology. The first-step method makes identification of hazards and impacts from such danger as well as anyone and anything that would be affected by such risks. The next step, namely to determine the frequency of the occurrence or likelihood of occurrence of such hazards and how often such events can occur, and the last step is performing risk evaluation. Risk assessment on pipeline system is done to evaluate the damaging impact of the pipeline to the community and to identify how to resolve risks more effectively.

Risk-based inspection is a method which is closely related to the discussion of risk assessment. This method uses risk as the basis to prioritize and manage inspection programs. The methodology of risk-based inspection allows the optimization of inspection and maintenance resources on the areas that have a higher risk. In addition, risk-based Inspection is an optimized combination of equipment inspection method, inspection scope, and frequency. The purpose of the Risk-based inspection is listed below [3-5].

a. Identify the area that is included in the high-risk classification.
b. Estimate the magnitude of the value of risk that exist on any equipment while operating based on a consistent methodology.
c. Prioritizing equipment based on the measurement of the enormity of risks.
d. Plan and design appropriate inspection programs.
e. Overcome risks systematically when a failure occurs.

The risk-based inspection will not eliminate the risk, the likelihood of the occurrence of an adverse incident (probability) and the impact of the incident (consequences) of equipment will always be there. Risk-based inspection is useful to assist and control the risks to a level that can still be received with prioritizing resources to the equipment that is known to have a higher risk [5-7]. American Petroleum Institute is an international organization that issued the guidelines for conducting the procedure of risk-based inspection. This guideline is in the form of codes and for pipeline system, codes API 570, API 580, and API 581 are used. Based on this codes, risk analysis of 20-inch natural gas transmission pipeline have been performed.

## 2. Experimental Method

### *2.1 Risk based inspection*

Risk-based inspection , are nowadays in a well-advanced stage of application and make a well-established part of modern practice. Risk-based inspection involves the programming of an inspection on the basis of information obtained from a risk assessment. Risk-based inspection allows people to view potential hazards that simultaneously accounts for both likelihood and consequences of an event. Risk based inspection is a systematic tool that helps users make informed business decisions regarding inspection and maintenance spending [5].

Risk-based inspection is a method of planning or testing and inspection programs and maintenance strategies using risk as a fundamental method. The risk is defined as a function of probability of Failure (PoF), and a function of the Consequences of Failure (CoF) is formulated with the equation below.

$$Risk = CoF \times PoF(t)$$

*2.2 Consequences of Failure*

Analysis of the consequences of failure due to release fluid representative in the semi-quantitative Risk-based inspection method consists of two parts. The consequences are regardless of combustible fluid and consequences from the release of toxic fluid. Analysis of the consequences of API Risk-based inspection assessment performed to aid in establishing ranking items of equipment on the basis of risk. The measures consequences presented are intended to be used to set priorities for inspection program. The main consequence categories are analyzed using different techniques as described below.

a. Flammable and explosive consequences are computed to determine the probability of event combined with computer modeling to determine the magnitude of consequences. The area can be determined based on consequences of a serious injury to personnel and damage to components of the thermal radiation and blast. Financial losses are also determined based on the area affected by the release.
b. Consequences of toxic calculated using a computer model to determine consequences area as a result of overexposure of personnel to toxic concentrations of a vapor cloud. Where flammable liquids and toxic, toxic event probability assumes that if the release ignited, the consequences of toxic ignored (i.e. toxins consumed in the fire). Financial losses are also determined based on the affected area by the release.
c. Non-flammable, non-toxic releases are also considered since it can still result in serious consequences. Consequences from chemical splashes and high-temperature steam burns are determined based on serious injuries to personnel. Physical explosions and BLEVEs can also cause serious personal injuries and component damage.
d. Financial consequences include losses due to business interruption and costs associated with environmental releases. Business interruption consequences are estimated as a function of the results of flammable and non-flammable consequence area. Environmental consequences are determined directly from the mass available for release or from the release rate.

## 3. RBI Case Study on 20 Inch Natural Gas Transmission Pipeline

Risk analysis with Risk-based inspection requires data such as sheets data, design and operational data, and inspection reports that have been done. After required data are collected, then perform risk analysis refers to API 581 Risk-based inspection Semi-Quantitative to find Probability of Failure (PoF) and Consequence of Failure (CoF). Furthermore, PoF and CoF values are combined to obtain the risk level [8-9].

*3.1 Release Rate Analysis*

The first step in analyzing the rate of leakage according to the workbook for semi-quantitative standard API 581 Appendix B is to determine the representative fluid and equipment category (Table 7.2 API 581 BRD) [1,2]. The equipment that is analyzed in this study is described in Table 1.

**Table 1.** Representative Fluid And Equipment Category

| Inventory | Representative Fluid | Inventory Value | |
|---|---|---|---|
| | | Detection | Isolation |
| Automatic Shutdown Valve Straight Pipe KP-0 Ø 20" | C1 – C2 | A | B |
| Automatic Shutdown Valve Straight Pipe KP-20 Ø 20" | C1 – C2 | A | B |
| Manual Shutdown Valve Straight Pipe KP-40 Ø 20" | C1 – C2 | A | C |

Based on the detection system and isolation systems Table 7.6 API 581, the components are categorized as A for detection systems, B, and C for insulation systems. Observation of changes or leakage of the fluid in the pipe visually and in the case of a leak was isolated by automatically and manually operated valve. Based on the detection system and the insulation system which is then adjusted by BRD 581 Table 7.7 API, the estimated duration of leakage shown in Table 2.

**Table 2.** Leak Durations Based on Detection and Isolation Systems

| Hole Size | ¼ inch | 1 inch | 4 inch | 16 inch |
|---|---|---|---|---|
| Leak Duration | 30 minutes | 20 minutes | 10 minutes | 0 |
| | 40 minutes | 30 minutes | 20 minutes | 0 |

Subsequently, calculates leakage rate of the fluid phase of natural gas contained in the pipe. To calculate the rate of fluid leakage by using the equations contained in the API BRD 581. Having the rate of the leak, the next is to calculate the duration of the leak of the amount (capacity) total fluid stored in it. Then analyzed the leak flow types, whether the kind of continuous flow or instantaneous. In accordance with the method of Risk-based inspection , to determine the type of leakage flow, the mass flow out within 3 minutes can be calculated. If within 3 minutes of outgoing mass flow exceeds 10,000 lbs., then the flow is categorized into the instantaneous flow and vice versa. Based on the calculation, the size of the holes ¼ inch and 1 inch is continuous flow while the size of the hole for 4-inch and 16-inch is the instantaneous flow. The last stage of this step is a comparison between the estimated duration of the leak detection system accordingly and isolation systems with a real leak. For instantaneous flow is considered 0 minutes, whereas for continuous flow compared and determined the smallest, which is then used for the duration of the leak. Determining the size of the leak hole that has been determined by API, the mass flow rate out of any hole's state can be determined. The rate of fluid flow out due to leakage for each hole is presented in the following Table 3.

**Table 3.** Rate of fluid flow out due to leakage for each hole

| Inventory | Fluid Phase | Hole Size | Release Rate | Release Type |
|---|---|---|---|---|
| Inlet Automatic Shutdown Valve Straight Pipe KP-0 Ø 20" | Gas | ¼ inch | 366.403197 | Continuous |
| | | 1 inch | 5862.451152 | Continuous |
| | | 4 inch | 93799.21844 | Instantaneous |
| | | 16 inch | 844192.9659 | Instantaneous |
| Inlet Automatic Shutdown Valve Straight Pipe KP-20 Ø 20" | Gas | ¼ inch | 385.3093463 | Continuous |
| | | 1 inch | 6164.94954 | Continuous |
| | | 4 inch | 98639.19264 | Instantaneous |
| | | 16 inch | 887752.7338 | Instantaneous |
| Inlet Manual Shutdown Valve Straight Pipe KP-40 Ø 20" | Gas | ¼ inch | 373.4385391 | Continuous |
| | | 1 inch | 5975.016626 | Continuous |
| | | 4 inch | 95600.26602 | Instantaneous |
| | | 16 inch | 860402.3942 | Instantaneous |

*3.2 Likelihood Analysis*

Analysis of Probability of Failure on the equipment was observed, with a semi-quantitative method Risk-based inspection is done through a process of TMSF (Technical Modules Sub-Factor) for each mechanism of the damage suffered. The mechanism for each damage caused to the equipment can be determined by scanning the operating conditions of the equipment. TSMF used in the damage analysis consists of several mechanisms of damage that can occur from a piece of equipment for the operating conditions and the type of working fluid. In this pipeline equipment only thinning factor that has probably caused by fluid services that are in it in the form of gas.

The thinning rate can be determined from available thickness data (which so far has shown that it contains different kinds of errors and human mistakes) or an alternative technique based upon estimated rates can be used from the API 581 Appendix G. Thinning technical module (Appendix G) includes 'Estimated Corrosion Rate Tables' for different kind of steels and alloys in various acidic and basic environments. API 581 suggests that this information can be used for RBI analysis whenever the potential thinning mechanism is known, and there is not any reliable data from inspections. In this research, enough data are available from thickness measurements of the pipeline. Here, the data is compared with the API 581 thinning rates for RBI analysis. After finding the thinning rate, the fraction of wall loss due to thinning and the number of 'highest effective' inspections will be used to determine the thinning technical module sub-factor. The fraction of wall loss due to thinning is calculated by the formula cited below (see page 9-9 API 581):

$$\text{Fraction of wall loss} = ar/t,$$

Where $a$ is the time (years) equipment age; $r$ the corrosion rate; $t$ the thickness. Based on the analysis, the result of TMSF thinning of inventory shown in Table 4.

**Table 4.** TMSF Thinning

| Inventory | $ar/t$ | Likelihood Category |
|---|---|---|
| Automatic Shutdown Valve Straight Pipe KP-0 Ø 20" | 0.04 | 2 |
| Automatic Shutdown Valve Straight Pipe KP-20 Ø 20" | 0.07 | 1 |
| Manual Shutdown Valve Straight Pipe KP-40 Ø 20" | 0.03 | 2 |

### *3.3 Consequence Analysis*

There are two analyses of the consequences of failure due to the release of a representative fluid in the semi-quantitative method API 581: the consequences of the release of flammable fluid representative and the consequences due to the release of toxic fluid representative.

*3.3.1 Detection and Isolation System.* Type of detection system for cases observed is the type A. Detection systems according to API 581 type *A* system is only performed visual observation to detect leakage of material out if the system exceeds the operating pressure. In the case of the observed type of isolation system, there are type B and C isolation systems according to API 581 type B, and C of this system depends on the isolation valves are operated manually or automatically if there is a leak material. According to API 581 BRD isolation and detection systems are 20% for KP-0 and KP-20 and 10% for a KP-40 reduction in the rate adjustment due to leak fluid.

*3.3.2 Mitigation System.* The next area has been determined as a result of leakage is reduced by mitigation system. Mitigation system conditions on a case observed in the case of leaks is inventory blow down system. The condition of the system is the area due to leakage can be reduced by 25%.
The consequence of failure is determined based on the area of damage or hazard due to toxicity. In this study analyzed the equipment is not toxic in its flow, the area the consequences of failure can use the value of the consequences of the damage. The consequences of the damage value equal to the value consequences of the fires which have been calculated by the analysis of the consequences of failure. The consequences of the fire are determined by the value of the area due to leakage of the area which

consists of extensive fire and hazardous areas. Value consequences of fires and the area of the consequences of the failure of each piece of equipment at Table 6.

**Table 5.** Area of Equipment Damage and Fatalities

| Inventory | Hole Size | Area of Equipment Damage ($ft^2$) | Area of Fatalities ($ft^2$) |
|---|---|---|---|
| Automatic Shutdown Valve Straight Pipe KP-0 Ø 20" | ¼ inch | 5.230.818.235 | 1.332.775.805 |
| | 1 inch | 791.784.942 | 1.908.588.527 |
| | 4 inch | 1.198.518.899 | 2.733.175.734 |
| | 16 inch | 1.032.291.919 | 2.252.891.869 |
| Automatic Shutdown Valve Straight Pipe KP-20 Ø 20" | ¼ inch | 5.495.192.156 | 1.398.728.259 |
| | 1 inch | 8.318.031.598 | 2.003.035.424 |
| | 4 inch | 1.259.093.939 | 2.868.426.975 |
| | 16 inch | 1.084.465.738 | 2.364.376.583 |
| Manual Shutdown Valve Straight Pipe KP-40 Ø 20" | ¼ inch | 5.981.277.376 | 1.519.823.525 |
| | 1 inch | 9.053.812.048 | 2.176.448.001 |
| | 4 inch | 1.370.468.398 | 3.116.760.617 |
| | 16 inch | 1.180.393.065 | 256.907.168 |

**Table 6.** Consequences of Area of Equipment Damage and Fatalities

| Inventory | Hole Size | Flammable Consequences ($ft^2$) | Area of Consequences of Failure ($ft^2$) | Total Area of Failure ($ft^2$) | Failure Consequence Category |
|---|---|---|---|---|---|
| Automatic Shutdown Valve Straight Pipe KP-0 Ø 20" | ¼ inch | 5,230,818.235 | 1,332,775,805 | 7,740,259,148 | D |
| | 1 inch | 791,784,942 | 1,908,588,527 | | |
| | 4 inch | 1,198,518,899 | 2,733,175,734 | | |
| | 16 inch | 1,032,291,919 | 2,252,891,869 | | |
| Automatic Shutdown Valve Straight Pipe KP-20 Ø 20" | ¼ inch | 5,495,192,156 | 1,398,728,259 | 7,740,259,148 | D |
| | 1 inch | 8,318,031,598 | 2,003,035,424 | | |
| | 4 inch | 1,259,093,939 | 2,868,426,975 | | |
| | 16 inch | 1,084,465,738 | 2,364,376,583 | | |
| Manual Shutdown Valve Straight Pipe KP-40 Ø 20" | ¼ inch | 5,981,277,376 | 1,519,823,525 | 7,740,259,148 | D |
| | 1 inch | 9,053,812,048 | 2,176,448,001 | | |
| | 4 inch | 1,370,468,398 | 3,116,760,617 | | |
| | 16 inch | 1,180,393,065 | 256,907,168 | | |

### *3.3.3 Risk Level*

The level of risk on a semi-quantitative method API 581 is a combination of categories the possibility of failure and the consequences of failure categories. Based on the analysis that has been done, then the value of the category of possible failures obtained from the calculation of the value category TMSF and the consequences of failure for each equipment are analyzed at Table 7.

**Table 7.** Risk Level

| Equipment | Likelihood Category | Failure Consequence Category | Risk Level |
|---|---|---|---|
| Automatic Shutdown Valve Straight Pipe KP-0 Ø 20" | 2 | D | Medium Risk |
| Automatic Shutdown Valve Straight Pipe KP-20 Ø 20" | 1 | D | Medium Risk |
| Manual Shutdown Valve Straight Pipe KP-40 Ø 20" | 2 | D | Medium Risk |

The final Risk Ranking is obtained by considering the probability of failure rating (1 to 5) on the Y-axis and the consequence rating (A to E) on the X-axis of the Risk Matrix. The risk rating is shown below in Figure 1.

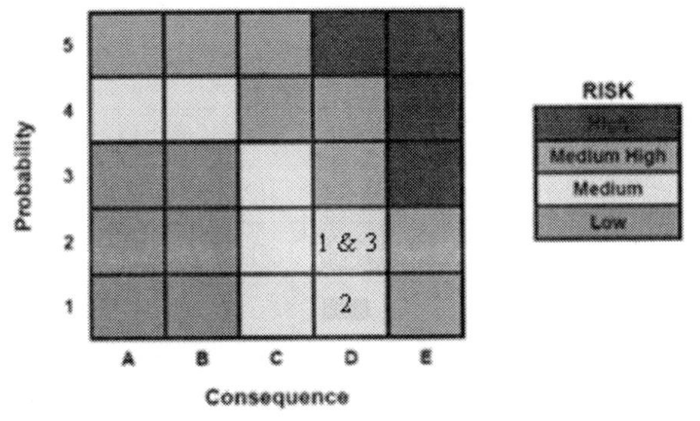


**Figure 1.** Risk Matrix

1. Automatic Shutdown Valve Straight Pipe KP-0 Ø 20"
2. Automatic Shutdown Valve Straight Pipe KP-20 Ø 20"
3. Inlet Manual Shutdown Valve Straight Pipe KP-40 Ø 20"

*3.4 Remaining Life Analysis*

Pipes are designed to flow the fluid with a certain pressure, the wall thickness of the pipe is determined based the pressure of working fluid therein. Besides the stress factors work in it, the pipe wall is also designed to accommodate the corrosion process when the pipeline is operated. Corrosion can cause thinning of the pipe wall, thinning process on the pipe will be directly proportional to the time. Therefore it needs to be regularly monitored in order to keep the pipeline wall thickness was observed. The wall thickness of the pipe which has thinning process can not be tolerated anymore if it has reached the minimum thickness of the pipe according to the calculation based on the otherwise require a minimum thickness of the pipe to be able to withstand the working pressure.

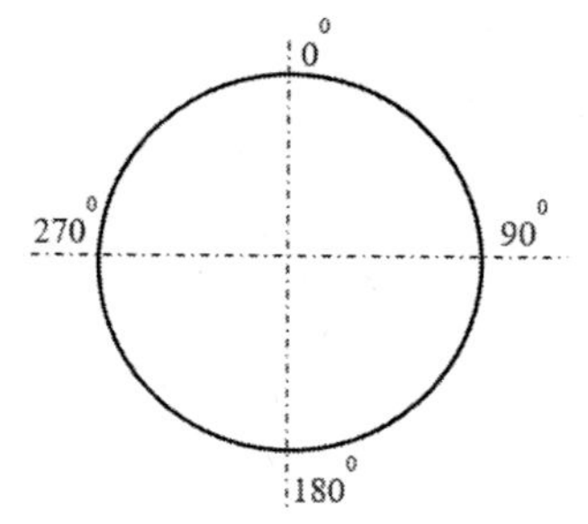


**Figure 2.** Position of Wall Thickness Inspection

**Table 8.** Wall Thickness of Pipe

| Location | Point | Thickness (inch) | | | | |
|---|---|---|---|---|---|---|
| | | Position (°) | | | | average |
| | | 0 | 90 | 180 | 270 | |
| Pipe on Automatic Shutdown Valve KP-0 Ø 20 Inch | 1 | 0.52 | 0.52 | 0.54 | 0.53 | 0.5275 |
| | 2 | 0.51 | 0.53 | 0.52 | 0.51 | 0.5175 |
| | 3 | 0.52 | 0.53 | 0.51 | 0.55 | 0.5275 |
| | 4 | 0.53 | 0.55 | 0.57 | 0.58 | 0.5425 |
| Pipe on Automatic Shutdown Valve KP-20 Ø 20 Inch | 1 | 0.54 | 0.53 | 0.56 | 0.52 | 0.5375 |
| | 2 | 0.52 | 0.51 | 0.59 | 0.52 | 0.5350 |
| | 3 | 0.54 | 0.52 | 0.55 | 0.51 | 0.5300 |
| | 4 | 0.53 | 0.52 | 0.55 | 0.52 | 0.5300 |
| Pipe on Manual Shutdown Valve KP-40 Ø 20 Inch | 1 | 0.55 | 0.54 | 0.55 | 0.56 | 0.5500 |
| | 2 | 0.52 | 0.53 | 0.53 | 0.51 | 0.5225 |
| | 3 | 0.56 | 0.57 | 0.56 | 0.56 | 0.5625 |
| | 4 | 0.54 | 0.52 | 0.52 | 0.53 | 0.5275 |

Periodic monitoring is performed to determine the thickness of the pipe at a certain time and also to determine the projected thickness of the pipe in the future. Projection thickness of the pipe can be use for predicting remaining life of the pipe.

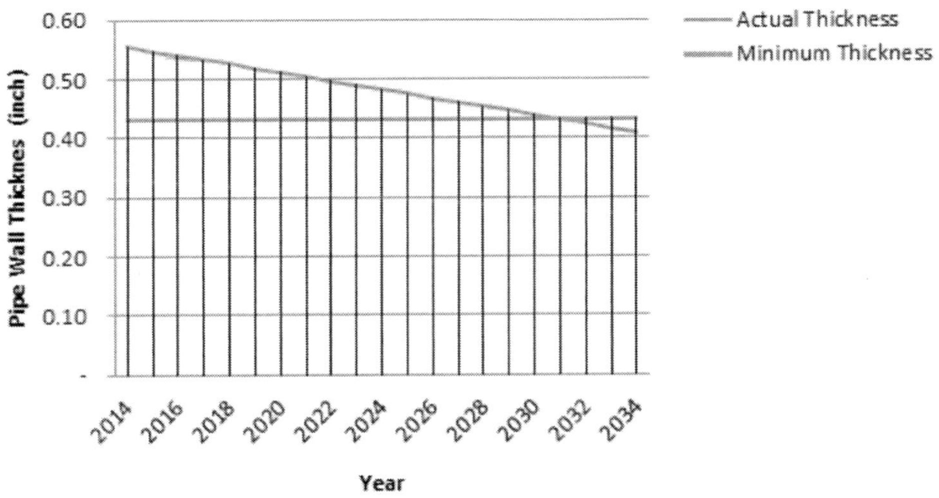


**Figure 3.** Remaining Life Prediction Charts

The minimum thickness of the pipe is calculated base on the requirement of thickness pipe to resist working pressure of the fluid. Refer to the basic design of this pipe; minimum thickness requirement is 0.432 inch. Base on the data inspection of actual pipe wall thickness, prediction of remaining life the pipe until 2032.

*3.5 Risk Evaluation*

Risk evaluation carried out aimed at reducing the risk to the optimization of the risk assessment events. In this context, it is also necessary that the optimal inspection programs to be performed are: Risk ranking, Risk reduction, and Optimization of inspection activities [8-10].

From the analysis that has been done, recommendations for lowering the risk are described as follow. Necessary maintenance and periodic inspections at intervals that are not too long. Need to be re-evaluated equipment operating conditions, the material conditions of the equipment, mitigation system, and still consider economic factors. Need inspection activities more effectively using a risk-based approach to the equipment being analyzed.

## 4. Conclusion

Based on the results of analysis using semi-quantitative method of API 581, the risk level of each pipeline equipment are analyzed: pipeline at KP-0 has a medium-high risk (2D), pipeline at gas station KP-20 has a medium risk (1D), and pipeline at gas station KP-40 has a medium risk (2D). The results of the risk assessment using the semi-quantitative method of standard API 581 based on the existing equipment at medium risk. Facts on the ground there is no critical problem in the equipment components. Damage mechanisms were prominent throughout the equipment is thinning mechanism. Evaluation of the risk approach is done with the aim of reducing risk by optimizing the risk assessment activities.

## 5. References

[1] American Petroleum Institute 2000 *Risk-based inspection, Base Resource Document API 581* API Publishing Services Washington DC

[2] American Petroleum Institute 2008 *Recommended Practice 581: Risk-Based Inspection Technology* API Publishing Services Washington DC

[3] American Petroleum Institute 2002 *Inspection, Repair, Alteration, and Rerating of In-service Piping Systems*. API Publishing Services Washington DC

[4] Perumal K E 2014 Corrosion Risk Analysis, Risk based inspection and a Case Study

Concerning a Condensate Pipeline *Procedia Engineering* **86**: 597–605
[5] Peters R W 2015 Understanding Risk-Based Maintenance by Using Risked-Based Planning with Risk-Based Inspections. *Reliable Maintenance Planning, Estimating, and Scheduling* **151**(151): 223–240
[6] Satmoko M E A et al 2015 Risk Assessment on Gas Piping Against Corrosion Using a Risk based inspection API 581 *Science and Engineering National Seminar* **1**(1): 64–71.
[7] Tien S W, Hwang W T dan Tsai C H 2007 Study of a Risk-Based Piping Inspection Guideline System. *ISA Transactions* **46**(1): 119–126
[8] Santosh et al 2006 Reliability Analysis of Pipelines Carrying H2S for Risk based inspection of Heavy Water Plants. *Reliability Engineering and System Safety* **91**(2): 163–170
[9] Jusoh I 1998 Onshore Pipeline Risk and Consequence Assesment. *Journal of Mechanical Engineering*(1): 1-17
[10] Seo J K et al 2015 A Risk based inspection Planning Method for Corroded Subsea Pipelines *Ocean Engineering* 109: 539–552

**Acknowledgments**
The author would like to acknowledge input and support from colleagues in EAEM Reliability Analysis Laboratory, Department of Mechanical Engineering, Diponegoro University, Semarang and Department of Mechanical & Automotive Engineering of Pukyong National University, South Korea.

IOP Conf. Series: Materials Science and Engineering **202** (2017) 012100 doi:10.1088/1757-899X/202/1/012100

# Vector Analysis of Ionic Collision on $CaCO_3$ Precipitation Based on Vibration Time History

**W Mangestiyono[1], S Muryanto[2], J Jamari[3] and A P Bayuseno[3]**

[1] Mechanical Engineering, Vocational Education, Diponegoro University, Semarang 50275, Indonesia
[2] Department of Chemical Engineering, UNTAG University, Semarang 50233, Indonesia
[3] Graduate Program, Mechanical Engineering, Diponegoro University, Semarang 50275, Indonesia

Email: rhombo_hedral@yahoo.co.id

**Abstract.** Vibration effects on the piping system can result from the internal factor of fluid or the external factor of the mechanical equipment operation. As the pipe vibrated, the precipitation process of $CaCO_3$ on the inner pipe could be affected. In the previous research, the effect of vibration on $CaCO_3$ precipitation in piping system was clearly verified. This increased the deposition rate and decreased the induction time. However, the mechanism of vibration control in $CaCO_3$ precipitation process as the presence of vibration has not been recognized yet. In the present research, the mechanism of vibration affecting the $CaCO_3$ precipitation was investigated through vector analysis of ionic collision. The ionic vector force was calculated based on the amount of the activation energy and the vibration force was calculated based on the vibration sensor data. The vector resultant of ionic collision based on the vibration time history was analyzed to prove that vibration brings ionic collision randomly to the planar horizontal direction and its collision model was suspected as the cause of the increasing deposition rate.

**Keywords:** $CaCO_3$; vector; ionic collision; vibration

## 1. Introduction

Mechanical vibration is often encountered in the piping system industries when the pipe transports some fluid from a point to other points during the industrial processes. The vibration can be caused by internal and external factors. The internal factor is usually generated by the flowing fluid [1-5] which is often called as fluid-induced vibration [1,5-6]. Moreover, the external factor can be caused by the operation of mechanical equipment such as an electric pump, blower, extruder, compressor [7] diesel engine, and a turbine which is called as external excitation [6-7].

In a certain condition, the internal and external factors can vibrate the piping system simultaneously and produce some forces, which have been investigated previously by Morison [8-9]. In the Morison equation, the relationships among the inertia force, drag force, and excitation force obviously described [7,10] the strength of the piping system structure to function without a rupture [7,10]. This Morison equation can still be applied by researchers for the time being [9] but other researchers have a different direction in particular crystallization and scaling process fields. As the fluid in the pipes is vibrated by internal and external forces, and affected by physical and chemical factors, it brings the curiosity to discover their influences in the precipitation process.

The influences of vibration on $CaCO_3$ scale formation in piping system include increasing the deposition rate, decreasing the induction time, and controlling the crystal polymorph and crystal phases [11]. To justify the origin of such phenomena, many factors are suspected as the mechanism of the vibration in affecting the precipitation. First, vibration force can promote the ionic collision through increasing ionic force when the ions collide with others. Second, the vibration might change the ionic displacement as vibration direction.

The collision of $CaCO_3$ ions during the precipitation process of the vibrating system may occur in a different manner since the hydrodynamic force affects the system [11]. Drag force, inertia force as part of the flowing force and excitation force as a function of vibration force [9-11] take in as the primary part of the precipitation process. The system will be different in the strength of the ionic force and the direction of the ionic displacement. An appropriate method to investigate this phenomenon is using vector analysis of ionic collision based on Vibration Time History (VTH). Furthermore, the present research was undertaken to understand the mechanism of deposition rate in $CaCO_3$ precipitation due to the presence of mechanical vibration.

## 2. Experimental Method

### *2.1 Experimental design*

The concept of the experimental design was here described firstly as guidance in the performance of the research. The calculated deposition rate was used to derive the activation energy (Ea) as suggested by Arrhenius [12] according to the Equation (1) below.

$$k = A\ \exp.[-Ea/RT] \qquad (1)$$

Where k is deposition rate (M.s-1); A is the frequency of effective collision; exp is an exponential number; Ea is activation energy (kJ/mol); R is ideal gas constant, and T is the temperature in K. Furthermore, the activation energy found was used to derive ionic force by dividing the amount by the number of one mol $CaCO_3$ ion, and by the time needed to produce one mol $CaCO_3$. The vibration force inserted to the system as also calculated in the N/ion unit based on the sensor data of the experiment. The Ionic shape was defined as spherical or nearly spherical as it had been investigated by Marcus [13] even for (Ca2+) or (CO32-) ion. Usually, the ionic collision may occur between monoatomic ions, diatomic ions, triatomic ions and polyatomic ions [14]. The collision also can occur between ion-atom, atom-atom or atom-molecule [15]. Even in this research, collision was investigated between ion-ion since the material had been diluted in the solvent. The Ionic collision was defined as a perfect bouncing collision in which the energy of the collision was bounced totally [12]. Four arrangement of ionic vector direction was chosen as the representation of randomness displacement i.e. by the angle of 900; 600; 300 and 00 to the horizontal axis. The resultant between ionic and vibration forces was calculated step by step in accordance with the vibration direction as it had been recorded through Vibration Time History until the ion direction is the same as the vibration direction.

### *2.2 Material*

$CaCO_3$ precipitation was carried out experimentally by mixing $CaCl_2$ and $Na_2CO_3$ solutions which were supplied by Merck® according to the reaction in equation (2).

$$CaCl_2\,(aq) + Na_2CO_3\,(aq) \rightarrow CaCO_3\,(s) + 2NaCl\,(aq) \qquad (2)$$

The concentration of calcium in the solution was determined to be 3.500 ppm, prepared by dissolving 77.175 gr of $CaCl_2$ into 6,000 mL of demineralized water. Sodium solution was also prepared by dissolving 55.65 gr of $Na_2CO_3$ into the 6,000 mL demineralized water. The solution was subsequently filtrated two times by 0.22 μm micropore® paper to ensure that the unwanted particles remained in the solution. Each of the solution was then stored in a covered vessel to avoid dust and other dirty materials. The chosen vessel was made from the material that was considered not to react with the solution and it must be strong enough when the solution was heated to 318 K.

*2.3 Experimental rig*

A built in-house experimental rig equipped with a computerized program control was employed to conduct the experiment as shown schematically in Figure 1. The rig consisted of two vessels (1) and (2) in which contained $CaCl_2$ and $Na_2CO_3$ solution respectively. A stirrer was attached in each vessel under a computer program control and automatically set at 30 rpm to ensure the homogeneous solution. All solution in each vessel was pumped by CHEM FEED Ca-92683 dosing pump (5) at a flow rate of 30 mL/min, and both solutions met in the test pipe section (6). The temperature was set by an electrical heater (3), which was immersed in the solution and controlled by a sensor (4) to keep it stable. When the various temperatures were set, a computer program sent a command to the pump to delay the operation for a couple of minutes until the right temperature condition was achieved.

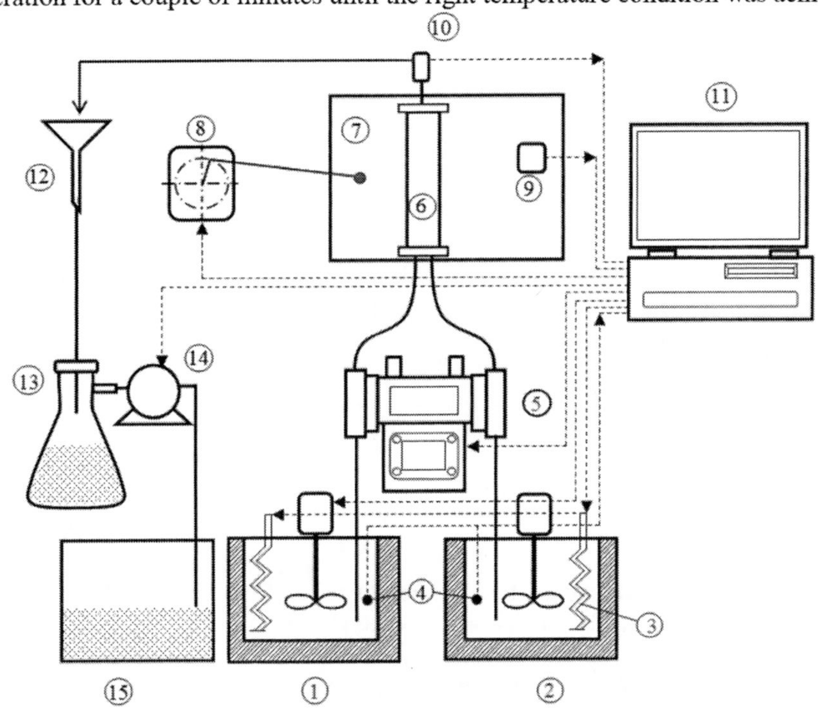


**Figure 1**. Experimental rig

In the test pipe section, four pairs of coupons were inserted in which a scale investigation was needed to form. The pipe was attached to the table (7) and vibrated mechanically by an electrical motor (8) which was set by a computer program (11). The vibration was operated at the frequency of 0.00; 4.00 and 8.00 Hz. Lutron VT-8204 Vibration Meter (9) was selected to record the vibration parameter such as peak to peak displacement (m), velocity (m/sec-RMS) and acceleration (m/sec$^2$-RMS), and the results were shown by the computer program.

The solution conductivity was measured by a TDS meter supplied by Komatsu (10) after leaving the coupon, and the data were recorded by a computer program (11). A filter paper of 0.22 μm (12) was used to screen the crystals brought by the flowing solution. A flask tube (13) and a vacuum pump

(14) were employed to help the capture of those crystals. Finally, the solution was sent to the waste vessel (15).

*2.4 Deposition rate*
$CaCO_3$ deposition rate in the flows and vibration system was described in the previous research [11] as the modification of the formula given by Sheikholeslami [16] as follows:

$$W_{net} = W_{precipitation} + W_{strainer} \quad (3)$$

Where $W_{net}$ is net deposition rate (gr/hr), $W_{precipitation}$ is deposition rate measured in coupon (g/hr) and $W_{strainer}$ is deposition rate measured in the strainer (g/hr). The deposition rate was measured in three various temperatures, i.e. 298, 308 and 318 K, and also three vibration frequency of 0.00, 4.00 and 8.00 Hz. The crystals found in the coupon and in the strainer were dried in an electrical heater at 60 °C for six hours. The weighing was done using Ohaus AR-2140 analytical balance when the crystals were still attached in the coupon or strainer. Crystals were kept in a dry cup after being removed from the coupon and strainer. The crystal mass of the precipitation was found by calculating the difference between the coupon mass with crystals and the coupon mass without crystal. Meanwhile, the crystal mass of the strainer was found by calculating the difference between strainer mass with crystals and the strainer mass without crystal.

## 3. Results and Discussion

*3.1. Deposition*
After running for 4 hours, the deposition rate was determined by the equation (3) for non-vibrated and vibrated system. The results of scale deposition rate calculation are given in Table 1. It shows that

**Table 1**. $CaCO_3$ deposition rate in various vibration frequency and temperatures (gr/hr)

| Frequency (Hz) | 298 K | 308 K | 318 K |
|---|---|---|---|
| 0.00 | 5.2157 | 6.3471 | 7.8786 |
| 4.00 | 6.1614 | 7.1799 | 8.3942 |
| 8.00 | 8.5598 | 9.4510 | 10.5422 |

the increase of the vibration frequency and temperature developed the $CaCO_3$ deposition rate. The influence of temperature on the deposition rate had been investigated by previous researchers [17-21] and is in the close agreement with this research in term of deposition rate at the frequency of 0.00 Hz. These deposition rates were subsequently used to calculate the activation energy.

*3.2. Vector force fields*
The vector component in this system could be identified in three forces, i.e. ionic force, vibration force and flowing force. Drag and inertia forces determined the flowing force to be 0.00209 x $10^{-23}$ N. These two forces were difficult to be drawn in a vector graph since their values were approximately 1/10,000 from other vectors. If drag and inertia forces were drawn in a vector system, their magnitude was atomic and therefore it would be better if these vectors were neglected.

The ionic collision force was calculated based on the activation energy of the reaction. On the other hand, the vibration force was calculated as the data were found by the sensor. The amount of the kinetic energy in order to be able to react should be at least similar to the amount of the activation energy (Ea). Referring to Arrhenius equation, the amount of the activation energy depends on the alteration of the solution temperatures [12]. Equation (4) expressed a linear model of Arrhenius equation in serving the calculation to be more simple and appropriate to the data found [12].

$$\ln k = -(Ea/R)(1/T) + \ln A \tag{4}$$

where $k$ is the reaction rate ($M.sec^{-1}$), A is the frequency of effective collision, R is ideal gas constant = 8,134 K, Ea is activation energy (kJ/mol), and T is temperature (K). Furthermore, the data listed in Table 2 were used in the calculation of the activation energy (Ea).

**Table 2**. Calculation of $\ln k - 1/T$

| $\ln k$ ($1/M.s^{-1}$) | $1/T$ ($K^{-1}$) |
|---|---|
| -11.1432 | $3.3355 \times 10^{-3}$ |
| -10.9469 | $3.2467 \times 10^{-3}$ |
| -10.7308 | $3.1446 \times 10^{-3}$ |

The experiment was conducted in three various temperatures, i.e. 298, 308 and 318 K for all vibration frequency. As the reaction of such equation (2) was assumed not to be in reversible term since the order of reaction was defined as one. The deposition rate was calculated via such equation (3) and then converted to $M.s^{-1}$ unit. To meet the equation (4), the deposition rate was converted to ln k and the temperature was also converted to 1/T as listed in Table 2. Through Excel program, the values of 1/T and ln k were plotted to determine the regression equation to find the gradient (m). The regression line of the Excel showed the equation of y line of which determinant $R^2$ value was one. The line showed that there is not a residue found in the y equation. The equation was:

$$y = -2.170.4\,x - 3.9156$$

As the equation of linear regression of $y = m\,x + n$, the value of m was -2,170.4 and n was -3.9156.

$$Ea = -\,m \times R = -(-2.170.4\ J/K.mol) \times (8.314\ K)$$
$$Ea = 18{,}044\ J/mol \text{ or } 18.044\ kJ/mol$$

The value of Ea = 18.044 kJ/mol was in good agreement with the work of Petrou [22]. Furthermore, the frequency of molecular collision was calculated as equation (5).

$$A = k/e[-Ea/RT] \tag{5}$$

Take the experiment of 318 K, of which reaction rate (k) was $21.8611 \times 10^{-6}$ mol/sec, Ea = 18.044 kJ/mol, R = 8.314 K, so the frequency of the collision (A) was found:

$$A = 21.8611 \times 10^{-6} \times e[-18.044/(8{,}314 \times 318)] = 21.8612 \times 10^{-6} \text{ mol collision}$$

As ion is an atom or a group of atoms that has a net positive or negative charge [12], so each $CaCO_3$ molecule has two ions i.e. Ca ion in positive charge and $CO_3$ ion in negative charge. Ionic effective collision (A) could be calculated as follow:

$$A = 21.8612 \times 10^{-6} \text{ mol collision} \times (6.022 \times 10^{23}) \text{ molecule/mol} \times 2 \text{ ion/molecule}$$
$$A = 263.2963 \times 10^{17} \text{ ionic collision}$$

As the effective collision occurs between the two ions in positive and negative charge, the number of ions was as much as two times of the number of the collision [12] = $2 \times 263.2963 \times 10^{17} = 526.5926 \times 10^{17}$ ions. To calculate the ionic force of the collision, take the experiment of 298 K, consider that ionic collision as a vector in trajectory Tr = 0.004 m as vibration displacement. Ionic collision force named as $F_{coll}$ and calculated using equation (6).

$$F_{coll} = Ea \times 1.000 \text{ N/kJ} / (NA \times \text{reactant ion} \times t \times Tr) \quad (6)$$

$$F_{coll} = (18.044 \text{ kJ/mol} \times 1.000 \text{ N/kJ})/(6.022 \times 10^{23} \text{ molecule} \times 2 \text{ ion} \times 45.693 \text{ sec} \times 0.004 \text{ m})$$

$$F_{coll} = 8.2 \times 10^{-23} \text{ N/ion}$$

Where number 2 represents the amount of ions per molecule i.e. ($Ca^{2+}$) and ($CO_3^{2-}$) ions, Notation t stands as the time needed to produce one mol of $CaCO_3$ which is calculated as a molar weight of $CaCO_3$ divided by the weight of product resulted from the reaction per hour by the experiment. The calculation of the time needed was t = (100 g/ 7.8786 g/hr) = 12.6926 hr or 45.693 sec.

In the current experiment of 8 Hz, the vibration sensor showed an acceleration of 22 m/sec$^2$ RMS. The solution mass in the pipe with a flow rate of 30 mL/min was 0.0005 kg in every second, so then the force given to the system by vibration ($F_{vib}$) in ionic term was calculated as Equation (7).

$$F_{vib} = (m \times a) / \Sigma \text{ ion} \quad (7)$$

$$F_{vib} = (0.0005 \text{ kg} \times 22 \text{ m/sec}^2) / (526.5926 \times 10^{17} \text{ ion})$$

$$F_{vib} = 21 \times 10^{-23} \text{ N/ion}$$

### *3.3. Vector analysis based on Vibration Time History (VTH)*

Take four arrays of ionic vectors with displacement direction i.e. ion at 90° (a), 60° (b), 30° (c), and 0° (d) to the horizontal axis, as represented in random directions as shown in Figure 2. It had been calculated that the ionic force was $8.2 \times 10^{-23}$ N/ion. When vibration affected the system, a new vector of $F_{vib} = 21 \times 10^{-23}$ N/ion acted as the direction of vibration displacement to the ion.

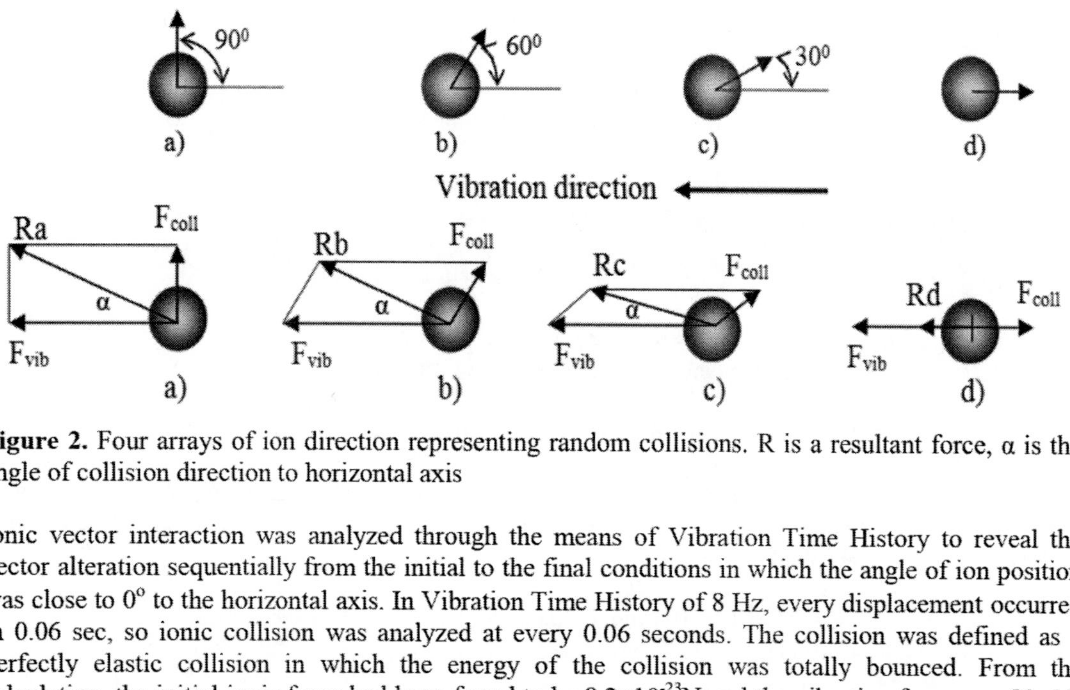


**Figure 2.** Four arrays of ion direction representing random collisions. R is a resultant force, α is the angle of collision direction to horizontal axis

Ionic vector interaction was analyzed through the means of Vibration Time History to reveal the vector alteration sequentially from the initial to the final conditions in which the angle of ion position was close to 0° to the horizontal axis. In Vibration Time History of 8 Hz, every displacement occurred in 0.06 sec, so ionic collision was analyzed at every 0.06 seconds. The collision was defined as a perfectly elastic collision in which the energy of the collision was totally bounced. From the calculation, the initial ionic force had been found to be $8.2 \times 10^{-23}$N and the vibration force was $21 \times 10^{-23}$ N.

Furthermore, the Ionic and vibration vector was analyzed through graphical method as shown in Figure 3, which based on Vibration Time History as presented in figure 4. Notation "α" represented the angle of ion direction to the horizontal axis. Ra, Rb, Rc and Rd show the resultant force of ion "a"; "b"; "c" and "d". After being analyzed in 1.18 seconds of Vibration Time History, the angle position of ion "a" changed from 90° to 1.27°. Ion "b" changed from 60° to 1.86°. Ion "c" changed from 30 ° to

0.88°. Otherwise, ion "d" remained in its position as 0°. The final direction of the ionic collision was in the horizontal direction. The analysis proved that the ionic collision in random direction had been affected by the vibration to planar-horizontally direction in a few seconds. Ion "a" needed 1.00 second to achieve horizontal direction from its initial direction, otherwise, ions "b"; "c" and "d" needed 1.18; 1.00 and 0.06 second as listed in Table 3.

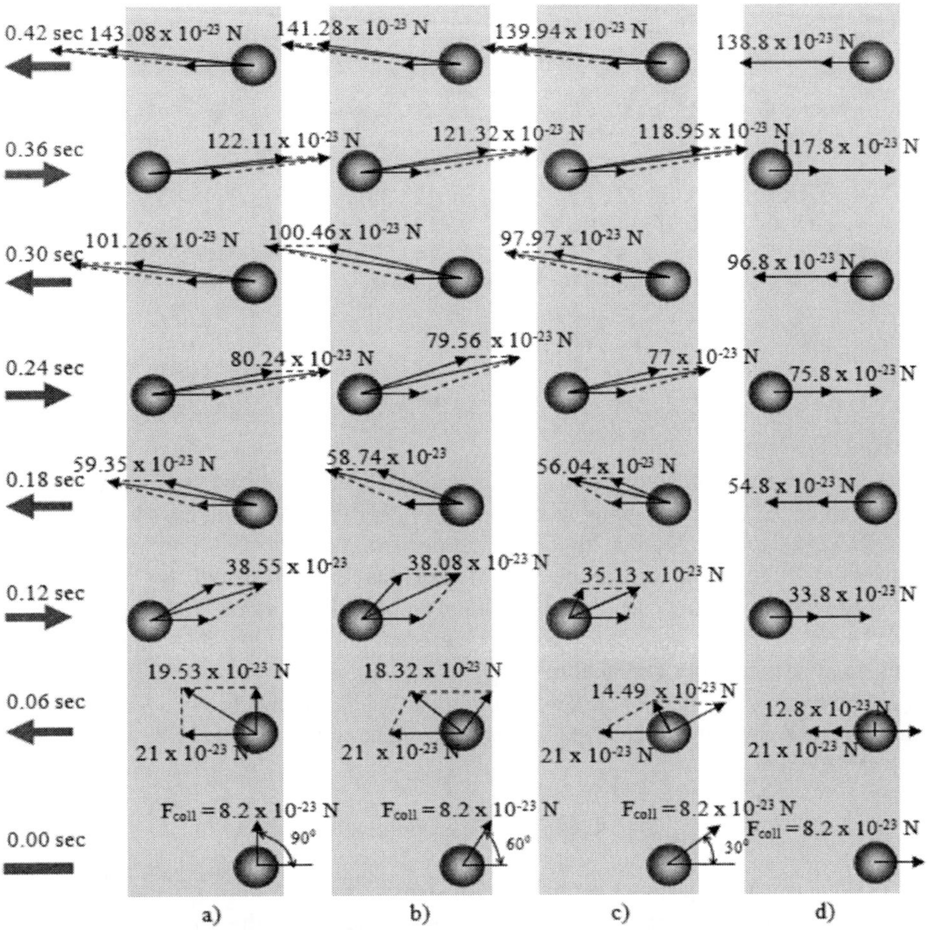


**Figure 3.** Vector analysis of four arrays of a), b), c), and d) ion directions as the representations of the randomic directions

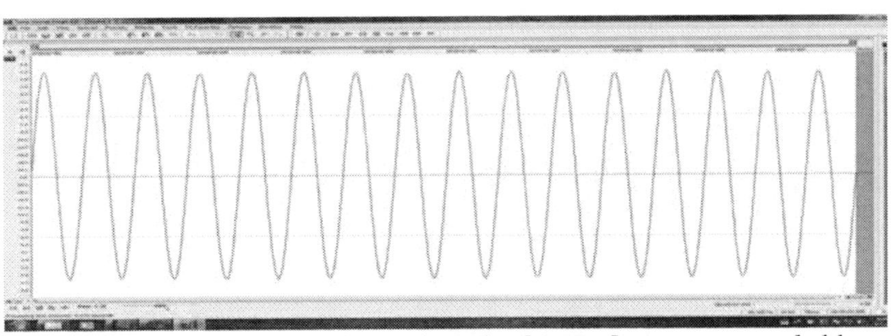

**Figure 4.** Vibration Time History (VTH) of 8 Hz frequency recorded by a computer program

IOP Conf. Series: Materials Science and Engineering **202** (2017) 012100 doi:10.1088/1757-899X/202/1/012100

**Table 3**. Resultant and direction of four ions after being analyzed based on VTH (R in x$10^{-23}$ N)

| Time | Vibration | ion a | | ion b | | ion c | | ion d | |
|---|---|---|---|---|---|---|---|---|---|
| (sec) | Direction | α (°) | Ra | α (°) | Rb | α (°) | Rc | α (°) | Rd |
| 0.06 | Left | 21.5 | 19.53 | 28.8 | 18.32 | 16.43 | 14.49 | 0 | 12.8 |
| 0.12 | Right | 10.2 | 38.55 | 13.4 | 38.08 | 6.7 | 35.13 | 0 | 33.8 |
| 0.18 | Left | 6.6 | 59.35 | 8.6 | 58.74 | 4.18 | 56.04 | 0 | 54.8 |
| 0.24 | Right | 4.9 | 80.24 | 6.4 | 79.56 | 3.04 | 77.00 | 0 | 75.8 |
| 0.30 | Left | 3.8 | 101.16 | 5.1 | 100.46 | 2.38 | 97.97 | 0 | 96.8 |
| 0.36 | Right | 3.1 | 122.11 | 4.2 | 121.32 | 1.95 | 118.95 | 0 | 117.8 |
| 0.42 | Left | 2.6 | 143.08 | 3.6 | 141.28 | 1.65 | 139.94 | 0 | 138.8 |
| 0.48 | Right | 2.3 | 164.04 | 3.1 | 161.24 | 1.43 | 160.94 | 0 | 159.8 |
| 0.54 | Left | 2.0 | 184.21 | 2.75 | 181.21 | 1.26 | 181.93 | 0 | 180.8 |
| 1.00 | Right | 1.8 | 204.32 | 2.45 | 202.20 | 1.13 | 202.92 | 0 | 201.8 |
| 1.06 | Left | 1.6 | 224.24 | 2.22 | 223.17 | 1.02 | 223.92 | 0 | 222.8 |
| 1.12 | Right | 1.4 | 265.07 | 2.03 | 244.17 | 0.93 | 244.91 | 0 | 243.8 |
| 1.18 | Left | 1.27 | 286.07 | 1.86 | 265.15 | 0.88 | 256.91 | 0 | 264.8 |

The force of ionic collision also increased dramatically from its initial strength as shown in Table 3. The initial force of those all ions were $8.2\times10^{-23}$ N and then increased to $286.07\times10^{-23}$ N; $265.15\times10^{-23}$ N; $256.91\times10^{-23}$ N and $264.8\times10^{-23}$ N for ion "a"; "b"; "c" and "d" respectively. This phenomenon showed the agreement to the data of previous research in which $CaCO_3$ deposition rate increased and the induction time decreased by the presence of vibration. The direction uniformity of the ionic collision in planar-horizontal was an unusual phenomenon and does not occur in the environment, so, it needed more explanation.

The data showed that this planar-horizontal collision produced much more $CaCO_3$ deposition. To confirm this unusual phenomenon, an equation of product probability [23] as in Equation (8) was used.

$$P(\theta) = (\sigma_R / 2\pi)\, G(\cos\theta) \qquad (8)$$

Here, P (θ) described the product probability of density distribution in the scatter angle with the angle of attack θ. $\sigma_R$ is the integral cross section. G is the density of the product. As the equation and as a function of cosine, the product probability of the collision depended on the value of cosθ of which the angle of θ could be scattered at 0° until 360°, as presented schematically in Figure 5.

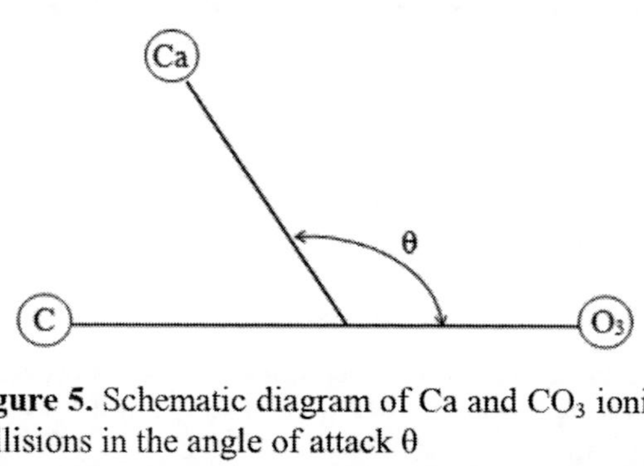


**Figure 5.** Schematic diagram of Ca and $CO_3$ ionic collisions in the angle of attack θ

The highest value of product probability was achieved when θ as 0° and 180° or in other temperatures was in horizontal position. When the cosθ achieved the maximum value, P (θ) as the product probability would reach the ultimate value. This would be the reason why ionic collision achieved maximally when the collision occurred in planar horizontally.

To distinguish the ionic collision between the vibrated and un-vibrated systems, a simple graph as shown in Figure 6 was used to express the ion distribution and ion direction. Figure 6a indicates that ion position in random configuration and collide in random direction. Figure 6b shows that ion position in a planar configuration and collide in horizontal direction. The ionic configuration of figure 6b which was plotted based on the vector analysis had been done and even the ionic configuration of Figure 6a was depicted as a usual phenomenon [12].

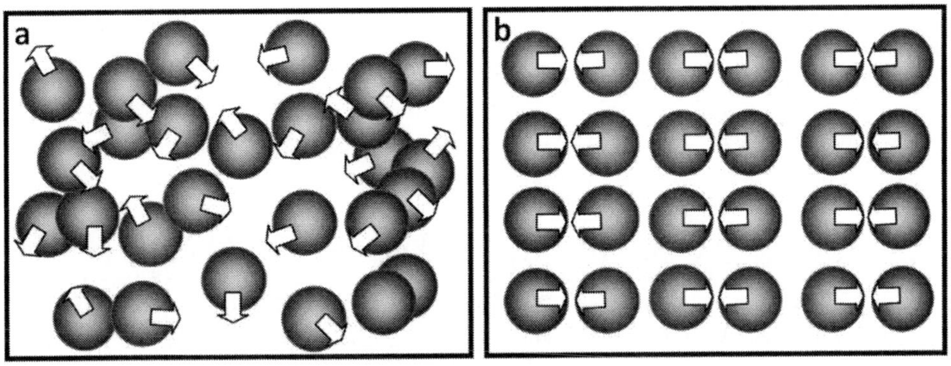


**Figure 6.** Ionic collision of a) un-vibrated, b) vibrated system

It is revealed that the planar horizontal direction as presented in Figure 6b brought the collision more effective than the random direction as shown in Figure 6a. At least there were two points of novelty of the research that had been found. First, the product probability of ionic collision was larger on the planar-horizontal direction. Second, the effective collision was easier to achieve as the data showed that the ionic force increased dramatically by the vibration. It only needed 1.18 sec to increase the ionic force from the initial stage until 35; 33; 32 and 33 times for ion "a"; "b"; "c" and "d".

## 4. Conclusion

The research of ionic collision vector analysis on $CaCO_3$ precipitation based on vibration time history was successfully conducted. The Ionic force was derived from its activation energy and there was $8.2\times10^{-23}$ N on every ion. Otherwise, the vibration force was calculated according to the data sent by the sensor and was found to be $21\times10^{-23}$ N. Normally, ion collides in a random direction, but vibration changed the ion collided in planar-horizontal direction. It was proved through vector analysis that the process needed only 1.18 sec. This study also confirmed that ionic force increased dramatically by the vibration and that is the reason why the vibration increased the deposition rate. Thus, the vibration mechanism in the affected $CaCO_3$ precipitation had been clearly justified.

## 5. References

[1] Kim B S, Kim Y K and Choi J 2008 Analysis of fluid-induced vibration of cryogenic pipes in consideration of the cooling effect *J. of Mech. Sci. and Tech.* **22** (2008) pp 2375-2385

[2] Sinir B G and Demir D D 2015 The Analysis of nonlinear vibration of pipe conveying an ideal fluid *J. of Mech. Fluids* **52** (2015) pp 38-44

[3] Oz H R 2001 Nonlinear vibrations and stability analysis of tensional pipes conveying fluid with variable velocity *J. Non-linear Mech.* **36**(2001) pp 1031-1039

[4] Han X, Lin W, Tang Y, Zhao C and Sammut K 2015 Effects of natural frequency ratio on vortex induced vibration of a cylindrical structure *J. Comp. & Fluids* **110**(2015) pp 62-76

[5] Li S J, Liu G and Kong W 2014 Vibration analysis of pipes conveying fluid by transfer matrix method *Nuclear Eng. and Design* **226**(2014) pp 78-88

[6] Liang F and Wen B 2014 Force vibrations with internal resonance of a pipe conveying fluid under external periodic excitation. *J. of Hydrodynamics*, DOI:10.1016/S101-5068 (14) 60086-X2014.

[7] Ruiz C, Jaramillo V H, Mba D, Ottewil J R and Cao Y 2015 Combination process and vibration data for improve condition monitoring of industrial system working under variable conditions. *Mech. Syst. and Signal Proc.* 66-67(2015) pp 699-714

[8] Andrew C, Hecimovich M and Nistor I 2015 Extrem wave loads on submerged water intakes in shallow water *J. of Hydr.* (2015)DOI:10.1016/S1001-6058(15)60454-1.2015,27(1)38-51

[9] Wolfram J and Naghipour M 1999 On the estimation of Morisson force coefficients and their predictive accuracy for very rough circular cylinders *J. App. Ocean Res.* **21**(1999) pp 311-328.

[10] Laskovski D, Stevenson P, Galvin K P 2009 Lift and drag forces on an isolated cubic particle in pipe flow. *Chem. Eng. Res. and Des,* **87**(2009): pp 1573-1581.

[11] Mangestiyono W, Muryanto S, Jamari J and Bayuseno A P 2015 The influence of vibration on CaCO3 scale formation in piping system *The 3rd Bali Int. Sem. on Sci. and Tech, October15' 2015. Denpasar Indonesia* DOI: http://dx.doi.org/10.1051/matecconf/ 20165801027

[12] Chang R and Overby J, 2011 *General Chemistry: The Essential Concepts*, 6th edition, Mc. Graw Hill, New York.

[13] Marcus Y 2006 Ionic volumes in solution *Biophy.l Chem.* **124** (2006) pp 200–207

[14] De S, Ghosh P N, Roy A and Safvan C P 2006 A setup for probing ion–molecule collision dynamics *Nuclear Inst. and Met. in Phys. Res. B* **243** (2006) pp 435–441

[15] Wang F, Xua X C, Honga X H, J. Wang H J and Goua B C 2011 A theoretical model for electron transfer in ion–atom collisions: Calculations for the collision of a proton with an argon atom *Physics Let. A* **375** (2011) pp 3290–3295

[16] Sheikoleslami 2000 Calcium sulfate fouling-Precipitation or particulate : A proposed Composite model *Heat Transfer Eng.* **21**-3(2000) pp 24-33

[17] Muryanto S, Bayuseno A P, Ma'mun H, Usamah M and Jotho 2014 Calcium carbonate scale formation in pipes: effect of flow rates, temperature, and malic acid as additives on the mass and morphology of the scale *Proc. Chem.* **9**(2014) pp 69-76.

[18] Bots P, Benning L G, Rodriguez-Blanco J D and Roncal-Herrero T, Shaw S 2012 Mechanistic Insights into the crystallization of amorphous calcium carbonate (ACC), *Cryst. Gr. and Des.* **12**(2012) pp 3806-3814

[19] Rodriguez-Blanco J D, Shaw S, Benning L G, 2015 A route for the direct crystallization of dolomite, *American Min.* **100**(2015) pp 1172-1181

[20] Ukrainczyk M, Kontrec J, Babić-Ivančić V, Brečević L and Damir-Kralj D 2007 Experimental design approach to calcium carbonate precipitation in a semicontinuous process *Powder Tech.* **171**(2007) pp 192-199

[21] Andritsos N and Karabelas A J 2003 Calcium carbonate scaling in a plate heat exchanger in the presence of particles *I. J. of Heat and Mass Trans.* **46**(2003) pp 4613-4627

[22] Petrou A L and Terzidaki A 2014 Calcium carbonate and calcium sulfate precipitation, crystallization and dissolution: Evidence for the activated steps and the mechanisms from the enthalpy and entropy of activation values *Chem. Geo.* **381** (2014) pp 144–153

[23] He X, Wu VWKC, Han K, Hao C, Zhang Y 2015 Mechanism of the collision energy and reagent vibration's effects on the collision time for the reaction Ca + HCl *Comp. and Th. Chem.* **1056** (2015) pp 1-10

**Acknowledgments**
This research was supported by the Ministry of Research, Technology and Higher Education of Republic of Indonesia. The author would like to thank to the staffs of Metrology and Instrumentation Laboratory of Industrial Technology Department, Vocational Education, Diponegoro University, for their help in the device calibration.

**Institute of Physics**
Dirac House, Temple Back
Bristol BS1 6BE UK

ISSN: 1757-8981
ISBN 978-1-5108-4176-5

CURRAN ASSOCIATES INC.
proceedings.com

# Mathematics, Informatics, Science and Education International Conference (MISEIC 2018)

## Emerging Trends of Research in Mathematics, Informatics, Sciences, and Education

Journal of Physics: Conference Series
Volume 1108

Surabaya, Indonesia
21 July 2018

IOP
Institute of Physics